Quantum Field Theory

Quantum Field Theory

An Integrated Approach

Eduardo Fradkin

PRINCETON UNIVERSITY PRESS
PRINCETON AND OXFORD

Published by Princeton University Press
41 William Street, Princeton, New Jersey 08540
6 Oxford Street, Woodstock, Oxfordshire OX20 1TR

press.princeton.edu

Library of Congress Cataloging-in-Publication Data
Names: Fradkin, Eduardo, author.
Title: Quantum field theory : an integrated approach / Eduardo Fradkin.
Description: Princeton : Princeton University Press, [2021] | Includes
bibliographical references and index.
Identifiers: LCCN 2020044826 (print) | LCCN 2020044827 (ebook) |
ISBN 9780691149080 (hardback) | ISBN 9780691189550 (ebook)
Subjects: LCSH: Quantum field theory.
Classification: LCC QC174.45 .F695 2021 (print) | LCC QC174.45 (ebook) |
DDC 530.14/3—dc23
LC record available at https://lccn.loc.gov/2020044826
LC ebook record available at https://lccn.loc.gov/2020044827

British Library Cataloging-in-Publication Data is available

Editorial: Jessica Yao and Arthur Werneck
Production Editorial: Nathan Carr
Text and Jacket/Cover Design: Pamela L. Schnitter
Production: Jacquie Poirier
Publicity: Matthew Taylor and Amy Stewart
Copyeditor: Cyd Westmoreland

This book has been composed in Minionpro

Printed on acid-free paper. ∞

Printed in the United States of America

10 9 8 7 6 5 4 3 2

to my parents
to Claudia

Contents

Preface and Acknowledgments

Quantum field theory (QFT) is the natural language and framework to study broad classes of physical systems with an infinite number of degrees of freedom. This is certainly the case in high-energy physics (and in nuclear physics), where QFT was born, but it is just as important in wide areas of condensed matter and statistical physics as well. This is particularly important in physical systems close to phase transitions, both quantum and classical, regimes in which perturbation theories fail. In the perspective developed by Kenneth Wilson, the renormalization group provides a framework for defining a QFT as a theory tuned to be close enough to a continuous phase transition, represented by critical fixed points. From this perspective, defining a QFT is the same as the problem of understanding continuous phase transitions. Moreover, both perspectives complement each other at the conceptual and the methodological levels. This point of view is the guiding principle behind this book.

Yet it is remarkable that, even though the renormalization group was developed almost 50 years ago, this perspective is seldom found in the many (and excellent) textbooks on the subject. Most textbooks on QFT focus, for good reasons, on relativistic field theory, and their main aim is to present the foundations of our understanding of the Standard Model of particle physics and to prepare students to do relevant computations. There are also many excellent books on a subject traditionally called "many-body physics," aimed primarily at nuclear and condensed matter physicists. Most books of this type focus on the computation of various Green functions and lay the groundwork for the perturbative computation of susceptibilities (for instance) using Feynman diagrams. Finally, there are also outstanding books on the theory of critical phenomena and phase transitions, both classical and quantum mechanical. It is my strongly held opinion that this fragmentation is not a good way to learn the subject. Moreover, it tends to make QFT rather formal and often physically obscure.

One salient example is the treatment of the renormalization group. In more traditional QFT textbooks, the renormalization group is presented almost entirely in the context of perturbation theory and of renormalized perturbation theory. This is a time-honored approach that has been highly successful. However, it mostly lacks physical transparency. Here, after presenting the problems that arise in field-theoretic perturbation theory, instead of proceeding directly to develop renormalized perturbation theory, I take a "detour" into the theory of phase transitions. So, rather than learn how to "hide infinities," one learns the concepts of the fixed point and universality, and how they provide a definition of a QFT. In this sense, this is not a book on methods but rather on a way of thinking about systems with an infinite number of degrees of freedom in the large-fluctuation regime. In class, and in this book, I discuss the main ideas and principles, using examples from high-energy, condensed matter, and statistical physics. The problem sets were designed so that students see how these concepts work out in different settings.

This book is based on lecture notes that I prepared for a course on quantum field theory that I have taught, off and on, at the Department of Physics of the University of Illinois at Urbana-Champaign since the 1980s. In most physics departments, at least in the United States, the teaching of QFT focuses primarily on training high-energy physics students. For this and historical reasons, a typical QFT course deals almost exclusively with relativistic field theory, focusing on quantum electrodynamics, quantum chromodynamics, and the Standard Model of particle physics. Other students, particularly those in condensed matter and statistical physics, are often taught a separate sequence of courses on many-body and statistical physics. It is my strongly held opinion that this separation is both artificial and detrimental to the understanding of the subject for both types of students (and others as well). Similarly, I also find the separation between "formal" and "phenomenological," common among my high-energy friends, both artificial and detrimental.

The lectures were developed for a two-semester course in QFT intended for students in high-energy, condensed-matter, and statistical physics, combined in a single course. After completing these QFT courses, students go on to take other classes that are more focused on their specific subfields. As can be expected, it is a challenge to cover the range of material of interest to such a wide audience of graduate students. The result is that I cover a fraction of the material in class, and its application to high-energy and condensed matter physics is done in the problem sets, which the students have endured stoically. During the first semester, I focus on basic material, mostly on free-field theories of different types, both relativistic and nonrelativistic. The Hamiltonian and path integral approaches are presented from the beginning. This basic material is covered in the first 11 chapters of the book. It includes an important chapter (chapter 10) on relating measurements with correlation functions, both in relativistic theories (the computation of cross sections) and in nonrelativistic systems (the computation of susceptibilities). The remaining eleven chapters cover the more advanced material taught in the second semester. Here, the focus is on the renormalization group, nonperturbative phenomena, and topology.

While teaching the QFT sequence, in addition to the lecture notes on which this book is based, I have recommended that students read several outstanding books. Prime examples include the classic book *An Introduction to Quantum Field Theory* by Michael Peskin and Daniel Schroeder (Peskin and Schroeder, 1995), which gives a modern perspective on high-energy physics. There are of course many outstanding books on relativistic QFT, such as Steven Weinberg's monumental three-volume book, *The Quantum Theory of Fields* (Weinberg, 2005). Two books that I really like are Alexander Polyakov's *Gauge Fields and Strings* (Polyakov, 1987), which has deep insights, and Sidney Coleman's *Aspects of Symmetry* (Coleman, 1985). On the condensed-matter side, I have often recommended *Green's Functions for Solid State Physicists* by Sebastian Doniach and E. Sondheimer (Doniach and Sondheimer, 1998); Richard Feynman's wonderful *Statistical Physics, A Set of Lectures* (Feynman, 1972); John Cardy's brilliant *Scaling and Renormalization in Statistical Physics* (Cardy, 1996); and the more technical *Field Theory, the Renormalization Group and Critical Phenomena* by Daniel Amit (Amit, 1980). I also often recommended *Methods of Quantum Field Theory in Statistical Physics* by Alexey Abrikosov, Lev Gorkov, and Igor Dzyaloshinskii (Abrikosov et al., 1963), and my *Field Theories of Condensed Matter Systems* (Fradkin, 2013).

Some comments about the organization of the book are in order. The first few chapters of the book cover fairly basic material, including classical field theory, with an emphasis on both relativistic theories and classical statistical mechanics. Here the emphasis is on the role of global and local symmetries at the classical level. The Aharonov-Bohm effect and Wilson loops are introduced here. The quantum theory of free scalar fields using canonical quantization is discussed in chapter 4, and path integral quantization both in quantum mechanics and in scalar fields is discussed in chapter 5. The connection between relativistic QFT (both at zero and finite temperatures) is discussed here as well. Chapter 6 is devoted to nonrelativistic (for both bosons and fermions) QFT. The quantum theory of the free Dirac field is discussed in chapter 7. Coherent-state path integrals for both bosons and fermions are discussed in chapter 8, focusing on quantization of the free Dirac field and the nonrelativistic Bose gas. The computation of functional determinants is discussed in this chapter. Chapter 9 is devoted to the quantization of gauge field theories, both abelian and non-abelian, using the Faddeev-Popov procedure.

The important problem of relating measurements to the computation of correlation functions (i.e., the relation between observables and propagators) is the subject of chapter 10, including the computation of scattering cross sections in relativistic theories (including the Lehmann, Symanzik, and Zimmermann (LSZ) formalism). Linear response theory and the computation of susceptibilities in nonrelativistic systems are also discussed here, emphasizing the role of conservation laws.

Feynman diagrams and Feynman rules are introduced in chapter 11. While the text focuses entirely on ϕ^4 scalar fields (both in Minkowski and Euclidean spacetimes), the perturbative approach to other field theories of interest is the subject of exercises. The subject of spontaneous breaking of a global symmetry is developed in chapter 12, including the theory of the effective potential, Ward identities, and Goldstone's theorem. Perturbative renormalizability and regularization schemes are presented in chapter 13, with special emphasis on dimensional regularization.

The connection between statistical mechanics and QFT is discussed in depth in chapter 14. The solution of the two-dimensional Ising model is discussed as an example of how a QFT is defined nonperturbatively. The Wilson-Kadanoff conceptual framework of the renormalization group is the subject of chapter 15, where the concepts of scaling, universality, and fixed points are introduced, including a heuristic discussion of scale invariance as an emergent symmetry, and renormalization group flows. These concepts are introduced using block-spin transformations, momentum-shell effective actions, and the operator product expansion. Armed with this physical framework, I return to the development of the perturbative renormalization group in chapter 16. Several theories are discussed, including ϕ^4 theory and the ϵ expansion, and asymptotically free theories, such as the nonlinear sigma model (and a proof of renormalizability) and Yang-Mills gauge theory.

The remaining chapters focus on more advanced topics. Several field theories in the large-N limit are discussed in chapter 17, including ϕ^4 theory, (vector) nonlinear sigma models, the Gross-Neveu model, and quantum electrodynamics. I also give an introduction to theories of planar diagrams (both with global and gauge symmetries). Chapter 18 is devoted to the introduction of the basic aspects of lattice gauge theory and discusses the phases of gauge theories, confinement, and the Higgs mechanism. Instantons and solitons are the subject of chapter 19. An intuitive (and highly nonrigorous) introduction to homotopy groups and the classification of topological excitations, drawing from examples in QFT and statistical physics, is presented here. An introduction to anomalies in QFT is presented in chapter 20. The concepts of conformal field theory are presented in chapter 21. The book ends with an introduction to the ideas of topological field theory in chapter 22.

Some comments about the relation between this book and the two-semester course that I teach are required. This book has 22 chapters, ranging from fairly basic aspects of QFT to more advanced material. During the first semester, I normally cover the first 11 chapters. The students taking the first part of the course include both theorists and experimentalists in high-energy, condensed matter, and statistical physics, and from time to time, some students in electrical engineering. In the second semester, which involves more advanced material, I typically cover nine of the remaining 11 chapters. The students taking the second half of the course are theorists. The problem sets are designed so that the students can go deeper into the subject, covering topics of interest in high-energy and condensed matter physics. The students typically are assigned seven problem sets (drawn from the problems I included in this book) during the first semester. In the second semester, they are assigned four problem sets and also have to write a term paper on a subject of their choice from a fairly long list of possible topics (with recommended/required bibliography). They also give an oral presentation on their paper, in a workshop format.

In designing the course (and the book), I had to make some significant choices. One is on the mathematical background that I assumed. At the University of Illinois, the students taking this class either had taken already, or were taking concurrently, a two-semester course, Mathematics for Physics, largely based on the outstanding book by my colleague Michael Stone and my (former) colleague Paul Goldbart (Stone and Goldbart, 2009). This is a thoroughly modern textbook, which I strongly recommend to my students (and to the readers of this book). This has allowed me to avoid being distracted by formal mathematical interludes on such subjects as group theory and many others. It has also allowed me to keep the discussion of field theory at a less formal level than otherwise would have been possible.

I also had to choose what topics to cover, at what level, and what not to cover. For pedagogical reasons, and to level the playing field, I have included some material that should also be covered in advanced quantum mechanics courses (and books). Another important choice was to present the Hamiltonian and the path-integral pictures from the outset on equal footing. In my opinion, this is essential. However, this approach has the downside that it takes longer to get to a discussion of subjects such as Feynman diagrams, which in some texts is done quite early on. As I noted above, chapter 10 on observables and correlation functions is quite important. It is where I introduce the computation of both scattering cross sections and susceptibilities, crucial to experiments. However, I cannot possibly discuss cross sections at the same level of detail as done in such books as Peskin and Schroeder (1995), or susceptibilities, as in Doniach and Sondheimer (1998). The result was that in the text (and in class) I covered cross sections for some simple cases, such as ϕ^4 theory and for Dirac fermions, and susceptibilities only for the computation of the electrical conductivity of a metal. The students did more detailed calculations in the exercises.

Some parts of this book have been available as informal lecture notes for my class. Apparently, these notes have been well received. The present book came to be at the (persistent) encouragement of Ingrid Gnerlich from Princeton University Press. She persuaded me to organize all my lecture notes as a book. Eventually, I (reluctantly) agreed to do it, even though I knew that it would take a lot of work. I am glad that the book is now in its final form, albeit several years after my originally intended completion date. During this period, which began in 2010, the people at Princeton University Press showed great patience and understanding.

I thought of several possible titles for this book, searching for one that would reflect its character while being concise. After several failed attempts (all long and boring titles), I settled on the current title, *An Integrated Approach to Quantum Field Theory*, which reflects my broad approach to the subject. I am grateful to Jessica Yao, my current Princeton University Press editor, for helping me converge on this title.

I must particularly thank Leonard Susskind, who was my PhD thesis advisor. Much of what I know on QFT is based on what I learned from him. I also learned from many other people—collaborators and friends. In particular, I thank Tom Faulkner, Leo Kadanoff, Steven Kivelson, John Kogut, Rob Leigh, Fidel Schaposnik, Stephen Shenker, and Michael Stone. Over the years, I had many enlightening discussions on QFT with Michael Stone, and several sections of this book reflect those discussions. I am particularly grateful for a discussion of Noether's theorem and its applications. The derivation of the classical energy-momentum tensor presented in chapter 3 follows Mike's very clear and elegant presentation rather than my original version.

Many students have helped me with my lecture notes over the years. Several of them have been teaching assistants in my class, and they had to write solutions to the exercises that I include in the book. It has been my policy that each teaching assistant should write his/her own solutions. This way I am sure they understand what they are doing (and learn!). Consistent with this policy, I have not included here the solutions to the exercises. Of the many students that helped me in this endeavor, I particularly thank Manuel Fuentes, Hart Goldman, Eun-Ah Kim, and Ramanjit Sohal, among others. I am also thankful for the patience shown by my collaborators, and my current students, with me during the past few months while I was completing this project.

Although I enjoy teaching QFT, writing this book was truly a challenge, much more than I had anticipated. It took a great deal of time, effort, and focus, particularly during the final stretch. This book would not have been finished without the love and support (and patience!) of Claudia, my love and companion.

Finally, I am thankful for the support I received from the Department of Physics of the University of Illinois at Urbana-Champaign. The superb atmosphere in the department, both friendly and intellectually challenging, with great communication across different fields, has made it a truly nurturing place.

Eduardo Fradkin
Urbana, Illinois, USA
April 2020

Quantum Field Theory

1

Introduction to Field Theory

The purpose of this book is twofold. Here I will introduce field theory as a framework for the study of systems with a very large number of degrees of freedom, $N \to \infty$. And I will also introduce and develop the tools that will allow us to treat such systems. Systems that involve a large (in fact, infinite) number of *coupled* degrees of freedom arise in many areas of physics, notably in high-energy and in condensed matter physics, among others. Although the physical meanings of these systems and their symmetries are quite different, they actually have much more in common than it may seem at first glance. Thus, we will discuss, on the same footing, the properties of relativistic quantum field theories, classical statistical mechanical systems, and condensed matter systems at finite temperature. This is a very broad field of study, and we will not be able to cover each area in great depth. Nevertheless, we will learn that it is often the case that what is clear in one context can be used to expand our knowledge in a different physical setting. We will focus on a few unifying themes, such as the construction of the ground state (the "vacuum"), the role of quantum fluctuations, collective behavior, and the response of these systems to weak external perturbations.

1.1 Examples of fields in physics

1.1.1 The electromagnetic field

Let us consider a very large box of linear size $L \to \infty$ and the electromagnetic field enclosed inside it. At each point in space $\boldsymbol{x}$, we can define a *vector* (which is a function of time as well) $\boldsymbol{A}(\boldsymbol{x}, t)$ and a *scalar* $A_0(\boldsymbol{x}, t)$. These are the vector and scalar potentials. The physically observable electric field $\boldsymbol{E}(\boldsymbol{x}, t)$ and the magnetic field $\boldsymbol{B}(\boldsymbol{x}, t)$ are defined in the usual way:

$$\boldsymbol{B}(\boldsymbol{x}, t) = \boldsymbol{\nabla} \times \boldsymbol{A}(\boldsymbol{x}, t), \qquad \boldsymbol{E}(\boldsymbol{x}, t) = -\frac{1}{c}\frac{\partial \boldsymbol{A}}{\partial t}(\boldsymbol{x}, t) - \boldsymbol{\nabla} A_0(\boldsymbol{x}, t) \tag{1.1}$$

The time evolution of this dynamical system is determined by a *local* Lagrangian density (which we will consider in section 2.6). The equations of motion are just the Maxwell equations. Let us define the 4-vector field

$$A^\mu(x) = \left(A^0(x), \boldsymbol{A}(x)\right), \qquad A^0 \equiv A_0 \tag{1.2}$$

where $\mu = 0, 1, 2, 3$ are the time and space components. Here x stands for the 4-vector

$$x^\mu = (ct, \boldsymbol{x}) \tag{1.3}$$

To every point x^μ of Minkowski spacetime $\mathcal{M}$ we associate a value of the vector potential A^μ. The vector potentials are ordered sets of four real numbers and hence are elements of $\mathbb{R}^4$. Thus a field configuration can be viewed as a mapping of the Minkowski spacetime $\mathcal{M}$ onto $\mathbb{R}^4$,

$$A^\mu : \mathcal{M} \mapsto \mathbb{R}^4 \tag{1.4}$$

Since spacetime is continuous, we need an infinite number of 4-vectors to specify a configuration of the electromagnetic field, even if the box were finite (which it is not). Thus we have a infinite number of degrees of freedom for two reasons: spacetime is both continuous and infinite.

1.1.2 The elastic field of a solid

Consider a three-dimensional crystal. A configuration of the system can be described by the set of positions of its atoms relative to their equilibrium state (i.e., the set of deformation vectors $\boldsymbol{d}$ at every time t). Lattices are labeled by ordered sets of three integers and are equivalent to the set

$$\mathbb{Z}^3 = \mathbb{Z} \times \mathbb{Z} \times \mathbb{Z} \tag{1.5}$$

whereas deformations are given by sets of three real numbers and are elements of $\mathbb{R}^3$. Hence a *crystal configuration* is a mapping

$$\boldsymbol{d} : \mathbb{Z}^3 \times \mathbb{R} \mapsto \mathbb{R}^3 \tag{1.6}$$

At length scales ℓ, which are large compared to the lattice spacing a but small compared to the linear size L of the system, we can replace the lattice $\mathbb{Z}^3$ by a *continuum* description, in which the crystal is replaced by a continuum three-dimensional Euclidean space $\mathbb{R}^3$. Thus the dynamics of the crystal requires a four-dimensional spacetime $\mathbb{R}^3 \times \mathbb{R} = \mathbb{R}^4$. Hence the configuration space becomes the set of continuous mappings

$$\boldsymbol{d} : \mathbb{R}^4 \mapsto \mathbb{R}^3 \tag{1.7}$$

In this continuum description, the dynamics of the crystal is specified in terms of the displacement vector field $\boldsymbol{d}(\boldsymbol{x}, t)$ and its time derivatives, the velocities $\frac{\partial \boldsymbol{d}}{\partial t}(\boldsymbol{x}, t)$, which define the mechanical state of the system. This is the starting point of the theory of elasticity. The displacement field $\boldsymbol{d}$ is the elastic field of the crystal.

1.1.3 The order-parameter field of a ferromagnet

Let us now consider a ferromagnet. This is a physical system, usually a solid, in which there is a local average magnetization field $\boldsymbol{M}(\boldsymbol{x})$ in the vicinity of a point $\boldsymbol{x}$. The local magnetization is simply the sum of the local magnetic moments of each atom in the neighborhood of $\boldsymbol{x}$. At scales long compared to microscopic distances (the interatomic spacing a), $\boldsymbol{M}(\boldsymbol{x})$ is a continuous real vector field. In some situations of interest, the magnitude of the local moment does not fluctuate, but its local orientation does. Hence, the local state of the system is specified locally by a three-component unit vector $\boldsymbol{n}$. Since the set of unit vectors is in one-to-one correspondence with the points on a sphere S^2, the configuration space is equivalent (isomorphic) to the sets of mappings of Euclidean three-dimensional space onto S^2,

$$\boldsymbol{n}:\mathbb{R}^3 \mapsto S^2 \tag{1.8}$$

In an ordered state, the individual magnetic moments become spontaneously oriented along some direction. For this reason, the field $\boldsymbol{n}$ is usually said to be an *order-parameter field*. In the theory of phase transitions, the order-parameter field represents the important degrees of freedom of the physical system (i.e., the degrees of freedom that drive the phase transition).

1.1.4 Hydrodynamics of a charged fluid

Charged fluids can be described in terms of hydrodynamics. In hydrodynamics, one specifies the charge density $\rho(\boldsymbol{x}, t)$ and the current density $\boldsymbol{j}(\boldsymbol{x}, t)$ near a spacetime point x^μ. The charge and current densities can be represented in terms of the 4-vector

$$j^\mu(x) = \big(c\rho(\boldsymbol{x}, t), \boldsymbol{j}(\boldsymbol{x}, t)\big) \tag{1.9}$$

where c is a suitably chosen speed (generally *not* the speed of light!). Clearly, the configuration space is the set of maps

$$j^\mu : \mathbb{R}^4 \mapsto \mathbb{R}^4 \tag{1.10}$$

In general, we will be interested both in the dynamical evolution of such systems and in their large-scale (thermodynamic) properties. Thus, we will need to determine how a system that, at some time t_0, is in some initial state, manages to evolve to some other state after time T. In classical mechanics, the dynamics of any physical system can be described in terms of a *Lagrangian*. The Lagrangian is a *local functional* of the field and of its space and time derivatives. "Local" here means that the equations of motion can be expressed in terms of partial differential equations. In other words, we do not allow for "action-at-a-distance," but only for local evolution. Similarly, the thermodynamic properties of these systems are governed by a local energy functional, the *Hamiltonian*. That the dynamics is determined by a Lagrangian means that the field itself is regarded as a mechanical system, to which the standard laws of classical mechanics apply. Here the wave equations of the fluid are the equations of motion of the field. This point of view will also tell us how to quantize a field theory.

1.2 Why quantum field theory?

From a historical point of view, quantum field theory (QFT) arose as an outgrowth of research in the fields of nuclear and particle physics. In particular, Dirac's theory of electrons and positrons was, perhaps, the first QFT. Nowadays, QFT is used, both as a picture and as a tool, in a wide range of areas of physics. In this book, I will not follow the historical path of the way QFT was developed. By and large, it was a process of trial and error in which the results had to be reinterpreted a posteriori. The introduction of QFT as the general framework of particle physics implied that the concept of *particle* had to be understood as an *excitation* of a *field*. Thus, *photons* become the quantized excitations of the electromagnetic field with particle-like properties (such as momentum), as anticipated by Einstein's 1905 paper on the photoelectric effect. Dirac's theory of the electron implied that even such "conventional" particles should also be understood as the *excitations of a field*.

The main motivation of these developments was the need to reconcile, or *unify*, quantum mechanics with special relativity. In addition, the experimental discoveries of the spin of the electron and of electron-positron creation by photons showed that not only was the Schrödinger equation inadequate to describe such physical phenomena, but the very notion of a particle itself had to be revised.

Indeed, let us consider the Schrödinger equation

$$H\Psi = i\hbar \frac{\partial \Psi}{\partial t} \tag{1.11}$$

where H is the Hamiltonian

$$H = \frac{\hat{\boldsymbol{p}}^2}{2m} + V(\boldsymbol{x}) \tag{1.12}$$

and $\hat{\boldsymbol{p}}$ is the momentum represented as a differential operator

$$\hat{\boldsymbol{p}} = \frac{\hbar}{i}\boldsymbol{\nabla} \tag{1.13}$$

acting on the Hilbert space of wave functions $\Psi(\boldsymbol{x})$.

The Schrödinger equation is invariant under Galilean transformations, provided the potential $V(\boldsymbol{x})$ is constant, but not under general Lorentz transformations. Hence, quantum mechanics, as described by the Schrödinger equation, is not compatible with the requirement that the description of physical phenomena must be identical for all inertial observers. In addition, it cannot describe pair-creation processes, since in the nonrelativistic Schrödinger equation, the number of particles is strictly conserved.

Back in the late 1920s, two apparently opposite approaches were proposed to solve these problems. We will see that these approaches actually do not exclude each other. The first approach was to stick to the basic structure of "particle" quantum mechanics and to write down a relativistically invariant version of the Schrödinger equation. Since in special relativity, the natural Lorentz scalar involving the energy E of a particle of mass m is $E^2 - (\boldsymbol{p}^2c^2 + m^2c^4)$, it was proposed that the "wave functions" should be solutions of the equation (the "square" of the energy)

$$\left[\left(i\hbar\frac{\partial}{\partial t}\right)^2 - \left(\left(\frac{\hbar c}{i}\boldsymbol{\nabla}\right)^2 + m^2c^4\right)\right]\Psi(\boldsymbol{x}, t) = 0 \tag{1.14}$$

This is the Klein-Gordon equation. This equation is invariant under the Lorentz transformations

$$x^\mu = \Lambda^{\mu,\nu}x'_\nu \qquad x^\mu = (x_0, \boldsymbol{x}) \tag{1.15}$$

provided that the "wave function" $\Psi(x)$ is also a scalar (i.e., invariant) under Lorentz transformations

$$\Psi(x) = \Psi'(x') \tag{1.16}$$

However, it soon became clear that the Klein-Gordon equation was not compatible with a particle interpretation. In addition, it cannot describe particles with spin. In particular, the solutions of the Klein-Gordon equation have the (expected) dispersion law

$$E^2 = \mathbf{p}^2 c^2 + m^2 c^4 \tag{1.17}$$

which implies that there are positive and negative energy solutions

$$E = \pm\sqrt{\mathbf{p}^2 c^2 + m^2 c^4} \tag{1.18}$$

From a "particle" point of view, negative energy states are unacceptable, since they would imply that there is no ground state. We will see in chapter 4 that in QFT, there is a natural and simple interpretation of these solutions that in no way make the system unstable. However, the meaning of the negative-energy solutions was unclear in the early 1930s.

To satisfy the requirement from special relativity that energy and momentum must be treated equally, and to avoid the negative-energy solutions that came from working with the square of the Hamiltonian H, Dirac proposed to look for an equation that was linear in derivatives (Dirac, 1928). To be compatible with special relativity, the equation must be *covariant* under Lorentz transformations (i.e., it should have the same form in all reference frames). Dirac proposed a *matrix* equation that is linear in derivatives with a "wave function" $\Psi(x)$ in the form of a four-component vector, a 4-spinor $\Psi_a(x)$ (with $a = 1, \ldots, 4$):

$$i\hbar\frac{\partial \Psi_a}{\partial t}(x) + \frac{\hbar c}{i}\sum_{j=1}^{3} \alpha_j^{ab}\partial_j \Psi_b(x) + mc^2\, \beta_{ab}\Psi_b(x) = 0 \tag{1.19}$$

where α_j and β are four 4×4 matrices. For this equation to be *covariant*, it is necessary that the 4-spinor field Ψ should transform as a *spinor* under Lorentz transformations

$$\Psi'_a(\Lambda x) = S_{ab}(\Lambda)\Psi_b(x) \tag{1.20}$$

where $S(\Lambda)$ is a suitable matrix. The matrix elements of the matrices α_j and β have to be pure numbers that are independent of the reference frame. By further requiring that the iterated form of this equation (i.e., the "square") satisfies the Klein-Gordon equation for each component separately, Dirac found that the matrices obey the (Clifford) algebra

$$\{\alpha_j, \alpha_k\} = 2\delta_{jk}\mathbf{1}, \qquad \{\alpha_j, \beta\} = 0, \qquad \alpha_j^2 = \beta^2 = \mathbf{1} \tag{1.21}$$

where $\mathbf{1}$ is the 4×4 identity matrix. The solutions are easily found to have the energy eigenvalues $E = \pm\sqrt{\mathbf{p}^2 c^2 + m^2 c^4}$. (We will come back to this in chapter 2.) It is also possible to show that the solutions are spin-1/2 particles and antiparticles (we will discuss this later on).

However, the particle interpretation of both the Klein-Gordon and the Dirac equations was problematic. Although spin 1/2 appeared now in a natural way, the meaning of the negative energy states remained unclear.

The resolution of all of these difficulties was the fundamental idea that these equations should *not* be regarded as the generalization of Schrödinger's equation for relativistic particles but, instead, as the *equations of motion of a field*, whose excitations are the particles, much in the same way as photons are the excitations of the electromagnetic field. In this picture, particle number is not conserved, but charge is. Thus, photons interacting with matter can create electron-positron pairs. Such processes do not violate charge conservation, but the notion is lost of a particle as an object that is a fundamental entity and has a distinct physical identity. Instead, the field becomes the fundamental object, and the particles become the excitations of the field.

Thus, the relativistic generalization of quantum mechanics is QFT. This concept is the starting point of QFT. The basic strategy is to seek a field theory with specific symmetry properties and whose equations of motion are Maxwell, Klein-Gordon, and Dirac equations, respectively. Notice that if the particles are to be regarded as the excitations of a field, there can be as many particles as we wish. Thus, the Hilbert space of a QFT has an arbitrary (and indefinite) number of particles. Such a Hilbert space is called a *Fock space.*

Therefore, in QFT, the field is *not* the wave function of anything. Instead the field represents an infinite number of degrees of freedom. In fact, the wave function in a QFT is a *functional* of the field configurations, which themselves specify the state of the system. We will see in chapter 4 that the states in Fock space are given either by specifying the number of particles and their quantum numbers or, alternatively, in terms of the amplitudes (or configurations) of some properly chosen fields.

Different fields transform differently under Lorentz transformations and constitute different representations of the Lorentz group. Consequently, their excitations are particles with different quantum numbers that label the representation. Thus,

1) The Klein-Gordon field $\phi(x)$ represents charge-neutral scalar spin-0 particles. Its configuration space is the set of mappings of Minkowski space onto the real numbers $\phi : \mathcal{M} \mapsto \mathbb{R}$, or complex numbers for charged spin-0 particles $\phi : \mathcal{M} \mapsto \mathbb{C}$.
2) The Dirac field represents charged spin-1/2 particles. It is a complex 4-spinor $\Psi_\alpha(x)$ ($\alpha = 1, \ldots, 4$), and its configuration space is the set of maps $\Psi_\alpha : \mathcal{M} \mapsto \mathbb{C}^4$, while it is real for neutral spin-1/2 particles (such as neutrinos).
3) The gauge field $A^\mu(x)$ represents the electromagnetic field, and its non-abelian generalizations for gluons (and so forth).

The description of relativistic quantum mechanics in terms of relativistic quantum fields solved essentially all problems that originated in its initial development. Moreover, QFT gives exceedingly accurate predictions of the behavior of quantized electromagnetic fields and charged particles, as described by quantum electrodynamics (QED). QFT also gives a detailed description of both the strong and weak interactions in terms of field theories known as quantum chromodynamics (QCD), based on Yang-Mills gauge field theories, and unified and grand unified gauge theories.

However, along with its successes, QFT also brought with it a completely new set of physical problems and questions. Essentially, any QFT of physical interest is necessarily a nonlinear theory, as it has to describe interactions. So even though the quantum numbers of the excitations (i.e., the "particle" spectrum) may be quite straightforward in the absence of interactions, the intrinsic nonlinearities of the theory may actually unravel much of this structure. Note that the equations of motion of QFT are nonlinear, as they also are in quantum mechanics. However, the wave functional of a QFT obeys a linear Schrödinger equation, just as the wave function does in nonrelativistic quantum mechanics.

In the early days of QFT, and indeed for some time thereafter, it was assumed that perturbation theory could be used in all cases to determine the actual spectrum. It was soon found out that while there are several cases of great physical interest in which some sort of perturbation theory yields an accurate description of the physics, in many more situations this is not the case. Early on it was found that, at every order in perturbation theory, there are singular contributions to many physical quantities. These singularities reflected the existence of an infinite number of degrees of freedom, both at short distances, since spacetime is a continuum (the *ultraviolet* (UV) domain), and at long distances, since spacetime is (essentially) infinite (the *infrared* (IR) domain). Qualitatively, divergent contributions in perturbation theory come about because degrees of freedom from a wide range of length scales (or wavelengths) and energy scales (or frequencies) contribute to the expectation values of physical observables.

Historically, the way these problems were dealt with was through the process of regularization (i.e., making the divergent contributions finite), and renormalization (i.e., defining a set of effective parameters which are functions of the energy and/or momentum scale at which the system is probed). Regularization required that the integrals be cut off at some high-energy scale (in the UV). Renormalization was then thought of as the process by which these arbitrarily introduced cutoffs were removed from the expressions for physical quantities. This was a physically obscure procedure, but it worked brilliantly in QED and, to a lesser extent, in QCD. Theories for which such a procedure can be implemented with the definition of only a *finite* number of renormalized parameters (the actual input parameters to be taken from experiment) are said to be *renormalizable* QFTs. QED and QCD are the most important examples of renormalizable QFTs, although there are many others.

Renormalization implies that the connection between the physical observables and the parameters in the Lagrangian of a QFT is highly nontrivial, and that the spectrum of the theory may have little to do with the predictions of perturbation theory. This is the case for QCD, whose "fundamental fields" involve quarks and gluons but the actual physical spectrum consists only of bound states whose quantum numbers are not those of either quarks or gluons. Renormalization also implies that the behavior of the physical observables depends on the scale at which the theory is probed. Moreover, a closer examination of these theories also revealed that they may exist in different *phases*, in which the observables have different behaviors with a specific particle spectrum in each phase. In this way, to understand what a given QFT predicted became very similar to the study of phases in problems in statistical physics. We will explore these connections in detail later in this book when we develop the machinery of the renormalization group in chapter 15. In this picture, the *vacuum* (or ground state) of a QFT corresponds to a *phase*, much in the same way as in statistical (or condensed matter) physics.

While the requirement of renormalizability works for the Standard Model of particle physics, it fails for gravity. The problem of unifying gravity with the rest of the forces of Nature remains a major problem in contemporary physics. A major program to solve this problem is string theory. String theory is the only known viable candidate to quantize gravity in a consistent manner. However, in string theory, QFT is seen as an effective low-energy (hydrodynamic) description of nature, and the QFT singularities are "regulated" by string theory in a natural way (but at the price of locality).

2

Classical Field Theory

In what follows, we will consider rather general field theories. The only guiding principles that we will use in constructing these theories are (a) symmetries and (b) a generalized least action principle.

2.1 Relativistic invariance

In chapter 1, we considered three examples of relativistic wave equations. They are the Maxwell equations for classical electromagnetism, the Klein-Gordon equation, and the Dirac equation. Maxwell's equations govern the dynamics of a *vector* field, the vector potentials $A^\mu(x) = (A^0(x), \boldsymbol{A}(x))$, whereas the Klein-Gordon equation describes excitations of a *scalar* field $\phi(x)$, and the Dirac equation governs the behavior of the four-component spinor field $\psi_\alpha(x)$, $(\alpha = 0, 1, 2, 3)$. Each of these fields transforms in a very definite way under the group of Lorentz transformations (the *Lorentz group*). The Lorentz group is defined as a group of linear transformations Λ of Minkowski spacetime $\mathcal{M}$ onto itself, $\Lambda : \mathcal{M} \mapsto \mathcal{M}$, such that the new coordinates are related to the old ones by a linear (Lorentz) transformation

$$x'^\mu = \Lambda^\mu_\nu x^\nu \tag{2.1}$$

The spacetime components of a Lorentz transformation, Λ^0_i, are the *Lorentz boosts*. Lorentz boosts relate inertial reference frames moving at relative velocity $\boldsymbol{v}$ with respect to each other. Lorentz boosts along the x^1-axis have the familiar form

$$\begin{aligned} x^{0\prime} &= \frac{x^0 + v x^1/c}{\sqrt{1 - v^2/c^2}} \\ x^{1\prime} &= \frac{x^1 + v x^0/c}{\sqrt{1 - v^2/c^2}} \\ x^{2\prime} &= x^2 \\ x^{3\prime} &= x^3 \end{aligned} \tag{2.2}$$

where $x^0 = ct$, $x^1 = x$, $x^2 = y$, and $x^3 = z$ (note: the superscripts indicate components, not powers!). If we use the notation $\gamma = (1 - v^2/c^2)^{-1/2} \equiv \cosh\alpha$, we can write the Lorentz boost as a matrix:

$$\begin{pmatrix} x^{0\prime} \\ x^{1\prime} \\ x^{2\prime} \\ x^{3\prime} \end{pmatrix} = \begin{pmatrix} \cosh\alpha & \sinh\alpha & 0 & 0 \\ \sinh\alpha & \cosh\alpha & 0 & 0 \\ 0 & 0 & 1 & 0 \\ 0 & 0 & 0 & 1 \end{pmatrix} \begin{pmatrix} x^0 \\ x^1 \\ x^2 \\ x^3 \end{pmatrix} \tag{2.3}$$

The space components of Λ^i_j are conventional rotations R of three-dimensional Euclidean space.

Infinitesimal Lorentz transformations are generated by the hermitian operators

$$L_{\mu\nu} = i(x_\mu \partial_\nu - x_\nu \partial_\mu) \tag{2.4}$$

where $\partial_\mu = \frac{\partial}{\partial x^\mu}$, and $\mu, \nu = 0, 1, 2, 3$. The infinitesimal generators $L_{\mu\nu}$ satisfy the algebra

$$[L_{\mu\nu}, L_{\rho\sigma}] = ig_{\nu\rho} L_{\mu\sigma} - ig_{\mu\rho} L_{\nu\sigma} - ig_{\nu\sigma} L_{\mu\rho} + ig_{\mu\sigma} L_{\nu\rho} \tag{2.5}$$

where $g_{\mu\nu}$ is the metric tensor for flat Minkowski spacetime (see eq. 2.8). This is the algebra of the Lie group $SO(3,1)$. Actually, any operator of the form

$$M_{\mu\nu} = L_{\mu\nu} + S_{\mu\nu} \tag{2.6}$$

where the $S_{\mu\nu}$ are 4×4 matrices satisfying the algebra of eq. (2.5) are also generators of $SO(3,1)$, defined in eq. (2.122).

Lorentz transformations form a *group*, since (a) the product of two Lorentz transformations is a Lorentz transformation, (b) there exists an identity transformation, and (c) Lorentz transformations are invertible. Notice, however, that in general, two transformations do not commute with each other. Hence, the Lorentz group is non-abelian.

The Lorentz group has the *defining property* of leaving invariant the relativistic interval

$$x^2 \equiv x_0^2 - \boldsymbol{x}^2 = c^2 t^2 - \boldsymbol{x}^2 \tag{2.7}$$

The group of Euclidean rotations leaves invariant the Euclidean distance $\boldsymbol{x}^2$ and is a subgroup of the Lorentz group. The rotation group is denoted by $SO(3)$, and the Lorentz group is denoted by $SO(3,1)$. This notation emphasizes that the signature of the metric has one $+$ sign and three $-$ signs.

The group $SO(3,1)$ of linear transformations is noncompact in the following sense. Let us consider first the group of rotations in three-dimensional space, $SO(3)$. The linear transformations in $SO(3)$ leave the Euclidean distance (squared) $R^2 = x_1^2 + x_2^2 + x_3^2$ invariant. The set of points with a fixed value of R is the two-dimensional surface of a sphere of radius R in three dimensions, which we will denote by S_2. The elements of the group of rotations $SO(3)$ are in one-to-one correspondence with the points on S_2. The area of a 2-sphere S_2 of unit radius is 4π. Then we say that the "volume" of the group $SO(3)$ is finite and equal to 4π. A group of linear transformations with finite volume is said to be *compact*. In contrast, the Lorentz group is the set of linear transformations, denoted by $SO(3,1)$, that leave the relativistic interval $x_\mu x^\mu$ invariant, which is not positive definite. As we well know, Lorentz boosts map points along hyperbolas of Minkowski spacetime. In this sense, the Lorentz group is noncompact, since its "volume" is infinite.

We will adopt the following conventions and definitions:

1) *Metric tensor*: We will use the standard ("Bjorken and Drell") metric for Minkowski spacetime, in which the metric tensor $g_{\mu\nu}$ is

$$g_{\mu\nu} = g^{\mu\nu} = \begin{pmatrix} 1 & 0 & 0 & 0 \\ 0 & -1 & 0 & 0 \\ 0 & 0 & -1 & 0 \\ 0 & 0 & 0 & -1 \end{pmatrix} \tag{2.8}$$

In this notation, the infinitesimal relativistic interval is

$$ds^2 = dx^\mu dx_\mu = g_{\mu\nu} dx^\mu dx^\nu = dx_0^2 - d\boldsymbol{x}^2 = c^2 dt^2 - d\boldsymbol{x}^2 \tag{2.9}$$

2) *4-vectors*:

   i) x^μ is a *contravariant* 4-vector, $x^\mu = (ct, \boldsymbol{x})$
   ii) x_μ is a *covariant* 4-vector, $x_\mu = (ct, -\boldsymbol{x})$
   iii) Covariant and contravariant vectors (and tensors) are related through the metric tensor $g_{\mu\nu}$

$$A^\mu = g^{\mu\nu} A_\nu \tag{2.10}$$

   iv) $\boldsymbol{x}$ is a vector in $\mathbb{R}^3$
   v) $p^\mu = \left(\frac{E}{c}, \boldsymbol{p}\right)$ is the energy-momentum 4-vector. Hence, $p_\mu p^\mu = \frac{E^2}{c^2} - \boldsymbol{p}^2$ is a Lorentz scalar.

3) *Scalar product*:

$$p \cdot q = p_\mu q^\mu = p_0 q_0 - \boldsymbol{p} \cdot \boldsymbol{q} \equiv p_\mu q_\nu g^{\mu\nu} \tag{2.11}$$

4) *Gradients*: $\partial_\mu \equiv \frac{\partial}{\partial x^\mu}$ and $\partial^\mu \equiv \frac{\partial}{\partial x_\mu}$. We define the d'Alambertian ∂^2 as

$$\partial^2 \equiv \partial^\mu \partial_\mu \equiv \frac{1}{c^2} \partial_t^2 - \nabla^2 \tag{2.12}$$

which is a Lorentz scalar. From now on, we will use units of time $[T]$ and length $[L]$ such that $\hbar = c = 1$. Thus, $[T] = [L]$, and we will use units like centimeters (or any other unit of length).

5) *Interval*: The interval in Minkowski spacetime is x^2:

$$x^2 = x_\mu x^\mu = x_0^2 - \boldsymbol{x}^2 \tag{2.13}$$

Time-like intervals have $x^2 > 0$, while space-like intervals have $x^2 < 0$ (see figure 2.1).

Since a field is a function (or mapping) of Minkowski space onto some other (properly chosen) space, it is natural to require that the fields should have simple transformation properties under Lorentz transformations. For example, the vector potential $A^\mu(x)$ transforms like a 4-vector under Lorentz transformations, that is, if $x'^\mu = \Lambda^\mu_\nu x^\nu$, then $A'^\mu(x') = \Lambda^\mu_\nu A^\nu(x)$. In other words, A^μ transforms like x^μ. Thus, it is a vector. All vector fields have this property. In contrast, a *scalar* field $\Phi(x)$ remains *invariant* under Lorentz transformations:

$$\Phi'(x') = \Phi(x) \tag{2.14}$$

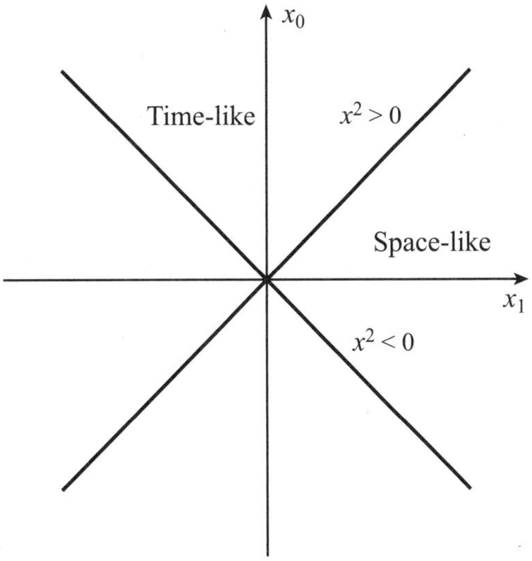

Figure 2.1 The Minkowski spacetime and its light cone. Events at a relativistic interval with $x^2 = x_0^2 - \boldsymbol{x}^2 > 0$ are time-like (and are causally connected with the origin), while events with $x^2 = x_0^2 - \boldsymbol{x}^2 < 0$ are space-like and are not causally connected with the origin.

A 4-spinor $\psi_\alpha(x)$ transforms under Lorentz transformations. Namely, there exists an induced 4×4 linear transformation matrix $S(\Lambda)$ such that

$$S(\Lambda^{-1}) = S^{-1}(\Lambda) \tag{2.15}$$

and

$$\Psi'(\Lambda x) = S(\Lambda)\Psi(x) \tag{2.16}$$

In section 2.5, we give an explicit expression for $S(\Lambda)$.

2.2 The Lagrangian, the action, and the least action principle

The evolution of any dynamical system is determined by its Lagrangian. In the classical mechanics of systems of particles described by the generalized coordinates q, the Lagrangian L is a differentiable function of the coordinates q and their time derivatives. L must be differentiable, since otherwise, the equations of motion would not be local in time (i.e., could not be written in terms of differential equations). An argument à la Landau and Lifshitz (1959a) enables us to "derive" the Lagrangian. For example, for a particle in free space, the homogeneity, uniformity, and isotropy of space and time require that L be only a function of the absolute value of the velocity $|\boldsymbol{v}|$. Since $|\boldsymbol{v}|$ is not a differentiable function of $\boldsymbol{v}$, the Lagrangian must be a function of $\boldsymbol{v}^2$. Thus, $L = L(\boldsymbol{v}^2)$. In principle, there is no reason to assume that L cannot be a function of the acceleration $\boldsymbol{a}$ (or rather, $\boldsymbol{a}^2$) or of its higher derivatives. Experiment tells us that in classical mechanics, it is sufficient to specify the initial position $\boldsymbol{x}(0)$ of a particle and its initial velocity $\boldsymbol{v}(0)$ to determine the time evolution of the system. Thus we have to choose

$$L(v^2) = \text{const} + \frac{1}{2}mv^2 \tag{2.17}$$

The additive constant is irrelevant in classical physics. Naturally, the coefficient of v^2 is just one-half of the inertial mass.

However, in special relativity, the natural invariant quantity to consider is not the Lagrangian but the *action* S. For a free particle, the relativistic invariant (i.e., Lorentz invariant) action must involve the invariant interval, the *proper length* $ds = c\sqrt{1-\frac{v^2}{c^2}}\,dt$. Hence, we write the action for a relativistic massive particle as

$$S = -mc\int_{s_i}^{s_f} ds = -mc^2\int_{t_i}^{t_f} dt\,\sqrt{1-\frac{v^2}{c^2}} \tag{2.18}$$

The relativistic Lagrangian is then

$$L = -mc^2\sqrt{1-\frac{v^2}{c^2}} \tag{2.19}$$

As a power series expansion, it contains all powers of v^2/c^2. It is elementary to see that, as expected, the canonical momentum $\boldsymbol{p}$ is

$$\boldsymbol{p} = \frac{\partial L}{\partial \boldsymbol{v}} = \frac{m\boldsymbol{v}}{\sqrt{1-\frac{v^2}{c^2}}} \tag{2.20}$$

from which it follows that the Hamiltonian (or energy) is given by

$$H = \frac{mc^2}{\sqrt{1-\frac{v^2}{c^2}}} = \sqrt{p^2c^2 + m^2c^4} \tag{2.21}$$

as it should be.

Once the Lagrangian is found, the classical equations of motion are determined by the *least action principle* (figure 2.2). Thus, we construct the action S

$$S = \int dt\, L\,(q, \dot{q}) \tag{2.22}$$

where $\dot{q} = \frac{dq}{dt}$, and demand that the physical trajectories $q(t)$ leave the action S stationary (i.e., $\delta S = 0$). The variation of S is

$$\delta S = \int_{t_i}^{t_f} dt\,\left(\frac{\partial L}{\partial q}\delta q + \frac{\partial L}{\partial \dot{q}}\delta \dot{q}\right) \tag{2.23}$$

Integrating by parts, we get

$$\delta S = \int_{t_i}^{t_f} dt\,\frac{d}{dt}\left(\frac{\partial L}{\partial \dot{q}}\delta q\right) + \int_{t_i}^{t_f} dt\,\delta q\left[\frac{\partial L}{\partial q} - \frac{d}{dt}\left(\frac{\partial L}{\partial \dot{q}}\right)\right] \tag{2.24}$$

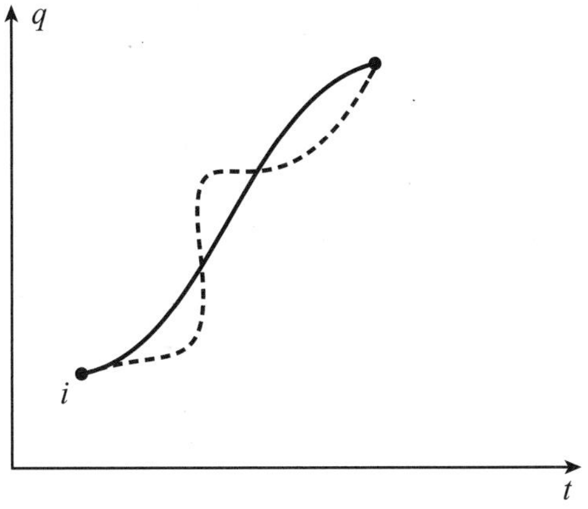

Figure 2.2 The least action principle: The solid curve is the classical trajectory and extremizes the classical action. The dashed curve represents a variation.

Hence, we get

$$\delta S = \frac{\partial L}{\partial \dot{q}} \delta q \bigg|_{t_i}^{t_f} + \int_{t_i}^{t_f} dt \, \delta q \left[\frac{\partial L}{\partial q} - \frac{d}{dt} \left(\frac{\partial L}{\partial \dot{q}} \right) \right] \tag{2.25}$$

If we assume that the variation δq is an arbitrary function of time that vanishes at the initial and final times t_i and t_f (i.e., $\delta q(t_i) = \delta q(t_f) = 0$), we find that $\delta S = 0$ if and only if the integrand of eq. (2.25) vanishes identically. Thus,

$$\frac{\partial L}{\partial q} - \frac{d}{dt} \left(\frac{\partial L}{\partial \dot{q}} \right) = 0 \tag{2.26}$$

These are the equations of motion or *Newton's equations*. In general, the equation determining the trajectories that leave the action stationary is called the *Euler-Lagrange equation*.

2.3 Scalar field theory

For the case of a field theory, we can proceed very much in the same way. Let us consider first the case of a scalar field $\Phi(x)$. The action S must be invariant under Lorentz transformations. Since we want to construct *local* theories, it is natural to assume that S is given in terms of a Lagrangian *density* $\mathcal{L}$:

$$S = \int d^4x \, \mathcal{L} \tag{2.27}$$

where $\mathcal{L}$ is a *local* differentiable function of the field and its derivatives. These assumptions in turn guarantee that the resulting equations of motion have the form of partial differential equations. In other words, the dynamics does not allow for action at a distance.

Since the volume element of Minkowski space d^4x is invariant under Lorentz transformations, the action S is invariant if $\mathcal{L}$ is a local, differentiable function of Lorentz invariants that can be constructed out of the field $\Phi(x)$. Simple invariants are $\Phi(x)$ itself and all of its powers. The gradient $\partial^\mu \Phi \equiv \frac{\partial \Phi}{\partial x_\mu}$ is not an invariant, but the d'Alambertian $\partial^2 \Phi$ is. Bilinears, such as $\partial_\mu \Phi \partial^\mu \Phi$, are also Lorentz invariant as well as under a change of the sign of Φ. So we can write the following simple expression for $\mathcal{L}$:

$$\mathcal{L} = \frac{1}{2}\partial_\mu \Phi \partial^\mu \Phi - V(\Phi) \tag{2.28}$$

where $V(\Phi)$ is some potential, which we can assume is a polynomial function of the field Φ. Let us consider the simple choice

$$V(\Phi) = \frac{1}{2}\bar{m}^2 \Phi^2 \tag{2.29}$$

where $\bar{m} = mc/\hbar$. Thus,

$$\mathcal{L} = \frac{1}{2}\partial_\mu \Phi \partial^\mu \Phi - \frac{1}{2}\bar{m}^2 \Phi^2 \tag{2.30}$$

This is the Lagrangian density for a *free scalar field*. Notice, in passing, that we could have added a term like $\partial^2 \Phi$. However, this term, in addition to being odd under $\Phi \to -\Phi$, is a *total divergence* and, as such, it has an effect only on the boundary conditions and not on the equations of motion. In what follows, unless stated to the contrary, we will not consider surface terms.

The least action principle requires that S be stationary under arbitrary variations of the field Φ and of its derivatives $\partial_\mu \Phi$. Thus, we get

$$\delta S = \int d^4x \left[\frac{\delta \mathcal{L}}{\delta \Phi} \delta \Phi + \frac{\delta \mathcal{L}}{\delta \partial_\mu \Phi} \delta \partial_\mu \Phi \right] \tag{2.31}$$

Notice that since $\mathcal{L}$ is a *functional* of Φ, we have to use functional derivatives (i.e., partial derivatives at each point of spacetime). On integrating by parts, we get

$$\delta S = \int d^4x\, \partial_\mu \left(\frac{\delta \mathcal{L}}{\delta \partial_\mu \Phi} \delta \Phi \right) + \int d^4x\, \delta \Phi \left[\frac{\delta \mathcal{L}}{\delta \Phi} - \partial_\mu \left(\frac{\delta \mathcal{L}}{\delta \partial_\mu \Phi} \right) \right] \tag{2.32}$$

Instead of considering initial and final conditions, we now have to imagine that the field Φ is contained in some very large box of spacetime. The term with the total divergence yields a surface contribution. We will consider field configurations such that $\delta \Phi = 0$ on that surface. Thus, the Euler-Lagrange equations are

$$\frac{\delta \mathcal{L}}{\delta \Phi} - \partial_\mu \left(\frac{\delta \mathcal{L}}{\delta \partial_\mu \Phi} \right) = 0 \tag{2.33}$$

More explicitly, we find

$$\frac{\delta \mathcal{L}}{\delta \Phi} = -\frac{\partial V}{\partial \Phi} \tag{2.34}$$

and, since

$$\frac{\partial \mathcal{L}}{\delta \partial_\mu \Phi} = \partial^\mu \Phi \tag{2.35}$$

then

$$\partial_\mu \frac{\delta \mathcal{L}}{\delta \partial_\mu \Phi} = \partial_\mu \partial^\mu \Phi \equiv \partial^2 \Phi \tag{2.36}$$

By direct substitution, we get the equation of motion (or *field equation*):

$$\partial^2 \Phi + \frac{\partial V}{\partial \Phi} = 0 \tag{2.37}$$

For the choice

$$V(\Phi) = \frac{\bar{m}^2}{2}\Phi^2, \quad \frac{\partial V}{\partial \Phi} = \bar{m}^2 \Phi \tag{2.38}$$

the field equation is

$$\left(\partial^2 + \bar{m}^2\right)\Phi = 0 \tag{2.39}$$

where $\partial^2 = \frac{1}{c^2}\frac{\partial^2}{\partial t^2} - \nabla^2$. Thus, we find that the equation of motion for the free massive scalar field Φ is

$$\frac{1}{c^2}\frac{\partial^2 \Phi}{\partial t^2} - \nabla^2 \Phi + \bar{m}^2 \Phi = 0 \tag{2.40}$$

This is precisely the Klein-Gordon equation if the constant $\bar{m}$ is identified with $\frac{mc}{\hbar}$. Indeed, the plane-wave solutions of these equations are

$$\Phi = \Phi_0 e^{i(p_0 x_0 - \boldsymbol{p}\cdot\boldsymbol{x})/\hbar} \tag{2.41}$$

where p_0 and $\boldsymbol{p}$ are related through the dispersion law:

$$p_0^2 = \boldsymbol{p}^2 c^2 + m^2 c^4 \tag{2.42}$$

Thus for each momentum $\boldsymbol{p}$, there are two solutions, one with *positive frequency* and one with *negative frequency*. We will see in section 4.3 that, in the quantized theory, the energy of the excitation is indeed equal to $|p_0|$. Notice that $\frac{1}{\bar{m}} = \frac{\hbar}{mc}$ has units of length and is equal to the Compton wavelength for a particle of mass m. From now on (unless it is stated to the contrary), we will use units in which $\hbar = c = 1$ and $m = \bar{m}$.

2.4 Classical field theory in the canonical formalism

In classical mechanics, it is often convenient to use the canonical formulation in terms of a Hamiltonian instead of the Lagrangian approach. For the case of a system of particles, the canonical formalism proceeds as follows. Given a Lagrangian $L(q, \dot{q})$, a canonical momentum p is defined to be

$$\frac{\partial L}{\partial \dot{q}} = p \tag{2.43}$$

The classical Hamiltonian $H(p, q)$ is defined by the Legendre transformation

$$H(p, q) = p\dot{q} - L(q, \dot{q}) \tag{2.44}$$

If the Lagrangian L is quadratic in the velocities $\dot{q}$ and separable, for example,

$$L = \frac{1}{2}m\dot{q}^2 - V(q) \tag{2.45}$$

then $H(p, q)$ is simply given by

$$H(p,q) = p\dot{q} - \left(\frac{m\dot{q}^2}{2} - V(q)\right) = \frac{p^2}{2m} + V(q) \tag{2.46}$$

where

$$p = \frac{\partial L}{\partial \dot{q}} = m\dot{q} \tag{2.47}$$

The (conserved) quantity H is then identified with the total energy of the system.

In this language, the least action principle becomes

$$\delta S = \delta \int L\, dt = \delta \int [p\dot{q} - H(p,q)]\, dt = 0 \tag{2.48}$$

Hence

$$\int dt \left(\delta p\, \dot{q} + p\, \delta\dot{q} - \delta p \frac{\partial H}{\partial p} - \delta q \frac{\partial H}{\partial q}\right) = 0 \tag{2.49}$$

On integration by parts, we get

$$\int dt \left[\delta p \left(\dot{q} - \frac{\partial H}{\partial p}\right) + \delta q \left(-\frac{\partial H}{\partial q} - \dot{p}\right)\right] = 0 \tag{2.50}$$

which can only be satisfied for arbitrary variations $\delta q(t)$ and $\delta p(t)$ if

$$\dot{q} = \frac{\partial H}{\partial p} \qquad \dot{p} = -\frac{\partial H}{\partial q} \tag{2.51}$$

These are Hamilton's equations.

Let us introduce the Poisson bracket $\{A, B\}_{PB}$ of two functions A and B of q and p by

$$\{A, B\}_{PB} \equiv \frac{\partial A}{\partial q}\frac{\partial B}{\partial p} - \frac{\partial A}{\partial p}\frac{\partial B}{\partial q} \tag{2.52}$$

Let $F(q, p, t)$ be some differentiable function of q, p, and t. Then the total time variation of F is

$$\frac{dF}{dt} = \frac{\partial F}{\partial t} + \frac{\partial F}{\partial q}\frac{dq}{dt} + \frac{\partial F}{\partial p}\frac{dp}{dt} \tag{2.53}$$

Using Hamilton's equations, we get the result

$$\frac{dF}{dt} = \frac{\partial F}{\partial t} + \frac{\partial F}{\partial q}\frac{\partial H}{\partial p} - \frac{\partial F}{\partial p}\frac{\partial H}{\partial q} \tag{2.54}$$

or, in terms of Poisson brackets:

$$\frac{dF}{dt} = \frac{\partial F}{\partial t} + \{F, H\}_{PB} \tag{2.55}$$

In particular,

$$\frac{dq}{dt} = \frac{\partial H}{\partial p} = \frac{\partial q}{\partial q}\frac{\partial H}{\partial p} - \frac{\partial q}{\partial p}\frac{\partial H}{\partial q} = \{q, H\}_{PB} \tag{2.56}$$

since

$$\frac{\partial q}{\partial p} = 0 \qquad \text{and} \qquad \frac{\partial q}{\partial q} = 1 \tag{2.57}$$

Also, the total rate of change of the canonical momentum p is

$$\frac{dp}{dt} = \frac{\partial p}{\partial q}\frac{\partial H}{\partial p} - \frac{\partial p}{\partial p}\frac{\partial H}{\partial q} \equiv -\frac{\partial H}{\partial q} \tag{2.58}$$

since $\frac{\partial p}{\partial q} = 0$ and $\frac{\partial p}{\partial p} = 1$. Thus,

$$\frac{dp}{dt} = \{p, H\}_{PB} \tag{2.59}$$

Notice that, for an isolated system, H is time independent. So,

$$\frac{\partial H}{\partial t} = 0 \tag{2.60}$$

and

$$\frac{dH}{dt} = \frac{\partial H}{\partial t} + \{H, H\}_{PB} = 0 \tag{2.61}$$

since

$$\{H, H\}_{PB} = 0 \tag{2.62}$$

Therefore, H can be regarded as the generator of infinitesimal time translations. Since it is conserved for an isolated system, for which $\frac{\partial H}{\partial t} = 0$, we can indeed identify H with the total energy. In passing, let us also notice that the above definition of the Poisson bracket implies that q and p satisfy

$$\{q, p\}_{PB} = 1 \tag{2.63}$$

This relation is fundamental for the quantization of these systems.

Much of this formulation can be generalized to the case of fields. Let us first discuss the canonical formalism for the case of a scalar field Φ with Lagrangian density $\mathcal{L}(\Phi, \partial_\mu, \Phi)$. We will *choose* $\Phi(x)$ to be the (infinite) set of *canonical coordinates*. The *canonical momentum* $\Pi(x)$ is defined by

$$\Pi(x) = \frac{\delta \mathcal{L}}{\delta \partial_0 \Phi(x)} \tag{2.64}$$

If the Lagrangian is *quadratic* in $\partial_\mu \Phi$, the canonical momentum $\Pi(x)$ is simply given by

$$\Pi(x) = \partial_0 \Phi(x) \equiv \dot{\Phi}(x) \tag{2.65}$$

The Hamiltonian *density* $\mathcal{H}(\Phi, \Pi)$ is a local function of $\Phi(x)$ and $\Pi(x)$; it is given by

$$\mathcal{H}(\Phi, \Pi) = \Pi(x)\, \partial_0 \Phi(x) - \mathcal{L}(\Phi, \partial_0 \Phi) \tag{2.66}$$

If the Lagrangian density $\mathcal{L}$ has the simple form

$$\mathcal{L} = \frac{1}{2}(\partial_\mu \Phi)^2 - V(\Phi) \tag{2.67}$$

then the Hamiltonian density $\mathcal{H}(\Phi, \Pi)$ is

$$\mathcal{H} = \Pi\dot{\Phi} - \mathcal{L}(\Phi, \dot{\Phi}, \partial_j\Phi) \equiv \frac{1}{2}\Pi^2(x) + \frac{1}{2}(\nabla\Phi(x))^2 + V(\Phi(x)) \tag{2.68}$$

which is explicitly a positive-definite quantity if the potential $V(\Phi) \geq 0$. This is the case for a free massive scalar field where $V(\Phi) = \frac{m^2}{2}\Phi^2$. Thus, the energy of a plane-wave solution of a massive scalar field theory (i.e., a solution of the Klein-Gordon equation) is always positive, no matter the sign of the *frequency*. In fact, the lowest energy state is simply $\Phi =$ constant. A solution made of linear superpositions of plane waves (i.e., a wave packet) has positive energy. Therefore, in *field theory, the energy is always positive*. We will see that, in the quantized theory, in the case of a complex field, the negative frequency solutions are identified with antiparticle states, and their existence does not signal a possible instability of the theory.

The canonical field $\Phi(x)$ and the canonical momentum $\Pi(x)$ satisfy the equal-time Poisson bracket (PB) relations:

$$\{\Phi(\boldsymbol{x}, x_0), \Pi(\boldsymbol{y}, x_0)\}_{PB} = \delta(\boldsymbol{x} - \boldsymbol{y}) \tag{2.69}$$

where $\delta(\boldsymbol{x})$ is the Dirac δ function and the Poisson bracket $\{A, B\}_{PB}$ is defined to be

$$\{A, B\}_{PB} = \int d^3x \left[\frac{\delta A}{\delta\Phi(\boldsymbol{x}, x_0)} \frac{\delta B}{\delta\Pi(\boldsymbol{x}, x_0)} - \frac{\delta A}{\delta\Pi(\boldsymbol{x}, x_0)} \frac{\delta B}{\delta\Phi(\boldsymbol{x}, x_0)} \right] \tag{2.70}$$

for any two functionals A and B of $\Phi(x)$ and $\Pi(x)$. This approach can be extended to theories other than that of a scalar field without too much difficulty. We will come back to these issues when considering the problem of quantization in section 4.2. Finally, note that while Lorentz invariance is apparent in the Lagrangian formulation, it is not so in the Hamiltonian formulation of a classical field.

2.5 Field theory of the Dirac equation

We now turn to the problem of a field theory for spinors. We will discuss the theory of spinors as a classical field theory. It turns out that this theory is not consistent unless it is properly quantized as a QFT of spinors. We will return to this problem in chapter 7.

Let us rewrite the Dirac equation

$$i\hbar\frac{\partial\Psi}{\partial t} = \frac{\hbar c}{i}\boldsymbol{\alpha}\cdot\nabla\Psi + \beta mc^2\,\Psi \equiv H_{\text{Dirac}}\Psi \tag{2.71}$$

in a manner so that relativistic covariance is apparent. The operator H_{Dirac} is the Dirac Hamiltonian.

We first recall that the 4×4 hermitian matrices $\boldsymbol{\alpha}$ and β should satisfy the algebra

$$\{\alpha_i, \alpha_j\} = 2\delta_{ij}\, I, \qquad \{\alpha_i, \beta\} = 0, \qquad \alpha_i^2 = \beta^2 = I \tag{2.72}$$

where I is the 4×4 identity matrix.

A simple representation of this algebra is the set of 2 × 2 block (Dirac) matrices

$$\alpha^i = \begin{pmatrix} 0 & \sigma^i \\ \sigma^i & 0 \end{pmatrix} \qquad \beta = \begin{pmatrix} I & 0 \\ 0 & -I \end{pmatrix} \tag{2.73}$$

where the σ^i matrices are the three 2 × 2 Pauli matrices,

$$\sigma^1 = \begin{pmatrix} 0 & 1 \\ 1 & 0 \end{pmatrix} \quad \sigma^2 = \begin{pmatrix} 0 & -i \\ i & 0 \end{pmatrix} \quad \sigma^3 = \begin{pmatrix} 1 & 0 \\ 0 & -1 \end{pmatrix} \tag{2.74}$$

and I is the 2 × 2 identity matrix. This is the *Dirac representation* of the Dirac algebra.

It is now convenient to introduce the Dirac gamma matrices:

$$\gamma^0 = \beta \qquad \gamma^i = \beta\alpha^i \tag{2.75}$$

The Dirac gamma matrices γ^μ have the block form

$$\gamma^0 = \beta = \begin{pmatrix} I & 0 \\ 0 & -I \end{pmatrix}, \qquad \gamma^i = \begin{pmatrix} 0 & \sigma^i \\ -\sigma^i & 0 \end{pmatrix} \tag{2.76}$$

and obey the Dirac algebra

$$\{\gamma^\mu, \gamma^\nu\} = 2g^{\mu\nu}\, I \tag{2.77}$$

where I is the 4 × 4 identity matrix.

In terms of the gamma matrices, the Dirac equation takes the much simpler, covariant, form

$$\left(i\gamma^\mu\partial_\mu - \frac{mc}{\hbar}\right)\Psi = 0 \tag{2.78}$$

where Ψ is a 4-spinor. It is also customary to introduce the notation (known as "Feynman's slash")

$$\not{a} \equiv a_\mu\gamma^\mu \tag{2.79}$$

Using Feynman's slash, we can write the Dirac equation in the form

$$\left(i\not{\partial} - \frac{mc}{\hbar}\right)\Psi = 0 \tag{2.80}$$

From now on, we will use units in which $\hbar = c = 1$. In these units, energy has units of length^{-1} and time has units of length.

Notice that, if Ψ satisfies the Dirac equation, then it also satisfies

$$(i\not{\partial} + m)(i\not{\partial} - m)\Psi = 0 \tag{2.81}$$

Also, we find that

$$\begin{aligned} \not{\partial} \cdot \not{\partial} &= \partial_\mu\partial_\nu\gamma^\mu\gamma^\nu = \partial_\mu\partial_\nu\left(\frac{1}{2}\{\gamma^\mu, \gamma^\nu\} + \frac{1}{2}[\gamma^\mu\gamma^\nu]\right) \\ &= \partial_\mu\partial_\nu g^{\mu\nu} = \partial^2 \end{aligned} \tag{2.82}$$

where we used that the commutator $[\gamma^\mu, \gamma^\nu]$ is antisymmetric in the indices μ and ν. As a result, we find that each component of the 4-spinor Ψ must also satisfy the Klein-Gordon equation

$$\left(\partial^2 + m^2\right) \Psi = 0 \tag{2.83}$$

2.5.1 Solutions of the Dirac equation

Let us briefly discuss the properties of the solutions of the Dirac equation. First consider solutions representing particles at rest. Thus, Ψ must be constant in space, and all its space derivatives must vanish. The Dirac equation becomes

$$i\gamma^0 \frac{\partial \Psi}{\partial t} = m\Psi \tag{2.84}$$

where $t = x_0$ $(c = 1)$. Let us introduce the *bispinors* ϕ and χ

$$\Psi = \begin{pmatrix} \phi \\ \chi \end{pmatrix} \tag{2.85}$$

We find that the Dirac equation reduces to a simple system of two 2×2 equations:

$$i\frac{\partial \phi}{\partial t} = +m\phi, \qquad i\frac{\partial \chi}{\partial t} = -m\chi \tag{2.86}$$

The four linearly independent solutions are

$$\phi_1 = e^{-imt} \begin{pmatrix} 1 \\ 0 \end{pmatrix} \quad \phi_2 = e^{-imt} \begin{pmatrix} 0 \\ 1 \end{pmatrix} \tag{2.87}$$

and

$$\chi_1 = e^{imt} \begin{pmatrix} 1 \\ 0 \end{pmatrix} \quad \chi_2 = e^{imt} \begin{pmatrix} 0 \\ 1 \end{pmatrix} \tag{2.88}$$

Thus, the upper component ϕ represents the solutions with positive energy $+m$, while χ represents the solutions with negative energy $-m$. The additional two-fold degeneracy of the solutions is related to the spin of the particle, as we will see below.

More generally, in terms of the bispinors ϕ and χ, the Dirac equation takes the form

$$i\frac{\partial \phi}{\partial t} = m\phi + \frac{1}{i}\boldsymbol{\sigma} \cdot \boldsymbol{\nabla}\chi \tag{2.89}$$

$$i\frac{\partial \chi}{\partial t} = -m\chi + \frac{1}{i}\boldsymbol{\sigma} \cdot \boldsymbol{\nabla}\phi \tag{2.90}$$

Furthermore, in the nonrelativistic limit, taking formally $c \to \infty$, it should reduce to the Schrödinger-Pauli equation. To see this, we define the slowly varying amplitudes $\tilde{\phi}$ and $\tilde{\chi}$:

$$\begin{aligned} \phi &= e^{-imt}\tilde{\phi} \\ \chi &= e^{-imt}\tilde{\chi} \end{aligned} \tag{2.91}$$

The field $\tilde{\chi}$ is small and nearly static. We will now see that the field $\tilde{\phi}$ describes solutions with positive energies close to $+m$. In terms of $\tilde{\phi}$ and $\tilde{\chi}$, the Dirac equation becomes

$$i\frac{\partial \tilde{\phi}}{\partial t} = \frac{1}{i}\boldsymbol{\sigma} \cdot \nabla \tilde{\chi} \tag{2.92}$$

$$i\frac{\partial \tilde{\chi}}{\partial t} = -2m\tilde{\chi} + \frac{1}{i}\boldsymbol{\sigma} \cdot \nabla \tilde{\phi} \tag{2.93}$$

Indeed, in this limit, the left side of eq. (2.93) is much smaller than its right side. Thus we can approximate:

$$2m\tilde{\chi} \approx \frac{1}{i}\boldsymbol{\sigma} \cdot \nabla \tilde{\phi} \tag{2.94}$$

We can now eliminate the "small component" $\tilde{\chi}$ from eq. (2.92) to find that $\tilde{\phi}$ satisfies

$$i\frac{\partial \tilde{\phi}}{\partial t} = -\frac{1}{2m}\nabla^2 \tilde{\phi} \tag{2.95}$$

which is indeed the Schrödinger-Pauli equation.

2.5.2 Conserved current

Let us introduce one last bit of useful notation. Define $\bar{\Psi}$ by

$$\bar{\Psi} = \Psi^\dagger \gamma^0 \tag{2.96}$$

in terms of which we can write down the 4-vector j^μ:

$$j^\mu = \bar{\Psi}\gamma^\mu \Psi \tag{2.97}$$

which is conserved, that is,

$$\partial_\mu j^\mu = 0 \tag{2.98}$$

Notice that the time component of j^μ is the density

$$j^0 = \bar{\Psi}\gamma^0\Psi \equiv \Psi^\dagger \Psi \tag{2.99}$$

and that the space components of j^μ are

$$\boldsymbol{j} = \bar{\Psi}\boldsymbol{\gamma}\Psi = \Psi^\dagger\gamma^0\boldsymbol{\gamma}\Psi = \Psi^\dagger\boldsymbol{\alpha}\Psi \tag{2.100}$$

Thus the Dirac equation has an associated 4-vector field, $j^\mu(x)$, which is conserved and obeys a local continuity equation:

$$\partial_0 j_0 + \nabla \cdot \boldsymbol{j} = 0 \tag{2.101}$$

However, it is easy to see that in general, the density j_0 can be positive or negative. Hence this current cannot be associated with a probability current (as in nonrelativistic quantum mechanics). Instead we will see that it is associated with the *charge* density and current.

2.5.3 Relativistic covariance

Let Λ be a Lorentz transformation. Let $\Psi(x)$ be a spinor field in an inertial frame, and $\Psi'(x')$ be the same Dirac spinor field in the transformed frame. The Dirac equation is *covariant* if the Lorentz transformation

$$x'_\mu = \Lambda^{\ \nu}_{\mu} x_\nu \tag{2.102}$$

induces a linear transformation $S(\Lambda)$ in *spinor space*,

$$\Psi'_\alpha(x') = S(\Lambda)_{\alpha\beta}\Psi_\beta(x) \tag{2.103}$$

such that the transformed Dirac equation has the same form as the original equation in the original frame. That is, we will require that if

$$\left(i\gamma^\mu \frac{\partial}{\partial x^\mu} - m\right)_{\alpha\beta} \Psi_\beta(x) = 0, \qquad \text{then} \qquad \left(i\gamma^\mu \frac{\partial}{\partial x'^\mu} - m\right)_{\alpha\beta} \Psi'_\beta(x') = 0 \tag{2.104}$$

Notice two important facts: (1) both the field Ψ and the coordinate x change under the action of the Lorentz transformation, and (2) the gamma matrices and the mass m do not change under a Lorentz transformation. Thus, the gamma matrices are independent of the choice of a reference frame. However, they do depend on the choice of the basis in spinor space.

What properties should the representation matrices $S(\Lambda)$ have? Let us first observe that if $x'^\mu = \Lambda^\mu_\nu x^\nu$, then

$$\frac{\partial}{\partial x'^\mu} = \frac{\partial x^\nu}{\partial x'^\mu}\frac{\partial}{\partial x^\nu} \equiv \left(\Lambda^{-1}\right)^\nu_{\ \mu}\frac{\partial}{\partial x^\nu} \tag{2.105}$$

Thus, $\frac{\partial}{\partial x^\mu}$ is a covariant vector. By substituting this transformation law back into the Dirac equation, we find

$$i\gamma^\mu \frac{\partial}{\partial x'^\mu}\Psi'(x') = i\gamma^\mu (\Lambda^{-1})^\nu{}_\mu \frac{\partial}{\partial x^\nu}(S(\Lambda)\Psi(x)) \tag{2.106}$$

Thus, the Dirac equation now reads

$$i\gamma^\mu \left(\Lambda^{-1}\right)_\mu^{\ \nu} S(\Lambda)\frac{\partial \Psi}{\partial x^\nu} - mS\left(\Lambda\right)\Psi = 0 \tag{2.107}$$

Or, equivalently,

$$S^{-1}\left(\Lambda\right) i\gamma^\mu \left(\Lambda^{-1}\right)_\mu{}^\nu S\left(\Lambda\right)\frac{\partial \Psi}{\partial x^\nu} - m\Psi = 0 \tag{2.108}$$

Therefore, covariance holds provided $S\left(\Lambda\right)$ satisfies the identity

$$S^{-1}\left(\Lambda\right)\gamma^\mu S\left(\Lambda\right)\left(\Lambda^{-1}\right)_\mu^{\ \nu} = \gamma^\nu \tag{2.109}$$

Since the set of Lorentz transformations forms a group, the representation matrices $S(\Lambda)$ should also form a group. In particular, it must be true that the property

$$S^{-1}\left(\Lambda\right) = S\left(\Lambda^{-1}\right) \tag{2.110}$$

holds. Recall that the invariance of the relativistic interval $x^2 = x_\mu x^\mu$ implies that Λ must obey

$$\Lambda^\nu{}_\mu \Lambda_\nu{}^\lambda = g_\mu{}^\lambda \equiv \delta_\mu{}^\lambda \tag{2.111}$$

Hence,

$$\Lambda_\mu{}^\nu = \left(\Lambda^{-1}\right)_\nu{}^\mu \tag{2.112}$$

So we can rewrite eq. (2.109) as

$$S(\Lambda)\, \gamma^\mu S(\Lambda)^{-1} = \left(\Lambda^{-1}\right)^\mu{}_\nu \gamma^\nu \tag{2.113}$$

Eq. (2.113) shows that a Lorentz transformation induces a *similarity* transformation on the gamma matrices, which is equivalent to (the inverse of) a Lorentz transformation. From this equation, it follows that, for the case of Lorentz boosts, eq. (2.113) shows that the matrices $S(\Lambda)$ are *hermitian*. Instead, for the subgroup $SO(3)$ of rotations about a fixed origin, the matrices $S(\Lambda)$ are *unitary*. These different properties follow from the fact that the matrices $S(\Lambda)$ are a representation of the Lorentz group $SO(3, 1)$, which is a noncompact Lie group.

We will now find the form of $S(\Lambda)$ for an infinitesimal Lorentz transformation. Since the identity transformation is $\Lambda^\mu_\nu = g^\mu_\nu$, a Lorentz transformation infinitesimally close to the identity should have the form

$$\Lambda^\mu_\nu = g^\mu_\nu + \omega^\mu_\nu, \quad \text{and} \quad \left(\Lambda^{-1}\right)^\mu_\nu = g^\mu_\nu - \omega^\mu_\nu \tag{2.114}$$

where $\omega^{\mu\nu}$ is infinitesimal and antisymmetric in its spacetime indices:

$$\omega^{\mu\nu} = -\omega^{\nu\mu} \tag{2.115}$$

Let us parametrize $S(\Lambda)$ in terms of a 4×4 matrix $\sigma_{\mu\nu}$, which is also antisymmetric in its indices (i.e., $\sigma_{\mu\nu} = -\sigma_{\nu\mu}$). Then we can write

$$\begin{aligned} S(\Lambda) &= I - \frac{i}{4}\sigma_{\mu\nu}\omega^{\mu\nu} + \cdots \\ S^{-1}(\Lambda) &= I + \frac{i}{4}\sigma_{\mu\nu}\omega^{\mu\nu} + \cdots \end{aligned} \tag{2.116}$$

where I stands for the 4×4 identity matrix. If we substitute back into eq. (2.113), we get

$$(I - \frac{i}{4}\sigma_{\mu\nu}\omega^{\mu\nu} + \cdots)\gamma^\lambda (I + \frac{i}{4}\sigma_{\alpha\beta}\omega^{\alpha\beta} + \cdots) = \gamma^\lambda - \omega^\lambda_\nu \gamma^\nu + \cdots \tag{2.117}$$

Collecting all the terms linear in ω, we obtain

$$\frac{i}{4}\left[\gamma^\lambda, \sigma_{\mu\nu}\right]\omega^{\mu\nu} = \omega^\lambda_\nu \gamma^\nu \tag{2.118}$$

Or equivalently, the matrices $\sigma_{\mu\nu}$ must obey

$$[\gamma^\mu, \sigma_{\nu\lambda}] = 2i(g^\mu_\nu \gamma_\lambda - g^\mu_\lambda \gamma_\nu) \tag{2.119}$$

This matrix equation has the solution

$$\sigma_{\mu\nu} = \frac{i}{2}[\gamma_\mu, \gamma_\nu] \tag{2.120}$$

Under a finite Lorentz transformation $x' = \Lambda x$, the 4-spinors transform as

$$\Psi'(x') = S(\Lambda)\,\Psi \tag{2.121}$$

with

$$S(\Lambda) = \exp\left(-\frac{i}{4}\sigma_{\mu\nu}\omega^{\mu\nu}\right) \tag{2.122}$$

The matrices $\sigma_{\mu\nu}$ are the *generators* of the group of Lorentz transformations in the spinor representation. From this solution, we see that the space components σ_{jk} are *hermitian* matrices, while the spacetime components σ_{0j} are *antihermitian.* This feature is telling us that the Lorentz group is not a *compact* unitary group, since in that case, all of its generators would be hermitian matrices. Instead, this result tells us that the Lorentz group is isomorphic to the *noncompact* group $SO(3,1)$. Thus, the representation matrices $S(\Lambda)$ are unitary only under space rotations with fixed origin.

The linear operator $S(\Lambda)$ gives the field in the *transformed frame* in terms of the coordinates of the *transformed frame*. However, we may also wish to ask for the transformation $U(\Lambda)$ that *compensates* for the effect of the coordinate transformation. In other words, we seek a matrix $U(\Lambda)$ such that

$$\Psi'(x) = U(\Lambda)\,\Psi(x) = S(\Lambda)\,\Psi(\Lambda^{-1}x) \tag{2.123}$$

For an infinitesimal Lorentz transformation, we seek a matrix $U(\Lambda)$ of the form

$$U(\Lambda) = I - \frac{i}{2}J_{\mu\nu}\omega^{\mu\nu} + \cdots \tag{2.124}$$

and we wish to find an expression for $J_{\mu\nu}$. We find

$$\begin{aligned}\left(I - \frac{i}{2}J_{\mu\nu}\omega^{\mu\nu} + \cdots\right)\Psi &= \left(I - \frac{i}{4}\sigma_{\mu\nu}\omega^{\mu\nu} + \cdots\right)\Psi\left(x^\rho - \omega^\rho_\nu x^\nu + \cdots\right) \\ &\cong \left(I - \frac{i}{4}\sigma_{\mu\nu}\omega^{\mu\nu} + \cdots\right)\left(\Psi - \partial_\rho\Psi\,\omega^\rho_\nu x^\nu + \cdots\right)\end{aligned} \tag{2.125}$$

Hence,

$$\Psi'(x) \cong \left(I - \frac{i}{4}\sigma_{\mu\nu}\omega^{\mu\nu} + x_\mu\omega^{\mu\nu}\partial_\nu + \cdots\right)\Psi(x) \tag{2.126}$$

From this expression, we see that $J_{\mu\nu}$ is given by the operator

$$J_{\mu\nu} = \frac{1}{2}\sigma_{\mu\nu} + i(x_\mu\partial_\nu - x_\nu\partial_\mu) \tag{2.127}$$

Note that the second term is the *orbital* angular momentum operator (we will come back to this issue shortly). The first term is then interpreted as the *spin*.

In fact, let us consider purely spatial rotations, whose infinitesimal generators are the space components of $J_{\mu\nu}$, that is,

$$J_{jk} = i(x_j\partial_k - x_k\partial_j) + \frac{1}{2}\sigma_{jk} \tag{2.128}$$

We can also define a three-component *vector* J_ℓ as the three-dimensional dual of J_{jk}:

$$J_{jk} = \epsilon_{jkl}J_\ell \tag{2.129}$$

where ϵ^{ijk} is the third-rank Levi-Civita tensor:

$$\epsilon^{ijk} = \begin{cases} 1 & \text{if } (ijk) \text{ is an even permutation of } (123) \\ -1 & \text{if } (ijk) \text{ is an odd permutation of } (123) \\ 0 & \text{otherwise} \end{cases} \tag{2.130}$$

Thus, we get (after restoring the factors of $\hbar$):

$$\begin{aligned} J_\ell &= \frac{i\hbar}{2}\epsilon_{\ell jk}(x_j\partial_k - x_k\partial_j) + \frac{\hbar}{4}\epsilon_{\ell jk}\sigma_{jk} \\ &= i\,\hbar\,\epsilon_{\ell jk}\,x_j\partial_k + \frac{\hbar}{2}\left(\frac{1}{2}\epsilon_{\ell jk}\sigma_{jk}\right) \\ J_\ell &\equiv \left(\boldsymbol{x}\times\hat{\boldsymbol{p}}\right)_\ell + \frac{\hbar}{2}\sigma_\ell \end{aligned} \tag{2.131}$$

The first term is clearly the orbital angular momentum, and the second term can be regarded as the spin. With this definition, it is straightforward to check that the spinors (ϕ, χ), which are solutions of the Dirac equation, carry spin 1/2.

2.5.4 Transformation properties of field bilinears in the Dirac theory

We will now consider the transformation properties of several physical observables of the Dirac theory under Lorentz transformations. Let Λ be a general Lorentz transformation, and $S(\Lambda)$ be the induced transformation for the Dirac spinors $\Psi_a(x)$ (with $a = 1, \ldots, 4$):

$$\Psi'_a(x') = S(\Lambda)_{ab}\,\Psi_b(x) \tag{2.132}$$

Using the properties of the induced Lorentz transformation $S(\Lambda)$ and of the Dirac gamma matrices, it is straightforward to verify that the Dirac bilinears obey the following transformation laws.

Scalar:

$$\bar{\Psi}'(x')\,\Psi'(x') = \bar{\Psi}(x)\,\Psi(x) \tag{2.133}$$

which transforms as a *scalar*.

Pseudoscalar: Define the Dirac matrix $\gamma_5 = i\gamma_0\gamma_1\gamma_2\gamma_3$. Then the bilinear

$$\bar{\Psi}'(x')\,\gamma_5\,\Psi'(x') = \det\Lambda\;\bar{\Psi}(x)\,\gamma_5\,\Psi(x) \tag{2.134}$$

transforms as a *pseudoscalar.*

Vector: Likewise,

$$\bar{\Psi}'(x')\,\gamma^\mu\,\Psi'(x') = \Lambda^\mu_\nu\;\bar{\Psi}(x)\,\gamma^\nu\,\Psi(x) \tag{2.135}$$

transforms as a *vector.*

Pseudovector:

$$\bar{\Psi}'(x')\,\gamma_5\gamma^\mu\,\Psi'(x') = \det\Lambda\;\Lambda^\mu_\nu\;\bar{\Psi}(x)\,\gamma_5\gamma^\nu\,\Psi(x) \tag{2.136}$$

transforms as a *pseudovector.*

Tensor: Finally,

$$\bar{\Psi}'(x')\,\sigma^{\mu\nu}\,\Psi'(x') = \Lambda^\mu_\alpha\,\Lambda^\nu_\beta\;\bar{\Psi}(x)\,\sigma^{\alpha\beta}\,\Psi(x) \tag{2.137}$$

transforms as a *tensor.*

In the above equations, Λ^μ_ν denotes a Lorentz transformation, and $\det\Lambda$ is its determinant. We have also used that

$$S^{-1}(\Lambda)\gamma_5 S(\Lambda) = \det\Lambda\;\gamma_5 \tag{2.138}$$

Also, note that the Dirac algebra provides a natural basis for the space of 4×4 matrices, which we denote by

$$\Gamma^S \equiv I, \quad \Gamma^V_\mu \equiv \gamma_\mu, \quad \Gamma^T_{\mu\nu} \equiv \sigma_{\mu\nu}, \quad \Gamma^A_\mu \equiv \gamma_5\gamma_\mu, \quad \Gamma^P = \gamma_5 \tag{2.139}$$

where S, V, T, A, and P stand for scalar, vector, tensor, axial vector (or pseudovector), and parity, respectively. For future reference, note the following useful trace identities obeyed by products of Dirac gamma matrices:

$$\operatorname{tr} I = 4, \quad \operatorname{tr}\gamma_\mu = \operatorname{tr}\gamma_5 = 0, \quad \operatorname{tr}\gamma_\mu\gamma_\nu = 4g_{\mu\nu} \tag{2.140}$$

Also, if we denote by a_μ and b_μ two arbitrary 4-vectors, then

$$\not{a}\,\not{b} = a_\mu b^\mu - i\sigma_{\mu\nu}\,a^\mu b^\nu, \qquad \text{and} \quad \operatorname{tr}\left(\not{a}\,\not{b}\right) = 4a\cdot b \tag{2.141}$$

2.5.5 The Dirac Lagrangian

We now seek a Lagrangian density $\mathcal{L}$ for the Dirac theory. It should be a local differentiable Lorentz-invariant functional of the spinor field Ψ. Since the Dirac equation is first order in derivatives and it is Lorentz covariant, the Lagrangian should be Lorentz invariant and first order in derivatives. A simple choice is

$$\mathcal{L} = \bar{\Psi}(i\not{\partial} - m)\Psi \equiv \frac{1}{2}\bar{\Psi}i\overleftrightarrow{\not{\partial}}\,\Psi - m\bar{\Psi}\Psi \tag{2.142}$$

where $\bar{\Psi}\overleftrightarrow{\not{\partial}}\,\Psi \equiv \bar{\Psi}(\not{\partial}\Psi) - (\partial_\mu\bar{\Psi})\gamma^\mu\Psi$. This choice satisfies all the requirements.

The equations of motion are derived in the usual manner, that is, by demanding that the action $S=\int d^4x\,\mathcal{L}$ be stationary:

$$\delta S=0=\int d^4x\left[\frac{\delta\mathcal{L}}{\delta\Psi_\alpha}\delta\Psi_\alpha+\frac{\delta\mathcal{L}}{\delta\partial_\mu\Psi_\alpha}\delta\partial_\mu\Psi_\alpha+(\Psi\leftrightarrow\bar{\Psi})\right] \tag{2.143}$$

The equations of motion are

$$\begin{aligned}\frac{\delta\mathcal{L}}{\delta\Psi_\alpha}-\partial_\mu\frac{\delta\mathcal{L}}{\delta\partial_\mu\Psi_\alpha}&=0\\ \frac{\delta\mathcal{L}}{\delta\bar{\Psi}_\alpha}-\partial_\mu\frac{\delta\mathcal{L}}{\delta\partial_\mu\bar{\Psi}_\alpha}&=0\end{aligned} \tag{2.144}$$

By direct substitution, we find

$$(i\partial\!\!\!/-m)\Psi=0,\quad\text{and}\quad\bar{\Psi}(i\overleftarrow{\partial\!\!\!/}+m)=0 \tag{2.145}$$

Here, $\overleftarrow{\partial\!\!\!/}$ indicates that the derivatives are acting *on the left*.

Finally, we can also write down the Hamiltonian density that follows from the Lagrangian of eq. (2.142). As usual, we need to determine the canonical momentum conjugate to the field Ψ:

$$\Pi(x)=\frac{\delta\mathcal{L}}{\delta\partial_0\Psi(x)}=i\bar{\Psi}(x)\gamma^0\equiv i\Psi^\dagger(x) \tag{2.146}$$

Thus the Hamiltonian density is

$$\begin{aligned}\mathcal{H}&=\Pi(x)\partial_0\Psi(x)-\mathcal{L}=i\bar{\Psi}\gamma^0\partial_0\Psi-\mathcal{L}\\ &=\bar{\Psi}i\boldsymbol{\gamma}\cdot\nabla\Psi+m\bar{\Psi}\Psi\\ &=\Psi^\dagger\left(i\boldsymbol{\alpha}\cdot\nabla+m\beta\right)\Psi\end{aligned} \tag{2.147}$$

Thus we find that the "one-particle" Dirac Hamiltonian H_{Dirac} of eq. (2.71) appears naturally in the field theory as well.

Since the Hamiltonian of eq. (2.147) is first order in derivatives, unlike its Klein-Gordon relative, it is not manifestly positive. Thus, there is a question of the stability of this theory. We will see below that the proper quantization of this theory as a QFT of *fermions* solves this problem. In other words, it will be necessary to *impose* the Pauli principle for this theory to describe a stable system with an energy spectrum that is bounded from below. In this way, we will see that there is a natural connection between the *spin* of the field and the *statistics*. This connection, which actually is an *axiom* of QFT, is known as the *spin-statistics theorem*.

2.6 Classical electromagnetism as a field theory

We now turn to the problem of the electromagnetic field generated by a set of sources. Let $\rho(x)$ and $\boldsymbol{j}(x)$ represent the charge and current densities, respectively, at a point x of spacetime. Charge conservation requires that a continuity equation has to be obeyed:

$$\frac{\partial \rho}{\partial t} + \nabla \cdot \boldsymbol{j} = 0 \tag{2.148}$$

Given an initial condition, which specifies the values of the electric field $\boldsymbol{E}(x)$ and the magnetic field $\boldsymbol{B}(x)$ at some t_0 in the past, the time evolution is governed by the Maxwell equations:

$$\nabla \cdot \boldsymbol{E} = \rho \qquad\qquad \nabla \cdot \boldsymbol{B} = 0 \tag{2.149}$$

$$\nabla \times \boldsymbol{B} - \frac{1}{c}\frac{\partial \boldsymbol{E}}{\partial t} = \boldsymbol{j} \qquad\qquad \nabla \times \boldsymbol{E} + \frac{1}{c}\frac{\partial \boldsymbol{B}}{\partial t} = 0 \tag{2.150}$$

It is possible to reformulate classical electrodynamics in a manner in which (1) relativistic covariance is apparent, and (2) the Maxwell equations follow from a least action principle. A convenient way to see the above is to define the electromagnetic field tensor $F^{\mu\nu}$, which is the (contravariant) antisymmetric real tensor whose components are given by

$$F^{\mu\nu} = -F^{\nu\mu} = \begin{pmatrix} 0 & -E^1 & -E^2 & -E^3 \\ E^1 & 0 & -B^3 & B^2 \\ E^2 & B^3 & 0 & -B^1 \\ E^3 & -B^2 & B^1 & 0 \end{pmatrix} \tag{2.151}$$

Or, equivalently,

$$\begin{aligned} F^{0i} &= -F^{i0} = -E^i \\ F^{ij} &= -F^{ji} = \epsilon^{ijk} B^k \end{aligned} \tag{2.152}$$

The *dual tensor* $\widetilde{F}_{\mu\nu}$ is defined as follows:

$$\widetilde{F}^{\mu\nu} = -\widetilde{F}^{\nu\mu} = \frac{1}{2}\epsilon^{\mu\nu\rho\sigma} F_{\rho\sigma} \tag{2.153}$$

where $\epsilon^{\mu\nu\rho\sigma}$ is the fourth-rank Levi-Civita tensor, defined similarly to the third-rank Levi-Civita tensor of eq. (2.130). In particular,

$$\widetilde{F}^{\mu\nu} = \begin{pmatrix} 0 & -B^1 & -B^2 & -B^3 \\ B^1 & 0 & E^3 & -E^2 \\ B^2 & -E^3 & 0 & E^2 \\ B^3 & E^2 & -E^2 & 0 \end{pmatrix} \tag{2.154}$$

With these notations, we can rewrite the left column of Maxwell's equations, eq. (2.150), in the more compact and manifestly covariant form

$$\partial_\mu F^{\mu\nu} = j^\nu \tag{2.155}$$

which we will interpret as the equations of motion of the electromagnetic field. The right column of the Maxwell equations, eq. (2.150), becomes the *constraint*

$$\partial_\mu \widetilde{F}^{\mu\nu} = 0 \tag{2.156}$$

which is known as the Bianchi identity. Then consistency of the Maxwell equations requires that the continuity equation

$$\partial_\mu j^\mu = 0 \tag{2.157}$$

be satisfied.

By inspection, we see that the field tensor $F^{\mu\nu}$ and the dual field tensor $\widetilde{F}^{\mu\nu}$ map into each other on exchanging the electric and magnetic fields. This *electromagnetic duality* would be an exact property of electrodynamics if in addition to the electric charge current j^μ, the Bianchi identity (eq. (2.156)) included a magnetic charge current (of magnetic monopoles).

At this point, it is convenient to introduce the vector potential A^μ, whose contravariant components are

$$A^\mu(x) = \left(\frac{A^0}{c}, \boldsymbol{A}\right) \tag{2.158}$$

The current 4-vector $j^\mu(x)$ is

$$j^\mu(x) = (\rho c, \boldsymbol{j}) \equiv (j^0, \boldsymbol{j}) \tag{2.159}$$

The electric field strength $\boldsymbol{E}$ and the magnetic field $\boldsymbol{B}$ are defined to be

$$\boldsymbol{E} = -\frac{1}{c}\nabla A^0 - \frac{1}{c}\frac{\partial \boldsymbol{A}}{\partial t}, \qquad \boldsymbol{B} = \nabla \times \boldsymbol{A} \tag{2.160}$$

In a more compact, relativistically covariant notation, we write

$$F^{\mu\nu} = \partial^\mu A^\nu - \partial^\nu A^\mu \tag{2.161}$$

In terms of the vector potential A^μ, the Maxwell equations remain unchanged under the (local) *gauge transformation*:

$$A^\mu(x) \mapsto A^\mu(x) + \partial^\mu \Phi(x) \tag{2.162}$$

where $\Phi(x)$ is an *arbitrary smooth* function of the spacetime coordinates x^μ. It is easy to check that, under the transformation of eq. (2.162), the field strength remains invariant (i.e., $F^{\mu\nu} \mapsto F^{\mu\nu}$). This property is called *gauge invariance*, and it plays a fundamental role in modern physics.

By directly substituting the definitions of the magnetic field $\boldsymbol{B}$ and the electric field $\boldsymbol{E}$ in terms of the 4-vector A^μ into the Maxwell equations, we obtain the wave equation. Indeed, the equation of motion

$$\partial_\mu F^{\mu\nu} = j^\nu \tag{2.163}$$

yields the equation for the vector potential:

$$\partial^2 A^\nu - \partial^\nu(\partial_\mu A^\mu) = j^\nu \tag{2.164}$$

which is the wave equation.

We can now use gauge invariance to further restrict the vector potential A^μ, which is not completely determined. These restrictions are known as the procedure of *fixing a gauge*. The choice

$$\partial_\mu A^\mu = 0 \tag{2.165}$$

known as the Lorentz gauge, yields the simpler and standard form of the wave equation:

$$\partial^2 A^\mu = j^\mu \tag{2.166}$$

Notice that the Lorentz gauge preserves Lorentz covariance.

Another popular choice is the radiation (or Coulomb) gauge:

$$\nabla \cdot \boldsymbol{A} = 0 \tag{2.167}$$

which yields (in units with $c = 1$)

$$\partial^2 A^\nu - \partial^\nu(\partial_0 A^0) = j^\nu \tag{2.168}$$

which is not Lorentz covariant. In the absence of external sources, $j^\nu = 0$, we can further impose the restriction that $A^0 = 0$. This choice reduces the set of three equations, one for each spatial component of $\boldsymbol{A}$, which satisfy

$$\partial^2 \boldsymbol{A} = 0, \quad \text{provided} \quad \nabla \cdot \boldsymbol{A} = 0 \tag{2.169}$$

The solutions are, as we well know, plane waves of the form

$$\boldsymbol{A}(x) = \boldsymbol{A}\, e^{i(p_0 x_0 - \boldsymbol{p}\cdot\boldsymbol{x})} \tag{2.170}$$

which are only consistent if $p_0^2 - \boldsymbol{p}^2 = 0$ and $\boldsymbol{p} \cdot \boldsymbol{A} = 0$. For the last reason, this choice is also known as the *transverse* gauge.

We can also regard the electromagnetic field as a dynamical system and construct a Lagrangian picture for it. Since the Maxwell equations are local, gauge invariant, and Lorentz covariant, we should demand that the Lagrangian density also be local, gauge invariant, and Lorentz invariant. Since, in terms of the vector potential A_μ, the Maxwell equations are second order in derivatives, we seek a local Lagrangian density that is also second order in derivatives of the vector potential. A simple choice that satisfies all the requirements is

$$\mathcal{L} = -\frac{1}{4} F_{\mu\nu} F^{\mu\nu} - j_\mu A^\mu \tag{2.171}$$

This Lagrangian density is manifestly Lorentz invariant. Gauge invariance is satisfied if and only if j_μ is a conserved current (i.e., if $\partial_\mu j^\mu = 0$) since under a gauge transformation $A_\mu \mapsto A_\mu + \partial_\mu \Phi(x)$, the field strength does not change. However, the source term changes as follows:

$$\begin{aligned}\int d^4x\, j_\mu A^\mu \mapsto & \int d^4x \left[j_\mu A^\mu + j_\mu \partial^\mu \Phi \right] \\ &= \int d^4x\, j_\mu A^\mu + \int d^4x\, \partial^\mu (j_\mu \Phi) - \int d^4x\, \partial^\mu j_\mu\, \Phi\end{aligned} \tag{2.172}$$

If the sources vanish at infinity ($\lim_{|x|\to\infty} j_\mu = 0$), then the surface term can be dropped. Thus the action $S = \int d^4x \mathcal{L}$ is gauge invariant if and only if the current j^μ is locally conserved,

$$\partial_\mu j^\mu = 0 \tag{2.173}$$

which is the continuity equation.

We can now derive the equations of motion by demanding that the action S be stationary:

$$\delta S = \int d^4x \left[\frac{\delta \mathcal{L}}{\delta A^\mu} \delta A^\mu + \frac{\delta \mathcal{L}}{\delta \partial^\nu A^\mu} \delta \partial^\nu A^\mu \right] = 0 \tag{2.174}$$

Much as we did before, we can now proceed to integrate by parts to get

$$\delta S = \int d^4x\, \partial^\nu \left[\frac{\delta \mathcal{L}}{\delta \partial^\nu A^\mu} \delta A^\mu \right] + \int d^4x\, \delta A^\mu \left[\frac{\delta \mathcal{L}}{\delta A^\mu} - \partial^\nu \left(\frac{\delta \mathcal{L}}{\delta \partial^\nu A^\mu} \right) \right] \tag{2.175}$$

By demanding that at the surface, the variation vanishes ($\delta A^\mu = 0$), we get

$$\frac{\delta \mathcal{L}}{\delta A^\mu} = \partial^\nu \left(\frac{\delta \mathcal{L}}{\delta \partial^\nu A^\mu} \right) \tag{2.176}$$

Explicitly, we find

$$\frac{\delta \mathcal{L}}{\delta A^\mu} = -j_\mu, \quad \text{and} \quad \frac{\delta \mathcal{L}}{\delta \partial^\nu A^\mu} = F^{\mu\nu} \tag{2.177}$$

Thus, we obtain

$$j^\mu = -\partial_\nu F^{\mu\nu} \tag{2.178}$$

or, equivalently,

$$j^\nu = \partial_\mu F^{\mu\nu} \tag{2.179}$$

Therefore, the least action principle *implies* the Maxwell equations.

2.7 The Landau theory of phase transitions as a field theory

We now turn to the problem of the statistical mechanics of a magnet. To be a little more specific, we consider the simplest *model* of a ferromagnet: the classical Ising model. In this model, one considers an array of atoms on some lattice (say, cubic). Each site is assumed to have a net spin magnetic moment $\mathbf{S}(i)$. From elementary quantum mechanics, we know that the simplest interaction among the spins is the Heisenberg exchange Hamiltonian,

$$H = -\sum_{<i,j>} J_{ij} \mathbf{S}(i) \cdot \mathbf{S}(j) \tag{2.180}$$

where $< i, j >$ are nearest neighboring sites on the lattice (see figure 2.3). In many situations, in which there is magnetic anisotropy, only the z-component of the spin operators plays a role. The Hamiltonian now reduces to that of the Ising model:

$$H_I = -J \sum_{<ij>} \sigma(i)\sigma(j) \equiv E[\sigma] \tag{2.181}$$

where $[\sigma]$ denotes a configuration of spins with $\sigma(i)$ being the z-projection of the spin at each site i.

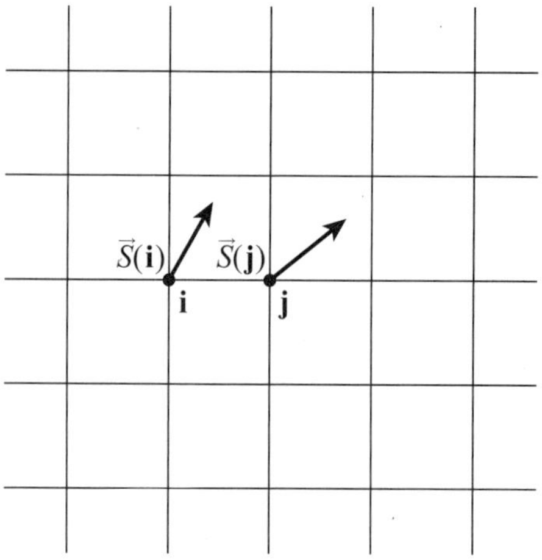

Figure 2.3 Spins on a lattice.

The equilibrium properties of the system are determined by the partition function Z, which is the sum over all spin configurations $[\sigma]$ of the Boltzmann weight for each state,

$$Z = \sum_{\{\sigma\}} \exp\left(-\frac{E[\sigma]}{T}\right) \tag{2.182}$$

where T is the temperature, and $\{\sigma\}$ is the set of all spin configurations.

In the 1950s, Landau developed a picture to study these types of problems, which in general are very difficult. Landau first proposed to work not with the microscopic spins but with a set of *coarse-grained* configurations (Landau, 1937). One way to do this, using the more modern version of the argument due to Kadanoff (Kadanoff, 1966) and Wilson (Wilson, 1971), is to partition a large system of linear size L into regions or *blocks* of smaller linear size ℓ such that $a_0 \ll \ell \ll L$, where a_0 is the lattice spacing. Each one of these regions will be centered around a site, say, $\boldsymbol{x}$. We will denote such a region by $\mathcal{A}(\boldsymbol{x})$. The idea is now to perform the sum (i.e., the partition function Z), while keeping the total magnetization of each region $\mathcal{A}(\boldsymbol{x})$ fixed at the values

$$M(\boldsymbol{x}) = \frac{1}{N[\mathcal{A}]} \sum_{y \in \mathcal{A}(\boldsymbol{x})} \sigma(\boldsymbol{y}) \tag{2.183}$$

where $N[\mathcal{A}]$ is the number of sites in $\mathcal{A}(\boldsymbol{x})$. The restricted partition function is now a *functional* of the coarse-grained local magnetizations $M(\boldsymbol{x})$:

$$Z[M] = \sum_{\{\sigma\}} \exp\left\{-\frac{E[\sigma]}{T}\right\} \prod_{x} \delta\left(M(\boldsymbol{x}) - \frac{1}{N(\mathcal{A})} \sum_{y \in \mathcal{A}(x)} \sigma(\boldsymbol{y})\right) \tag{2.184}$$

The variables $M(\boldsymbol{x})$ have the property that, for $N(\mathcal{A})$ very large, they effectively take values on the real numbers. Also, the coarse-grained configurations $\{M(\boldsymbol{x})\}$ are smoother than the microscopic configurations $\{\sigma\}$.

At very high temperatures, the average magnetization $\langle M \rangle = 0$ and the system is in a paramagnetic phase. In contrast, at low temperatures, the average magnetization can be nonzero, and the system may be in a ferromagnetic phase. Thus, at high temperatures, the partition function Z is dominated by configurations that have $\langle M \rangle = 0$, while at very low temperatures, the most frequent configurations have $\langle M \rangle \neq 0$. Landau proceeded to write down an approximate form for the partition function in terms of sums over *smooth, continuous* configurations $M(\boldsymbol{x})$, which formally, can be represented in the form

$$Z \approx \int \mathcal{D}M(\boldsymbol{x}) \exp\left(-\frac{E\left[M(\boldsymbol{x}), T\right]}{T} \right) \tag{2.185}$$

where $\mathcal{D}M(\boldsymbol{x})$ is an integration measure that means "sum over all configurations." We will define this more properly later on, in chapter 15.

Assume that for the dominant configurations in the sum (integral!) shown in eq. (2.185), the local averages $M(\boldsymbol{x})$ are smooth functions of $\boldsymbol{x}$ and are small. Then the energy functional $E[M]$ can be written as an expansion in powers of $M(\boldsymbol{x})$ and of its space derivatives. With these assumptions, the free energy of the magnet in D dimensions can be approximated by the Ginzburg-Landau form (Landau and Lifshitz, 1959a):

$$E(M) = \int d^D x \left(\frac{1}{2} K(T) |\nabla M(\boldsymbol{x})|^2 + \frac{1}{2} a(T) M^2(\boldsymbol{x}) + \frac{1}{4!} b(T) M^4(\boldsymbol{x}) + \cdots \right) \tag{2.186}$$

Ginzburg and Landau made the additional (and drastic) assumption that there is a *single configuration* $M(\boldsymbol{x})$ that dominates the full partition function. Under this assumption, which can be regarded as a mean-field approximation, the free energy $F = -T \ln Z$ takes the same form as eq. (2.186). In chapter 12, we will develop the theory of the effective action and return to this point.

Thermodynamic stability requires that the stiffness $K(T)$ and the coefficient $b(T)$ of the quartic term must be positive. The second term has a coefficient $a(T)$, which can have either sign. A simple choice of these parameters is

$$K(T) \simeq K_0, \qquad b(T) \simeq b_0, \qquad a(T) \simeq \bar{a}\,(T - T_c) \tag{2.187}$$

where T_c is an approximate value of the critical temperature.

The *free energy* $F(M)$ defines a *classical*, or *Euclidean*, field theory. In fact, by rescaling the field $M(x)$ in the form

$$\Phi(x) = \sqrt{K} M(x) \tag{2.188}$$

we can write the free energy as

$$F(\Phi) = \int d^d x \left\{ \frac{1}{2} (\nabla \Phi)^2 + U(\Phi) \right\} \tag{2.189}$$

where the potential $U(\Phi)$ is

$$U(\Phi) = \frac{\bar{m}^2}{2} \Phi^2 + \frac{\lambda}{4!} \Phi^4 + \cdots \tag{2.190}$$

where $\bar{m}^2 = \frac{a(T)}{K}$, and $\lambda = \frac{b}{K^2}$. Except for the absence of the term involving the canonical momentum $\Pi^2(x)$, $F(\Phi)$ has a striking resemblance to the Hamiltonian of a scalar field

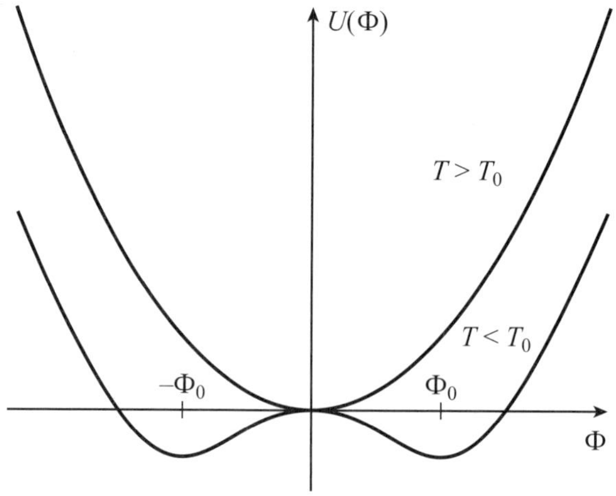

Figure 2.4 The Landau free energy for the order-parameter field Φ: For $T > T_0$, the free energy has a unique minimum at $\Phi = 0$, while for $T < T_0$, there are two minima at $\Phi = \pm\Phi_0$.

in Minkowski space! We will see in section 2.8 that this is not an accident and that the Ginzburg-Landau theory is closely related to a QFT of a scalar field with a Φ^4 potential.

Let us now ask: Is there a configuration $\Phi_c(\boldsymbol{x})$ that gives the *dominant* contribution to the partition function Z? If so, we should be able to approximate

$$Z = \int \mathcal{D}\Phi \ \exp\{-F(\Phi)\} \approx \exp\{-F(\Phi_c)\}\{1 + \cdots\} \tag{2.191}$$

This statement is usually called the *mean field approximation*. Since the integrand is an exponential, the dominant configuration Φ_c must be such that F has a (local) minimum at Φ_c. Thus, configurations Φ_c that leave $F(\Phi)$ *stationary* are good candidates (we actually need local minima!). The problem of finding extrema is simply the condition $\delta F = 0$. This is the same problem we solved for classical field theory in Minkowski spacetime. Notice that in the derivation of F, we have invoked essentially the same type of arguments as before: (1) invariance and (2) locality (differentiability).

The Euler-Lagrange equations can be derived by using the same arguments that we employed in the context of a scalar field theory. In the case at hand, they become

$$-\frac{\delta F}{\delta\Phi(\boldsymbol{x})} + \partial_j\left(\frac{\delta F}{\delta\partial_j\Phi(\boldsymbol{x})}\right) = 0 \tag{2.192}$$

For the case of the Landau theory, the Euler-Lagrange equation becomes the Ginzburg-Landau equation:

$$0 = -\nabla^2\,\Phi_c(\boldsymbol{x}) + \bar{m}^2\Phi_c(\boldsymbol{x}) + \frac{\lambda}{3!}\Phi_c^3(\boldsymbol{x}) \tag{2.193}$$

The solution $\Phi_c(\boldsymbol{x})$ that *minimizes* the energy is uniform in space and thus has $\partial_j\Phi_c = 0$. Hence, Φ_c is the solution of the very simple equation

$$\bar{m}^2\Phi_c + \frac{\lambda}{3!}\Phi_c^3 = 0 \tag{2.194}$$

Since λ is positive and $\bar{m}^2$ may have either sign (depending on whether $T > T_c$ or $T < T_c$), we have to explore both cases (see figure 2.4).

For $T > T_c$, $\bar{m}^2$ is also positive, and the only real solution is $\Phi_c = 0$. This is the paramagnetic phase. But, for $T < T_c$, $\bar{m}^2$ is negative, and two new solutions are available:

$$\Phi_c = \pm\sqrt{\frac{6 \mid \bar{m}^2 \mid}{\lambda}} \tag{2.195}$$

These are the solutions with lowest energy, and they are *degenerate*. They both represent the *magnetized* (or ferromagnetic) phase.

We now must ask whether this procedure is correct, or rather, when we can expect this approximation to work. The answer to this question is the central problem of the theory of phase transitions, which describes the behavior of statistical systems in the vicinity of a continuous (or second-order) phase transition. It turns out that this problem is also connected with a central problem of QFT, namely, when and how it is possible to remove the singular behavior of perturbation theory and in the process, remove all dependence on the short distance (or high energy) cutoff from physical observables. In QFT, this procedure amounts to a definition of the continuum limit. The answer to these questions motivated the development of the renormalization group, which solved both problems simultaneously.

2.8 Field theory and statistical mechanics

Let us now discuss a mathematical procedure that will allow us to connect QFT to classical statistical mechanics. We will use a mapping called a *Wick rotation*.

Go back to the action for a real scalar field $\Phi(x)$ in $D = d + 1$ spacetime dimensions:

$$S = \int d^D x\, \mathcal{L}(\Phi, \partial_\mu \Psi) \tag{2.196}$$

where $d^D x$ is

$$d^D x \equiv dx_0 d^d x \tag{2.197}$$

Let us *formally* carry out the *analytic continuation* of the time component x_0 of x_μ from real to imaginary time x_D:

$$x_0 \mapsto -ix_D \tag{2.198}$$

under which

$$\Phi(x_0, \boldsymbol{x}) \mapsto \Phi(\boldsymbol{x}, x_D) \equiv \Phi(x) \tag{2.199}$$

where $x = (\boldsymbol{x}, x_D)$. Under this transformation, the action (or rather, i times the action) becomes

$$iS \equiv i \int dx_0\, d^d x\, \mathcal{L}(\Phi, \partial_0 \Psi, \partial_j \Phi) \mapsto \int d^D x\, \mathcal{L}(\Phi, -i\partial_D \Phi, \partial_j \Phi) \tag{2.200}$$

If $\mathcal{L}$ has the form

$$\mathcal{L} = \frac{1}{2}(\partial_\mu \Phi)^2 - V(\Phi) \equiv \frac{1}{2}(\partial_0 \Phi)^2 - \frac{1}{2}(\boldsymbol{\nabla}\Phi)^2 - V(\Phi) \tag{2.201}$$

then the analytic continuation yields

$$\mathcal{L}(\Phi, -i\partial_D \Psi, \nabla\Phi) = -\frac{1}{2}(\partial_D \Phi)^2 - \frac{1}{2}(\boldsymbol{\nabla}\Phi)^2 - V(\Phi) \tag{2.202}$$

Then we can write

$$iS(\Phi, \partial_\mu \Phi) \xrightarrow[x_0 \to -ix_D]{} -\int d^D x \left[\frac{1}{2}(\partial_D \Phi)^2 + \frac{1}{2}(\nabla \Phi)^2 + V(\Phi) \right] \tag{2.203}$$

This expression has the same form as (minus) the potential energy $E(\Phi)$ for a classical field Φ in $D = d + 1$ *space* dimensions. However, it is also the same as the energy for a classical statistical mechanics problem in the same number of dimensions (i.e., the Landau-Ginzburg free energy of section 2.7.)

In classical statistical mechanics, the equilibrium properties of a system are determined by the *partition function*. For the case of the Landau theory of phase transitions, the partition function is

$$Z = \int \mathcal{D}\Phi \, e^{-E(\Phi)/T} \tag{2.204}$$

where the symbol "$\int \mathcal{D}\Phi$" means *sum over all configurations*. We will discuss the meaning of this expression and the definition of the "measure" $\mathcal{D}\Phi$ in section 5.3. If we choose for the energy functional $E(\Phi)$ the expression

$$E(\Phi) = \int d^D x \left[\frac{1}{2}(\partial \Phi)^2 + V(\Phi) \right] \tag{2.205}$$

where

$$(\partial \Phi)^2 \equiv (\partial_D \Phi)^2 + (\nabla \Phi)^2 \tag{2.206}$$

we see that the partition function Z is formally the analytic continuation of

$$\mathcal{Z} = \int \mathcal{D}\Phi \, e^{iS(\Phi, \partial_\mu \Phi)/\hbar} \tag{2.207}$$

where we have used $\hbar$, which has units of action (instead of temperature).

What is the physical meaning of $\mathcal{Z}$? This expression suggests that $\mathcal{Z}$ should have the interpretation of a sum of all possible functions $\Phi(\boldsymbol{x}, t)$ (i.e., the *histories* of the configurations of the field Φ) weighed by the phase factor $\exp\{\frac{i}{\hbar} S(\Phi, \partial_\mu \Phi)\}$. We will discover in section 5.3 that if T is formally identified with the Planck constant $\hbar$, then $\mathcal{Z}$ represents the *path-integral quantization* of the field theory! Notice that the semiclassical limit $\hbar \to 0$ is formally equivalent to the low-temperature limit of the statistical mechanical system.

The analytic continuation procedure that we just discussed is called a *Wick rotation*. It amounts to a passage from $D = d + 1$-dimensional Minkowski space to a D-dimensional Euclidean space. We will find that this analytic continuation is a very powerful tool. As we will see, various difficulties will arise when the theory is defined directly in Minkowski space. Primarily, the problem is the presence of ill-defined integrals, which are given precise meaning by a deformation of the integration contours from the real time (or frequency) axis to the imaginary time (or frequency) axis. The deformation of the contour amounts to a definition of the theory in Euclidean rather than in Minkowski space. The underlying assumption is that the analytic continuation can be carried out without problems. Namely, the assumption is that the result of this procedure is unique and that, whatever singularities may be present in the complex plane, they do not affect the result. It is important to stress that the success of this procedure is not guaranteed. However, in almost all theories that we know of, this assumption holds. The only case in which problems are known to exist is the theory of quantum gravity (which we will not discuss in this book).

Exercises

2.1 The Dirac equation

1) Use the Dirac equation to show that the 4-current $j^\mu = \bar{\psi}\gamma^\mu\psi$ is conserved.
2) Show that if ψ is a 4-spinor that satisfies the Dirac equation, then ψ also satisfies the Klein-Gordon equation.
3) Verify that the following identities hold:

i)

$$\not{A}\not{B} = A \cdot B - i\sigma_{\mu\nu} A^\mu B^\nu \tag{2.208}$$

where A^μ and B^ν are two arbitrary 4-vectors.

ii)

$$tr\not{A}\not{B} = 4\, A \cdot B \tag{2.209}$$

iii)

$$\gamma^\lambda \gamma^\mu \gamma_\lambda = -2\gamma^\mu \tag{2.210}$$

2.2 Transformation properties of field bilinears in the Dirac theory

In this problem, you will consider again the Dirac theory and study the transformation properties of its physical observable under Lorentz transformations. Let

$$x'_\mu = \Lambda_\mu^{\ \nu}\, x_\nu \tag{2.211}$$

be a general Lorentz transformation, and $S(\Lambda)$ be the induced transformation for the Dirac spinors $\psi_a(x)$ (with $a = 1, \ldots, 4$):

$$\psi'_a(x') = S(\Lambda)_{ab}\, \psi_b(x) \tag{2.212}$$

Verify that the Dirac bilinears listed obey the following transformation laws:

1)

$$\bar{\psi}'(x')\, \psi'(x') = \bar{\psi}(x)\, \psi(x) \tag{2.213}$$

2)

$$\bar{\psi}'(x')\, \gamma_5\, \psi'(x') = \det\Lambda\, \bar{\psi}(x)\, \gamma_5\, \psi(x) \tag{2.214}$$

3)

$$\bar{\psi}'(x')\, \gamma^\mu\, \psi'(x') = \Lambda^\mu_\nu\, \bar{\psi}(x)\, \gamma^\nu\, \psi(x) \tag{2.215}$$

4)

$$\bar{\psi}'(x')\, \gamma_5\gamma^\mu\, \psi'(x') = \det\Lambda\ \Lambda^\mu_\nu\, \bar{\psi}(x)\, \gamma_5\gamma^\nu\, \psi(x) \tag{2.216}$$

5)

$$\bar{\psi}'(x')\, \sigma^{\mu\nu}\, \psi'(x') = \Lambda^\mu_\alpha\, \Lambda^\nu_\beta\, \bar{\psi}(x)\, \sigma^{\alpha\beta}\, \psi(x) \tag{2.217}$$

where Λ^μ_ν is a Lorentz transformation, and $\det\Lambda$ is its determinant.

2.3 Chiral symmetry

Let us again consider the Dirac equation

$$(i\partial\!\!\!/ - m)\,\psi = 0 \tag{2.218}$$

but this time, in the *chiral* representation for the Dirac γ-matrices:

$$\gamma^0 = -\sigma^1 \otimes I = \begin{pmatrix} 0 & -I \\ -I & 0 \end{pmatrix}$$

$$\vec{\gamma} = i\sigma^2 \otimes \vec{\sigma} = \begin{pmatrix} 0 & \vec{\sigma} \\ -\vec{\sigma} & 0 \end{pmatrix}$$

$$\gamma_5 = \gamma^5 = \begin{pmatrix} I & 0 \\ 0 & -I \end{pmatrix}$$

$$\sigma^{0i} = i\begin{pmatrix} \sigma^i & 0 \\ 0 & -\sigma^i \end{pmatrix}$$

$$\sigma^{ij} = \epsilon^{ijk}\begin{pmatrix} \sigma^k & 0 \\ 0 & \sigma^k \end{pmatrix}$$

where σ^i are the three Pauli matrices, and I is the 2×2 identity matrix. Recall the definition of the matrix $\gamma_5 = i\gamma^0\gamma^1\gamma^2\gamma^3$.

1) Using the Dirac matrices in the chiral representation, write down the Dirac equation in terms of the 2-spinors ϕ and χ, where

$$\psi = \begin{pmatrix} \phi \\ \chi \end{pmatrix} \tag{2.219}$$

2) Show that if the excitations have zero mass (i.e., $m = 0$), the Dirac equation, written in the chiral basis, decouples into two 2×2 equations. Find the plane-wave solutions of these equations, and calculate their dispersion law (i.e., energy-momentum relation). Assign a chirality (γ_5) quantum number to each solution.
3) Consider the chiral transformation (CT)

$$\psi' = e^{i\gamma_5\theta}\,\psi \tag{2.220}$$

i) Find out how the 2-spinors ϕ and χ transform under a CT.
ii) Find out how $\bar{\psi}$ transforms under a CT.
iii) Find the transformation laws under a CT of the bilinears $\bar{\psi}\psi$ and $\bar{\psi}\gamma^\mu\psi$.
iv) Is the Dirac equation covariant under a CT if $m \neq 0$? Find the form of the Dirac equation, in terms of 4-spinors ψ, after a CT with angle θ has been carried out. What new terms do you find?

2.4 The Landau theory of phase transitions as a classical field theory

In the Landau-Ginzburg approach to the theory of phase transitions, the thermodynamic properties of a one-component classical ferromagnet in thermal equilibrium are described by a *free energy functional* of an order-parameter field $\phi(\boldsymbol{x})$ (the local magnetization). This functional contains, in addition to gradient terms, contributions proportional to various powers of the local order parameter. Under some circumstances the coefficient λ of the ϕ^4 term of the energy functional may become negative. This happens when the local magnetic moments have spin 1 rather than spin 1/2. In this case, we have to include, in the energy functional, a term with a higher power of ϕ (e.g., ϕ^6) to ensure the thermodynamic stability of the system.

The (free) energy density $\mathcal{E}$ for this system has the form

$$\mathcal{E} = \frac{1}{2}\left(\nabla\phi(\boldsymbol{x})\right)^2 + U(\phi(\boldsymbol{x})) \tag{2.221}$$

where the potential $U(\phi(\vec{x}))$ is

$$U(\phi(\boldsymbol{x})) = \frac{m_0^2}{2}\phi^2(\boldsymbol{x}) + \frac{\lambda_4}{4!}\phi^4(\boldsymbol{x}) + \frac{\lambda_6}{6!}\phi^6(\boldsymbol{x}) \tag{2.222}$$

with $m_0^2 = a(T - T_0)$, and $\lambda_4 < 0$, $\lambda_6 > 0$.

1) Use a variational principle to derive the saddle-point equations (i.e., the Landau-Ginzburg equations) for this system.
2) Plot the potential $U(\phi)$ for a constant field $\bar{\phi}$ for $\lambda_4 < 0$ (and fixed) at several temperatures. Show that, as the temperature T is lowered, there exists a temperature $T^* > T_0$ at which the state with lowest energy has $\langle\phi\rangle \neq 0$ (for fixed $\lambda_4 < 0$ and $\lambda_6 > 0$). Plot the qualitative behavior of $\langle\phi\rangle$ as a function of T. Is this a continuous function? Is this a first-order or a second-order transition? Find the value of the energy of the system in the ordered state.
3) Consider now the case $\lambda_4 > 0$, and show that the transition now takes place at T_0. Plot the qualitative behavior of $\langle\phi\rangle$ as a function of T for this case. Is this a continuous function? Is this a first-order or a second-order transition?
4) Summarize your results for the previous parts of this exercise in the form of a plot of λ_4 as a function of $T - T_0$. Indicate on the graph the areas in which the system is ordered and those in which it is disordered. Indicate where the transition is first order and where it is second order. Find an analytic expression for the phase boundary (i.e., the curve that separates the ordered and disordered states).

2.5 Scalar electrodynamics

The dynamics of a *charged* (complex) scalar field $\phi(x)$ coupled to the electromagnetic field $A_\mu(x)$ is governed by the Lagrangian density,

$$\mathcal{L} = \left(D_\mu\phi(x)\right)^* \left(D^\mu\phi(x)\right) - m_0^2|\phi(x)|^2 - \frac{\lambda}{2}\left(|\phi(x)|^2\right)^2 - \frac{1}{4}F^{\mu\nu}F_{\mu\nu} \tag{2.223}$$

where D_μ is the *covariant derivative*

$$D_\mu \equiv \partial_\mu + ieA_\mu \tag{2.224}$$

e is the electric charge, and * denotes complex conjugation.

1) Show that this Lagrangian density is invariant under the local gauge transformations

$$\phi'(x) = \phi(x)\, e^{-ie\Lambda(x)}$$
$$\phi'^*(x) = \phi^*(x)\, e^{+ie\Lambda(x)}$$
$$A'_\mu(x) = A_\mu(x) + \partial_\mu \Lambda(x)$$

2) Derive the classical equations of motion in a manifestly relativistically covariant form.
3) Find the Hamiltonian density for this system.
4) Write the complex field $\phi(x)$ in its polar components

$$\phi(x) = \rho(x)\, e^{i\theta(x)} \tag{2.225}$$

and find the equations of motion obeyed by the *real* fields ρ and θ. Write these equations of motion in the gauge $\theta = 0$, known as the London or unitary gauge. Find the Lagrangian for the field ρ.
5) Show that, if $m_0^2 < 0$, the equation of motion for the field ρ in the London gauge (derived in part 4 of this exercise) has a solution with $\rho = \bar{\rho} = \text{constant} > 0$. Freeze the field ρ at the value $\bar{\rho}$, and find the effective Lagrangian for the remaining degrees of freedom A_μ. Show that this Lagrangian has a term that is *quadratic* in A_μ, and calculate its coefficient. By solving the equations of motion for A_μ (derived from this effective Lagrangian), show that the coefficient of this quadratic term can be interpreted as a photon mass. *Note:* This phenomenon is known as the Meissner effect. This theory represents a system in a superconducting state.

2.6 Fluctuations of the surface of a crystal

In this problem in classical statistical mechanics, the field $\phi(\boldsymbol{x})$ represents the local vertical displacement (or height) of a lattice of atoms in d = two-dimensional space from their equilibrium positions. This model is used to describe the statistical fluctuations of a two-dimensional surface of a three-dimensional cubic crystal (known as a *surface-roughening problem*). The energy of a field configuration $\{\phi(\boldsymbol{x})\}$ is made up of the sum of two terms. The first term represents the elastic part

$$E_{\text{elastic}}[\phi(\boldsymbol{x})] = \frac{1}{2}\int d^2x\, (\boldsymbol{\nabla}\phi(\boldsymbol{x}))^2 \tag{2.226}$$

which arises from small deviations of the positions of the atoms from their equilibrium positions. The second part of the energy, which we denote by $V[\phi(\boldsymbol{x})]$, represents the interactions that tend to lock the height field $\phi(\boldsymbol{x})$ to some definite periodically arranged values. As a simple model, take the potential to be

$$V[\phi(\boldsymbol{x})] = -g\int d^2x\, \cos(\beta\phi(\boldsymbol{x})) \tag{2.227}$$

where g is a coupling constant, and $2\pi/\beta$ is the lock-in period. The total energy of a configuration is then given by

$$E[\phi(\boldsymbol{x})] = \int d^2x \left[\frac{K}{2} (\nabla\phi(\boldsymbol{x}))^2 - g\cos(\beta\phi(\boldsymbol{x})) \right] \tag{2.228}$$

1) What are the dimensions of: the field ϕ, the coupling constant g, and the parameter β? Assume that we have normalized the units, so that the energy is dimensionless.
2) Find the differential equation satisfied by the field configurations that extremize the energy, that is, the field equations.
3) By properly rescaling the field $\phi(\boldsymbol{x})$ and the space coordinate $\boldsymbol{x}$, write the field equations in a dimensionless form without any free parameters. What combination of parameters has dimensions of length?
4) Find a solution of the field equations that satisfies the following: the rescaled (dimensionless) $\phi(\boldsymbol{x})$ is a function of only one coordinate (say, x), and it satisfies the boundary conditions

$$\lim_{x\to+\infty} \phi(x) = 2\pi, \qquad \text{and} \qquad \lim_{x\to-\infty} \phi(x) = 0 \tag{2.229}$$

Hint: This problem reduces to a differential equation that can be solved by regarding it as a classical mechanics problem of a particle with coordinate $q = \phi$ as a function of time labeled by x. Note: This solution is known as a kink, a soliton, or a discommensuration.
5) Is this solution unique?
6) Estimate the width of the kink.
7) A strain in the displacement field ϕ can be represented by modifying the elastic energy as follows:

$$E_{\text{elastic}} = \frac{1}{2} \int d^2x \left(\nabla\phi(\boldsymbol{x}) - \boldsymbol{p} \right)^2 \tag{2.230}$$

How does the strain $\boldsymbol{p}$ modify the field equations?
8) Perform an analytic continuation of Minkowski spacetime (a Wick rotation) by regarding the y direction as imaginary time.
   i) Write down the resulting Lagrangian density of the relativistic (1+1)-dimensional field theory that you have found (known as the sine-Gordon field theory).
   ii) Write down the action of a field configuration.
   iii) Compare the potential terms of the original problem in classical statistical mechanics with that of the relativistic field theory you just derived. Make sure to write your expressions in a manifestly relativistically invariant form.
9) Write down the classical equations of motion in a manifestly covariant form.

3

Classical Symmetries and Conservation Laws

We have used the existence of symmetries in a physical system as a guiding principle for the construction of their Lagrangians and energy functionals. We will show now that these symmetries imply the existence of conservation laws.

There are different types of symmetries, which can be classified roughly into two classes: (1) spacetime symmetries and (2) internal symmetries. Some symmetries involve discrete operations and hence are called *discrete* symmetries, while others are *continuous* symmetries. Furthermore, in some theories, these are *global symmetries*, while in others, they are *local symmetries*. The latter class of symmetries go under the name of *gauge symmetries*. We will see that, in the fully quantized theory, global and local symmetries play different roles.

Spacetime symmetries are the most common examples of symmetries that are encountered in physics. They include translation invariance and rotation invariance. If the system is isolated, then time translation is also a symmetry. A nonrelativistic system is in general invariant under Galilean transformations, while relativistic systems, are instead Lorentz invariant. Other spacetime symmetries include *time-reversal* (T), *parity* (P), and *charge conjugation* (C). These symmetries are *discrete*.

In classical mechanics, the existence of symmetries has important consequences. Thus, translation invariance, which is a consequence of uniformity of space, implies the conservation of the total momentum $\boldsymbol{P}$ of the system. Similarly, *isotropy* implies the conservation of the total angular momentum $\boldsymbol{L}$, and *time-translation invariance* implies the conservation of the total energy E.

All of these concepts have analogs in field theory. However, in field theory, new symmetries also appear that do not have an analog in the classical mechanics of particles. These are the *internal symmetries*, which will be discussed below in detail.

3.1 Continuous symmetries and Noether's theorem

We will show now that the existence of continuous symmetries has very profound implications, such as the existence of conservation laws. One important feature of these conservation laws is the existence of *locally conserved currents*. This is the content of the following theorem, due to Emmy Noether.

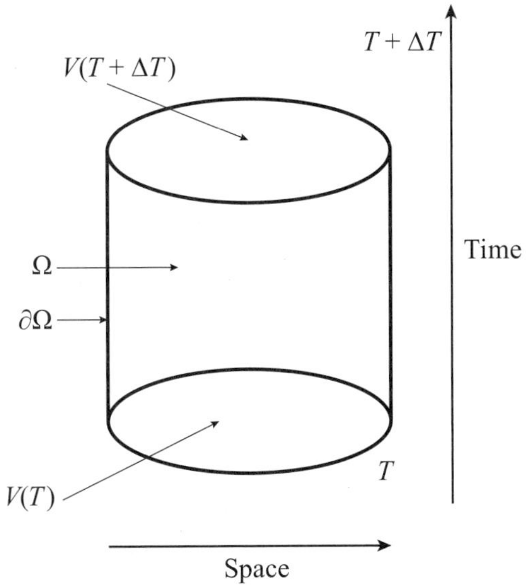

Figure 3.1 A spacetime 4-volume.

Noether's theorem: *For every continuous global symmetry there exists a global conservation law.*

Before we prove this statement, let us discuss the connection between locally conserved currents and constants of motion. In particular, let us show that for every locally conserved current, there exists a globally conserved quantity (i.e., a constant of motion). Let $j^\mu(x)$ be some locally conserved current (i.e., $j_\mu(x)$) that satisfies the local constraint

$$\partial_\mu j^\mu(x) = 0 \tag{3.1}$$

Let Ω be a bounded 4-volume in spacetime, with boundary $\partial\Omega$. Then the divergence (Gauss) theorem tells us that

$$0 = \int_\Omega d^4x\, \partial_\mu j^\mu(x) = \oint_{\partial\Omega} dS_\mu j^\mu(x) \tag{3.2}$$

where the r.h.s. is a surface integral on the oriented closed surface $\partial\Omega$ (a 3-volume). Suppose that the 4-volume Ω extends all the way to infinity in space and has a finite extent in time ΔT.

If there are no currents at *spatial* infinity (i.e., $\lim_{|\boldsymbol{x}|\to\infty} j^\mu(\boldsymbol{x}, x_0) = 0$), then only the top (at time $T + \Delta T$) and the bottom (at time T) of the boundary $\partial\Omega$ (shown in figure 3.1) will contribute to the surface (boundary) integral. Hence, the r.h.s. of eq. (3.2) becomes

$$0 = \int_{V(T+\Delta T)} dS_0\, j^0(\boldsymbol{x}, T + \Delta T) - \int_{V(T)} dS_0\, j^0(\boldsymbol{x}, T) \tag{3.3}$$

Since $dS_0 \equiv d^3x$, the boundary contributions reduce to two oriented 3-volume integrals:

$$0 = \int_{V(T+\Delta T)} d^3x\, j^0(\boldsymbol{x}, T + \Delta T) - \int_{V(T)} d^3x\, j^0(\boldsymbol{x}, T) \tag{3.4}$$

Thus, the quantity $Q(T)$

$$Q(T) \equiv \int_{V(T)} d^3x\, j^0(\boldsymbol{x}, T) \tag{3.5}$$

is a *constant of motion*, that is,

$$Q(T + \Delta T) = Q(T) \qquad \forall\, \Delta T \tag{3.6}$$

Hence, the existence of a locally conserved current, satisfying $\partial_\mu j^\mu = 0$, implies the existence of a globally conserved charge (or *Noether charge*) $Q = \int d^3x\, j^0(\boldsymbol{x}, T)$, which is a constant of motion. Thus, the proof of Noether's theorem reduces to proving the existence of a locally conserved current. In the following sections, we will prove Noether's theorem for internal and spacetime symmetries.

3.2 Internal symmetries

Let us begin, for simplicity, with the case of the complex scalar field $\phi(x) \neq \phi^*(x)$. The arguments that follow below are easily generalized to other cases. Let $\mathcal{L}(\phi, \partial_\mu\phi, \phi^*, \partial_\mu\phi^*)$ be the Lagrangian density. We will assume that the Lagrangian is invariant under the continuous global internal symmetry transformation

$$\begin{aligned} \phi(x) &\mapsto \phi'(x) = e^{i\alpha}\phi(x) \\ \phi^*(x) &\mapsto \phi'^*(x) = e^{-i\alpha}\phi^*(x) \end{aligned} \tag{3.7}$$

where α is an arbitrary real number (not a function!). The system is invariant under the transformation of eq. (3.7) if the Lagrangian $\mathcal{L}$ satisfies

$$\mathcal{L}(\phi', \partial_\mu\phi', \phi'^*, \partial_\mu\phi'^*) \equiv \mathcal{L}(\phi, \partial_\mu\phi, \phi^*, \partial_\mu\phi^*) \tag{3.8}$$

Then we say that the transformation shown in eq. (3.7) is a global symmetry of the system.
In particular, for an infinitesimal transformation, we have

$$\phi'(x) = \phi(x) + \delta\phi(x) + \cdots, \qquad \phi'^*(x) = \phi^*(x) + \delta\phi^*(x) + \cdots \tag{3.9}$$

where $\delta\phi(x) = i\alpha\phi(x)$. Since $\mathcal{L}$ is invariant, its variation must be identically equal to zero. The variation $\delta\mathcal{L}$ is

$$\delta\mathcal{L} = \frac{\delta\mathcal{L}}{\delta\phi}\delta\phi + \frac{\delta\mathcal{L}}{\delta\partial_\mu\phi}\delta\partial_\mu\phi + \frac{\delta\mathcal{L}}{\delta\phi^*}\delta\phi^* + \frac{\delta\mathcal{L}}{\delta\partial_\mu\phi^*}\delta\partial_\mu\phi^* \tag{3.10}$$

Using the equation of motion

$$\frac{\delta\mathcal{L}}{\delta\phi} - \partial_\mu\left(\frac{\delta\mathcal{L}}{\delta\partial_\mu\phi}\right) = 0 \tag{3.11}$$

and its complex conjugate, we can write the variation $\delta\mathcal{L}$ in the form of a total divergence:

$$\delta\mathcal{L} = \partial_\mu\left[\frac{\delta\mathcal{L}}{\delta\partial_\mu\phi}\delta\phi + \frac{\delta\mathcal{L}}{\delta\partial_\mu\phi^*}\delta\phi^*\right] \tag{3.12}$$

Thus, since $\delta\phi = i\alpha\phi$ and $\delta\phi^* = -i\alpha\phi^*$, we get

$$\delta\mathcal{L} = \partial_\mu\left[i\left(\frac{\delta\mathcal{L}}{\delta\partial_\mu\phi}\phi - \frac{\delta\mathcal{L}}{\delta\partial_\mu\phi^*}\phi^*\right)\alpha\right] \tag{3.13}$$

Hence, since α is arbitrary, $\delta\mathcal{L}$ will vanish identically if and only if the 4-vector j^μ, defined by

$$j^\mu = i\left(\frac{\delta\mathcal{L}}{\delta\partial_\mu\phi}\phi - \frac{\delta\mathcal{L}}{\delta\partial_\mu\phi^*}\phi^*\right) \tag{3.14}$$

is locally conserved, that is,

$$\delta\mathcal{L} = 0 \qquad \text{iff} \qquad \partial_\mu j^\mu = 0 \tag{3.15}$$

In particular, if $\mathcal{L}$ has the form

$$\mathcal{L} = (\partial_\mu\phi)^*(\partial^\mu\phi) - V(|\phi|^2) \tag{3.16}$$

which is manifestly invariant under the symmetry transformation of eq. (3.7), then the current j^μ is given by

$$j^\mu = i\left(\partial^\mu\phi^*\phi - \phi^*\partial^\mu\phi\right) \equiv i\phi^* \overleftrightarrow{\partial_\mu}\, \phi \tag{3.17}$$

Thus, the presence of a continuous internal symmetry implies the existence of a locally conserved current.

Furthermore, the conserved charge $\mathcal{Q}$ is given by

$$\mathcal{Q} = \int d^3x\, j^0(\boldsymbol{x}, x_0) = \int d^3x\, i\phi^* \overleftrightarrow{\partial_0}\, \phi \tag{3.18}$$

In terms of the canonical momentum $\Pi(x)$, the globally conserved charge $\mathcal{Q}$ of the charged scalar field is

$$\mathcal{Q} = \int d^3x\, i(\phi^*\Pi - \phi\Pi^*) \tag{3.19}$$

3.3 Global symmetries and group representations

Let us generalize the result of section 3.2. Consider a scalar field ϕ^a that transforms irreducibly under a certain representation of a Lie group G. In the case considered in section 3.2, the group G is the group of complex numbers of unit length, the group $U(1)$. The elements of this group, $g \in U(1)$, have the form $g = e^{i\alpha}$.

This set of complex numbers forms a group in the sense that

1) It is closed under complex multiplication, i.e.,

$$g = e^{i\alpha} \in U(1), \quad \text{and} \quad g' = e^{i\beta} \in U(1) \Rightarrow g * g' = e^{i(\alpha+\beta)} \in U(1) \tag{3.20}$$

2) There is an identity element (i.e., $g = 1$).
3) For every element $g = e^{i\alpha} \in U(1)$, there is a unique inverse element $g^{-1} = e^{-i\alpha} \in U(1)$.

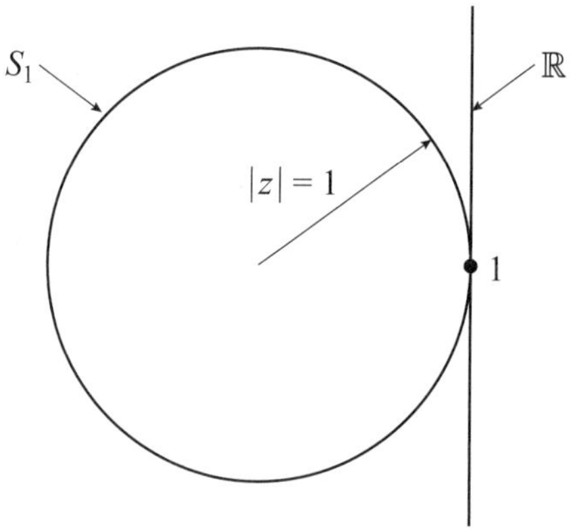

Figure 3.2 The $U(1)$ group is isomorphic to the unit circle, while the real numbers $\mathbb{R}$ are isomorphic to a tangent line.

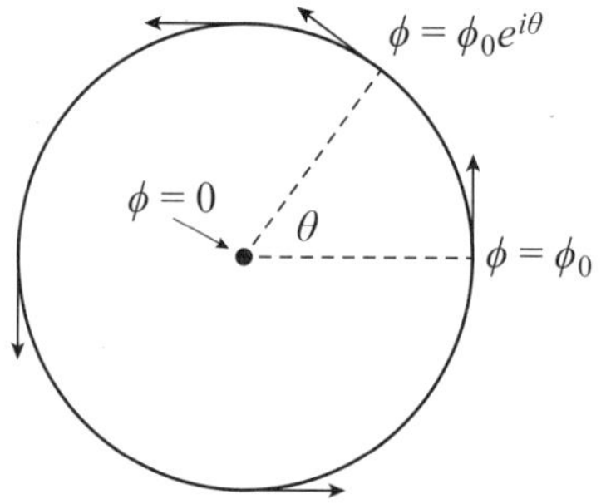

Figure 3.3 A vortex.

The elements of the group $U(1)$ are in one-to-one correspondence with the points of the unit circle S_1. Consequently, the parameter α that labels the transformation (or element of this group) is defined modulo 2π, and it should be restricted to the interval $(0, 2\pi]$. However, transformations infinitesimally close to the identity element, 1, lie essentially on the line tangent to the circle at 1 and so are isomorphic to the group of real numbers $\mathbb{R}$. The group $U(1)$ is compact, in the sense that the length of the natural parametrization of its elements is 2π, which is finite. In contrast, the group $\mathbb{R}$ of real numbers is not compact (see figure 3.2). Subsequently we will almost always work with internal symmetries with a compact Lie group.

For infinitesimal transformations, the groups $U(1)$ and $\mathbb{R}$ are essentially identical. There are, however, field configurations for which they are not. A typical case is the *vortex* configuration in two dimensions. For a vortex, the phase of the field on a large circle of radius $R \to \infty$ winds by 2π (figure 3.3). Such configurations would not exist if the symmetry group was $\mathbb{R}$ instead of $U(1)$. (Note that analyticity requires that $\phi \to 0$ as $x \to 0$.)

Another example is the N-component real scalar field $\phi^a(x)$, with $a = 1, \ldots, N$. In this case, the symmetry is the group of rotations in N-dimensional Euclidean space:

$$\phi'^a(x) = R^{ab} \phi^b(x) \tag{3.21}$$

The field ϕ^a is said to transform like the N-dimensional (vector) representation of the orthogonal group $O(N)$.

The elements of the orthogonal group, $R \in O(N)$, satisfy

1) if $R_1 \in O(N)$ and $R_2 \in O(N)$, then $R_1 R_2 \in O(N)$,
2) $\exists I \in O(N)$ such that $\forall R \in O(N)$, $RI = IR = R$,
3) $\forall R \in O(N)$ $\exists R^{-1} \in O(N)$ such that $R^{-1} = R^t$,

where R^t is the transpose of the matrix R.

Similarly, if the N-component vector $\phi^a(x)$ is a *complex field*, it transforms under the group of $N \times N$ unitary transformations U:

$$\phi'^a(x) = U^{ab}\phi^b(x) \tag{3.22}$$

The complex $N \times N$ matrices U are elements of the unitary group $U(N)$ and satisfy

1) if $U_1 \in U(N)$ and $U_2 \in U(N)$, then $U_1 U_2 \in U(N)$,
2) $\exists I \in U(N)$ such that $\forall U \in U(N)$, $UI = IU = U$,
3) $\forall U \in U(N)$, $\exists U^{-1} \in U(N)$ such that $U^{-1} = U^\dagger$, where $U^\dagger = (U^t)^*$.

In the particular case discussed above, ϕ^a transforms like the fundamental (spinor) representation of $U(N)$. If we impose the further restriction that $|\det U| = 1$, the group becomes $SU(N)$. For instance, if $N = 2$, the group is $SU(2)$, and

$$\boldsymbol{\phi} = \begin{pmatrix} \phi_1 \\ \phi_2 \end{pmatrix} \tag{3.23}$$

it transforms like the spin-1/2 representation of $SU(2)$.

In general, for an arbitrary continuous Lie group G, the field transforms like

$$\phi'_a(x) = \left(\exp\left[i\lambda^k\theta^k\right]\right)_{ab}\phi_b(x) \tag{3.24}$$

where the vector $\boldsymbol{\theta}$ is arbitrary and constant (i.e., independent of x). The matrices λ^k are a set of $N \times N$ linearly independent matrices that span the algebra of the Lie group G. For a given Lie group G, the number of such matrices is $D(G)$, and it is independent of the dimension N of the representation that was chosen. $D(G)$ is called the *rank* of the group. The matrices λ^k_{ab} are the *generators* of the group in this representation.

In general, from a symmetry point of view, the field ϕ does not have to be a vector, as it can also be a tensor or, for that matter, transform under any representation of the group. For simplicity, we will only consider the case of the vector representation of $O(N)$, and the fundamental (spinor) and adjoint (vector) (see later in this section and section 3.6) representations of $SU(N)$.

For an arbitrary compact Lie group G, the generators $\{\lambda^j\}$, $j = 1, \ldots, D(G)$, are a set of hermitian ($\lambda_j^\dagger = \lambda_j$), traceless ($\mathrm{tr}\lambda_j = 0$) matrices, which obey the commutation relations

$$[\lambda^j, \lambda^k] = if^{jkl}\lambda^l \tag{3.25}$$

The numerical constants f^{jkl} are known as the *structure constants* of the Lie group and are the same in all its representations. In addition, the generators have to be normalized. It is standard to require the normalization condition

$$\mathrm{tr}\lambda^a\lambda^b = \frac{1}{2}\delta^{ab} \tag{3.26}$$

In the case considered above, the complex scalar field $\phi(x)$, the symmetry group is the group of unit-length complex numbers of the form $e^{i\alpha}$. This group is known as the group $U(1)$. All its representations are one-dimensional, and it has only one generator.

A commonly used group is $SU(2)$. This group, which is familiar from nonrelativistic quantum mechanics, has three generators, J_1, J_2, and J_3, that obey the angular momentum algebra

$$[J_i, J_j] = i\epsilon_{ijk}J_k \tag{3.27}$$

with

$$\mathrm{tr}(J_iJ_j) = \frac{1}{2}\delta_{ij} \quad \text{and} \quad \mathrm{tr}J_i = 0 \tag{3.28}$$

The representations of $SU(2)$ are labeled by the angular momentum quantum number J. Each representation J is a $2J+1$–fold degenerate multiplet (i.e., the dimension of the representation is $2J+1$).

The lowest nontrivial representation of $SU(2)$ (i.e., $J \neq 0$) is the spinor representation, which has $J = \frac{1}{2}$ and is two-dimensional. In this representation, the field $\phi_a(x)$ is a two-component complex field, and the generators J_1, J_2, and J_3 are given by the set of 2×2 Pauli matrices $J_j = \frac{1}{2}\sigma_j$.

The vector (or spin-1) representation is three-dimensional, and ϕ_a is a three-component vector. In this representation, the generators are very simple:

$$(J_j)_{kl} = \epsilon_{jkl} \tag{3.29}$$

Notice that the dimension of this representation (3) is the same as the rank (3) of the group $SU(2)$. In this representation, known as the *adjoint* representation, the matrix elements of the generators are the structure constants. This is a general property of all Lie groups. In particular, the group $SU(N)$, whose rank is N^2-1, has N^2-1 infinitesimal generators, and the dimension of its adjoint (vector) representation is N^2-1. For instance, for $SU(3)$, the number of generators is eight.

Another important case is the group of rotations of N-dimensional Euclidean space, $O(N)$. In this case, the group has $N(N-1)/2$ generators, which can be labeled by the matrices L^{ij} $(i,j=1,\ldots,N)$. The fundamental (vector) representation of $O(N)$ is N-dimensional, and in this representation, the generators are

$$(L^{ij})_{kl} = -i(\delta_{ik}\delta_{j\ell} - \delta_{i\ell}\delta_{jk}) \tag{3.30}$$

It is easy to see that the L^{ij}s generate infinitesimal rotations in N-dimensional space.

Quite generally, in a given representation, an element of a Lie group is labeled by a set of Euler angles denoted by $\boldsymbol{\theta}$. If the Euler angles $\boldsymbol{\theta}$ are infinitesimal, then the representation matrix $\exp(i\boldsymbol{\lambda}\cdot\boldsymbol{\theta})$ is close to the identity and can be expanded in powers of $\boldsymbol{\theta}$. To leading order in $\boldsymbol{\theta}$, the change in ϕ^a is

$$\delta\phi^a(x) = i(\boldsymbol{\lambda}\cdot\boldsymbol{\theta})^{ab}\phi^b(x) + \cdots \tag{3.31}$$

If ϕ_a is real, the conserved current j^μ is

$$j_\mu^k(x) = \frac{\delta \mathcal{L}}{\delta \partial_\mu \phi^a(x)} \lambda_{ab}^k \phi_b(x) \tag{3.32}$$

where $k = 1, \ldots, D(G)$. Here, the generators λ^k are real hermitian matrices. In contrast, for a complex field ϕ_a, the conserved currents are

$$j_\mu^k(x) = i\left(\frac{\delta \mathcal{L}}{\delta \partial_\mu \phi^a(x)} \lambda_{ab}^k \phi_b(x) - \frac{\delta \mathcal{L}}{\delta \partial_\mu \phi^a(x)^*} \lambda_{ab}^k \phi_b(x)^*\right) \tag{3.33}$$

Here the generators λ^k are hermitian matrices (but are not all real).

Thus, we conclude that *the number of conserved currents is equal to the number of generators of the group*. For the particular choice

$$\mathcal{L} = (\partial_\mu \phi_a)^*(\partial^\mu \phi_a) - V(\phi_a^* \phi_a) \tag{3.34}$$

the conserved current is

$$j_\mu^k = i\lambda_{ab}^k \phi_a^* \overleftrightarrow{\partial_\mu} \phi_b \tag{3.35}$$

and the conserved charges are

$$Q^k = \int_V d^3x \, i\lambda_{ab}^k \phi_a^* \overleftrightarrow{\partial_0} \phi_b \tag{3.36}$$

where V is the volume of space.

3.4 Global and local symmetries: Gauge invariance

The existence of global symmetries assumes that, at least in principle, we can measure and change all components of a field $\phi^a(x)$ at all points $\boldsymbol{x}$ in space at the same time. Relativistic invariance tells us that, although the theory may possess this global symmetry, in principle, this experiment cannot be carried out. One is then led to consider theories that are invariant if the symmetry operations are performed locally. Namely, we should require that the Lagrangian be invariant under *local transformations*

$$\phi_a(x) \to \phi_a'(x) = \left(\exp\left[i\lambda^k \theta^k(x)\right]\right)_{ab} \phi_b(x) \tag{3.37}$$

For instance, we can demand that the theory of a complex scalar field $\phi(x)$ be invariant under local changes of phase

$$\phi(x) \to \phi'(x) = e^{i\theta(x)} \phi(x) \tag{3.38}$$

The standard local Lagrangian $\mathcal{L}$

$$\mathcal{L} = (\partial_\mu \phi)^*(\partial^\mu \phi) - V(|\phi|^2) \tag{3.39}$$

is invariant under *global* transformations with $\theta = $ constant, but it is not invariant under arbitrary smooth *local* transformations $\theta(x)$. The main problem is that since the derivative of the field does not transform like the field itself, the kinetic energy term is no longer invariant. Indeed, under a local transformation, we find

$$\partial_\mu \phi(x) \to \partial_\mu \phi'(x) = \partial_\mu \left[e^{i\theta(x)}\phi(x)\right] = e^{i\theta(x)}\left[\partial_\mu \phi + i\phi\partial_\mu \theta\right] \tag{3.40}$$

To make $\mathcal{L}$ *locally* invariant, we must find a new derivative operator D_μ, the *covariant derivative*, which transforms in the same way as the field $\phi(x)$ under local phase transformations:

$$D_\mu \phi \to D'_\mu \phi' = e^{i\theta(x)} D_\mu \phi \tag{3.41}$$

From a geometric point of view, we can picture the situation as follows. To define the phase of $\phi(x)$ locally, we have to define a local frame (or fiducial field) with respect to which the phase of the field is measured. Local gauge invariance is then the statement that the physical properties of the system must be independent of the particular choice of frame. From this point of view, local gauge invariance is an extension of the principle of relativity to the case of internal symmetries.

Now, if we wish to make phase transformations that differ from point to point in spacetime, we have to specify how the phase changes as we go from one point x in spacetime to another one y. In other words, we have to define a *connection* that will tell us how to *parallel transport* the phase of ϕ from x to y as we travel along some path Γ. Let us consider the situation in which x and y are arbitrarily close to each other (i.e., $y_\mu = x_\mu + dx_\mu$, where dx_μ is an infinitesimal 4-vector). The change in ϕ is

$$\phi(x+dx) - \phi(x) = \delta\phi(x) \tag{3.42}$$

If the *transport* of ϕ along some path going from x to $x + dx$ is to correspond to a phase transformation, then $\delta\phi$ must be proportional to ϕ. So we are led to define

$$\delta\phi(x) = iA_\mu(x)dx^\mu \phi(x) \tag{3.43}$$

where $A_\mu(x)$ is a suitably chosen vector field. Clearly, this implies that the covariant derivative D_μ must be defined to be

$$D_\mu \phi \equiv \partial_\mu \phi(x) - ieA_\mu(x)\phi(x) \equiv \left(\partial_\mu - ieA_\mu\right)\phi \tag{3.44}$$

where e is a parameter, which we will give the physical interpretation of a coupling constant.

How should $A_\mu(x)$ transform? We must choose its transformation law in such a way that $D_\mu \phi$ transforms like $\phi(x)$ itself. Thus, if $\phi \to e^{i\theta}\phi$, we have

$$D'\phi' = (\partial_\mu - ieA'_\mu)(e^{i\theta}\phi) \equiv e^{i\theta} D_\mu \phi \tag{3.45}$$

This requirement can be met if

$$i\partial_\mu \theta - ieA'_\mu = -ieA_\mu \tag{3.46}$$

Hence, A_μ should transform like

$$A_\mu \to A'_\mu = A_\mu + \frac{1}{e}\partial_\mu \theta \tag{3.47}$$

But this is nothing but a gauge transformation! Indeed, if we define the gauge transformation $\Phi(x)$

$$\Phi(x) \equiv \frac{1}{e}\theta(x) \tag{3.48}$$

we see that the vector field A_μ transforms like the vector potential of Maxwell's electromagnetism.

We conclude that we can promote a global symmetry to a local (i.e., gauge symmetry) by replacing the derivative operator by the covariant derivative. Thus, we can make a system invariant under local gauge transformations at the price of introducing a vector field A_μ, the gauge field, which plays the role of a connection. From a physical point of view, this result means that the impossibility of making a comparison at a distance of the phase of the field $\phi(x)$ requires that a physical gauge field $A_\mu(x)$ must be present. This procedure, which relates the matter and gauge fields through the covariant derivative, is known as *minimal coupling*.

There is a set of configurations of $\phi(x)$ that changes only because of the presence of the gauge field. These are the *geodesic* configurations $\phi_c(x)$. They satisfy the equation

$$D_\mu \phi_c = (\partial_\mu - ieA_\mu)\phi_c \equiv 0 \tag{3.49}$$

which is equivalent to the linear equation (using eq. (3.43))

$$\partial_\mu \phi_c = ieA_\mu \phi_c \tag{3.50}$$

For example, let us consider two points x and y in spacetime at the ends of a path $\Gamma(x,y)$. For a given path $\Gamma(x,y)$, the solution of eq. (3.49) is the path-ordered exponential of a line integral:

$$\phi_c(x) = e^{-ie\int_{\Gamma(x,y)} dz_\mu A^\mu(z)}\, \phi_c(y) \tag{3.51}$$

Indeed, under a gauge transformation, the line integral transforms like

$$\begin{aligned} e\int_{\Gamma(x,y)} dz_\mu A^\mu &\mapsto e\int_{\Gamma(x,y)} dz_\mu A^\mu + e\int_{\Gamma(x,y)} dz_\mu \frac{1}{e}\partial^\mu\theta \\ &= e\int_{\Gamma(x,y)} dz_\mu A^\mu(z) + \theta(y) - \theta(x) \end{aligned} \tag{3.52}$$

Hence, we get

$$\begin{aligned} \phi_c(y)\, e^{-ie\int_\Gamma dz_\mu A^\mu} &\mapsto \phi_c(y)\, e^{-ie\int_\Gamma dz_\mu A^\mu}\, e^{-i\theta(y)} e^{i\theta(x)} \\ &\equiv e^{i\theta(x)}\phi_c(x) \end{aligned} \tag{3.53}$$

as it should be.

However, we may now want to know how the change of phase of ϕ_c depends on the choice of the path Γ. Let $\phi_c^{\Gamma_1}(y)$ and $\phi_c^{\Gamma_2}(y)$ be solutions of the geodesic equations for two different paths Γ_1 and Γ_2 with the same end points, x and y. Clearly, we have that the change of phase $\Delta\gamma$ is given by

$$\Delta\gamma = -e\int_{\Gamma_1} dz_\mu A^\mu + e\int_{\Gamma_2} dz_\mu A^\mu \equiv -e\oint_{\Gamma+} dz_\mu A^\mu \tag{3.54}$$

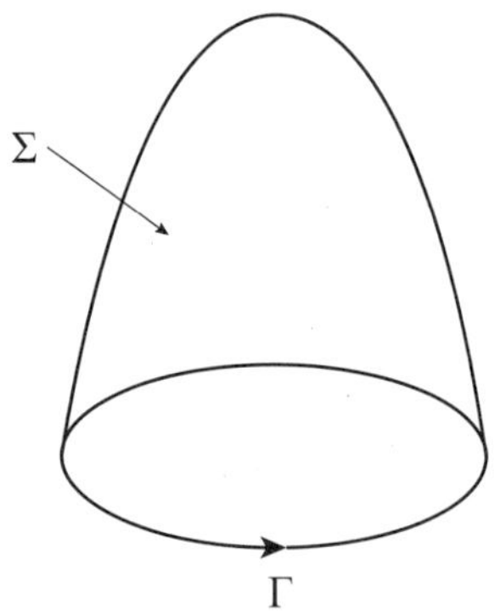

Figure 3.4 A closed path Γ is the boundary of the open surface Σ.

Here Γ^+ is the *closed oriented path*

$$\Gamma^+ = \Gamma_1^+ \cup \Gamma_2^- \tag{3.55}$$

and

$$\int_{\Gamma_2^-} dz_\mu A_\mu = -\int_{\Gamma_2^+} dz_\mu A_\mu \tag{3.56}$$

Using Stokes' theorem, we see that, if Σ^+ is an *oriented surface* whose boundary is the oriented closed path Γ^+, $\partial\Sigma^+ \equiv \Gamma^+$ (see figure 3.4), then $\Delta\gamma$ is given by the flux $\Phi(\Sigma)$ of the curl of the vector field A_μ through the surface Σ^+:

$$\Delta\gamma = -\frac{e}{2}\int_{\Sigma^+} dS_{\mu\nu} F^{\mu\nu} = -e\,\Phi(\Sigma) \tag{3.57}$$

where $F^{\mu\nu}$ is the field tensor

$$F^{\mu\nu} = \partial^\mu A^\nu - \partial^\nu A^\mu \tag{3.58}$$

$dS_{\mu\nu}$ is the *oriented* area element, and $\Phi(\Sigma)$ is the flux through the surface Σ. Both $F^{\mu\nu}$ and $dS_{\mu\nu}$ are antisymmetric in their spacetime indices. In particular, $F^{\mu\nu}$ can also be written as a commutator of two covariant derivatives:

$$F^{\mu\nu} = \frac{i}{e}[D^\mu, D^\mu] \tag{3.59}$$

Thus, $F^{\mu\nu}$ measures the (infinitesimal) incompatibility of displacements along two independent directions. In other words, $F^{\mu\nu}$ is a *curvature* tensor. These results show very clearly that if $F^{\mu\nu}$ is nonzero in some region of spacetime, then the *phase* of ϕ cannot be uniquely determined: The *phase* of ϕ_c depends on the path Γ along which it is measured.

3.5 The Aharonov-Bohm effect

The path dependence of the phase of ϕ_c is closely related to the *Aharonov-Bohm effect*. This effect is subtle. It was first discovered in the context of elementary quantum mechanics and plays a fundamental role in (quantum) field theory as well.

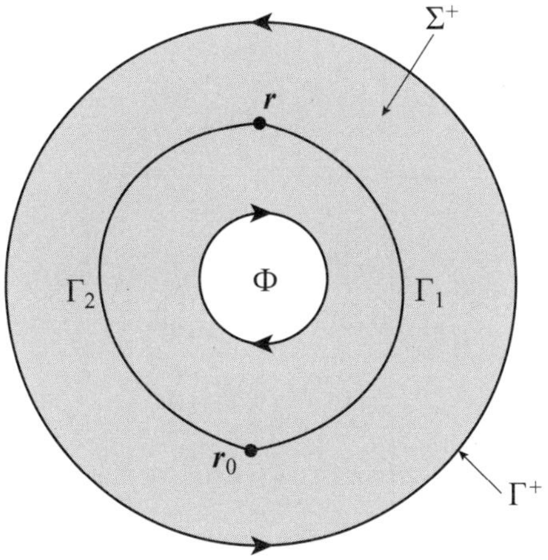

Figure 3.5 Geometric setup of the Aharonov-Bohm effect: a magnetic flux $\Phi \neq 0$ is threading through the small hole in the punctured plane Σ^+. Here Γ^+ represents the sum of the oriented outer and inner edges of the punctured plane Σ^+; Γ_1 and Γ_2 are two inequivalent paths from $\boldsymbol{r}_0$ to $\boldsymbol{r}$ described in the text.

Consider a quantum mechanical particle of charge e and mass m moving on a plane. The particle is coupled to an external electromagnetic field A_μ (here $\mu = 0, 1, 2$ only, since there is no motion out of the plane). Let us consider the geometry shown in figure 3.5, in which an infinitesimally thin solenoid pierces the plane in the vicinity of some point $\boldsymbol{r} = 0$. The Schrödinger equation for this problem is

$$H\Psi = i\hbar \frac{\partial \Psi}{\partial t} \tag{3.60}$$

where

$$H = \frac{1}{2m}\left(\frac{\hbar}{i}\nabla + \frac{e}{c}\boldsymbol{A}\right)^2 \tag{3.61}$$

is the Hamiltonian. The magnetic field $\boldsymbol{B} = B\hat{z}$ vanishes everywhere except at $\boldsymbol{r} = 0$,

$$B = \Phi_0\, \delta(\boldsymbol{r}) \tag{3.62}$$

Using Stokes' theorem, we see that the flux of $\boldsymbol{B}$ through an arbitrary region Σ^+ with boundary Γ^+ is

$$\Phi = \int_{\Sigma^+} d\boldsymbol{S} \cdot \boldsymbol{B} = \oint_{\Gamma^+} d\boldsymbol{\ell} \cdot \boldsymbol{A} \tag{3.63}$$

Hence, $\Phi = \Phi_0$ for all surfaces Σ^+ that enclose the point $\boldsymbol{r} = 0$, and it is equal to zero otherwise. Therefore, although the magnetic field is zero for $\boldsymbol{r} \neq 0$, the vector potential does not (and cannot) vanish.

The wave function $\Psi(\boldsymbol{r})$ can be calculated in a very simple way. Let us define

$$\Psi(\boldsymbol{r}) = e^{i\theta(\boldsymbol{r})}\, \Psi_0(\boldsymbol{r}) \tag{3.64}$$

where $\Psi_0(\boldsymbol{r})$ satisfies the Schrödinger equation in the absence of the field, that is,

$$H_0\Psi_0 = i\hbar\frac{\partial\Psi_0}{\partial t} \tag{3.65}$$

with

$$H_0 = -\frac{\hbar^2}{2m}\dot{\nabla}^2 \tag{3.66}$$

Since the wave function Ψ has to be differentiable and Ψ_0 is single valued, we must also satisfy the boundary condition:

$$\lim_{r\to 0}\Psi_0(\boldsymbol{r},t) = 0 \tag{3.67}$$

The wave function $\Psi = \Psi_0 e^{i\theta}$ looks like a gauge transformation. But there is a subtlety here. Indeed, θ can be determined as follows. By direct substitution, we get

$$\left(\frac{\hbar}{i}\nabla + \frac{e}{c}\boldsymbol{A}\right)(e^{i\theta}\Psi_0) = e^{i\theta}\left(\hbar\,\nabla\theta + \frac{e}{c}\boldsymbol{A} + \frac{\hbar}{i}\nabla\right)\Psi_0 \tag{3.68}$$

Thus, to succeed in our task, we only have to require that $\boldsymbol{A}$ and θ must obey the relation

$$\hbar\nabla\theta + \frac{e}{c}\boldsymbol{A} \equiv 0 \tag{3.69}$$

Or equivalently,

$$\nabla\theta(\boldsymbol{r}) = -\frac{e}{\hbar c}\boldsymbol{A}(\boldsymbol{r}) \tag{3.70}$$

However, if this relation holds, θ cannot be a smooth function of $\boldsymbol{r}$. In fact, the line integral of $\nabla\theta$ on an arbitrary closed path Γ^+ is given by

$$\int_{\Gamma^+} d\boldsymbol{\ell}\cdot\nabla\theta = \Delta\theta \tag{3.71}$$

where $\Delta\theta$ is the total change of θ in one full counterclockwise turn around the path Γ. It follows immediately that $\Delta\theta$ is given by

$$\Delta\theta = -\frac{e}{\hbar c}\oint_{\Gamma^+} d\boldsymbol{\ell}\cdot\boldsymbol{A} \tag{3.72}$$

Thus in general, $\theta(\boldsymbol{r})$ must be a multivalued function of $\boldsymbol{r}$ that has a branch cut going from $\boldsymbol{r}=0$ out to some arbitrary point at infinity. The actual position and shape of the branch cut is irrelevant, but the discontinuity $\Delta\theta$ of θ across the cut is not irrelevant.

Hence, Ψ_0 is chosen to be a smooth, single-valued solution of the Schrödinger equation in the absence of the solenoid, satisfying the boundary condition eq. (3.67). Such wave functions are (almost) plane waves.

Since the function $\theta(\boldsymbol{r})$ is multivalued and, hence, path dependent, the wave function Ψ is also multivalued and path dependent. In particular, let $\boldsymbol{r}_0$ be some arbitrary point on the plane, and $\Gamma(\boldsymbol{r}_0,\boldsymbol{r})$ be a path that begins at $\boldsymbol{r}_0$ and ends at $\boldsymbol{r}$. The phase $\theta(\boldsymbol{r})$ is, for that choice of path, given by

$$\theta(\boldsymbol{r}) = \theta(\boldsymbol{r}_0) - \frac{e}{\hbar c}\int_{\Gamma(\boldsymbol{r}_0,\boldsymbol{r})} d\boldsymbol{x}\cdot\boldsymbol{A}(\boldsymbol{x}) \tag{3.73}$$

The overlap of two wave functions that are defined by two different paths $\Gamma_1(\boldsymbol{r}_0, \boldsymbol{r})$ and $\Gamma_2(\boldsymbol{r}_0, \boldsymbol{r})$ is (with $\boldsymbol{r}_0$ fixed)

$$\begin{aligned}\langle\Gamma_1|\Gamma_2\rangle &= \int d^2\boldsymbol{r}\, \Psi^*_{\Gamma_1}(\boldsymbol{r})\, \Psi_{\Gamma_2}(\boldsymbol{r}) \\ &\equiv \int d^2\boldsymbol{r}\, |\Psi_0(\boldsymbol{r})|^2 \exp\left\{+\frac{ie}{\hbar c}\left(\int_{\Gamma_1(\boldsymbol{r}_0,\boldsymbol{r})} d\boldsymbol{\ell}\cdot\boldsymbol{A} - \int_{\Gamma_2(\boldsymbol{r}_0,\boldsymbol{r})} d\boldsymbol{\ell}\cdot\boldsymbol{A}\right)\right\}\end{aligned} \tag{3.74}$$

If Γ_1 and Γ_2 are chosen in such a way that the origin (where the solenoid is piercing the plane) is always to the left of Γ_1 but also always to the right of Γ_2, the difference of the two line integrals is the circulation of $\boldsymbol{A}$

$$\int_{\Gamma_1(\boldsymbol{r}_0,\boldsymbol{r})} d\boldsymbol{\ell}\cdot\boldsymbol{A} - \int_{\Gamma_2(\boldsymbol{r}_0,\boldsymbol{r})} d\boldsymbol{\ell}\cdot\boldsymbol{A} \equiv \oint_{\Gamma^+_{(\boldsymbol{r}_0)}} d\boldsymbol{\ell}\cdot\boldsymbol{A} \tag{3.75}$$

on the closed, positively oriented, contour $\Gamma^+ = \Gamma_1(\boldsymbol{r}_0, \boldsymbol{r}) \cup \Gamma_2(\boldsymbol{r}, \boldsymbol{r}_0)$. Since this circulation is constant and equal to the flux Φ, we find that the overlap $\langle\Gamma_1|\Gamma_2\rangle$ is

$$\langle\Gamma_1|\Gamma_2\rangle = \exp\left\{\frac{ie}{\hbar c}\Phi\right\} \tag{3.76}$$

where we have taken Ψ_0 to be normalized to unity. Eq. (3.76) is known as the *Aharonov-Bohm effect*.

We find that the overlap is a pure phase factor, which in general is different from 1. Notice that, although the wave function is always defined up to a *constant* arbitrary phase factor, *phase changes* are physical effects. In addition, for some special choices of Φ, the wave function becomes single valued. These values correspond to the choice

$$\frac{e}{\hbar c}\Phi = 2\pi n \tag{3.77}$$

where n is an arbitrary integer. This requirement amounts to a quantization condition for the magnetic flux Φ:

$$\Phi = n\left(\frac{hc}{e}\right) \equiv n\Phi_0 \tag{3.78}$$

where Φ_0 is the *flux quantum*, $\Phi_0 = \frac{hc}{e}$.

In 1931, Dirac considered the effects of a monopole configuration of magnetic fields on the quantum mechanical wave functions of charged particles (Dirac, 1931). In Dirac's construction, a magnetic monopole is represented as a long, thin solenoid in three-dimensional space. The magnetic field near the end of the solenoid is the same as that of a magnetic charge m equal to the magnetic flux going through the solenoid. Dirac argued that for the solenoid (the "Dirac string") to be unobservable, the wave function must be single valued. This requirement leads to the Dirac quantization condition for the smallest magnetic charge:

$$me = 2\pi\hbar c \tag{3.79}$$

which is the same as the flux quantization condition of eq. (3.78).

3.6 Non-abelian gauge invariance

Let us now consider systems with a non-abelian global symmetry. This means that the field ϕ transforms like some representation of a Lie group G,

$$\phi'_a(x) = U_{ab}\phi_b(x) \tag{3.80}$$

where U is a matrix that represents the action of a group element. The local Lagrangian density,

$$\mathcal{L} = \partial_\mu \phi^*_a \partial^\mu \phi^a - V(|\phi|^2) \tag{3.81}$$

is invariant under global transformations.

Suppose now that we want to promote this global symmetry to a local one. However, it is also true for this general case that although the potential term $V(|\phi|^2)$ is invariant even under local transformations $U(x)$, the first term of the Lagrangian in eq. (3.81) is not. Indeed, the gradient of ϕ does not transform properly (i.e., covariantly) under the action of the Lie group G:

$$\begin{aligned}\partial_\mu \phi'(x) &= \partial_\mu[U(x)\phi(x)] \\ &= \left(\partial_\mu U(x)\right)\phi(x) + U(x)\partial_\mu \phi(x) \\ &= U(x)[\partial_\mu \phi(x) + U^{-1}(x)\partial_\mu U(x)\phi(x)]\end{aligned} \tag{3.82}$$

Hence $\partial_\mu \phi$ does not transform in the same way as the ϕ field.

We can now follow the same approach that we used in the abelian case and define a *covariant derivative* operator D_μ, which should have the property that $D_\mu \phi$ obeys the same transformation law as the field ϕ:

$$\left(D_\mu \phi(x)\right)' = U(x)\left(D_\mu \phi(x)\right) \tag{3.83}$$

It is clear that D_μ is now both a differential operator and a matrix acting on the field ϕ. Thus, D_μ depends on the representation that was chosen for the field ϕ. We can now proceed in analogy with what we did in the case of electrodynamics, and guess that the covariant derivative D_μ should be of the form

$$D_\mu = I\,\partial_\mu - igA_\mu(x) \tag{3.84}$$

where g is a coupling constant, I is the $N \times N$ identity matrix, and A_μ is a matrix-valued vector field. If ϕ has N components, the vector field $A_\mu(x)$ is an $N \times N$ hermitian matrix that can be expanded in the basis of the group generators λ^k_{ab} (with $k = 1, \ldots, D(G)$, and $a, b = 1, \ldots, N$), which span the *algebra* of the Lie group G:

$$\left(A_\mu(x)\right)_{ab} = A^k_\mu(x)\lambda^k_{ab} \tag{3.85}$$

Thus, the vector field $A_\mu(x)$ is parametrized by the $D(G)$-component 4-vectors $A^k_\mu(x)$. Let us choose the transformation properties of $A_\mu(x)$ in such a way that $D_\mu \phi$ transforms covariantly under gauge transformations:

$$\begin{aligned} D'_\mu\phi(x)' &\equiv D'_\mu\left(U(x)\phi(x)\right) = \left(\partial_\mu - igA'_\mu(x)\right)\left(U\phi(x)\right) \\ &= U(x)[\partial_\mu\phi(x) + U^{-1}(x)\partial_\mu U(x)\phi(x) - igU^{-1}(x)A'_\mu(x)U(x)\phi(x)] \\ &\equiv U(x)\,D_\mu\phi(x) \end{aligned} \tag{3.86}$$

This condition is met if we require that

$$U^{-1}(x)igA'_\mu(x)U(x) = igA_\mu(x) + U^{-1}(x)\partial_\mu U(x) \tag{3.87}$$

or, equivalently, that A_μ obey the transformation law

$$A'_\mu(x) = U(x)A_\mu(x)U^{-1}(x) - \frac{i}{g}\left(\partial_\mu U(x)\right)U^{-1}(x) \tag{3.88}$$

Since the matrices $U(x)$ are unitary and invertible, we have

$$U^{-1}(x)U(x) = I \tag{3.89}$$

and we can equivalently write the transformed vector field $A'_\mu(x)$ in the form

$$A'_\mu(x) = U(x)A_\mu(x)U^{-1}(x) + \frac{i}{g}U(x)\left(\partial_\mu U^{-1}(x)\right) \tag{3.90}$$

This is the general form of a gauge transformation for a non-abelian Lie group G.

In the case of an abelian symmetry group, such as the group $U(1)$, the matrix reduces to a simple phase factor, $U(x) = e^{i\theta(x)}$, and the field $A_\mu(x)$ is a real-number vector field. It is easy to check that, in this case, A_μ transforms as follows:

$$\begin{aligned} A'_\mu(x) &= e^{i\theta(x)}A_\mu(x)e^{-i\theta(x)} + \frac{i}{g}e^{i\theta(x)}\partial_\mu\left(e^{-i\theta(x)}\right) \\ &\equiv A_\mu(x) + \frac{1}{g}\partial_\mu\theta(x) \end{aligned} \tag{3.91}$$

which recovers the correct form for an abelian gauge transformation.

Returning now to the non-abelian case, we see that under an infinitesimal transformation $U(x)$,

$$(U(x))_{ab} = \left[\exp\left(i\lambda^k\theta^k(x)\right)\right]_{ab} \cong \delta_{ab} + i\lambda^k_{ab}\,\theta^k(x) + \cdots \tag{3.92}$$

the scalar field $\phi(x)$ transforms as

$$\delta\phi_a(x) \cong i\lambda^k_{ab}\phi_b(x)\,\theta^k(x) + \cdots \tag{3.93}$$

while the vector field A^k_μ now transforms as

$$\delta A^k_\mu(x) \cong if^{ksj}A^j_\mu(x)\,\theta^s(x) + \frac{1}{g}\partial_\mu\theta^k(x) + \cdots \tag{3.94}$$

Thus, $A^k_\mu(x)$ transforms as a vector in the adjoint representation of the Lie group G, since, in that representation, the matrix elements of the generators are the group structure constants f^{ksj}. Notice that A^k_μ is always in the adjoint representation of the group G, regardless of the representation in which $\phi(x)$ happens to be.

From this discussion, it is clear that the field $A_\mu(x)$ can be interpreted as a generalization of the vector potential of electromagnetism. Furthermore, A_μ provides for a natural connection which tells us how the "internal coordinate system," in reference to which the field $\phi(x)$ is defined, changes from one point x_μ of spacetime to a neighboring point $x_\mu + dx_\mu$. In particular, the configurations $\phi^a(x)$, which are solutions of the geodesic equation

$$D^{ab}_\mu \phi_b(x) = 0 \tag{3.95}$$

correspond to the *parallel transport* of ϕ from some point x to some point y. This equation can be written in the equivalent form

$$\partial_\mu \phi_a(x) = igA^k_\mu(x)\,\lambda^k_{ab}\,\phi_b(x) \tag{3.96}$$

This linear partial differential equation can be solved as follows. Let x_μ and y_μ be two arbitrary points in spacetime and $\Gamma(x, y)$ be a fixed path with endpoints at x and y. This path is parametrized by a mapping z_μ from the real interval $[0, 1]$ to Minkowski space $\mathcal{M}$ (or any other space), $z_\mu : [0, 1] \mapsto \mathcal{M}$, of the form

$$z_\mu = z_\mu(t), \qquad t \in [0, 1] \tag{3.97}$$

with the boundary conditions

$$z_\mu(0) = x_\mu \quad \text{and} \quad z_\mu(1) = y_\mu \tag{3.98}$$

By integrating the geodesic equation, eq. (3.95), along the path Γ, we obtain

$$\int_{\Gamma(x,y)} dz_\mu \frac{\partial \phi_a(z)}{\partial z_\mu} = ig \int_{\Gamma(x,y)} dz_\mu\, A^\mu_{ab}(z)\,\phi_b(z) \tag{3.99}$$

Hence, we find that $\phi(x)$ must be the solution of the integral equation

$$\phi(y) = \phi(x) + ig \int_{\Gamma(x,y)} dz_\mu\, A^\mu(z)\phi(z) \tag{3.100}$$

where all the indices have been omitted to simplify the notation. In terms of the parametrization $z_\mu(t)$ of the path $\Gamma(x, y)$, we can write

$$\phi(y) = \phi(x) + ig \int_0^1 dt\, \frac{dz_\mu}{dt}\, A^\mu\,(z(t))\; \phi\,(z(t)) \tag{3.101}$$

We will solve this equation by means of an iteration procedure, similar to what is used for the evolution operator in quantum theory. By substituting repeatedly the l.h.s. of this equation into its r.h.s., we get the series

$$\begin{aligned}
\phi(y) = {} & \phi(x) + ig \int_0^1 dt \, \frac{dz_\mu(t)}{dt} \, A^\mu \,(z(t))\, \phi(x) \\
& + (ig)^2 \int_0^1 dt_1 \int_0^{t_1} dt_2 \frac{dz_{\mu_1}(t_1)}{dt_1} \, \frac{dz_{\mu_2}(t_2)}{dt_2} A^{\mu_1}\,(z(t_1))\, A^{\mu_2}\,(z(t_2))\, \phi(x) \\
& + \cdots + (ig)^n \int_0^1 dt_1 \int_0^{t_1} dt_2 \cdots \int_0^{t_{n-1}} dt_n \prod_{j=1}^{n} \left(\frac{dz_{\mu_j}(t_j)}{dt_j} A^{\mu_j}\left(z(t_j)\right) \right) \phi(x) \\
& + \cdots
\end{aligned} \tag{3.102}$$

Here we need to keep in mind that the A^μs are matrix-valued fields that are ordered from left to right.

The nested integrals in eq. (3.102) can be written in the form

$$\begin{aligned}
I_n = {} & (ig)^n \int_0^1 dt_1 \int_0^{t_1} dt_2 \cdots \int_0^{t_{n-1}} dt_n \, F(t_1) \cdots F(t_n) \\
\equiv {} & \frac{(ig)^n}{n!} \widehat{P} \left[\left(\int_0^1 dt \, F(t) \right)^n \right]
\end{aligned} \tag{3.103}$$

where the Fs are matrices, and the operator $\widehat{P}$ means the path-ordered product of the objects sitting to its right. If we formally define the exponential of an operator to be equal to its power series expansion,

$$e^A \equiv \sum_{n=0}^{\infty} \frac{1}{n!} A^n \tag{3.104}$$

where A is some arbitrary matrix, we see that the geodesic equation has the formal solution

$$\phi(y) = \widehat{P} \left[\exp \left(+ig \int_0^1 dt \, \frac{dz_\mu}{dt} \, A^\mu(z(t)) \right) \right] \phi(x) \tag{3.105}$$

or equivalently,

$$\phi(y) = \widehat{P} \left[\exp \left(ig \int_{\Gamma(x,y)} dz_\mu \, A^\mu(z) \right) \right] \phi(x) \tag{3.106}$$

Thus, $\phi(y)$ is given by an operator, the *path-ordered* exponential of the line integral of the vector potential A^μ, acting on $\phi(x)$.

By expanding the exponential in a power series, it is easy to check that, under an arbitrary local gauge transformation $U(z)$, the path-ordered exponential transforms as follows:

$$\begin{aligned}
& \widehat{P} \left[\exp\left(ig \int_{\Gamma(x,y)} dz^\mu \, A'_\mu(z(t)) \right) \right] \\
& \qquad \equiv U(y) \, \hat{P} \left[\exp \left(ig \int_{\Gamma(x,y)} dz^\mu \, A_\mu(z) \right) \right] U^{-1}(x)
\end{aligned} \tag{3.107}$$

In particular, we can consider the case of a closed path $\Gamma(x, x)$, where x is an arbitrary point on Γ. The path-ordered exponential $\widehat{W}_{\Gamma(x,x)}$

$$\widehat{W}_{\Gamma(x,x)} = \widehat{P}\left[\exp\left(ig\int_{\Gamma(x,x)} dz^\mu\, A_\mu(z)\right)\right] \tag{3.108}$$

is not gauge invariant, since under a gauge transformation, it transforms as

$$\begin{aligned}\widehat{W}'_{\Gamma(x,x)} &= \hat{P}\left[\exp\left(ig\int_{\Gamma(x,x)} dz^\mu\, A'_\mu(z)\right)\right]\\ &= U(x)\,\hat{P}\left[\exp\left(ig\int_{\Gamma(x,x)} dz^\mu\, A_\mu(t)\right)\right] U^{-1}(x)\end{aligned} \tag{3.109}$$

Therefore, $\widehat{W}_{\Gamma(x,x)}$ transforms like a group element:

$$\widehat{W}_{\Gamma(x,x)} = U(x)\,\widehat{W}_{\Gamma(x,x)}\,U^{-1}(x) \tag{3.110}$$

However, the trace of $\widehat{W}_{\Gamma(x,x)}$, which we denote by

$$W_\Gamma = \operatorname{tr}\widehat{W}_{\Gamma(x,x)} \equiv \operatorname{tr}\widehat{P}\left[\exp\left(ig\int_{\Gamma(x,x)} dz^\mu\, A_\mu(z)\right)\right] \tag{3.111}$$

not only is gauge-invariant but is also independent of the choice of the point x. However, it is a functional of the path Γ. The quantity W_Γ, which is known as the *Wilson loop*, plays a crucial role in gauge theories. In quantum theory, this object becomes the Wilson loop operator.

Let us now consider the case of a small closed path Γ. If Γ is small, then the minimal area $a(\Gamma)$ enclosed by Γ and its length $\ell(\Gamma)$ are both infinitesimal. In this case, we can expand the exponential in powers and retain only the leading terms. We get

$$\widehat{W}_\Gamma \approx I + ig\widehat{P}\oint_\Gamma dz^\mu A_\mu(z) + \frac{(ig)^2}{2!}\widehat{P}\left(\oint_\Gamma dz^\mu A_\mu(z)\right)^2 + \cdots \tag{3.112}$$

Stokes' theorem says that the first integral (the circulation of the vector field A_μ on the closed path Γ) is given by

$$\oint_\Gamma dz_\mu A^\mu(z) = \iint_\Sigma dx^\mu \wedge dx^\nu\, \frac{1}{2}(\partial_\mu A_\nu - \partial_\nu A_\mu) \tag{3.113}$$

where $\partial\Sigma = \Gamma$ is the infinitesimal area element bounded by Γ, and $dx^\mu \wedge dx^\nu$ is the oriented infinitesimal area element. Furthermore, the quadratic term in eq. (3.112) can be expressed as follows:

$$\frac{1}{2!}\widehat{P}\left(\oint_\Gamma dz^\mu A_\mu(z)\right)^2 \equiv \frac{1}{2}\iint_\Sigma dx^\mu \wedge dx^\nu\,(-[A_\mu, A_\nu]) + \cdots \tag{3.114}$$

Therefore, for an infinitesimally small loop, we get

$$\widehat{W}_{\Gamma(x,x')} \approx I + \frac{ig}{2} \iint_{\Sigma} dx^{\mu} \wedge dx^{\nu} \, F_{\mu\nu} + O(a(\Sigma)^2) \tag{3.115}$$

where $F_{\mu\nu}$ is the field tensor, defined by

$$F_{\mu\nu} \equiv \partial_{\mu} A_{\nu} - \partial_{\nu} A_{\mu} - ig[A_{\mu}, A_{\nu}] = i\left[D_{\mu}, D_{\nu}\right] \tag{3.116}$$

Keep in mind that since the fields A_{μ} are matrices, the field tensor $F_{\mu\nu}$ is also a matrix.

Notice also that now $F_{\mu\nu}$ is not gauge invariant. Indeed, under a local gauge transformation $U(x)$, $F_{\mu\nu}$ transforms as a similarity transformation:

$$F'_{\mu\nu}(x) = U(x) F_{\mu\nu}(x) U^{-1}(x) \tag{3.117}$$

This property follows from the transformation properties of A_{μ}. However, although $F_{\mu\nu}$ itself is not gauge invariant, other quantities, such as $\mathrm{tr}(F_{\mu\nu}F^{\mu\nu})$, are gauge invariant.

Let us finally consider the form of $F_{\mu\nu}$ in components. By expanding $F_{\mu\nu}$ in the basis of the group generators λ^k (hence, in the algebra of the gauge group),

$$F_{\mu\nu} = F^k_{\mu\nu} \lambda^k \tag{3.118}$$

we find that the components, $F^k_{\mu\nu}$, are

$$F^k_{\mu\nu} = \partial_{\mu} A^k_{\nu} - \partial_{\nu} A^k_{\mu} + g f^{k\ell m} A^{\ell}_{\mu} A^m_{\nu} \tag{3.119}$$

The natural local, gauge-invariant theory for a non-abelian gauge group is the Yang-Mills Lagrangian

$$\mathcal{L} = -\frac{1}{4} \mathrm{tr} F_{\mu\nu} F^{\mu\nu} \tag{3.120}$$

for a general compact Lie group G, which we will call the *gauge group*. Notice that the apparent similarity with the Maxwell Lagrangian is only superficial, since in this theory, it is not quadratic in the vector potentials. We will see in chapter 22 that other Lagrangians are possible in other dimensions if some symmetry (e.g., time reversal) is violated.

3.7 Gauge invariance and minimal coupling

We are now in a position to give a general prescription for the coupling of matter and gauge fields. Since the issue here is local gauge invariance, this prescription is valid for both relativistic and nonrelativistic theories.

So far, we have considered two cases: (1) fields that describe the dynamics of matter and (2) gauge fields that describe electromagnetism and chromodynamics. In our description of Maxwell's electrodynamics, we saw that, if the Lagrangian is required to respect local gauge invariance, then only conserved currents can couple to the gauge field. However, we have also seen that the presence of a global symmetry is a sufficient condition for the existence of a locally conserved current. This is not only a necessary condition, since a local symmetry also requires the existence of a conserved current.

We now consider more general Lagrangians that include both matter and gauge fields. In sections 3.4–3.7, we saw that if a system with Lagrangian $\mathcal{L}(\phi, \partial_\mu \phi)$ has a global symmetry $\phi \to U\phi$, then by replacing all derivatives by covariant derivatives, we promote a global symmetry to a local (or gauge) symmetry. We will proceed with our general philosophy and write down gauge-invariant Lagrangians for systems that contain both matter and gauge fields. Next I give a few explicit examples.

3.7.1 Quantum electrodynamics

Quantum electrodynamics (QED) is a theory of electrons and photons. The electrons are described by *Dirac spinor fields* $\psi_\alpha(x)$. The reason for this choice will become clear when we discuss quantum theory and the spin-statistics theorem. Photons are described by a $U(1)$ gauge field A_μ. The Lagrangian for *free electrons* is just the *free Dirac Lagrangian*, $\mathcal{L}_{\text{Dirac}}$:

$$\mathcal{L}_{\text{Dirac}}(\psi, \bar{\psi}) = \bar{\psi}(i\partial\!\!\!/ - m)\psi \tag{3.121}$$

The Lagrangian for the gauge field is the free Maxwell Lagrangian, $\mathcal{L}_{\text{gauge}}(A)$:

$$\mathcal{L}_{\text{gauge}}(A) = -\frac{1}{4} F_{\mu\nu} F^{\mu\nu} \equiv -\frac{1}{4} F^2 \tag{3.122}$$

The prescription that we will adopt, known as *minimal coupling*, consists of requiring that the total Lagrangian be invariant under local gauge transformations.

The free Dirac Lagrangian is invariant under the global phase transformation (i.e., with the same phase factor for all the Dirac components),

$$\psi_\alpha(x) \to \psi'_\alpha(x) = e^{i\theta} \psi_\alpha(x) \tag{3.123}$$

if θ is a constant, arbitrary phase, but it is not invariant under the *local phase* transformation:

$$\psi_\alpha(x) \to \psi'_\alpha(x) = e^{i\theta(x)} \psi_\alpha(x) \tag{3.124}$$

As we saw before, the matter part of the Lagrangian can be made invariant under the local transformations,

$$\begin{aligned} \psi_\alpha(x) &\to \psi'_\alpha(x) = e^{i\theta(x)} \psi_\alpha(x) \\ A_\mu(x) &\to A'_\mu(x) = A_\mu(x) + \frac{1}{e} \partial_\mu \theta(x) \end{aligned} \tag{3.125}$$

if the derivative $\partial_\mu \psi$ is replaced by the covariant derivative D_μ:

$$D_\mu = \partial_\mu - ieA_\mu(x) \tag{3.126}$$

The total Lagrangian is now given by the sum of two terms,

$$\mathcal{L} = \mathcal{L}_{\text{matter}}(\psi, \bar{\psi}, A) + \mathcal{L}_{\text{gauge}}(A) \tag{3.127}$$

where $\mathcal{L}_{\text{matter}}(\psi, \bar{\psi}, A)$ is the gauge-invariant extension of the Dirac Lagrangian:

$$\begin{aligned}\mathcal{L}_{\text{matter}}(\psi, \bar{\psi}, A) &= \bar{\psi}(i\not{D} - m)\psi \\ &= \bar{\psi}\left(i\not{\partial} - m\right)\psi + e\bar{\psi}\gamma_\mu\psi A^\mu\end{aligned} \tag{3.128}$$

$\mathcal{L}_{\text{gauge}}(A)$ is the usual Maxwell term, and $\not{D}$ is a shorthand for $\gamma^\mu D_\mu$. Thus, the total Lagrangian for QED is

$$\mathcal{L}_{\text{QED}} = \bar{\psi}(i\not{D} - m)\psi - \frac{1}{4}F^2 \tag{3.129}$$

Notice that now both matter and gauge fields are dynamical degrees of freedom.

The QED Lagrangian has a local gauge invariance. Hence, it also has a locally conserved current. In fact, the argument that we used in section 3.1 to show that there are conserved (Noether) currents if there is a continuous global symmetry, is also applicable to gauge-invariant Lagrangians. As a matter of fact, under an arbitrary infinitesimal gauge transformation

$$\delta\psi = i\theta\psi, \quad \delta\bar{\psi} = -i\theta\bar{\psi}, \quad \delta A_\mu = \tfrac{1}{e}\partial_\mu\theta \tag{3.130}$$

the QED Lagrangian remains invariant (i.e., $\delta\mathcal{L} = 0$). An arbitrary variation of $\mathcal{L}$ is

$$\delta\mathcal{L} = \frac{\delta\mathcal{L}}{\delta\psi}\delta\psi + \frac{\delta\mathcal{L}}{\delta\partial_\mu\psi}\delta\partial_\mu\psi + (\psi \leftrightarrow \bar{\psi}) + \frac{\delta\mathcal{L}}{\delta\partial_\mu A_\nu}\delta\partial_\mu A_\nu + \frac{\delta\mathcal{L}}{\delta A_\mu}\delta A_\mu \tag{3.131}$$

After using the equations of motion and the form of the gauge transformation, $\delta\mathcal{L}$ can be written in the form

$$\delta\mathcal{L} = \partial_\mu[j^\mu(x)\theta(x)] - \frac{1}{e}F^{\mu\nu}(x)\partial_\mu\partial_\nu\theta(x) + \frac{\delta\mathcal{L}}{\delta A_\mu}\frac{1}{e}\partial_\mu\theta(x) \tag{3.132}$$

where $j^\mu(x)$ is the *electron number current*:

$$j^\mu = i\left(\frac{\partial\mathcal{L}}{\delta\partial_\mu\psi}\psi - \bar{\psi}\frac{\delta\mathcal{L}}{\delta\partial_\mu\bar{\psi}}\right) \tag{3.133}$$

For smooth gauge transformations $\theta(x)$, the term $F^{\mu\nu}\partial_\mu\partial_\nu\theta$ in eq. (3.132) vanishes because of the antisymmetry of the field tensor $F_{\mu\nu}$. Hence we can write

$$\delta\mathcal{L} = \theta(x)\partial_\mu j^\mu(x) + \partial_\mu\theta(x)\left[j^\mu(x) + \frac{1}{e}\frac{\delta\mathcal{L}}{\delta A_\mu(x)}\right] \tag{3.134}$$

The first term tells us that since the infinitesimal gauge transformation $\theta(x)$ is arbitrary, the *Dirac current $j^\mu(x)$ locally is conserved* (i.e., $\partial_\mu j^\mu = 0$).

Let us define the *charge* (or *gauge*) current $J^\mu(x)$ by the relation

$$J^\mu(x) \equiv \frac{\delta\mathcal{L}}{\delta A_\mu(x)} \tag{3.135}$$

which is the current that enters in the equations of motion for the gauge field A_μ (i.e., the Maxwell equations). The vanishing of the second term in eq. (3.134), required since the changes of the infinitesimal gauge transformations are also arbitrary, tells us that the charge current and the number current are related by

$$J_\mu(x) = -e j_\mu(x) = -e\,\bar{\psi}\gamma_\mu\psi \tag{3.136}$$

This relation tells us that since $j^\mu(x)$ is locally conserved, the global conservation of Q_0,

$$Q_0 = \int d^3x j_0(x) \equiv \int d^3x\, \psi^\dagger(x)\psi(x) \tag{3.137}$$

implies the global conservation of the electric charge Q:

$$Q \equiv -eQ_0 = -e\int d^3x\,\psi^\dagger(x)\psi(x) \tag{3.138}$$

This property justifies the interpretation of the coupling constant e as the electric charge. In particular, the gauge transformation of eq. (3.125) tells us that the matter field $\psi(x)$ represents excitations that carry the unit of charge, $\pm e$. From this point of view, the electric charge can be regarded as a *quantum number*. This point of view becomes very useful in quantum theory in the strong coupling limit. In this case, under special circumstances, the excitations may acquire unusual quantum numbers. This is not the case for QED, but it is the case for a number of theories in one and two space dimensions, with applications in condensed matter systems, such as polyacetylene, or the two-dimensional electron gas in high magnetic fields (i.e., the fractional quantum Hall effect), or in gauge theories with magnetic monopoles.

3.7.2 Quantum chromodynamics

Quantum chromodynamics (QCD) is the gauge field theory of strong interactions in hadron physics. In this theory, the elementary constituents of hadrons, the *quarks*, are represented by the Dirac spinor field $\psi^i_\alpha(x)$. The theory also contains a set of gauge fields $A^a_\mu(x)$ that represent the *gluons*. The quark fields have both Dirac indices $\alpha = 1, \ldots, 4$ and *color* indices $i = 1, \ldots, N_c$, where N_c is the number of colors. In the Standard Model of weak, strong, and electromagnetic interactions in particle physics, and in QCD, there are in addition $N_f = 6$ flavors of quarks, grouped into three generations, each labeled by a flavor index, and six flavors of leptons, also grouped into three generations. The flavor symmetry is a global symmetry of the theory.

Quarks are assumed to transform under the *fundamental representation* of the gauge (or color) group G, say, $SU(N_c)$. The theory is invariant under the group of gauge transformations. In QCD, the color group is $SU(3)$, and so $N_c = 3$. The color symmetry is a non-abelian gauge symmetry. The gauge field A_μ is needed to enforce local gauge invariance. In components, we get $A_\mu = A^a_\mu \lambda^a$, where the λ^a are the generators of $SU(N_c)$. Thus, $a = 1, \ldots, D(SU(N_c))$, and $D(SU(N_c)) = N_c^2 - 1$. So A_μ is an $N_c^2 - 1$-dimensional vector in the adjoint representation of G. For $SU(3), N_c^2 - 1 = 8$, and so there are eight generators.

The gauge-invariant matter term of the Lagrangian, $\mathcal{L}_{\text{matter}}$, is

$$\mathcal{L}_{\text{matter}}(\psi, \bar{\psi}, A) = \bar{\psi}(i\not{D} - m)\psi \tag{3.139}$$

where $\not{D} = \not{\partial} - ig\not{A} \equiv \not{\partial} - ig\not{A}^a\lambda^a$ is the covariant derivative. The gauge field term of the Lagrangian $\mathcal{L}_{\text{gauge}}$,

$$\mathcal{L}_{\text{gauge}}(A) = -\frac{1}{4}\text{tr}F_{\mu\nu}F^{\mu\nu} \equiv -\frac{1}{4}F^a_{\mu\nu}F^{\mu\nu}_a \tag{3.140}$$

is the Yang-Mills Lagrangian. The total Lagrangian for QCD is

$$\mathcal{L}_{\text{QCD}} = \mathcal{L}_{\text{matter}}(\psi, \bar{\psi}, A) + \mathcal{L}_{\text{gauge}}(A) \tag{3.141}$$

Can we define a *color charge*? Since the color group is non-abelian, it has more than one generator. In section 3.1, we showed before that there are as many conserved currents as there are generators in the group. Now, in general, the group generators do not commute with each other. For instance, in $SU(2)$, there is only one diagonal generator, J_3, while $SU(3)$ has only two diagonal generators, and so forth. Can all the global *charges* Q^a

$$Q^a \equiv \int d^3x\, \psi^\dagger(x)\lambda^a\psi(x) \tag{3.142}$$

be defined simultaneously? It is straightforward to show that the Poisson brackets of any pair of charges are, in general, different from zero. In section 4.4 we will show, when we quantize the theory, that the charges Q^a obey the same commutation relations as the group generators themselves do. So, in the quantum theory, the only charges that can be assigned to *states* are precisely the same as the quantum numbers that label the representations. Thus, if the group is $SU(2)$, we can only assign to the states the values of the quadratic Casimir operator $\boldsymbol{J}^2$ and of the projection J_3. Similar restrictions apply to the case of $SU(3)$ and to other Lie groups.

3.8 Spacetime symmetries and the energy-momentum tensor

Until now we have considered only the role of internal symmetries. We now turn to spacetime symmetries and consider the role of coordinate transformations. In this more general setting, we will have to require the invariance of the action rather than only of the Lagrangian, as we did for internal symmetries.

There are three continuous spacetime symmetries that will be important to us: (1) translation invariance, (2) rotation invariance, and (3) homogeneity of time. While rotations are a subgroup of Lorentz transformation, space and time translations are examples of inhomogeneous Lorentz transformations (in the relativistic case) and of Galilean transformations (in the nonrelativistic case). Inhomogeneous Lorentz transformations also form a group, known as the *Poincaré group*. Note that the transformations discussed in section 3.7 are particular cases of more general coordinate transformations. However, it is important to keep in mind that, in most cases, general coordinate transformations are not symmetries of an arbitrary system. They are the symmetries of general relativity.

In what follows, we are going to consider the response of a system to infinitesimal coordinate transformations of the form

$$x'_\mu = x_\mu + \delta x_\mu \tag{3.143}$$

where δx_μ can be a function of the spacetime point x_μ. Under a coordinate transformation the fields change as

$$\phi(x) \to \phi'(x') = \phi(x) + \delta\phi(x) + \partial_\mu \phi\, \delta x^\mu \tag{3.144}$$

where $\delta\phi$ is the variation of ϕ in the absence of a change of coordinates (i.e., a functional change). In this notation, a uniform infinitesimal translation by a constant vector a_μ is $\delta x_\mu = a_\mu$, and an infinitesimal rotation of the space axes is $\delta x_0 = 0$ and $\delta x_i = \epsilon_{ijk}\theta_j x_k$.

In general, the action of the system is not invariant under arbitrary changes in both coordinates and fields. Indeed, under an arbitrary change of coordinates $x_\mu \to x'_\mu(x_\mu)$, the volume element d^4x is not $x_\mu \to x'_\mu(x_\mu)$, the volume element d^4x is not invariant and changes by a multiplicative factor of the form

$$d^4x' = d^4x\, J \tag{3.145}$$

where J is the Jacobian of the coordinate transformation

$$J = \frac{\partial x'_1 \cdots x'_4}{\partial x_1 \cdots x_4} \equiv \left| \det\left(\frac{\partial x'_\mu}{\partial x_\nu} \right) \right| \tag{3.146}$$

For an infinitesimal transformation $x'_\mu = x_\mu + \delta x_\mu(x)$, we get

$$\frac{\partial x'_\mu}{\partial x_\nu} = g^\nu_\mu + \partial^\nu \delta x_\mu \tag{3.147}$$

Since δx_μ is small, the Jacobian can be approximated by

$$J = \left| \det\left(\frac{\partial x'_\mu}{\partial x_\nu} \right) \right| = \left| \det\left(g^\nu_\mu + \partial^\nu \delta x_\mu \right) \right| \approx 1 + \mathrm{tr}(\partial^\nu \delta x_\mu) + O(\delta x^2) \tag{3.148}$$

Thus

$$J \approx 1 + \partial^\mu \delta x_\mu + O(\delta x^2) \tag{3.149}$$

The Lagrangian itself is in general not invariant. For instance, even though we will always be interested in systems whose Lagrangians are not an explicit function of x, still they are not in general invariant under the given transformation of coordinates. Also, under a coordinate change, the fields might also transform. Thus, in general, $\delta\mathcal{L}$ does not vanish.

The most general variation of $\mathcal{L}$ is

$$\delta\mathcal{L} = \partial_\mu \mathcal{L}\, \delta x^\mu + \frac{\delta\mathcal{L}}{\delta\phi}\delta\phi + \frac{\delta\mathcal{L}}{\delta\partial_\mu\phi}\delta\partial_\mu\phi \tag{3.150}$$

If ϕ obeys the equations of motion,

$$\frac{\delta\mathcal{L}}{\delta\phi} = \partial_\mu \frac{\delta\mathcal{L}}{\delta\partial_\mu\phi} \tag{3.151}$$

then the general change $\delta\mathcal{L}$ obeyed by solutions of the equations of motion is

$$\delta\mathcal{L} = \delta x^\mu\, \partial_\mu \mathcal{L} + \partial_\mu \left(\frac{\delta\mathcal{L}}{\delta\partial_\mu\phi}\delta\phi \right) \tag{3.152}$$

The total change in the action is a sum of two terms

$$\delta S = \delta \int d^4x\, \mathcal{L} = \int \delta d^4x\, \mathcal{L} + \int d^4x\, \delta \mathcal{L} \tag{3.153}$$

where the change in the integration measure is due to the Jacobian factor,

$$\delta d^4x = d^4x\, \partial_\mu \delta x^\mu \tag{3.154}$$

Hence, δS is given by

$$\delta S = \int d^4x \left[(\partial_\mu \delta x^\mu)\, \mathcal{L} + \delta x^\mu\, \partial_\mu \mathcal{L} + \partial_\mu \left(\frac{\delta \mathcal{L}}{\delta \partial_\mu \phi} \delta\phi \right) \right] \tag{3.155}$$

Since the total variation of ϕ, $\delta_T \phi$, is the sum of the functional change of the fields plus the changes in the fields caused by the coordinate transformation,

$$\delta_T \phi \equiv \delta\phi + \partial_\mu \phi \delta x^\mu \tag{3.156}$$

we can write δS as a sum of two contributions: one due to change of coordinates and another due to functional changes of the fields:

$$\delta S = \int d^4x \left[(\partial_\mu \delta x^\mu)\, \mathcal{L} + \delta x^\mu \partial_\mu \mathcal{L} + \partial_\mu \left(\frac{\delta \mathcal{L}}{\delta \partial_\mu \phi} (\delta_T \phi - \partial_\nu \phi \delta x^\nu) \right) \right] \tag{3.157}$$

Therefore, for the change of the action, we get

$$\delta S = \int d^4x \left\{ \partial_\mu \left[\left(g^\mu_\nu \mathcal{L} - \frac{\delta \mathcal{L}}{\delta \partial_\mu \phi} \partial_\nu \phi \right) \delta x^\nu \right] + \partial_\mu \left[\frac{\delta \mathcal{L}}{\delta \partial_\mu \phi} \delta_T \phi \right] \right\} \tag{3.158}$$

We have already encountered the second term when we discussed the case of internal symmetries. The first term represents the change of the action S as a result of a change of coordinates.

Let us now consider a few explicit examples. To simplify matters, we consider the effects of coordinate transformations alone. For simplicity, here we restrict our discussion to the case of the scalar field.

3.8.1 Spacetime translations

Under a uniform infinitesimal translation $\delta x_\mu = a_\mu$, the field ϕ does not change:

$$\delta_T \phi = 0 \tag{3.159}$$

The change of the action now is

$$\delta S = \int d^4x\, \partial_\mu \left(g^{\mu\nu} \mathcal{L} - \frac{\delta \mathcal{L}}{\delta \partial_\mu \phi} \partial^\nu \phi \right) a^\nu \tag{3.160}$$

For a system that is isolated and translationally invariant, the action must not change under a redefinition of the origin of the coordinate system. Thus, $\delta S = 0$. Since a_μ is arbitrary, it follows that the tensor $T^{\mu\nu}$

$$T^{\mu\nu} \equiv -g^{\mu\nu}\mathcal{L} + \frac{\delta\mathcal{L}}{\delta\partial_\mu\phi}\partial^\nu\phi \tag{3.161}$$

is conserved:

$$\partial_\mu T^{\mu\nu} = 0 \tag{3.162}$$

The tensor $T^{\mu\nu}$ is known as the *energy-momentum tensor*. The reason for this name is the following. Given that $T^{\mu\nu}$ is locally conserved, by Noether's theorem, we can define the 4-vector P^ν

$$P^\nu = \int d^3x\, T^{0\nu}(\boldsymbol{x}, x_0) \tag{3.163}$$

which is a constant of motion. In particular, P^0 is given by

$$P^0 = \int d^3x\, T^{00}(\boldsymbol{x}, x_0) \equiv \int d^3x \left[-\mathcal{L} + \frac{\delta\mathcal{L}}{\delta\partial_0\phi}\partial^0\phi \right] \tag{3.164}$$

But $\frac{\delta\mathcal{L}}{\delta\partial_0\phi}$ is just the canonical momentum $\Pi(x)$:

$$\Pi(x) = \frac{\delta\mathcal{L}}{\delta\partial_0\phi} \tag{3.165}$$

Then we can easily recognize that the quantity in brackets in eq. (3.164) in the definition of P^0 is just the Hamiltonian density $\mathcal{H}$:

$$\mathcal{H} = \Pi\, \partial^0\phi - \mathcal{L} \tag{3.166}$$

Therefore, the time component of P^ν is just the total energy of the system:

$$P^0 = \int d^3x\, \mathcal{H} \tag{3.167}$$

The space components P^j are

$$P^j = \int d^3x\, T^{0j} = \int d^3x \left[-g^{0j}\mathcal{L} + \frac{\delta\mathcal{L}}{\delta\partial_j\phi}\partial^j\phi \right] \tag{3.168}$$

Thus, since $g^{0j} = 0$, we get

$$\boldsymbol{P} = \int d^3x\, \Pi(x)\, \boldsymbol{\partial}\phi(x) \tag{3.169}$$

The vector $\boldsymbol{P}$ is identified with the total linear momentum, since (1) it is a constant of motion, and (2) it is the generator of infinitesimal space translations. For the same reasons, we will identify the component $T^{0j}(x)$ with the linear momentum density $\mathcal{P}^j(x)$. It is important to stress that the *canonical momentum* $\Pi(x)$ and the *total linear momentum density* $\mathcal{P}^j$ are obviously completely different physical quantities. While the canonical momentum is a field that is canonically conjugate to the field ϕ, the total momentum is the linear momentum stored in the field (i.e., the linear momentum of the center of mass).

3.8.2 Rotations

If the action is invariant under global infinitesimal Lorentz transformations, of which spatial rotations are a particular case,

$$\delta x_\mu = \omega_\mu^\nu \, x_\nu \tag{3.170}$$

where $\omega^{\mu\nu}$ is infinitesimal and antisymmetric, then the variation of the action is zero: $\delta S = 0$. If ϕ is a scalar field, then $\delta_T \phi$ is also zero. This is not the case for spinor or vector fields, which transform under Lorentz transformations. Because of the transformation properties of these fields, the angular momentum tensor that we define below will be missing the contribution representing the spin of the field (which vanishes for a scalar field). Here we consider only the case of scalars.

Then, since for a scalar field $\delta_T \phi = 0$, we find

$$\delta S = 0 = \int d^4x \, \partial_\mu \left[\left(g^{\mu\nu} \mathcal{L} - \frac{\delta \mathcal{L}}{\delta \partial_\mu \phi} \partial^\nu \phi \right) \omega^{\nu\rho} \, x_\rho \right] \tag{3.171}$$

Since $\omega^{\nu\rho}$ is constant and arbitrary, the quantity in brackets must also define a conserved current.

Let $M^{\mu\nu\rho}$ be the tensor defined by

$$M^{\mu\nu\rho} \equiv T^{\mu\nu} x^\rho - T^{\mu\rho} x^\nu \tag{3.172}$$

in terms of which the quantity in brackets in eq. (3.171) becomes $\frac{1}{2}\omega_{\nu\rho} M^{\mu\nu\rho}$. Thus, for an arbitrary constant ω, we find that the tensor $M^{\mu\nu\rho}$ is locally conserved:

$$\partial_\mu M^{\mu\nu\rho} = 0 \tag{3.173}$$

In particular, the transformation

$$\delta x_0 = 0, \qquad \delta x_j = \omega_{jk} x_k \tag{3.174}$$

represents an infinitesimal rotation of the spatial axes with

$$\omega_{jk} = \epsilon_{jkl} \theta_l \tag{3.175}$$

where the θ_l are three infinitesimal Euler angles. Thus, we suspect that $M^{\mu\nu\rho}$ must be related to the total angular momentum. Indeed, the local conservation of the current $M^{\mu\nu\rho}$ leads to the global conservation of the tensorial quantity $L^{\nu\rho}$:

$$L^{\nu\rho} \equiv \int d^3x \, M^{0\nu\rho}(\boldsymbol{x}, x_0) \tag{3.176}$$

In particular, the space components of $L_{\nu\rho}$ are

$$L_{jk} = \int d^3x \, (T_{0j}(x) \, x_k - T_{0k}(x) \, x_j)$$
$$= \int d^3x \, (\mathcal{P}_j(x) \, x_j - \mathcal{P}_k(x) \, x_j) \tag{3.177}$$

If we denote by L_j the (pseudo) vector

$$L_j \equiv \frac{1}{2} \, \epsilon_{jkl} \, L_{kl} \tag{3.178}$$

we get

$$L_j \equiv \int d^3x \, \epsilon_{jkl} \, x_k \, \mathcal{P}_l(x) \equiv \int d^3x \, \ell_j(x) \tag{3.179}$$

The vector L_j is the generator of infinitesimal rotations and is thus identified with the total angular momentum, whereas $\ell_j(x)$ is the corresponding (spatial) angular momentum density. Notice that, since we are dealing with a scalar field, there is no spin contribution to the angular momentum density.

The generalized angular momentum tensor $L^{\nu\rho}$ of eq. (3.176) is not translationally invariant, since under a displacement of the origin of the coordinate system by a^μ, $L^{\nu\rho}$ changes by an amount $a^\nu P^\rho - a^\rho P^\nu$. A truly intrinsic angular momentum is given by the Pauli-Lubanski vector W^μ,

$$W^\mu = -\frac{1}{2} \epsilon^{\mu\nu\lambda\rho} \frac{L^{\nu\lambda} P^\rho}{\sqrt{P^2}} \tag{3.180}$$

which, in the rest frame $\boldsymbol{P} = 0$, reduces to the angular momentum.

Finally, we find that if the angular momentum tensor $M^{\mu\nu\lambda}$ has the form

$$M^{\mu\nu\lambda} = T^{\mu\nu} x^\lambda - T^{\mu\lambda} x^\nu \tag{3.181}$$

then the conservation of the energy-momentum tensor $T^{\mu\nu}$ and of the angular momentum tensor $M^{\mu\nu\lambda}$ together lead to the condition that the energy-momentum tensor should be a symmetric second rank tensor:

$$T^{\mu\nu} = T^{\nu\mu} \tag{3.182}$$

Thus, we conclude that the conservation of angular momentum requires that the energy-momentum tensor $T^{\mu\nu}$ for a scalar field be symmetric.

The expression for $T^{\mu\nu}$ that we derived in eq. (3.161) is not manifestly symmetric. However, if $T^{\mu\nu}$ is conserved, then the "improved" tensor $\widetilde{T}^{\mu\nu}$

$$\widetilde{T}^{\mu\nu} = T^{\mu\nu} + \partial_\lambda K^{\mu\nu\lambda} \tag{3.183}$$

is also conserved, provided the tensor $K^{\mu\nu\lambda}$ is antisymmetric in (μ, λ) and (ν, λ). It is always possible to find such a tensor $K^{\mu\nu\lambda}$ to make $\widetilde{T}^{\mu\nu}$ symmetric. The improved, symmetric, energy-momentum tensor is known as the *Belinfante energy-momentum tensor.*

In particular, for the scalar field $\phi(x)$ whose Lagrangian density $\mathcal{L}$ is

$$\mathcal{L}=\frac{1}{2}(\partial_\mu\phi)^2-V(\phi) \tag{3.184}$$

the locally conserved energy-momentum tensor $T^{\mu\nu}$ is

$$T^{\mu\nu}=-g^{\mu\nu}\mathcal{L}+\frac{\delta\mathcal{L}}{\delta\partial_\mu\phi}\partial^\nu\phi\equiv-g^{\mu\nu}\mathcal{L}+\partial^\mu\phi\partial^\nu\phi \tag{3.185}$$

which is symmetric. The conserved energy-momentum 4-vector is

$$P^\mu=\int d^3x\,(-g^{0\mu}\mathcal{L}+\partial^0\phi\partial^\mu\phi) \tag{3.186}$$

Thus, we find that

$$P^0=\int d^3x\,(\Pi\,\partial_0\phi-\mathcal{L})=\int d^3x\left[\frac{1}{2}\Pi^2+\frac{1}{2}(\nabla\phi)^2+V(\phi)\right] \tag{3.187}$$

is the total energy of the field, and

$$\boldsymbol{P}=\int d^3x\,\Pi(x)\,\nabla\phi(x) \tag{3.188}$$

is the linear momentum $\boldsymbol{P}$ of the field. Both are constants of motion.

3.9 The energy-momentum tensor for the electromagnetic field

For the case of the Maxwell field A_μ, a straightforward application of these methods yields an energy-momentum tensor $T^{\mu\nu}$ of the form

$$\begin{aligned}T^{\mu\nu}&=-g^{\mu\nu}\mathcal{L}+\frac{\delta\mathcal{L}}{\delta\partial_\nu A_\lambda}\partial^\mu A_\lambda\\&=\frac{1}{4}g^{\mu\nu}F^{\alpha\beta}F_{\alpha\beta}-F^{\nu\lambda}\partial^\mu A_\lambda\end{aligned} \tag{3.189}$$

It obeys $\partial_\mu T^{\mu\nu}=0$, and hence is locally conserved. However, this tensor is not gauge invariant. We can construct a gauge-invariant and conserved energy-momentum tensor by exploiting the ambiguity in the definition of $T^{\mu\nu}$. Thus, if we choose $K_{\mu\nu\lambda}=F_{\nu\lambda}A_\mu$, which is antisymmetric in the indices ν and λ, we can construct the required gauge-invariant and conserved energy-momentum tensor,

$$\widetilde{T}^{\mu\nu}=\frac{1}{4}g^{\mu\nu}F^2-F^\nu_\lambda F^{\mu\lambda} \tag{3.190}$$

where we used the equation of motion of the free electromagnetic field, $\partial^\lambda F_{\nu\lambda}=0$. Notice that this "improved" energy-momentum tensor is both gauge invariant and symmetric.

From here we find that the 4-vector

$$P^{\mu} = \int_{x_0 \text{ fixed}} d^3x \, \widetilde{T}^{\mu 0} \tag{3.191}$$

is a constant of motion. Thus, we identify

$$P^0 = \int_{x_0 \text{ fixed}} d^3x \, \widetilde{T}^{00} = \int_{x_0 \text{ fixed}} d^3x \, \frac{1}{2} \left(\boldsymbol{E}^2 + \boldsymbol{B}^2 \right) \tag{3.192}$$

with the Hamiltonian, and

$$P^i = \int_{x_0 \text{ fixed}} d^3x \, \widetilde{T}^{i0} = \int_{x_0 \text{ fixed}} d^3x \, (\boldsymbol{E} \times \boldsymbol{B})_i \tag{3.193}$$

with the linear momentum (or Poynting vector) of the electromagnetic field.

3.10 The energy-momentum tensor and changes in the geometry

The energy-momentum tensor $T^{\mu\nu}$ appears in classical field theory as a result of the translation invariance, in both space and time, of the physical system. We have seen in section 3.9 that for a scalar field, $T^{\mu\nu}$ is a symmetric tensor as a consequence of the conservation of angular momentum. The definition we found does not require that $T^{\mu\nu}$ should have any definite symmetry. However, we found that it is always possible to modify $T^{\mu\nu}$ by adding a suitably chosen antisymmetric conserved tensor to find a symmetric version of $T^{\mu\nu}$. Given this fact, it is natural to ask whether there is a way to define the energy-momentum tensor in a such a way that it is always symmetric. This issue becomes important if we want to consider theories for systems containing fields that are not scalars.

It turns out that it is possible to regard $T^{\mu\nu}$ as the change in the action due to a change of the geometry in which the system lives. From classical physics, we know that when a body is distorted in some manner, in general its energy increases, since we have to perform some work against the body to deform it. A deformation of a body is a change of the geometry, in which its component parts evolve. Examples of such changes of geometry are shear distortions, dilatations, bending, and twists. In contrast, some changes do not cost any energy, since they are symmetry operations. Examples of symmetry operations are translations and rotations. These symmetry operations can be viewed as simple changes in the coordinates of the parts of the body, which do not change its geometrical properties (e.g., the distances and angles between different points). Thus, coordinate transformations do not alter the energy of the system. The same type of arguments apply to any dynamical system. In the most general case, we have to consider transformations that leave the action invariant. This leads us to consider how changes in the geometry of the spacetime affect the action of a dynamical system.

The information about the geometry in which a system evolves is encoded in the *metric tensor* of the space (and spacetime). The metric tensor is a symmetric tensor that specifies how to measure the distance $|ds|$ between a pair of nearby points x and $x + dx$:

$$ds^2 = g^{\mu\nu}(x) \, dx_{\mu} \, dx_{\nu} \tag{3.194}$$

Under an arbitrary local change of coordinates $x_\mu \to x_\mu + \delta x_\mu$, the metric tensor changes as follows:

$$g'_{\mu\nu}(x') = g_{\lambda\rho}(x) \frac{\partial x^\lambda}{\partial x'^\mu} \frac{\partial x^\rho}{\partial x'^\nu} \tag{3.195}$$

For an infinitesimal change, the functional change in the metric tensor $\delta g_{\mu\nu}$ is then

$$\delta g_{\mu\nu} = -\frac{1}{2}\left(g_{\mu\lambda}\partial^\nu \delta x_\lambda + g_{\lambda\nu}\partial^\mu \delta x_\lambda + \partial^\lambda g_{\mu\nu}\delta x_\lambda\right) \tag{3.196}$$

The volume element, invariant under coordinate transformations, is $d^4x\,\sqrt{g}$, where g is the determinant of the metric tensor.

Coordinate transformations change the metric of spacetime but do not change the action. *Physical* or geometric changes are changes in the metric tensor that are not due to coordinate transformations. For a system in a space with metric tensor $g^{\mu\nu}(x)$, not necessarily the Minkowski (or Euclidean) metric, the *change* in the action is a linear function of the infinitesimal change in the metric $\delta g^{\mu\nu}(x)$ (i.e., "Hooke's law"). Thus, we can write the change of the action due to an arbitrary infinitesimal change of the metric in the form

$$\delta S = \int d^4x\,\sqrt{g}\; T^{\mu\nu}(x)\,\delta g_{\mu\nu}(x) \tag{3.197}$$

Below we will identify the proportionality constant, the tensor $T^{\mu\nu}$, with the conserved energy-momentum tensor. This definition implies that $T^{\mu\nu}$ can be regarded as the derivative of the action with respect to the metric

$$T^{\mu\nu}(x) \equiv \frac{\delta S}{\delta g_{\mu\nu}(x)} \tag{3.198}$$

Since the metric tensor is symmetric, this definition always yields a symmetric energy-momentum tensor.

To prove that this definition of the energy-momentum tensor agrees with the one we obtained in section 3.9 (which was not unique!), we have to prove that the form of $T^{\mu\nu}$ just defined is a conserved current for coordinate transformations. Under an arbitrary local change of coordinates, which leave the distance ds unchanged, the metric tensor changes by the $\delta g_{\mu\nu}$ given in eq. (3.196). The change of the action δS must be zero in this case. If we substitute the expression for $\delta g_{\mu\nu}$ in δS, then an integration by parts will yield a conservation law. Indeed, for the particular case of a flat metric, such as the Minkowski or Euclidean metrics, the change δS is

$$\delta S = -\frac{1}{2}\int d^4x\, T^{\mu\nu}(x)\,\left(g_{\mu\lambda}\partial_\nu \delta x^\lambda + g_{\lambda\nu}\partial_\mu \delta x^\lambda\right) \tag{3.199}$$

since for a global coordinate transformation, the Jacobian factor $\sqrt{g}$ and the measure of spacetime d^4x are constant. Thus, if δS is to vanish for an arbitrary change δx, the tensor $T^{\mu\nu}(x)$ has to be a locally conserved current: that is, $\partial_\mu T^{\mu\nu}(x) = 0$. This definition can be extended to the case of more general spaces. Note, however, that the energy-momentum tensor can be made symmetric only if the space does not have a property known as *torsion*.

Finally, note that this definition of the energy-momentum tensor also allows us to identify the spatial components of $T^{\mu\nu}$ with the stress-energy tensor of the system.

Indeed, for a change in geometry that does not vary with time, the change in the action reduces to a change in the total energy of the system. Hence, the space components of the energy-momentum tensor tell us how much the total energy changes for a specific deformation of the geometry. But this is precisely what the stress energy tensor is!

Exercises

3.1 Symmetries and conservation laws of the abelian Higgs model

Here you will look again at a *charged* (complex) scalar field $\phi(x)$ coupled to the electromagnetic field $A_\mu(x)$. Recall that the Lagrangian density $\mathcal{L}$ for this system is

$$\mathcal{L} = \left(D_\mu \phi(x)\right)^* \left(D_\mu \phi(x)\right) - m_0^2 |\phi(x)|^2 - \frac{\lambda}{2} \left(|\phi(x)|^2\right)^2 - \frac{1}{4} F^{\mu\nu} F_{\mu\nu} \tag{3.200}$$

where D_μ is the *covariant derivative*,

$$D_\mu \equiv \partial_\mu + ieA_\mu \tag{3.201}$$

e is the electric charge, and * denotes complex conjugation. In this problem set, you will determine several important properties of this field theory at the classical level.

1) Derive an expression for the *locally conserved current* $j_\mu(x)$, associated with the global symmetry

$$\begin{aligned} \phi(x) &\to \phi'(x) = e^{i\theta} \phi(x) \\ \phi^*(x) &\to \phi'^*(x) = e^{-i\theta} \phi^*(x) \\ A_\mu(x) &\to A'_\mu(x) = A_\mu(x) \end{aligned} \tag{3.202}$$

in terms of the fields of the theory.

2) Show that the conservation of the current j_μ implies the existence of a *constant of motion*. Find an explicit form for this constant of motion.

3) Consider now the case of the *local* (or *gauge*) transformation

$$\begin{aligned} \phi(x) &\to \phi'(x) = e^{i\theta(x)} \phi(x) \\ \phi^*(x) &\to \phi'^*(x) = e^{-i\theta(x)} \phi^*(x) \\ A_\mu(x) &\to A'_\mu(x) = A_\mu(x) + \partial_\mu \Lambda(x) \end{aligned} \tag{3.203}$$

where $\theta(x)$ and $\Lambda(x)$ are two functions. What should be the relation between $\theta(x)$ and $\Lambda(x)$ for this transformation to be a symmetry of the Lagrangian of the system?

4) Show that, if the system has the local symmetry of problem 3, there is a locally conserved gauge current $J_\mu(x)$. Find an explicit expression for J_μ and discuss in what way it is different from the current j_μ of problem 1). Find an explicit expression for the associated constant of motion, and discuss its physical meaning.

5) Find the energy-momentum tensor $T^{\mu\nu}$ for this system. Show that it can be written as the sum of two terms,

$$T^{\mu\nu} = T^{\mu\nu}(A) + T^{\mu\nu}(\phi, A) \tag{3.204}$$

where $T^{\mu\nu}(A)$ is the energy-momentum tensor for the free electromagnetic field, and $T^{\mu\nu}(\phi, A)$ is the tensor that results from modifying the energy-momentum tensor for the decoupled complex scalar field ϕ by the *minimal coupling* procedure.

6) Find explicit expressions for the Hamiltonian $\mathcal{H}(x)$ and the linear momentum $\mathcal{P}^j(x)$ densities for this system. Give a physical interpretation for all terms that you have found for each quantity.

7) Consider now the case of an infinitesimal Lorentz transformation

$$x_\mu \to x'_\mu + \omega_{\mu\nu} x^\nu \tag{3.205}$$

where $\omega_{\mu\nu}$ is infinitesimal and antisymmetric. Show that the invariance of the Lagrangian of this system under these Lorentz transformations leads to the existence of a conserved tensor $M_{\mu\nu\lambda}$. Find the explicit form of this tensor. Give a physical interpretation for its spatial components. Does the conservation of this tensor impose any restriction on the properties of the energy-momentum tensor $T^{\mu\nu}$? Explain.

8) In this problem, you will consider again the same system but in a *polar* representation for the scalar field ϕ:

$$\phi(x) = \rho(x)\, e^{i\omega(x)} \tag{3.206}$$

In exercise 2.5, you showed that for $m_0^2 < 0$, the lowest energy states of the system can be well approximated by freezing the amplitude mode ρ to a constant value ρ_0, which you obtained by an energy minimization argument. In the present problem, find the form of (i) the conserved gauge current J_μ, (ii) the total energy E, and (iii) the total linear momentum $\boldsymbol{P}$ in this limit.

9) Consider now the analytic continuation to imaginary time of this theory. Find the energy functional of the equivalent system in classical statistical mechanics. Give a physical interpretation for each of the terms of this energy functional. If D is the dimensionality of spacetime for the original system, what is the dimensionality of *space* for the equivalent classical problem? Warning: Be very careful in the way you continue the components of the vector potential.

4

Canonical Quantization

We now begin the discussion of our main subject of interest: the role of quantum mechanical fluctuations in systems with infinitely many degrees of freedom. We begin with a brief overview of quantum mechanics of a single particle.

4.1 Elementary quantum mechanics

Elementary quantum mechanics describes the quantum dynamics of systems with a finite number of degrees of freedom. Two axioms are involved in the standard procedure for quantizing a classical system. Let $L(q, \dot{q})$ be the Lagrangian of an abstract dynamical system described by the generalized coordinate q. In chapter 2, we recalled that the canonical formalism of classical mechanics is based on the concept of canonical pairs of dynamical variables. So the canonical coordinate q has for its partner the canonical momentum p:

$$p = \frac{\partial L}{\partial \dot{q}} \tag{4.1}$$

In the canonical formalism, the dynamics of the system is governed by the classical Hamiltonian

$$H(q, p) = p\dot{q} - L(q, \dot{q}) \tag{4.2}$$

which is the Legendre transform of the Lagrangian. In the canonical (Hamiltonian) formalism, the equations of motion are just Hamilton's equations:

$$\dot{p} = -\frac{\partial H}{\partial q}, \qquad \dot{q} = \frac{\partial H}{\partial p} \tag{4.3}$$

The dynamical state of the system is defined by the values of the canonical coordinates and momenta at any given time t. As a result of these definitions, the coordinates and momenta satisfy a set of Poisson bracket relations,

$$\{q, p\}_{PB} = 1, \quad \{q, q\}_{PB} = \{p, p\}_{PB} = 0 \tag{4.4}$$

where

$$\{A, B\}_{PB} \equiv \frac{\partial A}{\partial q}\frac{\partial B}{\partial p} - \frac{\partial A}{\partial p}\frac{\partial B}{\partial q} \tag{4.5}$$

In quantum mechanics, the primitive (or fundamental) notion is the concept of a *physical state*. A physical state of a system is represented by a *state vector* in an abstract vector space, which is called the Hilbert space $\mathcal{H}$ of quantum states. The space $\mathcal{H}$ is a vector space in the sense that if two vectors, $|\Psi\rangle \in \mathcal{H}$ and $|\Phi\rangle \in \mathcal{H}$, represent physical states, then the linear superposition $|a\Psi + b\Phi\rangle = a|\Psi\rangle + b|\Phi\rangle$, where a and b are two arbitrary complex numbers, also represents a physical state, and thus it is an element of the Hilbert space (i.e., $|a\Psi + b\Phi\rangle \in \mathcal{H}$). The *superposition principle* is an axiom of quantum mechanics.

In quantum mechanics, the dynamical variables (i.e., the generalized coordinates $\hat{q}$ and the associated canonical momenta $\hat{p}$), the Hamiltonian H, and so forth are represented by operators that act linearly on the Hilbert space of states. Hence, quantum mechanics is a linear theory, even though the physical observables obey nonlinear Heisenberg equations of motion. Let us denote by $\hat{A}$ an arbitrary operator acting on the Hilbert space $\mathcal{H}$. The result of acting on the state $|\Psi\rangle \in \mathcal{H}$ with the operator $\hat{A}$ is another state $|\Phi\rangle \in \mathcal{H}$:

$$\hat{A}|\Psi\rangle = |\Phi\rangle \tag{4.6}$$

The Hilbert space $\mathcal{H}$ is endowed with an *inner product*. An inner product is an operation that assigns a complex number $\langle\Phi|\Psi\rangle$ to a given pair of states $|\Phi\rangle \in \mathcal{H}$ and $|\Psi\rangle \in \mathcal{H}$.

Since $\mathcal{H}$ is a vector space, there exists a set of linearly independent states $\{|\lambda\rangle\}$, called a *basis*, that spans the entire Hilbert space. Thus, an arbitrary state $|\Psi\rangle$ can be expanded as a linear combination of a complete set of states that form a basis of $\mathcal{H}$,

$$|\Psi\rangle = \sum_{\lambda} \Psi_{\lambda}\, |\lambda\rangle \tag{4.7}$$

which is unique for a fixed set of basis states. The basis states can be chosen to be *orthonormal* with respect to the inner product:

$$\langle\lambda|\mu\rangle = \delta_{\lambda\mu} \tag{4.8}$$

In general, if $|\Psi\rangle$ and $|\Phi\rangle$ are normalized states,

$$\langle\Psi|\Psi\rangle = \langle\Phi|\Phi\rangle = 1 \tag{4.9}$$

then the action of an operator $\hat{A}$ on a state $|\Psi\rangle$ is proportional to a (generally different) state $|\Phi\rangle$:

$$\hat{A}|\Psi\rangle = \alpha|\Phi\rangle \tag{4.10}$$

The coefficient α is a complex number that depends on the pair of states and on the operator $\hat{A}$. This coefficient is the *matrix element* of $\hat{A}$ between the states $|\Psi\rangle$ and $|\Phi\rangle$, which we write with the notation

$$\alpha = \langle\Phi|\hat{A}|\Psi\rangle \tag{4.11}$$

Operators that act on a Hilbert space do not generally commute with each other. One of the axioms of quantum mechanics is the *correspondence principle*, which states that in the classical limit, $\hbar \to 0$, the operators should effectively become numbers and commute with each other in the classical limit.

The procedure of *canonical quantization* consists of demanding that to the classical canonical pair (q, p) that satisfies the Poisson bracket $\{q, p\}_{PB} = 1$, we associate a pair of operators $\hat{q}$ and $\hat{p}$, acting on the Hilbert space of states $\mathcal{H}$, which obey the *canonical commutation relations*:

$$[\hat{q}, \hat{p}] = i\hbar \qquad [\hat{q}, \hat{q}] = [\hat{p}, \hat{p}] = 0 \tag{4.12}$$

Here $[\hat{A}, \hat{B}]$ is the commutator of the operators $\hat{A}$ and $\hat{B}$:

$$[\hat{A}, \hat{B}] = \hat{A}\hat{B} - \hat{B}\hat{A} \tag{4.13}$$

In particular, two operators that do not commute with each other cannot be diagonalized simultaneously. Hence, it is not possible to measure simultaneously with arbitrary precision in the same physical state two noncommuting observables. This is the *uncertainty principle.*

By following this prescription, we assign to the classical Hamiltonian $H(q, p)$, which is a function of the dynamical variables q and p, an operator $\hat{H}(\hat{q}, \hat{p})$ obtained by replacing the dynamical variables with the corresponding operators. Other classical dynamical quantities are similarly associated in quantum mechanics with quantum operators that act on the Hilbert space of states. Moreover, in quantum mechanics, all operators associated with classical physical quantities in quantum mechanics are *hermitian operators* relative to the inner product defined in the Hilbert space $\mathcal{H}$. Namely, if $\hat{A}$ is an operator and $\hat{A}^\dagger$ is the *adjoint* of $\hat{A}$,

$$\langle \Psi | \hat{A}^\dagger | \Phi \rangle \equiv \langle \Phi | \hat{A} | \Psi \rangle^* \tag{4.14}$$

then $\hat{A}$ is hermitian iff $\hat{A} = \hat{A}^\dagger$ (with suitable boundary conditions).

The quantum mechanical state of the system at time t, $|\Psi(t)\rangle$, obeys the Schrödinger equation:

$$i\hbar \frac{\partial}{\partial t} |\Phi(t)\rangle = \hat{H}(\hat{q}, \hat{p})\, |\Psi(t)\rangle \tag{4.15}$$

The state $|\Psi(t)\rangle$ is uniquely determined by the initial state $|\Psi(0)\rangle$. Thus, in quantum mechanics, just as in classical mechanics, the Hamiltonian is the generator of the (infinitesimal) time evolution of the state of a physical system.

It is always possible to choose a basis in which a particular operator is diagonal. For instance, if the operator is the canonical coordinate $\hat{q}$, a possible set of basis states are the eigenstates of $\hat{q}$, labeled by q, that is,

$$\hat{q}|q\rangle = q\, |q\rangle \tag{4.16}$$

The basis states $\{|q\rangle\}$ are orthonormal and complete:

$$\langle q | q' \rangle = \delta(q - q'), \qquad \hat{I} = \int dq\, |q\rangle \langle q| \tag{4.17}$$

A state vector $|\Psi\rangle$ can be expanded in an arbitrary basis. If the basis of states is $\{|q\rangle\}$, the expansion is

$$|\Psi\rangle = \sum_q \Psi(q)\, |q\rangle \equiv \int_{-\infty}^{+\infty} dq\, \Psi(q)\, |q\rangle \tag{4.18}$$

where we used the property that the eigenvalues of the coordinate q are the real numbers. The coefficients $\Psi(q)$ of this expansion,

$$\Psi(q) = \langle q|\Psi\rangle \tag{4.19}$$

(i.e., the amplitude required to find the system at coordinate q in this state), are the values of the wave function of the state $|\Psi\rangle$ in the *coordinate representation.*

Since the canonical momentum $\hat{p}$ does not commute with $\hat{q}$, it is not diagonal in this representation. Just as in classical mechanics, in quantum mechanics the momentum operator $\hat{p}$ is the generator of infinitesimal displacements. Consider the states $|q\rangle$ and $\exp(-\frac{i}{\hbar}a\hat{p})\,|q\rangle$. It is easy to prove that the latter is the state $|q+a\rangle$, since

$$\hat{q}\,\exp\left(-\frac{i}{\hbar}a\hat{p}\right)|q\rangle \equiv \hat{q}\sum_{n=0}^{\infty}\frac{1}{n!}\left(\frac{-ia}{\hbar}\right)^{n}\hat{p}^{n}\,|q\rangle \tag{4.20}$$

Using the commutation relation $[\hat{q},\hat{p}] = i\hbar$, it is easy to show that

$$[\hat{q},\hat{p}^{n}] = i\hbar n\hat{p}^{n-1} \tag{4.21}$$

Hence, we can write

$$\hat{q}\,\exp\left(-\frac{i}{\hbar}a\hat{p}\right)|q\rangle = (q+a)\,\exp\left(-\frac{i}{\hbar}a\hat{p}\right)|q\rangle \tag{4.22}$$

Thus,

$$\exp\left(-\frac{i}{\hbar}a\hat{p}\right)|q\rangle = |q+a\rangle \tag{4.23}$$

This is a unitary transformation, that is,

$$\left(\exp\left(-\frac{i}{\hbar}a\hat{p}\right)\right)^{-1} = \left(\exp\left(-\frac{i}{\hbar}a\hat{p}\right)\right)^{\dagger} \tag{4.24}$$

We can now use this property to compute the matrix element

$$\langle q|\,\exp\left(\frac{i}{\hbar}a\hat{p}\right)|\Psi\rangle \equiv \Psi(q+a) \tag{4.25}$$

For a infinitesimally small, it can be approximated by

$$\Psi(q+a) \approx \Psi(q) + \frac{i}{\hbar}a\langle q|\hat{p}|\Psi\rangle + \cdots \tag{4.26}$$

We find that the matrix element for $\hat{p}$ has to satisfy

$$\langle q|\hat{p}|\Psi\rangle = \frac{\hbar}{i}\lim_{a\to 0}\frac{\Psi(q+a)-\Psi(q)}{a} \tag{4.27}$$

Thus, the operator $\hat{p}$ is represented by a differential operator:

$$\langle q|\hat{p}|\Psi\rangle \equiv \frac{\hbar}{i}\frac{\partial}{\partial q}\Psi(q) = \frac{\hbar}{i}\frac{\partial}{\partial q}\langle q|\Psi\rangle \tag{4.28}$$

It is easy to check that the coordinate representation of the operator

$$\hat{p} = \frac{\hbar}{i}\frac{\partial}{\partial q} \tag{4.29}$$

and the coordinate operator $\hat{q}$ satisfy the commutation relation $[\hat{q}, \hat{p}] = i\hbar$.

4.2 Canonical quantization in field theory

We will now apply the axioms of quantum mechanics to a classical field theory. The result will be a QFT. For the sake of simplicity, we first consider the case of a scalar field $\phi(x)$. We have seen before that, given a Lagrangian density $\mathcal{L}(\phi, \partial_\mu\phi)$, the Hamiltonian can be found once the canonical momentum $\Pi(x)$ is defined:

$$\Pi(x) = \frac{\delta\mathcal{L}}{\delta\partial_0\phi(x)} \tag{4.30}$$

On a given time surface x_0, the classical Hamiltonian is

$$H = \int d^3x \left[\Pi(\boldsymbol{x}, x_0)\partial_0\phi(\boldsymbol{x}, x_0) - \mathcal{L}(\phi, \partial_\mu\phi)\right] \tag{4.31}$$

We quantize this theory by assigning to each dynamical variable of the classical theory a hermitian operator that acts on the Hilbert space of the quantum states of the system. Thus, the field $\hat{\phi}(\boldsymbol{x})$ and the canonical momentum $\widehat{\Pi}(\boldsymbol{x})$ are operators acting on a Hilbert space. These operators obey equal-time canonical commutation relations:

$$\left[\hat{\phi}(\boldsymbol{x}), \widehat{\Pi}(\boldsymbol{y})\right] = i\hbar\delta(\boldsymbol{x} - \boldsymbol{y}) \tag{4.32}$$

In the field representation, the Hilbert space is the vector space of wave functions Ψ, which are functionals of the configurations of the field (at a fixed time), which we denote by $\{\phi(\boldsymbol{x})\}$. In this notation, the wave functionals (i.e., the amplitude to find the state of the field in a given configuration) are

$$\Psi[\{\phi(\boldsymbol{x})\}] \equiv \langle\{\phi(\boldsymbol{x})\}|\Psi\rangle \tag{4.33}$$

In this representation, the field is a diagonal operator:

$$\langle\{\phi\}|\hat{\phi}(\boldsymbol{x})|\Psi\rangle \equiv \phi(\boldsymbol{x})\,\langle\{\phi(\boldsymbol{x})\}|\Psi\rangle = \phi(\boldsymbol{x})\,\Psi[\{\phi\}] \tag{4.34}$$

The canonical momentum $\widehat{\Pi}(\boldsymbol{x})$ is not diagonal in this representation, but it acts on the wave functionals as a functional differential operator:

$$\langle\{\phi\}|\widehat{\Pi}(\boldsymbol{x})|\Psi\rangle \equiv \frac{\hbar}{i}\frac{\delta}{\delta\phi(\boldsymbol{x})}\Psi[\{\phi\}] \tag{4.35}$$

What we just described is the Schrödinger picture of QFT. In this picture, as usual, the operators are time independent but the states are time dependent and satisfy the Schrödinger equation:

$$i\hbar\partial_0\Psi[\{\phi\},x_0]=\widehat{H}\Psi[\{\phi\},x_0] \tag{4.36}$$

For the particular case of a scalar field ϕ with the classical Lagrangian $\mathcal{L}$,

$$\mathcal{L}=\frac{1}{2}(\partial_\mu\phi)^2-V(\phi) \tag{4.37}$$

the quantum mechanical Hamiltonian operator $\hat{H}$ is

$$\widehat{H}=\int d^3x\left\{\frac{1}{2}\widehat{\Pi}^2(\boldsymbol{x})+\frac{1}{2}(\nabla\hat{\phi}(\boldsymbol{x}))^2+V(\boldsymbol{\phi}(\boldsymbol{x}))\right\} \tag{4.38}$$

The stationary states are the eigenstates of the Hamiltonian $\widehat{H}$.

While it is possible to proceed further with the Schrödinger picture, the manipulation of wave functionals becomes very cumbersome rather quickly. For this reason, the Heisenberg picture is commonly used.

In the Schrödinger picture, the time evolution of the system is encoded in the time dependence of the states. In contrast, in the Heisenberg picture, the operators are time dependent, while the states are time independent. The operators of the Heisenberg picture obey quantum mechanical equations of motion.

Let $\hat{A}$ be some operator that acts on the Hilbert space of states. Let us denote by $\hat{A}_H(x_0)$ the Heisenberg operator at time x_0, defined by

$$\hat{A}_H(x_0)=e^{\frac{i}{\hbar}\hat{H}x_0}\,\hat{A}\,e^{-\frac{i}{\hbar}\hat{H}x_0} \tag{4.39}$$

for a system with a time-independent Hamiltonian $\widehat{H}$. It is straightforward to check that $\hat{A}_H(x_0)$ obeys the equation of motion:

$$i\hbar\partial_0\hat{A}_H(x_0)=[\hat{A}_H(x_0),\widehat{H}] \tag{4.40}$$

Notice that in the classical limit, the dynamical variable $A(x_0)$ obeys the classical equation of motion

$$\partial_0A(x_0)=\{A(x_0),H\}_{PB} \tag{4.41}$$

where it is assumed that all the time dependence in A comes from the time dependence of the field (the "coordinates") and of their canonical momenta.

In the Heisenberg picture, both $\hat{\phi}(\boldsymbol{x},x_0)$ and $\widehat{\Pi}(\boldsymbol{x},x_0)$ are time-dependent operators that obey the quantum mechanical equations of motion:

$$i\hbar\partial_0\hat{\phi}(\boldsymbol{x},x_0)=\left[\hat{\phi}(\boldsymbol{x},x_0),\widehat{H}\right],\quad i\hbar\partial_0\widehat{\Pi}(\boldsymbol{x},x_0)=[\widehat{\Pi}(\boldsymbol{x},x_0),\widehat{H}] \tag{4.42}$$

The Heisenberg field operators $\hat{\phi}(\boldsymbol{x},x_0)$ and $\widehat{\Pi}(\boldsymbol{x},x_0)$ (we will omit the subindex "H" from now on) obey *equal-time commutation relations*:

$$\left[\hat{\phi}(\boldsymbol{x},x_0),\widehat{\Pi}(\boldsymbol{y},x_0)\right]=i\hbar\delta(\boldsymbol{x}-\boldsymbol{y}) \tag{4.43}$$

4.3 Quantization of the free scalar field theory

We will now quantize the theory of a relativistic scalar field $\phi(x)$. In particular, we consider a *free real* scalar field ϕ whose Lagrangian density is

$$\mathcal{L} = \frac{1}{2}(\partial_\mu \phi)(\partial^\mu \phi) - \frac{1}{2} m^2 \phi^2 \tag{4.44}$$

The quantum mechanical Hamiltonian $\widehat{H}$ for a free real scalar field is

$$\widehat{H} = \int d^3x \left[\frac{1}{2} \widehat{\Pi}^2(\boldsymbol{x}) + \frac{1}{2} \left(\nabla \hat{\phi}(\boldsymbol{x}) \right)^2 + \frac{1}{2} m^2 \hat{\phi}^2(\boldsymbol{x}) \right] \tag{4.45}$$

where $\hat{\phi}$ and $\widehat{\Pi}$ satisfy the equal-time commutation relations (in units with $\hbar = c = 1$):

$$[\hat{\phi}(\boldsymbol{x}, x_0), \widehat{\Pi}(\boldsymbol{y}, x_0)] = i\delta(\boldsymbol{x} - \boldsymbol{y}) \tag{4.46}$$

In the Heisenberg representation, $\hat{\phi}$ and $\widehat{\Pi}$ are time-dependent operators, while the states are time independent. The field operators obey the equations of motion:

$$i\partial_0 \hat{\phi}(\boldsymbol{x}, x_0) = [\hat{\phi}(\boldsymbol{x}, x_0), \widehat{H}], \qquad i\partial_0 \widehat{\Pi}(\boldsymbol{x}, x_0) = [\widehat{\Pi}(\boldsymbol{x}, x_0), \widehat{H}] \tag{4.47}$$

These are operator equations. After some algebra, we find that the Heisenberg equations of motion for the field and for the canonical momentum are

$$\partial_0 \hat{\phi}(\boldsymbol{x}, x_0) = \widehat{\Pi}(\boldsymbol{x}, x_0) \tag{4.48}$$

$$\partial_0 \widehat{\Pi}(\boldsymbol{x}, x_0) = \nabla^2 \hat{\phi}(\boldsymbol{x}, x_0) - m^2 \hat{\phi}(\boldsymbol{x}, x_0) \tag{4.49}$$

Upon substitution, we derive the field equation for the scalar field operator:

$$\left(\partial^2 + m^2 \right) \hat{\phi}(x) = 0 \tag{4.50}$$

Thus, the field operators $\hat{\phi}(x)$ satisfy the Klein-Gordon equation as their Heisenberg equation of motion.

4.3.1 Field expansions

Let us solve the field equation of motion by a Fourier transform,

$$\hat{\phi}(x) = \int \frac{d^3k}{(2\pi)^3} \hat{\phi}(\boldsymbol{k}, x_0)\, e^{i\boldsymbol{k}\cdot\boldsymbol{x}} \tag{4.51}$$

where the $\hat{\phi}(\boldsymbol{k}, x_0)$ are the Fourier amplitudes of $\hat{\phi}(x)$. By demanding that $\hat{\phi}(x)$ satisfies the Klein-Gordon equation, we find that $\hat{\phi}(\boldsymbol{k}, x_0)$ should satisfy

$$\partial_0^2 \hat{\phi}(\boldsymbol{k}, x_0) + (\boldsymbol{k}^2 + m^2)\hat{\phi}(\boldsymbol{k}, x_0) = 0 \tag{4.52}$$

Also, since $\hat{\phi}(\boldsymbol{x}, x_0)$ is a real hermitian field operator, $\hat{\phi}(\boldsymbol{k}, x_0)$ must satisfy

$$\hat{\phi}^\dagger(\boldsymbol{k}, x_0) = \hat{\phi}(-\boldsymbol{k}, x_0) \tag{4.53}$$

The time dependence of $\hat{\phi}(\boldsymbol{k}, x_0)$ is trivial. Let us write $\hat{\phi}(\boldsymbol{k}, x_0)$ as the sum of two terms:

$$\hat{\phi}(\boldsymbol{k}, x_0) = \hat{\phi}_+(\boldsymbol{k}) e^{i\omega(\boldsymbol{k})x_0} + \hat{\phi}_-(\boldsymbol{k}) e^{-i\omega(\boldsymbol{k})x_0} \tag{4.54}$$

The operators $\hat{\phi}_+(\boldsymbol{k})$ and $\hat{\phi}_+^\dagger(\boldsymbol{k})$ are not independent, since the reality condition of the field $\hat{\phi}(\boldsymbol{x}, x_0)$ implies that

$$\hat{\phi}_+(\boldsymbol{k}) = \hat{\phi}_-^\dagger(-\boldsymbol{k}) \quad \hat{\phi}_+^\dagger(\boldsymbol{k}) = \hat{\phi}_-(-\boldsymbol{k}) \tag{4.55}$$

This expansion is a solution of the equation of motion (the Klein-Gordon equation), provided $\omega(\boldsymbol{k})$ is given by

$$\omega(\boldsymbol{k}) = \sqrt{\boldsymbol{k}^2 + m^2} \tag{4.56}$$

Let us define the operator $\hat{a}(\boldsymbol{k})$ and its adjoint $\hat{a}^\dagger(\boldsymbol{k})$ by

$$\hat{a}(\boldsymbol{k}) = 2\omega(\boldsymbol{k})\hat{\phi}_-(\boldsymbol{k}) \quad \hat{a}^\dagger(\boldsymbol{k}) = 2\omega(\boldsymbol{k})\hat{\phi}_-^\dagger(\boldsymbol{k}) \tag{4.57}$$

The operators $\hat{a}^\dagger(\boldsymbol{k})$ and $\hat{a}(\boldsymbol{k})$ obey the (generalized) creation-annihilation operator algebra:

$$[\hat{a}(\boldsymbol{k}), \hat{a}^\dagger(\boldsymbol{k}')] = (2\pi)^3 2\omega(\boldsymbol{k})\, \delta^3(\boldsymbol{k} - \boldsymbol{k}') \tag{4.58}$$

In terms of the operators $\hat{a}^\dagger(\boldsymbol{k})$ and $\hat{a}(\boldsymbol{k})$, the field operator becomes

$$\hat{\phi}(x) = \int \frac{d^3k}{(2\pi)^3 2\omega(\boldsymbol{k})} \left[\hat{a}(\boldsymbol{k}) e^{-i\omega(\boldsymbol{k})x_0 + i\boldsymbol{k}\cdot\boldsymbol{x}} + \hat{a}^\dagger(\boldsymbol{k}) e^{i\omega(\boldsymbol{k})x_0 - i\boldsymbol{k}\cdot\boldsymbol{x}}\right] \tag{4.59}$$

where we have chosen to normalize the operators in such a way that the phase-space factor takes the Lorentz-invariant form $\dfrac{d^3k}{2\omega(\boldsymbol{k})}$.

The canonical momentum can also be expanded in a similar way:

$$\widehat{\Pi}(x) = -i \int \frac{d^3k}{(2\pi)^3 2\omega(k)} \omega(k) \left[\hat{a}(\boldsymbol{k}) e^{-i\omega(\boldsymbol{k})x_0 + i\boldsymbol{k}\cdot\boldsymbol{x}} - \hat{a}^\dagger(\boldsymbol{k}) e^{i\omega(\boldsymbol{k})x_0 - i\boldsymbol{k}\cdot\boldsymbol{x}}\right] \tag{4.60}$$

Notice that, in both expansions, there are terms with positive and negative frequency, and that the terms with positive frequency have *creation* operators $\hat{a}^\dagger(\boldsymbol{k})$, while the terms with negative frequency have *annihilation* operators $\hat{a}(\boldsymbol{k})$. This observation motivates the notation

$$\hat{\phi}(x) = \hat{\phi}_+(x) + \hat{\phi}_-(x) \tag{4.61}$$

where the expansion of $\hat{\phi}_+$ has only positive frequency terms and the expansion of $\hat{\phi}_-$ has only negative frequency terms. This decomposition will turn out to be very useful.

4.3.2 The Hamiltonian and its spectrum

Let us now write the Hamiltonian in terms of the operators $\hat{a}(\boldsymbol{k})$ and $\hat{a}^\dagger(\boldsymbol{k})$. The result is

$$\widehat{H} = \int \frac{d^3k}{(2\pi)^3 2\omega(\boldsymbol{k})} \frac{1}{2}\omega(\boldsymbol{k}) \left(\hat{a}(\boldsymbol{k})\hat{a}^\dagger(\boldsymbol{k}) + \hat{a}^\dagger(\boldsymbol{k})\hat{a}(\boldsymbol{k})\right) \tag{4.62}$$

This Hamiltonian needs to be normal-ordered relative to a ground state, which we will now define.

A: Vacuum state Let $|0\rangle$ be the state that is annihilated by all the operators $\hat{a}(\boldsymbol{k})$:

$$\hat{a}(\boldsymbol{k})|0\rangle = 0 \tag{4.63}$$

Relative to this state, which we will call the *vacuum* state, the Hamiltonian can be written in the form

$$\widehat{H} =: \widehat{H}: + E_0 \tag{4.64}$$

where $:\widehat{H}:$ is normal ordered relative to the state $|0\rangle$. In other words, in $:\widehat{H}:$, all the destruction operators appear to the right of all the creation operators. Therefore, the normal-ordered operator $:\widehat{H}:$ annihilates the vacuum state

$$:\widehat{H}: |0\rangle = 0 \tag{4.65}$$

The real number E_0 is the ground state energy. In this case, it is equal to

$$E_0 = \int d^3k \frac{1}{2}\omega(\boldsymbol{k})\, \delta(0) \tag{4.66}$$

where $\delta(0)$, the delta function at zero momentum, is the IR divergent quantity

$$\delta(0) = \lim_{p\to 0} \delta^3(\boldsymbol{p}) = \lim_{p\to 0} \int \frac{d^3x}{(2\pi)^3} e^{i\boldsymbol{p}\cdot\boldsymbol{x}} = \frac{V}{(2\pi)^3} \tag{4.67}$$

where V is the (infinite) volume of space. Thus, E_0 is extensive and can be written as $E_0 = \varepsilon_0 V$, where ε_0 is the ground state energy density. We find

$$\varepsilon_0 = \int \frac{d^3k}{(2\pi)^3} \frac{\omega(\boldsymbol{k})}{2} \equiv \frac{1}{2} \int \frac{d^3k}{(2\pi)^3} \sqrt{\boldsymbol{k}^2 + m^2} \tag{4.68}$$

Eq. (4.68) is the sum of the zero-point energies of all the oscillators. This quantity is formally divergent, since the integral is dominated by the contributions with large momenta or, what is the same, short distances. This is a *UV divergence*. It is divergent because the system has an infinite number of degrees of freedom even in a finite volume. We will encounter other examples of similar divergences in field theory. It is important to keep in mind that they are not artifacts of our scheme: They result from the fact that the system is in continuous spacetime and has an infinite number of degrees of freedom.

We can take two different points of view with respect to this problem. One possibility is simply to say that the ground state energy is not a physically observable quantity, since any experiment will only yield information on excitation energies, and in this theory, they are finite. Thus, we can simply redefine the zero of the energy by dropping this term. Normal ordering is then just the mathematical statement that all energies are measured relative to that of the ground state. As far as free field theory is concerned, this subtraction is sufficient, since it makes the theory finite.

However, once interactions are considered, divergences will show up in the formal computation of physical quantities. This procedure then requires further subtractions. An alternative approach consists of introducing a regulator or cutoff. The theory is now finite, but one is left with the task of proving that the physics is independent of the cutoff procedure. This is the program of the renormalization group. It is not presently known whether there should be a fundamental cutoff in these theories (i.e., whether there is a more fundamental description of nature at short distances and high energies, as postulated by string theory). However, it is clear that if QFTs are to be regarded as effective *hydrodynamic* theories valid below some high energy scale, then a cutoff is actually natural.

B: Hilbert space We can construct the spectrum of states by inspecting the normal-ordered Hamiltonian:

$$:\widehat{H}:=\int \frac{d^3k}{(2\pi)^3 2\omega(\boldsymbol{k})}\,\omega(\boldsymbol{k})\,\hat{a}^\dagger(\boldsymbol{k})\hat{a}(\boldsymbol{k}) \tag{4.69}$$

This Hamiltonian commutes with the linear momentum $\widehat{\boldsymbol{P}}$

$$\widehat{\boldsymbol{P}}=\int_{x_0 \text{ fixed}} d^3x\,\widehat{\Pi}(\boldsymbol{x},x_0)\nabla\hat{\phi}(\boldsymbol{x},x_0) \tag{4.70}$$

which, up to operator ordering ambiguities, is the quantum mechanical version of the classical linear momentum $\boldsymbol{P}$:

$$\boldsymbol{P}=\int_{x_0} d^3x\,T^{0j}\equiv\int_{x_0} d^3x\,\Pi(\boldsymbol{x},x_0)\nabla\phi(\boldsymbol{x},x_0) \tag{4.71}$$

In Fourier space, $\widehat{\boldsymbol{P}}$ becomes

$$\widehat{\boldsymbol{P}}=\int \frac{d^3k}{(2\pi)^3 2\omega(\boldsymbol{k})}\boldsymbol{k}\,\hat{a}^\dagger(\boldsymbol{k})\hat{a}(\boldsymbol{k}) \tag{4.72}$$

The normal-ordered Hamiltonian $:\hat{H}:$ also commutes with the occupation numbers of the oscillators, $\hat{n}(\boldsymbol{k})$, defined by

$$\hat{n}(\boldsymbol{k})\equiv\hat{a}^\dagger(\boldsymbol{k})\hat{a}(\boldsymbol{k}) \tag{4.73}$$

Since $\{\hat{n}(\boldsymbol{k})\}$ and the Hamiltonian $\hat{H}$ commute with each other, we can use a complete set of eigenstates of $\{\hat{n}(\boldsymbol{k})\}$ to span the Hilbert space. We will regard the excitations counted by $\hat{n}(k)$ as particles that have energy and momentum (in more general theories, they will also have other quantum numbers). Their Hilbert space has an indefinite number of particles, and it is called a *Fock space*. The states $\{|\{n(\boldsymbol{k})\}\rangle\}$ of the Fock space, defined by

$$|\{n(\boldsymbol{k})\}\rangle=\prod_{\boldsymbol{k}}\mathcal{N}(\boldsymbol{k})[\hat{a}^\dagger(\boldsymbol{k})]^{n(\boldsymbol{k})}|0\rangle \tag{4.74}$$

with $\mathcal{N}(\boldsymbol{k})$ being normalization constants, are eigenstates of the operator $\hat{n}(\boldsymbol{k})$:

$$\hat{n}(\boldsymbol{k})|\{n(\boldsymbol{k})\}\rangle = (2\pi)^3 2\omega(\boldsymbol{k})n(\boldsymbol{k})|\{n(\boldsymbol{k})\}\rangle \tag{4.75}$$

These states span the *occupation-number basis* of the Fock space.

The *total number operator* $\widehat{N}$,

$$\widehat{N} \equiv \int \frac{d^3k}{(2\pi)^3 2\omega(\boldsymbol{k})} \hat{n}(\boldsymbol{k}) \tag{4.76}$$

commutes with the Hamiltonian $\widehat{H}$, and it is diagonal in this basis.

$$\widehat{N}|\{n(\boldsymbol{k})\}\rangle = \int d^3k\, n(\boldsymbol{k})\, |\{n(\boldsymbol{k})\}\rangle \tag{4.77}$$

The energy of these states is

$$\widehat{H}|\{n(\boldsymbol{k})\}\rangle = \left[\int d^3k\, n(\boldsymbol{k})\omega(\boldsymbol{k}) + E_0\right] |\{n(\boldsymbol{k})\}\rangle \tag{4.78}$$

Thus, the *excitation energy* $\varepsilon(\{n(\boldsymbol{k})\})$ of this state is

$$\varepsilon(\boldsymbol{k}) = \int d^3k\, n(\boldsymbol{k})\omega(\boldsymbol{k}) \tag{4.79}$$

The total linear momentum operator $\widehat{\boldsymbol{P}}$ has an operator ordering ambiguity. It will be fixed by requiring that the vacuum state $|0\rangle$ be translationally invariant:

$$\widehat{\boldsymbol{P}}|0\rangle = 0 \tag{4.80}$$

In terms of creation and annihilation operators, the total momentum operator is

$$\widehat{\boldsymbol{P}} = \int \frac{d^3k}{(2\pi)^3 2\omega(\boldsymbol{k})} \boldsymbol{k}\,\hat{n}(\boldsymbol{k}) \tag{4.81}$$

Thus, $\widehat{\boldsymbol{P}}$ is diagonal in the basis $|\{n(\boldsymbol{k})\}\rangle$, since

$$\widehat{\boldsymbol{P}}|\{n(\boldsymbol{k})\}\rangle = \left[\int d^3k\, \boldsymbol{k}\, n(\boldsymbol{k})\right] |\{n(\boldsymbol{k})\}\rangle \tag{4.82}$$

The state with lowest energy, the vacuum state $|0\rangle$, has $n(\boldsymbol{k}) = 0$ for all $\boldsymbol{k}$. Thus the vacuum state has zero momentum, and it is translationally invariant.

The states $|\boldsymbol{k}\rangle$, defined by

$$|\boldsymbol{k}\rangle \equiv \hat{a}^\dagger(\boldsymbol{k})|0\rangle \tag{4.83}$$

have *excitation energy* $\omega(\boldsymbol{k})$ and total *linear momentum* $\boldsymbol{k}$. Thus, the states $\{|\boldsymbol{k}\rangle\}$ are particle-like excitations that have an energy dispersion curve

$$E = \sqrt{\boldsymbol{k}^2 + m^2} \tag{4.84}$$

characteristic of a relativistic particle of momentum $\boldsymbol{k}$ and mass m. Thus, the excitations of the ground state of this field theory are particle-like and have positive energy (relative to the vacuum state). From this discussion, we can see that these particles are free, since their energies and momenta are additive.

4.3.3 Causality

The starting point of the quantization procedure was to impose equal-time commutation relations on the canonical fields $\hat{\phi}(x)$ and their canonical momenta $\widehat{\Pi}(x)$. In particular, two field operators on different spatial locations *commute* at the same times. But do they commute at different times?

To address this question, let us calculate the commutator $\Delta(x-y)$:

$$i\Delta(x-y) = [\hat{\phi}(x), \hat{\phi}(y)] \tag{4.85}$$

where $\hat{\phi}(x)$ and $\hat{\phi}(y)$ are Heisenberg field operators for spacetime points x and y, respectively. From the Fourier expansion of the fields, we know that the field operator can be split into a sum of two terms:

$$\hat{\phi}(x) = \hat{\phi}_+(x) + \hat{\phi}_-(x) \tag{4.86}$$

where $\hat{\phi}_+$ contains only creation operators and positive frequencies, and $\hat{\phi}_-$ contains only annihilation operators and negative frequencies. Thus the commutator is

$$\begin{aligned} i\Delta(x-y) = {} & [\hat{\phi}_+(x), \hat{\phi}_+(y)] + [\hat{\phi}_-(x), \hat{\phi}_-(y)] \\ & + [\hat{\phi}_+(x), \hat{\phi}_-(y)] + [\hat{\phi}_-(x), \hat{\phi}_+(y)] \end{aligned} \tag{4.87}$$

The first two terms always vanish, since the $\hat{\phi}_+$ operators commute among themselves and so do the operators $\hat{\phi}_-$. Thus, the only nonvanishing contributions are

$$\begin{aligned} i\Delta(x-y) &= [\hat{\phi}_+(x), \hat{\phi}_-(y)] + [\hat{\phi}_-(x), \hat{\phi}_+(y)] \\ &= \int d\bar{k} \int d\bar{k}' \Big\{ [\hat{a}^\dagger(\boldsymbol{k}), \hat{a}(\boldsymbol{k}')] \exp\left(-i\omega(k)x_0 + i\boldsymbol{k}\cdot\boldsymbol{x} + i\omega(k')y_0 - i\boldsymbol{k}'\cdot\boldsymbol{y}\right) \\ &\quad + [\hat{a}(\boldsymbol{k}), \hat{a}^\dagger(\boldsymbol{k}')] \exp\left(i\omega(k)x_0 - i\boldsymbol{k}\cdot\boldsymbol{x} - i\omega(k')y_0 + i\boldsymbol{k}'\cdot\boldsymbol{y}\right) \Big\} \end{aligned} \tag{4.88}$$

where

$$\int d\bar{k} \equiv \int \frac{d^3k}{(2\pi)^3 2\omega(\boldsymbol{k})} \tag{4.89}$$

Furthermore, using the commutation relations of the creation and annihilation operators, we find that the operator $\Delta(x-y)$ is proportional to the identity operator, and hence, it is actually a function. It is given by

$$i\Delta(x-y) = \int d\bar{k}\, \left[e^{i\omega(\boldsymbol{k})(x_0-y_0) - i\boldsymbol{k}\cdot(\boldsymbol{x}-\boldsymbol{y})} - e^{-i\omega(\boldsymbol{k})(x_0-y_0) + i\boldsymbol{k}\cdot(\boldsymbol{x}-\boldsymbol{y})}\right] \tag{4.90}$$

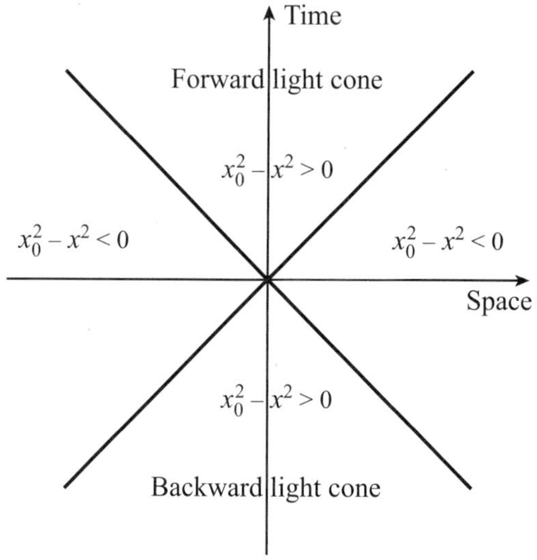

Figure 4.1 Minkowski spacetime.

With the help of the Lorentz-invariant function $\epsilon(k^0)$, defined by

$$\epsilon(k^0) = \frac{k^0}{|k^0|} \equiv \text{sign}(k^0) \tag{4.91}$$

we can write $\Delta(x-y)$ in the manifestly Lorentz invariant form:

$$i\Delta(x-y) = \int \frac{d^4k}{(2\pi)^3} \delta(k^2 - m^2)\epsilon(k^0) e^{-ik\cdot(x-y)} \tag{4.92}$$

The integrand of eq. (4.92) vanishes unless the *mass shell condition*

$$k^2 - m^2 = 0 \tag{4.93}$$

is satisfied.

Notice that $\Delta(x-y)$ satisfies the initial condition

$$\partial_0 \Delta|_{x_0=y_0} = -\delta^3(\boldsymbol{x} - \boldsymbol{y}) \tag{4.94}$$

At equal times $x_0 = y_0$ the commutator vanishes:

$$\Delta(\boldsymbol{x} - \boldsymbol{y}, 0) = 0 \tag{4.95}$$

Furthermore, by Lorentz invariance, $\Delta(x-y)$ also vanishes if the spacetime points x and y are separated by a *space-like* interval, $(x-y)^2 < 0$. This must be the case, since $\Delta(x-y)$ is manifestly Lorentz invariant. Thus if it vanishes at equal times, where $(x-y)^2 = (x_0 - y_0)^2 - (\boldsymbol{x} - \boldsymbol{y})^2 = -(\boldsymbol{x} - \boldsymbol{y}^2)^2 < 0$, it must vanish for all events with negative values of $(x-y)^2$. This implies that, for events x and y that *are not* causally connected, $\Delta(x-y) = 0$, and that $\Delta(x-y)$ is nonzero only for causally connected events (i.e., in the forward light cone shown in figure 4.1).

4.4 Symmetries of the quantum theory

In our discussion of classical field theory, we discovered that the presence of continuous global symmetries implied the existence of constants of motion. In addition, the constants of motion were the generators of infinitesimal symmetry transformations. Let us now explore what role symmetries play in the quantized theory.

In the quantized theory, all physical quantities are represented by operators that act on the Hilbert space of states. The classical statement that a quantity A is conserved if its Poisson bracket with the Hamiltonian vanishes

$$\frac{dA}{dt} = \{A, H\}_{PB} = 0 \tag{4.96}$$

in quantum theory becomes the operator identity

$$i\frac{d\widehat{A}_H}{dt} = [\widehat{A}_H, \widehat{H}] = 0 \tag{4.97}$$

We are using the Heisenberg representation. Then the constants of motion of the quantum theory are operators that commute with the Hamiltonian: $[\widehat{A}_H, \widehat{H}] = 0$.

Therefore, the quantum theory has a symmetry iff the charge $\widehat{Q}$, which is a hermitian operator associated with a classically conserved current $j^\mu(x)$ via the correspondence principle,

$$\widehat{Q} = \int_{x_0 \text{ fixed}} d^3x\, \widehat{j}^0(\boldsymbol{x}, x_0) \tag{4.98}$$

is an operator that commutes with the Hamiltonian $\widehat{H}$:

$$[\widehat{Q}, \widehat{H}] = 0 \tag{4.99}$$

If so, then the charges $\widehat{Q}$ constitute a representation of the generators of the algebra of the Lie group of the symmetry transformations in the Hilbert space of the theory. The transformations $\widehat{U}(\alpha)$ associated with the symmetry

$$\widehat{U}(\alpha) = \exp(i\alpha\widehat{Q}) \tag{4.100}$$

are unitary transformations that act on the Hilbert space of the system.

For instance, we saw that for a translationally invariant system, the classical energy-momentum four-vector P^μ,

$$P^\mu = \int_{x_0} d^3x\, T^{0\mu} \tag{4.101}$$

is conserved. In the quantum theory, P^0 becomes the Hamiltonian operator $\widehat{H}$, and P^i becomes the total momentum operator $\widehat{\boldsymbol{P}}$. In the case of a free scalar field, we saw before that these operators commute with each other: $[\widehat{\boldsymbol{P}}, \widehat{H}] = 0$. Thus, the eigenstates of the system have well-defined total energy and total momentum. Since $\boldsymbol{P}$ is the generator of infinitesimal translations of the classical theory, it is easy to check that its equal-time Poisson bracket with the field $\phi(x)$ is

$$\{\phi(\boldsymbol{x}, x_0), P^j\}_{PB} = \partial_x^j \phi \tag{4.102}$$

In quantum theory, the equivalent statement is that the field operator $\hat{\phi}(x)$ and the total momentum operator $\widehat{\boldsymbol{P}}$ satisfy the equal-time commutation relation:

$$[\hat{\phi}(\boldsymbol{x}, x_0), \widehat{P}^j] = i\partial_x^j \hat{\phi}(\boldsymbol{x}, x_0) \tag{4.103}$$

Consequently, the field operators $\hat{\phi}(\boldsymbol{x}+\boldsymbol{a}, x_0)$ and $\hat{\phi}(\boldsymbol{x}, x_0)$ are related by

$$\hat{\phi}(\boldsymbol{x}+\boldsymbol{a}, x_0) = e^{i\boldsymbol{a}\cdot\widehat{\boldsymbol{P}}} \hat{\phi}(\boldsymbol{x}, x_0) e^{-i\boldsymbol{a}\cdot\widehat{\boldsymbol{P}}} \tag{4.104}$$

Translational invariance of the ground state $|0\rangle$ implies that it is a state with zero total momentum: $\widehat{\boldsymbol{P}}|0\rangle = 0$. For a finite displacement $\boldsymbol{a}$, we get

$$e^{i\boldsymbol{a}\cdot\widehat{\boldsymbol{P}}}|0\rangle = |0\rangle \tag{4.105}$$

which says that the state $|0\rangle$ is invariant under translations and belongs to a one-dimensional representation of the group of global translations.

Let us now discuss what happens to global internal symmetries. The simplest case is the *free complex scalar field* $\phi(x)$, whose Lagrangian $\mathcal{L}$ is invariant under global phase transformations. If ϕ is a complex field, it can be decomposed into its real and imaginary parts:

$$\phi = \frac{1}{\sqrt{2}}(\phi_1 + i\phi_2) \tag{4.106}$$

The classical Lagrangian for a free complex scalar field ϕ is

$$\mathcal{L} = \partial_\mu \phi^* \partial^\mu \phi - m^2 \phi^* \phi \tag{4.107}$$

which now splits into a sum of two independent terms,

$$\mathcal{L}(\phi) = \mathcal{L}(\phi_1) + \mathcal{L}(\phi_2) \tag{4.108}$$

where $\mathcal{L}(\phi_1)$ and $\mathcal{L}(\phi_2)$ are the Lagrangians for the free scalar real fields ϕ_1 and ϕ_2, respectively. Likewise, the canonical momenta $\Pi(x)$ and $\Pi^*(x)$ decompose into

$$\Pi(x) = \frac{\delta\mathcal{L}}{\delta\partial_0\phi} = \frac{1}{\sqrt{2}}(\dot{\phi}_1 - i\dot{\phi}_2) \qquad \Pi^*(x) = \frac{1}{\sqrt{2}}(\dot{\phi}_1 + i\dot{\phi}_2) \tag{4.109}$$

In quantum theory, the operators $\hat{\phi}$ and $\hat{\phi}^\dagger$ are no longer equal to each other, and neither are $\widehat{\Pi}$ and $\widehat{\Pi}^\dagger$. Still, the canonical quantization procedure tells us that $\hat{\phi}$ and $\widehat{\Pi}$ (and $\hat{\phi}^\dagger$ and $\widehat{\Pi}^\dagger$) satisfy the equal-time canonical commutation relations:

$$[\hat{\phi}(\boldsymbol{x}, x_0), \widehat{\Pi}(\boldsymbol{y}, x_0)] = i\delta^3(\boldsymbol{x}-\boldsymbol{y}) \tag{4.110}$$

The theory of the free complex scalar field is solvable by the same methods that we used for a free real scalar field. Instead of a single creation-annihilation algebra, we must introduce now two algebras, with operators $\hat{a}_1(\mathbf{k})$ and $\hat{a}_1^\dagger(\mathbf{k})$, and $\hat{a}_2(\mathbf{k})$ and $\hat{a}_2^\dagger(\mathbf{k})$. Let $\hat{a}(\mathbf{k})$ and $\hat{b}(\mathbf{k})$ be defined by

$$\hat{a}(\mathbf{k}) = \frac{1}{\sqrt{2}}\left(\hat{a}_1(\mathbf{k}) + i\hat{a}_2(\mathbf{k})\right), \quad \hat{a}^\dagger(\mathbf{k}) = \frac{1}{\sqrt{2}}\left(\hat{a}_1^\dagger(\mathbf{k}) - i\hat{a}_2^\dagger(\mathbf{k})\right)$$
$$\hat{b}(\mathbf{k}) = \frac{1}{\sqrt{2}}\left(\hat{a}_1(\mathbf{k}) - i\hat{a}_2(\mathbf{k})\right), \quad \hat{b}^\dagger(\mathbf{k}) = \frac{1}{\sqrt{2}}\left(\hat{a}_1^\dagger(\mathbf{k}) + i\hat{a}_2^\dagger(\mathbf{k})\right) \tag{4.111}$$

which satisfy the algebra

$$[\hat{a}(\mathbf{k}), \hat{a}^\dagger(\mathbf{k}')] = [\hat{b}(\mathbf{k}), \hat{b}^\dagger(\mathbf{k}')] = (2\pi)^3 2\omega(\mathbf{k})\, \delta^3(\mathbf{k} - \mathbf{k}') \tag{4.112}$$

while all other commutators vanish.

The Fourier expansion for the fields now is

$$\hat{\phi}(x) = \int \frac{d^3k}{(2\pi)^3 2\omega(\mathbf{k})} \left(\hat{a}(\mathbf{k})e^{-ik\cdot x} + \hat{b}^\dagger(\mathbf{k})e^{ik\cdot x}\right)$$
$$\hat{\phi}^\dagger(x) = \int \frac{d^3k}{(2\pi)^3 2\omega(\mathbf{k})} \left(\hat{b}(\mathbf{k})e^{-ik\cdot x} + \hat{a}^\dagger(\mathbf{k})e^{ik\cdot x}\right) \tag{4.113}$$

where $\omega(\mathbf{k}) = \sqrt{\mathbf{k}^2 + m^2}$, and $k_0 = \omega(\mathbf{k})$. In this notation, the normal-ordered Hamiltonian is

$$:\widehat{H}: = \int \frac{d^3k}{(2\pi)^3 2\omega(\mathbf{k})}\, \omega(\mathbf{k}) \left(\hat{a}^\dagger(\mathbf{k})\hat{a}(\mathbf{k}) + \hat{b}^\dagger(\mathbf{k})\hat{b}(\mathbf{k})\right) \tag{4.114}$$

The normal-ordered total linear momentum $\widehat{\mathbf{P}}$ is given by a similar expression:

$$\widehat{\mathbf{P}} = \int \frac{d^3k}{(2\pi)^3 2\omega(\mathbf{k})}\, \mathbf{k} \left(\hat{a}^\dagger(\mathbf{k})\hat{a}(\mathbf{k}) + \hat{b}^\dagger(\mathbf{k})\hat{b}(\mathbf{k})\right) \tag{4.115}$$

We see that there are two types of quanta, a and b. The field $\hat{\phi}$ creates b-quanta, and it destroys a-quanta. The vacuum state has no quanta and is annihilated by both operators: $\hat{a}(\mathbf{k})|0\rangle = \hat{b}(\mathbf{k})|0\rangle = 0$. The one-particle states now have a twofold degeneracy, since the states $\hat{a}^\dagger(\mathbf{k})|0\rangle$ and $\hat{b}^\dagger(\mathbf{k})|0\rangle$ have one particle of type a and one of type b, respectively. These states have exactly the same energy $\omega(\mathbf{k})$ and the same momentum $\mathbf{k}$. Thus, for each value of the energy and of momentum, we have a two-dimensional space of possible states. This degeneracy is a consequence of the symmetry: The states form multiplets.

What is the quantum operator that generates this symmetry? The classically conserved current is

$$j_\mu = i\phi^* \overleftrightarrow{\partial_\mu} \phi \tag{4.116}$$

In the quantum theory, j_μ becomes the normal-ordered operator $:\hat{j}_\mu:$. The corresponding *global charge* $\widehat{Q}$ is

$$\begin{aligned}\widehat{Q} &= :\int d^3x\, i\left(\hat{\phi}^\dagger \partial_0 \hat{\phi} - (\partial_0 \hat{\phi}^\dagger)\hat{\phi}\right): \\ &= \int \frac{d^3k}{(2\pi)^3 2\omega(\boldsymbol{k})} \left(\hat{a}^\dagger(\boldsymbol{k})\hat{a}(\boldsymbol{k}) - \hat{b}^\dagger(\boldsymbol{k})\hat{b}(\boldsymbol{k})\right) \\ &= \hat{N}_a - \hat{N}_b \end{aligned} \tag{4.117}$$

where $\hat{N}_a$ and $\hat{N}_b$ are the number operators for quanta of types a and b, respectively. Since $[\widehat{Q}, \widehat{H}] = 0$, the difference $\hat{N}_a - \hat{N}_b$ is conserved. Since this property is the consequence of a symmetry, it is expected to hold in more general theories than the simple free-field case that we are discussing here, provided that $[\widehat{Q}, \widehat{H}] = 0$. Thus, although $\hat{N}_a$ and $\hat{N}_b$ in general may not be conserved separately, the difference $\hat{N}_a - \hat{N}_b$ will be conserved if the symmetry is exact.

Let us now briefly discuss how this symmetry is realized in the spectrum of states.

4.4.1 The vacuum state

The vacuum state has $N_a = N_b = 0$. Thus, the generator $\widehat{Q}$ annihilates the vacuum

$$\widehat{Q}|0\rangle = 0 \tag{4.118}$$

Therefore, the vacuum state is invariant (i.e., is a *singlet*) under the symmetry:

$$|0\rangle' = e^{i\widehat{Q}\alpha}|0\rangle = |0\rangle \tag{4.119}$$

Because the state $|0\rangle$ is always defined up to an overall phase factor, it spans a one-dimensional subspace of states that are invariant under the symmetry. This is the vacuum sector and, for this problem, is trivial.

4.4.2 One-particle states

There are two linearly independent one-particle states, $|+, \boldsymbol{k}\rangle$ and $|-, \boldsymbol{k}\rangle$, defined by

$$|+, \boldsymbol{k}\rangle = \hat{a}^\dagger(\boldsymbol{k})\,|0\rangle, \qquad |-, \boldsymbol{k}\rangle = \hat{b}^\dagger(\boldsymbol{k})\,|0\rangle \tag{4.120}$$

Both states have the same momentum $\boldsymbol{k}$ and energy $\omega(\boldsymbol{k})$. The $\widehat{Q}$-quantum numbers of these states, which we will refer to as their *charge*, are

$$\begin{aligned}\widehat{Q}|+, \boldsymbol{k}\rangle &= (\hat{N}_a - \hat{N}_b)\hat{a}^\dagger(\boldsymbol{k})|0\rangle = \hat{N}_a\,\hat{a}^\dagger(\boldsymbol{k})|0\rangle = +|+, \boldsymbol{k}\rangle \\ \widehat{Q}|-, \boldsymbol{k}\rangle &= (\hat{N}_a - \hat{N}_b)\,\hat{b}^\dagger(\boldsymbol{k})|0\rangle = -|-, \boldsymbol{k}\rangle \end{aligned} \tag{4.121}$$

Hence

$$\widehat{Q}|\sigma, \boldsymbol{k}\rangle = \sigma\,|\sigma, \boldsymbol{k}\rangle \tag{4.122}$$

where $\sigma = \pm 1$. Thus, the state $\hat{a}^\dagger(\boldsymbol{k})|0\rangle$ has *positive* charge, while $\hat{b}^\dagger(\boldsymbol{k})|0\rangle$ has *negative* charge.

Under a finite transformation $\widehat{U}(\alpha) = \exp(i\alpha\widehat{Q})$, the states $|\pm, \boldsymbol{k}\rangle$ transform as follows:

$$|+,\boldsymbol{k}\rangle' = \widehat{U}(\alpha)\,|+,\boldsymbol{k}\rangle = \exp(i\alpha\hat{Q})\,|+,\boldsymbol{k}\rangle = e^{i\alpha}\,|+,\boldsymbol{k}\rangle$$
$$|-,\boldsymbol{k}\rangle' = \widehat{U}(\alpha)\,|-,\boldsymbol{k}\rangle = \exp(i\alpha\hat{Q})\,|-,\boldsymbol{k}\rangle = e^{-i\alpha}\,|-,\boldsymbol{k}\rangle \tag{4.123}$$

The field $\hat{\phi}(x)$ itself transforms as

$$\hat{\phi}'(x) = \exp(-i\alpha\widehat{Q})\,\hat{\phi}(x)\,\exp(i\alpha\widehat{Q}) = e^{i\alpha}\hat{\phi}(x) \tag{4.124}$$

since

$$[\widehat{Q}, \hat{\phi}(x)] = -\hat{\phi}(x), \qquad [\widehat{Q}, \hat{\phi}^\dagger(x)] = \hat{\phi}^\dagger(x) \tag{4.125}$$

Thus the one-particle states are doubly degenerate, and each state transforms nontrivially under the symmetry group.

By inspection of the Fourier expansion for the complex field $\hat{\phi}$, we see that $\hat{\phi}$ is a sum of two terms: a set of positive frequency terms, symbolized by $\hat{\phi}_+$, and a set of negative frequency terms, $\hat{\phi}_-$. In this case, all positive frequency terms create particles of type b (which carry *negative* charge) while the negative frequency terms annihilate particles of type a (which carry *positive* charge). The states $|\pm, \boldsymbol{k}\rangle$ are commonly referred to as *particles* and *antiparticles*: Particles have rest mass m, momentum $\boldsymbol{k}$, and charge $+1$, while the antiparticles have the same mass and momentum but carry charge -1. This charge is measured in units of the electromagnetic charge $-e$ (see the discussion on the gauge current in section 3.7).

Finally, note that this theory contains an additional operator: the charge conjugation operator $\widehat{C}$, which maps particles into antiparticles and vice versa. This operator commutes with the Hamiltonian: $[\widehat{C}, \widehat{H}] = 0$. This property ensures that the *spectrum* is invariant under charge conjugation. In other words, for every state of charge Q, there exists a state with charge $-Q$, all other quantum numbers being the same.

Our analysis of the free complex scalar field can be easily extended to systems that are invariant under a more general symmetry group G. In all cases, the classically conserved charges become operators of the quantum theory. Thus, there are as many charge operators $\widehat{Q}^a$ as there are generators in the group. The charge operators represent the generators of the group in the Hilbert (or Fock) space of the system, and obey the same commutation relations as the generators themselves do. A simple generalization of the arguments that we have used here tells us that the states of the spectrum of the theory must transform under the irreducible representations of the symmetry group.

However, one important caveat should be made. Our discussion of the *free* complex scalar field showed us that, in that case, the ground state is invariant under the symmetry. In general, the only possible invariant state is the singlet state. All other states are not invariant and transform nontrivially.

But should the ground state always be invariant? In elementary quantum mechanics, there is a theorem, due to Wigner and Weyl, which states that for a finite system, the ground state is always a singlet under the action of the symmetry group. However, there are many systems in nature (e.g., magnets, Higgs phases, and superconductors) that have ground states that are not invariant under the symmetries of the Hamiltonian. This phenomenon, known as *spontaneous symmetry breaking*, does not occur in simple free-field theories, but it does happen in interacting-field theories. We will return to this important issue in chapter 12.

Exercises

4.1 Two-component complex scalar field

Consider the theory of a two-component complex scalar field $\phi_a(x)$ $(a=1,2)$, which has the Lagrangian

$$\mathcal{L}=(\partial_\mu \phi_a(x))^* \, (\partial^\mu \phi_a(x)) - V(\phi(x)) \tag{4.126}$$

where the potential $V(\phi(x))$ is

$$V(\phi(x)) = m_0^2 \, \phi_a^*(x)\, \phi_a(x) \tag{4.127}$$

As you know from the discussion in chapter 3, this theory is invariant under the classical global symmetry:

$$\begin{aligned}
\phi_a(x) &\to \phi'_a(x) = U_{ab}\, \phi_b(x)\\
\phi_a^*(x) &\to \phi'^*_a(x) = U^{-1}_{ab}\, \phi_b^*(x)\\
\mathcal{L}(\phi') &= \mathcal{L}(\phi)
\end{aligned} \tag{4.128}$$

where U is a 2×2 unitary matrix (i.e., $U^{-1}=U^\dagger$). This is the symmetry group $SU(2)$. Thus, ϕ transforms like the fundamental (spinor) representation of $SU(2)$. The matrices U can be expanded in the basis of 2×2 Pauli matrices $(\sigma_k)_{ab}$ and are parametrized by three Euler angles θ_k $(k=1,2,3)$:

$$\begin{aligned}
U_{ab} &= \left[\exp\left(i\theta_k\, \sigma_k\right)\right]_{ab}\\
&= \cos(|\boldsymbol{\theta}|)\delta_{ab} + i\frac{\boldsymbol{\theta}}{|\boldsymbol{\theta}|}\cdot \boldsymbol{\sigma}_{ab}\, \sin(|\boldsymbol{\theta}|)
\end{aligned} \tag{4.129}$$

1) Use the classical canonical formalism to find (a) the canonical momentum Π_a, conjugate to the field ϕ_a; (b) the Hamiltonian H; and (c) the total momentum P_j.
2) Derive the classical constants of motion associated with the global symmetry $SU(2)$. Relate these constants of motion to the generators of infinitesimal $SU(2)$ transformations. How many constants of motion do you find? Explain your results.
3) Quantize this theory by imposing canonical commutation relations. Write an expression for the quantum mechanical Hamiltonian and total momentum operators in terms of the field and canonical momentum operators.
4) Derive an expression for the *quantum mechanical* generators of global infinitesimal $SU(2)$ transformations in the Hilbert space of states of the system. Explain what relation, if any, they have with the conserved charges of the classical theory.
5) Derive the quantum mechanical equations of motion of the Heisenberg representation operators.

6) Find an expansion of the field and canonical momentum operators in terms of a suitable set of creation and annihilation operators. How many species of creation and annihilation operators do you need? Justify your results.
7) Find an expression for the $SU(2)$ generators in terms of creation and annihilation operators.
8) Find the ground state of the system and its quantum numbers. Find the Hamiltonian, the total momentum, and the $SU(2)$ generators normal-ordered relative to the ground state.
9) Find the spectrum of single particle states. Derive an expression for their energies, and assign quantum numbers to these states. Do you find any degeneracies? What is the degree of this degeneracy and why?

5

Path Integrals in Quantum Mechanics and Quantum Field Theory

In chapter 4, we discussed the Hilbert space picture of quantum mechanics and QFT for the case of free relativistic scalar fields. Here we consider the *path integral* picture of quantum mechanics and of relativistic scalar field theories.

The path integral picture is important for two reasons. First, it offers an alternative, complementary picture of quantum mechanics in which the role of the classical limit is apparent. Second, it gives a direct route to the study regimes where perturbation theory is either inadequate or fails completely. In quantum mechanics, a standard approach to such problems is the Wentzel, Kramers, and Brillouin (WKB) approximation. However, as it happens, it is extremely difficult (if not impossible) to generalize the WKB approximation to a QFT. Instead, the nonperturbative treatment of the Feynman path integral, which in quantum mechanics is equivalent to WKB, is generalizable to nonperturbative problems in QFT. In this chapter, we use path integrals only for bosonic systems, such as scalar fields. In subsequent chapters, we will also give a full treatment of the path integral, including its applications to fermionic fields, abelian and non-abelian gauge fields, classical statistical mechanics, and nonrelativistic many-body systems.

There is a huge literature on path integrals, going back to the original papers by Dirac (1933), and particularly Feynman's 1942 PhD thesis (Feynman, 2005) and his review paper (Feynman, 1948). Popular textbooks on path integrals include the classic by Feynman and Hibbs (1965), and Schulman's book (Schulman, 1981), among many others.

5.1 Path integrals and quantum mechanics

Consider a simple quantum mechanical system whose dynamics can be described by a generalized coordinate operator $\hat{q}$. We want to compute the amplitude

$$F(q_f, t_f | q_i, t_i) = \langle q_f, t_f | q_i, t_i \rangle \tag{5.1}$$

known as the Wightman function. This function represents the amplitude of finding the system at coordinate q_f at the final time t_f, knowing that it was at coordinate q_i at the initial time t_i. The amplitude $F(q_f, t_f | q_i, t_i)$ is just a matrix element of the evolution operator

$$F(q_f, t_f|q_i, t_i) = \langle q_f|e^{i\hat{H}(t_i-t_f)/\hbar}|q_i\rangle \tag{5.2}$$

For simplicity, let us set $|q_i, t_i\rangle = |0, 0\rangle$ and $|q_f, t_f\rangle = |q, t\rangle$. Then, from the definition of this matrix element, we find out that it obeys

$$\lim_{t\to 0} F(q, t|0, 0) = \langle q|0\rangle = \delta(q) \tag{5.3}$$

Furthermore, after some algebra, we also find that

$$\begin{aligned} i\hbar\frac{\partial F}{\partial t} &= i\hbar\frac{\partial}{\partial t}\langle q, t|0, 0\rangle = i\hbar\frac{\partial}{\partial t}\langle q|e^{-i\hat{H}t/\hbar}|0\rangle \\ &= \langle q|\hat{H}e^{-i\hat{H}t/\hbar}|0\rangle \\ &= \int dq' \langle q|\hat{H}|q'\rangle\langle q'|e^{-i\hat{H}t/\hbar}|0\rangle \end{aligned} \tag{5.4}$$

where we have used that, since $\{|q\rangle\}$ is a complete set of states, the identity operator I has the following expansion (called the *resolution of the identity*):

$$I = \int dq'|q'\rangle\langle q'| \tag{5.5}$$

Here we have assumed that the states are orthonormal:

$$\langle q|q'\rangle = \delta(q - q') \tag{5.6}$$

Hence,

$$i\hbar\frac{\partial}{\partial t}F(q, t|0, 0) = \int dq' \langle q|\hat{H}|q'\rangle F(q', t|0, 0) \equiv \hat{H}_q F(q, t|0, 0) \tag{5.7}$$

In other words, $F(q, t|0, 0)$ is the solution of the Schrödinger equation that satisfies the initial condition, eq. (5.3). For this reason, the amplitude $F(q, t|0, 0)$ is called the *Schrödinger propagator.*

The superposition principle tells us that the amplitude to find the system in the final state at the final time is the sum of amplitudes of the form

$$F(q_f, t_f|q_i, t_i) = \int dq' \langle q_f, t_f|q', t'\rangle\langle q', t'|q_i, t_i\rangle \tag{5.8}$$

where the system is in an arbitrary set of states at an intermediate time t'. Here we have represented this situation by inserting the identity operator I at the intermediate time t' in the form of the resolution of the identity in eq. (5.8) (see figure 5.1).

Let us next define a *partition* of the time interval $[t_i, t_f]$ into N subintervals, each of length Δt:

$$t_f - t_i = N\Delta t \tag{5.9}$$

Let $\{t_j\}$, with $j = 0, \ldots, N + 1$, denote a set of points in the interval $[t_i, t_f]$, such that

$$t_i = t_0 \le t_1 \le \cdots \le t_N \le t_{N+1} = t_f \tag{5.10}$$

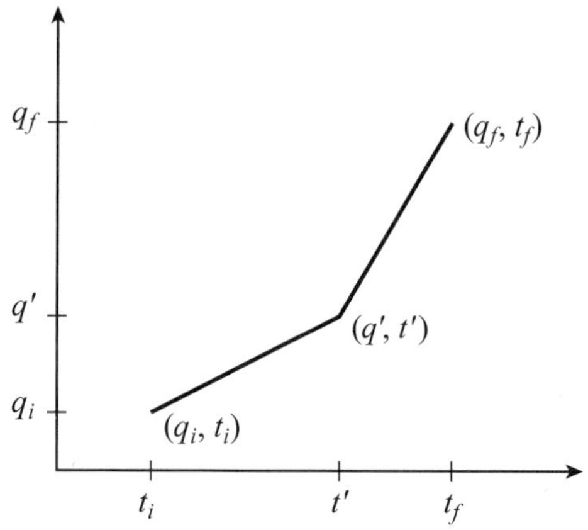

Figure 5.1 The amplitude to go from $|q_i, t_i\rangle$ to $|q_f, t_f\rangle$ is a sum of products of amplitudes through the intermediate states $|q', t'\rangle$.

Clearly, $t_k = t_0 + k\Delta t$, for $k = 1, \ldots, N+1$. By repeating the procedure used in eq. (5.8) of inserting the resolution of the identity at the intermediate times $\{t_k\}$, we find

$$F(q_f, t_f | q_i, t_i) = \int dq_1 \cdots dq_N \langle q_f, t_f | q_N, t_N \rangle \langle q_N, t_N | q_{N-1}, t_{N-1} \rangle \times \cdots$$
$$\times \cdots \langle q_j, t_j | q_{j-1}, t_{j-1} \rangle \cdots \langle q_1, t_1 | q_i, t_i \rangle \tag{5.11}$$

Each factor $\langle q_j, t_j | q_{j-1}, t_{j-1} \rangle$ in eq. (5.11) has the form

$$\langle q_j, t_j | q_{j-1}, t_{j-1} \rangle = \langle q_j | e^{-i\hat{H}(t_j - t_{j-1})/\hbar} | q_{j-1} \rangle \equiv \langle q_j | e^{-i\hat{H}\Delta t/\hbar} | q_{j-1} \rangle \tag{5.12}$$

In the limit $N \to \infty$, with $|t_f - t_i|$ fixed and finite, the interval Δt becomes infinitesimally small, and $\Delta t \to 0$. Hence, as $N \to \infty$, we can approximate the expression for $\langle q_j, t_j | q_{j-1}, t_{j-1} \rangle$ in eq. (5.12) as follows (see figure 5.2):

$$\begin{aligned} \langle q_j, t_j | q_{j-1}, t_{j-1} \rangle &= \langle q_j | e^{-iH\Delta t/\hbar} | q_{j-1} \rangle \\ &= \langle q_j | \left\{ I - i\frac{\Delta t}{\hbar} \hat{H} + O((\Delta t)^2) \right\} | q_{j-1} \rangle \\ &= \delta(q_j - q_{j-1}) - i\frac{\Delta t}{\hbar} \langle q_j | \hat{H} | q_{j-1} \rangle + O((\Delta t)^2) \end{aligned} \tag{5.13}$$

which becomes asymptotically exact as $N \to \infty$.

We can also introduce at each intermediate time t_j a complete set of momentum eigenstates $\{|p\rangle\}$, using their resolution of the identity

$$I = \int_{-\infty}^{\infty} dp \, |p\rangle \langle p| \tag{5.14}$$

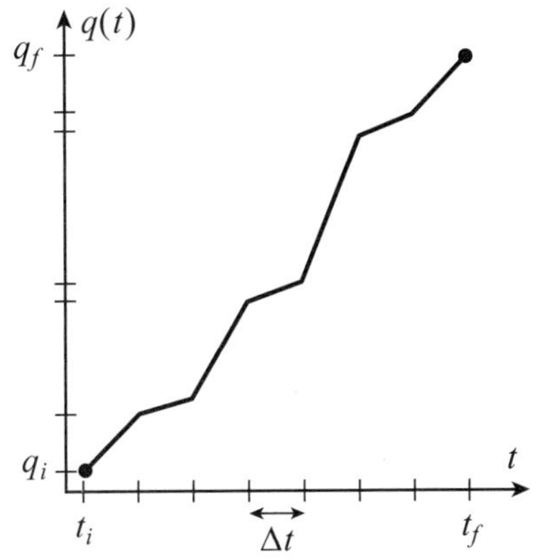

Figure 5.2 A history $q(t)$ of the system.

Recall that the overlap between the states $|q\rangle$ and $|p\rangle$ is

$$\langle q|p\rangle = \frac{1}{\sqrt{2\pi\hbar}}\, e^{ipq/\hbar} \tag{5.15}$$

For a typical Hamiltonian of the form

$$\hat{H} = \frac{\hat{p}^2}{2m} + V(\hat{q}) \tag{5.16}$$

its matrix elements are

$$\langle q_j|\hat{H}|q_{j-1}\rangle = \int_{-\infty}^{\infty} \frac{dp_j}{2\pi\hbar}\, e^{ip_j(q_j-q_{j-1})/\hbar} \left[\frac{p_j^2}{2m} + V(q_j)\right] \tag{5.17}$$

To the same level of approximation, we can also write

$$\langle q_j, t_j|q_{j-1}, t_{j-1}\rangle \approx \int \frac{dp_j}{2\pi\hbar} \exp\left[i\left(\frac{p_j}{\hbar}(q_j - q_{j-1}) - \Delta t\, H\left(p_j, \frac{q_j + q_{j-1}}{2}\right)\right)\right] \tag{5.18}$$

where we have introduced the "mid-point rule," which amounts to the replacement $q_j \to \frac{1}{2}\left(q_j + q_{j-1}\right)$ inside the Hamiltonian $H(p, q)$. Putting everything together, we find that the matrix element $\langle q_f, t_f|q_i, t_i\rangle$ becomes

$$\langle q_f, t_f|q_i, t_i\rangle = \lim_{N\to\infty} \int \prod_{j=1}^{N} dq_j \int_{-\infty}^{\infty} \prod_{j=1}^{N+1} \frac{dp_j}{2\pi\hbar} \exp\left\{\frac{i}{\hbar}\sum_{j=1}^{N+1}\left[p_j(q_j - q_{j-1}) - \Delta t\, H\left(p_j, \frac{q_j + q_{j-1}}{2}\right)\right]\right\} \tag{5.19}$$

Therefore, in the limit $N \to \infty$, holding $|t_i - t_f|$ fixed, the amplitude $\langle q_f, t_f | q_i, t_i \rangle$ is given by the (formal) expression

$$\langle q_f, t_f | q_i, t_i \rangle = \int \mathcal{D}p\mathcal{D}q \; e^{\frac{i}{\hbar}\int_{t_i}^{t_f} dt\, [p\dot{q} - H(p,q)]} \tag{5.20}$$

where we have used the notation

$$\mathcal{D}p\mathcal{D}q \equiv \lim_{N\to\infty} \prod_{j=1}^{N} \frac{dp_j dq_j}{2\pi\hbar} \tag{5.21}$$

which defines the integration measure. The functions, or *configurations*, $(q(t), p(t))$, must satisfy the initial and final conditions:

$$q(t_i) = q_i, \qquad q(t_f) = q_f \tag{5.22}$$

Thus the matrix element $\langle q_f, t_f | q_i, t_i \rangle$ can be expressed as a sum over histories in *phase space*. The weight of each history is the exponential factor in eq. (5.20). Notice that the quantity in brackets is just the Lagrangian,

$$L = p\dot{q} - H(p,q) \tag{5.23}$$

Thus the matrix element is just

$$\langle q_f, t_f | q, t \rangle = \int \mathcal{D}p\mathcal{D}q \, e^{\frac{i}{\hbar}S(q,p)} \tag{5.24}$$

where $S(q,p)$ is the action of each history $(q(t), p(t))$. Also notice that the sum (or integral) runs over *independent* functions $q(t)$ and $p(t)$, which are not required to satisfy any constraint (apart from the initial and final conditions), and in particular, they are not the solution of the equations of motion. Expressions of this type are known as *path integrals.* They are also called "functional integrals," since the integration measure is a sum over a space of functions, instead of a field of numbers, as in a conventional integral.

Using a Gaussian integral of the form (which involves an analytic continuation),

$$\int_{-\infty}^{\infty} \frac{dp}{2\pi\hbar} \, e^{i\left(p\dot{q} - \frac{p^2}{2m}\right)\frac{\Delta t}{\hbar}} = \sqrt{\frac{m}{2\pi i\hbar\Delta t}} \, e^{i\frac{\Delta t}{2\hbar}\dot{q}^2} \tag{5.25}$$

we can integrate out explicitly the momenta in the path integral and find a formula that involves only the histories of the coordinate alone. Notice that there are no initial and final conditions on the momenta, since the initial and final states have well-defined positions. The result is

$$\langle q_f, t_f | q_i, t_i \rangle = \int \mathcal{D}q \; e^{\frac{i}{\hbar}\int_{t_i}^{t_f} dt\, L(q,\dot{q})} \tag{5.26}$$

which is known as the *Feynman path integral* (Feynman, 1948, 2005). Here $L(q, \dot{q})$ is the Lagrangian,

$$L(q, \dot{q}) = \frac{1}{2} m\dot{q}^2 - V(q) \tag{5.27}$$

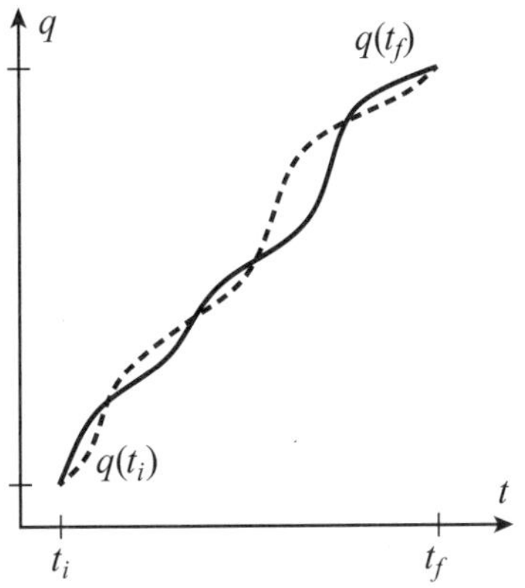

Figure 5.3 Two histories with the same initial and final states.

and the sum over histories $q(t)$ is restricted by the boundary conditions $q(t_i) = q_i$ and $q(t_f) = q_f$.

The Feynman path integral tells us that in the correspondence limit, $\hbar \to 0$, the only history (or possibly, histories, such as those shown in figure 5.3) that contributes significantly to the path integral must be one that leaves the action S stationary, since otherwise, the contributions of the rapidly oscillating exponential would add up to zero. In other words, in the classical limit, there is only one history $q_c(t)$ that contributes. For this history, $q_c(t)$, the action S is stationary, $\delta S = 0$, and $q_c(t)$ is the solution of the classical equation of motion:

$$\frac{\partial L}{\partial q} - \frac{d}{dt}\frac{\partial L}{\partial \dot{q}} = 0 \tag{5.28}$$

In other terms, in the correspondence limit $\hbar \to 0$, the evaluation of the Feynman path integral reduces to the requirement that the least action principle should hold. This is the classical limit.

5.2 Evaluating path integrals in quantum mechanics

Let us first discuss the following problem. We wish to know how to compute the amplitude $\langle q_f, t_f | q_i, t_i \rangle$ for a dynamical system whose Lagrangian has the standard form of eq. (5.27). For simplicity, we begin with a linear harmonic oscillator.

The Hamiltonian for a linear harmonic oscillator is

$$H = \frac{p^2}{2m} + \frac{m\omega^2}{2} q^2 \tag{5.29}$$

and the associated Lagrangian is

$$L = \frac{m}{2}\dot{q}^2 - \frac{m\omega^2}{2} q^2 \tag{5.30}$$

Let $q_c(t)$ be the classical trajectory. It is the solution of the classical equation of motion:

$$\frac{d^2 q_c}{dt^2} + \omega^2 q_c = 0 \tag{5.31}$$

Let us denote by $q(t)$ an arbitrary history of the system and by $\xi(t)$ its deviation from the classical solution $q_c(t)$. Since all the histories, including the classical trajectory $q_c(t)$, obey the *same* initial and final conditions,

$$q(t_i) = q_i, \qquad q(t_f) = q_f \tag{5.32}$$

it follows that $\xi(t)$ obeys instead *vanishing* initial and final conditions:

$$\xi(t_i) = \xi(t_f) = 0 \tag{5.33}$$

After some trivial algebra, it is easy to show that the action S for an arbitrary history $q(t)$ becomes

$$S(q, \dot{q}) = S(q_c, \dot{q}_c) + S(\xi, \dot{\xi}) + \int_{t_i}^{t_f} dt\, \frac{d}{dt}\left[m\xi \frac{dq_c}{dt}\right] + \int_{t_i}^{t_f} dt\, m\xi \left(\frac{d^2 q_c}{dt^2} + \omega^2 q_c\right). \tag{5.34}$$

The third term vanishes due to the boundary conditions obeyed by the fluctuations $\xi(t)$, eq. (5.33). The last term also vanishes, since q_c is a solution of the classical equation of motion, eq. (5.31). These two features hold for all systems, even if they are not harmonic. However, the Lagrangian (and hence the action) for ξ, the second term in eq. (5.34), in general is not the same as the action for the classical trajectory (the first term). Only the harmonic oscillator $S(\xi, \dot{\xi})$ has the same form as $S(q_c, \dot{q}_c)$.

Hence, for a harmonic oscillator, we get the path integral

$$\langle q_f, t_f | q_i, t_i \rangle = e^{\frac{i}{\hbar} S(q_c, \dot{q}_c)} \int_{\xi(t_i)=\xi(t_f)=0} \mathcal{D}\xi\, e^{\frac{i}{\hbar}\int_{t_i}^{t_f} dt L(\xi, \dot{\xi})} \tag{5.35}$$

Notice that the information on the initial and final states enters only through the factor associated with the classical trajectory. For the linear harmonic oscillator, the quantum mechanical contribution is independent of the initial and final states. Thus, we need to do two things: (1) we need an explicit solution $q_c(t)$ of the equation of motion, for which we will compute $S(q_c, \dot{q}_c)$, and (2) we need to compute the quantum mechanical correction, the last factor in eq. (5.35), which measures the strength of the quantum fluctuations.

For a general dynamical system whose Lagrangian has the form of eq. (5.27), the action of eq. (5.34) takes the form

$$\begin{aligned} S(q, \dot{q}) &= S(q_c, \dot{q}_c) + S_{\text{eff}}(\xi, \dot{\xi}; q_c) \\ &\quad + \int_{t_i}^{t_f} dt\, \frac{d}{dt}\left[m\xi \frac{dq_c}{dt}\right] + \int_{t_i}^{t_f} dt \left(m \frac{d^2 q_c}{dt^2} + \left.\frac{\partial V}{\partial q}\right|_{q_c}\right) \xi(t) \end{aligned} \tag{5.36}$$

where S_{eff} is the effective action for the fluctuations $\xi(t)$, which has the form

$$S_{\text{eff}}(\xi, \dot{\xi}) = \int_{t_i}^{t_f} dt \frac{1}{2} m\dot{\xi}^2 - \frac{1}{2}\int_{t_i}^{t_f} dt \int_{t_i}^{t_f} dt' \left.\frac{\partial^2 V}{\partial q(t) \partial q(t')}\right|_{q_c} \xi(t)\xi(t') + O(\xi^3) \tag{5.37}$$

Once again, the boundary conditions $\xi(t_i)=\xi(t_f)=0$ and the fact that $q_c(t)$ is a solution of the equation of motion together imply that the last two terms of eq. (5.36) vanish identically.

Thus, to the extent that we are allowed to neglect the $O(\xi^3)$ (and higher) corrections, the effective action S_{eff} can be approximated by an action that is quadratic in the fluctuation ξ. In general, this effective action will depend on the actual classical trajectory, since in general, $V''(q_c)$ is not constant but is a function of time determined by $q_c(t)$. However, if one is interested in the quantum fluctuations about a minimum of the potential $V(q)$, then $q_c(t)$ is constant (and equal to the minimum). Below we will discuss this case in detail.

Before we embark on an actual computation, it is worthwhile to ask when it would be a good approximation to neglect the terms $O(\xi^3)$ (and higher). Since we are expanding about the classical path q_c, we expect that this approximation should be correct as we formally take the limit $\hbar\to 0$. In the path integral, the effective action always appears in the combination $S_{\text{eff}}/\hbar$. Hence, for an effective action that is quadratic in ξ, we can eliminate the dependence on $\hbar$ by the rescaling

$$\xi=\sqrt{\hbar}\,\tilde{\xi} \tag{5.38}$$

This rescaling leaves the classical contribution $S(q_c)/\hbar$ unaffected. However, terms with powers higher than quadratic in ξ (say, $O(\tilde{\xi}^n)$) scale like $\hbar^{n/2}$. Thus the action (divided by $\hbar$) has an expansion of the form

$$\frac{S}{\hbar}=\frac{1}{\hbar}S^{(0)}(q_c)+S^{(2)}(\tilde{\xi};q_c)+\sum_{n=3}^{\infty}\hbar^{n/2-1}S^{(n)}(\tilde{\xi};q_c) \tag{5.39}$$

Thus, in the limit $\hbar\to 0$, we can formally expand the weight of the path integral in powers of $\hbar$. The matrix element we are calculating then takes the form

$$\langle q_f,t_f|q_i,t_i\rangle=e^{iS^{(0)}(q_c)/\hbar}\,\mathcal{Z}^{(2)}(q_c)\;(1+O(\hbar)) \tag{5.40}$$

The quantity $\mathcal{Z}^{(2)}(q_c)$ is the result of keeping only the quadratic approximation. The higher-order terms are a power series expansion in $\hbar$ and are analytic functions of $\hbar$. Here I have used the fact that, by symmetry, in most cases of interest the odd powers in ξ in general do not contribute, although there are some cases where they do.

Let us now calculate the effect of the quantum fluctuations to quadratic order. This is equivalent to the WKB approximation. Denote this factor by $\mathcal{Z}$:

$$\mathcal{Z}^{(2)}(q_c)=\int_{\tilde{\xi}(t_i)=\tilde{\xi}(t_f)=0}\mathcal{D}\xi\; e^{iS^{(2)}_{\text{eff}}(\tilde{\xi},\dot{\tilde{\xi}};q_c)} \tag{5.41}$$

It is elementary to show that, due to the boundary conditions, the action $S_{\text{eff}}(\xi,\dot{\xi})$ becomes

$$S_{\text{eff}}(\tilde{\xi},\dot{\tilde{\xi}})=\frac{1}{2}\int_{t_i}^{t_f}dt\,\tilde{\xi}(t)\left[-m\frac{d^2}{dt^2}-V''(q_c(t))\right]\tilde{\xi}(t) \tag{5.42}$$

The differential operator

$$\hat{A}=-m\frac{d^2}{dt^2}-V''(q_c(t)) \tag{5.43}$$

has the form of a Schrödinger operator for a particle on a "coordinate" t in a potential $-V''(q_c(t))$. Let $\psi_n(t)$ be a complete set of eigenfunctions of $\hat{A}$ satisfying the boundary conditions $\psi(t_i)=\psi(t_f)=0$. Completeness and orthonormality imply that the eigenfunctions $\{\psi_n(t)\}$ satisfy

$$\sum_n \psi_n^*(t)\psi_n(t')=\delta(t-t'), \qquad \int_{t_i}^{t_f} dt\, \psi_n^*(t)\psi_m(t)=\delta_{n,m} \tag{5.44}$$

An arbitrary function $\tilde{\xi}(t)$ that satisfies the vanishing boundary conditions of eq. (5.33) can be expanded as a linear combination of the basis eigenfunctions $\{\psi_n(t)\}$:

$$\tilde{\xi}(t)=\sum_n c_n\psi_n(t) \tag{5.45}$$

Clearly, we have $\tilde{\xi}(t_i)=\tilde{\xi}(t_f)=0$, as we should.

For the special case of $q_i=q_f=q_0$, where q_0 is a minimum of the potential $V(q)$, we have that $V''(q_0)=\omega_{\text{eff}}>0$ is a constant, and the eigenvectors of the Schrödinger operator are just plane waves. For a linear harmonic oscillator, $\omega_{\text{eff}}=\omega$. Thus, in this case the eigenvectors are

$$\psi_n(t)=b_n\sin(k_n(t-t_i)) \tag{5.46}$$

where

$$k_n=\frac{\pi n}{t_f-t_i} \qquad n=1,2,3,\ldots \tag{5.47}$$

and $b_n=1/\sqrt{t_f-t_i}$. The eigenvalues of $\hat{A}$ are

$$A_n=k_n^2-\omega_{\text{eff}}^2=\frac{\pi^2}{(t_f-t_i)^2}n^2-\omega_{\text{eff}}^2 \tag{5.48}$$

By using the expansion of eq. (5.45), we find that the action $S^{(2)}$ takes the form

$$S^{(2)}=\frac{1}{2}\int_{t_i}^{t_f} dt\,\tilde{\xi}(t)\,\hat{A}\,\tilde{\xi}(t)=\frac{1}{2}\sum_n A_n c_n^2 \tag{5.49}$$

where we have used the completeness and orthonormality of the basis functions $\{\psi_n(t)\}$.

The expansion in eq. (5.45) is a canonical transformation $\tilde{\xi}(t)\to c_n$. More to the point, the expansion is actually a *parametrization* of the possible histories in terms of a set of orthonormal functions, and it can be used to define the integration measure to be

$$\mathcal{D}\tilde{\xi}=\mathcal{N}\prod_n \frac{dc_n}{\sqrt{2\pi}} \tag{5.50}$$

with unit Jacobian. Here $\mathcal{N}$ is an irrelevant normalization constant that will be defined shortly.

Finally, the (formal) Gaussian integral, which is defined by a suitable analytic continuation procedure, is

$$\int_{-\infty}^{\infty}\frac{dc_n}{\sqrt{2\pi}}\,e^{\frac{i}{2}A_nc_n^2}=[-iA_n]^{-1/2} \tag{5.51}$$

and can be used to write the amplitude as

$$\mathcal{Z}^{(2)} = \mathcal{N} \prod_n A_n^{-1/2} \equiv \mathcal{N}\,(\mathrm{Det}\,\hat{A})^{-1/2} \tag{5.52}$$

where we have used the definition that the determinant of an operator is equal to the product of its eigenvalues. Therefore, up to a normalization constant, we obtain the result

$$\mathcal{Z}^{(2)} = (\mathrm{Det}\,\hat{A})^{-1/2} \tag{5.53}$$

We have thus reduced the problem of the computation of the leading (Gaussian) fluctuations in the path integral to the computation of a determinant of the fluctuation operator, a differential operator defined by the choice of classical trajectory. Next we will see how this is done.

5.2.1 Analytic continuation to imaginary time

It is useful to consider the related problem obtained by an analytic continuation to *imaginary time*, $t \to -i\tau$. We saw in section 2.8 that this problem is related to statistical physics. We will now work out one example that will be very instructive.

Formally, upon analytic continuation $t \to -i\tau$, the matrix element of the time evolution operator becomes

$$\langle q_f | e^{-\frac{i}{\hbar} H(t_f - t_i)} | q_i \rangle \to \langle q_f | e^{-\frac{1}{\hbar} H(\tau_f - \tau_i)} | q_i \rangle \tag{5.54}$$

Let us choose

$$\tau_i = 0 \qquad \tau_f = \beta\hbar \tag{5.55}$$

where $\beta = 1/T$, and T is the temperature (in units of $k_B = 1$). Hence, we find that

$$\langle q_f, -i\beta/\hbar | q_i, 0 \rangle = \langle q_f | e^{-\beta H} | q_i \rangle \tag{5.56}$$

The operator $\hat{\rho}$

$$\hat{\rho} = e^{-\beta H} \tag{5.57}$$

is the *density matrix* in the canonical ensemble of statistical mechanics for a system with Hamiltonian H in thermal equilibrium at temperature T.

It is customary to define the *partition function* $\mathcal{Z}$,

$$\mathcal{Z} = \mathrm{tr} e^{-\beta H} \equiv \int dq\, \langle q | e^{-\beta H} | q \rangle \tag{5.58}$$

where I inserted a complete set of eigenstates of $\hat{q}$. Using the results that were derived above, we see that the partition function $\mathcal{Z}$ can be written as a (Euclidean) *Feynman path integral in imaginary time*, of the form

$$\begin{aligned} \mathcal{Z} &= \int \mathcal{D}q[\tau] \exp\left\{ -\frac{1}{\hbar} \int_0^{\beta\hbar} d\tau \left[\frac{1}{2} m \left(\frac{\partial q}{\partial \tau} \right)^2 + V(q) \right] \right\} \\ &\equiv \int \mathcal{D}q[\tau] \exp\left\{ - \int_0^{\beta} d\tau \left[\frac{m}{2\hbar^2} \left(\frac{\partial q}{\partial \tau} \right)^2 + V(q) \right] \right\} \end{aligned} \tag{5.59}$$

where, in the last equality, we have rescaled $\tau \to \tau/\hbar$.

Since the partition function is a *trace* over states, we must use boundary conditions such that the initial and final states are the same state, and to sum over all such states. In other words, we must have *periodic* boundary conditions in imaginary time:

$$q(\tau) = q(\tau + \beta) \tag{5.60}$$

Therefore, a quantum mechanical system at finite temperature T can be described in terms of an equivalent system in classical statistical mechanics with Hamiltonian (or energy)

$$\mathcal{H} = \frac{m}{2\hbar^2}\left(\frac{\partial q}{\partial \tau}\right)^2 + V(q) \tag{5.61}$$

on a segment of length $1/T$ and obeying periodic boundary conditions. This effectively means that the segment is actually a ring of length $\beta = 1/T$.

Alternatively, on inserting a complete set of eigenstates of the Hamiltonian, it is easy to see that an arbitrary matrix element of the density matrix has the form

$$\begin{aligned}\langle q'|e^{-\beta H}|q\rangle &= \sum_{n=0}^{\infty} \langle q'|n\rangle\langle n|q\rangle e^{-\beta E_n} \\ &= \sum_{n=0}^{\infty} e^{-\beta E_n} \psi_n^*(q')\psi_n(q) \xrightarrow[\beta\to\infty]{} e^{-\beta E_0}\psi_0^*(q')\psi_0(q)\end{aligned} \tag{5.62}$$

where $\{E_n\}$ are the eigenvalues of the Hamiltonian, E_0 is the ground state energy, and $\psi_0(q)$ is the ground state wave function.

Therefore, we can calculate both the ground state energy E_0 and the ground state wave function from the density matrix and, consequently, from the (imaginary time) path integral. For example, from the identity

$$E_0 = -\lim_{\beta\to\infty} \frac{1}{\beta} \ln \operatorname{tr} e^{-\beta H} \tag{5.63}$$

we see that the ground state energy is given by

$$E_0 = -\lim_{\beta\to\infty} \frac{1}{\beta} \ln \int_{q(0)=q(\beta)} \mathcal{D}q \exp\left\{-\int_0^\beta d\tau \left[\frac{m}{2\hbar^2}\left(\frac{\partial q}{\partial \tau}\right)^2 + V(q)\right]\right\} \tag{5.64}$$

Mathematically, the imaginary time path integral is a better-behaved object than its real time counterpart, since it is a sum of positive quantities (the statistical weights). In contrast, the Feynman path integral (in real time) is a sum of phases and, as such, is an ill-defined object. It is actually conditionally convergent, and to make sense of it, convergence factors (or regulators) have to be introduced. The effect of these convergence factors is actually to produce an analytic continuation to imaginary time. We will encounter the same problem in the calculation of propagators. Thus, the imaginary time path integral, often referred to as the "Euclidean" path integral (as opposed to Minkowski), can be used to describe both a quantum system and a statistical mechanics system.

Finally, notice that at low temperatures $T \to 0$, the Euclidean path integral can be approximated using methods similar to the ones we discussed for the (real time) Feynman path integral. The main difference is that we must sum over trajectories that are periodic in imaginary time with period $\beta = 1/T$. In practice, this sum can only be done exactly for simple systems, such as the harmonic oscillator; for more general systems, one has to resort to some form of perturbation theory. Here we consider a physical system described by a dynamical variable q and a potential energy $V(q)$ that has a minimum at $q_0 = 0$. For simplicity, take $V(0) = 0$, and make the identfication $m\omega^2 = V''(0)$ (in other words, treat the system as an effective harmonic oscillator). The partition function is given by the Euclidean path integral

$$\mathcal{Z} = \int \mathcal{D}q[\tau] \exp\left(-\frac{1}{2}\int_0^\beta \xi(\tau)\hat{A}_E\xi(\tau)d\tau\right) \tag{5.65}$$

where $\hat{A}_E$ is the imaginary time (or Euclidean) version of the operator $\hat{A}$, and it is given by

$$\hat{A}_E = -\frac{m}{\hbar^2}\frac{d^2}{d\tau^2} + V''(q_c(\tau)) \tag{5.66}$$

The functions this operator acts on obey periodic boundary conditions with period β. Notice the important change in the sign of the term of the potential. Hence, once again, we will need to compute a functional determinant, although the operator now acts on functions obeying periodic boundary conditions. In chapter 8, we will see that in the case of *fermionic* theories, the boundary conditions become *antiperiodic.*

5.2.2 The functional determinant

Let us now compute the determinant in $\mathcal{Z}^{(2)}$ given in eq. (5.52). We will do the calculation in imaginary time and then carry out the analytic continuation to real time. We follow closely the method as explained in detail in Sidney Coleman's book (Coleman, 1985).

We want to compute

$$D = \text{Det}\left[-\frac{m}{\hbar^2}\frac{d^2}{d\tau^2} + V''(q_c(\tau))\right] \tag{5.67}$$

subject to the requirement that the space of functions that the operator acts on obeys specific boundary conditions in (imaginary) time. We are interested in two cases: (a) vanishing boundary conditions, which are useful for studying quantum mechanics at $T = 0$, and (b) periodic boundary conditions with period $\beta = 1/T$. The approach is somewhat different in the two situations.

A: Vanishing boundary conditions We define the (real) variable $x = \frac{\hbar}{m}\tau$. The range of x is the interval $[0, L]$, with $L = \hbar\beta/\sqrt{m}$. Let us consider the following eigenvalue problem for the Schrödinger operator $-\partial^2 + W(x)$,

$$\left(-\partial^2 + W(x)\right)\psi(x) = \lambda\psi(x) \tag{5.68}$$

subject to the boundary conditions $\psi(0) = \psi(L) = 0$. Formally, the determinant is given by

$$D = \prod_n \lambda_n \tag{5.69}$$

where $\{\lambda_n\}$ is the spectrum of eigenvalues of the operator $-\partial^2 + W(x)$ for a space of functions satisfying a given boundary condition.

Let us define an auxiliary function $\psi_\lambda(x)$, with λ a real number not necessarily in the spectrum of the operator, such that the following requirements are met:

1) $\psi_\lambda(x)$ is a solution of eq. (5.68), and
2) ψ_λ obeys the *initial* conditions, $\psi_\lambda(0) = 0$ and $\partial_x \psi_\lambda(0) = 1$.

It is easy to see that $-\partial^2 + W(x)$ has an eigenvalue at λ_n if and only if $\psi_{\lambda_n}(L) = 0$. (Because of this property, this procedure is known as the *shooting method*.) Hence, the determinant D of eq. (5.69) is equal to the product of the zeros of $\psi_\lambda(x)$ at $x = L$.

Consider now two potentials $W^{(1)}$ and $W^{(2)}$, and the associated functions, $\psi_\lambda^{(1)}(x)$ and $\psi_\lambda^{(2)}(x)$. Let us show that

$$\frac{\mathrm{Det}\left[-\partial^2 + W^{(1)}(x) - \lambda\right]}{\mathrm{Det}\left[-\partial^2 + W^{(2)}(x) - \lambda\right]} = \frac{\psi_\lambda^{(1)}(L)}{\psi_\lambda^{(2)}(L)} \tag{5.70}$$

The left-hand side of eq. (5.70) is a meromorphic function of λ in the complex plane and has simple zeros at the eigenvalues of $-\partial^2 + W^{(1)}(x)$ and simple poles at the eigenvalues of $-\partial^2 + W^{(2)}(x)$. Also, the left-hand side of eq. (5.70) approaches 1 as $|\lambda| \to \infty$, except along the positive real axis, which is where the spectrum of eigenvalues of both operators is located. Here we have assumed that the eigenvalues of the operators are nondegenerate, which is the general case. Similarly, the right-hand side of eq. (5.70) is also a meromorphic function of λ, which has *exactly the same zeros and the same poles* as the left-hand side. It also goes to 1 as $|\lambda| \to \infty$ (again, except along the positive real axis), since the wave functions ψ_λ asymptotically are plane waves in this limit. Therefore, the function formed by taking the ratio of the right- to the left-hand side is an analytic function on the entire complex plane, and it approaches 1 as $|\lambda| \to \infty$. Then general theorems of the theory of functions of a complex variable tell us that this function is equal to 1 everywhere.

From these considerations, we conclude that the following ratio is independent of $W(x)$:

$$\frac{\mathrm{Det}\left(-\partial^2 + W(x) - \lambda\right)}{\psi_\lambda(L)} \tag{5.71}$$

We now define a constant $\mathcal{N}$ such that

$$\frac{\mathrm{Det}\left(-\partial^2 + W(x)\right)}{\psi_0(L)} = \pi \hbar \mathcal{N}^2 \tag{5.72}$$

Then we can write

$$\mathcal{N}\left[\mathrm{Det}\left(-\partial^2 + W\right)\right]^{-1/2} = \left[\pi \hbar \psi_0(L)\right]^{-1/2} \tag{5.73}$$

Thus we have reduced the computation of the determinant, including the normalization constant, to finding the function $\psi_0(L)$. For the case of the linear harmonic oscillator, this function is the solution of

$$\left[-\frac{\partial^2}{\partial x^2} + m\omega^2\right] \psi_0(x) = 0 \tag{5.74}$$

with the initial conditions $\psi_0(0) = 0$ and $\psi_0'(0) = 1$. The solution is

$$\psi_0(x) = \frac{1}{\sqrt{m\omega}} \sinh(\sqrt{m}\omega x) \tag{5.75}$$

Hence,

$$\mathcal{Z} = \mathcal{N}\left[\text{Det}\left(-\frac{\partial^2}{\partial x^2} + m\omega^2\right)\right]^{-1/2} = [\pi\hbar\psi_0(L)]^{-1/2} \tag{5.76}$$

and we find

$$\mathcal{Z} = \left[\frac{\pi\hbar}{\sqrt{m\omega}} \sinh(\beta\omega)\right]^{-1/2} \tag{5.77}$$

where we have used $L = \hbar\beta/\sqrt{m}$. From this result, we find that the ground state energy is

$$E_0 = \lim_{\beta\to\infty} \frac{-1}{\beta} \ln \mathcal{Z} = \frac{\hbar\omega}{2} \tag{5.78}$$

as it should be.

Finally, by means of an analytic continuation back to real time, we can use these results to find, for instance, the amplitude to return to the origin after some time T. Thus, for $t_f - t_i = T$ and $q_f = q_i = 0$, we get

$$\langle 0, T|0, 0\rangle = \left[\frac{i\pi\hbar}{\sqrt{m\omega}} \sin(\omega T)\right]^{-1/2} \tag{5.79}$$

B: Periodic boundary conditions Periodic boundary conditions imply that the histories satisfy $q(\tau) = q(\tau + \beta)$. Hence, these functions can be expanded in a Fourier series of the form

$$q(\tau) = \sum_{n=-\infty}^{\infty} e^{i\omega_n\tau} q_n \tag{5.80}$$

where $\omega_n = 2\pi n/\beta$. Since $q(\tau)$ is real, we have the constraint $q_{-n} = q_n^*$. For such configurations (or histories), the action becomes

$$\begin{aligned} S &= \int_0^\beta d\tau \left[\frac{m}{2\hbar^2}\left(\frac{\partial q}{\partial \tau}\right)^2 + \frac{1}{2}V''(0)q^2\right] \\ &= \frac{\beta}{2}V''(0)q_0^2 + \beta\sum_{n\geq 1}\left[\frac{m}{\hbar^2}\omega_n^2 + V''(0)\right]|q_n|^2 \end{aligned} \tag{5.81}$$

The integration measure is now

$$\mathcal{D}q[\tau] = \mathcal{N}\frac{dq_0}{\sqrt{2\pi}} \prod_{n\geq 1} \frac{d\,\text{Re}\, q_n\, d\,\text{Im}\, q_n}{2\pi} \tag{5.82}$$

where $\mathcal{N}$ is a normalization constant that will be discussed below. After doing the Gaussian integrals, the partition function becomes

$$\mathcal{Z} = \mathcal{N} \frac{1}{\sqrt{\beta V''(0)}} \prod_{n \geq 1} \frac{1}{\frac{\beta m}{\hbar^2}\omega_n^2 + \beta V''(0)} = \mathcal{N} \left[\prod_{n=-\infty}^{\infty} \frac{1}{\frac{\beta m}{\hbar^2}\omega_n^2 + \beta V''(0)} \right]^{1/2} \tag{5.83}$$

Formally, the infinite product in this equation is divergent. The normalization constant $\mathcal{N}$ eliminates this divergence. This is an example of what is called a *regularization*. The regularized partition function is

$$\mathcal{Z} = \sqrt{\frac{m}{\hbar^2 \beta}} \frac{1}{\sqrt{\beta V''(0)}} \prod_{n \geq 1} \left[1 + \frac{\hbar^2 V''(0)}{m\omega_n^2} \right]^{-1} \tag{5.84}$$

Using the identity

$$\prod_{n \geq 1} \left(1 + \frac{a^2}{n^2 \pi^2} \right) = \frac{\sinh a}{a} \tag{5.85}$$

we find

$$\mathcal{Z} = \frac{1}{2 \sinh \left(\frac{\beta \hbar}{2} \left(\frac{V''(0)}{m} \right)^{1/2} \right)} \tag{5.86}$$

which is the partition function for a linear harmonic oscillator; see *Statistical Physics* (Landau and Lifshitz, 1959b).

5.3 Path integrals for a scalar field theory

Let us now develop the path-integral quantization picture for a scalar field theory. Our starting point is the canonically quantized scalar field. As we have seen, in canonical quantization, the scalar field $\hat{\phi}(x)$ is an operator that acts on a Hilbert space of states. We will use the field representation, which is the analog of the conventional coordinate representation in quantum mechanics.

Thus, the basis states are labeled by the field configuration at some fixed time x_0, a set of states of the form $\{ |\{\phi(\boldsymbol{x}, x_0)\}\rangle \}$. The field operator $\hat{\phi}(\boldsymbol{x}, x_0)$ acts trivially on these states:

$$\hat{\phi}(\boldsymbol{x}, x_0) |\{\phi(\boldsymbol{x}, x_0)\}\rangle = \phi(\boldsymbol{x}, x_0) |\{\phi(\boldsymbol{x}, x_0)\}\rangle \tag{5.87}$$

The set of states $\{|\{\phi(\boldsymbol{x}, x_0)\}\rangle\}$ is both complete and orthonormal. Completeness here means that these states span the entire Hilbert space. Consequently, the identity operator $\hat{\mathcal{I}}$ in the full Hilbert space can be expanded in a complete basis in the usual manner, which for this basis means

$$\hat{\mathcal{I}} = \int \mathcal{D}\phi(\boldsymbol{x}, x_0) \, |\{\phi(\boldsymbol{x}, x_0)\}\rangle\langle\{\phi(\boldsymbol{x}, x_0)\}| \tag{5.88}$$

Since the completeness condition is a sum over all the states in the basis, and since this basis is the set of field configurations at a given time x_0, we will need to define the integration measure that represents the sums over the field configurations. In this case, the definition of the integration measure is trivial:

$$\mathcal{D}\phi(\boldsymbol{x}, x_0) = \prod_{\boldsymbol{x}} d\phi(\boldsymbol{x}, x_0) \tag{5.89}$$

Likewise, orthonormality of the basis states is the condition

$$\langle \phi(\boldsymbol{x}, x_0) | \phi'(\boldsymbol{x}, x_0) \rangle = \prod_{\boldsymbol{x}} \delta\left(\phi(\boldsymbol{x}, x_0) - \phi'(\boldsymbol{x}, x_0)\right) \tag{5.90}$$

Thus, we have a working definition of the Hilbert space for a real scalar field.

In canonical quantization, the classical canonical momentum $\Pi(\boldsymbol{x}, x_0)$, defined as

$$\Pi(\boldsymbol{x}, x_0) \equiv \frac{\delta \mathcal{L}}{\delta \partial_0 \phi(\boldsymbol{x}, x_0)} = \partial_0 \phi(\boldsymbol{x}, x_0) \tag{5.91}$$

becomes an operator that acts on the same Hilbert space as the field ϕ itself does. The field operator $\hat{\phi}(x)$ and the canonical momentum operator $\hat{\Pi}(x)$ satisfy *equal-time canonical commutation relations*:

$$\left[\hat{\phi}(\boldsymbol{x}, x_0), \hat{\Pi}(\boldsymbol{y}, x_0)\right] = i\hbar\delta^3(\boldsymbol{x} - \boldsymbol{y}) \tag{5.92}$$

Here we consider a real scalar field whose Lagrangian density is

$$\mathcal{L} = \frac{1}{2}\left(\partial_\mu \phi\right)^2 - V(\phi) \tag{5.93}$$

It is a simple matter to generalize what follows to more general cases, such as complex fields and/or several components. Let us also recall that the Hamiltonian for a scalar field is given by

$$\hat{H} = \int d^3x \left[\frac{1}{2}\hat{\Pi}^2(x) + \frac{1}{2}\left(\nabla\hat{\phi}(x)\right)^2 + V(\hat{\phi}(x))\right] \tag{5.94}$$

For reasons that will become clear soon, it is convenient to add an extra term to the Lagrangian density of the scalar field, eq. (5.93), of the form

$$\mathcal{L}_{\text{source}} = J(x)\,\phi(x) \tag{5.95}$$

The field $J(x)$ is called an *external source*. The field $J(x)$ is the analog of external *forces* acting on a system of classical particles. Here we will always assume that the sources $J(x)$ vanish both at spatial infinity (at all times) and everywhere in both the remote past and in the remote future:

$$\lim_{|\boldsymbol{x}|\to\infty} J(\boldsymbol{x}, x_0) = 0 \qquad \lim_{x_0\to\pm\infty} J(\boldsymbol{x}, x_0) = 0 \tag{5.96}$$

The total Lagrangian density is

$$\mathcal{L}(\phi, J) = \mathcal{L} + \mathcal{L}_{\text{source}} \tag{5.97}$$

Since the source $J(x)$ is in general a function of space and time, the Hamiltonian that follows from this Lagrangian is formally time dependent.

Let us derive the path integral for this QFT by following the same procedure we used for the case of a finite quantum mechanical system. Hence we begin by considering the Wightman function, defined as the amplitude

$$_J\langle\{\phi(\mathbf{x}, x_0)\}|\{\phi'(\mathbf{y}, y_0)\}\rangle_J \tag{5.98}$$

In other words, we want the transition amplitude in the background of the sources $J(x)$. We will be interested in situations where x_0 is in the remote future and y_0 is in the remote past. It turns out that this amplitude is intimately related to the computation of ground state (or *vacuum*) expectation values of *time-ordered* products of field operators in the Heisenberg representation

$$G^{(N)}(x_1, \ldots, x_N) \equiv \langle 0|T[\hat{\phi}(x_1)\cdots\hat{\phi}(x_N)]|0\rangle \tag{5.99}$$

which are known as the N-point functions (or correlators). In particular, the 2-point function

$$G^{(2)}(x_1 - x_2) \equiv -i\langle 0|T[\hat{\phi}(x_1)\hat{\phi}(x_2)]|0\rangle \tag{5.100}$$

is called the *Feynman propagator* for this theory. We will see later on that all quantities of physical interest can be obtained from a suitable correlation function of the type of eq. (5.99).

In eq. (5.99), we have used the notation $T[\hat{\phi}(x_1)\ldots\hat{\phi}(x_N)]$ for the *time-ordered product* of Heisenberg field operators. For any pair of Heisenberg operators $\hat{A}(x)$ and $\hat{B}(y)$ that commute for space-like separations, their time-ordered product is defined to be

$$T[\hat{A}(x)\hat{B}(y)] = \theta(x_0 - y_0)\hat{A}(x)\hat{B}(y) + \theta(y_0 - x_0)\hat{B}(y)\hat{A}(x) \tag{5.101}$$

where $\theta(x)$ is the step (or Heaviside) function:

$$\theta(x) = \begin{cases} 1 & \text{if } x \geq 0, \\ 0 & \text{otherwise} \end{cases} \tag{5.102}$$

This definition is generalized by induction to the product of any number of operators. Notice that inside a time-ordered product, the Heisenberg operators behave as if they were c-numbers.

Let us now recall the structure of the derivation that we gave of the path integral in quantum mechanics in section 5.1. Let us paraphrase that derivation for this field theory. We consider the amplitude of eq. (5.98), and realize that this amplitude is actually a matrix element of the evolution operator,

$$_J\langle\{\phi(\mathbf{x}, x_0)\}|\{\phi'(\mathbf{y}, y_0)\}\rangle_J = \langle\{\phi(\mathbf{x})\}|T\, e^{-\frac{i}{\hbar}\int_{y_0}^{x_0} dx_0'\, \widehat{H}(x_0')}|\{\phi'(\mathbf{y})\}\rangle \tag{5.103}$$

where T stands for the time-ordering symbol (not temperature!), and $\widehat{H}(x_0')$ is the time-dependent Hamiltonian, whose Hamiltonian density is

$$\widehat{\mathcal{H}}(x_0) = \frac{1}{2}\widehat{\Pi}^2(\mathbf{x}, x_0) + \frac{1}{2}\left(\nabla\hat{\phi}(\mathbf{x}, x_0)\right)^2 + V(\hat{\phi}(\mathbf{x}, x_0)) - J(\mathbf{x}, x_0)\hat{\phi}(\mathbf{x}, x_0) \tag{5.104}$$

Paraphrasing the construction used in the case of quantum mechanics of a particle, we first partition the time interval into a large number of steps N, each of width Δt, and then insert a complete set of eigenstates of the field operator $\hat{\phi}$, since the field plays the role of the coordinate. Here too, we also have to insert complete sets of eigenstates of the canonical momentum operator, which here means the canonical field operator $\hat{\Pi}(\boldsymbol{x})$. On formally taking the time-continuum limit, $N \to \infty$ and $\Delta t \to 0$, while keeping $N\Delta t$ fixed, we obtain the result that the phase-space path integral of the field theory is

$$_J\langle\{\phi(\boldsymbol{x},x_0)\}|\{\phi'(\boldsymbol{y},y_0)\}\rangle_J = \int_{\text{b.c.}} \mathcal{D}\phi\mathcal{D}\Pi\, e^{\frac{i}{\hbar}\int d^4x\left[\dot{\phi}\Pi - \mathcal{H}(\phi,\Pi) + J\phi\right]} \tag{5.105}$$

where "b.c." indicates the boundary conditions specified by the requirement that the initial and final states be $|\{\phi(\boldsymbol{x},x_0)\}\rangle$ and $|\{\phi'(\boldsymbol{y},y_0)\}\rangle$, respectively.

Exactly as in the case of the path integral for a particle, the Hamiltonian of this theory is quadratic in the canonical momenta $\Pi(x)$. Hence, we can further integrate out the field $\Pi(x)$ and obtain the Feynman path integral for the scalar field theory in the form of a *sum over histories* of field configurations:

$$_J\langle\{\phi(\boldsymbol{x},x_0)\}|\{\phi'(\boldsymbol{y},y_0)\}\rangle_J = \mathcal{N}\int_{\text{b.c.}} \mathcal{D}\phi\, e^{\frac{i}{\hbar}S(\phi,\partial_\mu\phi,J)} \tag{5.106}$$

where $\mathcal{N}$ is an (unimportant) normalization constant, and $S(\phi,\partial_\mu\phi,J)$ is the action for a real scalar field $\phi(x)$ coupled to a source $J(x)$:

$$S(\phi,\partial_\mu\phi,J) = \int d^4x\left[\frac{1}{2}\left(\partial_\mu\phi\right)^2 - V(\phi) + J\phi\right] \tag{5.107}$$

5.4 Path integrals and propagators

In QFT, we are interested in calculating vacuum (ground state) expectation values of field operators at various spacetime locations. Thus, instead of the amplitude $_J\langle\{\phi(\boldsymbol{x},x_0)\}|\{\phi'(\boldsymbol{y},y_0)\}\rangle_J$, we might be interested in a transition between an initial state, at $y_0 \to -\infty$, which is the *vacuum state* $|0\rangle$ (i.e., the ground state of the scalar field in the absence of the source $J(x)$), and a final state at $x_0 \to \infty$, which is also the vacuum state of the theory in the absence of sources. Denote this matrix element by

$$Z[J] = {}_J\langle 0|0\rangle_J \tag{5.108}$$

This matrix element is called the *vacuum persistence amplitude.*

Now let us see how the vacuum persistence amplitude is related to the Feynman path integral for a scalar field, eq. (5.106). To do this, assume that the source $J(x)$ is "on" between times $t < t'$ and that we watch the system on a much longer time interval $T < t < t' < T'$. For this interval, we can now use the superposition principle to insert complete sets of states at intermediate times t and t', and write the amplitude in the form:

$$\begin{aligned}&_J\langle\{\Phi'(\boldsymbol{x},T')\}|\{\Phi(\boldsymbol{x},T)\}\rangle_J = \\ &\int \mathcal{D}\phi(\boldsymbol{x},t)\,\mathcal{D}\phi'(\boldsymbol{x},t')\langle\{\Phi'(\boldsymbol{x},T')\}|\{\phi'(\boldsymbol{x},t')\}\rangle \\ &\qquad \times {}_J\langle\{\phi'(\boldsymbol{x},t')\}|\{\phi(\boldsymbol{x},t)\}\rangle_J\langle\{\phi(\boldsymbol{x},t)\}|\{\Phi(\boldsymbol{x},T)\}\rangle\end{aligned} \tag{5.109}$$

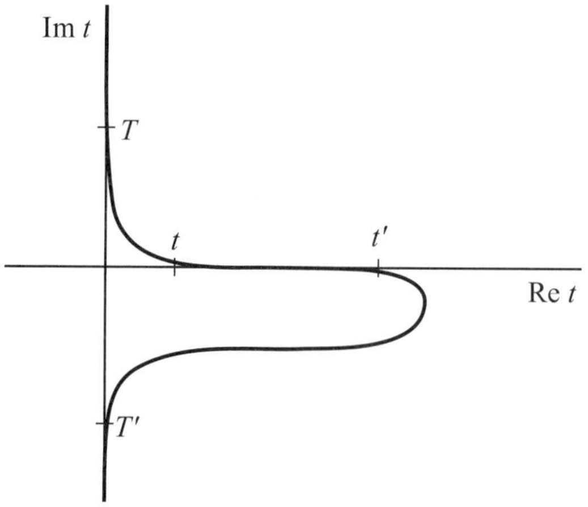

Figure 5.4 Analytic continuation.

The matrix elements $\langle\{\phi(\boldsymbol{x},t)\}|\{\Phi(\boldsymbol{x},T)\}\rangle$ and $\langle\{\Phi'(\boldsymbol{x},T')\}|\{\phi'(\boldsymbol{x},t')\}\rangle$ are given by

$$\langle\{\phi(\boldsymbol{x},t)\}|\{\Phi(\boldsymbol{x},T)\}\rangle=\sum_n \Psi_n[\{\phi(\boldsymbol{x})\}]\Psi_n^*[\{\Phi(\boldsymbol{x})\}]\,e^{-iE_n(t-T)/\hbar}$$

$$\langle\{\Phi'(\boldsymbol{x},T')\}|\{\phi'(\boldsymbol{x},t')\}\rangle=\sum_m \Psi_m[\{\Phi'(\boldsymbol{x})\}]\Psi_m^*[\{\phi'(\boldsymbol{x})\}]\,e^{-iE_m(T'-t')/\hbar} \tag{5.110}$$

where we have introduced complete sets of eigenstates $|\{\Psi_n\}\rangle$ of the Hamiltonian of the scalar field (without sources) and the corresponding wave functions, $\{\Psi_n[\Phi(\boldsymbol{x})]\}$.

At long times T and T', these series expansions oscillate very rapidly, and a definition must be provided to make sense of these expressions. To this end, let us now analytically continue T along the *positive* imaginary time axis, and T' along the *negative* imaginary time axis, as shown in figure 5.4. After carrying out the analytic continuation, we find that the following identities hold:

$$\lim_{T\to+i\infty} e^{-iE_0T/\hbar}\langle\{\phi(\boldsymbol{x},t)\}|\{\Phi(\boldsymbol{x},T)\}\rangle=\Psi_0[\{\phi\}]\,\Psi_0^*[\{\Phi\}]\,e^{-iE_0t/\hbar}$$

$$\lim_{T'\to-i\infty} e^{iE_0T'/\hbar}\langle\{\Phi'(\boldsymbol{x},T')\}|\{\phi(\boldsymbol{x},t')\}\rangle=\Psi_0[\{\Phi'\}]\,\Psi_0^*[\{\phi'\}]\,e^{iE_0t'/\hbar} \tag{5.111}$$

This result is known as the Gell-Mann-Low theorem (Gell-Mann and Low, 1951). In this limit, the contributions from excited states drop out, provided the vacuum state $|0\rangle$ is nondegenerate. This procedure is equivalent to the standard adiabatic turning on and off of the external sources. The restriction to a nondegenerate vacuum state can be done by lifting a possible degeneracy by means of an infinitesimally weak external perturbation, which is switched off after the infinite time limit is taken. We will encounter similar issues when discussing spontaneous symmetry breaking in chapter 12.

Hence, in the same limit, we also find that the following relation holds:

$$\lim_{T\to+i\infty}\lim_{T'\to-i\infty}\frac{\langle\{\Phi'(\boldsymbol{x},T')\}|\{\Phi(\boldsymbol{x},T)\}\rangle}{\exp\left[-iE_0(T'-T)/\hbar\right]\ \Psi_0^*[\{\Phi\}]\ \Psi_0[\{\Phi'\}]}$$
$$=\int\mathcal{D}\Phi\mathcal{D}\Phi'\ \Psi_0^*[\{\phi'(\boldsymbol{x},t')\}]\ \Psi_0[\{\phi(\boldsymbol{x},t)\}]\ _J\langle\{\phi'(\boldsymbol{x},t')\}|\{\phi(\boldsymbol{x},t)\}\rangle_J$$
$$\equiv_J\langle 0|0\rangle_J \tag{5.112}$$

Eq. (5.112) gives us a direct relation between the Feynman path integral and the vacuum persistence amplitude of the form:

$$Z[J]={}_J\langle 0|0\rangle_J=\mathcal{N}\lim_{T\to+i\infty}\lim_{T'\to-i\infty}\int\mathcal{D}\phi\, e^{\frac{i}{\hbar}\int_T^{T'}d^4x\left[\mathcal{L}(\phi,\partial_\mu\phi)+J\phi\right]} \tag{5.113}$$

In other words, in this asymptotically long-time limit, the amplitude of eq. (5.98) becomes identical to the vacuum persistence amplitude ${}_J\langle 0|0\rangle_J$, regardless of the choice of the initial and final states.

Hence we find a direct relation between the vacuum persistence function $Z[J]$ and the Feynman path integral, given by eq. (5.113). Notice that, in this limit, we can ignore the "hard" boundary condition and work instead with free boundary conditions. Or equivalently, physical properties become independent of the initial and final conditions imposed.

For these reasons, from now on, we will work with the simpler expression

$$Z[J]={}_J\langle 0|0\rangle_J=\mathcal{N}\int\mathcal{D}\phi\, e^{\frac{i}{\hbar}\int d^4x\left[\mathcal{L}(\phi,\partial_\mu\phi)+J\phi\right]} \tag{5.114}$$

This is a very useful relation. We will see now that $Z[J]$ is the generating function(al) of all the vacuum expectation values of time-ordered products of fields (i.e., the correlators of the theory).

In particular, let us compute the expression

$$\frac{1}{Z[0]}\frac{\delta^2 Z[J]}{\delta J(x)\delta J(x')}\bigg|_{J=0}=\frac{1}{\langle 0|0\rangle}\frac{\delta^2\,{}_J\langle 0|0\rangle_J}{\delta J(x)\delta J(x')}\bigg|_{J=0}=\left(\frac{i}{\hbar}\right)^2\langle 0|T[\phi(x)\phi(x')]|0\rangle \tag{5.115}$$

where T is the time-ordering symbol. Thus, the 2-point function (i.e., the *Feynman propagator* or *propagator* of the scalar field $\phi(x)$) becomes

$$\langle 0|T[\phi(x)\phi(x')]|0\rangle=-i\frac{1}{\langle 0|0\rangle}\int\mathcal{D}\phi\,\phi(x)\,\phi(x')\,\exp\left(\frac{i}{\hbar}S[\phi,\partial_\mu\phi]\right) \tag{5.116}$$

Similarly, the N-point function $\langle 0|T[\phi(x_1)\cdots\phi(x_N)]|0\rangle$ becomes

$$\langle 0|T[\phi(x_1)\cdots\phi(x_N)]|0\rangle=(-i\hbar)^N\frac{1}{\langle 0|0\rangle}\frac{\delta^N\,{}_J\langle 0|0\rangle_J}{\delta J(x_1)\cdots\delta J(x_N)}\bigg|_{J=0}$$
$$=\frac{1}{\langle 0|0\rangle}\int\mathcal{D}\phi\,\phi(x_1)\cdots\phi(x_N)\,\exp\left(\frac{i}{\hbar}S[\phi,\partial_\mu\phi]\right) \tag{5.117}$$

where

$$Z[0] = \langle 0|0\rangle = \int \mathcal{D}\phi \, \exp\left(\frac{i}{\hbar} S[\phi, \partial_\mu \phi]\right) \tag{5.118}$$

Therefore, we find that the path integral always yields vacuum expectation values of time-ordered products of operators. The quantity $Z[J]$ can thus be viewed as the generating functional of the correlation functions of this theory. These are actually general results that hold for the path integrals of all theories.

5.5 Path integrals in Euclidean spacetime and statistical physics

In section 5.4, we saw how to relate the computation of transition amplitudes to path integrals in Minkowski spacetime with specific boundary conditions dictated by the nature of the initial and final states. In particular, we derived explicit expressions for the case of fixed boundary conditions.

However, we could have chosen other boundary conditions. For instance, we could have chosen the amplitude to begin in any state at the initial time and to go back to the *same* state at the final time, but summing over all states. This is the same as to ask for the *trace*,

$$\begin{aligned} Z'[J] &= \int \mathcal{D}\Phi_J \langle\{\Phi(\boldsymbol{x}, t')\}|\{\Phi(\boldsymbol{x}, t)\}\rangle_J \\ &\equiv \operatorname{Tr} T\, e^{-\frac{i}{\hbar}\int d^4x\, (\mathcal{H} - J\phi)} \\ &\equiv \int_{\text{PBC}} \mathcal{D}\phi \; e^{\frac{i}{\hbar}\int d^4x\, (\mathcal{L} + J\phi)} \end{aligned} \tag{5.119}$$

where PBC stands for "periodic boundary conditions" on some generally *finite* time interval $t' - t$, and T is the time-ordering symbol.

Let us now carry the analytic continuation to imaginary time $t \to -i\tau$ (i.e., known as a Wick rotation). Upon Wick rotation, the theory has Euclidean invariance (i.e., rotations and translations in $D = d + 1$-dimensional space). Imaginary time plays the same role as the other d spatial dimensions. Hereafter, we denote imaginary time by x_D, and all vectors will have indices μ that run from 1 to D.

Consider two cases: an infinite imaginary time interval and a finite imaginary time interval.

5.5.1 Infinite imaginary time interval

In this case, the path integral becomes

$$Z'[J] = \int \mathcal{D}\phi \; e^{-\int d^Dx\, (\mathcal{L}_E - J\phi)} \tag{5.120}$$

where D is the total number of spacetime dimensions. For the sake of definiteness, here we discuss the four-dimensional case, but the results are obviously valid more generally. Here $\mathcal{L}_E$ is the Euclidean Lagrangian:

$$\mathcal{L}_E = \frac{1}{2}(\partial_0\phi)^2 + \frac{1}{2}(\nabla\phi)^2 + V(\phi) \tag{5.121}$$

The path integral in eq. (5.120) has two interpretations.

The first interpretation is simply that the time limit (in imaginary time) is infinite and therefore, the integral must be identical to the vacuum persistence amplitude ${}_J\langle 0|0\rangle_J$. The only difference is that from this interpretation, we get all the N-point functions in Euclidean spacetime (imaginary time). Therefore, the relativistic interval is

$$x_0^2 - \boldsymbol{x}^2 \to -\tau^2 - \boldsymbol{x}^2 < 0 \tag{5.122}$$

which is always space-like. Hence, with this procedure, we get the correlation functions for space-like separations of its arguments. To get to time-like separations, we need to do an analytic continuation back to real time. This we will do in section 5.6.4.

The second interpretation is that the path integral in eq. (5.120) is the *partition function* of a system in classical statistical mechanics in D dimensions with energy density (divided by T) equal to $\mathcal{L}_E - J\phi$. This will turn out to be a very useful connection (both ways!).

5.5.2 Finite imaginary time interval

In this case, we have

$$0 \le x_0 = \tau \le \beta = 1/T \tag{5.123}$$

where T will be interpreted as the temperature. Indeed, in this case, the path integral is

$$Z'[0] = \operatorname{Tr} e^{-\beta H} \tag{5.124}$$

and we are effectively looking at a problem of the same QFT but at *finite temperature* $T = 1/\beta$. The path integral is once again the partition function but of a system in quantum statistical physics! The partition function thus is (after setting $\hbar = 1$)

$$Z'[J] = \int \mathcal{D}\phi \, e^{-\int_0^\beta d\tau \, (\mathcal{L}_E - J\phi)} \tag{5.125}$$

where the field $\phi(\boldsymbol{x}, \tau)$ obeys *periodic boundary conditions* in imaginary time:

$$\phi(\boldsymbol{x}, \tau) = \phi(\boldsymbol{x}, \tau + \beta) \tag{5.126}$$

This boundary condition will hold for all *bosonic* theories. We will see in section 8.5 that theories with fermions obey instead *antiperiodic* boundary conditions in imaginary time.

Hence, QFT at finite temperature T is just QFT on a Euclidean spacetime that is periodic (and finite) in one direction: imaginary time. In other words, we have wrapped (or *compactified*) Euclidean spacetime into a cylinder with perimeter (circumference) $\beta = 1/T$ (in units of $\hbar = k_B = 1$), as shown in figure 5.5.

The correlation functions in imaginary time (which we will call the "Euclidean correlation functions") are given by

$$\frac{1}{Z'[J]} \frac{\delta^N Z'[J]}{\delta J(x_1) \cdots J(x_N)}\bigg|_{J=0} = \langle \phi(x_1) \cdots \phi(x_N) \rangle \tag{5.127}$$

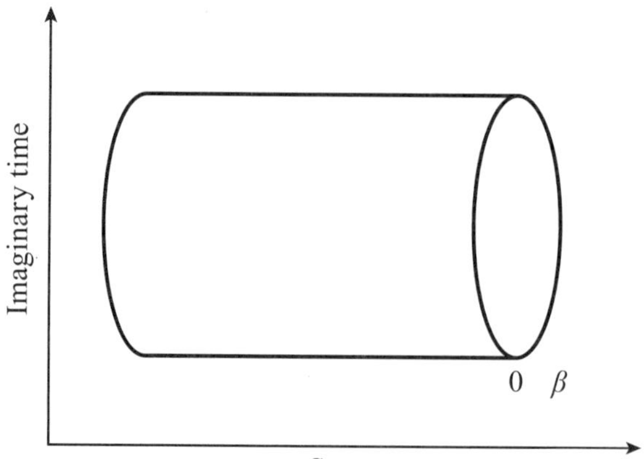

Figure 5.5 Periodic boundary conditions wrap spacetime into a cylinder.

which are just the correlation functions in the equivalent problem in statistical mechanics. Upon analytic continuation, the Euclidean correlation functions $\langle\phi(x_1)\ldots\phi(x_N)\rangle$ and the N-point functions of the QFT are related by

$$\langle\phi(x_1)\cdots\phi(x_N)\rangle \leftrightarrow (i\hbar)^N\langle 0|T\phi(x_1)\cdots\phi(x_N)|0\rangle \tag{5.128}$$

For the case of a QFT at finite temperature T, the path integral yields the correlation functions of the Heisenberg field operators in imaginary time. These correlation functions are often called the *thermal correlation functions* (or propagators). They are functions of the spatial positions of the fields, $\boldsymbol{x}_1,\ldots,\boldsymbol{x}_N$, and of their *imaginary time coordinates*, $x_{D1},\ldots,x_{DN}$ (here $x_D\equiv\tau$). To obtain the correlation functions as a function of the *real time coordinates* $x_{01},\ldots,x_{0N}$ at finite temperature T, it is necessary to do an analytic continuation. We will discuss how this is done in section 5.8.

5.6 Path integrals for the free scalar field

Let us now consider the case of a *free scalar field*. We consider *Euclidean spacetime* (i.e., in imaginary time) and then will do the relevant analytic continuation back to real time at the end of the calculation.

The Euclidean Lagrangian $\mathcal{L}_E$ for a free field ϕ coupled to a source J is

$$\mathcal{L}_E=\frac{1}{2}\left(\partial_\mu\phi\right)^2+\frac{1}{2}m^2\phi^2-J\phi \tag{5.129}$$

where we are using the notation

$$\left(\partial_\mu\phi\right)^2=\partial_\mu\phi\partial_\mu\phi \tag{5.130}$$

Here the index is $\mu=1,\ldots,D$ for an Euclidean spacetime of $D=d+1$ dimensions. For the most part (but not always), we will be interested in the case of $d=3$, and so the Euclidean space has four dimensions. Notice the way the Euclidean spacetime indices are placed in eq. (5.130). This is not a misprint!

We will compute the Euclidean path integral (or partition function) $\mathcal{Z}_E[J]$ exactly. The Euclidean path integral for a free field has the form

$$\mathcal{Z}_E[J]=\mathcal{N}\int \mathcal{D}\phi \; e^{-\int d^Dx \left[\frac{1}{2}\left(\partial_\mu\phi\right)^2+\frac{1}{2}m^2\phi^2-J\phi\right]} \tag{5.131}$$

In classical statistical mechanics, this theory is known as the Gaussian model.

In what follows, let us assume that the boundary conditions of the field ϕ (and the source J) at infinity are either vanishing or periodic, and that the source J also either vanishes at spatial infinity or is periodic. With these assumptions, all terms that are total derivatives drop out identically. Therefore, upon integration by parts and after dropping boundary terms, the Euclidean Lagrangian becomes

$$\mathcal{L}_E=\frac{1}{2}\phi\left[-\partial^2+m^2\right]\phi-J\phi \tag{5.132}$$

Since this action is a quadratic form of the field ϕ, this path integral can be calculated exactly. It has terms that are quadratic (or rather, bilinear) in ϕ and a term linear in ϕ, the source term. By means of the following shift of the field ϕ,

$$\phi(x)=\bar{\phi}(x)+\xi(x) \tag{5.133}$$

the Lagrangian becomes

$$\begin{aligned}\mathcal{L}_E&=\frac{1}{2}\phi\left[-\partial^2+m^2\right]\phi-J\phi\\ &=\frac{1}{2}\bar{\phi}\left[-\partial^2+m^2\right]\bar{\phi}-J\bar{\phi}+\frac{1}{2}\xi\left[-\partial^2+m^2\right]\xi+\xi\left[-\partial^2+m^2\right]\bar{\phi}-J\xi\end{aligned} \tag{5.134}$$

Hence, we can decouple the source $J(x)$ by requiring that the shift $\bar{\phi}$ be such that the terms linear in ξ cancel each other exactly. This requirement leads to the condition that the classical field $\bar{\phi}$ be the solution of the following inhomogeneous partial differential equation:

$$\left[-\partial^2+m^2\right]\bar{\phi}=J(x) \tag{5.135}$$

Equivalently, we can write the classical field $\bar{\phi}$ in terms of the source $J(x)$ through the action of the inverse of the operator $-\partial^2+m^2$:

$$\bar{\phi}=\frac{1}{-\partial^2+m^2}J \tag{5.136}$$

The solution of eq. (5.135) is

$$\bar{\phi}(x)=\int d^Dx'\, G_0^E(x-x')\, J(x') \tag{5.137}$$

where

$$G_0^E(x-x')=\left\langle x\left|\frac{1}{-\partial^2+m^2}\right|x'\right\rangle \tag{5.138}$$

is the correlation function of the linear partial differential operator $-\partial^2 + m^2$. Thus, $G_0^E(x-x')$ is the solution of

$$\left[-\partial_x^2 + m^2\right] G_0^E(x-x') = \delta^D(x-x') \tag{5.139}$$

In terms of $G_0^E(x-x')$, the terms of the shifted action become

$$\begin{aligned}\int d^Dx\Big(\frac{1}{2}\bar{\phi}(x)\left[-\partial^2 + m^2\right]\bar{\phi}(x) - J\bar{\phi}(x)\Big) \\ &= -\frac{1}{2}\int d^Dx\bar{\phi}(x)J(x) \\ &= -\frac{1}{2}\int d^Dx\, d^Dx'\, J(x)\, G_0^E(x-x')\, J(x')\end{aligned} \tag{5.140}$$

Therefore, the path integral for the generating function of the free Euclidean scalar field $Z_E[J]$, defined in eq. (5.131), is given by

$$Z_E[J] = Z_E[0]\; e^{\frac{1}{2}\int d^Dx \int d^Dx'\, J(x)\, G_0^E(x-x')\, J(x')} \tag{5.141}$$

where $Z_E[0]$ is

$$Z_E[0] = \int \mathcal{D}\xi\; e^{-\frac{1}{2}\int d^Dx\, \xi(x)\left[-\partial^2 + m^2\right]\xi(x)} \tag{5.142}$$

Eq. (5.141) shows that, after the decoupling, $Z_E[J]$ is a product of two factors: (1) a factor that is a function of a bilinear form in the source J, and (2) a path integral, $Z_E[0]$, that is independent of the sources.

5.6.1 Calculation of $Z_E[0]$

The path integral $Z_E[0]$ is analogous to the fluctuation factor that we found in section 5.2 for the path integral for a harmonic oscillator in elementary quantum mechanics. There we saw that the analogous factor can be written as a determinant of a differential operator, the kernel of the bilinear form that entered in the action. The same result holds here as well. The only difference is that the kernel is now the *partial differential* operator $\hat{A} = -\partial^2 + m^2$, whereas in quantum mechanics, it is an ordinary differential operator. Here too, the operator $\hat{A}$ has a set of eigenstates $\{\Psi_n(x)\}$ which, once the boundary conditions in spacetime are specified, are both complete and orthonormal. The associated spectrum of eigenvalues A_n is

$$\begin{aligned}&\left[-\partial^2 + m^2\right]\Psi_n(x) = A_n \Psi_n(x) \\ &\int d^Dx\, \Psi_n(x)\, \Psi_m(x) = \delta_{n,m} \\ &\sum_n \Psi_n(x)\, \Psi_n(x') = \delta(x-x')\end{aligned} \tag{5.143}$$

Hence, once again we can expand the field $\phi(x)$ in the complete set of states $\{\Psi_n(x)\}$:

$$\phi(x) = \sum_n c_n\, \Psi_n(x) \tag{5.144}$$

The field configurations are thus parametrized by the coefficients $\{c_n\}$.

The action now becomes

$$S = \int d^D x \, \mathcal{L}_E(\phi, \partial\phi) = \frac{1}{2} \sum_n A_n c_n^2 \tag{5.145}$$

Thus, up to a normalization factor, we find that $Z_E[0]$ is given by

$$Z_E[0] = \prod_n A_n^{-1/2} \equiv \left(\text{Det}\left[-\partial^2 + m^2\right]\right)^{-1/2} \tag{5.146}$$

and we have reduced the calculation of $Z_E[0]$ to the computation of the determinant of a differential operator, $\text{Det}\left[-\partial^2 + m^2\right]$.

In chapter 8, we will discuss efficient methods to compute such determinants. For the moment, it is sufficient to notice that there is a simple, but formal, way to compute this determinant. First, notice that if we are interested in the behavior of an infinite system at $T = 0$, then the eigenstates of the operator $-\partial^2 + m^2$ are simply suitably normalized plane waves. Let L be the linear size of the system, with $L \to \infty$. Then the eigenfunctions are labeled by a D-dimensional momentum p_μ (with $\mu = 0, 1, \ldots, d$)

$$\Psi_p(x) = \frac{1}{(2\pi L)^{D/2}} e^{i p_\mu x_\mu} \tag{5.147}$$

with eigenvalues,

$$A_p = p^2 + m^2 \tag{5.148}$$

Hence the *logarithm* of the determinant is

$$\begin{aligned} \ln \text{Det}\left[-\partial^2 + m^2\right] &= \text{Tr} \ln\left[-\partial^2 + m^2\right] \\ &= \sum_p \ln(p^2 + m^2) \\ &= V \int \frac{d^D p}{(2\pi)^D} \ln(p^2 + m^2) \end{aligned} \tag{5.149}$$

where $V = L^D$ is the volume of Euclidean spacetime. Hence,

$$\ln Z_E[0] = -\frac{V}{2} \int \frac{d^D p}{(2\pi)^D} \ln(p^2 + m^2) \tag{5.150}$$

This expression has two singularities: an *infrared divergence* and a *UV divergence*. As shown in eq. (5.150), $\ln Z[0]$ diverges as $V \to \infty$. This IR singularity actually is not a problem, since $\ln Z_E[0]$ should be an *extensive* quantity that must scale with the volume of spacetime. In other words, this is how it should behave. However, the integral in eq. (5.150) diverges at large momenta unless there is an upper bound (or *cutoff*) for the allowed momenta. This is a UV singularity. It has the same origin as the UV divergence of the ground state energy. In fact, $Z_E[0]$ is closely related to the ground state (vacuum) energy, since

$$Z_E[0] = \lim_{\beta \to \infty} \sum_n e^{-\beta E_n} \sim e^{-\beta E_0} + \cdots \tag{5.151}$$

Thus,

$$E_0 = -\lim_{\beta\to\infty} \frac{1}{\beta} \ln Z_E[0] = \frac{1}{2} L^d \int \frac{d^D p}{(2\pi)^D} \ln(p^2+m^2) \tag{5.152}$$

where L^d is the volume of space, and $V = L^d \beta$. Notice that eq. (5.152) is UV divergent. In section 8.8, we will discuss how to compute expressions of the form of eq. (5.152).

5.6.2 Propagators and correlators

Some interesting results are found immediately by direct inspection of eq. (5.141). We can easily see that, once we set $J=0$, the correlation function $G_E^{(0)}(x-x')$,

$$G_E^{(0)}(x-x') = \left\langle x \left| \frac{1}{-\partial^2+m^2} \right| x' \right\rangle \tag{5.153}$$

is equal to the 2-point correlation function for this theory (at $J=0$):

$$\langle \phi(x)\phi(x')\rangle = \frac{1}{Z_E[0]} \frac{\delta^2 Z_E[J]}{\delta J(x)\delta J(x')}\bigg|_{J=0} = G_E^{(0)}(x-x') \tag{5.154}$$

Likewise, we find that, for a free-field theory, the N-point correlation function $\langle \phi(x_1)\cdots \phi(x_N)\rangle$ is equal to

$$\begin{aligned}\langle \phi(x_1)\cdots\phi(x_N)\rangle =& \frac{1}{Z_E[0]} \frac{\delta^N Z_E[J]}{\delta J(x_1)\cdots\delta J(x_N)}\bigg|_{J=0} \\ =& \langle\phi(x_1)\phi(x_2)\rangle \cdots \langle \phi(x_{N-1})\phi(x_N)\rangle + \text{permutations}\end{aligned} \tag{5.155}$$

Therefore, for a free field, up to permutations of the coordinates $x_1, \ldots, x_N$, the N-point function reduces to a sum of products of 2-point functions. Hence, N must be a positive *even* integer. This result, eq. (5.155), which we derived in the context of a theory for a free scalar field, is actually much more general. It is known as *Wick's theorem*. It applies to all free theories, theories whose Lagrangians are bilinear in the fields. And it is independent of the statistics and on whether there is relativistic invariance or not. The only caveat is that (as we will see in chapter 8) for the case of fermionic theories, a sign is associated with each term in this sum.

It is easy to see that, for $N=2k$, the total number of terms in the sum is

$$(2k-1)(2k-3)\cdots = \frac{(2k)!}{2^k k!} \tag{5.156}$$

In eq. (5.155) each of a 2-point function $\langle\phi(x_1)\phi(x_2)\rangle$ is a free propagator. It is also called a *contraction*. It is common to use the notation

$$\langle\phi(x_1)\phi(x_2)\rangle = \overbracket{\phi(x_1)\phi}(x_2) \tag{5.157}$$

to denote a contraction or propagator.

5.6.3 Calculation of the propagator

Let us now calculate the 2-point function, or propagator, $G_E^{(0)}(x-x')$ for infinite Euclidean space. This is the case of interest in QFT at $T=0$. In section 5.8, we will calculate the propagator at finite temperature.

Eq. (5.139) tells us that $G_E^{(0)}(x-x')$ is the Green function of the operator $-\partial^2+m^2$. We will use Fourier transform methods and write $G_E^{(0)}(x-x')$ in the form

$$G_E^{(0)}(x-x') = \int \frac{d^D p}{(2\pi)^D}\, G_0^E(p)\, e^{i p_\mu (x_\mu - x'_\mu)} \tag{5.158}$$

which is a solution of eq. (5.139) if

$$G_E^{(0)}(p) = \frac{1}{p^2+m^2} \tag{5.159}$$

Therefore, the correlation function in real (Euclidean!) space is the integral

$$G_E^{(0)}(x-x') = \int \frac{d^D p}{(2\pi)^D} \frac{e^{i p_\mu (x_\mu - x'_\mu)}}{p^2+m^2} \tag{5.160}$$

We will often encounter integrals of this type and for that reason, let us evaluate this one in some detail. We begin by using the identity

$$\frac{1}{A} = \frac{1}{2}\int_0^\infty d\alpha\, e^{-\frac{A}{2}\alpha} \tag{5.161}$$

where $A>0$ is a positive real number. The variable α is called a "Feynman-Schwinger parameter."

Now choose $A=p^2+m^2$, and substitute this expression back in eq. (5.160), which takes the form

$$G_E^{(0)}(x-x') = \frac{1}{2}\int_0^\infty d\alpha \int \frac{d^D p}{(2\pi)^D}\, e^{-\frac{\alpha}{2}(p^2+m^2) + i p_\mu (x_\mu - x'_\mu)} \tag{5.162}$$

The integrand is a Gaussian, and the integral can be calculated by a shift of the integration variables p_μ (i.e., by completing squares)

$$\frac{\alpha}{2}(p^2+m^2) - i p_\mu (x_\mu - x'_\mu) = \frac{1}{2}\left(\sqrt{\alpha}\, p_\mu - i\frac{x_\mu - x'_\mu}{\sqrt{\alpha}}\right)^2 - \frac{1}{2}\left(\frac{x_\mu - x'_\mu}{\sqrt{\alpha}}\right)^2 \tag{5.163}$$

and by using the Gaussian integral

$$\int \frac{d^D p}{(2\pi)^D}\, e^{-\frac{1}{2}\left(\sqrt{\alpha}\, p_\mu - i\frac{x_\mu - x'_\mu}{\sqrt{\alpha}}\right)^2} = (2\pi\alpha)^{-D/2} \tag{5.164}$$

After all of this is done, we find the formula

$$G_E^{(0)}(x-x') = \frac{1}{2(2\pi)^{D/2}} \int_0^\infty d\alpha\, \alpha^{-D/2}\, e^{-\frac{|x-x'|^2}{2\alpha} - \frac{1}{2}m^2\alpha} \tag{5.165}$$

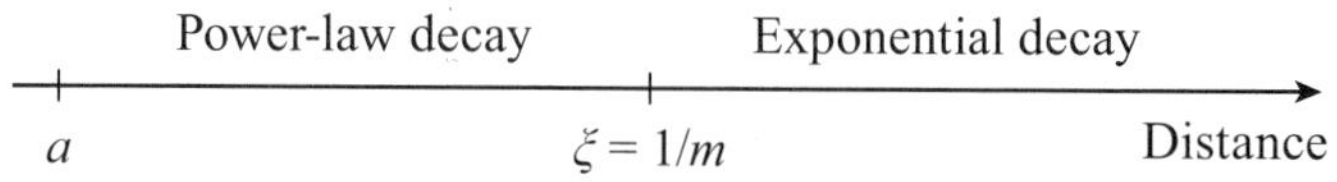

Figure 5.6 Behaviors of the Euclidean propagator.

Let us now define a rescaling of the variable α,

$$\alpha = \lambda t \tag{5.166}$$

by which

$$\frac{|x-x'|^2}{2\alpha} + \frac{1}{2}m^2\alpha = \frac{|x-x'|^2}{2\lambda t} + \frac{1}{2}m^2\lambda t \tag{5.167}$$

If we choose

$$\lambda = \frac{|x-x'|}{m} \tag{5.168}$$

then the exponent becomes

$$\frac{|x-x'|^2}{2\alpha} + \frac{1}{2}m^2\alpha = \frac{m|x-x'|}{2}\left(t+\frac{1}{t}\right) \tag{5.169}$$

After this final change of variables, we find that the correlation function is

$$G_E^{(0)}(x-x') = \frac{1}{(2\pi)^{D/2}}\left(\frac{m}{|x-x'|}\right)^{\frac{D}{2}-1} K_{\frac{D}{2}-1}(m|x-x'|) \tag{5.170}$$

where $K_\nu(z)$ is the modified Bessel function, which has the integral representation

$$K_\nu(z) = \frac{1}{2}\int_0^\infty dt\, t^{\nu-1}\, e^{-\frac{z}{2}\left(t+\frac{1}{t}\right)} \tag{5.171}$$

where $\nu = \frac{D}{2} - 1$, and $z = m|x-x'|$.

There are two interesting regimes: (1) long distances, $m|x-x'| \gg 1$, and (2) short distances, $m|x-x'| \ll 1$. (See figure 5.6.)

A: Long-distance behavior In this regime, $z = m|x-x'| \gg 1$, a saddle-point calculation shows that the Bessel function $K_\nu(z)$ has the asymptotic behavior,

$$K_\nu(z) = \sqrt{\frac{\pi}{2z}}e^{-z}\left[1 + O(1/z)\right] \tag{5.172}$$

Thus, in this regime, the Euclidean propagator (or correlation function) behaves like

$$G_E^{(0)}(x-x') = \frac{\sqrt{\pi/2}\, m^{D-2}\, e^{-m|x-x'|}}{(2\pi)^{D/2}\,(m|x-x'|)^{\frac{D-1}{2}}}\left[1 + O\left(\frac{1}{m|x-x'|}\right)\right] \tag{5.173}$$

Therefore, at long distances, the Euclidean (or imaginary time) propagator has an exponential decay with distance (and imaginary time). The length scale for this decay is

$1/m$, which is natural, since it is the only quantity with units of length in the theory. In real time, and in conventional units, this length scale is just the Compton wavelength, $\hbar/mc$. In statistical physics, this length scale is known as the *correlation length* ξ.

B: Short-distance behavior In this regime, we must use the behavior of the Bessel function for small values of the argument:

$$K_\nu(z) = \frac{\Gamma(\nu)}{2\left(\frac{z}{2}\right)^\nu} + O(1/z^{\nu-2}) \tag{5.174}$$

The correlation function now behaves instead like

$$G_E^{(0)}(x-x') = \frac{\Gamma\left(\frac{D}{2}-1\right)}{4\pi^{D/2}|x-x'|^{D-2}} + \cdots \tag{5.175}$$

where $\cdots$ are terms that vanish as $m|x-x'| \to 0$. Notice that the leading term is independent of the mass m. This is the behavior of the free *massless* theory.

5.6.4 Behavior of the propagator in Minkowski spacetime

Let us find the behavior of the propagator in real time. Now we must do the analytic continuation back to real time.

Recall that in going from Minkowski to Euclidean space, we continued $x_0 \to -ix_4$. In addition, there is also a factor of i difference in the definition of the propagator. Thus, the propagator in Minkowski spacetime, $G^{(0)}(x-x')$, is the expression that results from the analytic continuation:

$$G^{(0)}(x-x') = iG_E^{(0)}(x-x')\Big|_{x_4 \to ix_0} \tag{5.176}$$

We can also obtain this result from the path integral formulation in Minkowski spacetime. Indeed, the generating functional for a free real massive scalar field $Z[J]$ in $D=(d+1)$-dimensional Minkowski spacetime is

$$Z[J] = \int \mathcal{D}\phi e^{i\int d^Dx\left[\frac{1}{2}(\partial\phi)^2 - \frac{m^2}{2}\phi^2 + J\phi\right]} \tag{5.177}$$

Hence, the expectation value of the time-ordered product of two fields is

$$\langle 0|T\phi(x)\phi(y)|0\rangle = -\frac{1}{Z[J]}\frac{\delta Z[J]}{\delta J(x)\delta J(y)}\Big|_{J=0} \tag{5.178}$$

In contrast, for a free field, the generating function is given by (up to a normalization constant $\mathcal{N}$)

$$Z[J] = \mathcal{N}\left[\text{Det}\left(\partial^2+m^2\right)\right]^{-1/2} e^{\frac{i}{2}\int d^Dx \int d^Dy J(x) G^{(0)}(x-y) J(y)} \tag{5.179}$$

where $G_0(x-y)$ is the Green function of the Klein-Gordon operator and satisfies

$$\left(\partial^2+m^2\right) G^{(0)}(x-y) = \delta^D(x-y) \tag{5.180}$$

Hence, we obtain the expected result:

$$\langle 0|T\phi(x)\phi(y)|0\rangle = -iG^{(0)}(x-y) \tag{5.181}$$

Let us compute the propagator in $D=4$ Minkowski spacetime by analytic continuation from the $D=4$ Euclidean propagator. The relativistic interval s is given by

$$s^2 = (x_0 - x_0')^2 - (\boldsymbol{x} - \boldsymbol{x}')^2 \tag{5.182}$$

The Euclidean interval (length) $|x-x'|$ and the relativistic interval s are related by

$$|x-x'| = \sqrt{(x-x')^2} \to \sqrt{-s^2} \tag{5.183}$$

Therefore, in $D=4$ spacetime dimensions, the Minkowski space propagator is

$$G^{(0)}(x-x') = \frac{i}{4\pi^2}\frac{m}{\sqrt{-s^2}}K_1(m\sqrt{-s^2}) \tag{5.184}$$

We will need the asymptotic behavior of the Bessel function $K_1(z)$,

$$\begin{aligned} K_1(z) &= \sqrt{\frac{\pi}{2z}}e^{-z}\left[1+\frac{3}{8z}+\cdots\right], \quad \text{for } z \gg 1 \\ K_1(z) &= \frac{1}{z}+\frac{z}{2}\left(\ln z + C - \frac{1}{2}\right)+\cdots, \text{ for } z \ll 1 \end{aligned} \tag{5.185}$$

where $C=0.577215\ldots$ is the Euler-Mascheroni constant. Let us examine the behavior of eq. (5.184) in the regimes: (1) space-like, $s^2<0$, and (2) time-like, $s^2>0$, intervals.

A: Space-like intervals: $(x-x')^2 = s^2 < 0$ This is the space-like domain. By inspecting eq. (5.184), we see that for space-like separations, the factor $\sqrt{-s^2}$ is a positive real number. Consequently, the argument of the Bessel function is real (and positive), and the propagator is pure imaginary. In particular, we see that for $s^2<0$, the Minkowski propagator is essentially the Euclidean correlation function:

$$G^{(0)}(x-x') = iG_E^{(0)}(x-x'), \qquad \text{for } s^2<0 \tag{5.186}$$

Hence, for $s^2<0$, we have the asymptotic behaviors:

$$\begin{aligned} G^{(0)}(x-x') &= i\frac{\sqrt{\pi/2}}{4\pi^2}\frac{m^2}{\left(m\sqrt{-s^2}\right)^{3/2}}e^{-m\sqrt{-s^2}}, \qquad \text{for } m\sqrt{-s^2}\gg 1 \\ G^{(0)}(x-x') &= \frac{i}{4\pi^2(-s^2)}, \qquad \text{for } m\sqrt{-s^2}\ll 1 \end{aligned} \tag{5.187}$$

(see figure 5.7).

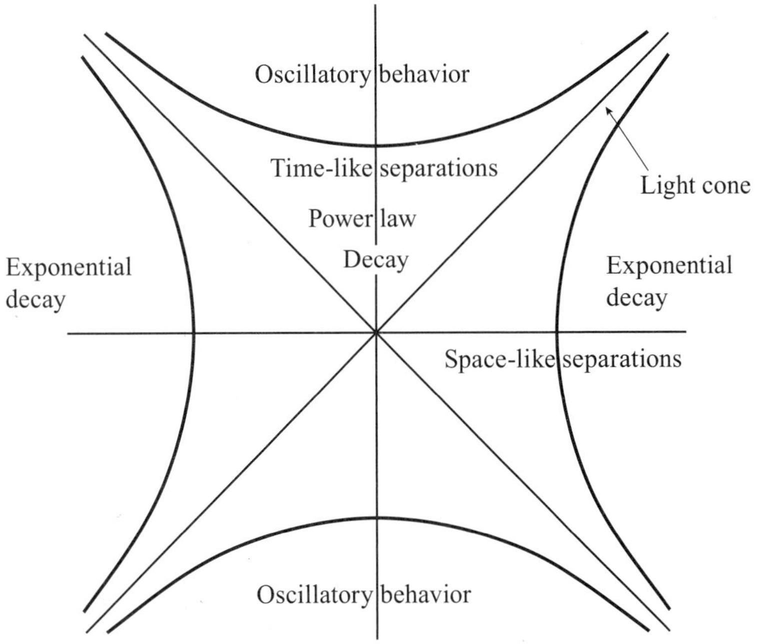

Figure 5.7 Behaviors of the propagator in Minkowski spacetime.

B: Time-like intervals: $(x-x')^2 = s^2 > 0$ This is the time-like domain. The analytic continuation yields

$$G^{(0)}(x-x') = \frac{m}{4\pi^2\sqrt{s^2}} K_1(im\sqrt{s^2}) \tag{5.188}$$

For pure imaginary arguments, the Bessel function $K_1(iz)$ is the analytic continuation of the Hankel function, $K_1(iz) = -\frac{\pi}{2}H_1^{(1)}(-z)$. This function is oscillatory for large values of its argument. Indeed, we now get the behaviors

$$\begin{aligned} G^{(0)}(x-x') &= \frac{\sqrt{\pi/2}}{4\pi^2} \frac{m^2}{\left(m\sqrt{s^2}\right)^{3/2}} e^{im\sqrt{s^2}}, && \text{for } m\sqrt{s^2} \gg 1 \\ G^{(0)}(x-x') &= \frac{1}{4\pi^2 s^2}, && \text{for } m\sqrt{s^2} \ll 1 \end{aligned} \tag{5.189}$$

Notice that, up to a factor of i, the short-distance behavior is the same for both time-like and space-like separations. The main difference is that at large time-like separations, we get oscillatory behavior instead of exponential decay. The length scale of the oscillations is, once again, set by the only scale in the theory: the Compton wavelength.

5.7 Exponential decays and mass gaps

The exponential decay at long space-like separations (and the oscillatory behavior at long time-like separations) is not a peculiarity of the free field theory. It is a general consequence of the existence of a *mass gap* in the spectrum. We can see this by considering the 2-point function of a generic theory (for simplicity, in imaginary time). The 2-point function is

$$G^{(2)}(\boldsymbol{x}-\boldsymbol{x}', \tau-\tau') = \langle 0|T\hat{\phi}(\boldsymbol{x},\tau)\hat{\phi}(\boldsymbol{x}',\tau')|0\rangle \tag{5.190}$$

where T is the imaginary time-ordering operator.

The Heisenberg representation of the operator $\hat{\phi}$ in imaginary time is ($\hbar=1$):

$$\hat{\phi}(\boldsymbol{x},\tau) = e^{H\tau}\,\hat{\phi}(\boldsymbol{x},0)\,e^{-H\tau} \tag{5.191}$$

Hence, we can write the 2-point function as

$$\begin{aligned}
G^{(2)}&(\boldsymbol{x}-\boldsymbol{x}', \tau-\tau') = \\
&= \theta(\tau-\tau')\langle 0|e^{H\tau}\,\hat{\phi}(\boldsymbol{x},0)\,e^{-H(\tau-\tau')}\,\hat{\phi}(\boldsymbol{x}',0)\,e^{-H\tau'}|0\rangle \\
&+ \theta(\tau'-\tau)\langle 0|e^{H\tau'}\,\hat{\phi}(\boldsymbol{x}',0)\,e^{-H(\tau'-\tau)}\,\hat{\phi}(\boldsymbol{x},0)\,e^{-H\tau}|0\rangle \\
&= \theta(\tau-\tau')\,e^{E_0(\tau-\tau')}\,\langle 0|\hat{\phi}(\boldsymbol{x},0)\,e^{-H(\tau-\tau')}\,\hat{\phi}(\boldsymbol{x}',0)|0\rangle \\
&+ \theta(\tau'-\tau)\,e^{E_0(\tau'-\tau)}\,\langle 0|\hat{\phi}(\boldsymbol{x}',0)\,e^{-H(\tau'-\tau)}\,\hat{\phi}(\boldsymbol{x},0)|0\rangle
\end{aligned} \tag{5.192}$$

We now insert a complete set of eigenstates $\{|n\rangle\}$ of the Hamiltonian $\hat{H}$, with eigenvalues $\{E_n\}$. The 2-point function now reads

$$\begin{aligned}
G^{(2)}&(\boldsymbol{x}-\boldsymbol{x}', \tau-\tau') = \\
&= \theta(\tau-\tau')\sum_n \langle 0|\hat{\phi}(\boldsymbol{x},0)|n\rangle\,\langle n|\hat{\phi}(\boldsymbol{x}',0)|0\rangle\,e^{-(E_n-E_0)(\tau-\tau')} \\
&+ \theta(\tau'-\tau)\sum_n \langle 0|\hat{\phi}(\boldsymbol{x}',0)|n\rangle\,\langle n|\hat{\phi}(\boldsymbol{x},0)|0\rangle\,e^{-(E_n-E_0)(\tau'-\tau)}
\end{aligned} \tag{5.193}$$

Since

$$\hat{\phi}(\boldsymbol{x},0) = e^{i\hat{\boldsymbol{P}}\cdot\boldsymbol{x}}\hat{\phi}(0,0)e^{-i\hat{\boldsymbol{P}}\cdot\boldsymbol{x}} \tag{5.194}$$

and, in a translation-invariant system, the eigenstates of the Hamiltonian are also eigenstates of the total momentum $\boldsymbol{P}$,

$$\hat{\boldsymbol{P}}|0\rangle = 0, \qquad \hat{\boldsymbol{P}}|n\rangle = \boldsymbol{P}_n|n\rangle \tag{5.195}$$

where $\boldsymbol{P}_n$ is the linear momentum of state $|n\rangle$, we can write

$$\langle 0|\hat{\phi}(\boldsymbol{x},0)|n\rangle\,\langle n|\hat{\phi}(\boldsymbol{x}',0)|0\rangle = |\langle 0|\hat{\phi}(0,0)|n\rangle|^2\,e^{-i\boldsymbol{P}_n\cdot(\boldsymbol{x}-\boldsymbol{x}')} \tag{5.196}$$

Using the above expressions, we can rewrite eq. (5.193) in the form

$$\begin{aligned}
G^{(2)}&(\boldsymbol{x}-\boldsymbol{x}', \tau-\tau') \\
&= \sum_n |\langle 0|\hat{\phi}(0,0)|n\rangle|^2\Big[\theta(\tau-\tau')e^{-i\boldsymbol{P}_n\cdot(\boldsymbol{x}-\boldsymbol{x}')}e^{-(E_n-E_0)(\tau-\tau')} \\
&\qquad + \theta(\tau'-\tau)e^{-i\boldsymbol{P}_n\cdot(\boldsymbol{x}'-\boldsymbol{x})}e^{-(E_n-E_0)(\tau'-\tau)}\Big]
\end{aligned} \tag{5.197}$$

Thus, at equal positions, $\boldsymbol{x}=\boldsymbol{x}'$, we obtain the following simpler expression in the imaginary time interval $\tau-\tau'$:

$$G^{(2)}(0,\tau-\tau')=\sum_n |\langle 0|\hat{\phi}(\mathbf{0},0)|n\rangle|^2 \times e^{-(E_n-E_0)|\tau-\tau'|} \tag{5.198}$$

In the limit of large imaginary time separation, $|\tau-\tau'|\to\infty$, there is always a largest nonvanishing term in the sums. This is the term for the state $|n_0\rangle$ that mixes with the vacuum state $|0\rangle$ through the field operator $\hat{\phi}$, and with the lowest excitation energy, the *mass gap* $E_{n_0}-E_0$. Hence, for large imaginary time separations, $|\tau-\tau'|\to\infty$, the 2-point function decays exponentially,

$$G^{(2)}(0,\tau-\tau')\simeq|\langle 0|\hat{\phi}(\mathbf{0},0)|n_0\rangle|^2\times e^{-(E_{n_0}-E_0)|\tau-\tau'|} \tag{5.199}$$

a result that we already derived for a free field in eq. (5.173). Therefore, if the spectrum has a gap, the correlation functions (or propagators) decay exponentially in imaginary time. In real time, we will get instead an oscillatory behavior. This is a very general result.

Finally, notice that Lorentz invariance in Minkowski spacetime (real time) implies rotational (Euclidean) invariance in imaginary time. Hence, exponential decay in imaginary times, at equal positions, must imply (in general) exponential decay in real space at equal imaginary times. Thus, in a Lorentz invariant system, the propagator at space-like separations is always equal to the propagator in imaginary time.

5.8 Scalar fields at finite temperature

Let us now discuss briefly the behavior of free scalar fields in thermal equilibrium at finite temperature T. We will give a more detailed discussion in chapter 10, where we discuss more extensively the relation between observables and propagators.

We saw in section 5.5 that the field theory is now defined on a Euclidean cylindrical spacetime, which is finite and periodic along the imaginary time direction with circumference $\beta=1/T$, where T is the temperature (and the Boltzmann constant is set to $k_B=1$). Hence the imaginary time dimension has been compactified.

5.8.1 The free energy

Let us begin by computing the free energy. We will work in $D=d+1$ Euclidean spacetime dimensions. The partition function $Z(T)$ is computed using the result of eq. (5.146) except that the differential operator is now

$$\hat{A}=-\partial_\tau^2-\partial^2+m^2 \tag{5.200}$$

In addition, the caveat is now that ∂^2 denotes the Laplacian operator that acts only on the spatial coordinates, $\mathbf{x}$, and that the imaginary time is periodic. The mode expansion for the field in this Euclidean (cylinder) space is

$$\phi(\boldsymbol{x},\tau)=\sum_{n=-\infty}^{\infty}\int\frac{d^dp}{(2\pi)^d}\phi(\omega_n,\boldsymbol{p})e^{i\omega_n\tau+i\boldsymbol{p}\cdot\boldsymbol{x}} \tag{5.201}$$

where $\omega_n = 2\pi Tn$ are the Matsubara frequencies, and $n \in \mathbb{Z}$. The field operator is periodic in imaginary time τ with period $\beta = 1/T$. The Euclidean action now is

$$
\begin{aligned}
S &= \int_0^\beta d\tau \int d^dx \left[\frac{1}{2}(\partial_\tau \phi)^2 + \frac{1}{2}(\partial\phi)^2 + \frac{1}{2}m^2\phi^2\right] \\
&= \frac{\beta}{2}\int \frac{d^dp}{(2\pi)^d}(\boldsymbol{p}^2 + m^2)|\phi_0(\boldsymbol{p})|^2 \\
&+ \beta \int \frac{d^dp}{(2\pi)^d} \sum_{n\geq 1} \left(\omega_n^2 + \boldsymbol{p}^2 + m^2\right) |\phi(\omega_n, \boldsymbol{p})|^2
\end{aligned}
\tag{5.202}
$$

where the action has been split into the sum of the contribution from the zero-frequency Matsubara mode, denoted by $\phi_0(\boldsymbol{p}) = \phi(0, \boldsymbol{p})$, and the contributions of the modes for the rest of the frequencies.

Since the free energy is given by $F(T) = -T \ln Z(T)$, we need to compute (again, up to the usual UV-divergent normalization constant)

$$
F(T) = \frac{T}{2} \ln \operatorname{Det}\left[-\partial_\tau^2 - \partial^2 + m^2\right]
\tag{5.203}
$$

We can now expand the determinant in the eigenvalues of the operator $-\partial_\tau^2 - \partial^2 + m^2$ to obtain the formally divergent expression

$$
F(T) = \frac{1}{2} VT \int \frac{d^dp}{(2\pi)^d} \sum_{n=-\infty}^{\infty} \ln\left(\beta\left[\omega_n^2 + \boldsymbol{p}^2 + m^2\right]\right)
\tag{5.204}
$$

where V is the spatial volume. This expression is formally divergent both in the momentum integrals and in the frequency sum and needs to be regularized. We already encountered this problem in our discussion of path integrals in quantum mechanics (see section 5.2.2). As in that case, recall that we have a formally divergent normalization constant, $\mathcal{N}$, which we have not made explicit here and that can be defined so as to cancel the divergence of the frequency sum (as we did in eq. (5.84)).

The regularized frequency sum can now be computed:

$$
F(T) = VT \int \frac{d^dp}{(2\pi)^d} \ln\left[\left(\beta\left(\boldsymbol{p}^2 + m^2\right)^{1/2}\right) \prod_{n=1}^{\infty}\left(1 + \frac{\boldsymbol{p}^2 + m^2}{\omega_n^2}\right)\right]
\tag{5.205}
$$

Using the identity in eq. (5.85), the free energy $F(T)$ becomes

$$
F(T) = VT \int \frac{d^dp}{(2\pi)^d} \ln\left[2\sinh\left(\frac{\sqrt{\boldsymbol{p}^2 + m^2}}{2T}\right)\right]
\tag{5.206}
$$

which can be written in the form

$$
F(T) = V\varepsilon_0 + VT \int \frac{d^dp}{(2\pi)^d} \ln\left(1 - e^{-\frac{\sqrt{\boldsymbol{p}^2 + m^2}}{T}}\right)
\tag{5.207}
$$

where

$$\varepsilon_0 = \frac{1}{2} \int \frac{d^d p}{(2\pi)^d} \sqrt{\boldsymbol{p}^2 + m^2} \tag{5.208}$$

is the (UV divergent) vacuum (ground state) energy density. Notice that the UV divergence is absent in the finite temperature contribution.

5.8.2 The thermal propagator

The thermal propagator is the time-ordered propagator in imaginary time. It is equivalent to the Euclidean correlation function in cylindrical geometry. Denote the thermal propagator by:

$$G_T^{(0)}(\boldsymbol{x}, \tau) = \langle \phi(\boldsymbol{x}, \tau) \phi(\boldsymbol{0}, 0) \rangle_T \tag{5.209}$$

It has the Fourier expansion

$$\langle \phi(\boldsymbol{x}, \tau) \phi(\boldsymbol{0}, 0) \rangle_T = \frac{1}{\beta} \sum_{n=-\infty}^{\infty} \int \frac{d^d p}{(2\pi)^d} \frac{e^{i\omega_n \tau + i\boldsymbol{p}\cdot\boldsymbol{x}}}{\omega_n^2 + \boldsymbol{p}^2 + m^2} \tag{5.210}$$

where, once again, $\omega_n = 2\pi T n$ are the Matsubara frequencies.

We will now obtain two useful expressions for the thermal propagator. The expressions follow from doing the momentum integrals first. The Matsubara frequencies act as mass terms of a field in one dimension lower. This observation allows us to identify the integrals in eq. (5.210) with the Euclidean propagators of an infinite number of fields, each labeled by an integer n, in d Euclidean dimensions with mass squared equal to

$$m_n^2 = m^2 + \omega_n^2 \tag{5.211}$$

We can now use the result of eq. (5.170) for the Euclidean correlator (now in d Euclidean dimensions) and write the thermal propagator as the following series:

$$G_T^{(0)}(\boldsymbol{x}, \tau) = \frac{1}{\beta} \sum_{n=-\infty}^{\infty} \frac{e^{i\omega_n \tau}}{(2\pi)^{d/2}} \left(\frac{m_n}{|x - x'|} \right)^{\frac{d}{2} - 1} K_{\frac{d}{2}-1}(m_n |\boldsymbol{x}|) \tag{5.212}$$

where m_n is given by eq. (5.211). Since the thermal propagator is expressed as an infinite series of massive propagators, each with increasing mass, it implies that at distances large compared with the length scale $\lambda_T = (2\pi T)^{-1}$ (known as the thermal wavelength), all the terms of the series become negligible compared with the term with vanishing Matsubara frequency. In this limit, the thermal propagator reduces to the correlator of the classical theory in d (spatial) Euclidean dimensions:

$$G_T^{(0)}(\boldsymbol{x}, \tau) \simeq \langle \phi(\boldsymbol{x}) \phi(\boldsymbol{0}) \rangle, \qquad \text{for } |\boldsymbol{x}| \gg \lambda_T \tag{5.213}$$

In other terms, at distances large compared with the circumference β of the cylindrical Euclidean spacetime, the theory becomes asymptotically equivalent to the Euclidean theory in one lower spacetime dimension.

Let us now find an alternative expression for the thermal propagator by doing the sum over Matsubara frequencies shown in eq. (5.210). We will use the residue theorem to

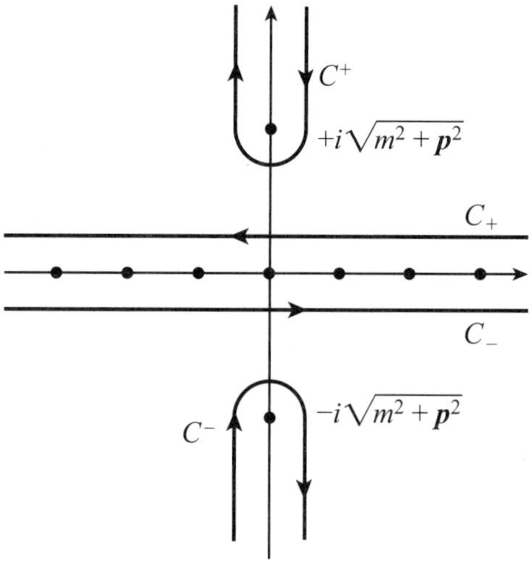

Figure 5.8 Contour integral for eq. (5.214).

represent the sum as a contour integral on the complex plane, as shown in figure 5.8,

$$\frac{1}{\beta}\sum_{n=-\infty}^{\infty}\frac{e^{i\omega_n\tau}}{\omega_n^2+\boldsymbol{p}^2+m^2}=\frac{1}{2}\oint_{C_+\cup C_-}\frac{dz}{2\pi i}\frac{e^{iz\tau}}{z^2+m^2+\boldsymbol{p}^2}\cot\left(\frac{z}{2T}\right) \tag{5.214}$$

where the (positively oriented) contour $C=C_-\cup C_+$ in the complex z plane is shown in figure 5.8. The black dots on the real axis represent the integers $z=n$, while the black dots on the imaginary axis represent the poles at $\pm i\sqrt{m^2+\boldsymbol{p}^2}$. Upon distorting the contour C_+ to the negatively oriented contour C^+ of the upper half-plane, and the contour C_- to the negatively oriented contour C^- of the lower half-plane, we can evaluate the integrals by using the residue theorem once again, but now at the poles on the imaginary axis.

This computation yields the following result for the thermal propagator:

$$G_T(\boldsymbol{x},\tau)=\int\frac{d^dp}{(2\pi)^d}\frac{\coth\left(\frac{\sqrt{\boldsymbol{p}^2+m^2}}{2T}\right)}{2\sqrt{\boldsymbol{p}^2+m^2}}e^{-|\tau|\sqrt{\boldsymbol{p}^2+m^2}}e^{i\boldsymbol{p}\cdot\boldsymbol{x}} \tag{5.215}$$

This expression applies to the regime $\tau\ll\beta=1/T$, in which quantum fluctuations play a dominant role. After a little algebra, we can now write the thermal propagator as

$$\begin{aligned}G_T(\boldsymbol{x},\tau)=&\int\frac{d^dp}{(2\pi)^d}\frac{e^{-|\tau|\sqrt{\boldsymbol{p}^2+m^2}}e^{i\boldsymbol{p}\cdot\boldsymbol{x}}}{2\sqrt{\boldsymbol{p}^2+m^2}}\\&+\int\frac{d^dp}{(2\pi)^d}\frac{1}{\exp\left(\frac{\sqrt{\boldsymbol{p}^2+m^2}}{T}\right)-1}\frac{e^{-|\tau|\sqrt{\boldsymbol{p}^2+m^2}}e^{i\boldsymbol{p}\cdot\boldsymbol{x}}}{\sqrt{\boldsymbol{p}^2+m^2}}\end{aligned} \tag{5.216}$$

By inspection of eq. (5.216), we see that the first term on the r.h.s. is the $T \to 0$ limit, and that it is just (as it should be!) the propagator in $D = d + 1$ Euclidean spacetime dimensions, $G_E^{(0)}(\boldsymbol{x}, \tau)$, after an integration over frequencies. The second term on the r.h.s. of eq. (5.216) describes the contributions to the thermal propagator from thermal fluctuations, shown in the form of the Bose occupation numbers, the Bose-Einstein distribution:

$$n(\boldsymbol{p}, T) = \frac{1}{\exp\left(\frac{\sqrt{\boldsymbol{p}^2 + m^2}}{T}\right) - 1} \tag{5.217}$$

The appearance of the Bose-Einstein distribution is to be expected (and, in fact, required), since the excitations of the scalar field are bosons.

Finally, we can find the time-ordered propagator, in real time x_0, at finite temperature T, which we denote by $G^{(0)}(\boldsymbol{x}, x_0; T)$. By means of the analytic continuation $\tau \to i x_0$ of the thermal propagator of eq. (5.216), we find

$$G^{(0)}(\boldsymbol{x}, x_0; T) = G_M^{(0)}(x) + \int \frac{d^d p}{(2\pi)^d} \frac{1}{\exp\left(\frac{\sqrt{\boldsymbol{p}^2 + m^2}}{T}\right) - 1} \frac{e^{-i|x_0|\sqrt{\boldsymbol{p}^2 + m^2}} e^{i\boldsymbol{p}\cdot\boldsymbol{x}}}{\sqrt{\boldsymbol{p}^2 + m^2}} \tag{5.218}$$

where $G_M^{(0)}(x)$ is the (Lorentz-invariant) Minkowski spacetime propagator in D dimensions (i.e., at zero temperature), given in section 5.6.4. Notice that the finite temperature contribution is not Lorentz invariant. This result is to be expected, since at finite temperatures space and time do not play equivalent roles.

Exercises

5.1 Path integral for a particle in a double-well potential

Consider a particle with coordinate q and mass m moving in the one-dimensional double-well potential $V(q)$:

$$V(q) = \lambda \left(q^2 - q_0^2\right)^2 \tag{5.219}$$

In this problem, you will use the path integral methods, in imaginary time, presented in this chapter, to calculate the matrix element

$$\langle q_0, \tfrac{T}{2} | -q_0, \tfrac{-T}{2}\rangle = \langle q_0 | e^{-\frac{1}{\hbar}HT} | -q_0\rangle \tag{5.220}$$

to leading order in the semiclassical expansion, in the limit $T \to \infty$.

1) Write down the expression of the imaginary time path integral that is appropriate for this problem. Write an explicit expression for the Euclidean Lagrangian (the Lagrangian in imaginary time). How does it differ from the Lagrangian in real

time? Make sure that you specify the initial and final conditions. Do not calculate anything yet!

2) Derive the Euler-Lagrange equation for this problem (always in imaginary time). Compare it with the equation of motion in real time. Find the explicit solution for the trajectory (in imaginary time) that satisfies the initial and final conditions. Is the solution unique? Explain. What is the physical interpretation of this trajectory and of the amplitude? A simple way to solve for the trajectory that you need in this problem is to think of this equation of motion as if imaginary time were real time, then to find the analog of the classical energy, and to use the conservation of energy to find the trajectory.
3) Compute the imaginary time action for the trajectory you found in part 2 of this exercise.
4) Expand around the solution you found in part 3. Write a formal expression for the amplitude to leading order. Find an explicit expression for the operator that enters into the fluctuation determinant.

5.2 Path integral for a charged particle moving on a plane in the presence of a perpendicular magnetic field

Consider a particle of mass m and charge $-e$ moving on a plane in the presence of an external uniform magnetic field perpendicular to the plane and with strength B. Let $\boldsymbol{r}=(x_1,x_2)$ and $\boldsymbol{p}=(p_1,p_2)$ represent the coordinates and momentum of the particle, respectively. The Hamiltonian of the system is

$$H(q,p)=\frac{1}{2m}\left(\boldsymbol{p}+\frac{e}{c}\boldsymbol{A}(\boldsymbol{r})\right)^2 \tag{5.221}$$

where $\boldsymbol{A}(\boldsymbol{r})$ is the vector potential. In the gauge $\nabla\cdot\boldsymbol{A}(\boldsymbol{r})=0$, the vector potential is given by

$$A_1(\boldsymbol{r})=-\frac{B}{2}x_2 \tag{5.222}$$

and

$$A_2(\boldsymbol{r})=\frac{B}{2}x_1 \tag{5.223}$$

1) Derive a path integral formula for the transition amplitude of the process in which the particle returns to its initial location $\boldsymbol{r}_0$ at time t_f, having left that point at t_i:

$$\langle \boldsymbol{r}_0, t_f|\boldsymbol{r}_0, t_i\rangle \tag{5.224}$$

where $\boldsymbol{r}_0$ is an arbitrary point of the plain and $|t_f-t_i|\to\infty$.
2) Consider now the "ultra-quantum" limit $m\to 0$. Find the form of the action S in this limit for a path that begins and ends at $\boldsymbol{r}_0$. Find a geometric interpretation for this formula. Hint: At some point you may have to use Stokes' theorem.
3) Consider now the case in which instead of a plane the charged particle moves on a torus of size $L\times L$. Are there any ambiguities involved in the evaluation of this formula? Think of the regions enclosed by the path and those left outside the path. What condition should satisfy the field strength B for a torus of dimensions $L\times L$ (with $L\to\infty$), so that the amplitude $e^{\frac{i}{\hbar}S}$ is free from any ambiguities?

5.3 Path integrals for a scalar field theory

Consider the problem of a charged (complex) free scalar field $\phi(x)$ in $D=4$ spacetime dimensions. The action of this theory coupled to complex sources $J(x)$ is

$$\mathcal{S}=\int d^4x\,\left(\partial_\mu\phi(x)^*\partial^\mu\phi(x)-m^2\phi(x)^*\phi(x)-J(x)^*\phi(x)-J(x)\phi(x)^*\right) \quad (5.225)$$

1) Find an explicit formula for the vacuum persistence amplitude ${}_J\langle 0|0\rangle_J$ for this theory in the form of a path integral in Minkowski space. Find the form of the path integral in Euclidean space (imaginary time).
2) Using the methods discussed in the text, evaluate the path integral you just wrote down, for a general set of complex sources $J(x)$, both in Minkowski and in Euclidean spacetimes.
3) Use the formulas you derived in part 2 to compute the following *two-point functions* (in the limit $J\to 0$):

$$G_2(x-x')=\langle 0|T\phi(x)\phi^*(x')|0\rangle \quad (5.226)$$

$$G_2^*(x-x')=\langle 0|T\phi^*(x)\phi(x')|0\rangle \quad (5.227)$$

$$G_2'(x-x')=\langle 0|T\phi(x)\phi(x')|0\rangle \quad (5.228)$$

$$G_2'^*(x-x')=\langle 0|T\phi^*(x)\phi^*(x')|0\rangle \quad (5.229)$$

In these equations, x and x' are two arbitrary points in spacetime. Write your solutions in terms of a propagator. Do not calculate the propagator yet.
4) Find the equation that is satisfied by the propagator in both Euclidean and Minkowski spacetimes. Solve the equation for the Euclidean case, and find the solution in Minkowski spacetime by analytic continuation. Discuss the different asymptotic behaviors of the two-point functions in both Euclidean and Minkowski spacetimes.
5) Find explicit expressions for the following *four-point functions*:

$$G_4^a(x_1,x_2,x_3,x_4)=\langle 0|T\phi(x_1)^*\phi(x_2)^*\phi(x_3)\phi(x_4)|0\rangle \quad (5.230)$$

$$G_2'(x_1,x_2,x_3,x_4)=\langle 0|T\phi(x_1)\phi(x_2)\phi(x_3)\phi(x_4)|0\rangle \quad (5.231)$$

$$G_2'^*(x_1,x_2,x_3,x_4)=\langle 0|T\phi(x_1)^*\phi(x_2)^*\phi(x_3)^*\phi(x_4)^*|0\rangle \quad (5.232)$$

$$G_2^b(x_1,x_2,x_3,x_4)=\langle 0|T\phi(x_1)^*\phi(x_2)\phi(x_3)^*\phi(x_4)|0\rangle \quad (5.233)$$

$$G_2^c(x_1,x_2,x_3,x_4)=\langle 0|T\phi(x_1)^*\phi(x_2)\phi(x_3)\phi(x_4)^*|0\rangle \quad (5.234)$$

in terms of the two-point functions. Find relations among these four-point functions.

6

Nonrelativistic Field Theory

In this chapter, we discuss the field-theory description of nonrelativistic systems. The material presented here is, for the most part, introductory, as this topic is covered in depth in many classic textbooks, such as *Methods of Quantum Field Theory in Statistical Physics* (Abrikosov et al., 1963) and *Statistical Mechanics, A Set of Lectures* (Feynman, 1972) (among many others). The discussion of "second quantization" is very standard and is similar to Feynman's presentation. It is presented here for completeness.

6.1 Second quantization and the many-body problem

Let us consider now the problem of a system of N identical nonrelativistic particles. For the sake of simplicity, assume that the physical state of each particle j is described by its position $\boldsymbol{x}_j$ relative to some reference frame. This case is easy to generalize.

The wave function for this system is $\Psi(\boldsymbol{x}_1, \ldots, \boldsymbol{x}_N)$. If the particles are identical, then the probability density, $|\Psi(\boldsymbol{x}_1, \ldots, \boldsymbol{x}_N)|^2$, must be invariant under arbitrary exchanges of the labels that we use to identify (or designate) the particles. In quantum mechanics, particles do not have well-defined trajectories. Only the states of a physical system are well defined. Thus, even though at some initial time t_0 the N particles may be localized to a set of well-defined positions $\boldsymbol{x}_1, \ldots, \boldsymbol{x}_N$, they will become delocalized as the system evolves. Furthermore, the Hamiltonian itself is invariant under a permutation of the particle labels. Hence, *permutations* constitute a *symmetry* of a many-particle quantum mechanical system. In other words, in quantum mechanics, identical particles are indistinguishable. In particular, the probability density of any eigenstate must remain invariant if the labels of any pair of particles are exchanged.

If we denote by P_{jk} the operator that exchanges the labels of particles j and k, the wave functions must satisfy

$$P_{jk}\Psi(\boldsymbol{x}_1, \ldots, \boldsymbol{x}_j, \ldots, \boldsymbol{x}_k, \ldots, \boldsymbol{x}_N) = e^{i\phi}\Psi(\boldsymbol{x}_1, \ldots, \boldsymbol{x}_j, \ldots, \boldsymbol{x}_k, \ldots, \boldsymbol{x}_N) \tag{6.1}$$

Under a further exchange operation, the particles return to their initial labels, and we recover the original state. This sample argument then requires that $\phi = 0, \pi$, since 2ϕ must not be an observable phase. We then conclude that there are two possibilities: either Ψ is *even* under permutation and $P\Psi = \Psi$, *or* Ψ is *odd* under permutation and $P\Psi = -\Psi$.

Systems of identical particles with wave functions that are even under a pairwise permutation of the particle labels are called *bosons*. In the other case of Ψ being *odd* under pairwise permutation, they are called *fermions*. It must be stressed that these arguments only show that the requirement that the state Ψ can be either even or odd is only a sufficient condition. It turns out that under special circumstances, the phase factor ϕ can take values that are different from 0 or π. These particles are called *anyons* and are discussed in chapter 22. It turns out that the only cases in which they can exist is if the particles are restricted to move on a line or in a plane.

In the case of relativistic QFTs, the requirement that the states have well-defined statistics (or symmetry) is demanded by a very deep and fundamental theorem linking the statistics of the states of spin of the field. This is the *spin-statistics* theorem and will be discussed in section 7.1.6.

6.1.1 Fock space

Let us now discuss a procedure, known as *second quantization*, which will enable us to keep track of the symmetry of the states in a simple way. Consider again a system of N particles. The wave functions in the coordinate representation are $\Psi(\boldsymbol{x}_1, \ldots, \boldsymbol{x}_N)$, where the labels $\boldsymbol{x}_1, \ldots, \boldsymbol{x}_N$ denote both the coordinates and the spin states of the particles in the state $|\Psi\rangle$. For the sake of definiteness, we will discuss physical systems describable by Hamiltonians $\hat{H}$ of the form:

$$\hat{H} = -\frac{\hbar^2}{2m}\sum_{j=1}^{N}\nabla_j^2 + \sum_{j=1}^{N} V(\boldsymbol{x}_j) + g\sum_{j,k} U(\boldsymbol{x}_j - \boldsymbol{x}_k) + \cdots \tag{6.2}$$

Let $\{\phi_n(x)\}$ be the wave functions for a complete set of one-particle states. Then an arbitrary N-particle state can be expanded in a basis of the tensor product of the one-particle states, namely,

$$\Psi(\boldsymbol{x}_1, \ldots, \boldsymbol{x}_N) = \sum_{\{nj\}} C(n_1, \ldots, n_N)\phi_{n1}(\boldsymbol{x}_1)\cdots\phi_{nN}(\boldsymbol{x}_N) \tag{6.3}$$

Thus, if Ψ is symmetric (antisymmetric) under an arbitrary exchange $\boldsymbol{x}_j \to \boldsymbol{x}_k$, then the coefficients $C(n_1, \ldots, n_N)$ must be symmetric (antisymmetric) under the exchange $n_j \leftrightarrow n_k$.

A set of N-particle basis states with well-defined permutation symmetry is the properly symmetrized or antisymmetrized tensor product

$$|\Psi_1, \ldots, \Psi_N\rangle \equiv |\Psi_1\rangle \times |\Psi_2\rangle \times \cdots \times |\Psi_N\rangle = \frac{1}{\sqrt{N!}}\sum_P \xi^P |\Psi_{P(1)}\rangle \times \cdots \times |\Psi_{P(N)}\rangle \tag{6.4}$$

where the sum runs over the set of all possible permutations P. The weight factor ξ is $+1$ for bosons and -1 for fermions. For fermions, the N-particle state vanishes if two particles are in the same one-particle state. This is the *exclusion principle*.

The inner product of two N-particle states is

$$\begin{aligned}\langle \chi_1, \ldots, \chi_N | \psi_1, \ldots, \psi_N\rangle &= \frac{1}{N!}\sum_{P,Q}\xi^{P+Q}\langle \chi_{Q(1)}|\psi_{P(1)}\rangle \cdots \langle \chi_{Q(N)}|\psi_{P(1)}\rangle = \\ &= \sum_{P'}\xi^{P'}\langle \chi_1|\psi_{P(1)}\rangle \cdots \langle \chi_N|\psi_{P(N)}\rangle \end{aligned} \tag{6.5}$$

which is nothing but the *permanent (determinant)* of the matrix $\langle \chi_j | \psi_k \rangle$ for symmetric (antisymmetric) states:

$$\langle \chi_1, \ldots, \chi_N | \psi_1, \ldots, \psi_N \rangle = \begin{vmatrix} \langle \chi_1 | \psi_1 \rangle & \ldots & \langle \chi_1 | \psi_N \rangle \\ \vdots & & \vdots \\ \langle \chi_N | \psi_1 \rangle & \ldots & \langle \chi_N | \psi_N \rangle \end{vmatrix}_{\xi} \tag{6.6}$$

In the case of antisymmetric states, the inner product is the familiar *Slater determinant*. Let us denote by $\{|\alpha\rangle\}$ the complete set of one-particle states satisfying

$$\langle \alpha | \beta \rangle = \delta_{\alpha\beta}, \qquad \sum_{\alpha} |\alpha\rangle\langle\alpha| = 1 \tag{6.7}$$

The N-particle states are given by $\{|\alpha_1, \ldots, \alpha_N\rangle\}$. Because of the symmetry requirements, the labels α_j can be arranged in the form of a monotonic sequence $\alpha_1 \leq \alpha_2 \leq \cdots \leq \alpha_N$ for bosons, or in the form of a strict monotonic sequence $\alpha_1 < \alpha_2 < \cdots < \alpha_N$ for fermions. Let n_j be an integer that counts how many particles are in the jth one-particle state. The boson states $|\alpha_1, \ldots, \alpha_N\rangle$ must be normalized by a factor of the form

$$\frac{1}{\sqrt{n_1! \ldots n_N!}} |\alpha_1, \ldots, \alpha_N\rangle, \qquad \text{with} \qquad \alpha_1 \leq \alpha_2 \leq \cdots \leq \alpha_N \tag{6.8}$$

where n_j are nonnegative integers. For fermions, the states are

$$|\alpha_1, \ldots, \alpha_N\rangle, \qquad \text{with} \qquad \alpha_1 < \alpha_2 < \cdots < \alpha_N \tag{6.9}$$

but now $n_j = 0, 1$. These N-particle states are complete and orthonormal:

$$\frac{1}{N!} \sum_{\alpha_1, \ldots, \alpha_N} |\alpha_1, \ldots, \alpha_N\rangle\langle\alpha_1, \ldots, \alpha_N| = \hat{I} \tag{6.10}$$

where the sum over the αs is unrestricted, and the operator $\hat{I}$ is the identity operator in the space of N-particle states.

Now consider the more general problem in which the number of particles N is not fixed a priori. Instead we consider an enlarged space of states in which the number of particles is allowed to fluctuate. In the language of statistical physics, what we are doing is going from the *canonical ensemble* to the *grand canonical ensemble*. Thus, denote by $\mathcal{H}_0$ the Hilbert space with no particles; $\mathcal{H}_1$ the Hilbert space with only one particle; and in general, $\mathcal{H}_N$ the Hilbert space for N particles. The direct sum $\mathcal{H}$ of these spaces

$$\mathcal{H} = \mathcal{H}_0 \oplus \mathcal{H}_1 \oplus \cdots \oplus \mathcal{H}_N \oplus \cdots \tag{6.11}$$

is called the *Fock space*. An arbitrary state $|\psi\rangle$ in Fock space is the sum over the subspaces $\mathcal{H}_N$:

$$|\psi\rangle = |\psi^{(0)}\rangle + |\psi^{(1)}\rangle + \cdots + |\psi^{(N)}\rangle + \cdots \tag{6.12}$$

The subspace with no particles is a one-dimensional space spanned by the vector $|0\rangle$, which we will call the *vacuum*. The subspaces with well-defined numbers of particles are defined to be orthogonal to one another in the sense that the inner product in the Fock space,

$$\langle \chi | \psi \rangle \equiv \sum_{j=0}^{\infty} \langle \chi^{(j)} | \psi^{(j)} \rangle \tag{6.13}$$

vanishes if $|\chi\rangle$ and $|\psi\rangle$ belong to different subspaces.

6.1.2 Creation and annihilation operators

Let $|\phi\rangle$ be an arbitrary one-particle state. Let us define the *creation operator* $\hat{a}^\dagger(\phi)$ by its action on an arbitrary state in Fock space:

$$\hat{a}^\dagger(\phi)|\psi_1, \ldots, \psi_N\rangle = |\phi, \psi_1, \ldots, \psi_N\rangle \tag{6.14}$$

Clearly, $\hat{a}^\dagger(\phi)$ maps the N-particle state with proper symmetry $|\psi_1, \ldots, \psi_N\rangle$ onto the $N+1$-particle state $|\phi, \psi, \ldots, \psi_N\rangle$, also with proper symmetry. The *destruction* or *annihilation operator* $\hat{a}(\phi)$ is defined to be the adjoint of $\hat{a}^\dagger(\phi)$:

$$\langle \chi_1, \ldots, \chi_{N-1}|\hat{a}(\phi)|\psi_1, \ldots, \psi_N\rangle = \langle \psi_1, \ldots, \psi_N|\hat{a}^\dagger(\phi)|\chi_1, \ldots, \chi_{N-1}\rangle^* \tag{6.15}$$

Hence

$$\langle \chi_1, \ldots, \chi_{N-1}|\hat{a}(\phi)|\psi_1, \ldots, \psi_N\rangle = \langle \psi_1, \ldots, \psi_N|\phi, \chi_1, \ldots, \chi_{N-1}\rangle^* =$$
$$= \begin{vmatrix} \langle\psi_1|\phi\rangle & \langle\psi_1|\chi_1\rangle & \cdots & \langle\psi_1|\chi_{N-1}\rangle \\ \vdots & \vdots & & \vdots \\ \langle\psi_N|\phi\rangle & \langle\psi_N|\chi_1\rangle & \cdots & \langle\psi_N|\chi_{N-1}\rangle \end{vmatrix}_\xi^* \tag{6.16}$$

We can now expand the permanent (or determinant) to get

$$\langle \chi_1, \ldots, \chi_{N-1}|\hat{a}(\phi)|\psi_1, \ldots, \psi_N\rangle =$$
$$= \sum_{k=1}^{N} \xi^{k-1} \langle\psi_k|\phi\rangle \begin{vmatrix} \langle\psi_1|\chi_1\rangle & \ldots & \langle\psi_1|\chi_{N-1}\rangle \\ \vdots & & \vdots \\ \ldots & (\text{no } \psi_k) & \ldots \\ \langle\psi_N|\chi_1\rangle & \ldots & \langle\psi_N|\chi_{N-1}\rangle \end{vmatrix}_\xi^*$$
$$= \sum_{k=1}^{N} \xi^{k-1} \langle\psi_k|\phi\rangle \langle \chi_1, \ldots, \chi_{N-1}|\psi_1, \ldots, \hat{\psi}_k, \ldots, \psi_N\rangle \tag{6.17}$$

where $\hat{\psi}_k$ denotes that the one-particle state ψ_k is absent. Thus, the destruction operator is given by

$$\hat{a}(\phi)|\psi_1, \ldots, \psi_N\rangle = \sum_{k=1}^{N} \xi^{k-1} \langle\phi|\psi_k\rangle |\psi_1, \ldots, \hat{\psi}_k, \ldots, \psi_N\rangle \tag{6.18}$$

With these definitions, we can easily see that the operators $\hat{a}^\dagger(\phi)$ and $\hat{a}(\phi)$ obey the commutation relations:

$$\hat{a}^\dagger(\phi_1)\hat{a}^\dagger(\phi_2) = \xi\, \hat{a}^\dagger(\phi_2)\hat{a}^\dagger(\phi_1) \tag{6.19}$$

Let us introduce the notation

$$\left[\hat{A}, \hat{B}\right]_\xi \equiv \hat{A}\hat{B} - \xi\hat{B}\hat{A} \tag{6.20}$$

where $\hat{A}$ and $\hat{B}$ are two arbitrary operators. For $\xi = +1$ (bosons), we have the *commutator*

$$\left[\hat{a}^\dagger(\phi_1), \hat{a}^\dagger(\phi_2)\right]_{+1} \equiv \left[\hat{a}^\dagger(\phi_1), \hat{a}^\dagger(\phi_2)\right] = 0 \tag{6.21}$$

while for $\xi = -1$, it is the *anticommutator*:

$$\left[\hat{a}^\dagger(\phi_1), \hat{a}^\dagger(\phi_2)\right]_{-1} \equiv \left\{\hat{a}^\dagger(\phi_1), \hat{a}^\dagger(\phi_2)\right\} = 0 \tag{6.22}$$

Similarly, for any pair of arbitrary one-particle states $|\phi_1\rangle$ and $|\phi_2\rangle$, we have

$$\left[\hat{a}(\phi_1), \hat{a}(\phi_2)\right]_{-\xi} = 0 \tag{6.23}$$

It is also easy to check that the following identity holds:

$$\left[\hat{a}(\phi_1), \hat{a}^\dagger(\phi_2)\right]_{-\xi} = \langle\phi_1|\phi_2\rangle \tag{6.24}$$

So far we have not picked any particular representation. Let us consider the occupation number representation, in which the states are labeled by the number of particles n_k in the single-particle state k. In this case, we have

$$|n_1, \ldots, n_k, \ldots\rangle \equiv \frac{1}{\sqrt{n_1! n_2! \cdots}} |1, \ldots, 1, 2, \ldots, 2, \ldots\rangle \tag{6.25}$$

where the state has n_1 particles in state 1, n_2 particles in state 2, and so on.

In the case of bosons, the n_js can be any nonnegative integer, while for fermions, they can only be equal to zero or one. In general, we have that if $|\alpha\rangle$ is the αth single-particle state, then

$$\begin{aligned} \hat{a}^\dagger_\alpha |n_1, \ldots, n_\alpha, \ldots\rangle &= \sqrt{n_\alpha + 1}\, |n_1, \ldots, n_\alpha + 1, \ldots\rangle \\ \hat{a}_\alpha |n_1, \ldots, n_\alpha, \ldots\rangle &= \sqrt{n_\alpha}\, |n_1, \ldots, n_\alpha - 1, \ldots\rangle \end{aligned} \tag{6.26}$$

Thus for both fermions and bosons, $\hat{a}_\alpha$ annihilates all states with $n_\alpha = 0$, while for fermions, $\hat{a}^\dagger_\alpha$ annihilates all states with $n_\alpha = 1$.

The commutation relations are

$$[\hat{a}_\alpha, \hat{a}_\beta] = [\hat{a}^\dagger_\alpha, \hat{a}^\dagger_\beta] = 0, \qquad [\hat{a}_\alpha, \hat{a}^\dagger_\beta] = \delta_{\alpha\beta} \tag{6.27}$$

for bosons, and

$$\{\hat{a}_\alpha, \hat{a}_\beta\} = \{\hat{a}^\dagger_\alpha, \hat{a}^\dagger_\beta\} = 0, \qquad \{\hat{a}_\alpha, \hat{a}^\dagger_\beta\} = \delta_{\alpha\beta} \tag{6.28}$$

for fermions. Here, $\{\hat{A}, \hat{B}\}$ is the anticommutator of the operators $\hat{A}$ and $\hat{B}$

$$\{\hat{A}, \hat{B}\} \equiv \hat{A}\hat{B} + \hat{B}\hat{A} \tag{6.29}$$

If a unitary transformation is performed in the space of one-particle state vectors, then a unitary transformation is *induced* in the space of the operators themselves; that is, if $|\chi\rangle = \alpha|\psi\rangle + \beta|\phi\rangle$, then

$$\begin{aligned} \hat{a}(\chi) &= \alpha^* \hat{a}(\psi) + \beta^* \hat{a}(\phi) \\ \hat{a}^\dagger(\chi) &= \alpha \hat{a}^\dagger(\psi) + \beta \hat{a}^\dagger(\phi) \end{aligned} \tag{6.30}$$

and we say that $\hat{a}^\dagger(\chi)$ transforms like the *ket* $|\chi\rangle$, while $\hat{a}(\chi)$ transforms like the *bra* $\langle\chi|$.

For example, we can pick as the complete set of one-particle states the momentum states $\{|\boldsymbol{p}\rangle\}$. This is momentum space. For this choice, the commutation relations are

$$\begin{aligned} \left[\hat{a}^\dagger(\boldsymbol{p}), \hat{a}^\dagger(\boldsymbol{q})\right]_\xi &= \left[\hat{a}(\boldsymbol{p}), \hat{a}(\boldsymbol{q})\right]_\xi = 0 \\ \left[\hat{a}(\boldsymbol{p}), \hat{a}^\dagger(\boldsymbol{q})\right]_\xi &= (2\pi)^d \delta^d(\boldsymbol{p} - \boldsymbol{q}) \end{aligned} \tag{6.31}$$

where d is the dimensionality of space. In this representation, an N-particle state is

$$|\boldsymbol{p}_1, \ldots, \boldsymbol{p}_N\rangle = \hat{a}^\dagger(\boldsymbol{p}_1) \cdots \hat{a}^\dagger(\boldsymbol{p}_N)|0\rangle \tag{6.32}$$

However, we can also pick the one-particle states to be eigenstates of the position operators:

$$|\boldsymbol{x}_1, \ldots, \boldsymbol{x}_N\rangle = \hat{a}^\dagger(\boldsymbol{x}_1) \cdots \hat{a}^\dagger(\boldsymbol{x}_N)|0\rangle \tag{6.33}$$

In position space, the operators satisfy

$$\begin{aligned} [\hat{a}^\dagger(\boldsymbol{x}_1), \hat{a}^\dagger(\boldsymbol{x}_2)]_\xi &= [\hat{a}(\boldsymbol{x}_1), \hat{a}(\boldsymbol{x}_2)]_\xi = 0 \\ [\hat{a}(\boldsymbol{x}_1), \hat{a}^\dagger(\boldsymbol{x}_2)]_\xi &= \delta^d(\boldsymbol{x}_1 - \boldsymbol{x}_2) \end{aligned} \tag{6.34}$$

This is the position space, or coordinate representation.

A transformation from position space to momentum space is the Fourier transform

$$|\boldsymbol{p}\rangle = \int d^d x\, |\boldsymbol{x}\rangle\langle\boldsymbol{x}|\boldsymbol{p}\rangle = \int d^d x\, |\boldsymbol{x}\rangle e^{i\boldsymbol{p}\cdot\boldsymbol{x}} \tag{6.35}$$

and, conversely,

$$|\boldsymbol{x}\rangle = \int \frac{d^d p}{(2\pi)^d} |\boldsymbol{p}\rangle e^{-i\boldsymbol{p}\cdot\boldsymbol{x}} \tag{6.36}$$

Then, the operators themselves obey

$$\hat{a}^\dagger(\boldsymbol{p}) = \int d^d x \, \hat{a}^\dagger(\boldsymbol{x}) e^{i\boldsymbol{p}\cdot\boldsymbol{x}}$$

$$\hat{a}^\dagger(\boldsymbol{x}) = \int \frac{d^d p}{(2\pi)^d} \hat{a}^\dagger(\boldsymbol{p}) e^{-i\boldsymbol{p}\cdot\boldsymbol{x}} \tag{6.37}$$

6.1.3 General operators in Fock space

Let $A^{(1)}$ be an operator acting on one-particle states. We can always define an *extension* of A acting on any arbitrary state $|\psi\rangle$ of the N-particle Hilbert space as follows:

$$\widehat{A}|\psi\rangle \equiv \sum_{j=1}^{N} |\psi_1\rangle \times \cdots \times A^{(1)}|\psi_j\rangle \times \cdots \times |\psi_N\rangle \tag{6.38}$$

For instance, if the one-particle basis states $\{|\psi_j\rangle\}$ are eigenstates of A with eigenvalues $\{a_j\}$, we get

$$\widehat{A}|\psi\rangle = \left(\sum_{j=1}^{N} a_j\right) |\psi\rangle \tag{6.39}$$

We wish to find an expression for an arbitrary operator A in terms of creation and annihilation operators. Let us first consider the operators $A^{(1)}_{\alpha\beta} = |\alpha\rangle\langle\beta|$, which act on one-particle states. The operators $A^{(1)}_{\alpha\beta}$ form a basis of the space of operators acting on one-particle states. Then, the N-particle extension of $A_{\alpha\beta}$ is

$$\widehat{A}_{\alpha\beta}|\psi\rangle = \sum_{j=1}^{N} |\psi_1\rangle \times \cdots \times |\alpha\rangle \times \cdots \times |\psi_N\rangle \langle\beta|\psi_j\rangle \tag{6.40}$$

Thus

$$\widehat{A}_{\alpha\beta}|\psi\rangle = \sum_{j=1}^{N} |\psi_1, \ldots, \overbrace{\alpha}^{j}, \ldots, \psi_N\rangle \, \langle\beta|\psi_j\rangle \tag{6.41}$$

In other words, we can replace the one-particle state $|\psi_j\rangle$ from the basis with the state $|\alpha\rangle$ at the price of a weight factor, the overlap $\langle\beta|\psi_j\rangle$. This operator has a very simple expression in terms of creation and annihilation operators. Indeed,

$$\hat{a}^\dagger(\alpha)\hat{a}(\beta)|\psi\rangle = \sum_{k=1}^{N} \xi^{k-1} \langle\beta|\psi_k\rangle \, |\alpha, \psi_1, \ldots, \psi_{k-1}, \psi_{k+1}, \ldots, \psi_N\rangle \tag{6.42}$$

We can now use the symmetry of the state to write

$$\xi^{k-1}|\alpha, \psi_1, \ldots, \psi_{k-1}, \psi_{k+1}, \ldots, \psi_N\rangle = |\psi_1, \ldots, \overbrace{\alpha}^{k}, \ldots, \psi_N\rangle \tag{6.43}$$

Thus the operator $A_{\alpha\beta}$, the extension of $|\alpha\rangle\langle\beta|$ to the N-particle space, coincides with $\hat{a}^\dagger(\alpha)\hat{a}(\beta)$:

$$\widehat{A}_{\alpha\beta} \equiv \hat{a}^\dagger(\alpha)\hat{a}(\beta) \tag{6.44}$$

We can use this result to find that the extension for an arbitrary operator $A^{(1)}$ of the form

$$A^{(1)} = \sum_{\alpha,\beta} |\alpha\rangle\langle\alpha|A^{(1)}|\beta\rangle\,\langle\beta| \tag{6.45}$$

is

$$\widehat{A} = \sum_{\alpha,\beta} \hat{a}^\dagger(\alpha)\hat{a}(\beta)\langle\alpha|A^{(1)}|\beta\rangle \tag{6.46}$$

Hence the coefficients of the expansion are the matrix elements of $A^{(1)}$ between arbitrary one-particle states. We now discuss a few operators of interest.

A: The identity operator The identity operator $\hat{1}$ of the one-particle Hilbert space

$$\hat{1} = \sum_{\alpha} |\alpha\rangle\langle\alpha| \tag{6.47}$$

becomes the *number operator* $\widehat{N}$

$$\widehat{N} = \sum_{\alpha} \hat{a}^\dagger(\alpha)\hat{a}(\alpha) \tag{6.48}$$

In position and in momentum space, we find

$$\widehat{N} = \int \frac{d^dp}{(2\pi)^d}\, \hat{a}^\dagger(\boldsymbol{p})\hat{a}(\boldsymbol{p}) = \int d^dx\, \hat{a}^\dagger(\boldsymbol{x})\hat{a}(\boldsymbol{x}) = \int d^dx\, \hat{\rho}(\boldsymbol{x}) \tag{6.49}$$

where $\hat{\rho}(\boldsymbol{x}) = \hat{a}^\dagger(\boldsymbol{x})\hat{a}(\boldsymbol{x})$ is the *particle density operator.*

B: The linear momentum operator In the space $\mathcal{H}_1$, the linear momentum operator is

$$\hat{p}_j^{(1)} = \int \frac{d^dp}{(2\pi)^d}\, p_j\,|\boldsymbol{p}\rangle\langle\boldsymbol{p}| = \int d^dx\,|\boldsymbol{x}\rangle\,\frac{\hbar}{i}\partial_j\,\langle\boldsymbol{x}| \tag{6.50}$$

Thus, we get that the total momentum operator $\widehat{P}_j$ is

$$\widehat{P}_j = \int \frac{d^dp}{(2\pi)^d}\, p_j\hat{a}^\dagger(\boldsymbol{p})\hat{a}(\boldsymbol{p}) = \int d^dx\, \hat{a}^\dagger(\boldsymbol{x})\frac{\hbar}{i}\partial_j\hat{a}(\boldsymbol{x}) \tag{6.51}$$

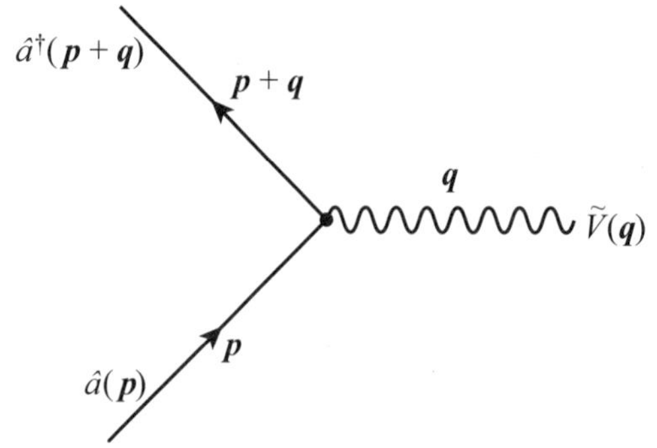

Figure 6.1 Feynman diagram of one-body scattering.

C: The Hamiltonian The one-particle Hamiltonian $H^{(1)}$

$$H^{(1)} = \frac{\boldsymbol{p}^2}{2m} + V(\boldsymbol{x}) \tag{6.52}$$

has the matrix elements

$$\langle \boldsymbol{x}|H^{(1)}|\boldsymbol{y}\rangle = -\frac{\hbar^2}{2m}\,\nabla^2\,\delta^d(\boldsymbol{x}-\boldsymbol{y}) + V(\boldsymbol{x})\delta^d(\boldsymbol{x}-\boldsymbol{y}) \tag{6.53}$$

Thus, in Fock space, we get

$$\widehat{H} = \int d^d x\, \hat{a}^\dagger(\boldsymbol{x}) \left(-\frac{\hbar^2}{2m}\,\nabla^2 + V(\boldsymbol{x}) \right) \hat{a}(\boldsymbol{x}) \tag{6.54}$$

in position space. In momentum space, we can define

$$\widetilde{V}(\boldsymbol{q}) = \int d^d x\, V(\boldsymbol{x}) e^{-i\boldsymbol{q}\cdot\boldsymbol{x}} \tag{6.55}$$

which is the Fourier transform of the potential $V(x)$, and we get

$$\widehat{H} = \int \frac{d^d p}{(2\pi)^d}\,\frac{\boldsymbol{p}^2}{2m}\,\hat{a}^\dagger(\boldsymbol{p})\hat{a}(\boldsymbol{p}) + \int \frac{d^d p}{(2\pi)^d} \int \frac{d^d q}{(2\pi)^d}\,\widetilde{V}(\boldsymbol{q})\hat{a}^\dagger(\boldsymbol{p}+\boldsymbol{q})\hat{a}(\boldsymbol{p}) \tag{6.56}$$

The last term has a very simple physical interpretation. When acting on a one-particle state with well-defined momentum, say $|\boldsymbol{p}\rangle$, the potential term yields another one-particle state with momentum $\boldsymbol{p}+\boldsymbol{q}$, where $\boldsymbol{q}$ is the *momentum transfer*, with amplitude $\widetilde{V}(\boldsymbol{q})$. This process is usually depicted by the *Feynman diagram* shown in figure 6.1.

D: Two-body interactions A two-particle interaction is an operator $\hat{V}^{(2)}$ that acts on the space of two-particle states $\mathcal{H}_2$ and has the form

$$V^{(2)} = \frac{1}{2} \sum_{\alpha,\beta} |\alpha,\beta\rangle V^{(2)}(\alpha,\beta)\langle\alpha,\beta| \tag{6.57}$$

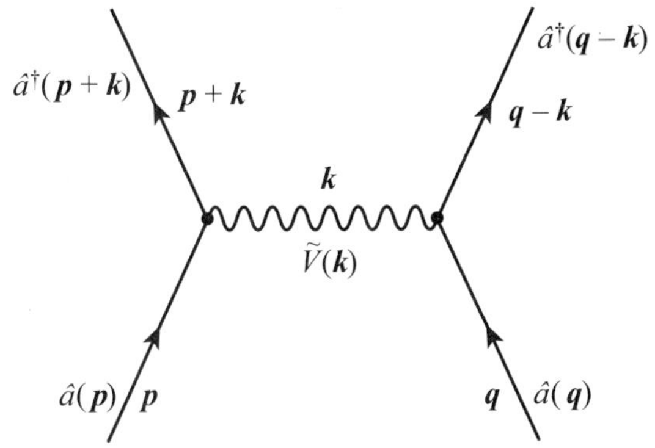

Figure 6.2 Feynman diagram of a two-body interaction.

The methods developed above yield an extension of $V^{(2)}$ to Fock space of the form

$$\widehat{V} = \frac{1}{2} \sum_{\alpha,\beta} \hat{a}^\dagger(\alpha)\hat{a}^\dagger(\beta)\hat{a}(\beta)\hat{a}(\alpha)\, V^{(2)}(\alpha,\beta) \tag{6.58}$$

In position space, ignoring spin, we get

$$\begin{aligned}\widehat{V} &= \frac{1}{2}\int d^d x \int d^d y\, \hat{a}^\dagger(\boldsymbol{x})\, \hat{a}^\dagger(\boldsymbol{y})\, \hat{a}(\boldsymbol{y})\, \hat{a}(\boldsymbol{x})\, V^{(2)}(\boldsymbol{x},\boldsymbol{y}) \\ &\equiv \frac{1}{2}\int d^d x \int d^d y\, \hat{\rho}(\boldsymbol{x}) V^{(2)}(\boldsymbol{x},\boldsymbol{y})\hat{\rho}(\boldsymbol{y}) + \frac{1}{2}\int d^d x\, V^{(2)}(\boldsymbol{x},\boldsymbol{x})\, \hat{\rho}(\boldsymbol{x})\end{aligned} \tag{6.59}$$

while in momentum space, we find

$$\widehat{V} = \frac{1}{2}\int \frac{d^d p}{(2\pi)^d} \int \frac{d^d q}{(2\pi)^d} \int \frac{d^d k}{(2\pi)^d}\, \widetilde{V}(\boldsymbol{k})\, \hat{a}^\dagger(\boldsymbol{p}+\boldsymbol{k})\hat{a}^\dagger(\boldsymbol{q}-\boldsymbol{k})\hat{a}(\boldsymbol{q})\hat{a}(\boldsymbol{p}) \tag{6.60}$$

where $\widetilde{V}(\boldsymbol{k})$ is a function of only the momentum transfer $\boldsymbol{k}$, represented in the Feynman diagram in figure 6.2. This is a consequence of translation invariance. In particular, for a Coulomb interaction,

$$V^{(2)}(\boldsymbol{x},\boldsymbol{y}) = \frac{e^2}{|\boldsymbol{x}-\boldsymbol{y}|} \tag{6.61}$$

for which we have

$$\widetilde{V}(\boldsymbol{k}) = \frac{4\pi e^2}{\boldsymbol{k}^2} \tag{6.62}$$

6.2 Nonrelativistic field theory and second quantization

We can now reformulate the problem of an N-particle system as a nonrelativistic field theory. The procedure described in section 6.1 is commonly known as second quantization. If the (identical) particles are bosons, the operators $\hat{a}(\phi)$ obey *canonical commutation relations.*

If the (identical) particles are fermions, the operators $\hat{a}(\phi)$ obey *canonical anticommutation relations*. In position space, it is customary to represent $\hat{a}(\phi)$ by the operator $\hat{\psi}(\boldsymbol{x})$, which obeys the equal-time algebra,

$$\left[\hat{\psi}(\boldsymbol{x}), \hat{\psi}^{\dagger}(\boldsymbol{y})\right]_{\xi} = \delta^{d}(\boldsymbol{x}-\boldsymbol{y})$$
$$\left[\hat{\psi}(\boldsymbol{x}), \hat{\psi}(\boldsymbol{y})\right]_{\xi} = \left[\hat{\psi}^{\dagger}(\boldsymbol{x}), \hat{\psi}^{\dagger}(\boldsymbol{y})\right]_{\xi} = 0 \tag{6.63}$$

where we have used the notation

$$\left[\hat{A}, \hat{B}\right]_{\xi} \equiv \hat{A}\hat{B} - \xi \hat{B}\hat{A} \tag{6.64}$$

with $\xi = 1$ for bosons and $\xi = -1$ for fermions.

In this framework, the one-particle Schrödinger equation becomes the classical field equation:

$$\left[i\hbar\frac{\partial}{\partial t} + \frac{\hbar^2}{2m}\nabla^2 - V(\boldsymbol{x})\right]\hat{\psi} = 0 \tag{6.65}$$

Can we find a Lagrangian density $\mathcal{L}$ from which the one-particle Schrödinger equation follows as its classical equation of motion? The answer is yes, and $\mathcal{L}$ is given by

$$\mathcal{L} = i\hbar\psi^{\dagger}\partial_t\psi - \frac{\hbar^2}{2m}\boldsymbol{\nabla}\psi^{\dagger}\cdot\nabla\psi - V(\boldsymbol{x})\psi^{\dagger}\psi \tag{6.66}$$

Its Euler-Lagrange equations are

$$\partial_t\frac{\delta\mathcal{L}}{\delta\partial_t\psi^{\dagger}} = -\boldsymbol{\nabla}\cdot\frac{\delta\mathcal{L}}{\delta\boldsymbol{\nabla}\psi^{\dagger}} + \frac{\delta\mathcal{L}}{\delta\psi^{\dagger}} \tag{6.67}$$

which are equivalent to the field equation, eq. (6.65). The canonical momenta Π_{ψ} and $\Pi_{\psi}^{\dagger}$ are

$$\Pi_{\psi} = \frac{\delta\mathcal{L}}{\delta\partial_t\psi} = i\hbar\psi^{\dagger} \tag{6.68}$$

and

$$\Pi_{\psi^{\dagger}} = \frac{\delta\mathcal{L}}{\delta\partial_t\psi^{\dagger}} = -i\hbar\psi \tag{6.69}$$

Thus, the (equal-time) canonical commutation relations are

$$\left[\hat{\psi}(\boldsymbol{x}), \hat{\Pi}_{\psi}(\boldsymbol{y})\right]_{\xi} = i\hbar\delta(\boldsymbol{x}-\boldsymbol{y}) \tag{6.70}$$

which require that

$$\left[\hat{\psi}(\boldsymbol{x}), \hat{\psi}^{\dagger}(\boldsymbol{y})\right]_{\xi} = \delta^{d}(\boldsymbol{x}-\boldsymbol{y}) \tag{6.71}$$

Notice that the canonical momenta are proportional to the field. This is a consequence of the Lagrangian being first order in time derivatives. It is also necessary for the equal-time commutation and anticommutation relations to be those of creation and annihilation operators.

6.3 Nonrelativistic fermions at zero temperature

The results of the previous sections tell us that the action for nonrelativistic fermions with two-body interactions is, in $D = d + 1$ spacetime dimensions,

$$S = \int d^D x \left[\hat{\psi}^\dagger i\hbar \partial_t \hat{\psi} - \frac{\hbar^2}{2m} \nabla \hat{\psi}^\dagger \cdot \nabla \hat{\psi} - V(\boldsymbol{x}) \hat{\psi}^\dagger(x) \hat{\psi}(x) \right]$$
$$- \frac{1}{2} \int d^D x \int d^D x' \hat{\psi}^\dagger(x) \hat{\psi}^\dagger(x') U(x - x') \hat{\psi}(x') \hat{\psi}(x) \tag{6.72}$$

where $U(x - x')$ represents instantaneous pair-interactions:

$$U(x - x') \equiv U(\boldsymbol{x} - \boldsymbol{x}') \delta(x_0 - x_0') \tag{6.73}$$

The Hamiltonian $\hat{H}$ for this system is

$$\hat{H} = \int d^d x \left[\frac{\hbar^2}{2m} \nabla \hat{\psi}^\dagger \cdot \nabla \hat{\psi} + V(\boldsymbol{x}) \hat{\psi}^\dagger(\boldsymbol{x}) \psi(\boldsymbol{x}) \right]$$
$$+ \frac{1}{2} \int d^d x \int d^d x' \hat{\psi}^\dagger(x) \hat{\psi}^\dagger(x') U(x - x') \hat{\psi}(x') \hat{\psi}(x) \tag{6.74}$$

For fermions, the fields $\hat{\psi}$ and $\hat{\psi}^\dagger$ satisfy equal-time canonical anticommutation relations

$$\{\hat{\psi}(\boldsymbol{x}), \hat{\psi}^\dagger(\boldsymbol{x})\} = \delta(\boldsymbol{x} - \boldsymbol{x}') \tag{6.75}$$

while for bosons, they satisfy

$$[\hat{\psi}(\boldsymbol{x}), \hat{\psi}^\dagger(\boldsymbol{x}')] = \delta(\boldsymbol{x} - \boldsymbol{x}') \tag{6.76}$$

In both cases, the Hamiltonian $\hat{H}$ commutes with the total number operator $\hat{N} = \int d^d x \hat{\psi}^\dagger(\boldsymbol{x}) \hat{\psi}(\boldsymbol{x})$, since $\hat{H}$ conserves the total number of particles.

The Fock space picture of the many-body problem is equivalent to the grand canonical ensemble of statistical mechanics. Thus, instead of fixing the number of particles, we can introduce a Lagrange multiplier μ, the chemical potential, to weigh contributions from different parts of the Fock space, and we define the operator $\tilde{H}$

$$\widetilde{H} \equiv \widehat{H} - \mu \widehat{N} \tag{6.77}$$

In a Hilbert space with fixed $\hat{N}$, this amounts to a shift of the energy by μN.

Now let us allow the system to choose the sector of the Fock space but with the requirement that the average number of particles $\langle \hat{N} \rangle$ is fixed to be some number $\bar{N}$. In the thermodynamic limit ($N \to \infty$), the chemical potential μ represents the difference of the ground state energies between two sectors with $N+1$ and N particles, respectively. The modified Hamiltonian $\widetilde{H}$ is

$$\widetilde{H}=\int d^dx \sum_{\sigma} \hat{\psi}_{\sigma}^{\dagger}(\boldsymbol{x}) \left(-\frac{\hbar^2}{2m}\nabla^2+V(\boldsymbol{x})-\mu\right)\hat{\psi}_{\sigma}(\boldsymbol{x})$$
$$+\frac{1}{2}\int d^dx \int d^dy \sum_{\sigma,\sigma'} \hat{\psi}_{\sigma}^{\dagger}(\boldsymbol{x})\hat{\psi}_{\sigma'}^{\dagger}(\boldsymbol{y})U(\boldsymbol{x}-\boldsymbol{y})\hat{\psi}_{\sigma'}(\boldsymbol{y})\hat{\psi}_{\sigma}(\boldsymbol{x}) \tag{6.78}$$

6.3.1 The ground state of free fermions

Let us discuss now the very simple problem of finding the ground state for a system of N spinless free fermions. In this case, the interaction potential $U(\boldsymbol{x}-\boldsymbol{y})$ vanishes, and if the system is isolated and translationally invariant, the external potential $V(\boldsymbol{x})$ also vanishes. In general, there is a complete set of one-particle states of eigenstates of the single-particle Hamiltonian $\{|\alpha\rangle\}$. In this basis, the second quantized Hamiltonian $\hat{H}$ is

$$\hat{H}=\sum_{\alpha} E_{\alpha}\hat{a}_{\alpha}^{\dagger}a_{\alpha} \tag{6.79}$$

where the index α labels the one-particle states by increasing order of their single-particle energies:

$$E_1 \leq E_2 \leq \cdots \leq E_n \leq \cdots \tag{6.80}$$

Since we are dealing with fermions, we cannot put more than one particle in each state. Thus the state with the lowest energy is obtained by filling up all the first N single-particle states. Let $|\text{gnd}\rangle$ denote this ground state

$$|\text{gnd}\rangle=\prod_{\alpha=1}^{N}\hat{a}_{\alpha}^{\dagger}|0\rangle \equiv \hat{a}_1^{\dagger}\cdots\hat{a}_N^{\dagger}|0\rangle=|1\cdots 1,00\cdots\rangle \tag{6.81}$$

where the ket on the extreme right side has the lowest N states occupied, and all other states are empty. The energy of this state is E_{gnd}, where

$$E_{\text{gnd}}=E_1+\cdots+E_N \tag{6.82}$$

The energy of the topmost occupied single particle state, E_N, is called the *Fermi energy*. The set of occupied states is called the *filled Fermi sea*, shown in figure 6.3.

6.3.2 Excited states

A state such as $|\psi\rangle$

$$|\psi\rangle=|1\cdots 1010\cdots\rangle \tag{6.83}$$

is an excited state. Here the ket denotes a state where the lowest $N-1$ states plus the $N+1$ state are occupied, and all other states are empty. It is obtained by removing one particle from the single-particle state N (thus leaving a *hole* behind), and putting the particle in the unoccupied single-particle state $N+1$. This is a state with one *particle-hole pair*, and it has the form

$$|1\cdots 1010\cdots\rangle=\hat{a}_{N+1}^{\dagger}\hat{a}_N|\text{gnd}\rangle \tag{6.84}$$

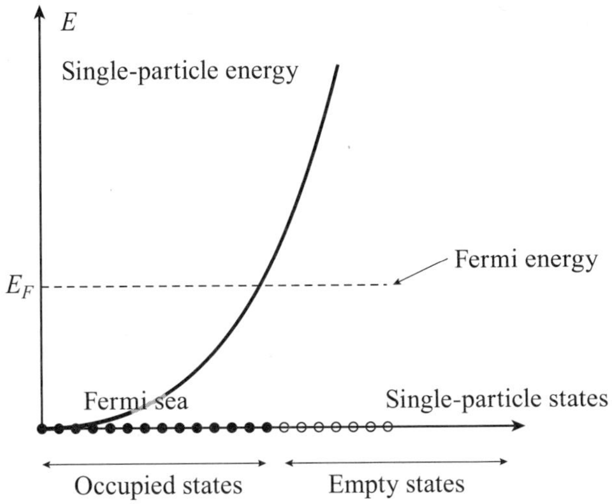

Figure 6.3 The Fermi sea.

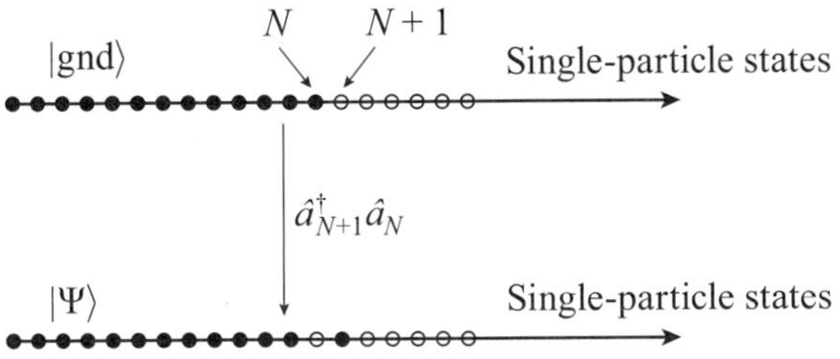

Figure 6.4 An excited (particle-hole) state.

See figure 6.4. The energy of this state is

$$E_\psi = E_1 + \cdots + E_{N-1} + E_{N+1} \tag{6.85}$$

Hence,

$$E_\psi = E_{\text{gnd}} + E_{N+1} - E_N \tag{6.86}$$

Since $E_{N+1} \geq E_N, E_\psi \geq E_{\text{gnd}}$, the *excitation energy* $\varepsilon_\psi = E_\psi - E_{\text{gnd}}$ is positive:

$$\varepsilon_\psi = E_{N+1} - E_N \geq 0 \tag{6.87}$$

6.3.3 Normal-ordering and particle-hole transformation: Construction of the physical Hilbert space

It is apparent that, instead of using the empty state $|0\rangle$ for the reference state, it is physically more reasonable to use instead the filled Fermi sea $|\text{gnd}\rangle$ as the physical reference state or *vacuum state*. Thus, this state is a vacuum in the sense of an absence of excitations.

These arguments motivate the introduction of the *particle-hole transformation*. Let us introduce the fermion operators b_α such that

$$\hat{b}_\alpha = \hat{a}^\dagger_\alpha, \qquad \text{for } \alpha \leq N \tag{6.88}$$

Since $\hat{a}^\dagger_\alpha|\text{gnd}\rangle = 0$ (for $\alpha \leq N$), the operators $\hat{b}_\alpha$ annihilate the ground state $|\text{gnd}\rangle$, that is,

$$\hat{b}_\alpha|\text{gnd}\rangle = 0 \tag{6.89}$$

The following canonical anticommutation relations hold:

$$\begin{gathered}\{\hat{a}_\alpha, \hat{a}'_\alpha\} = \left\{\hat{a}_\alpha, \hat{b}_\beta\right\} = \left\{\hat{b}_\beta, \hat{b}'_\beta\right\} = \left\{\hat{a}_\alpha, \hat{b}^\dagger_\beta\right\} = 0 \\ \left\{\hat{a}_\alpha, \hat{a}^\dagger_{\alpha'}\right\} = \delta_{\alpha\alpha'}, \qquad \left\{\hat{b}_\beta, \hat{b}^\dagger_{\beta'}\right\} = \delta_{\beta\beta'}\end{gathered} \tag{6.90}$$

where $\alpha, \alpha' > N$, and $\beta, \beta' \leq N$. Thus, relative to the state $|\text{gnd}\rangle$, $\hat{a}^\dagger_\alpha$ and $\hat{b}^\dagger_\beta$ behave like creation operators. An arbitrary excited state has the form

$$|\alpha_1 \cdots \alpha_m, \beta_1 \cdots \beta_n; \text{gnd}\rangle \equiv \hat{a}^\dagger_{\alpha_1} \cdots \hat{a}^\dagger_{\alpha_m} \hat{b}^\dagger_{\beta_1} \cdots \hat{b}^\dagger_{\beta_n}|\text{gnd}\rangle \tag{6.91}$$

This state has m particles, in the single-particle states $\alpha_1, \ldots, \alpha_m$, and n holes, in the single-particle states $\beta_1, \ldots, \beta_n$. The ground state is annihilated by the operators $\hat{a}_\alpha$ and $\hat{b}_\beta$:

$$\hat{a}_\alpha|\text{gnd}\rangle = \hat{b}_\beta|\text{gnd}\rangle = 0, \qquad \text{with } \alpha > N \text{ and } \beta \leq N \tag{6.92}$$

The Hamiltonian $\hat{H}$ is normal ordered relative to the *empty state* $|0\rangle$ (i.e., $\hat{H}|0\rangle = 0$), but it is not normal ordered relative to the actual ground state $|\text{gnd}\rangle$. The particle-hole transformation enables us to normal order $\hat{H}$ relative to $|\text{gnd}\rangle$,

$$\widehat{H} = \sum_\alpha E_\alpha \hat{a}^\dagger_\alpha \hat{a}_\alpha = \sum_{\alpha \leq N} E_\alpha + \sum_{\alpha > N} E_\alpha \hat{a}^\dagger_\alpha \hat{a}_\alpha - \sum_{\beta \leq N} E_\beta \hat{b}^\dagger_\beta \hat{b}_\beta \tag{6.93}$$

where the minus sign in the last term reflects the Fermi statistics.

Thus

$$\widehat{H} = E_{\text{gnd}} + :\widehat{H}: \tag{6.94}$$

where

$$E_{\text{gnd}} = \sum_{\alpha=1}^{N} E_\alpha \tag{6.95}$$

is the ground state energy, and the normal-ordered Hamiltonian is

$$:\widehat{H}: = \sum_{\alpha > N} E_\alpha \hat{a}^\dagger_\alpha \hat{a}_\alpha - \sum_{\beta \leq N} \hat{b}^\dagger_\beta \hat{b}_\beta E_\beta \tag{6.96}$$

The number operator $\widehat{N}$ is also not normal-ordered relative to $|\text{gnd}\rangle$. Thus, we write

$$\widehat{N} = \sum_{\alpha} \hat{a}_{\alpha}^{\dagger}\hat{a}_{\alpha} = N + \sum_{\alpha>N} \hat{a}_{\alpha}^{\dagger} a_{\alpha} - \sum_{\beta\leq N} \hat{b}_{\beta}^{\dagger}\hat{b}_{\beta} \tag{6.97}$$

We see that particles raise the energy, while holes reduce it. However, if we deal with Hamiltonians that conserve the particle number $\widehat{N}$ (i.e., $[\widehat{N}, \hat{H}] = 0$), then for every particle that is removed a hole must be created. Hence particles and holes can only be created in pairs. A particle-hole state $|\alpha, \beta\ \text{gnd}\rangle$ is

$$|\alpha, \beta\ \text{gnd}\rangle \equiv \hat{a}_{\alpha}^{\dagger}\hat{b}_{\beta}^{\dagger}|\text{gnd}\rangle \tag{6.98}$$

It is an eigenstate with energy

$$\begin{aligned}\widehat{H}|\alpha, \beta\ \text{gnd}\rangle &= \left(E_{\text{gnd}} + :\widehat{H}:\right) \hat{a}_{\alpha}^{\dagger}\hat{b}_{\beta}^{\dagger}|\text{gnd}\rangle \\ &= \left(E_{\text{gnd}} + E_{\alpha} - E_{\beta}\right)|\alpha, \beta\ \text{gnd}\rangle\end{aligned} \tag{6.99}$$

and the excitation energy is $E_{\alpha} - E_{\beta} \geq 0$. Hence the ground state is stable to the creation of particle-hole pairs.

This state has exactly N particles, since

$$\widehat{N}|\alpha, \beta\ \text{gnd}\rangle = (N + 1 - 1)|\alpha, \beta\ \text{gnd}\rangle = N|\alpha, \beta\ \text{gnd}\rangle \tag{6.100}$$

Let us finally notice that the field operator $\hat{\psi}^{\dagger}(\boldsymbol{x})$ in position space is

$$\hat{\psi}^{\dagger}(\boldsymbol{x}) = \sum_{\alpha} \langle \boldsymbol{x}|\alpha\rangle \hat{a}_{\alpha}^{\dagger} = \sum_{\alpha>N} \phi_{\alpha}(\boldsymbol{x})\hat{a}_{\alpha}^{\dagger} + \sum_{\beta\leq N} \phi_{\beta}(\boldsymbol{x})\hat{b}_{\beta} \tag{6.101}$$

where $\{\phi_{\alpha}(\boldsymbol{x})\}$ are the single-particle wave functions.

The procedure of normal ordering allows us to define the physical Hilbert space. The physical meaning of this approach becomes more transparent in the *thermodynamic limit* $N \to \infty$ and $V \to \infty$ at constant density ρ. In this limit, the Hilbert space is the set of states that is obtained by acting finitely with creation and annihilation operators on the ground state (the vacuum). The spectrum of states that results from this approach consists of the set of states with finite excitation energy. Hilbert spaces that are built on reference states with macroscopically different numbers of particles are effectively disconnected from one another. Thus, the normal ordering of a Hamiltonian of a system with an infinite number of degrees of freedom amounts to a choice of the Hilbert space. This restriction becomes of fundamental importance when interactions are taken into account.

6.3.4 The free Fermi gas

Let us consider the case of free spin-1/2 electrons moving in free space. The Hamiltonian for this system is

$$\tilde{H} = \int d^d x \sum_{\sigma=\uparrow,\downarrow} \hat{\psi}_{\sigma}^{\dagger}(\boldsymbol{x}) \left[-\frac{\hbar^2}{2m} \nabla^2 - \mu \right] \hat{\psi}_{\sigma}(\boldsymbol{x}) \tag{6.102}$$

where $\sigma = \uparrow, \downarrow$ indicates the z-projection of the spin of the electron. The value of the chemical potential μ will be determined once we satisfy the condition that the electron density is equal to some fixed value $\bar{\rho}$.

In momentum space, we get

$$\hat{\psi}_\sigma(\boldsymbol{x}) = \int \frac{d^d p}{(2\pi)^d}\, \hat{\psi}_\sigma(\boldsymbol{p})\, e^{-i\boldsymbol{p}\cdot\boldsymbol{x}/\hbar} \tag{6.103}$$

where the operators $\hat{\psi}_\sigma(\boldsymbol{p})$ and $\hat{\psi}^\dagger_\sigma(\boldsymbol{p})$ satisfy the canonical equal-time anticommutation relations:

$$\begin{aligned} \left\{\hat{\psi}_\sigma(\boldsymbol{p}), \hat{\psi}^\dagger_{\sigma'}(\boldsymbol{p})\right\} &= (2\pi)^d \delta_{\sigma\sigma'} \delta^d(\boldsymbol{p}-\boldsymbol{p}') \\ \left\{\hat{\psi}^\dagger_\sigma(\boldsymbol{p}), \hat{\psi}^\dagger_{\sigma'}(\boldsymbol{p})\right\} &= \{\hat{\psi}(\boldsymbol{p}), \hat{\psi}_{\sigma'}(\boldsymbol{p}')\} = 0 \end{aligned} \tag{6.104}$$

The Hamiltonian has the very simple form

$$\tilde{H} = \int \frac{d^d p}{(2\pi)^d} \sum_{\sigma=\hat{\imath},\downarrow} \left(\varepsilon(\boldsymbol{p}) - \mu\right) \hat{\psi}^\dagger_\sigma(\boldsymbol{p}) \hat{\psi}_\sigma(\boldsymbol{p}) \tag{6.105}$$

where $\varepsilon(\boldsymbol{p})$ is given by

$$\varepsilon(\boldsymbol{p}) = \frac{\boldsymbol{p}^2}{2m} \tag{6.106}$$

For this simple case, $\varepsilon(\boldsymbol{p})$ is independent of the spin orientation.

It is convenient to measure the energy relative to the chemical potential (i.e., the *Fermi energy*) $\mu = E_F$. The relative energy $E(\boldsymbol{p})$ is

$$E(\boldsymbol{p}) = \varepsilon(\boldsymbol{p}) - \mu \tag{6.107}$$

Hence, $E(\boldsymbol{p})$ is the excitation energy measured from the Fermi energy $E_F = \mu$. The energy $E(\boldsymbol{p})$ does not have a definite sign, since there are states with $\varepsilon(\boldsymbol{p}) > \mu$ as well as those with $\varepsilon(\boldsymbol{p}) < \mu$. Let us define p_F as the value of $|\boldsymbol{p}|$ for which

$$E(p_F) = \varepsilon(p_F) - \mu = 0 \tag{6.108}$$

This is the Fermi momentum. Thus, for $|\boldsymbol{p}| < p_F$, $E(\boldsymbol{p})$ is negative, while for $|\boldsymbol{p}| > p_F$, $E(\boldsymbol{p})$ is positive.

We can construct the ground state of the system by finding the state with the lowest energy at fixed μ. Since $E(\boldsymbol{p})$ is negative for $|\boldsymbol{p}| \leq p_F$, we see that by filling up all of those states, we get the lowest possible energy. It is then natural to normal order the system relative to a state in which all one-particle states with $|\boldsymbol{p}| \leq p_F$ are occupied. Hence we make the particle-hole transformation

$$\begin{aligned} \hat{b}_\sigma(\boldsymbol{p}) &= \hat{\psi}^\dagger_\sigma(\boldsymbol{p}) \quad \text{for } |\boldsymbol{p}| \leq p_F \\ \hat{a}_\sigma(\boldsymbol{p}) &= \hat{\psi}_\sigma(\boldsymbol{p}) \quad \text{for } |\boldsymbol{p}| > p_F \end{aligned} \tag{6.109}$$

In terms of the operators $\hat{a}_\sigma$ and $\hat{b}_\sigma$, the Hamiltonian is

$$\tilde{H} = \sum_{\sigma=\uparrow,\downarrow} \int \frac{d^d p}{(2\pi)^d} [E(\boldsymbol{p})\theta(|\boldsymbol{p}| - p_F)\hat{a}^\dagger_\sigma(\boldsymbol{p})\hat{a}_\sigma(\boldsymbol{p}) + \theta(p_F - |\boldsymbol{p}|)E(\boldsymbol{p})\hat{b}_\sigma(\boldsymbol{p})\hat{b}^\dagger_\sigma(\boldsymbol{p})] \tag{6.110}$$

where $\theta(x)$ is the step function:

$$\theta(x) = \begin{cases} 1, & x > 0 \\ 0, & x \leq 0 \end{cases} \tag{6.111}$$

Using the anticommutation relations in the last term, we get

$$\tilde{H} = \sum_{\sigma=\uparrow,\downarrow} \int \frac{d^d p}{(2\pi)^d} E(\boldsymbol{p})[\theta(|\boldsymbol{p}| - p_F)\hat{a}^\dagger_\sigma(\boldsymbol{p})\hat{a}_\sigma(\boldsymbol{p}) - \theta(p_F - |(\boldsymbol{p})|)\hat{b}^\dagger_\sigma(\boldsymbol{p})\hat{b}_\sigma(\boldsymbol{p})] + \tilde{E}_{\text{gnd}} \tag{6.112}$$

where $\tilde{E}_{\text{gnd}}$, the ground state energy measured from the chemical potential μ, is given by

$$\tilde{E}_{\text{gnd}} = \sum_{\sigma=\uparrow,\downarrow} \int \frac{d^d p}{(2\pi)^d} \theta(p_F - |\boldsymbol{p}|)E(\boldsymbol{p})(2\pi)^d\delta^d(0) = E_{\text{gnd}} - \mu N \tag{6.113}$$

Recall that

$$(2\pi)^d\delta^d(0) = \lim_{\boldsymbol{p}\to 0}(2\pi)^d\delta^d(\boldsymbol{p}) = \lim_{\boldsymbol{p}\to 0} \int d^d x e^{i\boldsymbol{p}\cdot\boldsymbol{x}} = V \tag{6.114}$$

where V is the volume of the system. Thus, $\tilde{E}_{\text{gnd}}$ is extensive,

$$\tilde{E}_{\text{gnd}} = V\tilde{\varepsilon}_{\text{gnd}} \tag{6.115}$$

and the ground state energy density $\tilde{\varepsilon}_{\text{gnd}}$ is

$$\tilde{\varepsilon}_{\text{gnd}} = 2\int_{|\boldsymbol{p}|\leq p_F} \frac{d^d p}{(2\pi)^d} E(\boldsymbol{p}) = \varepsilon_{\text{gnd}} - \mu\bar{\rho} \tag{6.116}$$

where the factor of 2 comes from the two spin orientations. Putting everything together, we get

$$\tilde{\varepsilon}_{\text{gnd}} = 2\int_{|\boldsymbol{p}|\leq p_F} \frac{d^d p}{(2\pi)^d} \left(\frac{\boldsymbol{p}^2}{2m} - \mu\right) = 2\int_0^{p_F} dp\, p^{d-1} \frac{S_d}{(2\pi)^d}\left(\frac{p^2}{2m} - \mu\right) \tag{6.117}$$

where S_d is the area of the d-dimensional hypersphere. Our definitions tell us that the chemical potential is $\mu = \frac{p_F^2}{2m} \equiv E_F$, where E_F is the Fermi energy. Thus the ground state energy density ε_{gnd} (measured from the empty state) is

$$\varepsilon_{\text{gnd}} = \frac{1}{m}\frac{S_d}{(2\pi)^d}\int_0^{p_F} dp\, p^{d+1} = \frac{p_F^{d+2}}{m(d+2)}\frac{S_d}{(2\pi)^d} = 2E_F\frac{p_F^d S_d}{(d+2)(2\pi)^d} \tag{6.118}$$

How many particles does this state have? To find that out, we need to look at the number operator. The number operator can also be normal-ordered with respect to this state:

$$\begin{aligned}\hat{N} &= \int \frac{d^d p}{(2\pi)^d} \sum_{\sigma=\uparrow,\downarrow} \hat{\psi}^\dagger_\sigma(\boldsymbol{p}) \hat{\psi}_\sigma(\boldsymbol{p}) = \\ &= \int \frac{d^d p}{(2\pi)^d} \sum_{\sigma=\uparrow,\downarrow} \{\theta(|\boldsymbol{p}| - p_F) \hat{a}^\dagger_\sigma(\boldsymbol{p}) \hat{a}_\sigma(\boldsymbol{p}) + \theta(p_F - |\boldsymbol{p}|) \hat{b}_\sigma(\boldsymbol{p}) \hat{b}^\dagger_\sigma(\boldsymbol{p})\}\end{aligned} \tag{6.119}$$

Hence, $\hat{N}$ can also be written in the form

$$\hat{N} =: \hat{N}: + N \tag{6.120}$$

where the normal-ordered number operator $: \hat{N}:$ is

$$: \hat{N}:= \int \frac{d^d p}{(2\pi)^d} \sum_{\sigma=\uparrow,\downarrow} [\theta(|\boldsymbol{p}| - p_F) \hat{a}^\dagger_\sigma(\boldsymbol{p}) \hat{a}_\sigma(\boldsymbol{p}) - \theta(p_F - |\boldsymbol{p}|) \hat{b}^\dagger_\sigma(\boldsymbol{p}) \hat{b}_\sigma(\boldsymbol{p})] \tag{6.121}$$

and N, the number of particles in the reference state $|\text{gnd}\rangle$, is

$$N = \int \frac{d^d p}{(2\pi)^d} \sum_{\sigma=\uparrow,\downarrow} \theta(p_F - |\boldsymbol{p}|)(2\pi)^d \delta^d(0) = \frac{2}{d} p_F^d \frac{S_d}{(2\pi)^d} V \tag{6.122}$$

Therefore, the particle density $\bar{\rho} = \frac{N}{V}$ is

$$\bar{\rho} = \frac{2}{d} \frac{S_d}{(2\pi)^d} p_F^d \tag{6.123}$$

This equation determines the *Fermi momentum* p_F in terms of the density $\bar{\rho}$. Similarly, we find that the ground state energy per particle is $\frac{E_{\text{gnd}}}{N} = \frac{2d}{d+2} E_F$.

The excited states can be constructed in a similar fashion. The state

$$|+, \sigma, \boldsymbol{p}\rangle = \hat{a}^\dagger_\alpha(\boldsymbol{p}) |\text{gnd}\rangle \tag{6.124}$$

is a state representing an electron with spin σ and momentum $\boldsymbol{p}$, while the state $|-, \sigma, \boldsymbol{p}\rangle$

$$|-, \sigma, \boldsymbol{p}\rangle = \hat{b}^\dagger_\sigma(\boldsymbol{p}) |\text{gnd}\rangle \tag{6.125}$$

represents a hole with spin σ and momentum $\boldsymbol{p}$. We see that electrons have momentum $\boldsymbol{p}$ with $|\boldsymbol{p}| > p_F$, while holes have momentum $\boldsymbol{p}$ with $|\boldsymbol{p}| < p_F$. The excitation energy of a one-electron state is $E(\boldsymbol{p}) \geq 0$ (for $|\boldsymbol{p}| > p_F$), while the excitation energy of a one-hole state is $-E(\boldsymbol{p}) \geq 0$ (for $|\boldsymbol{p}| < p_F$).

Similarly, an electron-hole *pair* is a state of the form

$$|\sigma \boldsymbol{p}, \sigma' \boldsymbol{p}'\rangle = \hat{a}^\dagger_\sigma(\boldsymbol{p}) \hat{b}^\dagger_{\sigma'}(\boldsymbol{p}') |\text{gnd}\rangle \tag{6.126}$$

with $|\boldsymbol{p}| > p_F$ and $|\boldsymbol{p}'| < p_F$. This state has excitation energy $E(\boldsymbol{p}) - E(\boldsymbol{p}')$, which is positive. Hence, states obtained from the ground state without changing the density can only increase the energy. This proves that $|\text{gnd}\rangle$ is indeed the ground state. However, if the density is allowed to change, we can always construct states with energy less than E_{gnd} by creating some number of holes without creating an equal number of particles.

Exercises

6.1 Spin waves in a quantum Heisenberg antiferromagnet

Consider the Heisenberg model of a one-dimensional quantum antiferromagnet. I first give you a brief summary of the Heisenberg model. You do not need to have any previous knowledge of magnetism (or the Heisenberg model) to do this problem. You will be able to solve this problem using only the methods discussed in this chapter.

The one-dimensional Heisenberg model is defined for a linear chain (a one-dimensional lattice) with N sites. The lattice spacing is taken to be equal to one (i.e., it is the unit of length). The quantum mechanical Hamiltonian for this system is

$$\hat{H} = J \sum_{j=-N/2+1}^{N/2} \hat{S}_k(j) \cdot \hat{S}_k(j+1) \tag{6.127}$$

where the exchange constant $J > 0$ (i.e., an antiferromagnet), and the operators $\hat{S}_k$ ($k = 1, 2, 3$) are the three angular momentum operators in the spin-S representation (S is integer or half-integer) that satisfy the commutation relations

$$[\hat{S}_j, \hat{S}_k] = i\epsilon_{jkl}\hat{S}_l \tag{6.128}$$

For simplicity, assume periodic boundary conditions: $\hat{S}_k(j) \equiv \hat{S}_k(j+N)$.

In the semiclassical limit, $S \to \infty$, the operators act like real numbers, since the commutators vanish. In this limit, the state with lowest energy has nearby spins that point in opposite (but arbitrary!) directions in spin space. This is the classical Néel state. In this state, the spins on one sublattice (say, the even sites) point up along some direction in space, while the spins on the other sublattice (the odd sites) point down. For finite values of S, the spins can only have a definite projection along one axis but not along all three at the same time. Thus we should expect to see some zero-point motion precessional effect that will depress the net projection of the spin along any axis. But if the state is stable, even sites will have predominantly up spins, while odd sites will have predominantly down spins. These observations motivate the following definition of a set of basis states for the full Hilbert space of this system.

The states $|\Psi\rangle$ of the Hilbert space of this chain are spanned by the tensor product of the Hilbert spaces of each individual jth spin $|\Psi_j\rangle$, $|\Psi\rangle = \prod_j \otimes |\Psi_j\rangle$. The latter are simply the $2S+1$ degenerate multiplet of states with angular momentum S of the form $\{|S, M(j)\rangle\}$, with $|M(j)| \leq S$, which satisfy

$$\begin{aligned} \mathbf{S}^2(j)|S, M(j)\rangle &= S(S+1)|S, M(j)\rangle \\ S_3(j)|S, M(j)\rangle &= M(j)|S, M(j)\rangle \end{aligned} \tag{6.129}$$

The states in this multiplet can be obtained from the highest-weight state $|S,S\rangle$ by using the lowering operator $\hat{S}^- = \hat{S}_1 - i\hat{S}_2$. Its adjoint is the raising operator $\hat{S}^+(j) = \hat{S}_1(j) + i\hat{S}_2(j)$. For reasons that will become clear below, it is convenient to define for j even the *spin-deviation* operator $\hat{n}(j) \equiv S - \hat{S}_3(j)$. For j odd, the spin-deviation operator is $\hat{n}(j) \equiv S + \hat{S}_3(j)$. For j even, the highest-weight state $|S,S\rangle$ is an eigenstate of $\hat{n}(j)$ with eigenvalue zero, while the state $|S,-S\rangle$ has eigenvalue $2S$:

$$\begin{aligned}\hat{n}(j)|S,S\rangle &= (S - \hat{S}_3(j))|S,S\rangle = 0\\ \hat{n}(j)|S,-S\rangle &= (S - \hat{S}_3(j))|S,-S\rangle = 2S\,|S,-S\rangle\end{aligned} \tag{6.130}$$

In contrast, for j odd, the state $|S,-S\rangle$ has eigenvalue zero while the state $|S,S\rangle$ has eigenvalue $2S$.

In terms of the operators $\hat{n}(j)$, the basis states are $\{|S,M(j)\rangle\} \equiv \{|n(j)\rangle\}$, where $M(j) = S \mp n(j)$. For even sites, the raising and lowering operators $\hat{S}(j)^{\pm}$ act on the states of this basis as follows:

$$\begin{aligned}\hat{S}^+|n\rangle &= \left[2S\left(1 - \frac{n-1}{2S}\right)n\right]^{\frac{1}{2}} |n-1\rangle\\ \hat{S}^-|n\rangle &= \left[2S(n+1)\left(1 - \frac{n}{2S}\right)\right]^{\frac{1}{2}} |n+1\rangle\end{aligned} \tag{6.131}$$

For odd sites, the action of the above two operators is interchanged.

The action of the operators $\hat{S}^{\pm}$ is somewhat similar to that of annihilation and creation operators in harmonic oscillator states. For this reason, we define a set of creation and annihilation operators $\hat{a}^\dagger$ and $\hat{a}$ such that

$$\begin{aligned}\hat{a}^\dagger|n\rangle &= \sqrt{n+1}|n+1\rangle\\ \hat{a}|n\rangle &= \sqrt{n}|n-1\rangle\end{aligned} \tag{6.132}$$

which satisfy the conventional algebra $[\hat{a}, \hat{a}^\dagger] = 1$. Since we have two sublattices and the operators $\hat{S}^{\pm}$ are different on each sublattice, it is useful to introduce two types of creation and annihilation operators: the operators $\hat{a}^\dagger(j)$ and $\hat{a}(j)$, which act on even sites; and $\hat{b}^\dagger(j)$ and $\hat{b}(j)$, which act on odd sites. They obey the commutation relations

$$\begin{aligned}\left[\hat{a}(j), \hat{a}^\dagger(k)\right] &= \left[\hat{b}(j), \hat{b}^\dagger(k)\right] = \delta_{jk}\\ \left[\hat{a}(j), \hat{a}(k)\right] &= \left[\hat{b}(j), \hat{b}(k)\right] = \left[\hat{a}(j), \hat{b}(k)\right] = 0\end{aligned} \tag{6.133}$$

and similar equations for their hermitian conjugates. It is easy to check that the action of raising and lowering operators on the states $\{|n\rangle\}$ is the same as the action of the following operators on the same states:

$$\hat{S}^+(j) = \sqrt{2S}\left[1 - \frac{\hat{n}(j)}{2S}\right]^{\frac{1}{2}} \hat{a}(j)$$

$$\hat{S}^-(j) = \sqrt{2S}\hat{a}^\dagger(j)\left[1 - \frac{\hat{n}(j)}{2S}\right]^{\frac{1}{2}}$$

$$\hat{S}_3(j) = S - \hat{n}(j), \qquad \hat{n}(j) = \hat{a}^\dagger(j)\hat{a}(j) \tag{6.134}$$

on even sites, and

$$\hat{S}^-(j) = \sqrt{2S}\left[1 - \frac{\hat{n}(j)}{2S}\right]^{\frac{1}{2}} \hat{b}(j)$$

$$\hat{S}^+(j) = \sqrt{2S}\hat{b}^\dagger(j)\left[1 - \frac{\hat{n}(j)}{2S}\right]^{\frac{1}{2}}$$

$$\hat{S}_3(j) = -S + \hat{n}(j), \qquad \hat{n}(j) = \hat{b}^\dagger(j)\hat{b}(j) \tag{6.135}$$

on odd sites. Notice that although the integers n can now range from 0 to infinity, the Hilbert space is still finite, since (for even sites) $\hat{S}^-|n=2S\rangle = 0$. Similarly, for odd sites, $\hat{S}^+|n=2S\rangle = 0$.

1) Derive the quantum mechanical equations of motion obeyed by the spin operators $\hat{S}^\pm(j), \hat{S}_3(j)$ in the Heisenberg representation, for both j even and j odd. Are these equations linear? Explain your result.
2) Verify that the definition for the operators $S^\pm$ and S_3 of eqs. (6.134) and (6.135) are consistent with those of eq. (6.131).
3) Use the definitions given above to show that the Heisenberg Hamiltonian can be written in terms of two sets of creation and annihilation operators $\hat{a}^\dagger(j)$ and $\hat{a}(j)$ (which act on even sites), and $\hat{b}^\dagger(j)$ and $\hat{b}(j)$ (which act on odd sites).
4) Find an approximate form for the Hamiltonian that is valid in the semiclassical limit $S \to \infty$ (or $\frac{1}{S} \to 0$). Include terms of order $\frac{1}{S}$ (relative to the leading-order term). Show that the approximate Hamiltonian is quadratic in the operators a and b.
5) Make the approximations from part 4 for the equations of motion of part 1. Show that the equations of motion are now linear. Of what order in $\frac{1}{S}$ are the terms that have been neglected?
6) Show that the Fourier transform

$$\hat{a}(q) = \sqrt{\frac{2}{N}} \sum_{j\,\text{even}} e^{iqj}\,\hat{a}(j)$$

$$\hat{b}(q) = \sqrt{\frac{2}{N}} \sum_{j\,\text{odd}} e^{-iqj}\,\hat{b}(j) \tag{6.136}$$

followed by the canonical (Bogoliubov) transformation

$$\hat{c}(q) = \cosh(\theta(q))\,\hat{a}(q) + \sinh(\theta(q))\,\hat{b}^\dagger(q)$$

$$\hat{d}(q) = \cosh(\theta(q))\,\hat{b}(q) + \sinh(\theta(q))\,\hat{a}^\dagger(q) \tag{6.137}$$

yields a diagonal Hamiltonian H_{SW} of the form

$$H_{\text{SW}} = E_0 + \int_{-\frac{\pi}{2}}^{+\frac{\pi}{2}} \frac{dq}{2\pi}\, \omega(q)(\hat{n}_c(q) + \hat{n}_d(q)) \qquad (6.138)$$

where $\hat{n}_c(q) = \hat{c}^\dagger(q)\hat{c}(q)$, $\hat{n}_d(q) = \hat{d}^\dagger(q)\hat{d}(q)$, provided that the angle $\theta(q)$ is chosen properly. The operators $\hat{c}(q)$ and $\hat{d}(q)$ and their hermitian conjugates obey the algebra of eq. (6.133). Derive an explicit expression for the angle $\theta(q)$ and for the frequency $\omega(q)$.

7) Find the ground state for this system in this approximation (usually called the *spin-wave* approximation).
8) Find the single-particle eigenstates in this approximation. Determine the quantum numbers of the excitations. Find their dispersion (or energy-momentum) relations. Find a set of values of the momentum q for which the energy of the excited states goes to zero. Show that the energy of these states vanishes linearly as the momentum approaches the special points, and determine the spin-wave velocity v_s at these points.

Note: This is the semiclassical or spin-wave approximation. The identities in eq. (6.134) and eq. (6.135) are known as the Holstein-Primakoff identities.

7

Quantization of the Free Dirac Field

7.1 The Dirac equation and quantum field theory

The Dirac equation is a relativistic wave equation that describes the quantum dynamics of spinors. We will see in this section that a consistent description of this theory cannot be done outside the framework of (local) relativistic QFT.

The Dirac equation

$$(i\partial\!\!\!/ - m)\psi = 0 \tag{7.1}$$

can be regarded as the equation of motion of a complex field ψ. Much as for the case of the scalar field, and also in close analogy to the theory of nonrelativistic many-particle systems discussed in chapter 6, we will regard the Dirac field as an operator that acts on a Fock space of states.

We have already discussed that the Dirac equation also follows from a least action principle. Indeed, the Lagrangian

$$\mathcal{L} = \frac{i}{2}[\bar{\psi}\partial\!\!\!/\psi - (\partial_\mu\bar{\psi})\gamma^\mu\psi] - m\bar{\psi}\psi \equiv \bar{\psi}(i\partial\!\!\!/ - m)\psi \tag{7.2}$$

has the Dirac equation for its equation of motion. Also, the momentum $\Pi_\alpha(x)$ canonically conjugate to $\psi_\alpha(x)$ is

$$\Pi^\psi_\alpha(x) = \frac{\delta\mathcal{L}}{\delta\partial_0\psi_\alpha(x)} = i\psi^\dagger_\alpha \tag{7.3}$$

Thus, they obey the equal-time Poisson brackets:

$$\{\psi_\alpha(\boldsymbol{x}), \Pi^\psi_\beta(\boldsymbol{y})\}_{PB} = \delta_{\alpha\beta}\delta^3(\boldsymbol{x}-\boldsymbol{y}) \tag{7.4}$$

Thus,

$$\{\psi_\alpha(\boldsymbol{x}), \psi^\dagger_\beta(\boldsymbol{y})\}_{PB} = i\delta_{\alpha\beta}\delta^3(\boldsymbol{x}-\boldsymbol{y}) \tag{7.5}$$

In other words, the field ψ_α and its adjoint $\psi^\dagger_\alpha$ are a canonical pair. This result follows from the fact that the Dirac Lagrangian is first order in time derivatives. Notice that the field theory of nonrelativistic many-particle systems (for both fermions and bosons) also has a Lagrangian that is first order in time derivatives. We will see that, because of this property, the QFT of both types of systems follows rather similar lines, at least at a formal level. As in the case of the many-particle systems, two types of statistics are available to us: Fermi-Dirac and Bose-Einstein. We will see that only the choice of Fermi statistics leads to a physically meaningful theory of the Dirac equation.

The Hamiltonian for the Dirac theory is

$$H = \int d^3x \, \bar{\psi}_\alpha(\boldsymbol{x}) \Big[-i\boldsymbol{\gamma} \cdot \nabla + m \Big]_{\alpha\beta} \psi_\beta(\boldsymbol{x}) \tag{7.6}$$

where the fields $\psi(x)$ and $\bar{\psi} = \psi^\dagger \gamma_0$ are operators that act on a Hilbert space to be specified below. Notice that the one-particle operator in eq. (7.6) is just the one-particle Dirac Hamiltonian obtained if we regard the Dirac equation as a Schrödinger equation for spinors. We will leave the issue of their commutation relations (i.e., Fermi or Bose) open for the time being. In any event, the equations of motion are *independent* of that choice (i.e., they do not depend on the statistics).

In the Heisenberg representation, we find

$$i\gamma_0 \partial_0 \psi = \Big[\gamma_0 \psi, H \Big] = (-i\boldsymbol{\gamma} \cdot \nabla + m)\psi \tag{7.7}$$

which is just the Dirac equation.

We will solve this equation by means of a Fourier expansion in modes of the form

$$\psi(x) = \int \frac{d^3p}{(2\pi)^3} \frac{m}{\omega(p)} \left(\tilde{\psi}_+(p) e^{-ip\cdot x} + \tilde{\psi}_-(p) e^{ip\cdot x} \right) \tag{7.8}$$

where $\omega(p)$ is a quantity with units of energy (which will turn out to be equal to $p_0 = \sqrt{\boldsymbol{p}^2 + m^2}$), and $p \cdot x = p_0 x_0 - \boldsymbol{p} \cdot \boldsymbol{x}$. In terms of $\tilde{\psi}_\pm(p)$, the Dirac equation becomes

$$(p_0\gamma_0 - \boldsymbol{\gamma} \cdot \boldsymbol{p} \pm m)\, \tilde{\psi}_\pm(p) = 0 \tag{7.9}$$

In other words, $\tilde{\psi}_\pm(p)$ creates one-particle states with energy $\pm p_0$. Let us make the substitution

$$\tilde{\psi}_\pm(p) = (\pm \not{p} + m)\tilde{\phi} \tag{7.10}$$

We get

$$(\not{p} \mp m)(\pm \not{p} + m)\tilde{\phi} = \pm(p^2 - m^2)\tilde{\phi} = 0 \tag{7.11}$$

This equation has nontrivial solutions only if the mass-shell condition is obeyed:

$$p^2 - m^2 = 0 \tag{7.12}$$

Thus, we can identify $p_0 = \omega(\boldsymbol{p}) = \sqrt{\boldsymbol{p}^2 + m^2}$. At zero momentum, these states become

$$\tilde{\psi}_\pm(p_0, \boldsymbol{p} = 0) = (\pm p_0 \gamma_0 + m)\tilde{\phi} \tag{7.13}$$

where $\tilde{\phi}$ is an arbitrary 4-spinor. Let us choose $\tilde{\phi}$ to be an eigenstate of γ_0. Recall that in the Dirac representation, γ_0 is diagonal:

$$\gamma_0 = \begin{pmatrix} I & 0 \\ 0 & -I \end{pmatrix} \tag{7.14}$$

Thus the spinors $u^{(1)}(m,\mathbf{0})$ and $u^{(2)}(m,\mathbf{0})$,

$$u^{(1)}(m,\mathbf{0}) = \begin{pmatrix} 1 \\ 0 \\ 0 \\ 0 \end{pmatrix}, \qquad u^{(2)}(m,\mathbf{0}) = \begin{pmatrix} 0 \\ 1 \\ 0 \\ 0 \end{pmatrix} \tag{7.15}$$

have γ_0-eigenvalue $+1$

$$\gamma_0 u^{(i)}(m,\mathbf{0}) = +u^{(\sigma)}(m,\mathbf{0}), \qquad \sigma = 1,2 \tag{7.16}$$

and the spinors $v^{(\sigma)}(m,\mathbf{0})(\sigma = 1,2)$

$$v^{(1)}(m,\mathbf{0}) = \begin{pmatrix} 0 \\ 0 \\ 1 \\ 0 \end{pmatrix}, \qquad v^{(2)}(m,\mathbf{0}) = \begin{pmatrix} 0 \\ 0 \\ 0 \\ 1 \end{pmatrix} \tag{7.17}$$

have γ_0-eigenvalue -1

$$\gamma_0 v^{(\sigma)}(m,\mathbf{0}) = -v^{(\sigma)}(m,\mathbf{0}), \qquad \sigma = 1,2 \tag{7.18}$$

Let $\varphi^{(i)}(m,0)$ be the 2-spinors $(\sigma = 1,2)$

$$\varphi^{(1)} = \begin{pmatrix} 1 \\ 0 \end{pmatrix}, \qquad \varphi^{(2)} = \begin{pmatrix} 0 \\ 1 \end{pmatrix} \tag{7.19}$$

In terms of $\varphi^{(i)}$, the solutions are

$$\tilde{\psi}_+(p) = u^{(\sigma)}(p) = \frac{(\not{p}+m)}{\sqrt{2m(p_0+m)}} u^{(\sigma)}(m,\mathbf{0}) = \begin{pmatrix} \sqrt{\frac{p_0+m}{2m}} & \varphi^{(\sigma)}(m,0) \\ \frac{\boldsymbol{\sigma}\cdot\mathbf{p}}{\sqrt{2m(p_0+m)}} & \varphi^{(\sigma)}(m,0) \end{pmatrix} \tag{7.20}$$

and

$$\tilde{\psi}_-(p) = v^{(\sigma)}(p) = \frac{(-\not{p}+m)}{\sqrt{2m(p_0+m)}} v^{(\sigma)}(m,\mathbf{0}) = \begin{pmatrix} \frac{\boldsymbol{\sigma}\cdot\mathbf{p}}{\sqrt{2m(p_0+m)}} & \varphi^{(\sigma)}(m,\mathbf{0}) \\ \sqrt{\frac{p_0+m}{2m}} & \varphi^{(\sigma)}(m,\mathbf{0}) \end{pmatrix} \tag{7.21}$$

where the two solutions $\tilde{\psi}_+(p)$ have energy $+p_0 = +\sqrt{\mathbf{p}^2+m^2}$, while the two solutions $\tilde{\psi}_-(p)$ have energy $-p_0 = -\sqrt{\mathbf{p}^2+m^2}$.

Therefore, the one-particle states of the Dirac theory can have both positive and negative energy and, as it stands, the spectrum of the one-particle Dirac Hamiltonian, shown schematically in figure 7.1, is not positive. In addition, each Dirac state has a twofold degeneracy due to spin.

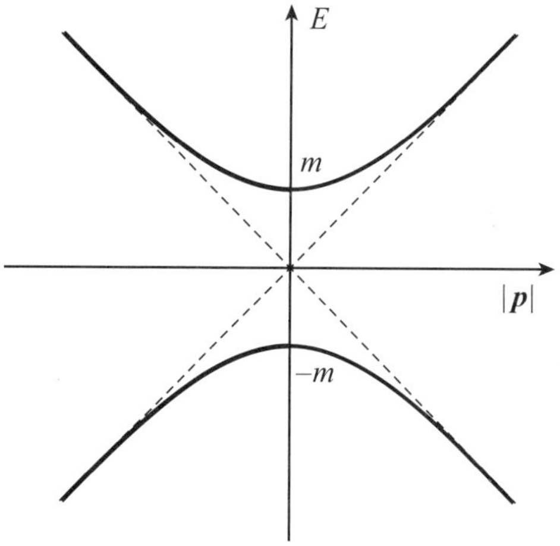

Figure 7.1 Single-particle spectrum of the Dirac theory.

The spinors $u^{(i)}$ and $v^{(i)}$ obey the orthogonality conditions

$$\begin{aligned} \bar{u}^{(\sigma)}(p)u^{(\nu)}(p) &= \delta_{\sigma\nu} \\ \bar{v}^{(\sigma)}(p)v^{(\nu)}(p) &= -\delta_{\sigma\nu} \\ \bar{u}^{(\sigma)}(p)v^{(\nu)}(p) &= 0 \end{aligned} \tag{7.22}$$

where $\bar{u} = u^\dagger \gamma_0$ and $\bar{v} = v^\dagger \gamma_0$. It is straightforward to check that the operators $\Lambda_\pm(p)$

$$\Lambda_\pm(p) = \frac{1}{2m}(\pm \not{p} + m) \tag{7.23}$$

are *projection operators* of the spinors onto the subspaces with positive (Λ_+) energy and negative (Λ_-) energy, respectively. These operators satisfy

$$\Lambda_\pm^2 = \Lambda_\pm, \quad \mathrm{Tr}\,\Lambda_\pm = 2, \quad \Lambda_+ + \Lambda_- = 1 \tag{7.24}$$

Hence, the four 4-spinors $u^{(\sigma)}$ and $v^{(\sigma)}$ are orthonormal and complete natural bases of the Hilbert space of single-particle states.

We can use these results to write the expansion of the field operator

$$\psi(x) = \int \frac{d^3p}{(2\pi)^3}\frac{m}{p_0}\sum_{\sigma=1,2}\left[a_{\sigma,+}(\boldsymbol{p})u_+^{(\sigma)}(p)e^{-ip\cdot x} + a_{\sigma,-}(\boldsymbol{p})v_-^{(\sigma)}(p)e^{ip\cdot x}\right] \tag{7.25}$$

where the coefficients $a_{\sigma,\pm}(\boldsymbol{p})$ are operators with as-yet unspecified commutation relations. The (formal) Hamiltonian for this system is

$$H = \int \frac{d^3p}{(2\pi)^3}\frac{m}{p_0}\sum_{\sigma=1,2} p_0[a_{\sigma,+}^\dagger(\boldsymbol{p})a_{\sigma,+}(p) - a_{\sigma,-}^\dagger(\boldsymbol{p})a_{\sigma,-}(\boldsymbol{p})] \tag{7.26}$$

Since the single-particle spectrum does not have a lower bound, any attempt to quantize the theory with canonical commutation relations will have the problem that the total energy of

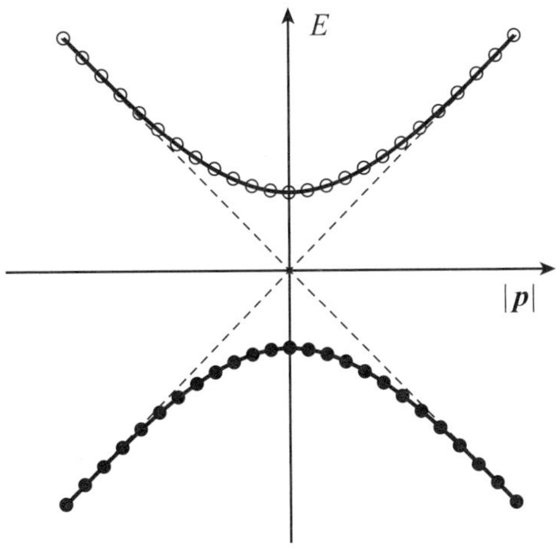

Figure 7.2 Ground state of the Dirac theory.

the system is not bounded from below. In other words, "Dirac bosons" do not have a ground state, and the system is unstable, since we can put as many bosons as we wish in states with arbitrarily large but negative energy.

Dirac realized that the simple and elegant way out of this problem was to require the electrons to obey the Pauli exclusion principle, since, in that case, there is a natural and stable ground state. However, this assumption implies that the Dirac theory must be quantized as a theory of fermions. Hence we are led to quantize the theory with *canonical anticommutation relations*

$$\begin{aligned}
&\{a_{s,\sigma}(\boldsymbol{p}), a_{s',\sigma'}(\boldsymbol{p}')\} = 0 \\
&\{a_{s,\sigma}(\boldsymbol{p}), a^{\dagger}_{s',\sigma'}(\boldsymbol{p}')\} = (2\pi)^3 \frac{p_0}{m} \delta^3(\boldsymbol{p}-\boldsymbol{p}')\delta_{ss'}\delta_{\sigma\sigma'}
\end{aligned} \tag{7.27}$$

where $s = \pm$. Let us denote by $|0\rangle$ the state annihilated by the operators $a_{s,\sigma}(\boldsymbol{p})$:

$$a_{s,\sigma}(\boldsymbol{p})|0\rangle = 0 \tag{7.28}$$

We will see now that this state is not the vacuum (or ground state) of the Dirac theory.

7.1.1 Ground state and normal ordering

We will show now that the *ground state* or *vacuum* $|\text{vac}\rangle$ is the state in which *all the negative energy states are filled* (as shown in figure 7.2):

$$|\text{vac}\rangle = \prod_{\sigma,\boldsymbol{p}} a^{\dagger}_{-,\sigma}(\boldsymbol{p})|0\rangle \tag{7.29}$$

Let us normal order all the operators relative to the vacuum state $|\text{vac}\rangle$. This amounts to a particle-hole transformation for the negative energy states. Thus, we define the fermion creation and annihilation operators $b_\sigma(\boldsymbol{p}), b^{\dagger}_\sigma(\boldsymbol{p})$ and $d_\sigma(\boldsymbol{p}), d^{\dagger}_\sigma(\boldsymbol{p})$ to be

$$b_\sigma(\mathbf{p}) = a_{\sigma,+}(\mathbf{p})$$
$$d_\sigma(\mathbf{p}) = a^\dagger_{\sigma,-}(\mathbf{p}) \tag{7.30}$$

which obey

$$b_\sigma(\mathbf{p})|\text{vac}\rangle = d_\sigma(\mathbf{p})|\text{vac}\rangle = 0 \tag{7.31}$$

The Hamiltonian now reads

$$H = \int \frac{d^3p}{(2\pi)^3}\frac{m}{p_0}p_0 \sum_{\sigma=1,2}[b^\dagger_\sigma(\mathbf{p})b_\sigma(\mathbf{p}) - d_\sigma(\mathbf{p})d^\dagger_\sigma(\mathbf{p})] \tag{7.32}$$

We now normal order $\hat{H}$ relative to the vacuum state

$$H = :H: + E_0 \tag{7.33}$$

with a normal-ordered Hamiltonian

$$:H: = \int \frac{d^3p}{(2\pi)^3}\frac{m}{p_0} \sum_{\sigma=1,2} p_0\left[b^\dagger_\sigma(\mathbf{p})b_\sigma(\mathbf{p}) + d^\dagger_\sigma(\mathbf{p})d_\sigma(\mathbf{p})\right] \tag{7.34}$$

The constant E_0 is the (negative and divergent) ground state energy,

$$E_0 = -2V \int d^3p\,\sqrt{\mathbf{p}^2 + m^2} \tag{7.35}$$

similar to the expression we already countered in the Klein-Gordon theory, but with opposite sign. The factor of 2 is due to spin.

In terms of the operators b_σ and d_σ, the Dirac field has the mode expansion

$$\psi_\alpha(x) = \int \frac{d^3p}{(2\pi)^3}\frac{m}{p_0} \sum_{\sigma=1,2}\left[b_\sigma(\mathbf{p})u^{(\sigma)}_\alpha(\mathbf{p})e^{-ip\cdot x} + d^\dagger_\sigma(\mathbf{p})v^{(\sigma)}_\alpha(\mathbf{p})e^{ip\cdot x}\right] \tag{7.36}$$

which satisfies equal-time canonical anticommutation relations

$$\{\psi_\alpha(\mathbf{x}), \psi^\dagger_\beta(\mathbf{x}')\} = \delta_{\alpha\beta}\delta^3(\mathbf{x} - \mathbf{x}')$$
$$\{\psi_\alpha(\mathbf{x}), \psi_\beta(\mathbf{x}')\} = \{\psi^\dagger_\alpha(\mathbf{x}), \psi^\dagger_\beta(\mathbf{x}')\} = 0 \tag{7.37}$$

7.1.2 One-particle states

The excitations of this theory can be constructed by using the same methods employed for nonrelativistic many-particle systems. Let us first construct the total four-momentum operator P^μ:

$$P^\mu = \int d^3x\, T^{0\mu} = \int \frac{d^3p}{(2\pi)^3}\frac{m}{p_0}p^\mu \sum_{\sigma=1,2} :b^\dagger_\sigma(\mathbf{p})b_\sigma(\mathbf{p}) - d_\sigma(\mathbf{p})d^\dagger_\sigma(\mathbf{p}): \tag{7.38}$$

Hence

$$:P^{\mu}:=\int \frac{d^3p}{(2\pi)^3}\frac{m}{p_0}p^{\mu}\sum_{\sigma=1,2}\left[b_{\sigma}^{\dagger}(\boldsymbol{p})b_{\sigma}(\boldsymbol{p})+d_{\sigma}^{\dagger}(\boldsymbol{p})d_{\sigma}(\boldsymbol{p})\right] \tag{7.39}$$

The states $b_{\sigma}^{\dagger}(\boldsymbol{p})|\text{vac}\rangle$ and $d_{\sigma}^{\dagger}(\boldsymbol{p})|\text{vac}\rangle$ have energy $p_0=\sqrt{\boldsymbol{p}^2+m^2}$ and momentum $\boldsymbol{p}$:

$$\begin{aligned}
:H:b_{\sigma}^{\dagger}(\boldsymbol{p})|\text{vac}\rangle &= p_0 b_{\sigma}^{\dagger}(\boldsymbol{p})|\text{vac}\rangle\\
:H:d_{\sigma}^{\dagger}(\boldsymbol{p})|\text{vac}\rangle &= p_0 d_{\sigma}^{\dagger}(\boldsymbol{p})|\text{vac}\rangle\\
:P^i:b_{\sigma}^{\dagger}(\boldsymbol{p})|\text{vac}\rangle &= p^i b_{\sigma}^{\dagger}(\boldsymbol{p})|\text{vac}\rangle\\
:P^i:d_{\sigma}^{\dagger}(\boldsymbol{p})|\text{vac}\rangle &= p^i d_{\sigma}^{\dagger}(\boldsymbol{p})|\text{vac}\rangle
\end{aligned} \tag{7.40}$$

We see that there are four different states that have the same energy and momentum. Let us find quantum numbers to classify these states.

7.1.3 Spin

The angular momentum tensor $\mathcal{M}_{\mu\nu\lambda}$ for the Dirac theory is

$$\mathcal{M}_{\mu\nu\lambda}=\int d^3x\,\bar{\psi}(x)\,\gamma^{\mu}\left[i\left(x^{\nu}\partial^{\lambda}-x^{\lambda}\partial^{\nu}\right)+\frac{1}{2}\sigma^{\nu\lambda}\right]\psi(x) \tag{7.41}$$

where $\sigma^{\nu\lambda}$ is the matrix

$$\sigma^{\nu\lambda}=\frac{i}{2}\left[\gamma^{\nu},\gamma^{\lambda}\right] \tag{7.42}$$

The conserved angular momentum $J^{\nu\lambda}$ is

$$J^{\nu\lambda}=\mathcal{M}^{0\nu\lambda}=\int d^3x\,\psi^{\dagger}(x)\left[i(x^{\nu}\partial^{\lambda}-x^{\lambda}\partial^{\nu})+\frac{1}{2}\sigma^{\nu\lambda}\right]\psi(x) \tag{7.43}$$

In particular, out of its space components J^{ij}, we can construct the total angular momentum three-vector $\boldsymbol{J}$:

$$J^i=\frac{1}{2}\epsilon^{ijk}J^{jk}=\int d^3x\,\psi^{\dagger}\left(i\epsilon^{ijk}x^j\partial^k+\frac{1}{2}\epsilon^{ijk}\sigma^{jk}\right)\psi \tag{7.44}$$

It is easy to check that, in the Dirac representation, the last term represents the spin.

In the quantized theory, the angular momentum operator is

$$\boldsymbol{J}=\boldsymbol{L}+\boldsymbol{S} \tag{7.45}$$

where $\boldsymbol{L}$ is the *orbital angular momentum*

$$\boldsymbol{L}=\int d^3x\,\psi^{\dagger}(\boldsymbol{x})\boldsymbol{x}\times i\boldsymbol{\partial}\psi(\boldsymbol{x}) \tag{7.46}$$

while $\boldsymbol{S}$ is the *spin*

$$\boldsymbol{S}=\int d^3x\,\psi^{\dagger}(\boldsymbol{x})\,\boldsymbol{\Sigma}\,\psi(\boldsymbol{x}) \tag{7.47}$$

where $\boldsymbol{\Sigma}$ is the 4×4 matrix

$$\boldsymbol{\Sigma}=\frac{1}{2}\begin{pmatrix}\boldsymbol{\sigma} & 0\\ 0 & \boldsymbol{\sigma}\end{pmatrix}\equiv\frac{1}{2}\boldsymbol{\sigma} \tag{7.48}$$

To measure the spin polarization of a state, we first go to the rest frame in which $\boldsymbol{p}=0$. In this frame, we can consider the four-vector W^μ:

$$W^\mu=(0,m\boldsymbol{\Sigma}) \tag{7.49}$$

Let n^μ be the space-like 4-vector $n^\mu=(0,\boldsymbol{n})$, where $\boldsymbol{n}$ has unit length. Thus, $n^\mu n_\mu=-1$. We will use n^μ to fix the direction of polarization in the rest frame.

The scalar product $W_\mu n^\mu$ is a Lorentz-invariant scalar, and hence, its value is independent of the choice of frame. In the rest frame, we have

$$W_\mu n^\mu=-m\boldsymbol{n}\cdot\boldsymbol{\Sigma}\equiv-\frac{m}{2}\vec{n}\cdot\boldsymbol{\sigma}=-\frac{m}{2}\begin{pmatrix}\boldsymbol{n}\cdot\boldsymbol{\sigma} & 0\\ 0 & \boldsymbol{n}\cdot\boldsymbol{\sigma}\end{pmatrix} \tag{7.50}$$

In particular, if $\boldsymbol{n}=e_z$, then $W_\mu n^\mu$ is

$$W_\mu n^\mu=-\frac{m}{2}\begin{pmatrix}\sigma_3 & 0\\ 0 & \sigma_3\end{pmatrix} \tag{7.51}$$

which is diagonal. The operator $-\frac{1}{m}W\cdot n$ is a Lorentz scalar that measures the spin polarization:

$$\begin{aligned}
-\frac{1}{m}W\cdot nu^{(1)}(p)&=+\frac{1}{2}u^{(1)}(p)\\
-\frac{1}{m}W\cdot nu^{(2)}(p)&=-\frac{1}{2}u^{(2)}(p)\\
-\frac{1}{m}W\cdot nv^{(1)}(p)&=+\frac{1}{2}v^{(1)}(p)\\
-\frac{1}{m}W\cdot nv^{(2)}(p)&=-\frac{1}{2}v^{(2)}(p)
\end{aligned} \tag{7.52}$$

It is straightforward to check that $-\frac{1}{m}W\cdot n$ is the Lorentz scalar

$$-\frac{1}{m}W\cdot n=\frac{1}{4m}\epsilon_{\mu\nu\lambda\rho}n^\mu p^\nu\sigma^{\lambda\rho}=\frac{1}{2m}\gamma_5\,\not{n}\,\not{p} \tag{7.53}$$

which enables us to write the spin projection operator $P(n)$,

$$P(n)=\frac{1}{2}(I+\gamma_5\,\not{n}) \tag{7.54}$$

where we used the following relations:

$$\begin{aligned}
\frac{1}{2m}\gamma_5\,\not{n}\,\not{p}\,u^{(\sigma)}(p)&=\frac{1}{2}\gamma_5\,\not{n}\,u^{(\sigma)}(p)=(-1)^\sigma\frac{1}{2}u^{(\sigma)}(p)\\
\frac{1}{2m}\gamma_5\,\not{n}\,\not{p}\,v^{(\sigma)}(p)&=-\frac{1}{2}\gamma_5\,\not{n}\,v^{(\sigma)}(p)=(-1)^\sigma\frac{1}{2}v^{(\sigma)}(p)
\end{aligned} \tag{7.55}$$

7.1.4 Charge

The Dirac Lagrangian is invariant under the global (phase) transformation

$$\begin{aligned} \psi &\to \psi' = e^{i\alpha}\psi \\ \bar{\psi} &\to \bar{\psi}' = e^{-i\alpha}\bar{\psi} \end{aligned} \tag{7.56}$$

Consequently, it has a locally conserved current j^μ,

$$j^\mu = \bar{\psi}\gamma^\mu\psi \tag{7.57}$$

which is also locally gauge invariant. As a result, it has a conserved total charge $Q = -e \int d^3x j^0(x)$. The corresponding operator in the quantized theory Q is

$$Q = -e\int d^3x\, j^0(x) = -e\int d^3x\, \psi^\dagger(x)\psi(x) \tag{7.58}$$

The total charge operator Q commutes with the Dirac Hamiltonian $\hat{H}$:

$$[Q, H] = 0 \tag{7.59}$$

Hence, the eigenstates of the Hamiltonian H have a well-defined charge.

In terms of the creation and annihilation operators, the total charge operator Q becomes

$$Q = -e\int \frac{d^3p}{(2\pi)^3}\frac{m}{p_0}\sum_{\sigma=1,2}\left(b_\sigma^\dagger(\mathbf{p})b_\sigma(\mathbf{p}) + d_\sigma(\mathbf{p})d_\sigma^\dagger(\mathbf{p})\right) \tag{7.60}$$

which is not normal-ordered relative to $|\text{vac}\rangle$. The normal-ordered charge operator $:Q:$ is

$$:Q: = -e\int \frac{d^3p}{(2\pi)^3}\frac{m}{p_0}\sum_{\sigma=1,2}\left[b_\sigma^\dagger(\mathbf{p})b_\sigma(\mathbf{p}) - d_\sigma^\dagger(\mathbf{p})d_\sigma(\mathbf{p})\right] \tag{7.61}$$

and we can write

$$Q = :Q: + Q_{\text{vac}} \tag{7.62}$$

where Q_{vac} is the unobservable (and divergent) vacuum charge

$$Q_{\text{vac}} = -eV\int \frac{d^3p}{(2\pi)^3} \tag{7.63}$$

V being the volume of space. From now on, we will *define* the charge to be the subtracted charge operator

$$:Q: = Q - Q_{\text{vac}} \tag{7.64}$$

which annihilates the vacuum state

$$:Q:\, |\text{vac}\rangle = 0 \tag{7.65}$$

(i.e., the vacuum is neutral). In other words, we measure the charge of a state relative to the vacuum charge, which we define to be zero. Equivalently, this amounts to a definition of the order of the operators in $:Q:$

$$:Q:=-e\int d^3x\frac{1}{2}[\psi^\dagger(x),\psi(x)] \tag{7.66}$$

The one-particle states $b_\sigma^\dagger|\text{vac}\rangle$ and $d_\sigma^\dagger|\text{vac}\rangle$ have well-defined charge:

$$\begin{aligned} :Q:b_\sigma^\dagger(\boldsymbol{p})|\text{vac}\rangle &= -\,eb_\sigma^\dagger(\boldsymbol{p})|\text{vac}\rangle \\ :Q:d_\sigma^\dagger(\boldsymbol{p})|\text{vac}\rangle &= +\,ed_\sigma^\dagger(\boldsymbol{p})|\text{vac}\rangle \end{aligned} \tag{7.67}$$

Hence we identify the state $b_\sigma^\dagger(\boldsymbol{p})|\text{vac}\rangle$ with an *electron* of charge $-e$, spin σ, momentum $\boldsymbol{p}$, and energy $p_0=\sqrt{\boldsymbol{p}^2+m^2}$. Similarly, the state $d_\sigma^\dagger(\boldsymbol{p})|\text{vac}\rangle$ is a *positron* with the same quantum numbers and energy of the electron but with *positive charge* $+e$.

7.1.5 The Dirac energy-momentum tensor

The energy-momentum tensor of the Dirac theory is obtained following the standard approach presented in section 3.8. On general grounds, we expect that the energy-momentum tensor should be given by

$$T^{\mu\nu}=i\bar\psi\gamma^\mu\partial^\nu\psi-\mathcal{L} \tag{7.68}$$

Using the equation of motion of the free Dirac field (i.e., the Dirac equation), we find that the energy-momentum tensor for the Dirac theory reduces to

$$T^{\mu\nu}=i\bar\psi\gamma^\mu\partial^\nu\psi \tag{7.69}$$

From this expression, it follows that the Hamiltonian is

$$H=\int d^3x\,T^{00}i\bar\psi\gamma^0\partial^0\psi=\int d^3x\psi^\dagger\gamma^0(-i\boldsymbol{\gamma}\cdot\boldsymbol{\partial}+m)\psi \tag{7.70}$$

which is indeed the Hamiltonian of the Dirac field.

7.1.6 Causality and the spin-statistics connection

Let us finally discuss the question of causality and the spin-statistics connection in the Dirac theory. To this end, we consider the *anticommutator* of two Dirac fields at different times:

$$i\Delta_{\alpha\beta}(x-y)=\{\psi_\alpha(x),\bar\psi_\beta(y)\} \tag{7.71}$$

By using the field expansion, we obtain the expression

$$i\Delta_{\alpha\beta}(x-y)=\int\frac{d^3p}{(2\pi)^3}\frac{m}{p_0}\sum_{\sigma=1,2}\left[e^{-ip\cdot(x-y)}u_\alpha^{(\sigma)}(p)\bar u_\beta^\sigma(p)+e^{ip\cdot(x-y)}v_\alpha^{(\sigma)}(p)\bar v_\beta^{(\sigma)}(p)\right] \tag{7.72}$$

By using the (completeness) identities,

$$\sum_{\sigma=1,2} u_\alpha^{(\sigma)}(p)\bar{u}_\beta^{(\sigma)}(p) = \left(\frac{\not{p}+m}{2m}\right)_{\alpha\beta}$$
$$\sum_{\sigma=1,2} v_\alpha^{(\sigma)}(p)\bar{v}_\beta^{(\sigma)}(p) = \left(\frac{\not{p}-m}{2m}\right)_{\alpha\beta} \tag{7.73}$$

we can write the anticommutator in the form

$$i\Delta_{\alpha\beta}(x-y) = \int \frac{d^3p}{(2\pi)^3}\frac{m}{p_0}\left[\left(\frac{\not{p}+m}{2m}\right)_{\alpha\beta} e^{-ip\cdot(x-y)} + \left(\frac{\not{p}-m}{2m}\right)_{\alpha\beta} e^{ip\cdot(x-y)}\right] \tag{7.74}$$

After some straightforward algebra, we obtain the result

$$\begin{aligned} i\Delta_{\alpha\beta}(x-y) &= \int \frac{d^3p}{(2\pi)^3}\frac{1}{2p_0}(i\not{\partial}_x + m)\left[e^{-ip\cdot(x-y)} - e^{ip\cdot(x-y)}\right] \\ &= (i\not{\partial}_x + m)_{\alpha\beta}\int \frac{d^3p}{(2\pi)^3 2p_0}\left[e^{-ip\cdot(x-y)} - e^{ip\cdot(x-y)}\right] \end{aligned} \tag{7.75}$$

We recognize the integral on the r.h.s. of eq. (7.75) to be the commutator of two free scalar (Klein-Gordon) fields, $\Delta_{KG}(x-y)$.

Hence, the anticommutator of two Dirac fields of the Dirac theory is

$$i\Delta_{\alpha\beta}(x-y) = (i\not{\partial} + m)_{\alpha\beta}\, i\Delta_{KG}(x-y) \tag{7.76}$$

Since $\Delta_{KG}(x-y)$ vanishes at space-like separations, so does $\Delta_{\alpha\beta}(x-y)$. Hence, the Dirac theory quantized with anticommutators obeys causality.

However, had we quantized the Dirac theory with commutators (which, as we saw, leads to a theory without a ground state), we would have also found a violation of causality. Indeed, we would have obtained instead the result

$$\Delta_{\alpha\beta}(x-y) = (i\not{\partial} + m)_{\alpha\beta}\tilde{\Delta}(x-y) \tag{7.77}$$

where $\tilde{\Delta}(x-y)$ is given by

$$\tilde{\Delta}(x-y) = \int \frac{d^3p}{(2\pi)^3 2p_0}\left(e^{-ip\cdot(x-y)} + e^{ip\cdot(x-y)}\right) \tag{7.78}$$

which does not vanish at space-like separations. Instead, at equal times and at long distances, $\tilde{\Delta}(x-y)$ decays as

$$\tilde{\Delta}(R,0) \simeq \frac{e^{-mR}}{R^2}, \qquad \text{for } mR \gg 1 \tag{7.79}$$

Thus, if the Dirac theory were to be quantized with commutators, the field operators would not commute at equal times at distances shorter than the Compton wavelength. This would be a violation of locality. The same result holds in the theory of the scalar field if it is quantized with anticommutators.

These results can be summarized in the spin-statistics theorem: Fields with half-integer spin must be quantized as fermions (i.e., obey canonical anticommutation relations), whereas fields with integer spin must be quantized as bosons (i.e., obey canonical commutation relations). If a field theory is quantized with the wrong spin-statistics connection, either the theory becomes nonlocal, with violations of causality, and/or it does not have a ground state, or it contains states in its spectrum with negative norms. Notice that the arguments we have used were derived for free local theories. It is a highly nontrivial task to prove that the spin-statistics connection also remains valid for interacting theories. Although this can be done by making sufficiently strong assumptions about the behavior of perturbation theory, in reality, the spin-statistics connection must be regarded as an axiom of local relativistic QFTs.

7.2 The propagator of the Dirac spinor field

Let us now compute the propagator for a spinor field $\psi_\alpha(x)$. We will find that it is essentially the Green function for the Dirac operator. The propagator is defined by

$$S_{\alpha\beta}(x-x') = -i\langle \text{vac}|T\psi_\alpha(x)\bar{\psi}_\beta(x')|\text{vac}\rangle \tag{7.80}$$

where we have used the *time-ordered product* of two fermionic field operators, which is defined by

$$T\psi_\alpha(x)\bar{\psi}_\beta(x') = \theta(x_0-x_0')\psi_\alpha(x)\bar{\psi}_\beta(x') - \theta(x_0'-x_0)\bar{\psi}_\beta(x')\psi_\alpha(x) \tag{7.81}$$

Notice the change in sign with respect to the time-ordered product of bosonic operators. The sign change reflects the anticommutation properties of the field. In other words, inside a time-ordered product, Fermi fields behave as if they were anticommuting c-numbers.

We will show now that this propagator is closely connected to the propagator of the free scalar field (i.e., the Green function for the Klein-Gordon operator ∂^2+m^2).

By acting with the Dirac operator on $S_{\alpha\beta}(x-x')$, we find

$$\left(i\partial\!\!\!/ - m\right)_{\alpha\beta} S_{\beta\lambda}(x-x') = -i\left(i\partial\!\!\!/ - m\right)_{\alpha\beta} \langle \text{vac}|T\psi_\beta(x)\bar{\psi}_\lambda(x')|\text{vac}\rangle \tag{7.82}$$

Let us now use

$$\frac{\partial}{\partial x_0}\theta(x_0-x_0') = \delta(x_0-x_0') \tag{7.83}$$

and the fact that the equation of motion of the Heisenberg field operators ψ_α is the Dirac equation,

$$\left(i\partial\!\!\!/ - m\right)_{\alpha\beta}\psi_\beta(x) = 0 \tag{7.84}$$

to show that

$$\begin{aligned}(i\partial\!\!\!/ - m)_{\alpha\beta} S_{\beta\lambda}(x-x') &= -i\langle \text{vac}|T\left(i\partial\!\!\!/ - m\right)_{\alpha\beta}\psi_\beta(x)\bar{\psi}_\lambda(x')|\text{vac}\rangle \\ &\quad + \delta(x_0 - x_0')\Big(\langle \text{vac}|\gamma^0_{\alpha\beta}\psi_\beta(x)\bar{\psi}_\lambda(x')|\text{vac}\rangle \\ &\qquad + \langle \text{vac}|\bar{\psi}_\lambda(x')\gamma^0_{\alpha\beta}\psi_\beta(x)|\text{vac}\rangle\Big) \\ &= \delta(x_0 - x_0')\gamma^0_{\alpha\beta}\langle \text{vac}|\left\{\psi_\beta(x), \psi^\dagger_\nu(x')\right\}|\text{vac}\rangle\gamma^0_{\nu\lambda} \\ &= \delta(x_0 - x_0')\delta^3(\boldsymbol{x} - \boldsymbol{x}')\delta_{\alpha\lambda}\end{aligned} \tag{7.85}$$

Therefore we find that $S_{\beta\lambda}(x-x')$ is the solution of the equation

$$\left(i\partial\!\!\!/ - m\right)_{\alpha\beta} S_{\beta\lambda}(x-x') = \delta^4(x-x')\delta_{\alpha\lambda} \tag{7.86}$$

Hence, $S_{\beta\lambda}(x-x') = -i\langle \text{vac}|T\psi_\beta(x)\bar{\psi}_\lambda(x')|\text{vac}\rangle$ is the Green function of the Dirac operator.

We saw before that there is a close connection between the Dirac and the Klein-Gordon operators. We will now use this connection to relate their propagators. Let us write the Green function $S_{\alpha\lambda}(x-x')$ in the form

$$S_{\alpha\lambda}(x-x') = \left(i\partial\!\!\!/ + m\right)_{\alpha\beta} G_{\beta\lambda}(x-x') \tag{7.87}$$

Since $S_{\alpha\lambda}(x-x')$ satisfies eq. (7.86), we find that

$$\left(i\partial\!\!\!/ - m\right)_{\alpha\beta} S_{\beta\lambda}(x-x') = \left(i\partial\!\!\!/ - m\right)_{\alpha\beta}\left(i\partial\!\!\!/ + m\right)_{\beta\nu} G_{\nu\lambda}(x-x') \tag{7.88}$$

But

$$\left(i\partial\!\!\!/ - m\right)_{\alpha\beta}\left(i\partial\!\!\!/ + m\right)_{\beta\nu} = -\left(\partial^2 + m^2\right)\delta_{\alpha\nu} \tag{7.89}$$

Hence, $G_{\alpha\nu}(x-x')$ must satisfy

$$-\left(\partial^2 + m^2\right) G_{\alpha\nu}(x-x') = \delta^4(x-x')\delta_{\alpha\nu} \tag{7.90}$$

Therefore, $G_{\alpha\nu}(x-x')$ is given by

$$G_{\alpha\nu}(x-x') = -G^{(0)}(x-x')\delta_{\alpha\nu} \tag{7.91}$$

where $G^{(0)}(x-x')$ is the propagator for a free massive scalar field (i.e., the Green function of the Klein-Gordon equation):

$$\left(\partial^2 + m^2\right) G^{(0)}(x-x') = \delta^4(x-x') \tag{7.92}$$

We then conclude that the Dirac propagator $S_{\alpha\beta}(x-x')$ and the Klein-Gordon propagator $G^{(0)}(x-x')$ are related by

$$S_{\alpha\beta}(x-x') = -\left(i\partial\!\!\!/ + m\right)_{\alpha\beta} G^{(0)}(x-x') \tag{7.93}$$

In particular, this relationship implies that they have essentially the same asymptotic behaviors that we discussed for the free scalar field: a power-law behavior at short distances (albeit with a different power) and exponential (or oscillatory) behavior at large distances. The spinor structure of the Dirac propagator is determined by the operator in front of $G^{(0)}(x-x')$ in eq. (7.93).

In momentum space, the Feynman propagator for the Dirac field, given by eq. (7.93), becomes

$$S_{\alpha\beta}(p) = \left(\frac{\not p + m}{p^2 - m^2 + i\epsilon}\right)_{\alpha\beta} \tag{7.94}$$

Hence we get the same pole structure in the time-ordered propagator as we did for the Klein-Gordon field.

7.3 Discrete symmetries of the Dirac theory

We now discuss three important discrete symmetries in relativistic field theories: *charge conjugation*, *parity*, and *time reversal*. These discrete symmetries have a different role, and a different standing, than the continuous symmetries discussed before. In a relativistic QFT, the ground state (i.e., the vacuum) must be invariant under continuous Lorentz transformations, but it may not be invariant under $\mathcal{C}$, $\mathcal{P}$, or $\mathcal{T}$. However, in a local relativistic QFT, the product $\mathcal{CPT}$ is always a good symmetry. This is in fact an axiom of relativistic local QFT. Thus, although $\mathcal{C}$, $\mathcal{P}$, or $\mathcal{T}$ may or may not be good symmetries of the vacuum state, $\mathcal{CPT}$ *must* be a good symmetry. As in the case of the symmetries discussed in section 4.4 (in the case of the free scalar field), these symmetries must also be realized in the Fock space of the QFT.

7.3.1 Charge conjugation

Charge conjugation is a symmetry that exchanges particles and antiparticles (or holes). Consider a Dirac field minimally coupled to an external electromagnetic field A_μ. The equation of motion for the Dirac field ψ is

$$\left(i\not\partial - e\not A - m\right)\psi = 0 \tag{7.95}$$

Let us define the charge conjugate field ψ^c

$$\psi^c(x) = \mathcal{C}\psi(x)\mathcal{C}^{-1} \tag{7.96}$$

where $\mathcal{C}$ is the (unitary) charge conjugation operator, $\mathcal{C}^{-1} = \mathcal{C}^\dagger$, such that ψ^c obeys

$$\left(i\not\partial + e\not A - m\right)\psi^c = 0 \tag{7.97}$$

The conjugate field $\bar\psi = \psi^\dagger\gamma^0$ obeys the equation

$$\bar\psi\left[\gamma^\mu\left(-i\overleftarrow{\partial}_\mu - eA_\mu\right) - m\right] = 0 \tag{7.98}$$

which, when transposed, becomes

$$\left[\gamma^{\mu\,T}\left(-i\partial_\mu - eA_\mu\right) - m\right]\bar\psi^{\,T} = 0 \tag{7.99}$$

where T is the transpose, and where we use the notation

$$\bar{\psi}^{\,T} = \gamma^{0\,T}\psi^* \tag{7.100}$$

Let C be an invertible 4×4 matrix, where C^{-1} is its inverse. Then we can write

$$C\left[\gamma^{\mu\,T}\left(-i\partial_\mu - eA_\mu\right) - m\right]C^{-1}C\bar{\psi}^{\,T} = 0 \tag{7.101}$$

such that

$$C\,(\gamma^{\mu})^T\,C^{-1} = -\gamma^{\mu} \tag{7.102}$$

Hence,

$$\left[\left(i\partial\!\!\!/ + eA\!\!\!/\right) - m\right]C\bar{\psi}^{\,T} = 0 \tag{7.103}$$

For eq. (7.103) to hold, we must have

$$C\psi C^{\dagger} = \psi^c = C\bar{\psi}^{\,T} = C\gamma^{0T}\psi^* \tag{7.104}$$

Hence the field ψ^c thus defined has positive charge $+e$.

We can find the charge conjugation matrix C explicitly:

$$C = i\gamma^2\gamma^0 = \begin{pmatrix} 0 & i\sigma^2 \\ -i\sigma^2 & 0 \end{pmatrix} = C^{-1} \tag{7.105}$$

In particular, this definition means that ψ^c is given by

$$\psi^c = C\bar{\psi}^{T} = C\gamma^{0t}\psi^* = i\gamma^2\psi^* \tag{7.106}$$

Eq. (7.106) provides us with a definition for a charge-neutral Dirac fermion (i.e., ψ represents a neutral fermion if $\psi = \psi^c$). Hence the condition is

$$\psi = i\gamma^2\psi^* \tag{7.107}$$

A Dirac fermion that satisfies the neutrality condition is known as a *Majorana fermion.*

To understand the action of C on physical states, we can look at the charge conjugate u^c of the positive energy, up spin, and charge $-e$ spinor in the rest frame

$$u = \begin{pmatrix} \varphi \\ 0 \end{pmatrix} e^{-imt} \tag{7.108}$$

which is

$$u^c = \begin{pmatrix} 0 \\ -i\gamma^2\varphi^* \end{pmatrix} e^{imt} \tag{7.109}$$

which has negative energy, down spin, and charge $+e$.

At the level of the full QFT, we will require the vacuum state $|\text{vac}\rangle$ to be invariant under charge conjugation:

$$\mathcal{C}|\text{vac}\rangle = |\text{vac}\rangle \tag{7.110}$$

How do one-particle states transform? To determine that, let us look at the action of charge conjugation $\mathcal{C}$ on the one-particle states, and demand that particle and antiparticle states be exchanged under charge conjugation

$$\begin{aligned} \mathcal{C}b_\sigma^\dagger(p)|\text{vac}\rangle &= \mathcal{C}b_\sigma^\dagger(p)\mathcal{C}^{-1}\mathcal{C}|\text{vac}\rangle \equiv d_\sigma^\dagger(p)|\text{vac}\rangle \\ \mathcal{C}d_\sigma^\dagger(p)|\text{vac}\rangle &= \mathcal{C}d_\sigma^\dagger(p)\mathcal{C}^{-1}\mathcal{C}|\text{vac}\rangle \equiv b_\sigma^\dagger(p)|\text{vac}\rangle \end{aligned} \tag{7.111}$$

Hence, for the one-particle states to satisfy these rules, it is sufficient to require that the field operators $b_\sigma(\mathbf{p})$ and $d_\sigma(\mathbf{p})$ satisfy

$$\mathcal{C}b_\sigma(\mathbf{p})\mathcal{C}^\dagger = d_\sigma(\mathbf{p}); \qquad \mathcal{C}d_\sigma(\mathbf{p})\mathcal{C}^\dagger = b_\sigma(\mathbf{p}) \tag{7.112}$$

Using

$$u_\sigma(\mathbf{p}) = -i\gamma^2\left(v_\sigma(\mathbf{p})\right)^*; \qquad v_\sigma(\mathbf{p}) = -i\gamma^2\left(u_\sigma(\mathbf{p})\right)^* \tag{7.113}$$

we find that the field operator $\psi(x)$ transforms as

$$\mathcal{C}\psi(x)\mathcal{C}^\dagger = \left(-i\bar{\psi}\gamma^0\gamma^2\right)^T \tag{7.114}$$

and

$$\mathcal{C}\bar{\psi}(x)\mathcal{C}^\dagger = \left(-i\gamma^0\gamma^2\psi\right)^T \tag{7.115}$$

In particular, the fermionic bilinears defined in section 2.5.4 satisfy the transformation laws:

$$\begin{aligned} \mathcal{C}\bar{\psi}\psi\mathcal{C}^\dagger &= +\bar{\psi}\psi, & \mathcal{C}i\bar{\psi}\gamma^5\psi\mathcal{C}^\dagger &= i\bar{\psi}\gamma^5\psi, \\ \mathcal{C}\bar{\psi}\gamma^\mu\psi\mathcal{C}^\dagger &= -\bar{\psi}\gamma^\mu\psi, & \mathcal{C}\bar{\psi}\gamma^\mu\gamma^5\psi\mathcal{C}^\dagger &= +\bar{\psi}\gamma^\mu\gamma^5\psi \end{aligned} \tag{7.116}$$

7.3.2 Parity

Define *parity* as the transformation $\mathcal{P} = \mathcal{P}^{-1}$, which reverses the momentum of a particle but not its spin. Once again, the vacuum state is invariant under parity. Thus, we must require:

$$\begin{aligned} \mathcal{P}b_\sigma^\dagger(\mathbf{p})|\text{vac}\rangle &= \mathcal{P}b_\sigma^\dagger(\mathbf{p})\mathcal{P}^{-1}\mathcal{P}|\text{vac}\rangle \equiv b_\sigma^\dagger(-\mathbf{p})|\text{vac}\rangle \\ \mathcal{P}d_\sigma^\dagger(\mathbf{p})|\text{vac}\rangle &= \mathcal{P}d_\sigma^\dagger(\mathbf{p})\mathcal{P}^{-1}\mathcal{P}|\text{vac}\rangle \equiv d_\sigma^\dagger(-\mathbf{p})|\text{vac}\rangle \end{aligned} \tag{7.117}$$

In real space, this transformation is equivalent to

$$\mathcal{P}\psi(\mathbf{x}, x_0)\mathcal{P}^{-1} = \gamma^0\psi(-\mathbf{x}, x_0), \quad \mathcal{P}\bar{\psi}(\mathbf{x}, x_0)\mathcal{P}^{-1} = \bar{\psi}(-\mathbf{x}, x_0)\gamma_0 \tag{7.118}$$

7.3.3 Time reversal

Finally, we discuss time reversal $\mathcal{T}$. Define $\mathcal{T}$ as the operator

$$\mathcal{T}e^{iHx_0}\mathcal{T}^{-1} = e^{-iHx_0} \tag{7.119}$$

We will require the time-reversal operator to be unitary in the sense that

$$\mathcal{T}^{-1} = \mathcal{T}^{\dagger} \tag{7.120}$$

However, we also require the operator to act on c-numbers as complex conjugation, that is,

$$\mathcal{T}\,(\text{c-number}) = (\text{c-number})^{*}\,\mathcal{T} \tag{7.121}$$

An operator with these properties is said to be *antilinear* (anti-unitary).

As a result, time reversal is the operator that reverses both the momentum and the spin of the particles:

$$\mathcal{T} b_{\sigma}(\boldsymbol{p})\mathcal{T}^{\dagger} = b_{-\sigma}(-\boldsymbol{p}), \qquad \mathcal{T} d_{\sigma}(\boldsymbol{p})\mathcal{T}^{\dagger} = d_{-\sigma}(-\boldsymbol{p}) \tag{7.122}$$

while leaving the vacuum state invariant:

$$\mathcal{T}|\text{vac}\rangle = |\text{vac}\rangle \tag{7.123}$$

In real space, this implies:

$$\mathcal{T}\psi(\boldsymbol{x}, x_0)\mathcal{T}^{\dagger} = -\gamma^1\gamma^3\,\psi^{*}(\boldsymbol{x}, -x_0) \tag{7.124}$$

7.4 Chiral symmetry

Let us now discuss a global symmetry specific to theories of spinors and known as *chiral symmetry*. Consider again the Dirac Lagrangian:

$$\mathcal{L} = \bar{\psi}\left(i\partial\!\!\!/ - m\right)\psi \tag{7.125}$$

Define the *chiral transformation*

$$\psi' = e^{i\gamma_5\theta}\,\psi \tag{7.126}$$

where θ is a *constant phase* in the range $0 \leq \theta < 2\pi$. We wish to find the Lagrangian for the new transformed field ψ. From

$$\mathcal{L} = \bar{\psi} e^{i\gamma_5\theta}\left(i\gamma^{\mu}\partial_{\mu} - m\right) e^{i\gamma_5\theta}\,\psi \tag{7.127}$$

and the fact that

$$\left\{\gamma_{\mu}, \gamma_5\right\} = 0 \tag{7.128}$$

after some simple algebra, which uses the identity

$$\gamma^{\mu} e^{i\gamma_5\theta} = e^{-i\gamma_5\theta}\gamma^{\mu} \tag{7.129}$$

we find that the field ψ satisfies a modified Dirac Lagrangian of the form

$$\mathcal{L} = \bar{\psi}\left(i\partial\!\!\!/ - m e^{2i\gamma_5\theta}\right)\psi \tag{7.130}$$

Thus for $m \neq 0$, the form of the Dirac Lagrangian changes under a chiral transformation. However, if the theory is massless, the Dirac theory has an exact global chiral symmetry.

It is also instructive to determine how various fermion bilinears transform under the chiral transformation. We find that the Dirac mass, $\bar{\psi}\psi$, and the axial mass, $i\bar{\psi}\gamma^5\psi$, transform as follows under a chiral transformation:

$$\begin{aligned} \bar{\psi}'\psi' &= \cos(2\theta)\bar{\psi}\psi + \sin(2\theta)i\bar{\psi}\gamma_5\psi \\ i\bar{\psi}'\gamma^5\psi' &= -\sin(2\theta)\bar{\psi}\psi + \cos(2\theta)i\bar{\psi}\gamma_5\psi \end{aligned} \tag{7.131}$$

and, hence, are not invariant under this transformation. Instead, the Dirac mass, $\bar{\psi}\psi$, and the γ_5 mass, $i\bar{\psi}\gamma_5\psi$, transform (rotate) into each other. Finally, note that, although the Dirac Lagrangian is not chiral invariant if $m \neq 0$, the Dirac current

$$\bar{\psi}'\gamma^\mu\psi' = \bar{\psi}\gamma^\mu\psi \tag{7.132}$$

is invariant under the chiral transformation, and so is the coupling of the Dirac field to a gauge field.

7.5 Massless fermions

Let us look at the massless limit of the Dirac equation in more detail. Historically, this problem grew out of the study of neutrinos (which we now know are not necessarily massless, at least not all of them). For an eigenstate of 4-momentum p_μ, the Dirac equation is

$$(\not{p} - m)\,\psi(p) = 0 \tag{7.133}$$

In the massless limit $m = 0$, the Dirac equation simply becomes

$$\not{p}\psi(p) = 0 \tag{7.134}$$

which is equivalent to

$$\gamma_5\gamma_0\not{p}\psi(p) = 0 \tag{7.135}$$

On expanding in components, we find

$$\gamma_5 p_0\psi(p) = \gamma_5\gamma_0\boldsymbol{\gamma}\cdot\boldsymbol{p}\psi(p) \tag{7.136}$$

However, since

$$\gamma_5\gamma_0\boldsymbol{\gamma} = \begin{bmatrix} \boldsymbol{\sigma} & 0 \\ 0 & \boldsymbol{\sigma} \end{bmatrix} = \boldsymbol{\Sigma} \tag{7.137}$$

we can write

$$\gamma_5 p_0\psi(p) = \boldsymbol{\Sigma}\cdot\boldsymbol{p}\psi(p) \tag{7.138}$$

Thus, the *chirality* γ_5 is equivalent to the *helicity* $\boldsymbol{\Sigma}\cdot\boldsymbol{p}$ of the state. (But only in the massless limit!) This suggests the introduction of the chiral basis in which γ_5 is diagonal:

$$\gamma_0 = \begin{pmatrix} 0 & -I \\ -I & 0 \end{pmatrix}, \qquad \gamma_5 = \begin{pmatrix} I & 0 \\ 0 & -I \end{pmatrix}, \qquad \boldsymbol{\gamma} = \begin{pmatrix} 0 & \boldsymbol{\sigma} \\ -\boldsymbol{\sigma} & 0 \end{pmatrix} \tag{7.139}$$

In this basis, the massless Dirac equation becomes

$$\begin{pmatrix} \boldsymbol{\sigma} & 0 \\ 0 & \boldsymbol{\sigma} \end{pmatrix} \cdot \boldsymbol{p}\psi(p) = \begin{pmatrix} I & 0 \\ 0 & -I \end{pmatrix} p_0 \psi(p) \tag{7.140}$$

Let us write the 4-spinor ψ in terms of two 2-spinors of the form

$$\psi = \begin{pmatrix} \psi_R \\ \psi_L \end{pmatrix} \tag{7.141}$$

in terms of which the Dirac equation decomposes into two separate equations for each chiral component, ψ_R and ψ_L. Thus the *right-handed* (positive chirality) component ψ_R satisfies the Weyl equation

$$\left(\boldsymbol{\sigma} \cdot \boldsymbol{p} - p_0\right) \psi_R = 0 \tag{7.142}$$

while the *left-handed* (negative chirality) component satisfies instead

$$\left(\boldsymbol{\sigma} \cdot \boldsymbol{p} + p_0\right) \psi_L = 0 \tag{7.143}$$

Hence, one massless Dirac spinor is equivalent to two Weyl spinors.

We conclude that a theory of free massless Dirac fermions has a global chiral symmetry. It is natural to assume that it must also have a locally conserved chiral (axial) current (as required by Noether's theorem), which we will denote by j^5_μ. An elementary calculation gives the result

$$j^5_\mu = i\bar{\psi}\gamma^5\gamma_\mu\psi \tag{7.144}$$

However, we also know that the Dirac theory has a global $U(1)$ phase symmetry which, when coupled to the vector field A_μ, becomes local $U(1)$ gauge invariance and has a locally conserved gauge current, $j_\mu = \bar{\psi}\gamma_\mu\psi$. We will see in chapter 20 that in an interacting theory, its UV divergences make it impossible for both currents to be simultaneously conserved and that, if the theory is to be gauge invariant, then the chiral current cannot be conserved. In other words, a classically conserved current may not be conserved at the quantum level. The phenomenon of nonconservation at the quantum level of a classically conserved current is known as a *quantum anomaly*. In this case, the nonconservation of the chiral current is known as the chiral (or axial) anomaly.

Exercises

7.1 The Dirac theory in a potential well

Consider the problem of trying to localize a Dirac fermion with an external electrostatic potential well of width a and depth U_0. To make matters simpler, assume that the fermions move on a line (say, the axis $x_3 \equiv x$) and that, as a result, the problem is effectively one dimensional.

1) Find the second quantized Hamiltonian for the problem of a Dirac field coupled to a static external electrostatic field $\boldsymbol{E} = -\nabla A_0(x)$, for the case of this one-dimensional geometry.

2) Find the single-particle states for the Dirac theory in the presence of the potential well

$$A_0(x) = \begin{cases} -U_0, & \text{for } |x| \leq \frac{a}{2} \\ 0, & \text{otherwise} \end{cases} \tag{7.145}$$

where $U_0 > 0$. Be careful to include both extended and bound states, and to discuss the behavior of these states as a function of the depth U_0 and of the mass m.

3) Construct the ground state and the excited state with lowest energy for the various regimes that you found in part 2 of this exercise and give their quantum numbers. Derive an expansion of the ground state in the presence of the potential in terms of the eigenstates of the same theory without the potential. Give a physical interpretation of the various terms in this expansion.

7.2 Currents in the Dirac theory

Consider a theory of massive Dirac fermions in four Minkowski spacetime dimensions.

1) Show that the gauge current $J_\mu(x) = \bar{\psi}(x)\gamma_\mu\psi(x)$ is locally conserved.
2) Show that the axial current $J^5_\mu(x) = i\bar{\psi}(x)\gamma_5\gamma_\mu\psi(x)$ is not conserved if the Dirac mass does not vanish.

7.3 Wick's theorem for Dirac fermions

This is an exercise on the use of Wick's theorem on a specific theory, the two-component free Dirac field $\psi_{\alpha,a}(x)$, with $\alpha = 1,\ldots,4$ being the components of the spinor, and $a = 1, 2$, with the global $SU(2)$ symmetry $\psi_a(x) \to U_{ab}\psi_b(x)$ (we are not making the spinor components explicit), where U_{ab} is an arbitrary 2×2 $SU(2)$ matrix.

1) Use symmetry arguments to determine which of the following vacuum expectation values are nonzero (sums over repeated indices are implied, and the spinor labels are not explicit):

a) $\langle 0|T\bar{\psi}_a(x)\psi_a(y)|0\rangle$
b) $\langle 0|T\bar{\psi}_a(x)\psi_b(y)\bar{\psi}_b(z)\psi_c(w)|0\rangle$
c) $\langle 0|T\bar{\psi}_a(x)\psi_b(y)\bar{\psi}_b(z)\psi_a(w)|0\rangle$

2) Use Wick's theorem to find expressions for the vacuum expectation values for (b) and (c) in (part 1) in terms of the value for (a).

7.4 Fermions in one dimension

Consider an application of the Dirac theory to a problem in condensed matter physics: polyacetylene. Polyacetylene is a long polymer chain $(CH)_n$. In the low-energy regime, the important electronic states are close to the Fermi energy E_F, with momentum p close to the two Fermi points $\pm p_F = \frac{\pi}{2a_0}$ (the band is half filled), the right- and left-moving components of a two-component spinor Fermi field $\psi_{R,\sigma}(x)$ and $\psi_{L,\sigma}(x)$; $\sigma = \uparrow, \downarrow$ are their spin polarizations. The important lattice vibrations (phonons) that matter have lattice momentum $2p_F = \pi/a_0$ and are described by a set of heavy classical bosons $\Delta(x)$. The low-energy Hamiltonian density is

$$\mathcal{H} = -iv_F\psi^\dagger_\sigma(x)\sigma_3\partial_x\psi_\sigma(x) + \frac{1}{2}\Delta^2(x) + g\Delta(x)\,\bar{\psi}_\sigma(x)\psi_\sigma(x) \tag{7.146}$$

In this form, this system is closely related to the Gross-Neveu model. Here g is the electron-phonon coupling constant, and σ_3 is the Pauli matrix. The fermions are a relativistic Dirac field with "speed of light" equal to the Fermi velocity, v_F, and they obey equal-time canonical anticommutation relations. Use the 2×2 Dirac matrices,

$$\gamma_0 = \sigma_2 = \begin{pmatrix} 0 & -i \\ i & 0 \end{pmatrix}, \quad \gamma_1 = i\sigma_1 = \begin{pmatrix} 0 & i \\ i & 0 \end{pmatrix}, \quad \gamma_5 = \gamma_0\gamma_1 = \sigma_3 \tag{7.147}$$

and the notation $\bar{\psi} = \psi^\dagger \gamma_0 = \psi^\dagger \sigma_2$.

1) Find the ground state vector $|\text{gnd}\rangle$ and energy of the system in this limit by calculating the constant value of Δ (it describes a staggered distortion of the lattice) for which the ground state energy is minimized. You will have to cut off some of the momentum integrals at a large cutoff momentum Λ. Note: The *spontaneous* staggered distortion of the lattice is known as *dimerization*, and this phenomenon is called the *Peierls instability.*
2) Determine the energy spectrum and quantum numbers of the single-particle electronic states in this approximation.
3) Show that the continuum Hamiltonian (and the associated Lagrangian) is invariant under the global discrete symmetry transformation

$$\psi(x) \to \gamma_5 \psi(x), \quad \Delta(x) \to -\Delta(x) \tag{7.148}$$

with $\gamma_5 = \sigma_3$. Show that, in terms of the lattice model, this transformation amounts to a shift of all fields by one lattice site. In that language, the symmetry corresponds to an ambiguity in the way the dimerized structure is placed on the lattice. Show that the operator $\bar{\psi}_\sigma(x)\psi_\sigma(x)$ is odd under the discrete symmetry, and hence, it is an order parameter.
4) Compute the ground state expectation value for the order parameter of the previous part,

$$\Delta(x) = \sum_\sigma \langle \text{gnd} | \bar{\psi}_\sigma(x) \psi_\sigma(x) | \text{gnd} \rangle \tag{7.149}$$

in the $M \to \infty$ approximation. Show that this order parameter has nonvanishing expectation value only if $\Delta \neq 0$, and establish a connection between both quantities.

8

Coherent-State Path-Integral Quantization of Quantum Field Theory

8.1 Coherent states and path-integral quantization

The path integral that we have used so far is a powerful tool, but it can only be used in theories based on canonical quantization. For example, it cannot be used in theories of fermions (relativistic or not). In this chapter, we discuss a more general approach based on the concept of coherent states. Coherent states and their application to path integrals have been widely discussed. Excellent references include the 1975 Les Houches lectures by Faddeev (1976) and the books by Perelomov (1986), Klauder (Klauder and Skagerstam, 1985), and Schulman (1981). The related concept of geometric quantization is insightfully presented in the work by Wiegmann (1989).

8.2 Coherent states

Consider a Hilbert space spanned by a complete set of harmonic oscillator states $\{|n\rangle\}$, with $n=0,\ldots,\infty$. Let $\hat{a}^\dagger$ and $\hat{a}$ be a pair of creation and annihilation operators acting on this Hilbert space and satisfying the commutation relations

$$[\hat{a}, \hat{a}^\dagger] = 1, \quad [\hat{a}^\dagger, \hat{a}^\dagger] = 0, \quad [\hat{a}, \hat{a}] = 0 \tag{8.1}$$

These operators generate the harmonic oscillator states $\{|n\rangle\}$ in the usual way,

$$|n\rangle = \frac{1}{\sqrt{n!}} \left(\hat{a}^\dagger\right)^n |0\rangle, \qquad \hat{a}|0\rangle = 0 \tag{8.2}$$

where $|0\rangle$ is the vacuum state of the oscillator.

Let us denote by $|z\rangle$ the *coherent state*

$$|z\rangle = e^{z\hat{a}^\dagger} |0\rangle, \qquad \langle z| = \langle 0| e^{\bar{z}\hat{a}} \tag{8.3}$$

where z is an arbitrary complex number, and $\bar{z}$ is its complex conjugate. The coherent state $|z\rangle$ has the defining property of being a wave packet with optimal spread (i.e., the Heisenberg uncertainty inequality is an equality for these coherent states).

How does $\hat{a}$ act on the coherent state $|z\rangle$? Consider

$$\hat{a}|z\rangle = \sum_{n=0}^{\infty} \frac{z^n}{n!} \hat{a} \left(\hat{a}^\dagger\right)^n |0\rangle \tag{8.4}$$

Since

$$\left[\hat{a}, \left(\hat{a}^\dagger\right)^n\right] = n \left(\hat{a}^\dagger\right)^{n-1} \tag{8.5}$$

we find

$$\hat{a}|z\rangle = \sum_{n=0}^{\infty} \frac{z^n}{n!} n \left(\hat{a}^\dagger\right)^{n-1} |0\rangle \equiv z\,|z\rangle \tag{8.6}$$

Therefore, $|z\rangle$ is a right eigenvector of $\hat{a}$, and z is the (right) eigenvalue.

Likewise, we get

$$\hat{a}^\dagger |z\rangle = \hat{a}^\dagger \sum_{n=0}^{\infty} \frac{z^n}{n!} \left(\hat{a}^\dagger\right)^n |0\rangle = \sum_{n=1}^{\infty} n \frac{z^{n-1}}{n!} \left(\hat{a}^\dagger\right)^n |0\rangle \tag{8.7}$$

Thus,

$$\hat{a}^\dagger |z\rangle = \partial_z |z\rangle \tag{8.8}$$

Therefore, the operators z and ∂_z provide a representation of the algebra of creation and annihilation operators.

Another quantity of interest is the overlap of two coherent states, $\langle z|z'\rangle$:

$$\langle z|z'\rangle = \langle 0|e^{\bar{z}\hat{a}}\, e^{z'\hat{a}^\dagger} |0\rangle \tag{8.9}$$

We will calculate this matrix element using the Baker-Hausdorff formulas

$$e^{\hat{A}}\, e^{\hat{B}} = e^{\hat{A}+\hat{B}+\frac{1}{2}\left[\hat{A},\hat{B}\right]} = e^{\left[\hat{A},\hat{B}\right]}\, e^{\hat{B}}\, e^{\hat{A}} \tag{8.10}$$

which hold provided the commutator $\left[\hat{A}, \hat{B}\right]$ is a c-number (i.e., it is proportional to the identity operator). Since $\left[\hat{a}, \hat{a}^\dagger\right] = 1$, we find

$$\langle z|z'\rangle = e^{\bar{z}z'}\, \langle 0|e^{z'\hat{a}^\dagger}\, e^{\bar{z}\hat{a}}|0\rangle \tag{8.11}$$

But

$$e^{\bar{z}\hat{a}}\, |0\rangle = |0\rangle, \qquad \langle 0|\, e^{z'\hat{a}^\dagger} = \langle 0| \tag{8.12}$$

Hence we get

$$\langle z|z'\rangle = e^{\bar{z}z'} \tag{8.13}$$

An arbitrary state $|\psi\rangle$ of this Hilbert space can be expanded in the harmonic oscillator basis states $\{|n\rangle\}$:

$$|\psi\rangle = \sum_{n=0}^{\infty} \frac{\psi_n}{\sqrt{n!}} |n\rangle = \sum_{n=0}^{\infty} \frac{\psi_n}{n!} \left(\hat{a}^\dagger\right)^n |0\rangle \tag{8.14}$$

The projection of the state $|\psi\rangle$ onto the coherent state $|z\rangle$ is

$$\langle z|\psi\rangle = \sum_{n=0}^{\infty} \frac{\psi_n}{n!} \langle z| \left(\hat{a}^\dagger\right)^n |0\rangle \tag{8.15}$$

Since

$$\langle z| \hat{a}^\dagger = \bar{z} \langle z| \tag{8.16}$$

we find

$$\langle z|\psi\rangle = \sum_{n=0}^{\infty} \frac{\psi_n}{n!} \bar{z}^n \equiv \psi(\bar{z}) \tag{8.17}$$

Therefore, the projection of $|\psi\rangle$ onto $|z\rangle$ is the *antiholomorphic* (i.e., *anti-analytic*) function $\psi(\bar{z})$. In other words, in this representation, the space of states $\{|\psi\rangle\}$ is in one-to-one correspondence with the space of anti-analytic functions.

In summary, the coherent states $\{|z\rangle\}$ satisfy the following properties:

$$\begin{aligned} \hat{a}|z\rangle &= z|z\rangle, & \langle z|\hat{a} &= \partial_{\bar{z}}\langle z| \\ \hat{a}^\dagger|z\rangle &= \partial_z|z\rangle, & \langle z|\hat{a}^\dagger &= \bar{z}\langle z| \\ \langle z|\psi\rangle &= \psi(\bar{z}), & \langle \psi|z\rangle &= \bar{\psi}(z) \end{aligned} \tag{8.18}$$

Next we will prove the *resolution of identity:*

$$\hat{I} = \int \frac{dz d\bar{z}}{2\pi i} e^{-z\bar{z}} |z\rangle \langle z| \tag{8.19}$$

Let $|\psi\rangle$ and $|\phi\rangle$ be two arbitrary states,

$$|\psi\rangle = \sum_{n=0}^{\infty} \frac{\psi_n}{\sqrt{n!}} |n\rangle, \qquad |\phi\rangle = \sum_{n=0}^{\infty} \frac{\phi_n}{\sqrt{n!}} |n\rangle \tag{8.20}$$

such that their inner product is

$$\langle \phi|\psi\rangle = \sum_{n=0}^{\infty} \frac{1}{n!} \bar{\phi}_n \psi_n \tag{8.21}$$

Let us compute the matrix element of the operator $\hat{I}$ given in eq. (8.19):

$$\langle \phi|\hat{I}|\psi\rangle = \sum_{m,n} \frac{\bar{\phi}_m \psi_n}{n!} \langle n|\hat{I}|m\rangle \tag{8.22}$$

Thus we need to compute

$$\langle n|\hat{I}|m\rangle = \int \frac{dz d\bar{z}}{2\pi i} e^{-|z|^2} \langle n|z\rangle \langle z|m\rangle \tag{8.23}$$

Recall that the integration measure is defined to be given by

$$\frac{dzd\bar{z}}{2\pi i} = \frac{d\,\text{Re}z\, d\,\text{Im}z}{\pi} \tag{8.24}$$

The overlaps are given by

$$\langle n|z\rangle = \frac{1}{\sqrt{n!}}\langle 0|\left(\hat{a}\right)^n|z\rangle = \frac{z^n}{\sqrt{n!}}\langle 0|z\rangle \tag{8.25}$$

and

$$\langle z|m\rangle = \frac{1}{\sqrt{m!}}\langle z|\left(\hat{a}^\dagger\right)^m|0\rangle = \frac{\bar{z}^m}{\sqrt{m!}}\langle z|0\rangle \tag{8.26}$$

Since $|\langle 0|z\rangle|^2 = 1$, we get

$$\begin{aligned}\langle n|\hat{I}|m\rangle &= \int \frac{dzd\bar{z}}{2\pi i}\frac{e^{-|z|^2}}{\sqrt{n!m!}} z^n \bar{z}^m \\ &= \int_0^\infty \rho d\rho \int_0^{2\pi} \frac{d\varphi}{2\pi}\frac{e^{-\rho^2}}{\sqrt{n!m!}}\rho^{n+m} e^{i(n-m)\varphi}\end{aligned} \tag{8.27}$$

Thus,

$$\langle n|\hat{I}|m\rangle = \frac{\delta_{n,m}}{n!}\int_0^\infty dx\, x^n e^{-x} = \langle n|m\rangle \tag{8.28}$$

Hence, we have found that

$$\langle \phi|\hat{I}|\psi\rangle = \langle \phi|\psi\rangle \tag{8.29}$$

for any pair of states $|\psi\rangle$ and $|\phi\rangle$. Therefore, $\hat{I}$ is the identity operator in this Hilbert space. We conclude that the set of coherent states $\{|z\rangle\}$ is an *overcomplete* set of states.

Furthermore, since

$$\langle z|\left(\hat{a}^\dagger\right)^n\left(\hat{a}\right)^m|z'\rangle = \bar{z}^n z'^m\langle z|z'\rangle = \bar{z}^n z'^m e^{\bar{z}z'} \tag{8.30}$$

we conclude that the matrix elements in generic coherent states $|z\rangle$ and $|z'\rangle$ of any arbitrary normal-ordered operator of the form

$$\hat{A} = \sum_{n,m} A_{n,m}\left(\hat{a}^\dagger\right)^n\left(\hat{a}\right)^m \tag{8.31}$$

are equal to

$$\langle z|\hat{A}|z'\rangle = \left(\sum_{n,m} A_{n,m}\bar{z}^n z'^m\right) e^{\bar{z}z'} \tag{8.32}$$

Therefore, if $\hat{A}(\hat{a},\hat{a}^\dagger)$ is an arbitrary normal-ordered operator (relative to the state $|0\rangle$), its matrix elements are given by

$$\langle z|\hat{A}(\hat{a},\hat{a}^\dagger)|z'\rangle = A(\bar{z},z')\, e^{\bar{z}z'} \tag{8.33}$$

where $A(\bar{z}, z')$ is a function of two complex variables, $\bar{z}$ and z', obtained from $\hat{A}$ by the formal replacement

$$\hat{a} \leftrightarrow z', \qquad \hat{a}^\dagger \leftrightarrow \bar{z} \tag{8.34}$$

The complex function $A(\bar{z}, z')$ is often called the *symbol* of the (normal-ordered) operator $\hat{A}$.

For example, the matrix elements of the number operator $\hat{N} = \hat{a}^\dagger \hat{a}$, which measures the number of excitations, is

$$\langle z|\hat{N}|z'\rangle = \langle z|\hat{a}^\dagger \hat{a}|z'\rangle = \bar{z}z' \, e^{\bar{z}z'} \tag{8.35}$$

8.3 Path integrals and coherent states

The concept of coherent states has been applied to broad areas of quantum mechanics (Klauder and Skagerstam, 1985; Perelomov, 1986). Here we will focus on its application to path integrals (Faddeev, 1976).

We want to compute the matrix elements of the evolution operator $\mathcal{U}$,

$$\mathcal{U} = e^{-i\frac{T}{\hbar}\hat{H}(\hat{a}^\dagger, \hat{a})} \tag{8.36}$$

where $\hat{H}(\hat{a}^\dagger, \hat{a})$ is a normal-ordered operator, and $T = t_f - t_i$ is the total time lapse. Thus, if $|i\rangle$ and $|f\rangle$ denote two arbitrary initial and final states, respectively, we can write the matrix element of $\mathcal{U}$ as

$$\langle f|e^{-i\frac{T}{\hbar}\hat{H}(\hat{a}^\dagger, \hat{a})}|i\rangle = \lim_{\epsilon \to 0, N \to \infty} \langle f|\left(1 - i\frac{\epsilon}{\hbar}\hat{H}(\hat{a}^\dagger, \hat{a})\right)^N |i\rangle \tag{8.37}$$

However, instead of inserting a complete set of states at each intermediate time t_j (with $j = 1, \ldots, N$), we will now insert an overcomplete set of coherent states $\{|z_j\rangle\}$ at each time t_j through the insertion of the resolution of the identity:

$$\begin{aligned}
&\langle f|\left(1 - i\frac{\epsilon}{\hbar}\hat{H}(\hat{a}^\dagger, \hat{a})\right)^N |i\rangle = \\
&= \int \left(\prod_{j=1}^{N} \frac{dz_j d\bar{z}_j}{2\pi i}\right) e^{-\sum_{j=1}^{N} |z_j|^2} \left[\prod_{k=1}^{N-1} \langle z_{k+1}|\left(1 - i\frac{\epsilon}{\hbar}\hat{H}(\hat{a}^\dagger, \hat{a})\right)|z_k\rangle\right] \\
&\quad \times \langle f|\left(1 - i\frac{\epsilon}{\hbar}\hat{H}(\hat{a}^\dagger, \hat{a})\right)|z_N\rangle\langle z_1|\left(1 - i\frac{\epsilon}{\hbar}\hat{H}(\hat{a}^\dagger, \hat{a})\right)|z_i\rangle
\end{aligned} \tag{8.38}$$

In the limit $\epsilon \to 0$, these matrix elements become

$$\begin{aligned}
\langle z_{k+1}|\left(1 - i\frac{\epsilon}{\hbar}\hat{H}(\hat{a}^\dagger, \hat{a})\right)|z_k\rangle &= \langle z_{k+1}|z_k\rangle - i\frac{\epsilon}{\hbar}\langle z_{k+1}|\hat{H}(\hat{a}^\dagger, \hat{a})|z_k\rangle \\
&= \langle z_{k+1}|z_k\rangle \left[1 - i\frac{\epsilon}{\hbar}H(\bar{z}_{k+1}, z_k)\right]
\end{aligned} \tag{8.39}$$

where $H(\bar{z}_{k+1}, z_k)$ is a function obtained from the normal-ordered Hamiltonian by performing the substitutions $\hat{a}^\dagger \to \bar{z}_{k+1}$ and $\hat{a} \to z_k$.

Hence, we can write the following expression for the matrix element of the evolution operator:

$$\langle f| e^{-i\frac{T}{\hbar}\hat{H}(\hat{a}^\dagger,\hat{a})}|i\rangle = \lim_{\epsilon\to 0, N\to\infty} \int \left(\prod_{j=1}^{N} \frac{dz_j d\bar{z}_j}{2\pi i}\right) e^{-\sum_{j=1}^{N}|z_j|^2} e^{\sum_{j=1}^{N-1}\bar{z}_{j+1}z_j} \prod_{j=1}^{N-1}$$

$$\times \left[1 - i\frac{\epsilon}{\hbar}H(\bar{z}_{k+1}, z_k)\right] \langle f|z_N\rangle\langle z_1|i\rangle \left[1 - i\frac{\epsilon}{\hbar}\frac{\langle f|\hat{H}|z_N\rangle}{\langle f|z_N\rangle}\right]$$

$$\times \left[1 - i\frac{\epsilon}{\hbar}\frac{\langle z_1|\hat{H}|i\rangle}{\langle z_1|i\rangle}\right] \tag{8.40}$$

By further expanding the initial and final states in coherent states,

$$\langle f| = \int \frac{dz_f d\bar{z}_f}{2\pi i} e^{-|z_f|^2} \bar{\psi}_f(z_f)\langle z_f|$$

$$|i\rangle = \int \frac{dz_i d\bar{z}_i}{2\pi i} e^{-|z_i|^2} \psi_i(\bar{z}_i)|z_i\rangle \tag{8.41}$$

we find the (formal) result:

$$\langle f| e^{-i\frac{T}{\hbar}\hat{H}(\hat{a}^\dagger,\hat{a})}|i\rangle =$$

$$= \int \mathcal{D}z\mathcal{D}\bar{z}\, e^{\frac{i}{\hbar}\int_{t_i}^{t_f} dt\left[\frac{\hbar}{2i}(z\partial_t\bar{z} - \bar{z}\partial_t z) - H(z,\bar{z})\right]}$$

$$\times e^{\frac{1}{2}(|z_i|^2 + |z_f|^2)} \bar{\psi}_f(z_f)\psi_i(\bar{z}_i) \tag{8.42}$$

This is the coherent-state form of the path integral. In this expression, we can identify the Lagrangian L as the quantity

$$L = \frac{\hbar}{2i}(z\partial_t\bar{z} - \bar{z}\partial_t z) - H(z,\bar{z}) \tag{8.43}$$

It is easy to check that this expression is equivalent to the phase-space path integral derived in section 5.1.

Notice that the Lagrangian for the coherent-state representation presented in eq. (8.43) is first order in time derivatives. Because of this feature, we are not guaranteed that the paths are differentiable. This property leads to all kinds of subtleties that for the most part we will ignore in what follows.

8.4 Path integral for a nonrelativistic Bose gas

The field-theoretic description of a gas of (spinless) nonrelativistic bosons is given in terms of the creation and annihilation field operators $\hat{\phi}^\dagger(\boldsymbol{x})$ and $\hat{\phi}(\boldsymbol{x})$, which satisfy the equal-time commutation relations (in d space dimensions):

$$\left[\hat{\phi}(\boldsymbol{x}), \hat{\phi}^\dagger(\boldsymbol{y})\right] = \delta^d(\boldsymbol{x}-\boldsymbol{y}) \tag{8.44}$$

Relative to the empty state $|0\rangle$, such that

$$\hat{\phi}(\boldsymbol{x})|0\rangle = 0 \tag{8.45}$$

the normal-ordered Hamiltonian (in the grand canonical ensemble) is

$$\begin{aligned}\hat{H} = &\int d^dx\, \hat{\phi}^\dagger(\boldsymbol{x})\left[-\frac{\hbar^2}{2m}\nabla^2 - \mu + V(\boldsymbol{x})\right]\hat{\phi}(\boldsymbol{x}) \\ &+ \frac{1}{2}\int d^dx \int d^dy\, \hat{\phi}^\dagger(\boldsymbol{x})\hat{\phi}^\dagger(\boldsymbol{y})U(\boldsymbol{x}-\boldsymbol{y})\hat{\phi}(\boldsymbol{y})\hat{\phi}(\boldsymbol{x})\end{aligned} \tag{8.46}$$

where m is the mass of the bosons, μ is the chemical potential, $V(\boldsymbol{x})$ is an external potential, and $U(\boldsymbol{x}-\boldsymbol{y})$ is the interaction potential between pairs of bosons.

Following our discussion of the coherent-state path integral, we see that we can immediately write down a path integral for a thermodynamically large system of bosons. The boson coherent states are now labeled by a complex field $\phi(\boldsymbol{x})$ and its complex conjugate $\bar{\phi}(\boldsymbol{x})$,

$$|\{\phi(\boldsymbol{x})\}\rangle = e^{\int dx\, \phi(\boldsymbol{x})\hat{\phi}^\dagger(\boldsymbol{x})}|0\rangle \tag{8.47}$$

which has the coherent state property of being a right eigenstate of the field operator $\hat{\phi}(\boldsymbol{x})$,

$$\hat{\phi}(\boldsymbol{x})|\{\phi\}\rangle = \phi(\boldsymbol{x})|\{\phi\}\rangle \tag{8.48}$$

as well as obeying the resolution of the identity in this space of states:

$$\mathcal{I} = \int \mathcal{D}\phi\mathcal{D}\phi^* e^{-\int dx\, |\phi(\boldsymbol{x})|^2}|\{\phi\}\rangle\langle\{\phi\}| \tag{8.49}$$

The matrix element of the evolution operator of this system between an arbitrary initial state $|i\rangle$ and an arbitrary final state $|f\rangle$, separated by a time lapse $T = t_f - t_i$ (not to be confused with the temperature!), now takes the form

$$\begin{aligned}&\langle f|e^{-\frac{i}{\hbar}\hat{H}T}|i\rangle = \\ &\int \mathcal{D}\phi\mathcal{D}\bar{\phi}\,\exp\left\{\frac{i}{\hbar}\int_{t_i}^{t_f} dt\left(\int d^dx\,\frac{\hbar}{i}\left[\phi(\boldsymbol{x},t)\partial_t\bar{\phi}(\boldsymbol{x},t) - \bar{\phi}(\boldsymbol{x},t)\partial_t\phi(\boldsymbol{x},t)\right] - H[\phi,\bar{\phi}]\right)\right\} \\ &\times \bar{\Psi}_f(\phi(\boldsymbol{x},t_f))\Psi_i(\bar{\phi}(\boldsymbol{x},t_i))\, e^{\frac{1}{2}\int d\boldsymbol{x}(|\phi(\boldsymbol{x},t_f)|^2 + |\phi(\boldsymbol{x},t_i)|^2)}\end{aligned} \tag{8.50}$$

where the functional $H[\phi,\bar{\phi}]$ is

$$H[\phi,\bar{\phi}] = \int d^d x\, \bar{\phi}(\boldsymbol{x}) \left[-\frac{\hbar^2}{2m} \boldsymbol{\nabla}^2 - \mu + V(\boldsymbol{x}) \right] \phi(\boldsymbol{x}) + \frac{1}{2} \int d^d x \int d^d y\, |\phi(\boldsymbol{x})|^2 |\phi(\boldsymbol{y})|^2 U(\boldsymbol{x}-\boldsymbol{y}) \tag{8.51}$$

It is also possible to write the action S in the less symmetric but simpler form (where $d^D x \equiv dt\, d^d x$)

$$S = \int d^D x\, \bar{\phi}(x) \left(i\hbar\partial_t + \frac{\hbar^2}{2m} \vec{\nabla}^2 + \mu - V(\boldsymbol{x}) \right) \phi(x) - \frac{1}{2} \int d^D x \int d^D y\, |\phi(x)|^2 |\phi(y)|^2 U(x-y) \tag{8.52}$$

where $U(x-y) = U(\boldsymbol{x}-\boldsymbol{y})\delta(t_x - t_y)$.

Therefore the path integral for a system of nonrelativistic bosons, with chemical potential μ, has the same form as the path integral of the charged scalar field (see section 5.6), except that the action is first order in time derivatives. The fact that the field is complex follows from the requirement that the number of bosons is a globally conserved quantity, which is why we can introduce a chemical potential.

This formulation is useful for studying superfluid helium and similar problems. Suppose, for instance, that we want to compute the partition function Z for this system of bosons at finite temperature T,

$$Z = \operatorname{tr} e^{-\beta \hat{H}} \tag{8.53}$$

where $\beta = 1/T$ (in units with $k_B = 1$). The coherent-state path integral representation of the partition function is obtained by (1) restricting the initial and final states to be the same ($|i\rangle = |f\rangle$) and arbitrary, (2) summing over all possible states, and (3) Wick rotating to imaginary time $t \to -i\tau$, with the time-span $T \to -i\beta\hbar$ and periodic boundary conditions in imaginary time.

The result is the (imaginary time) path integral

$$Z = \int \mathcal{D}\phi \mathcal{D}\bar{\phi}\, e^{-S_E(\phi,\bar{\phi})} \tag{8.54}$$

where S_E is the Euclidean action

$$S_E(\phi,\bar{\phi}) = \frac{1}{\hbar} \int_0^\beta d\tau \int d\boldsymbol{x}\, \bar{\phi} \left[\hbar\partial_\tau - \mu - \frac{\hbar^2}{2m} \boldsymbol{\nabla}^2 + V(x) \right] \phi + \frac{1}{2\hbar} \int_0^\beta d\tau \int d\boldsymbol{x} \int d\boldsymbol{y}\, U(x-y)\, |\phi(x)|^2 |\phi(y)|^2 \tag{8.55}$$

The fields $\phi(x) = \phi(\boldsymbol{x},\tau)$ satisfy periodic boundary conditions in imaginary time:

$$\phi(\boldsymbol{x},\tau) = \phi(\boldsymbol{x},\tau + \beta\hbar) \tag{8.56}$$

This requirement suggests an expansion of the field $\phi(x)$ in Fourier modes of the form

$$\phi(\boldsymbol{x}, \tau) = \sum_{n=-\infty}^{\infty} e^{i\omega_n \tau} \phi(\vec{x}, \omega_n) \tag{8.57}$$

where the frequencies ω_n (the *Matsubara* frequencies) must be chosen so that ϕ obeys the required periodic boundary conditions. We find

$$\omega_n = \frac{2\pi}{\beta\hbar} n = \frac{2\pi T}{\hbar} n, \qquad n \in \mathbb{Z} \tag{8.58}$$

where n is an arbitrary integer. In exercise 8.5 at the end of this chapter, you will evaluate this path integral using a semiclassical approximation.

8.5 Fermion coherent states

In this section, we develop a formalism for fermions that follows closely what we have done for bosons while taking into account the anticommuting nature of fermionic operators.

Let $\{c_i^\dagger\}$ be a set of fermion creation operators, with $i = 1, \ldots, N$, and $\{c_i\}$ the set of their N adjoint operators, the associated annihilation operators. The number operator for the ith fermion is $n_i = c_i^\dagger c_i$. Let us define the basis states of the ith fermion by the kets $|0_i\rangle$ and $|1_i\rangle$, which obey the obvious definitions:

$$c_i|0_i\rangle = 0, \qquad c_i^\dagger|0_i\rangle = |1_i\rangle, \qquad c_i^\dagger c_i|0_i\rangle = 0, \qquad c_i^\dagger c_i|1_i\rangle = |1_i\rangle \tag{8.59}$$

For N fermions, the Hilbert space is spanned by the antisymmetrized states $|n_1, \ldots, n_N\rangle$. Let

$$|0\rangle \equiv |0_1, \ldots, 0_N\rangle \tag{8.60}$$

be the empty state. A general state in this Hilbert space is

$$|n_1, \ldots, n_N\rangle = \left(c_1^\dagger\right)^{n_1} \ldots \left(c_N^\dagger\right)^{n_N} |0\rangle \tag{8.61}$$

As we saw before, the wave function $\langle n_1, \ldots, n_N|\Psi\rangle$ is fully antisymmetric. If the state $|\Psi\rangle$ is a product state, then the wave function is a Slater determinant.

8.5.1 Definition of fermion coherent states

We now define fermion coherent states. Let $\{\bar{\xi}_i, \xi_i\}$, with $i = 1, \ldots, N$, be a set of $2N$ *Grassmann variables*, a set of symbols also known as the *generators* of a Grassmann algebra. By definition, Grassmann variables satisfy the following properties:

$$\{\xi_i, \xi_j\} = \{\bar{\xi}_i, \bar{\xi}_j\} = \{\xi_i, \bar{\xi}_j\} = \xi_i^2 = \bar{\xi}_i^2 = 0 \tag{8.62}$$

Therefore, Grassmann variables behave like a set of time-ordered fermion operators.

We also require that the Grassmann variables anticommute with the fermion operators:

$$\{\xi_i, c_j\} = \left\{\xi_i, c_j^\dagger\right\} = \{\bar{\xi}_i, c_j\} = \left\{\bar{\xi}_i, c_j^\dagger\right\} = 0 \tag{8.63}$$

Let us define the *fermion coherent states* to be

$$|\xi\rangle \equiv e^{-\xi c^{\dagger}}|0\rangle \tag{8.64}$$

$$\langle\xi| \equiv \langle 0|\, e^{\bar{\xi}c} \tag{8.65}$$

As a consequence of these definitions, we have:

$$e^{-\xi c^{\dagger}} = 1 - \xi c^{\dagger} \tag{8.66}$$

Similarly, if ψ is a Grassmann variable, then

$$\langle\xi|\psi\rangle = \langle 0|e^{\bar{\xi}c}e^{-\psi c^{\dagger}}|0\rangle = 1 + \bar{\xi}\psi = e^{\bar{\xi}\psi} \tag{8.67}$$

For N fermions, we have

$$|\xi\rangle \equiv |\xi_1, \ldots, \xi_N\rangle = \Pi_{i=1}^{N} e^{-\xi_i c_i^{\dagger}}|0\rangle \equiv e^{-\sum_{i=1}^{N}\xi_i c_i^{\dagger}}|0\rangle \tag{8.68}$$

since the following commutator vanishes:

$$\left[\xi_i c_i^{\dagger}, \xi_j c_j^{\dagger}\right] = 0 \tag{8.69}$$

8.5.2 Analytic functions of Grassmann variables

We define $\psi(\xi)$ to be an *analytic* function of the Grassmann variable if it has a power series expansion in ξ,

$$\psi(\xi) = \psi_0 + \psi_1\xi + \psi_2\xi^2 + \cdots \tag{8.70}$$

where $\psi_n \in \mathbb{C}$. Since

$$\xi^n = 0, \qquad \forall n \geq 2 \tag{8.71}$$

then all analytic functions of a Grassmann variable reduce to a first-degree polynomial,

$$\psi(\xi) \equiv \psi_0 + \psi_1\xi \tag{8.72}$$

Similarly, we define complex conjugation by

$$\overline{\psi(\xi)} \equiv \bar{\psi}_0 + \bar{\psi}_1\bar{\xi} \tag{8.73}$$

where $\bar{\psi}_0$ and $\bar{\psi}_1$ are the complex conjugates of ψ_0 and ψ_1, respectively.

We can also define functions of two Grassmann variables ξ and $\bar{\xi}$,

$$A(\bar{\xi}, \xi) = a_0 + a_1\xi + \bar{a}_1\bar{\xi} + a_{12}\bar{\xi}\xi \tag{8.74}$$

where a_1, $\bar{a}_1$, and a_{12} are complex numbers; a_1 and $\bar{a}_1$ are not necessarily complex conjugates of each other.

8.5.3 Differentiation over Grassmann variables

Since analytic functions of Grassmann variables have such a simple structure, differentiation is just as simple. Indeed, we define the derivative as the coefficient of the linear term:

$$\partial_\xi \psi(\xi) \equiv \psi_1 \tag{8.75}$$

Likewise, we also have

$$\partial_{\bar\xi} \overline{\psi(\xi)} \equiv \bar\psi_1 \tag{8.76}$$

Clearly, using this rule, we can write

$$\partial_\xi (\bar\xi \xi) = -\partial_\xi (\xi \bar\xi) = -\bar\xi \tag{8.77}$$

A similar argument shows that

$$\partial_\xi A(\bar\xi, \xi) = a_1 - a_{12}\,\bar\xi \tag{8.78}$$

$$\partial_{\bar\xi} A(\bar\xi, \xi) = \bar a_1 + a_{12}\,\xi \tag{8.79}$$

$$\partial_{\bar\xi} \partial_\xi A(\bar\xi, \xi) = -a_{12} = -\partial_\xi \partial_{\bar\xi} A(\bar\xi, \xi) \tag{8.80}$$

from which we conclude that ∂_ξ and $\partial_{\bar\xi}$ *anticommute*:

$$\left\{\partial_{\bar\xi}, \partial_\xi\right\} = 0, \qquad \text{and} \qquad \partial_\xi \partial_\xi = \partial_{\bar\xi} \partial_{\bar\xi} = 0 \tag{8.81}$$

8.5.4 Integration over Grassmann variables

The basic differentiation rule of eq. (8.75) implies that

$$1 = \partial_\xi \xi \tag{8.82}$$

which suggests the following definitions:

$$\int d\xi\ 1 = 0, \qquad \int d\xi\ \partial_\xi \xi = 0, \qquad \int d\xi\ \xi = 1 \tag{8.83}$$

Analogous rules also apply for the conjugate variables $\bar\xi$.

It is instructive to compare the differentiation and integration rules:

$$\begin{aligned} \textstyle\int d\xi\ 1 = 0 \quad &\leftrightarrow \quad \partial_\xi 1 = 0 \\ \textstyle\int d\xi\ \xi = 1 \quad &\leftrightarrow \quad \partial_\xi \xi = 1 \end{aligned} \tag{8.84}$$

Thus, for Grassmann variables, differentiation and integration are exactly equivalent:

$$\partial_\xi \Longleftrightarrow \int d\xi \tag{8.85}$$

These rules imply that the integral of an analytic function $f(\xi)$ is

$$\int d\xi\, f(\xi) = \int d\xi\ \left(f_0 + f_1 \xi\right) = f_1 \tag{8.86}$$

and

$$\int d\xi A(\bar{\xi},\xi)=\int d\xi \left(a_0+a_1\xi+\bar{a}_1\bar{\xi}+a_{12}\bar{\xi}\xi\right)=a_1-a_{12}\bar{\xi}$$

$$\int d\bar{\xi} A(\bar{\xi},\xi)=\int d\xi \left(a_0+a_1\xi+\bar{a}_1\bar{\xi}+a_{12}\bar{\xi}\xi\right)=\bar{a}_1+a_{12}\xi$$

$$\int d\bar{\xi}d\xi A(\bar{\xi},\xi)=-\int d\xi d\bar{\xi} A(\bar{\xi},\xi)=-a_{12} \tag{8.87}$$

It is straightforward to show that with these definitions, the following expression is a consistent definition of a delta function:

$$\delta(\xi',\xi)=\int d\eta\, e^{-\eta(\xi-\xi')} \tag{8.88}$$

where ξ, ξ', and η are Grassmann variables.

Finally, given that we have a vector space of analytic functions, we can define an *inner product* as follows:

$$\langle f|g\rangle=\int d\bar{\xi}d\xi\, e^{-\bar{\xi}\xi}\, \bar{f}(\xi)\, g(\bar{\xi})=\bar{f}_0 g_0+\bar{f}_1 g_1 \tag{8.89}$$

as expected.

8.5.5 Properties of fermion coherent states

In section 8.5.1 we defined above the fermion bra and ket coherent states:

$$|\{\xi_j\}\rangle=e^{-\sum\limits_j \xi_j c_j^\dagger}|0\rangle, \qquad \langle\{\xi_j\}|=\langle 0|\, e^{\sum\limits_j \bar{\xi}_j c_j} \tag{8.90}$$

After a little algebra, using the rules defined above, it is easy to see that the following identities hold:

$$c_i|\{\xi_j\}\rangle=\xi_i|\{\xi_j\}\rangle, \qquad c_i^\dagger|\{\xi_j\}\rangle=-\partial_{\xi_i}|\{\xi_j\}\rangle \tag{8.91}$$

$$\langle\{\xi_j\}|\, c_i=\partial_{\bar{\xi}_i}\langle\{\xi_j\}|, \qquad \langle\{\xi_j\}|\, c_i^\dagger=\bar{\xi}_i\langle\{\xi_j\}| \tag{8.92}$$

The inner product of two coherent states, $|\{\xi_j\}\rangle$ and $|\{\xi_j'\}\rangle$, is

$$\langle\{\xi_j\}|\{\xi_j'\}\rangle=e^{\sum\limits_j \bar{\xi}_j\xi_j} \tag{8.93}$$

Similarly, we also have the *resolution of the identity* (which is easy to prove):

$$I=\int \left(\Pi_{i=1}^N d\bar{\xi}_i d\xi_i\right)\, e^{-\sum\limits_{i=1}^N \bar{\xi}_i\xi_i}\, |\{\xi_i\}\rangle\langle\{\xi_i\}| \tag{8.94}$$

Let $|\Psi\rangle$ be some state. Then we can use eq. (8.94) to expand the state $|\Psi\rangle$ in fermion coherent states $|\xi\rangle$,

$$|\Psi\rangle = \int \left(\Pi_{i=1}^{N} d\bar{\xi}_i d\xi_i\right) e^{-\sum_{i=1}^{N} \bar{\xi}_i \xi_i} \Psi(\xi)|\{\xi_i\}\rangle \tag{8.95}$$

where

$$\Psi(\bar{\xi}) \equiv \Psi(\bar{\xi}_1, \ldots, \bar{\xi}_N) \tag{8.96}$$

We can use the rules derived above to compute the following matrix elements

$$\langle \xi | c_j | \Psi\rangle = \partial_{\bar{\xi}_j} \Psi(\bar{\xi}), \qquad \langle \xi | c_j^\dagger | \Psi\rangle = \bar{\xi}_j \, \Psi(\bar{\xi}) \tag{8.97}$$

which are consistent with what we concluded in section 8.5.1.

Let $|0\rangle$ be the "empty state." We will not call it the "vacuum," since it is not in the sector of the ground state for the systems of interest. Let $A\left(\{c_j^\dagger\}, \{c_j\}\right)$ be a *normal-ordered operator* (with respect to the state $|0\rangle$). By using the formalism worked out above, one can show without difficulty that its matrix elements in the coherent states $|\xi\rangle$ and $|\xi'\rangle$ are

$$\langle \xi | A\left(\{c_j^\dagger\}, \{c_j\}\right) |\xi'\rangle = e^{\sum_i \bar{\xi}_i \xi_i'} A\left(\{\bar{\xi}_j\}, \{\xi_j'\}\right) \tag{8.98}$$

For example, the expectation value of the fermion number operator $\hat{N}$,

$$\hat{N} = \sum_j c_j^\dagger c_j \tag{8.99}$$

in the coherent state $|\xi\rangle$ is

$$\frac{\langle \xi | \hat{N} | \xi\rangle}{\langle \xi | \xi\rangle} = \sum_j \bar{\xi}_j \xi_j \tag{8.100}$$

8.5.6 Grassmann Gaussian integrals

Let us consider a Gaussian integral over Grassmann variables of the form

$$\mathcal{Z}[\bar{\zeta}, \zeta] = \int \left(\prod_{i=1}^{N} d\bar{\xi}_i d\xi_i\right) e^{-\sum_{i,j} \bar{\xi}_i M_{ij} \xi_j + \bar{\xi}_i \zeta_i + \bar{\zeta}_i \xi_i} \tag{8.101}$$

where $\{\zeta_i\}$ and $\{\bar{\zeta}_i\}$ are a set of $2N$ Grassmann variables, and the matrix M_{ij} is a complex hermitian matrix. We now show that

$$\mathcal{Z}[\bar{\zeta}, \zeta] = (\det M)\, e^{\sum_{i,j} \bar{\zeta}_i \left(M^{-1}\right)_{ij} \zeta_j} \tag{8.102}$$

Before showing that eq. (8.102) is correct, let us make a few observations:

1) Eq. (8.102) looks like the familiar expression for Gaussian integrals for bosons, except that instead of a factor of $(\det M)^{-1/2}$, we have a factor of $\det M$ in the

numerator. Except for the absence of a square root, the fluctuation determinant appearing in the numerator is the main effect of the Fermi statistics!

2) Moreover, if we had considered a system of N Grassmann variables (instead of $2N$), we would have obtained instead a factor of

$$(\det M)^{1/2} = \mathrm{Pf}(M) \tag{8.103}$$

where M would now be an $N \times N$ *real anti-symmetric* matrix, and $\mathrm{Pf}(M)$ denotes the Pfaffian of the matrix M.

To prove that eq. (8.102) is correct, it will be sufficient to consider the case $\zeta_i = \bar{\zeta}_i = 0$, since the contribution from these sources is identical to the bosonic case (except for the ordering of the Grassmann variables). Using the Grassmann identities, we can write the exponential factor as

$$e^{-\sum_{i,j} \bar{\xi}_i M_{ij} \xi_j} = \prod_{i,j} \left(1 - \bar{\xi}_i M_{ij} \xi_j\right) \tag{8.104}$$

The integral that we need to do is

$$\mathcal{Z}[0,0] = \int \left(\prod_{i=1}^{N} d\bar{\xi}_i \xi_i \right) \prod_{i,j} \left(1 - \bar{\xi}_i M_{ij} \xi_j\right) \tag{8.105}$$

From the integration rules, we can easily see that the only nonvanishing terms in this expression are those that have just one ξ_i and one $\bar{\xi}_i$ (for each i). Hence we can write

$$\begin{aligned} \mathcal{Z}[0,0] &= (-1)^N \int \left(\prod_{i=1}^{N} d\bar{\xi}_i \, d\xi_i \right) \bar{\xi}_1 M_{12} \xi_2 \bar{\xi}_2 M_{23} \xi_3 \cdots + \text{permutations} \\ &= (-1)^N M_{12} M_{23} M_{34} \cdots \int \left(\prod_{i=1}^{N} d\bar{\xi}_i \, d\xi_i \right) \bar{\xi}_1 \xi_2 \bar{\xi}_2 \xi_3 \bar{\xi}_3 \cdots \bar{\xi}_N \xi_1 + \text{permutations} \\ &= (-1)^{2N} M_{12} M_{23} M_{34} \cdots M_{N,1} + \text{permutations} \end{aligned} \tag{8.106}$$

What is the contribution of the terms labeled "permutations"? It is easy to see that if we permute any pair of labels, say, 2 and 3, we will get a contribution of the form

$$(-1)^{2N} (-1) M_{13} M_{32} M_{24} \cdots \tag{8.107}$$

Hence we conclude that the Gaussian Grassmann integral is just the determinant of the matrix M:

$$\mathcal{Z}[0,0] = \int \left(\prod_{i=1}^{N} d\bar{\xi}_i d\xi_i \right) e^{-\sum_{i,j} \bar{\xi}_i M_{ij} \xi_j} = \det M \tag{8.108}$$

Alternatively, we can diagonalize the quadratic form and notice that the Jacobian is "upside-down" compared to the result we found for bosons.

8.6 Path integrals for fermions

We are now ready to give a prescription for the construction of a fermion path integral in a general system. Let H be a normal-ordered Hamiltonian, with respect to some reference state $|0\rangle$, of a system of fermions. Let $|\Psi_i\rangle$ be the ket at the initial time t_i and $|\Psi_f\rangle$ be the final state at time t_f. The matrix element of the evolution operator can be written as a Grassmann path integral,

$$\begin{aligned}\langle \Psi_f, t_f | \Psi_i, t_i \rangle &= \langle \Psi_f | e^{-\frac{i}{\hbar} H(t_f - t_i)} | \Psi_i \rangle \\ &\equiv \int \mathcal{D}\bar{\psi}\mathcal{D}\psi \, e^{\frac{i}{\hbar} S(\bar{\psi}, \psi)} \times \text{projection operators}\end{aligned} \tag{8.109}$$

where we have not written down the explicit form of the projection operators onto the initial and final states. The action $S(\bar{\psi}, \psi)$ is

$$S(\bar{\psi}, \psi) = \int_{t_i}^{t_f} dt \, \left[i\hbar\bar{\psi}\partial_t\psi - H(\bar{\psi}, \psi) \right] \tag{8.110}$$

This expression of the fermion path integral holds for any theory of fermions, relativistic or not. Notice that it has the same form as the bosonic path integral. The only change is that for fermions, the determinant appears in the numerator, while for bosons, it is in the denominator!

8.7 Path-integral quantization of the Dirac field

Let us now apply the methods just developed to the case of the Dirac theory. Let $\psi_\alpha(x)$, with $\alpha = 1, \ldots, 4$ be a free massive Dirac field in 3+1 spacetime dimensions. This field satisfies the Dirac equation as an equation of motion,

$$\left(i\partial\!\!\!/ - m\right)\psi = 0 \tag{8.111}$$

where ψ is a 4-spinor and $\partial\!\!\!/ = \gamma^\mu \partial_\mu$. Recall that the Dirac γ-matrices satisfy the algebra

$$\{\gamma^\mu, \gamma^\nu\} = 2g^{\mu\nu} \tag{8.112}$$

where $g^{\mu\nu}$ is the Minkowski space metric tensor (in the Bjorken-Drell form).

We saw before that in the QFT description of the Dirac theory, ψ is an operator acting on the Fock space of (fermionic) states. We also saw that the Dirac equation can be regarded as the classical equation of motion of the Lagrangian density,

$$\mathcal{L} = \bar{\psi}\left(i\partial\!\!\!/ - m\right)\psi \tag{8.113}$$

where $\bar{\psi} = \psi^{\dagger}\gamma^0$. We also noted that the momentum canonically conjugate to the field ψ is $i\psi^{\dagger}$, from which the standard fermionic equal-time anticommutation relations follow:

$$\left\{\psi_\alpha(\boldsymbol{x}, x_0), \psi^{\dagger}_\beta(\boldsymbol{y}, x_0)\right\} = \delta_{\alpha\beta}\delta^3(\boldsymbol{x} - \boldsymbol{y}) \tag{8.114}$$

The Lagrangian density $\mathcal{L}$ for a Dirac fermion coupled to sources η_α and $\bar{\eta}_\alpha$ is

$$\mathcal{L} = \bar{\psi}\left(i\partial\!\!\!/ - m\right)\psi + \bar{\psi}\eta + \bar{\eta}\psi \tag{8.115}$$

where the sources are "classical" (Grassmann) anticommuting fields.

We can follow the same steps described in the preceding sections to find the following expression for the path integral of the Dirac field in terms of Grassmann fields $\psi(x)$ and $\bar{\psi}(x)$ (which here are independent variables!)

$$\begin{aligned} \mathcal{Z}[\bar{\eta}, \eta] &= \frac{1}{\langle 0|0\rangle}\langle 0|T\, e^{i\int d^4x\,(\bar{\psi}\eta + \bar{\eta}\psi)}|0\rangle \\ &\equiv \int \mathcal{D}\bar{\psi}\mathcal{D}\psi\, e^{iS + i\int d^4x(\bar{\psi}\eta + \bar{\eta}\psi)} \end{aligned} \tag{8.116}$$

where $S = \int d^4x\mathcal{L}$.

From this result, it follows that the Dirac propagator is given by

$$\begin{aligned} iS_{\alpha\beta}(x-y) &= \langle 0|T\psi_\alpha(x)\bar{\psi}_\beta(y)|0\rangle \\ &= \frac{(-i)^2}{\mathcal{Z}[0,0]}\left.\frac{\delta^2\mathcal{Z}[\bar{\eta},\eta]}{\delta\bar{\eta}_\alpha(x)\delta\eta_\beta(y)}\right|_{\bar{\eta}=\eta=0} \\ &= \langle x, \alpha|\frac{1}{i\partial\!\!\!/ - m}|y, \beta\rangle \end{aligned} \tag{8.117}$$

Similarly, the partition function for a theory of free Dirac fermions is

$$\mathcal{Z}_{\text{Dirac}} = \text{Det}\left(i\partial\!\!\!/ - m\right) \tag{8.118}$$

In contrast, in the case of a free real scalar field, we found

$$\mathcal{Z}_{\text{scalar}} = \left[\text{Det}\left(\partial^2 + m^2\right)\right]^{-1/2} \tag{8.119}$$

Therefore for the case of the Dirac field, the partition function is a determinant, whereas for the scalar field, it is the inverse of a determinant (actually, of its square root). The fact that one result is the inverse of the other is a consequence of Dirac fields being quantized as fermions, whereas scalar fields are quantized as bosons. As we saw, this is a general result for Fermi and Bose fields, regardless of whether they are relativistic. Moreover, from this result, it follows that the vacuum (ground state) energy of a Dirac field is negative, whereas the vacuum energy for a scalar field is positive. We will see that in interacting field theories, eq. (8.118) leads to the Feynman diagram rule that each fermion loop carries a minus sign that reflects the Fermi-Dirac statistics; hence this rule holds even in the absence of relativistic invariance.

Note that, due to the charge-conjugation symmetry of the Dirac theory, the spectrum of the Dirac operator is symmetric. That is, for every positive eigenvalue of the Dirac operator, there is a negative eigenvalue of equal magnitude. More formally, since $\{\gamma^5, \gamma^\mu\} = 0$,

$$\gamma_5 \left(i\partial\!\!\!/ - m\right) \gamma_5 = \left(-i\partial\!\!\!/ - m\right) \tag{8.120}$$

and $\gamma_5^2 = I$ (the 4×4 identity matrix), it follows that

$$\text{Det}\left(i\partial\!\!\!/ - m\right) = \text{Det}\left(i\partial\!\!\!/ + m\right) \tag{8.121}$$

Hence

$$\begin{aligned} \text{Det}\left(i\partial\!\!\!/ - m\right) &= \left[\text{Det}\left(i\partial\!\!\!/ - m\right) \text{Det}\left(i\partial\!\!\!/ + m\right)\right]^{1/2} \\ &= \left[\text{Det}\left(\partial^2 + m^2\right)\right]^2 \end{aligned} \tag{8.122}$$

where we used that the "square" of the Dirac operator is the Klein-Gordon operator multiplied by the 4×4 identity matrix, which is why the exponent on the right side of eq. (8.122) is $2 = 4 \times \frac{1}{2}$. It is easy to see that this result implies that the vacuum energy for the Dirac fermion E_0^{Dirac} and the vacuum energy of a scalar field E_0^{scalar}, with the same mass m, are related by

$$E_0^{\text{Dirac}} = -4E_0^{\text{scalar}} \tag{8.123}$$

Here we have ignored the fact that both the left and right sides of this equation are divergent, as we saw before. However, since they have the same divergence (or equivalently, they are regularized in the same way), the comparison is meaningful.

In contrast, if instead of Dirac fermions (which are charged and hence complex fields), we consider Majorana fermions (which are charge neutral and hence are real fields), we would have obtained instead the results

$$\mathcal{Z}_{\text{Majorana}} = \text{Pf}\left(i\partial\!\!\!/ - m\right) = \left[\text{Det}\left(i\partial\!\!\!/ - m\right)\right]^{1/2} \tag{8.124}$$

where Pf is the Pfaffian, given by the square root of the determinant. Thus the vacuum energy of a Majorana fermion is half the vacuum energy of a Dirac fermion.

Finally, since a massless Dirac fermion is equivalent to two Weyl fermions (one for each chirality), it follows that the vacuum energy of a Majorana Weyl fermion is equal and opposite to the vacuum energy of a scalar field. Moreover, this relation holds for all the states of their spectra, which are identical. This observation is the origin of the concept of supersymmetry, in which the fermionic states and the bosonic states are precisely matched. As a result, the vacuum energy of a supersymmetric theory is zero.

We end this section by discussing briefly the theory of Dirac fermions in Euclidean spacetime. We focus on the theory in four dimensions, but this can be done in any dimension. There are two equivalent ways to do this analytic continuation. One option is to define a set of four antihermitian Dirac gamma matrices

$$\gamma_j = \gamma^j, \qquad \gamma_4 = -i\gamma^0 \tag{8.125}$$

with $j = 1, 2, 3$, that satisfy the algebra

$$\{\gamma_\mu, \gamma_\nu\} = -2\delta_{\mu\nu} \tag{8.126}$$

with $\mu = 1, \ldots, 4$. Similarly, define the hermitian γ_5 matrix:

$$\gamma_5 = \gamma^5 = \gamma_1\gamma_2\gamma_3\gamma_4 \tag{8.127}$$

In this notation, the partition function is

$$Z = \int \mathcal{D}\psi\mathcal{D}\bar{\psi} \exp(-S_E(\psi, \bar{\psi})) \tag{8.128}$$

where S_E is the Euclidean for a Dirac spinor, which, coupled to an abelian gauge field, becomes

$$S_E = \int d^4x\bar{\psi}(i\not{D} + m)\psi \tag{8.129}$$

where $D_\mu = \partial_\mu + iA_\mu$, again for $\mu = 1, 2, 3, 4$, with $A_4 = -iA_0$. The Euclidean Dirac propagator in momentum space is given by

$$S(p) = \frac{1}{-\not{p} + m} = \frac{\not{p} + m}{p^2 + m^2} \tag{8.130}$$

(with $p_0 = -ip_4$). Alternatively, we can define the four gamma matrices to be hermitian and satisfy the Clifford algebra $\{\gamma_\mu, \gamma_\nu\} = 2\delta_{\mu nu}$. In this notation, the Euclidean action is

$$S_E = \int d^4x\bar{\psi}(\not{D} + m)\psi \tag{8.131}$$

where the operator $\not{\partial}$ is antihermitian, and the propagator is

$$S(p) = \frac{1}{-i\not{p} + m} = \frac{i\not{p} + m}{p^2 + m^2} \tag{8.132}$$

8.8 Functional determinants

We now face the problem of how to compute functional determinants. We have already discussed how to do this for path integrals with a few degrees of freedom (i.e., in quantum mechanics). Let us now generalize these ideas to QFT. We begin by discussing some simple determinants that show up in systems of fermions and bosons at finite temperature and density.

8.8.1 Functional determinants for coherent states

Consider a system of fermions (or bosons) with one-body Hamiltonian $\hat{h}$ at nonzero temperature T and chemical potential μ. The partition function is

$$Z = \operatorname{tr} e^{-\beta\left(\hat{H} - \mu\hat{N}\right)} \tag{8.133}$$

where $\beta = 1/k_B T$,

$$\hat{H} = \int dx\, \hat{\psi}^\dagger(x)\, \hat{h}\, \hat{\psi}(x) \tag{8.134}$$

and

$$\hat{N} = \int dx\, \hat{\psi}^\dagger(x)\, \hat{\psi}(x) \tag{8.135}$$

is the number operator. Here x denotes both spatial and internal (spin) labels.

The functional (or path) integral expression for the partition function is

$$Z = \int \mathcal{D}\psi^* \mathcal{D}\psi\, e^{\frac{i}{\hbar} \int dt\, \psi^* \left(i\hbar\partial_t - \hat{h} + \mu\right) \psi} \tag{8.136}$$

In imaginary time, set $t \to -i\tau$, with $0 \leq \tau \leq \beta\hbar$:

$$Z = \int \mathcal{D}\psi^* \mathcal{D}\psi\, e^{-\int_0^{\beta\hbar} d\tau\, \psi^* \left(\partial_\tau + \hat{h} - \mu\right) \psi} \tag{8.137}$$

The fields $\psi(\tau)$ can represent either bosons (in which case they are just complex functions of x and τ), or fermions (in which case they are complex Grassmann functions of x and τ). The only subtlety resides in the choice of boundary conditions.

1) *Bosons.* Since the partition function is a trace, in this case, the fields (complex or real) must obey the usual *periodic boundary conditions* in imaginary time:

$$\psi(\tau) = \psi(\tau + \beta\hbar) \tag{8.138}$$

2) *Fermions.* In the case of fermions, the fields are complex Grassmann variables. However, if we want to compute a trace, because of the anticommutation rules, it is necessary to require the fields to obey instead *antiperiodic boundary conditions*:

$$\psi(\tau) = -\psi(\tau + \beta\hbar) \tag{8.139}$$

Let $\{|\lambda\rangle\}$ be a complete set of eigenstates of the one-body Hamiltonian $\hat{h}$, and let $\{\varepsilon_\lambda\}$ be its eigenvalue spectrum with λ being a spectral parameter. Hence, we have a suitable set of quantum numbers spanning the spectrum of $\hat{h}$. Also, let $\{\phi_\lambda(\tau)\}$ be the associated complete set of eigenfunctions. We now expand the field configurations in the basis of eigenfunctions of $\hat{h}$:

$$\psi(\tau) = \sum_\lambda \psi_\lambda\, \phi_\lambda(\tau) \tag{8.140}$$

The eigenfunctions of $\hat{h}$ are complete and orthonormal.

Thus, if we expand the fields, the path integral in eq. (8.136) becomes (with $\hbar = 1$)

$$Z = \int \left(\prod_\lambda d\psi_\lambda^* d\psi_\lambda\right) e^{-\int d\tau \sum_\lambda \psi_\lambda^* \left(-\partial_\tau - \varepsilon_\lambda + \mu\right) \psi_\lambda} \tag{8.141}$$

which becomes (after absorbing all uninteresting constant factors in the integration measure)

$$Z=\prod_{\lambda}[\text{Det}\,(-\partial_\tau-\varepsilon_\lambda+\mu)]^\sigma \tag{8.142}$$

where $\sigma=+1$ for fermions, and $\sigma=-1$ for bosons.

Let $\psi_n^\lambda(\tau)$ be the solution of the equation

$$(-\partial_\tau-\varepsilon_\lambda+\mu)\,\psi_n^\lambda(\tau)=\alpha_n\psi_n^\lambda(\tau) \tag{8.143}$$

where α_n is the (generally complex) eigenvalue. The eigenfunctions $\psi_n^\lambda(\tau)$ will be required to satisfy either periodic or antiperiodic boundary conditions,

$$\psi_n^\lambda(\tau)=-\sigma\,\psi_n^\lambda(\tau+\beta) \tag{8.144}$$

where, once again, $\sigma=\pm1$.

The eigenvalue condition, eq. (8.143), is solved by

$$\psi_n(\tau)=\psi_n\,e^{i\alpha_n\tau} \tag{8.145}$$

provided α_n satisfies

$$\alpha_n=-i\omega_n+\mu-\varepsilon_\lambda \tag{8.146}$$

where the Matsubara frequencies are given by (with $k_B=1$):

$$\omega_n=\begin{cases}2\pi T\left(n+\frac{1}{2}\right), & \text{for fermions}\\ 2\pi Tn, & \text{for bosons}\end{cases} \tag{8.147}$$

Let us consider now the function $\varphi_\alpha(\tau)$, which is an eigenfunction of $-\partial_\tau-\varepsilon_\lambda+\mu$,

$$(-\partial_\tau-\varepsilon_\lambda+\mu)\,\varphi_\alpha(\tau)=\alpha\,\varphi_\alpha(\tau) \tag{8.148}$$

which satisfies only an initial condition for $\varphi_\alpha^\lambda(0)$, such as

$$\varphi_\alpha^\lambda(0)=1 \tag{8.149}$$

Notice that since the operator is linear in ∂_τ, we cannot impose additional conditions on the derivative of φ_α.

The solution of

$$\partial_\tau\ln\varphi_\alpha^\lambda(\tau)=\mu-\varepsilon-\alpha \tag{8.150}$$

is

$$\varphi_\alpha^\lambda(\tau)=\varphi_\alpha^\lambda(0)\,e^{(\mu-\varepsilon_\lambda-\alpha)\,\tau} \tag{8.151}$$

After imposing the initial condition eq. (8.149), we find

$$\varphi_\alpha^\lambda(\tau)=e^{-(\alpha+\varepsilon_\lambda-\mu)\,\tau} \tag{8.152}$$

But even though this function $\varphi_\alpha^\lambda(\tau)$ satisfies all the requirements, it does not have the same zeros as the determinant $\mathrm{Det}(-\partial_\tau + \mu - \varepsilon_\lambda - \alpha)$. However, the function

$$ {}_\sigma F_\alpha^\lambda(\tau) = 1 + \sigma\, \varphi_\alpha^\lambda(\tau) \tag{8.153}$$

does satisfy all the requirements. Indeed,

$$ {}_\sigma F_\alpha^\lambda(\beta) = 1 + \sigma\, e^{-(\alpha+\varepsilon_\lambda-\mu)\beta} \tag{8.154}$$

which vanishes for $\alpha = \alpha_n$. Then, a version of Coleman's argument, discussed in section 5.2.2, tells us that

$$ \frac{\mathrm{Det}\,(-\partial_\tau + \mu - \varepsilon_\lambda - \alpha)}{{}_\sigma F_\alpha^\lambda(\beta)} = \text{constant} \tag{8.155}$$

where the right side is a constant in the sense that it does not depend on the choice of the eigenvalues $\{\varepsilon_\lambda\}$.

Hence,

$$ \mathrm{Det}\,(-\partial_\tau + \mu - \varepsilon_\lambda) = \text{const.}\; {}_\sigma F_0^\lambda(\beta) \tag{8.156}$$

The partition function is

$$ Z = e^{-\beta F} = \prod_\lambda \left[\mathrm{Det}\,(-\partial_\tau + \mu - \varepsilon_\lambda)\right]^\sigma \tag{8.157}$$

where F is the free energy, which is given by

$$ \begin{aligned} F &= -\sigma T \sum_\lambda \ln \mathrm{Det}\,(-\partial_\tau + \mu - \varepsilon_\lambda - \alpha) \\ &= -\sigma T \sum_\lambda \ln\left(1 + \sigma e^{\beta(\mu - \varepsilon_\lambda)}\right) + f(\beta\mu) \end{aligned} \tag{8.158}$$

which is the correct result for noninteracting fermions and bosons. Here we have set

$$ f(\beta\mu) = \begin{cases} 0, & \text{fermions} \\ -2T\mathcal{N}\,\ln\left(1 - e^{\beta\mu}\right), & \text{bosons} \end{cases} \tag{8.159}$$

where $\mathcal{N}$ is the number of states in the spectrum $\{\lambda\}$.

In some cases, the spectrum has the symmetry $\varepsilon_\lambda = -\varepsilon_{-\lambda}$ (e.g., the Dirac theory, whose spectrum is $\varepsilon_\pm = \pm\sqrt{p^2 + m^2}$), and these expressions can be simplified further,

$$ \begin{aligned} \prod_\lambda \mathrm{Det}\,(-\partial_t + i\varepsilon_\lambda) &= \prod_{\lambda>0} \left[\mathrm{Det}\,(-\partial_t + i\varepsilon_\lambda)\,\mathrm{Det}\,(-\partial_t - i\varepsilon_\lambda)\right] \\ &= \prod_{\lambda>0} \mathrm{Det}\left(\partial_t^2 + \varepsilon_\lambda^2\right) \\ &\equiv \prod_{\lambda>0} \mathrm{Det}\left(-\partial_\tau^2 + \varepsilon_\lambda^2\right) \end{aligned} \tag{8.160}$$

where, in the last step, we performed a Wick rotation. This last expression we have encountered before. The result is

$$\prod_{\lambda>0} \text{Det}\left(-\partial_\tau^2 + \varepsilon_\lambda^2\right) = \text{const.}\, \psi_0(\beta) \tag{8.161}$$

where $\psi_0(\tau)$ is the solution of the differential equation

$$\left(-\partial_\tau^2 + \varepsilon_\lambda^2\right) \psi_0(\tau) = 0 \tag{8.162}$$

which satisfies the initial conditions

$$\psi_0(0) = 0, \qquad \partial_\tau \psi_0(0) = 1 \tag{8.163}$$

The solution is

$$\psi_0(\tau) = \frac{\sinh(|\varepsilon_\lambda|\tau)}{|\varepsilon_\lambda|} \tag{8.164}$$

Hence

$$\psi_0(\beta) = \frac{\sinh(|\varepsilon_\lambda|\beta)}{|\varepsilon_\lambda|} \longrightarrow \frac{e^{|\varepsilon_\lambda|\beta}}{2|\varepsilon_\lambda|}, \qquad \text{as } \beta \to \infty \tag{8.165}$$

In particular, since

$$\prod_{\lambda>0} \frac{e^{|\varepsilon_\lambda|\beta}}{2|\varepsilon_\lambda|} = e^{\beta \sum_{\lambda>0} |\varepsilon_\lambda|} = e^{-\beta \sum_{\lambda<0} \varepsilon_\lambda} \tag{8.166}$$

we get that the ground state energy E_G is the sum of the single-particle energies of the occupied (negative energy) states:

$$E_G = \sum_{\lambda<0} \varepsilon_\lambda \tag{8.167}$$

8.8.2 Functional determinants, heat kernels, and ζ-function regularization

We have seen before that the evaluation of the effects of quantum fluctuations involves the calculation of the determinant of a differential operator. In the case of nonrelativistic single-particle quantum mechanics, in section 5.2.2, we discussed in detail how to calculate a functional determinant of the form $\text{Det}\left[-\partial_t^2 + W(t)\right]$. However, the method we used for that purpose becomes unmanageably cumbersome if applied to the calculation of determinants of partial differential operators of the form $\text{Det}\left[-D^2 + W(x)\right]$, where $x \equiv x_\mu$ and D is some differential operator. Fortunately, there are better and more efficient ways of doing such calculations.

Let $\hat{A}$ be an operator, and $\left\{f_n(x)\right\}$ be a complete set of eigenstates of $\hat{A}$, with the eigenvalue spectrum $\{a_n\}$, such that

$$\hat{A} f_n(x) = a_n f_n(x) \tag{8.168}$$

We will assume that $\hat{A}$ has a discrete spectrum of real positive eigenvalues, and hence it is bounded from below. For the case of a continuous spectrum, we will put the system in a finite box, which makes the spectrum discrete, and take limits at the end of the calculation.

The function $\zeta(s)$,

$$\zeta(s) = \sum_{n=1}^{\infty} \frac{1}{n^s}, \qquad \text{for Re}\, s > 1 \tag{8.169}$$

is the well-known Riemann ζ-function. We will now use the eigenvalue spectrum of the operator $\hat{A}$ to define the *generalized ζ-function*

$$\zeta_A(s) = \sum_n \frac{1}{a_n^s} \tag{8.170}$$

where the sum runs over the labels (here denoted by n) of the spectrum of the operator $\widehat{A}$. Assume that the sum (infinite series) is convergent, which, in practice, will require that we introduce some sort of regularization at the high end (high energies) of the spectrum.

On differentiation, we find

$$\frac{d\zeta_A}{ds} = \sum_n \frac{d}{ds} e^{-s \ln a_n} = -\sum_n \frac{\ln a_n}{a_n^s} \tag{8.171}$$

where we have assumed convergence. Then, in the limit $s \to 0^+$, we find that the following identity holds:

$$\lim_{s\to 0^+} \frac{d\zeta_A}{ds} = -\sum_n \ln a_n = -\ln \prod_n a_n \equiv -\ln \operatorname{Det} A \tag{8.172}$$

Hence, we can formally relate the generalized ζ-function, ζ_A, to the functional determinant of the operator A:

$$\left.\frac{d\zeta_A}{ds}\right|_{s\to 0^+} = -\ln \operatorname{Det} A \tag{8.173}$$

We have thus reduced the computation of a determinant to the computation of a function with specific properties.

Let us define now the generalized *heat kernel*,

$$G_A(x, y; \tau) = \sum_n e^{-a_n \tau} f_n(x) f_n^*(y) \equiv \langle x | e^{-\tau \hat{A}} | y \rangle \tag{8.174}$$

where $\tau > 0$. The heat kernel $G_A(x, y; \tau)$ clearly obeys the differential equation

$$-\partial_\tau G_A(x, y; \tau) = \hat{A} G_A(x, y; \tau) \tag{8.175}$$

which can be regarded as a generalized heat equation. Indeed, for $\hat{A} = -D\nabla^2$, this is the regular heat equation (where D is the diffusion constant) and, in this case, τ represents time. In general, we will refer to τ as *proper time.*

The heat kernel $G_A(x, y; \tau)$ satisfies the initial condition

$$\lim_{\tau\to 0^+} G_A(x, y; \tau) = \sum_n f_n(x) f_n^*(y) = \delta(x - y) \tag{8.176}$$

where we have used the completeness relation of the eigenfunctions $\{f_n(x)\}$. Hence, $G_A(x, y; \tau)$ is the solution of a generalized heat equation with kernel $\hat{A}$. It defines a generalized random walk or Markov process.

We now show that $G_A(x, y; \tau)$ is related to the generalized ζ-function $\zeta_A(s)$. Indeed, let us consider the heat kernel $G_A(x, y; \tau)$ at *short distances*, $y \to x$, and compute the integral

(below, we denote by D the dimensionality of spacetime)

$$\int d^D x \lim_{y \to x} G_A(x, y; \tau) = \sum_n e^{-a_n \tau} \int d^D x f_n(x) f_n^*(x)$$
$$= \sum_n e^{-a_n \tau} \equiv \operatorname{tr} e^{-\tau \hat{A}} \tag{8.177}$$

where we assumed that the eigenfunctions are normalized to unity:

$$\int d^D x \, |f_n(x)|^2 = 1 \tag{8.178}$$

(i.e., normalized inside a box).

We will now use that, for $s > 0$ and $a_n > 0$ (or at least, it has a positive real part), we can write

$$\int_0^\infty d\tau \, \tau^{s-1} \, e^{-a_n \tau} = \frac{\Gamma(s)}{a_n^s} \tag{8.179}$$

where $\Gamma(s)$ is the Euler gamma function:

$$\Gamma(s) = \int_0^\infty d\tau \, \tau^{s-1} \, e^{-\tau} \tag{8.180}$$

Then, we obtain the identity

$$\int_0^\infty d\tau \, \tau^{s-1} \int d^D x \lim_{y \to x} G_A(x, y; \tau) = \sum_n \frac{\Gamma(s)}{a_n^s} \tag{8.181}$$

Therefore, we find that the generalized ζ-function, $\zeta_A(s)$, can be obtained from the generalized heat kernel $G_A(x, y; \tau)$:

$$\zeta_A(s) = \frac{1}{\Gamma(s)} \int_0^\infty d\tau \, \tau^{s-1} \int d^D x \lim_{y \to x} G_A(x, y; \tau) \tag{8.182}$$

This result suggests the following strategy for the computation of determinants. Given the hermitian operator $\hat{A}$, we solve the generalized heat equation

$$\hat{A} \, G_A = -\partial_\tau \, G_A \tag{8.183}$$

subject to the initial condition

$$\lim_{\tau \to 0^+} G_A(x, y; \tau) = \delta^D(x - y) \tag{8.184}$$

Next we find the associated ζ-function, $\zeta_A(s)$, using the expression

$$\zeta_A(s) = \frac{1}{\Gamma(s)} \int_0^\infty d\tau \, \tau^{s-1} \int d^D x \lim_{y \to x} G_A(x, y; \tau) \tag{8.185}$$

where we recognize the identity

$$\text{tr}\, e^{-\tau \hat{A}} \equiv \int d^D x \lim_{y \to x} G_A(x, y; \tau) \tag{8.186}$$

We next take the limit $s \to 0^+$ to relate the generalized ζ-function to the determinant:

$$\lim_{s \to 0^+} \frac{d\zeta_A(s)}{ds} = -\ln \det \hat{A} \tag{8.187}$$

In practice, we will have to exercise some care in this step, since singularities arise as we take this limit. Most often, we will keep the points x and y apart by a small but finite distance a, which we will eventually attempt to take to zero. Hence, we need to understand in detail the short-distance behavior of the heat kernel.

Furthermore, the propagator of the theory

$$S_A(x, y) = \langle x | \hat{A}^{-1} | y \rangle \tag{8.188}$$

can also be related to the heat kernel. Indeed, by expanding eq. (8.188) in the eigenstates of $\hat{A}$, we find

$$S_A(x, y) = \sum_n \frac{\langle x|n\rangle \, \langle n|y\rangle}{a_n} = \sum_n \frac{f_n(x) f_n^*(y)}{a_n} \tag{8.189}$$

We can now write the integral of the heat kernel as

$$\int_0^\infty d\tau \; G_A(x, y; \tau) = \sum_n f_n(x) f_n^*(y) \int_0^\infty d\tau \; e^{-a_n \tau} = \sum_n \frac{f_n(x) f_n^*(y)}{a_n} \tag{8.190}$$

Hence, the propagator $S_A(x, y)$ can be expressed as an integral of the heat kernel:

$$S_A(x, y) = \int_0^\infty d\tau \; G_A(x, y; \tau) \tag{8.191}$$

Equivalently, since $G_A(x, y; \tau)$ satisfies the heat equation

$$\hat{A} G_A = -\partial_\tau G_A, \qquad \text{then} \qquad G_A(x, y; \tau) = \langle x | e^{-\tau \hat{A}} | y \rangle \tag{8.192}$$

thus,

$$S_A(x, y) = \int_0^\infty d\tau \; \langle x | e^{-\tau \hat{A}} | y \rangle = \langle x | \hat{A}^{-1} | y \rangle \tag{8.193}$$

and which is indeed the Green function of $\hat{A}$,

$$\hat{A}_x S(x, y) = \delta(x - y) \tag{8.194}$$

It is worth noting that the heat kernel $G_A(x, y; \tau)$, as can be seen from eq. (8.192), is also the *Gibbs density matrix* of the bounded Hermitian operator $\hat{A}$. As such, it has an imaginary time (τ!) path-integral representation. Here, to actually ensure convergence, we must also require that the spectrum of $\hat{A}$ be positive. In that picture, we view $G_A(x, y; \tau)$ as the amplitude for the imaginary time (proper time) evolution from the initial state $|y\rangle$ to the final state $|x\rangle$. In other words, we picture $S_A(x, y)$ as the amplitude needed to go from y to x in an arbitrary time.

8.9 The determinant of the Euclidean Klein-Gordon operator

As an example of the use of the heat kernel method, let us use it to compute the determinant of the Euclidean Klein-Gordon operator. Thus, we take the hermitian operator $\hat{A}$ to be

$$\hat{A} = -\nabla^2 + m^2 \tag{8.195}$$

in D Euclidean spacetime dimensions. This operator has a bounded positive spectrum. Here we are interested in a system with infinite size $L \to \infty$, and a large volume $V = L^D$. We will follow the steps outlined above.

We begin by constructing the heat kernel $G(x, y; \tau)$. By definition, it is the solution of the partial differential equation

$$\left(-\nabla^2 + m^2\right) G(x, y; \tau) = -\partial_\tau G(x, y; \tau) \tag{8.196}$$

satisfying the initial condition

$$\lim_{\tau \to 0^+} G(x, y; \tau) = \delta^D(x - y) \tag{8.197}$$

We will find $G(x, y; \tau)$ by Fourier transforms:

$$G(x, y; \tau) = \int \frac{d^D p}{(2\pi)^D} \, G(\boldsymbol{p}, \tau) \, e^{i\boldsymbol{p} \cdot (\boldsymbol{x} - \boldsymbol{y})} \tag{8.198}$$

For $G(x, y; \tau)$ to satisfy eq. (8.196), its Fourier transform $G(\boldsymbol{p}; \tau)$ must satisfy the differential equation

$$-\partial_\tau G(\boldsymbol{p}; \tau) = \left(\boldsymbol{p}^2 + m^2\right) G(\boldsymbol{p}; \tau) \tag{8.199}$$

The solution of this equation, consistent with the initial condition of eq. (8.197), is

$$G(\boldsymbol{p}; \tau) = e^{-\left(\boldsymbol{p}^2 + m^2\right)\tau} \tag{8.200}$$

We can now easily find $G(x, y; \tau)$ by simply finding the inverse transform of $G(\boldsymbol{p}; \tau)$:

$$\begin{aligned} G(x, y; \tau) &= \int \frac{d^D p}{(2\pi)^D} \, e^{-\left(\boldsymbol{p}^2 + m^2\right)\tau + i\boldsymbol{p} \cdot (\boldsymbol{x} - \boldsymbol{y})} \\ &= \frac{1}{(4\pi\tau)^{D/2}} \, e^{-\left(m^2 \tau + \frac{|\boldsymbol{x} - \boldsymbol{y}|^2}{4\tau}\right)} \end{aligned} \tag{8.201}$$

Notice that for $m \to 0$, $G(x, y; \tau)$ reduces to the usual diffusion kernel (with unit diffusion constant)

$$\lim_{m\to 0} G(x, y; \tau) = \frac{1}{(4\pi\tau)^{D/2}} e^{-\frac{|x-y|^2}{4\tau}} \tag{8.202}$$

Next we construct the ζ-function

$$\zeta_{-\nabla^2+m^2}(s) = \frac{1}{\Gamma(s)} \int_0^\infty d\tau\, \tau^{s-1} \int d^D x \lim_{y\to x} G(x, y; \tau) \tag{8.203}$$

We first do the integral

$$\begin{aligned} &\int_0^\infty d\tau\, \tau^{s-1} \int d^D x\, G(x, y; \tau) \\ &\qquad = \frac{V}{(4\pi)^{D/2}} \int_0^\infty d\tau\, \tau^{s-1-D/2}\, e^{-\left(m^2\tau + \frac{R^2}{4\tau}\right)} \end{aligned} \tag{8.204}$$

where $R = |\boldsymbol{x} - \boldsymbol{y}|$. Upon scaling the variable $\tau = \lambda t$, with $\lambda = R/2m$, we find that

$$\int_0^\infty d\tau\, \tau^{s-1} G(x, y; \tau) = \frac{2}{(4\pi)^{D/2}} \left(\frac{R}{2m}\right)^{s-\frac{D}{2}} K_{\frac{D}{2}-s}(mR) \tag{8.205}$$

where $K_\nu(z)$,

$$K_\nu(z) = \frac{1}{2} \int_0^\infty dt\, t^{\nu-1} e^{-\frac{z}{2}\left(t+\frac{1}{t}\right)} \tag{8.206}$$

is a modified Bessel function. Its short-argument behavior is

$$K_\nu(z) \sim \frac{\Gamma(\nu)}{2} \left(\frac{2}{z}\right)^\nu + \cdots \tag{8.207}$$

As a check, notice that for $s = 1$, the integral of eq. (8.204) does reproduce the Euclidean Klein-Gordon propagator discussed in section 5.6.3.

The next step is to take the short-distance limit:

$$\begin{aligned} \lim_{R\to 0} \int_0^\infty d\tau\, \tau^{s-1} G(\boldsymbol{x}, \boldsymbol{y}; \tau) &= \lim_{R\to 0} \frac{2^{1-s}}{(2\pi)^{D/2}} \frac{m^{D-2s}}{(mR)^{\frac{D}{2}-s}} K_{\frac{D}{2}-s}(mR) \\ &= \frac{\Gamma\left(s-\frac{D}{2}\right)}{(4\pi)^{D/2} m^{2s-D}} \end{aligned} \tag{8.208}$$

Notice that the order of the limit and the integration have been exchanged. Also, after taking the short-distance limit $R \to 0$, the expression above acquired a factor of $\Gamma(s - D/2)$, which is singular as $s - D/2$ approaches zero (or any negative integer). Thus, a small but finite R smears this singularity.

Finally we find the ζ-function by doing the (trivial) integration over space

$$\begin{aligned}\zeta(s) &= \frac{1}{\Gamma(s)} \int_0^\infty d\tau \int d^D x \lim_{y \to x} G(\boldsymbol{x}, \boldsymbol{y}; \tau) \\ &= V \mu^{-2s} \frac{m^D}{(4\pi)^{D/2}} \frac{\Gamma\left(s - \frac{D}{2}\right)}{\Gamma(s)} \left(\frac{m}{\mu}\right)^{-2s}\end{aligned} \tag{8.209}$$

where $\mu = 1/R$ plays the role of a cutoff mass (or momentum) scale that we will need to make some quantities dimensionless. The appearance of this quantity is also a consequence of the singularities.

Let us now consider the specific case of $D=4$ dimensions. For $D=4$, the ζ-function is

$$\zeta(s) = V \frac{m^4}{16\pi^2} \frac{\mu^{-2s}}{(s-1)(s-2)} \left(\frac{m}{\mu}\right)^{-2s} \tag{8.210}$$

We can now compute the desired (logarithm of the) determinant for $D=4$ dimensions:

$$\ln \operatorname{Det}\left[-\nabla^2 + m^2\right] = -\lim_{s \to 0^+} \frac{d\zeta}{ds} = \frac{m^4}{16\pi^2} \left[\ln \frac{m}{\mu} - \frac{3}{4}\right] V \tag{8.211}$$

where $V = L^4$. A similar calculation for $D=2$ yields the result

$$\ln \operatorname{Det}\left[-\nabla^2 + m^2\right] = -\frac{m^2}{2\pi} \left[\ln \frac{m}{\mu} - \frac{1}{2}\right] V \tag{8.212}$$

where $V = L^2$.

8.10 Path integral for spin

We now discuss the use of path-integral methods to describe a quantum mechanical spin. Consider a quantum mechanical system that consists of a spin in the spin-S representation of the group $SU(2)$. The space of states of the spin-S representation is $2S+1$-dimensional, and it is spanned by the basis $\{|S, M\rangle\}$ which consists of the eigenstates of the operators $\boldsymbol{S}^2$ and S_3, that is,

$$\begin{aligned}\boldsymbol{S}^2 \, |S, M\rangle &= S(S+1) \, |S, M\rangle \\ S_3 \, |S, M\rangle &= M \, |S, M\rangle\end{aligned} \tag{8.213}$$

with $|M| \leq S$ (in integer-spaced intervals). This set of states is complete, and it forms a basis of this Hilbert space. The operators S_1, S_2, and S_3 obey the $SU(2)$ algebra,

$$[S_a, S_b] = i\epsilon_{abc} S_c \tag{8.214}$$

where $a, b, c = 1, 2, 3$.

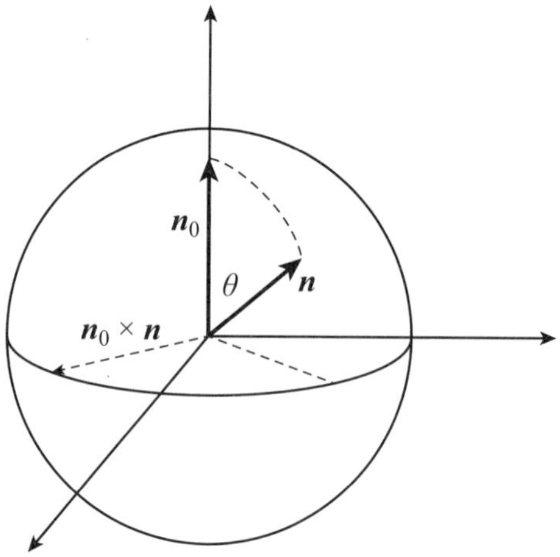

Figure 8.1 Geometry for a spin coherent state $|\boldsymbol{n}\rangle$.

The simplest physical problem involving spin is the coupling to an external magnetic field $\boldsymbol{B}$ through the Zeeman interaction

$$H_{\text{Zeeman}} = \mu\, \boldsymbol{B} \cdot \boldsymbol{S} \tag{8.215}$$

where μ is the Zeeman coupling constant (i.e., the product of the Bohr magneton and the gyromagnetic factor).

Let us denote by $|0\rangle$ the *highest weight state* $|S, S\rangle$. Define the spin raising and lowering operators $S^{\pm}$ as

$$S^{\pm} = S_1 \pm i S_2 \tag{8.216}$$

The highest weight state $|0\rangle$ is annihilated by S^+:

$$S^+|0\rangle = S^+|S, S\rangle = 0 \tag{8.217}$$

Clearly, we also have

$$\begin{aligned} \boldsymbol{S}^2|0\rangle &= S(S+1)|0\rangle \\ S_3|0\rangle &= S|0\rangle \end{aligned} \tag{8.218}$$

Let us consider now the spin coherent state $|\boldsymbol{n}\rangle$ (Perelomov, 1986),

$$|\boldsymbol{n}\rangle = e^{i\theta \boldsymbol{n}_0 \times \boldsymbol{n} \cdot \boldsymbol{S}}\,|0\rangle \tag{8.219}$$

where $\boldsymbol{n}$ is a three-dimensional unit vector ($\boldsymbol{n}^2 = 1$), $\boldsymbol{n}_0$ is a unit vector pointing along the direction of the quantization axis (i.e., the "north pole" of the unit sphere), and θ is the *colatitude* (see figure 8.1),

$$\boldsymbol{n} \cdot \boldsymbol{n}_0 = \cos\theta \tag{8.220}$$

As we will see, the state $|\boldsymbol{n}\rangle$ is a coherent spin state that represents a spin polarized along the $\boldsymbol{n}$-axis. The state $|\boldsymbol{n}\rangle$ can be expanded in the basis $|S, M\rangle$:

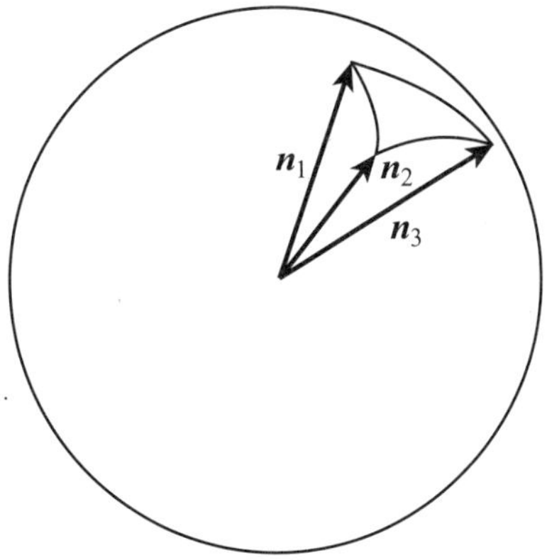

Figure 8.2 Spherical triangle with vertices at the unit vectors, $\boldsymbol{n}_1$, $\boldsymbol{n}_2$, and $\boldsymbol{n}_3$.

$$|\boldsymbol{n}\rangle = \sum_{M=-S}^{S} D^{(S)}_{MS}(\boldsymbol{n})\,|S, M\rangle \tag{8.221}$$

Here the $D^{(S)}_{MS}(\boldsymbol{n})$ are the representation matrices in the spin-S representation.

It is important to note that many rotations lead to the same state $|\boldsymbol{n}\rangle$ from the highest weight $|0\rangle$. For example, any rotation along the direction $\boldsymbol{n}$ results only in a change in the phase of the state $|\boldsymbol{n}\rangle$. These rotations are equivalent to a multiplication on the right by a rotation about the z-axis. However, in quantum mechanics, this phase has no physically observable consequence. Hence we will regard all of these states as being physically equivalent.

In other words, the states form *equivalence classes* (or *rays*), and we must pick one and only one state from each class. These rotations are generated by S_3, the (only) diagonal generator of $SU(2)$. Hence, the physical states are not in one-to-one correspondence with the elements of $SU(2)$ but instead with the elements of the right *coset* $SU(2)/U(1)$, with the $U(1)$ generated by S_3. In the case of the coherent state of a more general Lie group, the coset is obtained by dividing out the maximal torus generated by all diagonal generators of the group. In mathematical language, if we consider all rotations at once, the spin coherent states are said to form a *hermitian line bundle.*

A consequence of these observations is that the D matrices do not form a group under matrix multiplication. Instead they satisfy

$$D^{(S)}(\boldsymbol{n}_1)D^{(S)}(\boldsymbol{n}_2) = D^{(S)}(\boldsymbol{n}_3)\, e^{i\Phi(\boldsymbol{n}_1, \boldsymbol{n}_2, \boldsymbol{n}_3)S_3} \tag{8.222}$$

where the phase factor is usually called a *cocycle*. Here $\Phi(\boldsymbol{n}_1, \boldsymbol{n}_2, \boldsymbol{n}_3)$ is the (oriented) area of the spherical triangle with vertices at $\boldsymbol{n}_1, \boldsymbol{n}_2, \boldsymbol{n}_3$ (see figure 8.2).

However, since the sphere is a closed surface, which area do we actually mean? “Inside” or “outside”? Thus, the phase factor is ambiguous by an amount determined by 4π, the total area of the sphere:

$$e^{i4\pi M} \tag{8.223}$$

However, since M is either an integer or a half-integer, this ambiguity in Φ has no consequence whatsoever:

$$e^{i4\pi M} = 1 \tag{8.224}$$

We can also regard this result as a requirement that M be quantized to be an integer or a half-integer (i.e., the representations of $SU(2)$).

The states $|\boldsymbol{n}\rangle$ are coherent states that satisfy the following properties (Perelomov, 1986). The overlap of two coherent states $|\vec{n}_1\rangle$ and $|\boldsymbol{n}_2\rangle$ is

$$\begin{aligned}\langle \boldsymbol{n}_1|\boldsymbol{n}_2\rangle &= \langle 0|D^{(S)}(\boldsymbol{n}_1)^\dagger D^{(S)}(\boldsymbol{n}_2)|0\rangle \\ &= \langle 0|D^{(S)}(\boldsymbol{n}_0)e^{i\Phi(\boldsymbol{n}_1,\boldsymbol{n}_2,\vec{n}_0)S_3}|0\rangle \\ &= \left(\frac{1+\boldsymbol{n}_1\cdot\boldsymbol{n}_2}{2}\right)^S e^{i\Phi(\boldsymbol{n}_1,\boldsymbol{n}_2,\boldsymbol{n}_0)S}\end{aligned} \tag{8.225}$$

The (diagonal) matrix element of the spin operator is

$$\langle \boldsymbol{n}|\boldsymbol{S}|\boldsymbol{n}\rangle = S\,\boldsymbol{n} \tag{8.226}$$

Finally, the (overcomplete) set of coherent states $\{|\boldsymbol{n}\rangle\}$ has a resolution of the identity of the form

$$\hat{I} = \int d\mu(\boldsymbol{n})\, |\boldsymbol{n}\rangle\langle \boldsymbol{n}| \tag{8.227}$$

where the integration measure $d\mu(\boldsymbol{n})$ is

$$d\mu(\boldsymbol{n}) = \left(\frac{2S+1}{4\pi}\right)\delta(\boldsymbol{n}^2-1)d^3n \tag{8.228}$$

Let us now use the coherent states $\{|\boldsymbol{n}\rangle\}$ to find the path integral for a spin. In imaginary time τ (and with periodic boundary conditions), the path integral is simply the partition function

$$Z = \text{tr}\, e^{-\beta H} \tag{8.229}$$

where $\beta = 1/T$ (T is the temperature), and H is the Hamiltonian. As usual, the path-integral form of the partition function is found by splitting up the imaginary time interval $0 \leq \tau \leq \beta$ into N_τ steps, each of length $\delta\tau$, such that $N_\tau \delta\tau = \beta$. Hence we have

$$Z = \lim_{N_\tau\to\infty,\delta\tau\to 0} \text{tr}\left(e^{-\delta\tau H}\right)^{N_\tau} \tag{8.230}$$

and must insert the resolution of the identity at every intermediate time step:

$$\begin{aligned}Z &= \lim_{N_\tau\to\infty,\delta\tau\to 0}\left(\prod_{j=1}^{N_\tau}\int d\mu(\boldsymbol{n}_j)\right)\left(\prod_{j=1}^{N_\tau}\langle \boldsymbol{n}(\tau_j)|e^{-\delta\tau H}|\boldsymbol{n}(\tau_{j+1})\rangle\right) \\ &\simeq \lim_{N_\tau\to\infty,\delta\tau\to 0}\left(\prod_{j=1}^{N_\tau}\int d\mu(\vec{n}_j)\right)\left(\prod_{j=1}^{N_\tau}\left[\langle \boldsymbol{n}(\tau_j)|\vec{n}(\tau_{j+1})\rangle - \delta\tau\,\langle \boldsymbol{n}(\tau_j)|H|\boldsymbol{n}(\tau_{j+1})\rangle\right]\right)\end{aligned} \tag{8.231}$$

However, since

$$\frac{\langle \boldsymbol{n}(\tau_j)|H|\boldsymbol{n}(\tau_{j+1})\rangle}{\langle \boldsymbol{n}(\tau_j)|\boldsymbol{n}(\tau_{j+1})\rangle} \simeq \langle \boldsymbol{n}(\tau_j)|H|\boldsymbol{n}(\tau_j)\rangle = \mu S\boldsymbol{B}\cdot\boldsymbol{n}(\tau_j) \tag{8.232}$$

and

$$\langle \boldsymbol{n}(\tau_j)|\boldsymbol{n}(\tau_{j+1})\rangle = \left(\frac{1+\boldsymbol{n}(\tau_j)\cdot\boldsymbol{n}(\tau_{j+1})}{2}\right)^S e^{i\Phi(\boldsymbol{n}(\tau_j),\boldsymbol{n}(\tau_{j+1}),\boldsymbol{n}_0)S} \tag{8.233}$$

we can write the partition function in the form

$$Z = \lim_{N_\tau\to\infty,\delta\tau\to 0}\int \mathcal{D}\boldsymbol{n}\, e^{-S_E[\boldsymbol{n}]} \tag{8.234}$$

where $S_E[\boldsymbol{n}]$ is given by

$$\begin{aligned} -S_E[\boldsymbol{n}] &= iS\sum_{j=1}^{N_\tau}\Phi(\boldsymbol{n}(\tau_j),\vec{n}(\tau_{j+1}),\boldsymbol{n}_0)\\ &+S\sum_{j=1}^{N_\tau}\ln\left(\frac{1+\boldsymbol{n}(\tau_j)\cdot\vec{n}(\tau_{j+1})}{2}\right)-\sum_{j=1}^{N_\tau}(\delta\tau)\mu S\,\boldsymbol{n}(\tau_j)\cdot\boldsymbol{B} \end{aligned} \tag{8.235}$$

The first term on the right side of eq. (8.235) contains the expression $\Phi(\boldsymbol{n}(\tau_j),\vec{n}(\tau_{j+1}),\boldsymbol{n}_0)$, which has a simple geometric interpretation: It is the sum of the areas of the N_τ contiguous spherical triangles. These triangles have the pole $\boldsymbol{n}_0$ as a common vertex, and their other pairs of vertices trace a spherical polygon with vertices at $\{\boldsymbol{n}(\tau_j)\}$. In the time-continuum limit, this spherical polygon becomes the *history* of the spin, which traces a closed oriented curve $\Gamma=\{\boldsymbol{n}(\tau)\}$ (with $0\le\tau\le\beta$). Let us denote by Ω^+ the region of the sphere whose boundary is Γ and which contains the pole $\boldsymbol{n}_0$. The complement of this region is Ω^-, and it contains the opposite pole $-\boldsymbol{n}_0$. Hence we find that

$$\lim_{N_\tau\to\infty,\delta\tau\to 0}\Phi(\boldsymbol{n}(\tau_j),\boldsymbol{n}(\tau_{j+1}),\boldsymbol{n}_0) = \mathcal{A}[\Omega^+] = 4\pi - \mathcal{A}[\Omega^-] \tag{8.236}$$

where $\mathcal{A}[\Omega]$ is the area of the region Ω. Once again, the ambiguity of the area leads to the requirement that S should be an integer or a half-integer.

There is a simple and elegant way to write the area enclosed by Γ. Let $\boldsymbol{n}(\tau)$ be a history and Γ be the set of points on the 2-sphere traced by $\boldsymbol{n}(\tau)$ for $0\le\tau\le\beta$. Define $\boldsymbol{n}(\tau,s)$ (with $0\le s\le 1$) to be an *arbitrary* extension of $\boldsymbol{n}(\tau)$ from the curve Γ to the interior of the upper cap Ω_+, as shown in figure 8.3, such that

$$\boldsymbol{n}(\tau,0)=\boldsymbol{n}(\tau),\quad \boldsymbol{n}(\tau,1)=\boldsymbol{n}_0,\qquad \boldsymbol{n}(\tau,0)=\boldsymbol{n}(\tau+\beta,0) \tag{8.237}$$

Then the area can be written in the compact form

$$\mathcal{A}[\Omega_+] = \int_0^1 ds\int_0^\beta d\tau\, \boldsymbol{n}(\tau,s)\cdot\partial_\tau\boldsymbol{n}(\tau,s)\times\partial_s\boldsymbol{n}(\tau,s) \equiv S_{\mathrm{WZ}}[\boldsymbol{n}] \tag{8.238}$$

In mathematics, this expression for the area is called the (simplectic) 2-form, and in the literature, it is usually called a Wess-Zumino action (Witten, 1984), S_{WZ}, or a *Berry phase*

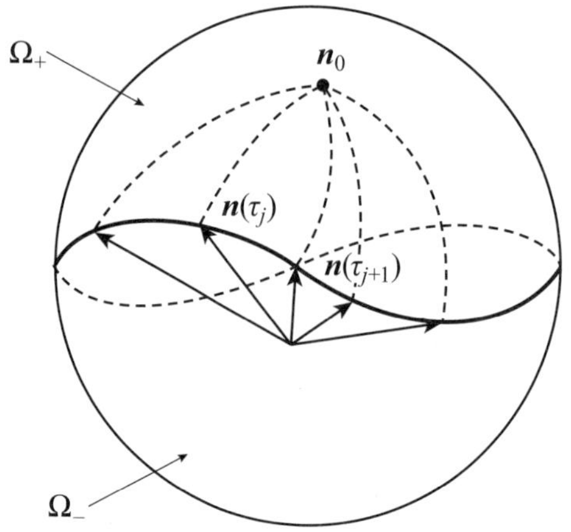

Figure 8.3 The function $n(\tau, s)$ as an arbitrary extension of the history $n(\tau)$ to the upper cap Ω_+ of the sphere S_2.

(Simon, 1983; Berry, 1984). The coherent-state path integral for spin is a special case of the method of geometric quantization (Wiegmann, 1989).

Thus, in the (formal) time continuum limit, the action S_E becomes (Fradkin and Stone, 1988)

$$\mathcal{S}_E = -iS\,\mathcal{S}_{WZ}[\boldsymbol{n}] + \frac{S\delta\tau}{2}\int_0^\beta d\tau\,(\partial_\tau \boldsymbol{n}(\tau))^2 + \int_0^\beta d\tau\,\mu S\,\boldsymbol{B}\cdot\boldsymbol{n}(\tau) \tag{8.239}$$

Notice that we have kept (temporarily) a term of order $\delta\tau$, which we will drop shortly.

How should we interpret eq. (8.239)? Since $\boldsymbol{n}(\tau)$ is constrained to be a point on the surface of the unit sphere (i.e., $\boldsymbol{n}^2 = 1$), the action $S_E[\boldsymbol{n}]$ can be interpreted as the action of a particle of mass $M = S\delta\tau \to 0$, and $\vec{n}(\tau)$ is the position vector of the particle at (imaginary) time τ. Thus, the second term is a (vanishingly small) kinetic energy term, and the last term in eq. (8.239) is a potential energy term.

What is the meaning of the first term? In eq. (8.238), we saw that $\mathcal{S}_{WZ}[\boldsymbol{n}]$, the so-called Wess-Zumino or Berry phase term in the action, is the area of the (positively oriented) region $\mathcal{A}[\Omega_+]$ "enclosed" by the "path" $\boldsymbol{n}(\tau)$. In fact,

$$\mathcal{S}_{WZ}[\boldsymbol{n}] = \int_0^1 ds \int_0^\beta d\tau\,\boldsymbol{n}\cdot\partial_\tau\boldsymbol{n}\times\partial_s\boldsymbol{n} \tag{8.240}$$

is the area of the oriented surface Ω_+, whose boundary is the oriented path $\Gamma = \partial\Omega_+$ (see figure 8.3). Using the Stokes' theorem, we can write the expression $S\mathcal{A}[\boldsymbol{n}]$ as the circulation of a vector field $\boldsymbol{A}[\boldsymbol{n}]$,

$$\oint_{\partial\Omega} d\boldsymbol{n}\cdot\boldsymbol{A}[\boldsymbol{n}(\tau)] = \iint_{\Omega_+} d\boldsymbol{S}\cdot\nabla_{\boldsymbol{n}}\times\boldsymbol{A}[\boldsymbol{n}(\tau)] \tag{8.241}$$

provided the "magnetic field" $\nabla_{\boldsymbol{n}}\times\boldsymbol{A}$ is "constant," namely,

$$\boldsymbol{B} = \nabla_{\boldsymbol{n}}\times\boldsymbol{A}[\boldsymbol{n}(\tau)] = S\,\boldsymbol{n}(\tau) \tag{8.242}$$

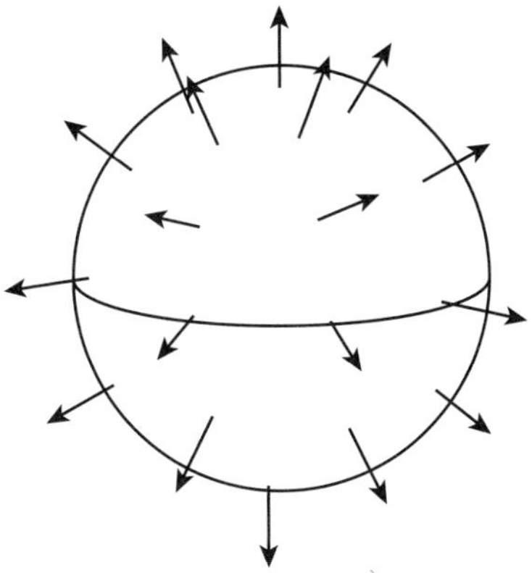

Figure 8.4 A hairy ball or monopole.

In other words, this is the magnetic field of a magnetic monopole located at the center of the sphere (see figure 8.4). What is the total flux Φ of this magnetic field?

$$\Phi = \int_{\text{sphere}} d\boldsymbol{S} \cdot \nabla_{\boldsymbol{n}} \times \boldsymbol{A}[\boldsymbol{n}] = S \int d\boldsymbol{S} \cdot \boldsymbol{n} \equiv 4\pi S \tag{8.243}$$

Thus, the total number of flux quanta N_ϕ piercing the unit sphere is

$$N_\phi = \frac{\Phi}{2\pi} = 2S = \text{magnetic charge} \tag{8.244}$$

We reach the conclusion that the magnetic charge is *quantized*, a result known as the *Dirac quantization condition*.

Is this result consistent with what we know about charged particles in magnetic fields? In particular, how is this result related to the physics of spin? To answer these questions, let us go back to real time and write the action

$$\mathcal{S}[\boldsymbol{n}] = \int_0^T dt \left[\frac{M}{2} \left(\frac{d\boldsymbol{n}}{dt} \right)^2 + \boldsymbol{A}[\boldsymbol{n}(t)] \cdot \frac{d\boldsymbol{n}}{dt} - \mu S \boldsymbol{n}(t) \cdot \boldsymbol{B} \right] \tag{8.245}$$

with the constraint $\boldsymbol{n}^2 = 1$ and where the limit $M \to 0$ is implied.

The classical Hamiltonian associated to the action in eq. (8.245) is

$$H = \frac{1}{2M} \left[\boldsymbol{n} \times \left(\boldsymbol{p} - \boldsymbol{A}[\boldsymbol{n}] \right) \right]^2 + \mu S \boldsymbol{n} \cdot \boldsymbol{B} \equiv H_0 + \mu S \boldsymbol{n} \cdot \boldsymbol{B} \tag{8.246}$$

It is easy to check that the vector $\boldsymbol{\Lambda}$,

$$\boldsymbol{\Lambda} = \boldsymbol{n} \times \left(\boldsymbol{p} - \boldsymbol{A} \right) \tag{8.247}$$

satisfies the algebra

$$[\Lambda_a, \Lambda_b] = i\hbar \epsilon_{abc} \left(\Lambda_c - \hbar S n_c \right) \tag{8.248}$$

where $a, b, c = 1, 2, 3$; ϵ_{abc} is the (third rank) Levi-Civita tensor; and with

$$\boldsymbol{\Lambda} \cdot \boldsymbol{n} = \boldsymbol{n} \cdot \boldsymbol{\Lambda} = 0 \tag{8.249}$$

the generators of rotations for this system are

$$\boldsymbol{L} = \boldsymbol{\Lambda} + \hbar S \boldsymbol{n} \tag{8.250}$$

The operators $\boldsymbol{L}$ and $\boldsymbol{\Lambda}$ satisfy the (joint) algebra

$$\begin{aligned} [L_a, L_b] &= -i\hbar\epsilon_{abc} L_s, \qquad & \left[L_a, \boldsymbol{L}^2\right] &= 0 \\ [L_a, n_b] &= i\hbar\epsilon_{abc} n_c, \qquad & [L_a, \Lambda_b] &= i\hbar\epsilon_{abc}\Lambda_c \end{aligned} \tag{8.251}$$

Hence

$$\left[L_a, \boldsymbol{\Lambda}^2\right] = 0 \Rightarrow [L_a, H] = 0 \tag{8.252}$$

Since the operators L_a satisfy the angular momentum algebra, we can diagonalize $\boldsymbol{L}^2$ and L_3 simultaneously. Let $|m, \ell\rangle$ be the simultaneous eigenstates of $\boldsymbol{L}^2$ and L_3,

$$\boldsymbol{L}^2|m, \ell\rangle = \hbar^2 \ell(\ell+1)|m, \ell\rangle \tag{8.253}$$

$$L_3|m, \ell\rangle = \hbar m|m, \ell\rangle \tag{8.254}$$

$$H_0|m, \ell\rangle = \frac{\hbar^2}{2MR^2}\left(\frac{\ell(\ell+1) - S}{2S}\right)|m, \ell\rangle \tag{8.255}$$

where $R = 1$ is the radius of the sphere. The eigenvalues ℓ are of the form $\ell = S + n$, $|m| \leq \ell$, with $n \in \mathbb{Z}^+ \cup \{0\}$ and $2S \in \mathbb{Z}^+ \cup \{0\}$. Hence each level is $(2\ell + 1)$-fold degenerate, or what is equivalent, $(2n + 1 + 2S)$-fold degenerate. Then we get

$$\boldsymbol{\Lambda}^2 = \boldsymbol{L}^2 - \boldsymbol{n}^2\hbar^2 S^2 = \boldsymbol{L}^2 - \hbar^2 S^2 \tag{8.256}$$

Since $M = S\delta t \to 0$, the lowest energies in the spectrum of H_0 are those with the smallest values of ℓ (i.e., states with $n = 0$ and $\ell = S$). The degeneracy of this "Landau" level is $2S + 1$, and the gap to the next excited states diverges as $M \to 0$. Thus, in the $M \to 0$ limit, the lowest energy states have the same degeneracy as the spin-S representation. Moreover, the operators $\boldsymbol{L}^2$ and L_3 become the corresponding spin operators. Thus the equivalency found is indeed correct.

Thus we have shown that the quantum states of a scalar (nonrelativistic) particle bound to a magnetic monopole of magnetic charge $2S$, obeying the Dirac quantization condition, are identical to those of a spinning particle (Wu and Yang, 1976)!

We close this section with some observations on the semiclassical motion. From the (real time) action (already in the $M \to 0$ limit),

$$\mathcal{S} = -\int_0^T dt\, \mu S\, \boldsymbol{n} \cdot \boldsymbol{B} + S\int_0^T dt \int_0^1 ds\, \boldsymbol{n} \cdot \partial_t \boldsymbol{n} \times \partial_s \boldsymbol{n} \tag{8.257}$$

we can derive a classical equation of motion by looking at the stationary configurations. The variation of the second term in eq. (8.257) is

$$\delta S = S\,\delta \int_0^T dt \int_0^1 ds\, \boldsymbol{n} \cdot \partial_t \boldsymbol{n} \times \partial_s \boldsymbol{n} = S \int_0^T dt \delta \boldsymbol{n}(t) \cdot \boldsymbol{n}(t) \times \partial_t \boldsymbol{n}(t) \tag{8.258}$$

and the variation of the first term in eq. (8.257) is

$$\delta \int_0^T dt\, \mu S \boldsymbol{n}(t) \cdot \boldsymbol{B} = \int_0^T dt\, \delta \boldsymbol{n}(t) \cdot \mu S \boldsymbol{B} \tag{8.259}$$

Hence,

$$\delta S = \int_0^T dt\, \delta \boldsymbol{n}(t) \cdot \Big(- \mu S \boldsymbol{B} + S \boldsymbol{n}(t) \times \partial_t \boldsymbol{n}(t) \Big) \tag{8.260}$$

which implies that the classical trajectories must satisfy the equation of motion

$$\mu \boldsymbol{B} = \boldsymbol{n} \times \partial_t \boldsymbol{n} \tag{8.261}$$

If we now use the vector identity

$$\boldsymbol{n} \times \boldsymbol{n} \times \partial_t \boldsymbol{n} = (\boldsymbol{n} \cdot \partial_t \boldsymbol{n})\, \boldsymbol{n} - \boldsymbol{n}^2 \partial_t \boldsymbol{n} \tag{8.262}$$

and

$$\boldsymbol{n} \cdot \partial_t \boldsymbol{n} = 0, \qquad \boldsymbol{n}^2 = 1 \tag{8.263}$$

we get the classical equation of motion:

$$\partial_t \boldsymbol{n} = \mu \boldsymbol{B} \times \boldsymbol{n} \tag{8.264}$$

Therefore, the classical motion is *precessional*, with an angular velocity $\boldsymbol{\Omega}_{\rm pr} = \mu \boldsymbol{B}$.

Exercises

8.1 Hilbert space of analytic functions

Let $\hat{a}$ and $\hat{a}^\dagger$ be a pair of creation and annihilation operators that act on a Hilbert space of states $\mathcal{H}$ to be specified below. The operators satisfy the algebra

$$\left[\hat{a}, \hat{a}^\dagger\right] = 1, \qquad \left[\hat{a}, \hat{a}\right] = \left[\hat{a}^\dagger, \hat{a}^\dagger\right] = 0$$

The Hilbert space $\mathcal{H}$ is defined to be the space of (locally) analytic functions $f(z)$ of a complex variable $z = x + iy$. The *inner product* in the space $\mathcal{H}$ for two analytic functions f and g is defined to be

$$(f, g) = \int \frac{dz d\bar{z}}{2\pi i}\, e^{-|z|^2} \bar{f}(z) g(z) \tag{8.265}$$

where $\bar{z}$ is the complex conjugate of z, $|z|^2 = z\bar{z}$, and the integration measure is

$$\frac{dzd\bar{z}}{2\pi i} \equiv \frac{dxdy}{\pi} \tag{8.266}$$

1) Show that $\hat{a}^\dagger \equiv z$ and $\hat{a} \equiv \frac{d}{dz}$ are a representation of the algebra in this space (i.e., show that they satisfy the algebra when acting on the space $\mathcal{H}$).
2) Show that the monomials $\psi_n(z) = \frac{z^n}{\sqrt{n!}}$ are orthonormal with the inner product defined above.
3) Show that $\psi_n(z)$ is an eigenfunction of the number operator $\hat{N} = \hat{a}^\dagger \hat{a}$. Find its eigenvalue.

8.2 Bose coherent states

Let $|z\rangle$ be the coherent state

$$|z\rangle = e^{z\hat{a}^\dagger}|0\rangle \tag{8.267}$$

where $|0\rangle$ is the state annihilated by $\hat{a}$.

1) Let $|\psi\rangle$ denote the state

$$|\psi\rangle = \sum_{n=0}^{\infty} \frac{\psi_n}{n!}(\hat{a}^\dagger)^n|0\rangle \tag{8.268}$$

Evaluate the inner product $\langle\psi|z\rangle$.

2) Compute the inner product $\langle u|w\rangle$, where u and w are two arbitrary complex numbers.
3) Let $\hat{A}$ be the normal-ordered operator

$$\hat{A} = \sum_{n,m=0}^{\infty} A_{n,m}(\hat{a}^\dagger)^n(\hat{a})^m \tag{8.269}$$

Prove the formula

$$\langle u|\hat{A}|w\rangle = A(\bar{u}, w)\langle u|w\rangle \tag{8.270}$$

where $A(\bar{u}, w)$ is obtained from $\hat{A}$ by means of the substitutions

$$\hat{a}^\dagger \to \bar{u}, \qquad \hat{a} \to w \tag{8.271}$$

4) Prove the formula ("resolution of the identity")

$$\hat{I} = \int \frac{dzd\bar{z}}{2\pi i}\, e^{-|z|^2}\, |z\rangle\langle z| \tag{8.272}$$

where $\hat{I}$ is the identity operator.

8.3 Grassmann variables

1) Let a and a^* be a pair of Grassmann variables. Let $g(a^*)$ be an analytic function of a single Grassmann variable a^*, that is,

$$g(a^*) = g_0 + g_1 a^* \tag{8.273}$$

and let $f(a)$ be another such function. Show that the inner product $\langle f|g\rangle$ defined by

$$\langle f|g\rangle = \int da^* da\, e^{-a^* a} f(a^*)^* g(a^*) \tag{8.274}$$

implies that

$$\langle f|g\rangle = \bar{f}_0 g_0 + \bar{f}_1 g_1 \tag{8.275}$$

where $\bar{x}$ stands for the complex conjugate of x.

2) Show that

$$(Af)(a^*) = \int d\alpha^* d\alpha A(a^*, \alpha) f(\alpha^*) e^{-\alpha^* \alpha} = g(a^*) \tag{8.276}$$

is equivalent to

$$\begin{pmatrix} g_0 \\ g_1 \end{pmatrix} = \begin{pmatrix} A_{00} & A_{10} \\ A_{01} & A_{11} \end{pmatrix} \begin{pmatrix} f_0 \\ f_1 \end{pmatrix} \tag{8.277}$$

and that

$$(A\,B)\,(a^*, a) = \int d\alpha^* d\alpha\; e^{-\alpha^* \alpha}\; A(a^*, \alpha)\, B(\alpha^*, a) = C(a^*, a) \tag{8.278}$$

is equivalent to the standard definition of the product of two 2×2 matrices.

3) Show that the operators $\hat{a}^*$ and $\hat{a}$, defined by

$$\hat{a}^* f(\alpha^*) = a^* f(a^*), \qquad \hat{a} f(a^*) = \frac{d}{da^*} f(a^*) \tag{8.279}$$

satisfy canonical anticommutation relations, that is, $\hat{a}^*\,\hat{a}^* = \hat{a}\,\hat{a} = 0$ and $\{\hat{a}^*, \hat{a}\} = 1$.

4) Show that, if $\{\xi_j\}$ is a set of N Grassmann variables ($j = 1, \ldots, N$), then the Grassmann integral is

$$\mathcal{Z} = \int \prod_{j=1}^{N} d\xi_j^* d\xi_j \,\exp\left\{-\sum \xi_k^* M_{kl} \xi_l\right\} = \mathrm{Det}\,\mathrm{M} \tag{8.280}$$

8.4 Path integral for Dirac fermions

The Lagrangian density $\mathcal{L}$ for the free massive Dirac field in four-dimensional Minkowski space is

$$\mathcal{L} = \bar{\psi}\,(i\partial\!\!\!/ - m)\,\psi \tag{8.281}$$

1) Consider the path integral for a free Dirac field in four spacetime dimensions, coupled to a set of Grassmann sources $\bar{\eta}_\alpha(x)$ and $\eta_\alpha(x)$. Derive an expression for this generating function in terms of the sources and a fermion determinant. Do not compute the determinant.

2) Use the results of the first part of this exercise to show that the Feynman propagator of the Dirac theory is given by

$$S_F^{\alpha\beta}(x-y) = \langle x, \alpha| \frac{1}{i\partial\!\!\!/ - m} |y, \beta\rangle \tag{8.282}$$

3) Use the results of the first part of this exercise to derive an expression for the four-point function

$$S_F^{(4)}(x_1, x_2, x_3, x_4)_{\alpha,\beta,\gamma,\delta} = \langle 0|\psi_\alpha(x_1)\psi_\beta(x_2)\bar{\psi}_\gamma(x_3)\bar{\psi}_\delta(x_4)|0\rangle \tag{8.283}$$

in terms of products of propagators. Beware of the signs!

8.5 The weakly interacting Bose gas
Consider a gas of nonrelativistic Bose particles at fixed density ρ inside a very large box of linear size L in three space dimensions. Let $\phi^\dagger(\boldsymbol{x})$ and $\phi(\boldsymbol{x})$ be a set of boson creation and annihilation operators, respectively. The second quantized Hamiltonian is

$$H = \int d^3x \left[\phi^\dagger(\boldsymbol{x}) \left(\frac{\hat{\boldsymbol{p}}^2}{2m} - \mu \right) \phi(\boldsymbol{x}) + \hat{n}(\boldsymbol{x}) V(\boldsymbol{x} - \boldsymbol{x}') \hat{n}(\boldsymbol{x}') \right] \tag{8.284}$$

where μ is the chemical potential, $\hat{n} = \phi^\dagger\phi$, and $V(r)$ is a rotationally invariant short-range interaction, which we will take to be equal to

$$V(\boldsymbol{x} - \boldsymbol{x}') = \lambda\delta^3(\boldsymbol{x} - \boldsymbol{x}') \tag{8.285}$$

The positive constant λ is the scattering amplitude and will play the role of a coupling constant for this system.

1) Use the method of Bose coherent states to find a path-integral formula for the partition function of this system at temperature T. Do not compute the path integral at this stage. Assume constant boundary conditions at spatial infinity (i.e., the field amplitude approaches a constant value at the boundaries). Carefully specify the boundary conditions in the imaginary time dimension. Write your answer down in the form

$$\mathcal{Z} = \int \mathcal{D}\phi^*\mathcal{D}\phi\, e^{-S_E(\phi^*, \phi)} \tag{8.286}$$

and give an explicit expression for the Euclidean action S_E.

2) Use the method of semiclassical quantization (i.e., the saddle point expansion) to determine the classical path at temperature T. What condition should be satisfied by $\phi(x)$ for it to be such a classical path? Find the relationship between the ground state of the system at $T = 0$ and this classical path in the limit $T \to 0$. Is the solution unique? Justify your answer. Hint: Think of the symmetries of the Lagrangian. This will give you an idea about the uniqueness of the classical path. You may find it convenient to write the classical path $\phi(x)$ in the form of an amplitude times a phase.

3) Compute the time-ordered Green function

$$G(x - y) = -i\langle \hat{T}\hat{\phi}(x)\hat{\phi}^\dagger(y)\rangle \tag{8.287}$$

at $T = 0$, in the semiclassical limit. What is the asymptotic value of G in the limit of equal times and large space separation? Give a physical interpretation of this result.

4) Consider small quantum fluctuations around the classical path found in part 2 of this exercise. Write an arbitrary configuration $\phi(x)$ in the form

$$\phi(x) = \sqrt{\rho_0 + \delta\rho(x)} \quad e^{i\theta(x)} \tag{8.288}$$

Expand the action in powers of $\delta\rho$ and θ up to second order in both variables. Check the cancellation of the linear terms. Integrate out the density fluctuations, and find an effective action for the phase variable

$$e^{-S_{\text{eff}}(\theta(x))} = \int \mathcal{D}\delta\rho \, e^{-S_E} \tag{8.289}$$

which is quadratic in θ.

5) Show that, for configurations $\{\theta(x)\}$ that are slowly varying, the effective action has the form

$$S_{\text{eff}} = \int d^4x \frac{1}{2} \left[a\,(\partial_\tau \theta)^2 + b\,(\nabla\theta(x))^2 \right] \tag{8.290}$$

and calculate the coefficients a and b. Find the analytic continuation of this expression back in real time. Find the time-ordered propagator of the phase field $\theta(x)$. What equation of motion does it satisfy? Draw an analogy between this equation and the equation of motion for a relativistic massless scalar field.

8.6 Functional determinants and the Casimir effect

Consider a free scalar field $\phi(x, t)$ in 1+1 spacetime dimensions. The Lagrangian density $\mathcal{L}$ is

$$\mathcal{L} = \frac{1}{2}\partial_\mu \phi(x)\partial^\mu \phi(x) - \frac{1}{2}m^2\phi(x)^2 \tag{8.291}$$

where $x \equiv (x, t)$. Consider the case in which the total length of the system along the space coordinate is equal to L, and assume periodic boundary conditions, that is,

$$\phi(x, t) = \phi(x + L, t) \tag{8.292}$$

for all times t.

1) Calculate the classical value of the ground state energy of the system with the boundary conditions specified above.
2) Use path integral methods to derive a formal expression for the total ground state energy density (i.e., energy per unit length). This formula should contain a determinant, which you should not compute for the moment.
3) Use the method of the ζ-function to compute the quantum correction to the ground state energy density. Consider the massless limit $m \to 0$ only. Write your answer in the form of an extensive term and a finite-sized term that vanishes as $L \to \infty$ like A/L^η, with $\eta > 0$. Find the value of this exponent η as well as that of the coefficient A, sign included. Note: You may have to keep a dependence on the mass in one of the two terms. Keep just the leading behavior in the small-mass

limit. Hint: At some point in the calculation, the following Poisson summation formula may be useful:

$$\sum_{n=-\infty}^{\infty} f(n) = \sum_{m=-\infty}^{\infty} \int_{-\infty}^{\infty} dx\, f(x)\, e^{2\pi imx} \tag{8.293}$$

4) If you interpret the dependence of the ground state energy on the linear size of the system L as a potential energy for the "walls" that confine the system, what can you say about the force that the zero-point fluctuations exert on these so-called walls? Note: To speak about walls, we should have used vanishing, instead of periodic boundary conditions, as we have done. The calculation is somewhat more complicated in that case. This effect (i.e., a force exerted on the walls of a system by the zero-point motion of a field) is known as the *Casimir effect*.

9

Quantization of Gauge Fields

We now turn to the problem of the quantization of gauge theories. We begin with the simplest gauge theory: the free electromagnetic field. This is an abelian gauge theory. After that we will discuss at length the quantization of non-abelian gauge fields. Unlike abelian theories, such as the free electromagnetic field, even in the absence of matter fields, non-abelian gauge theories are not free fields and have highly nontrivial dynamics.

9.1 Canonical quantization of the free electromagnetic field

Maxwell's theory was the first field theory to be quantized. The quantization procedure of a gauge theory, even for a free field, involves various subtleties not shared by the other problems we have considered so far. The issue is that this theory has a local gauge invariance. Unlike systems that only have global symmetries, not all the classical configurations of vector potentials represent physically distinct states. It could be argued that one should abandon the picture based on the vector potential and go back to a picture based on electric and magnetic fields instead. However, no local Lagrangian can describe the time evolution of the system in that representation. Furthermore, is not clear which field, $\boldsymbol{E}$ or $\boldsymbol{B}$ (or some other field), plays the role of coordinates and which can play the role of momentum. For that reason (and others), one sticks with the Lagrangian formulation with the vector potential A_μ as its independent coordinate-like variable.

The Lagrangian for Maxwell's theory,

$$\mathcal{L} = -\frac{1}{4} F_{\mu\nu} F^{\mu\nu} \tag{9.1}$$

where $F_{\mu\nu} = \partial_\mu A_\nu - \partial_\nu A_\mu$, can be written in the form

$$\mathcal{L} = \frac{1}{2}(\boldsymbol{E}^2 - \boldsymbol{B}^2) \tag{9.2}$$

where

$$E_j = -\partial_0 A_j - \partial_j A_0, \qquad B_j = -\epsilon_{jk\ell} \partial_k A_\ell \tag{9.3}$$

The electric field E_j and the space components of the vector potential A_j form a canonical pair, since by definition, the momentum Π_j that is conjugate to A_j is

$$\Pi_j(x) = \frac{\delta \mathcal{L}}{\delta \partial_0 A_j(x)} = \partial_0 A_j + \partial_j A_0 = -E_j \tag{9.4}$$

Notice that since $\mathcal{L}$ does not contain any terms that include $\partial_0 A_0$, the momentum Π_0, conjugate to A_0, vanishes:

$$\Pi_0 = \frac{\delta \mathcal{L}}{\delta \partial_0 A_0} = 0 \tag{9.5}$$

A consequence of this result is that A_0 is essentially arbitrary, and it plays the role of a Lagrange multiplier. Indeed, it is always possible to find a gauge transformation ϕ

$$A_0' = A_0 + \partial_0 \phi, \qquad A_j' = A_j - \partial_j \phi \tag{9.6}$$

such that $A_0' = 0$. The solution is

$$\partial_0 \phi = -A_0 \tag{9.7}$$

which is consistent, provided that A_0 vanishes both in the remote past and in the remote future, $x_0 \to \pm\infty$.

The canonical formalism can be applied to Maxwellian electrodynamics by noticing that the fields $A_j(\boldsymbol{x})$ and $\Pi_{j'}(\boldsymbol{x}')$ obey the equal-time Poisson brackets,

$$\{A_j(\boldsymbol{x}), \Pi_{j'}(\boldsymbol{x}')\}_{PB} = \delta_{jj'} \delta^3(\boldsymbol{x} - \boldsymbol{x}') \tag{9.8}$$

or, in terms of the electric field $\boldsymbol{E}$,

$$\{A_j(\boldsymbol{x}), E_{j'}(\boldsymbol{x}')\}_{PB} = -\delta_{jj'} \delta^3(\boldsymbol{x} - \boldsymbol{x}') \tag{9.9}$$

Thus, the spatial components of the vector potential and the components of the electric field are canonical pairs. However, the time component of the vector field, A_0, does not have a canonical pair. Thus, the quantization procedure treats it separately, as a Lagrange multiplier field that imposes a constraint, which we will see is Gauss's law. However, at the operator level, the condition $\Pi_0 = 0$ must then be imposed as a constraint. This fact led Dirac to formulate the theory of quantization of systems with constraints (Dirac, 1966). There is, however, another approach, also initiated by Dirac, consisting of setting $A_0 = 0$ and imposing Gauss's law as a constraint on the space of quantum states. As we will see, this approach amounts to fixing the gauge first (at the price of manifest Lorentz invariance).

The classical Hamiltonian density is defined in the usual manner:

$$\mathcal{H} = \Pi_j \partial_0 A_j - \mathcal{L} \tag{9.10}$$

We find

$$\mathcal{H}(x) = \frac{1}{2}(\boldsymbol{E}^2 + \boldsymbol{B}^2) - A_0(x) \boldsymbol{\nabla} \cdot \boldsymbol{E}(x) \tag{9.11}$$

Except for the last term, this is the usual form. It is easy to see that the last term is a constant of motion. Indeed, the equal-time Poisson bracket between the Hamiltonian density $\mathcal{H}(\boldsymbol{x})$

and $\nabla \cdot \boldsymbol{E}(\boldsymbol{y})$ is zero. By explicit calculation, we get

$$\{\mathcal{H}(\boldsymbol{x}), \nabla \cdot \boldsymbol{E}(\boldsymbol{y})\}_{PB} = \int d^3z \Big[-\frac{\delta \mathcal{H}(\boldsymbol{x})}{\delta A_j(\boldsymbol{z})} \frac{\delta \nabla \cdot \boldsymbol{E}(\boldsymbol{y})}{\delta E_j(\boldsymbol{z})} + \frac{\delta \mathcal{H}(\boldsymbol{x})}{\delta E_j(\boldsymbol{z})} \frac{\delta \nabla \cdot \boldsymbol{E}(\boldsymbol{y})}{\delta A_j(\boldsymbol{z})} \Big] \tag{9.12}$$

But

$$\frac{\delta \mathcal{H}(\boldsymbol{x})}{\delta A_j(\boldsymbol{z})} = \int d^3w \frac{\delta \mathcal{H}(\boldsymbol{x})}{\delta B_k(\boldsymbol{w})} \frac{\delta B_k(\boldsymbol{w})}{\delta A_j(\boldsymbol{z})} = \int d^3w B_k(\boldsymbol{w}) \delta(\boldsymbol{x}-\boldsymbol{w}) \epsilon_{k\ell j} \nabla_\ell^w \delta(\boldsymbol{w}-\boldsymbol{z})$$
$$= -\epsilon_{k\ell j} \nabla_\ell^z \int d^3w B_k(\boldsymbol{w}) \delta(\boldsymbol{x}-\boldsymbol{w}) \delta(\boldsymbol{w}-\boldsymbol{z}) \tag{9.13}$$

Hence,

$$\frac{\delta \mathcal{H}(\boldsymbol{x})}{\delta A_j(\boldsymbol{z})} = \epsilon_{j\ell k} \nabla_\ell^z (B_k(\boldsymbol{x}) \delta(\boldsymbol{x}-\boldsymbol{z})) = \epsilon_{j\ell k} B_k(\boldsymbol{x}) \nabla_\ell^x \delta(\boldsymbol{x}-\boldsymbol{z}) \tag{9.14}$$

Similarly, we get

$$\frac{\delta \nabla \cdot \boldsymbol{E}(\boldsymbol{y})}{\delta E_j(\boldsymbol{z})} = \nabla_j^y \delta(\boldsymbol{y}-\boldsymbol{z}), \qquad \frac{\delta \nabla \cdot \boldsymbol{E}(\boldsymbol{y})}{\delta A_j(\boldsymbol{z})} = 0 \tag{9.15}$$

Thus, the Poisson bracket is

$$\{\mathcal{H}(\boldsymbol{x}), \nabla \cdot \boldsymbol{E}(\boldsymbol{y})\}_{PB} = \int d^3z [-\epsilon_{j\ell k} B_k(\boldsymbol{x}) \nabla_\ell^x \delta(\boldsymbol{x}-\boldsymbol{z}) \nabla_j^y \delta(\boldsymbol{y}-\boldsymbol{z})]$$
$$= -\epsilon_{j\ell k} B_k(\boldsymbol{x}) \nabla_\ell^x \nabla_j^y \delta(\boldsymbol{x}-\boldsymbol{y})$$
$$= \epsilon_{j\ell k} B_k(\boldsymbol{x}) \nabla_\ell^x \nabla_j^x \delta(\boldsymbol{x}-\boldsymbol{y}) = 0 \tag{9.16}$$

provided that $\boldsymbol{B}(\boldsymbol{x})$ is nonsingular. Thus, $\nabla \cdot \boldsymbol{E}(\boldsymbol{x})$ is a constant of motion. It is easy to check that $\nabla \cdot \boldsymbol{E}$ generates infinitesimal gauge transformations. We will prove this statement directly in the quantum theory.

Since $\nabla \cdot \boldsymbol{E}(\boldsymbol{x})$ is a constant of motion, if we pick a value for it at some initial time x_0 it will remain constant in time. Thus we can write

$$\nabla \cdot \boldsymbol{E}(\boldsymbol{x}) = \rho(\boldsymbol{x}) \tag{9.17}$$

which we recognize to be Gauss's law. Naturally, an external charge distribution can be explicitly time dependent, and then we have

$$\frac{d}{dx_0}(\nabla \cdot \boldsymbol{E}) = \frac{\partial}{\partial x_0}(\nabla \cdot \boldsymbol{E}) = \frac{\partial}{\partial x_0} \rho_{ext}(\boldsymbol{x}, x_0) \tag{9.18}$$

Before turning to the quantization of this theory, notice that A_0 plays the role of a Lagrange multiplier field, whose variation yields Gauss's law, $\nabla \cdot \boldsymbol{E} = 0$. Hence, Gauss's law should be regarded as a *constraint* rather than as an equation of motion. This issue becomes very important in quantum theory. Indeed, without the constraint $\nabla \cdot \boldsymbol{E} = 0$, the theory is both absolutely trivial and wrong.

Constraints impose very tight restrictions on the allowed states of a quantum theory. For instance, consider a particle of mass m moving freely in three-dimensional space. Its stationary states have plane wave functions $\Psi_{\boldsymbol{p}}(\boldsymbol{r}, x_0)$, with energy $E(\boldsymbol{p}) = \frac{\boldsymbol{p}^2}{2m}$. If we constrain the particle to move only on the surface of a sphere of radius R, it becomes equivalent to a rigid rotor of moment of inertia $I = mR^2$ and energy eigenvalues $\epsilon_{\ell m} = \frac{\hbar^2}{2I}\ell(\ell+1)$ where $\ell = 0, 1, 2, \ldots$, and $|m| \leq \ell$. Thus, even the simple constraint $\boldsymbol{r}^2 = R^2$ has nontrivial effects.

Unlike the case of a particle forced to move on the surface of a sphere, the constraints that we have to impose when quantizing Maxwellian electrodynamics do not change the energy spectrum. This is because we can reduce the number of degrees of freedom to be quantized by taking advantage of the gauge invariance of the classical theory. This procedure is called *gauge fixing*. For example, the classical equation of motion

$$\partial^2 A^\mu - \partial^\mu(\partial_\nu A^\nu) = 0 \tag{9.19}$$

in the Coulomb gauge, $A_0 = 0$ and $\nabla \cdot \boldsymbol{A} = 0$, becomes

$$\partial^2 A_j = 0 \tag{9.20}$$

However, the Coulomb gauge is not compatible with the Poisson bracket,

$$\left\{A_j(\boldsymbol{x}), \Pi_{j'}(\boldsymbol{x}')\right\}_{PB} = \delta_{jj'}\delta(\boldsymbol{x} - \boldsymbol{x}') \tag{9.21}$$

since the spatial divergence of the delta function does not vanish. It follows that the quantization of the theory in the Coulomb gauge is achieved at the price of a modification of the commutation relations.

Since the classical theory is gauge invariant, we can always fix the gauge without any loss of physical content. The procedure of gauge fixing is attractive, because the number of independent variables is greatly reduced. A standard approach to the quantization of a gauge theory is to fix the gauge first, at the classical level, and to quantize later.

However, some problems arise immediately. For instance, in most gauges (e.g., the Coulomb gauge), Lorentz invariance is lost, or at least it is manifestly so. Thus, although the Coulomb gauge (also known as the radiation or transverse gauge) spoils Lorentz invariance, it has the attractive feature that the nature of the physical states (the photons) is quite transparent. In section 9.2, we will see that the quantization of the theory in this gauge has some peculiarities.

Another standard choice is the Lorentz gauge

$$\partial_\mu A^\mu = 0 \tag{9.22}$$

whose main appeal is its manifest covariance. The quantization of the system in this gauge follows the method developed by Suraj Gupta and Konrad Bleuer. While highly successful, it requires the introduction of states with negative norms (known as ghosts), which cancel all the gauge-dependent contributions to physical quantities. This approach is described in detail in the book by Itzykson and Zuber (1980).

More general covariant gauges can also be defined. A general approach consists not of imposing a rigid restriction on the degrees of freedom, but of adding new terms to the Lagrangian that eliminate the gauge freedom. For instance, the modified Lagrangian

$$\mathcal{L} = -\frac{1}{4}F_{\mu\nu}^2 + \frac{1}{2\alpha}(\partial_\mu A^\mu(x))^2 \tag{9.23}$$

is not gauge invariant because of the presence of the last term. We can easily see that this term weighs gauge-equivalent configurations differently, and the parameter $1/\alpha$ plays the role of a Lagrange multiplier field. In fact, in the limit $\alpha \to 0$, we recover the Lorentz gauge condition. In the path integral quantization of Maxwell's theory, it is proven that this approach is equivalent to averaging physical quantities over gauges. If $\alpha = 1$, the equations of motion become very simple (i.e., $\partial^2 A_\mu = 0$). This is the Feynman gauge. In this gauge, the calculations are simplest, although here too, the quantization of the theory has subtleties (such as ghosts).

Still, in the Hamiltonian or canonical quantization procedure, a third approach has been developed. In this approach, one fixes the gauge $A_0 = 0$. This condition is not enough to eliminate completely the gauge freedom. In this gauge, a residual set of time-independent gauge transformations are still allowed. In this approach, quantization is achieved by replacing the Poisson brackets by commutators, and Gauss's law now becomes a constraint on the space of physical quantum states. So, we quantize first and constrain later.

In general, it is a nontrivial task to prove that all the different quantizations yield a theory with the same physical properties. In practice, what one has to prove is that these different gauge choices yield theories whose states differ from one another by, at most, a unitary transformation. Otherwise, the quantized theories would be physically inequivalent. In addition, the recovery of Lorentz invariance may be a bit tedious in some cases. There is, however, an alternative, complementary approach to the quantum theory, in which most of these issues become transparent. This is the path-integral approach. This method has the advantage that all symmetries are taken care of from the outset. In addition, the canonical methods encounter very serious difficulties in the treatment of the non-abelian generalizations of Maxwellian electrodynamics.

We will consider here two canonical approaches: 1) quantization in the Coulomb gauge and 2) canonical quantization in the $A_0 = 0$ gauge in the Schrödinger picture.

9.2 Coulomb gauge

Quantization in the Coulomb gauge follows very closely the methods developed for the scalar field. Indeed, the classical constraints $A_0 = 0$ and $\nabla \cdot \boldsymbol{A} = 0$ allow for a Fourier expansion of the vector potential $\boldsymbol{A}(\boldsymbol{x}, x_0)$. In Fourier space, we write

$$\boldsymbol{A}(\boldsymbol{x}, x_0) = \int \frac{d^3p}{(2\pi)^3 2p_0} \boldsymbol{A}(\boldsymbol{p}, x_0) \exp(i\boldsymbol{p} \cdot \boldsymbol{x}) \tag{9.24}$$

where $\boldsymbol{A}(\boldsymbol{p}, x_0) = \boldsymbol{A}^*(-\boldsymbol{p}, x_0)$. Maxwell's equations yield the classical equation of motion, the wave equation:

$$\partial^2 \boldsymbol{A}(\boldsymbol{x}, x_0) = 0 \tag{9.25}$$

The Fourier expansion is consistent only if the amplitude $\boldsymbol{A}(\boldsymbol{p}, x_0)$ satisfies

$$\partial_0^2 \boldsymbol{A}(\boldsymbol{p}, x_0) + \boldsymbol{p}^2 \boldsymbol{A}(\boldsymbol{p}, x_0) = 0 \tag{9.26}$$

The constraint $\nabla \cdot \boldsymbol{A} = 0$ in turn becomes the transversality condition:

$$\boldsymbol{p} \cdot \boldsymbol{A}(\boldsymbol{p}, x_0) = 0 \tag{9.27}$$

Hence, $\boldsymbol{A}(\boldsymbol{p}, x_0)$ has the time dependence

$$\boldsymbol{A}(\boldsymbol{p}, x_0) = \boldsymbol{A}(\boldsymbol{p}) e^{ip_0 x_0} + \boldsymbol{A}(-\boldsymbol{p}) e^{-ip_0 x_0} \tag{9.28}$$

where $p_0 = |\boldsymbol{p}|$. Then, the mode expansion takes the form

$$\boldsymbol{A}(\boldsymbol{x}, x_0) = \int \frac{d^3p}{(2\pi)^3 2p_0} \left[\boldsymbol{A}^*(\boldsymbol{p}) e^{ip\cdot x} + \boldsymbol{A}(\boldsymbol{p}) e^{-ip\cdot x} \right] \tag{9.29}$$

where $p \cdot x = p_\mu x^\mu$. The transversality condition, eq. (9.27), is satisfied by introducing two polarization unit vectors, $\boldsymbol{\epsilon}_1(\boldsymbol{p})$ and $\boldsymbol{\epsilon}_2(\boldsymbol{p})$, such that $\boldsymbol{\epsilon}_1 \cdot \boldsymbol{\epsilon}_2 = \boldsymbol{\epsilon}_1 \cdot \boldsymbol{p} = \boldsymbol{\epsilon}_2 \cdot \boldsymbol{p} = 0$, and $\boldsymbol{\epsilon}_1^2 = \boldsymbol{\epsilon}_2^2 = 1$. Hence, if the amplitude $\boldsymbol{A}$ has to be orthogonal to $\boldsymbol{p}$, it must be a linear combination of $\boldsymbol{\epsilon}_1$ and $\boldsymbol{\epsilon}_2$,

$$\boldsymbol{A}(\boldsymbol{p}) = \sum_{\alpha=1,2} \boldsymbol{\epsilon}_\alpha(\boldsymbol{p}) a_\alpha(\boldsymbol{p}) \tag{9.30}$$

where the factors $a_\alpha(\boldsymbol{p})$ are complex amplitudes. In terms of $a_\alpha(\boldsymbol{p})$ and $a_\alpha^*(\boldsymbol{p})$, the Hamiltonian looks like a sum of oscillators.

In the Coulomb gauge, the passage to the quantum theory is achieved by assigning to each amplitude $a_\alpha(\boldsymbol{p})$ a Heisenberg annihilation operator $\hat{a}_\alpha(\boldsymbol{p})$. Similarly, $a_\alpha^*(\boldsymbol{p})$ maps onto the adjoint operator, the creation operator $\hat{a}_\alpha^\dagger(\boldsymbol{p})$. The expansion of the vector potential in modes is now

$$\hat{\boldsymbol{A}}(x) = \int \frac{d^3p}{(2\pi)^3 2p_0} \sum_{\alpha=1,2} \boldsymbol{\epsilon}_\alpha(p) \left[\hat{a}_\alpha(\boldsymbol{p}) e^{-ip\cdot x} + \hat{a}_\alpha^\dagger(\boldsymbol{p}) e^{ip\cdot x} \right] \tag{9.31}$$

with $p^2 = 0$ and $p_0 = |\boldsymbol{p}|$. The operators $\hat{a}_\alpha(\boldsymbol{p})$ and $\hat{a}_\alpha^\dagger(\boldsymbol{p})$ satisfy canonical commutation relations:

$$\begin{aligned} [\hat{a}_\alpha(\boldsymbol{p}), \hat{a}_{\alpha'}^\dagger(\boldsymbol{p}')] &= 2p_0 (2\pi)^3 \delta(\boldsymbol{p} - \boldsymbol{p}') \\ [\hat{a}_\alpha(\boldsymbol{p}), \hat{a}_{\alpha'}(\boldsymbol{p}')] &= [\hat{a}_\alpha^\dagger(\boldsymbol{p}), \hat{a}_{\alpha'}^\dagger(\boldsymbol{p}')] = 0 \end{aligned} \tag{9.32}$$

It is straightforward to check that the vector potential $\boldsymbol{A}(\boldsymbol{x})$ and the electric field $\boldsymbol{E}(\boldsymbol{x})$ obey the (unconventional) equal-time commutation relation,

$$[A_j(\boldsymbol{x}), E_{j'}(\boldsymbol{x}')] = -i \left(\delta_{jj'} - \frac{\nabla_j \nabla_{j'}}{\nabla^2} \right) \delta^3(\boldsymbol{x} - \boldsymbol{x}') \tag{9.33}$$

where the symbol $1/\nabla^2$ represents the inverse of the Laplacian (i.e., the Laplacian Green function). In the derivation of this relation, the following identity was used:

$$\sum_{\alpha=1,2} \epsilon_\alpha^j(\boldsymbol{p}) \epsilon_\alpha^{j'}(\boldsymbol{p}) = \delta_{jj'} - \frac{p_j p_{j'}}{\boldsymbol{p}^2} \tag{9.34}$$

These commutation relations are an extension of the canonical commutation relation and are a consequence of the transversality condition, $\nabla \cdot \boldsymbol{A} = 0$.

In this gauge, the (normal-ordered) Hamiltonian is

$$\hat{H} = \int \frac{d^3p}{(2\pi)^3 2p_0} \, p_0 \sum_{\alpha=1,2} \hat{a}^\dagger_\alpha(\boldsymbol{p})\hat{a}_\alpha(\boldsymbol{p}) \tag{9.35}$$

The ground state (i.e., the vacuum state $|0\rangle$) is annihilated by both polarizations $\hat{a}_\alpha(\boldsymbol{p})|0\rangle = 0$. The single-particle states are $\hat{a}^\dagger_\alpha(\boldsymbol{p})|0\rangle$ and represent transverse photons with momentum $\boldsymbol{p}$, energy $p_0 = |\boldsymbol{p}|$, and with the two possible linear polarizations labeled by $\alpha = 1, 2$. Circularly polarized photons can be constructed in the usual manner.

The Coulomb gauge has the advantage that, in this picture, the electromagnetic field can be regarded as a collection of linear harmonic oscillators, which are then quantized. Of course, this is a simple reflection of the fact that Maxwell electrodynamics is a free field theory. It has, however, several problems. One is that Lorentz invariance is violated from the outset and has to be recovered afterward in the computation of observables. The other is that (as we will discuss below) in non-abelian theories, the Coulomb gauge does not exist globally. For these reasons, its usefulness is essentially limited to Maxwell's theory.

9.3 The gauge $A_0 = 0$

In this gauge, we directly apply the canonical formalism. In what follows, we fix $A_0 = 0$ and associate to the three spatial components A_j of the vector potential an operator, $\hat{A}_j$, which acts on a Hilbert space of states. Similarly, to the canonical momentum $\Pi_j = -E_j$, we assign an operator $\widehat{\Pi}_j$. These operators obey the equal-time commutation relations

$$[\widehat{A}_j(\boldsymbol{x}), \widehat{\Pi}_{j'}(\boldsymbol{x}')] = i\delta(\boldsymbol{x} - \boldsymbol{x}')\delta_{jj'} \tag{9.36}$$

Hence, the vector potential $\boldsymbol{A}$ and the electric field $\boldsymbol{E}$ are not canonically conjugate operators and do not commute with each other:

$$[\hat{A}_j(\boldsymbol{x}), \hat{E}_{j'}(\boldsymbol{x}')] = -i\delta_{jj'}\delta(\boldsymbol{x} - \boldsymbol{x}') \tag{9.37}$$

Let us now specify the Hilbert space to be the space of states $|\Psi\rangle$ with wave functions that, in the field representation, have the form $\Psi(\{A_j(\boldsymbol{x})\})$. When acting on these states, the electric field is the functional differential operator:

$$\hat{E}_j(\boldsymbol{x}) \equiv i \frac{\delta}{\delta A_j(\boldsymbol{x})} \tag{9.38}$$

In this Hilbert space, the inner product is

$$\langle \{A_j(\boldsymbol{x})\} | \{A_j(\boldsymbol{x})\} \rangle \equiv \Pi_{\boldsymbol{x},j} \delta\left(A_j(\boldsymbol{x}) - A_j(\boldsymbol{x})\right) \tag{9.39}$$

This Hilbert space is actually much too large. Indeed, states with wave functions that differ by time-independent gauge transformations

$$\Psi_\phi(\{A_j(\boldsymbol{x})\}) \equiv \Psi(\{A_j(\boldsymbol{x}) - \nabla_j\phi(\boldsymbol{x})\}) \tag{9.40}$$

are *physically equivalent*, since the matrix elements of the electric field operator $\hat{E}_j(\boldsymbol{x})$ and magnetic field operator $\hat{B}_j(\boldsymbol{x}) = \epsilon_{jk\ell} \nabla_k \hat{A}_\ell(\boldsymbol{x})$ are the same for all gauge-equivalent states:

$$\begin{aligned}
\langle \Psi'_{\phi'}(\{A_j(\boldsymbol{x})\})|\hat{E}_j(\boldsymbol{x})|\Psi_\phi(\{A_j(\boldsymbol{x})\})\rangle &= \langle \Psi'(\{A_j(\boldsymbol{x})\})|\hat{E}_j(\boldsymbol{x})|\Psi(\{A_j(\boldsymbol{x})\})\rangle \\
\langle \Psi'_{\phi'}(\{A_j(\boldsymbol{x})\})|\hat{B}_j(\boldsymbol{x})|\Psi_\phi(\{A_j(\boldsymbol{x})\})\rangle &= \langle \Psi'(\{A_j(\boldsymbol{x})\})|\hat{B}_j(\boldsymbol{x})|\Psi(\{A_j(\boldsymbol{x})\})\rangle
\end{aligned} \tag{9.41}$$

The (local) operators $\hat{Q}(\boldsymbol{x})$,

$$\hat{Q}(\boldsymbol{x}) = \boldsymbol{\nabla} \cdot \hat{\boldsymbol{E}}(\boldsymbol{x}) \tag{9.42}$$

commute locally with the Hamiltonian and with each other:

$$[\hat{Q}(\boldsymbol{x}), \hat{H}] = 0, \qquad [\hat{Q}(\boldsymbol{x}), \hat{Q}(\boldsymbol{y})] = 0 \tag{9.43}$$

Hence, all the local operators $\hat{Q}(\boldsymbol{x})$ can be diagonalized simultaneously with $\hat{H}$.

Let us show now that $\hat{Q}(\boldsymbol{x})$ generates local infinitesimal time-independent gauge transformations. From the canonical commutation relation,

$$[\hat{A}_j(\boldsymbol{x}), \hat{E}_{j'}(\boldsymbol{x}')] = -i\delta_{jj'}\delta(\boldsymbol{x} - \boldsymbol{x}') \tag{9.44}$$

we get (by differentiation):

$$[\hat{A}_j(\boldsymbol{x}), \hat{Q}(\boldsymbol{x}')] = [\hat{A}_j(\boldsymbol{x}), \nabla_j \hat{E}_{j'}(\boldsymbol{x}')] = i \nabla_j^x \delta(\boldsymbol{x} - \boldsymbol{x}') \tag{9.45}$$

Hence, we also find

$$[i \int dz \phi(z)\hat{Q}(z), \hat{A}_j(\boldsymbol{x})] = -\int dz \phi(z) \nabla_j^z \delta(\boldsymbol{z} - \boldsymbol{x}) = \nabla_j \phi(\boldsymbol{x}) \tag{9.46}$$

and

$$\begin{aligned}
& e^{i \int dz \phi(z)\hat{Q}(z)} \hat{A}_j(\boldsymbol{x}) e^{-i \int dz \phi(z)\hat{Q}(z)} \\
&= e^{-i \int dz \nabla_k \phi(z)\hat{E}_k(z)} \hat{A}_j(\boldsymbol{x})\, e^{i \int dz \nabla_k \phi(z)\hat{E}_k(z)} \\
&= \hat{A}_j(\boldsymbol{x}) + \nabla_j \phi(\boldsymbol{x})
\end{aligned} \tag{9.47}$$

The physical requirement that states differing by time-independent gauge transformations be equivalent to one another leads to the demand that we should restrict the Hilbert space to the space of gauge-invariant states. These states, which we will denote by $|\text{Phys}\rangle$, satisfy

$$\hat{Q}(\boldsymbol{x})|\text{Phys}\rangle \equiv \boldsymbol{\nabla} \cdot \hat{\boldsymbol{E}}(\boldsymbol{x})|\text{Phys}\rangle = 0 \tag{9.48}$$

Thus, the constraint means that only states that obey Gauss's law are in the *physical Hilbert space*. Unlike the quantization in the Coulomb gauge, in the $A_0 = 0$ gauge, the commutators are canonical, and the states are constrained to obey Gauss's law.

In the Schrödinger picture, the eigenstates of the system obey the Schrödinger equation

$$\int dx \frac{1}{2}\Big[-\frac{\delta^2}{\delta A_j(\boldsymbol{x})^2}+B_j(\boldsymbol{x})^2\Big]\Psi[A]=\mathcal{E}\Psi[A] \tag{9.49}$$

where $\Psi[A]$ is a shorthand for the wave functional $\Psi(\{A_j(\boldsymbol{x})\})$. In this notation, the constraint of Gauss's law is

$$\nabla_j^x \hat{E}_j(\boldsymbol{x})\Psi[A]\equiv i\,\nabla_j^x \frac{\delta}{\delta A_j(\boldsymbol{x})}\Psi[A]=0 \tag{9.50}$$

This constraint can be satisfied by separating the real field $A_j(\boldsymbol{x})$ into longitudinal $A_j^L(\boldsymbol{x})$ and transverse $A_j^T(\boldsymbol{x})$ parts,

$$A_j(\boldsymbol{x})=A_j^L(\boldsymbol{x})+A_j^T(\boldsymbol{x})=\int \frac{d^3p}{(2\pi)^3}\left(A_j^L(\boldsymbol{p})+A_j^T(\boldsymbol{p})\right)e^{i\boldsymbol{p}\cdot\boldsymbol{x}} \tag{9.51}$$

where $A_j^L(\boldsymbol{x})$ and $A_j^T(\boldsymbol{x})$ satisfy

$$\nabla_j A_j^T(\boldsymbol{x})=0, \quad A_j^L(\boldsymbol{x})=\nabla_j\phi(\boldsymbol{x}) \tag{9.52}$$

and $\phi(\boldsymbol{x})$ is, for the moment, arbitrary. In terms of A_j^L and A_j^T, the constraint of Gauss's law simply becomes

$$\nabla_j^x \frac{\delta}{\delta A_j^L(\boldsymbol{x})}\Psi[A]=0 \tag{9.53}$$

and the Hamiltonian now is

$$\hat{H}=\int d^3p\,\frac{1}{2}\Big[-\frac{\delta^2}{\delta A_j^T(\boldsymbol{p})\delta A_j^T(-\boldsymbol{p})}-\frac{\delta^2}{\delta A_j^L(\boldsymbol{p})\delta A_j^L(-\boldsymbol{p})}+\boldsymbol{p}^2A_j^T(\boldsymbol{p})A_j^T(-\boldsymbol{p})\Big] \tag{9.54}$$

We satisfy the constraint by looking only at gauge-invariant states. Their wave functions do not depend on the longitudinal components of $\boldsymbol{A}(\boldsymbol{x})$. Hence, $\Psi[A]=\Psi[A^T]$. When acting on those states, the Hamiltonian is

$$H\Psi=\int d^3p\frac{1}{2}\Big[-\frac{\delta^2}{\delta A_j^T(\boldsymbol{p})\delta A_j^T(-\boldsymbol{p})}+\boldsymbol{p}^2A_j^T(\boldsymbol{p})A_j^T(-\boldsymbol{p})\Big]\Psi=\mathcal{E}\Psi \tag{9.55}$$

Let $\boldsymbol{\epsilon}_1(\boldsymbol{p})$ and $\boldsymbol{\epsilon}_2(\boldsymbol{p})$ be two vectors that, together with the unit vector $\boldsymbol{n}_p=\boldsymbol{p}/|\boldsymbol{p}|$, form an orthonormal basis. Let us define the operators ($\alpha=1,2; j=1,2,3$):

$$\hat{a}(\boldsymbol{p},\alpha)=\frac{1}{\sqrt{2|\boldsymbol{p}|}}\epsilon_j^\alpha(\boldsymbol{p})\left[\frac{\delta}{\delta A_j^T(-\boldsymbol{p})}+|\boldsymbol{p}|A_j^T(\boldsymbol{p})\right]$$

$$\hat{a}^\dagger(\boldsymbol{p},\alpha)=\frac{1}{\sqrt{2|\boldsymbol{p}|}}\epsilon_j^\alpha(\boldsymbol{p})\left[-\frac{\delta}{\delta A_j^T(\boldsymbol{p})}+|\boldsymbol{p}|A_j^T(-\boldsymbol{p})\right] \tag{9.56}$$

These operators satisfy the commutation relations

$$[\hat{a}(\boldsymbol{p},\alpha),\hat{a}^{\dagger}(\boldsymbol{p}',\alpha')]=\delta_{\alpha\alpha'}\delta^3(\boldsymbol{p}-\boldsymbol{p}') \tag{9.57}$$

In terms of these operators, the Hamiltonian $\hat{H}$ and the expansion of the transverse part of the vector potential are

$$\hat{H}=\int d^3p\frac{|\boldsymbol{p}|}{2}\sum_{\alpha=1,2}[\hat{a}^{\dagger}(\boldsymbol{p},\alpha)\hat{a}(\boldsymbol{p},\alpha)+\hat{a}(\boldsymbol{p},\alpha)\hat{a}^{\dagger}(\boldsymbol{p},\alpha)]$$

$$A_j^T(\boldsymbol{x})=\int\frac{d^3p}{\sqrt{(2\pi)^3 2|\boldsymbol{p}|}}\sum_{\alpha=1,2}\epsilon_\alpha^j(\boldsymbol{p})[\hat{a}(\boldsymbol{p},\alpha)e^{i\boldsymbol{p}\cdot\boldsymbol{x}}+\hat{a}^{\dagger}(\boldsymbol{p},\alpha)e^{-i\boldsymbol{p}\cdot\boldsymbol{x}}] \tag{9.58}$$

These expressions are the same ones that we obtained before in the Coulomb gauge (except for the normalization factors).

It is instructive to derive the wave functional for the ground state. The ground state $|0\rangle$ is the state annihilated by all the oscillators $\hat{a}(\boldsymbol{p},\alpha)$. Hence its wave function $\Psi_0[A]$ satisfies

$$\langle\{A_j(\boldsymbol{x})\}|\hat{a}(\boldsymbol{p},\alpha)|0\rangle=0 \tag{9.59}$$

This equation is the functional differential equation

$$\epsilon_\alpha^j(\boldsymbol{p})\left[\frac{\delta}{\delta A_j^T(-\boldsymbol{p})}+|\boldsymbol{p}|A_j^T(\boldsymbol{p})\right]\Psi_0(\{A_j^T(\boldsymbol{p})\})=0 \tag{9.60}$$

It is easy to check that the unique solution of this equation is

$$\Psi_0[A]=N\exp\left[-\frac{1}{2}\int d^3p\,|\boldsymbol{p}|A_j^T(\boldsymbol{p})A_j^T(-\boldsymbol{p})\right] \tag{9.61}$$

Since the transverse components of $A_j(\boldsymbol{p})$ satisfy

$$A_j^T(\boldsymbol{p})=\epsilon_{jk\ell}\frac{p_k A_\ell(\boldsymbol{p})}{|\boldsymbol{p}|}=\left(\frac{\boldsymbol{p}\times\boldsymbol{A}(\boldsymbol{p})}{|\boldsymbol{p}|}\right)_j \tag{9.62}$$

we can write $\Psi_0[A]$ in the form

$$\Psi_0[A]=\mathcal{N}\exp\left[-\frac{1}{2}\int\frac{d^3p}{|\boldsymbol{p}|}\left(\boldsymbol{p}\times\boldsymbol{A}(\boldsymbol{p})\right)\cdot\left(\boldsymbol{p}\times\boldsymbol{A}(-\boldsymbol{p})\right)\right] \tag{9.63}$$

It is instructive to write this wave function in position space (i.e., as a functional of the configuration of magnetic fields $\{\boldsymbol{B}(\boldsymbol{x})\}$). Clearly, we have

$$\boldsymbol{p}\times\boldsymbol{A}(\boldsymbol{p})=-i\int\frac{d^3x}{(2\pi)^{3/2}}\left(\nabla_x\times\boldsymbol{A}(\boldsymbol{x})\right)e^{-i\boldsymbol{p}\cdot\boldsymbol{x}}$$

$$\boldsymbol{p}\times\boldsymbol{A}(-\boldsymbol{p})=i\int\frac{d^3x}{(2\pi)^{3/2}}\left(\nabla_x\times\boldsymbol{A}(\boldsymbol{x})\right)e^{i\boldsymbol{p}\cdot\boldsymbol{x}} \tag{9.64}$$

By substitution of these identities back into the exponent of the wave function, we get

$$\Psi_0[A] = \mathcal{N} \exp\Big(-\frac{1}{2} \int d^3x \int d^3x' \boldsymbol{B}(\boldsymbol{x}) \cdot \boldsymbol{B}(\boldsymbol{x}') G(\boldsymbol{x}-\boldsymbol{x}') \Big) \tag{9.65}$$

where $G(\boldsymbol{x}-\boldsymbol{x}')$ is given by

$$G(\boldsymbol{x}-\boldsymbol{x}') = \int \frac{d^3p}{(2\pi)^3} \frac{e^{-i\boldsymbol{p}\cdot(\boldsymbol{x}-\boldsymbol{x}')}}{|\boldsymbol{p}|} \tag{9.66}$$

This function has singular behavior at large values of $|\boldsymbol{p}|$. Define a *smoothed* version $G_\Lambda(\boldsymbol{x}-\boldsymbol{x}')$ to be

$$G_\Lambda(\boldsymbol{x}-\boldsymbol{x}') = \int \frac{d^3p}{(2\pi)^3} \frac{e^{-i\boldsymbol{p}\cdot(\boldsymbol{x}-\boldsymbol{x}')}}{|\boldsymbol{p}|} e^{-|\boldsymbol{p}|/\Lambda} \tag{9.67}$$

which cuts off the contributions for $|\boldsymbol{p}| \gg \Lambda$. Also, $G_\Lambda(\boldsymbol{x}-\boldsymbol{x}')$ formally goes back to $G(\boldsymbol{x}-\boldsymbol{x}')$ as $\Lambda \to \infty$. $G_\Lambda(\boldsymbol{x}-\boldsymbol{x}')$ can be evaluated explicitly to give

$$\begin{aligned} G_\Lambda(\boldsymbol{x}-\boldsymbol{x}') &= \frac{1}{2\pi^2|\boldsymbol{x}-\boldsymbol{x}'|^2} \int_0^\infty dt \, \sin t \, e^{-t/\Lambda|\boldsymbol{x}-\boldsymbol{x}'|} \\ &= \frac{1}{2\pi^2|\boldsymbol{x}-\boldsymbol{x}'|^2} \mathrm{Im} \left[\frac{1}{\frac{1}{\Lambda|\boldsymbol{x}-\boldsymbol{x}'|} - i} \right] \end{aligned} \tag{9.68}$$

Thus,

$$\lim_{\Lambda\to\infty} G_\Lambda(\boldsymbol{x}-\boldsymbol{x}') = \frac{1}{2\pi^2|\boldsymbol{x}-\boldsymbol{x}'|^2} \tag{9.69}$$

Hence, the ground state wave functional $\Psi_0[A]$ is

$$\Psi_0[A] = \mathcal{N} \exp\Big(-\frac{1}{4\pi^2} \int d^3x \int d^3x \frac{\boldsymbol{B}(\boldsymbol{x}) \cdot \boldsymbol{B}(\boldsymbol{x}')}{|\boldsymbol{x}-\boldsymbol{x}'|^2} \Big) \tag{9.70}$$

which is only a functional of the configuration of magnetic fields.

9.4 Path-integral quantization of gauge theories

We have discussed at length the quantization of the abelian gauge theory (i.e., Maxwell electromagnetism), within canonical quantization in the $A_0 = 0$ gauge and a modified canonical formalism in the Coulomb gauge. Although conceptually what we have done is correct, it poses some questions.

The canonical formalism is natural in the gauge $A_0 = 0$ and can be generalized to other gauge theories. However, this gauge is highly noncovariant, and it is necessary to prove covariance of physical observables in the end. In addition, the gauge field propagator in this gauge is very complicated.

The particle spectrum is most transparent in the transverse (or Coulomb) gauge. However, in addition to being noncovariant, it is not possible to generalize this gauge to non-abelian gauge theories (or even to abelian gauge theories on a compact gauge group) due to subtle topological problems known as Gribov ambiguities (or Gribov "copies"). The propagator is equally awful in this gauge. The commutation relations in real space look quite different from those in scalar field theory. In addition, for non-abelian gauge groups, even in the absence of matter fields, the theory is already nonlinear and needs to be regularized in a manner that preserves gauge invariance. Although it is possible to use covariant gauges, such as the Lorentz gauge $\partial_\mu A^\mu = 0$, the quantization of the theory in these gauges require an approach, known as Gupta-Bleuer quantization, that is difficult to generalize.

At the root of this problem is the issue of quantizing a theory that has a local (or gauge) invariance in which both Lorentz and gauge invariance are kept explicitly. It turns out that path-integral quantization is the most direct approach to deal with these problems.

Let us now construct the path integral for the free electromagnetic field. However, the procedure that we develop holds, at least formally, for any gauge theory.

We begin with the theory quantized canonically in the gauge $A_0 = 0$ (Dirac, 1966). We saw in section 9.3 that, in the gauge $A_0 = 0$, the electric field $\boldsymbol{E}$ is (minus) the momentum canonically conjugate to the vector potential $\boldsymbol{A}$ (the spatial components of the gauge field), and both fields obey equal-time canonical commutation relations:

$$\left[E_j(\boldsymbol{x}), A_k(\boldsymbol{x}')\right] = i\delta^3(\boldsymbol{x} - \boldsymbol{x}') \tag{9.71}$$

In addition, in this gauge, Gauss's law becomes a constraint on the space of states,

$$\nabla \cdot \boldsymbol{E}(\boldsymbol{x})|\text{Phys}\rangle = J_0(\boldsymbol{x})|\text{Phys}\rangle \tag{9.72}$$

which defines the physical Hilbert space. Here $J_0(x)$ is a charge density distribution. In the presence of a set of conserved sources, $J_\mu(x)$, that satisfy $\partial_\mu J^\mu = 0$, the Hamiltonian of the free-field theory is

$$\hat{H} = \int d^3x \, \frac{1}{2}\left(\boldsymbol{E}^2 + \boldsymbol{B}^2\right) + \int d^3x \, \boldsymbol{J} \cdot \boldsymbol{A} \tag{9.73}$$

Let us construct the path integral in the Hilbert space of gauge-invariant states defined by the condition eq. (9.72).

Denote by $Z[J_\mu]$ the partition function

$$Z[J] = \text{tr}' \, T e^{-i\int dx_0 \hat{H}} \equiv \text{tr}\left(T e^{-i\int dx_0 \hat{H}} \, \hat{P} \right) \tag{9.74}$$

where tr' means a trace (or sum) over the space of states that satisfy Gauss's law, eq. (9.72). We implement this constraint by means of the operator $\hat{P}$ that projects onto the gauge-invariant states:

$$\hat{P} = \prod_{\boldsymbol{x}} \delta\left(\nabla \cdot \hat{E}(\boldsymbol{x}) - J_0(\boldsymbol{x})\right) \tag{9.75}$$

We now follow the standard construction of the path integral, while making sure that we only sum over histories that are consistent with the constraint. In principle, all we need to do is to insert complete sets of states that are eigenstates of the field operator $\hat{\boldsymbol{A}}(x)$ at all intermediate times. These states, denoted by $|\{\boldsymbol{A}(\boldsymbol{x}, x_0)\}\rangle$, are not gauge invariant, and they do not satisfy the constraint. However, the projection operator $\hat{P}$ projects out the unphysical components of these states.

Hence, if the projection operator is included in the evolution operator, the inserted states actually are gauge invariant. Thus, to insert at every intermediate time x_0^k ($k=1,\ldots,N$ with $N\to\infty$ and $\Delta x_0\to 0$) a complete set of gauge-invariant states amounts to writing $Z[J]$ as

$$Z[J]=\prod_{k=1}^{N}\int \mathcal{D}A_j(\boldsymbol{x},x_0^k)$$
$$\times \langle\{A_j(\boldsymbol{x},x_0^k)\}|\left(1-i\Delta x_0\hat{H}\right)\prod_{\boldsymbol{x}}\delta\left(\nabla\cdot \boldsymbol{E}(\boldsymbol{x},x_0^k)-J_0(\boldsymbol{x},x_0^k)\right)|\{A_j(\boldsymbol{x},x_0^{k+1})\}\rangle \tag{9.76}$$

As an operator, the projection operator $\hat{P}$ is naturally spanned by the eigenstates of the electric field operator $|\{\boldsymbol{E}(\boldsymbol{x},x_0)\}\rangle$:

$$\prod_{\boldsymbol{x}}\delta\left(\nabla\cdot \boldsymbol{E}(\boldsymbol{x},x_0)-J_0(\boldsymbol{x},x_0)\right)$$
$$\equiv\int \mathcal{D}\boldsymbol{E}(\boldsymbol{x},x_0)\,|\{\boldsymbol{E}(\boldsymbol{x},x_0)\}\rangle\langle\{\boldsymbol{E}(\boldsymbol{x},x_0)\}|\prod_{\boldsymbol{x}}\delta\left(\nabla\cdot \boldsymbol{E}(\boldsymbol{x},x_0)-J_0(\boldsymbol{x},x_0)\right) \tag{9.77}$$

The delta function has the integral representation

$$\prod_{\boldsymbol{x}}\delta\left(\nabla\cdot \boldsymbol{E}(\boldsymbol{x},x_0)-J_0(\boldsymbol{x},x_0)\right)$$
$$=\mathcal{N}\int \mathcal{D}A_0(\boldsymbol{x},x_0)e^{i\Delta x_0\int d^3x\, A_0(\boldsymbol{x},x_0)\left(\nabla\cdot \boldsymbol{E}(\boldsymbol{x},x_0)-J_0(\boldsymbol{x},x_0)\right)} \tag{9.78}$$

Hence, the matrix elements of interest become

$$\int \mathcal{D}\boldsymbol{A}\prod_{x_0}\langle\{\boldsymbol{A}(\boldsymbol{x},x_0)\}|\left(1-i\Delta x_0\hat{H}\right)\prod_{\boldsymbol{x}}\delta\left(\nabla_j\hat{E}_j-J_0\right)|\{\boldsymbol{A}(\boldsymbol{x},x_0+\Delta x_0)\}\rangle$$
$$=\int \mathcal{D}A_0\mathcal{D}\boldsymbol{A}\mathcal{D}\boldsymbol{E}\prod_{x_0}\langle\{\boldsymbol{A}(\boldsymbol{x},x_0)\}|\{\boldsymbol{E}(\boldsymbol{x},x_0)\}\rangle\langle\{\boldsymbol{E}(\boldsymbol{x},x_0)\}|\{\boldsymbol{A}(\boldsymbol{x},x_0+\Delta x_0)\}\rangle$$
$$\times\exp\left[i\Delta x_0\int d^3x\, A_0(\boldsymbol{x},x_0)\left(\nabla\cdot \boldsymbol{E}(\boldsymbol{x},x_0)-J_0(\boldsymbol{x},x_0)\right)\right]\exp\left[-\frac{\langle\{\boldsymbol{A}(\boldsymbol{x},x_0)\}|\hat{H}|\{\boldsymbol{E}(\boldsymbol{x},x_0)\}\rangle}{\langle\{\boldsymbol{A}(\boldsymbol{x},x_0)\}|\{\boldsymbol{E}(\boldsymbol{x},x_0)\}\rangle}\right] \tag{9.79}$$

The overlaps are equal to

$$\langle\{\boldsymbol{A}(\boldsymbol{x},x_0)\}|\{\boldsymbol{E}(\boldsymbol{x},x_0)\}\rangle=e^{i\int d^3x\,\boldsymbol{A}(\boldsymbol{x},x_0)\cdot \boldsymbol{E}(\boldsymbol{x},x_0)} \tag{9.80}$$

Hence, we find that the product of the overlaps is given by

$$\prod_{x_0} \langle \{\boldsymbol{A}(\boldsymbol{x},x_0)\} | \{\boldsymbol{E}(\boldsymbol{x},x_0)\}\rangle \langle \{\boldsymbol{E}(\boldsymbol{x},x_0)\} | \{\boldsymbol{A}(\boldsymbol{x},x_0+\Delta x_0)\}\rangle =$$

$$= e^{-i \int dx_0 \int d^3x\, \boldsymbol{E}(\boldsymbol{x},x_0)\cdot \partial_0 \boldsymbol{A}(\boldsymbol{x},x_0)} \tag{9.81}$$

The matrix elements of the Hamiltonian are

$$\frac{\langle \{\boldsymbol{A}(\boldsymbol{x},x_0)\} | \hat{H} | \{\boldsymbol{E}(\boldsymbol{x},x_0)\}\rangle}{\langle \{\boldsymbol{A}(\boldsymbol{x},x_0)\} | \{\boldsymbol{E}(\boldsymbol{x},x_0)\}\rangle} = \int d^3x \left[\frac{1}{2}\left(\boldsymbol{E}^2+\boldsymbol{B}^2\right)+\boldsymbol{J}\cdot\boldsymbol{A}\right] \tag{9.82}$$

Putting everything together, we find that the path-integral expression for $Z[J]$ has the form

$$Z[J] = \int \mathcal{D}A_\mu \mathcal{D}\boldsymbol{E}\, e^{iS[A_\mu,\boldsymbol{E}]} \tag{9.83}$$

where

$$\mathcal{D}A_\mu = \mathcal{D}\boldsymbol{A}\mathcal{D}A_0 \tag{9.84}$$

and the action $S[A_\mu, \boldsymbol{E}]$ is given by

$$S[A_\mu,\boldsymbol{E}] = \int d^4x \left[-\boldsymbol{E}\cdot\partial_0\boldsymbol{A} - \frac{1}{2}\left(\boldsymbol{E}^2+\boldsymbol{B}^2\right) - \boldsymbol{J}\cdot\boldsymbol{A} + A_0\left(\boldsymbol{\nabla}\cdot\boldsymbol{E} - J_0\right)\right] \tag{9.85}$$

Notice that the Lagrange multiplier field A_0, which appeared when we introduced the integral representation of the delta function, has become the time component of the vector potential.

Since the action is quadratic in the electric fields, we can integrate them out explicitly to find

$$\int \mathcal{D}\boldsymbol{E}\, e^{i\int d^4x \left(-\frac{1}{2}\boldsymbol{E}^2 - \boldsymbol{E}\cdot\left(\partial_0\boldsymbol{A}+\boldsymbol{\nabla}A_0\right)\right)} = \text{const.}\, e^{i\int d^4x \frac{1}{2}\left(-\partial_0\boldsymbol{A}-\boldsymbol{\nabla}A_0\right)^2} \tag{9.86}$$

Collecting everything, we find that the path integral is

$$Z[J] = \int \mathcal{D}A_\mu\, e^{i\int d^4x \mathcal{L}} \tag{9.87}$$

where the Lagrangian is

$$\mathcal{L} = -\frac{1}{4}F_{\mu\nu}F^{\mu\nu} + J_\mu A^\mu \tag{9.88}$$

which is what we should have expected. Note that this formal argument is valid for *all* gauge theories, abelian or non-abelian. In other words, the path integral is *always* the sum over the histories of the field A_μ with a weight factor that equals the exponential of $i/\hbar$ times the action S of the gauge theory.

Therefore, we found that, at least formally, we can write a functional integral that will play the role of the generating functional of the N-point functions of these theories:

$$\langle 0|TA_{\mu_1}(x_1)\cdots A_{\mu_N}(x_N)|0\rangle \tag{9.89}$$

9.5 Path integrals and gauge fixing

The expression for the path integral in eq. (9.87) is formal, because we are summing over all histories of the field without restriction. In fact, since the action S and the integration measure $\mathcal{D}A_\mu$ are both gauge invariant, histories that differ by gauge transformations have the same weight in the path integral, and the partition function has an apparent divergence of the form $v(G)^V$, where $v(G)$ is the volume of the gauge group G, and V is the (infinite) volume of spacetime.

To avoid this problem, we must implement a procedure that restricts the sum over configurations in such a way that configurations differing by local gauge transformations are counted only once. This procedure is known as gauge fixing. We will follow that approach, introduced by L. Faddeev and V. Popov (Faddeev and Popov, 1967; Faddeev, 1976). Although the method works for all gauge theories, the non-abelian theories have subtleties and technical issues that we will address here (and in section 9.8). Let us begin with a general discussion of the method, and then we will specialize it first for the case of Maxwell's theory (the $U(1)$ gauge theory without matter fields) and later to the case of a general compact gauge group.

Let the vector potential A_μ be a field that takes values in the *algebra* of a gauge group G (i.e., A_μ is a linear combination of the group generators). Let $U(x)$ be a unitary-matrix field that takes values on a representation of the group G (recall our discussion on this subject in section 3.6). For the abelian group $U(1)$, we have

$$U(x) = e^{i\phi(x)} \tag{9.90}$$

where $\phi(x)$ is a real (scalar) field. A gauge transformation is, for a group G,

$$A_\mu^U = UA_\mu U^\dagger - iU\partial_\mu U^\dagger \tag{9.91}$$

For the abelian group $U(1)$, we have

$$A_\mu^U = A_\mu + \partial_\mu \phi \tag{9.92}$$

To avoid infinities in $Z[J]$, we must impose restrictions on the sum over histories such that histories that are related via a gauge transformation are *counted exactly once*. To do this, we must find a way to classify the configurations of the vector field A_μ into *classes*. Let us do this by defining gauge-fixing conditions. Each class is labeled by a *representative* configuration, and other elements in the class are related to it by smooth gauge transformations. Hence, all configurations in a given class are characterized by a set of gauge-invariant data, such as field strengths in the case of the abelian theory. The set of configurations that differ from one another by a local gauge transformation belong to the same class. We can think of the class as a set obtained by the action on some reference configuration by the gauge group, and the elements of a class constitute an orbit of the gauge

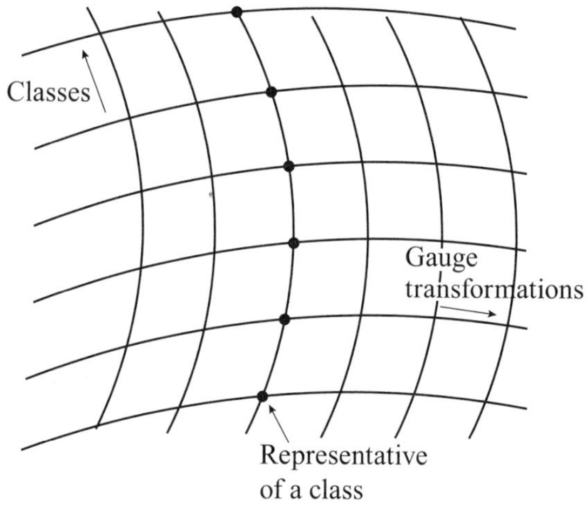

Figure 9.1 The gauge-fixing condition selects a manifold of configurations.

group. Mathematically, the elements of the gauge class form a vector bundle. The geometry of the configurations of gauge fields is shown in figure 9.1.

We must choose gauge conditions such that the theory remains local and, if possible, Lorentz covariant. It is essential that, whatever gauge condition we use, *each class is counted exactly once* by the gauge condition. It turns out that for the Maxwell gauge theory, this is always (and trivially) the case. However, in non-abelian theories, and in gauge theories with an abelian compact gauge group, there are many gauges in which a class can be counted more than once. The origin of this problem is a topological obstruction first shown by I. Singer. This issue is known as the Gribov problem. The Coulomb gauge is well known to always have this problem, except for the trivial case of Maxwell's theory.

Finally, we must also keep in mind that we are only fixing the local gauge invariance, but we should not alter the boundary conditions, since they represent physical degrees of freedom. In particular, if the theory is defined on a closed manifold (e.g., a sphere or tori) large gauge transformations, which wrap around the manifold, represent global degrees of freedom (or states). Large gauge transformations play a key role in gauge theories at finite temperature, where the transformation wraps around the (finite and periodic) imaginary time direction. Also, there is a class of gauge theories, known as topological field theories, whose only physical degrees of freedom are represented by large gauge transformations on closed manifolds. We will discuss these theories in chapter 22.

How do we impose a gauge condition consistently? We will do it in the following way. Let us denote the gauge condition that we wish to impose by

$$g(A_\mu) = 0 \tag{9.93}$$

where $g(A_\mu)$ is a local differentiable function of the gauge fields and/or of their derivatives. Examples of such local conditions are $g(A_\mu) = \partial_\mu A^\mu$ for the Lorentz gauge, and $g(A_\mu) = n_\mu A^\mu$ for an axial gauge.

The discussion that follows is valid for all compact Lie groups G of volume $v(G)$. For the special case of the Maxwell gauge theory, the gauge group is $U(1)$. Up to topological considerations, the group $U(1)$ is isomorphic to the real numbers $\mathbb{R}$, even though the volume of the compact $U(1)$ group is finite ($v(U(1)) = 2\pi$), while for the noncompact case, the group is the real numbers $\mathbb{R}$, whose "volume" is infinite ($v(\mathbb{R}) = \infty$).

Naively, to impose a gauge condition would mean to restrict the path integral by inserting eq. (9.93) as a delta function in the integrand:

$$Z[J] \sim \int \mathcal{D}A_\mu \, \delta\left(g(A_\mu)\right) \, e^{iS[A,J]} \tag{9.94}$$

In general, this is an inconsistent (and wrong) prescription, as now shown. Following Faddeev and Popov, we begin by considering the expression defined by the integral

$$\Delta_g^{-1}[A_\mu] \equiv \int \mathcal{D}U \, \delta\left(g(A_\mu^U)\right) \tag{9.95}$$

where the $A_\mu^U(x)$ are the configurations of gauge fields related by the gauge transformation $U(x)$ to the configuration $A_\mu(x)$ (i.e., we move *inside one class*). In other words, the integral in eq. (9.95) is a sum over the orbit of the gauge group. Thus, by construction, $\Delta_g[A_\mu]$ depends only on the class defined by the gauge-fixing condition g or, what is the same, it is gauge invariant.

Let us show that $\Delta_g^{-1}[A_\mu]$ is gauge invariant. We first observe that the integration measure $\mathcal{D}U$, called the "Haar measure," is invariant under the composition rule $U \to UU'$,

$$\mathcal{D}U = \mathcal{D}(UU') \tag{9.96}$$

where U' is an arbitrary but fixed element of G. For the case of $G = U(1)$, $U = \exp(i\phi)$ and $\mathcal{D}U \equiv \mathcal{D}\phi$.

Using the invariance of the measure, eq. (9.96), we can write

$$\Delta_g^{-1}[A_\mu^{U'}] = \int \mathcal{D}U \, \delta\left(g(A_\mu^{U'U})\right) = \int \mathcal{D}U'' \, \delta\left(g(A_\mu^{U''})\right) = \Delta_g^{-1}[A_\mu] \tag{9.97}$$

where we have set $U'U = U''$. Therefore, $\Delta_g^{-1}[A_\mu]$ is gauge invariant (i.e., it is a function of the *class* and not of the configuration A_μ itself). Obviously, we can also write eq. (9.95) in the form

$$1 = \Delta_g[A_\mu] \int \mathcal{D}U \, \delta\left(g(A_\mu^U)\right) \tag{9.98}$$

Let us now insert the number 1, as given by eq. (9.98), in the path integral for a general gauge theory to find

$$\begin{aligned} Z[J] &= \int \mathcal{D}A_\mu \times 1 \times e^{iS[A,J]} \\ &= \int \mathcal{D}A_\mu \, \Delta_g[A_\mu] \int \mathcal{D}U \, \delta\left(g(A_\mu^U)\right) \, e^{iS[A,J]} \end{aligned} \tag{9.99}$$

Now make the change of variables

$$A_\mu \to A_\mu^{U'} \tag{9.100}$$

where $U' = U'(x)$ is an arbitrary gauge transformation, to find

$$Z[J] = \int \mathcal{D}U \int \mathcal{D}A_\mu^{U'} \, e^{iS[A^{U'},J]} \, \Delta_g[A_\mu^{U'}] \, \delta\left(g(A_\mu^{U'U})\right) \tag{9.101}$$

(Notice the order of integration has been changed.) Now choose $U' = U^{-1}$, and use the gauge invariance of the action $S[A, J]$, of the measure $\mathcal{D}A_\mu$, and of $\Delta_g[A]$ to write the partition as

$$Z[J] = \left[\int \mathcal{D}U\right] \int \mathcal{D}A_\mu \, \Delta_g[A_\mu] \, \delta\left(g(A_\mu)\right) \, e^{iS[A,J]} \tag{9.102}$$

The factor in brackets in eq. (9.102) is the infinite constant

$$\int \mathcal{D}U = v(G)^V \tag{9.103}$$

where $v(G)$ is the volume of the gauge group, and V is the (infinite) volume of spacetime. This infinite constant is nothing but the result of summing over gauge-equivalent states inside each class.

Thus, provided the quantity $\Delta_g[A_\mu]$ is finite, and that it does not vanish identically, we find that the consistent rule for fixing the gauge consists of dividing out the (infinite) factor of the volume of the gauge group, and more importantly, of inserting, together with the constraint $\delta\left(g(A_\mu)\right)$, the factor $\Delta_g[A_\mu]$ in the integrand of $Z[J]$:

$$Z[J] \sim \int \mathcal{D}A_\mu \, \Delta_g[A_\mu] \, \delta\left(g(A_\mu)\right) \, e^{iS[A,J]} \tag{9.104}$$

Therefore, the measure $\mathcal{D}A_\mu$ has to be understood as a sum over classes of configurations of the gauge fields and not over all possible configurations.

We have only to compute $\Delta_g[A_\mu]$. Let us first show that $\Delta_g[A_\mu]$ is a determinant of a certain operator and is known as the Faddeev-Popov determinant. We first compute only this determinant for the case of the abelian theory $U(1)$. The non-abelian case, relevant for Yang-Mills gauge theories, is discussed in section 9.8.

Let us compute $\Delta_g[A_\mu]$ by using the fact that $g[A_\mu^U]$ can be regarded as a function of $U(x)$ (for $A_\mu(x)$ fixed). We now change variables from U to g. The price we pay is a Jacobian factor, since

$$\mathcal{D}U = \mathcal{D}g \, \text{Det}\left|\frac{\delta U}{\delta g}\right| \tag{9.105}$$

where the determinant is the Jacobian of the change of variables. Since this is a nonlinear change of variables, we expect a nontrivial Jacobian. Therefore, we can write

$$\Delta_g^{-1}[A_\mu] = \int \mathcal{D}U \, \delta\left(g(A_\mu^U)\right) = \int \mathcal{D}g \, \text{Det}\left|\frac{\delta U}{\delta g}\right| \delta(g) \tag{9.106}$$

to obtain

$$\Delta_g^{-1}[A_\mu] = \text{Det}\left|\frac{\delta U}{\delta g}\right|_{g=0} \tag{9.107}$$

or, conversely,

$$\Delta_g[A_\mu] = \text{Det}\left|\frac{\delta g}{\delta U}\right|_{g=0} \tag{9.108}$$

The results obtained thus far hold for all gauge theories with a compact gauge group. We specialize our discussion first to the case of the $U(1)$ gauge theory, Maxwell electromagnetism. We then discuss how this applies to non-abelian Yang-Mills gauge theories.

For example, for the particular case of the abelian $U(1)$ gauge theory, the Lorentz gauge condition is obtained by the choice $g(A_\mu) = \partial_\mu A^\mu$. Then, for a general $U(1)$ gauge transformation, $U(x) = \exp(i\phi(x))$, we get

$$g(A_\mu^U) = \partial_\mu\,(A^\mu + \partial^\mu \phi) = \partial_\mu A^\mu + \partial^2 \phi \tag{9.109}$$

Hence,

$$\frac{\delta g(x)}{\delta \phi(y)} = \partial^2 \delta(x-y) \tag{9.110}$$

Thus, for the Lorentz gauge of the abelian theory, the Faddeev-Popov determinant is given by

$$\Delta_g[A_\mu] = \mathrm{Det}\partial^2 \tag{9.111}$$

which is a constant independent of A_μ. This is a peculiarity of the abelian theory and, as shown below, it is not true in the non-abelian case.

Let us return momentarily to the general case of eq. (9.104) and modify the gauge condition from $g(A_\mu) = 0$ to $g(A_\mu) = c(x)$, where $c(x)$ is some arbitrary function of x. The partition function now reads

$$Z[J] \sim \int \mathcal{D}A_\mu\, \Delta_g[A_\mu]\, \delta\left(g(A_\mu) - c(x)\right)\, e^{iS[A,J]} \tag{9.112}$$

Averaging over the arbitrary functions with a Gaussian weight (properly normalized to unity) yields

$$\begin{aligned} Z_\alpha[J] &= \mathcal{N} \int \mathcal{D}A_\mu\, \mathcal{D}c\, e^{-i\int d^4x \frac{c(x)^2}{2\alpha}}\, \Delta_g[A_\mu]\, \delta\left(g(A_\mu) - c(x)\right)\, e^{iS[A,J]} \\ &= \mathcal{N} \int \mathcal{D}A_\mu\, \Delta_g(A_\mu)\, e^{i\int d^4x \left[\mathcal{L}[A,J] - \frac{1}{2\alpha}\left(g(A_\mu)\right)^2\right]} \end{aligned} \tag{9.113}$$

From now the discussion is restricted to the $U(1)$ abelian gauge theory (the electromagnetic field) and $g(A_\mu) = \partial_\mu A^\mu$. From eq. (9.113), we find that in this gauge, the Lagrangian is

$$\mathcal{L}_\alpha = -\frac{1}{4} F_{\mu\nu}^2 + J_\mu A^\mu - \frac{1}{2\alpha}\left(\partial_\mu A^\mu\right)^2 \tag{9.114}$$

The parameter α labels a family of gauge-fixing conditions known as the *Feynman–'t Hooft gauges*. For $\alpha \to 0$, we recover the strong constraint $\partial_\mu A^\mu = 0$, the Lorentz gauge. From the point of view of doing calculations, the simplest gauge is $\alpha = 1$, the Feynman gauge, as we will see now.

In this family of gauges parametrized by α, it is straightforward to see that, up to surface terms, the Lagrangian is

$$\mathcal{L}_\alpha = \frac{1}{2} A_\mu \left[g^{\mu\nu}\, \partial^2 - \frac{\alpha - 1}{\alpha}\, \partial^\mu \partial^\nu\right] A_\nu + J_\mu A^\mu \tag{9.115}$$

and the partition function reduces to

$$Z[J] = \mathcal{N}\,\mathrm{Det}\left[\partial^2\right] \int \mathcal{D}A_\mu\, e^{i\int d^4x\, \mathcal{L}_\alpha[A, J]} \tag{9.116}$$

Hence, in a general gauge labeled by α, we get

$$Z[J] = \mathcal{N}\,\mathrm{Det}\left[\partial^2\right]\,\mathrm{Det}\left[g^{\mu\nu}\,\partial^2 - \frac{\alpha-1}{\alpha}\,\partial^\mu\partial^\nu\right]^{-1/2} \times \exp\left(\frac{i}{2}\int d^4x \int d^4y\, J_\mu(x)\, G^{\mu\nu}(x-y)\, J_\nu(y)\right) \tag{9.117}$$

where

$$G^{\mu\nu}(x-y) = -\langle x|\left(g^{\mu\nu}\,\partial^2 - \frac{\alpha-1}{\alpha}\,\partial^\mu\partial^\nu\right)^{-1}|y\rangle \tag{9.118}$$

is the propagator in this gauge, parametrized by α. By inspection, we see that the propagator of the gauge field $G_{\mu\nu}(x-y)$ is related to the vacuum expectation value of the gauge fields by

$$G_{\mu\nu}(x-y) = i\langle 0|TA_\mu(x)A_\nu(y)|0\rangle \tag{9.119}$$

The form of eq. (9.117) may seem to imply that $Z[J]$ depends on the choice of gauge. However, this cannot be correct, since the path integral is, by construction, gauge invariant. We will see in section 9.7 that gauge invariance is indeed protected. This result comes about because J_μ is a conserved current, and so it satisfies the continuity equation $\partial_\mu J^\mu = 0$.

For the Feynman–'t Hooft family of gauges, the propagator takes the form

$$G_{\mu\nu}(x-y) = -\left[g^{\mu\nu} + (\alpha-1)\frac{\partial^\mu\partial^\nu}{\partial^2}\right] G^{(0)}(x-y) \tag{9.120}$$

where $G^{(0)}(x-y)$ is the propagator of the free *massless* scalar field and hence satisfies the Green function equation

$$\partial^2 G^{(0)}(x-y) = \delta^4(x-y) \tag{9.121}$$

where we set the mass of the scalar field to zero.

Thus, as expected for a free-field theory, $Z[J]$ is a product of two factors: a functional (or fluctuation) determinant, and a factor that depends solely on the sources J_μ and contains all the information on the correlation functions. For the case of a single scalar field, we also found a contribution in the form of a determinant factor, but its power was $-1/2$. In the present case, there are two such factors. The first one is the Faddeev-Popov determinant. The second one is the determinant of the fluctuation operator for the gauge field. However, in the Feynman gauge, $\alpha = 1$, this operator is just $g^{\mu\nu}\partial^2$, and its determinant has the same form as the Faddeev-Popov determinant except that it has a power of $-4/2$. This is what one would have expected for a theory with four independent fields (one for each component of A_μ). The Faddeev-Popov determinant has power $+1$. Thus the total power is just $1 - 4/2 = -1$, which is the correct answer for a theory with only two independent (real) fields.

9.6 The propagator

For general α, $G_{\mu\nu}(x-y)$ is the solution of the Green function equation

$$-\left[g^{\mu\nu}\,\partial^2 - \frac{\alpha-1}{\alpha}\partial^\mu\partial^\nu\right] G_{\nu\lambda}(x-y) = g^\mu_\lambda\,\delta^4(x-y) \tag{9.122}$$

Notice that in the special case of the Feynman gauge, $\alpha=1$, this equation becomes

$$-\partial^2 G^{\mu\nu}(x-y) = g^{\mu\nu}\delta^4(x-y) \tag{9.123}$$

Hence, in the Feynman gauge, $G_{\mu\nu}(x-y)$ takes the form

$$G^{\mu\nu}(x-y) = -g^{\mu\nu}\,G^{(0)}(x-y) \tag{9.124}$$

where $G^{(0)}(x-y)$ is just the propagator of a free massless scalar field, that is,

$$\partial^2 G^{(0)}(x-y) = \delta^4(x-y) \tag{9.125}$$

However, in a general gauge, the propagator of the gauge fields

$$G_{\mu\nu}(x-y) = i\langle 0|TA_\mu(x)A_\nu(y)|0\rangle \tag{9.126}$$

does not coincide with the propagator of a scalar field. Therefore, $G_{\mu\nu}(x-y)$, as expected, is a gauge-dependent quantity.

In spite of being gauge dependent, the propagator does contain physical information. Let us examine this issue by calculating the propagator in a general gauge α. The Fourier transform of $G_{\mu\nu}(x-y)$ in D spacetime dimensions is

$$G_{\mu\nu}(x-y) = \int \frac{d^Dp}{(2\pi)^D}\,\widetilde{G}_{\mu\nu}(p)\,e^{-ip\cdot(x-y)} \tag{9.127}$$

This is a solution of eq. (9.122), provided $\widetilde{G}_{\mu\nu}(p)$ satisfies

$$\left[g^{\mu\nu}p^2 - \frac{\alpha-1}{\alpha}p^\mu p^\nu\right]\widetilde{G}_{\nu\lambda}(p) = g^\mu_\lambda \tag{9.128}$$

The formal solution is

$$\widetilde{G}_{\mu\nu}(p) = \frac{1}{p^2}\left[g^{\mu\nu} + (\alpha-1)\frac{p^\mu p^\nu}{p^2}\right] \tag{9.129}$$

In spacetime, the form of this (still formal) solution is given by eq. (9.120).

In particular, in the Feynman gauge $\alpha=1$, (formally) we get

$$\widetilde{G}^F_{\mu\nu}(p) = \frac{g^{\mu\nu}}{p^2} \tag{9.130}$$

whereas in the Lorentz gauge, we find instead

$$\widetilde{G}^{L}_{\mu\nu}(p) = \frac{1}{p^2}\left[g^{\mu\nu} - \frac{p^\mu p^\nu}{p^2}\right] \tag{9.131}$$

Hence, in all cases, there is a pole in p^2 in front of the propagator and a matrix structure that depends on the gauge choice. Notice that the matrix in brackets in the Lorentz gauge, $\alpha \to 0$, becomes the transverse projection operator, which satisfies

$$p_\mu\left[g^{\mu\nu} - \frac{p^\mu p^\nu}{p^2}\right] = 0 \tag{9.132}$$

which follows from the gauge condition $\partial_\mu A^\mu = 0$.

The physical information for this propagator is contained in its analytic structure. It has a pole at $p^2 = 0$, which implies that $p_0 = \sqrt{\boldsymbol{p}^2} = |\boldsymbol{p}|$ is the singularity of $\widetilde{G}_{\mu\nu}(p)$. Hence the pole in the propagator tells us that this theory has a massless particle, the photon.

To actually compute from $\widetilde{G}_{\mu\nu}(p)$ the propagator in spacetime requires that we define the integrals in momentum space carefully. As it stands, the Fourier integral eq. (9.127) is ill defined, due to the pole in $\widetilde{G}_{\mu\nu}(p)$ at $p^2 = 0$. A proper definition requires that we move the pole into the complex plane by shifting $p^2 \to p^2 + i\epsilon$, where ϵ is real and $\epsilon \to 0^+$. This prescription yields the *Feynman propagator*. We will see in chapter 10 that this rule applies to any theory, and that it always yields the vacuum expectation value of the time-ordered product of fields. For the rest of this chapter, we will use the propagator in the Feynman gauge, which reduces to the propagator of a scalar field. This is a quantity we know quite well, both in Euclidean and Minkowski spacetimes.

9.7 Physical meaning of $Z[J]$ and the Wilson loop operator

We have already discussed that a general property of the path integral of any theory is that, in Euclidean spacetime, $Z[0]$ is just

$$Z[0] = \langle 0|0\rangle \sim e^{-TE_0} \tag{9.133}$$

where T is the time span, which in general is such that $T \to \infty$ (beware: Here T is not the temperature!), and E_0 is the vacuum energy. Thus, if the sources J_μ are static (or quasi-static), we get instead

$$\frac{Z[J]}{Z[0]} \sim e^{-T\,[E_0(J) - E_0]} \tag{9.134}$$

Thus, the change in the vacuum energy due to the presence of the sources is

$$U(J) = E_0(J) - E_0 = -\lim_{T\to\infty} \frac{1}{T} \ln \frac{Z[J]}{Z[0]} \tag{9.135}$$

As we will see, the behavior of this quantity has a lot of information about the physical properties of the vacuum (i.e., the ground state) of a theory. Quite generally, if the

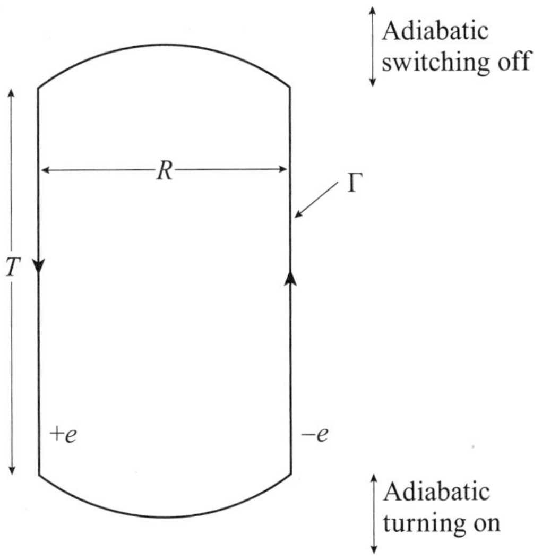

Figure 9.2 The Wilson loop operator can be viewed as representing a pair of quasi-static sources of charge $\pm e$ separated by a distance R from each other.

quasi-static sources J_μ are well separated from each other, $U(J)$ can be split into two terms: a self-energy of the sources, and an interaction energy,

$$U(J) = E_{\text{self-energy}}[J] + V_{\text{int}}[J] \tag{9.136}$$

As an example, let us now compute the expectation value of the Wilson loop operator,

$$W_\Gamma = \langle 0| P e^{ie \oint_\Gamma dx_\mu A^\mu} |0\rangle \tag{9.137}$$

where Γ is the closed path in spacetime shown in figure 9.2, and P is the path-ordering symbol. Physically, what we are doing is looking at the electromagnetic field created by the current

$$J_\mu(x) = e\delta(x_\mu - s_\mu)\,\hat{s}_\mu \tag{9.138}$$

where s_μ is the set of points in spacetime on the loop Γ, and $\hat{s}_\mu$ is a unit vector field tangent to Γ. The loop Γ has time span T and spatial size R.

We will be interested in loops such that $T \gg R$, so that the sources are turned on adiabatically in the remote past and switched off (also adiabatically) in the remote future. By current conservation, the loop must be oriented. Thus, at a fixed time x_0, the loop looks like a pair of static sources with charges $\pm e$ at $\pm R/2$. In other words, we are looking for the effects of a particle-antiparticle pair that is created at rest in the remote past. The members of the pair are then slowly separated (to avoid bremsstrahlung radiation) and live happily apart from each other, at a prudent distance R, for a long time T. And finally, they are adiabatically annihilated in the remote future. Thus, we are in the quasi-static regime described above, and $Z[J]/Z[0]$ should tell us what the effective interaction is between this pair of sources (or "electrodes").

What are the possible behaviors of the Wilson loop operator in general (i.e., for any gauge theory)? The answer to this depends on the nature of the vacuum state. In chapter 18, we will see that a given theory may have different *vacua* or *phases* (as in thermodynamic phases), and that the behavior of the physical observables is different in different vacua (or phases). Here we will do an explicit computation for the case of the simple Maxwell $U(1)$ gauge theory. However, the behavior that we find only holds for a free field, and it is not generic.

What are the possible behaviors, then? A loop is an extended object. In contrast to a local operator, the Wilson loop expectation value is characterized by its geometric properties: its area, perimeter, aspect ratio, and so on. We will see in chapter 18 that these geometric properties of the loop characterize the behavior of the Wilson loop operator. Here are the generic cases (Wilson, 1974; Kogut and Susskind, 1975a):

1) *Area law*: Let $A = RT$ be the minimal area of a surface bounded by the loop. One possible behavior of the Wilson loop operator is the *area law*:

$$W_\Gamma \sim e^{-\sigma RT} \tag{9.139}$$

We will see that this is the *fastest possible decay* of the Wilson loop operator as a function of size. If the area law is obeyed, the effective potential for R large (but still small compared to the time span T) behaves as

$$V_{\text{int}}(R) = \lim_{T\to\infty} \frac{-1}{T} \ln W_\Gamma = \sigma R \tag{9.140}$$

Hence, in this case, the energy to separate a pair of sources grows linearly with distance, and the sources are *confined*. We will say that in this case, the theory is in the *confined phase*. The quantity σ is known as the *string tension*.

2) *Perimeter law*: Another possible decay behavior, weaker than the area law, is a perimeter law,

$$W_\Gamma \sim e^{-\rho(R+T)} + O\left(e^{-R/\xi}\right) \tag{9.141}$$

where ρ is a constant with units of energy, and ξ is a length scale. This decay law implies that in this case,

$$V_{\text{int}} \sim \rho + \text{const.}\, e^{-R/\xi} \tag{9.142}$$

Thus the energy to separate two sources to infinite distance is *finite*. This is a *deconfined* phase. However, since it is massive, with a mass scale $m \sim \xi^{-1}$, there are no long-range gauge bosons. We will see that this phase can also be regarded as a Higgs phase. Since the gauge bosons are massive, this phase bears a close analogy with a superconductor.

3) *Scale invariant law*: Yet another possibility is that the Wilson loop behavior is determined by the aspect ratio R/T or T/R, for example,

$$W_\Gamma \sim e^{-\alpha\left(\frac{R}{T} + \frac{T}{R}\right)} \tag{9.143}$$

where α is a dimensionless constant. This behavior leads to an interaction

$$V_{\text{int}} \sim -\frac{\alpha}{R} \tag{9.144}$$

which coincides with the Coulomb law in four dimensions. We will see that this is a deconfined phase with massless gauge bosons (photons).

Let us now compute the expectation value of the Wilson loop operator in the Maxwell $U(1)$ gauge theory. We will return to the general problem when we discuss the strong coupling behavior of gauge theories in chapter 18. We begin by using the analytic continuation of eq. (9.117) to imaginary time,

$$\begin{aligned} Z[J] &= \mathcal{N}\,\mathrm{Det}\left[\partial^2\right]^{-1} e^{-\frac{1}{2}\int d^4x \int d^4y\, J_\mu(x)\, \langle A_\mu(x) A_\nu(y)\rangle\, J_\nu(y)} \\ &= \mathcal{N}\,\mathrm{Det}\left[\partial^2\right]^{-1} e^{-\frac{e^2}{2}\oint_\Gamma dx_\mu \oint_\Gamma dy_\nu\, \langle A_\mu(x) A_\nu(y)\rangle} \end{aligned} \tag{9.145}$$

where $\langle A_\mu(x) A_\nu(y)\rangle$ is the Euclidean propagator of the gauge fields in the family of gauges labeled by α. Here we have also analytically continued the temporal component of the gauge field $A_0 \to iA_D$, so that the inner products, such as $A_\mu A^\mu \to -A_\mu^2$ (where now $\mu = 1, \ldots, D$), behave as they should in D-dimensional Euclidean spacetime.

In the Feynman gauge $\alpha = 1$, the propagator is given by the expression

$$\langle A_\mu(x) A_\nu(y)\rangle = \delta_{\mu\nu} \int \frac{d^Dp}{(2\pi)^D} \frac{1}{p^2}\, e^{ip_\mu \cdot (x_\mu - y_\mu)} \tag{9.146}$$

where $\mu = 1, \ldots, D$. After doing the integral, we find that the Euclidean propagator (the correlation function) in the Feynman gauge is

$$\langle A_\mu(x) A_\nu(y)\rangle = \delta_{\mu\nu} \frac{\Gamma\left(\frac{D}{2} - 1\right)}{4\pi^{D/2} \left|x - y\right|^{D-2}} \tag{9.147}$$

Notice that the propagator has a short-distance singularity $\sim R^{-(D-2)}$, where R is a length scale. This singularity can be easily understood from dimensional analysis. Indeed, since the Lagrangian density must have units of inverse spacetime volume, $[\mathcal{L}] = L^{-D}$, it follows that the gauge field has units of $[A_\mu] = L^{-(D-2)/2}$, just as in the case of the scalar field. Thus, the circulation of the gauge field has units of $L^{(D-4)/2}$, and the electric charge has units of $[e] = L^{(D-4)/2}$. We will see below that this scaling is consistent.

To carry out this calculation, let us assume that the time span T of the Wilson loop is much larger than its spatial extent R, as shown in figure 9.2. We further assume that the loop is everywhere smooth and that both at long times in the past and in the future, the loop was turned on and off arbitrarily slowly (adiabatically). These assumptions are needed to avoid singularities that have the physical interpretation of the production of a large number of soft photons in the form of bremsstrahlung radiation, as noted above. With these assumptions, the contributions to the expectation value of the Wilson loop operator from the top and bottom of the loop in figure 9.2 can be neglected. Therefore, $E[J]-E_0$ is equal to

$$\begin{aligned} E[J]-E_0 &= \lim_{T\to\infty} \frac{e^2}{2T} \oint_\Gamma \oint_\Gamma dx\cdot dy \, \frac{\Gamma\left(\frac{D}{2}-1\right)}{4\pi^{D/2}\left|x-y\right|^{D-2}} \\ &= 2\times \text{self-energy} - \frac{e^2}{2T}\, 2\int_{-T/2}^{+T/2} dx_D \int_{-T/2}^{+T/2} dy_D \, \frac{\Gamma\left(\frac{D}{2}-1\right)}{4\pi^{D/2}\left|x-y\right|^{D-2}} \end{aligned} \tag{9.148}$$

where $\left|x-y\right|^2=(x_D-y_D)^2+R^2$. The integral in eq. (9.148) is equal to

$$\begin{aligned} &\int_{-T/2}^{+T/2} dx_D \int_{-T/2}^{+T/2} dy_D \, \frac{\Gamma\left(\frac{D}{2}-1\right)}{4\pi^{D/2}\left|x-y\right|^{D-2}} \\ &\quad= \int_{-T/2}^{+T/2} ds \int_{-(T/2+s)/R}^{+(T/2-s)/R} \frac{dt}{(t^2+1)^{(D-2)/2}} \frac{1}{R^{D-3}} \times \frac{\Gamma\left(\frac{D-2}{2}\right)}{4\pi^{D/2}} \\ &\quad\simeq \frac{1}{R^{D-3}} \int_{-T/2}^{+T/2} ds \int_{-\infty}^{+\infty} \frac{dt}{(t^2+1)^{(D-2)/2}} \times \frac{\Gamma\left(\frac{D-2}{2}\right)}{4\pi^{D/2}} \\ &\quad= \frac{T\sqrt{\pi}}{R^{D-3}} \frac{\Gamma\left(\frac{D-3}{2}\right)}{\Gamma\left(\frac{D-2}{2}\right)} \times \frac{\Gamma\left(\frac{D-2}{2}\right)}{4\pi^{D/2}} \end{aligned} \tag{9.149}$$

where

$$\Gamma(\nu)=\int_0^\infty dt\, t^{\nu-1}e^{-t} \tag{9.150}$$

is the Euler gamma function. In eq. (9.149), we have already taken the limit $T/R\to\infty$. Putting it all together, the *interaction energy* of a pair of static sources of charges $\pm e$ separated a distance R in D-dimensional spacetime is given by

$$V_{\text{int}}(R) = -\frac{\Gamma(\frac{D-1}{2})}{2\pi^{(D-1)/2}(D-3)} \frac{e^2}{R^{D-3}} \tag{9.151}$$

This is the Coulomb potential in D spacetime dimensions. It is straightforward to see that this result is consistent (as it should be) with our dimensional analysis. In the particular case of $D=4$ dimensions, we find

$$V_{\text{int}}(R) = -\left(\frac{e^2}{4\pi}\right)\frac{1}{R} \tag{9.152}$$

where the (dimensionless) quantity $\alpha=e^2/(4\pi)$ is the *fine structure constant*. Notice that in $D=4$ spacetime dimensions, the charge e is dimensionless. This fact plays a key role in the perturbative analysis of quantum electrodynamics. In contrast, in $D=1+1$ dimensions, this result implies that the Coulomb interaction is a linear function of the separation R between the sources (i.e., the charged sources are confined).

Therefore we find that, even at the quantum level, the effective interaction between a pair of static sources is the Coulomb interaction. This is true because Maxwell's theory is a free-field theory. It is also true in quantum electrodynamics (QED), the QFT of electrons and photons, at distances R much greater than the Compton wavelength of the electron. However, it is not true at short distances, where the effective charge is screened by fluctuations of the Dirac field, and the potential becomes exponentially suppressed. In quantum chromodynamics (QCD), the situation is quite different: Even in the absence of a matter field, for R large compared with a scale ξ determined by the dynamics of Yang-Mills theory, the effective potential $V(R)$ grows linearly with R. This long-distance behavior is known as *confinement*. The existence of the nontrivial scale ξ, known as the *confinement scale*, cannot be obtained in perturbation theory. Conversely, the potential is Coulomb-like at short distances, a behavior known as *asymptotic freedom*.

9.8 Path-integral quantization of non-abelian gauge theories

In this section, we discuss the general properties of the path-integral quantization of non-abelian gauge theories. Most of what we did for the abelian case carries over to non-abelian gauge theories, where, as we will see, it plays a much more central role. However, here we will not deal with the nonlinearities, which ultimately require the use of the ideas of the renormalization group and a nonperturbative treatment. We will do this in chapter 15. A more detailed presentation can be found in the classic books by Claude Itzykson and Jean-Bernard Zuber (1980) and Michael Peskin and Daniel Schroeder (1995).

The path integral $Z[J]$ for a non-abelian gauge field A_μ with gauge condition(s) $g^a[A]$ is

$$Z[J] = \int \mathcal{D}A_\mu^a \, e^{iS[A,J]} \, \delta(g[A]) \, \Delta_{\text{FP}}[A] \tag{9.153}$$

where $\Delta_{\text{FP}}[A]$ is the Faddeev-Popov determinant, and $A_\mu = A_\mu^a \lambda^a$ is in the algebra of a simply connected compact Lie group G, whose generators are the Hermitian matrices λ^a. We will use the family of covariant gauge conditions

$$g^a[A] = \partial^\mu A_\mu^a(x) + c^a(x) = 0 \tag{9.154}$$

Notice that one gauge condition is imposed for each direction in the algebra of the gauge group G. We will proceed as we did in the abelian case and consider an average over gauges. In other words, we will work in the manifestly covariant Feynman–'t Hooft gauges. Notice that in the partition function of eq. (9.153), we have dropped the overall divergent factor $v(G)^V$ (or, rather, that we defined the integration measure $\mathcal{D}A_\mu$ so that this factor is explicitly canceled).

Let us work out the structure of the Faddeev-Popov determinant for a general gauge-fixing condition $g^a[A]$. Let U be an infinitesimal gauge transformation,

$$U \simeq 1 + i\epsilon^a(x)\lambda^a + \cdots \tag{9.155}$$

Under a gauge transformation, the vector field A_μ transforms as

$$A_\mu^U = UA_\mu U^{-1} + i\left(\partial_\mu U\right) U^{-1} \equiv A_\mu + \delta A_\mu \tag{9.156}$$

For an infinitesimal transformation, the change in A_μ is

$$\delta A_\mu = i\epsilon^a \left[\lambda^a, A_\mu\right] - \partial_\mu \epsilon^a \lambda^a + O(\epsilon^2) \tag{9.157}$$

where the λ^a are the generators of the algebra of the gauge group G.

In components, we can also write

$$\begin{aligned}
\delta A_\mu^c &= 2i\epsilon^b \, \mathrm{tr}\left(\lambda^c \left[\lambda^b, A_\mu\right]\right) - 2\,\partial_\mu \epsilon^b \, \mathrm{tr}\left(\lambda^c \lambda^b\right) + O(\epsilon^2) \\
&= i\epsilon^b \mathrm{tr}\left(\lambda^c \left[\lambda^b, \lambda^d\right]\right) A_\mu^d - \partial_\mu \epsilon^b \delta_{bc} + O(\epsilon^2) \\
&= -2 f^{bde}\, \epsilon^b \mathrm{tr}\left(\lambda^c \lambda^e\right) A_\mu^d - \partial_\mu \epsilon^b \, \delta_{bc} + O(\epsilon^2) \\
&= -f^{bdc}\, \epsilon^b \, A_\mu^d - \partial_\mu \epsilon^b \, \delta_{bc} + O(\epsilon^2)
\end{aligned} \tag{9.158}$$

where f^{abc} are the structure constants of the Lie group G.

Therefore, we find

$$\frac{\delta A_\mu^c(x)}{\delta \epsilon^b(y)} = -\left[\partial_\mu \delta_{bc} + f^{bcd}\, A_\mu^d\right] \delta(x-y) \equiv -D_\mu^{cd}[A]\delta(x-y) \tag{9.159}$$

where we have denoted by $D_\mu[A]$ the covariant derivative in the *adjoint* representation, which in components is given by

$$D_\mu^{ab}[A] = \delta_{ab}\, \partial_\mu - f^{abc}\, A_\mu^c \tag{9.160}$$

Using these results, we can put the Faddeev-Popov determinant (or Jacobian) in the form

$$\Delta_{\mathrm{FP}}[A] = \mathrm{Det}\left(\frac{\delta g}{\delta \epsilon}\right) = \mathrm{Det}\left(\frac{\partial g^a}{\partial A_\mu^c}\frac{\delta A_\mu^c}{\delta \epsilon^b}\right) \tag{9.161}$$

where we used the expression

$$\frac{\delta g^a}{\delta \epsilon^b} = \frac{\partial g^a}{\partial A_\mu^c}\frac{\delta A_\mu^c}{\delta \epsilon^b} \tag{9.162}$$

We will now define an operator M_{FP}, whose matrix elements are

$$\begin{aligned}
\langle x, a | M_{\mathrm{FP}} | y, b\rangle &= \langle x, a | \frac{\partial g}{\partial A_\mu^c}\frac{\delta A_\mu^c}{\delta \epsilon} | y, b\rangle \\
&= \int_z \frac{\partial g^a(x)}{\partial A_\mu^c(z)}\frac{\delta A_\mu^c(z)}{\delta \epsilon^b(y)} \\
&= -\int_z \frac{\partial g^a(x)}{\partial A_\mu^c(z)} D_\mu^{cb} \delta(z-y)
\end{aligned} \tag{9.163}$$

For the case of $g^a[A] = \partial^\mu A_\mu^a(x) - c^a(c)$, appropriate for the Feynman–'t Hooft gauges, we have

$$\frac{\partial g^a(x)}{\partial A_\mu^c(z)} = \delta_{ac}\, \partial^\mu \delta(x-z) \tag{9.164}$$

and also

$$\begin{aligned}\langle x, a|M_{\rm FP}|y, b\rangle &= -\int_z \delta_{ac}\partial_x^\mu \delta(x-z) D_\mu^{cb}[A]\delta(z-y) \\ &= -\int_z \delta_{ac}\delta(x-z)\, \partial_z^\mu D_\mu^{cb}\delta(x-y) \\ &= -\,\partial^\mu D_\mu^{ab}\delta(x-y)\end{aligned} \tag{9.165}$$

Thus, the Faddeev-Popov determinant is now

$$\Delta_{\rm FP} = {\rm Det}\left(\partial^\mu D_\mu[A]\right) \tag{9.166}$$

Notice that in the non-abelian case, this determinant is an explicit function of the gauge field A_μ.

Since $\Delta_{\rm FP}[A]$ is a determinant, it can be written as a path integral over a set of fermionic fields, denoted by $\eta_a(x)$ and $\bar\eta_a(x)$ (and known as *ghosts*), with one per gauge condition (i.e., one per generator):

$${\rm Det}\left[\partial^\mu D_\mu\right] = \int \mathcal{D}\eta_a \mathcal{D}\bar\eta_a\, e^{\displaystyle i\int d^Dx\, \bar\eta_a(x)\, \partial^\mu D_\mu^{ab}[A]\, \eta_b(x)} \tag{9.167}$$

Notice that these ghost fields are not spinors, and hence are quantized with the "wrong" statistics. In other words, these "particles" do not satisfy the general conditions for causality and unitarity. Hence, ghosts cannot create physical states (thus their ghostly character).

The full form of the path integral of a Yang-Mills gauge theory with coupling constant g, in the Feynman–'t Hooft covariant gauges with gauge parameter λ, is given by

$$Z = \int \mathcal{D}A\mathcal{D}\eta\mathcal{D}\bar\eta\, e^{\displaystyle i\int d^Dx\, \mathcal{L}_{\rm YM}[A,\eta,\bar\eta]} \tag{9.168}$$

where $\mathcal{L}_{\rm YM}$ is the effective Lagrangian density (defined in section 3.7.2)

$$\mathcal{L}_{\rm YM}[A,\eta,\bar\eta] = -\frac{1}{4g^2}{\rm tr}\left(F_{\mu\nu}F^{\mu\nu}\right) + \frac{\lambda}{2g^2}\left(\partial_\mu A^\mu\right)^2 - \bar\eta\, \partial_\mu D^\mu[A]\, \eta \tag{9.169}$$

Thus the pure gauge theory, even in the absence of matter fields, is nonlinear. We will return to this problem in chapters 16–18, when we look at both the perturbative and nonperturbative aspects of Yang-Mills gauge theories.

9.9 BRST invariance

In section 9.8, we developed in detail the path-integral quantization of non-abelian Yang-Mills gauge theories. We paid close attention to the role of gauge invariance and how to consistently fix the gauge to define the path integral. Here we will show that the effective Lagrangian of a Yang-Mills gauge field, eq. (9.169), has an extended symmetry, closely related to supersymmetry. This extended symmetry plays a crucial role in proving the renormalizability of non-abelian gauge theories.

Let us consider the QCD Lagrangian in the Feynman–'t Hooft covariant gauges (with gauge parameter λ and coupling constant g). The Lagrangian density $\mathcal{L}$ of this theory is

$$\mathcal{L} = \bar{\psi}\,(i\not{D} - m)\,\psi - \frac{1}{4} F^a_{\mu\nu} F^{\mu\nu}_a - \frac{1}{2\lambda} B_a B_a + B_a \partial^\mu A^a_\mu - \bar{\eta}^a \partial^\mu D^{ab}_\mu \eta^b \tag{9.170}$$

Here ψ is a Dirac Fermi field that represents quarks and transforms under the fundamental representation of the gauge group G. The "Hubbard-Stratonovich" field B_a is an auxiliary field that has no dynamics of its own; it transforms as a vector in the adjoint representation of G.

Becchi, Rouet, and Stora (Becchi et al., 1974, 1976) and Tyutin (2008) realized that this gauge-fixed Lagrangian has the following ("BRST") symmetry, where ϵ is an infinitesimal anticommuting parameter:

$$\delta A^a_\mu = \epsilon D^{ab}_\mu \eta_b \tag{9.171}$$

$$\delta\psi = ig\epsilon\eta^a t^a \psi \tag{9.172}$$

$$\delta\eta^a = -\frac{1}{2} g\epsilon f^{abc} \eta_b \eta_c \tag{9.173}$$

$$\delta\bar{\eta}^a = \epsilon B^a \tag{9.174}$$

$$\delta B^a = 0 \tag{9.175}$$

Equation (9.171) and eq. (9.172) are local gauge transformations, and as such, leave invariant the first two terms of the effective Lagrangian $\mathcal{L}$ of eq. (9.170). The third term in eq. (9.170) is trivial. The invariance of the fourth and fifth terms holds, because the change of δA in the fourth term cancels the change of $\bar{\eta}$ in the fifth term. Finally, it remains to determine whether the changes of the fields A_μ and η in the fifth term of eq. (9.170) cancel out. To see that this is the case, check that

$$\begin{aligned}\delta\left(D^{ab}_\mu \eta^b\right) &= D^{ab}_\mu \delta\eta^b + g f^{abc} \delta A^b_\mu \eta^c \\ &= -\frac{1}{2} g^2 \epsilon f^{abc} f^{cde} \left(A^b_\mu \eta^d \eta^e + A^d_\mu \eta^e \eta^b + A^e_\mu \eta^b \eta^d\right)\end{aligned} \tag{9.176}$$

which vanishes due to the Jacobi identity for the structure constants,

$$f^{ade} f^{bcd} + f^{bde} f^{cad} + f^{cde} f^{abd} = 0 \tag{9.177}$$

or, equivalently, from the nested commutators of the generators t^a:

$$\left[t^a, \left[t^b, t^c\right]\right] + \left[t^b, \left[t^c, t^a\right]\right] + \left[t^c, \left[t^a, t^b\right]\right] = 0 \tag{9.178}$$

Hence, BRST is at least a global symmetry of the gauge-fixed action with gauge-fixing parameter λ.

This symmetry has a remarkable property that follows from its fermionic nature. Let ϕ be any of the fields of the Lagrangian and $Q\phi$ be the BRST transformation of the field:

$$\delta\phi = \epsilon Q\phi \tag{9.179}$$

For instance,

$$Q^a A_\mu^a = D_\mu^{ab}\eta^b \tag{9.180}$$

and so on. It follows that for *any* field ϕ

$$Q^2\phi = 0 \tag{9.181}$$

(i.e., the BRST transformation of $Q\phi$ vanishes). This rule works for the field A_μ due to the transformation property of $\delta(D_\mu^{ab}\eta^b)$. It also holds for the ghosts, since

$$Q^2\eta^a = \frac{1}{2}g^2 f^{abc} f^{bde}\eta^c\eta^d\eta^e = 0 \tag{9.182}$$

which holds due to the Jacobi identity.

What are the implications of the existence of BRST as a continuous symmetry? To begin with, it implies that there is a conserved self-adjoint charge Q that must necessarily commute with the Hamiltonian H of the Yang-Mills gauge theory. In eq. (9.181), we saw how Q acts on the fields, $Q^2\phi = 0$, for all the fields in the Lagrangian. Hence, as an operator, $Q^2 = 0$; that is, the BRST charge Q is nilpotent, and it commutes with H.

Let us now show that Q divides the Hilbert space of the eigenstates of H into three sectors:

1) For $Q^2 = 0$ to hold, many eigenstates of H must be annihilated by Q. Let $\mathcal{H}_1$ be the set of eigenstates of H that are *not* annihilated by Q. Hence, if $|\psi_1\rangle \in \mathcal{H}_1$, then $Q|\psi_1\rangle \neq 0$. Thus, the states in $\mathcal{H}_1$ are not BRST invariant.
2) Let us consider the subspace of states $\mathcal{H}_2$ of the form $|\psi_2\rangle = Q|\psi_1\rangle$; that is, $\mathcal{H}_2 = Q\mathcal{H}_1$. Then for these states, $Q|\psi_2\rangle = Q^2|\psi_1\rangle = 0$. Hence, the states in $\mathcal{H}_2$ are BRST invariant but are the BRST transform of states in $\mathcal{H}_1$.
3) Finally, let $\mathcal{H}_0$ be the set of eigenstates of H that are annihilated by Q, $Q|\psi_0\rangle = 0$, but are *not* in $\mathcal{H}_2$ (i.e., $|\psi_0\rangle \neq Q|\psi_1\rangle$). Hence, the states in $\mathcal{H}_0$ are BRST invariant and are not the BRST transform of any other state. This is the *physical space of states.*

It follows from the above classification that any pair of states in $\mathcal{H}_2$, $|\psi_2\rangle$ and $|\psi_2'\rangle$, has a zero inner product:

$$\langle\psi_2|\psi_2'\rangle = \langle\psi_1|Q|\psi_2'\rangle = 0 \tag{9.183}$$

where we used the property that $|\psi_2\rangle$ is the BRST transform of a state in $\mathcal{H}_1$, $|\psi_1\rangle$. Similarly, one can show that if $|\psi_0\rangle \in \mathcal{H}_0$, then $\langle\psi_2|\psi_0\rangle = 0$.

What is the physical meaning of BRST and of this classification? Peskin and Schroeder (1995) give a simple argument. Consider the weak coupling limit of the theory, $g \to 0$. In this limit, we can find out what BRST does by looking at the transformation properties of the fields that appear in the Lagrangian of eq. (9.170). In particular, Q transforms a forward polarized (i.e., longitudinal) component of A_μ into a ghost. At $g = 0$, we see that $Q\eta = 0$ and that the anti-ghost $\bar{\eta}$ transforms into the auxiliary field B. Also, at the classical level, $B = \lambda\partial^\mu A_\mu$. Hence, the auxiliary fields B are backward (longitudinally) polarized quanta of A_μ. Thus, forward polarized gauge bosons and anti-ghosts are in $\mathcal{H}_1$, since they are not the BRST transform of states created by other fields. Ghosts and backward polarized gauge bosons are in $\mathcal{H}_2$, since they are the BRST transform of the former. Finally, transverse gauge bosons are in $\mathcal{H}_0$. Hence, in general, states with ghosts, anti-ghosts, and gauge bosons with unphysical polarization belong to either $\mathcal{H}_1$ or $\mathcal{H}_2$. Only the physical states belong to $\mathcal{H}_0$. It turns out that the S-matrix, when restricted to the physical space $\mathcal{H}_0$, is unitary (as it should be).

Finally, note that BRST symmetry appears in all theories with constraints (Henneaux and Teitelboim, 1992). For example, it plays a key role in the study of critical dynamics in the path-integral approach to the Langevin equation description of systems out of equilibrium, and in the statistical mechanics of disordered systems (Martin et al., 1973; Parisi, 1988; Hertz et al., 2016).

Exercises

9.1 Path-integral quantization of the free electromagnetic field
Use the path-integral quantization of the free electromagnetic (Maxwell) gauge field, in the Feynman–'t Hooft family of gauges, with gauge-fixing parameter α (see the text for definitions).

1) Find an expression, in momentum space, for the Feynman propagator for the free electromagnetic field, $D^F_{\mu\nu}(x,x')$:

$$D^F_{\mu\nu}(x,x') = -i\langle 0|T\, A_\mu(x)\, A_\nu(x')|0\rangle \tag{9.184}$$

Make sure to specify your $i\epsilon$ prescription.

2) Use your results from part 1 to derive an inhomogeneous partial differential equation obeyed by the propagator in this family of gauges.

3) Show that the expression that you derived in part 2 contains a Lorentz covariant term of the form

$$D^F_{\mu\nu}(x,x') = g_{\mu\nu}\, D_F(x,x') + \cdots \tag{9.185}$$

where $\cdots$ are the other terms. Find a relation between $D_F(x,x')$ and the Feynman propagator for a free real scalar field. Use this relationship to calculate an explicit formula for the dependence of $D_F(x,x')$ on the coordinates.

4) Derive an explicit expression for the path integral for a Maxwell gauge field, in the Feynman–'t Hooft family of gauges, coupled to a set of classical (i.e., not quantized) currents $J_\mu(x)$.

a) Derive an explicit formula for the Faddeev-Popov determinant for this family of gauges, and show that for a Maxwell field, it does not depend on the gauge field configuration.

b) Show that the explicit expression for the path integral does not depend on the choice of gauge if the currents are conserved.

c) Show the explicit dependence of the path integral on the propagator you computed in part 2. Show that this expression is independent of the gauge-fixing parameter α.

10

Observables and Propagators

In earlier chapters, we considered the properties of several field theories that describe physical systems isolated from the outside world. However, the only way to investigate the properties of a physical system is to interact with it. Thus, we must consider physical systems that are somehow coupled to external fields (or sources).

We should consider two cases. In one case, we will look at the problem of the interaction of states of an isolated system (i.e., a scattering problem). In this case, we prepare states (or wave packets) that are sufficiently far apart so that their mutual interactions can be neglected. The prototype is the scattering of particles off a target or each other in a particle accelerator experiment. Here the physical properties are encoded in a suitable set of cross sections.

In the second case, we want to understand properties of a large system "from outside," and we will consider the role of small external perturbations. Here we will develop a general approach, known as linear response theory. Our results will be couched in terms of a suitable set of susceptibilities. This is the typical situation of interest in many experiments in condensed matter physics.

In both cases, all quantities of physical interest will be derived from a suitably defined correlation function or propagator. Our task is twofold. First, we must determine the general expected properties of the propagators (in particular, their analytic properties). Second, we will show that their analytic properties largely determine the behavior of cross sections and susceptibilities.

10.1 The propagator in classical electrodynamics

In a classical field theory, such as classical electrodynamics, we can investigate the properties of the electromagnetic field by considering the effect of a set of highly localized external sources. These can be *electric charges* or, more generally, some well-defined distribution of *electric currents*. The result is familiar: The external currents set up a radiation field that propagates in space—a propagating electromagnetic field. In Maxwell's electrodynamics, these effects are described by Maxwell's equations, the equations of motion of the electromagnetic field generated by a current distribution $j^\mu(x)$:

$$\partial^2 A^\mu(x) = j^\mu(x) \tag{10.1}$$

where we have imposed the Lorentz gauge condition, $\partial_\mu A^\mu = 0$, and hence the equation of motion reduces to the wave equation.

In classical electrodynamics, the solutions to this equation are found using the *Green function* $G(x, x')$,

$$\partial_x^2 G(x, x') = \delta^4(x - x') \tag{10.2}$$

which satisfies the boundary condition (or, rather, initial condition) that

$$G(x, x') = 0, \quad \text{if} \quad x_0 < x'_0 \tag{10.3}$$

This is the *retarded Green function*, denoted by $G_R(x - x')$, which vanishes for events in the past, $x_0 < x'_0$.

The wave equation in the presence of a set of currents $j^\mu(x)$ is an inhomogeneous partial differential equation. For times in the remote past, before any currents were present, there should be no electromagnetic field present. The choice of retarded boundary conditions guarantees that the system obeys *causality*. The solution to the inhomogeneous partial differential equation is, as usual, the sum of an *arbitrary* solution of the homogeneous equation, A^μ_{in}, which represents a preexisting electromagnetic field, and a *particular* solution of the inhomogeneous field equation. We write the general solution in the form

$$A^\mu(x) = A^\mu_{in}(x) + \int d^4x' G_R(x - x') j^\mu(x') \tag{10.4}$$

where $A^\mu_{in}(x)$ is a solution of the wave equation in free space (in the absence of sources):

$$\partial^2 A^\mu_{in}(x) = 0 \tag{10.5}$$

Thus, all we need to know is the retarded Green function, $G_R(x - x')$. Notice that the choice of retarded boundary conditions ensures that, for $x_0 < x'_0$, $A^\mu(x) = A^\mu_{in}(x)$, since $G_R(x - x') = 0$ for $x_0 < x'_0$.

Let us solve for the Green function $G(x, x')$. This is most easily done in terms of the Fourier transform of $G(x, x')$,

$$G(x, x') = \int \frac{d^4p}{(2\pi)^4} e^{-ip \cdot (x - x')} \widetilde{G}(p) \tag{10.6}$$

where $p \equiv p^\mu$, and $p \cdot x \equiv p_0 x_0 - \boldsymbol{p} \cdot \boldsymbol{x}$. It is easy to check that, formally, the Fourier transform $\widetilde{G}(p)$ should be given by

$$\widetilde{G}(p) = -\frac{1}{p^2} \tag{10.7}$$

Now we have two problems. One is that $\widetilde{G}(p)$ has a singularity at $p^2 \equiv p_0^2 - \boldsymbol{p}^2 = 0$ (i.e., at the eigenfrequencies of the normal modes of the free electromagnetic field $p_0 = \pm|\boldsymbol{p}|$). Thus, the integral is ill defined, and we must specify what to do with the singularity. The other problem is that this function $G(x, x')$ does not satisfy, at least not in any obvious way, the boundary conditions.

We will solve both problems simultaneously. Let us define the *retarded Green function* by

$$G_R(x, x') \equiv \Theta(x_0 - x'_0) G(x, x') \tag{10.8}$$

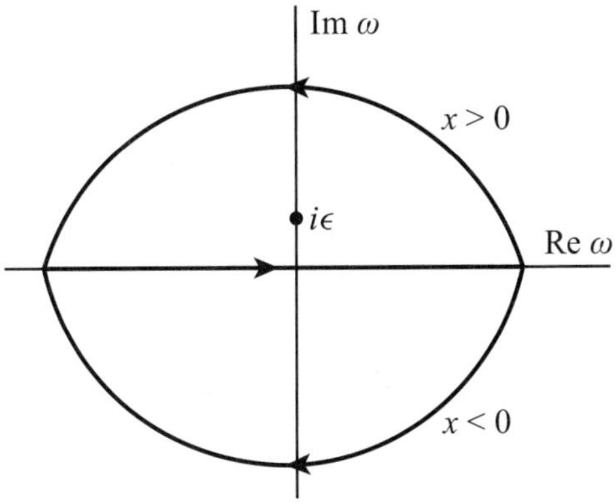

Figure 10.1 Contour in the complex plane that defines the function $\Theta(x)$.

where $\Theta(x)$ is the step (or Heaviside) function:

$$\Theta(x) = \begin{cases} 1 & x \geq 0, \\ 0 & x < 0 \end{cases} \tag{10.9}$$

Therefore, $G_R(x, x')$ vanishes for $x_0 < x_0'$ and is equal to $G(x - x')$ for $x_0 > x_0'$.

The step function $\Theta(x)$ has the formal integral representation (a Fourier transform):

$$\Theta(x) = \lim_{\epsilon \to 0^+} \int_{-\infty}^{+\infty} \frac{d\omega}{2\pi i} \frac{e^{i\omega x}}{\omega - i\epsilon} \tag{10.10}$$

where the integral is interpreted as an integral over the contour shown in figure 10.1. Thus, when closing the contour on the lower half-plane, as in the case where $x < 0$, shown in figure 10.1, the closed contour does not contain a pole, and the integral vanishes by the Cauchy theorem. Conversely, for $x > 0$, we close the contour on a large arc in the upper half-plane and pick up a contribution from the enclosed pole equal to the residue, $e^{-\epsilon x}$, which converges to 1 as $\epsilon \to 0^+$. Notice that the integral on the large arc in the upper half-plane converges to zero, for arcs with radius $R \to \infty$, only if $x > 0$.

Define the retarded Green function, $G_R(x - x')$, by the following expression (see figure 10.2):

$$G_R(x - x') = - \lim_{\epsilon \to 0^+} \int \frac{d^4p}{(2\pi)^4} \frac{e^{-ip \cdot (x - x')}}{(p_0 + i\epsilon)^2 - \boldsymbol{p}^2} \tag{10.11}$$

which satisfies all the requirements.

We can also define the *advanced Green function*, $G_A(x - x')$, which vanishes in the future but not in the past,

$$G_A(x - x') = 0, \qquad \text{for} \quad x_0 - x_0' > 0 \tag{10.12}$$

by changing the sign of ϵ:

$$G_A(x - x') = - \lim_{\epsilon \to 0^+} \int \frac{d^4p}{(2\pi)^4} \frac{e^{-ip \cdot (x - x')}}{(p_0 - i\epsilon)^2 - \boldsymbol{p}^2} \tag{10.13}$$

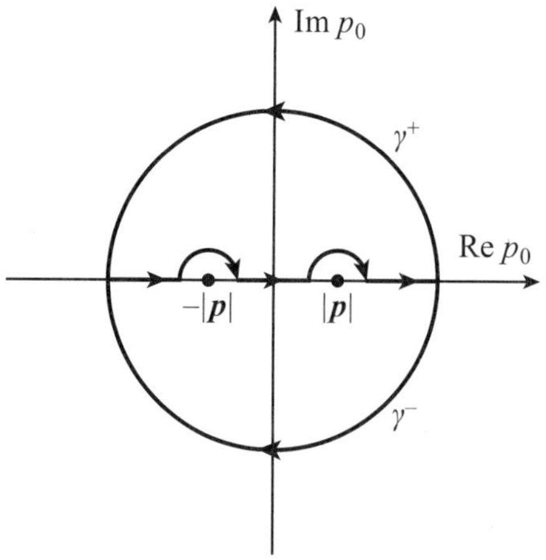

Figure 10.2 Contour in the complex plane that defines the retarded Green function.

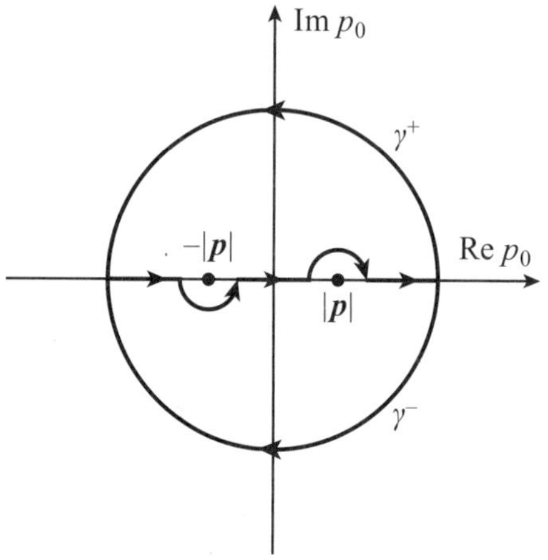

Figure 10.3 Contour in the complex plane that defines the Feynman propagator, or time-ordered Green function.

Now the poles are in the upper half-plane, and the integral vanishes for $x_0 - x_0' > 0$. However, for $x_0 - x_0' < 0$, we pick up the poles when we close on the lower half-plane, and this contribution does not vanish.

There are still two other possible choices of contours, such as the one shown in figure 10.3. In this case, when we close on the contour γ^-, in the lower half-plane, we pick up the contribution of the positive frequency pole at $+|\boldsymbol{p}|$. The resulting frequency integral is

$$\oint_{\gamma^-} \frac{dp_0}{2\pi} \frac{e^{-ip \cdot (x-x')}}{p_0^2 - \boldsymbol{p}^2 + i\epsilon} = -\frac{i}{2|\boldsymbol{p}|} e^{i\boldsymbol{p} \cdot (\boldsymbol{x}-\boldsymbol{x}') - i|\boldsymbol{p}|(x_0 - x_0')}, \qquad \text{for } x_0 > x_0' \quad (10.14)$$

In the opposite case, we close on the contour γ^+ in the upper half-plane, which encloses the negative frequency pole at $-|\boldsymbol{p}|$. The integral now becomes

$$\oint_{\gamma^+} \frac{dp_0}{2\pi i} \frac{e^{-ip\cdot(x-x')}}{p_0^2 - \boldsymbol{p}^2 + i\epsilon} = \frac{i}{-2|\boldsymbol{p}|} e^{i\boldsymbol{p}\cdot(\boldsymbol{x}-\boldsymbol{x}') - i|\boldsymbol{p}|(x_0 - x_0')}, \qquad \text{for } x_0 < x_0' \tag{10.15}$$

With this choice, the contour integral yields the *time-ordered Green function*

$$\begin{aligned} G_F(x-x') &= -\int \frac{d^4p}{(2\pi)^4} \frac{e^{-ip\cdot(x-x')}}{p^2 + i\epsilon} \\ &= i\int \frac{d^3p}{(2\pi)^3 2|\boldsymbol{p}|} \Big\{ \Theta(x_0 - x_0')\, e^{-i|\boldsymbol{p}|(x_0 - x_0') + i\boldsymbol{p}\cdot(\boldsymbol{x}-\boldsymbol{x}')} \\ &\qquad + \Theta(x_0' - x_0) e^{+i|\boldsymbol{p}|(x_0 - x_0') - i\boldsymbol{p}\cdot(\boldsymbol{x}-\boldsymbol{x}')} \Big\} \end{aligned} \tag{10.16}$$

which is known as the *Feynman propagator* and bears a close formal resemblance to the mode expansions of a free-field theory. Here, too, the integration measure is Lorentz invariant. Also notice that this Feynman propagator or Green function propagates the positive frequency modes forward in time and the negative frequency modes backward in time. The alternative choice of contour simply yields the negative of $G_F(x-x')$.

10.2 The propagator in nonrelativistic quantum mechanics

In nonrelativistic quantum mechanics, the evolution of quantum states is governed by the Schrödinger equation:

$$(i\hbar\partial_t - H)\psi = 0 \tag{10.17}$$

Let $H = H_0 + V$, and V be some position- and time-dependent potential that vanishes (*very slowly*) both in the remote past ($t \to -\infty$) and in the remote future ($t \to +\infty$). In this case, the eigenstates of the system are, in both limits, eigenstates of H_0. If $V(x,t)$ varies slowly with time, the states of H evolve smoothly, or adiabatically. Thus, we are describing *scattering processes* between free particle states (see figure 10.4). Let $F(\boldsymbol{x}', t'|\boldsymbol{x}, t)$ denote the amplitude

$$F(\boldsymbol{x}', t'|\boldsymbol{x}, t) \equiv \langle \boldsymbol{x}', t'|\boldsymbol{x}, t\rangle \tag{10.18}$$

We have already seen this amplitude when we discussed the path integral picture of quantum mechanics.

Let us suppose that, at some time t, the system is in the state $|\psi(t)\rangle = |\boldsymbol{x}, t\rangle$. At some time $t' > t$, the state of the system is $|\psi, t'\rangle$, which is obtained by solving the Schrödinger equation

$$i\hbar\partial_t|\psi\rangle = H|\psi\rangle \tag{10.19}$$

where H is generally time dependent. The formal solution of this equation is

$$|\psi(t')\rangle = T\, e^{-\frac{i}{\hbar}\int_t^{t'} dt'' H(t'')} |\psi(t)\rangle \tag{10.20}$$

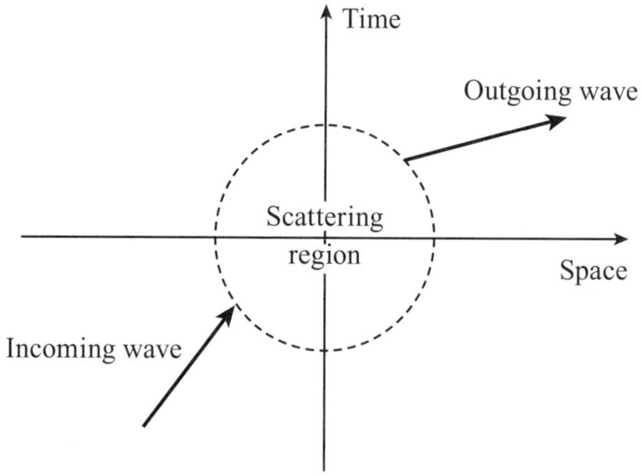

Figure 10.4 A scattering process.

where T is the time ordering symbol, that is,

$$T\, e^{\frac{-i}{\hbar}\int_t^{t'} dt'' H(t'')} \equiv \sum_{n=0}^{\infty} \frac{1}{n!}\left(\frac{-i}{\hbar}\right)^n \int_t^{t'} dt_1 \cdots \int_t^{t_{n-1}} dt_n H(t_1)\cdots H(t_n) \tag{10.21}$$

Thus, the amplitude $F(\boldsymbol{x}', t'|\boldsymbol{x}, t)$ is

$$F(\boldsymbol{x}', t'|\boldsymbol{x}, t) = \langle \boldsymbol{x}', t'|\boldsymbol{x}, t\rangle \equiv \langle \boldsymbol{x}'|\psi(t')\rangle \tag{10.22}$$

Hence, if the initial state is $|\psi(t)\rangle = |\boldsymbol{x}\rangle$, then we get

$$F(\boldsymbol{x}', t'|\boldsymbol{x}, t) = \langle \boldsymbol{x}'|T\, e^{-\frac{i}{\hbar}\int_t^{t'} dt'' H(t'')}|\boldsymbol{x}\rangle \tag{10.23}$$

For an arbitrary initial state $|\psi(t)\rangle$, we get

$$\langle \boldsymbol{x}'|\psi(t')\rangle = \langle \boldsymbol{x}'|\, e^{-\frac{i}{\hbar}\int_t^{t'} dt'' H(t'')}|\psi(t)\rangle \tag{10.24}$$

Since the states $\{|\boldsymbol{x}\rangle\}$ are complete, we have the completeness relation

$$1 = \int d\boldsymbol{x} |\boldsymbol{x}\rangle\langle \boldsymbol{x}| \tag{10.25}$$

which allows us to write

$$\langle \boldsymbol{x}'|\psi(t')\rangle = \int_{-\infty}^{+\infty} d\boldsymbol{x} \langle \boldsymbol{x}'|\, e^{-\frac{i}{\hbar}\int_t^{t'} dt'' H(t'')}|\boldsymbol{x}\rangle\langle \boldsymbol{x}|\psi(t)\rangle \tag{10.26}$$

so we get

$$\psi(\boldsymbol{x}', t') = \int_{-\infty}^{+\infty} d\boldsymbol{x} \langle \boldsymbol{x}', t' | \boldsymbol{x}, t \rangle \psi(\boldsymbol{x}, t) \tag{10.27}$$

In other words, the amplitude $F(\boldsymbol{x}', t'|\boldsymbol{x}, t)$ is the kernel of the time evolution for arbitrary states. The amplitude $F(\boldsymbol{x}', t'|\boldsymbol{x}, t)$ is known as the *Schwinger function*.

The initial state $|\psi(t)\rangle$ and the final state $|\psi(t')\rangle$ are connected by the *evolution operator* $U(t', t)$

$$U(t', t) \equiv T\, e^{-\frac{i}{\hbar}\int_t^{t'} dt'' H(t'')} \tag{10.28}$$

Here T is the time ordering symbol. $U(t, t')$ is *unitary*, since, as a result of the hermiticity of the Hamiltonian,

$$\begin{aligned} U^{\dagger}(t', t) &= T\, e^{\frac{i}{\hbar}\int_t^{t'} dt'' H(t'')} \\ &= T\, e^{-\frac{i}{\hbar}\int_{t'}^{t} dt'' H(t'')} = U(t, t') \end{aligned} \tag{10.29}$$

By definition, $U(t, t')$ is the inverse of $U(t', t)$, since it evolves the states backward in time. In addition, the operator $U(t', t)$ obeys the initial condition

$$\lim_{t' \to t} U(t', t) = I \tag{10.30}$$

where I is the identity operator. If the Hamiltonian is time independent, the evolution operator is

$$U(t', t) = T\, e^{-\frac{i}{\hbar}\int_t^{t'} dt'' H(t'')} = e^{-\frac{i}{\hbar}H(t'-t)} \tag{10.31}$$

where the last step holds only for a time-independent Hamiltonian.

These ideas will also allow us to introduce the *scattering matrix* (or *S-matrix*). If $\psi_i(\boldsymbol{x}, t)$ is some initial state and $\psi_f(\boldsymbol{x}', t')$ is some final state, then the matrix elements of the S-matrix between states ψ_i and ψ_f, S_{fi}, are obtained by evolving the state ψ_i up to time t' and projecting it onto the state ψ_f:

$$S_{fi} = \lim_{t' \to +\infty} \lim_{t \to -\infty} \int d\boldsymbol{x} \int d\boldsymbol{x}'\, \psi_f^*(\boldsymbol{x}', t') \langle \boldsymbol{x}', t' | \boldsymbol{x}, t \rangle \psi_i(\boldsymbol{x}, t) \tag{10.32}$$

Let us define the *Green function* or *propagator* $G(\boldsymbol{x}', t'|\boldsymbol{x}, t)$:

$$G(\boldsymbol{x}', t'|\boldsymbol{x}, t) \equiv -\frac{i}{\hbar}\Theta(t' - t)\langle \boldsymbol{x}', t' | \boldsymbol{x}, t \rangle = -\frac{i}{\hbar}\Theta(t' - t) F(\boldsymbol{x}, t'|\boldsymbol{x}, t) \tag{10.33}$$

It satisfies the equation

$$(i\hbar \partial_{t'} - H)\, G(\boldsymbol{x}', t'|\boldsymbol{x}, t) = \delta(\boldsymbol{x} - \boldsymbol{x}')\delta(t - t') \tag{10.34}$$

with the boundary condition

$$G(\boldsymbol{x}', t'|\boldsymbol{x}, t) = 0, \quad \text{if}\, t' < t \tag{10.35}$$

Hence, $G(\boldsymbol{x}', t'|\boldsymbol{x}, t)$ is known as the *retarded Schrödinger propagator*.

In terms of G, the S-matrix is given by (recall that $t' > t$):

$$S_{fi} = i \lim_{t' \to +\infty} \lim_{t \to -\infty} \int dx \int dx' \psi_f^*(\boldsymbol{x}', t') G(\boldsymbol{x}', t' | \boldsymbol{x}, t) \psi_i(\boldsymbol{x}, t) \tag{10.36}$$

Let us consider now the case of a free particle with Hamiltonian H_0 that is coupled to an external perturbation represented by a potential $V(\boldsymbol{x}, t)$. The free Green function, G_0, satisfies the equation

$$(i\hbar\partial_{t'} - H_0)\, G_0 = \delta(\boldsymbol{x}' - \boldsymbol{x})\delta(t' - t) \tag{10.37}$$

G_0 can be regarded as the matrix elements of the following operator:

$$G_0(\boldsymbol{x}', t' | \boldsymbol{x}, t) = \langle \boldsymbol{x}', t' | (i\hbar\partial_t - H)^{-1} | \boldsymbol{x}, t \rangle \tag{10.38}$$

Clearly, G satisfies the same equation but with the full H:

$$(i\hbar\partial_t - H)\, G = 1 \tag{10.39}$$

Hence, we can write

$$[(i\hbar\partial_t - H_0) - V]\, G = 1 \tag{10.40}$$

By using the definition of G_0, we get the operator equation

$$(G_0^{-1} - V)G = 1 \tag{10.41}$$

Thus, G satisfies the integral equation

$$G(\boldsymbol{x}', t' | \boldsymbol{x}, t) = G_0(\boldsymbol{x}', t' | \boldsymbol{x}, t) + \int dx'' \int dt'' G_0(\boldsymbol{x}', t' | \boldsymbol{x}'', t'') V(\boldsymbol{x}'', t'') G(\boldsymbol{x}'', t'' | \boldsymbol{x}, t) \tag{10.42}$$

which is known as the *Dyson equation*. It has the formal operator solution

$$G^{-1} = G_0^{-1} - V \tag{10.43}$$

The integral equation can be solved by an iterative procedure, which amounts to a perturbative expansion in powers of V. The result is the *Born series*. Using an obvious matrix notation, we get

$$G = G_0 + G_0 V G_0 + G_0 V G_0 V G_0 + \cdots \tag{10.44}$$

We can represent this series by a set of diagrams. Let us consider the first term, to which we assign the diagram in figure 10.5. The oriented arrow ranging from $(\boldsymbol{x}, t)$ to $(\boldsymbol{x}', t')$ represents the unperturbed propagator $G_0(\boldsymbol{x}' t' | \boldsymbol{x}, t)$. Because of the causal boundary conditions obeyed by G_0, it can only propagate forward in time. The second term of the series, the *Born approximation*, $\delta G^{(1)}$,

$$\delta G^{(1)}(\boldsymbol{x}', t' | \boldsymbol{x}, t) = \int dx' \int dt'' G_0(\boldsymbol{x}', t' | \boldsymbol{x}'', t'') V(\boldsymbol{x}'', t'') G_0(\boldsymbol{x}'', t'' | \boldsymbol{x}, t) \tag{10.45}$$

is represented by figure 10.6a, where the shaded circle represents the action of the potential V. In general, we get a diagram of the form of figure 10.6b, which represents a multiple scattering process. Notice that all contributions propagate strictly forward in time.

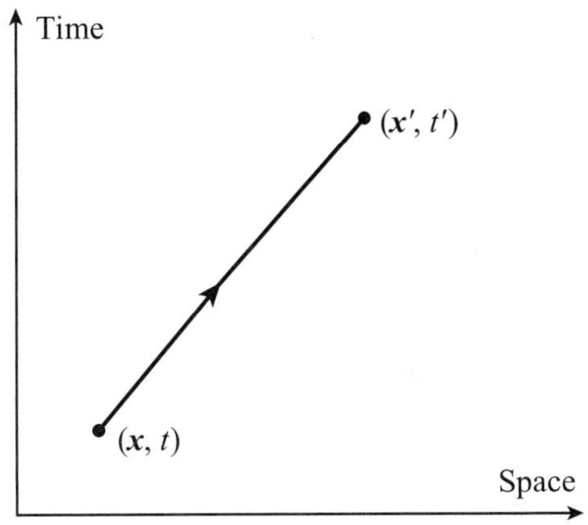

Figure 10.5 The zeroth-order term.

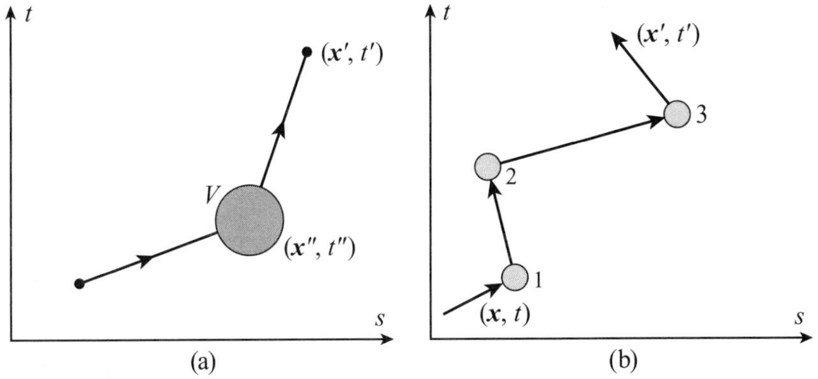

Figure 10.6 (a) The first-order term: the Born approximation. (b) A multiple scattering process.

Let us compute the propagator $G_0(\boldsymbol{x}'t'|\boldsymbol{x}t)$ for a free spinless particle in three-dimensional space. The Hamiltonian H is just $H = -\frac{\hbar^2}{2m}\nabla^2$. Thus, G_0 obeys the equation

$$\left(i\hbar\partial_{t'} + \frac{\hbar^2}{2m}\nabla_x^2\right) G_0(\boldsymbol{x}'t'|\boldsymbol{x}t) = \delta^3(\boldsymbol{x}' - \boldsymbol{x})\delta(t' - t) \tag{10.46}$$

with *causal* boundary conditions (i.e., $G(\boldsymbol{x}'t'|\boldsymbol{x}t) = 0$ for $t' < t$). Given the symmetries of this very simple system, we can Fourier expand G_0:

$$G_0(\boldsymbol{x}'t'|\boldsymbol{x}t) = \int \frac{d^3p}{(2\pi)^3} \int \frac{d\omega}{2\pi} \widetilde{G}_0(\boldsymbol{p}, \omega) e^{-\frac{i}{\hbar}\boldsymbol{p}\cdot(\boldsymbol{x}'-\boldsymbol{x}) + i\omega(t'-t)} \tag{10.47}$$

By direct substitution, we find that $\widetilde{G}_0(\boldsymbol{p}, \omega)$ is given by

$$\widetilde{G}_0(\boldsymbol{p}, \omega) = \frac{1}{\hbar\omega - \frac{\boldsymbol{p}^2}{2m}} \tag{10.48}$$

Notice that, once again, $\widetilde{G}_0(\boldsymbol{p}, \omega)$ has a pole at $\hbar\omega = \frac{p^2}{2m}$, on the real frequency axis, which is the dispersion law or the "mass shell" condition.

Apart from being singular, this "solution" does not obey the causal boundary condition. We will enforce the boundary condition by deforming the integration contour in the complex frequency plane. Following our discussion of the Green function for classical electrodynamics, we move the pole by an infinitesimal positive amount $i\epsilon$ into the upper half of the complex frequency plane. Write the *retarded* propagator as

$$G_0^{\text{ret}}(\boldsymbol{p}, \omega) = \frac{1}{\hbar\omega - \dfrac{\boldsymbol{p}^2}{2m} - i\epsilon} \tag{10.49}$$

and we will take the limit $\epsilon \to 0^+$ at the end of our calculations.

The frequency integral is equal to

$$\int_{-\infty}^{+\infty} \frac{d\omega}{2\pi} \frac{e^{i\omega(t'-t)}}{\hbar\omega - \dfrac{\boldsymbol{p}^2}{2m} - i\epsilon} = \frac{i}{\hbar}\Theta(t'-t) e^{\frac{i}{\hbar}\frac{\boldsymbol{p}^2}{2m}(t'-t)} \tag{10.50}$$

Hence, the Green function $G_0(\boldsymbol{x}', t'|\boldsymbol{x}, t)$ is

$$G_0(\boldsymbol{x}', t'|\boldsymbol{x}, t) = \frac{i}{\hbar}\Theta(t'-t) \int \frac{d^3p}{(2\pi)^3} e^{\frac{i}{\hbar}\frac{\boldsymbol{p}^2}{2m}(t'-t) - \frac{i}{\hbar}\boldsymbol{p}\cdot(\boldsymbol{x}'-\boldsymbol{x})} \tag{10.51}$$

Let $\psi_{\boldsymbol{p}}(\boldsymbol{x}, t)$ denote the wave functions for the *stationary states* $|\boldsymbol{p}\rangle$,

$$\psi_{\boldsymbol{p}}(\boldsymbol{x}, t) = \frac{1}{(2\pi)^{3/2}} e^{-\frac{i}{\hbar}\boldsymbol{p}\cdot\boldsymbol{x} + \frac{i}{\hbar}E(\boldsymbol{p})t} \tag{10.52}$$

where $E(\boldsymbol{p}) = \dfrac{\boldsymbol{p}^2}{2m}$. We see that $G_0(\boldsymbol{x}', t'|\boldsymbol{x}, t)$ can be written in the form

$$G_0(\boldsymbol{x}', t'|\boldsymbol{x}, t) = \frac{i}{\hbar}\,\Theta(t'-t) \int \frac{d^3p}{(2\pi)^3}\, \psi_{\boldsymbol{p}}(\boldsymbol{x}', t') \psi_{\boldsymbol{p}}^*(\boldsymbol{x}, t) \tag{10.53}$$

In general, if the Hamiltonian has a complete set of stationary states $\{|n\rangle\}$ with wave functions $\psi_n(x)$ and eigenvalues E_n, the Green function is

$$G_0^{\text{ret}}(\boldsymbol{x}', t'|\boldsymbol{x}, t) = \frac{i}{\hbar}\Theta(t'-t) \sum_n \psi_n(\boldsymbol{x}', t') \psi_n^*(\boldsymbol{x}, t) \tag{10.54}$$

where

$$\psi_n(\boldsymbol{x}, t) = \psi_n(\boldsymbol{x}) e^{-\frac{i}{\hbar}E_n t} \tag{10.55}$$

If the system is isolated, the Hamiltonian is time independent, and the Green function is a function of $t' - t$. In this case, it is convenient to consider the Fourier transform

$$G_0^{\text{ret}}(\boldsymbol{x}', \boldsymbol{x}; t'-t) = \int_{-\infty}^{\infty} \frac{d\omega}{2\pi}\, G_0^{\text{ret}}(\boldsymbol{x}', \boldsymbol{x}; \omega)\, e^{i\omega(t'-t)/\hbar} \tag{10.56}$$

where we have to pick the correct integration contour so that G_0^{ret} is retarded.

Quite explicitly, we find

$$G_0^{\text{ret}}(\boldsymbol{x}', \boldsymbol{x}; \omega) = \lim_{\epsilon \to 0^+} \sum_n \frac{\psi_n(\boldsymbol{x}')\psi_n^*(\boldsymbol{x})}{\hbar\omega - E_n - i\epsilon} \tag{10.57}$$

Here too, the denominators in this equation have zeros on the real frequency axis as $\epsilon \to 0^+$. Thus, the Green function is a series of *distributions*. In the limit $\epsilon \to 0^+$, we can write

$$\lim_{\epsilon \to 0^+} \frac{1}{x - i\epsilon} = \mathcal{P}\frac{1}{x} + i\pi\,\delta(x) \tag{10.58}$$

where $\mathcal{P}$ denotes the principal value, and $\delta(x)$ is the Dirac delta function. Using these results, we can write the following expressions for the real and imaginary parts of the Green function:

$$\begin{aligned}
\operatorname{Re} G_0^{\text{ret}}(\boldsymbol{x}', \boldsymbol{x}; \omega) &= \sum_n \mathcal{P}\frac{\psi_n(\boldsymbol{x}')\psi_n^*(\boldsymbol{x})}{\hbar\omega - E_n} \\
\operatorname{Im} G_0^{\text{ret}}(\boldsymbol{x}', \boldsymbol{x}; \omega) &= \pi \sum_n \psi_n(\boldsymbol{x}')\psi_n^*(\boldsymbol{x})\delta(\hbar\omega - E_n)
\end{aligned} \tag{10.59}$$

We can use these results to write an expression for the *density of states* $\rho(\omega)$

$$\rho(\omega) = \sum_n \delta(\hbar\omega - E_n) \tag{10.60}$$

in terms of the retarded Green function of the form

$$\rho(\omega) = \frac{1}{\pi} \operatorname{Im} \int d\boldsymbol{x}\, G_0^{\text{ret}}(\boldsymbol{x}, \boldsymbol{x}; \omega) \tag{10.61}$$

In other words, the spectral density is determined by the imaginary part of the Green function. In section 10.4, we will find similar relationships for other quantities of physical interest.

Let us close by noting that there is a close connection between the Green function in quantum mechanics and the kernel of the diffusion equation. In d-dimensional space, the Green function is

$$G_0(\boldsymbol{x}' - \boldsymbol{x}, t' - t) = -\frac{i}{\hbar}\Theta(t' - t) \int \frac{d^d p}{(2\pi)^d} e^{-\frac{i}{\hbar}\frac{\boldsymbol{p}^2}{2m}(t'-t) + \frac{i}{\hbar}\boldsymbol{p}\cdot(\boldsymbol{x}'-\boldsymbol{x})} \tag{10.62}$$

By completing squares inside the exponent,

$$\begin{aligned}
\frac{\boldsymbol{p}^2}{2m}(t' - t) - \boldsymbol{p}' \cdot (\boldsymbol{x}' - \boldsymbol{x}) &= \frac{(t' - t)}{2m}\left[\boldsymbol{p} - 2m\boldsymbol{p} \cdot \frac{\boldsymbol{x}' - \boldsymbol{x}}{(t' - t)}\right] \\
&= \left(\frac{t' - t}{2m}\right)\left[\boldsymbol{p} - \left(\frac{\boldsymbol{x}' - \boldsymbol{x}}{t' - t}\right) m\right]^2 - \frac{m|\boldsymbol{x}' - \boldsymbol{x}|^2}{2|t' - t|}
\end{aligned} \tag{10.63}$$

we can write

$$G_0(\boldsymbol{x}' - \boldsymbol{x}, t' - t) =$$

$$-\frac{i}{\hbar}\Theta\left(t'-t\right)\int \frac{d^d p}{(2\pi)^d}\, e^{-\frac{i}{\hbar}\left(\frac{t'-t}{2m}\right)\left(\boldsymbol{p}-\left(\frac{\boldsymbol{x}'-\boldsymbol{x}}{t'-t}\right)m\right)^2+\frac{i}{2\hbar}m\frac{|\boldsymbol{x}'-\boldsymbol{x}|^2}{|t'-t|}} \tag{10.64}$$

After a straightforward integration, we find that G_0 is equal to

$$G_0(\boldsymbol{x}' - \boldsymbol{x}, t' - t) = -\frac{i}{\hbar}\Theta(t'-t)\left(\frac{m\hbar}{2\pi(t'-t)}\right)^{d/2} e^{\frac{im}{2\hbar}\frac{|\boldsymbol{x}'-\boldsymbol{x}|^2}{(t'-t)}} \tag{10.65}$$

This formula is strongly reminiscent of the kernel for the heat equation (or diffusion equation) that was discussed in section 8.8.2:

$$\partial_\tau \psi = D\,\nabla^2\,\psi \tag{10.66}$$

where D is the diffusion constant, and τ is the diffusion time. Indeed, after an analytic continuation to *imaginary time*, $t \to i\tau$, the Schrödinger equation becomes a diffusion equation with a diffusion constant $D = \frac{\hbar}{2m}$. The Green function (or heat kernel) for the heat equation is

$$K(\boldsymbol{x}' - \boldsymbol{x}, \tau' - \tau) = \Theta(\tau' - \tau)\frac{1}{(4\pi D\tau)^{d/2}}\, e^{-\frac{|\boldsymbol{x}'-\boldsymbol{x}|^2}{4D(\tau'-\tau)}} \tag{10.67}$$

which, of course, agrees with the analytic continuation of the Feynman propagator $G_0(\boldsymbol{x}' - \boldsymbol{x}, t' - t)$. This connection between quantum mechanics and diffusion processes is central to the path integral picture of quantum mechanics.

10.3 Analytic properties of the propagators of free relativistic fields

10.3.1 Properties of the propagator of the real scalar field

The propagator for a free real relativistic field,

$$G^{(0)}(x - x') = i\langle 0|T\phi(x)\phi(x')|0\rangle \tag{10.68}$$

is the solution of the partial differential equation,

$$(\partial^2 + m_0^2)\, G^{(0)}(x - x') = \delta^4(x - x') \tag{10.69}$$

which we discussed and solved in section 5.6.3. $G^{(0)}(x - x')$ can be calculated by the usual Fourier expansion methods:

$$G^{(0)}(x - x') = \int \frac{d^4 p}{(2\pi)^4}\, \widetilde{G}^{(0)}(p)\, e^{-ip\cdot(x-x')} \tag{10.70}$$

where $\widetilde{G}^{(0)}(p)$ is given by

$$\widetilde{G}^{(0)}(p) = \frac{-1}{p^2 - m^2} \tag{10.71}$$

where $p^2 = p_\mu p^\mu$.

Once again, we have to give a prescription for going around the poles of $\widetilde{G}_0(p)$ which yields the correct boundary conditions. We will adopt the same conventions used in section 10.1. For the Feynman (or time-ordered) propagator, we shift the denominator by $i\epsilon$:

$$\widetilde{G}^{(0)}(p) = \frac{-1}{p^2 - m^2 + i\epsilon} \tag{10.72}$$

The poles of $\widetilde{G}^{(0)}(p)$ are located at $p_0 = \pm\sqrt{\mathbf{p}^2 + m^2} \mp i\epsilon$. Thus, the positive frequency pole is in the lower half-plane and the negative frequency pole is in the upper half-plane.

We will use the integration paths shown in figure 10.3. For $x_0 > x_0'$, we close the integration contour on the path γ^- and pick up the contribution from the positive frequency pole at $+\sqrt{\mathbf{p}^2 + m^2} - i\epsilon$. Thus, for $x_0 > x_0'$, we find:

$$\oint_{\gamma^+} \frac{dp_0}{2\pi} \frac{e^{-ip_0(x_0 - x_0')}}{p_0^2 - (\mathbf{p}^2 + m^2) + i\epsilon} \xrightarrow[\epsilon \to 0^+]{} -i\frac{e^{-i\sqrt{\mathbf{p}^2+m^2}(x_0 - x_0')}}{2\sqrt{\mathbf{p}^2 + m^2}} \tag{10.73}$$

Similarly, for $x_0 < x_0'$, we close the integration contour on the path γ^+ on the upper half-plane, where we pick up the contribution from the negative frequency pole at $-\sqrt{\mathbf{p}^2 + m^2} + i\epsilon$. Thus, for $x_0 < x_0'$, we obtain

$$\oint_{\gamma^-} \frac{dp_0}{2\pi} \frac{e^{-ip_0(x_0 - x_0')}}{p_0^2 - (\mathbf{p}^2 + m^2) + i\epsilon} \xrightarrow[\epsilon \to 0^+]{} i\frac{e^{i\sqrt{\mathbf{p}^2+m^2}(x_0 - x_0')}}{-2\sqrt{\mathbf{p}^2 + m_0^2}} \tag{10.74}$$

By collecting terms, we get

$$\begin{aligned} G^{(0)}(x - x') = {} & i\Theta\left(x_0 - x_0'\right) \int \frac{d^3p}{(2\pi)^3 2\omega(\mathbf{p})} e^{+i\mathbf{p}\cdot(\mathbf{x}-\mathbf{x}') - i\omega(\mathbf{p})\left(x_0 - x_0'\right)} \\ & + i\Theta(x_0' - x_0) \int \frac{d^3p}{(2\pi)^3 2\omega(\mathbf{p})} e^{-i\mathbf{p}\cdot(\mathbf{x}-\mathbf{x}') + i\omega(\mathbf{p})(x_0 - x_0')} \end{aligned} \tag{10.75}$$

This result shows that $G^{(0)}(x - x')$ does satisfy the required boundary condition. It also shows that the positive frequency components of the field propagate forward in time, while the negative frequency components propagate backward in time.

The simplest way to compute $G_0(x - x')$ is by means of an analytic continuation (or Wick rotation) to imaginary time, $x_0 \to ix_4$. This amounts to a rotation of the integration contour from the real p_0 axis to the imaginary p_0 axis, namely, $p_0 \to ip_4$, as shown in figure 10.7. The Wick-rotated or Euclidean correlation function $G_0^E(x - x')$, the Euclidean propagator that we calculated before, is given by

$$G^{(0)}(x - x') = i \int \frac{d^4p}{(2\pi)^4} \frac{e^{-ip\cdot(x - x')}}{p^2 + m^2} \equiv -iG_0^E(x, x') \tag{10.76}$$

where $p^2 = -\sum_{i=1}^4 p_i p_i$, and $p \cdot x = -\sum_{i=1}^4 p_i x_i$.

The time-ordered (or Feynman) propagator does not obey causality, since it does not vanish for space-like separated events, $s^2 < 0$. We can define a *causal*, or *retarded*, propagator that obeys the causal boundary condition (i.e., $G^{(0)}(x - x') = 0$) except inside the *forward* light-cone. Similarly, we can also define as *advanced* that which vanishes outside the *backward* light-cone. We will discuss only the retarded propagator $G_{\text{ret}}^{(0)}(x - x')$.

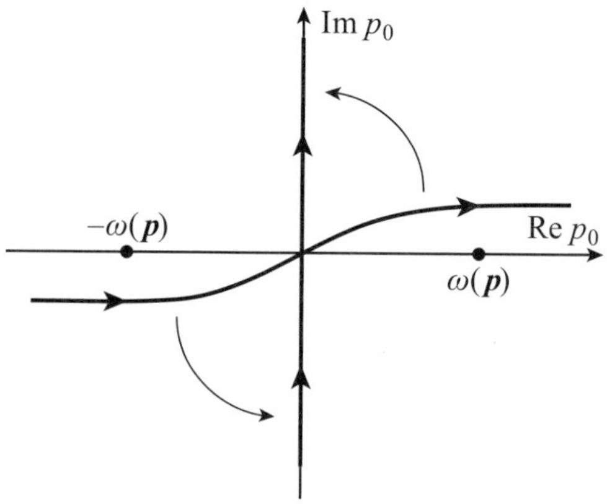

Figure 10.7 Wick rotation.

The retarded propagator is defined by computing the frequency integral on the path shown in figure 10.2 (using the replacement $|\boldsymbol{p}| \to \omega(\boldsymbol{p})$). The retarded propagator $G^{(0)}_{\text{ret}}(x-x')$ is given by

$$\begin{aligned}
G^{(0)}_{\text{ret}}(x-x') &= -\int \frac{d^3p}{(2\pi)^3} e^{i\boldsymbol{p}\cdot(\boldsymbol{x}-\boldsymbol{x}')} \int_{-\infty}^{+\infty} \frac{dp_0}{2\pi} \frac{e^{-ip_0(x_0-x'_0)}}{p_0^2-\omega^2(\boldsymbol{p})} \\
&\equiv -i\int \frac{d^3p}{(2\pi)^3 2\omega(\boldsymbol{p})} e^{i\boldsymbol{p}\cdot(\boldsymbol{x}-\boldsymbol{x}')} \int_{-\infty}^{+\infty} \frac{dp_0}{2\pi i} e^{-ip_0(x_0-x'_0)} \\
&\quad \times \left[\frac{1}{p_0-\omega(\boldsymbol{p})+i\epsilon} - \frac{1}{p_0+\omega(\boldsymbol{p})+i\epsilon}\right]
\end{aligned} \tag{10.77}$$

Hence, we get

$$\begin{aligned}
G^{(0)}_{\text{ret}}(x-x') &= -i\Theta(x_0-x'_0)\int \frac{d^3p}{(2\pi)^3 2\omega(\boldsymbol{p})} \\
&\quad \times \left[e^{i\omega(\boldsymbol{p})(x_0-x'_0)-i\boldsymbol{p}\cdot(\boldsymbol{x}-\boldsymbol{x}')} - e^{-i\omega(\boldsymbol{p})(x_0-x'_0)+i\boldsymbol{p}\cdot(\boldsymbol{x}-\boldsymbol{x}')}\right]
\end{aligned} \tag{10.78}$$

The integral over the momentum variables is just the quantity $i\Delta(x-x')$ that we have encountered before (cf. eq. (4.90)),

$$i\Delta(x-x') = \left[\phi(x), \phi(x')\right] \equiv \left\langle 0\middle|\left[\phi(x), \phi(x')\right]\middle|0\right\rangle \tag{10.79}$$

which allows us to write

$$G^{(0)}_{\text{ret}}(x-x') = \Theta(x_0-x'_0)\Delta(x-x') = -i\Theta(x_0-x'_0)\left\langle 0\middle|\left[\phi(x), \phi(x')\right]\middle|0\right\rangle \tag{10.80}$$

10.3.2 Properties of the propagator of the Dirac field

The same line of argument we have used for the scalar field can be used for the Dirac field. The vacuum state of the Dirac theory is defined by filling up all negative-energy states.

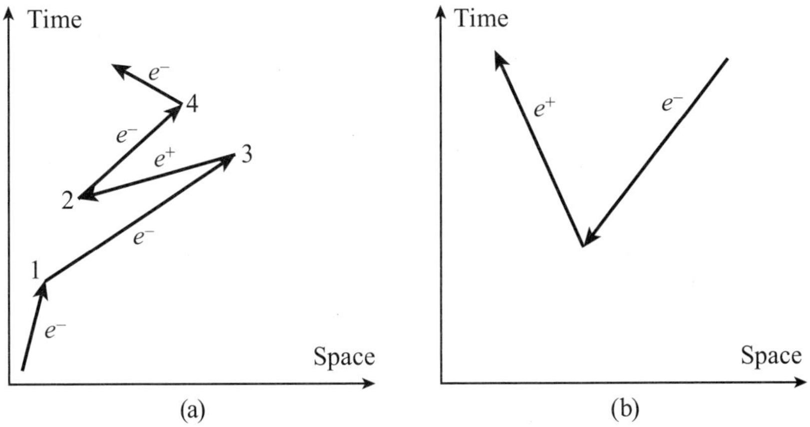

Figure 10.8 (a) Scattering processes of a Dirac field. (b) A pair-creation process.

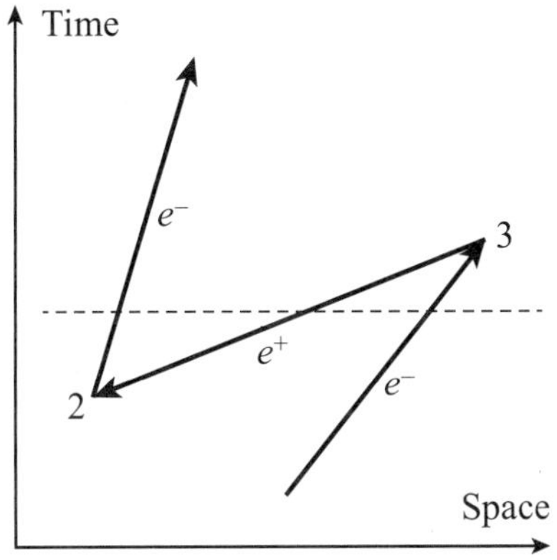

Figure 10.9 An intermediate state with an electron-positron pair.

Now imagine that an electron is propagating in free space, and then an external potential is adiabatically switched on. If the potential is not too strong, we can still describe its effects by means of a Born series of multiple scattering processes, as shown in figure 10.8a.

The electron scatters off the potential at point 1 in figure 10.8a, and as a result, it propagates up to point 3. If the potential has the correct matrix elements, at point 3, the positive energy state may turn into a negative energy state. In general, this won't be allowed, since all negative energy states are filled, unless a negative energy state was emptied (by the action of the potential) *before the electron became scattered into that state*. Indeed, the potential can *create an electron-positron pair* out of the vacuum, as in the process shown in figure 10.8b.

Thus, the process can remove an electron from an occupied negative energy state and promote it to a previously empty positive energy state. Hence, if the time t lies between t_2 and t_3, there are three different states propagating in the system (see figure 10.9):

1) a positive energy state that disappears at point 3,
2) a negative energy state that propagates *backwards* in time from point 3 to point 2,
3) a positive energy state that appeared at point 2.

An alternative interpretation is that an electron-positron pair was created at point 2 and that the positron annihilated the original electron at point 3. This process clearly shows that the Dirac theory is a *quantum field theory*, and it cannot be described in the framework of quantum mechanics with a fixed number of particles, as in the nonrelativistic case. Thus, the Fock-space description is essential to the relativistic case.

These arguments suggest that we may want to seek a propagator that propagates positive energy states forward in time, while negative energy states propagate backward in time. This is the Feynman, or time-ordered, propagator, $S_F^{\alpha\alpha'}(x-x')$. It is straightforward to see that these requirements are met by the following expression:

$$S_F^{\alpha\alpha'}(x-x') = -i\langle 0|T\psi_\alpha(x)\bar{\psi}_{\alpha'}(x')|0\rangle \tag{10.81}$$

which satisfies the equation of motion

$$(i\partial\!\!\!/ - m)S_F(x-x') = \delta^4(x-x') \tag{10.82}$$

The same methods that we used for the scalar field yield the solution (dropping the spinor indices)

$$S_F(x-x') = \int \frac{d^4p}{(2\pi)^4} \frac{p\!\!\!/ + m}{p^2 - m^2 + i\epsilon} e^{-ip\cdot(x-x')} \tag{10.83}$$

where an $i\epsilon$ has been introduced in order to get the correct boundary conditions. We have also shown that $S_F(x-x')$ satisfies

$$S_F(x-x') = (i\partial\!\!\!/ + m)\int \frac{d^4p}{(2\pi)^4} \frac{e^{-ip\cdot(x-x')}}{p^2 - m^2 + i\epsilon} = -(i\partial\!\!\!/ + m)G^{(0)}(x-x') \tag{10.84}$$

where $G^{(0)}(x-x')$ is the Feynman or time-ordered propagator for the free massive scalar field. The $i\epsilon$ prescription ensures that positive energy states propagate forward in time and that negative energy states propagate backward in time.

10.4 The propagator of the nonrelativistic electron gas

Let us now discuss the propagator, or *one-particle Green function*, for a nonrelativistic free electron gas at finite density and zero temperature. It is defined in the usual way:

$$G_0^{\alpha\alpha'}(x,x') = -i\langle \text{gnd}|T\psi_\alpha(x)\psi^\dagger_{\alpha'}(x')|\text{gnd}\rangle \tag{10.85}$$

where α and α' are spin indices. Here, $|\text{gnd}\rangle$ is the ground state of the system. To simplify the notation, in what follows, all expectation values are taken with respect to the ground state (but will not be denoted explicitly). This propagator can be used to compute various quantities of physical interest. For example, the average *electron density* $\langle \hat{n}(x)\rangle$ is

$$\langle n(\boldsymbol{x})\rangle = \langle \psi^\dagger(\boldsymbol{x})\psi(\boldsymbol{x})\rangle = -i \lim_{\boldsymbol{x}'\to\boldsymbol{x}} \lim_{x_0'\to x_0} \text{tr}\, G(x,x') \tag{10.86}$$

Likewise, the *current density* $\langle \boldsymbol{j}(\boldsymbol{x})\rangle$ is

$$\begin{aligned}\langle \boldsymbol{j}(\boldsymbol{x})\rangle =& \frac{\hbar}{2mi}\mathrm{tr}\,\langle \psi^\dagger(\boldsymbol{x})\left(\nabla_x\psi(\boldsymbol{x})\right) - \left(\nabla_x\psi^\dagger(\boldsymbol{x})\right)\psi(\boldsymbol{x})\rangle \\ =& -\frac{1}{m}\lim_{\boldsymbol{x}'\to\boldsymbol{x}}\lim_{x_0'\to x_0}\left(\nabla_{\boldsymbol{x}} - \nabla_{\boldsymbol{x}'}\right)\mathrm{tr}\,G(x,x')\end{aligned} \tag{10.87}$$

and the *magnetization density*, $\boldsymbol{M}(\boldsymbol{x}) = \langle \psi^\dagger_\alpha(\boldsymbol{x})\boldsymbol{\sigma}_{\alpha\beta}\psi_\beta(\boldsymbol{x})\rangle$, is

$$\boldsymbol{M}(\boldsymbol{x}) = -i\lim_{\boldsymbol{x}'\to\boldsymbol{x}}\lim_{x_0'\to x_0}\mathrm{tr}\,[G(x,x')\boldsymbol{\sigma}] \tag{10.88}$$

Let us compute $G(x,x')$ by the standard method of Fourier transforms:

$$\begin{aligned}G(x,x') =& \int\frac{d^3p}{(2\pi)^3}\int\frac{d\omega}{2\pi}\,\widetilde{G}(\boldsymbol{p},\omega)\,e^{i\left[\boldsymbol{p}\cdot(\boldsymbol{x}-\boldsymbol{x}')-\omega(t-t')\right]} \\ \equiv& \int\frac{d^4p}{(2\pi)^4}\,\widetilde{G}(p)\,e^{ip\cdot(x-x')}\end{aligned} \tag{10.89}$$

where $p_0=\omega$ and $t=x_0$. Let $E(\boldsymbol{p})$ be the single-particle energies measured from the chemical potential μ, $E(\boldsymbol{p}) = \frac{p^2}{2m} - \mu$. Then we get

$$\begin{aligned}G(\boldsymbol{x},t) =& -i\int\frac{d^3p}{(2\pi)^3}\Theta(t)\langle\mathrm{gnd}|\psi(\boldsymbol{p})\psi^\dagger(\boldsymbol{p})|\mathrm{gnd}\rangle e^{i(\boldsymbol{p}\cdot\boldsymbol{x}-E(\boldsymbol{p})\,t)} \\ &+ i\int\frac{d^3p}{(2\pi)^3}\Theta(-t)\langle\mathrm{gnd}|\psi^\dagger(\boldsymbol{p})\psi(\boldsymbol{p})|\mathrm{gnd}\rangle e^{i(\boldsymbol{p}\cdot\boldsymbol{x}-E(\boldsymbol{p})t)}\end{aligned} \tag{10.90}$$

Recall that $|\mathrm{gnd}\rangle$ is the state in which all negative energy states, with $E(\boldsymbol{p})<0$ (or, equivalently, $\epsilon(p)<\mu$) are filled. Thus, the Green function becomes

$$G(\boldsymbol{x},t) = -i\int\frac{d^3p}{(2\pi)^3}\left[\Theta(t)\left(1-n(\boldsymbol{p})\right) - \Theta(-t)n(\boldsymbol{p})\right]e^{i(\boldsymbol{p}\cdot\boldsymbol{x}-E(\boldsymbol{p})t)} \tag{10.91}$$

where $n(\boldsymbol{p}) = \langle\mathrm{gnd}|\psi^\dagger(\boldsymbol{p})\psi(\boldsymbol{p})|\mathrm{gnd}\rangle$ is the Fermi-Dirac distribution (or "Fermi function" for short) at zero temperature:

$$n(\boldsymbol{p}) = \begin{cases}1, & |\boldsymbol{p}|\le p_F \\ 0, & \text{otherwise}\end{cases} \tag{10.92}$$

We can write the Fourier transform $G(\boldsymbol{p},\omega)$ in the form

$$G(\boldsymbol{p},\omega) = -i\left\{\Theta(|\boldsymbol{p}|-p_F)\int_0^\infty e^{i(\omega-E(\boldsymbol{p}))t}dt - \Theta(p_F-|\boldsymbol{p}|)\int_0^\infty dt e^{-i(\omega-E(\boldsymbol{p}))t}\right\} \tag{10.93}$$

The integrals in this expression define distributions of the form

$$\int_0^\infty dt\,e^{ist} = \lim_{\epsilon\to0^+}\int_0^\infty dt\,e^{ist-\epsilon t} = i\lim_{\epsilon\to0^+}\frac{1}{s+i\epsilon} = i\left(\mathcal{P}\frac{1}{s} - i\pi\delta(s)\right) \tag{10.94}$$

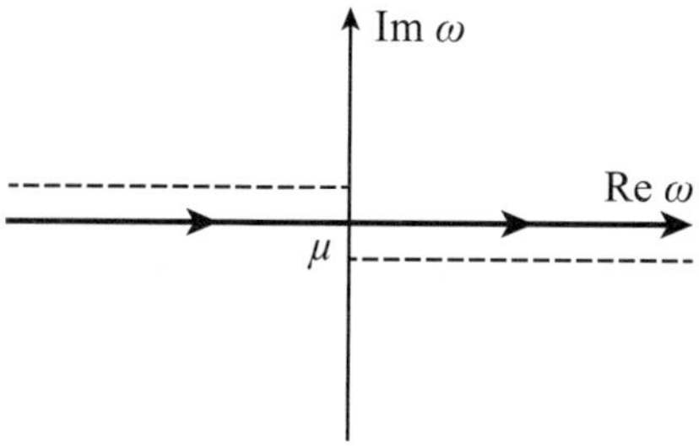

Figure 10.10 Analytic structure of the nonrelativistic fermion propagator at finite chemical potential μ. The broken lines show the location of the poles for different momenta $\boldsymbol{p}$ on the upper and lower half-planes.

where $\mathcal{P}\frac{1}{s}$ is the *principal value*,

$$\mathcal{P}\frac{1}{s} = \lim_{\epsilon\to 0} \frac{s}{s^2+\epsilon^2} \tag{10.95}$$

and $\delta(s)$ is the Dirac δ-function,

$$\delta(s) = \frac{1}{\pi} \lim_{\epsilon\to 0} \frac{\epsilon}{s^2+\epsilon^2} \tag{10.96}$$

We can use these results to write $\widetilde{G}(\boldsymbol{p},\omega)$ as

$$\widetilde{G}(\boldsymbol{p},\omega) = \frac{\Theta(|\boldsymbol{p}|-p_F)}{\omega - E(\boldsymbol{p}) + i\epsilon} + \frac{\Theta(p_F - |\boldsymbol{p}|)}{\omega - E(\boldsymbol{p}) - i\epsilon} \tag{10.97}$$

where $E(\boldsymbol{p}) = \epsilon(\boldsymbol{p}) - \mu$. An equivalent, and more compact, expression is

$$\widetilde{G}(\boldsymbol{p},\omega) = \frac{1}{\omega - E(\boldsymbol{p}) + i\epsilon\, \text{sign}(|\boldsymbol{p}| - p_F)} \tag{10.98}$$

Notice that

$$\begin{aligned} \text{Im}\, \widetilde{G}(\boldsymbol{p},\omega) &= -\pi\Theta(|\boldsymbol{p}|-p_F)\delta(\omega - \epsilon(\boldsymbol{p}) + \mu) + \pi\Theta(p_F - |\boldsymbol{p}|)\delta(\omega - \epsilon(\boldsymbol{p}) + \mu) \\ &= -\pi\delta(\omega - \epsilon(\boldsymbol{p}) + \mu)\Big[\Theta\left(|\boldsymbol{p}| - p_F\right) - \Theta(p_F - |\boldsymbol{p}|)\Big] \end{aligned} \tag{10.99}$$

The last identity shows that

$$\text{sign Im}\widetilde{G}(\boldsymbol{p},\omega) = -\text{sign}\,\omega \tag{10.100}$$

Hence, we can also write $\widetilde{G}(\boldsymbol{p},\omega)$ as

$$\widetilde{G}(\boldsymbol{p},\omega) = \frac{1}{\omega - E(\boldsymbol{p}) + i\epsilon\, \text{sign}\,\omega} \tag{10.101}$$

In this expression, $\widetilde{G}(\boldsymbol{p},\omega)$ has poles at $\omega = E(\boldsymbol{p})$, and all the poles with $\omega > 0$ are infinitesimally shifted down to the lower half-plane, while the poles with $\omega < 0$ are raised to the upper half-plane by the same amount. Since $E(\boldsymbol{p}) = \epsilon(\boldsymbol{p}) - \mu$, all poles with $\epsilon(\boldsymbol{p}) > \mu$ are shifted down, while all poles with $\epsilon(\boldsymbol{p}) < \mu$ are shifted up (figure 10.10).

10.5 The scattering matrix

The problems of real physical interest very rarely involve free fields. In general, we have to deal with interacting fields. For the sake of definiteness, let us consider a scalar field, but the ideas discussed have general applicability.

The field $\phi(\boldsymbol{x}, t)$ in the Heisenberg representation is related to the Schrödinger operator $\phi(\boldsymbol{x})$ through the time-evolution operator generated by the Hamiltonian

$$\phi(\boldsymbol{x}, t) \equiv \phi(x) = e^{iHt}\, \phi(\boldsymbol{x}, 0)\, e^{-iHt} \tag{10.102}$$

(for $\hbar = 1$). Recall that in the Schrödinger representation, the *fields are fixed* but the *states evolve* according to the Schrödinger equation

$$H|\Phi\rangle = i\partial_t|\Phi\rangle \tag{10.103}$$

whereas, in the Heisenberg representation, the *states are fixed* but the *fields evolve* according to the equations of motion

$$i\partial_t\phi(x) = [\phi(x), H] \tag{10.104}$$

In an interacting system, the problem is precisely how to determine the evolution operator e^{iHt}. Thus, the Heisenberg representation cannot be constructed a priori.

Let us assume that the Hamiltonian can be split into a sum of two terms:

$$H = H_0 + H_{int} \tag{10.105}$$

where H_0 represents a system whose states are fully known to us (a problem that we know how to solve), and in this sense are "free." H_{int} represents the interactions. For technical reasons, we have to assume that $H_{int}(t)$, as a function of time, vanishes (very smoothly) both in the remote past and in the remote future.

We now define the *interaction representation*. In this representation, one defines the *fields* ϕ_{in}, which evolve as Heisenberg fields with Hamiltonian H_0:

$$i\partial_t\phi_{\text{in}}(x) = [\phi_{\text{in}}, H_0] \tag{10.106}$$

These operators create and destroy free incoming states. We call these states *incoming*, since as $t \to -\infty$, there are no interactions. In the absence of interactions, the states do not evolve; but if interactions are present, they do. A unitary operator $U(t)$ governs the time evolution of the states and the S-matrix.

We want to find an operator $U(t)$ such that

$$\phi(x, t) = U^{-1}\phi_{\text{in}}(x, t)U(t) \tag{10.107}$$

where $\phi(x, t)$ is the Heisenberg field operator for the full Hamiltonian H. The operator $U(t)$ must be *unitary* and satisfy the initial condition

$$\lim_{t\to-\infty} U(t) = I \tag{10.108}$$

Since $U(t)$ is unitary and invertible, it must satisfy the condition $U^{-1}(t) = U^{\dagger}(t)$. Thus

$$\partial_t U(t) U^{-1}(t) + U(t) \partial_t U^{-1}(t) = 0 \tag{10.109}$$

Since the operators $\phi_{\rm in}$ and ϕ are defined by the Heisenberg evolution equation, eq. (10.107), we get

$$\begin{aligned} \partial_t \phi_{\rm in} &= \partial_t U \,\phi\, U^{-1} + U\, \partial_t \phi \, U^{-1} + U\,\phi\, \partial_t U^{-1} \\ &= \partial_t U\, \phi\, U^{-1} + iU\,[H, \phi]\, U^{-1} + U\,\phi\, U^{-1}\, U\, \partial_t U^{-1} \end{aligned} \tag{10.110}$$

In other words,

$$\partial_t \phi_{\rm in} = iU(t)[H(\phi), \phi] U^{-1}(t) + \partial_t U U^{-1} \phi_{\rm in} + \phi_{\rm in} U \partial_t U^{-1} \tag{10.111}$$

Similarly, the Hamiltonian must obey the identity

$$H(\phi_{\rm in}) = U(t) H(\phi) U^{-1}(t) \tag{10.112}$$

which implies that $\phi_{\rm in}$ should obey

$$\partial_t \phi_{\rm in} = i[H(\phi_{\rm in}), \phi_{\rm in}] + [(\partial_t U)\; U^{-1}, \phi_{\rm in}] \tag{10.113}$$

Since $\phi_{\rm in}$ obeys the free equation of motion, eq. (10.106), we find that the evolution operator $U(t)$ must satisfy the condition

$$[iH_{\rm int}(\phi_{\rm in}) + (\partial_t U)\; U^{-1}, \phi_{\rm in}] = 0 \tag{10.114}$$

for *all* operators $\phi_{\rm in}$. Therefore, the operator in the left argument of the commutator must be a c-number (i.e., it is proportional to the identity operator). However, since $\lim_{t\to -\infty} H_{\rm int}(\phi_{\rm in}) = 0$, and $\lim_{t\to -\infty} U(t) = I$, this c-number must be equal to zero.

We thus arrive at an operator equation for $U(t)$:

$$i\partial_t U = H_{\rm int}(\phi_{\rm in})\, U(t) \tag{10.115}$$

The operator U governs the time evolution of the states in the interaction representation, since the state $|\Phi\rangle_{\rm in}$ becomes

$$U(t)|\Phi\rangle_{\rm in} = |\Phi(t)\rangle \tag{10.116}$$

In particular, the *outgoing states* $|\Phi\rangle_{\rm out}$ (i.e., the states at $t \to +\infty$, which are also *free states*) are related to the in-states by the operator $U(t)$ in the limit $t \to +\infty$

$$|\Phi\rangle_{\rm out} = \lim_{t\to +\infty} U(t)|\Phi\rangle_{\rm in} \equiv S|\Phi\rangle_{\rm in} \tag{10.117}$$

where $S = \lim_{t\to +\infty} U(t)$ is the S-matrix.

The solution is

$$U(t) = Te^{-i\int_{-\infty}^{t} dt' H_{\rm int}(\phi_{\rm in}(t'))} \tag{10.118}$$

where T is the time-ordering operator. In terms of the interaction part $\mathcal{L}_{\text{int}}$ of the Lagrangian, we find that the S-matrix is given by the expression

$$S = \lim_{t\to+\infty} U(t) = T\, e^{-i\int_{-\infty}^{+\infty} dt' H_{\text{int}}\left(\phi_{\text{in}}(t')\right)} = T\, e^{i\int d^4x\, \mathcal{L}_{\text{int}}(\phi_{\text{in}})} \tag{10.119}$$

This result is the starting point for the computation of the S-matrix using perturbation theory in H_{int}.

10.6 Physical information contained in the *S*-matrix

Let us compute transition matrix elements between arbitrary *in* and *out* states. Let $|i, \text{in}\rangle$ be the initial incoming state, and $|f, \text{out}\rangle$ be the final outgoing state. The transition probability $W_{i\to f}$ is then given by a matrix element of the S-matrix,

$$W_{i\to f} = |\langle f, \text{out}|i, \text{in}\rangle|^2 \equiv |\langle f, \text{in}|S|i, \text{in}\rangle|^2 \tag{10.120}$$

since $\langle f, \text{out}|S = \langle f, \text{in}|$. We can split S into noninteracting and interacting parts

$$S = I + iT \tag{10.121}$$

where I, the identity operator, represents the free part, and the T-matrix (not to be confused with the time-ordering symbol!) represents the interactions. In terms of the T-matrix, the transition probability is

$$W_{i\to f} = |\langle f|i\rangle + i\langle f|T|i\rangle|^2 \tag{10.122}$$

From now on we will discuss the case of a scalar field, but the arguments can be generalized to all other problems of interest with only minor modifications.

Let us consider the situation in which the initial state $|i, \text{in}\rangle$ consists of two wave packets with only positive frequency components:

$$|i, \text{in}\rangle = \int \frac{d^3p_1}{2p_1^0(2\pi)^3} \int \frac{d^3p_2}{2p_2^0(2\pi)^3} f_1(p_1) f_2(p_2) |p_1, p_2; \text{in}\rangle \tag{10.123}$$

The incoming flux is equal to $\int \frac{d^3p}{2p_0(2\pi)^3} |f(p)|^2$. Each component $|p_1, p_2; \text{in}\rangle$ will have a matrix element with the final state $|f, \text{out}\rangle$.

Since we have translation invariance, the total 4-momentum should be conserved. If we denote by P_f the momentum of the state $|f, \text{out}\rangle$, we can write the matrix element of the T-matrix as

$$\langle f|T|p_1 p_2\rangle = (2\pi)^4 \delta^4(P_f - p_1 - p_2)\langle f|\mathcal{T}|p_1 p_2\rangle \tag{10.124}$$

where $\mathcal{T}$ is the reduced operator, which, as we see, acts only on the energy shell, $p^2 = m^2$.

If we neglect the forward-scattering contribution, the transition probability $W_{i\to f}$ becomes

$$W_{i\to f} = \int \frac{d^3p_1}{2p_1^0(2\pi)^3} \int \frac{d^3p_2}{2p_2^0(2\pi)^3} \int \frac{d^3q_1}{2q_1^0(2\pi)^3} \int \frac{d^3q_2}{2q_2^0(2\pi)^3}$$
$$\times (2\pi)^4\delta^4(p_1+p_2-q_1-q_2)\,(2\pi)^4\delta^4(P_f-p_1-p_2)$$
$$\times f_1^*(p_1)f_2^*(p_2)f_1(q_1)f_2(q_2)\langle f|\mathcal{T}|p_1p_2\rangle^*\langle f|\mathcal{T}|q_1q_2\rangle \qquad (10.125)$$

The incoming states are assumed to be sharply peaked around some momenta $\bar{p}_1$ and $\bar{p}_2$ with a spread Δp, such that we can approximate the matrix element as follows:

$$\langle f|\mathcal{T}|p_1p_2\rangle \approx \langle f|\mathcal{T}|q_1q_2\rangle \approx \langle f|\mathcal{T}|\bar{p}_1\bar{p}_2\rangle \qquad (10.126)$$

Under these assumptions, the form for the transition probability is

$$W_{i\to f} = \int d^4x |\widetilde{f}_1(x)|^2 |\widetilde{f}_2(x)|^2 (2\pi)^4\delta^4(P_f-\bar{p}_1-\bar{p}_2)|\langle f|\mathcal{T}|\bar{p}_1\bar{p}_2\rangle|^2 \qquad (10.127)$$

where $\widetilde{f}(x)$ is the Fourier transform of $f(p)$. The integrand of eq. (10.127) is the transition probability per unit time and volume:

$$\frac{dW_{i\to f}}{dtdV} = |\widetilde{f}_1(x)|^2|\widetilde{f}(x)|^2(2\pi)^4\delta^4(P_f-\bar{p}_1-\bar{p}_2)|\langle f|\mathcal{T}|\bar{p}_1\bar{p}_2\rangle|^2 \qquad (10.128)$$

In position space, the flux is $i\int d^3x\widetilde{f}^*(x)\overleftrightarrow{\partial_0}\widetilde{f}(x)$. If $\widetilde{f}(x)$ is sufficiently smooth, we can make the following approximation:

$$i\widetilde{f}^*(x)\,\overleftrightarrow{\partial_0}\widetilde{f}(x) \approx 2\bar{p}_0|\widetilde{f}(x)|^2 \qquad (10.129)$$

Let us assume that particle 1 is incident in the laboratory reference frame, and that in the laboratory frame, particle 2 is at rest. The density of particles in the target is

$$\frac{dn_2}{dV} = 2\bar{p}_2^0|\widetilde{f}_2(x)|^2 \qquad (10.130)$$

where $\bar{p}_2^0 = m_2$, since particle 2 is at rest. The incident flux is the velocity multiplied by the density of particles in the beam:

$$\Phi_{\text{in}} = \frac{|\bar{p}_1|}{|\bar{p}_0^1|}2\bar{p}_1^0|\widetilde{f}_1(x)|^2 = 2|\bar{p}_1||\widetilde{f}_1(x)|^2 \qquad (10.131)$$

Then the *differential cross section* $d\sigma$ is related to the transition probability by the relation

$$\frac{dW_{i\to f}}{dtdV} = \frac{dn_2}{dV}\cdot\Phi_{\text{in}}\cdot d\sigma \qquad (10.132)$$

Hence

$$2m_2|\widetilde{f}_2(x)|^2 2|\bar{p}_1||\widetilde{f}_1(x)|^2 d\sigma =$$
$$|\widetilde{f}_1(x)|^2|\widetilde{f}_2(x)|^2(2\pi)^4\delta^4(P_f-\bar{p}_1-\bar{p}_2)|\langle f|\mathcal{T}|\bar{p}_1\bar{p}_2\rangle|^2 \qquad (10.133)$$

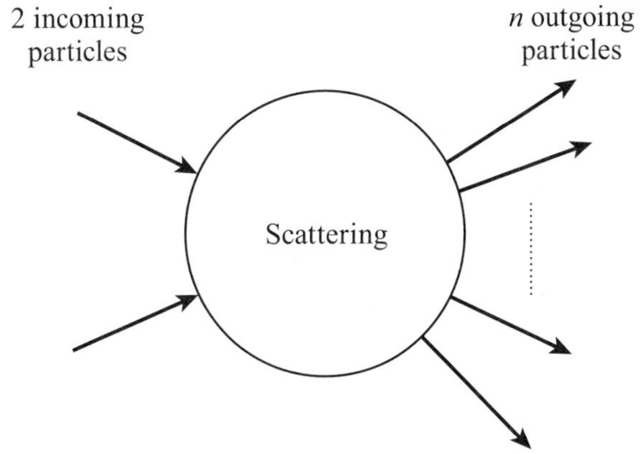

Figure 10.11 A $2 \rightarrow n$ scattering process.

Therefore, the differential cross section $d\sigma$ is

$$d\sigma = (2\pi)^4 \delta^4(P_f - \bar{p}_1 - \bar{p}_2) \frac{1}{4m_2|\bar{p}_1|} |\langle f|\mathcal{T}|\bar{p}_1\bar{p}_2\rangle|^2 \tag{10.134}$$

The quantity in the denominator, $m_2|\bar{p}_1|$, can be written in the relativistic invariant way:

$$m_2|\bar{p}_1| = m_2\sqrt{\bar{p}_1^{02} - m_1^2} = \left[(\bar{p}_2 \cdot \bar{p}_1)^2 - m_1^2 m_2^2\right]^{1/2} \tag{10.135}$$

Thus far, we have not made any assumptions about the nature of the final state. If the process that we consider involves two particles going in and n particles coming out (see figure 10.11), the total differential cross section becomes

$$\begin{aligned} d\sigma = & \frac{1}{4\left[(\bar{p}_2 \cdot \bar{p}_1)^2 - m_1^2 m_2^2\right]^{1/2}} \int_\Delta \frac{d^3 p_3}{(2\pi)^3 2p_3^0} \cdots \int_\Delta \frac{d^3 p_{n+2}}{(2\pi)^3 2p_{n+2}^0} \\ & \times |\langle p_3, \ldots, p_{n+2}|\mathcal{T}|p_1 p_2\rangle|^2 (2\pi)^4 \delta\left(p_1 + p_2 - \sum_{i=3}^{n+2} p_i\right) \end{aligned} \tag{10.136}$$

where Δ is an energy-momentum resolution. This expression shows that the central issue is to compute matrix elements of the reduced operator $\mathcal{T}$.

10.7 Asymptotic states and the analytic properties of the propagator

Let us now show that the S-matrix elements can be calculated if we know the vacuum expectation value (v.e.v.) of time-ordered products of field operators $\phi(x)$ in the Heisenberg representation. This is the Lehmann, Symanzik, and Zimmermann (LSZ) approach.

To make this connection, it is necessary to relate the interacting fields $\phi(x)$ to a set of fields that create (or destroy) the *actual states* in the spectrum of the system, representing particles that are *far from one another.* In any scattering experiment, the initial states are sharply peaked wave packets that can be constructed to be arbitrarily close to the eigenstates of the system. Of course, the true eigenstates are plane waves, and any two such states will necessarily have a nonvanishing overlap in space. But wave packets that are essentially made of just one state will not overlap if the wave packets (the "particles") are sufficiently far apart from each other at the initial time. Since they do not overlap, they do not interact. In this sense, the spectrum of incoming states can be generated by a set of free fields. Let us make the further assumption that the *in* and *out* states, the so-called *asymptotic states*, are created by such a set of free fields. We denote these free fields by $\phi_{\text{in}}(x)$.

These assumptions amount to saying that the states of the fully interacting theory are in one-to-one correspondence with the states of a noninteracting theory. In some loose sense, this hypothesis implies that the information that we can obtain from perturbation theory is always qualitatively correct. In other words, it is assumed that the states of the interacting theory are adiabatically connected to those of a noninteracting theory.

However, these assumptions can fail in several possible ways. A mild failure would be the appearance of *bound states*, which, of course, are not present in the unperturbed theory. This situation is actually rather common, and it can be remedied without too much difficulty. Two examples of this case are positronium states in QED and the collective modes of the Landau theory of the Fermi liquid.

However, there are several ways in which this picture can fail in a rather serious way. One case is when the fields of the Lagrangian do not describe any of the asymptotic states of the theory. An example of this case is quantum chromodynamics (QCD), whose Lagrangian describes the dynamics of *quarks* and *gluons*, which are not present in the asymptotic states, since quarks are confined and gluons are screened. The asymptotic states of QCD are mesons, baryons, and glue-balls, which are bound states of quarks and gluons.

Another possible failure of this hypothesis is the case in which the states created by the fields of the Lagrangian are not the true elementary excitations; instead they behave like an effective composite object of more elementary states. In such cases, the true one-particle states may be orthogonal to the states created by the fields of the Lagrangian. This is a rather common situation in theories in $1+1$ dimensions, whose spectrum is generated by a set of soliton-like states, which are extended objects in terms of the bare fields of the Lagrangian. Something very similar happens in the theory of the fractional quantum Hall states of two-dimensional electron fluids in strong magnetic fields.

A third way in which the hypothesis may fail is that the quantum fluctuations may be so strong that the particle interpretation simply fails. We will see that this happens if the field theory is at a nontrivial fixed point of the renormalization group. In general, at a fixed point, most (but not all) observables develop *anomalous dimensions*. In this case, the result of a scattering process is not a set of well-defined particles. Then we will see that the scattering amplitude does not have poles (which are associated with particle states) but instead has branch cuts. Hence, in this case, the conventional particle interpretation of QFT fails.

In this chapter, we do not consider these very interesting situations (which will be discussed in other chapters). Instead, we assume that the adiabatic hypothesis (or scenario) actually holds. For the sake of concreteness, we will deal with scalar fields, and we require ϕ_{in} to be a free massive field that obeys the Klein-Gordon equation

$$\left(\partial^2 + m^2\right)\phi_{\text{in}} = 0 \tag{10.137}$$

where m is the physical mass. In general, the physical mass is different from the mass parameter m_0 that enters in the Lagrangian of the interacting theory.

As the initial state evolves in time, the particles approach each other and begin to overlap. Interactions take place, and after some time, the system evolves to some final state, consisting of a set of well-defined particles, the out state. The unitary operator that connects the in and out states is precisely the S-matrix of the interaction representation. The only difference here is that the in and out states are not eigenstates of some unperturbed system but are the actual eigenstates of the full theory.

This picture assumes that there is a stable vacuum state $|0\rangle$ such that the observed particles are the elementary excitations of this vacuum. The free fields $\phi_{\text{in}}(x)$ are just a device to generate the spectrum and have no real connection with actual dynamics. However, the interacting field $\phi(x)$ creates not only one-particle states but also many-particle states. This is so because its equations of motion are nonlinear. Hence the matrix elements of $\phi(x)$ and $\phi_{\text{in}}(x)$ between the vacuum $|0\rangle$ and one-particle in-states $|1\rangle$ are generally different, since ϕ_{in} creates only one-particle states. This must be true even as $t \to -\infty$. We state this difference by writing

$$\langle 1|\phi(x)|0\rangle = Z^{1/2}\langle 1|\phi_{\text{in}}|0\rangle \tag{10.138}$$

The proportionality constant Z is known as the wave-function renormalization. If $Z \neq 1$, the operator $\phi(x)$ must have a nonzero multiparticle projection. Notice that this is only an identity of these matrix elements and is not an identity between the fields themselves.

In the interaction representation, it is possible to derive a similar looking identity that originates because the unperturbed and perturbed states do not have the same normalization. It is important to stress that this approach makes the *essential* assumption that the states that are reached through perturbation theory in the interaction representation can approximate, with arbitrary precision, *all* of the *exact* states of the theory. This assumption is the hypothesis that the asymptotic states are generated by free fields.

The operators that create the physical asymptotic states satisfy canonical commutation relations, and the commutator of a pair of such fields is

$$\langle 0|[\phi_{\text{in}}(x), \phi_{\text{in}}(x')]|0\rangle = i\Delta(x - x'; m) \tag{10.139}$$

where m is the physical mass. In contrast, the interacting fields satisfy

$$\langle 0|[\phi(x), \phi(x')]|0\rangle = \sum_n \left[\langle 0|\phi(x)|n\rangle\langle n|\phi(x')|0\rangle - (x \leftrightarrow x')\right] \tag{10.140}$$

where $\{|n\rangle\}$ is a complete set of physical (in) states. The operators $\phi(x)$ are related to the operator $\phi(0)$ at the origin at some time $x_0 = 0$ by

$$\phi(x) = e^{iP\cdot x}\phi(0)e^{-iP\cdot x} \tag{10.141}$$

where P_μ is the *total* 4-momentum operator. If P_n^μ is the 4-momentum of the state $|n\rangle$, we can write

$$\langle 0|[\phi(x), \phi(x')]|0\rangle = \sum_n \left[\langle 0|\phi(0)|n\rangle e^{-iP_n\cdot(x-x')}\langle n|\phi(0)|0\rangle - (x \leftrightarrow x')\right] \tag{10.142}$$

Let us now insert the identity

$$1 = \int d^4Q\, \delta^4(Q - P_n) \tag{10.143}$$

to get

$$\langle 0|[\phi(x), \phi(x')]|0\rangle = \int d^4Q \sum_n \delta^4(Q - P_n)|\langle 0|\phi(0)|n\rangle|^2 \left(e^{-iQ\cdot(x-x')} - e^{iQ\cdot(x-x')}\right) \tag{10.144}$$

We can rewrite this expression in terms of a *spectral density* $\rho(Q)$:

$$\langle 0|[\phi(x), \phi(x')]|0\rangle = \int \frac{d^4Q}{(2\pi)^3}\rho(Q)\left(e^{-iQ\cdot(x-x')} - e^{iQ\cdot(x-x')}\right) \tag{10.145}$$

where $\rho(Q)$ is given by

$$\rho(Q) = (2\pi)^3 \sum_n \delta^4(Q - P_n)|\langle 0|\phi(0)|n\rangle|^2 \tag{10.146}$$

Recall that $\Delta(x - x'; m)$ is given by

$$\begin{aligned} i\Delta(x - x'; m) &= \int \frac{d^3Q}{(2\pi)^3 2Q_0}\left(e^{-iQ\cdot(x-x')} - e^{iQ\cdot(x-x')}\right) \\ &= \int \frac{d^4Q}{(2\pi)^3}\epsilon(Q^0)\delta^4(Q^2 - m^2)e^{-iQ\cdot(x-x')} \end{aligned} \tag{10.147}$$

where $\epsilon(Q^0) = \text{sign}(Q^0)$. Thus we can write

$$\langle 0|[\phi(x), \phi(x')]|0\rangle = \int \frac{d^4Q}{(2\pi)^3}\rho(Q)\epsilon(Q^0)e^{-iQ\cdot(x-x')} \tag{10.148}$$

Since $\rho(Q)$ is Lorentz invariant, by construction, it can only be a real and positive function of Q^2:

$$\rho(Q) = \sigma(Q^2) > 0 \tag{10.149}$$

Hence

$$\langle 0|[\phi(x), \phi(x')]|0\rangle = \int \frac{d^4Q}{(2\pi)^3}\sigma(Q^2)\epsilon(Q^0)e^{-iQ\cdot(x-x')} \tag{10.150}$$

where $\sigma(Q)^2$ is known as the spectral function.

Let us now rewrite this expression in the form of an integral over the spectrum. Insert the identity

$$1 = \int_0^\infty d\mu^2 \delta(Q^2 - \mu^2) \tag{10.151}$$

where μ^2 is a spectral parameter, to obtain

$$\begin{aligned} \langle 0|[\phi(x), \phi(x')]|0\rangle &= \int \frac{d^4Q}{(2\pi)^3}\left[\int_0^\infty d\mu^2 \delta(Q^2 - \mu^2)\right]\sigma(Q^2)\epsilon(Q_0)e^{-iQ\cdot(x-x')} \\ &= \int_0^\infty d\mu^2 \sigma(\mu^2)\left[\int \frac{d^4Q}{(2\pi)^3}\epsilon(Q_0)\delta(Q^2 - \mu^2)e^{-iQ\cdot(x-x')}\right] \end{aligned} \tag{10.152}$$

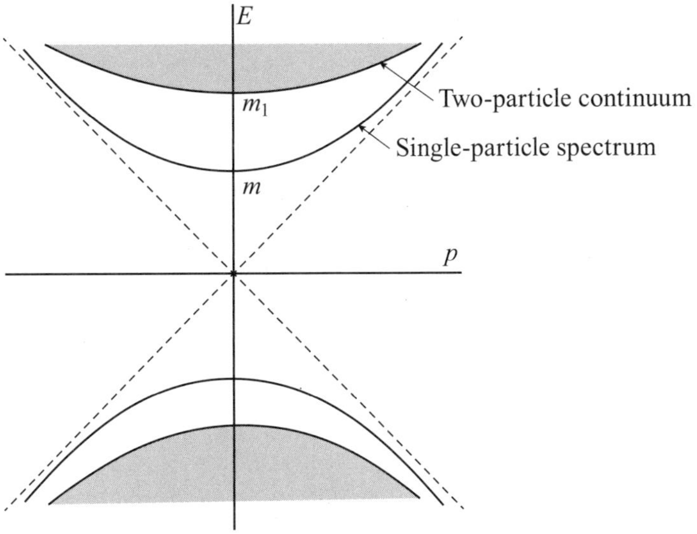

Figure 10.12 Spectrum of the propagator.

where $\sigma(\mu^2)$ is

$$\sigma(\mu^2) = (2\pi)^3 \sum_n \delta(P_n^2 - \mu^2)|\langle 0|\phi(0)|n\rangle|^2 \tag{10.153}$$

Thus, we can write the v.e.v. of the commutator of the interacting fields as

$$\langle 0|[\phi(x), \phi(x')]|0\rangle = i \int_0^\infty d\mu^2 \sigma(\mu^2) \Delta(x - x'; \mu^2) \tag{10.154}$$

If we assume that the theory has a physical particle with mass m and one-particle states of mass m, we can write the final expression:

$$-i\langle 0|[\phi(x), \phi(x')]|0\rangle = Z\Delta(x - x'; m^2) + \int_{m_1^2}^\infty d\mu^2 \sigma(\mu^2) \Delta(x - x'; \mu^2) \tag{10.155}$$

where the first term represents the one-particle states, and the integral represents the continuum of multiparticle states with a threshold at m_1, as shown in figure 10.12. In other words, if there is a stable particle with mass m, the spectral function must have a δ-function at $\mu^2 = m^2$ with strength Z, the spectral weight of the one-particle state.

Since the field $\phi(x)$ obeys equal-time canonical commutation relations with the canonical momentum $\Pi(x) = \partial_0 \phi(x)$, we get

$$-i\langle 0|[\Pi(\boldsymbol{x}, x_0), \phi(\boldsymbol{x}', x_0)]|0\rangle = Z \lim_{x_0' \to x_0} \partial_0 \Delta(x - x'; m^2) + \int_{m_1^2}^\infty d\mu^2 \sigma(\mu^2) \lim_{x_0' \to x_0} \partial_0 \Delta(x - x'; \mu^2) \tag{10.156}$$

In contrast, the free-field commutator $\Delta(x-x';m^2)$ obeys the initial condition

$$\lim_{x_0'\to x_0} \partial_0 \Delta(x-x';m^2) = \lim_{x_0'\to x_0} [\Pi(x), \phi(x')] = -i\delta^3(\boldsymbol{x}-\boldsymbol{x}') \tag{10.157}$$

Hence, we find that the spectral function $\sigma(\mu^2)$ obeys the *spectral sum rule*:

$$1 = Z + \int_{m_1^2}^{\infty} d\mu^2 \sigma(\mu^2) \tag{10.158}$$

Since $\sigma(\mu^2) > 0$, we find that $0 \leq Z \leq 1$. The lower end of the integration range, the threshold for multiparticle production m_1^2, is equal to $4m^2$, since we must create at least two elementary excitations.

A similar analysis can be done for the Feynman (time-ordered) propagator

$$G_F(x-x';m) = -i\langle 0|T\phi(x)\phi(x')|0\rangle \tag{10.159}$$

which has the spectral representation

$$G_F(x-x';m) = ZG_0(x-x';m) + \int_{m_1^2}^{\infty} d\mu^2 \sigma(\mu^2) G_0(x-x';\mu^2) \tag{10.160}$$

This decomposition is known as the *Lehmann representation.*

In eq. (10.160), $G_0(x-x';m^2)$ is the Feynman propagator for a free field:

$$G_0(x-x') = -\int \frac{d^4p}{(2\pi)^4} \frac{e^{-ip\cdot(x-x')}}{p^2-m^2+i\epsilon} \tag{10.161}$$

In the limit $\epsilon \to 0^+$, the poles of the integrand can be manipulated to give

$$\lim_{\epsilon\to 0^+} \frac{1}{p^2-m^2+i\epsilon} = \mathcal{P}\frac{1}{p^2-m^2} - i\pi\delta(p^2-m^2) \tag{10.162}$$

Using this identity, we get that, in momentum space, the propagator is

$$G_F(p;m) = -\frac{Z}{p^2-m^2+i\epsilon} - \int_{m_1}^{\infty} d\mu^2 \frac{\sigma(\mu^2)}{p^2-\mu^2+i\epsilon} \tag{10.163}$$

Its imaginary part is given by

$$\text{Im}\, G_F(p;m) = \pi Z\delta(p^2-m^2) + \pi\int_{m_1^2}^{\infty} \sigma(\mu^2)\delta(p^2-\mu^2) \tag{10.164}$$

Hence

$$\frac{1}{\pi}\text{Im}\, G_F(p;m) = Z\delta(p^2-m^2) + \sigma(p^2)\Theta(p^2-m_1^2) \tag{10.165}$$

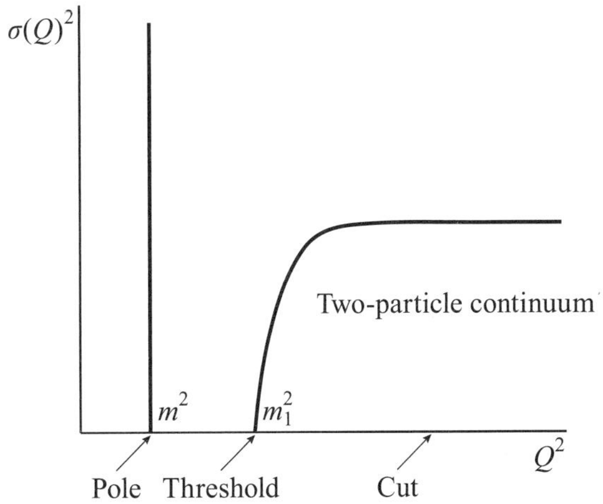

Figure 10.13 The analytic structure of the propagator is encoded in the spectral density $\sigma(Q^2)$.

Once the imaginary part is known, the real part can be found by using the *Kramers-Krönig* or *dispersion relation*:

$$\mathrm{Re}\, G_F(p, m^2) = \frac{1}{\pi} \mathcal{P} \int_0^\infty d\mu^2 \frac{\mathrm{Im}\, G_F(p, \mu^2)}{\mu^2 - p^2 - i\epsilon} \tag{10.166}$$

We see that, in general, there are two contributions to $\mathrm{Im}\, G_F(p; m)$. The first term is the contribution from the single-particle states. In addition, there is a smooth contribution (the second term), which results from multiparticle production. While the single-particle states contribute with an isolated pole (a δ-function in the imaginary part), the multiple particle states (or continuum) are represented by a branch cut (see figure 10.13).

There is a simple and natural physical interpretation of these results. If the incoming state has $Q^2 < m^2$, it cannot propagate, since the allowed value must be at least m^2 (the physical mass). If $Q^2 > m_1^2$, and if there are no bound states, the incoming state can decay into at least two single-particle states. Hence $m_1^2 = 4m^2$. These states should form a continuum, since given the initial momentum P_i, there are many multiparticle states with the same total momentum. Thus, those processes are incoherent. Notice that without interactions, the incoming state would not have been able to decay into several single-particle states.

Note that the propagators of all theories that we have previously discussed have the same type of analytic structure as discussed here.

10.8 The S-matrix and the expectation value of time-ordered products

We are now in a position to find the connection between S-matrix elements and the v.e.v. of time-ordered fields. For simplicity, keep in mind the case of scalar fields, but the results are easily generalizable. The actual derivation is rather lengthy and unilluminating. We will discuss its meaning and refer to standard textbooks for details.

Let's assume that we want to evaluate the matrix element

$$\langle p_1, \ldots, p_n; \mathrm{out}|q_1, \ldots, q_m; \mathrm{in}\rangle = \langle p_1, \ldots, p_n|S|q_1, \ldots, q_m\rangle \tag{10.167}$$

Assume that all incoming and outgoing momenta are different. This matrix element is given by the *reduction formula*

$$\begin{aligned}\langle p_1, \ldots, p_n|S|q_1, \ldots, q_m\rangle =& \frac{i}{Z^{(n+m)/2}} \int d^4y_1 \cdots d^4y_n d^4x_1 \cdots d^4x_n \\ \times \exp\left[i \left(\sum_{\ell=1}^{n} p_\ell \cdot y_\ell - \sum_{k=1}^{m} q_k \cdot x_k \right)\right] & \prod_{\ell=1}^{n} \left(\partial_{y_\ell}^2 + m^2\right) \prod_{k=1}^{m} \left(\partial_{x_k}^2 + m^2\right) \\ & \times \langle 0|T\left(\phi(y_1)\ldots\phi(y_n)\phi(x_1)\ldots\phi(x_m)\right)|0\rangle\end{aligned} \tag{10.168}$$

where m^2 is the physical mass, and the external momenta p and q are on the mass shell: $p^2 = q^2 = m^2$.

Let us consider, for example, the $2 \to 2$ process:

$$\langle p_1, p_2; \mathrm{out}|q_1, q_2; \mathrm{in}\rangle = \langle p_1 p_2, \mathrm{in}|S|q_1, q_2; \mathrm{in}\rangle = \langle p_1 p_2, \mathrm{out}|a_{\mathrm{in}}^\dagger(q_1)|q_2, \mathrm{in}\rangle \tag{10.169}$$

where $a_{\mathrm{in}}^\dagger(q_1)$ is (using the notation introduced in eq. (3.17))

$$a_{\mathrm{in}}^\dagger(q_1) = -i \int_{\text{fixed } t} d^3x \, e^{-iq_1 \cdot x} \overleftrightarrow{\partial_0} \phi_{\mathrm{in}}(x) \tag{10.170}$$

Hence, the matrix element is

$$\begin{aligned}\langle p_1 p_2; \mathrm{out}|q_1 q_2; \mathrm{in}\rangle &= -i \lim_{t\to-\infty} \int_t d^3x \, e^{-iq_1 \cdot x} \overleftrightarrow{\partial_0} \langle p_1 p_2; \mathrm{out}|\phi_{\mathrm{in}}(x)|q_2; \mathrm{in}\rangle \\ &\equiv -i \lim_{t\to-\infty} \frac{1}{Z^{1/2}} \int_t d^3x \, e^{-iq_1 \cdot x} \overleftrightarrow{\partial_0} \langle p_1 p_2; \mathrm{out}|\phi(x)|q_2; \mathrm{in}\rangle\end{aligned} \tag{10.171}$$

where we have made the replacement of ϕ_{in} by $\frac{1}{Z^{1/2}}\phi$ inside the matrix element ($t \to -\infty$).

But

$$\langle 0|\phi(x)|1\rangle = Z^{1/2}\langle 0|\phi_{\mathrm{in}}(x)|1\rangle = Z^{1/2}\langle 0|\phi_{\mathrm{out}}(x)|1\rangle \tag{10.172}$$

and

$$\left(\lim_{t_f\to+\infty} - \lim_{t_i\to-\infty}\right) \int d^3x \, F(\mathbf{x}, t) = \lim_{t_f\to+\infty} \lim_{t_i\to-\infty} \int_{t_i}^{t_f} dt \, \frac{\partial}{\partial t} \int d^3x \, F(\mathbf{x}, t) \tag{10.173}$$

These formulas allow us to write

$$
\begin{aligned}
&\lim_{t_f\to+\infty}\int d^3x\,\langle p_1p_2,\text{out}|a^\dagger_{\text{in}}(q_1)|q_2,\text{in}\rangle\\
&=\lim_{t_i\to-\infty}\int d^3x\,\langle p_1p_2,\text{out}|a^\dagger_{\text{in}}(q_1)|q_2,\text{in}\rangle+\int d^4x\,\partial_0\langle p_1p_2,\text{out}|a^\dagger_{\text{in}}(q_1)|q_2,\text{in}\rangle\\
&=\langle p_1p_2,\text{out}|a^\dagger_{\text{out}}(q_1)|q_2,\text{in}\rangle
\end{aligned}
\tag{10.174}
$$

Thus, the matrix element is

$$
\begin{aligned}
\langle p_1p_2,\text{out}|q_1q_2,\text{in}\rangle=&\langle p_1p_2,\text{out}|a^\dagger_{\text{out}}(q_1)|q_2,\text{in}\rangle\\
&+\frac{i}{Z^{1/2}}\int d^4x\,\partial_0\left[e^{-iq_1\cdot x}\overleftrightarrow{\partial_0}\langle p_1p_2,\text{out}|\phi(x)|q_2,\text{in}\rangle\right]
\end{aligned}
\tag{10.175}
$$

The first contribution is a *disconnected term*, and it is given by

$$
\begin{aligned}
&\langle p_1p_2,\text{out}|a^\dagger_{\text{out}}(q_1)|q_2,\text{in}\rangle\\
&=(2\pi)^3 2p_1^0\delta^3(p_1-q_1)\langle p_2,\text{out}|q_1,\text{in}\rangle+(2\pi)^3 2p_2^0\delta^3(p_2-q_1)\langle p_1,\text{out}|q_1,\text{in}\rangle
\end{aligned}
\tag{10.176}
$$

Notice that q_1 is on the mass shell, $q_1^2=m^2$, and e^{-iq_1x} is a solution of the Klein-Gordon equation

$$
(\partial^2+m^2)e^{iq_1\cdot x}=0\qquad (q_1^2=m^2)
\tag{10.177}
$$

The second contribution to the matrix element can also be evaluated (for arbitrary states α and β):

$$
\begin{aligned}
&\int d^4x\,\partial_0\left[e^{-iq_1\cdot x}\overleftrightarrow{\partial_0}\langle\beta,\text{out}|\phi(x)|\alpha,\text{in}\rangle\right]=\\
&=\int d^4x\left[e^{-iq_1\cdot x}\partial_0^2\langle\beta,\text{out}|\phi(x)|\alpha,\text{in}\rangle+(-\partial_0^2e^{-iq_1\cdot x})\langle\beta,\text{out}|\phi(x)|\alpha,\text{in}\rangle\right]\\
&=\int d^4x\left[e^{-iq_1\cdot x}\partial_0^2\langle\beta,\text{out}|\phi(x)|\alpha,\text{in}\rangle+\left[(-\nabla^2+m^2)\,e^{-iq_1\cdot x}\right]\langle\beta,\text{out}|\phi(x)|\alpha,\text{in}\rangle\right]\\
&=\int d^4x e^{-iq_1\cdot x}(\partial^2+m^2)\langle\beta,\text{out}|\phi(x)|\alpha,\text{in}\rangle
\end{aligned}
\tag{10.178}
$$

where we have integrated by parts. Hence, the matrix element is

$$
\begin{aligned}
\langle p_1p_2,\text{out}|q_1q_2,\text{in}\rangle=&\frac{i}{Z^{1/2}}\int d^4x_1e^{-iq_1\cdot x_1}(\partial^2+m^2)\langle p_1p_2,\text{out}|\phi(x_1)|q_2,\text{in}\rangle\\
&+(2\pi)^3 2p_1^0\delta^3(p_1-q_1)\langle p_2,\text{out}|q_2,\text{in}\rangle+(2\pi)^3 2p_2^0\delta^3(p_2-q_1)\langle p_1,\text{out}|q_2,\text{in}\rangle
\end{aligned}
\tag{10.179}
$$

The matrix element inside the integrand of the first term is equal to

$$\langle p_1 p_2, \text{out}|\phi(x_1)|q_2, \text{in}\rangle = \lim_{y_1^0 \to +\infty} \frac{i}{Z^{1/2}} \int d^3 y_1 e^{ip_1 \cdot y_1} \overleftrightarrow{\partial}_{y_1^0} \langle p_2, \text{out}|\phi(y_1)\phi(x_1)|q_2, \text{in}\rangle \tag{10.180}$$

where (by definition) $y_1^0 > x_1^0$. This expression is also equal to

$$\begin{aligned}\langle p_1 p_2, \text{out}|\phi(x_1)|q_2, \text{in}\rangle =& \langle p_2, \text{out}|\phi(x_1) a_{\text{in}}(p_1)|q_2, \text{in}\rangle \\ &+ \frac{i}{Z^{1/2}} \int d^4 y_1 e^{ip_1 \cdot y_1} (\partial_{y_1}^2 + m^2) \langle p_2, \text{out}|T\phi(y_1)\phi(x_1)|q_2, \text{in}\rangle\end{aligned} \tag{10.181}$$

By substituting back into the expression from the matrix element, we find that the latter is equal to

$$\begin{aligned}&\langle p_1 p_2, \text{out}|q_1 q_2, \text{in}\rangle = \\ &\quad = (2\pi)^3 2p_1^0 \delta^3(p_1 - q_1)\langle p_2, \text{out}|q_2, \text{in}\rangle + (2\pi)^3 2p_2^0 \delta^3(p_2 - q_1)\langle p_1, \text{out}|q_2, \text{in}\rangle \\ &+ \frac{i}{Z^{1/2}} \int d^4 x_1 e^{-iq_1 \cdot x_1} (\partial_{x_1}^2 + m^2) \langle p_2, \text{out}|\phi(x_1)|0, \text{in}\rangle (2\pi)^3 2q_2^0 \delta^3(q_2 - p_1) \\ &+ \left(\frac{1}{Z^{1/2}}\right)^2 \int d^4 x_1 d^4 y_1 e^{i(p_1 \cdot y_1 - q_1 \cdot x_1)} (\partial_{x_1}^2 + m^2)(\partial_{y_1}^2 + m^2) \\ &\qquad \times \langle p_2, \text{out}|T\phi(y_1)\phi(x_1)|q_2, \text{in}\rangle\end{aligned} \tag{10.182}$$

By iterating this process once more, we obtain the reduction formula of eq. (10.168) plus disconnected terms.

The reduction formula provides the connection between the on-shell S-matrix elements and the v.e.v. of time-ordered products. Notice that the reduction formula implies that the v.e.v. of time-ordered products of the field operators must have poles in the variables p_1^2 (where p_i is conjugate to x_i) and that the matrix element is the residue of this pole.

The reduction formula shows that all scattering data can be understood in terms of an appropriate v.e.v. of a time-ordered product of field operators. The problem that we are left to solve is the computation of these v.e.v.s. In chapter 11, we will use perturbation theory to compute these expectation values.

10.9 Linear response theory

In addition to the problem of evaluating S-matrix elements, it is of interest to consider the response of a system to weak localized external perturbations. These responses will tell us much about the nature of both the ground state and of the low-lying states of the system. This method is of great importance for the study of systems in condensed matter physics.

Let H be the full Hamiltonian of a system. We will consider the coupling of the system to weak external sources. Let $\hat{O}(x,t)$ be a hermitian operator representing a *local observable*, such as the charge density, the current density, or the local magnetic moment. Let us represent the coupling to the external source by an extra term $H_{\text{ext}}(t)$ in the Hamiltonian. The total Hamiltonian is now

$$H_T = H + H_{\text{ext}} \tag{10.183}$$

If the source is adiabatically switched on and off, then the Heisenberg representation for the isolated system becomes the interaction representation for the full system. Hence, exactly as in the interaction representation, all the observables obey the Heisenberg equations of motion of the system in the absence of the external source, while the states will follow the external source in their evolution.

Let $|\text{gnd}\rangle$ be the *exact* ground state (or vacuum) of the system in the absence of any external sources. The external sources perturb this ground state and cause the v.e.v. of the local observable $\hat{\mathcal{O}}(x,t)$ to change:

$$\langle\text{gnd}|\hat{\mathcal{O}}(\boldsymbol{x},t)|\text{gnd}\rangle \to \langle\text{gnd}|U^{-1}(t)\,\hat{\mathcal{O}}(\boldsymbol{x},t)\,U(t)|\text{gnd}\rangle \tag{10.184}$$

where the time-evolution operator $U(t)$ is now given by

$$U(t) = T\exp\left\{-\frac{i}{\hbar}\int_{-\infty}^{t} dt' H_{\text{ext}}(t')\right\} \tag{10.185}$$

Linear response theory consists of evaluating the changes in the expectation values of the observables to leading order in the external perturbation. Thus, to leading order in the external sources, the change of the v.e.v. is

$$\delta\langle\text{gnd}|\hat{\mathcal{O}}(\boldsymbol{x},t)|\text{gnd}\rangle = \frac{i}{\hbar}\int_{-\infty}^{t} dt'\;\langle\text{gnd}|[H_{\text{ext}}(t'),\hat{\mathcal{O}}(\boldsymbol{x},t)]|\text{gnd}\rangle + \cdots \tag{10.186}$$

Quite generally, we will be interested in the case where H_{ext} represents the local coupling of the system to an external source $f(\boldsymbol{x},t)$ through the observable $\hat{\mathcal{O}}(\boldsymbol{x},t)$. Thus, let us choose the perturbation $H_{\text{ext}}(t)$ to have the form

$$H_{\text{ext}}(t) = \int d^3x\, f(\boldsymbol{x},t)\,\hat{\mathcal{O}}(\boldsymbol{x},t) \tag{10.187}$$

The function $f(\boldsymbol{x},t)$ is usually called the *force*.

If the observable is normal ordered relative to the ground state of the isolated system, it must have a vanishing expectation value in the ground state: $\langle\text{gnd}|\hat{\mathcal{O}}(\boldsymbol{x},t)|\text{gnd}\rangle = 0$. Hence, the change of its expectation value will be equal to the final value and is given by

$$\delta\langle\text{gnd}|\hat{\mathcal{O}}(\boldsymbol{x},t)|\text{gnd}\rangle =$$
$$= \frac{i}{\hbar}\int_{-\infty}^{t} dt' \int d^3x'\;\langle\text{gnd}|[\hat{\mathcal{O}}(\boldsymbol{x},t'),\hat{\mathcal{O}}(\boldsymbol{x},t)]|\text{gnd}\rangle\, f(\boldsymbol{x}',t') + \cdots \tag{10.188}$$

The main assumption of linear response theory is that the response is proportional to the force. The proportionality constant is interpreted as a *generalized susceptibility* χ. Thus, we write the change in the v.e.v. in the form

$$\delta\langle\text{gnd}|\hat{\mathcal{O}}(\boldsymbol{x},t)|\text{gnd}\rangle = \chi\cdot f \equiv \int d^3x' \int_{-\infty}^{t} dt'\;\chi(x,x')\,f(x') + \cdots \tag{10.189}$$

where $\chi(x,x')$ is the susceptibility.

Let $D^R(x,x')$ represent the *retarded correlation function* of the observable $\hat{\mathcal{O}}(\boldsymbol{x},t)$:

$$D^R(x,x') = -i\Theta(x_0 - x_0')\langle \text{gnd}|[\hat{\mathcal{O}}(x), \hat{\mathcal{O}}(x')]|\text{gnd}\rangle \qquad (10.190)$$

We see that $\langle \mathcal{O}(x)\rangle$ is determined by $D^R(x,x')$, since

$$\delta\langle \text{gnd}|\mathcal{O}(x)|\text{gnd}\rangle = \frac{1}{\hbar}\int d^4x' D^R(x,x')f(x') + \cdots \qquad (10.191)$$

Therefore, the responses and the susceptibilities are given by retarded correlation functions, not by time-ordered ones. However, since the retarded and time-ordered correlation functions are related by an analytic continuation, the knowledge of the latter gives the information about the former.

Let us Fourier transform the time dependence of the ground state expectation value $\langle \mathcal{O}(\boldsymbol{x},t)\rangle$. The Fourier transform, $\langle \mathcal{O}(\boldsymbol{x},\omega)\rangle$, is given by the expression

$$\delta\langle \text{gnd}|\mathcal{O}(\boldsymbol{x},\omega)|\text{gnd}\rangle =$$

$$\int d^3x' \left\{ -\frac{i}{\hbar}\int_{-\infty}^{0} d\tau \,\langle \text{gnd}|[\mathcal{O}(\boldsymbol{x},t), \mathcal{O}(\boldsymbol{x}',t+\tau)]|\text{gnd}\rangle \, e^{i\omega\tau}\right\} f(\boldsymbol{x}',\omega) \quad (10.192)$$

where $f(\boldsymbol{x},\omega)$ is the Fourier transform of $f(\boldsymbol{x},t)$. Thus, the Fourier transform of the generalized susceptibility $\chi(\boldsymbol{x},\boldsymbol{x}';\omega)$ is given by

$$\chi(\boldsymbol{x},\boldsymbol{x}';\omega) = -\frac{i}{\hbar}\int_{-\infty}^{0} d\tau \, e^{i\omega\tau}\,\langle \text{gnd}|[\mathcal{O}(\boldsymbol{x},0), \mathcal{O}(\boldsymbol{x}',\tau)]|\text{gnd}\rangle \qquad (10.193)$$

which is known as the *Kubo formula*. Hence

$$\chi(\boldsymbol{x},\boldsymbol{x}';\omega) = \frac{1}{\hbar}\int_{-\infty}^{+\infty} d\tau \, e^{i\omega\tau} D^R(x,x') \qquad (10.194)$$

If we also Fourier transform the space dependence, we get

$$\langle \mathcal{O}(\boldsymbol{p},\omega)\rangle = \frac{1}{\hbar}D^R(\boldsymbol{p},\omega)f(\boldsymbol{p},\omega) \qquad (10.195)$$

The generalized susceptibility $\chi(\boldsymbol{p},\omega)$ now becomes

$$\chi(\boldsymbol{p},\omega) = \frac{\langle \mathcal{O}(\boldsymbol{p},\omega)\rangle}{f(\boldsymbol{p},\omega)} = \frac{1}{\hbar}D^R(\boldsymbol{p},\omega) \qquad (10.196)$$

In practice we will compute the *time-ordered* correlation function $D(x,x')$. Recalling our discussion of the propagator, we expect $D(\boldsymbol{p},\omega)$ to have poles on the real frequency axis. For $D(x,x')$ to be time ordered, all poles with $\omega < 0$ should be moved (infinitesimally) into the upper half of the complex frequency plane, while all poles with $\omega > 0$ should be moved into lower half-plane. Thus, $D(\boldsymbol{p},\omega)$ is not analytic on either half-plane. However, the retarded correlation function $D_R(\boldsymbol{p},\omega)$ is (given our conventions for Fourier transforms) analytic in the lower half-plane. Thus, we can relate the time-ordered correlation function $D(\boldsymbol{p},\omega)$ to the retarded correlation function $D^R(\boldsymbol{p},\omega)$ by

$$\operatorname{Re} D^R(\boldsymbol{p},\omega) = \operatorname{Re} D(\boldsymbol{p},\omega) \equiv \hbar \operatorname{Re}\chi(\boldsymbol{p},\omega) \tag{10.197}$$

$$\operatorname{Im} D^R(\boldsymbol{p},\omega) = \operatorname{Im} D(\boldsymbol{p},\omega)\ \operatorname{sign}\omega \equiv \hbar \operatorname{Im}\chi(\boldsymbol{p},\omega) \tag{10.198}$$

The time-ordered correlation function $D(\boldsymbol{p},\omega)$ (i.e., the propagator for the observable $\hat{\mathcal{O}}(\boldsymbol{x},t)$) admits a spectral (or Lehmann) representation similar to that of the propagator for the relativistic scalar field of eq. (10.160). Similarly, we can define the *spectral function* $A(\boldsymbol{p},\omega)$ of the observable to be

$$A(\boldsymbol{p},\omega) = \operatorname{Im} D^{\text{ret}}(\boldsymbol{p},\omega) \tag{10.199}$$

The relations of eq. (10.197) and eq. (10.198) imply that the susceptibility $\chi(\boldsymbol{p},\omega)$ obeys the Kramers-Krönig (or dispersion) relation

$$\operatorname{Re}\chi(p,\omega) = \frac{1}{\pi}\mathcal{P}\int_{-\infty}^{+\infty} d\omega'\ \frac{\operatorname{Im}\chi(\boldsymbol{p},\omega')}{\omega'-\omega} \tag{10.200}$$

where $\mathcal{P}$ denotes the principal value of the expression.

Finally, let us recast the formulas for a general change of an arbitrary operator into a more compact form. We can apply the formulas derived for the interaction representation just to the part of the Hamiltonian that involves the coupling to the external sources $H_{\text{ext}}(t)$. The interaction representation S-matrix is

$$S = \lim_{t\to+\infty} U(t) = T\, e^{-\frac{i}{\hbar}\int_{-\infty}^{+\infty} dt\, H_{\text{ext}}(t)} \tag{10.201}$$

Let $\langle \text{gnd, out}|\text{gnd, in}\rangle$ be the *vacuum persistence amplitude*

$$\begin{aligned}\langle \text{gnd, out}|\text{gnd, in}\rangle &= \langle \text{gnd}|S|\text{gnd}\rangle \\ &= \langle \text{gnd}|T\, e^{-\frac{i}{\hbar}\int_{-\infty}^{+\infty} dt\, H_{\text{ext}}(t)}|\text{gnd}\rangle\end{aligned} \tag{10.202}$$

which is a functional of the forces (or sources) $f(\boldsymbol{x},t)$. We will denote the vacuum persistence amplitude by $Z[f]$:

$$Z[f] = \langle \text{gnd}|T\, e^{-\frac{i}{\hbar}\int d^4x f(x)\,\mathcal{O}(x)}|\text{gnd}\rangle \tag{10.203}$$

In the limit in which the sources are weak, we can expand $Z[f]$ in powers of $f(x)$ to obtain

$$\begin{aligned}Z[f] =& 1 - \frac{i}{\hbar}\int d^4x f(x)\langle \text{gnd}|\mathcal{O}(x)|\text{gnd}\rangle \\ &+ \frac{1}{2!}\left(-\frac{i}{\hbar}\right)^2 \int d^4x \int d^4x' f(x) f(x')\langle \text{gnd}|T\,\mathcal{O}(x)\mathcal{O}(x')|\text{gnd}\rangle + \cdots\end{aligned} \tag{10.204}$$

The second term vanishes if the operator $\mathcal{O}(\boldsymbol{x}, t)$ is normal-ordered. With the same degree of precision, we can re-exponentiate the resulting expression:

$$Z[f] = e^{\frac{i}{2\hbar}\int d^4x \int d^4x' f(x) K(x,x') f(x') + O(f^3)} \tag{10.205}$$

where the kernel $K(x, x')$ is the time-ordered correlation function of the observable $A(x)$:

$$K(x, x') = \frac{i}{\hbar}\langle \text{gnd}| T\mathcal{O}(x)\mathcal{O}(x')|\text{gnd}\rangle \tag{10.206}$$

In general, the observables of physical interest are, at least, bilinear functions of the fields. Thus, the kernels $K(x, x')$ represent not one-particle propagators, but, in general, propagators for two or more excitations (i.e., they involve four-point functions of the field operators, or higher).

We can learn a lot from a physical system if the spectral functions of the kernels $K(x, x')$ are known. In general, we expect that the spectral function will have a structure similar to that of the propagator: one (or more) delta-function contributions and a branch cut. The delta functions are two (or more) particle bound states, which are known as the collective modes. The branch cuts originate from the two or multi-particle continuum. Examples of collective modes are plasmons (or sound waves) in electron liquids, spin waves in magnets, and phase modes in superconductors and superfluids.

10.10 The Kubo formula and the electrical conductivity of a metal

As an application, let us now consider the response of an electron gas to weak external electromagnetic fields $A_\mu(x)$. The formalism can be generalized easily to other systems (relativistic or not) and responses. In particular, we discuss how to relate response functions to the electrical conductivity of a metal. More general applications can be found in the classic text by Martin (1968).

There are three effects (and couplings) that we need to take into consideration: (1) electrostatic, (2) diamagnetic (or orbital), and (3) paramagnetic. The electrostatic coupling is simply the coupling of the local charge density of the electron fluid to an external (scalar) potential. In this case, H_{ext} is given by

$$H_{\text{ext}} = \sum_{\sigma=\uparrow,\downarrow} \int d^3x\, e\,\phi(\boldsymbol{x}, t)\, \psi_\sigma^\dagger(\boldsymbol{x}, t)\psi_\sigma(\boldsymbol{x}, t) \tag{10.207}$$

where $\phi(\boldsymbol{x}, t) \equiv A_0(\boldsymbol{x}, t)$ is the scalar potential (or time component) of the vector potential $A_\mu(x)$.

The diamagnetic (or orbital) coupling follows from the minimal coupling to the space components of the external vector potential $\boldsymbol{A}(x)$. The kinetic energy term H_{Kin} is modified following the minimal coupling prescription to become

$$H_{\text{Kin}}(A) = \int d^3x \sum_{\sigma=\uparrow,\downarrow} \frac{\hbar^2}{2m}\left(\boldsymbol{\nabla} + \frac{ie}{\hbar c}\boldsymbol{A}(x)\right)\psi_\sigma^\dagger(x) \cdot \left(\boldsymbol{\nabla} - \frac{ie}{\hbar c}\boldsymbol{A}(x)\right)\psi_\sigma(x) \tag{10.208}$$

which can be written as a sum of two terms

$$H_{\text{Kin}}(A) = H_{\text{Kin}}(0) + H_{\text{ext}}(A) \tag{10.209}$$

where $H_{\text{Kin}}(0)$ is the Hamiltonian in the absence of the field and $H_{\text{ext}}(A)$ is the perturbation:

$$H_{\text{ext}}(A) = \int d^3x \left[\boldsymbol{J}(\boldsymbol{x},t) \cdot \boldsymbol{A}(\boldsymbol{x},t) - \frac{e^2}{2mc^2} \boldsymbol{A}^2(\boldsymbol{x},t) \sum_\sigma \psi_\sigma^\dagger(\boldsymbol{x},t)\psi_\sigma(\boldsymbol{x},t) \right] \tag{10.210}$$

Here $\boldsymbol{J}(x)$ is the gauge-invariant charge current

$$\begin{aligned} \boldsymbol{J}(x) &= \frac{ie\hbar}{2mc} \sum_\sigma \left[\psi_\sigma^\dagger(x) \boldsymbol{\nabla} \psi_\sigma(x) - \boldsymbol{\nabla} \psi_\sigma^\dagger(x) \psi_\sigma(x) \right] - \frac{e^2}{mc^2} \boldsymbol{A}(x) \sum_\sigma \psi_\sigma^\dagger(x)\psi_\sigma(x) \\ &\equiv \frac{ie\hbar}{2mc} \sum_\sigma \left[\psi_\sigma^\dagger(x) \boldsymbol{D} \psi_\sigma(x) - (\boldsymbol{D}\psi_\sigma(x))^\dagger \psi_\sigma(x) \right] \end{aligned} \tag{10.211}$$

where $\boldsymbol{D} = \boldsymbol{\nabla} + i\frac{e}{\hbar c}\boldsymbol{A}(x)$ is the covariant derivative (in space). Clearly $\boldsymbol{J}(x)$ is the sum of two terms, the mass current and the diamagnetic term, $\frac{e^2}{mc^2}\boldsymbol{A}^2 \sum_\sigma \psi_\sigma^\dagger \psi_\sigma$.

We can write the total perturbation, including the scalar potential A_0, in the form

$$H_{\text{ext}} = \int d^3x \left[J_\mu(x) A^\mu(x) - \frac{e^2}{2mc^2} \boldsymbol{A}^2 \sum_\sigma \psi_\sigma^\dagger \psi_\sigma \right] \tag{10.212}$$

Finally, consider the paramagnetic coupling of the external magnetic field to the spin degrees of freedom of the system, which has the form of a Zeeman interaction

$$H_{\text{ext}}^{\text{Zeeman}} = \int d^3x\, g\boldsymbol{B}(x) \cdot \sum_{\sigma,\sigma'} \psi_\sigma^\dagger(x) \boldsymbol{S}_{\sigma\sigma'} \psi_{\sigma'}(x) \tag{10.213}$$

where g is typically of the order of the Bohr magneton μ_B, and $\boldsymbol{S} = \frac{\hbar}{2}\boldsymbol{\sigma}$ for spin-1/2 systems.

A straightforward application of the linear response formulas derived above yields the following expression for the expectation value of the current $\langle J_\mu \rangle'$ in the presence of the perturbation:

$$\langle J_\mu(x) \rangle' = \langle J_\mu(x) \rangle_{\text{gnd}} - \frac{i}{\hbar} \int_{-\infty}^{t} dt' \int d^3x' \langle \text{gnd} | [J_\nu(x'), J_\mu(x)] | \text{gnd} \rangle A_\nu(x') \tag{10.214}$$

This formula suggests that we should define the *retarded current correlation function* $\mathcal{D}_{\mu\nu}^{\text{ret}}(x,x')$:

$$\mathcal{D}_{\mu\nu}^{\text{ret}}(x,x') = -i\Theta(x_0 - x_0') \langle \text{gnd} | [J_\mu(x), J_\nu(x')] | \text{gnd} \rangle \tag{10.215}$$

The *induced current* $\langle J_\mu \rangle_{\text{ind}}$

$$\langle J_\mu \rangle_{\text{ind}} = \langle J_\mu \rangle' - \langle j_\mu \rangle_{\text{gnd}} \tag{10.216}$$

(where j_μ is the mass current) has a very simple form in terms of the retarded correlation function $\mathcal{D}_{\mu\nu}^{\text{ret}}(x,x')$:

$$\langle J_\mu(x) \rangle_{\text{ind}} = \frac{1}{\hbar} \int d^4x' \mathcal{D}_{\mu\nu}^{\text{Ret}}(x,x') A^\nu(x') - \frac{e^2}{mc^2} A_k(x) \langle n(x) \rangle \delta_{\mu k} + O(A^2) \tag{10.217}$$

Below we will see that $\langle J_\mu(x) \rangle_{\text{ind}}$ is conserved (i.e., $\partial_\mu^x \langle J^\mu(x) \rangle_{\text{ind}} = 0$) and gauge invariant.

We can express these results in terms of an effective action for the external electromagnetic field A_μ,

$$Z_{\rm eff}[A_\mu] = \mathcal{N} e^{\frac{i}{2\hbar}\int d^4x \int d^4y\, A_\mu(x)\Pi^{\mu\nu}(x-y)A_\nu(y)} \tag{10.218}$$

such that

$$\langle J_\mu(x)\rangle_{\rm ind} = \frac{\hbar}{i}\frac{1}{Z_{\rm eff}[A_\mu]}\frac{\delta Z_{\rm eff}[A_\mu]}{\delta A_\mu(x)} = \int d^4y\, \Pi_{\mu\nu}(x-y)A^\nu(y) \tag{10.219}$$

By inspection, we see that the effective *polarization tensor* $\Pi_{\mu\nu}(x-y)$ is related to the current correlation function $\mathcal{D}_{\mu\nu}(x-y)$,

$$\Pi_{\mu\nu}(x-y) = \frac{1}{\hbar}\mathcal{D}_{\mu\nu}(x-y) + \frac{e\rho}{mc^2}\delta(x-y)\tilde{g}_{\mu\nu} \tag{10.220}$$

where we introduced the diagonal tensor

$$\tilde{g}_{\mu\nu} = \begin{cases} 0, & \text{if } \mu=0 \text{ and/or } \nu=0 \\ -\delta_{ij}, & \text{if } \mu,\nu = i,j \end{cases} \tag{10.221}$$

Since $\langle J_\mu(x)\rangle_{\rm ind}$ is gauge invariant, we can compute its form in any gauge. In the gauge $A_0=0$, the spatial components of $\langle J_\mu(x)\rangle_{\rm ind}$ are

$$\langle J_k(x)\rangle_{\rm ind} = -\frac{e\rho}{mc^2}A_k(x) + \int d^4x' \mathcal{D}^{\rm ret}_{k\ell}(x-x')A_\ell(x') + O(A^2) \tag{10.222}$$

where $\rho = e\langle n\rangle$ is the expectation value of the charge density, and we will assume that it is uniform.

In the $A_0=0$ gauge, the external electric field $\boldsymbol{E}_{\rm ext}$ and magnetic field $\boldsymbol{H}$ are

$$\boldsymbol{E}_{\rm ext} = -\partial_0 \boldsymbol{A}, \qquad \boldsymbol{H} = \nabla\times\boldsymbol{A} \tag{10.223}$$

Now, in Fourier space, we can write

$$\begin{aligned}\langle J_k(\boldsymbol{p},\omega)\rangle_{\rm ind} &= -\frac{e^2\langle n\rangle}{mc^2}A_k(\boldsymbol{p},\omega) + \mathcal{D}^{\rm ret}_{k\ell}(\boldsymbol{p},\omega)A_\ell(\boldsymbol{p},\omega)\\ &\equiv \left(\mathcal{D}^{\rm ret}_{k\ell}(\boldsymbol{p},\omega) - \frac{e^2\langle n\rangle}{mc^2}\delta_{k\ell}\right)\frac{E^{\rm ext}_\ell}{i\omega}(\boldsymbol{p},\omega)\end{aligned} \tag{10.224}$$

This expression is almost the conductivity. It is not quite that, since the conductivity is a relation between the total current $\boldsymbol{J} = \boldsymbol{J}_{\rm ind} + \boldsymbol{J}_{\rm ext}$ and the total electric field $\boldsymbol{E}$. To take these electromagnetic effects into account, we must use Maxwell's equations in a medium, which involves the vector fields $\boldsymbol{E}$, $\boldsymbol{D}$, $\boldsymbol{B}$, and $\boldsymbol{H}$

$$\begin{aligned} \nabla\cdot\boldsymbol{D} &= \rho, & \nabla\times\boldsymbol{E} &= -\frac{\partial \boldsymbol{H}}{\partial t}\\ \nabla\cdot\boldsymbol{B} &= 0, & \nabla\times\boldsymbol{H} &= \frac{\partial \boldsymbol{E}}{\partial t} + \boldsymbol{J}\end{aligned} \tag{10.225}$$

where

$$\boldsymbol{B}=\boldsymbol{H}+\boldsymbol{M}, \qquad \boldsymbol{E}=\boldsymbol{E}^{\text{ext}}+\boldsymbol{E}^{\text{ind}} \tag{10.226}$$

Here $\boldsymbol{M}$ and $\boldsymbol{E}^{\text{ind}}$ are the magnetic and electric polarization vectors, respectively. In particular,

$$\boldsymbol{J}^{\text{ind}}=\partial_t \boldsymbol{E}^{\text{ind}} \tag{10.227}$$

and

$$\partial_t \boldsymbol{D}=\partial_t \boldsymbol{E}+\boldsymbol{J}^{\text{ind}} \tag{10.228}$$

Linear response theory is the statement that $\boldsymbol{D}$ (not to be confused with the covariant derivative!) must be proportional to $\boldsymbol{E}$,

$$D_j=\varepsilon_{jk}E_k \tag{10.229}$$

where ε_{jk} is the dielectric tensor.

$\boldsymbol{E}$ and $\boldsymbol{E}^{\text{ext}}$ satisfy similar equations:

$$\begin{aligned} -\boldsymbol{\nabla}\times\boldsymbol{\nabla}\times\boldsymbol{E}&=\partial_t^2\boldsymbol{E}+\partial_t\boldsymbol{J}\\ -\boldsymbol{\nabla}\times\boldsymbol{\nabla}\times\boldsymbol{E}^{\text{ext}}&=\partial_t^2\boldsymbol{E}^{\text{ext}}+\partial_t\boldsymbol{J}^{\text{ext}} \end{aligned} \tag{10.230}$$

Since $\boldsymbol{\nabla}\times\boldsymbol{\nabla}\times\boldsymbol{E}=\boldsymbol{\nabla}(\boldsymbol{\nabla}\cdot\boldsymbol{E})-\nabla^2\boldsymbol{E}$, we can write, in Fourier space,

$$\begin{aligned} p_ip_jE_j(\boldsymbol{p},\omega)-\boldsymbol{p}^2E_i(\boldsymbol{p},\omega)&=-\,\omega^2E_j(\boldsymbol{p},\omega)-i\omega J_i(\boldsymbol{p},\omega)\\ p_ip_jE_j^{\text{ext}}(\boldsymbol{p},\omega)-\boldsymbol{p}^2E_i^{\text{ext}}(\boldsymbol{p},\omega)&=-\,\omega^2E_i^{\text{ext}}(\boldsymbol{p},\omega)-i\omega J_i^{\text{ext}}(\boldsymbol{p},\omega) \end{aligned} \tag{10.231}$$

Thus, we get

$$\begin{aligned} p_ip_jE_j(\boldsymbol{p},\omega)-\boldsymbol{p}^2E_i(\boldsymbol{p},\omega)+\omega^2E_i(\boldsymbol{p},\omega)&=-i\omega J_i^{\text{ind}}(\boldsymbol{p},\omega)\\ &\quad+p_ip_jE_j^{\text{ext}}(\boldsymbol{p},\omega)-\boldsymbol{p}^2E_i^{\text{ext}}(\boldsymbol{p},\omega)+\omega^2E_i^{\text{ext}}(\boldsymbol{p},\omega) \end{aligned} \tag{10.232}$$

and

$$-\,i\omega J_i^{\text{ind}}(\boldsymbol{p},\omega)=\left(\delta_{ij}\frac{e^2\langle n\rangle}{mc^2}-\mathcal{D}_{ij}^{\text{ret}}(\boldsymbol{p},\omega)\right)E_j^{\text{ext}}(\boldsymbol{p},\omega) \tag{10.233}$$

From these results we conclude that

$$\begin{aligned} &(p_ip_j-\boldsymbol{p}^2\delta_{ij}+\omega^2\delta_{ij})\,E_j(\boldsymbol{p},\omega)=\\ &\qquad\left(\delta_{ij}\frac{e^2\langle n\rangle}{mc^2}-\mathcal{D}_{ij}^{\text{ret}}(\boldsymbol{p},\omega)+p_ip_j-\boldsymbol{p}^2\delta_{ij}+\omega^2\delta_{ij}\right)E_j^{\text{ext}}(\boldsymbol{p},\omega) \end{aligned} \tag{10.234}$$

In matrix form, these equations have the simpler form:

$$(\boldsymbol{p}\otimes\boldsymbol{p}-\boldsymbol{p}^2I+\omega^2I)\boldsymbol{E}(\boldsymbol{p},\omega)=\left(\frac{e^2\langle n\rangle}{mc^2}I-\mathcal{D}^{\text{ret}}+\boldsymbol{p}\otimes\boldsymbol{p}-\boldsymbol{p}^2I+\omega^2I\right)\boldsymbol{E}_{\text{ext}}(\boldsymbol{p},\omega) \tag{10.235}$$

This equation allows us to write $\boldsymbol{E}_{\text{ext}}$ in terms of $\boldsymbol{E}$. We find that the induced current is

$$i\omega \boldsymbol{J}_{\text{ind}}(\boldsymbol{p},\omega) =$$
$$\left(\mathcal{D}^{\text{ret}} - \frac{e^2\langle n\rangle}{mc^2} I\right)\left[\frac{e^2\langle n\rangle}{mc^2} I - \mathcal{D}^{\text{ret}} + \boldsymbol{p}\otimes\boldsymbol{p} - \boldsymbol{p}^2 I + \omega^2 I\right]^{-1} (\boldsymbol{p}\otimes\boldsymbol{p} - \boldsymbol{p}^2 I + \omega^2 I)\, \boldsymbol{E}(\boldsymbol{p},\omega) \tag{10.236}$$

and we find that the *conductivity tensor* σ_{jk} is

$$\begin{aligned} i\omega\sigma(\boldsymbol{p},\omega) = &\left(\mathcal{D}^{\text{ret}}(\boldsymbol{p},\omega) - \frac{e^2\langle n\rangle}{mc^2} I\right) + \left(\mathcal{D}^{\text{ret}}(\boldsymbol{p},\omega) - \frac{e^2\langle n\rangle}{mc^2} I\right) \\ &\times \left[\frac{e^2\langle n\rangle}{mc^2} I - \mathcal{D}^{\text{ret}}(\boldsymbol{p},\omega) + p\otimes p - \boldsymbol{p}^2 I + \omega^2 I\right]^{-1} \left(\mathcal{D}^{\text{ret}}(\boldsymbol{p},\omega) - \frac{e^2\langle n\rangle}{mc^2} I\right) \end{aligned} \tag{10.237}$$

Also, since $\boldsymbol{D} = \varepsilon \boldsymbol{E}$, the dielectric tensor ε_{jk} and the conductivity tensor σ_{jk} are related by

$$\varepsilon = I + \frac{i}{\omega}\sigma \tag{10.238}$$

10.11 Correlation functions and conservation laws

In the problem discussed in section 10.10, we saw that we had to consider a correlation function of currents. Since the currents are conserved,

$$\partial_\mu J^\mu = 0 \tag{10.239}$$

we expect that the correlation function $\mathcal{D}_{\mu\nu}(x,x')$ should obey a similar equation. Let us compute the divergence of the retarded correlation function:

$$\partial_x^\mu \mathcal{D}^{\text{ret}}_{\mu\nu}(x,x') = \partial_x^\mu \left(-i\Theta(x_0 - x_0')\langle \text{gnd}|[J_\mu(x), J_\nu(x')]|\text{gnd}\rangle\right) \tag{10.240}$$

Except for the contribution coming from the step function, it is apparent that we can operate with the derivative inside the expectation value to get

$$\begin{aligned} \partial_x^\mu \mathcal{D}^{\text{ret}}_{\mu\nu}(x,x') = &- i\left(\partial_x^\mu \Theta(x_0 - x_0')\right)\, \langle \text{gnd}|[J_\mu(x), J_\nu(x')]|\text{gnd}\rangle \\ &- i\Theta(x_0 - x_0')\langle \text{gnd}|[\partial_x^\mu J_\mu(x), J_\nu(x')]|\text{gnd}\rangle \end{aligned} \tag{10.241}$$

The second term vanishes since $J_\mu(x)$ is a conserved current, and the first term is nonzero only if $\mu = 0$. Hence we obtain the identity

$$\partial_x^\mu \mathcal{D}^{\text{ret}}_{\mu\nu}(x,x') = -i\delta(x_0 - x_0')\langle \text{gnd}|\left[J_0(x), J_\nu(x')\right]|\text{gnd}\rangle \tag{10.242}$$

which is the v.e.v. of an equal-time commutator.

These commutators are given by (Kadanoff and Martin, 1961)

$$\begin{aligned}
&\langle \text{gnd}| \left[J^0(\boldsymbol{x}, x_0), J^0(\boldsymbol{x}, x_0)\right] |\text{gnd}\rangle = 0 \\
&\langle \text{gnd}| \left[J^0(\boldsymbol{x}, x_0), J^i(\boldsymbol{x}', x_0)\right] |\text{gnd}\rangle = \frac{ie^2}{mc^2} \partial_k^x \left[\delta(\boldsymbol{x}-\boldsymbol{x}')\langle n(\boldsymbol{x})\rangle\right]
\end{aligned} \tag{10.243}$$

Identities of this type also arise in relativistic systems (e.g., the theory of Dirac fermions), where they are known as Schwinger terms. However, the Schwinger terms of relativistic QFT have a subtle origin (Schwinger, 1959). They are related to *anomalies*, and we will discuss them in chapter 20. In the nonrelativistic case, they arise because in these systems, the Fermi sea has a bottom, and hence there is always a high-energy regulator (or cutoff) that preserves gauge invariance.

Hence, the divergence of $\mathcal{D}^{\text{ret}}_{\mu\nu}$ is

$$\begin{aligned}
&\partial_x^\mu \mathcal{D}^{\text{ret}}_{\mu k}(x, x') = \frac{e^2}{mc^2} \partial_k^x \left[\delta^4(x-x')\langle n(x)\rangle\right], \qquad \partial_{x'}^\mu \mathcal{D}^{\text{ret}}_{0\mu}(x, x') = 0 \\
&\partial_{x'}^\nu \mathcal{D}^{\text{ret}}_{k\nu}(x, x') = -\frac{e^2}{mc^2} \partial_k^x \left[\delta^4(x-x')\langle n(x')\rangle\right], \qquad \partial_{x'}^\mu \mathcal{D}^{\text{ret}}_{0\mu}(x, x') = 0
\end{aligned} \tag{10.244}$$

Notice that the time-ordered functions also satisfy the same identities. These identities can be used to prove that $\langle \boldsymbol{J}^{\text{ind}}\rangle$ is indeed gauge invariant and conserved.

However, by comparing eq. (10.220) (which relates the polarization tensor $\Pi_{\mu\nu}$ with the current correlation function $\mathcal{D}_{\mu\nu}$) with eq. (10.244), we see that the presence of the Schwinger terms in the divergence of $\mathcal{D}_{\mu\nu}$ ensures that the polarization tensor $\Pi_{\mu\nu}$ is conserved and hence transverse,

$$\partial^\mu \Pi_{\mu\nu} = 0 \tag{10.245}$$

as required by gauge invariance. We will return to this issue in section 12.6, where we discuss the relation between Ward identities and gauge invariance.

Furthermore, in momentum and frequency space, the identities become

$$\begin{aligned}
&-i\omega \mathcal{D}^{\text{ret}}_{00}(\boldsymbol{p}, \omega) - ip_k \mathcal{D}^{\text{ret}}_{k0}(\boldsymbol{p}, \omega) = 0 \\
&-i\omega \mathcal{D}^{\text{ret}}_{0k}(\boldsymbol{p}, \omega) - ip_\ell \mathcal{D}^{\text{ret}}_{\ell k}(\boldsymbol{p}, \omega) = -\frac{e^2\langle n\rangle}{mc^2} ip_k \\
&-i\omega \mathcal{D}^{\text{ret}}_{00}(\boldsymbol{p}, \omega) - ip_k \mathcal{D}^{\text{ret}}_{0k}(\boldsymbol{p}, \omega) = 0 \\
&-i\omega \mathcal{D}^{\text{ret}}_{k0}(\boldsymbol{p}, \omega) - ip_\ell \mathcal{D}^{\text{ret}}_{k\ell}(\boldsymbol{p}, \omega) = -\frac{e^2\langle n\rangle}{mc^2} ip_k
\end{aligned} \tag{10.246}$$

We can combine them to obtain the result

$$\omega^2 \mathcal{D}^{\text{ret}}_{00}(\boldsymbol{p}, \omega) - p_\ell p_k \mathcal{D}^{\text{ret}}_{\ell k}(\boldsymbol{p}, \omega) = -\frac{e^2\langle n\rangle}{mc^2} \boldsymbol{p}^2 \tag{10.247}$$

Hence, the density-density and the current-current correlation functions are not independent.

A number of interesting identities follow from this equation. In particular, if we take the static limit $\omega \to 0$ at fixed momentum $\boldsymbol{p}$, we get

$$\lim_{\omega\to 0} p_\ell p_k \mathcal{D}^{\text{ret}}_{\ell k}(\boldsymbol{p},\omega) = \frac{e^2\bar{n}}{mc^2}\boldsymbol{p}^2 \tag{10.248}$$

provided that $\lim_{\omega\to 0}\mathcal{D}^{\text{ret}}_{00}(\boldsymbol{p},\omega)$ is not singular for $\boldsymbol{p}\neq 0$. Also, from the equal-time commutator,

$$\langle \text{gnd}|[J_k(\boldsymbol{x},x_0),J_0(\boldsymbol{x},x_0)]|\text{gnd}\rangle = \frac{ie^2}{mc^2}\partial^x_k\left(\delta(\boldsymbol{x}-\boldsymbol{x}')\langle n(\boldsymbol{x})\rangle\right) \tag{10.249}$$

we get

$$\lim_{x_0'\to x_0}\partial^x_k\mathcal{D}^{\text{ret}}_{k0}(x,x') = \frac{e^2}{mc^2}\nabla^2_x\left(\delta(\boldsymbol{x}-\boldsymbol{x}')\langle n(\boldsymbol{x})\rangle\right) \tag{10.250}$$

If the system is uniform, $\langle n(\boldsymbol{x})\rangle = \bar{n}$, we can Fourier transform this last identity to obtain

$$\int_{-\infty}^{+\infty}\frac{d\omega}{2\pi} ip_k\mathcal{D}^{\text{ret}}_{k0}(\boldsymbol{p},\omega) = -\frac{e^2\bar{n}}{mc^2}\boldsymbol{p}^2 \tag{10.251}$$

Using the conservation laws, we find the identity

$$\int_{-\infty}^{+\infty}\frac{d\omega}{2\pi} i\omega\mathcal{D}^{\text{ret}}_{00}(\boldsymbol{p},\omega) = \frac{e^2\bar{n}}{mc^2}\boldsymbol{p}^2 \tag{10.252}$$

This identity is known as the *f-sum rule*.

If the system is isotropic, these relations can be used to yield a simpler form for the conductivity tensor. Indeed, for an isotropic system, $\mathcal{D}^{\text{ret}}_{k\ell}(\boldsymbol{p},\omega)$ can only have the form of a sum of a longitudinal part $\mathcal{D}^{\text{ret}}_{\parallel}$ and a transverse part $\mathcal{D}^{\text{ret}}_{\perp}$:

$$\mathcal{D}^{\text{ret}}_{\ell k}(\boldsymbol{p},\omega) = \mathcal{D}^{\text{ret}}_{\parallel}(\boldsymbol{p},\omega)\frac{p_\ell p_k}{\boldsymbol{p}^2} + \mathcal{D}^{\text{ret}}_{\perp}(\boldsymbol{p},\omega)\left(\frac{p_\ell p_k}{\boldsymbol{p}^2} - \delta_{\ell k}\right) \tag{10.253}$$

Thus, we get a relation between $\mathcal{D}^{\text{ret}}_{00}$ and $\mathcal{D}^{\text{ret}}_{\parallel}$:

$$\omega^2\mathcal{D}^{\text{ret}}_{00}(\boldsymbol{p},\omega) - \boldsymbol{p}^2\mathcal{D}^{\text{ret}}_{\parallel}(\boldsymbol{p},\omega) = -\frac{e^2\bar{n}}{mc^2}\boldsymbol{p}^2 \tag{10.254}$$

Hence

$$\mathcal{D}^{\text{ret}}_{00}(\boldsymbol{p},\omega) = \frac{\boldsymbol{p}^2}{\omega^2}\left(\mathcal{D}^{\text{ret}}_{\parallel}(\boldsymbol{p},\omega) - \frac{e^2\bar{n}}{mc^2}\right) \tag{10.255}$$

and

$$\lim_{\omega\to 0}\mathcal{D}^{\text{ret}}_{\parallel}(\boldsymbol{p},\omega) = \frac{e^2\bar{n}}{mc^2} \tag{10.256}$$

for all $\boldsymbol{p}$.

The conductivity tensor can also be separated into longitudinal ($\sigma_\parallel$) and transverse ($\sigma_\perp$) pieces:

$$\sigma_{ij} = \sigma_\parallel\frac{p_i p_j}{\boldsymbol{p}^2} + \sigma_\perp\left(\frac{p_i p_j}{\boldsymbol{p}^2} - \delta_{ij}\right) \tag{10.257}$$

We find

$$\sigma_{\parallel} = i\omega \left[\frac{\mathcal{D}_{\parallel}^{\text{ret}} - \frac{e^2 \bar{n}}{mc^2}}{-\mathcal{D}_{\parallel}^{\text{ret}} + \frac{e^2 \bar{n}}{mc^2} + \omega^2} \right] \tag{10.258}$$

and

$$\sigma_{\perp} = \frac{1}{i\omega} \left(\mathcal{D}_{\perp}^{\text{ret}} - \frac{e^2 \bar{n}}{mc^2} \right) \left[1 + \frac{\mathcal{D}_{\perp}^{\text{ret}} - \frac{e^2 \bar{n}}{mc^2}}{\frac{e^2 \bar{n}}{mc^2} - \mathcal{D}_{\perp}^{\text{ret}} + \omega^2 - \boldsymbol{p}^2} \right] \tag{10.259}$$

These relations tell us that the real part of $\sigma_{\parallel}$ is determined by the imaginary part of $\mathcal{D}_{\parallel}^{\text{ret}}$. Thus, the *resistive part* (the real part) of the longitudinal conductivity $\sigma_{\parallel}$, which reflects the dissipation in the system, is determined by the imaginary part of a response function. This general result is known as the fluctuation-dissipation theorem.

10.12 The Dirac propagator in a background electromagnetic field

Let us consider briefly the Dirac propagator in a background electromagnetic field and use it to compute the S-matrix for Coulomb scattering. By a "background field," we mean a classical (fixed but possibly time-dependent) electromagnetic field $A_\mu(x)$. Denote by $S_F(x, x'|A)$ the propagator for a free Dirac field in a background gauge field A_μ.

$S_F(x, x'|A)$ obeys the Green function equation

$$(i\partial\!\!\!/ - eA\!\!\!/ - m)\, S_F(x, x'|A) = \delta^4(x - x') \tag{10.260}$$

In the absence of a background field, the Dirac propagator $S_F(x, x')$ obeys instead

$$(i\partial\!\!\!/ - m)\, S_F(x, x') = \delta^4(x - x') \tag{10.261}$$

Thus, we can also write eq. (10.260) as

$$\left(S_F^{-1} - eA\!\!\!/\right) S_F(A) = 1 \tag{10.262}$$

Hence

$$S_F(x, x'|A) = S_F(x - x') + e \int d^4y\, S_F(x - y) A\!\!\!/(y)\, S_F(y, x'|A) \tag{10.263}$$

or, in components,

$$S_F^{\alpha\beta}(x, x'|A) = S_F^{\alpha\beta}(x - x') + e \int d^4y\, S_F^{\alpha\lambda}(x - y) \left[A_\mu(y)\gamma^\mu\right]^{\lambda\sigma} S_F^{\sigma\beta}(y, x'|A) \tag{10.264}$$

As an explicit application, consider the case of Coulomb scattering of (free) Dirac electrons from a fixed nucleus with positive electric charge Ze. Let us now compute the S-matrix for this problem in the Born approximation. As in nonrelativistic quantum mechanics, in this approximation, we replace the propagator in the integrand of eqs. (10.263) and (10.264) by the free Dirac propagator, $S_F(x - x')$.

Consider now an incoming state, a spinor denoted by $\Psi_i(x)$, with a particle with positive energy (an electron) and spin up (say, in the z direction), and momentum $\boldsymbol{p}_i$. This incoming (initial) state is (for $x_0 \to -\infty$)

$$\Psi_i(x) = \frac{1}{\sqrt{V}}\, u^{(\alpha)}(p_i)\,\sqrt{\frac{m}{E_i}}\; e^{-ip_i \cdot x} \tag{10.265}$$

The outgoing (final) state $\Psi_f(x)$ is a spinor representing a particle with positive energy (an electron) with spin up (also in the z direction) and momentum $\boldsymbol{p}_f$, and it is given by

$$\Psi_f(y) = \frac{1}{\sqrt{V}}\, u^{(\beta)}(p_f)\,\sqrt{\frac{m}{E_f}}\; e^{-ip_f \cdot y} \tag{10.266}$$

The S-matrix is

$$S_{fi} = i \lim_{x_0\to-\infty}\,\lim_{y_0\to+\infty} \int d^3x \int d^3y\, \overline{\Psi}_f(\boldsymbol{y}, y_0)\, S_F(y, x|A)\, \Psi_i(\boldsymbol{x}, x_0) \tag{10.267}$$

At the level of the Born approximation, we can write

$$\begin{aligned} S_{fi} = {} & i \lim_{x_0\to-\infty}\,\lim_{y_0\to+\infty} \int d^3x \int d^3y\, \overline{\Psi}_f(\boldsymbol{y}, y_0)\, S_F(y, x)\, \Psi_i(\boldsymbol{x}, x_0) \\ & + i \lim_{x_0\to-\infty}\,\lim_{y_0\to+\infty} \int d^3x \int d^3y \int d^4z\, \overline{\Psi}_f(\boldsymbol{y}, y_0)\, S_F(y, z)\, A\!\!\!/(z)\, S_F(z, x)\, \Psi_i(\boldsymbol{x}, x_0) \\ & + \cdots \end{aligned} \tag{10.268}$$

Recall the expression for the free Dirac propagator:

$$\begin{aligned} S_F(x - x') = {} & -i\langle 0|T\psi_\alpha(x)\bar{\psi}_{\alpha'}(x')|0\rangle \\ = {} & -i \int \frac{d^3p}{(2\pi)^3} \left(\frac{m}{E(p)}\right) \Big(\Theta(x_0' - x_0)e^{-ip\cdot(x'-x)}\Lambda_+(p) \\ & + \Theta(x_0 - x_0')e^{-ip\cdot(x-x')}\Lambda_-(p)\Big) \end{aligned} \tag{10.269}$$

where $\Lambda_\pm(p)$ are projection operators onto positive (particle) and negative (antiparticle) energy states:

$$\Lambda_\pm(p) = \frac{1}{2m}\left(\pm p\!\!\!/ + m\right) \tag{10.270}$$

Alternatively, we can express the propagator in terms of the basis spinors $u^\sigma(p)$ (which span the positive energy states) and $v^\sigma(p)$ (which span the negative energy states), as

$$\begin{aligned} S_F(x' - x) = {} & -i\Theta(x_0' - x_0) \int d^3p \sum_{\sigma=1,2} u_p^{(\sigma)}(x')\bar{u}_p^{(\sigma)}(x) \\ & + i\Theta(x_0 - x_0') \int d^3p \sum_{\sigma=1,2} v_p^{(\sigma)}(x')\bar{v}_p^{(\sigma)}(x) \end{aligned} \tag{10.271}$$

where we have used the notation

$$u_p^{(\sigma)}(x) \equiv u^{(\sigma)}(p)\, e^{-ip\cdot x}, \quad v_p^{(\sigma)}(x) = v^{(\sigma)}(p)\, e^{-ip\cdot x} \tag{10.272}$$

Let us begin by computing the first term in eq. (10.268), the projection of the free propagator onto the initial and final states. By expanding the propagator, we find

$$\begin{aligned}\int d^3x \int d^3y\, \bar{\psi}_f(y) S_F(y-x)\psi_i(x) = \\ -i\Theta(y_0-x_0)\int d^3p \sum_{\sigma=1,2}\int d^3x d^3y\, \bar{\psi}_f(y) u_p^{(\sigma)}(y)\bar{u}_p^{(\sigma)}(x)\psi_i(x) \\ +i\Theta(x_0-y_0)\int d^3p \sum_{\sigma=1,2}\int d^3x d^3y\, \bar{\psi}_f(y) v_p^{(\sigma)}(y)\bar{v}_p^{(\sigma)}(x)\psi_i(x)\end{aligned} \tag{10.273}$$

We now use the orthogonality relations of the Dirac basis spinors to find

$$\begin{aligned}\int d^3y\, \bar{\psi}_f^{(\beta)}(y) u_p^{(\sigma)}(y) =& \delta^{\beta\sigma}\delta^3(\mathbf{p}-\mathbf{p}_f) \\ \int d^3x\, \bar{u}_p^{(\sigma)}(x)\psi_i^{(\alpha)}(x) =& \delta^{\sigma\alpha}\delta^3(\mathbf{p}-\mathbf{p}_i)\end{aligned} \tag{10.274}$$

Hence, to leading order, the matrix element S_{fi} of the S-matrix is

$$S_{fi} = \delta^3(\mathbf{p}_f - \mathbf{p}_i)\delta^{\alpha\beta} + \text{Born term} \tag{10.275}$$

Let us now compute the Born term (the first Born approximation). We first need to compute an expression for

$$\int d^3y\, \bar{\psi}_f(y) S_F(y,z) \tag{10.276}$$

and for

$$\int d^3x\, S_F(z,x)\psi_i(x) \tag{10.277}$$

Using once again the expansion of the propagator, we find that eq. (10.276) is

$$\begin{aligned}&\int d^3y\, \bar{\psi}_f(y) S_F(y,z) = \\ &\int d^3y\, \bar{\psi}_f^{(\beta)}(y)(-i)\Theta(y_0-z_0)\sum_{\sigma=1,2}\int d^3p\, u_p^{(\sigma)}(y)\bar{u}_p^{(\sigma)}(z) \\ &+\int d^3y\, \bar{\psi}_f^{(\beta)}(i)\Theta(z_0-y_0)\sum_{\sigma=1,2} v_p^{(\sigma)}(y)\bar{v}_p^{(\sigma)}(z)\end{aligned}$$

$$
\begin{aligned}
&= -\,i\Theta(y_0 - z_0) \sum_{\sigma=1,2} \int d^3p \left(\int d^3y\, \bar{\psi}_f^{(\beta)}(y) u_p^{(\sigma)}(y) \right) \bar{u}_p^{(\sigma)}(z) \\
&\quad + i\Theta(z_0 - y_0) \sum_{\sigma=1,2} \int d^3p \left(\int d^3y\, \bar{\psi}_f^{(\beta)}(y) v_p^{(\sigma)}(y) \right) \bar{v}_p^{(\sigma)}(z) \\
&= \begin{cases} -i\Theta(y_0 - z_0)\, \bar{\psi}_f^{(\beta)}(z), & \text{if the final state is a particle} \\ +i\Theta(z_0 - y_0)\, \bar{\psi}_f^{(\beta)}(z), & \text{if the final state is an antiparticle} \end{cases}
\end{aligned}
\tag{10.278}
$$

The other expression, eq. (10.277), can be computed similarly. Putting it all together, we find that the Born term is

$$
\text{Born term} = -ie \int d^4z\, \bar{\psi}_f^{(\beta)}(z) \not{A} \psi_i^{(\alpha)}(z)\, \Theta(y_0 - z_0)\, \Theta(z_0 - x_0) \tag{10.279}
$$

corresponding to an electron propagating forward in time.

Let us evaluate this expression for the case of a Coulomb potential,

$$
A^\mu = (A_0, 0), \quad A_0 = \frac{-Ze}{4\pi r} \tag{10.280}
$$

with $r = |\boldsymbol{z}|$. The Born term now becomes

$$
\begin{aligned}
\text{Born term} &= -\,ie \int_{-\infty}^{\infty} dz_0 \int d^3z\, \bar{\psi}_f^{(\beta)}(z) \gamma_0 \psi_i^{(\alpha)}(z)\, \Theta(y_0 - z_0) \left(\frac{-Ze}{4\pi r} \right) \\
&= \frac{ie}{V} \frac{m}{\sqrt{E_i E_f}} \frac{Ze}{4\pi} \int_{-\infty}^{\infty} dz_0 \int d^3z\, e^{i(p_f - p_i)\cdot z}\, \frac{1}{r}\, \bar{u}^{(\beta)}(p_f) \gamma_0 u^{(\alpha)}(p_i)
\end{aligned}
\tag{10.281}
$$

where V is the volume. Using that

$$
\int_{-\infty}^{\infty} dz_0\, e^{i(E_f - E_i) z_0} = 2\pi\, \delta(E_f - E_i) \tag{10.282}
$$

we can write the Born term as

$$
\text{Born term} = \frac{iZ\alpha}{V} \frac{m}{\sqrt{E_i E_f}\, 2\pi\, \delta(E_i - E_f)} \int d^3r\, \frac{e^{-i\boldsymbol{q}\cdot \boldsymbol{r}}}{r}\, \bar{u}^{(\beta)}(p_f) \gamma_0 u^{(\alpha)}(p_i) \tag{10.283}
$$

where $\alpha = \frac{e^2}{4\pi}$ is the fine structure constant, $\boldsymbol{q} = \boldsymbol{p}_f - \boldsymbol{p}_i$ is the momentum transfer, and

$$
\int d^3r\, \frac{e^{-i\boldsymbol{q}\cdot \boldsymbol{r}}}{r} = \frac{4\pi}{|\boldsymbol{q}|} \tag{10.284}
$$

The matrix element of the S-matrix, in the Born approximation, is then equal to

$$
S_{fi} = \delta^{\alpha\beta} \delta^3(\boldsymbol{p}_f - \boldsymbol{p}_i) + i \frac{Z\alpha}{V} \frac{M}{\sqrt{E_i E_f}}\, 2\pi\, \delta(E_i - E_f)\, \frac{4\pi}{q^2}\, \bar{u}^{(\beta)}(p_f) \gamma_0 u^{(\alpha)}(p_i) \tag{10.285}
$$

Since

$$\# \text{ states with } \boldsymbol{p}_f \text{ within } d^3p_f = V\frac{d^3p_f}{(2\pi)^3} \tag{10.286}$$

we can write the transition probability per particle into these final states as

$$|S_{fi}|^2\, V\,\frac{d^3p_f}{(2\pi)^3} = Z^2\frac{(4\pi\alpha)^2}{E_iV}\, m^2\,\frac{|\bar{u}^{(\beta)}(p_f)\gamma_0 u^{(\alpha)}(p_i)|^2}{|\boldsymbol{q}|^4}\,\frac{d^3p_f}{(2\pi)^3E_f}\, 2\pi\,\delta(E_f-E_i)\,T \tag{10.287}$$

where T is the time of measurement (this is Fermi's golden rule).

Thus, the number of transitions per particle and unit time is

$$\frac{dP_{fi}}{dt} = \int \left|\frac{iZ\alpha}{V}\,\frac{m}{\sqrt{E_fE_i}}\,\frac{4\pi}{|\boldsymbol{q}|^2}\,\bar{u}^{(\beta)}(p_f)\gamma_0 u^{(\alpha)}(p_i)\right|^2 2\pi\,\delta(E_f-E_i)\,V\frac{d^3p_f}{(2\pi)^3} \tag{10.288}$$

Dividing out this expression by the incoming flux, $\frac{1}{V}\frac{|\boldsymbol{p}_i|}{E_i}$, we obtain an expression for the differential cross section:

$$d\sigma_{fi} = \left(\int dp_f p_f^2\,\frac{4Z^2\alpha^2m^2}{|\boldsymbol{p}_i|E_f|\boldsymbol{q}|^4}|\bar{u}^{(\beta)}(p_f)\gamma_0 u^{(\alpha)}(p_i)|^2\delta(E_f-E_i)\right)^2 d\Omega_f \tag{10.289}$$

For elastic scattering, $|\boldsymbol{p}_i| = |\boldsymbol{p}_f| = p_f$ and $EdE = p_f dp_f$, we obtain that the differential cross section is

$$d\sigma_{fi} = \frac{4Z^2\alpha^2m^2}{|\boldsymbol{q}|^4}|\bar{u}^{(\beta)}(p_f)\gamma_0 u^{(\alpha)}(p_i)|^2 d\Omega_f \tag{10.290}$$

For an unpolarized beam, we get

$$\left.\frac{d\sigma_{fi}}{d\Omega}\right|_{\text{unpolarized}} = \frac{Z^2\alpha^2}{4|\boldsymbol{p}|^2\beta^2\sin^4(\Theta/2)}\left(1-\beta^2\sin^2\frac{\Theta}{2}\right) \tag{10.291}$$

where $\beta = v/c$.

Exercises

10.1 Spectral function for the Dirac propagator

1) Derive a formal expression for the spectral function $\rho_{\alpha\alpha'}(p)$ of the Feynman propagator for the Dirac theory,

$$S_F^{\alpha\alpha'}(p) = \int_0^\infty dm'^2\,\frac{\rho_{\alpha\alpha'}(p)}{p^2-m'^2+i\epsilon} \tag{10.292}$$

in terms of matrix elements of the field operators. Recall that the Dirac propagator is

$$S_F^{\alpha\alpha'}(x,x') = -i\langle 0|T\,\psi_\alpha(x)\,\bar{\psi}_{\alpha'}(x')|0\rangle \tag{10.293}$$

2) Show that, if the vacuum is invariant under parity $\mathcal{P}$,

$$\mathcal{P}\psi(x)\mathcal{P}^{-1} \equiv \gamma_0 \psi \tag{10.294}$$

then $\rho_{\alpha\alpha'}$ has the simpler form

$$\rho_{\alpha\alpha'} = \rho_1(p^2) \not{p}_{\alpha\alpha'} + \rho_2(p^2)\, \delta_{\alpha\alpha'} \tag{10.295}$$

Compute explicitly $\rho_1(p^2)$ and $\rho_2(p^2)$ for the free Dirac theory.

10.2 Wick's theorem for the scalar field

This is an exercise on the application of Wick's theorem for a specific theory, the 3-component free scalar field $\phi_a(x)$, with $a = 1, 2, 3$, with the global $O(3)$ invariance $\phi_a(x) \to O_{ab}\phi_b(x)$, where O_{ab} is an arbitrary 3×3 rotation matrix.

1) Use symmetry arguments to determine which of the following v.e.v.s are nonzero (summation over repeated indices is implied):

a) $\langle 0|T\phi_a(x)\phi_a(x')|0\rangle$
b) $\langle 0|T\phi_a(x)\phi_b(x')\phi_b(x'')|0\rangle$
c) $\langle 0|T\phi_a(x)\phi_b(x')\phi_b(x'')\phi_a(x''')|0\rangle$

2) Use Wick's theorem to find expressions for the v.e.v.s of (b) and (c) in (part 1) in terms of the v.e.v. of (a).

10.3 Reduction formulas

Consider a theory of a complex scalar field $\phi(x)$ ("pions" $\pi^{\pm}$) coupled to the quantized electromagnetic field $A_\mu(x)$. Consider the process $\gamma \to \pi^+ + \pi^-$ (pair creation) with a photon of 4-momentum p_i and polarization α in the initial state and pions of momenta p_+ and p_- in the final state. The S-matrix element is

$$\langle p_+ p_- | \hat{S} | p_i, \alpha\rangle \tag{10.296}$$

Find a reduction formula that relates this matrix element to the v.e.v. of a set of time-ordered fields for this particular process.

10.4 Cross sections for a real scalar field

Find the analog for this problem of the reduction formula shown in eq. (10.168) for the real scalar field, and derive an explicit formula that relates the differential cross section for a $2 \to 2$ process to a suitable four-point function, where the *in* state has two particles of positive charge and momenta p_1 and p_2 (with $p_1 - p_2 \to 0$), and the *out* state also has two particles with positive charge and momenta q_1 and q_2 (with $q_1 - q_2 \to 0$). Calculate the cross section to lowest order in perturbation theory (i.e., to order λ).

10.5 Complex scalar field coupled to a classical electromagnetic field

Consider a complex scalar field coupled to an external classical (i.e., not quantized) electromagnetic field $A_\mu(x)$.

1) Write the Lagrangian density of this theory coupled to the gauge field.
2) Verify that the theory is gauge invariant.
3) Derive an expression for the classical current $\mathcal{J}_\mu$, and show that the current is conserved and gauge invariant.

4) Show that, in the Heisenberg representation of the full theory, the current operator $\hat{\mathcal{J}}_\mu$ obeys the local conservation equation $\partial^\mu \hat{\mathcal{J}}_\mu = 0$.
5) Consider the current-current correlation function

$$\mathcal{D}_{\mu\nu}(x, x') = \langle 0 | T\, \hat{\mathcal{J}}_\mu(x) \hat{\mathcal{J}}_\nu(x') | 0 \rangle \tag{10.297}$$

(where $|0\rangle$ is the exact ground state of the full theory). Use your results to find an expression for $\partial_x^\mu \mathcal{D}_{\mu\nu}(x, x')$. When is it equal to zero?

10.6 Propagators and correlation functions for the one-dimensional quantum Heisenberg antiferromagnet.
In exercise 6.1 of chapter 6, you studied the one-dimensional quantum Heisenberg antiferromagnet in the spin wave (or semiclassical) approximation. In that problem, you constructed the ground state and the single-particle excitations, the spin waves. In this problem, you will study the response of that system to an external space- and time-dependent magnetic field. This external field will be represented by an additional term to the Hamiltonian of the form

$$H_{\text{ext}} = \sum_{n=-N/2+1}^{N/2} B_k(n, t)\, \hat{S}_k(n, t) \tag{10.298}$$

which we will regard as a perturbation.

1) Consider for the moment that the external field has been switched off. Derive an expression for the following propagators:

$$D_{33}(nt, n't') = -i\langle \text{gnd} | T\, \hat{S}_3(n, t) \hat{S}_3(n', t') | \text{gnd} \rangle \tag{10.299}$$

$$D_{+-}(nt, n't') = -i\langle \text{gnd} | T\, \hat{S}^+(n, t) \hat{S}^-(n', t') | \text{gnd} \rangle \tag{10.300}$$

in momentum and frequency space. Be very careful and very explicit in the way you treat the poles of these propagators. Show that your choice of frequency integration contour yields a propagator that satisfies the correct boundary conditions.
2) Use linear response theory to derive an expression for the magnetic susceptibilities χ_{33} and χ_{+-} (in position space) of this system in terms of correlation (or retarded) functions of the system.
3) Use Wick's theorem to find an expression for the corresponding time-ordered functions in the spin-wave approximation in momentum and frequency space.
4) Use the results of the previous parts of this exercise to show that $\chi_{+-}(p, \omega)$ has, in the limit $\omega \to 0$, a pole at $p = \pi$. Calculate the residue of this pole. The residue is the square of the order parameter of the system in this approximation.

11

Perturbation Theory and Feynman Diagrams

Let us now turn our attention to interacting quantum field theories. All of the results derived in this chapter apply equally to both relativistic and nonrelativistic theories with only minor changes. Here we will use the path-integral approach developed in previous chapters.

The properties of any field theory can be understood if the N-point correlation functions are known:

$$G_N(x_1, \ldots, x_N) = \langle 0|T\phi(x_1)\cdots\phi(x_N)|0\rangle \tag{11.1}$$

Much of what we will do below can be adapted to any field theory of interest. We will discuss in detail the simplest case, the relativistic self-interacting scalar field theory. It is straightforward to generalize this case to other theories of interest. Only a summary of results will be given for the other cases.

11.1 The generating functional in perturbation theory

The N-point function of a scalar field theory,

$$G_N(x_1, \ldots, x_N) = \langle 0|T\phi(x_1)\cdots\phi(x_N)|0\rangle \tag{11.2}$$

can be computed from the generating functional $Z[J]$:

$$Z[J] = \langle 0|Te^{i\int d^Dx J(x)\phi(x)}|0\rangle \tag{11.3}$$

In $D=d+1$-dimensional Minkowski spacetime, $Z[J]$ is given by the path integral

$$Z[J] = \int \mathcal{D}\phi\, e^{iS[\phi]+i\int d^Dx J(x)\phi(x)} \tag{11.4}$$

where the action $S[\phi]$ is the action for a relativistic scalar field. The N-point function, eq. (11.1), is obtained by functional differentiation:

$$G_N(x_1, \ldots, x_N) = (-i)^N \frac{1}{Z[J]} \frac{\delta^N Z[J]}{\delta J(x_1)\cdots\delta J(x_N)}\bigg|_{J=0} \tag{11.5}$$

Similarly, the Feynman propagator $G_F(x_1 - x_2)$, which is essentially the 2-point function, is given by

$$G_F(x_1 - x_2) = i\langle 0|T\phi(x_1)\phi(x_2)|0\rangle = -i\,\frac{1}{Z[J]}\,\frac{\delta^2 Z[J]}{\delta J(x_1)\delta J(x_2)}\bigg|_{J=0} \tag{11.6}$$

Thus, in principle, all we need to do is to compute $Z[J]$, and once it is determined, to use it to compute the N-point functions.

We will derive an expression for $Z[J]$ in the simplest theory, the relativistic real scalar field with a ϕ^4 interaction, but the methods are very general. We will work in Euclidean spacetime (i.e., in imaginary time), where the generating function takes the form

$$Z[J] = \int \mathcal{D}\phi\, e^{-S[\phi] + \int d^D x J(x)\phi(x)} \tag{11.7}$$

where $S[\phi]$ now is

$$S[\phi] = \int d^D x \left[\frac{1}{2}(\partial\phi)^2 + \frac{m^2}{2}\phi^2 + \frac{\lambda}{4!}\phi^4\right] \tag{11.8}$$

In the Euclidean theory, the N-point functions are

$$G_N(x_1, \ldots, x_N) = \langle \phi(x_1)\cdots\phi(x_N)\rangle = \frac{1}{Z[J]}\,\frac{\delta^N Z[J]}{\delta J(x_1)\cdots\delta J(x_N)}\bigg|_{J=0} \tag{11.9}$$

Let us denote by $Z_0[J]$ the generating action for the free scalar field, with action $S_0[\phi]$. In chapter 5, we already calculated $Z_0[J]$ and obtained the result (see eq. (5.141) and eq. (5.146)):

$$\begin{aligned} Z_0[J] &= \int \mathcal{D}\phi\, e^{-S_0[\phi] + \int d^D x J(x)\phi(x)} \\ &= \left[\operatorname{Det}\left(-\partial^2 + m^2\right)\right]^{-1/2} e^{\frac{1}{2}\int d^D x \int d^D y J(x) G_0(x-y) J(y)} \end{aligned} \tag{11.10}$$

where ∂^2 is the Laplacian operator in D-dimensional Euclidean space, and $G_0(x-y)$ is the free-field Euclidean propagator (i.e., the two-point correlation function):

$$G_0(x-y) = \langle \phi(x)\phi(y)\rangle_0 = \langle x|\frac{1}{-\partial^2 + m^2}|y\rangle \tag{11.11}$$

and the sub-index label 0 denotes a free-field expectation value.

We can write the full generating function $Z[J]$ in terms of the free-field generating function $Z_0[J]$ by noting that the interaction part of the action contributes with a weight of the path integral, which, on expanding in powers of the coupling constant λ, takes the form

$$\begin{aligned} e^{-S_{\text{int}}[\phi]} &= e^{-\int d^D x \frac{\lambda}{4!}\phi^4(x)} \\ &= \sum_{n=0}^{\infty} \frac{(-1)^n}{n!}\left(\frac{\lambda}{4!}\right)^n \int d^D x_1 \cdots \int d^D x_n\, \phi^4(x_1)\cdots\phi^4(x_n) \end{aligned} \tag{11.12}$$

Hence, on expanding in powers of λ, the generating function $Z[J]$ is

$$
\begin{aligned}
Z[J] &= \int \mathcal{D}\phi \, e^{-S_0[\phi] + \int d^D x J(x)\phi(x) - S_{\text{int}}[\phi]} \\
&= \sum_{n=0}^{\infty} \frac{(-1)^n}{n!} \left(\frac{\lambda}{4!}\right)^n \\
&\times \int d^D x_1 \cdots \int d^D x_n \int \mathcal{D}\phi \, \phi^4(x_1) \cdots \phi^4(x_n) \, e^{-S_0[\phi] + \int d^D x J(x)\phi(x)} \\
&= \sum_{n=0}^{\infty} \frac{(-1)^n}{n!} \left(\frac{\lambda}{4!}\right)^n \int d^D x_1 \cdots \int d^D x_n \frac{\delta^4}{\delta J(x_1)^4} \cdots \frac{\delta^4}{\delta J(x_n)^4} Z_0[J] \\
&\equiv e^{-\frac{\lambda}{4!} \int d^D x \frac{\delta^4}{\delta J(x)^4}} Z_0[J]
\end{aligned}
\tag{11.13}
$$

where the operator of the last line is defined by its power series expansion. We see that this amounts to the formal replacement

$$
S_{\text{int}}[\phi] \mapsto S_{\text{int}}\left(\frac{\delta}{\delta J}\right) \tag{11.14}
$$

This expression allows us to write the generating function of the full theory $Z[J]$ in terms of $Z_0[J]$, the generating function of the free-field theory:

$$
Z[J] = e^{-S_{\text{int}}\left(\frac{\delta}{\delta J}\right)} Z_0[J] \tag{11.15}
$$

Notice that this expression holds for any theory, not just a ϕ^4 interaction. This result is the starting point of perturbation theory.

Before we embark on explicit calculations in perturbation theory, it is worthwhile to see what assumptions have been made along the way. We assumed (a) that the fields obey Bose commutation relations, and (b) that the vacuum (or ground state) is nondegenerate.

The restriction to Bose statistics was made at the level of the path integral of the scalar field. This approach is, however, of general validity, and it also applies to theories with Fermi fields, which are path integrals over Grassmann fields. As we saw before, in all cases, the generating functional yields vacuum (or ground state) expectation values of time-ordered products of fields. The assumption of relativistic invariance is also not essential, although it simplifies the calculations significantly.

The restriction to a nondegenerate vacuum state has more subtle physical consequences. We have noted that, in some cases, the vacuum may be degenerate if a global symmetry is spontaneously broken. Here, the thermodynamic limit plays an essential role. For example, if the vacuum is doubly degenerate, we can do perturbation theory on one of the two vacuum states. If they are related by a global symmetry, the number of orders in perturbation theory that are necessary to have a mixing with its degenerate partner is proportional to the total number of degrees of freedom. Thus, in the thermodynamic limit, they will not mix unless the vacuum is unstable. Such an instability usually shows up in the form of IR divergent contributions in the perturbation expansion. However, this is not a sickness of the expansion, but the consequence of having an inadequate starting point for the ground state! In general, the "safe procedure" to deal with degeneracies that are due to symmetry is to add an additional term (like a source) to the Lagrangian that breaks the symmetry and to do all the calculations in the presence of such a symmetry-breaking term. This term should be removed only after the thermodynamic limit is taken.

The case of *gauge symmetries* has other and important subtleties. In the case of Maxwell's electrodynamics, in chapter 9 we saw that the ground state is locally gauge invariant and that, as a result, it is unique. It turns out that this is a generic feature of theories that are locally gauge invariant for all symmetry groups and in all dimensions. There is a very powerful theorem (due to S. Elitzur, and which will be discussed in section 18.6), which states that not only the ground state of theories with gauge invariance is unique and gauge invariant, but also that this restriction extends to the entire spectrum of the system. Thus only gauge-invariant operators have nonzero expectation values, and only such operators can generate the physical states. To put it differently, a local symmetry cannot be broken spontaneously even in the thermodynamic limit. The physical reason behind this statement is that if the symmetry is local, it takes only a finite order in perturbation theory to mix all symmetry-related states. Thus, whatever may happen at the boundaries of the system has no consequence for what happens in the interior, and the thermodynamic limit no longer plays a role. This is why we can fix the gauge and remove the enormous redundancy of the description of the states.

Nevertheless, we have to be very careful about two issues. First, the gauge-fixing procedure must select one and only one state from each gauge class. Second, the perturbation theory is based on the propagator of the gauge fields, $G_{\mu\nu}(x-x')=-i\langle 0|TA_\mu(x)A_\nu(x')|0\rangle$, which is not gauge invariant and, unless a gauge is fixed, it vanishes. If a gauge is fixed, this propagator has contributions that depend on the choice of gauge, but the poles of this propagator do not depend on the choice of gauge since they describe physical excitations (e.g., photons). Furthermore, although the propagator is gauge dependent, it will only appear in combination with matter currents, which are conserved. Thus, the gauge-dependent terms of the propagator do not contribute to physical processes.

Except for these caveats, we can now proceed to do perturbation theory for all field theories of interest.

11.2 Perturbative expansion for the two-point function

Let us discuss the perturbative computation of the two-point function in ϕ^4 field theory in D-dimensional Euclidean spacetime. Recall that, under a Wick rotation, the analytic continuation to imaginary time $ix_0 \mapsto x_D$, the two-point function in D-dimensional Minkowski spacetime, $\langle 0|T\phi(x_1)\phi(x_2)|0\rangle$, maps onto the correlation function of two fields in D-dimensional Euclidean spacetime:

$$\langle 0|T\phi(x_1)\phi(x_2)|0\rangle \mapsto \langle \phi(x_1)\phi(x_2)\rangle \tag{11.16}$$

Let us formally write the two-point function $G^{(2)}(x_1-x_2)$ as a power series in the coupling constant λ:

$$G^{(2)}(x_1-x_2)=\sum_{n=0}^{\infty}\frac{\lambda^n}{n!}\,G_n^{(2)}(x_1-x_2) \tag{11.17}$$

Using the generating functional $Z[J]$, we can write

$$G^{(2)}(x_1 - x_2) = \frac{1}{Z[J]} \frac{\delta^2 Z[J]}{\delta J(x_1)\delta J(x_2)}\Big|_{J=0} \tag{11.18}$$

where

$$Z[J] = e^{-S_{\text{int}}\left(\frac{\delta}{\delta J}\right)} Z_0[J] \tag{11.19}$$

Hence, the two-point function can be expressed as a ratio of two series expansions in powers of the coupling constant. The numerator of eq. (11.18) is given by an expansion in powers of the coupling constant λ, which can formally be written as

$$\frac{\delta^2 Z[J]}{\delta J(x_1)\delta J(x_2)}\Big|_{J=0} = \sum_{n=0}^{\infty} \frac{(-1)^n}{n!} \left(\frac{\lambda}{4!}\right)^n$$
$$\times \int d^D y_1 \cdots \int d^D y_n \frac{\delta^2}{\delta J(x_1)\delta J(x_2)} \frac{\delta^4}{\delta J(y_1)^4} \cdots \frac{\delta^4}{\delta J(y_n)^4} Z_0[J]\Big|_{J=0} \tag{11.20}$$

Likewise, the denominator of eq. (11.18) is given by the expansion of $Z[0]$ in powers of λ, which has a similar expression but without a contribution corresponding to the functional derivatives with respect to the source at the external points x_1 and x_2. The equivalent expansions in Minkowski spacetime are obtained, term by term, by the replacement

$$-\lambda \mapsto i\lambda \tag{11.21}$$

at every order in the expansion.

Let us now look at the form of the first few terms of the expansion of the two-point function in perturbation theory.

11.2.1 Zeroth order in λ

To zeroth order in λ, $O(\lambda^0)$, the numerator of eq. (11.18) reduces to

$$\frac{\delta}{\delta J(x_1)} \frac{\delta}{\delta J(x_2)} Z[J]\Big|_{J=0} = \frac{\delta}{\delta J(x_1)} \frac{\delta}{\delta J(x_2)} Z_0[J]\Big|_{J=0} + O(\lambda)$$
$$= G_0(x_1 - x_2) + O(\lambda) \tag{11.22}$$

while the denominator of eq. (11.18) is simply equal to one:

$$Z[0] = Z_0[0] + O(\lambda) = 1 + O(\lambda) \tag{11.23}$$

where the contribution of the determinant in $Z_0[0]$ has been dropped, since it does not contribute to the expectation values of the local observables. Hence, we find that to zeroth order, the two-point function is simply given by

$$G^{(2)}(x_1 - x_2) = G_0(x_1 - x_2) + O(\lambda) \tag{11.24}$$

11.2.2 First order in λ

To first order in λ, the denominator $Z[0]$ of eq. (11.18) is given by

$$Z[0] = 1 + \frac{-1}{1!}\left(\frac{\lambda}{4!}\right)\int dy \frac{\delta^4}{\delta J(y)^4} Z_0[J]\Big|_{J=0} + O(\lambda^2) \tag{11.25}$$

The expression in the integrand can be calculated from the Taylor expansion of $Z_0[J]$ in powers of $J(x)$. To find a nonzero contribution, we need to bring down from the exponent enough factors of J so that they can be canceled by the functional derivatives. Since the argument of the exponential factor in $Z_0[J]$ is a *bilinear* functional of $J(x)$,

$$Z_0[J] \propto e^{\frac{1}{2}\int d^Dx \int d^Dy J(x) G_0(x-y) J(y)} \tag{11.26}$$

only an even number of derivatives in $J(x)$ can be canceled to a given order. In particular, to first order in λ, we have to cancel four derivatives. Thus we need to expand the exponential in $Z_0[J]$ to second order in its argument to obtain the only nonvanishing contribution to first order in λ to $Z[J]$ at $J=0$:

$$\begin{aligned} Z[0] = 1 &+ \frac{(-1)}{1!}\left(\frac{\lambda}{4!}\right) \\ &\times \int d^Dx \frac{\delta^4}{\delta J(x)^4} \frac{1}{2!}\left(\frac{1}{2}\int d^Dy_1 \int d^Dy_2 J(y_1) G_0(y_1-y_2) J(y_2)\right)^2 \Big|_{J=0} \\ &+ O(\lambda^2) \end{aligned} \tag{11.27}$$

The derivatives yield a set of δ functions

$$\frac{\delta^4}{\delta J(x)^4}[J(y_1)\cdots J(y_4)]\Big|_{J=0} = \sum_P \prod_{j=1}^{4} \delta(y_{Pj} - x) \tag{11.28}$$

where P runs over the 4! permutations of the arguments y_1, y_2, y_3, and y_4.

We can now write the first-order correction to $Z[0]$ in the form

$$\begin{aligned} Z[0] = 1 &+ \left(\frac{-1}{1!}\right)\left(\frac{\lambda}{4!}\right)\frac{1}{2!}\left(\frac{1}{2}\right)^2 \\ &\times \int d^Dx \int d^Dy_1 \cdots d^Dy_4 G_0(y_1-y_2) G_0(y_3-y_4) \sum_P \prod_{j=1}^{4} \delta(y_{Pj}-x) \\ &+ O(\lambda^2) \\ = 1 &+ \frac{(-1)}{1!}\left(\frac{\lambda}{4!}\right) S \int d^Dx\, G_0(x,x)\, G_0(x,x) + O(\lambda^2) \end{aligned} \tag{11.29}$$

where $S=3$ is the multiplicity factor, which counts the number of times this picture is obtained by contracting all the lines in a pairwise fashion without changing the topology of this term.

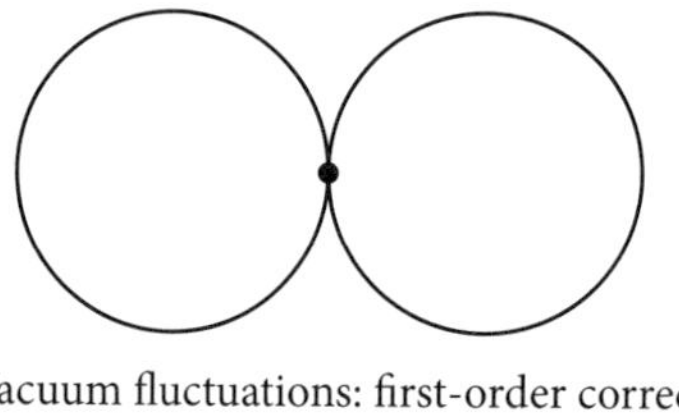

Figure 11.1 Vacuum fluctuations: first-order correction to $Z[0]$.

It is useful to introduce a picture or Feynman diagram to represent this contribution. Let us mark four points $y_1, \ldots, y_4$ and an additional point at x (which we call a *vertex*) with four legs coming out of it. Let us join y_1 and y_2 by a line and y_3 with y_4 by another line. To each line, we assign a factor of $G_0(y_1 - y_2)$ and $G_0(y_3 - y_4)$, respectively:

$$G_0(x-y) = \quad \overset{x \qquad\qquad\qquad\qquad y}{\overline{\qquad\qquad\qquad\qquad\qquad}} \tag{11.30}$$

Next, because of the δ functions, we have to identify each of the points $y_1, \ldots, y_4$ with each one of the legs attached to y in all possible ways. The resulting expression has to be integrated over all values of the coordinates and of x. The result is

$$Z[0] = 1 - \frac{\lambda}{8} \int d^D x \, (G_0(x,x))^2 + O(\lambda^2) \tag{11.31}$$

Physically, the first-order contribution represents corrections to the ground state energy due to vacuum fluctuations. This expression can be represented more simply by the Feynman diagram shown in figure 11.1.

Here, and below, a solid line denotes the bare propagator $G_0(x-y)$.

Let us now compute the first-order corrections to $\left.\dfrac{\delta^2 Z[J]}{\delta J(x_1)\delta J(x_2)}\right|_{J=0}$:

$$\begin{aligned}
&\frac{(-1)}{1!}\frac{\lambda}{4!}\int d^D y \frac{\delta^2}{\delta J(x_1)\delta J(x_2)} \frac{\delta^4}{\delta J(y)^4} Z[J]\Big|_{J=0} \\
&= \frac{(-1)}{1!}\frac{\lambda}{4!}\int d^D y \frac{\delta^2}{\delta J(x_1)\delta J(x_2)} \frac{\delta^4}{\delta J(y)^4} \\
&\quad \times \frac{1}{3!}\left(\frac{1}{2}\int d^D z_1 \int d^D z_2 J(z_1) G_0(z_1 - z_2) J(z_2)\right)^3 \Bigg|_{J=0}
\end{aligned} \tag{11.32}$$

The nonvanishing contributions are obtained by matching the derivatives in eq. (11.32) with an equal number of powers of J. We see that there are six factors of the source J at points $z_1, \ldots, z_6$ and six derivatives, one at x_1 and at x_2, and four at y. To match derivatives with powers amounts to finding all possible pairwise contractions of these two sets of points. Once the δ functions have acted, we are left with just one integral over the position y of the internal vertex. Hence, the result amounts to finding all possible contractions among the external legs at x_1 and x_2, with each other and/or with the internal vertex at y. Notice that for each contraction, we get a power of the bare (unperturbed) propagator G_0.

The only nonvanishing terms resulting from this process are represented by the Feynman diagrams in figure 11.2.

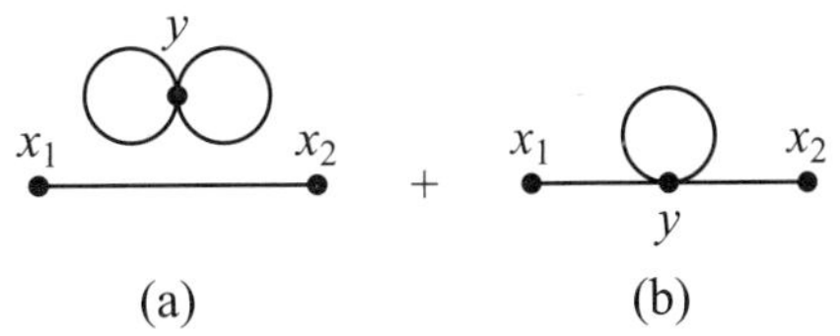

Figure 11.2 First-order contributions to the two-point function.

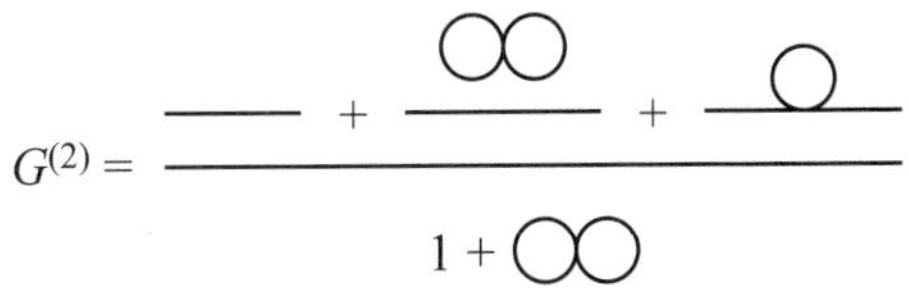

Figure 11.3 Diagrams for the two-point function to first order in λ.

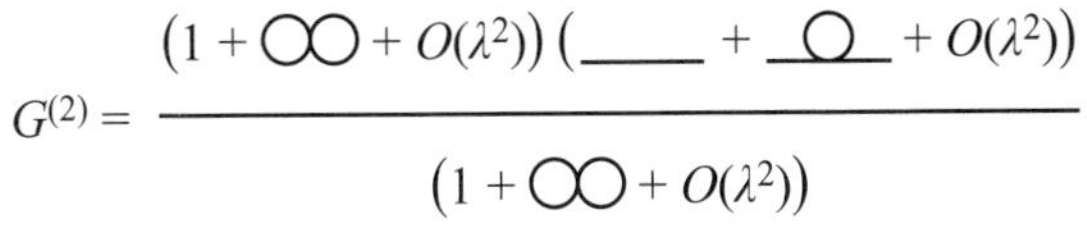

Figure 11.4 Factorization of Feynman diagrams for the two-point function to first order in λ.

The first contribution, figure 11.2a, is the product of the bare propagator $G_0(x_1 - x_2)$ between the external points and the first-order correction of the vacuum diagrams:

$$(a) = G_0(x_1 - x_2) \times \left[-\left(\frac{\lambda}{8}\right) \int d^D y \left(G_0(y, y)\right)^2 \right] \tag{11.33}$$

The second term, the "tadpole" diagram of figure 11.2b, is given by the expression

$$(b) = -\left(\frac{\lambda}{4!}\right) S \int d^D y \, G_0(x_1, y) G_0(y, y) G_0(y, x_2) \tag{11.34}$$

where the multiplicity factor is $S = 4 \times 3$. It counts the number of ways (12) of contracting the external points to the internal vertex: There are four different ways of contracting one external point to one of the four lines attached to the internal vertex at y, and three different ways of contracting the remaining external point to one of the three remaining lines of the internal vertex. There is only one way to contract the two leftover internal lines attached to the vertex at y.

By collecting terms, we get the result shown in figure 11.3.

To first order in λ, the expansion of the two-point function can also be written in the form shown in figure 11.4.

At least to this order in perturbation theory, some of the diagrams that contribute to the expansion of the numerator get exactly canceled by the expansion of the denominator, the vacuum diagrams. The diagrams that get canceled are unlinked in the sense that one can split a diagram in two by drawing a line that does not cut any of the propagator lines. These diagrams contain a factor consisting of terms of the vacuum diagrams. We will see in section 11.3 that this is a feature of this expansion to all orders in λ.

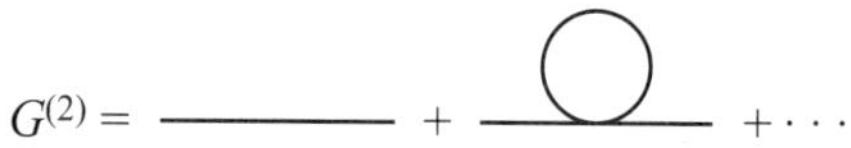

Figure 11.5 The two-point function to first order in λ.

Thus, to first order in λ, the two-point function is given by

$$G^{(2)}(x_1,x_2) = G_0^{(2)}(x_1,x_2) - \frac{\lambda}{2}\int d^Dy\, G_0^{(2)}(x_1,y)G_0^{(2)}(y,y)G_0^{(2)}(y,x_2) + O(\lambda^2) \tag{11.35}$$

as shown in figure 11.5.

11.3 Cancellation of the vacuum diagrams

The cancellation of the vacuum diagrams is a general feature of perturbation theory. Let us reexamine this issue in more general terms. We will go through the arguments for the case of the two-point function, but they are trivial to generalize to any N-point function. This feature also holds in all theories, relativistic or not, bosonic or fermionic, provided the fields satisfy local canonical commutation (or anticommutation) relations.

The expansion of the two-point function has the form

$$\langle\phi(x_1)\phi(x_2)\rangle = \frac{1}{Z[0]}\sum_{n=0}^{\infty}\int d^Dy_1\cdots d^Dy_n\,\frac{(-1)^n}{n!}\left\langle\phi(x_1)\phi(x_2)\prod_{j=1}^{n}\mathcal{L}_{\text{int}}\left(\phi(y_j)\right)\right\rangle_0 . \tag{11.36}$$

The denominator factor $Z[0]$ has a similar expansion

$$Z[0] = \sum_{n=0}^{\infty}\frac{(-1)^n}{n!}\int d^Dy_1\cdots d^Dy_n\left\langle\prod_{j=1}^{N}\mathcal{L}_{\text{int}}\left(\phi(y_j)\right)\right\rangle_0 \tag{11.37}$$

where $\langle\mathcal{O}[\phi]\rangle_0$ denotes the expectation value of the operator $\mathcal{O}[\phi]$ in free-field theory.

Consider first the numerator. Each expectation value involves a sum of products of pairwise contractions. If we assign a Feynman diagram to each contribution, it is clear that we can classify these terms into two classes: (a) linked and (b) unlinked diagrams. A diagram is said to be *unlinked* if it contains a *sub-diagram*, in which a set of internal vertices are linked with each other but not to an external vertex. The *linked* diagrams satisfy the opposite property. Since the vacuum diagrams by definition do not contain any external vertices, they are unlinked.

All the expectation values that appear in the numerator of eq. (11.36) can be written as a sum of terms, each of the form of a linked diagram times a vacuum graph, that is,

$$\langle\phi(x_1)\phi(x_2)\mathcal{L}_{\text{int}}\left(\phi(y_1)\right)\cdots\mathcal{L}_{\text{int}}\left(\phi(y_n)\right)\rangle_0 =$$

$$= \sum_{k=0}^{n}\binom{n}{k}\left\langle\phi(x_1)\phi(x_2)\prod_{j=1}^{k}\mathcal{L}_{\text{int}}\left(\phi(y_j)\right)\right\rangle_0\left\langle\prod_{j=k+1}^{n}\mathcal{L}_{\text{int}}\left(\phi(y_j)\right)\right\rangle_0 \tag{11.38}$$

where the super-index ℓ denotes a *linked* factor (i.e., a factor that does not contain any vacuum sub-diagram). Thus, the numerator has the form

$$\sum_{n=0}^{\infty}\sum_{k=0}^{n}\frac{(-1)^n}{n!}\binom{n}{k}\left\langle \phi(x_1)\phi(x_2)\prod_{j=1}^{k}\mathcal{L}_{\text{int}}(y_k)\right\rangle_0^{\ell}\left\langle \prod_{j=k+1}^{N}\mathcal{L}_{\text{int}}\left(\phi(y_j)\right)\right\rangle_0 \tag{11.39}$$

which factorizes into

$$\left(\sum_{k=0}^{\infty}\frac{(-1)^k}{k!}\int d^Dy_1\cdots d^Dy_k\left\langle \phi(x_1)\phi(x_2)\prod_{j=1}^{k}\mathcal{L}_{\text{int}}\left(\phi(y_j)\right)\right\rangle_0^{\ell}\right)$$
$$\times\left(\sum_{n=0}^{\infty}\frac{(-1)^n}{n!}\int d^Dy_1\cdots d^Dy_n\left\langle \prod_{j=1}^{n}\mathcal{L}_{\text{int}}\left(\phi(y_j)\right)\right\rangle_0\right) \tag{11.40}$$

We can clearly recognize that the second factor is identically equal to the denominator $Z[0]$.

Hence we can write the two-point function as a sum of linked Feynman diagrams:

$$\langle \phi(x_1)\phi(x_2)\rangle = \sum_{n=0}^{\infty}\frac{(-1)^n}{n!}\int d^Dy_1\cdots d^Dy_n\,\langle \phi(x_1)\phi(x_2)\prod_{j=1}^{n}\mathcal{L}_{\text{int}}\left(\phi(y_n)\right)\rangle_0^{\ell} \tag{11.41}$$

This result is known as the *linked-cluster theorem*. This theorem, which proves that the vacuum diagrams cancel out exactly to all orders in perturbation theory, is valid for all N-point functions (not just for two-point functions) and for any local theory. It also holds in Minkowski spacetime on the replacement $(-1)^n \leftrightarrow i^n$. It holds for all theories with a local canonical structure, relativistic or not.

11.4 Summary of Feynman rules for ϕ^4 theory

11.4.1 Position space

The general rules to construct the diagrams in position space for the N-point function $\langle 0|T\phi(x_1)\cdots\phi(x_N)|0\rangle$ in ϕ^4 theory in Minkowski space and $\langle \phi(x_1)\cdots\phi(x_N)\rangle$ in Euclidean space are

1) A general graph for the N-point function has N external points and n internal interaction vertices, where n is the order in perturbation theory. Each vertex is a point with a coordinate label and four lines (for a ϕ^4 theory) coming out of it.
2) Draw all topologically distinct graphs obtained by connecting the external points and the internal vertices in all possible ways. Discard all graphs that contain sub-diagrams not linked to at least one external point.
3) The following weights are assigned to each graph:

   i) For every vertex, a factor of $-i\frac{\lambda}{4!}$ in Minkowski space and $-\frac{\lambda}{4!}$ in Euclidean space.
   ii) For every line connecting a pair of points z_1 and z_2, a propagator factor of $\langle 0|T\phi(z_1)\phi(z_2)|0\rangle_0 = -iG_0^{(2)}(z_1, z_2)$ in Minkowski space, or $\langle \phi(z_1)\phi(z_2)\rangle_0 = G_0(z_1 - z_2)$ in Euclidean space.
   iii) An overall factor of $\frac{1}{n!}$.

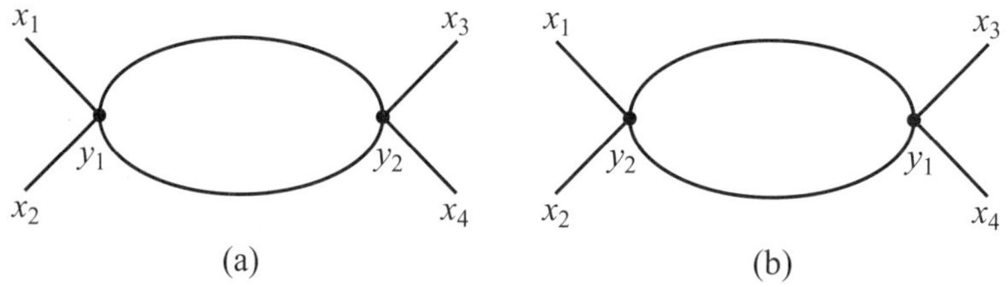

Figure 11.6 Two topologically distinct contributions to the four-point function to order λ^2 that have the same analytic expression.

4) There is a multiplicity factor that counts the number of ways in which the lines can be joined without changing the topology of the graph.
5) Integrate over all internal coordinates $\{z_i\}$.

For example, the 4-point function

$$G^{(4)}(x_1, x_2, x_3, x_4) = \langle \phi(x_1)\phi(x_2)\phi(x_3)\phi(x_4) \rangle \tag{11.42}$$

has the two contributions to order λ^2 shown in figure 11.6. These two diagrams have exactly the same weight (if $G_0^{(2)}(1, 2) = G_0^{(2)}(2, 1)$), and their total contribution to the 4-point function is

$$\frac{1}{2!}\left(\frac{-\lambda}{4!}\right)^2 S \int d^D y_1 \int d^D y_2\, G_0^{(2)}(x_1, y_1)\, G_0^{(2)}(x_2, y_1) \times \left[G_0^{(2)}(y_1, y_2)\right]^2 G_0^{(2)}(x_3, y_2)\, G_0^{(2)}(x_4, y_2) \tag{11.43}$$

but they are topologically distinct and thus count as separate contributions. The multiplicity factor is $S = (4 \times 3)^2 \times 2 \times 2$.

11.4.2 Momentum space

The N-point functions can also be computed in momentum space. The rules for constructing Feynman diagrams in momentum space are:

1) A graph has N external legs, labeled by a set of external momenta $k_1, \ldots, k_N$, flowing into the diagram and n internal vertices (for order n in perturbation theory). Each vertex has four lines (for ϕ^4 theory), each carrying a momentum $q_1, \ldots, q_4$ (flowing out of the vertex). All lines must be connected in pairs. All vacuum terms have to be discarded.
2) All topologically different graphs must be drawn.
3) The weight of each diagram is determined as follows:
   - Each vertex contributes a factor of $(\frac{-\lambda}{4!})(2\pi)^D \delta^D(\sum_{i=1}^{4} q_i)$.
   - Each line carries a momentum p_μ and contributes to the weight with a factor $G_0^{(2)}(p) = \frac{1}{p^2+m^2}$, in Euclidean space. In Minkowski space, the factor becomes $-iG_0^{(2)} = \frac{-i}{p^2-m^2+i\epsilon}$.
   - All the numerical factors are the same as in position space.
   - All the internal momenta must be integrated over.

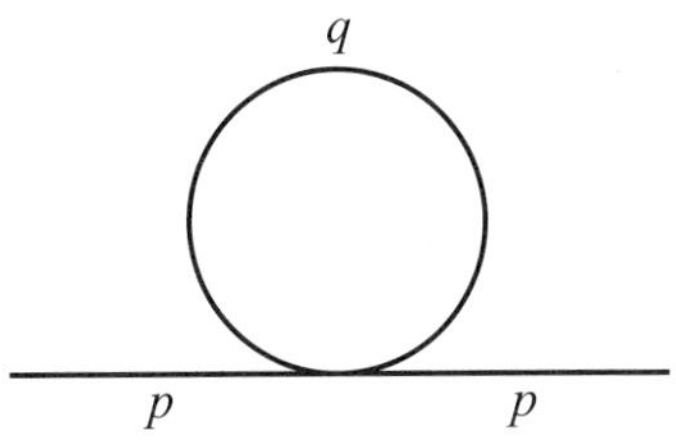

Figure 11.7 First-order contribution to the two-point function in momentum space.

For example, the first-order contribution to the two-point function is the tadpole diagram shown in figure 11.7. It has the algebraic weight

$$\left(\frac{-\lambda}{4!}\right)\frac{1}{1!}(4\times 3)\int\frac{d^Dq}{(2\pi)^D}\,\frac{1}{q^2+m^2}\left(\frac{1}{p^2+m^2}\right)^2 \tag{11.44}$$

11.5 Feynman rules for theories with fermions and gauge fields

The Feynman rules that we derived for ϕ^4 theory can be easily extended to other theories, relativistic or not. An important difference occurs in theories with fermions, in which the rules need to be modified to account for the Fermi-Dirac statistics. As an example, let us consider the case of quantum electrodynamics (QED), the theory of electrons and photons. Similar rules are formulated in quantum chromodynamics (QCD), which we will discuss in section 16.8. Feynman rules for self-interacting fermions are also devised in the case of the Gross-Neveu model (or the Nambu-Jona-Lasinio model) and in the case of a nonrelativistic electron gas.

The Lagrangian density of QED, in the Feynman– 't Hooft gauges, is

$$\mathcal{L}_{\text{QED}} = \bar{\psi}\, i\not{D}\psi - \frac{1}{4}F_{\mu\nu}^2 - \frac{1}{2\alpha}(\partial_\mu A^\mu)^2 \tag{11.45}$$

In the presence of separate sources $\bar{\eta}(x)$ for the Dirac (spinor) field $\psi(x)$ (and $\eta(x)$ for the adjoint Dirac field $\bar{\psi}(x)$, and $J_\mu(x)$ for the electromagnetic gauge field $A_\mu(x)$), the partition function is

$$Z[\eta,\bar{\eta},J_\mu] = \int \mathcal{D}\psi\mathcal{D}\bar{\psi}\mathcal{D}A_\mu\, e^{\,i\int d^4x\left[\mathcal{L}_{\text{QED}} + J_\mu(x)A^\mu(x) + \bar{\eta}(x)\psi(x) + \bar{\psi}(x)\eta(x)\right]} \tag{11.46}$$

where the Faddeev-Popov determinant has been dropped, since it is trivial in this abelian gauge theory (see section 9.5). In what follows, we drop the spinor labels to simplify the notation.

As in the case of ϕ^4 theory, the partition function $Z[\eta,\bar{\eta},J_\mu]$ in eq. (11.46) can be expressed in terms of the partition function of a free-field theory with sources. In this case, we have two partition functions of free-field theories: the partition function for free Dirac fields with sources,

$$Z_{\text{Dirac}}[\eta,\bar{\eta}] = \int \mathcal{D}\psi\mathcal{D}\bar{\psi}\,\exp\left[i\int d^4x\left(\bar{\psi}(i\not{\partial}-m)\psi + \bar{\eta}\psi + \bar{\psi}\eta\right)\right] \tag{11.47}$$

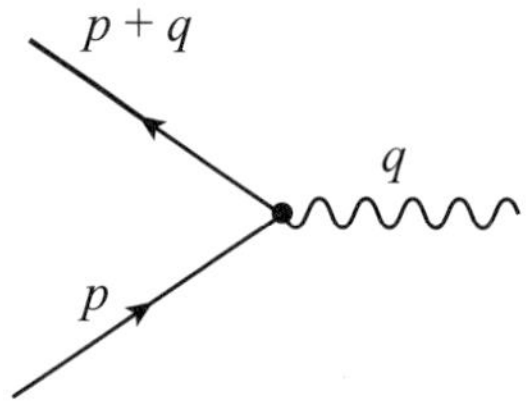

Figure 11.8 The bare interaction vertex of QED: The solid lines are Dirac fermion propagators, and the wavy line is a photon propagator. The bare vertex has a weight $e\gamma_\mu$.

and the partition function of the free Maxwell field, in the Feynman–'t Hooft gauges, with sources J_μ given by

$$Z_{\text{Maxwell}}[J_\mu] = \int \mathcal{D}A_\mu \, \exp\left[i \int d^4x \left(-\frac{1}{4}F_{\mu\nu}^2 - \frac{1}{2\alpha}\left(\partial_\mu A^\mu\right)^2 \right) + J_\mu A^\mu \right] \tag{11.48}$$

where, here too, the Faddeev-Popov determinant has been dropped.

In QED, the interaction term of the Lagrangian is

$$\mathcal{L}_{\text{int}}[\psi, \bar{\psi}, A_\mu] = eA_\mu \bar{\psi}\gamma^\mu \psi \tag{11.49}$$

Then the full generating function of eq. (11.46) is expressed as

$$Z[\eta, \bar{\eta}, J_\mu] = \exp\left(i \int d^4x \mathcal{L}_{\text{int}} \left[-i\frac{\delta}{\delta\bar{\eta}}, -i\frac{\delta}{\delta\eta}, -i\frac{\delta}{\delta J_\mu} \right] \right) Z_{\text{Dirac}}[\eta, \bar{\eta}] Z_{\text{Maxwell}}[J_\mu] \tag{11.50}$$

where the replacements

$$\psi(x) \to -i\frac{\delta}{\delta\bar{\eta}(x)}, \quad \bar{\psi}(x) \to -i\frac{\delta}{\delta\eta(x)}, \quad A_\mu(x) \to -i\frac{\delta}{\delta J^\mu(x)} \tag{11.51}$$

have been made inside the interaction term of the Lagrangian.

The Feynman rules then follow by the same line of argument that we used in ϕ^4 theory. The main differences are

1) Now we have two propagators: the propagator of the free massive Dirac field, $S_F(x-y) = -i\langle 0|T\psi(x)\bar{\psi}(y)|0\rangle$, which was given explicitly in section 7.2, and the propagator of the Maxwell field (in the Feynman–'t Hooft gauges) $G_{\mu\nu}(x-y,\alpha) = i\langle 0|TA_\mu(x)A_\nu(y)|0\rangle$, given explicitly in section 9.6.
2) The interaction vertex involves two Dirac fermions and a gauge field, shown in figure 11.8. Each bare vertex has a weight in a Feynman diagram of $e\gamma_\mu$.
3) Each closed Dirac fermion loop contributes with a minus sign, and each diagram now has a weight $(-1)^F$, where F is the number of fermion loops. This rule follows from the form of the path integral for fermions (see chapter 7). It applies to all theories of fermions, relativistic or not.

Aside from these changes, the Feynman rules are the same.

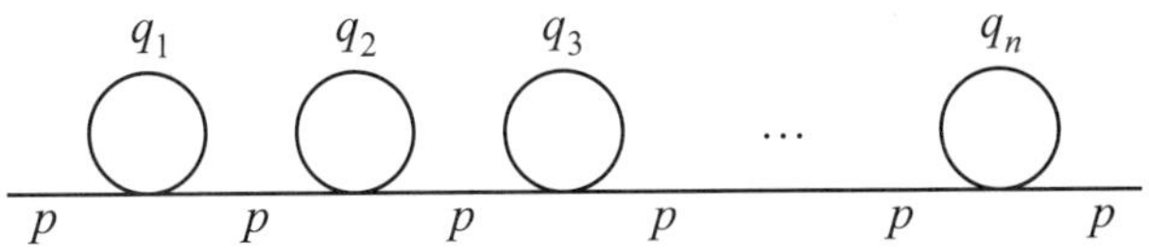

Figure 11.9 The set of all one-particle reducible diagrams of the two-point function to leading order in λ.

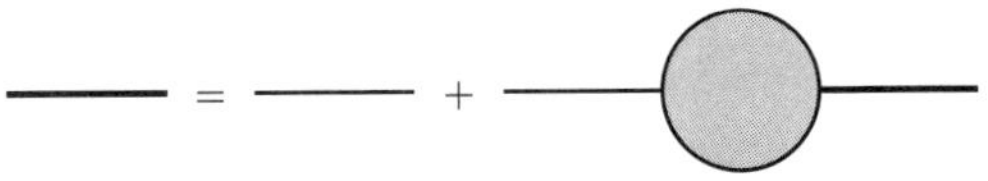

Figure 11.10 The Dyson equation for the two-point function.

11.6 The two-point function and the self-energy in ϕ^4 theory

We now return to ϕ^4 theory. To first order in λ, and in momentum space, the two-point function in Euclidean space is

$$G^{(2)}(p) = \frac{1}{p^2+m^2} + \frac{-\lambda}{4!}(4\times 3)\left(\int \frac{d^Dq}{(2\pi)^D}\frac{1}{q^2+m^2}\right)\left(\frac{1}{p^2+m^2}\right)^2 + O(\lambda^2) \tag{11.52}$$

Let us define by μ^2 the *effective* or *renormalized mass* (squared) such that

$$\frac{1}{p^2+\mu^2} = \frac{1}{p^2+m^2}\left\{1 - \frac{\lambda}{2}\left(\int \frac{d^Dq}{(2\pi)^D}\frac{1}{q^2+m^2}\right)\left(\frac{1}{p^2+m^2}\right) + O(\lambda^2)\right\} \tag{11.53}$$

Again, to first order in λ, we can write the equivalent expression

$$G^{(2)}(p) = \frac{1}{p^2+m^2+\frac{\lambda}{2}\int \frac{d^Dq}{(2\pi)^D}\frac{1}{q^2+m^2}} + O(\lambda^2) \tag{11.54}$$

This expression leads us to define μ^2 to be

$$\mu^2 = m^2 + \frac{\lambda}{2}\int \frac{d^Dq}{(2\pi)^D}\frac{1}{q^2+m^2} + \cdots \tag{11.55}$$

This equation is equivalent to a sum of a large number of diagrams with higher order in λ. How do we know that it is consistent? First note that we have summed diagrams of the form shown in figure 11.9. These diagrams have the very special feature that it is possible to split the diagram into two sub-diagrams by cutting only a single internal line. Momentum conservation requires that the momentum of that line be equal to the momentum on the incoming external leg. Thus, once again, we have two types of diagrams: (a) one-particle *reducible* diagrams (which satisfy the property defined above) and (b) the one-particle *irreducible* graphs that do not. Hence, the total contribution to the two-point function is the solution of the equation shown in figure 11.10.

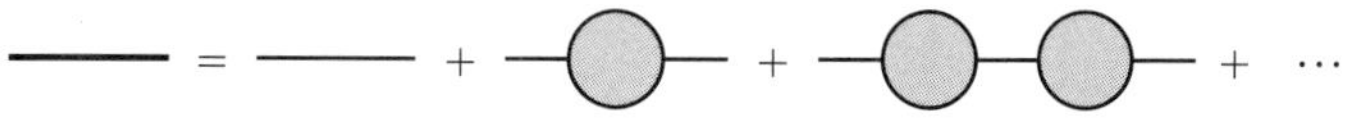

Figure 11.11 Feynman diagrams summed by the Dyson equation.

Here the thick lines are the *full* propagator, the thin line is the *bare* propagator, and the shaded circle represents all the irreducible diagrams (i.e., diagrams with amputated external legs). We represent the shaded circle by the *self-energy* operator $\Sigma(p)$, shown in figure 11.11.

Thus, the total sum satisfies the Dyson equation

$$G^{(2)}(p) = G_0^{(2)}(p) + G_0^{(2)}(p)\,\Sigma(p)\,G^{(2)}(p) \tag{11.56}$$

The inverse of $G^{(2)}(p)$, $\Gamma^{(2)}(p)$, satisfies

$$\Gamma^{(2)}(p) = G_0^{(2)}(p)^{-1} - \Sigma(p) = p^2 + m^2 - \Sigma(p) \tag{11.57}$$

To first order in λ, $\Sigma(p)$ is just the tadpole term

$$\Sigma(p) = -\frac{\lambda}{2}\int \frac{d^Dq}{(2\pi)^D}\frac{1}{q^2+m^2} + O(\lambda^2) \tag{11.58}$$

which, to this order in λ, happens to be independent of the external momentum p_μ. Of course, the higher-order terms in general will be functions of p_μ.

In terms of the renormalized mass μ^2, to order one-loop (i.e., $O(\lambda)$), we get

$$\mu^2 = m^2 - \Sigma(p) = m^2 + \frac{\lambda}{2}\int \frac{d^Dq}{(2\pi)^D}\frac{1}{q^2+m^2} + O(\lambda^2) \tag{11.59}$$

Thus, we conclude that, to $O(\lambda)$, vacuum fluctuations renormalize the mass. However, a quick look at eq. (11.59) reveals that this is a very large renormalization. Indeed, fluctuations of all momenta, ranging from long wavelengths (and low energies) with $q \sim 0$, to short wavelengths (or high energies), contribute to the mass renormalization. In fact, the high-energy fluctuations, with $q^2 \gg m^2$, yield the largest contributions to eq. (11.59), since the mass effectively cuts off the contributions in the infrared (IR), $q \to 0$. Moreover, for all dimensions $D \geq 2$, the high-energy (or ultraviolet, (UV), $q \to \infty$) contribution is divergent. If we were to cut off the integral at a high-momentum scale Λ in general spacetime dimension D, the contribution from this diagram will diverge as Λ^{D-2}. In particular, the tadpole contribution to the mass renormalization is logarithmically divergent for $D = 2$ (i.e., $1+1$) dimensions, and it is quadratically divergent for $D = 4$ dimensions.

Thus, although it is consistent (and quite physical) to regard the leading effect of fluctuations as a mass renormalization, they amount to a divergent change. The reason for this divergence is that all wavelengths contribute, from the IR to the UV. This happens because spacetime is continuous, and we assumed that there is no intrinsic short-distance scale below which local field theory should not be valid.

There is a way to think about this problem. The problem of how to understand the physics of these singular contributions, indeed, of how the *continuum limit* (a theory without cutoff) of QFT is defined is the central purpose of the renormalization group (RG). We will study this approach in detail in chapters 15 and 16. Here we discuss some qualitative features. From the point of view of the RG, the problem is that the continuum theory (i.e., defining a

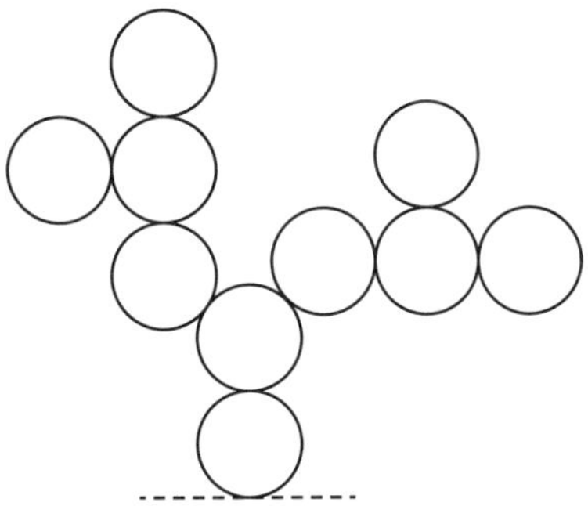

Figure 11.12 A typical tree diagram.

theory without a UV cutoff) cannot be defined naively. We will see that for such a procedure to work, it is necessary to be able to define the theory in a regime in which there is no scale (i.e., in a scale-invariant regime). This requirement means that one should look at a regime in which the renormalized mass becomes arbitrarily small, $\mu^2 \to 0$. As we will see in this section, this requires fine-tuning the bare coupling constant and the bare mass to some pre-determined critical values. It turns out that, near such a critical point, a continuum field theory (without a UV cutoff) can be defined. The RG point of view relates the problem of the definition of a QFT to that of finding a continuous phase transition, which is a central problem in statistical physics.

However, there are alternative descriptions, such as string theory, that postulate that local field theory is not the correct description at short distances, typically near the *Planck scale*, $\ell_{\text{Planck}} = \sqrt{\frac{\hbar G}{c^3}} \sim 10^{-33}\text{cm}$ (!), where G is Newton's gravitational constant. From this viewpoint, these singular contributions at high energies signal a breakdown of the theory at those scales.

Before we try to compute $\Sigma(p)$, it is worth mentioning the *Hartree approximation*. It consists of summing up all tadpole diagrams (and only the tadpole diagrams) to all orders in λ. A typical graph is shown in figure 11.12.

The sum of all the tadpole diagrams can be done by means of a very simple trick. Let us modify the expression for the self-energy to make it self-consistent, that is,

$$\Sigma_0(p) = -\frac{\lambda}{2} \int \frac{d^D q}{(2\pi)^D} \frac{1}{q^2 + m^2 - \Sigma_0(q)} \tag{11.60}$$

This formula is equivalent to a Dyson equation in which the internal propagator is replaced by the full propagator, as in figure 11.13. This approximation becomes exact for a theory of an N-component real scalar field $\phi_a(x)$ $(a = 1, \ldots, N)$, with $O(N)$ symmetry, and interaction (shown in figure 11.14)

$$\mathcal{L}_{\text{int}}[\phi] = \frac{\lambda}{4!} \left((\boldsymbol{\phi})^2\right)^2 = \frac{\lambda}{4!} \left(\sum_{a=1}^{N} \phi_a(x)\phi_a(x) \right)^2 \tag{11.61}$$

in the large-N limit, $N \to \infty$. Otherwise, the solution of this integral equation just yields the leading correction. The reason this works can be seen by directly counting the powers of N in the expansion of the two-point function to order one-loop, as shown in figure 11.15.

Figure 11.13 The self-energy in the one-loop (Hartree) approximation.

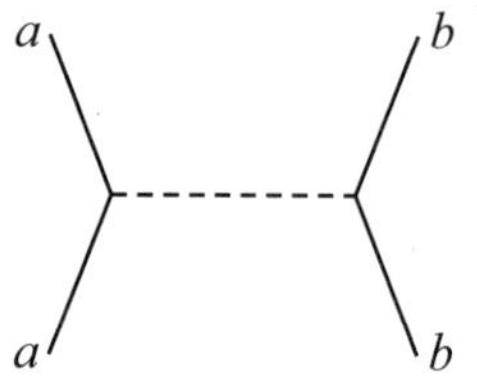

Figure 11.14 The vertex for the real scalar field with $O(N)$ symmetry; here $a, b = 1, \ldots, N$.

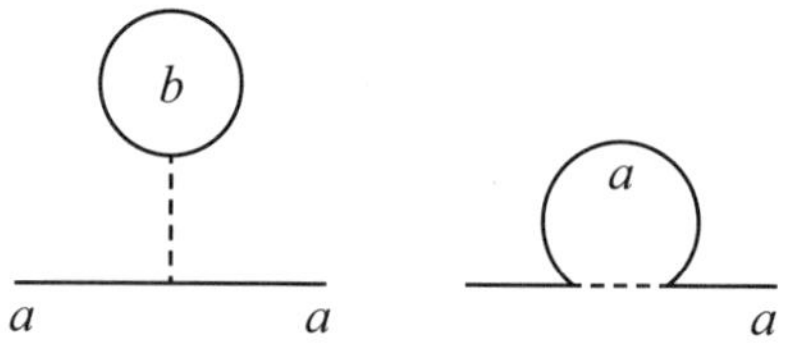

Figure 11.15 The two-point function to one-loop order in the $O(N)$ theory; here $a, b = 1, \ldots, N$. The diagram on the left is of order N, while the diagram on the right is of order 1.

Equation (11.60) is an integral nonlinear equation for $\Sigma_0(p)$. Equations of this type are common in many-body physics. For example, the gap equation of the Bardeen-Cooper-Schrieffer (BCS) theory of superconductivity has a similar form (although in BCS theory, one sums ladder diagrams instead of bubble diagrams).

Let us now evaluate the integral in the equation for $\Sigma_0(p)$, eq. (11.60). Clearly, $\Sigma_0(p)$ is a correction due to virtual fluctuations with momenta q_μ ranging from zero to infinity. These "off-shell" fluctuations do not obey the mass shell condition $p^2 = m^2$. Notice that, at this level of approximation, $\Sigma(p)$ is *independent of the momentum*. This is only correct to order one-loop.

Before computing the integral, let us rewrite eq. (11.60) in terms of the effective or renormalized mass μ^2:

$$\mu^2 = m^2 - \Sigma(p) = m^2 + \frac{\lambda}{2} \int \frac{d^D q}{(2\pi)^D} \frac{1}{q^2 + \mu^2} \tag{11.62}$$

Let us denote by m_c^2 the value of the bare mass such that the renormalized mass vanishes, $\mu^2 = 0$:

$$0 = m_c^2 + \frac{\lambda}{2} \int \frac{d^D q}{(2\pi)^D} \frac{1}{q^2} \tag{11.63}$$

Clearly, m_c^2 is IR divergent for $D \leq 2$ and UV divergent for $D \geq 2$. Let us now express the renormalized mass μ^2 in terms of m_c^2, and define $\delta m^2 = m^2 - m_c^2$. We find

$$
\begin{aligned}
\mu^2 &= \delta m^2 + m_c^2 + \frac{\lambda}{2}\int \frac{d^D q}{(2\pi)^D}\frac{1}{q^2+\mu^2} \\
&= \delta m^2 + \frac{\lambda}{2}\int \frac{d^D q}{(2\pi)^D}\left(\frac{1}{q^2+\mu^2}-\frac{1}{q^2}\right) \\
&= \delta m^2 - \mu^2\frac{\lambda}{2}\int \frac{d^D q}{(2\pi)^D}\frac{1}{q^2(q^2+\mu^2)}
\end{aligned}
\tag{11.64}
$$

In $D=4$ dimensions, the integral contributes with a logarithmic divergence (cf. eq. (11.99) of section 11.8):

$$
\mu^2 = \delta m^2 - \frac{\lambda}{2}\mu^2\left(\frac{1}{8\pi^2}\ln\left(\frac{\Lambda}{\mu}\right)-\frac{\gamma}{16\pi^2}\right)
\tag{11.65}
$$

Thus, although the stronger, quadratic divergence was absorbed by a renormalization of the mass, a weaker logarithmic singularity remains. It turns out, as we discuss in chapter 16, that this remaining singular contribution can be absorbed only by a renormalization of the coupling constant λ. In any case, what is clear is that, already at the lowest order in perturbation theory, the leading corrections can (and do) yield a larger contribution to the behavior of physical quantities than the bare, unperturbed values, and these corrections are not small.

Let us end this section with a discussion of what these large perturbative corrections mean in the context of the theory of phase transitions, since, after all, these transitions are also described by a theory with the same form. We saw in section 2.7 that in the Landau theory of phase transitions, the mass (squared) parameter (i.e., the coefficient of the ϕ^2 term in the action) is related to the difference $m^2 \equiv T - T_0$, where the temperature T is measured from the mean field critical temperature, T_0. We can think of $m_c^2 \equiv T_c - T_0$ as defining a corrected critical temperature T_c:

$$
T_c = T_0 - \frac{\lambda}{2}\int \frac{d^D q}{(2\pi)^D}\frac{1}{q^2}
\tag{11.66}
$$

This shows that fluctuations suppress T_c downward from T_0 due to (precisely) the large contributions from fluctuations at short distances. Thus, the UV singular effects can be absorbed in a new (and lower) T_c. The *subtracted* mass, $\delta m^2 \equiv T - T_c$, is now the new control parameter, as shown in the last line of eq. (11.64).

Returning to the mass renormalization of eq. (11.64), we can use the integral of eq. (11.98) (of section 11.8) to find

$$
\mu^2 = \delta m^2 - \lambda\frac{(\mu^2)^{\frac{D}{2}-1}}{(4\pi)^{D/2}}\frac{\Gamma\left(2-\frac{D}{2}\right)}{D-2}
\tag{11.67}
$$

Since $\delta m^2 = T - T_c$, we can write this result in the suggestive form

$$
\delta m^2 = T - T_c = \mu^2 + \left[\frac{\lambda}{(4\pi)^{D/2}}\frac{\Gamma\left(2-\frac{D}{2}\right)}{D-2}\right](\mu^2)^{\frac{D}{2}-1}
\tag{11.68}
$$

Let us now look at the IR behavior, where the renormalized mass $\mu^2 \to 0$. We see that if $D>4$, the exponent of μ^2 in the second term of the r.h.s. of eq. (11.68) is always larger than 1, and hence this contribution is small compared to the first term. However, for $D < 4$, the reverse is true and for μ^2 small enough, the fluctuations dominate. Hence, perturbation theory breaks down for some small value of μ^2 where the two contributions are comparable. As a result, perturbation theory is reliable (in the IR) for

$$\delta T \gtrsim \mu^2 \approx \left(\frac{\lambda}{(4\pi)^{D/2}} \frac{\Gamma\left(2-\frac{D}{2}\right)}{D-2} \right)^{2/(4-D)} \tag{11.69}$$

and breaks down for δT (or μ^2) smaller than that value. This is known as the Ginzburg criterion, which gives an estimate of the width of the critical region: the range temperatures (or renormalized mass) for which the perturbation theory breaks down in the IR and does not describe the true behavior.

Recall that the relation between the susceptibility and the effective (or renormalized) mass is $\mu^2 = \chi^{-1}$. Hence, at one-loop order, we would predict that

$$\chi(T) \propto \begin{cases} (T-T_c)^{-2/(D-2)}, & D<4 \\ (T-T_c)^{-1}, & D>4 \\ (T-T_c)^{-1} \times \text{small logarithmic corrections}, & D=4 \end{cases} \tag{11.70}$$

Thus one effect of these fluctuations can be to change the dependence of a physical quantity, such as the susceptibility, on the control parameter, $T - T_c$, which sets how close the theory is to the *massless* or *critical* regime. A key purpose of the RG program is the prediction of *critical exponents*, such as the one we found in eq. (11.70). We will see that this result is actually the beginning of a set of controlled approximations to the exact values. Just as important will be the fact that the RG will give a deeper interpretation of the meaning of renormalization.

11.7 The four-point function and the effective coupling constant

Let us now discuss briefly the perturbative contributions to the four-point function,

$$G^{(4)}(x_1, x_2, x_3, x_4) = \langle \phi(x_1)\phi(x_2)\phi(x_3)\phi(x_4) \rangle \tag{11.71}$$

which is also known as the "two-particle" correlation function. We will discuss its connection with the effective (or renormalized) coupling constant.

To zeroth order in perturbation theory, $O(\lambda^0)$, the four-point function factorizes into a product of all (three) possible two-point functions obtained by pairwise contractions of the four field operators (shown in figure 11.16):

$$\begin{aligned} &G^{(4)}(x_1, x_2, x_3, x_4) \\ &\quad = G_0(x_1, x_3)G_0(x_2, x_4) + G_0(x_1, x_2)G_0(x_3, x_4) + G_0(x_1, x_4)G_0(x_2, x_3) + O(\lambda) \end{aligned} \tag{11.72}$$

As it is apparent, to zeroth order in λ, the four-point function reduces to just products of bare two-point functions and hence, nothing new is learned from it. We will show in the

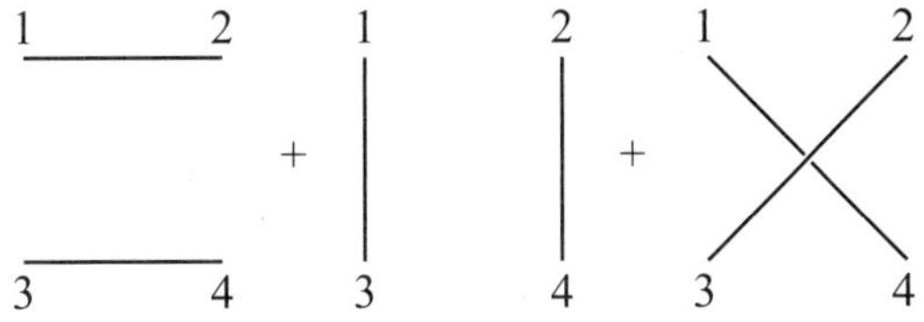

Figure 11.16 The four-point function $G^{(4)}(1,2,3,4)$ to order $O(\lambda^0)$.

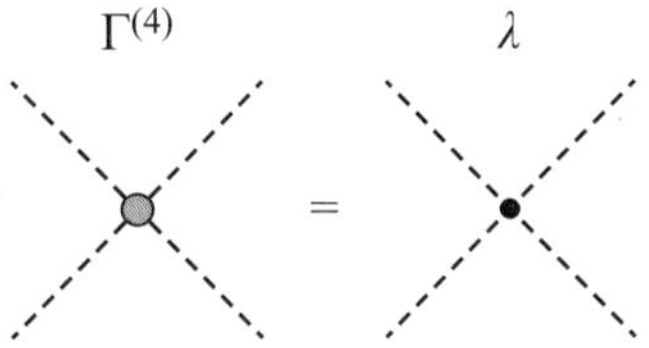

Figure 11.17 The one-particle irreducible four-point (vertex) function at the tree level.

next chapter (eq. (12.40)) that to all orders of perturbation theory, the four-point function has the following structure:

$$\begin{aligned}G^{(4)}(x_1,x_2,x_3,x_4) &= G^{(2)}(x_1,x_3)G^{(2)}(x_2,x_4)\\ &\quad + G^{(2)}(x_1,x_2)G^{(2)}(x_3,x_4) + G^{(2)}(x_1,x_4)G^{(2)}(x_2,x_3)\\ &\quad - \int d^D y_1 \cdots \int d^D y_4\, G^{(2)}(x_1,y_1)G^{(2)}(x_2,y_2)G^{(2)}(x_3,y_3)G^{(2)}(x_4,y_4)\\ &\quad \times \Gamma^{(4)}(y_1,y_2,y_3,y_4)\end{aligned} \tag{11.73}$$

where the factors of $G^{(2)}(x,x')$ represent the *exact two-point function*, and the new four-point function, $\Gamma^{(4)}(y_1,y_2,y_3,y_4)$, is known as the four-point *vertex function*. The vertex function is defined as the set of one-particle irreducible (1-PI) Feynman diagrams (i.e., diagrams that cannot be split in two by cutting a single propagator line), with the external lines "amputated" (they are already accounted for in the propagator factors).

In momentum space, due to momentum conservation at the vertex, $\Gamma^{(4)}$ has the form

$$\Gamma^{(4)}(p_1,\ldots,p_4) = (2\pi)^D\,\delta^D\left(\sum_{i=1}^{4} p_i\right)\overline{\Gamma^{(4)}}(p_1,\ldots,p_4) \tag{11.74}$$

The lowest-order contribution to $\Gamma^{(4)}(y_1,y_2,y_3,y_4)$ appears at order λ

$$\Gamma^{(4)}(y_1,y_2,y_3,y_4) = \lambda + O(\lambda^2) \tag{11.75}$$

depicted by the tree-level diagrams in figure 11.17. In momentum space, the one-particle irreducible four-point function is

$$\overline{\Gamma^{(4)}}(p_1,\ldots,p_4) = \lambda + O(\lambda^2) \tag{11.76}$$

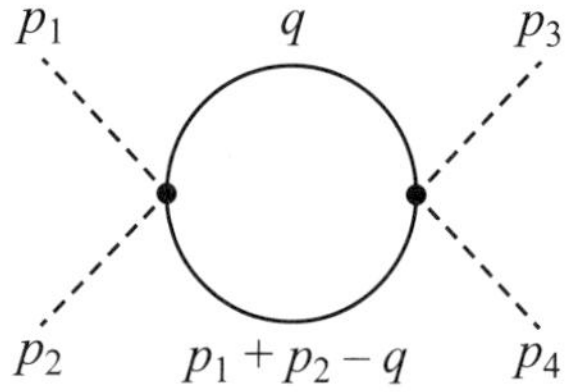

Figure 11.18 The one-loop contribution to the one-particle irreducible four-point (vertex) function.

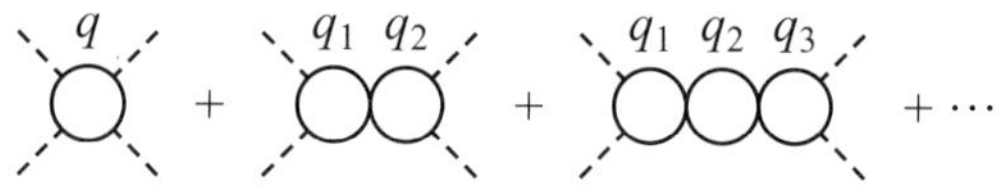

Figure 11.19 The sum of bubble diagrams.

To one-loop order, $O(\lambda^2)$, the four-point vertex function is a sum of (three) Feynman diagrams of the form shown in figure 11.18. The total contribution to the vertex function $\overline{\Gamma^{(4)}}$, to order one-loop, is

$$\overline{\Gamma^{(4)}}(p_1,\ldots,p_4)=$$
$$\lambda-\frac{\lambda^2}{2}\left\{\int\frac{d^Dq}{(2\pi)^D}\frac{1}{\left(q^2+m^2\right)\left((p_1+p_2-q)^2+m^2\right)}+\text{two permutations}\right\}$$
$$+O(\lambda^3) \tag{11.77}$$

This expression has a logarithmic UV divergence in $D=4$, and more severe divergences for $D>4$. To address this problem, let us proceed by analogy with the mass renormalization and define the physical or *renormalized* coupling constant g by the value of $\overline{\Gamma^{(4)}}(p_1,\ldots,p_4)$ at zero external momenta, $p_1=\cdots=p_4=0$ (it is up to us to define it at any momentum scale we wish):

$$g\equiv\lim_{p_i\to 0}\overline{\Gamma^{(4)}}(p_1,\ldots,p_4)=\overline{\Gamma^{(4)}}(0,\ldots,0) \tag{11.78}$$

This definition is convenient and simple, but it is problematic if the renormalized mass μ^2 vanishes (i.e., in the massless or critical theory). To order one-loop, the renormalized coupling constant g is

$$g=\lambda-3\frac{\lambda^2}{2}\int\frac{d^Dq}{(2\pi)^D}\frac{1}{\left(q^2+m^2\right)^2}+O(\lambda^3) \tag{11.79}$$

Using the same line of argument we used to define the self-energy Σ, let us now sum all the one-loop diagrams, as shown in figure 11.19. This "bubble" sum is a geometric series, and it is equivalent to the replacement of eq. (11.79) by

$$g=\lambda-3\frac{g^2}{2}\int\frac{d^Dq}{(2\pi)^D}\frac{1}{\left(q^2+m^2\right)^2}+O(\lambda^3) \tag{11.80}$$

or, alternatively, to write the bare coupling constant λ in terms of the renormalized coupling g:

$$\lambda = g + \frac{3}{2}g^2 \int \frac{d^D q}{(2\pi)^D} \frac{1}{\left(q^2+\mu^2\right)^2} + O(g^3) \tag{11.81}$$

where we have replaced the bare mass m^2 with the renormalized mass μ^2. This replacement amounts to adding tadpole diagrams in the internal propagators. This result is consistent at this order in perturbation theory.

Written in terms of the renormalized coupling constant g and of the renormalized mass μ^2, the vertex function becomes

$$\begin{aligned}\overline{\Gamma^{(4)}}(p_1,\ldots,p_4) &= g - \frac{g^2}{2}\int \frac{d^D q}{(2\pi)^D}\left[\frac{1}{(q^2+\mu^2)\left((p_1+p_2-q)^2+\mu^2\right)} - \frac{1}{\left(q^2+\mu^2\right)^2}\right] \\ &\quad + \text{two permutations} + O(g^3)\end{aligned} \tag{11.82}$$

which is UV finite for $D < 6$. Thus, the renormalization of the coupling constant leads to a subtraction of the singular expression for the vertex function.

After the renormalization of the coupling constant, the singular behavior of the integral now appears only in the relation between the bare coupling constant λ and the renormalized coupling constant g, given in eq. (11.81). Clearly there are several ways of interpreting this relation. One interpretation is to say that at a fixed value of the bare coupling constant λ, eq. (11.81) relates the regulator Λ needed to make the integral finite (which is a momentum scale) and the renormalized coupling constant. In other terms, the effective or renormalized coupling constant g has become a function of a momentum (or energy) scale! Conversely, we can fix the renormalized coupling and ask how we have to change the bare coupling constant λ as we send the regulator to infinity.

It will be useful to work with dimensionless quantities. Since the bare and the renormalized coupling constants have units of Λ^{4-D}, we define the dimensionless bare coupling constant u by

$$\lambda = \Lambda^{4-D} u \tag{11.83}$$

Then

$$\begin{aligned} u &= \Lambda^{D-4}\left(g + \frac{3}{2}g^2 \int \frac{d^D q}{(2\pi)^D}\frac{1}{\left(q^2\right)^2} + O(g^3)\right) \\ &= \Lambda^{D-4}\left(g + \frac{1}{(4\pi)^{D/2}}\frac{6}{(D-2)(D-4)}\Lambda^{D-4} g^2 + O(g^3)\right)\end{aligned} \tag{11.84}$$

We will vary the momentum scale Λ and the dimensionless bare coupling constant u at fixed g. The differential change of the dimensionless coupling constant is known as the RG *beta function*,

$$\beta(u) = -\Lambda \frac{\partial u}{\partial \Lambda}\bigg|_g \tag{11.85}$$

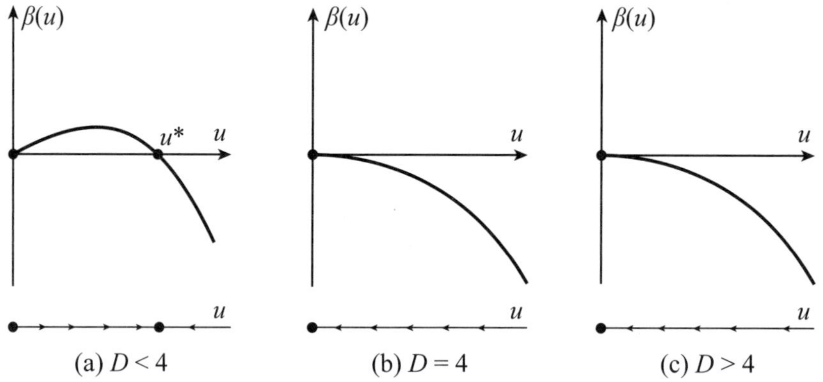

Figure 11.20 The beta function (top) and the IR RG flows (bottom) for (a) $D < 4$, (b) $D = 4$, and (c) $D > 4$. The UV RG flows are the reverse of their IR counterparts.

where I have used the sign convention standard in statistical physics (opposite to that in high-energy physics!). Hence (for $D \to 4$),

$$\beta(u) = (4 - D)u - \frac{3}{16\pi^2}u^2 + O(u^3) \tag{11.86}$$

The behavior of the RG beta function (or Gell-Mann-Low function) for $D < 4$, $D = 4$, and $D > 4$ is shown in figure 11.20. The IR RG flows (i.e., the flow of the dimensionless coupling constant u as the momentum scale λ is decreased) is also shown in figure 11.20 (bottom). The UV flows, i.e., the flow of u as the momentum scale is *increased*, is obtained by reversing the direction of the RG flows.

Clearly for $D \geq 4$, the dimensionless coupling constant flows to 0 in the IR (at low energies and long distances). Thus in the IR regime, $\Lambda \to 0$, the theory becomes *weakly coupled*, and perturbation theory becomes reliable in that regime. However, at short distances (or at high energies), $\Lambda \to \infty$, the opposite happens: The dimensionless coupling becomes large, and perturbation theory breaks down. Four dimensions is special in that the approach and departure from the decoupled limit, $u = 0$, is very slow and leads to logarithmic corrections to the free-field values.

However, for $D < 4$, something new happens: There is a nontrivial fixed point at u^* where $\beta(u^*) = 0$. At the fixed point, the coupling constant does not flow as the momentum scale changes. Hence, at a fixed point, it is possible to send the momentum scale $\Lambda \to \infty$ and effectively have a theory without a cutoff. Notice that even infinitesimally away from the fixed point, the UV flows are unstable. Also (and for the same reason) at the fixed point, it is possible to go into the deep IR regime and have a theory with a finite coupling constant u^*. It turns out that this behavior is central to the theory of phase transitions. We will come back to the problem of the RG in chapter 16, where we will develop it in detail and discuss its application to different theories.

11.8 One-loop integrals

Let us now sketch the computation of one-loop integrals (in Euclidean spacetime). We introduce a momentum cutoff Λ and to suppress the contributions at large momenta, $q \gg \Lambda$

of integrals of the form

$$I_D\left(\frac{\mu^2}{\Lambda^2}\right)=\int \frac{d^Dq}{(2\pi)^D}\frac{1}{q^2+\mu^2}\, e^{-\frac{q^2}{\Lambda^2}} \tag{11.87}$$

where we have used a Gaussian cutoff function (or *regulator*). We will only be interested in the regime $\mu^2 \ll \Lambda^2$. Using a Feynman-Schwinger parametrization, we can write

$$\begin{aligned} I_D\left(\frac{\mu^2}{\Lambda^2}\right)&=\int_0^\infty d\alpha \int \frac{d^Dq}{(2\pi)^D}\, e^{-\frac{q^2}{\Lambda^2}-\alpha(q^2+\mu^2)}\\ &=\int_0^\infty d\alpha\, e^{-\alpha\mu^2}\int \frac{d^Dq}{(2\pi)^D}\, e^{-\left(\frac{1}{\Lambda^2}+\alpha\right)q^2}\\ &=e^{\frac{\mu^2}{\Lambda^2}}\frac{(\mu^2)^{\frac{D}{2}-1}}{(4\pi)^{D/2}}\int_{\mu^2/\Lambda^2}^\infty dt\, t^{-\frac{D}{2}}e^{-t} \end{aligned} \tag{11.88}$$

Hence

$$I_D\left(\frac{\mu^2}{\Lambda^2}\right)=\frac{(\mu^2)^{\frac{D}{2}-1}}{(4\pi)^{D/2}}\,\Gamma\left(1-\frac{D}{2},\frac{\mu^2}{\Lambda^2}\right)e^{\frac{\mu^2}{\Lambda^2}} \tag{11.89}$$

where $\Gamma(\nu,z)$ is the incomplete gamma function, with $z=\frac{\mu^2}{\Lambda^2}$ and $\nu=1-\frac{D}{2}$,

$$\Gamma(\nu,z)=\int_z^\infty dt\, t^{\nu-1}e^{-t} \tag{11.90}$$

and $\Gamma(\nu,0)=\Gamma(\nu)$ is the gamma function

$$\Gamma(\nu)=\int_0^\infty dt\, t^{\nu-1}e^{-t} \tag{11.91}$$

If the regulator Λ is removed (i.e., if we take the limit $\Lambda\to\infty$), $I_D(\mu^2)$ formally becomes:

$$I_D(\mu^2)=\frac{(\mu^2)^{\frac{D}{2}-1}}{(4\pi)^{D/2}}\,\Gamma\left(1-\frac{D}{2}\right) \tag{11.92}$$

For general D, $\Gamma(1-D/2)$ is a meromorphic function of the complex variable D, and has simple poles for $\nu=0$ or any negative integer. Hence, $\Gamma(1-D/2)$ has poles for $D=2,4,6,\ldots$.

In $D=4$ dimensions, $\nu=-1$, where $\Gamma(\nu)$ has a pole, the incomplete gamma function at $\nu=-1$ is (as $z\to 0$)

$$\Gamma(-1,z)=\left(\frac{1}{z}-\ln\frac{1}{z}\right)e^{-z}+\gamma \tag{11.93}$$

where γ is the Euler-Mascheroni constant

$$\gamma=-\int_0^\infty dt\, e^{-t}\,\ln t=0.5772\ldots \tag{11.94}$$

Hence, for $\mu^2 \ll \Lambda^2$, I_4 is

$$I_4\left(\frac{\mu^2}{\Lambda^2}\right) = \frac{\Lambda^2}{16\pi^2} - \frac{\mu^2}{8\pi^2}\ln\left(\frac{\Lambda}{\mu}\right) + \frac{\gamma}{16\pi^2}\mu^2 \tag{11.95}$$

Here we see that the leading singularity is quadratic in the regulator Λ, with a sub-leading logarithmic component.

In two dimensions, I_2 has instead a logarithmic singularity for $\mu^2 \ll \Lambda^2$:

$$\begin{aligned} I_2\left(\frac{\mu^2}{\Lambda^2}\right) &= \frac{1}{4\pi}\Gamma\left(0, \frac{\mu^2}{\Lambda^2}\right) \\ &= \frac{1}{2\pi}\ln\left(\frac{\Lambda}{\mu}\right) - \frac{\gamma}{4\pi} \end{aligned} \tag{11.96}$$

A second integral of interest is

$$\begin{aligned} J_D\left(\frac{\mu^2}{\Lambda^2}\right) &= \int \frac{d^Dq}{(2\pi)^D}\frac{1}{q^2\,(q^2+\mu^2)} \\ &= \frac{1}{\mu^2}\left(I_D(0) - I_D\left(\frac{\mu^2}{\Lambda^2}\right)\right) \end{aligned} \tag{11.97}$$

$J_D(\mu^2/\Lambda^2)$ is UV finite if $D < 4$, where it is given by

$$J_D\left(\mu^2\right) = \frac{(\mu^2)^{\frac{D}{2}-2}}{(4\pi)^{D/2}}\frac{\Gamma\left(\frac{D}{2}-2\right)}{\frac{D}{2}-1} \tag{11.98}$$

In four dimensions, J_4 has a logarithmic divergence:

$$J_4\left(\frac{\mu^2}{\Lambda^2}\right) = \frac{1}{8\pi^2}\ln\left(\frac{\Lambda}{\mu}\right) - \frac{\gamma}{16\pi^2} \tag{11.99}$$

A third useful integral is

$$I'_D\left(\frac{\mu^2}{\Lambda^2}\right) = \int \frac{d^Dq}{(2\pi)^D}\frac{1}{\left(q^2+\mu^2\right)^2}e^{-\frac{q^2}{\Lambda^2}} = -\frac{\partial I_D}{\partial \mu^2} \tag{11.100}$$

In the massless limit, it becomes

$$I'_D(0) = \frac{1}{(4\pi)^{D/2}}\frac{4}{(D-2)(D-4)}\Lambda^{D-4} \tag{11.101}$$

In four dimensions, it becomes

$$I'_4\left(\frac{\mu^2}{\Lambda^2}\right) = \frac{1}{8\pi^2}\ln\left(\frac{\Lambda}{\mu}\right) - \left(\frac{\gamma+1}{16\pi^2}\right) \tag{11.102}$$

Exercises

11.1 Feynman diagrams for a complex scalar field

In this problem, you will study the properties of a theory of a self-interacting complex scalar field $\phi(x)$, in perturbation theory. The Lagrangian density for this theory in four-dimensional Minkowski spacetime is

$$\mathcal{L} = \partial_\mu \phi^*(x)\, \partial_\mu \phi(x) - m_0^2\, \phi^*(x)\phi(x) \\ - \lambda \left(\phi^*(x)\, \phi(x)\right)^2 + J(x)^* \phi(x) + J(x)\phi^*(x) \tag{11.103}$$

Assume that $m_0^2 > 0$ and that the *coupling constant* $\lambda > 0$. Here the $J(x)$ are a set of (complex) sources. You are asked to do calculations using both canonical (operator) and path integral methods.

1) Use path integral methods to derive an explicit expression for the generating function for the vacuum expectation values (v.e.v.s) of time-ordered products $\mathcal{Z}_0[J, J^*]$ for the free field theory both in Minkowski spacetime and in Euclidean spacetime (imaginary time). Assume that the spacetime has infinite extent and that the sources (and the fields) vanish at infinity. Write your answer as an expression involving the sources $J(x)$, and a suitable v.e.v. of a time-ordered product of two free fields.
2) Use the path integral method to calculate, at the level of the free field theory (i.e., $\lambda = 0$) the following time-ordered products of the field $\phi(x)$:

   a) $\langle 0|T\, \phi(x)\phi(x')|0\rangle$,
   b) $\langle 0|T\, \phi^*(x)\phi(x')|0\rangle$,
   c) $\langle 0|T\, \phi(x)\phi^*(x')|0\rangle$,
   d) $\langle 0|T\, \phi^*(x)\phi^*(x')|0\rangle$.

   Determine which of these expectation values are different from zero, and explain why. Use this result to define a set of Feynman propagators for this system. You can give your answer in terms of the Feynman propagator for a real scalar field, $G_0(x, x')$, which was discussed in this chapter.
3) Use path integral methods to derive a formula for the generating function $\mathcal{Z}[J, J^*]$ of the full interacting theory in terms of $\mathcal{Z}_0[J, J^*]$ and of the interaction part of the Lagrangian. Use this formula to derive the Feynman rules for this theory both in Minkowski spacetime and in Euclidean spacetime.
4) Give a general argument to prove that the vacuum diagrams cancel out from the computation of v.e.v.s of any number of field operators.
5) Use the Feynman rules that you derived in part 1 of this exercise to obtain the perturbation theory expansion in Euclidean spacetime for

   a) the vacuum diagrams,
   b) the two-point functions, and
   c) the four-point functions

   in position space and up to and including all terms up to second order in the coupling constant λ. Assign a Feynman diagram to each term. Give an explicit formula for each term, including the multiplicity factors. Do not do the integrals! Verify the cancellation of the vacuum diagrams.

6) Calculate the two-point functions and the four-point functions in Euclidean spacetime in momentum space to order λ for the two-point functions and to order λ^2 for the four-point functions. Draw a Feynman diagram for each term in momentum space. Do not do the integrals!
7) Derive an expression for the self-energy to first order in λ in Euclidean spacetime. Use this result to get a formula for the effective (or renormalized) mass m to first order in λ. Does the mass get renormalized up or down? Can you give a simple explanation for this result? Hint: think of the shift of the energy levels in quantum mechanics.
8) Use a procedure analogous to the one you employed in section 11.7 to define an effective or renormalized coupling constant g in terms of the bare coupling constant λ and of the renormalized mass m, to leading order in λ.
9) Consider the free propagators in Minkowski spacetime. Rotate the integration contour to the imaginary frequency axis. Explain the relation between this procedure and the analytic continuation to imaginary time. What domain of spacetime events is actually described by this procedure? Show that the integrals that you get are equivalent to integrals in Euclidean spacetime with dimension $D=4$.
10) Calculate the momentum integrals of part 7 and part 8 for the domain indicated in section 11.8. Here, assume that the spacetime Euclidean dimension D is arbitrary. Show that the corrections to the two-point functions diverge for $D \geq 2$, and for the four-point functions, they diverge for $D \geq 4$. Use the formulas given at the end of this problem to show that, in the complex D plane, the integrals have isolated poles in D. Assume that the external momenta in the case of the four-point functions are taken to be at the symmetry point $p_j \cdot p_k = -\frac{1}{4}\kappa^2$, $p_j^2 = \kappa^2$, where κ is an arbitrary momentum scale.

    You will find it useful to know the following integrals in Euclidean momentum space:

$$\frac{1}{A^n B^m} = \frac{\Gamma(n+m)}{\Gamma(n)\Gamma(m)} \int_0^1 dx \, \frac{x^{n-1}(1-x)^{m-1}}{(xA+(1-x)B)^{n+m}} \tag{11.104}$$

$$I_{D,n} = \int \frac{d^D p}{(2\pi)^D} \frac{1}{\left(p^2+2p\cdot q+m_0^2\right)^n} = \frac{1}{2}\frac{S_D}{(2\pi)^D}\frac{\Gamma(\frac{D}{2})\Gamma(n-\frac{D}{2})}{\Gamma(n)}\left(m_0^2-q^2\right)^{\frac{D}{2}-n} \tag{11.105}$$

where S_D is the volume of the D-dimensional unit hypersphere

$$S_D = \frac{2\pi^{\frac{D}{2}}}{\Gamma\left(\frac{D}{2}\right)} \tag{11.106}$$

and $\Gamma(s)$ is the Γ-function

$$\Gamma(s) = \int_0^\infty dt \, t^{s-1} e^{-t} \tag{11.107}$$

For $s \to 0$, the Γ-function has the asymptotic behavior

$$\Gamma(s) = \frac{\Gamma(s+1)}{s} = \frac{1}{s} - \gamma + O(s) \tag{11.108}$$

where γ is the Euler constant

$$\gamma = 0.57721... = \lim_{m\to\infty} \left[1 + \frac{1}{2} + \frac{1}{3} + \cdots + \frac{1}{m} - \ln m\right] \tag{11.109}$$

11.2 Perturbative $SU(2)$ Yang-Mills gauge theory

Here you will consider the $SU(2)$ Yang-Mills gauge field $A_\mu(x)$ in a four-dimensional Minkowski spacetime, coupled to a Fermi field $\psi(x)$. Under local gauge transformations, the Fermi field (e.g., quarks) is assumed to transform like the spinor (fundamental) representation of $SU(2)$. The Lagrangian density for this $SU(2)$ analog of QCD (with one species of quark) Yang-Mills gauge theory in the Feynman–'t Hooft gauge is

$$\mathcal{L} = \bar{\psi} i \not{D} \psi - \frac{1}{2}\text{tr}\left(F_{\mu\nu}F^{\mu\nu}\right) + \frac{\lambda}{2}\text{tr}\left(\partial_\mu A^\mu\right)^2 - \bar{\eta}_a \partial_\mu D^\mu_{ab} \eta_b \tag{11.110}$$

where ψ is a 4-spinor Dirac fermion, A_μ is the gauge field, and η and $\bar{\eta}$ are Faddeev-Popov ghosts. Recall the following definitions:

$$\not{D} = \gamma^\mu D_\mu \tag{11.111}$$

where

$$D_\mu = \partial_\mu + igA_\mu = \partial_\mu + it^a A^a_\mu \tag{11.112}$$

is the covariant derivative in the fundamental representation,

$$D^\mu_{ab} = \partial^\mu \delta_{ab} + g f_{abc} A^\mu_c \tag{11.113}$$

is the covariant derivative in the adjoint representation, and

$$F_{\mu\nu} = t^a F^a_{\mu\nu} \tag{11.114}$$

is the field strength, where

$$F^a_{\mu\nu} = \partial_\mu A^a_\nu - \partial_\nu A^a_\mu + g f^{abc} A^b_\mu A^c_\nu \tag{11.115}$$

Here the f^{abc} are $SU(2)$ structure constants (i.e., $f^{abc} = \epsilon^{abc}$), and the t^a (with $a = 1, 2, 3$) are three generators of $SU(2)$ that satisfy

$$\left[t^a, t^b\right] = i f^{abc} t^c, \qquad \text{tr}\left(t^a t^b\right) = \frac{1}{2}\delta_{ab} \tag{11.116}$$

For $SU(2)$ in the fundamental (spinor) representation, the generators are simply related to the three 2×2 hermitian Pauli matrices $\{\sigma^a\}$:

$$t^a = \frac{1}{2}\sigma^a \tag{11.117}$$

1) Write down the Feynman propagators in momentum space for the quark fields ψ, the gluons (gauge fields) A^a_μ, and the Faddeev-Popov ghosts η^a and $\bar{\eta}^a$.

2) Use the Lagrangian density of eq. (11.110) to derive the Feynman rules for this theory in position space. Draw diagrams that you feel may be necessary.
3) Use the (Feynman gauge) Feynman rules derived in the previous part to find the one-loop correction in a perturbation expansion in powers of the gauge coupling constant g for

   a) the fermion propagator
   b) the gauge field propagator
   c) the fermion-gauge field vertex
   d) the three- and four-gauge field vertices

   in this gauge. Do not do the integrals. Draw a Feynman diagram for each contribution.

12

Vertex Functions, the Effective Potential, and Symmetry Breaking

12.1 Connected, disconnected, and irreducible propagators

In this chapter, we return to the structure of perturbation theory in a canonical local field theory that we discussed in chapter 11. The results derived here for the simpler case of a scalar field theory apply, with some changes, to any local field theory, relativistic or not.

Let us suppose that we want to compute the four-point function in ϕ^4 theory of a scalar field, $G_4(x_1, x_2, x_3, x_4)$. Obviously, there is a set of graphs in which the four-point function is reduced to products of two-point functions:

$$\begin{aligned} G_4(x_1, x_2, x_3, x_4) &= \langle \phi(x_1)\phi(x_2)\phi(x_3)\phi(x_4)\rangle \\ &= G_2(x_1, x_2)G_2(x_3, x_4) + \text{permutations} + \text{other terms} \end{aligned} \tag{12.1}$$

An example of such diagrams is shown in figure 12.1. This graph is linked (i.e., it has no vacuum part), but it is disconnected, since we can split the graph into two pieces by drawing a line without cutting any propagator line.

However, as we have already seen, the N-point function can be computed from the generating functional $Z[J]$ by functional differentiation with respect to the sources $J(x)$, that is,

$$G_N(x_1, \ldots, x_N) = \frac{1}{Z[J]} \frac{\delta^N Z[J]}{\delta J(x_1) \cdots \delta J(x_N)}\bigg|_{J=0} \tag{12.2}$$

Let us now compute instead the following expression:

$$G_N^c(x_1, \ldots, x_N) = \frac{\delta^N \ln Z[J]}{\delta J(x_1) \cdots \delta J(x_N)}\bigg|_{J=0} \tag{12.3}$$

We will now see that $G_N^{(c)}(x_1, \ldots, x_N)$ is an N-point function that contains only connected Feynman diagrams.

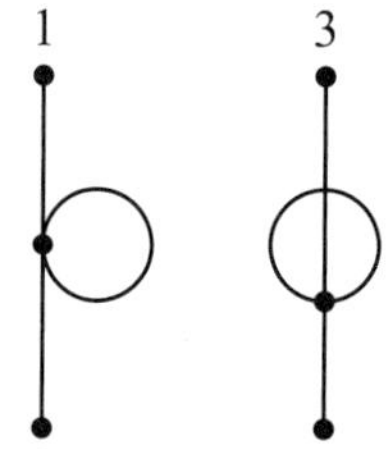

Figure 12.1 A factorized contribution to the four-point function.

As an example, let us consider first the two-point function $G_2^c(x_1, x_2)$, which is formally given by the expression:

$$G_2^c(x_1,x_2) = \left.\frac{\delta^2 \ln Z[J]}{\delta J(x_1)\delta J(x_2)}\right|_{J=0} \tag{12.4}$$

$$= \left.\frac{\delta}{\delta J(x_1)}\frac{1}{Z[J]}\frac{\delta Z[J]}{\delta J(x_2)}\right|_{J=0} \tag{12.5}$$

$$= \left.\frac{1}{Z[J]}\frac{\delta^2 Z[J]}{\delta J(x_1)\delta J(x_2)}\right|_{J=0} - \left.\frac{1}{Z[J]}\frac{\delta Z[J]}{\delta J(x_1)}\right|_{J=0}\left.\frac{1}{Z[J]}\frac{\delta Z[J]}{\delta J(x_2)}\right|_{J=0} \tag{12.6}$$

Thus, we find that the connected two-point function can be expressed in terms of the two-point function and the one-point functions,

$$G_2^c(x_1,x_2) = G_2(x_1,x_2) - G_1(x_1)G_1(x_2) \tag{12.7}$$

We can express this result equivalently in the form

$$\langle \phi(x_1)\phi(x_2)\rangle_c = \langle \phi(x_1)\phi(x_2)\rangle - \langle \phi(x_1)\rangle\langle \phi(x_2)\rangle \tag{12.8}$$

and the quantity

$$G_2^c(x_1,x_2) = \langle \phi(x_1)\phi(x_2)\rangle_c = \langle [\phi(x_1) - \langle \phi(x_1)\rangle]\,[\phi(x_2) - \langle \phi(x_2)\rangle]\rangle \tag{12.9}$$

is called the *connected* two-point function. Hence, it is the two-point function of the field $\phi(x) - \langle \phi(x)\rangle$ that has been normal ordered with respect to the true vacuum. It is straightforward to show that the same identification holds for the connected N-point functions.

The generating functional of the connected N-point functions $F[J]$,

$$F[J] = \ln Z[J] \tag{12.10}$$

is identified with the free energy (or *vacuum energy*) of the system. The connected N-point functions are obtained from the free energy by

$$G_N^c(x_1,\ldots,x_N) = \left.\frac{\delta^N F[J]}{\delta J(x_1)\cdots\delta J(x_N)}\right|_{J=0} \tag{12.11}$$

Notice that the connected N-point functions play a role analogous to the cumulants (or moments) of a probability distribution.

Recall that in the theory of phase transitions, the source $J(x)$ plays the role of the symmetry-breaking field $H(x)$ that breaks the global symmetry of the scalar field theory (e.g., the external magnetic field in the Landau theory of magnetism). For a uniform external field $H(x) = H$, we find

$$\frac{\delta F}{\delta J} = \frac{dF}{dH} = \int d^d x \langle \phi(x) \rangle = V \langle \phi \rangle \tag{12.12}$$

In terms of the free-energy density $f = \frac{F}{V}$, we can write

$$\frac{df}{dH} = \langle \phi \rangle = m \tag{12.13}$$

where m is the magnetization density. Similarly, the magnetic susceptibility χ,

$$\chi = \frac{dm}{dH} = \frac{d^2 f}{dH^2} \tag{12.14}$$

is given by an integral of the two-point function:

$$\begin{aligned} \chi &= \frac{1}{V} \frac{d}{dH} \left\langle \int d^d x_1 \phi(x_1) \right\rangle \\ &= \frac{1}{V} \left\langle \int d^d x_1 \int d^d x_2 \phi(x_1) \phi(x_2) \right\rangle - \frac{1}{V} \left\langle \int d^d x_1 \phi(x_1) \right\rangle \left\langle \int d^d x_2 \phi(x_2) \right\rangle \end{aligned} \tag{12.15}$$

Using the fact that in a translation-invariant system, the expectation value of the field is constant, and that the two-point function is only a function of distance, we find

$$\begin{aligned} \chi &= \frac{1}{V} \int d^d x_1 \int d^d x_2 \, G_2(x_1, x_2) - V \langle \phi \rangle^2 \\ &= \int d^d y \, G_2(|\mathbf{y}|) - V \langle \phi \rangle^2 \\ &= \int d^d y \, G_2^c(|\mathbf{y}|) = \lim_{\boldsymbol{k} \to 0} G_2^c(\boldsymbol{k}) \end{aligned} \tag{12.16}$$

Hence, the magnetic susceptibility is the integral of the connected two-point function, which is also known as the correlation function.

12.2 Vertex functions

So far, we have been able to reduce the number of diagrams to be considered by:

1) showing that vacuum parts do not contribute to $G_N(x_1, \ldots, x_N)$, and
2) showing that disconnected parts need not be considered by working instead with the connected N-point function, $G_N^c(x_1, \ldots, x_N)$.

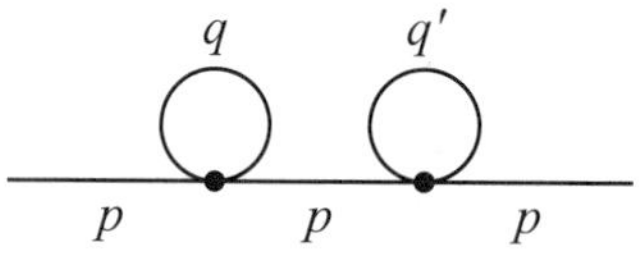

Figure 12.2 A reducible contribution to the two-point function.

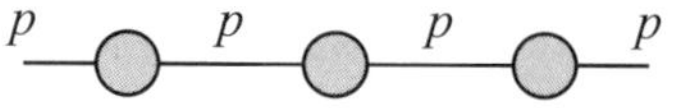

Figure 12.3 Three blobs.

There is still another set of graphs that can be handled easily. Consider the second-order contribution to the connected two-point function $G_2^c(x_1, x_2)$ shown in figure 12.2. The explicit form of this contribution is, in momentum space, given by the following expression:

$$\left(-\frac{\lambda}{4!}\right)^2 \frac{1}{2!}(4\times 3)\cdot(4\times 3)\left(G_0(p)\right)^3 \int \frac{d^dq}{(2\pi)^d} G_0(q) \int \frac{d^dq'}{(2\pi)^d} G_0(q') \tag{12.17}$$

Note two features of this contribution. One is that the momentum in the middle propagator line is the same as the momentum p of the external line. This follows from momentum conservation. The other feature is that this graph can be split in two by a line that cuts either the middle propagator line or any of the two external propagator lines. A graph that can be split into two disjoint parts by cutting a single propagator line is said to be one-particle reducible. No matter how complicated the graph is, that line must have the same momentum as the momentum on an incoming leg (again, by momentum conservation).

In general, we need to do a sum of diagrams with the structure shown in figure 12.3, which represents the expression

$$G_0^4(p)(\Sigma(p))^3 \tag{12.18}$$

and where the shaded circles in figure 12.3 represent the self-energy $\Sigma(p)$ (i.e., the sum of one-particle irreducible diagrams of the two-point function). In fact, we can do this sum to all orders and obtain

$$\begin{aligned} G_2(p) &= G_0(p) + G_0(p)\Sigma(p)G_0(p) + G_0^3(p)(\Sigma(p))^2 + \cdots \\ &= G_0(p)\sum_{n=0}^{\infty}(\Sigma(p)G_0(p))^n \\ &= \frac{G_0(p)}{1-\Sigma(p)G_0(p)} \end{aligned} \tag{12.19}$$

We can write this result in the equivalent form

$$G_2^{-1}(p) = G_0^{-1}(p) - \Sigma(p) \tag{12.20}$$

Armed with this result, we can express the relation between the bare and the full two-point function as the *Dyson equation*

$$G_2(p) = G_0(p) + G_0(p)\Sigma(p)G_2(p) \tag{12.21}$$

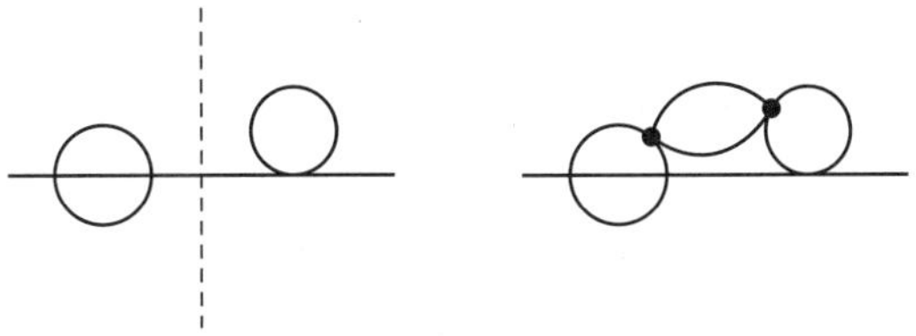

Figure 12.4 One-particle reducible and one-particle irreducible diagrams.

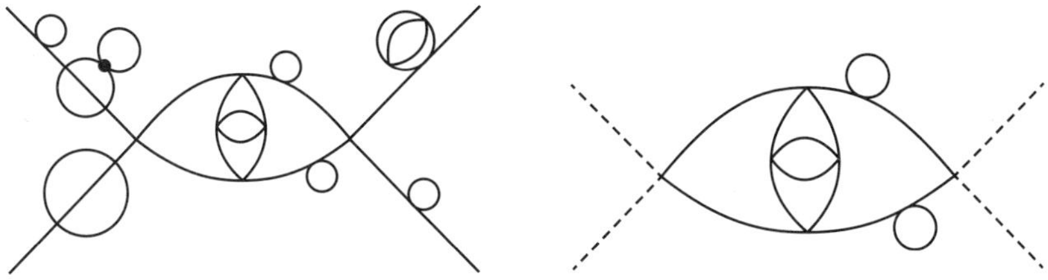

Figure 12.5 A contribution (a) to a one-particle reducible vertex function, and (b) to a one-particle irreducible vertex function.

where, as before, $\Sigma(p)$ represents the set of all possible connected one-particle irreducible graphs with their external legs amputated as in the example shown in figure 12.4.

The one-particle irreducible two-point function $\Sigma(p)$ is known as the mass operator or as the self-energy (or two-point vertex). Why? In the limit $p \to 0$, the inverse bare propagator reduces to the bare mass (squared):

$$G_0^{-1}(0) = m_0^2 \tag{12.22}$$

Similarly, also in the zero-momentum limit, the inverse full propagator takes the value of the effective (or renormalized) mass (squared)

$$G_2^{-1}(0) = m_0^2 - \Sigma(0) = m^2 \tag{12.23}$$

Thus, $\Sigma(0)$ represents a renormalization of the mass.

Let us now extend the concept of the sum of one-particle irreducible diagrams to a general N-point function, the *vertex functions*. Examples of reducible and one-particle irreducible contributions to the four-point vertex function are shown in figure 12.5. To this end, we will need to find a suitable generating functional for these correlators.

In chapter 11, we considered the free energy $F[J]$ and showed that it is the generating functional of the connected N-point functions. $F[J]$ is a function of the external sources J. However, in many cases, this is inconvenient, since in systems that exhibit *spontaneous symmetry breaking*, as $J \to 0$, we may still have $\langle \phi \rangle \neq 0$. Thus, it will be desirable to have a quantity that is a functional of the expectation values of the observables instead of the sources. So we will seek instead a functional of the expectation values rather than of the external sources. We will find this functional by means of a Legendre transformation from the sources J to the expectation values $\langle \phi \rangle$. This procedure is closely analogous to the relation in thermodynamics between the Helmholtz free energy and the Gibbs free energy.

The local expectation value of the field, $\langle\phi(x)\rangle \equiv \bar{\phi}(x)$, is related to the functional $F[J]$ by

$$\langle\phi(x)\rangle = \frac{\delta F}{\delta J(x)} \tag{12.24}$$

The Legendre transform of $F[J]$, denoted by $\Gamma\left[\bar{\phi}\right]$, is defined by

$$\Gamma[\bar{\phi}] = \int d^d x\, \bar{\phi}(x) J(x) - F[J] \tag{12.25}$$

where, for simplicity, we have omitted all other indices (e.g., components of the scalar field). What we do below can be easily extended to theories with other types of fields and symmetries.

Let us now compute the functional derivative of the generating functional $\Gamma[\bar{\phi}]$ with respect to $\bar{\phi}(x)$. After some simple algebra, we find

$$\begin{aligned}\frac{\delta\Gamma[\bar{\phi}]}{\delta\bar{\phi}(x)} &= \int d^d y\, J(y)\delta(y-x) + \int d^d y\, \bar{\phi}(y)\frac{\delta J(y)}{\delta\bar{\phi}(x)} - \int d^d y\, \frac{\delta F}{\delta J(y)}\frac{\delta J(y)}{\delta\bar{\phi}(x)} \\ &= J(x) + \int d^d y\, \bar{\phi}(y)\frac{\delta J(y)}{\delta\bar{\phi}(x)} - \int d^d y\, \bar{\phi}(y)\frac{\delta J(y)}{\delta\bar{\phi}(x)}\end{aligned} \tag{12.26}$$

Since the last two terms on the right-hand side cancel each other, we find

$$\frac{\delta\Gamma[\bar{\phi}]}{\delta\bar{\phi}(x)} = J(x) \tag{12.27}$$

However, let us consider a theory in which, even in the limit $J \to 0$, $\bar{\phi}(x)$ may still be nonzero. On symmetry grounds, one expects that if the source vanishes, then the expectation value of the field should also vanish. However, in many situations, the expectation value of the field does not vanish in the limit of a vanishing source. In this case, we say that a symmetry is *spontaneously broken* if $\bar{\phi}(x) \neq 0$ as $J \to 0$. An example is a magnet, where ϕ is the local magnetization, and J is the external magnetic field. Another example, from the theory of Dirac fermions, is the bilinear $\bar{\psi}\psi$ which is the order parameter for chiral symmetry breaking, and the fermion mass is the symmetry-breaking field.

Returning to the general case, since the expectation value $\bar{\phi}(x)$ satisfies the extremal condition $\frac{\delta\Gamma[\bar{\phi}]}{\delta\bar{\phi}(x)} = 0$, this state is an extremum of the potential Γ. Naturally, for the state to be stable, it must also be a minimum, not just an extremum. The value of $\bar{\phi}(x)$ is known as the *classical field*.

The functional $\Gamma[\bar{\phi}]$ can be formally expanded in a Taylor series expansion of the form

$$\Gamma[\bar{\phi}] = \sum_{N=0}^{\infty} \frac{1}{N!}\int_{dz_1\ldots dz_N} \Gamma^{(N)}(z_1,\ldots,z_N)\bar{\phi}(z_1)\cdots\bar{\phi}(z_N) \tag{12.28}$$

The coefficients

$$\Gamma^{(N)}(z_1,\ldots,z_N) = \frac{\delta\Gamma[\phi]}{\delta\bar{\phi}(z_1)\cdots\delta\bar{\phi}(z_N)} \tag{12.29}$$

are the N-point vertex functions.

To find relations between the vertex functions and the connected functions, differentiate the classical field $\bar{\phi}(x)$ by $\bar{\phi}(y)$, resulting in

$$\delta(x-y) = \frac{\delta^2 F}{\delta J(x)\delta\bar{\phi}(y)} = \int_z \frac{\delta^2 F}{\delta J(x)\delta J(z)} \frac{\delta J(z)}{\delta\bar{\phi}(y)}$$
$$= \int_z \frac{\delta^2 F}{\delta J(x)\delta J(z)} \frac{\delta^2 \Gamma}{\delta\bar{\phi}(z)\delta\bar{\phi}(y)} \tag{12.30}$$

Since the connected two-point function $G_2^c(x-z)$ is given by

$$G_2^c(x-z) = \left.\frac{\delta^2 F}{\delta J(x)\delta J(z)}\right|_{J=0} \tag{12.31}$$

we see that the operator

$$\Gamma^{(2)}(x-y) = \left.\frac{\delta^2 \Gamma}{\delta\bar{\phi}(x)\bar{\phi}(y)}\right|_{J=0} \tag{12.32}$$

is the inverse of $G_2^c(x-y)$ (as an operator).

We can gain further insight by passing to momentum space, where we find

$$\Gamma^{(2)}(p) = \left[G_2^c(p)\right]^{-1} = p^2 + m_0^2 - \Sigma(p) \tag{12.33}$$

Thus, $\Gamma^{(2)}(p)$ is essentially the negative of the self-energy, and it is the sum of all 1-PI graphs of the two-point function.

To find relations of this type for more general N-point functions, differentiate eq. (12.30) by $J(u)$ to obtain

$$\frac{\delta}{\delta J(u)}\delta(x-y) = 0$$
$$= \int_z \left[\frac{\delta^3 F}{\delta J(x)\delta J(z)\delta J(u)} \frac{\delta J(y)}{\delta\bar{\phi}(y)} + \frac{\delta^2 F}{\delta J(x)\delta J(z)} \frac{\delta^2 J(z)}{\delta J(u)\delta\bar{\phi}(y)}\right] \tag{12.34}$$

But since

$$\frac{\delta^2 J(z)}{\delta J(u)\delta\bar{\phi}(y)} = \int_w \frac{\delta^3 \Gamma}{\delta\bar{\phi}(w)\delta\bar{\phi}(z)\delta\bar{\phi}(y)} \frac{\delta\bar{\phi}(w)}{\delta J(u)}$$
$$= \int_w \frac{\delta^3 \Gamma}{\delta\bar{\phi}(w)\delta\bar{\phi}(z)\delta\bar{\phi}(y)} \frac{\delta^2 F}{\delta J(u)\delta J(w)} \tag{12.35}$$

we get

$$0 = \int_z G_3^c(x,z,y)\Gamma^{(2)}(z-y) + \int_{z,w} G_2^c(x-z)G_2^c(u-w)\Gamma^{(3)}(w,z,y) \tag{12.36}$$

where $\Gamma^{(2)} = [G_2]^{-1}$. Hence, we find the expression for the connected three-point function at points x_1, x_2, and x_3:

$$G_3^c(x_1,x_2,x_3) = -G_2^c(x_1,y_1)G_2^c(x_2,y_2)G_2^c(x_3,y_3)\Gamma^{(3)}(y_1,y_2,y_3) \tag{12.37}$$

where repeated labels are integrated over.

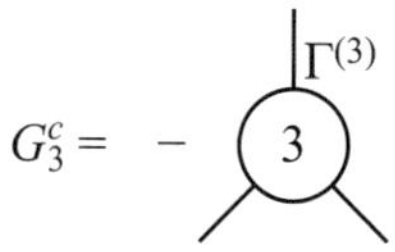

Figure 12.6 The three-point vertex function.

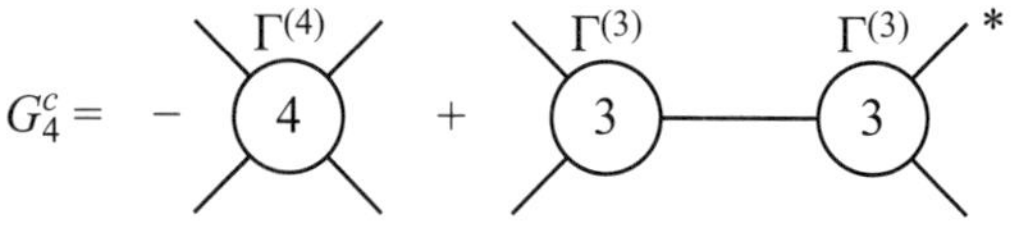

Figure 12.7 The four-point vertex function.

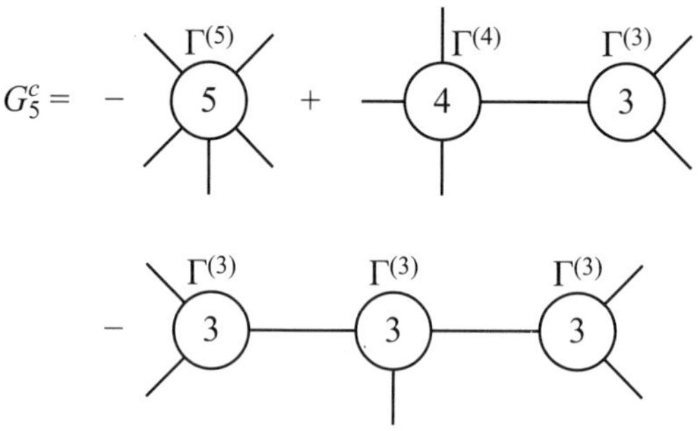

Figure 12.8 The five-point vertex function.

Notice in passing that we can write the two-point function in a similar fashion:

$$G_2^c(x_1, x_2) = G_2^c(x_1, y_1)G_2^c(x_2, y_2)\Gamma^{(2)}(y_1, y_2) \tag{12.38}$$

(again with repeated labels being integrated over), since $G_2^c = [\Gamma^{(2)}]^{-1}$.

Thus, $\Gamma^{(3)}$ is the 1-PI three-point vertex function. Eq. (12.37) has the pictorial representation shown in figure 12.6, where the circle is the three-point vertex function, and the sticks are connected two-point functions.

If we now further differentiate eq. (12.34) with respect to additional fields $\bar{\phi}$, we obtain relations between the four-point function (and the lower-point functions) shown in figure 12.7. Here the circles are four- and three-point vertex functions, and the sticks are again connected two-point functions.

This procedure generalizes to the higher-point functions. Figure 12.8 and figure 12.9 show the pictorial representation for the connected five- and six-point functions in terms of the corresponding vertex functions and connected two-point functions. In figure 12.9, the symbol (*) means that the respective diagram is one-particle reducible by a body cut. In each diagram, the summation over all possible equivalent combinations is implied. Clearly, a graph can be reducible either by a cut of only an external line or by a body cut.

In general, from the definition of the vertex function $\Gamma^{(N)}$,

$$\Gamma^{(N)}(1, \ldots, N) = \frac{\delta^N \Gamma(\bar{\phi})}{\delta\bar{\phi}(1) \cdots \delta\bar{\phi}(N)}\Big|_{J=0} \tag{12.39}$$

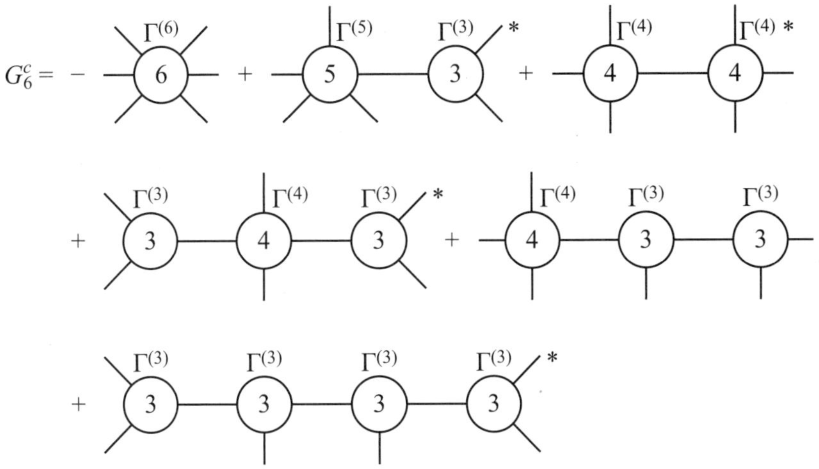

Figure 12.9 The six-point vertex function.

we find that the connected N-point function, for $N > 2$, is related to the vertex function by (repeated labels are again integrated over)

$$G_N^c(1,\ldots,N) = -G_2^c(1,1')\cdots G_2^c(N,N')\Gamma^{(N)}(1',\ldots,N') + Q^{(N)}(1,\ldots,N) \tag{12.40}$$

where the first term is one-particle reducible only via cuts of the external legs and the second by body cuts. Notice that for the r-point function in a ϕ^r theory, this second term vanishes.

In momentum space, these expressions become simpler. Thus, the connected two-point function obeys

$$G_2^c(k_1,k_2) = (2\pi)^d\delta^d(\mathbf{k}_1+\mathbf{k}_2)G_2^c(k_1) \tag{12.41}$$

By using eq. (12.33), we can write the two-point vertex function as

$$\Gamma^{(2)}(k_1,k_2) = (2\pi)^d\delta^d(\mathbf{k}_1+\mathbf{k}_2)\Gamma^{(2)}(k_1) \tag{12.42}$$

In the general case, $N > 2$, we have

$$G_N^c(k_1,\ldots,k_N) = -G_2^c(k_1)\cdots G_2^c(k_N)\Gamma^{(N)}(k_1,\ldots,k_N) + Q^{(N)}(k_1,\ldots,k_N) \tag{12.43}$$

In what follows, we focus on the vertex functions.

12.3 The effective potential and spontaneous symmetry breaking

Let $v = \bar{\phi} = \langle\phi\rangle$. Then, with the above definition for the vertex functions $\Gamma^{(N)}$, we can write the generating function $\Gamma[\bar{\phi}]$ as a power series expansion of the form

$$\Gamma[\bar{\phi}] = \sum_{N=1}^{\infty}\frac{1}{N!}\int d^dx_1\cdots\int d^dx_N\Gamma^{(N)}(x_1,\ldots,x_N|v)[\bar{\phi}(x_1)-v]\cdots[\bar{\phi}(x_N)-v] \tag{12.44}$$

If $J \to 0$, then the sum starts at $N = 2$. Here $v = \lim_{J\to 0} \bar{\phi}$, which is a local minimum of $\Gamma[\bar{\phi}]$ since

$$\left.\frac{\delta \Gamma}{\delta \bar{\phi}}\right|_{\bar{\phi}=v} = J \mapsto 0, \quad \text{and} \quad \Gamma^{(2)}\Big|_{\bar{\phi}=v} \geq 0 \tag{12.45}$$

In the symmetric phase of the theory, the generating function $\Gamma[\bar{\phi}]$ has the form

$$\Gamma\left[\bar{\phi}\right] = \sum_{N=1}^{\infty} \frac{1}{N!} \int d^d x_1 \cdots \int d^d x_N \Gamma^{(N)}(x_1, \ldots, x_N) \bar{\phi}(x_1) \cdots \bar{\phi}(x_N) \tag{12.46}$$

The classical field $\bar{\phi} = v$ is defined by the condition $\frac{\delta \Gamma}{\delta \bar{\phi}} = 0$. If $\bar{\phi} \neq 0$, then the global symmetry $\phi \leftrightarrow -\phi$ is spontaneously broken. Moreover, for $\langle \phi \rangle = \bar{\phi} = \text{const.}$, the generating functional Γ becomes

$$\Gamma\left[\bar{\phi}\right] = \sum_{N=2}^{\infty} \frac{1}{N!} \left[\left(\int d^d x_1 \cdots \int d^d x_N \Gamma^{(N)}(x_1, \ldots, x_N)\right)\right] \bar{\phi}^N \tag{12.47}$$

The Fourier transform of $\Gamma^{(N)}(x_1, \ldots, x_N)$ is

$$\Gamma^{(N)}(x_1, \ldots, x_N) = \int \frac{d^d k_1}{(2\pi)^d} \cdots \int \frac{d^d k_N}{(2\pi)^d} \Gamma^{(N)}(k_1, \ldots, k_N) e^{-i\Sigma_{j=1}^N \mathbf{k}_j \cdot \mathbf{x}_j} \tag{12.48}$$

Momentum conservation requires that $\Gamma^{(N)}(k_1, \ldots, k_N)$ should take the form

$$\Gamma^{(N)}(k_1, \ldots, k_N) = (2\pi)^d \delta^d \left(\sum_j \mathbf{k}_j\right) \widetilde{\Gamma}^{(N)}(k_1, \ldots, k_N) \tag{12.49}$$

So we find that $\Gamma(\bar{\phi})$ is given by the expression

$$\Gamma(\bar{\phi}) = V \sum_{N=2}^{\infty} \frac{1}{N!} \widetilde{\Gamma}^{(N)}(0, \ldots, 0)\, \bar{\phi}^N \tag{12.50}$$

where V is the volume of Euclidean spacetime. Clearly, we can also write $\Gamma(\bar{\phi}) = VU(\bar{\phi})$, where

$$U(\bar{\phi}) = \sum_{N=2}^{\infty} \frac{1}{N!} \widetilde{\Gamma}^{(N)}(0, \ldots, 0)\, \bar{\phi}^N \tag{12.51}$$

is the *effective potential*.

Note that the $\widetilde{\Gamma}^{(N)}(0, \ldots, 0)$ terms are computed in the symmetric theory. In this framework, if U has a minimum at $\bar{\phi} \neq 0$ for $J = 0$, then the vacuum state (i.e., the ground state) is not invariant under the global symmetry of the theory: We have a *spontaneously broken global symmetry* (or spontaneous symmetry breaking). If we identify $J(x) \equiv H$ with the external physical field, then it follows from eq. (12.27) that $\frac{dU}{d\bar{\phi}} = H$. From this relation, the equation of state follows:

$$H = \sum_{N=1}^{\infty} \frac{1}{N!} \widetilde{\Gamma}^{(N+1)}(0, \ldots, 0)\, \bar{\phi}^N \tag{12.52}$$

These results provide the following strategy. Compute the effective potential and from it the vacuum (ground state). Next, compute the full vertex functions, either in the symmetric or broken symmetry state, by identifying in $\Gamma[\bar{\phi}]$ the coefficients of the products $\prod_i(\bar{\phi}(x_i) - v)$, where v is the classical field that minimizes the effective potential $U(\bar{\phi})$, that is,

$$\Gamma^{(N)}(1,\ldots,N|v) = \left.\frac{\delta^N\Gamma[\bar{\phi}]}{\delta\bar{\phi}(1)\cdots\delta\bar{\phi}(N)}\right|_{\bar{\phi}=v} \tag{12.53}$$

12.4 Ward identities

Let us now discuss the consequences of the existence of a continuous global symmetry G. We begin with a discussion of the simpler case, in which the symmetry group is $G = O(2) \simeq U(1)$. Consider the case of a two-component real scalar field $\boldsymbol{\phi}(x) = (\phi_\pi(x), \phi_\sigma(x))$, whose Euclidean Lagrangian is

$$\mathcal{L}(\boldsymbol{\phi}) = \frac{1}{2}\left[(\partial\boldsymbol{\phi})^2 + m_0^2\boldsymbol{\phi}^2\right] + \frac{\lambda}{4!}(\boldsymbol{\phi}^2)^2 + \boldsymbol{J}(x)\cdot\boldsymbol{\phi}(x) \tag{12.54}$$

where $\boldsymbol{J}(x)$ is a set of sources. In the absence of such sources, $\boldsymbol{J} = 0$, the Lagrangian $\mathcal{L}$ is invariant under global $O(2)$ transformations:

$$\boldsymbol{\phi}' = \exp(i\theta\sigma_2)\boldsymbol{\phi} = \begin{pmatrix} \cos\theta & \sin\theta \\ -\sin\theta & \cos\theta \end{pmatrix}\boldsymbol{\phi} \equiv T\boldsymbol{\phi} \tag{12.55}$$

For an infinitesimal angle θ, we can approximate

$$T = I + \epsilon\begin{pmatrix} 0 & 1 \\ -1 & 0 \end{pmatrix} + \cdots \tag{12.56}$$

The partition function $Z[\boldsymbol{J}]$

$$Z[\boldsymbol{J}] = \int \mathcal{D}\boldsymbol{\phi}\,\exp\left(-\int d^dx\mathcal{L}[\boldsymbol{\phi},\boldsymbol{J}]\right) \tag{12.57}$$

is invariant under the global symmetry $\boldsymbol{\phi}' = T\boldsymbol{\phi}$ if the sources transform accordingly: $\boldsymbol{J}' = T\boldsymbol{J}$. Then $\boldsymbol{J}(x)\cdot\boldsymbol{\phi}(x)$ is also invariant. Provided the integration measure of the path integral is invariant, $\mathcal{D}\boldsymbol{\phi} = \mathcal{D}\boldsymbol{\phi}'$, it follows that

$$Z[\boldsymbol{J}'] = Z[\boldsymbol{J}] \tag{12.58}$$

Thus, the partition function $Z[\boldsymbol{J}]$, the generating function of the connected correlators $F[\boldsymbol{J}]$, and the generating functional of the vertex (one-particle irreducible) functions $\Gamma[\boldsymbol{\phi}]$ are invariant under the action of the global symmetry.

In our discussion of classical field theory in section 3.1, we proved Noether's theorem, which states that a system with a global continuous symmetry has a locally conserved current and a globally conserved charge. However, this result only held at the classical level, since the derivation required the use of the classical equations of motion. We will now show that in the full quantum theory, the correlators of the fields obey a set of identities, known as

Ward identities, that follow from the existence of a global continuous symmetry. Moreover, these identities will also allow us to find consequences that hold if the global continuous symmetry is spontaneously broken.

To derive the Ward identities, consider the action of infinitesimal global transformations on the generating functionals. For an infinitesimal transformation T, the sources transform as

$$\boldsymbol{J}' = \boldsymbol{J} + \epsilon \begin{pmatrix} 0 & 1 \\ -1 & 0 \end{pmatrix} \boldsymbol{J} \tag{12.59}$$

In terms of the components of the source, $\boldsymbol{J} = (J_\sigma, J_\pi)$, the infinitesimal transformation is

$$J'_\sigma = J_\sigma + \epsilon J_\pi \tag{12.60}$$

$$J'_\pi = J_\pi - \epsilon J_\sigma \tag{12.61}$$

Or, equivalently, $\delta J_\sigma = \epsilon J_\pi$ and $\delta J_\pi = -\epsilon J_\sigma$. Since the generating functional $F[\boldsymbol{J}]$ is invariant under the global symmetry, we find

$$\begin{aligned} \delta F &= \int d^d x \left[\frac{\delta F[\boldsymbol{J}]}{\delta J_\sigma(x)} \delta J_\sigma(x) + \frac{\delta F[\boldsymbol{J}]}{\delta J_\pi(x)} \delta J_\pi(x) \right] = 0 \\ &= \int d^d x \, \epsilon \left[\frac{\delta F[\boldsymbol{J}]}{\delta J_\sigma(x)} J_\pi(x) - \frac{\delta F[\boldsymbol{J}]}{\delta J_\pi(x)} J_\sigma(x) \right] = 0 \end{aligned} \tag{12.62}$$

which implies that

$$\int d^d x \left[\bar{\phi}_\sigma(x) J_\pi(x) - \bar{\phi}_\pi(x) J_\sigma(x) \right] = 0 \tag{12.63}$$

Therefore, the generating functional $\Gamma[\bar{\boldsymbol{\phi}}]$ satisfies the identity

$$\int d^d x \left[\bar{\phi}_\sigma(x) \frac{\delta \Gamma[\bar{\boldsymbol{\phi}}]}{\delta \bar{\phi}_\pi(x)} - \bar{\phi}_\pi(x) \frac{\delta \Gamma[\bar{\boldsymbol{\phi}}]}{\delta \bar{\phi}_\sigma(x)} \right] = 0 \tag{12.64}$$

Eq. (12.64) is called the *Ward identity* for the generating functional $\Gamma[\boldsymbol{\phi}]$. It states that $\Gamma[\boldsymbol{\phi}]$ is invariant under the global transformation $\boldsymbol{\phi} \to T\boldsymbol{\phi}$. This identity is always valid (i.e., to all orders in perturbation theory).

We will now find several (many!) Ward identities that follow by differentiation of the Ward identity of eq. (12.64). By differentiating eq. (12.64) with respect to $\bar{\phi}_\pi(y)$, we find

$$0 = \int d^d x \left\{ \frac{\delta^2 \Gamma}{\delta \bar{\phi}_\pi(y) \delta \bar{\phi}_\pi(x)} \bar{\phi}_\sigma(x) - \frac{\delta^2 \Gamma}{\delta \bar{\phi}_\pi(y) \delta \bar{\phi}_\sigma(x)} \bar{\phi}_\pi(x) - \frac{\delta \Gamma}{\delta \bar{\phi}_\sigma(x)} \delta^d(x-y) \right\} \tag{12.65}$$

From this equation, it follows that

$$\frac{\delta \Gamma}{\delta \bar{\phi}_\sigma(y)} = \int d^d x \left[\frac{\delta^2 \Gamma}{\delta \bar{\phi}_\pi(x) \delta \bar{\phi}_\pi(y)} \bar{\phi}_\sigma(x) - \frac{\delta^2 \Gamma}{\delta \bar{\phi}_\sigma(x) \delta \bar{\phi}_\pi(y)} \bar{\phi}_\pi(x) \right] \tag{12.66}$$

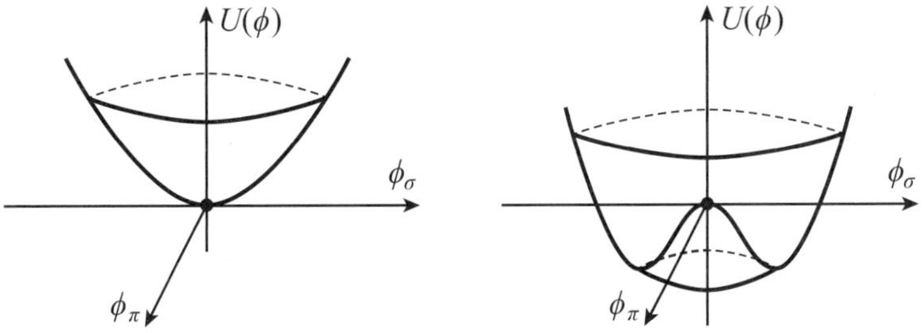

(a) (b)

Figure 12.10 The effective potential with $O(2)$ symmetry (a) in the symmetric phase, and (b) in the broken symmetry phase.

If the $O(2)$ symmetry is spontaneously broken, and the minimum of the effective potential $U(\boldsymbol{\phi})$ is on the circle shown in figure 12.10b, say $\bar{\boldsymbol{\phi}} = (v, 0)$, then eq. (12.66) becomes

$$v \int d^d x \frac{\delta^2 \Gamma}{\delta \bar{\phi}_\pi(x) \delta \bar{\phi}_\pi(y)} = H \tag{12.67}$$

where H denotes the uniform component of $\boldsymbol{J}$ along the direction of symmetry breaking, $J_\sigma = H$. Eq. (12.67) can be recast as

$$v \int d^d x \Gamma^{(2)}_{\pi\pi}(x - y) = H \tag{12.68}$$

or, equivalently,

$$\lim_{p \to 0} v \widetilde{\Gamma}^{(2)}_{\pi\pi}(\boldsymbol{p}) = H \tag{12.69}$$

So if the $O(2)$ symmetry is broken spontaneously, then the vacuum expectation value is nonzero ($v \neq 0$) as the symmetry-breaking field is removed ($H \to 0$). Then the 1-PI two-point function of the transverse components vanishes at long wavelengths, $\lim_{\boldsymbol{p} \to 0} \widetilde{\Gamma}^{(2)}_{\pi\pi}(\boldsymbol{p}) \to 0$, as $H \to 0$. Notice the important order of limits: first $\boldsymbol{p} \to 0$ and then $H \to 0$. Therefore, the connected transverse two-point function $\widetilde{G}^c_{2,\pi\pi}(\boldsymbol{p})$ has a pole at zero momentum with zero energy in the spontaneously broken phase. In other terms, in the broken symmetry state, the transverse components of the field, ϕ_π, describe a massless excitation known as the *Goldstone boson*.

Conversely, in the symmetric phase, in which $v \to 0$ as $H \to 0$, we find instead that

$$\lim_{H \to 0} \lim_{\boldsymbol{p} \to 0} \widetilde{\Gamma}^{(2)}_{\pi\pi}(\boldsymbol{p}) = \lim_{H \to 0} \frac{H}{v} \neq 0 \tag{12.70}$$

and the "transverse" modes are also massive. We will see shortly that in the symmetric phase, all masses are equal (as they should be!). In fact, the limiting value of $\frac{H}{v}$ in the symmetric phase is just equal to the inverse susceptibility $\chi^{-1}_{\pi\pi}$.

Thus we conclude that there is an alternative: either (a) the theory is in the symmetric phase (i.e., $v = 0$), or (b) the symmetry is spontaneously broken ($v \neq 0$ with $J \to 0$) and there are massless excitations (Goldstone bosons). This result, known as the Goldstone theorem, is actually generally valid in a broken symmetry state of a system with a global continuous symmetry.

Let us now consider the more general case of a global symmetry with group $O(N)$. The analysis for other Lie groups (e.g., $U(N)$) is similar. In the $O(N)$ case, the group has $N(N-1)/2$ generators

$$(L_{ij})_{kl} = -i\left[\delta_{ik}\delta_{jl} - \delta_{il}\delta_{jk}\right] \tag{12.71}$$

where $i, j, k, l = 1, \ldots, N$. Assume that the $O(N)$ symmetry is spontaneously broken along the direction $\bar{\boldsymbol{\phi}} = (v, \mathbf{0})$, where $\mathbf{0}$ has $N-1$ components. More generally, we can write the field as $\bar{\boldsymbol{\phi}} = (\bar{\phi}_\sigma, \bar{\boldsymbol{\phi}}_\pi)$. In the broken symmetry state, only the $\bar{\phi}_\sigma$ component has a nonvanishing expectation value. In this state there is a residual, unbroken, $O(N-1)$ symmetry obtained by rotating transverse components among one another. Thus, the symmetry, which is actually broken, rather than $O(N)$, is in the coset $O(N)/O(N-1)$, which is isomorphic to the N-dimensional sphere S_N.

Under the action of $O(N)$, the field $\bar{\boldsymbol{\phi}}$ transforms as follows:

$$\bar{\phi}'_a = \left(e^{i\vec{\lambda}\cdot\vec{L}}\right)_{ab} \bar{\phi}_b \tag{12.72}$$

where $\lambda_{ij} = -\lambda_{ji}$ (with $i, j = 1, \ldots, N$) are the Euler angles of the S_N sphere. The broken generators (the generators that mix with the direction of the broken symmetry) are L_{i1}, and the L_{ij} (with $i, j \neq 1$) are the generators of the unbroken $O(N-1)$ symmetry.

Let us perform an infinitesimal transformation away from the direction of symmetry breaking with the L_{i1} (with $i = 2, \ldots, N$) generators:

$$\delta\bar{\phi}_a = i\lambda_{i1}\,(L_{i1})_{ab}\,\bar{\phi}_b = \lambda_{i1}[\delta_{ia}\delta_{1b} - \delta_{ib}\delta_{1a}]\bar{\phi}_b = \lambda_{a1}\bar{\phi}_1 - \lambda_{b1}\delta_{1a}\bar{\phi}_b \tag{12.73}$$

or, what is the same,

$$\delta\bar{\phi}_\sigma = \lambda_{1b}\bar{\phi}_{\pi,b}, \qquad \delta\bar{\phi}_{\pi,a} = -\lambda_{1a}\bar{\phi}_\sigma \tag{12.74}$$

Likewise,

$$\delta J_\sigma = \lambda_{1b} J_{\pi_b}, \qquad \delta J_{\pi_a} = -\lambda_{1a} J_\sigma \tag{12.75}$$

Thus,

$$\begin{aligned}\delta F &= \int d^d x \left[\frac{\delta F}{\delta J_\sigma(x)}\delta J_\sigma(x) + \frac{\delta F}{\delta J_{\pi,a}(x)}\delta J_{\pi,a}(x)\right] \\ &= \int d^d x \lambda_{1a}\left[-\frac{\delta\Gamma}{\delta\bar{\phi}_\sigma(x)}\bar{\phi}_{\pi,a}(x) + \frac{\delta\Gamma}{\delta\bar{\phi}_{\pi,a}(x)}\bar{\phi}_\sigma(x)\right] = 0\end{aligned} \tag{12.76}$$

Since the infinitesimal Euler angles λ_{1a} are arbitrary, we have a Ward identity for each component ($a = 2, \ldots, N$):

$$\int d^d x \left[\frac{\delta\Gamma}{\delta\bar{\phi}_\sigma(x)}\bar{\phi}_{\pi,a}(x) - \frac{\delta\Gamma}{\delta\bar{\phi}_{\pi,a}(x)}\bar{\phi}_\sigma(x)\right] = 0 \tag{12.77}$$

We will proceed as in the $O(2)$ theory and differentiate eq. (12.76) with respect to a transverse field component, $\bar{\phi}_{\pi,b}$, to obtain

$$\int d^d x \left[\frac{\delta^2 \Gamma}{\delta \bar{\phi}_\sigma(x) \delta \bar{\phi}_{\pi,b}(y)} \bar{\phi}_{\pi,a}(x) + \frac{\delta \Gamma}{\delta \bar{\phi}_\sigma(x)} \delta_{ab} \delta(x-y) - \frac{\delta^2 \Gamma}{\delta \bar{\phi}_{\pi,a}(x) \delta \bar{\phi}_{\pi,b}(y)} \bar{\phi}_\sigma(x) \right] = 0 \tag{12.78}$$

Let us now assume that the $O(N)$ symmetry is spontaneously broken and that the field has the expectation value $\bar{\boldsymbol{\phi}} = (v, \mathbf{0})$. Then eq. (12.78) becomes

$$\delta_{ab} J_\sigma(y) = v \int d^d x \Gamma^{(2)}_{\pi_a \pi_b}(x-y) = v \lim_{p \to 0} \Gamma^{(2)}_{\pi_a \pi_b}(p) \tag{12.79}$$

which requires that J_σ be uniform. From this equation, the following conclusions can be drawn:

1) $\Gamma^{(2)}_{\pi_a \pi_b}(0)$ must be diagonal: $\Gamma^{(2)}_{\pi_a \pi_b}(0) \equiv \delta_{ab} \Gamma_{\pi\pi}(0)$. Hence the masses of the transverse components, $\bar{\phi}_a$, are equal, $m^2_{\pi_a} = m^2_{\pi_b}$, and we have a degenerate multiplet.
2) In the limit in which the source along the symmetry-breaking direction is removed, $J_\sigma \to 0$, it must hold that $v \lim_{p \to} \Gamma^{(2)}_{\pi_a \pi_a}(p) = 0$. Thus, we find that there are two possible cases:
   - 2a) if $v \neq 0$, all transverse $\bar{\phi}_{\pi,a}$ excitations are massless, and there are $N-1$ massless excitations (Goldstone bosons);
   - 2b) if $v = 0$, the theory is in the symmetric phase, all excitations are massive, and $m^2_\sigma = \Gamma^{(2)}_{\sigma\sigma}(0) \neq 0$.

These results are exact identities, and hence are valid order by order in perturbation theory.

Following the same procedure, we can get, in fact, an infinite set of identities. For example, in the $O(2)$ theory (for simplicity), by differentiating eq. (12.65) with respect to the field ϕ_σ, we find the relation

$$\Gamma^{(2)}_{\sigma\sigma}(p) - \Gamma^{(2)}_{\pi\pi}(p) = v\, \Gamma^{(3)}_{\sigma\pi\pi}(0, p, -p) \tag{12.80}$$

In the symmetric phase ($v = 0$), eq. (12.80) implies that the irreducible two-point functions for the ϕ_σ and ϕ_π components must be equal: $\Gamma^{(2)}_{\sigma\sigma}(p) = \Gamma^{(2)}_{\pi\pi}(p)$, and that in particular, the masses must be equal: $m^2_\sigma = m^2_\pi$. However, in the broken symmetry phase, where $v \neq 0$, this equation and the requirement that the ϕ_π field must be a Goldstone boson (and hence massless) implies that the mass of the ϕ_σ field must be related to the three-point function (with all momenta equal to zero): $\Gamma^{(2)}_{\sigma\sigma}(0) = v \Gamma^{(3)}_{\sigma\pi\pi}(0,0,0)$. This result implies that in the broken symmetry phase, the ϕ_σ field is massive. In the Minkowski space interpretation, this result also implies that the ϕ_σ particle has an amplitude to decay into two Goldstone modes (at zero momentum) and hence that, as a state, it must have a finite width (or lifetime) determined by its mass and by the vacuum expectation value v.

Also, by further differentiating the identity eq. (12.65) with respect to the ϕ_σ field two more times, and using the identity eq. (12.80), we can derive one more identity relating the four-point functions of the ϕ_σ and the ϕ_π fields:

$$\Gamma^{(4)}_{\pi\pi\sigma\sigma}(z, y, t, w) + \Gamma^{(4)}_{\pi\pi\sigma\sigma}(w, y, z, t) + \Gamma^{(4)}_{\pi\pi\sigma\sigma}(t, y, z, w) = \Gamma^{(4)}_{\sigma\sigma\sigma\sigma}(y, z, t, w) \tag{12.81}$$

The Fourier transforms of the above identity at a symmetric point of the four incoming momenta (e.g., for $p \to 0$) satisfy the relation $3\Gamma^{(4)}_{\pi\pi\sigma\sigma} = \Gamma^{(4)}_{\sigma\sigma\sigma\sigma}$, which ensures that $O(2)$ invariance holds.

12.5 The low-energy effective action and the nonlinear sigma model

In section 12.4, we showed that if a continuous global symmetry is spontaneously broken, then the theory has exactly massless excitations known as Goldstone bosons. These states constitute the low-energy manifold of states. It makes sense to find the possible forms the effective action takes for these low-energy states. We will now see that the form of their effective action is completely determined by the global symmetry.

To find the effective low-energy action, rather than using the Cartesian decomposition $\boldsymbol{\phi}(x) = (\phi_\pi(x), \phi_\sigma(x))$ into longitudinal and transverse fields, it will be more instructive to use instead a nonlinear representation. In the case of the $O(N)$ theory, let us write the field $\boldsymbol{\phi}(x)$ in terms of an amplitude field $\rho(x)$ and an N-component unit-vector field $\boldsymbol{n}(x)$,

$$\boldsymbol{\phi}(x) = \rho(x)\, \boldsymbol{n}(x), \qquad \text{provided } \boldsymbol{n}^2(x) = 1 \tag{12.82}$$

where the unit-length condition has been imposed as a constraint. This constraint ensures that we have not changed the number of degrees of freedom in the factorization of eq. (12.82). We will see that the constrained field $\boldsymbol{n}(x)$ describes the manifold of Goldstone states. Clearly, the target space of the field $\boldsymbol{n}$ is the sphere S_{N-1}.

Formally, the partition function now becomes

$$Z[J] = \int \mathcal{D}\rho\mathcal{D}\boldsymbol{n} \prod_x \delta(\boldsymbol{n}^2(x) - 1) \exp\left(- \int d^dx \mathcal{L}[\rho, \boldsymbol{n}] + \int d^dx \rho(x)\boldsymbol{n}(x) \cdot \boldsymbol{J}(x) \right) \tag{12.83}$$

where the Lagrangian $\mathcal{L}[\rho, \boldsymbol{n}]$ is

$$\mathcal{L}[\rho, \boldsymbol{n}] = \frac{1}{2}\left(\partial_\mu \rho\right)^2 + \frac{1}{2}\rho^2 \left(\partial_\mu \boldsymbol{n}\right)^2 + \frac{m_0^2}{2}\rho^2 + \frac{\lambda}{4!}\rho^4 \tag{12.84}$$

Notice that, due to the constraint on the $\boldsymbol{n}$ field, the integration measure of the path integral has changed:

$$\mathcal{D}\boldsymbol{\phi} \to \mathcal{D}\rho\mathcal{D}\boldsymbol{n} \prod_x \delta(\boldsymbol{n}^2(x) - 1) \tag{12.85}$$

to preserve the number of degrees of freedom.

It is clear that in this nonlinear representation, the global $O(N)$ transformations leave the amplitude field $\rho(x)$ invariant and only act on the nonlinear field $\boldsymbol{n}(x)$. This field represents the Goldstone manifold, which only enters in the second term of the Lagrangian of eq. (12.84). In fact, the field $\rho(x)$ represents the fluctuations of the amplitude about the minimum of the potential shown in figure 12.10(b). In contrast, the field $\boldsymbol{n}$ represents the Goldstone bosons: the field fluctuations along the "flat" directions of the potential at the bottom of the sombrero (or of a wine bottle). The other important observation is that only the derivatives of the field $\boldsymbol{n}$ enter in the Lagrangian and that, in particular, there is no mass term for this field. This is a consequence of the symmetry and of the Ward identities.

Let us look closer at what happens in the broken-symmetry state, where $m_0^2 < 0$. In this phase, the field ρ has a v.e.v. equal to $\bar{\rho} = v = \sqrt{|m_0^2|/(6\lambda)}$. Let us represent the amplitude fluctuations in terms of the field $\eta = \rho - v$. The Lagrangian for the fields η and $\boldsymbol{n}$ reads

$$\mathcal{L}[\eta, \boldsymbol{n}] = \frac{1}{2}\left(\partial_\mu \eta\right)^2 + \frac{m_{\text{eff}}^2}{2}\eta^2 + \frac{\lambda}{6} v\eta^3 + \frac{\lambda}{4!}\eta^4 + \frac{v^2}{2}\left(\partial_\mu \boldsymbol{n}\right)^2 + v\eta\left(\partial_\mu \boldsymbol{n}\right)^2 + \frac{1}{2}\eta^2\left(\partial_\mu \boldsymbol{n}\right)^2 \tag{12.86}$$

where $m_{\text{eff}}^2 = \lambda v^2/3 = 2|m_0^2|$ is the effective mass (squared) of the amplitude field ρ. (Here we have ignored the downward shift of the classical vacuum energy, $-\lambda v^4/24$, in the broken-symmetry phase.)

Several comments are now in order. One is that the amplitude field η is massive, and its mass grows parametrically larger when moving deeper into the broken symmetry state. Thus its effects should become weak at low energies (and long distances). The other comment is that in the broken-symmetry state, there is a trilinear coupling term (the next-to-last term in eq. (12.86)). This term is linear in the amplitude field η and quadratic in the Goldstone field $\boldsymbol{n}$, and its coupling constant is the expectation value v. This is precisely what follows from the Ward identities (cf. eq. (12.80)). For this reason, v^2 plays the role of the decay rate of the (massive) amplitude mode (into massless Goldstone modes).

Let us consider the effective Lagrangian of the massless Goldstone field $\boldsymbol{n}$. We can deduce what it is by integrating out the amplitude fluctuations—the massive field η—in perturbation theory. By carrying out this elementary calculation, we find that, to lowest order, the effective Lagrangian for the Goldstone field $\boldsymbol{n}$ has the form (recall the constraint $\boldsymbol{n}^2 = 1$):

$$\mathcal{L}_{\text{eff}}[\boldsymbol{n}] = \frac{1}{2g^2}\left(\partial_\mu \boldsymbol{n}\right)^2 + \frac{1}{2g^2 m_{\text{eff}}^2}\left(\partial_\mu \boldsymbol{n}\right)^4 + \cdots \tag{12.87}$$

where $g^2 = 1/v^2$ plays the role of the coupling constant. The first term of this effective Lagrangian is known, in this case, as the $O(N)$ nonlinear sigma model. This theory seems free, since it is quadratic in $\boldsymbol{n}$, but it is actually nonlinear due to the constraint $\boldsymbol{n}^2 = 1$. The correction term is clearly small in the low-energy regime: $\left(\partial_\mu \boldsymbol{n}\right)^2 \ll m_{\text{eff}}^2$. So this is actually a gradient expansion, which is accurate in the asymptotic low-energy regime.

In the low-energy (and long-wavelength) regime, the quartic term of eq. (12.87) can be neglected (in chapter 15, we will see that this is an example of an irrelevant operator). In this IR regime, the effective action becomes

$$\mathcal{L}_{\text{NLSM}}[\boldsymbol{n}] = \frac{1}{2g^2}\left(\partial_\mu \boldsymbol{n}\right)^2 \tag{12.88}$$

where the field $\boldsymbol{n}(x)$ obeys the local constraint $\boldsymbol{n}^2(x) = 1$. This theory is the $O(3)$ nonlinear sigma model.

The nonlinear sigma model is of interest in many areas of physics. It was originally introduced as a model for pion physics and chiral symmetry breaking in high-energy physics. In this context, the coupling constant g^2 is identified with the inverse of the pion decay constant. It is also of wide interest in classical statistical mechanics, where it is a model of the long-wavelength behavior of the free energy in the classical Heisenberg model of a ferromagnet, and in quantum magnets, where it is the effective Lagrangian for a quantum antiferromagnet. We will return to a detailed discussion of the behavior of the nonlinear sigma model in section 16.4, where we discuss the perturbative renormalization group, and in section 17.2, where we discuss the large-N limit.

12.6 Ward identities, Schwinger-Dyson equations, and gauge invariance

Let us now take a somewhat different look at Ward identities, focusing on the implications of the global symmetry on the correlators of a theory. We will derive a set of equations obeyed by the correlators as a consequence of the global symmetry. These equations take the form of Schwinger-Dyson equations. We will discuss two cases. The first is a scalar field theory, which, for simplicity, is that of a single complex scalar field with the global symmetry $U(1)$. The other case is the global $U(1)$ symmetry of QED. Both cases can be easily generalized to other continuous symmetry groups.

12.6.1 Schwinger-Dyson equation for the complex scalar field

Consider a complex scalar field with Lagrangian

$$\mathcal{L}(\phi, \phi^*) = |\partial_\mu \phi|^2 - V(|\phi|^2) \tag{12.89}$$

which has the global $U(1)$ symmetry

$$\phi(x) \mapsto e^{i\theta}\phi(x), \qquad \phi^*(x) \mapsto e^{-i\theta}\phi^*(x) \tag{12.90}$$

Or, in infinitesimal form,

$$\delta\phi(x) = i\theta\phi(x), \qquad \delta\phi^*(x) = -i\theta\phi^*(x) \tag{12.91}$$

The (Euclidean) two-point function

$$G^{(2)}(x-y) = \langle \phi(x)\phi^*(y)\rangle = \frac{1}{Z}\int \mathcal{D}\phi\mathcal{D}\phi^*\, \phi(x)\phi^*(y)\, e^{-\int d^Dx\mathcal{L}(\phi,\phi^*)} \tag{12.92}$$

and the partition function Z

$$Z = \int \mathcal{D}\phi\mathcal{D}\phi^*\, e^{-\int d^Dx\mathcal{L}(\phi,\phi^*)} \tag{12.93}$$

are invariant under the global $U(1)$ symmetry transformation.

Next we make the (infinitesimal) change of variables,

$$\phi(x) \to \phi'(x) = \phi(x) + i\theta(x)\phi(x), \qquad \phi(x)^* \to \phi'(x)^* = \phi(x)^* - i\theta(x)\phi(x)^* \tag{12.94}$$

where $\theta(x)$ is an arbitrary (and small) function of the coordinate x, which vanishes at infinity. The path integral of eq. (12.92) does not change under a change of the integration variable (the field), since the integration measure is invariant, so we can readily derive the identity

$$\partial_\mu^z \langle j^\mu(z)\phi(x)\phi^*(y)\rangle = \delta(z-x)\langle\phi(x)\phi^*(y)\rangle - \delta(z-y)\langle\phi(x)\phi^*(y)\rangle \tag{12.95}$$

where $j_\mu(z) = i(\phi(z)\partial_\mu\phi^*(z) - \phi^*(z)\partial_\mu\phi(z))$ is the Noether current for this global symmetry. An expression of this type is known as a Schwinger-Dyson equation. It has an obvious generalization to a general N-point function for any continuous global symmetry group. Notice that, at the classical level, we would have predicted that the left-hand side of eq. (12.95) should be identically zero. The so-called *contact terms* that appear on the right-hand side of this equation have a quantum origin.

12.6.2 Ward identity and gauge invariance

The result just presented can be easily generalized to the case of a gauge theory, such as QED. The QED Lagrangian

$$\mathcal{L}_{\text{QED}}[\phi, \bar{\psi}, A_\mu] = \bar{\psi} i\gamma^\mu \partial_\mu \psi - eA_\mu \bar{\psi}\gamma^\mu \psi - \frac{1}{4}F_{\mu\nu}^2 \tag{12.96}$$

is invariant under the local $U(1)$ gauge transformations $\psi(x) \to e^{ie\theta(x)}\psi(x)$, $\bar{\psi}(x) \to e^{-ie\theta(x)}\bar{\psi}(x)$, and $A_\mu(x) \to A_\mu(x) + \partial_\mu \theta(x)$.

However, the QED Lagrangian is also invariant under the global symmetry $\psi(x) \to e^{i\theta}\psi(x)$, $\bar{\psi}(x) \to e^{-i\theta}\bar{\psi}(x)$, and $A_\mu(x) \to A_\mu(x)$. This is a global symmetry of the Dirac sector of the theory. We can now derive a Ward identity (or Schwinger-Dyson equation) for the global $U(1)$ symmetry of QED. On repeating the same arguments used in the derivation of the Ward identity of eq. (12.95), we now obtain a similar-looking relation for the two-point function of the Dirac field:

$$\partial_\mu^z \langle j^\mu(z)\psi(x)\bar{\psi}(y)\rangle = -e\delta(z-x)\langle\psi(x)\bar{\psi}(y)\rangle + e\delta(z-y)\langle\psi(x)\bar{\psi}(y)\rangle \tag{12.97}$$

where $j_\mu = e\bar{\psi}\gamma_\mu\psi$ is the (gauge-invariant) Dirac current. Equivalently, we can write

$$-i\partial_\mu^z \langle j^\mu(z)\psi(x)\bar{\psi}(y)\rangle = -e\delta(z-x)S_F(x-y) + e\delta(z-y)S_F(x-y) \tag{12.98}$$

where $S_F(x-y) = -i\langle\psi(x)\bar{\psi}(y)\rangle$ is the (Feynman) Dirac propagator. This Ward identity relates the change of the Dirac propagator to the vertex function resulting from the insertion of a current (i.e., the coupling to the gauge field). In momentum space, this Ward identity can be brought into the form

$$-iq_\mu \Gamma^\mu(p+q, p, q) = S_F^{-1}(p+q) - S_F^{-1}(p) \tag{12.99}$$

where $\Gamma^\mu(p+q, p)$ is the vertex function, and $S_F(p)$ is the Feynman propagator for the Dirac field. Pictorially, we can represent the Ward identity of eq. (12.99) in terms of Feynman diagrams as follows:

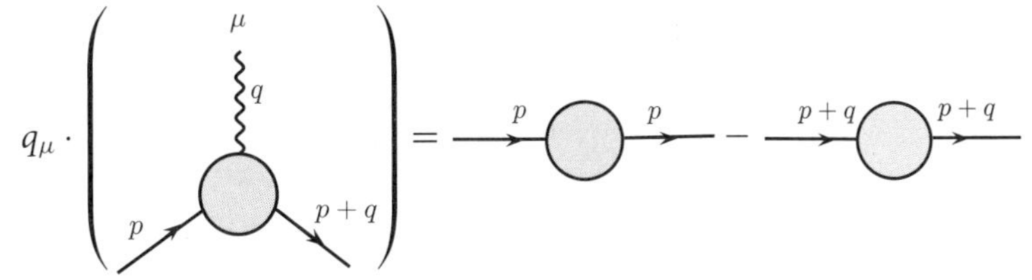

(12.100)

which relates the insertion of a photon of momentum q to the change of the fermion propagator with the momentum of the photon.

Let us now discuss a different derivation of the Ward identities, following an approach similar to what we did in section 12.4 for ϕ^4 theory with $O(N)$ symmetry. Consider the QED

Lagrangian in the Feynman–'t Hooft gauges parametrized by α, coupled to a conserved external current $J_\mu(x)$ and to Dirac sources $\eta(x)$ and $\bar{\eta}(x)$. The full gauge-fixed Lagrangian is now

$$\mathcal{L} = \bar{\psi} i \not{D} \psi - \frac{1}{4} F_{\mu\nu}^2 - \frac{1}{2\alpha}(\partial_\mu A^\mu)^2 + J_\mu(x) A^\mu(x) + \bar{\eta}(x)\psi(x) + \bar{\psi}(x)\eta(x) \tag{12.101}$$

where, as usual, $\not{D} = \gamma^\mu(\partial_\mu + ieA_\mu(x))$. The partition function is now

$$Z[J_\mu, \eta, \bar{\eta}] = \int \mathcal{D}\bar{\psi}\mathcal{D}\psi\mathcal{D}A_\mu e^{i\int d^4x \mathcal{L}} \tag{12.102}$$

where $\mathcal{L}$ is the Lagrangian of QED with sources (eq. (12.101)). Here we ignored the Faddeev-Popov determinant factor which, as we have seen, is unimportant in QED.

As in other theories, we focus now on $F = i \ln Z[J_\mu, \eta, \bar{\eta}]$, the generating function of connected N-point functions of Dirac fermions and gauge fields, which satisfies

$$\frac{\delta F}{\delta J_\mu(x)} = A_\mu(x), \qquad \frac{\delta F}{\delta \eta(x)} = \bar{\psi}(x), \qquad \frac{\delta F}{\delta \bar{\eta}(x)} = \psi(x) \tag{12.103}$$

where the expectation values have been omitted to simplify the notation. Let us now define $\Gamma[A_\mu, \psi, \bar{\psi}]$, the generating function of the one-particle irreducible vertex functions, obtained from F by the Legendre transform

$$\Gamma[A_\mu, \psi, \bar{\psi}] = \int d^4x (J_\mu(x) A^\mu(x) + \bar{\eta}(x)\psi(x) + \bar{\psi}(x)\eta(x)) - F \tag{12.104}$$

such that

$$\frac{\delta \Gamma}{\delta A_\mu(x)} = J_\mu(x), \qquad \frac{\delta \Gamma}{\delta \psi(x)} = \bar{\eta}(x), \qquad \frac{\delta \Gamma}{\delta \bar{\psi}(x)} = \eta(x) \tag{12.105}$$

We next make a change of variables in the partition function $A_\mu \to A_\mu + \delta A_\mu$, $\psi \to \psi + \delta\psi$, and $\bar{\psi} \to \bar{\psi} + \delta\bar{\psi}$, which corresponds to an infinitesimal gauge transformation

$$\delta A_\mu(x) = \partial_\mu \Lambda(x), \qquad \delta\psi(x) = -ie\Lambda(x)\psi(x), \qquad \delta\bar{\psi}(x) = ie\Lambda(x)\bar{\psi}(x) \tag{12.106}$$

where $\Lambda(x)$ is infinitesimal and local. From this we readily find that the generating function $\Gamma[A_\mu, \psi, \bar{\psi}]$ satisfies the generation function of all Ward identities:

$$\partial_\mu \frac{\delta \Gamma}{\delta A_\mu(x)} - ie\left(\psi(x)\frac{\delta \Gamma}{\delta \psi(x)} - \bar{\psi}(x)\frac{\delta \Gamma}{\delta \bar{\psi}(x)}\right) = \frac{1}{\alpha}\partial^2 \partial_\mu A^\mu(x) \tag{12.107}$$

where the right-hand side arises from the gauge-fixing condition.

Now, on functional differentiation of eq. (12.107) with respect to $\psi(x)$ and $\bar{\psi}(x)$, we obtain the identity

$$\partial_z^\mu \frac{\delta^3 \Gamma}{\delta\bar{\psi}(x)\delta\psi(y)\delta A_\mu(z)} + ie\left(\frac{\delta^2 \Gamma}{\delta\bar{\psi}(x)\delta\psi(y)}\delta(x-z) - \frac{\delta^2 \Gamma}{\delta\bar{\psi}(x)\delta\psi(y)}\delta(y-z)\right) = 0 \tag{12.108}$$

where we readily identify the first term with the one-particle irreducible three-point vertex function $\langle\psi(x)\bar{\psi}(y)A_\mu(z)\rangle$ and the second and third terms with the one-particle irreducible propagators of the Dirac fields. The Fourier transform of this identity is just eq. (12.99).

By functional differentiation of eq. (12.107) with respect to $A_\nu(y)$, we obtain the Ward identity:

$$\partial_\mu^x \frac{\delta^2\Gamma}{\delta A_\mu(x)\delta A_\nu(y)} = \partial_x^\mu \Gamma_{\mu\nu}^{(2)}(x-y) = \frac{1}{\alpha}\partial_x^2\partial_x^\mu\delta(x-y) \tag{12.109}$$

In contrast, the 1-PI photon two-point function $\Gamma_{\mu\nu}^{(2)}(x-y)$ is given by

$$\Gamma_{\mu\nu}^{(2)}(x-y) = G_{\mu\nu}^{-1}(x-y) - \Pi_{\mu\nu}(x-y) \tag{12.110}$$

where $G_{\mu\nu}(x-y)$ is the bare photon propagator (i.e., in free Maxwell theory) in the 't Hooft–Feynman gauges. Since the Ward identity eq. (12.109) is also obeyed by the free-field propagator $G_{\mu\nu}^{(2)}(x-y)$, we find that the QED photon self-energy, the polarization function $\Pi_{\mu\nu}(x-y)$, must satisfy the exact identity

$$\partial_x^\mu \Pi_{\mu\nu}(x-y) = 0 \tag{12.111}$$

Note that, although we derived it in the context of QED, all that this Ward identity requires is global symmetry. In fact, it also applies to nonrelativistic systems, such as the electron gas and Fermi liquids (see section 10.11). Naturally, the propagators that enter in the expressions above will be different in these theories. This identity ensures that gauge invariance is exactly respected, order by order, in perturbation theory.

Exercises

12.1 Vertex functions, effective potential, and Ward identities

In this problem, you will study the effects of interactions on the physical properties of a system of interacting relativistic fermions in 1+1 dimensions, known as the chiral Gross-Neveu model. This model is a reasonable description of the physics in quasi-one-dimensional systems, and it is also of interest for investigating the behavior of quantum field theories of relativistic Fermi fields.

In 1+1 dimensions, Fermi fields $\psi_a(x)$ are two-component spinors of the form

$$\psi_a = \begin{pmatrix} \psi_{R,a} \\ \psi_{L,a} \end{pmatrix} \tag{12.112}$$

The upper component, $\psi_{R,a}$ is the right-moving component of the fermion, and the lower component $\psi_{L,a}$ is the left-moving component of the fermion. In what follows, consider the case in which there are N species of fermions labeled by an index $a = 1, \ldots, N$.

Next, define the following set of two-dimensional Dirac γ matrices

$$\gamma_0 = \sigma_1 = \begin{pmatrix} 0 & 1 \\ 1 & 0 \end{pmatrix} \quad \gamma_1 = i\sigma_2 = \begin{pmatrix} 0 & 1 \\ -1 & 0 \end{pmatrix} \quad \gamma_5 = \gamma_0\gamma_1 = -\sigma_3 = \begin{pmatrix} -1 & 0 \\ 0 & 1 \end{pmatrix} \tag{12.113}$$

with the notation

$$\partial\!\!\!/ = \partial_\mu \gamma^\mu = \gamma_0 \partial_0 - \gamma_1 \partial_1 \tag{12.114}$$

where $\mu = 0, 1$ denotes the indices of a 1+1-dimensional Minkowski spacetime.

Introduce the operators

$$\bar{\psi}\psi = \psi^\dagger \gamma_0 \psi \equiv \psi_R^\dagger \psi_L + \psi_L^\dagger \psi_R \tag{12.115}$$

and

$$\bar{\psi}\gamma_5\psi = \psi^\dagger \gamma_1 \psi \equiv \psi_R^\dagger \psi_L - \psi_L^\dagger \psi_R \tag{12.116}$$

Using this notation, the Lagrangian density of the chiral Gross-Neveu model is (setting the speed of light to $c = 1$)

$$\mathcal{L} = \bar{\psi}_a(x) i\partial\!\!\!/ \psi_a(x) + \frac{g}{2}\left[(\bar{\psi}_a(x)\psi_a(x))^2 - (\bar{\psi}_a(x)\gamma^5\psi_a(x))^2\right] \tag{12.117}$$

where we have not written down the spinor indices explicitly. In all the parts of this exercise, you have to use path-integral methods.

1) Derive an expression for the free-fermion propagator in momentum space.
2) Derive the Feynman rules for a perturbation expansion of the fermion two-point function

$$S_{a,b}^{\alpha,\beta}(p) \tag{12.118}$$

(with $p = (p^0, p^1)$) in powers of the coupling constant g.
3) Derive an expression for the quantities listed in parts (a) and (b) up to and including their second-order corrections (i.e., $O(g^2)$). Draw a Feynman diagram for each contribution. Check the cancellation of the vacuum diagrams. Give a consistent sign for each contribution. Do *not* do the integrals!

   a) The fermion two-point function.
   b) The effective coupling constant g. What correlation function should you consider? Is it a connected Green function or a one-particle irreducible vertex function? Justify your answer.

4) a) Show that the Lagrangian of the chiral Gross-Neveu model is invariant under the continuous global chiral symmetry

$$\psi_{\alpha a}(x) \to \left(e^{i\theta\gamma_5}\right)_{\alpha\beta} \psi_{\beta a} \tag{12.119}$$

   b) Find the transformation law obeyed by the operators $\hat{\Delta}_0 \equiv \bar{\psi}_a\psi_a$ and $\hat{\Delta}_5 \equiv i\bar{\psi}_a\gamma_5\psi_a$ under this symmetry.
   c) Give a physical interpretation of this symmetry. What is the meaning of the operators $\hat{\Delta}_0$ and $\hat{\Delta}_5$ in terms of the right- and left-moving components of the fermions?

5) Now let us add chiral symmetry-breaking terms to the Lagrangian of the form

$$\mathcal{L}_{\text{chiral}} = H_0(x)\hat{\Delta}_0(x) + H_5(x)\hat{\Delta}_5(x) \tag{12.120}$$

   where $H_0(x)$ and $H_5(x)$ are the symmetry-breaking fields. Consider the path integral for this problem in the presence of the symmetry-breaking terms.

Assume that the operators $\hat{\Delta}_0$ and $\hat{\Delta}_5$ have uniform expectation values given by $\bar{\Delta}_0$ and $\bar{\Delta}_5$, respectively. Find the transformation law obeyed by the symmetry-breaking fields H_0 and H_5 under a global chiral transformation of the Fermi fields by an angle θ.

6) Derive a formal expression for the effective potential $U(\bar{\Delta}_0, \bar{\Delta}_5)$ in terms of a series expansion in powers of the expectation values. Relate the coefficients of this expansion in terms of vertex functions. What is the meaning of these vertex functions in terms of fermion operators?
7) Find a perturbation theory expression for all coefficients of the effective potential U to leading order (i.e., the first nonvanishing term) in an expansion in terms of the coupling constant g. Do not do the integrals! Draw the Feynman diagram associated with each contribution.
8) Derive the chiral Ward identity for the generating functional of the vertex functions for the operators $\bar{\Delta}_0$ and $\bar{\Delta}_5$.
9) Derive a Ward identity that relates the vertex functions Γ_{00} and Γ_{55} in the limit $p \to 0$ $(p \equiv p_\mu)$. Assume that, as $H_0 \to 0$ and $H_5 \to 0$, only $\bar{\Delta}_0$ has a nonzero expectation value (i.e., that the chiral symmetry is *spontaneously broken*).
10) Is there a Goldstone boson in this system? Justify your answer. Explain the significance of your answer for the original fermion problem. What does it say about its two-particle spectrum? And of the one-particle spectrum?

13

Perturbation Theory, Regularization, and Renormalization

13.1 The loop expansion

We now return to perturbation theory. For the sake of definiteness, we will focus on the theory of the real scalar field. But this approach is general. The perturbative expansion developed in chapters 11 and 12 is limited to situations in which the coupling constant λ is small. This is a serious limitation. The perturbative expansion presented in those chapters is also ambiguous, even when the theory has many coupling constants and/or several fields that are coupled. In these cases, the naive expansion does not provide a criterion for how to organize the expansion. There are, however, other expansion schemes that, in principle, do not have these limitations. One such scheme is the familiar semiclassical WKB method in quantum mechanics. The problem is that WKB is really difficult to generalize to a field theory, and it is not a feasible option. We will see in chapter 19 that semiclassical methods also play an important role in QFT. Another nonperturbative approach, to be discussed in chapter 17, involves taking the limit in which the rank of the symmetry group of the theory is large. These limits have proven very instructive for elucidating the behavior of many theories beyond perturbation theory.

Implicit in the use of perturbation theory (and even more so when *defining* the theory by its perturbative expansion, as is often done) is the assumption that this expansion is, in some sense, convergent in the weak-coupling regime. It is obvious that this cannot be the case. Thus, in ϕ^4 theory, changing the sign of the coupling constant λ from positive to negative turns a theory with a stable classical vacuum state into one with a metastable vacuum state. This argument, originally formulated by Dyson in the context of QED, implies that all perturbative expansions have at most a vanishing radius of convergence and that the perturbation-theory series is, at most, an asymptotic series. This fact is also true even in the simpler problem of the quantum anharmonic oscillator. Nevertheless, it is often the case (and the anharmonic oscillator is a useful example in this sense) that the nonanalyticities may be of the form of an essential singularity that cannot be detected to any finite order in perturbation theory. Another example is QED, which is to date the most precise theory in physics. In this case, changing the sign of the fine structure constant turns repulsion between like charges into an attraction, leading to a massive instability of the ground state.

For these (and other) reasons, it would be desirable to have other types of expansions. Within the framework of perturbation theory, there is a way to organize the expansion to

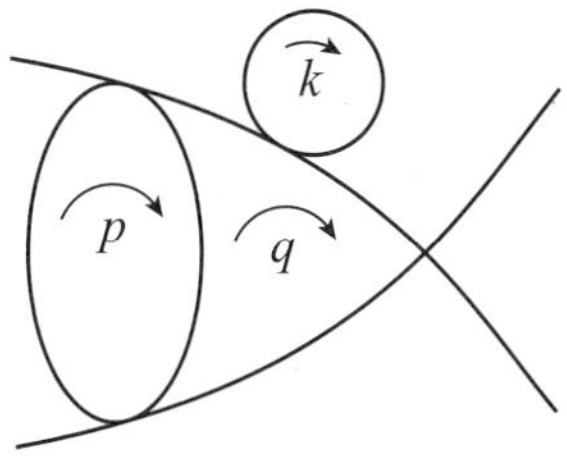

Figure 13.1 A three-loop contribution, $L=3$, to the $N=4$-point function at order $n=4$ in perturbation theory. The three internal momentum integrals run over the momenta labeled by p, q, and k.

make it more general and effective: the loop expansion. Although the loop expansion does not escape from these problems, nevertheless, it is a great formal tool. In this chapter, we follow in spirit (and in some detail) the presentation in the excellent text by Amit (1980) and that by Zinn-Justin (2002).

The loop expansion is essentially an expansion in powers of the number of internal momentum integrals ("loops") that appear in the perturbative expansion. The loop expansion involves introducing a formal expansion parameter, which we will call a, and organizing perturbation theory as a series expansion in powers of a. Naturally, this procedure is formal, since in the theory, we must have $a=1$, and it is far from obvious that this can be set consistently. Here, for simplicity, we will work with ϕ^4 theory for a one-component real scalar field in D Euclidean dimensions, but this procedure generalizes to any theory we may be interested in. Let us introduce the parameter a in the partition function in much the same way as $\hbar$ enters in the expression of the path integral:

$$Z[J]=\int \mathcal{D}\phi\, e^{-\frac{1}{a}S[\phi]}$$

$$=e^{-\frac{1}{a}\int d^Dx\mathcal{L}_{\text{int}}[\frac{\delta}{\delta J}]}\times\int \mathcal{D}\phi\, e^{-\frac{1}{a}\int d^Dx\mathcal{L}_0[\phi]+\int d^dxJ(x)\phi(x)} \tag{13.1}$$

$$=e^{-\frac{1}{a}\int d^Dx\mathcal{L}_{\text{int}}[\frac{\delta}{\delta J}]}\times e^{\frac{a}{2}\int d^Dx\int d^dx'J(x)G_0(x,x')J(x')} \tag{13.2}$$

Thus the Feynman diagrammatic rules are the same as in chapter 11, with the following changes: (1) every vertex now acquires a weight $\frac{1}{a}$, and (2) every propagator acquires a factor a. Therefore, at order n in perturbation theory, a graph with N external points and I internal lines will have a weight a^{I-n}. We now ask: How many momentum integrations does a Feynman diagram have? In each diagram, there are n δ functions (which enforce momentum conservation at each vertex). However, each diagram must have one δ function for overall momentum conservation. Since each internal propagator line carries momentum, and total momentum is conserved at each vertex, the number of independent momentum integrations L in each diagram is $L=I-(n-1)$. Thus the weight of a Feynman diagram is $a^{I-n}=a^{L-1}$. Hence the expansion in powers of a is really an expansion in powers of the number of independent integrals or *loops*.

An example is presented in figure 13.1, which is a Feynman diagram for $N=4$-point function to $n=4$ order in perturbation theory. This diagram has $I=6$ internal propagator lines. Our rules imply that the number of loops is $L=3$. In this case, the loop expansion

Figure 13.2 Tree-level contribution to the four-point vertex function $\Gamma^{(4)}$.

coincides with an expansion in powers of λ. However, if we had a theory with several coupling constants, the situation would be different.

13.1.1 The tree-level approximation

Let us compute the generating function $\Gamma[\bar{\phi}]$ to the tree-level order, $L=0$ (no loops).

$\Gamma^{(2)}$: At the tree level, the 1-PI two-point function $\Gamma_0^{(2)}$ is just

$$\Gamma_0^{(2)}(k_1, k_2) = (2\pi)^D \delta^D(k_1 + k_2)(k_1^2 + m_0^2) \tag{13.3}$$

since at the tree level, all 1-PI graphs to the self-energy Σ contain at least one loop, and hence, $\Sigma_0 = 0$.

$\Gamma^{(3)}$: The three-point function $\Gamma^{(3)}$ vanishes by symmetry at the tree level (and in fact to all orders in the loop expansion): $\Gamma^{(3)} = 0$.

$\Gamma^{(4)}$: At the tree level, the four-point function $\Gamma_0^{(4)}$ is just given by (see figure 13.2)

$$\Gamma_0^{(4)}(k_1, \ldots, k_4) = \lambda (2\pi)^D \delta^D(k_1 + \cdots + k_4) \tag{13.4}$$

$\Gamma^{(N)}$: In ϕ^4 theory, all vertex functions with $N > 4$ vanish at the tree level.

So, at the tree level, the generating function $\Gamma[\phi]$ is

$$\begin{aligned}\Gamma[\bar{\phi}] &= \sum_{N=1}^{\infty} \frac{1}{N!} \int \frac{d^D q_1}{(2\pi)^D} \cdots \int \frac{d^D q_N}{(2\pi)^D} \Gamma^{(N)}(q_1, \ldots, q_N) \bar{\phi}(-q_1) \cdots \bar{\phi}(-q_N) \\ &= \frac{1}{2!} \int \frac{d^D q_1}{(2\pi)^D} \int \frac{d^D q_2}{(2\pi)^D} (2\pi)^d \delta^d(q_1 + q_2)(q_1^2 + m_0^2) \bar{\phi}(-q_1) \bar{\phi}(-q_2) \\ &+ \frac{1}{4!} \lambda \int \frac{d^D q_1}{(2\pi)^D} \cdots \int \frac{d^D q_4}{(2\pi)^D} (2\pi)^d \delta^d(q_1 + \cdots + q_4) \bar{\phi}(-q_1) \cdots \bar{\phi}(-q_4) + O(\lambda^2)\end{aligned} \tag{13.5}$$

On Fourier transformation back to real space, we find that the tree-level effective action is just the classical action of ϕ^4 theory:

$$\Gamma_0(\bar{\phi}) = \int d^D x \left[\frac{1}{2} (\partial_\mu \bar{\phi})^2 + \frac{m_0^2}{2} \bar{\phi}^2 + \frac{\lambda}{4!} \bar{\phi}^4 \right] \tag{13.6}$$

Then, at the tree level, we find two phases. If $m_0^2 > 0$, then $\bar{\phi} \equiv \Phi = 0$, and we are in the symmetric phase (with an unbroken symmetry), while for $m_0^2 < 0$, $\Phi = \pm \sqrt{\frac{6|m_0^2|}{\lambda}}$, we are in the phase in which the symmetry is spontaneously broken. Here and below, Φ denotes the physical expectation value.

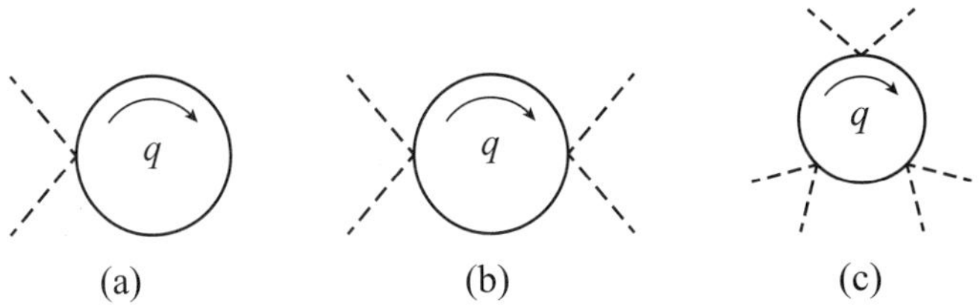

Figure 13.3 One-loop contributions to (a) $\Gamma^{(2)}$, (b) $\Gamma^{(4)}$, and (c) $\Gamma^{(6)}$; q is the internal momentum of the loop.

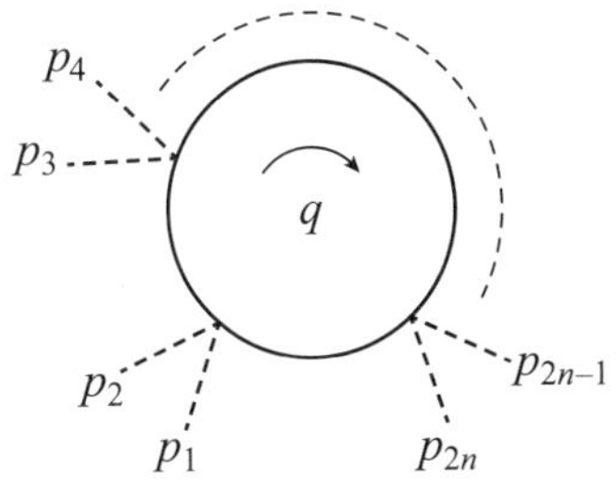

Figure 13.4 One-loop contribution to $\Gamma^{(N)}$, with $N=2n$; q is the internal momentum of the loop, and the p_i (with $i=1,\ldots,2n$) are the $2n$ external momenta.

13.1.2 One loop

Let us now examine the role of quantum fluctuations by computing the effective potential at the one-loop level. At one loop, $L=1$, which implies that the Feynman diagrams that contribute must have order n in perturbation theory equal to the number I of internal lines: $I=n$. In ϕ^4 theory, a graph with N external points, I internal lines, and order n must satisfy the identity

$$4n = N + 2I \tag{13.7}$$

since all the $4n$ lines that are emerging from the n vertices must either be contracted together pairwise or be attached to the N external points. From this we deduce that the one-loop diagrams must be such that $N=2n$. The one-loop corrections to $\Gamma^{(2)}$, $\Gamma^{(4)}$, and $\Gamma^{(6)}$ are shown in figure 13.3.

To compute the one-loop contribution to effective potential, $U_1(\Phi)$, we need to compute the one-loop contribution to the N-point function $\Gamma_1^{(N)}(0,\ldots,0)$, and with $N=2n$, and with all the external momenta set to zero: $p_1=p_2=\cdots=p_{2n-1}=p_{2n}=0$. A one-loop contribution to $\Gamma^{(N)}$ is shown in figure 13.4. We find:

$$\Gamma_1^{(N)}(0,\ldots,0) = -\left(-\frac{\lambda}{4!}\right)^n \frac{1}{n!} S_n \int \frac{d^D q}{(2\pi)^D} \left(\frac{1}{q^2+m_0^2}\right)^n \times (N-1)! \tag{13.8}$$

where the multiplicity factor is $S_n = (4\times 3)^n \times n!$, where $n!$ is the number of ways of reordering the vertices, and $(N-1)!$ is the number of ways of assigning the external momenta, p_i $(i=1,\ldots,2n)$, to the external vertices.

We can compute the corrections to the effective potential:

$$U_1[\Phi]=\sum_{N=1}^{\infty}\frac{1}{N!}\Phi^N\Gamma_1^{(N)}(0,\ldots,0)$$
$$=-\sum_{n=1}^{\infty}\frac{1}{(2n)!}\left(-\frac{\lambda\Phi^2}{4!}\right)^n\frac{(4\times 3)^n}{n!}n!(2n-1)!\int\frac{d^Dq}{(2\pi)^D}\left(\frac{1}{q^2+m_0^2}\right)^n$$
$$=-\sum_{n=1}^{\infty}\frac{1}{2n}\int\frac{d^Dq}{(2\pi)^D}\left(-\frac{\lambda\Phi^2/2}{q^2+m_0^2}\right)^n \tag{13.9}$$

Using the power series expansion of the logarithm, $\ln(1+x)=-\sum_{n=1}^{\infty}\frac{1}{n}(-x)^n$, we obtain

$$U_1[\Phi]=\frac{1}{2}\int\frac{d^Dq}{(2\pi)^D}\ln\left(1+\frac{\lambda\Phi^2/2}{q^2+m_0^2}\right) \tag{13.10}$$

Therefore, the effective potential $U(\Phi)$, including the one-loop corrections, is:

$$U[\Phi]=\frac{m_0^2}{2}\Phi^2+\frac{\lambda}{4!}\Phi^4+\frac{1}{2}\int\frac{d^Dq}{(2\pi)^D}\ln\left(q^2+m_0^2+\frac{\lambda\Phi^2}{2}\right) \tag{13.11}$$

where we canceled a constant, Φ-independent, term in eq. (13.10) against the contribution of the functional determinant to the effective potential for the free massive scalar field theory.

Let us examine what we have done more carefully. In going from eq. (13.9) to eq. (13.10), we switched the order of the momentum integral with the sum over the index n of the order of perturbation theory and, in so doing, we used the Taylor series expansion of the logarithm. This is valid if the terms of the series are smaller than 1 and the series is convergent. The problem is that, for dimensionality D, the first $D/2$ terms of this series are divergent, since the momentum integrals diverge. In particular, for $D=4$, the first two terms diverge in the UV. These are, of course, the quadratic UV divergence of the self-energy and the logarithmic UV divergence of the effective coupling constant, which we discussed in sections 11.6 and 11.7.

The solution to this problem, as we have already seen, is to define a renormalized mass μ^2 and a renormalized coupling constant g. The renormalized mass is defined by the condition

$$\mu^2=\Gamma^{(2)}(0)=m_0^2+\frac{\lambda}{2}\int\frac{d^Dq}{(2\pi)^D}\frac{1}{q^2+m_0^2}+\cdots \tag{13.12}$$

where we used the one-loop result. To one-loop order, we can invert the series as an expression of the bare mass m_0^2 in terms of the renormalized mass μ^2:

$$m_0^2=\mu^2-\frac{\lambda}{2}\int\frac{d^Dq}{(2\pi)^D}\frac{1}{q^2+\mu^2}+\cdots \tag{13.13}$$

We will see shortly that the error made in replacing the bare mass with the renormalized mass in the integral is canceled by a two-loop correction. Furthermore, to one-loop order, the renormalized two-point function $\Gamma_R^{(2)}(p)$ is finite:

$$\Gamma_R^{(2)}(p) = p^2 + \mu^2 \tag{13.14}$$

Similarly, we define a renormalized coupling constant g by the value of the four-point vertex function at zero external momenta, $p_i = 0$:

$$g = \Gamma^{(4)}(0) = \lambda - \frac{3}{2}\lambda^2 \int \frac{d^D q}{(2\pi)^D} \left(\frac{1}{q^2 + m_0^2}\right)^2 + \cdots \tag{13.15}$$

where we have used the result at one-loop order. At this order in the loop expansion, we can express the bare coupling λ as a power series of the renormalized coupling g:

$$\lambda = g + \frac{3}{2}g^2 \int \frac{d^d q}{(2\pi)^d} \left(\frac{1}{q^2 + \mu^2}\right)^2 + \cdots \tag{13.16}$$

where we have replaced the bare mass with the renormalized mass, which is consistent to one-loop order. Similarly, in the expression of the bare mass, it is consistent to replace the bare coupling constant with the renormalized coupling constant. We will see that these replacements amount to taking into account two-loop corrections.

Thus the one-loop relation between the bare and the renormalized mass becomes

$$m_0^2 = \mu^2 - \frac{g}{2} \int \frac{d^D q}{(2\pi)^D} \frac{1}{q^2 + \mu^2} + \cdots \tag{13.17}$$

To one-loop order, the renormalized four-point function, $\Gamma_R^{(4)}(p_1, \ldots, p_4)$, is

$$\Gamma_R^{(4)}(p_1, \ldots, p_4) = g - \frac{g^2}{2} \int \frac{d^D q}{(2\pi)^D} \left[\frac{1}{(q^2 + \mu^2)((p_1 + p_2 - q)^2 + \mu^2)} - \left(\frac{1}{q^2 + \mu^2}\right)^2\right]$$
$$+ 2 \text{ permutations} \tag{13.18}$$

where the two permutations involve terms with the external momenta p_1 and p_3, and p_1 and p_4, respectively.

Finally, to one-loop order, the (renormalized) effective potential $U_R[\Phi]$ is

$$U_R[\Phi] = \frac{\mu^2}{2}\Phi^2 + \frac{g}{4!}\Phi^4 + \frac{1}{2} \int \frac{d^D q}{(2\pi)^D} \ln\left(q^2 + \mu^2 + \frac{g\Phi^2}{2}\right)$$
$$- \frac{g}{2}\Phi^2 \int \frac{d^D q}{(2\pi)^D} \frac{1}{q^2 + \mu^2} + \frac{g^2}{16}\Phi^4 \int \frac{d^D q}{(2\pi)^D} \left(\frac{1}{q^2 + \mu^2}\right)^2 \tag{13.19}$$

which is finite for $D < 6$ dimensions. The renormalized mass μ^2 and the renormalized coupling constant are defined in terms of the renormalized effective potential $U_R[\Phi]$ by the *renormalization conditions:*

$$\mu^2 = \Gamma_R^{(2)}(0) = \left.\frac{\partial^2 U_R}{\partial \Phi^2}\right|_{\Phi=0}, \qquad g = \Gamma_R^{(4)}(0) = \left.\frac{\partial^4 U_R}{\partial \Phi^4}\right|_{\Phi=0} \tag{13.20}$$

Let us end this one-loop discussion with several observations. One is that we have formally manipulated divergent integrals. In what follows, we will see that it is necessary to go through a procedure of making them finite known *regularizations*. This amounts to defining the theory at short distances, which can be done in several possible ways. Another observation is that we have chosen to define the renormalized mass and coupling constant as the values of the two- and four-point vertex functions at zero external momenta. While this choice is intuitive, it is nevertheless arbitrary. One problem with this choice of renormalization conditions is that if we were to be interested in the renormalized massless theory (which defines a critical system), now the integrals contain IR divergences, which will play an important role and will require a change in the definition of the renormalized parameters. This is an important physical question that involves an operational definition of what the mass and the coupling constant are and how are they measured.

Last, but not least, is the following observation. To one-loop order, we defined two renormalized quantities: the mass and the coupling constant. Is this sufficient to all orders in perturbation theory? We will see next that already at two-loop order, a new renormalization condition is needed: the wave function renormalization. But, how do we know that the number of renormalized (or effective) parameters does not grow (or even explode) with the order in perturbation theory? If that were to be the case, this theory would not have much predictive power! We will see that theories with a finite number of renormalized parameters are what is called *renormalizable field theories,* which play a key role in physics.

13.2 Perturbative renormalization to two-loop order

We have just seen that, to one-loop order, the corrections to the two- and four-point functions amount to a redefinition of the bare mass m_0^2 and of the bare coupling constant λ in terms of a renormalized mass μ^2 and a renormalized coupling constant g. All of the singular behavior (i.e., the sensitive dependence on the UV cutoff) is contained in the relation between these bare and renormalized quantities. We will now look at what happens to next order in perturbation theory. Thus, we will consider the two-loop diagrams and examine whether this program works at the two-loop level or if new physical effects appear.

13.2.1 Mass renormalization for two loops

The two-loop contributions to the one-particle irreducible two-point function $\Gamma^{(2)}$ are the last two terms of the following Feynman diagrams:

$\Gamma^{(2)}(p) =$ 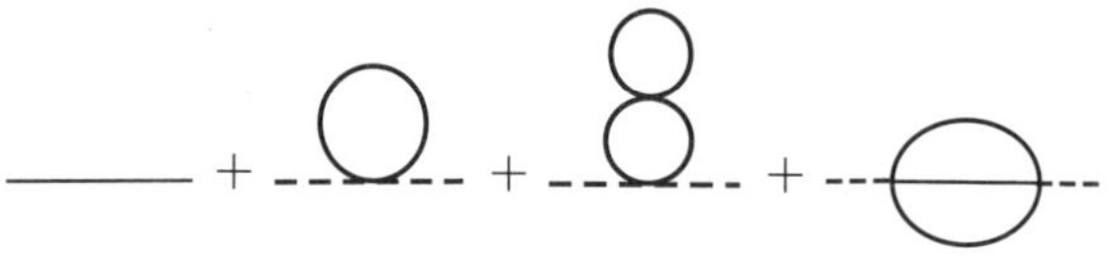 (13.21)

At two-loop level, the renormalized mass μ^2 is given by zero momentum limit of $\Gamma^{(2)}$ and has the formal expression

$$\mu^2 \equiv \Gamma^{(0)}(0) = m_0^2 + \frac{\lambda}{2} E_1(m_0^2) - \frac{\lambda^2}{4} E_2(m_0^2) E_1(m_0^2) - \frac{\lambda^2}{6} E_3(0, m_0^2) + 0(\lambda^2) \qquad (13.22)$$

where we used the notation

$$E_1(m_0^2, \Lambda) = \int^{\Lambda} \frac{d^D q}{(2\pi)^D} \frac{1}{q^2 + m_0^2}$$

$$E_2(m_0^2, \Lambda) = \int^{\Lambda} \frac{d^D q}{(2\pi)^D} \left(\frac{1}{q^2 + m_0^2} \right)^2$$

$$E_3(p, m_0^2, \Lambda) = \int^{\Lambda} \frac{d^D q_1}{(2\pi)^D} \int^{\Lambda} \frac{d^D q_2}{(2\pi)^D} \frac{1}{(q_1^2 + m_0^2)(q_2^2 + m_0^2)((p - q_1 - q_2)^2 + m_0^2)} \tag{13.23}$$

where Λ is a UV cutoff scale (or regulator). Clearly, for UV scale Λ, the degree of UV divergence of E_1 is Λ^{D-2}, of E_2 is Λ^{D-4}, and of $E_3(0)$ is Λ^{2D-6}. In particular, in $D=4$ dimensions, E_1 and E_3 are quadratically divergent, while E_2 is logarithmically divergent. Hence, some UV regularization (or rather, *definition*) must be supplied. We will do this shortly below.

By inspection, we see that the first two-loop contribution to $\Gamma^{(2)}(p)$, the third term in eq. (13.21), is just an insertion of the one-loop digram (the second term) inside the propagator. Indeed, if we carry the definition of the bare mass m_0^2 in terms of the renormalized mass μ^2 of eq. (13.17) beyond the leading term (i.e., the terms denoted by the ellipsis), we find

$$E_1(m_0^2) = \int^{\Lambda} \frac{d^D q}{(2\pi)^D} \frac{1}{q^2 + m_0^2} = \int^{\Lambda} \frac{d^D q}{(2\pi)^D} \frac{1}{q^2 + \mu^2 - \frac{\lambda}{2} E_1(\mu^2) + O(\lambda^2)}$$
$$= E_1(\mu^2) + \frac{\lambda}{2} E_1(\mu^2) E_2(\mu^2) + O(\lambda^2) \tag{13.24}$$

Thus, the expression of the renormalized mass at two-loop order becomes

$$\mu^2 = m_0^2 + \frac{\lambda}{2} \left(E_1(\mu^2) + \frac{\lambda}{2} E_1(\mu^2) E_2(\mu^2) \right) - \frac{\lambda^2}{4} E_2(m_0^2) E_1(m_0^2) - \frac{\lambda^2}{6} E_3(0, m_0^2)$$
$$= m_0^2 - \frac{\lambda}{2} E_1(\mu^2) - \frac{\lambda^2}{6} E_3(\mu^2) + O(\lambda^3) \tag{13.25}$$

Equivalently, at two-loop order, the expression of the bare mass m_0^2 in terms of the renormalized mass μ^2 is

$$m_0^2 = \mu^2 - \frac{\lambda}{2} E_1(\mu^2) + \frac{\lambda^2}{6} E_3(0, \mu^2) + O(\lambda^3) \tag{13.26}$$

Notice that the one-loop renormalization has partially canceled the two-loop contribution.

13.2.2 Coupling constant renormalization for two loops

Let us now turn to the renormalization of the coupling constant at two loops. At the two-loop level, the bare one-particle irreducible four-point function is given by the following Feynman diagrams:

$\Gamma^{(4)}(p_1, \ldots, p_4) =$ 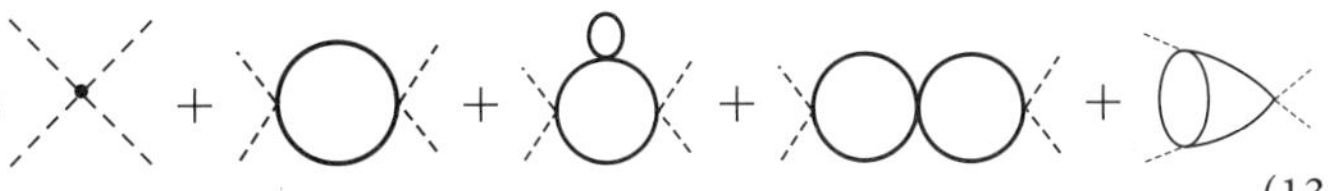 (13.27)

The actual expression is

$$\begin{aligned}\Gamma^{(4)}(p_1,\ldots,p_4) = \lambda &- \frac{\lambda^2}{2}\left[I(p_1+p_2;m_0^2) + 2\,\text{permutations}\right]\\ &+ \frac{\lambda^3}{4}\left[I(p_1+p_2;m_0^2)^2 + 2\,\text{permutations}\right]\\ &+ \frac{\lambda^3}{2}\left[I_3(p_1+p_2;m_0^2)E_1(m_0^2) + 2\,\text{permutations}\right]\\ &+ \frac{\lambda^3}{2}\left[I_4(p_1,\ldots,p_4;m_0^2) + 5\,\text{permutations}\right] + O(\lambda^4)\end{aligned} \tag{13.28}$$

where the (singular) integrals are given by

$$I(p;m_0^2) = \int_q \frac{1}{(q^2+m_0^2)((p-q)^2+m_0^2)} \tag{13.29}$$

$$I_3(p;m_0^2) = \int_q \frac{1}{(q^2+m_0^2)^2((p-q)^2+m_0^2)} \tag{13.30}$$

$$I_4(\{p_i\};m_0^2) = \int\int_{q_1,q_2} \frac{1}{(q_1^2+m_0^2)(q_2^2+m_0^2)((p_1+p_2-q_1)^2+m_0^2)((p_3+q_1-q_2)^2+m_0^2)} \tag{13.31}$$

and we used the notation

$$\int_q \equiv \int^\Lambda \frac{d^Dq}{(2\pi)^D} \tag{13.32}$$

By counting powers, we see that $I(p;m_0^2)$ scales with the UV cutoff as Λ^{D-4} (and it has a $\ln\Lambda$ divergence for $D=4$), whereas I_3 scales as Λ^{D-6} (and is finite for $D=4$ dimensions), and I_4 scales as $\Lambda^{2(D-4)}$ (and has a $\ln^2\Lambda$ divergence for $D=4$).

Let us first carry out the mass renormalization of the expressions involved in eq. (13.28), which lead to a partial cancellation. After that is done, eq. (13.28) becomes

$$\begin{aligned}\Gamma^{(4)}(p_1,\ldots,p_4) = \lambda &- \frac{\lambda^2}{2}\left[I(p_1+p_2;\mu^2) + 2\,\text{permutations}\right]\\ &+ \frac{\lambda^3}{4}\left[I(p_1+p_2;\mu^2)^2 + 2\,\text{permutations}\right]\\ &+ \frac{\lambda^3}{2}\left[I_4(p_1,\ldots,p_4;\mu^2) + 5\,\text{permutations}\right] + O(\lambda^4)\end{aligned} \tag{13.33}$$

Next, define the renormalized coupling constant at the two-loop level:

$$g = \Gamma^{(4)}([p_i=0]) = \lambda - \frac{3}{2}\lambda^2 E_2(\mu^2) + \frac{3}{4}\lambda^3 E_2^2(\mu^2) + 3\lambda^3 I_4([p_i=0];\mu^2) \tag{13.34}$$

which is inverted as

$$\lambda = g + \frac{3}{2}g^2 E_2(\mu^2) + \frac{15}{4}g^3 E_2^2(\mu^2) - 3g^3 I_4([p_i=0];\mu^2) + O(g^4) \tag{13.35}$$

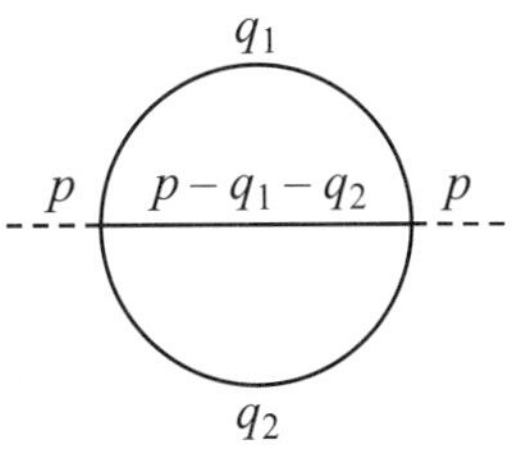

Figure 13.5 Feynman diagram with the leading contribution to the wave-function renormalization.

After this renormalization, the four-point function becomes

$$\begin{aligned}\Gamma^{(4)}(p_1,\ldots,p_4) = & g - \frac{g^2}{2}\left\{[I(p_1+p_2;\mu^2) - E_2(\mu^2)] + 2\,\text{permutations}\right\}\\ & + \frac{g^3}{4}\left\{[I(p_1+p_2;\mu^2) - E_2(\mu^2)]^2 + 2\,\text{permutations}\right\}\\ & + \frac{g^3}{2}\left\{[I_4(p_1,\ldots,p_4;\mu^2) - I_4([p_i=0];\mu^2)]\right.\\ & \left. - E_2(\mu^2)[I(p_1+p_2;\mu^2) - E_2(\mu^2)] + 5\,\text{permutations}\right\} + O(g^4) \end{aligned} \tag{13.36}$$

The coupling-constant renormalization changes the two-point function to

$$\Gamma^{(2)}(p;\mu^2) = p^2 + \mu^2 - \frac{g^2}{6}\left[E_3(p;\mu^2) - E_3(0;\mu^2)\right] + O(g^3) \tag{13.37}$$

and the relation between the bare and the renormalized mass is:

$$m_0^2 = \mu^2 - \frac{g}{2}E_1(\mu^2) - \frac{3}{4}g^2 E_1(\mu^2)E_2(\mu^2) + \frac{g^2}{6}E_3(0;\mu^2) + O(g^3) \tag{13.38}$$

13.2.3 Wave-function renormalization

The preceding results now tell us that, after mass renormalization at the two-loop level, the one-particle irreducible two-point function $\Gamma^{(2)}(p)$ becomes

$$\Gamma^{(2)}(p) = p^2 + \mu^2 - \frac{g^2}{6}\left[E_3(p,\mu^2) - E_3(0,\mu^2)\right] + O(g^3) \tag{13.39}$$

There is still, however, the momentum-dependent contribution of $\Gamma^{(2)}(p)$, the third term of the right-hand side of eq. (13.39). The subtracted expression, $E_3(p,\mu^2) - E_3(0,\mu^2)$, is logarithmically divergent in $D=4$ dimensions. Clearly, this behavior is unaffected by the mass renormalization. It is also obvious that it cannot be taken care of by the coupling-constant renormalization. Hence, we need a new renormalization, which, for historical reasons, is known as the *wave-function renormalization* (see figure 13.5). We now discuss this renormalization.

Since the remaining singular behavior in $\Gamma^{(2)}(p)$ comes from its momentum dependence, it enters as a factor in front of p^2. Equivalently, this new renormalization is a change in the prefactor of the gradient term of the action (as discussed below). This suggests that we define a renormalized one-particle irreducible two-point function by a rescaling of the form

$$\Gamma_R^{(2)}(p,\mu_R^2)=Z_\phi(g,\mu^2,\Lambda)\Gamma^{(2)}(p,\mu^2,\Lambda) \tag{13.40}$$

such that

$$\Gamma_R^{(2)}(p,\mu_R^2)=p^2+\mu_R^2 \tag{13.41}$$

Notice that the rescaling of the two-point function is forcing us to change the definition of the renormalized mass from μ^2 to μ_R^2. The introduction of the new renormalization "constant" $Z_\phi(g,\mu^2,\Lambda)$ is equivalent to a rescaling of the field ϕ by $Z_\phi^{1/2}$. For this reason, this procedure is known as the wave-function renormalization. This name, invented in the context of QED in the late 1940s, is still retained, even though it is a highly misleading term.

The wave-function renormalization constant Z_ϕ has the expansion

$$Z_\phi=1+z_1g+z_2g^2+\cdots \tag{13.42}$$

Since the wave-function renormalization only appears (in ϕ^4 theory) at the two-loop level, we see that we must set $z_1=0$. Furthermore, demanding that $\Gamma_R^{(2)}$ has the form of eq. (13.41) or, what is the same, that

$$\Gamma_R^{(2)}(0)=\mu_R^2, \qquad \text{and} \qquad \left.\frac{\partial\Gamma_R^{(2)}(p)}{\partial p^2}\right|_{p=0}=1 \tag{13.43}$$

we find

$$1=Z_\phi\left(1-\frac{g^2}{6}\left.\frac{\partial E_3(p,\mu^2)}{\partial p^2}\right|_{p=0}\right) \tag{13.44}$$

from which we conclude that the (logarithmically divergent) quantity z_2 is

$$z_2=\frac{1}{6}\left.\frac{\partial E_3(p,\mu^2)}{\partial p^2}\right|_{p=0} \tag{13.45}$$

However, this rescaling changes the new renormalized mass to μ_R^2:

$$\mu_R^2=(1+z_2g^2)\mu^2 \tag{13.46}$$

The introduction of the wave-function renormalization, in turn, also affects the definition of the renormalized coupling. The renormalized one-particle irreducible four-point function is defined as

$$\Gamma_R^{(4)}(\{p_i\};g_R,\mu_R^2)=Z_\phi^2\Gamma^{(4)}(\{p_i\};g,\mu^2) \tag{13.47}$$

This definition accounts for the contributions of two-loop corrections in the internal propagators. Similarly, the renormalized one-particle irreducible N-point functions are defined to be

$$\Gamma_R^{(N)}(\{p_i\};g_R,\mu_R^2)=Z_\phi^{N/2}\Gamma^{(N)}(\{p_i\};g,\mu^2) \tag{13.48}$$

13.2.4 Renormalization conditions

Let us summarize the procedure of perturbative renormalization outlined so far, which we carried out to the two-loop level:

1) We computed the bare two-point and four-point one-particle irreducible functions $\Gamma^{(2)}$ and $\Gamma^{(4)}$ as functions of the bare mass m_0^2 and the bare coupling constant λ.
2) We replaced the bare mass m_0^2 first by μ^2 and later by $\mu_R^2 = \mu^2 Z_\phi$. The renormalized mass was defined by the renormalization condition

$$\lim_{p\to 0} \Gamma_R^{(2)}(p) = \mu_R^2 \tag{13.49}$$

3) We replaced the bare coupling constant λ first by g and later by the renormalized coupling constant $g_R = gZ_\phi^2$. The renormalized coupling constant was defined by the renormalization condition

$$\lim_{\{p_i\}\to 0} \Gamma_R^{(4)}(p_1, \ldots, p_4) = g_R \tag{13.50}$$

4) The wave-function renormalization Z_ϕ was obtained by demanding the renormalization condition

$$\left.\frac{\partial \Gamma_R^{(2)}(p)}{\partial p^2}\right|_{p=0} = 1 \tag{13.51}$$

5) We defined the renormalized functions

$$\Gamma_R^{(N)} = Z_\phi^{N/2} \Gamma^{(N)}, \tag{13.52}$$

which are UV-finite functions of the renormalized mass μ_R^2 and of the renormalized coupling constant g_R.

Notice that these are definitions that we chose to make, and that many such definitions are possible. In particular, all of the singular (divergent) behavior is "hidden" in the relation between the bare and the renormalized quantities. By this procedure, at least up to two loops, we succeeded in removing the strong dependence on the UV definition of the theory. However, how do we know if this procedure will suffice to all orders in perturbation theory? In other words, how do we know if the number of renormalized parameters does not grow with the order of the perturbation theory? Clearly, a theory with an infinite number of arbitrary parameters would not be a theory at all! We will address this problem shortly.

Formally, the renormalized mass μ_R^2, the renormalized coupling constant g_R, and the wave function renormalization Z_ϕ are functions of the bare mass m_0^2, the bare coupling constant λ, and the UV regulator (or cutoff scale) Λ, of the form

$$\begin{aligned}
\mu_R^2 &= Z_\phi \mu^2(m_0^2, \lambda, \Lambda) \\
g_R &= Z_\phi^2 g(m_0^2, \lambda, \Lambda) \\
Z_\phi &= Z_\phi(m_0^2, \lambda, \Lambda)
\end{aligned} \tag{13.53}$$

which are given by their expressions in the perturbation theory expansion in powers of the coupling constant λ. Alternatively, we can invert these relations and write expressions for the bare parameters in terms of the renormalized ones:

$$m_0^2 = Z_\phi m_0^2(\mu_R^2, g_R, \Lambda)$$

$$\lambda = Z_\phi^2 g(\mu_R^2, g_R, \Lambda)$$

$$Z_\phi = Z_\phi(\mu_R^2, g_R, \Lambda) \tag{13.54}$$

Then the renormalized vertex functions

$$\Gamma_R^{(N)}(\{p_i\}, \mu_R^2, g_R, \Lambda) = Z_\phi^{N/2} \Gamma^{(N)}(m_0^2, \lambda, \Lambda) \tag{13.55}$$

have a finite limit as the regulator $\Lambda \to \infty$ to every order in an expansion in powers of the renormalized coupling constant g_R.

We will now consider the computation of the renormalization constants to two-loop order in ϕ^4 theory in $D=4$ dimensions. In general, in ϕ^4 theory in $D=4$ dimensions, we have three renormalization conditions

$$\Gamma_R^{(2)}(0, \mu_R^2, g_R) = \mu_R^2 \tag{13.56}$$

$$\frac{\partial}{\partial p^2} \Gamma_R^{(2)}(p, \mu_R^2, g_R)\Big|_{p=0} = 1 \tag{13.57}$$

$$\Gamma_R^{(4)}(\{p_i = 0\}, \mu_R^2, g_R) = g_R \tag{13.58}$$

where eq. (13.56) fixes the renormalized mass, eq. (13.57) fixes the wave-function renormalization, and eq. (13.58) fixes the coupling-constant renormalization.

For simplicity, we will discuss only the massless case, and hence we require that the normalized mass $\mu_R^2 = 0$. However, the renormalization conditions used so far are fine for the massive case, but not for the massless case, $\mu_R^2 = 0$, due to the IR divergences present in the expressions of the Feynman diagrams. For this reason, in the massless case, we instead impose renormalization conditions at a fixed momentum scale κ:

$$\Gamma_R^{(2)}(0, g_R) = 0$$

$$\frac{\partial}{\partial p^2} \Gamma_R^{(2)}(p, g_R)\Big|_{p^2=\kappa^2} = 1$$

$$\Gamma_R^{(4)}(\{p_i\}, g_R)\Big|_{SP} = g_R \tag{13.59}$$

where SP denotes the symmetric arrangement of the four external momenta $\{p_i\}$ (with $i = 1, \ldots, 4$), such that $p_i \cdot p_j = \frac{\kappa^2}{4}(4\delta_{ij} - 1)$. With this choice, we have $P^2 = (p_i + p_j)^2 = \kappa^2$. Since the renormalized quantities are defined at a fixed momentum scale κ, the renormalization constants will also be functions of that scale.

To proceed, we first need to find the value of the bare mass m_0^2 for which the renormalized mass vanishes, $\mu_R^2 = 0$. Let us call this value of the bare mass $m_c^2(\lambda, \Lambda)$, which is a function of the bare coupling constant and of the UV regulator. We have already done something like this at the one-loop level in section 11.6, where we observed that this approach is equivalent to finding the correction due to fluctuations to the critical temperature for the phase transition in the Landau theory of phase transitions. In that case, we identified $m_c^2 = T_c - T_0$, with T_0 being the bare (or mean field) value of the critical temperature, and T_c being its value corrected by fluctuations.

At two-loop order, we find that $m_c^2(\lambda, \Lambda)$ is the solution of the equation

$$0 = m_c^2 + \frac{\lambda}{2} D_1(m_c^2, \Lambda) - \frac{\lambda^2}{4} D_1(m_c^2, \Lambda) E_2(m_c^2, \Lambda) - \frac{\lambda^2}{6} E_3(0, m_c^2, \Lambda) + O(\lambda^3) \qquad (13.60)$$

The solution to this equation as a power series expansion in powers of the bare coupling constant λ is

$$m_c^2(\lambda, \Lambda) = -\frac{\lambda}{2} E_1(0, \Lambda) + \frac{\lambda^2}{6} E_3(0, 0, \Lambda) + O(\lambda^3) \qquad (13.61)$$

where, as before, we have used the notation

$$E_1(0, \Lambda) = \int^{\Lambda} \frac{d^D q}{(2\pi)^D} \frac{1}{q^2}$$
$$E_3(0, 0, \Lambda) = \int^{\Lambda} \frac{d^D q_1}{(2\pi)^D} \int^{\Lambda} \frac{d^D q_2}{(2\pi)^D} \frac{1}{q_1^2 q_2^2 (q_1 + q_2)^2} \qquad (13.62)$$

where Λ is an unspecified UV regulator (or cutoff). Notice that both integrals diverge like Λ^2 in the UV in $D = 4$ dimensions and are finite in the IR.

Next we need to do the wave-function and the coupling constant renormalizations. In each case, let us write the bare coupling constant λ and the wave-function renormalization Z_ϕ as power series in the renormalized coupling constant g_R, of the form

$$\lambda = g_R + \lambda_2 g_R^2 + \lambda_3 g_R^3 + O(g_R^4)$$
$$Z_\phi = 1 + z_2 g_R^2 + O(g_R^3) \qquad (13.63)$$

These expressions use the fact that no wave-function renormalization occurs at the one-loop level and hence, $z_1 = 0$.

The renormalization condition eq. (13.57) dictates that Z_ϕ must obey the condition

$$1 = Z_\phi \left[1 - \frac{\lambda^2}{6} \frac{\partial}{\partial p^2} \right.$$ 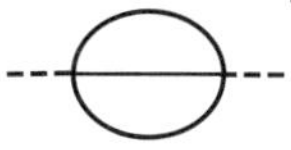 $$\left. \right]_{p^2 = \kappa^2} \qquad (13.64)$$

From this, it follows that z_2 is given by the expression

$$z_2 = \frac{1}{6} \frac{\partial}{\partial p^2} E_3(p, 0, \Lambda) \Big|_{p^2 = \kappa^2} \qquad (13.65)$$

Note that, although the integral $E_3(p, 0, \Lambda)$ diverges quadratically in the UV as Λ^2 in $D = 4$ dimensions, the derivative $\frac{\partial E_3}{\partial p^2}$ diverges logarithmically with the UV cutoff Λ.

Likewise, eq. (13.58) leads to the requirement that the renormalized coupling constant g_R should obey

$$g_R = Z_\phi^2 \left[\lambda - \frac{3}{2} \lambda^2 \right.$$ 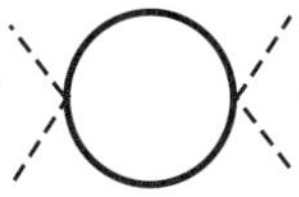 $$+ \frac{3}{4} \lambda^3$$ 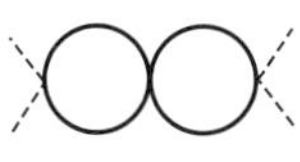 $$+ 3\lambda^3$$ 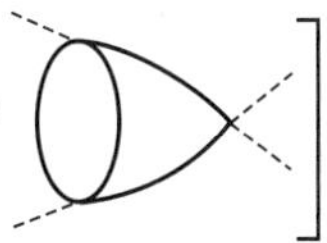 $$\left. \right]_{SP} \qquad (13.66)$$

Collecting terms and expanding order by order, we find that the constants λ_2 and λ_3 (defined in eq. (13.63)) are given by

$$\lambda_2 = \frac{3}{2} I_{SP}$$

$$\lambda_3 = \frac{15}{4} I_{SP}^2 - 3I_{4SP} - 2z_2 \tag{13.67}$$

where I_{SP} and I_{4SP} are the integrals for the bubble diagrams defined in eqs. (13.29), (13.30), and (13.31) and computed in the massless theory at the symmetric point of the external momenta with momentum scale κ.

Thus we have reduced the problem of computing the renormalization constants to the evaluation of singular integrals, which themselves are ill-defined unless we supply a definition of the theory of short distances. This is done in section 13.6, where we discuss different regularization schemes.

13.3 Subtractions, counterterms, and renormalized Lagrangians

The procedure of renormalized perturbation theory relates the bare connected N-point functions, $G_c^{(N)}(\{p_i\}; m_0^2, \lambda, \Lambda)$, which depend on N external momenta $\{p_i\}$ (with $i = 1, \ldots, N$), the bare mass m_0^2, and the bare coupling constant λ (and some as-yet unspecified UV regulator, or cutoff, Λ), to a renormalized connected N-point function $G_{cR}^{(N)}(\{p_i\}; \mu_R^2, g_R)$, which depends on the N external momenta, the renormalized mass μ_R^2, and the renormalized coupling constant g_R:

$$G_{cR}^{(N)}(\{p_i\}; \mu_R^2, g_R) = Z_\phi^{-N/2} G_c^{(N)}(\{p_i\}; m_0^2, \lambda, \Lambda) \tag{13.68}$$

In this way, the renormalized N-point functions formally do not depend on the regulator scale Λ, although (as we will see later on) they do depend on the renormalization procedure. At least at a formal level, the renormalized N-point functions describe a continuum QFT (i.e., without a cutoff). All the strong dependence on the UV definition of the theory (i.e., the regulator) is encoded in the relation between the bare and the renormalized quantities.

The bare connected N-point functions are determined by the generating functional $F[J] = \ln Z[J]$, where $Z[J]$ is the partition function for a theory with the bare Lagrangian (density) $\mathcal{L}_B$:

$$\mathcal{L}_B = \frac{1}{2}\left(\partial_\mu \phi\right)^2 + \frac{1}{2} m_0^2 \phi^2 + \frac{\lambda}{4!}\phi^4 - J\phi \tag{13.69}$$

Formally, we should expect that the renormalized connected N-point functions should be determined from a generating functional $F_R[J]$ for a renormalized Lagrangian $\mathcal{L}_R$ of the same form: that is,

$$\mathcal{L}_R = \frac{1}{2}\left(\partial_\mu \phi_R\right)^2 + \frac{1}{2}\mu_R^2 \phi^2 + \frac{g_R}{4!}\phi_R^4 - J\phi_R \tag{13.70}$$

which depends on the renormalized mass μ_R^2 and the renormalized coupling constant g_R. Here ϕ_R is the renormalized field,

$$\phi_R \equiv Z_\phi^{-1/2} \phi \tag{13.71}$$

which differs from the field ϕ by a multiplicative rescaling factor $Z_\phi^{-1/2}$.

On rescaling the field as in eq. (13.71), we can write the bare Lagrangian, eq (13.69), as

$$\mathcal{L}_B = \frac{1}{2} Z_\phi \left(\partial_\mu \phi_R\right)^2 + \frac{1}{2} Z_\phi m_0^2 \phi_R^2 + \frac{1}{4!} \lambda Z_\phi^2 \phi_R^4 - J \phi_R \sqrt{Z_\phi} \tag{13.72}$$

Then the difference $\Delta\mathcal{L} = \mathcal{L}_B - \mathcal{L}_R$ between the bare and the renormalized Lagrangians is

$$\Delta\mathcal{L} = \frac{1}{2}(Z_\phi - 1)\left(\partial_\mu \phi_R\right)^2 + \frac{1}{2}(Z_\phi m_0^2 - \mu_R^2)\phi_R^2 + \frac{1}{4!}(\lambda Z_\phi^2 - g_R)\phi_R^4 - J\phi_R(\sqrt{Z_\phi} - 1) \tag{13.73}$$

The procedure of renormalizing a theory can then be regarded as a series of steps in which the bare Lagrangian $\mathcal{L}_B$ is subtracted by a set of *counterterms*, such as those shown in eq. (13.73), which have the *same form* as the bare Lagrangian.

Thus, the program of renormalizing a QFT, defined in terms of its perturbation theory, can be recast as a systematic classification of the possible counterterms needed to make the perturbation theory UV finite. Here we have assumed that the bare and the renormalized Lagrangians have the same structure (and not just the same symmetries), and hence, so do the counterterms. If additional operators were to arise at some order in perturbation theory, these operators and their counterterms must be added to the Lagrangian to ensure consistency, or *renormalizability*.

This program traces its origins to the work by Feynman, Schwinger, and Tomonaga, later expanded by Bogoliubov, Symanzik, and many others, in QED, which, to date, is the most precise and the most successful QFT. But despite its many notable successes, this program has several drawbacks. First, it relies entirely on the perturbative definition of the theory. As we discussed, the perturbation series is at best an asymptotic series with a vanishing radius of convergence. Hence, it cannot be an analytic function of the bare coupling, even when a UV regulator is defined, since changing the sign of the coupling constant λ from positive to negative turns a theory with a stable vacuum state to one without a stable vacuum state (unless operators with higher powers of the field are added explicitly to the action to ensure stability).

Second, and more seriously, the procedure that we followed is physically obscure. A lot of important physics is hidden away in the relation between bare and renormalized quantities. We will clarify these problems when we discuss the renormalization group in chapter 15. There we will see that, at a price, we can formulate a nonperturbative definition of the theory beyond a definition in terms of its perturbation series. One important concept that we will encounter is that the coupling constants (and for that matter, all the parameters of the Lagrangian) are not really fixed quantities but depend on the energy (and momentum) scale at which they are defined (or measured).

13.4 Dimensional analysis and perturbative renormalizability

The previous sections showed that, at least to second order in the loop expansion, it is possible to recast the effects of fluctuations in a renormalization of a set of parameters (i.e., the coupling constant, the mass, and the wave function renormalization). Here we have made the implicit assumption that there is a finite number of such renormalized parameters.

When this is the case, we will say that a theory, such as ϕ^4 theory in $D=4$ dimensions, is "renormalizable." Or, to be more precise, we should say "perturbatively renormalizable," since it is a statement about the perturbative expansion. In chapter 15, we discuss the renormalization group, and we will see that several possible renormalized theories are defined by a fixed point of the renormalization group.

For now, let us discuss perturbative renormalization in the weak coupling regime, in several theories focusing specifically on the case of ϕ^4 theory. We begin by doing dimensional analysis in several theories of physical interest that we have discussed in other chapters. Here the term "dimensional analysis" always implies that this analysis is done at the level of the free field theory. We will see in chapters 15 and 16 that interactions can (and do) change this analysis in profound ways. Here we introduce two key concepts: scaling dimensions of operators and critical dimensions of spacetime. We will see that an intimate relation exists between critical dimensions and perturbative renormalizability.

13.4.1 Scalar fields

Let us begin with a theory of a scalar field ϕ, which, for simplicity, we take to have only one real component. The extension to many components (real or complex) will not change the essence of our analysis. The Euclidean action has the general form

$$S=\int d^Dx\,\mathcal{L}=\int d^Dx\left[\frac{1}{2}(\partial_\mu\phi)^2+\frac{m_0^2}{2}\phi^2+\sum_n\frac{\lambda_n}{n!}\phi^n\right] \tag{13.74}$$

Since the action S should be dimensionless, the Euclidean Lagrangian (density) must scale as the inverse volume, $[\mathcal{L}]=L^{-D}$, where L is a length scale. It then follows that the field ϕ must have units of $[\phi]=L^{-(D-2)/2}$. Or, in terms of a momentum scale Λ, the field scales as $[\phi]=\Lambda^{(D-2)/2}$. By consistency, we find that the mass has units of $[m_0]=L^{-1}=\Lambda$. Similarly, the operators ϕ^n have units of $[\phi^n]=L^{-n(D-2)/2}$, and the coupling constants λ_n, defined in the action of eq. (13.74), scale as $[\lambda_n]=L^{\frac{nD}{2}-D-n}=\Lambda^{D+n-nD/2}$.

We will say that an operator $\mathcal{O}$ has *scaling dimension* $\Delta_{\mathcal{O}}$ if its units are $[\mathcal{O}]=L^{-\Delta_{\mathcal{O}}}$. Clearly, the free scalar field ϕ has scaling dimension $\Delta_\phi=(D-2)/2$, and ϕ^n has (free-field) scaling dimension $\Delta_n=n\Delta_\phi$.

When can the coupling constant λ_n be *dimensionless*? For this to happen, the scaling dimension of the operator ϕ^n must be equal to the dimensionality D of Euclidean space. This can only happen at a certain *critical dimension* $D_n^c=2n/(n-2)$. Hence, the critical dimension for ϕ^3 is 6, for ϕ^4 is 4, for ϕ^6 is 3, and so forth. Also, we see that as $n\to\infty$, the critical dimension decreases: $D_n^c\to 2$. Hence in $D=2$ dimensions, the field ϕ and all its powers are dimensionless and their scaling dimensions are zero.

This dimensional analysis also tells us that the connected N-point functions of the scalar field, $G_N(x_1,\ldots,x_N)=\langle\phi(x_1)\cdots\phi(x_n)\rangle$, have units $[G_N]=[\phi]^N=L^{-N(D-2)/2}=\Lambda^{N(D-2)/2}$. Their Fourier transforms $\tilde{G}_N(p_1,\ldots,p_N)$ have units $[\tilde{G}_N]\Lambda^{-ND}=\Lambda^{-N(D+2)/2}$. It is easy to see that the Fourier transforms of the one-particle irreducible N-point vertex functions $\bar{\Gamma}^N$ (where we have factored out the delta function for momentum conservation) have units $[\bar{\Gamma}^N]=\Lambda^{N+D-ND/2}$, which are the same units as those of the coupling constants λ_N.

13.4.2 Nonlinear sigma models

Nonlinear sigma models are a class of scalar field theories in which the field obeys certain local constraints. In such theories, the global symmetry is realized nonlinearly.

The prototype nonlinear sigma model is an N-component real scalar field $n^a(x)$ (with $a=1,\ldots,N$) that satisfies the local constraint

$$\boldsymbol{n}^2(x)=1 \tag{13.75}$$

The global symmetry in this theory is $O(N)$. The Euclidean action of the nonlinear sigma model is

$$S=\frac{1}{2g}\int d^Dx\,\left(\partial_\mu \boldsymbol{n}(x)\right)^2 \tag{13.76}$$

where g is a coupling constant.

The constraint of the nonlinear sigma model, eq. (13.75), requires that the field $\boldsymbol{n}$ be dimensionless. Hence, the coupling constant g must have units $[g]=L^{D-2}$. It follows that the critical dimension of the nonlinear sigma model is $D_c=2$. We will come back to nonlinear sigma models in chapter 16, where we will find that this analysis holds for all such models.

13.4.3 Fermi fields

Let us now discuss the theory of an N-component relativistic Fermi field ψ_a, with $a=1,\ldots,N$ (here we drop the Dirac indices). The action has a free-field Dirac term plus some local interaction terms:

$$S=\int d^Dx\,\left[\bar{\psi}_a\left(i\gamma^\mu\partial_\mu-m_0\right)\psi_a+g(\bar{\psi}_a\psi_a)^2\right] \tag{13.77}$$

This is the Gross-Neveu model with a global $SU(N)$ symmetry and a discrete chiral symmetry. Here, as before, g is the coupling constant.

Once again, by requiring that the action S be dimensionless, we find the scaling dimension of the field, which is now $\Delta_\psi=(D-1)/2$, and the mass has units $[m_0]=L^{-1}=\Lambda$ (as it should). The scaling dimension of the interaction operator $(\bar{\psi}\psi)^2$ is $\Delta=4\Delta_\psi=2(D-1)$. It follows that the coupling constant g must have units $[g]=-D+2(D-1)=D-2$. Hence, g is dimensionless in $D_c=2$, which is the critical dimension of the Gross-Neveu model. Notice that this dimension is the same as the critical dimension of the nonlinear sigma model.

Another case of interest is a theory of a Dirac field ψ and a scalar field ϕ coupled through a Yukawa coupling of the form $g_Y\bar{\psi}\psi\phi$. Since the scaling dimension of the Dirac field is $\Delta_\psi=(D-1)/2$ and the scaling dimension of the scalar field is $\Delta_\phi=(D-2)/2$, the units of the Yukawa coupling are $[g_Y]=(D-4)/2$. Hence, the critical dimension for the Yukawa coupling is $D_c=4$.

13.4.4 Gauge theories

We now turn to the case of gauge theories. Consider the general case of a non-abelian Yang-Mills theory. This analysis also holds for the special case of the abelian (Maxwell) gauge theory. Here the gauge field A_μ is a connection and takes values in the algebra of a gauge group G. The covariant derivative is $D_\mu=\partial_\mu-iA_\mu$. In this case, dimensional analysis is useful in the asymptotically weak coupling regime, $g\to 0$, where the gauge fields might be expected to essentially behave classically. We will see in the next section that, here too, this analysis can also serve to assess the importance of low-order quantum fluctuations (i.e., one loop). We will see that our expectations are correct in some cases (e.g., QED) but incorrect in other cases (e.g., Yang-Mills).

From this definition, it follows that the gauge field A_μ has units of inverse length, and hence its scaling dimension is $\Delta_A = 1$, regardless of the dimension of spacetime. Notice that we have not included the coupling constant in the definition of the covariant derivative. Since the field tensor is the commutator of two covariant derivatives, $F_{\mu\nu} = i[D_\mu, D_\nu]$, it has scaling dimension $\Delta_F = 2$.

In contrast, the Yang-Mills action is

$$S = \frac{1}{4g^2} \int d^D x \, \mathrm{tr}(F_{\mu\nu} F^{\mu\nu}) \tag{13.78}$$

where the Yang-Mills coupling constant g has been introduced. Again, since S is dimensionless, it follows that the Yang-Mills coupling constant g has units $[g] = L^{-(D-4)/2}$.

Hence, the critical dimension of all gauge theories (with a continuous gauge group) is $D_c = 4$. This analysis also holds for gauge theories (minimally) coupled to matter fields, which is the case for Dirac fermions (as in QED and QCD) and for complex scalar fields (as in scalar electrodynamics and in Higgs models). In all cases, gauge invariance requires that there should be only one coupling constant. In chapter 18, we discuss gauge theories with a discrete gauge group, and we will see that their critical dimension is $D_c = 2$.

13.5 Criterion for perturbative renormalizability

In section 13.2, we discussed in detail the program of perturbative renormalization in ϕ^4 field theory. There we worked laboriously to repackage the results, up to two-loop order, in *the same form* as the classical theory by defining a finite number of renormalized parameters (e.g., the renormalized mass and the renormalized coupling constant). We also saw that at the two-loop level, we needed to introduce a new concept, the wave-function (or field) renormalization. Although we have not yet discussed it, products of fields at short distances become composite operators, which have their own renormalizations.

With some variants, this program has been carried out for all the theories mentioned in section 13.4. For instance, in the case of nonlinear sigma models and Yang-Mills gauge theories, the expansion in powers of the coupling constant is an expansion about the classical vacuum state. In all cases, one makes the (explicit or implicit) assumption that the theory has a small parameter that, at least qualitatively, might control the use of a perturbative expansion. This is the case in QED, whose expansion parameter is the fine structure constant $\alpha = 1/137$, but it is not the case in most other theories, particularly in Yang-Mills.

Theories that require a finite number of renormalizations to account for their UV divergences are said to be (perturbatively) renormalizable field theories. This criterion implicitly always uses free-field (or classical) theory as a reference theory. In chapters 15 and 16, when we discuss the renormalization group, we will see that one can define theories with respect to other *fixed points*, where the theory is not free (or classical). From now on, when we use the term "renormalizable," it will be understood to mean "perturbatively renormalizable." Examples of renormalizable field theories are ϕ^4 theory in $D=4$ dimensions, nonlinear sigma models in $D=2$ dimensions, Gross-Neveu models in $D=2$ dimensions, gauge theories in $D=4$ dimensions, QED and QCD in $D=4$ dimensions, and scalar field theories in $D=4$ dimensions.

We will see that the theories mentioned here are renormalizable at their critical dimensions, where their coupling constants are dimensionless. Let us consider why this is the case. To make the argument concrete, we examine perturbative ϕ^4 field theory and

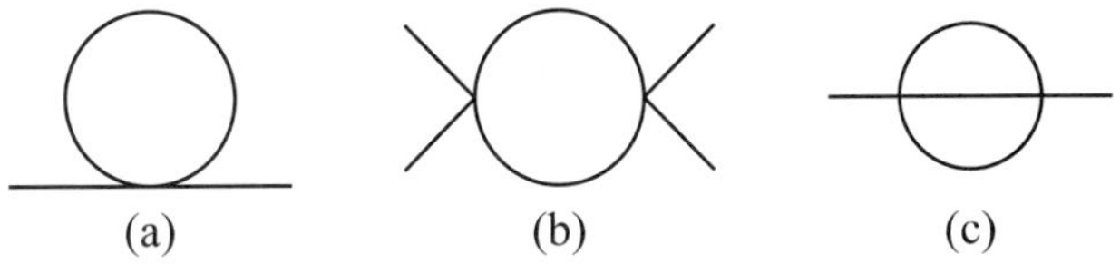

Figure 13.6 Primitively UV-divergent diagrams in ϕ^4 theory: (a) the tadpole diagram, (b) the bubble diagram, and (c) the watermelon diagram.

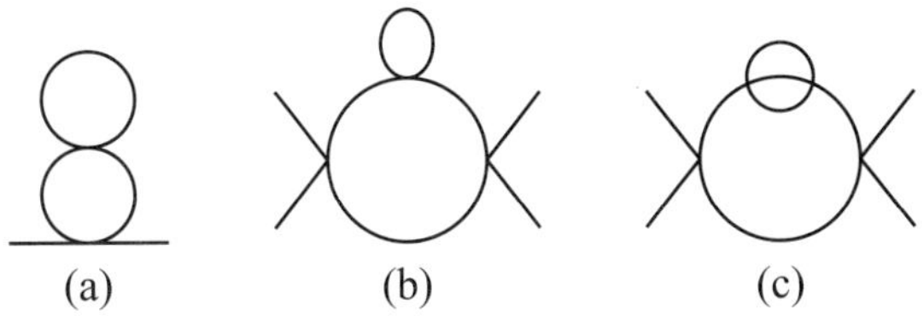

Figure 13.7 Examples of nonprimitively UV-divergent diagrams in ϕ^4 theory: (a) the tadpole insertion in a tadpole diagram, (b) the tadpole insertion in the bubble diagram, and (c) the watermelon insertion in the bubble diagram.

its divergent Feynman diagrams. The same framework works, with little change, for other theories.

There are two types of divergent diagrams, such as the ones shown in figure 13.6. At higher orders in perturbation theory, we will find more complex divergent diagrams, such as those shown in figure 13.7. However, this class of diagrams results from insertions of lower-order diagrams into themselves. Divergent diagrams that do not arise from lower-order insertions (i.e., those of figure 13.6) are said to be *primitively divergent.* In ϕ^4 theory in $D=4$ dimensions, these are the only primitively divergent diagrams. Clearly, the number and type of primitively divergent diagrams depend on the theory and on the dimension D.

When we say that a diagram is "divergent," we use implicitly its superficial degree of divergence, which follows from a simple power-counting argument. For instance, consider a scalar field theory with a ϕ^r interaction. Let us consider the perturbative contributions to the N-point vertex function to nth order in perturbation theory. As a function of the UV regulator Λ, these contributions scale as $\Lambda^{\delta(r,D,N,n)}$, where the exponent $\delta(r,D,N,n)$ is the superficial degree of divergence of a Feynman diagram. For a diagram with L internal loops, the quantity $\delta(r,D,N,n)$ must be just the difference between the phase-space contribution and the number I of propagators in the integrand of the diagram. Thus, we must have

$$\delta(r,D,N,n)=LD-2I, \quad \text{where} \quad L=I-(n-1) \tag{13.79}$$

However, in a diagram for the N-point function, the propagator lines must either connect the internal vertices with themselves or with the external points. Hence, we must have

$$nr=N+2I \tag{13.80}$$

We then find that the superficial degree of divergence $\delta(r,D,N,n)$ is

$$\delta(r,D,N,n)=\left(r+D-\frac{ND}{2}\right)-n\delta_r \tag{13.81}$$

where

$$\delta_r = r + D - \frac{rD}{2} \equiv D - \Delta_r \tag{13.82}$$

and we have introduced the (free-field) scaling dimension $\Delta_r = r\Delta_\phi = r(D-2)/2$ of the operator ϕ^r in spacetime dimension D, defined in section 13.4. Thus the quantity δ_r gives the units of the coupling constant λ_r of the operator ϕ^r.

This dimensional analysis tells us that the superficial degree of UV divergence of the diagrams that contribute to a vertex function depends on the canonical scaling dimension Δ_r of the operator ϕ^r. Eq. (13.81) depends linearly on the order n of the perturbation theory. We have three cases:

1) If $\delta(r, D, N, n) < 0$, then the superficial degree of UV divergence of the N-point function will increase as the order n of the perturbation theory increases. Clearly, if the degree of UV divergence increases with the order of the perturbation theory, we will have to introduce an infinite number of parameters (couplings) to account for this singular behavior. For this reason, a theory with $\delta_r < 0$ is said to be "nonrenormalizable."
2) If $\delta_r > 0$, the superficial degree of UV divergence decreases as the order of perturbation theory increases. In this case, the theory is said to be "super-renormalizable."
3) If $\delta_r = 0$, the superficial degree of divergence does not depend on the order n of the perturbation theory. However, from eq. (13.82), we see that $\delta_r = 0$ only if the scaling dimension satisfies $\Delta_r = D$. Hence, the dimension D must be equal to the critical dimension for ϕ^r, where the coupling constant λ_r is dimensionless.

We can now change the question somewhat and ask: What vertex functions $\Gamma^{(N)}$ have primitive divergences in ϕ^r theory at its critical dimension $D_c(r) = 2r/(r-2)$? The answer is those vertex functions for which $\delta(D_c(r)) \geq 0$. This yields the condition $N + D_c(r) - ND_c(r)/2 \geq 0$. Hence, vertex functions $\Gamma^{(N)}$ with $N \leq 2D_c(r)/(D_c(r)-2) = r$ have primitively divergent diagrams. In particular, in ϕ^4 theory in $D=4$ dimensions, $\Gamma^{(2)}$ and $\Gamma^{(4)}$ vertex functions have, respectively, quadratic and logarithmically primitively divergent diagrams. But in ϕ^6 theory in $D=3$ dimensions, $\Gamma^{(2)}$, $\Gamma^{(4)}$, and $\Gamma^{(6)}$ vertex functions have quadratic, linear, and logarithmically divergent primitive diagrams, and so forth.

This analysis only sets the stage for the proof of renormalizability at the critical dimension. In addition, a more sophisticated analysis is needed to prove renormalizability. This analysis is rather technical, and we will not do it here. In chapter 16, we will prove the renormalizability of the $O(N)$ nonlinear sigma model, where the proof makes heavy use of a Ward identity.

In this section, we have focused exclusively on the UV behavior of a theory of the type of ϕ^4. However, the IR behavior is just as interesting, and in some sense, it is physically more important, because IR divergences signal a breakdown of perturbation theory, indicating that the perturbative ground state may be unstable. IR divergences are suppressed (or, rather, controlled) by a finite renormalized mass, μ_R. However, IR divergences come back, with a vengeance, in the massless limit, $\mu_R \to 0$. So it is natural to ask: How does perturbation theory behave in the IR? It is easy to see that if a theory is nonrenormalizable, $\delta_r < 0$, then the vertex functions are finite in the IR. Conversely, super-renormalizable theories, with $\delta_r > 0$, are nontrivial in the IR, and their degree of IR divergence increases with the order in perturbation theory. Only in the renormalizable case, for which $\delta_r = 0$, is the degree of UV and IR divergence independent of the order of the perturbation theory. In particular, at the critical dimension, the theory has logarithmic divergences that blow up at both the IR and the UV limits.

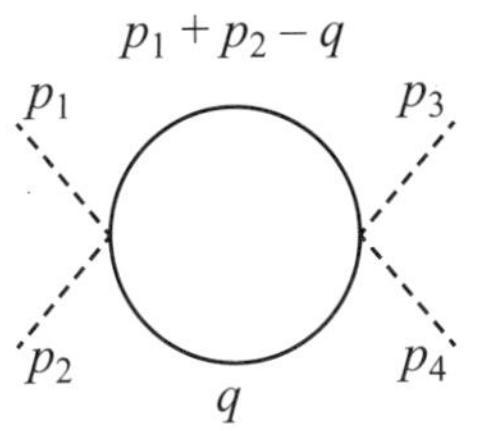

Figure 13.8 Feynman diagram with the leading contribution to the four-point vertex function.

In chapters 15 and 16, when we discuss the theory of the renormalization group, operators with $\delta < 0$ are called *irrelevant* operators (because their effect is negligible in the IR), those with $\delta > 0$ will be called *relevant* operators (which dominate in the IR), and those with $\delta = 0$ will be called *marginal* operators. In the framework of the renormalization group, a renormalizable theory is one whose Lagrangian has only marginal and relevant operators.

13.6 Regularization

We now must face the UV divergences of the Feynman diagrams and devise some procedure to define the theory in the UV. These procedures are known as regularizations. In principle there are many ways to regularize the theory, and we will consider a few. Most regularization procedures break one symmetry or another, such as Lorentz/Euclidean invariance or some internal symmetry. Hence the effects of the regularization must be considered with care.

13.6.1 Momentum cutoffs

The simplest and most intuitive approach is to define some sort of cutoff procedure that suppresses the contributions from very large momenta to each Feynman diagram. This can be done in momentum space, invoking a UV momentum space cutoff Λ, or in position space, by invoking a short-distance cutoff a. For the sake of definiteness, let us consider the one-loop contribution to the one-particle irreducible four-point function in ϕ^4 theory, shown in figure 13.8.

We can regularize the diagram in figure 13.8 by using a sharp momentum cutoff Λ to make the following expression UV finite:

$$I_{\text{reg}}(p) = \frac{\lambda^2}{2} \int \frac{d^D q}{(2\pi)^D} \frac{1}{(q^2+m_0^2)((q-p)^2+m_0^2)} f_\Lambda(p) + 2\,\text{permutations} \tag{13.83}$$

where $p = p_1 + p_2$ is the momentum transfer, $f_\Lambda(p) = \theta(\Lambda - |p|)$, and Λ is the UV momentum cutoff. Here $\theta(x)$ is the Heaviside (step) function: $\theta(x) = 1$ if $x > 0$, and zero otherwise.

A sharp momentum cutoff is a simple procedure that is useful only at one-loop order. However, multi-loop integrals with momentum cutoff are cumbersome analytically, to say the least, and are not compatible with full Lorentz (and Euclidean) invariance. More importantly, the cutoff procedure is not compatible with gauge invariance, which makes it ineffective outside the boundaries of theories with only global symmetries.

To improve the analytic tractability, other momentum cutoff regularizations are used. Thus, instead of cutting off the momentum integrals at a definite momentum scale Λ, as in eq. (13.83), smooth cutoff procedures have been introduced. They amount to multiplying the integrand of the Feynman diagrams by a smooth cutoff function $f_\Lambda(p)$, such as a Gaussian regulator, $f_\Lambda(p) = \exp(-p^2/\Lambda^2)$. More common is the use of rational functions, for example,

$$f_\Lambda(p) = \left(\frac{\Lambda^2}{p^2 + \Lambda^2}\right)^n \tag{13.84}$$

where the integer n is chosen to make the diagram UV finite.

13.6.2 Lattice regularization

Another possible regularization is a lattice cutoff. This entails defining the (Euclidean) theory on a lattice, normally a hypercubic lattice of dimension D with lattice spacing a. In other words, the Euclidean field theory then becomes identical to a system in classical statistical mechanics. In the case of ϕ^4 theory, the degrees of freedom are real fields defined on the sites of the lattice. Once this is done, the momenta of the fields is restricted to the first Brillouin zone of a hypercubic lattice, which, in the thermodynamic limit (where the linear size of the lattice is infinite), is defined on the D-dimensional torus $|q_\mu| \leq \pi/a$ (with $\mu = 1, \ldots, D$).

In this approach, the bare propagators become

$$G(p, m_0^2) = \frac{1}{\frac{2}{a^2}\sum_{\mu=1}^{D}[1 - \cos(q_\mu a)] + m_0^2} \tag{13.85}$$

Much as for a momentum cutoff, to compute the loop integrals (even one-loop!) is a challenging task. In practice, theories regularized on a lattice can be studied numerically using classical Monte Carlo simulation techniques, which are outside the framework of perturbation theory. However, the lattice regularization breaks the continuous rotational and translational Euclidean invariance down to the point-group symmetry and discrete translation symmetries of the lattice. The recovery of these continuous symmetries can only be achieved by tuning the theory close enough to a critical point, where the lattice effects are encoded in "irrelevant operators" of the renormalization group, as will be discussed in chapter 15. Note that, in contrast with naive cutoff procedures, lattice regularization is (or can be) compatible with local gauge invariance, as shown by Wilson (1974) and Kogut and Susskind (1975a). We will discuss these questions in chapter 18.

13.6.3 Pauli-Villars regularization

Given the drawbacks of sharp momentum cutoff procedures, other approaches have been devised. A procedure widely applied to the regularization of Feynman diagrams in QED is known as Pauli-Villars (PV) regularization (Pauli and Villars, 1949). A detailed presentation of PV regularization can be found in chapter 7 of Itzykson and Zuber (1980). In the Pauli-Villars approach, the bare propagator $G_0(p; m_0)$ is replaced by a regularized propagator $G_0^{\text{reg}}(p; m_0)$, which differs from the bare propagator by enough subtractions to render the Feynman diagrams UV finite. This is done by introducing a set of (unobservable) very massive fields with masses M_i. The regularized propagator is defined to be

$$G_0^{\text{reg}}(p; m_0) = G_0(p; m_0) + \sum_i c_i G_0(p; M_i) \tag{13.86}$$

where the $G_0(p; M_i)$ are the propagators of the regulating heavy fields. The coefficients, c_i in this sum, are chosen in such a way that the Feynman diagrams are finite in the UV and the regularized propagator is smooth. Let us consider the case of a single heavy field with mass $M \gg m_0$. To suppress the strong divergent behavior in the UV, we need to subtract the behavior at large momenta. In this case, the regularized propagator is

$$\begin{aligned} G_0^{\text{reg}}(p; m_0) &= G_0(p; m_0) - G_0(p; M) \\ &= \frac{1}{p^2 + m_0^2} - \frac{1}{p^2 + M^2} \\ &= \frac{M^2 - m_0^2}{(p^2 + m_0^2)(p^2 + M^2)} \end{aligned} \tag{13.87}$$

Hence, the regularized propagator behaves as $1/p^4$ for large moment. With this prescription, clearly the one-loop diagram of figure 13.8 is now finite for $D < 8$ dimensions. However, eq. (13.87) shows that the propagator of the heavy regulator fields is negative. In the Minkowski signature, this implies a violation of unitarity. The result is that in a theory regulated à la Pauli-Villars, it is necessary to prove that unitarity is preserved. Nevertheless, for some theories, such as those with anomalies, Pauli-Villars regularization is the only practical regularization method available.

13.6.4 Dimensional regularization

The complex analytic structure of multi-loop Feynman diagrams regularized with Pauli-Villars regulators (and, for that matter, with any soft cutoff as well) motivated the introduction of so-called analytic regularization methods. In section 8.8.2, we encountered one such method, the ζ-function regularization approach for the calculation of functional determinants (Hawking, 1977).

In the case of Feynman diagrams, an early analytic regularization approach consisted of replacing the Euclidean propagators as follows:

$$\frac{1}{p^2 + m_0^2} \mapsto \lim_{\eta \to 1} \frac{1}{(p^2 + m_0^2)^\eta} \tag{13.88}$$

Thus, since the Feynman diagram shown in figure 13.8 is logarithmically divergent in $D = 4$ dimensions, it becomes finite for any $\eta > 1$. However, this regularization has the serious drawback that it changes the analytic structure of the Feynman diagrams, since for any value of η, the poles of the propagators are replaced by branch cuts. In fact, the free-field theory of such propagators is nonlocal.

The most powerful and widely used regularization procedure is *dimensional regularization*. In this regularization, introduced in 1972 by Gerard 't Hooft and Martinus Veltman (1972) and, independently and simultaneously, by Carlos Bollini and Juan José Giambiagi (1972a,b), the Feynman diagrams are computed in a general dimension $D > D_c$ (where D_c is typically 4), where they are convergent. The resulting expressions are then analytically continued to the complex D plane. In these expressions, the UV divergences are replaced by poles as functions of $D - D_c$. The regularized expressions of Feynman diagrams are obtained by subtracting these poles, a procedure known as *minimal subtraction*.

One key advantage of dimensional regularization is that it is manifestly compatible with gauge invariance. For this reason, dimensional regularization was the key tool in the proof of renormalizability of non-abelian Yang-Mills gauge theories and had (and still does have) a huge impact on our understanding of modern-day particle physics. Dimensional regularization is also the key analytic tool in high-precision computation of critical exponents in the theory of phase transitions.

However, for all its successes and its great power, dimensional regularization has limitations. By construction, it relies on the assumption that the quantities of interest depend smoothly on the dimensionality D, and it cannot be used for quantities that cannot be continued in dimension. This is a problem in relativistic QFTs that involve fermions (Dirac or Majorana). Indeed, while many properties of spinors can be continued in dimension, some cannot. One such problem is chiral symmetry and the associated chiral anomaly, which exists only in even spacetime dimensions D and whose expression involves the Levi-Civita totally antisymmetric tensor. Similarly, in odd spacetime dimensions, fermionic theories have parity (or time-reversal) anomalies, which also involve the Levi-Civita tensor. The expressions for these anomalies are specific for a given dimension and cannot be unambiguously analytically continued in dimensionality. Consequently, dimensional regularization does not work for such problems.

13.7 Computation of regularized Feynman diagrams

Let us now compute the Feynman diagram shown in figure 13.8 in different regularization schemes and compare the results. More specifically, we will do the computation using (a) a Pauli-Villars regularization and (b) dimensional regularization.

Let us regularize one of the terms for the Feynman diagram given in eq. (13.83) with a cutoff function roughly of the form eq. (13.84) with $n=2$, resulting in the expression

$$I_D^{\text{reg}}(p^2) = \frac{\lambda^2}{2}\int \frac{d^Dq}{(2\pi)^D}\frac{1}{(q^2+m_0^2)((q-p)^2+m_0^2)}\left(\frac{\Lambda^2}{q^2+\Lambda^2}\right)\left(\frac{\Lambda^2}{(q-p)^2+\Lambda^2}\right) \tag{13.89}$$

By counting powers, we see that this regularized expression is finite for dimension $D<8$. Using partial fractions in pairs of factors, we obtain

$$\begin{aligned} I_D^{\text{reg}}(p^2) &= \frac{\lambda^2}{2} \\ &\times \int \frac{d^Dq}{(2\pi)^D}\left[\frac{1}{(q^2+m_0^2)}-\frac{1}{q^2+\Lambda^2}\right]\left[\frac{1}{((q-p)^2+m_0^2)}-\frac{1}{(q-p)^2+\Lambda^2}\right] \end{aligned} \tag{13.90}$$

where we have omitted a prefactor $[\Lambda^2/(\Lambda^2-m_0^2)]^2$, which approaches 1 for $\Lambda \gg m_0$. Hence, we have obtained essentially the same expression as we would have found with Pauli-Villars regularization with a large regulator mass $M=\Lambda$. Notice that instead we could have chosen to use a single smooth cutoff function $f_\Lambda(p)$ with $n=1$, which would have rendered the diagram finite for $D<6$. Clearly there is some degree of leeway in how one chooses to regularize the Feynman diagram.

Let us now proceed to compute the regularized expression $I_D^{\text{reg}}(p^2)$, shown in eq. (13.90). To this end, let A be a positive real number. Then introduce the Feynman-Schwinger parameter x through the integral

$$\frac{1}{A} = \frac{1}{2} \int_0^\infty dx\, e^{-Ax/2} \tag{13.91}$$

to raise all the denominator factors in eq. (13.90) to the arguments of exponentials. Since eq. (13.90) has two factors inside the momentum integral, we will need to introduce an expression of the form of eq. (13.91) for each factor, resulting in the expression:

$$I_D^{\text{reg}}(p^2) = \frac{\lambda^2}{8} \int_0^\infty dx \int_0^\infty dy \int \frac{d^D q}{(2\pi)^D} \left(e^{-x(q^2+m_0^2)/2} - e^{-x(q^2+\Lambda^2)/2}\right) \times \left(e^{-x((q-p)^2+m_0^2)/2} - e^{-x((q-p)^2+\Lambda^2)/2}\right) \tag{13.92}$$

Using the Gaussian integral identity (with $A > 0$),

$$\int \frac{d^D q}{(2\pi)^D} e^{-\frac{A}{2}q^2 - Bq\cdot p} = \frac{1}{(2\pi A)^{D/2}} e^{+\frac{B}{2A}p^2} \tag{13.93}$$

eq. (13.92) becomes

$$I_D^{\text{reg}}(p^2) = \frac{\lambda^2}{8} \int_0^\infty dx \int_0^\infty dy \frac{e^{-\frac{xy}{2(x+y)}p^2}}{(2\pi(x+y))^{D/2}} \left(e^{-xm_0^2/2} - e^{-x\Lambda^2/2}\right)\left(e^{-ym_0^2/2} - e^{-y\Lambda^2/2}\right) \tag{13.94}$$

Next, after the change of variables

$$x = uv, \qquad y = (1-u)v \tag{13.95}$$

with $0 \leq v < \infty$ and $0 \leq u \leq 1$, eq. (13.94) becomes

$$I_D^{\text{reg}}(p^2) = \frac{\lambda^2}{8(2\pi)^{D/2}} \int_0^1 du \int_0^\infty dv v^{(2-D)/2} e^{-uv(1-u)p^2/2} \times \left(e^{-vm_0^2/2} + e^{-v\Lambda^2/2} - e^{-\frac{1}{2}uvm_0^2 - \frac{1}{2}(1-u)v\Lambda^2} - e^{-\frac{1}{2}uv\Lambda^2 - \frac{1}{2}(1-u)vm_0^2}\right) \tag{13.96}$$

We can now explicitly carry out the integral over the variable v using the result

$$\int_0^\infty dv\, v^{(2-D)/2} e^{-\gamma v} = \gamma^{(D-4)/2}\Gamma(2 - D/2) \tag{13.97}$$

where $\Gamma(z)$ is the Euler gamma function

$$\Gamma(z) = \int_0^\infty dt\, t^{z-1} e^{-t} \tag{13.98}$$

which is well defined in the domain $\mathrm{Re}\, z > 0$. We obtain

$$I_D^{\text{reg}}(p^2) = \frac{\lambda^2}{2}\frac{\Gamma(2-D/2)}{4(2\pi)^{D/2}}\int_0^1 du \left\{\left[\frac{2}{u(1-u)p^2+m_0^2}\right]^{2-D/2} + \left[\frac{2}{u(1-u)p^2+\Lambda^2}\right]^{2-D/2} - \left[\frac{2}{u(1-u)p^2+um_0^2+(1-u)\Lambda^2}\right]^{2-D/2} - \left[\frac{2}{u(1-u)p^2+u\Lambda^2+(1-u)m_0^2}\right]^{2-D/2}\right\} \tag{13.99}$$

In $D=4$ dimensions, the regularized expression for $I_4^{\text{reg}}(p^2)$ becomes (for $\Lambda \gg m_0$)

$$I_4^{\text{reg}}(p^2) = \frac{\lambda^2}{32\pi^2}\ln\left(\frac{\Lambda^2}{m_0^2}\right) - \frac{\lambda^2}{16\pi^2} - \frac{\lambda^2}{32\pi^2}\int_0^1 du\, \ln\left[1+u(1-u)\frac{p^2}{m_0^2}\right] \tag{13.100}$$

where a logarithmically divergent part has been separated from a finite part. This regularized result is manifestly rotationally invariant in Euclidean space and is Lorentz invariant when continued to Minkowski spacetime.

A similar expression is found for the massless case, $m_0 \to 0$,

$$I_4^{\text{reg}}(p^2) = \frac{\lambda^2}{32\pi^2}\ln\left(\frac{\Lambda^2}{\mu^2}\right) - \frac{\lambda^2}{16\pi^2} - \frac{\lambda^2}{32\pi^2}\int_0^1 du\, \ln\left[u(1-u)\frac{p^2}{\mu^2}\right] \tag{13.101}$$

where μ is an arbitrary mass scale needed for dimensional reasons. Doing the integral in eq. (13.101), we find the simpler result for the massless $m_0 = 0$ case:

$$I_4^{\text{reg}}(p^2) = \frac{\lambda^2}{32\pi^2}\ln\left(\frac{\Lambda^2}{p^2}\right) \tag{13.102}$$

This result shows that in the massless case, there is an IR logarithmic divergence as $p \to 0$. In both cases, massive and massless, the UV regulator Λ enters in an essential and singular way.

13.8 Computation of Feynman diagrams with dimensional regularization

Let us now compute the same Feynman diagram (see figure 13.8 and eq. (13.83)) using dimensional regularization. Thus, as explained above, we will compute the Feynman diagram without any explicit UV regularization as a function of the dimension of the Euclidean spacetime D. We will seek the domain of dimensions D for which the diagram is finite and define its value at the dimension of physical interest (say, $D=4$) by means of an analytic continuation. The UV singular behavior will appear in the form of poles in the dependence on the dimension D, regarded as a complex variable.

The expression for the Feynman diagram of eq. (13.83) in general dimension D is obtained by setting the regulator $\Lambda \to \infty$ in eq. (13.99). The result is

$$I_D(p^2) = \frac{\lambda^2}{2}\frac{\Gamma(2-D/2)}{4(2\pi)^{D/2}}\int_0^1 du \left[\frac{2}{u(1-u)p^2+m_0^2}\right]^{2-D/2} \tag{13.103}$$

Except for the singularities of the gamma function, this expression is finite.

The Euler gamma function, $\Gamma(z)$, is an analytic function of z in the domain $\operatorname{Re} z > 0$. In the complex plane, the gamma function has simple poles at the negative integers, $z = -n$, with $n \in \mathbb{N}$, and $z = 0$. More explicitly, this can be seen using the Weierstrass representation of the gamma function:

$$\Gamma(z) = \int_0^\infty dt\, t^{z-1} e^{-t} = \int_0^1 dt\, t^{z-1} e^{-t} + \int_1^\infty dt\, t^{z-1} e^{-t} \tag{13.104}$$

The second integral is clearly finite, since the integration range does not reach down to $t = 0$. The first integral has a finite integration range, the interval $[0, 1]$, and hence we can expand the exponential in its power series expansion and integrate term by term. The result is the Weierstrass representation

$$\Gamma(z) = \sum_{n=0}^{\infty} \frac{(-1)^n}{n!(z+n)} + \Gamma_{\text{reg}}(z) \tag{13.105}$$

where $\Gamma_{\text{reg}}(z)$, the regularized gamma function, is

$$\Gamma_{\text{reg}}(z) = \int_1^\infty dt\, t^{z-1} e^{-t} \tag{13.106}$$

Thus, the Weierstrass representation of eq. (13.105) expresses the gamma function as a sum of a regularized function $\Gamma_{\text{reg}}(z)$ and a series of simple poles on the negative real axis at the negative integers and zero. It also tells us how to analytically continue the gamma function from the domain $\operatorname{Re} z > 0$ to the complex plane $\mathbb{C}$.

Furthermore, in the vicinity of its leading pole at $z = 0$, the gamma function has the asymptotic behavior

$$\Gamma(z) = \frac{1}{z} - \gamma + O(z), \qquad \text{as } z \to 0 \tag{13.107}$$

where

$$\gamma = \lim_{n \to \infty} \left(-\ln n + \sum_{k=1}^{n} \frac{1}{k} \right) = 0.5772\ldots \tag{13.108}$$

is the Euler-Mascheroni constant.

Armed with these results, we can write eq. (13.103), setting $\epsilon = 4 - D$, in the form

$$I_D(p^2) = \frac{\lambda^2}{2} \frac{\mu^{-\epsilon}}{4(2\pi)^{D/2}} 2^{\epsilon/2} \Gamma\left(\frac{\epsilon}{2}\right) \int_0^1 du \left[\frac{\mu^2}{u(1-u)p^2 + m_0^2}\right]^{\epsilon/2} \tag{13.109}$$

where once again, μ is an arbitrary mass scale.

In the limit $D \to 4$ dimensions, eq. (13.109) becomes

$$I_D(p^2) = \frac{\lambda^2}{32\pi^2} \left[\frac{2}{\epsilon} + \ln(4\pi) - \gamma + \int_0^1 du \ln\left(\frac{\mu^2}{u(1-u)p^2 + m_0^2}\right)\right] \tag{13.110}$$

Thus, the expression of this Feynman diagram is split into a singular term, with a pole in ϵ, and a finite regular term in $D = 4$ dimensions. Dimensional regularization defines the value of this Feynman diagram as this expression with the pole term subtracted. Clearly, one also could have subtracted some piece of the finite term, and that prescription would have been equally correct. The procedure of subtracting just the contribution of the pole is known as dimensional regularization with minimal subtraction (or MS).

In the massless limit, $m_0 \to 0$, eq. (13.110) becomes

$$I_D(p^2) = \frac{\lambda^2}{32\pi^2}\left[\frac{2}{\epsilon} + \ln(4\pi) - \gamma + 2 + \ln\left(\frac{\mu^2}{p^2}\right)\right] \tag{13.111}$$

which should be compared with the same result using Pauli-Villars regularization, eq. (13.102). Clearly, the factor $\ln(\Lambda^2/\mu^2)$ in the Pauli-Villars result corresponds to the pole $2/\epsilon$ in dimensional regularization. The same comparison holds for the massive case. Notice, however, that the finite parts are different in different regularizations.

That the finite parts of Feynman diagrams depend on the regularization should have been expected. Although superficially it may seem surprising, finite parts depend on different definitions of the theory at short distances. However, if we compute the difference of the diagram for two different momentum scales, computed with the same regularization, the regularization-dependent contributions should cancel each other out. Indeed, we can check explicitly that the dimensional regularization result of eq. (13.111) yields

$$I_D(p) - I_D(p^*) = \frac{\lambda^2}{16\pi^2}\ln\left(\frac{p^*}{p}\right) \tag{13.112}$$

which agrees with the result obtained using the Pauli-Villars regularization of eq. (13.102). This is a general result.

Exercises

13.1 Renormalization of the $O(N)$ scalar ϕ^4 theory

Consider the ϕ^4 theory in a regime in which the renormalized mass is nonzero for an N-component real scalar field ϕ_a, where $a = 1, \ldots, N$, with an $O(N)$-invariant Lagrangian density in D Euclidean dimensions of the form

$$\mathcal{L} = \frac{1}{2}\left(\partial_\mu \phi_a\right)^2 + \frac{m_0^2}{2}\phi_a^2 + \frac{\lambda}{4!}\left(\phi_a^2\right)^2 \tag{13.113}$$

As usual, repeated indices are summed over.

In the symmetric theory (i.e., in the absence of spontaneous symmetry breaking), the two-point and four-point 1-PI vertex functions $\Gamma^{(2)}_{ab}(p)$ and $\Gamma^{(4)}_{abcd}(p_1, \ldots, p_4)$ take the symmetric form

$$\Gamma^{(2)}_{ab}(p) = \delta_{ab}\,\bar{\Gamma}^{(2)}(p) \tag{13.114}$$

and

$$\Gamma^{(4)}_{abcd}(p_1, \ldots, p_4) = S_{abcd}\,\bar{\Gamma}^{(4)}(p_1, \ldots, p_4) \tag{13.115}$$

where

$$S_{abcd} = \frac{1}{3}\left(\delta_{ab}\delta_{cd} + \delta_{ac}\delta_{bd} + \delta_{ad}\delta_{bc}\right) \tag{13.116}$$

1) Find all the contributions to $\Gamma^{(2)}$ and $\Gamma^{(4)}$ to one-loop order for this N-component theory. Write down and draw all Feynman diagrams and their associated analytic expressions. Write your results in terms of the integrals discussed in section 11.8 (for the one-component theory). Do not do the integrals. Derive the explicit dependence of each diagram on the number of components N.
2) Define a set of renormalization conditions, at zero external momentum, for the vertex functions $\bar{\Gamma}^{(2)}(p)$ and $\bar{\Gamma}^{(4)}(p_1, \ldots, p_4)$ in the symmetric massive theory.
3) Determine m_0^2 and λ in terms of the renormalized mass μ and coupling constant g at fixed momentum cutoff Λ, to one-loop order. Express your answers in terms of the integrals defined in this chapter. Do not do the integrals!
4) Show that the renormalization conditions of part 2 and the renormalization constants obtained in part 3 yield finite vertex functions at arbitrary values of the external momentum to one-loop order in perturbation theory.

The following are useful identities:

$$\sum_c S_{abcc} = \frac{N+2}{3} \delta_{ab}$$

$$\sum_{ij} S_{abij} S_{ijcd} = \frac{2}{3} S_{abcd} + \frac{N+2}{3} \delta_{ab} \delta_{cd}$$

14

Quantum Field Theory and Statistical Mechanics

Until now we have been following the standard approach of defining QFT by its perturbative Feynman diagrammatic expansion around the free field theory. In chapter 13, we saw that each Feynman diagram needs to be regularized in the UV. In other words, the theory needs a definition in the UV. In this chapter, we will take a detour into statistical mechanics, which has a natural UV definition (as it is often defined on a lattice). We will see that the perspective of statistical mechanics offers a way to define QFT without the use of perturbation theory. In chapter 15, we will introduce the framework of the RG, which will make precise how to define a continuum field theory. We will return to renormalized perturbation theory in chapter 16.

We have seen in earlier chapters that there is a connection between QFT in D spacetime dimensions (at $T = 0$) and classical statistical mechanics in D Euclidean dimensions. We will now explore this correspondence further. This correspondence has profound consequences for the physics of both QFT and statistical mechanics, particularly in the vicinity of a continuous phase transition.

In section 2.8, we used heuristic arguments to show that the partition function of an Ising model in D dimensions can be represented by the path integral of a continuum ϕ^4 scalar field theory in a D-dimensional Euclidean spacetime. There are, however, some important differences. Since the configurations of the Ising model cannot vary on scales smaller than the lattice spacing, the expectation values of the physical observables do not have the UV divergences discussed in chapter 11. In other words, the Ising model can be regarded as a Euclidean scalar field theory with a lattice regularization. In addition, the microscopic degrees of freedom of the Ising model, the spins, obey the constraint $\sigma^2 = 1$. Instead, the fields of a scalar field theory are unconstrained, but in the path integral, its configurations are weighted by the potential $V(\phi) = \frac{g}{4!}\phi^4$.

The purpose of this chapter is not to present a complete theory of phase transitions, for which there are many excellent texts, such as the books by Cardy (1996), Goldenfeld (1992), and Amit (1980). The goal here is to present the perspective that QFTs can be defined in terms of regularized theories near a phase transition. This perspective was formulated and developed by K. G. Wilson (1983). The connection between statistical mechanics and QFT is the key to this framework. To this end, in this chapter, we revisit the Ising model and discuss its phase transition in some detail. We will see that its behavior near its phase transition provides an example of the definition of a QFT.

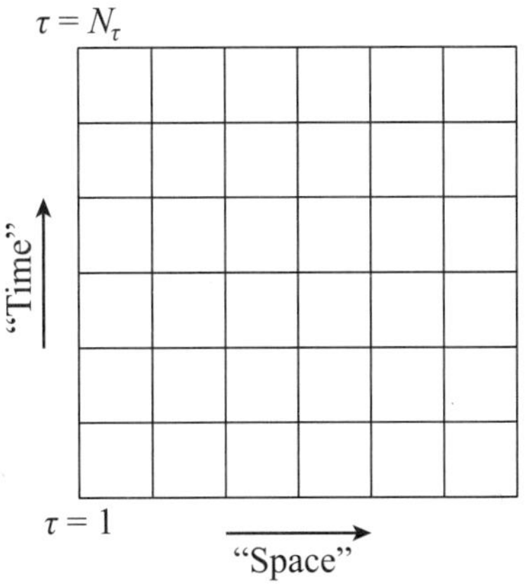

Figure 14.1 The Euclidean spacetime lattice.

14.1 The classical Ising model as a path integral

Let us consider a typical problem in equilibrium statistical mechanics. For the sake of simplicity, consider the Ising model in D dimensions. The arguments given below are straightforward to generalize to other cases.

Consider a D-dimensional hypercubic lattice of unit spacing. At each site $\boldsymbol{r}$, there is an Ising spin variable $\sigma(\boldsymbol{r})$, which can take one of two possible values, ± 1. We will refer to the state of this lattice (or *configuration*) $[\sigma]$ to a particular set of values of this collection of variables. Let $H[\sigma]$ be the classical energy ($j = 1, \ldots, D$)

$$H[\sigma] = -J \sum_{\boldsymbol{r}, j=1,\ldots,D} \sigma(\boldsymbol{r})\sigma(\boldsymbol{r} + \hat{\boldsymbol{e}}_j) - h \sum_{\boldsymbol{r}} \sigma(\boldsymbol{r}) \tag{14.1}$$

where J is the exchange constant, and h represents the Zeeman coupling to an external uniform magnetic field of the Ising spins. The partition function Z is (on setting $k_B = 1$) the sum over the configuration space $[\sigma]$ of the Gibbs weight for each configuration:

$$Z = \sum_{[\sigma]} e^{-H[\sigma]/T} \tag{14.2}$$

where T is the temperature.

Assume that the system is on a hypercubic lattice and obeys periodic boundary conditions, and hence that the Euclidean spacetime is a torus in D dimensions. Also assume that the interactions are ferromagnetic, and hence $J > 0$. Clearly the partition function is a function of the ratios J/T and h/T. The discussion that follows can be easily extended to any other lattice provided the interactions are ferromagnetic, and it applies to antiferromagnetic ($J < 0$) interactions if the lattice is bipartite. We will not consider the interesting problem of frustrated lattices.

Let us regard one of the spatial dimensions labeled by D as time, or more precisely, as imaginary time. Then we have $D - 1$ space dimensions and one time dimension. This is clearly an arbitrary choice, given the symmetries of this problem. Thus, each hyperplane of the hypercubic lattice will be labeled by an integer-valued variable $\tau = 1, \ldots, N_\tau$, where N_τ is the linear size of the system along the direction τ (see figure 14.1).

In this picture, we can regard each configuration $[\sigma]$ as the imaginary-time evolution of an initial configuration at some initial imaginary time $\tau = 1$ (i.e., the configuration on the first "row" or hyperplane, in time τ, as we go from row to row along the time direction, all the way to $\tau = N$). The integers τ will be regarded as a discretized imaginary time. The partition function is thus interpreted as a sum over the histories of these spin configurations, and hence, it is a path integral with a discretized imaginary (Euclidean) time.

14.2 The transfer matrix

Let us now show that the partition function can be written in the form of the trace of a matrix, known as the *transfer matrix* (see, e.g., Schultz et al. 1964),

$$Z = \sum_{[\sigma]} e^{-H[\sigma]/T} \equiv \operatorname{tr} \hat{T}^N \tag{14.3}$$

where N is the number of rows or, in general, hyperplanes. To this end, we define a complete set of states $\{|\sigma\rangle_\tau\}$ on each row (or hyperplane) labeled by $\tau = 1, \ldots, N$. Since each row contains N^{D-1} sites, and for the Ising model, we have two states per site (the two spin projections), the number of states in this basis is $2^{N^{D-1}}$. This concept is easily generalizable to systems with other degrees of freedom.

Let us find an explicit expression for the matrix $\hat{T}$. In particular, we will look for a transfer matrix $\hat{T}$ with the factorized form

$$\hat{T} = \hat{T}_1^{1/2}\, \hat{T}_2\, \hat{T}_1^{1/2} \tag{14.4}$$

It turns out that for a large number of interesting problems in equilibrium classical statistical mechanics, the transfer matrix $\hat{T}$ can be chosen to be a hermitian matrix. From the point of view of classical statistical mechanics, this follows (in part) because the Boltzmann weights are positive real numbers. However, this condition is not sufficient. We will show below that there is a sufficient condition known as *reflection positivity*, which is the Euclidean equivalent of unitarity in quantum mechanics and QFT.

For the particular case of the Ising model, it is possible to write the matrices $\hat{T}_1$ and $\hat{T}_2$ in terms of a set of real Pauli matrices $\hat{\sigma}_1(\boldsymbol{r})$ and $\hat{\sigma}_2(\boldsymbol{r})$ defined on each site $\boldsymbol{r}$ of the row, which act on the states defined on each row. For the case at hand, after some algebra, we find

$$\hat{T}_2 = \exp\left\{ \frac{J_s}{T} \sum_{\boldsymbol{r},j} \hat{\sigma}_3(\boldsymbol{r})\hat{\sigma}_3(\boldsymbol{r} + \hat{\boldsymbol{e}}_j) + \frac{h}{T} \sum_{\boldsymbol{r}} \hat{\sigma}_3(\boldsymbol{r}) \right\} \tag{14.5}$$

$$\hat{T}_1 = \left[\frac{1}{2} \sinh\left(\frac{2J_\tau}{T} \right) \right]^{N_s/2} \exp\left\{ b \sum_{\boldsymbol{r}} \hat{\sigma}_1(\boldsymbol{r}) \right\} \tag{14.6}$$

where N_s is the number of sites in each hyperplane (row), and $j = 1, \ldots, D-1$. Here we have assumed that J_s and J_τ, the coupling constants along the space and time directions, are not necessarily equal to each other, and the parameter b is given by

$$e^{-2b} = \tanh(J_\tau/T) \tag{14.7}$$

The correlation functions of the Ising model, for spins located at sites $R=(\boldsymbol{R},\tau)$ and $R'=(\boldsymbol{R}',\tau')$,

$$\left\langle\sigma(R)\,\sigma(R')\right\rangle=\frac{1}{Z}\sum_{[\sigma]}\sigma(R)\sigma(R')e^{-H[\sigma]} \tag{14.8}$$

can be expressed in terms of the transfer $\hat{T}$ in the suggestive form

$$\begin{aligned}\left\langle\sigma(R)\;\;\sigma(R')\right\rangle&=\frac{1}{Z}\,\mathrm{tr}\left(\hat{T}^{\tau}\hat{\sigma}_3(\boldsymbol{R})\hat{T}^{\tau'-\tau}\hat{\sigma}_3(\boldsymbol{R}')\hat{T}^{\tau-\tau'}\right)\\&\equiv\langle T[\hat{\sigma}_3(\boldsymbol{R},\tau)\,\hat{\sigma}_3(\boldsymbol{R}',\tau')]\rangle\end{aligned} \tag{14.9}$$

where T is the imaginary-time time ordering symbol. Here the (pseudo) Heisenberg representation of the spin operators has been introduced:

$$\hat{\sigma}_3(\boldsymbol{R},\tau)\equiv\hat{T}^{\tau}\hat{\sigma}_3(\boldsymbol{R})\hat{T}^{-\tau} \tag{14.10}$$

Let $\{\lambda_n\}$ be the eigenvalues of the eigenstates $|n\rangle$ of the (hermitian) transfer matrix $\hat{T}$

$$\hat{T}|n\rangle=\lambda_n|n\rangle \tag{14.11}$$

Since the transfer matrix is hermitian, its eigenvalues are real numbers. Moreover, since the transfer matrix is also real and symmetric, its eigenvalues are positive real numbers, $\lambda_n>0$ (for all n).

14.3 Reflection positivity

Let us now consider the case in which the correlation function is computed along the imaginary time coordinate only, and the spatial coordinates are set to be equal: $\boldsymbol{R}=\boldsymbol{R}'$. By following the line of argument used in section 5.7 for the computation of the correlators in QFT, we can formally compute the correlation function as a sum over the matrix elements of the operators on the eigenstates of the transfer matrix. In the thermodynamic limit, $N_\tau\to\infty$ and $N_s\to\infty$, the result simplifies to the following expression:

$$\begin{aligned}\langle\sigma(\boldsymbol{R},\tau)\sigma(\boldsymbol{R},\tau')\rangle&=\lim_{N_t\to\infty}\frac{1}{Z}\mathrm{tr}\Big(T[\hat{\sigma}_3(\boldsymbol{R},\tau)\hat{\sigma}_3(\boldsymbol{R},\tau')]\Big)\\&=\sum_n\left|\langle G|\hat{\sigma}_3|n\rangle\right|^2\left(\frac{\lambda_n}{\lambda_{\max}}\right)^{\tau'-\tau}\end{aligned} \tag{14.12}$$

where $|G\rangle$ is the eigenstate of the transfer matrix with largest eigenvalue $g_{\max}$. A comparison with the analogous expression of eq. (5.198) suggests that the quantity ξ defined by

$$\xi^{-1}=\ln\left(\frac{\lambda_{\max}}{\lambda_{n_0}}\right) \tag{14.13}$$

is the *correlation length* of the system. Here $|n_0\rangle$ is the eigenstate of the transfer matrix with the largest possible eigenvalue, λ_{n_0}, that is mixed with the state $|G\rangle$ by the spin operator $\hat{\sigma}_3(\boldsymbol{r})$.

From eq. (14.12) it is apparent that the correlation function is real and positive. A similar result can be derived for the correlation function on two general coordinates (not necessarily along a given axis). Hence, if the matrix $\hat{T}$ is hermitian, the correlation functions are positive.

There is a natural and important interpretation of this result in terms of a Hilbert space that will allow us to find a more precise connection between classical statistical mechanics and QFT. The positivity requirement for the correlators can be relaxed to the condition of *reflection positivity*, which states the following. Let $A_r[\sigma]$ be some arbitrary local operator localized near $r = (\boldsymbol{r}, \tau)$, and define the operation of the *reflection* across a hyperplane $\mathcal{P}$. Let $A_{\mathcal{P}r}[\sigma]$ be the same operator after reflection across the hyperplane $\mathcal{P}$. Then the positivity requirement

$$\langle A_r[\sigma]\, A_{\mathcal{P}_r}[\sigma]\rangle \geq 0 \tag{14.14}$$

implies that the transfer matrix $\hat{T}$, defined in a direction normal to the hyperplane $\mathcal{P}$, must be a hermitian operator for all its eigenstates to have positive norm. This is the analog of unitarity in QFT (i.e., the requirement that all states in the spectrum must have positive norm), which is a requirement for the probabilistic interpretation of quantum mechanics and of QFT. This condition is clearly obeyed in the ferromagnetic Ising model and in many systems of physical interest. However, many interesting problems in classical statistical mechanics do not satisfy the condition of reflection positivity.

14.4 The Ising model in the limit of extreme spatial anisotropy

Let us now go back to the Ising model and consider the spatial symmetries of the correlation functions. If the coupling constants are equal, $J_\tau = J_s$, then the system is invariant under the point-group symmetries of the hypercubic lattice. This means that the correlation functions must have these symmetries as well. Now, if the correlation length is very large (much larger than the lattice constant), it should be possible to approximate the point-group symmetry by the symmetries of D-dimensional Euclidean rotations. Hence, in this limit (which, as we will see, means that the system is close to a continuous phase transition) the hypersurfaces of equal correlation, on which the correlators take constant values, are, to an excellent approximation, the boundary of a D-dimensional hypersphere, S_D.

However, this line of reasoning also means that it should be possible to change the system by smoothly deforming it away from isotropy to a spatially anisotropic system, by changing J_τ and J_s away from the isotropy condition, without changing its properties in any essential way. Thus if we shear the lattice by increasing the coupling J_τ and decreasing the coupling J_s in a suitable way, all that must change is that the hypersurfaces of equal correlation must become hyper-ellipsoids. In such a process of deformation, the correlation length in lattice units increases along the direction we agreed to call time, and it decreases along the orthogonal (space) directions. We can now imagine compensating for this deformation by adding more hyperplanes, so that the hypersurfaces of equal correlation become, to a desired degree of approximation, spherical once again (see figure 14.2).

If this process of deformation is repeated ad infinitum, the coupling constant along the time direction becomes very large, $J_\tau \to \infty$, and the coupling constant along the space direction becomes very small, $J_s \to 0$. In this limit, the time direction becomes continuous. This limit also requires that we tune down the magnetic field: $h \to 0$.

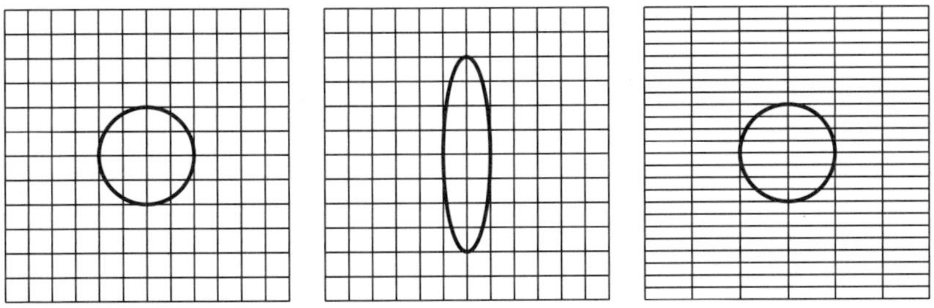

(a) (b) (c)

Figure 14.2 Curves of constant correlation and the time continuum limit: (a) isotropic lattice with $J_\tau = J_s$; (b) anisotropic lattice with $J'_\tau > J_\tau$ and $J'_s < J_s$, such that the correlation length is doubled along the (vertical) time axis and halved along the (horizontal) space axis; (c) lattice with twice as many rows in the time direction such that the correlation lengths along the two orthogonal directions are now equal to each other.

In this limit, the transfer matrix takes a much simpler form

$$\hat{T}_1^{1/2}\,\hat{T}_2\,\hat{T}_1^{1/2} \approx e^{-\epsilon \hat{H} + O(\epsilon^2)} \tag{14.15}$$

where $\hat{H}$ is a local hermitian operator, and ϵ is a suitably chosen small parameter. Also, since $\tanh b = e^{-2J_\tau/T}$, the limit of b small is the limit of $\frac{J_\tau}{T}$ large. Thus we write

$$\frac{J_s}{T} = g\, e^{-2J_\tau/T}, \qquad \frac{h}{T} = \bar{h}\, e^{-2J_\tau/T} \tag{14.16}$$

By choosing

$$\epsilon = e^{-2J_\tau/T} \tag{14.17}$$

we find that the operator $\hat{H}$ of eq. (14.15) is given by

$$\hat{H} = -\sum_{r} \hat{\sigma}_1(\boldsymbol{r}) - g\sum_{\boldsymbol{r},j} \hat{\sigma}_3(\boldsymbol{r})\hat{\sigma}_3(\boldsymbol{r}+\hat{\boldsymbol{e}}_j) - \bar{h}\sum_{\boldsymbol{r}} \hat{\sigma}_3(\boldsymbol{r}) \tag{14.18}$$

where g can be regarded as a coupling constant and $\bar{h}$ as a uniform longitudinal magnetic field.

In this picture, "time" is continuous. The equivalence of the sequence of deformed systems holds, provided that we require

$$N_\tau e^{-2J_\tau/T} = \text{fixed} = \bar{\beta} \tag{14.19}$$

Thus, as $\frac{J_\tau}{T} \to \infty$, we should let $N_\tau \to \infty$. The fixed number $\bar{\beta}$ should be infinite if the original system is thermodynamically large along that direction. Thus

$$Z = \operatorname{tr} \hat{T}^{N_\tau} \underset{N_\tau \to \infty}{\simeq} \operatorname{tr} e^{-\bar{\beta}\hat{H}} \tag{14.20}$$

and we recognize $\bar{\beta}$ as the effective inverse temperature of a quantum mechanical system in $D-1$ dimensions whose Hamiltonian is given by eq. (14.18). This quantum spin system is known as the Ising model in a transverse field (and a longitudinal field h). The reader should recognize that what we have done is the reverse of the procedure that we did to define a path integral: instead of discretizing the time variable, we have made it continuous.

Further taking the parameter $\bar{\beta} \to \infty$ (to restore full symmetry) amounts to taking the effective temperature of the equivalent system to zero. Hence, in the thermodynamic limit, we get

$$Z = \lim_{N_\tau \infty} \operatorname{tr} \hat{T}^{N_\tau} \to \lambda_{\max}^{N_\tau} = e^{-\bar{\beta} E_0} \tag{14.21}$$

where $\lambda_{\max}$ is the largest eigenvalue of the transfer matrix, and E_0 is the lowest eigenvalue of $\hat{H}$ (i.e., its ground state).

We conclude that, if we know the ground-state energy of $\hat{H}$ (or equivalently, the largest eigenvalue of the transfer matrix $\hat{T}$), we know the partition function Z in the thermodynamic limit. This observation is the key (for one of the very many methods) to solving the Ising model in $D=2$ dimensions.

In summary, we showed that the classical statistical mechanics of the Ising model in D dimensions is equivalent to the quantum mechanics of the Ising model in a transverse field in $D-1$ space dimensions (Fradkin and Susskind, 1978). This mapping also holds for any classical models, with global or local symmetries, and with real and positive Gibbs weight. And it satisfies reflection positivity. This mapping is simply the lattice version of the path integral of quantum theory.

14.5 Symmetries and symmetry breaking

14.5.1 Global $\mathbb{Z}_2$ symmetry

Let us now discuss the symmetries of this system, which are the same in all dimensions. The classical Ising model has (if the external field is set to zero, $h=0$) the global spin-flip symmetry $[\sigma] \to [-\sigma]$. Thus, the classical Hamiltonian $H[\sigma]$ is invariant under the operations

$$\begin{aligned} I &: [\sigma] \mapsto [\sigma], && \text{(the identity)} \\ R &: [\sigma] \mapsto [-\sigma], && \text{(global spin flip)} \end{aligned} \tag{14.22}$$

These operations form a group, since

$$I \star I = I, \qquad R \star R = I, \qquad I \star R = R \star I = R \tag{14.23}$$

where $\star$ denotes the composition of the two operations. This set of operations defines the group $\mathbb{Z}_2$ of permutations of two elements.

The equivalent quantum problem has, as it should, exactly the same symmetries. The operators $\hat{I}$ and $\hat{R}$ act on the quantum states defined on the rows (or hyperplanes) precisely in the same way as in eq. (14.23). We can construct these symmetry operators from the underlying operators of the lattice model quite explicitly. Let $\hat{I}(\boldsymbol{r})$ and $\hat{R}(\boldsymbol{r}) \equiv \hat{\sigma}_1(\boldsymbol{r})$ be the 2×2 identity matrix and the σ_1 Pauli matrix acting on the spin state at site $\boldsymbol{r}$. Then we can write for $\hat{I}$ and $\hat{R}$ the expressions

$$\hat{I} = \prod_{r} \otimes \hat{I}(r), \qquad \hat{R} = \prod_{r} \otimes \hat{\sigma}_1(r) \tag{14.24}$$

where $\otimes$ denotes the tensor product. These operators satisfy the obvious properties

$$\begin{aligned} \hat{I}^{-1} &= \hat{I}, & \hat{R}^{-1} &= \hat{R} \\ \hat{I}\hat{\sigma}_1(r)\hat{I}^{-1} &= \hat{\sigma}_1(r), & \hat{R}\hat{\sigma}_1(r)\hat{R}^{-1} &= \hat{\sigma}_1(r) \\ \hat{I}\hat{\sigma}_3(r)\hat{I}^{-1} &= \hat{\sigma}_3(r), & \hat{R}\hat{\sigma}_3(r)R^{-1} &= -\hat{\sigma}_3(r) \end{aligned} \tag{14.25}$$

From these properties, it follows that, if $\bar{h} = 0$, the Hamiltonian is invariant under the $\mathbb{Z}_2$ symmetry:

$$[\hat{I}, \hat{H}] = [\hat{R}, \hat{H}] = 0 \tag{14.26}$$

14.5.2 Qualitative behavior of the ground state and spontaneous symmetry breaking

We can write $\hat{H}$ as a sum of two terms, $\hat{H} = \hat{H}_0 + \hat{V}$, with

$$\hat{H}_0 = -\sum_{r} \hat{\sigma}_1(r), \qquad \hat{V} = -g \sum_{\mathrm{r},j} \hat{\sigma}_3(r)\hat{\sigma}_3(r + \hat{e}_j) \tag{14.27}$$

where $j = 1, \ldots, D - 1$.

A: The symmetric phase, $g \ll 1$ Let us first consider the limit $g \ll 1$, which corresponds to the high-temperature limit in the classical model in one higher dimension. If $g \ll 1$, we can study the properties of the ground state in perturbation theory in powers of g. The unperturbed ground state $|\Psi_0\rangle_0$ is

$$|\Psi_0\rangle_0 = \prod_{r} \otimes |+, r\rangle \equiv |+\rangle \tag{14.28}$$

Here $|+, r\rangle$ is the eigenstate of $\hat{\sigma}_1(r)$ with eigenvalue $+1$. Since it is an eigenstate of σ_1 at every site r, this ground state is nondegenerate and is invariant under global $\mathbb{Z}_2$ transformations

$$\hat{R}|\Psi_0\rangle_0 = |\Psi_0\rangle_0 \tag{14.29}$$

In this state,

$${}_0\langle\Psi_0|\hat{\sigma}_3(r)|\Psi_0\rangle_0 = 0 \tag{14.30}$$

since $\hat{\sigma}_3(r)$ is off-diagonal in the basis of eigenvectors of $\hat{\sigma}_1$. Similarly, the equal-time correlation function, which is equivalent to the correlation function on a fixed row in the classical system in D dimensions, also vanishes in this state:

$${}_0\langle\Psi_0|\hat{\sigma}_3(r)\, \hat{\sigma}_3(\mathbf{R}')|\Psi_0\rangle_0 = 0 \tag{14.31}$$

In perturbation theory in powers of g, the ground state gets corrected order-by-order by virtual processes that consist of σ_1 spin flips. Using Brillouin-Wigner perturbation theory, the perturbed ground state to lowest order is given by

$$|\Psi\rangle_0 = |\Psi_0\rangle_0 + \hat{P}\frac{\hat{V}}{E_0 - \hat{H}_0}|\Psi\rangle_0 + \cdots \tag{14.32}$$

where $\hat{P}$ projects the state $|\Psi_0\rangle_0$ out of the sum. For instance, for $D=2$, the state $|++\cdots+\rangle$ gets perturbed by pairs of σ_1 spin flips

$$\hat{V}|\Psi_0\rangle = -g\sum_R \mathcal{A}(R)|++\cdots+--+\cdots+\rangle \tag{14.33}$$

where the flipped pair of spins is located at R, and $\mathcal{A}$ is an easily calculable prefactor.

At higher orders in g, we will get more pairs in flipped states. Moreover, at higher orders, the individual spins of a pair get separated from each other. Hence in this picture, we can regard the individual spin flips as particles that are created in pairs from the vacuum (the ground state). The perturbative expansion can then be pictured as a set of worldlines of these pairs of particles that are, eventually, also annihilated in pairs and must be regarded as a set of *loops*. This picture of the expansion in powers of g is equivalent to the high-temperature expansion of classical statistical mechanics of the Ising model, which is also a theory of loops on the D-dimensional lattice.

What does this structure of the ground state imply for the following correlation function?

$$C(\boldsymbol{R}, \boldsymbol{R}') = \frac{\langle\Psi_0|\hat{\sigma}_3(\boldsymbol{R})\,\hat{\sigma}_3(\boldsymbol{R}')|\Psi_0\rangle}{\langle\Psi_0|\Psi_0\rangle} \tag{14.34}$$

Since $\sigma_3^2 = I$, we can write the product of two such spin operators at points $\boldsymbol{R}$ and $\boldsymbol{R}'$ as

$$\hat{\sigma}_3(\boldsymbol{R})\,\hat{\sigma}_3(\boldsymbol{R}') = \prod_{(\boldsymbol{r},\boldsymbol{r}')\in\Gamma_{\boldsymbol{R},\boldsymbol{R}'}} \hat{\sigma}_3(\boldsymbol{r})\,\hat{\sigma}_3(\boldsymbol{r}') \tag{14.35}$$

where $\boldsymbol{r}$ and $\boldsymbol{r}'$ are pairs of neighboring sites on an arbitrary path $\Gamma_{\boldsymbol{R},\boldsymbol{R}'}$ that goes from $\boldsymbol{R}$ to $\boldsymbol{R}'$. Thus, we can also write the correlation function as

$$C(\boldsymbol{R}, \boldsymbol{R}') = \frac{\langle\Psi_0|\prod_{(\boldsymbol{r},\boldsymbol{r}')\in\Gamma_{\boldsymbol{R},\boldsymbol{R}'}} \hat{\sigma}_3(\boldsymbol{r})\,\hat{\sigma}_3(\boldsymbol{r}')|\Psi_0\rangle}{\langle\Psi_0|\Psi_0\rangle} \tag{14.36}$$

which only picks up a nonzero contribution in perturbation theory in g if we go to an order n sufficiently high so that $n \gtrsim |\boldsymbol{R}-\boldsymbol{R}'|$, where $|\boldsymbol{R}-\boldsymbol{R}'|$ is the number of bonds on the shortest path $\Gamma_{\boldsymbol{R},\boldsymbol{R}'}$. This argument shows that the correlation function decays exponentially with distance,

$$C(\boldsymbol{R}, \boldsymbol{R}') \approx \text{const } \times e^{-|\boldsymbol{R}-\boldsymbol{R}'|/\xi} \tag{14.37}$$

where ξ is the correlation length. To lowest order in g, we find

$$\xi^{-1} \simeq \ln(1/g + \cdots) \tag{14.38}$$

It turns out that the expansion in powers of g has a finite radius of convergence in all space dimensions, $D-1$. Within the radius of convergence, this approximation is a strict upper bound for the correlation function $C(\boldsymbol{R}, \boldsymbol{R}')$. Hence, for all values of g small enough to be in the nonvanishing and finite radius of convergence, this system is in a phase that is smoothly related to the unperturbed state found at $g=0$. In this *entire phase*, the correlation functions decay exponentially with distance, characterized by a finite correlation length ξ.

This result suggests that in this phase, there is a finite energy (mass) gap $G \sim 2 + O(g)$, which can be easily checked by calculating the gap in perturbation theory. We will see in section 14.6 that the proportionality of the mass gap and the inverse correlation length become asymptotically exact near the phase transition, where this system becomes effectively Lorentz invariant, with a suitably defined speed of light.

B: The broken symmetry phase, $g \gg 1$ For $g \gg 1$, which is equivalent to the low-temperature regime of the classical problem in one extra dimension, the roles of the unperturbed and perturbation terms of the Hamiltonian are switched. Hence we rewrite the Hamiltonian as

$$\hat{H}_0 = -g \sum_{\boldsymbol{r},j} \hat{\sigma}_3(\boldsymbol{r}) \hat{\sigma}_3(\boldsymbol{r} + \hat{\boldsymbol{e}}_j), \qquad \hat{V} = -\sum_{\boldsymbol{r}} \hat{\sigma}_1(\boldsymbol{r}) \tag{14.39}$$

The unperturbed ground state now must be an eigenstate of $\hat{\sigma}_3(\boldsymbol{r})$. The leading-order ground state is *doubly degenerate*,

$$|\Psi_\uparrow\rangle_0 \equiv |\uparrow, \ldots, \uparrow\rangle, \qquad |\Psi_\downarrow\rangle_0 \equiv |\downarrow, \ldots, \downarrow\rangle \tag{14.40}$$

where, for all $\boldsymbol{r}$,

$$\hat{\sigma}_3(\boldsymbol{r}) |\Psi_\uparrow\rangle_0 = |\Psi_\uparrow\rangle_0, \qquad \hat{\sigma}_3(\boldsymbol{r}) |\Psi_\downarrow\rangle_0 = -|\Psi_\downarrow\rangle_0 \tag{14.41}$$

If the system is finite, these states will mix in perturbation theory. Indeed, if the lattice has N_s spatial sites, the mixing occurs to order N_s in perturbation theory. But, if we take the thermodynamic limit $N_s \to \infty$ first, then there is no mixing to any finite order in perturbation theory. The obvious exception is when the spacetime dimension is $D=1$, for which the number of sites in the "row" is just one, $N_s = 1$. Thus, for $D > 1$, or $D \geq 2$, these two states belong to two essentially decoupled pieces of the Hilbert space. In the classical system, the property just described is known as broken ergodicity, in the sense that the system only explores half of the total number of possible configurations.

This is an example of *spontaneous symmetry breaking*. Although the Hamiltonian is invariant under $\mathbb{Z}_2$ global transformations, in the thermodynamic limit, the ground state is not $\mathbb{Z}_2$ invariant. It is possible to prove rigorously that for $D \geq 2$, the expansion in powers in $1/g$ is convergent and that it has a finite radius of convergence. Therefore, the behavior of all quantities obtained in the limit $g \to \infty$ will also hold beyond some large but finite value of the coupling constant g. Other terms, for large enough g, have a stable phase with a spontaneously broken $\mathbb{Z}_2$ global symmetry. For instance, it is easy to see that in this phase, the energy gap between the ground state and the first excited state is $G(g) = \frac{2}{g} + O(1/g^2)$.

To illustrate the concept of spontaneous symmetry breaking, let us calculate, to lowest orders in $1/g$, the local spontaneous magnetization, which is the expectation value of the local spin operator:

$$m = \langle \Psi_\uparrow | \hat{\sigma}_3(\boldsymbol{r}) | \Psi_\uparrow \rangle \tag{14.42}$$

To lowest order in $1/g$, we have

$$_0\langle \Psi_\uparrow | \hat{\sigma}_3(\boldsymbol{r}) | \Psi_\uparrow \rangle_0 = +1, \qquad _0\langle \Psi_\downarrow | \hat{\sigma}_3(\boldsymbol{r}) | \Psi_\downarrow \rangle_0 = -1 \tag{14.43}$$

Clearly, in general, the expectation value is a function of the coupling constant g: $m = f(g)$. In general, the correlation function has the form

$$C(\boldsymbol{R}, \boldsymbol{R}') = \langle \Psi_\uparrow | \hat{\sigma}_3(\boldsymbol{R}) \hat{\sigma}_3(\boldsymbol{R}') | \Psi_\uparrow \rangle = m^2(g) + C_{\text{conn}}(\boldsymbol{R} - \boldsymbol{R}') \tag{14.44}$$

where the connected part, denoted by $C_{\text{conn}}(\boldsymbol{R} - \boldsymbol{R}')$, decays exponentially at long distances. Hence, as $|\boldsymbol{R} - \boldsymbol{R}'| \to \infty$, the correlation function approaches (exponentially fast) the constant value $m^2(g)$, which is the square of the expectation value of the order parameter (the local magnetization).

14.6 Solution of the two-dimensional Ising model

The solution of the two-dimensional Ising model on a square lattice by Lars Onsager in 1944 is one of the triumphs of theoretical physics of the twentieth century (Onsager, 1944). We will see that it will also help us understand how to define a QFT.

The arguments given in section 14.5 tell us that the Ising model on a square lattice is equivalent to the one-dimensional quantum system with Hamiltonian

$$H = -\sum_{n=-\frac{L}{2}+1}^{L/2} \sigma_1(n) - g \sum_{n=-\frac{L}{2}+1}^{L/2} \hat{\sigma}_3(n) \hat{\sigma}_3(n) \tag{14.45}$$

where $L = N_s$ is the number of sites along the one-dimensional space. We will consider here a system with periodic boundary conditions.

14.6.1 The Jordan-Wigner transformation

A naive look at the Hamiltonian leaves us with a puzzle. We have been brought up to think that quadratic Hamiltonians are trivial. So, why the fuss? Isn't $\hat{H}$ bilinear in $\hat{\sigma}$? The problem is that, even though H is a bilinear form in $\hat{\sigma}$, it is not a free theory. It is trivial to check that the equation of motion for $\hat{\sigma}_3(r)$ is not linear, since

$$i\partial_t \hat{\sigma}_3(\boldsymbol{r}) = [\hat{\sigma}_3(\boldsymbol{r}), \hat{H}] = -2i\hat{\sigma}_2(\boldsymbol{r}) = -2\hat{\sigma}_3(\boldsymbol{r})\hat{\sigma}_1(\boldsymbol{r}) \tag{14.46}$$

and

$$i\partial_t \hat{\sigma}_1(\boldsymbol{r}) = [\hat{\sigma}_1(\boldsymbol{r}), \hat{H}] = +g\hat{\sigma}_3(\boldsymbol{r})\hat{\sigma}_1(\boldsymbol{r}) \sum_{j=1,\ldots,D-1} \left(\hat{\sigma}_3(r + \hat{\boldsymbol{e}}_j) + \hat{\sigma}_3(r - \hat{\boldsymbol{e}}_j)\right) \tag{14.47}$$

The reason behind the nonlinearity is the form of the (equal-time) commutation relations

$$[\hat{\sigma}_3(\boldsymbol{r}),\hat{\sigma}_3(\boldsymbol{r}')]=[\hat{\sigma}_1(\boldsymbol{r}),\hat{\sigma}_1(\boldsymbol{r}')]=0$$
$$[\hat{\sigma}_3(\boldsymbol{r}),\hat{\sigma}_1(\boldsymbol{r}')]=0, \qquad \boldsymbol{r}\neq\boldsymbol{r}' \tag{14.48}$$
$$\{\hat{\sigma}_l(\boldsymbol{r}),\hat{\sigma}_k(\boldsymbol{r})\}=2\delta_{lk}, \qquad l,k=1,3$$

Hence, the commutation relations are not canonical. Instead, they seem to describe objects that are bosons on different sites but fermions on the same site. Alternatively, they can be regarded as bosons with hard cores. Indeed, the raising and lowering operators

$$\hat{\sigma}^{\pm}=\frac{1}{2}\left(\hat{\sigma}_3\mp i\hat{\sigma}_2\right) \tag{14.49}$$

act on the states $|\pm\rangle$, the eigenstates of $\hat{\sigma}_1$, as follows:

$$\hat{\sigma}^+|+\rangle=0, \qquad \hat{\sigma}^+|-\rangle=|+\rangle \tag{14.50}$$
$$\hat{\sigma}^-|+\rangle=|-\rangle, \qquad \hat{\sigma}^-|-\rangle=0$$

They can be regarded as the creation and annihilation operators of some oscillator, but with the hard-core constraint that the boson occupation number $\hat{n}=\hat{\sigma}^+\hat{\sigma}^-$ should only have eigenvalues 0 and 1, since

$$\hat{\sigma}^+\hat{\sigma}^-|+\rangle=|+\rangle, \qquad \hat{\sigma}^+\hat{\sigma}^-|-\rangle=0 \tag{14.51}$$

If $D=2$, the quantum problem has $d=1$ space dimensions. It turns out that there is a very neat and useful transformation that will enable us to deal with this problem. This is the Jordan-Wigner transformation (Jordan and Wigner, 1928; Lieb et al., 1961). The key idea behind this transformation is that, in one dimension only, hard-core bosons are equivalent to fermions! Qualitatively, this is easy to understand. If the particles live on a line, then bosons cannot exchange their positions by purely dynamical effects, since the hard-core condition forbids that possibility. Similarly, one-dimensional fermions cannot change their relative ordering as a result of their dynamics due to the Pauli principle. Thus, the strategy is to show that our problem is secretly a *fermion problem*. From now on, we will restrict our discussion to one (space) dimension.

Let us consider the kink creation operator $\hat{K}(n)$:

$$\hat{K}(n)=\prod_{j=-\frac{L}{2}+1}^{n}\left(-\hat{\sigma}_1(j)\right) \tag{14.52}$$

The operator flips all spins to the left of site $n+1$. Clearly, when acting on the "high-temperature" (or $g\ll 1$) ground state $|+\rangle$, we get

$$\hat{K}(n)|+\rangle=(-1)^n|+\rangle \tag{14.53}$$

That is, it acts as a counting operator in the phase with $g\ll 1$. But, in the "low-temperature" ground state (i.e., for $g\gg 1$), we get instead

$$\hat{K}(n)|\uparrow\cdots\uparrow\rangle = |\downarrow\cdots\underset{n}{\downarrow}\underset{n+1}{\uparrow}\cdots\uparrow\rangle \tag{14.54}$$

This state is called a *kink* (or domain wall) or a *topological soliton*. Clearly, the operator $\hat{K}_n$ disturbs the boundary conditions for $g > g_c$, but it does not for $g < g_c$, where

$$\langle\Psi_0|\hat{K}(n)|\Psi_0\rangle \neq 0 \qquad (g < g_c) \tag{14.55}$$

Hence, the high-temperature phase (disordered) is a condensate of kinks. The operator $\hat{K}$ is also known as a disorder operator (Kadanoff and Ceva, 1971; Fradkin and Susskind, 1978), since it takes a nonzero expectation value in a disordered phase.

We will now see that a clever combination of order (i.e., $\hat{\sigma}_3$) and disorder ($\hat{K}$) operators yields a Fermi field. Let us consider the operators

$$\hat{\chi}_1(j) = \hat{K}(j-1)\hat{\sigma}_3(j), \qquad \hat{\chi}_2(j) = i\hat{K}(j)\hat{\sigma}_3(j) \tag{14.56}$$

Since these operators are products of Pauli matrices, they trivially obey the following identities (for all j):

$$\hat{\chi}_1^\dagger(j) = \hat{\chi}_1(j), \qquad \hat{\chi}_2^\dagger(j) = \hat{\chi}_2(j), \qquad \hat{\chi}_1(j)^2 = \hat{\chi}_2(j)^2 = 1 \tag{14.57}$$

More generally, it is straightforward to show that they obey the algebra

$$\{\hat{\chi}_1(j), \hat{\chi}_1(j')\} = \{\hat{\chi}_2(j), \hat{\chi}_2(j')\} = 2\delta_{jj'}, \qquad \{\hat{\chi}_1(j), \hat{\chi}_2(j')\} = 0 \tag{14.58}$$

Therefore, the fields $\hat{\chi}_1(j)$ and $\hat{\chi}_2(j)$ are self-adjoint (i.e., real) fermions, known as Majorana fermions.

Let us define the complex (or Dirac) fermions $\hat{\psi}(j)$ and $\hat{\psi}^\dagger(j)$

$$\hat{\psi}^\dagger(j) \equiv \hat{K}(j-1)\hat{\sigma}^+(j), \qquad \hat{\psi}(j) \equiv \hat{K}(j-1)\hat{\sigma}^-(j) \tag{14.59}$$

where we used the projection operators $\hat{\sigma}^\pm$ (see eq. (14.49)). Equivalently, we can write

$$\hat{\psi} = \frac{1}{2}\left(\hat{\chi}_1 + i\hat{\chi}_2\right), \qquad \hat{\psi}^\dagger = \frac{1}{2}\left(\hat{\chi}_1 - i\hat{\chi}_2\right) \tag{14.60}$$

It is straightforward to check that the operators $\psi^\dagger(j)$ and $\psi(j)$ obey the canonical fermion algebra

$$\{\hat{\psi}(j), \hat{\psi}^\dagger(j')\} = \delta_{jj'}, \qquad \{\hat{\psi}(j), \hat{\psi}(j')\} = 0 \tag{14.61}$$

It is easy to invert the transformation of eq. (14.59). Observe that the local fermion number operator, $\hat{n}(j) = \hat{\psi}^\dagger(j)\hat{\psi}(j)$, is given by

$$\psi^\dagger(j)\psi(j) = \frac{1}{2}\left(1 + \hat{\sigma}_1(j)\right) \tag{14.62}$$

Hence,

$$-\hat{\sigma}_1(j) = -2\hat{\psi}^+(j)\hat{\psi}(j) + 1 \equiv 1 - 2\hat{n}(j) \tag{14.63}$$

Since the operator $\hat{n}(j)$ satisfies $\hat{n}^2(j) = \hat{n}(j)$, we can also write

$$-\hat{\sigma}_1(j) = e^{i\pi\hat{n}(j)} = e^{i\pi\hat{\psi}^\dagger(j)\psi(j)} \tag{14.64}$$

Therefore, the inverse Jordan-Wigner transformation is

$$\hat{\sigma}^{+}(j) = e^{i\pi \sum_{l<j} \hat{\psi}^{\dagger}(l)\hat{\psi}(l)} \hat{\psi}^{\dagger}(j) \tag{14.65}$$

$$\hat{\sigma}^{-}(j) = e^{i\pi \sum_{l<j} \hat{\psi}^{\dagger}(l)\hat{\psi}(l)} \hat{\psi}(j)$$

and, similarly,

$$\hat{\sigma}_3(j) = e^{i\pi \sum_{l<j} \hat{\psi}^{\dagger}(l)\hat{\psi}(l)} (\hat{\psi}^{\dagger}(j) + \psi(j)) \tag{14.66}$$

$$\hat{\sigma}_2(j) = e^{i\pi \sum_{l<j} \hat{\psi}^{\dagger}(l)\hat{\psi}(l)} \frac{1}{i}(-\hat{\psi}^{\dagger}(j) + \psi(j))$$

We can use these results to map the Hamiltonian of the spin system to a Hamiltonian for the equivalent Fermi system. For L even, we find

$$\hat{H} = -L + \sum_{j=-\frac{L}{2}+1}^{\frac{L}{2}} 2\hat{\psi}^{\dagger}(j)\hat{\psi}(j) + g \sum_{j=-\frac{L}{2}+1}^{\frac{L}{2}} (\hat{\psi}^{\dagger}(j) - \hat{\psi}(j))(\hat{\psi}^{\dagger}(j+1) + \hat{\psi}(j+1))$$
$$+ \text{boundary term} \tag{14.67}$$

The boundary term is given by

$$-g\eta\hat{\sigma}_3\left(\frac{L}{2}\right)\hat{\sigma}_3\left(-\frac{L}{2}+1\right) = \tag{14.68}$$
$$-g\eta\hat{Q}\left(\hat{\psi}^{\dagger}\left(\frac{L}{2}\right) - \hat{\psi}\left(\frac{L}{2}\right)\right)\left(\hat{\psi}^{\dagger}\left(-\frac{L}{2}+1\right) + \hat{\psi}\left(-\frac{L}{2}+1\right)\right)$$

where $\eta = \pm 1$ for periodic or antiperiodic boundary conditions for the spins, respectively. Here, $\hat{Q}$ is the operator

$$\hat{Q} = \hat{R} = \prod_{-\frac{L}{2}+1}^{\frac{L}{2}} (-\hat{\sigma}_1(j)) = e^{i\pi \hat{N}} \tag{14.69}$$

where $\hat{N}$ is the total number of fermions.

However, $\hat{N}$ does not commute with the Hamiltonian $\hat{H}$ and, hence, the total number of complex (Dirac) fermions is not conserved. This is apparent, since the fermion Hamiltonian has terms that create and destroy fermions in pairs. Nevertheless, the fermion *parity* operator $(-1)^F \equiv e^{i\pi \hat{N}}$ is conserved, since it commutes with the Hamiltonian, $[(-1)^F, \hat{H}] = 0$. Thus, we can only specify whether the fermion number is even or odd. This is natural in a system of Majorana fermions, which, as such, do not have a conserved charge (since they are charge neutral); instead, the fermion parity is conserved.

If the spin operators obey periodic (or antiperiodic) boundary conditions,

$$\hat{\sigma}_3\left(\frac{L}{2}+1\right)=\eta\hat{\sigma}_3\left(-\frac{L}{2}+1\right) \tag{14.70}$$

with $\eta=\pm 1$, then the fermions obey the boundary conditions

$$\hat{\psi}\left(\frac{L}{2}+1\right)=\hat{Q}\,\eta\,\hat{\psi}\left(-\frac{L}{2}+1\right) \tag{14.71}$$

In particular, for periodic boundary conditions for the spins ($\eta=+1$), the fermions obey the boundary conditions

$$\hat{\psi}\left(\frac{L}{2}+1\right)=\hat{Q}\hat{\psi}\left(-\frac{L}{2}+1\right) \tag{14.72}$$

Hence, for an even number of fermions, $\hat{N}$ even ($Q=+1$), the fermions obey periodic boundary conditions, but for $\hat{N}$ odd ($Q=-1$), they obey antiperiodic boundary conditions. Nevertheless, since the ground state energy for N odd is greater than that for N even, $E_0^- > E_0^+$, we can work in the even sector. In this sense, the fermion-boson mapping is 2 to 1.

14.6.2 Diagonalization

The fermion Hamiltonian is a bilinear form in Fermi fields. Thus, it should be diagonalizable by a suitable canonical transformation. Since fermion number is not conserved, this transformation cannot be just a Fourier transform. In the even sector,

$$\hat{Q}|\Psi\rangle=|\Psi\rangle \tag{14.73}$$

the Fourier transform, for L even, is

$$\hat{\psi}(j)=\frac{1}{L}\sum_{k=-\frac{L}{2}+1}^{\frac{L}{2}} e^{i2\pi\frac{kj}{L}}\tilde{a}(k), \qquad \tilde{a}(k)=\sum_{j=-\frac{L}{2}+1}^{\frac{L}{2}} e^{-i2\pi\frac{kj}{L}}\hat{\psi}(j) \tag{14.74}$$

such that

$$\{\tilde{a}(k),\tilde{a}^\dagger(k')\}=L\delta_{k,k'}, \qquad \{\tilde{a}(k),\tilde{a}(k')\}=\{\tilde{a}^\dagger(k),\tilde{a}^\dagger(k')\}=0 \tag{14.75}$$

In the thermodynamic limit, $L\to\infty$, the momenta $\frac{2\pi}{L}k\equiv k$ densely fill up the interval $(-\pi,\pi]$ (the first Brillouin zone). In this limit, we find a representation of the Dirac delta function,

$$\delta(k)\equiv\lim_{L\to\infty}\frac{1}{2\pi}\sum_{j=-\frac{L}{2}+1}^{\frac{L}{2}} e^{ikj}=\lim_{L\to\infty}\frac{L}{2\pi}\delta_{k,0} \tag{14.76}$$

and $\lim_{k\to 0}\delta(k)=\frac{L}{2\pi}$. Also, in the thermodynamic limit, we identify the operators $\tilde{a}_k\equiv\hat{a}(k)$, which obey the anticommutation relations

$$\{\hat{a}(k),\hat{a}^\dagger(k')\}=2\pi\delta(k-k') \tag{14.77}$$

These definitions can be used to derive the following identities:

$$\lim_{L\to\infty} \sum_{j=-\frac{L}{2}+1}^{\frac{L}{2}} \hat{\psi}^\dagger(j)\hat{\psi}(j) = \int_{-\pi}^{\pi} \frac{dk}{2\pi} \hat{a}^\dagger(k)\hat{a}(k)$$

$$\lim_{L\to\infty} \sum_{j=-\frac{L}{2}+1}^{\frac{L}{2}} \hat{\psi}^\dagger(j)\hat{\psi}(j\pm 1) = \int_{-\pi}^{\pi} \frac{dk}{2\pi} e^{\pm ik}\hat{a}^\dagger(k)\hat{a}(k) \tag{14.78}$$

and

$$\lim_{L\to\infty} \sum_{j=-\frac{L}{2}+1}^{\frac{L}{2}} \hat{\psi}^\dagger(j)\hat{\psi}^\dagger(j+1) = \int_{-\pi}^{\pi} \frac{dk}{2\pi} e^{ik}\hat{a}^\dagger(k)\hat{a}^\dagger(-k)$$

$$\lim_{L\to\infty} \sum_{j=-\frac{L}{2}+1}^{\frac{L}{2}} \hat{\psi}(j)\hat{\psi}(j+1) = \int_{-\pi}^{\pi} \frac{dk}{2\pi} e^{-ik}\hat{a}(k)\hat{a}(-k) \tag{14.79}$$

By collecting terms, we find that the Hamiltonian becomes

$$\begin{aligned} H = -L + 2 &\int_{-\pi}^{\pi} \frac{dk}{2\pi} (1 + g\cos k)\hat{a}^\dagger(k)\hat{a}(k) \\ +g &\int_{-\pi}^{\pi} \frac{dk}{2\pi} (e^{ik}\hat{a}^\dagger(k)\hat{a}^\dagger(-k) - e^{-ik}\hat{a}(k)\hat{a}(-k)) \end{aligned} \tag{14.80}$$

This Hamiltonian has the same form as the pairing Hamiltonian of the Bardeen-Cooper-Schrieffer (BCS) theory of superconductivity (Schrieffer, 1964).

It is possible to rewrite H in eq. (14.80) in terms of the spinor field $\hat{\Psi}(k)$

$$\hat{\Psi}(k) = \begin{pmatrix} \hat{a}^\dagger(k) \\ \hat{a}(-k) \end{pmatrix}, \qquad \hat{\Psi}^\dagger(k) = \begin{pmatrix} \hat{a}(k), & \hat{a}^\dagger(-k) \end{pmatrix} \tag{14.81}$$

which, in the theory of superconductivity, is known as the Nambu representation.

Notice that the two components of the spinor $\hat{\Psi}$ are not independent. Indeed if we denote by $\hat{C}$ the matrix

$$\hat{C} = \begin{pmatrix} 0 & 1 \\ 1 & 0 \end{pmatrix} \tag{14.82}$$

we find that the spinor field $\hat{\Psi}(k)$ must obey

$$\hat{\Psi}^\dagger(k) = \left(\hat{C}\,\hat{\Psi}(-k)\right)^T = \hat{\Psi}^T(-k)\,\hat{C} \tag{14.83}$$

This is a real spinor field, or what is the same, a Majorana fermion.

In terms of the spinor fields $\hat{\Psi}(k)$, the Hamiltonian is

$$H = +L\left(-1 + \int_0^{\pi} \frac{dk}{2\pi} 2(1 + g\cos k)\right) - \int_0^{\pi} \frac{dk}{2\pi} \hat{\Psi}^\dagger(k) \begin{pmatrix} 2(1+g\cos k) & -2ig\sin k \\ 2ig\sin k & -2(1+g\cos k) \end{pmatrix} \hat{\Psi}(k) \tag{14.84}$$

We will see shortly that, in the low-energy limit, this Hamiltonian reduces to the Dirac Hamiltonian in 1+1 dimensions.

To diagonalize this system, we finally perform a Bogoliubov transformation to a new set of fermion operators $\hat{\eta}(k)$,

$$\begin{aligned} \hat{a}(k) &= u(k)\hat{\eta}(k) - iv(k)\hat{\eta}^\dagger(-k) \\ \hat{a}(-k) &= u(k)\hat{\eta}(-k) + iv(k)\hat{\eta}^\dagger(k) \end{aligned} \tag{14.85}$$

where $u(k)$ and $v(k)$ are two real (even) functions of k, which will be determined below. The inverse transformation is

$$\begin{aligned} \hat{\eta}(k) &= u(k)\,\hat{a}(k) + iv(k)\,\hat{a}^\dagger(-k) \\ \hat{\eta}(-k) &= u(k)\,\hat{a}(-k) - iv(k)\,\hat{a}^\dagger(k) \end{aligned} \tag{14.86}$$

The transformation is canonical, since it preserves the anticommutation relations,

$$\{\hat{a}(k),\, a^\dagger(k')\} = 2\pi\delta(k-k') \Longrightarrow \{\hat{\eta}(k),\, \eta^\dagger(k')\} = 2\pi\delta(k-k') \tag{14.87}$$

which can be achieved provided the functions $u(k)$ and $v(k)$ satisfy the relation

$$u^2(k) + v^2(k) = 1 \tag{14.88}$$

This condition is satisfied by the choice

$$u(k) = \cos\theta(k), \qquad v(k) = \sin\theta(k) \tag{14.89}$$

We will determine the function $\theta(k)$ by demanding that the fermion-number nonconserving terms cancel out exactly in terms of the fields $\hat{\eta}(k)$. Let us now define the functions $\alpha(k)$ and $\beta(k)$:

$$\alpha(k) = 2(1 + g\cos k), \qquad \beta(k) = 2g\sin k \tag{14.90}$$

The condition that the terms that do not conserve the fermions $\hat{\eta}(k)$ cancel is given by

$$-2\alpha(k)u(k)v(k) + \beta(k)\left(u^2(k) - v^2(k)\right) = 0 \tag{14.91}$$

The solution for the functions $u(k)$ and $v(k)$ of eq. (14.89) implies that $\theta(k)$ must satisfy the simple relation

$$\tan(2\theta(k)) = \frac{\beta(k)}{\alpha(k)} = \frac{g\sin k}{1 + g\cos k} \tag{14.92}$$

With this choice, the Hamiltonian H takes the form

$$H = \int_0^{\pi} \frac{dk}{2\pi}\, \omega(k) \left(\hat{\eta}^{\dagger}(k)\hat{\eta}(k) + \hat{\eta}^{\dagger}(-k)\hat{\eta}(-k)\right) + \varepsilon_0(g)L \tag{14.93}$$

where $\varepsilon_0(g)$ is given by

$$\begin{aligned}\varepsilon_0(g) &= -1 + 2\int_0^{\pi} \frac{dk}{2\pi} \left(v^2(k)\, 2g \sin k - u(k)v(k)\, 2(1+g\cos k)\right)\\ &= -1 + \int_0^{\pi} \frac{dk}{2\pi} \left[4g \sin k \sin^2(\theta(k)) - 2\sin(2\theta(k))(1+g\cos k)\right]\end{aligned} \tag{14.94}$$

and $\omega(k)$ by

$$\begin{aligned}\omega(k) &= \alpha(k)\cos(2\theta(k)) + \beta(k)\sin(2\theta(k))\\ &= 2\left[(1+g\cos k)\cos(2\theta(k)) + g\sin k \sin(2\theta(k))\right]\end{aligned} \tag{14.95}$$

We will choose the range of $\theta(k)$ and the signs of $\cos(2\theta(k))$ and $\sin(2\theta(k))$ to be

$$\text{sgn}\cos(2\theta(k)) = \text{sgn}\alpha(k), \qquad \text{sgn}\sin(2\theta(k)) = \text{sgn}\beta(k) \tag{14.96}$$

which guarantee that $\omega(k) \geq 0$. Using these results, we can write $\omega(k)$ in the form

$$\omega(k) = |\alpha(k)||\cos 2\theta(k)| + |\beta(k)||\sin 2\theta(k)| \tag{14.97}$$

Hence, we find

$$\omega(k) = 2\sqrt{1+g^2+2g\cos k} \geq 0 \tag{14.98}$$

Similar algebraic manipulations lead to the simpler expression for $\varepsilon_0(g)$,

$$\varepsilon_0(g) = -\frac{1}{2}\int_{-\pi}^{\pi} \frac{dk}{2\pi}\, \omega(k) < 0 \tag{14.99}$$

which is clearly negative.

14.6.3 Energy spectrum and critical behavior

With these above choices, it is now elementary to find the spectrum of the Hamiltonian.

A: The ground state The ground state $|G\rangle$ is simply the state annihilated by all the destruction operators $\hat{\eta}(k)$:

$$\hat{\eta}(k)|0\rangle = 0, \qquad \hat{\eta}(-k)|0\rangle = 0 \tag{14.100}$$

The ground state energy density is equal to $\varepsilon_0(g)$, and it is negative.

By retracing our steps, it is easy to see that the free energy of the two-dimensional classical Ising model is related to the ground state energy of the one-dimensional quantum Ising model,

$$\lim_{N_\tau\to\infty} Z = \exp\left(-\frac{N_\tau L f}{T}\right) = \lim_{N_\tau\to\infty} \text{tr}\, \hat{T}^{N_\tau} = \lim_{\beta\to\infty} \text{tr}\, e^{-\beta\hat{H}} \tag{14.101}$$

where $\hat{T}$ is the transfer matrix of the two-dimensional classical Ising model at temperature T, and $\hat{H}$ is the Hamiltonian of the one-dimensional quantum Ising model with coupling constant g. Hence, the free energy density f of the two-dimensional classical model and the ground state energy density ε_0 of the one-dimensional quantum model are related by the identification

$$f \equiv \varepsilon_0(g) \tag{14.102}$$

From our results for the ground state energy density, we find the explicit result

$$\begin{aligned}\varepsilon_0(g) &= -2\int_0^{\pi} \frac{dk}{2\pi}\sqrt{(1+g)^2 - 4g\sin^2\left(\frac{k}{2}\right)} \\ &= -\frac{2|1+g|}{\pi}\int_0^{\pi/2} dx\sqrt{1-(1-\gamma^2)\sin^2 x} \\ &= -\frac{2|1+g|}{\pi} E\left(\frac{\pi}{2}, \sqrt{1-\gamma^2}\right)\end{aligned} \tag{14.103}$$

where $E(\frac{\pi}{2}, \gamma)$ is the complete elliptic integral of the second kind, and $\gamma = \left|\frac{1-g}{1+g}\right|$ is the modulus of the elliptic integral. Using the expansion of the elliptic integral in the limit $\gamma \to 0$ (Gradshteyn and Ryzhik, 2015),

$$E\left(\frac{\pi}{2}, \sqrt{1-\gamma^2}\right) \underset{\gamma\to 0}{\approx} 1 + \frac{\gamma^2}{4}\left(\ln\frac{16}{\gamma^2} - 1\right) + O(\gamma^4) \tag{14.104}$$

we can write an expression for $\varepsilon_0(g)$ valid for $g \sim 1$:

$$\varepsilon_0(g) = -2\left(\frac{1-g}{\pi}\right)\left\{1 + \frac{\gamma^4}{4}\left(\ln\frac{16}{\gamma^2} - 1\right) + \cdots\right\} \tag{14.105}$$

We can write the ground state energy density $\varepsilon_0(g)$ as a sum of two terms, $\varepsilon_0^{\text{sing}}$ (for the singular behavior for $g \approx 1$) and $\varepsilon_0^{\text{reg}}$ (for the nonsingular behavior away from $g = 1$). The singular piece, $\varepsilon_0^{\text{sing}}(g)$, is

$$\varepsilon_0^{\text{sing}} = -\frac{4}{\pi}\left[1 + \frac{t^2}{8}\left(\ln\left(\frac{8}{|t|}\right) - \frac{1}{2}\right) + \cdots\right] \tag{14.106}$$

where $t = |1-g|$ plays the role of the "reduced temperature" of the two-dimensional classical model, $t = (T - T_c)/T_c$ (where T is the temperature of the classical problem, and T_c is the critical temperature).

Let us now use these results to compute the behavior of the specific heat of the classical two-dimensional Ising model close to the phase transition. This follows from the identification of the free energy of the classical model with the ground state energy of the quantum model. Since the specific heat is obtained by twice differentiating the free energy with respect to temperature, it can be obtained equivalently by twice differentiating the ground state energy of the quantum model with respect to the coupling constant g. Hence, if we write the specific heat as a sum of a singular and a regular part, $C = C_{\text{sing}} + C_{\text{reg}}$, the singular part is readily found to be

$$C_{\text{sing}} \approx -\frac{\partial^2 \varepsilon_0^{\text{sing}}}{\partial t^2} \approx +\frac{1}{2\pi} \ln\left(\frac{8}{|t|}\right) - \frac{3}{4\pi} + \cdots \tag{14.107}$$

This is the famous logarithmic divergence of the specific heat of the two-dimensional classical Ising model, first derived by Onsager (1944). This result also identifies the critical coupling as $g_c = 1$, corresponding to the Onsager critical temperature.

In this section, we have worked consistently in the thermodynamic limit, $L \to \infty$, and obtained the expression for the ground state energy density ε_0 in eq. (14.99). We will see in chapter 21 that the Casimir energy, the leading finite correction to the ground state energy, has a special significance in conformal field theory (CFT). There it will be shown that at a fixed point, the Casimir energy, E_{Casimir}, is

$$E_{\text{Casimir}} = -\frac{\pi c v}{6L} \tag{14.108}$$

where L is the linear size of the system, v is the speed of light, and c is the central charge of the Virasoro algebra of the CFT. We will see below that at $g = 1$, the energy gap of the spectrum of the Ising model vanishes, and the system is at its phase transition point. Using the results of this section, we can compute the Casimir energy by computing the leading finite correction to the ground state energy. The result that one readily finds is that in the Ising model, the central charge is $c = 1/2$.

B: Excitation spectrum The first excited state is the fermion state $|k\rangle$ with momentum k:

$$|k\rangle = \hat{\eta}^\dagger(k)|0\rangle, \qquad H|k\rangle = (E_0 + \omega(k))|k\rangle \tag{14.109}$$

Thus, it is an exact eigenstate of the Hamiltonian with excitation energy $\omega(k)$, shown in figure 14.3. The excitation energy is positive for all momenta k in the first Brillouin zone, $-\pi < k \leq \pi$. The excitation energy is smallest at $k = \pi$, where it is equal to $2|1-g|$.

Therefore, we find that the fermion spectrum has an energy gap $G(g) = \min \omega(k)$. By writing $\omega(k)$ in the form

$$\omega(k) = 2\sqrt{1 + g^2 + 2g\cos k} = 2\sqrt{(1-g)^2 + 4g\cos^2\left(\frac{k}{2}\right)} \tag{14.110}$$

we find that, with our conventions, the gap is reached at $k = \pi$ (for $g > 0$). Thus for all $g \neq 1$, the gap does not vanish, and only exactly at $g = 1$ does it vanish. But this is precisely the phase transition point!

Let's take a closer look at the low-energy part of the spectrum. To do that, consider momenta k close to π by setting $k = \pi - q$, with $|q| \ll \pi$. In this regime, we can approximate the excitation energy to be

$$\omega(q) \simeq 2((1-g)^2 + gq^2)^{1/2} \tag{14.111}$$

In the limit $g \to 1$, the spectrum is that of a relativistic fermion with mass $m = 2|1-g|/(2g)$ and "speed of light" $v = 2$. We will see shortly that this relativistic behavior is not an accident.

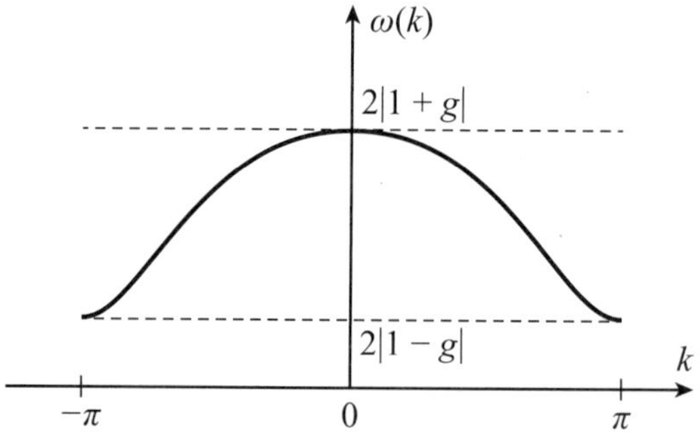

Figure 14.3 The excitation spectrum of the one-dimensional quantum Ising model for $g \neq 1$.

In addition, right at $g = 1$, the spectrum is a linear function of q (i.e., $\omega(q) = v|q|$) and behaves as a relativistic theory of massless fermions.

As $g \to g_c = 1$, the spectral gap $G(g)$ vanishes following a power law

$$G(g) = \mathcal{A}\left|g - g_c\right|^{\nu} \tag{14.112}$$

where $g_c = 1$, $\mathcal{A} = 2$. We can now identify the critical exponent $\nu = 1$ with the gap exponent. Likewise, the inverse of the mass gap G is the correlation length $\xi = G^{-1}$ of the Ising model (or the Compton wavelength of the relativistic fermion). Hence, the correlation length diverges as the critical point is approached, $g \to 1$, as $\xi \sim |g - g_c|^{-\nu}$, with $\nu = 1$.

What is the physical meaning here of a fermionic excitation? As we have seen, a fermion is essentially a kink or a domain wall. Hence, the excitation energy of a fermion is essentially the free energy needed to create a domain wall. In the one-dimensional quantum problem, as the phase transition is approached, the domain wall energy vanishes, and it becomes a massless excitation. Thus, the excitations that become massless are effectively topological solitons. However, the two-dimensional classical picture is equivalent to the Euclidean path-integral picture. In this framework, the transition consists of domain walls whose free-energy density vanishes. Hence the domain walls proliferate at the phase transition and span the size of the system.

14.7 Continuum limit and the two-dimensional Ising universality class

Let us now show that this model defines a nontrivial QFT by taking the continuum limit. We can do this since as $g \to g_c$, the correlation length diverges, $\xi \to \infty$, and becomes much larger than the lattice spacing. We will construct the continuum limit by looking at the equations of motion of the Majorana fermion fields.

Unlike the Ising spin operators, whose equations of motion are nonlinear, the Majorana fermions $\chi_1(j)$ and χ_2 have simple (linear!) equations of motion. Indeed, using the definition of the Majorana fermion operators of eq. (14.56), one readily finds that the Hamiltonian is

$$H = i\sum_j \chi_1(j)\chi_2(j) - ig\sum_j \chi_2(j)\chi_1(j+1) \tag{14.113}$$

where we have assumed periodic boundary conditions (for a chain with an even number of sites). The equations of motion of the Majorana fermions are

$$
\begin{aligned}
i\partial_t \chi_1(j) &= 2i\chi_2(j) + 2ig\chi_2(j-1) \\
i\partial_t \chi_2(j) &= -2i\chi_1(j) - 2ig\chi_1(j+1)
\end{aligned}
\tag{14.114}
$$

It is easy to check that the eigenstates (i.e., the Fourier modes of $\chi_1(j)$ and $\chi_2(j)$) have the spectrum of eq. (14.110).

As shown in figure 14.3, the low-energy states have wave vector $k \simeq \pi$. To take the continuum limit, let us focus on these low-energy states. To this end, it is convenient to redefine the Majorana fermion operators as

$$
\chi_1(j) = (-1)^j \tilde{\chi}_1(j), \quad \chi_2(j) = (-1)^j \tilde{\chi}_2(j)
\tag{14.115}
$$

whose equations of motion are obtained from those given in eq. (14.114) by replacing $g \leftrightarrow -g$. In doing so, the low-energy Fourier modes of the operators $\tilde{\chi}_i(j)$ are now near $k=0$. To proceed to the continuum limit, we now restore a lattice constant a_0 and assign to a lattice site labeled by j a coordinate $x_j = ja_0$ with units of length. Since we will be interested only in field configurations that vary slowly on the scale of the lattice spacing, we can use the approximations

$$
\begin{aligned}
\tilde{\chi}_2(j-1) &\approx \tilde{\chi}_2(x_j) - a_0\partial_x \tilde{\chi}_2(x_j) + O(a_0^2) \\
\tilde{\chi}_1(j+1) &\approx \tilde{\chi}_1(x_j) + a_0\partial_x \tilde{\chi}_1(x_j) + O(a_0^2)
\end{aligned}
\tag{14.116}
$$

Therefore, in this regime, we can rewrite the equations of motion as of two partial differential equations of the form

$$
\begin{aligned}
\frac{1}{2a_0 g} i\partial_t \tilde{\chi}_1 &\simeq i\frac{(1-g)}{a_0 g}\tilde{\chi}_2 + i\partial_x \tilde{\chi}_2 \\
\frac{1}{2a_0 g} i\partial_t \tilde{\chi}_2 &\simeq -i\frac{(1-g)}{a_0 g}\tilde{\chi}_1 + i\partial_x \tilde{\chi}_1
\end{aligned}
\tag{14.117}
$$

On rescaling and relabeling the time coordinate $t2a_0\sqrt{g} \mapsto x_0$, and the space coordinate $x \mapsto x_1\sqrt{g}$ (thus setting the "speed of light" to 1), we can write the equations of motion as

$$
\begin{aligned}
i\partial_0 \tilde{\chi}_1 - i\partial_1 \tilde{\chi}_2 + i\left(\frac{(1-g)}{a_0 g}\right)\tilde{\chi}_2 &= 0 \\
i\partial_0 \tilde{\chi}_2 - i\partial_1 \tilde{\chi}_1 - i\left(\frac{(1-g)}{a_0 g}\right)\tilde{\chi}_1 &= 0
\end{aligned}
\tag{14.118}
$$

Define the *scaling limit*, in which $g \to g_c = 1$ and simultaneously $a_0 \to 0$ while keeping finite the following quantity:

$$
m = \lim_{\substack{a_0 \to 0 \\ g \to 1}} \left(\frac{(1-g)}{a_0\sqrt{g}}\right)
\tag{14.119}
$$

As we will see, m will be identified with a mass scale. In this limit, the equations of motion take the form of Dirac equations:

$$i\partial_0\tilde{\chi}_1 - i\partial_1\tilde{\chi}_2 + im\tilde{\chi}_2 = 0$$
$$i\partial_0\tilde{\chi}_2 - i\partial_1\tilde{\chi}_1 - im\tilde{\chi}_1 = 0 \tag{14.120}$$

Let us define the 2×2 gamma matrices

$$\gamma_0 = -\sigma_2, \qquad \gamma_1 = i\sigma_3, \qquad \gamma_5 = \sigma_1 \tag{14.121}$$

in terms of which we find that the spinor field $\tilde{\chi} = (\tilde{\chi}_1, \tilde{\chi}_2)$ obeys the 1+1-dimensional Dirac equation

$$(i\partial\!\!\!/ - m)\tilde{\chi} = 0 \tag{14.122}$$

where the spinor field $\tilde{\chi}$ obeys the reality condition

$$\tilde{\chi}^\dagger \equiv \tilde{\chi}^T \tag{14.123}$$

and, hence, it is a massive relativistic Majorana fermion. From now on, to simplify the notation, we define $\tilde{\chi} \equiv \chi$.

The Lagrangian of the theory of free Majorana spinor χ (in the Minkowski signature) is

$$\mathcal{L} = \bar{\chi} i\partial\!\!\!/ \chi - \frac{1}{2} m \bar{\chi}\chi \tag{14.124}$$

Here $\bar{\chi} = \chi^T\gamma_0$, where we used the condition that the Majorana spinor is a real Fermi field. Notice that the (Majorana) mass term is the hermitian operator

$$\frac{1}{2}\bar{\chi}\chi = \frac{1}{2}\chi^T\gamma_0\chi = \frac{i}{2}\epsilon_{\alpha\beta}\chi_\alpha\chi_\beta = i\chi_1\chi_2 \tag{14.125}$$

where $\alpha, \beta = 1, 2$ label the two components of the Majorana spinor (in 1+1 dimensions).

Thus, in the *scaling limit* of $g \to g_c = 1$ (i.e., asymptotically close to the phase transition) and at distances large compared to the lattice spacing $a_0 \to 0$ (the "continuum limit"), the two-dimensional classical Ising model and the one-dimensional quantum Ising model are equivalent to (or rather, define) the continuum field theory of free Majorana fermions. Notice that this works only at distances long compared with the lattice constant (i.e., $a_0 \to 0$) but comparable to the diverging correlation length $\xi \approx \frac{1}{|m|}$.

The results of the past few sections show that the two-dimensional Ising model can be understood in terms of a theory of Majorana fermions that are nonlocal objects in terms of the spins of the lattice model. It should not be a surprise that the computation of the correlators of the spin operators themselves may be a nontrivial task in the Majorana fermion basis. In fact, it is possible to compute the correlation function of two spins, although the computation is rather technical and somewhat outside the scope of this book. The results of such a computation are nevertheless instructive. When far from the critical point, the results are what is expected (i.e., exponential decay of correlations in the symmetric phase and long-range order in the broken symmetry state). It is also found that, at the critical point, the two-point function decays as a power law: $G(R) \sim R^{-1/4}$. This result is very different from what would have been guessed using dimensional analysis in a free scalar field theory. We will see in chapter 21 that the two-dimensional Ising model is in a different universality class that is characterized by a fixed point of free Majorana fermions (which, qualitatively, are the domain walls of the Ising spins). Since domain walls are nonlocal objects in the language of local spins, it should be no surprise that the results are actually so different. We will return to these questions in chapter 21, where we discuss conformal field theory.

Let us summarize what we have done in this chapter. The two-dimensional classical Ising model is a nontrivial theory that happens to be integrable in terms of a theory of Majorana fermions. While it is great to have an exact solution, we should not lose sight of the important lessons that we can draw from a special case. In fact, most QFTs of interest are not solvable. The important lesson that we should draw is that interacting QFTs have properties that largely cannot be guessed by using perturbation theory. In the case of the two-dimensional Ising model, the fluctuations of the individual spin fields are not simple and cannot be accessed perturbatively. In earlier chapters, we used symmetry arguments to argue that the Ising model should be equivalent to ϕ^4 field theory. Although this is true, we will see in later chapters that the perturbative construction of ϕ^4 theory only works near four spacetime dimensions. What we did in this chapter shows that below four dimensions, we cannot rely on perturbation theory.

Exercises

14.1 Transfer matrices

In this chapter, we used the concept of the transfer matrix to relate the partition function of a classical system in D dimensions to a quantum system in $D-1$ dimensions. In this problem, you will consider a one-dimensional classical system with a global $U(1)$ symmetry. At each site $j=1,\ldots,N$ of a chain of N sites, define a degree of freedom $\boldsymbol{n}_j=(\cos\theta_j, \sin\theta_j)$, with $\theta\in(0,2\pi]$. The partition function (with periodic boundary conditions) is

$$Z=\prod_{j=1}^{N}\exp\left(\frac{J}{T}\sum_{j=1}^{N}\cos(\theta_{j+1}-\theta_j)+\frac{H}{T}\sum_{j=1}^{N}\cos\theta_j\right)$$

Assume here that we are at low temperatures, $J/T\gg 1$. In this regime, we can make the following useful approximation:

$$\exp(K\cos\theta)\simeq\frac{1}{\sqrt{K}}\sum_{n=-\infty}^{\infty}\exp\left(-\frac{n^2}{2K}\right)\exp(in\theta)$$

Also assume that the external field is weak: $H/T\ll 1$.

1) Show that, quite generally, the partition function Z can be written as

$$Z=\mathrm{tr}((T_1T_2)^N)$$

where T_1 is diagonal in the basis of the angular momentum operator $L=-i\partial_\theta$, and T_2 is diagonal in the basis of the angle variable θ.

2) Show that if $J/T \gg 1$ and $H/T \ll 1$, then $T_1 = \text{const} \cdot e^{-aL^2}$, and $T_2 = e^{b\cos\theta}$. Determine the value of the constants a and b in terms of the ratios J/T and H/T. Find the expression of the quantum Hamiltonian for the variable θ, and provide a physical interpretation.

14.2 The one-dimensional quantum Ising model at finite temperature
In section 14.4, we used a Jordan-Wigner transformation to show that the Hamiltonian of the one-dimensional quantum model on a chain of L sites,

$$H = -\sum_{n=1}^{L} \sigma_1(n) - \lambda \sum_{n=1}^{N} \sigma_3(n)\sigma_3(n+1) + \text{boundary term}$$

is equivalent to a theory of fermions using a Jordan-Wigner transformation, which we diagonalized. Here, as in the text, let us work with periodic boundary conditions for a chain with L an even number. Assume that the lattice spacing is $a = 1$.

1) Consider first the case in which the system is at the thermodynamic limit, $L \to \infty$. Use standard methods of quantum statistical mechanics to show that at the critical point, $g = 1$, the low-temperature behavior of the free energy per unit length $f(T)$ of the chain has the Stefan-Boltzmann form

$$f(T) = \varepsilon_0 - \frac{\pi c T^2}{6v} + O(T^4)$$

where $v = 2$ is the speed of the excitations at the critical point, and T is the temperature. Explain physically why this law should be obeyed at the critical point. Determine the value of the constant c. *Note:* A useful integral is

$$\int_0^\infty dx \frac{x}{e^x + 1} = \frac{\pi^2}{12}$$

2) Compute the Casimir energy of a chain of L spins in the large-L limit. In the text, it is shown that in the thermodynamic limit, $L \to \infty$, the ground state energy density is given by eq. (14.99). Consider a long but *finite* chain of L sites (with periodic boundary conditions). Again we will consider a system at the critical point $g = 1$. Show that for a finite system of length L, the ground state energy per unit length is

$$\varepsilon(L, g) = -\frac{1}{L} \sum_{n=-\frac{L}{2}+1}^{\frac{L}{2}} \frac{1}{2}\omega(k_n)$$

where $k_n = 2\pi n/L$, and $\omega(k)$ is given in eq. (14.98) with $k \equiv 2\pi n/L$. Use the Euler-Maclaurin expansion to show that, at $g = 1$, the ground state energy per unit length has, in the large L limit, the following form:

$$\varepsilon = \varepsilon_0 - \frac{\pi c v}{6L} + O(1/L^2)$$

where $v = 2$ is the "speed of light." Calculate the constant c, and show that it has the same value as in problem 1. Compare this result to what you found in the exercises of chapter 8 for a real scalar field in 1+1 dimensions. In chapter 21, we will see that the constant c is the conformal anomaly. The Euler-Maclaurin formula is useful:

$$\sum_{n=0}^{L} F(a+nh) = \frac{1}{h}\int_a^b F(t)dt + \frac{1}{2}(F(a)+F(b)) + \frac{h}{2!}B_2(F'(b)-F'(a)) + O(h^2)$$

where $F(x)$ is a differentiable function in the interval $[a,b]$, $h \ll |a-b|$ is small, and $B_2 = 1/6$ is the second Bernoulli number.

14.3 Majorana fermions and the Ising model in 1+1 dimensions
In this problem, you will work with the theory of Majorana fermions discussed in section 14.7.

1) Show that in 1+1 dimensions, the propagators for the massive Majorana and Dirac fermions are the same. Find an explicit expression for the propagator $S_{\alpha\beta}(x-y)$ in the massless case, $m=0$, as a function of $x_\mu - y_\mu$ $(\nu = 0,1)$ in (1+1)-dimensional Minkowski spacetime.
2) Derive an explicit expression for the correlation function of the Majorana mass operator $i\chi_1(x)\chi_2(x)$ for the massless theory, $m=0$, as a function of $x_\mu - y_\mu$.
3) Now consider the one-dimensional quantum Ising model, whose Hamiltonian is given in eq. (14.45). Here you will use the Jordan-Wigner transformation to compute the equal-time correlation function $G(n,n') = \langle G|\sigma_1(n)\sigma_1(n')|G\rangle$ by mapping this expectation value to a correlation function involving the lattice fermion operators. Here $|G\rangle$ is the exact ground state of the Hamiltonian of the quantum Ising model with coupling constant g.
4) Now consider the expression you obtained in part 3 at the critical point $g=1$. Find the behavior of the correlation function at long distances. Show that the long-distance behavior is described by the correlation function of the Majorana mass operators in the massless theory. Use this calculation to identify the operator of the quantum Ising chain in terms of the operators of the theory of massless Majorana fermions.

15

The Renormalization Group

15.1 Scale dependence in quantum field theory and in statistical physics

In earlier chapters, we used perturbation theory to compute the N-point functions in field theory, focusing on ϕ^4 theory. There we used a time-honored approach that has been successful in such theories as quantum electrodynamics. Even at low orders in this expansion, we found difficulties in the form of divergent contributions at every order. In comparison with quantum mechanics, this problem may seem unusual and troubling. In quantum mechanics, Rayleigh-Schrödinger perturbation theory is finite term by term, although the resulting series may not be convergent (e.g., an expansion in powers of the coupling constant of the nonlinear term in an anharmonic oscillator). Instead, in QFT, one generally has divergent contributions, at every order in perturbation theory. To handle these divergent contributions, one introduces a set of effective (or renormalized) quantities. In this approach, the singular behavior is hidden in the relation between bare and renormalized quantities. This approach looks like it is sweeping the problem under a rug, and in fact, it is doing just that.

Quite early on (in 1954), Murray Gell-Mann and Francis Low (1954) proposed to take a somewhat different look at the renormalization process in QFT (QED in their case), by recasting the renormalization of the coupling constant as the solution of a differential equation. The renormalization process tells us that we can change the bare coupling λ and the UV cutoff Λ while, at the same time, keeping the renormalized coupling constant fixed at some value g. This process defines a function, the Gell-Mann-Low beta function $\beta(\lambda)$,

$$\beta(\lambda) = \Lambda \frac{\partial \lambda}{\partial \Lambda}\bigg|_g \tag{15.1}$$

which can be calculated from the perturbation theory diagrams. In fact, integrating this equation amounts to summing over a large class of diagrams at a given order of the loop expansion. The physical significance of this approach, which was subsequently and substantially extended by Nikolay Bogoliubov and Dmitry Shirkov (1959), remained obscure for quite some time. Things changed in a fundamental way with the work of Kenneth Wilson in the late 1960s.

Wilson proposed to take a different view on the meaning of the divergences found in QFT (Wilson, 1983). He argued that these divergent contributions have a physical origin. In a classical field theory, the theory is defined by a partial differential equation, such as Maxwell's equations. In this case, every dimensional quantity is determined by the dimensional parameters present in the partial differential equation. In other words, in a classical theory, the dimensional parameters of the equation of motion are the only physical scales. In contrast, in a QFT and in statistical mechanics, we cannot isolate a single scale as being responsible for the physical behavior of macroscopic quantities. For example, when we consider the one-loop corrections to the four-point vertex function $\Gamma^{(4)}$, we find that it contains an integral of the form

$$\int^{\Lambda} \frac{d^D p}{(2\pi)^D} \frac{1}{(p^2+m_0^2)^2} \tag{15.2}$$

where we introduced a regulator Λ to cut off the contributions from field configurations with momenta $|p| > \Lambda$. By inspection of this integral, we see that the correction to the effective (or renormalized) coupling constant has contributions not from a single momentum scale $|p|$ (or length scale $|p|^{-1}$) but from the entire range of momenta

$$\xi^{-1} = m_0 \leq |p| \leq \Lambda \tag{15.3}$$

or, equivalently, in terms of length scales, from the range of distances

$$a = \Lambda^{-1} \leq |x| \leq \xi = m_0^{-1} \tag{15.4}$$

What the integral is telling us is that we must add up all the contributions in this range. However, on one hand, clearly the contributions at small momenta ($|p| \approx 0$) are important, since in that range, the denominator of the integrand is smallest and, hence, that is where the integrand is largest. On the other hand, the contributions from momentum scales close to the cutoff, $|p| \approx \Lambda$, are even bigger, since the phase space is growing by a factor of Λ^D. Thus, we should expect divergences to occur, since there is no physical mechanism to stop fluctuations from happening on length scales between a short-distance cutoff a and a macroscopic scale ξ. It is important to note that this sensitivity to the definition of the theory at short distances will still be present even if local field theory is replaced at even shorter distances by some other theory (perhaps one more fundamental, such as string theory). At any rate, in all cases, there will remain a sensitivity to the definition of the theory at some short-distance scale used to define the local field theory. We will see shortly that these very same issues arise in the theory of phase transitions in statistical physics.

In QFT in Euclidean spacetime, the N-point functions are given in terms of the path integral, which for a scalar field ϕ has the form

$$Z[J] = \int \mathcal{D}\phi e^{-S_\Lambda[\phi] + \int d^D x J\phi} \tag{15.5}$$

where $S_\Lambda[\phi]$ is the Euclidean action for a theory defined with a regulator $\Lambda = a^{-1}$. To define a theory without a regulator amounts to taking the limit $\Lambda \to \infty$ or, equivalently, to taking the limit in which the short-distance cutoff $a \to 0$. If one regards the short-distance cutoff as a lattice spacing, this is the same as defining a *continuum limit*.

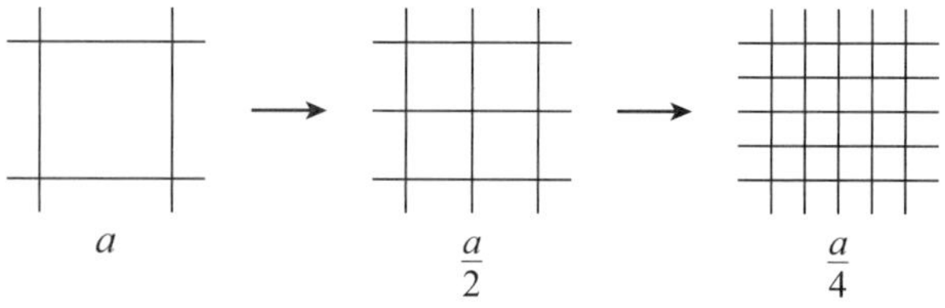

Figure 15.1 Taking the continuum limit. At each step, the number of degrees of freedom is doubled, the UV scale a is halved, and the grid becomes progressively denser.

In some sense, the procedure of removing the UV regulator from a field theory is reminiscent of the definition of the integral of a real function on a finite interval as the limit of a Riemann sum. In the case of the integral, the limit exists (provided the function has bounded variation). However, in the case of a functional integral, the problem is more complex. The problem of defining a field theory without a regulator is then reduced to knowing when, and how, it is possible to take such a limit. One can regard the field theory as being defined with a UV cutoff a (e.g., the "lattice spacing") and define a process of taking the continuum limit as the limit of a sequence of theories defined at progressively smaller UV length scales (e.g., from scale $a=\Lambda^{-1}$, to $\frac{a}{2}=\frac{2}{\Lambda}$, to $\frac{a}{4}=\frac{4}{\Lambda}$, etc). At each step, the number of local degrees of freedom is doubled, and the grid on which the degrees of freedom are defined becomes progressively more dense. In the limit of infinitely many such steps, it becomes a continuum (see figure 15.1). Kenneth Wilson had the deep insight to realize that for this continuum limit of a QFT to exist, it is necessary to *tune* the field theory to a particular value of its coupling constants at which the physical length scale of the theory (i.e., the correlation length ξ) should diverge in units of the short-distance UV cutoff a, $\xi \gg a$ (Wilson, 1983). Or, what is the same, we need to find a regime in which the mass gap $M=\xi^{-1}$ is much smaller than the UV momentum cutoff Λ: $M \ll \Lambda$. Phrased in this fashion, the problem of defining a QFT is reduced to finding a regime in which the physical correlation length is divergent. However, as we have seen, this is the same as the problem of finding a continuous phase transition in statistical physics! Thus, the two problems are equivalent.

15.2 RG flows, fixed points, and universality

In parallel, but independently from the development of these ideas in QFT, the problem of the behavior of physical systems near a continuous phase transition was being reexamined and developed by several people, notably by Widom (1965) and Fisher (1967), Patashinskii and Pokrovskii (1966), and, particularly, by Kadanoff (1966). Near a continuous phase transition ("second order" in Landau's terminology), certain physical quantities, such as the specific heat and the magnetic susceptibility, for a phase transition in a magnet should generally be divergent as the critical temperature T_c is approached. At T_c, the correlation length should be divergent. Using phenomenological arguments, these researchers realized that these properties can be described by a singular contribution to the free-energy density f_{sing} that has a *scaling form* (i.e., it is a homogeneous function of the reduced temperature, $t=(T-T_c)/T_c$, and of the external magnetic field H in dimensionless units). Implicit in these assumptions was that at the critical temperature T_c, where the correlation length diverges, the system exhibits scaling, or what is the same, the system acquires an emergent symmetry: *scale invariance*.

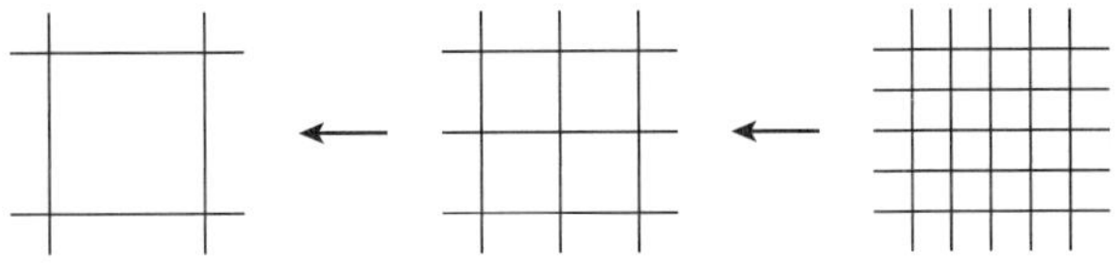

Figure 15.2 Coarse-graining the short-distance physics. In each step, the number of degrees of freedom is halved, the UV scale a is doubled, and the grid becomes progressively more sparse.

Kadanoff (1966) formulated these ideas in terms of the following physically intuitive picture. Consider for instance the simplest model of a magnet, the Ising model on a square lattice of spacing a. At each site, one has an Ising spin: a degree of freedom that can take only two values, $\sigma = \pm 1$. The partition function of this problem is a sum over all the configurations of spins, $[\sigma]$, weighed by the Boltzmann probability for each configuration at temperature T:

$$Z = \sum_{[\sigma]} \exp(-S[\sigma]/T) \tag{15.6}$$

In the Ising model, the Euclidean action $S[\sigma]$ is just the interaction energy of the spins,

$$S[\sigma] = -J \sum_{<i,j>} \sigma(i)\sigma(j) \tag{15.7}$$

where $< i, j >$ are nearest neighboring sites of the lattice, J is an energy scale (the exchange constant), and T is the temperature. In what follows, the energy of the classical statistical mechanical system will be called the "Euclidean action."

In a system with N sites, there are 2^N configurations of spins. The configurations that contribute predominantly at large distance scales, $|x| \gg a$, should be smooth at those scales. In particular, the physics at long distances should not depend sensitively on the physics at short distances. In fact, configurations that vary rapidly at short distances should tend to average out, and hence should make small contributions to the long-distance behavior. To make this picture concrete, Kadanoff proposed an iterative procedure for the computation of the partition function, which progressively sums over the short-distance degrees of freedom, resulting in an effective, renormalized, theory for the long-distance degrees of freedom. This procedure is an example of a renormalization group (RG) transformation.

15.2.1 Block-spin transformations

Beginning with a system with lattice spacing a, let us divide it into a set of block spins, so that the new, coarse-grained system will have lattice spacing $2a$. We then define a new effective spin for each block and compute its effective Hamiltonian by tracing over (integrating out) the rapidly varying configurations at scale a. In effect, this procedure is the same as that for the construction of the continuum field theory shown in figure 15.1, except that now we run the procedure backward, from short distances to long ones (see figure 15.2). To this end, let us divide the system into cells (the block spins), and let $\mathcal{A}$ be one of these cells. Define an effective degree of freedom μ for each cell

$$\mu = \frac{\sum_{i \in \mathcal{A}} \sigma(i)}{|| \sum_{i \in \mathcal{A}} \sigma(i) ||} \tag{15.8}$$

which represents the average over the configurations $\{\sigma\}$ that vary rapidly on the scale of the cell. Then the configurations on the block spins $\{\mu\}$, defined on a coarse-grained system with a new larger lattice spacing (say, $2a$), vary smoothly on short scales but not on long ones.

The next step is to define an effective Hamiltonian for the block spins $\{\mu\}$ through a block-spin transformation $T[\mu|\sigma]$ such that

$$\sum_{\{\mu\}} T[\mu|\sigma] = 1 \tag{15.9}$$

For example, one could take

$$T[\mu|\sigma] = \delta\left(\mu - \frac{\sum_{i\in\mathcal{A}} \sigma(i)}{||\sum_{i\in\mathcal{A}} \sigma(i)||}\right) \tag{15.10}$$

Once a specific transformation is chosen, the effective theory of the block spins is obtained by inserting eq. (15.9) in the partition function:

$$Z = \sum_{\{\sigma\}} e^{-S[\sigma]} = \sum_{\{\mu\}} \sum_{\{\sigma\}} T[\mu|\sigma]\, e^{-S[\sigma]} \tag{15.11}$$

which leads to the definition of an effective action, $S_{\text{eff}}[\mu]$, of the block spins:

$$e^{-S_{\text{eff}}[\mu]} = \sum_{\{\sigma\}} T[\mu|\sigma]\, e^{-S[\sigma]} \tag{15.12}$$

The partition function of the coarse-grained system is obviously the same as that of the original system

$$Z = \sum_{\{\mu\}} e^{-S_{\text{eff}}[\mu]} \tag{15.13}$$

even though, in general, the actions are not: $S_{\text{eff}}[\mu] \neq S[\sigma]$. As is apparent from this formal discussion, the form of the effective action S_{eff} depends on the block transformation chosen. Nevertheless, since S_{eff} results from summing over the contributions of a finite subset of degrees of freedom, it is a finite expression that can be written in terms of a set of local operators with effective (renormalized) coefficients.

Thus this coarse-graining procedure maps a system with lattice spacing a (the UV cutoff) with degrees of freedom σ to another system with a larger lattice spacing (say, $2a$) for the new degrees of freedom μ, while keeping the partition functions the same. As a result, the two-point correlation function of the coarse-grained degrees of freedom is related to the correlators of the old degrees of freedom by an expression of the form

$$\langle \mu(\boldsymbol{R})\mu(\boldsymbol{R}')\rangle_{S_{\text{eff}}} = \left\langle \frac{\sum_{i\in\mathcal{A}(\boldsymbol{R})} \sigma(i)}{||\sum_{i\in\mathcal{A}(\boldsymbol{R})} \sigma(i)||} \frac{\sum_{i\in\mathcal{A}(\boldsymbol{R}')} \sigma(i)}{||\sum_{i\in\mathcal{A}(\boldsymbol{R}')} \sigma(i)||} \right\rangle_S \tag{15.14}$$

Here $\boldsymbol{R}$ and $\boldsymbol{R}'$ are the locations of the two block spins. The right-hand side of eq. (15.14) is computed in a theory with lattice spacing a, and the left-hand side has lattice spacing $a' = ba$ ($b = 2$ in the above example). Now, suppose that the degrees of freedom σ of the theory of the right-hand side have correlation length $\xi[\sigma]$, and the correlation length of the coarse-grained degrees of freedom μ is $\xi[\mu]$. Since the theories are equivalent, they must be related simply by a change in scale

$$\xi[\mu] = \frac{1}{b}\xi[\sigma] \tag{15.15}$$

because the theory has b times fewer degrees of freedom. It is necessary to account for the change in scale in order to be able to compare the units of both theories (i.e., measurements must be made with the same ruler!).

15.2.2 RG flows and fixed points

Let us now assume that it will always be possible to define a complete set of (conveniently normalized) local operators, which we denote by $O_\alpha[\sigma]$, in terms of which a general action is

$$S[\sigma] = \sum_\alpha h_\alpha O_\alpha[\sigma] \tag{15.16}$$

where h_α are the coupling constants. For example, in the case of an Ising model, the local operators include the identity operator, the spin operator, the nearest-neighbor interactions, next-nearest-neighbor interactions, three- and four-spin interactions, and so forth. We will assume that the operators $O_\alpha[\sigma]$ are complete in the sense specified below.

We will also assume that the theory of the coarse-grained degrees of freedom μ is defined by the same set of local operators $O_\alpha[\mu]$,

$$S_{\text{eff}}[\mu] = \sum_\alpha h_\alpha^{\text{eff}}(b) O_\alpha[\mu] \tag{15.17}$$

where the $h_\alpha^{\text{eff}}(b)$ are a set of renormalized interactions that depend on $b = a'/a$, the change in the scale of the UV cutoff. The set of operators $\{O_\alpha[\sigma]\}$ is complete in the sense that its elements are the set of possible local operators generated under the RG transformation.

A consequence of this construction is that the coupling constants are no longer fixed but are different at different scales. This result implies that the RG transformation induces a mapping in the space of coupling constants. The repeated action of the RG can then be viewed as a *flow* in this space: the RG flow.

Let us suppose, for the moment, that we have been clever enough to choose a transformation such that the renormalized couplings are related to the old ones by a *homogeneous* transformation of the scale, that is,

$$h_\alpha^{\text{eff}}(b) = b^{y_\alpha} h_\alpha \tag{15.18}$$

which implicitly includes a change in length scale. The exponents y_α describe how the couplings transform under the change of scale from a to ba. We now see that if the exponent $y_\alpha > 0$, then the coupling will be larger in the coarse-grained system (with scale ba): $h_\alpha^{\text{eff}} > h_\alpha$. Conversely, if the exponent $y_\alpha < 0$, then the coupling will be smaller in the coarse-grained system: $h_\alpha^{\text{eff}} < h_\alpha$. We will say that an operator for which $y_\alpha > 0$ is a *relevant* operator, while one with $y_\alpha < 0$ is an *irrelevant* operator. The RG flow of the coupling constant h_α is given by the beta function $\beta(h_\alpha)$. In this context, the beta function is defined as

$$\beta(h_\alpha) \equiv (h_\alpha^{\text{eff}}(b) - h_\alpha)/\ln b \tag{15.19}$$

which measures the rate of change of the coupling h_α for a logarithmic change in the UV scale: $\ln b = \ln(a'/a)$. For $b \to 1$, the beta function reduces to

$$\beta(h_\alpha) = \frac{\partial h_\alpha}{\partial \ln b} = y_\alpha h_\alpha + O(h_\alpha^2) \tag{15.20}$$

which we will call the "tree-level" beta function. The special case of $y_\alpha = 0$ defines a *marginal* operator, and the beta function is given by higher-order terms that we have not yet included.

It is now clear that if we keep repeating this transformation n times, then in the limit $n \to \infty$, the contribution of the irrelevant operators disappears from the renormalized theory. In that limit, the Euclidean action of the effective theory contains only marginal and relevant operators. Following Wilson, we are led to conjecture that there should be theories that are invariant under the action of these transformations, called RG transformations. Such theories are said to be fixed points of the RG, and we denote their action by S^*. Then, a generic renormalized theory will have a Euclidean action of the form

$$S = S^* + \sum_\alpha h_\alpha \int d^D x\, O_\alpha[\sigma(x)] \tag{15.21}$$

which includes the fixed-point action, S^* plus the set of marginal and relevant operators.

The existence of theories described by such fixed points of the RG, conjectured by Wilson, is the key to understanding how to define a QFT. A theory described by such a fixed-point action S^* cannot depend on any microscopic scales, such as the UV cutoff (the lattice spacing a). Thus, at a fixed point, there are no scales left in the theory, and the theory becomes invariant under changes of scale (i.e., invariant under scale transformations). To achieve such a fixed point, it is necessary to rescale the lattice spacing ba of the renormalized theory described by S_{eff} back to what it was in the "bare" theory described by S. Hence, all lengths must be divided by b.

The construction of an RG transformation is summarized by the following two steps: (1) a block-spin (or coarse-graining) transformation that eliminates a finite fraction of local degrees of freedom, followed by (2) a rescaling of all lengths. Note that an RG transformation is for us to choose, and different choices (i.e., different ways of averaging over the short-distance physics) can always be made.

As noted, there are two types of fixed points, depending on whether the correlation length is zero or infinite. Clearly, the short-distance physics cannot be eliminated from the theory if the fixed point has zero correlation length. We will see that, in spite of this, such fixed points are important, since they turn out to label the different possible *phases* (or types of vacua, in field-theory jargon) of a theory. One such example is a theory with a global symmetry in its spontaneously broken symmetry phase: The effective field theory of such a phase contains dimensional quantities determined by the physics at some short-distance scale.

However, if the fixed point has a divergent (infinite) correlation length, the definition of the theory at short distances should become irrelevant at long distances, and the information on the short-distance physics disappears from a set of dimensionless quantities, which, in this sense, become *universal*. This observation leads to the concept of the existence of *universality classes* of theories, which are characterized by fixed points with the same universal properties. Once a fixed point of this type can be reached, the theory has a new, emergent symmetry: *scale invariance*. We will see shortly that, under most circumstances of physical interest, this symmetry can be extended to a larger emergent symmetry: *conformal invariance*. Importantly, a theory at a nontrivial fixed point and its set of marginal and

Figure 15.3 IR RG flow for one relevant operator. This is an IRU or, equivalently, a UV fixed point.

Figure 15.4 IR RG flow for one irrelevant operator. This is an IR stable fixed point, or equivalently, a UV unstable fixed point.

relevant operators defines a renormalizable QFT. From this perspective, each fixed point (and, hence, each universality class) defines a different QFT.

Let us consider now some simple examples of RG flows and fixed points.

Consider first a theory with a fixed point S^* with one relevant perturbation:

$$S = S^* + h \int d^D x \, O[\sigma(x)] \tag{15.22}$$

Under the RG with change of scale $b = a'/a$, we obtain

$$S' = S^* + b^y h \int d^D x \, O[\mu(x)] \tag{15.23}$$

with $y > 0$, for a relevant operator. The resulting IR RG flow is shown in figure 15.3. Here the fixed point at $h = 0$ is destabilized by the action of the relevant operator $O[\sigma(x)]$. Since we are flowing to long length scales, let us call this an IR unstable fixed point (IRU), known as a critical fixed point. Conversely, if we reverse the action of the RG and flow instead to short distances, we will say (equivalently!) that this is a UV stable fixed point (known as a UV fixed point).

The opposite flow is obtained if the operator is irrelevant and $y < 0$. Now the fixed point at $h = 0$ is stable in the IR (it is an IR stable fixed point), but it is unstable in the UV. This RG flow is shown in figure 15.4. Now the theory flows in the IR toward the fixed point at $h = 0$ for all $h \neq 0$. The *basin of attraction* of the fixed points is the set of values of the coupling for which the IR flow is toward the fixed point. In this sense, an IR stable fixed point defines a phase of the theory, while the IR unstable fixed point describes a continuous phase transition between two phases (to the right and to the left of the fixed point of figure 15.3). The IR flows of the two phases converge to IR stable fixed points located at $h = \pm\infty$, respectively.

Now consider a theory with two perturbations:

$$S = S^* + h_1 \int d^D x \, O_1[\sigma(x)] + h_2 \int d^D x \, O_2[\sigma(x)] \tag{15.24}$$

We have to consider three cases: (1) both operators are relevant, and hence, $y_1 > 0$ and $y_2 > 0$ (shown in figure 15.5a), (2) both are irrelevant, with $y_1 < 0$ and $y_2 < 0$ (shown in figure 15.5b), and (3) one operator has $y_1 > 0$ and is relevant, and the other has $y_2 < 0$ and is irrelevant (shown in figure 15.5c). While the cases in figure 15.5a and 15.5b are simple extensions of the examples of one-dimensional flows, the flows in figure 15.5c are richer. In this case, we see a manifold of values of the coupling constants that flows into the fixed point (driven by the flow of the irrelevant operator), while for other values, it flows away from this fixed point (driven now by the relevant operator). In particular, the manifold that

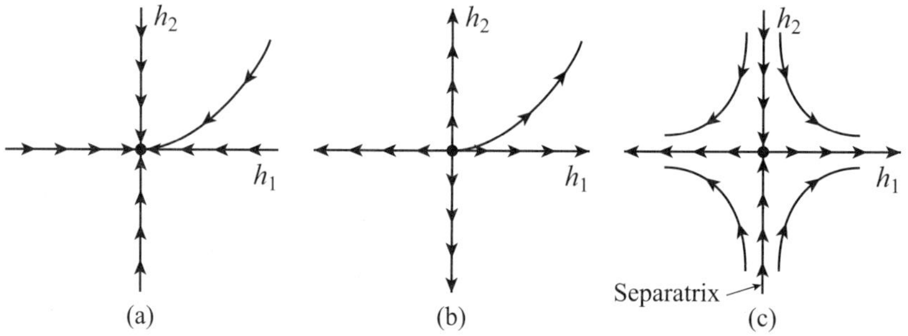

Figure 15.5 IR RG flows for (a) two irrelevant operators, (b) two relevant operators, and (c) one relevant and one irrelevant operator.

flows into the fixed point is a separatrix of the flow: Systems with bare values to the left and to the right of the separatrix flow to fixed points far away from the fixed point at the origin. Thus, a separatrix of the RG flow is a boundary between two different phases of the system, and it is the locus of phase transitions between them.

15.2.3 Simple examples of block-spin transformations

To make this discussion more concrete, let us now do some simple RG transformations, again using the Ising model as an example. The first example will be the trivial case of the Ising model in one dimension, a chain with N sites (with N even) and periodic boundary conditions. In this case, the block-spin transformation is quite simple: we will integrate out the spin degrees of freedom on the odd sub-lattice, and the spins on the even sub-lattice (with $N/2$ sites) are the block spins. This RG transformation is known as *decimation*. Thus, we begin with the partition function

$$Z = \sum_{\{\sigma\}} e^{-S[\sigma]} = \sum_{\{\sigma\}} e^{\sum_{j=1}^{N} \frac{1}{T}\sigma_j\sigma_{j+1}} \tag{15.25}$$

where we set the energy scale $J = 1$, such that $\sigma_{N+1} = \sigma_1$. Let us denote the spins on the even sub-lattice as $\sigma_{2r} \equiv \mu_r$ (with $r = 1, \ldots, N/2$) as the block spins. Thus, we begin with a system with N sites, lattice spacing a, and coupling constant $1/T$. We next integrate out the degrees of freedom on the odd sites, using

$$\sum_{\sigma_{2r+1}=\pm 1} e^{\frac{1}{T}\sigma_{2r+1}(\mu_r + \mu_{r+1})} = e^{\alpha + \frac{1}{T'}\mu_r\mu_{r+1}} \tag{15.26}$$

where

$$\alpha = \ln 2 + \frac{1}{2}\ln\cosh\left(\frac{2}{T}\right), \qquad T' = \frac{2}{\ln\cosh\left(\frac{2}{T}\right)} \tag{15.27}$$

Therefore, the effective action of the block spins $\{\mu_r\}$ is

$$S' = -\frac{N}{2}\alpha - \frac{1}{T'}\sum_{r=1}^{N/2} \mu_r\mu_{r+1} \tag{15.28}$$

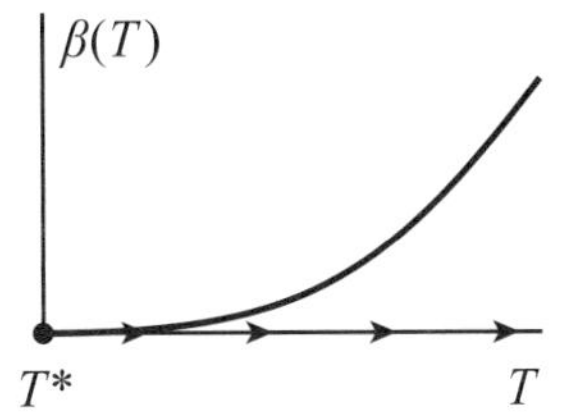

Figure 15.6 The beta function and the IR RG flow of the one-dimensional Ising model.

Thus, after the RG transformation, we obtain an action with the same form as S (up to a term proportional to the identity operator) but with a renormalized coupling T'. The change in the coupling (the temperature), $\delta T = T' - T$, for low values of $T \to 0$, is given by the beta function $\beta(T)$:

$$\beta(T) = \frac{\delta T}{\ln 2} = \frac{T^2}{2} + O(T^3) \qquad (15.29)$$

where $\ln 2 = \ln(a'/a)$ is the log of the change in the UV scale.

At a fixed point T^*, we must have $T'^* = T^*$, or what is the same, $\delta T^* = 0$. This clearly happens only at $T = T^* = 0$. Hence, the only finite fixed point (i.e., aside from $T^* \to \infty$) is at $T^* = 0$. Moreover, since the beta function of eq. (15.29) is always positive, the $T^* = 0$ fixed point is unstable, and the theory flows to the strong coupling fixed point at $T^* \to \infty$ for all $T > 0$. This RG flow is shown in figure 15.6. Notice that in this case, the beta function does not have a term linear in the coupling constant (T in this case). Thus, in this case, the operator perturbing the fixed point $T^* = 0$ is marginal. Since the coefficient of the leading term (quadratic, in this case) is positive, we say that this is a marginally relevant perturbation. Thus, in the one-dimensional problem, the RG flows in the IR to strong coupling, in this case $T \to \infty$, and there is no phase transition (as is well known).

Let us now look at a less trivial case and consider the Ising model in dimension $D > 1$. This problem is in general not exactly solvable (although it is in $D = 2$ dimensions; see chapter 14). The naive application of the decimation procedure we just used in one dimension leads to the generation of additional operators in the effective action of the form of next-nearest neighbor interactions, four-spin interactions, and so forth. However, there is a simple and instructive approximate approach, introduced by Migdal (1975a,b) and Kadanoff (1977), which makes the physics very apparent. It is a good approximation if the system is almost ordered and hence, for $T \ll 1$. The approximation consists of moving some of the bonds, as shown in figure 15.7, so that some of the spins can be integrated out without generating additional interactions. This is done in three steps: (a) move first half of the vertical bonds to the left, thus doubling the strength of the remaining interactions on these bonds; (b) integrate out the "middle spins" shown in figure 15.7b; and (c) move the resulting effective interaction along the vertical direction. This procedure is shown in figure 15.7. Let β_1 and β_2 be initial interactions along the vertical and horizontal bonds, respectively, where we have used the notation $\beta_1 = J_1/T$ and $\beta_2 = J_2/T$. After these three steps, the new interactions in both directions become

$$\beta_1' = \frac{1}{2} \ln \cosh(2\beta_1), \qquad \beta_2' = 2\beta_2 \qquad (15.30)$$

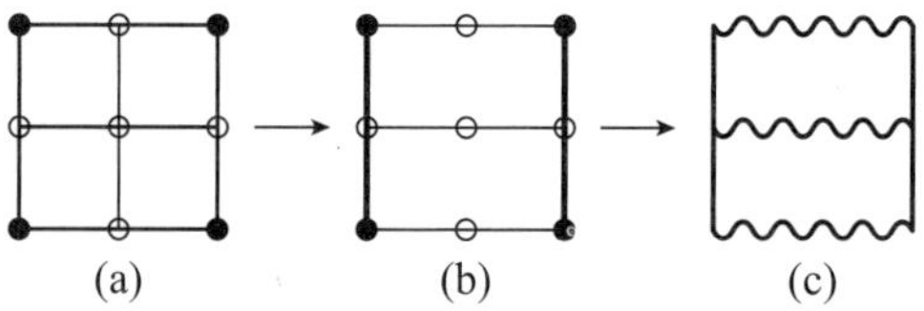

Figure 15.7 The Migdal-Kadanoff RG transformation for the two-dimensional Ising model. (a) The open circles are spins that will be integrated over. (b) The vertical bonds are moved to the left, and the strength of the remaining vertical bonds is doubled. (c) The middle spins on the horizontal bonds are integrated out, leading to the effective interactions represented by the wiggly lines. For details, see the text.

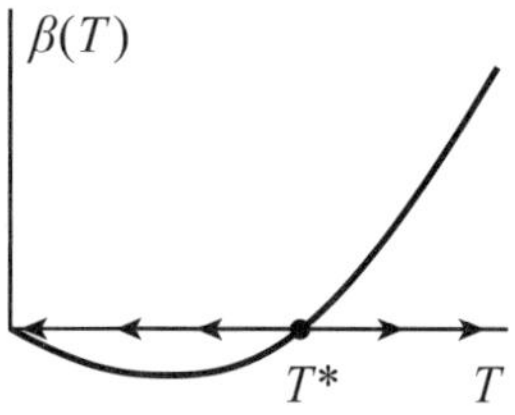

Figure 15.8 The beta function and the IR RG flow of the $d=(1+\epsilon)$-dimensional Ising model. This RG has two finite fixed points: the IR stable fixed point at $T=0$, and the IR unstable fixed point at $T^*=2\epsilon$, which labels the phase transition between the disordered phase at $T>T^*$ to the ordered (broken-symmetry) phase at $T<T^*$.

In D dimensions, this leads to a renormalized coupling

$$\frac{1}{T'} = 2^{D-2} \ln \cosh\left(\frac{2}{T}\right) \tag{15.31}$$

where we have ignored the artificial asymmetry induced by this procedure. This expression is very similar to the one-dimensional result of eq. (15.27). Let us now perform an analytic continuation in the dimensionality d and find an expression for the beta function in dimension $D=1+\epsilon$. The result is

$$\beta(T) = \frac{T'-T}{\ln 2} = -\epsilon T + \frac{T^2}{2} + O(T^3) \tag{15.32}$$

which is a generalization of eq. (15.27).

This beta function is shown in figure 15.8. The beta function has two zeros, at $T=0$ and at T^*, the two finite fixed points of this RG flow. These two fixed points have different characters and physical meanings. From the slope of the beta function in eq. (15.32) at the IR-unstable fixed point at T^*, we find that for a change of scale $b=a'/a=2$,

$$\delta T' = b^{\epsilon}\, \delta T \tag{15.33}$$

and hence, that here the exponent is $y=\epsilon=d-1>0$, which implies that the perturbation away from this fixed point is relevant. Conversely, at the IR-stable fixed point at $T=0$, the exponent is instead $y=-\epsilon$, and the perturbation is now irrelevant. As can be seen in this example, the notion of relevancy or irrelevancy of an operator depends on the fixed point and is not absolute.

Let us now determine the behavior of the correlation length ξ close to the nontrivial fixed point at T^*. After one action of this RG transformation, the correlation length becomes $\xi' = \xi/2$. Hence, after n iterations, we obtain $\xi_n = \xi/2^n$. Since $\delta T' = 2^y \delta T$, then after n iterations, it becomes $\delta T^{(n)} = 2^{ny}\delta T$. Thus, we can write

$$2^n = \left(\frac{\delta T^{(n)}}{\delta T}\right)^{1/y} = \frac{\xi}{\xi_n} \tag{15.34}$$

So if we begin the RG flow very close to T^*, such that $\delta T \ll T^*$, after n iterations, $\delta T^{(n)} \simeq T^*$ and $\xi_n \simeq a$, which is the UV cutoff (the lattice spacing). Hence, we deduce that the correlation length ξ must diverge as $\delta T \to 0$ and obey the scaling law

$$\xi(\delta T) = a \left(\frac{T^*}{\delta T}\right)^{\nu} \tag{15.35}$$

where the *critical exponent* ν is given by

$$\nu = \frac{1}{y} = \frac{1}{\epsilon} = \frac{1}{D-1} + O(\epsilon) \tag{15.36}$$

In other words, we now have an example of an IR unstable fixed point where the theory has a diverging correlation length (in units of the UV cutoff). Hence in a system of this type, it should be possible to construct a continuum field theory by tuning the theory to the scaling regime of its nontrivial fixed point. This is why fixed points of this type are also known as UV fixed points.

15.2.4 The Wilson-Fisher momentum-shell RG

We now look at a different type of RG transformation: the momentum shell RG introduced by Kenneth Wilson and Michael Fisher (1972). Instead of blocks of local degrees of freedom, we focus on momentum space and progressively integrate out the high-momentum modes of the field configurations (Wilson and Kogut, 1974). Here we focus on the case of Euclidean ϕ^4 theory, which at the classical level is equivalent to the Landau-Ginzburg theory of phase transitions. The action in D Euclidean dimensions is

$$S = \int d^Dx \left[\frac{1}{2}\left(\partial_\mu \phi\right)^2 + \frac{t}{2}\Lambda^2\phi^2 + \frac{u}{4!}\Lambda^\epsilon \phi^4\right] \tag{15.37}$$

where we have defined the mass m_0^2 and the coupling constant λ in terms of the UV cutoff Λ as $m_0^2 = t\Lambda^2$ and $\lambda = u\Lambda^\epsilon$, where t and u are dimensionless, and $\epsilon = 4 - D$.

This approach begins by splitting the field into slow and fast components, denoted by $\phi_<$ and $\phi_>$, respectively:

$$\phi(x) = \phi_<(x) + \phi_>(x) \tag{15.38}$$

For a theory with a UV cutoff Λ (which we assume is imposed using a smooth cutoff procedure), we split the momentum space into a (thin) shell of momenta $b\Lambda < |p| < \Lambda$ (with $b < 1$) and the rest, $|p| < b\Lambda$ (as shown in figure 15.9). The fast fields $\phi_>$ have support only on the momentum shell, while the slow fields $\phi_<$ have support on the rest of the momentum sphere:

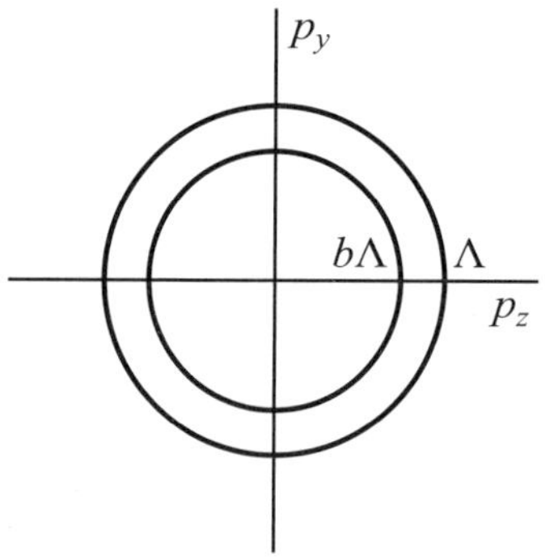

Figure 15.9 The momentum shell $b\Lambda < |p| < \Lambda$, with $b < 1$.

$$\phi_<(x) = \int_{|p|<b\Lambda} \frac{d^D p}{(2\pi)^D} \phi(p) e^{ip.x} \tag{15.39}$$

$$\phi_>(x) = \int_{b\Lambda<|p|<\Lambda} \frac{d^D p}{(2\pi)^D} \phi(p) e^{ip.x} \tag{15.40}$$

Written in terms of the fast and slow fields, the action of eq. (15.37) takes the form

$$S[\phi] = S_<[\phi_< +]S_>[\phi_>] + S_{\text{int}}[\phi_<, \phi_>] \tag{15.41}$$

where we have defined

$$S_<[\phi_<] \equiv \int d^D x \left[\frac{1}{2} \left(\partial_\mu \phi_<\right)^2 + \frac{t}{2} \Lambda^2 \phi_<^2 + \frac{u}{4!} \Lambda^\epsilon \phi_<^4 \right], \tag{15.42}$$

$$S_<[\phi_>] \equiv \int d^D x \left[\frac{1}{2} \left(\partial_\mu \phi_>\right)^2 + \frac{t}{2} \Lambda^2 \phi_>^2 + \frac{u}{4!} \Lambda^\epsilon \phi_>^4 \right], \tag{15.43}$$

$$S_{\text{int}}[\phi_<, \phi_>] \equiv \int d^D x \frac{u}{4!} \Lambda^\epsilon \left[4\phi_> \phi_<^3 + 6\phi_>^2 \phi_<^2 + 4\phi_>^3 \phi_< \right]. \tag{15.44}$$

Notice that the fast and slow fields do not mix at the quadratic (free-field) level, since

$$\begin{aligned} \int d^D x \frac{1}{2} \left[(\partial_\mu \phi)^2 + t\Lambda^2 \phi^2 \right] &= \int_{|p|<\Lambda} \frac{d^D p}{(2\pi)^D} \frac{1}{2} (p^2 + t\Lambda^2) \phi(p) \phi(-p) \\ &= S_<[\phi_<] + S_>[\phi_>] \end{aligned} \tag{15.45}$$

The next step is to obtain an effective action for the slow fields (with UV cutoff $b\Lambda$), which we will denote by $S_{\text{eff},<}^{b\Lambda}[\phi_<]$. We do this by integrating out the fast fields $\phi_>$ as follows:

$$\begin{aligned} Z &= \int \mathcal{D}\phi \, e^{-S^\Lambda[\phi]} = \int \mathcal{D}\phi_< \mathcal{D}\phi_> \, e^{-S^\Lambda[\phi_>, \phi_<]} \\ &= \int \mathcal{D}\phi_< \, e^{-S_<^{b\Lambda}[\phi_<]} \int \mathcal{D}\phi_> \, e^{-S_>^\Lambda[\phi_>] - S_{\text{int}}^\Lambda[\phi_>, \phi_<]} \end{aligned} \tag{15.46}$$

which defines the effective action $S^{b\Lambda}_{\text{eff},<}[\phi_<]$ of the slow fields in a theory with cutoff $b\Lambda$:

$$e^{-S^{b\Lambda}_{\text{eff},<}[\phi_<]} \equiv e^{-S^{\Lambda}_{<}[\phi_<]} \int \mathcal{D}\phi_> \, e^{-S^{\Lambda}_{>}[\phi_>] - S^{\Lambda}_{\text{int}}[\phi_>, \phi_<]} \tag{15.47}$$

The path integral in this expression now has to be computed. This we do by using perturbation theory in u, which is equivalent to working at one-loop order in the fluctuations of the fast fields. To this end, write the action of the fast fields as a sum of a massive free fast field action and an interaction term,

$$S^{\Lambda}_{>}[\phi_>] = S^{\Lambda}_{0,>}[\phi_>] + S^{\Lambda}_{\text{int},>}[\phi_>] \tag{15.48}$$

We find

$$\int \mathcal{D}\phi_> \, e^{-S^{\Lambda}_{>}[\phi_>] - S^{\Lambda}_{\text{int}}[\phi_>, \phi_<]} = Z_{0,>} \sum_{n=0}^{\infty} \frac{(-1)^n}{n!} \left(-\frac{u\Lambda^{\epsilon}}{4!} \right)^n I_n[\phi_<] \tag{15.49}$$

where $Z_{0,>} = \exp(-F^0_>)$ is the partition function for free fast fields defined in the momentum shell, and where we define $I_n[\phi_<]$ as

$$I_n[\phi_<] = \int_{\{x_j\}} \left\langle \prod_{j=1}^{n} \left[\phi^4_>(x_j) + 4\phi^3_>(x_j)\phi_<(x_j) + 6\phi^2_>(x_j)\phi^2_<(x_j) \right.\right.$$
$$\left.\left. + 4\phi_>(x_j)\phi^3_<(x_j) + \phi^4_<(x_j) \right] \right\rangle_{0,>} \tag{15.50}$$

where $\langle \cdots \rangle_{0,>}$ represents an expectation value on the free fast fields. Below we will use that, by symmetry, $\langle \phi^{2k+1}_>(x) \rangle_{0,>} = 0$.

To first (lowest) order in u, we find

$$I_1 = \int d^D x \left[\langle \phi^4_>(x) \rangle_{0,>} + 6\langle \phi^2_>(x) \rangle_{0,>} \phi^2_<(x) \right] \tag{15.51}$$

Similarly, I_2 is given by the expression

$$I_2[\phi_<] = \int d^D x_1 \int d^D x_2 \Big[\langle \phi^4_>(x_1)\phi^4_>(x_2) \rangle_{0,>} + 12\langle \phi^4_>(x_1)\phi^2_>(x_2) \rangle_{0,>} \phi^2_<(x_2)$$
$$+ 16\langle \phi^3_>(x_1)\phi^3_>(x_2) \rangle_{0,>} \phi_<(x_1)\phi_<(x_2)$$
$$+ 36\langle \phi^2_>(x_1)\phi^2_>(x_2) \rangle_{0,>} \phi^2_<(x_1)\phi^2_<(x_2)$$
$$+ 16\langle \phi_>(x_1)\phi_>(x_2) \rangle_{0,>} \phi^3_<(x_1)\phi^3_<(x_2)$$
$$+ 32\langle \phi^3_>(x_1)\phi_>(x_2) \rangle_{0,>} \phi_<(x_1)\phi^3_<(x_2) \Big] \tag{15.52}$$

Using these results, we find that, up to cubic order in the dimensionless coupling constant u, the effective action $S^{b\Lambda}_{\text{eff},<}[\phi_<]$ is

$$S^{b\Lambda}_{\text{eff},<}[\phi_<] = S^{b\Lambda}_{<}[\phi_<] + \left(\frac{u\Lambda^\epsilon}{4!}\right) I_1 - \frac{1}{2!}\left(\frac{u\Lambda^\epsilon}{4!}\right)^2 (I_2 - I_1^2) + O(u^3) \tag{15.53}$$

where I_1 and I_2 are given by eq. (15.51) and eq. (15.52), respectively. The explicit form of the effective action is

$$\begin{aligned}
S^{b\Lambda}_{\text{eff},<}[\phi_<] = & \int d^Dx \left[\frac{1}{2}(\partial_\mu \phi_<)^2 + \frac{t}{2}\Lambda^2 \phi_<^2 + \frac{u}{4!}\Lambda^\epsilon \phi_<^4\right] \\
& + \left[F^0_> + \left(\frac{u\Lambda^\epsilon}{4!}\right) \int d^Dx \langle \phi_>^4(x)\rangle_{0,>} - \frac{1}{2}\left(\frac{u\Lambda^\epsilon}{4!}\right)^2 \int_{x_1,x_2} \langle :\phi_>^4(x_1) :: \phi_>^4(x_2):\rangle_{>,0}\right] \\
& + \Bigg\{ \left(\frac{u\Lambda^\epsilon}{4!}\right) \int d^Dx\, 6 \langle \phi_>^2(x)\rangle_{>,0} \phi_<^2(x) \\
& - \frac{1}{2}\left(\frac{u\Lambda^\epsilon}{4!}\right)^2 \int_{x_1,x_2} \Big[36 \langle :\phi_>^2(x_1) :: \phi_>^2(x_2):\rangle_{>,0} \phi_<^2(x_1)\phi_<^2(x_2) \\
& + 12 \langle :\phi_>^4(x_1) :: \phi_>^2(x_2):\rangle_{>,0} \phi_<^2(x_2) + 16 \langle \phi_>(x_1)\phi_>(x_2)\rangle_{>,0} \phi_<^3(x_1)\phi_<^3(x_2) \\
& + 16 \langle \phi_>^3(x_1)\phi_>^3(x_2)\rangle_{>,0} \phi_<(x_1)\phi_<(x_2) + 32 \langle \phi_>^3(x_1)\phi_>(x_2)\rangle_{>,0} \phi_<(x_1)\phi_<^3(x_2) \Big] \\
& + O(u^3) \Bigg\}
\end{aligned} \tag{15.54}$$

where $F^0_> = -\ln Z_{0,>}$, and the notation $:A: = A - \langle A\rangle_{0,>}$ is used for the normal-ordered operator A with respect to the fast free-field theory.

Let us write the expression for the effective action for the slow fields in terms of a set of local operators of the form $\phi_<^n$ and derivatives (e.g., $\partial_i\partial_j\phi_<^n$). We can do this since the kernels that appear in eq. (15.54) involve only the correlators of the fast fields, which are rapidly decaying functions of the coordinates (provided the cutoff function used to restrict the fields to the momentum shell is smooth).

Let us consider, for example, the contribution

$$16 \int_{x_1,x_2} \langle \phi_>^3(x_1)\phi_>^3(x_2)\rangle_{>,0} \phi_<(x_1)\phi_<(x_2) =$$

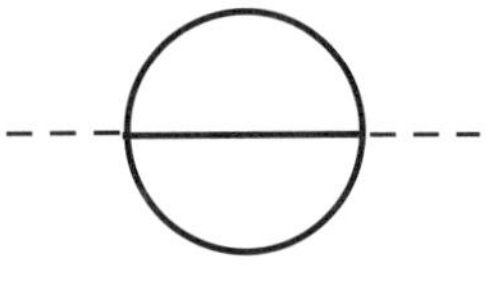

$$= 96 \int dx_1^D dx_2^D \phi_<(x_1)\phi_<(x_2) \left[G_{0,>}(x_1 - x_2)\right]^3 \tag{15.55}$$

where the internal momenta of the Feynman diagram are in the momentum shell (the solid line), and the external momenta (the broken lines) are outside the momentum shell. Here, $G_{0,>}(x_1 - x_2)$ is the Euclidean propagator of the fast free fields:

$$G_{0,>}(x_1 - x_2) = \int_{b\Lambda<|p|<\Lambda} \frac{d^Dp}{(2\pi)^D} \frac{e^{ip\cdot(x_1-x_2)}}{p^2 + t\Lambda^2} \tag{15.56}$$

We will assume that the cutoff implied by the restriction to the momentum shell is smooth enough so that this propagator decays exponentially with distance. Under these

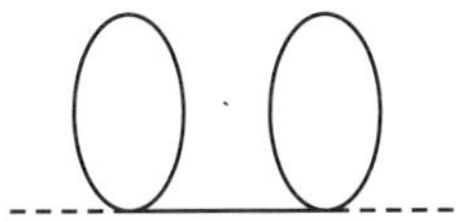

Figure 15.10 Reducible Feynman diagrams with internal momenta in the momentum shell vanish identically by momentum conservation.

assumptions, it is legitimate to perform inside the integral shown in eq. (15.55) a Taylor expansion of the slow field $\phi_<(x_2)$ about the coordinate x_1:

$$\phi_<(x_2) = \phi_<(x_2 + \ell) = \phi_<(x_1) + \ell_i \partial_i \phi_<(x_1) + \frac{1}{2}\ell_i \ell_j \partial_i \partial_j \phi_<(x_1) + \cdots \tag{15.57}$$

We can then write the integral on the right-hand side of eq. (15.55) as

$$\begin{aligned}&\int dx_1^D dx_2^D \, \phi_<(x_1)\phi_<(x_2) \left[G_{0,>}(x_1 - x_2)\right]^3 = \\ &= \int d^D x \, \phi_<^2(x) \int d^D \ell \left[G_{0,>}(\ell)\right]^3 - \frac{1}{D}\left[\int d^D \ell \, \ell^2 \left[G_{0,>}(\ell)\right]^3\right] \frac{1}{2}\left(\partial_\mu \phi_<(x)\right)^2 + \cdots\end{aligned} \tag{15.58}$$

Therefore, the contribution of eq. (15.55) is part of the mass renormalization and the wave-function renormalization.

Notice that reducible diagrams cannot appear in the effective actions for the slow fields: They vanish automatically, because the internal momenta are inside the momentum shell and the external momenta are not (as in the example shown in figure 15.10).

Thus, up to this order in an expansion in powers of the dimensionless coupling constant u, the effective action of the slow fields for a theory with UV cutoff $b\Lambda$ becomes

$$S_{\text{eff},<}^{b\Lambda}[\phi_<] = \int d^D x \left[\frac{A}{2}\left(\partial_\mu \phi_<\right)^2 + B\phi_<^2(x) + C\phi_<^4(x) + \text{const.} + \cdots\right] \tag{15.59}$$

where $\cdots$ denotes operators with higher powers of $\phi_<$ and/or higher derivatives (only even powers and even derivatives will appear by symmetry). The coefficients A, B, and C are

$$\begin{aligned}A &= 1 + 48\left(\frac{u\Lambda^\epsilon}{4!}\right)^2 \frac{1}{D}\int d^D \ell \left[G_{0,>}(\ell)\right]^3 \ell^2 + O(u^3) \\ B &= \frac{t}{2}\Lambda^2 + 6\left(\frac{u\Lambda^\epsilon}{4!}\right) G_{0,>}(0) - 72\left(\frac{u\Lambda^\epsilon}{4!}\right)^2 G_{0,>}(0) \int d^D \ell \left[G_{0,>}(\ell)\right]^2 \\ &\quad - 48\left(\frac{u\Lambda^\epsilon}{4!}\right)^2 \int d^D \ell \left[G_{0,>}(\ell)\right]^3 + O(u^3) \\ C &= \frac{u\Lambda^\epsilon}{4!} - 36\left(\frac{u\Lambda^\epsilon}{4!}\right)^2 \int d^D \ell \left[G_{0,>}(\ell)\right]^2 - 48\left(\frac{u\Lambda^\epsilon}{4!}\right)^2 G_{0,>}(0) \int d^D \ell \, G_{0,>}(\ell) \\ &\quad + O(u^3)\end{aligned} \tag{15.60}$$

Here we have ignored the constant terms, which contribute to the renormalization of the identity operator.

The remaining integrals can be done quite easily. In the limit of a very thin shell (compared to the cutoff scale Λ) and writing $b = e^{-\delta s} \to 1^-$ as $\delta s \to 0^+$, the integrals become

$$G_{0,>}(0) = \int_{\text{shell}} \frac{d^D p}{(2\pi)^D} \frac{1}{p^2 + t\Lambda^2} = \frac{\Lambda^{D-2}}{1+t} \frac{S_D}{(2\pi)^D} \delta s$$

$$\int d^D \ell \, G_{0,>}(\ell) = \int_{\text{shell}} \frac{d^D p}{(2\pi)^D} \frac{(2\pi)^D \delta^D(p)}{p^2 + t\Lambda^2} = 0$$

$$\int d^D \ell \left[G_{0,>}(\ell)\right]^2 = \frac{\Lambda^{D-4}}{(1+t)^2} \frac{S_D}{(2\pi)^D} \delta s$$

$$\int d^D \ell \left[G_{0,>}(\ell)\right]^3 = \frac{\Lambda^{2D-6}}{(1+t)^3} \frac{S_D}{(2\pi)^D} \delta s$$

$$\int d^D \ell \left[G_{0,>}(\ell)\right]^3 \ell^2 = \Lambda^{2D-8} \left[\frac{2D}{(1+t)^4} - \frac{8}{(1+t)^5}\right] \frac{S_D}{(2\pi)^D} \delta s \tag{15.61}$$

where

$$S_D = \frac{2\pi^{D/2}}{\Gamma(D/2)} \tag{15.62}$$

is the area of the hypersphere, and $\Gamma(x)$ is the gamma function.

The last step is to define a rescaled field

$$\phi' = Z_\phi^{-1/2} \phi_< \tag{15.63}$$

which we readily identify with the wave-function renormalization, and to rescale the coordinates,

$$x' = bx \tag{15.64}$$

so that the cutoff scale in the new rescaled coordinates is back to Λ. The resulting action for the (rescaled) field ϕ' in the rescaled coordinates x' is

$$\begin{aligned} S'(\phi') &= S_{\text{eff},<}^{b\Lambda}(\phi_<) \\ &= \int d^D x' b^{-D} \left[\frac{A}{2} Z_\phi b^2 \left(\partial'_\mu \phi'\right)^2 + B Z_\phi \phi'^2 + C Z_\phi^2 \phi'^4\right] \end{aligned} \tag{15.65}$$

and we require that the renormalization conditions

$$1 = b^{2-D} A Z_\phi \tag{15.66}$$

$$\frac{t'}{2} \Lambda^2 = b^{-D} B Z_\phi \tag{15.67}$$

$$\frac{u'}{4!} \Lambda^\epsilon = b^{-D} C Z_\phi^2 \tag{15.68}$$

be satisfied. By keeping only the leading corrections to t and u, we obtain

$$Z_\phi = b^{D-2}A^{-1} = b^{D-2}\left(1 + O(u^2)\right) \tag{15.69}$$

$$t' = b^{-2}\left[t + \frac{1}{2}(u - ut)\frac{S_D}{(2\pi)^D}\delta s + O(u^2, t^2)\right] \tag{15.70}$$

$$u' = b^{-\epsilon}\left[u - \frac{3}{2}u^2\frac{S_D}{(2\pi)^D}\delta s + O(u^3, t^2)\right] \tag{15.71}$$

Notice that, to this leading order, the wave-function renormalization is trivial. Also notice the important fact that all dependence from the momentum cutoff scale Λ has canceled exactly. This is a generic feature of RG transformations. Finally, these equations simplify if we absorb the phase-space factors in a new dimensionless coupling constant v:

$$v = u\frac{S_D}{(2\pi)^D} \tag{15.72}$$

The (Gell-Mann-Low) beta functions for the dimensionless couplings t and v are defined to be

$$\begin{aligned}\beta_t &= \frac{dt}{ds} = a\frac{dt}{da} = -\Lambda\frac{dt}{d\Lambda}\\ \beta_v &= \frac{dv}{ds} = a\frac{dv}{da} = -\Lambda\frac{dv}{d\Lambda}\end{aligned} \tag{15.73}$$

Using the results just derived, we find that the beta functions are

$$\beta_t = 2t + \frac{v}{2} - \frac{1}{2}vt + \cdots \tag{15.74}$$

$$\beta_v = \epsilon v - \frac{3}{2}v^2 + \cdots \tag{15.75}$$

to leading orders in the couplings t and v, and in $\epsilon = 4 - D$. Therefore, this *perturbative* RG transformation is accurate only to order ϵ and is the beginning of a series in powers of ϵ. This is the ϵ-expansion (Wilson and Kogut, 1974).

In $D < 4$ dimensions, the RG equations (eq. (15.74) and eq. (15.75)) define an RG flow. In $D < 4$ dimensions, this RG flow has two fixed points: (1) the free-field (or Gaussian) fixed point at $t = u = 0$, and (2) the nontrivial fixed point, the Wilson-Fisher fixed point, at $u^* = \frac{2}{3}\epsilon$ and $t^* = -\frac{\epsilon}{6}$. The RG flow of the coupling constant is shown in figure 15.11a, b. The full RG flow is shown in figure 15.11c. If $D < 4$, the free-field fixed point is completely unstable in the IR. In contrast, the Wilson-Fisher fixed point, which only exists if $D < 4$, is bistable: there is a trajectory from the free-field fixed point flows to the Wilson-Fisher fixed point, and another trajectory on which the IR flows away from the Wilson-Fisher fixed point toward the $t \to \pm\infty$ regimes. In these regimes, the theory describes a symmetric phase (for $t \to +\infty$) and a broken symmetry phase (for $t \to -\infty$). Hence, for $D < 4$, the free-field fixed point has two relevant operators (with coupling constants t and u), whereas the Wilson-Fisher fixed point has one irrelevant and one relevant operator. Notice that the Wilson-Fisher fixed point is at a finite value of $t^* < 0$, which reflects the mass renormalization at the one-loop level. But as $D \to 4$ dimensions, the Wilson-Fisher fixed point coalesces with the free-field fixed

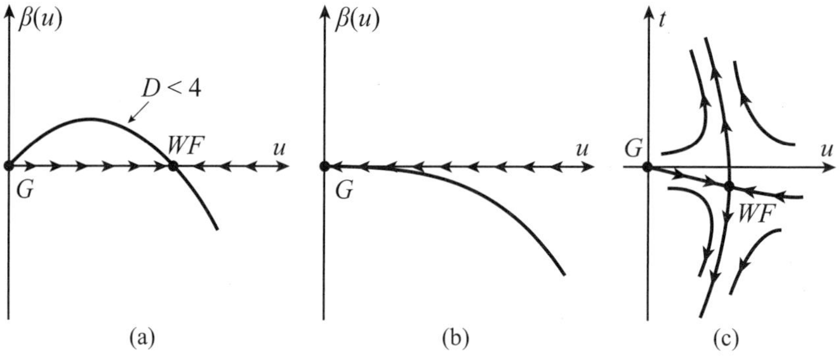

Figure 15.11 RG beta function for the coupling constant v in (a) $D<4$ dimensions and (b) $D=4$ dimensions. The full RG flow is shown in (c). Here t is the dimensionless mass, $m^2=t\Lambda^2$, u is the dimensionless coupling constant, $\lambda=u\Lambda^{\epsilon}$, and WF is the Wilson-Fisher fixed point.

point, which becomes the only remaining fixed point. As a result, in $D=4$ dimensions, the free-field fixed point is bistable in the IR: It is *marginally stable* in the coupling constant u and unstable in t.

The RG flow for the coupling constants determines the behavior of the correlation length ξ. By dimensional analysis, we can always write ξ in the form

$$\xi=\Lambda^{-1}f(t,u) \tag{15.76}$$

where $f(t,u)$ is a dimensionless function of t and u. Consider now two physical systems defined with different values of the UV scale Λ but with the same value of the physical scale ξ. Since ξ is fixed, we can write

$$0=\frac{\partial\xi}{\partial\Lambda}=-\Lambda^{-2}f(t,u)+\Lambda^{-1}\frac{\partial f}{\partial\Lambda} \tag{15.77}$$

However, $f(t,u)$ does not depend explicitly on Λ, but it depends implicitly on the scale through the RG flow. Hence,

$$\frac{\partial f}{\partial\Lambda}=\frac{\partial f}{\partial u}\frac{\partial u}{\partial\Lambda}+\frac{\partial f}{\partial t}\frac{\partial t}{\partial\Lambda} \tag{15.78}$$

By using the definitions of the beta functions, we find that the function $f(t,u)$ must obey the partial differential equation

$$f+\frac{\partial f}{\partial u}\beta_u+\frac{\partial f}{\partial t}\beta_t=0 \tag{15.79}$$

This is an example of a *Callan-Symanzik equation*.

Let us solve eq. (15.79) for the unstable trajectory of the Wilson-Fisher fixed point. Close to the fixed point, we can work with the linearized flow expressed in terms of the variables x and y,

$$x=4(t-t^*)+\left(1-\frac{\epsilon}{6}\right)(v-v^*),\qquad y=v-v^* \tag{15.80}$$

Figure 15.12 RG flows along the scaled variable x with two initial values, x_1 and x_2, and the same final value $\bar{x}$.

where (v^*, t^*) is the location of the Wilson-Fisher fixed point. The linearized beta functions are

$$\beta_x = \left(2 - \frac{\epsilon}{3}\right) x + \ldots, \qquad \beta_y = -\epsilon y + \cdots \tag{15.81}$$

In these coordinates, the fixed point is at $(0,0)$. Since the flow in y is irrelevant (it flows into the fixed point), we set $y = 0$ and focus on the flow away from the fixed point along x.

The Callan-Symanzik equation to be solved is now

$$0 = f + \frac{\partial f}{\partial x} \beta_x \tag{15.82}$$

The general solution of this equation is

$$\ln f = \text{const} - \int \frac{dx}{\beta_x} \tag{15.83}$$

For a flow beginning at some small value $x_0 \to 0$ and ending at some finite value x, of $O(1)$, this solution is

$$f(x) = f(x_0) \exp\left[- \int_{x_0}^{x} \frac{dx'}{\beta_x(x')} \right] \tag{15.84}$$

Let us consider two flows along x with different starting points, x_1 and x_2, and with the same final point $\bar{x}$, with the *same* UV cutoff Λ_0 (shown in figure 15.12). The value of the correlation length ξ at x_1 and x_2 is

$$\xi(x_1) = \Lambda_0^{-1} f(x_1), \qquad \xi(x_2) = \Lambda_0^{-1} f(x_2) \tag{15.85}$$

where

$$\begin{aligned} f(x_1) &= f(\bar{x}) \exp\left(\int_{x_1}^{\bar{x}} \frac{dx}{\beta(x)} \right) \\ f(x_2) &= f(\bar{x}) \exp\left(\int_{x_2}^{\bar{x}} \frac{dx}{\beta(x)} \right) \end{aligned} \tag{15.86}$$

Hence, we find

$$\xi(x_1) = \xi(x_2) \exp\left(\int_{x_1}^{x_2} \frac{dx}{\beta(x)} \right) \tag{15.87}$$

Suppose now that x_1 is very close to the fixed point ($x_1 \to 0$) and that x_2 is far enough away so that $\xi(x_1) \simeq a$ (the short-distance cutoff). Using the linearized beta function, we find that

$$\xi(x_1) \simeq a \exp\left(\int_{x_1}^{x_2} \frac{dx}{(2 - \frac{\epsilon}{3})x} \right) = a \left| \frac{x_1}{x_2} \right|^{-\nu} \tag{15.88}$$

where

$$\nu = \frac{1}{\beta'(0)} = \frac{1}{2 - \frac{\epsilon}{3}} = \frac{1}{2} + \frac{\epsilon}{12} + O(\epsilon^2) \tag{15.89}$$

Hence, the value of the critical exponent ν of the correlation length is determined by the slope of the beta function at the Wilson-Fisher fixed point, which at the present level of approximation is $\beta'(0) = 2 - \frac{\epsilon}{3} + O(\epsilon^2)$.

15.3 General properties of a fixed-point theory

At a fixed point of an RG transformation, the theory has an emergent symmetry: scale invariance. This is a new, emergent symmetry that is operative at length scales that are long compared to a short-distance cutoff a (the UV regulator) but short compared to the linear size L of the system (the IR regulator). No other scales are present at the fixed point. In chapter 21, we will see that, in most cases of interest, in QFT, scale invariance can be extended to a larger emergent symmetry: conformal invariance (i.e., invariance under conformal coordinate transformations $x' = f(x)$ such that angles between vectors are preserved). In chapter 21, we will see that most of the statements made in this section follow from conformal invariance. In this section, we will explore the consequences of these assumptions for the properties of theories at a fixed point. The discussion presented here follows two excellent sources: Paul Ginsparg's 1988 Les Houches lectures (Ginsparg, 1989) and John Cardy's book on scaling and renormalization (Cardy, 1996).

In a scale-invariant theory, the action (or, more precisely, the partition function) is invariant under scale transformations, $x \mapsto x' = \lambda x$. In such a theory, the expectation values of the observables transform under dilations as *homogeneous functions*. A function $F(x)$ is a homogeneous function of degree k if $F(\lambda x) = \lambda^k F(x)$.

From the scale invariance of the partition function, it follows that at a general fixed point (whose action we denote by S^*), the correlators of a physical observable $\mathcal{O}(x)$ must be homogeneous functions of the distance $|x - y|$ and should obey power laws of the form

$$\langle \mathcal{O}(x)\mathcal{O}(y)\rangle^* = \frac{\text{const.}}{|x-y|^{2\Delta_{\mathcal{O}}}} \tag{15.90}$$

Here $\langle A\rangle^*$ denotes the expectation value of the observable A at the fixed point S^*. The quantity $\Delta_{\mathcal{O}}$ is called the *scaling dimension* of the local operator $\mathcal{O}$. Hence, at the fixed point, the operator $\mathcal{O}$ has units of

$$[\mathcal{O}] = \ell^{-\Delta_{\mathcal{O}}} \tag{15.91}$$

where ℓ is a length scale.

The scaling dimension of the observables is one of the universal quantities that define a fixed point. A universal quantity here means a quantity whose value is independent of the short-distance definition of the theory (i.e., of the particular definition in the UV). Clearly, dimensional quantities are by definition not universal, since a microscopic scale is needed to define the units, and different choices will lead to different values. Hence, only dimensionless quantities, such as the scaling dimensions, can be universal.

At a general fixed point, the scaling dimensions of the operators must be positive real numbers, so that the correlators of local observables obey cluster decomposition and must decay (as a power law in this case) at large separations. In exceptional cases, the scaling dimensions take rational values. This is the case for scale-invariant free-field theories and also for integrable systems, such as the minimal models of $D=2$ dimensional conformal field theories, which will be discussed in chapter 21. There we will see that a theory with conformal invariance has the following properties:

1) The theory has a conformally invariant vacuum state $|0\rangle$.
2) It has a set of operators, which we will denote by $\{\phi_j\}$, called quasi-primary, that transform irreducibly under conformal transformations, meaning that they obey the transformation law

$$\phi_j(x) \mapsto \left|\frac{\partial x'}{\partial x}\right|^{\Delta_j/D} \phi_j(x') \tag{15.92}$$

where

$$J = \left|\frac{\partial x'}{\partial x}\right| \tag{15.93}$$

is the Jacobian of the conformal transformation, D is the spacetime dimension, and the numbers $\{\Delta_j\}$ are the scaling dimensions of the operators. The scaling dimensions are the quantum numbers that label the representations of the conformal group.
3) Under a conformal transformation, the correlators of the fields transform as

$$\langle\phi_1(x_1)\cdots\phi_N(x_N)\rangle^* = \left|\frac{\partial x'}{\partial x}\right|_{x=x_1}^{\Delta_1/D} \cdots \left|\frac{\partial x'}{\partial x}\right|_{x=x_N}^{\Delta_N/D} \langle\phi_1(x_1')\cdots\phi_N(x_N')\rangle^* \tag{15.94}$$

4) All other operators in the theory are linear combinations of the quasi-primary fields and of their derivatives.

The symmetries of the fixed point strongly constrain the behavior of the correlators. Translation and rotation (or Lorentz) invariance require that the N-point functions depend only on the pairwise distances, $|x_{12}|=|x_1-x_2|$. Scale invariance then requires that they depend only on scale-invariant ratios, $|x_{12}|/|x_{34}|$. In addition, conformal invariance requires that they depend only on the cross ratios $|x_{12}||x_{34}|/(|x_{13}||x_{24}|)$.

Translation, rotation, and scale invariance restrict the form of the two-point functions, $\langle\phi_1(x_1)\phi_2(x_2)\rangle^*$, to have the form

$$\langle\phi_1(x_1)\phi_2(x_2)\rangle^* = \frac{C_{12}}{|x_{12}|^{\Delta_1+\Delta_2}} \tag{15.95}$$

However, in chapter 21, we will see that conformal invariance further restricts the form of the two-point functions, requiring that $C_{12}=0$ if $\Delta_1 \neq \Delta_2$. Hence

$$\langle\phi_1(x_1)\phi_2(x_2)\rangle^* = \begin{cases} \dfrac{C_{12}}{|x_{12}|^{2\Delta}}, & \Delta=\Delta_1=\Delta_2 \\ 0, & \Delta_1 \neq \Delta_2 \end{cases} \tag{15.96}$$

Hence, primary fields with different scaling dimensions (i.e., in different representations) are "orthogonal" to each other. Clearly, by a suitable redefinition of the operators, we can always set the nonvanishing coefficient $C_{12}=1$.

The form of the three-point functions is also completely determined by translation, rotation, scale, and conformal invariance. In chapter 21, we will see that the three-point function has the form (Polyakov, 1974)

$$\langle \phi_1(x_1)\phi_2(x_2)\phi_3(x_3)\rangle^* = \frac{C_{123}}{|x_{12}|^{\Delta_{12}}|x_{23}|^{\Delta_{23}}|x_{31}|^{\Delta_{31}}} \tag{15.97}$$

where $\Delta_{ij} = \Delta_i + \Delta_j - \Delta_k$ (with $i, j, k = 1, 2, 3$). Provided the primary fields are normalized so that the coefficient of the two-point function is $C_{ij} = \delta_{ij}$, the coefficients C_{ijk} of the three-point functions are universal numbers that further characterize the fixed point. We will see that they play a key role.

These symmetries also restrict the form of the higher-point functions but not as completely. For instance, the four-point function must have the form

$$\langle \phi_1(x_1)\phi_2(x_2)\phi_3(x_3)\phi_4(x_4)\rangle^* = F\left(\frac{|x_{12}||x_{34}|}{|x_{13}||x_{24}|}\right)\prod_{i<j}\frac{1}{|x_{ij}|^{\Delta_i+\Delta_j-\bar{\Delta}/3}} \tag{15.98}$$

where $\bar{\Delta} = \sum_{i=1}^{4}\Delta_i$. The function $F(z)$, where $z = |x_{12}||x_{34}|/(|x_{13}||x_{24}|)$ is the cross-ratio of the coordinates of the four operators, is a universal scaling function. Similar but more complicated expressions apply for the higher-order N-point functions (with $N > 4$).

15.4 The operator product expansion

Let us consider an N-point function of a theory at a general fixed point $\langle \phi_1(x_1)\cdots\phi_N(x_N)\rangle^*$. Consider the case in which two of the operators, say, $\phi_1(x_1)$ and $\phi_2(x_2)$, approach each other, and hence, they are closer to each other than to the insertions of any of the other operators (i.e., $a \ll |x_{12}| \ll |x_{ij}|$, where $i = 1, 2$ and $j = 3, \ldots, N$; but they are still farther apart than the scale of the UV cutoff a). The form of the N-point functions in a fixed-point theory (cf. eq. (15.98)) tells us that the N-point function will be singular as the coordinates of any pair of fields approach each other, and furthermore, that the singularity is determined by the scaling dimensions of the fields involved.

This property suggests that this behavior should be the same as if the product of these two fields is replaced by some suitable combination of all the operators in the theory. This observation suggests that, inside expectation values, the product of fields can be replaced by an expansion involving all other fields. The precise statement is as follows. Given that the set of operators $\{\phi_j\}$ and their descendants (involving derivatives) are a complete set, it follows that the product of two such fields (say, $\phi_i(x)$ and $\phi_j(y)$) at short distances should be equivalent to a suitable linear combination of the fields $\{\phi_k\}$ and their descendants (Kadanoff, 1969; Wilson, 1969; Polyakov, 1970):

$$\lim_{y\to x}\phi_i(x)\phi_j(y) = \lim_{y\to x}\sum_k \frac{\widetilde{C}_{ijk}}{|x-y|^{\Delta_i+\Delta_j-\Delta_k}}\phi_k\left(\frac{x+y}{2}\right) \tag{15.99}$$

This statement is the content of the *operator product expansion* (OPE). Notice that this operator identity is a weak identity, as it is only meant to hold inside an expectation value in the fixed-point theory. Also notice that the only operators ϕ_k that contribute to the singular behavior must have scaling dimensions such that $\Delta_k \leq \Delta_i + \Delta_j$.

We will now see that the coefficients of the OPE are the same as the coefficients of the three-point functions. To this effect, let us consider the three-point function $\langle \phi_i(x)\phi_j(y)\phi_l(z)\rangle^*$ and use the OPE of eq. (15.99) to write

$$\lim_{y \to x} \langle \phi_i(x)\phi_j(y)\phi_l(z)\rangle^* = \lim_{y \to x} \sum_k \frac{\widetilde{C}_{ijk}}{|x-y|^{\Delta_i+\Delta_j-\Delta_k}} \left\langle \phi_k\left(\frac{x+y}{2}\right)\phi_l(z)\right\rangle^* \tag{15.100}$$

We can now use the expressions for the two-point functions of eq. (15.96) to find that

$$\begin{aligned}\lim_{y \to x} \langle \phi_i(x)\phi_j(y)\phi_l(z)\rangle^* &= \sum_k \widetilde{C}_{ijk} \lim_{y \to x} \frac{1}{|x-y|^{\Delta_i+\Delta_j-\Delta_k}} \left\langle \phi_k\left(\frac{x+y}{2}\right)\phi_l(z)\right\rangle^* \\ &= \widetilde{C}_{ijl} \lim_{y \to x} \frac{1}{|x-y|^{\Delta_i+\Delta_j-\Delta_l}} \frac{1}{|x-z|^{2\Delta_l}}\end{aligned} \tag{15.101}$$

Using the expression for the three-point functions, eq. (15.97), we readily find that

$$\widetilde{C}_{ijk} = C_{ijk} \tag{15.102}$$

In other terms, the coefficients of the OPE are the same as the coefficients of the three-point functions (for normalized operators).

These results are also interpreted as meaning that at a fixed point, as they approach each other, the operators (or fields) satisfy a *fusion algebra*, which is conventionally denoted as

$$[\phi_i]\cdot[\phi_j] = \sum_k C_{ijk}[\phi_k] \tag{15.103}$$

Here the possible multiplicities of these fusion channels have not been explicitly written down. We will see below that the fusion algebra encodes the properties of the one-loop RG beta functions in the neighborhood of the fixed point.

15.5 Simple examples of fixed points

Let us now discuss several simple examples of fixed points in free-field theories.

15.5.1 Free massless scalar field

Let $\phi(x)$ be a free massless scalar field in D Euclidean dimensions. The Lagrangian is

$$\mathcal{L} = \frac{1}{2}\left(\partial_\mu \phi\right)^2 \tag{15.104}$$

From dimensional analysis and the condition that action be dimensionless, we see that the Lagrangian has units of $[\mathcal{L}] = \ell^{-D}$. Hence the free field ϕ has units of $[\phi] = \ell^{-(D-2)/2}$. Thus, the scaling dimension of the free field $\phi(x)$ is

$$\Delta_\phi = \frac{1}{2}(D-2) \tag{15.105}$$

which is consistent with the fact that the two-point function of a free scalar field is

$$\langle \phi(x)\phi(y)\rangle = \frac{\text{const.}}{|x-y|^{2\Delta_\phi}} \tag{15.106}$$

where $2\Delta_\phi = D-2$.

In a free field, it is trivial to find the scaling dimensions of all other operators. Thus, the operator $\phi^n(x)$ has dimension

$$\Delta_n \equiv \Delta[\phi^n] = n\Delta_\phi = \frac{n}{2}(D-2) \tag{15.107}$$

which is consistent with the two-point function of the normal-ordered operator $:\phi^n(x):$

$$\langle :\phi^n(x)::\phi^n(y):\rangle = \frac{\text{const.}}{|x-y|^{2\Delta_n}} \tag{15.108}$$

Here normal ordering means $:A: = A - \langle A\rangle$. Another equivalent way to see that eq. (15.108) is correct is to use Wick's theorem and show that the correlator of ϕ^n reduces to a sum of products of n two-point correlators.

These results are also consistent with a simple dimensional analysis of the Lagrangian. Indeed, under a scale transformation $x \mapsto \lambda x$, the scalar field transforms as $\phi(\lambda x) \mapsto \lambda^{-\Delta_\phi}\phi(x)$, where $\Delta_\phi = \frac{1}{2}(D-2)$, which leaves the action invariant.

15.5.2 Free massless Dirac field

The Lagrangian of a free massless Dirac field in D dimensions is

$$\mathcal{L} = \bar{\psi}(x) i\gamma^\mu \partial_\mu \psi(x) \tag{15.109}$$

where $\psi(x)$ is a spinor Fermi field, with two components in $D=2,3$ dimensions, four components in $D=4,5$ dimensions, and so forth, and γ_μ are the Dirac matrices in D dimensions.

By dimensional analysis, we see that the Dirac field has units of $[\psi] = \ell^{-(D-1)/2}$. Hence, its scaling dimension is $\Delta_\psi = \frac{1}{2}(D-1)$, and the two-point function is

$$\langle \bar{\psi}(x)\psi(y)\rangle = \frac{\text{const.}}{|x-y|^{2\Delta_\psi}} \tag{15.110}$$

Similarly, the scaling dimension of a composite operator $(\bar{\psi}\psi)^n$ is

$$\Delta((\bar{\psi}\psi)^n) = 2n\Delta_\psi = n(D-1) \tag{15.111}$$

In particular, the current operator $j_\mu = \bar{\psi}\gamma_\mu\psi$ has scaling dimension $D-1$.

15.5.3 Gauge theories

The Lagrangian for a gauge theory is

$$\mathcal{L} = \frac{1}{4g^2}\operatorname{tr} F_{\mu\nu}^2 \tag{15.112}$$

where the field strength is $F_{\mu\nu} = i[D_\mu, D_\nu] = \partial_\mu A_\nu - \partial_\nu A_\mu + [A_\mu, A_\nu]$, and $D_\mu = \partial_\mu + iA_\mu$ is the covariant derivative. From the fact that the gauge field is a one-form, it follows that it has units of $[A] = \ell^{-1}$, and hence its scaling dimension is $\Delta[A] = 1$. Notice that here the coupling constant is in the prefactor of the Lagrangian. Thus the field strength has scaling dimension $\Delta[F]=2$. The action is dimensionless if we assign the coupling constant units of $[g]=(D-4)/2$.

15.6 Perturbing a fixed-point theory

Let us now consider a theory close to a fixed point. We will use perturbation theory in powers of the coupling constants of the theory. This framework is used to derive the general form of the RG beta functions close to a general fixed-point theory. This approach is known as conformal perturbation theory. It is a generalization of the Kosterlitz RG (Kosterlitz, 1974) used to describe the Kosterlitz-Thouless phase transition in two-dimensional classical statistical mechanics (Kosterlitz and Thouless, 1973). An insightful discussion of this RG is presented in Cardy (1996), whose approach we will follow closely.

Let S^* be the action of a fixed-point theory in D Euclidean dimensions. Let δS be a set of general perturbations parametrized by the complete set of primary fields $\{\phi_j\}$ of the fixed point. The total action S is

$$S=S^*+\delta S=S^*+\sum_j \int dx^D\, g_j\, a^{\Delta_j-D}\phi_j(x) \tag{15.113}$$

where the g_j are the dimensionless coupling constants, the short-distance scale a is the UV regulator, and the Δ_j are the scaling dimensions of the fields ϕ_j.

Let Z be the path integral for this theory, which we formally denote by

$$\begin{aligned} Z &= \operatorname{tr}\exp(-S^*-\delta S)\\ &= \operatorname{tr}\exp\left(-S^*-\sum_j \int d^Dx\, g_j\, a^{\Delta_j-D}\phi_j(x)\right)\\ &= Z^*\times\left\langle \exp\left(-\sum_j \int d^Dx\, g_j\, a^{\Delta_j-D}\phi_j(x)\right)\right\rangle^* \end{aligned} \tag{15.114}$$

where $Z^*=\operatorname{tr}\exp(-S^*)$ is the partition function of the fixed-point theory. Here, as before, we represent expectation values in the fixed-point theory by the notation

$$\langle A\rangle^*=\frac{1}{Z^*}\operatorname{tr}\left(A\exp(-S^*)\right) \tag{15.115}$$

Our next step is to write the expansion of the partition function Z in powers of the coupling constants $\{g_j\}$:

$$\begin{aligned} \frac{Z}{Z^*} &= \left\langle \exp\left(-\sum_j \int d^Dx\, g_j\, a^{\Delta_j-D}\phi_j(x)\right)\right\rangle^*\\ &= 1-\sum_j \int d^Dx\, g_j\, a^{\Delta_j-D}\langle\phi_j(x)\rangle^* \end{aligned} \tag{15.116}$$

$$+\frac{1}{2!}\sum_{j,k}\int d^D x_j \int d^D x_k \, g_j g_k \, a^{\Delta_j+\Delta_k-2D}\langle\phi_j(x_j)\phi_k(x_k)\rangle^* \tag{15.117}$$

$$-\frac{1}{3!}\sum_{j,k,l}\int d^D x_j \int d^D x_k \int d^D x_l \, g_j g_k g_l \, a^{\Delta_j+\Delta_k+\Delta_l-3D}\langle\phi_j(x_j)\phi_k(x_k)\phi_l(x_l)\rangle^* \tag{15.118}$$

$$+\cdots$$

At a formal level, we can reinterpret this expression as the partition function of a gas of particles of different types in a grand canonical ensemble, with each species labeled by j at coordinates x_j. In this perspective, the coupling constants g_j play the role of the fugacity of each type of particle, and the interactions are given by the (negative of the) logarithm of the correlators at the fixed point. Since S^* is a fixed-point action, the integrals of the correlators present in eqs. (15.116), (15.117), and (15.118) will generally contain IR divergences. Assume that the theory is defined in a box of linear size L to cut off the IR behavior. But, precisely because S^* is a fixed-point action and hence is scale invariant, we will need essentially the short-distance UV singular behavior, which is controlled by the OPE. Let us control the short-distance singularities by cutting off the integrals at some short distance a. Hence the span of the integrals will be the range $L \gg |x_j - x_k| \gg a$. In other words, the "particles" of this gas have a "hard-core" a.

Let us analyze this expansion using the following RG transformation. We will attempt to change the UV cutoff a by some small amount $\delta a = a\delta\ell$ (with $\delta\ell \ll 1$) and then compute the change of the action needed to compensate for that UV-cutoff change, while requiring that the full partition function Z be left unchanged in a box of fixed linear size L. Instead of integrating out modes in momentum shells, as we did in section 15.2.4, let us integrate out contributions to the integrals appearing in eqs. (15.116), (15.117), and (15.118) in a shell in real space between the UV cutoff a and the dilated cutoff ba, with $b = 1 + \delta\ell$. We will proceed as follows.

Rescale the UV cutoff $a \mapsto ba$, with $b = 1 + \delta\ell > 1$. Hence, we are increasing the UV cutoff. We then need to rescale the coupling constants $\{g_j\}$ to compensate for this change, while keeping Z unchanged. How do we do this? The UV cutoff a appears in three places:

1) in factors of the form $a^{\Delta_j - D}$, where Δ_j is the scaling dimension of the operators,
2) as a UV cutoff of the integrals, and
3) through the dependence of the integrals on the IR cutoff L, which by scale invariance must enter in the form L/a.

If the relative change of UV scale is small (i.e., $\delta\ell \ll 1$), we need to look only for linear changes in $\delta\ell$. Since these factors always appear multiplying the dimensionless coupling constants g_j,

$$g_j a^{\Delta_j - D} \mapsto g_j a^{\Delta_j - D} b^{\Delta_j - D} \tag{15.119}$$

we can compensate for this change by a change in the coupling constants:

$$g_j \mapsto g_j b^{D-\Delta_j} = g_j(i + (D - \Delta_j)\delta\ell) \tag{15.120}$$

Hence, we will change the coupling constants by

$$g_j \mapsto g_j + (D - \Delta_j) g_j \delta\ell \tag{15.121}$$

Next, let us compute the result of a change to the cutoff in the integrals. This amounts to computing the contributions to the integrals in the shell in real space $a(1+\delta\ell) > |x_i - x_j| > a$ (instead of a shell in momentum space). Thus we split the integrals as follows:

$$\int_{|x_i-x_j|>a(1+\delta\ell)} = \int_{|x_i-x_j|>a} - \int_{a(1+\delta\ell)>|x_i-x_j|>a} \tag{15.122}$$

The application of this analysis to terms linear in the coupling constants, eq. (15.116), yields a purely numerical contribution to the partition function Z (and hence it amounts to a renormalization of the identity operator). The contribution of the terms quadratic in the coupling constants, eq. (15.117), can be computed using the OPE. This is legitimate, since we will look at contributions only inside the infinitesimal shell in real space. Using the OPE, we find that this contribution is

$$\frac{1}{2!}\sum_{j,k}\sum_{l} C_{jkl}\, g_j g_k\, a^{\Delta_l-\Delta_j-\Delta_k} \int_{\text{shell}} d^D x_j d^D x_k \langle \phi_l((x_j+x_k)/2)\rangle^* a^{\Delta_j+\Delta_k-2D}$$

$$= \frac{1}{2!}\sum_{j,k}\sum_{l} C_{jkl}\, g_j g_k\, a^{\Delta_l-D} \int d^D x \langle \phi_l(x)\rangle^* S_D \delta\ell \tag{15.123}$$

where

$$S_D = \frac{2\pi^D}{\Gamma(D/2)} \tag{15.124}$$

is the area of the hypersphere in D dimensions, and $\Gamma(x)$ is the Euler gamma function. The "one-loop" contribution, eq. (15.123), can be compensated by a change in the coupling constants

$$g_l \mapsto g_l - \frac{1}{2} S_D \sum_{j,k} g_j g_k C_{jkl} \delta\ell \tag{15.125}$$

In fact, the structure of the OPEs implies that similar changes happen at all orders in the expansion.

Finally, there is the dependence in a due to the dependence in L. However, to keep the system unchanged, the IR cutoff L must be held fixed. Hence this dependence is trivial.

Let us now collect all changes in the coupling constants in the differential equation

$$\frac{dg_l}{d\ell} = (D-\Delta_l) g_l - \sum_{j,k} C_{jkl} g_j g_k + \cdots \tag{15.126}$$

where we have absorbed a factor of $S_D/2$ in a (multiplicative) redefinition of the coupling constants.

In summary, we derived the beta functions for all the coupling constants using conformal perturbation theory to order one-loop. The beta functions depend on the data of the fixed point: the scaling dimensions Δ_j and the coefficients C_{ijk} of the OPEs of the primary fields. This is a general result that holds for all fixed-point theories.

15.7 Example of operator product expansions: ϕ^4 theory

Let us now apply the ideas of the preceding sections to the free-field fixed point of ϕ^4 theory. The Euclidean Lagrangian is

$$\mathcal{L} = \frac{1}{2}(\partial_\mu \phi)^2 + ta^{-2}\phi^2 + ua^{D-4}\phi^4 + ha^{-(1+D/2)}\phi \tag{15.127}$$

where we defined the mass m^2, the coupling constant λ, and the symmetry-breaking field J in terms of the dimensionless couplings, t, u, and h:

$$m^2 = ta^{-2}, \quad \lambda = ua^{D-4}, \quad J = ha^{-(1+D/2)} \tag{15.128}$$

Let us now define the normal-ordered composite operators

$$:\phi^2:= \phi^2 - \langle\phi^2\rangle, \qquad :\phi^4:= \phi^4 - 3\langle\phi^2\rangle\phi^2, \qquad \text{etc.} \tag{15.129}$$

In what follows, we will use the notations $\phi_n =: \phi^n :$. Clearly, $\phi_1 = \phi$, since $\langle\phi\rangle^* = 0$ at the free-field fixed point.

Let us begin by computing the OPE of two $\phi \equiv \phi_1$ fields:

$$\begin{aligned}\lim_{y\to x} \phi_1(x)\phi_1(y) &= \lim_{y\to x} :\phi(x)::\phi(y): \\ &= \lim_{y\to x}\left[\frac{1}{|x-y|^{D-2}} + :\phi^2\left((x+y)/2\right): + \cdots\right]\end{aligned} \tag{15.130}$$

Notice that the first term, proportional to the identity operator, is the most singular term in the expansion, and that the next term is finite (as $|x-y| \to 0$). The next possible operator would be $:(\partial\phi)^2:$. However, its coefficient will vanish as $|x-y|^2$. Hence, in the limit, it does not contribute. Only singular and finite terms will be kept in the expansion.

We record this equation as the OPE of two ϕ_1 fields:

$$[\phi_1]\cdot[\phi_1] = [1] + [\phi_2] \tag{15.131}$$

where "1" denotes the identity field.

Next we compute the OPE of a ϕ_1 field (i.e., the field ϕ itself) and the $\phi_2 \equiv: \phi(x)^2 :$ field. Using Wick's theorem, we find

$$[\phi_1]\cdot[\phi_2] = 2[\phi_1] + [\phi_3] \tag{15.132}$$

where $\phi_3 =: \phi(x)^3 :$.

We then compute the OPE of two $\phi_2 =: \phi^2 :$ fields. Again, using Wick's theorem, we obtain

$$\lim_{y\to x} :\phi^2(x)::\phi^2(y):= \lim_{y\to x}\left[\frac{2}{|x-y|^{2(D-2)}} + \frac{4}{|x-y|^{D-2}}:\phi^2((x+y)/2): + \cdots\right] \tag{15.133}$$

which we record as

$$[\phi_2]\cdot[\phi_2] = 2 + 4[\phi_2] + [\phi_4] \tag{15.134}$$

The following OPEs are derived in a similar fashion:

$$[\phi_1]\cdot[\phi_4]=4[\phi_3] \tag{15.135}$$

$$[\phi_2]\cdot[\phi_4]=12[\phi_2]+8[\phi_4] \tag{15.136}$$

$$[\phi_4]\cdot[\phi_4]=24+96[\phi_2]+72[\phi_4] \tag{15.137}$$

Notice that we have neglected possible descendant operators, such as $:(\partial_\mu\phi)^2:$, $:(\partial^2\phi)^2:$, $:(\partial_\mu\phi)^4:$. These higher-derivative operators are typically leading irrelevant operators. However, this is not always the case. For instance, in the OPE of $:\phi^2:$ with itself, the operator $:(\partial_\mu\phi)^2:$ can appear. By matching scaling dimensions, we see that the prefactor of $:(\partial_\mu\phi)^2:$ must be proportional to $|x-y|^{-(D-4)}$, which is singular if $D>4$ and has a logarithmic divergence if $D=4$. These contributions, however, enter in higher-order loops (as we will see in chapter 16).

Using these results, we can now write the beta functions for ϕ^4 theory. With the identification $g_1=h$ (the symmetry-breaking field), $g_2=t$ (the dimensionless mass squared), and $g_4=u$ (the coupling constant), we find that their one-loop beta functions are

$$\frac{dh}{d\ell}=\left(\frac{D}{2}+1\right)h-4ht+\cdots \tag{15.138}$$

$$\frac{dt}{d\ell}=2t-4h^2-4t^2-24tu-96u^2+\cdots \tag{15.139}$$

$$\frac{du}{d\ell}=(4-D)u-t^2-16tu-72u^2+\cdots \tag{15.140}$$

For $D=4-\epsilon$, these beta functions have two fixed points: (1) the trivial fixed point at $h^*=t^*=u^*=0$, and (2) the nontrivial Wilson-Fisher fixed point at $h^*=0$, $t^*=O(\epsilon^2)$, and $u^*=\frac{\epsilon}{72}+O(\epsilon^2)$. The critical exponent ν of the correlation length is the inverse of the slope of the beta function for t at the Wilson-Fisher fixed point. We obtain $\nu=\frac{1}{2}+\frac{\epsilon}{12}+O(\epsilon^2)$, consistent with what we found in section 15.2.4. Also notice that since $t^*=O(\epsilon^2)$, the anomalous dimension of the field ϕ is zero at one-loop order. We will see in chapter 16 that it is nonzero at two-loop order.

Exercises

15.1 A simple block-spin transformation

The Ashkin-Teller model is a simple generalization of the Ising model in which there are two Ising degrees of freedom, σ and τ, each taking values ±1, defined on the sites of a lattice. Here we will consider a D-dimensional hypercubic lattice whose sites are labeled by the two-component lattice vectors $\boldsymbol{r}$. The interactions are restricted to nearest neighboring sites, and the classical Hamiltonian H is

$$H=-K_2\sum_{\langle \boldsymbol{r},\boldsymbol{r}'\rangle}\big(\sigma(\boldsymbol{r})\sigma(\boldsymbol{r}')+\tau(\boldsymbol{r})\tau(\boldsymbol{r}')\big)-K_4\sum_{\langle \boldsymbol{r},\boldsymbol{r}'\rangle}\sigma(\boldsymbol{r})\sigma(\boldsymbol{r}')\tau(\boldsymbol{r})\tau(\boldsymbol{r}') \tag{15.141}$$

where $\langle \boldsymbol{r}, \boldsymbol{r}' \rangle$ denotes nearest neighboring sites, and $K_2 > 0$ and $K_4 > 0$ are two coupling constants. Clearly for $K_4 = 0$, this system is equivalent to two decoupled Ising models.

1) Show that the following statements are true:

   a) If $K_2 \to \infty$, the system is in its ground state. Find the possible ground state(s).
   b) If $K_4 \to \infty$, the system is equivalent to a single Ising model with coupling constant $K = 2K_2$.
   c) If $K_2 \to 0$, the system is also equivalent to a single Ising model with coupling constant K_{eff}. Find K_{eff}.

2) Consider now the Ashkin-Teller model on a one-dimensional lattice. Use a decimation method (analogous to what was done in this chapter for the one-dimensional Ising model) to derive an exact RG transformation. Find the fixed points, and calculate their eigenvalues. Draw the RG flows explicitly. Discuss the similarities and differences with the Ising model.
3) Consider now the Ashkin-Teller model on a D-dimensional hypercubic lattice. Use the Migdal-Kadanoff bond-moving procedure to derive a block-spin transformation for $D = 1 + \varepsilon$, where $\varepsilon > 0$ and "small."

   a) Find all the fixed points, both stable and unstable. Calculate all the eigenvalues of the RG transformation for each fixed point, and the correlation length exponent ν for each fixed point.
   b) Sketch the qualitative RG flows on the $K_2 - K_4$ plane. Use the flows to derive a phase diagram. What is the order parameter(s) which labels each phase? Justify your arguments.

   Warning: The Midgal-Kadanoff bond-moving procedure yields an unphysical phase transition at $K_2 \to \infty$, with $K_4 < \infty$. It is easy to see that the associated stable fixed points are physically equivalent.

15.2 Momentum-shell RG for the nonlinear σ-model
In this problem, you are asked to consider the $O(N)$ nonlinear σ-model, in $D = 2 + \varepsilon$ dimensions, coupled to an external field. Since we consider this theory in Euclidean space, it is equivalent to a classical Heisenberg model. Parametrize the configurations by means of an N-component field $\boldsymbol{n}(\boldsymbol{x})$, which satisfies the constraint $\boldsymbol{n}^2 = 1$. The Euclidean Lagrangian (or classical Hamiltonian) is

$$\mathcal{L} = \frac{1}{2g}(\partial_\mu \boldsymbol{n}(\boldsymbol{x}))^2 + \frac{1}{2g}\boldsymbol{H}(\boldsymbol{x}) \cdot \boldsymbol{n}(\boldsymbol{x}) \tag{15.142}$$

The partition function is

$$Z = \int \mathcal{D}\boldsymbol{n}(\boldsymbol{x}) \prod_{\boldsymbol{x}} \delta(\boldsymbol{n}^2(\boldsymbol{x}) - 1)\, e^{-\int d^d x\, \mathcal{L}} \tag{15.143}$$

Follow the procedure to decompose the field $\boldsymbol{n}(\boldsymbol{x})$ into its slow and fast components. Let $\boldsymbol{n}_0(\boldsymbol{x})$ be an N-component slowly varying configuration of the σ-model, which is also a solution of the classical equations of motion (i.e., it extremizes the Euclidean action). Let $\phi_i(\boldsymbol{x})$ $(i = 1, \ldots, N-1)$ be an (essentially unrestricted) $N-1$-component field, which we use to parametrize the fluctuations of the field as follows. Let $\{\boldsymbol{e}_i(\boldsymbol{x})\}$ be a set of $N-1$ unit-length vectors defined at each point in space. They are required to form, together with $\boldsymbol{n}_0(\boldsymbol{x})$, an orthonormal basis for the local configurations,

$$n_0^a(\boldsymbol{x})e_i^a(\boldsymbol{x}) = 0; \qquad e_i^a(\boldsymbol{x})e_j^a(\boldsymbol{x}) = \delta_{i,j} \tag{15.144}$$

for $i,j = 1,\ldots,N-1$ and $a = 1,\ldots,n$ (repeated indices are summed over). Thus, we can write

$$n^a(\boldsymbol{x}) = \sqrt{1-\boldsymbol{\phi}^2(\boldsymbol{x})}\; n_0^a(\boldsymbol{x}) + \sum_{i=1}^{N-1}\phi_i(\boldsymbol{x})e_i^a(\boldsymbol{x}) \tag{15.145}$$

In other words, we have defined a *local frame*. The local changes of the slow field $n_0^a(x)$ and of the unit vectors e_i^a (i.e., of the local frame) can also be expanded in that basis:

$$\partial_\mu n_0^a(\boldsymbol{x}) = \sum_{i=1}^{N-1} B_\mu^i(\boldsymbol{x})e_i^a(\boldsymbol{x}) \tag{15.146}$$

$$\partial_\mu e_i^a(\boldsymbol{x}) = \sum_{j=1}^{N-1} A_\mu^{ij}(\boldsymbol{x})e_j^a(\boldsymbol{x}) - B_\mu^i(\boldsymbol{x})n_0^a(\boldsymbol{x}) \tag{15.147}$$

where

$$B_\mu^i(\boldsymbol{x}) = \boldsymbol{e}_i(x)\cdot\partial_\mu \boldsymbol{n}_0(x), \qquad A_\mu^{ij}(x) = -A_\mu^{ji}(\boldsymbol{x}) = -\boldsymbol{e}_i(x)\cdot\partial_\mu \boldsymbol{e}_j(x) \tag{15.148}$$

1) Write the coupling constant g and the magnetic field $\boldsymbol{H}$ in terms of the dimensionless coupling constant u and dimensionless field $\boldsymbol{h}$, by using dimensional analysis in terms of a momentum cutoff Λ.
2) Write the action as a function(al) of the fluctuating field $\boldsymbol{\phi}(x)$ in the background of the slow field $\boldsymbol{n}_0(x)$, parametrized by the fields $B_\mu^i(x)$ and $A_\mu^{ij}(x)$. Expand the Euclidean action as a power series expansion of the field $\boldsymbol{\phi}(x)$, and write down the explicit form of the linear and quadratic terms in $\phi_i(x)$.
3) Show that for the configuration $\boldsymbol{n}_0(\boldsymbol{x})$ to be stationary, the background fields $B_\mu^i(x)$ and $A_\mu^{ij}(x)$ must satisfy the equation of motion

$$D_\mu^{ij}B_\mu^j(x) \equiv \left[\partial_\mu\delta_{ij} - A_\mu^{ij}(x)\right]B_\mu^j(x) = 0 \tag{15.149}$$

   Hint: Demand that the action be stationary with respect to a variation of the fluctuation field ϕ_i.
4) Use the momentum-shell integration technique to integrate out the fast variables of the field $\phi_i(x)$ inside the momentum shell $b\Lambda \le |\boldsymbol{p}| \le \Lambda$, with $b = 1-\delta\ell \lesssim 1$ and $\delta\ell$ small. Obtain a set of differential equations that govern the (differential) renormalization for u and $h = |\boldsymbol{h}|$.
5) Find the fixed points and draw the qualitative flows for dimension $D = 2+\varepsilon$. Calculate the eigenvalues of u and h at each fixed point, and determine when u and h are relevant, irrelevant, or marginal. Show that precisely at $D = 2$, there is a marginal operator.

6) Use an RG argument to find an analytic expression for the singular dependence of the correlation length as a function of u for $u > u_c$ and $h = 0$. Express your result in terms of the critical exponent ν.
7) Use an RG argument to find the behavior of the order parameter, the spontaneous magnetization $\mathcal{M}^a = \langle n^a(\boldsymbol{x}) \rangle$, as a function of u at $h = 0$ and close to u_c. Show that it obeys a power law of the form $\mathcal{M} \sim (u_c - u)^\beta$, and compute the critical exponent β to lowest order in ε. Use a similar RG argument to find the field dependence of the magnetization *at* u_c, and determine the value of the critical exponent δ of the h dependence of $\mathcal{M}$ as $h \to 0$, $\mathcal{M} \sim |h|^{1/\delta}$.
8) Consider the special (and important) case of $D = 2$ dimensions.

   a) Solve the RG flow equation for the dimensional coupling constant $u = u(\Lambda)$ as an explicit function of the cutoff Λ in terms of its value at a reference scale μ.
   b) Discuss the behavior of this effective (or *running*) coupling constant in the UV regime $\Lambda \gg \mu$. Does the running coupling constant get big or small? How fast? Why is this behavior referred to as *asymptotic freedom*?
   c) Find the behavior of the correlation length ξ as a function of the dimensional coupling constant u for $D = 2$. Be explicit about the assumptions you make and state their physical justifications.

Note: In this problem, you have to do the momentum-shell integral. You may find it useful to express your angular integrations in terms of the area of the unit hypersphere in D dimensions

$$S_D = \frac{2\pi^{D/2}}{\Gamma(D/2)}$$

where $\Gamma(x)$ is the Euler gamma function,

$$\Gamma(z) = \int_0^\infty dt\, t^{z-1} e^{-t}$$

which is convergent for Re $z > 0$.

15.3 The free massless scalar field fixed point in $D = 2$ spacetime dimensions
Consider a free massless scalar field $\phi(x)$ in $D = 2$ Euclidean dimensions, $\boldsymbol{x} = (x_1, x_2)$. This system is known in statistical mechanics as the Gaussian model. The (Euclidean) Lagrangian density is

$$\mathcal{L} = \frac{K}{2} (\nabla\phi)^2 \tag{15.150}$$

where K is a positive real parameter (a coupling constant). The Euclidean propagator for this system in a region of linear size $L \equiv m^{-1}$ (a very large disk) is

$$G(x - x') = \langle \phi(x)\phi(x') \rangle = \frac{1}{K} \int \frac{d^2p}{(2\pi)^2} \frac{e^{ip\cdot(x-x')}}{p^2 + m^2} \tag{15.151}$$

This formula is valid for $|\boldsymbol{x} - \boldsymbol{x}'| \ll m^{-1} = L$, where the propagator takes the form

$$G(x-x') = -\frac{1}{4\pi K} \ln\left(\frac{|x-x'|^2 + a^2}{L^2}\right) \tag{15.152}$$

where a is a short-distance cutoff. You will use this propagator in the rest of this exercise.

1) *Scale invariance*: Show that $S[\phi]$ is invariant under scale transformations (dilations)

$$x \to x' = \lambda x \tag{15.153}$$

where $\lambda > 0$ is a positive real number.

2) *Operators*: Consider the vertex operators $V_q[x] = e^{iq\phi(x)}$. Add a source term to the action of the form

$$J(x) = i \sum_{j=1}^{N} q_j \delta^{(2)}(x - x_j) \tag{15.154}$$

to show that the correlation function of N vertex operators

$$\langle V_{q_1}(\boldsymbol{x}_1) \cdots V_{q_N}(\boldsymbol{x}_N)\rangle \tag{15.155}$$

is invariant under the shift $\phi(x) \to \phi(x) + \alpha$ (with α arbitrary) only if $\sum_{j=1}^{N} q_j = 0$.

3) *Correlators*: Show that the correlation functions of N vertex operators with "charges" $q_1, \ldots, q_N$ are given by

$$\begin{aligned}\langle V_{q_1}(x_1) \cdots V_{q_N}(x_N)\rangle =& e^{-\left(\sum_{j=1}^{N} q_j\right)^2 G(0)} \\ & \times e^{-\sum_{j>j'=1}^{N} q_j q_{j'} \left[G(x_j - x_{j'}) - G(0)\right]}\end{aligned} \tag{15.156}$$

Show that, in the $L \to \infty$ limit, these correlation functions vanish identically unless the charge neutrality condition, $\sum_{j=1}^{N} q_j = 0$, is satisfied.

4) *Two-point function*: Use the expression you just derived to compute the explicit form of the correlation function of two vertex operators:

$$G_q^{(2)}(x - x') = \langle V_q(x) V_{-q}(x')\rangle \tag{15.157}$$

Show that the scaling dimension Δ_q of the vertex operator V_q is

$$\Delta_q = \frac{q^2}{4\pi K} \tag{15.158}$$

5) *Three-point function*: Show that the correlation function of three vertex operators $V_{q_1}(\boldsymbol{x}_1)$, $V_{q_2}(\boldsymbol{x}_2)$, and $V_{q_3}(\boldsymbol{x}_3)$ is given by

$$\langle V_{q_1}(x_1) V_{q_2}(x_2) V_{q_3}(x_3)\rangle = \frac{1}{|x_1 - x_2|^{\Delta_{12}} |x_2 - x_3|^{\Delta_{23}} |x_3 - x_1|^{\Delta_{31}}} \tag{15.159}$$

with $q_1 + q_2 + q_3 = 0$. Show that $\Delta_{12} = \Delta_{q_1} + \Delta_{q_2} - \Delta_{q_3}$, $\Delta_{23} = \Delta_{q_2} + \Delta_{q_3} - \Delta_{q_1}$, and $\Delta_{31} = \Delta_{q_3} + \Delta_{q_1} - \Delta_{q_2}$.

6) *Scaling dimensions*: Suppose we were to add to the free-field fixed-point Lagrangian a perturbation of the form

$$\mathcal{L}_{\text{int}} = g\cos(q\phi(x)) = \frac{g}{2}\left(V_q(x) + V_{-q}(x)\right) \tag{15.160}$$

where g is a coupling constant. Determine the values of K for which this perturbation is (a) *marginal*, (b) relevant, and (c) irrelevant. Use these results and the arguments given in this chapter to write down the RG β-function $\beta(g)$ to linear order in the coupling constant g.

7) *Operator product expansion* (OPE):

a) Show that the operators $\{V_{q_i}(x)\}$ obey an OPE of the form

$$V_{q_1}(x_1)V_{q_2}(x_2) \sim \frac{C_{q_1,q_2,q_1+q_2}}{|x_1 - x_2|^{\mu_{q_1,q_2}}} V_{q_1+q_2}\left(\frac{x_1+x_2}{2}\right)$$

as $x_1 \to x_2$. Use the results of the above problems to find the coefficient C_{q_1,q_2,q_1+q_2} and the exponent μ_{q_1,q_2}.

b) Show that the operators $V_q(x)$ and $V_{-q}(y)$ obey an OPE of the form

$$V_q(x)V_{-q}(y) \sim \frac{1}{|x-y|^{\eta_q}} + \frac{C_q}{|x-y|^{\mu_q}} :(\nabla\phi)^2\left(\frac{x+y}{2}\right):$$

as $x \to y$, where $:(\nabla\phi)^2(x): = (\nabla\phi)^2(x) - \langle(\nabla\phi)^2(x)\rangle$. Find the coefficient C_q and the exponents η_q and μ_q.

c) Show that the OPE of the operator $E(x) =: (\nabla\phi)^2(x):$ with itself does not contain the operator $E(x)$ (i.e., $C_{EEE} = 0$).

8) *RG:* Now consider the sine-Gordon theory, whose (Euclidean) Lagrangian density is

$$\mathcal{L} = \frac{K}{2}(\nabla\phi)^2 + \frac{g}{2}\left(V_q(x) + V_{-q}(x)\right) \tag{15.161}$$

where K and g play the role of the coupling constants. Use the results of the OPE of part 7(c) to construct the RG beta functions for the “stiffness” K (this effect can also be regarded as a wave function renormalization) and the coupling constant g, up to quadratic order in g.

16

The Perturbative Renormalization Group

In this chapter, we reexamine the renormalized perturbation theory discussed in chapters 12 and 13 from the perspective of the RG. We discuss ϕ^4 theory, the $O(N)$ nonlinear sigma model, and Yang-Mills gauge theories. More detailed presentations of some of this material can be found in Amit (1980), Zinn-Justin (2002), and Peskin and Schroeder (1995).

16.1 The perturbative renormalization group

We begin by setting up the perturbative RG to ϕ^4 theory. In section 13.2, we used renormalized perturbation theory to show that, to two-loop order, the theory can be made finite by defining a set of renormalization constants such that the renormalized one-particle irreducible N-point vertex functions are related to the bare functions by

$$\Gamma_R^{(N)}(\{p_i\}; m_R, g_R, \kappa) = Z_\phi^{N/2} \Gamma^{(N)}(\{p_i\}, m_0, \lambda, \Lambda) \tag{16.1}$$

where Λ is the UV momentum regulator, and κ is the renormalization scale. The relation between the bare and the renormalized theory is encoded in the renormalization constants (the wave-function renormalization Z_ϕ, the renormalized mass m_R, and the renormalized coupling constant g_R) such that as the UV cutoff is removed ($\Lambda \to \infty$), the renormalized vertex functions have a finite limit.

Here we focus attention on the massless theory, defined by the condition that the renormalized mass vanishes, $m_R = 0$. Thus, we express the wave-function renormalization Z_ϕ and the bare coupling constant λ as functions of the renormalized coupling constant and of the renormalization scale. Similarly, the bare mass will be tuned to a value m_c such that $m_R = 0$. Thus we write

$$Z_\phi = Z_\phi(g_R(\kappa), \kappa, \Lambda) \tag{16.2}$$

$$\lambda = \lambda(g_R(\kappa), \kappa, \Lambda) \tag{16.3}$$

$$m_c^2 = m_c^2(g_R(\kappa), \kappa, \Lambda) \tag{16.4}$$

which makes explicit the fact that the renormalized coupling constant g_R is actually not a constant but depends on the value we choose for the renormalization scale κ.

Since the value of the renormalization scale κ is arbitrary, the vertex functions defined at two different scales κ_1 and κ_2 must be related to each other, since they correspond to the same bare theory:

$$\Gamma_R^{(N)}(\{p_i\}; g_R(\kappa_1), \kappa_1) = Z_\phi^{N/2}(g_R(\kappa_1), \kappa_1)\Gamma^{(N)}(\{p_i\}, m_c, \lambda, \Lambda)$$

$$\Gamma_R^{(N)}(\{p_i\}; g_R(\kappa_2), \kappa_2) = Z_\phi^{N/2}(g_R(\kappa_2), \kappa_2)\Gamma^{(N)}(\{p_i\}, m_c, \lambda, \Lambda) \tag{16.5}$$

We see that the renormalized vertex functions at the two scales are related by an expression of the form

$$\Gamma_R^{(N)}(\{p_i\}; g_R(\kappa_1), \kappa_1) = Z^{N/2}(g_R(\kappa_2), \kappa_2; g_R(\kappa_1), \kappa_1)\Gamma_R^{(N)}(\{p_i\}; g_R(\kappa_2), \kappa_2) \tag{16.6}$$

where by definition,

$$Z(g_R(\kappa_2), \kappa_2; g_R(\kappa_1), \kappa_1) = \frac{Z_\phi(g_R(\kappa_1), \kappa_1, \Lambda)}{Z_\phi(g_R(\kappa_2), \kappa_2, \Lambda)} \tag{16.7}$$

$$= \frac{\partial}{\partial p^2}\Gamma_R^{(2)}(p, g_R(\kappa_1), \kappa_1)\Big|_{p^2=\kappa_2^2} \tag{16.8}$$

which has a finite limit as the UV cutoff is removed ($\Lambda \to \infty$).

Eq. (16.6) is a relation between finite quantities at different scales, and it is a finite quantity. It implies that a change in the renormalization scale κ is equivalent to a rescaling of the fields by $Z_\phi^{1/2}$ and a change of the renormalized coupling constant $g_1 = g_R(\kappa_1) \mapsto g_2 = g_R(\kappa_2)$, with

$$g_2 \equiv F(\kappa_2, \kappa_1, g_1) = Z_\phi^{-2}\Gamma_R^{(4)}(\{p_i\}, g_1, \kappa_1)\Big|_{SP(\kappa_2)} \tag{16.9}$$

where $SP(\kappa)$ is the symmetric point of the four momenta $\{p_i\}$, with each momentum being at the scale κ. The mapping $g_2 = F(\kappa_2, \kappa_1, g_1)$, such that $g = F(\kappa, \kappa, g)$ defines a *flow* in the space of coupling constants (i.e., an RG flow).

These relations apply to the full generating functional of renormalized vertex functions, which then obeys

$$\Gamma_R\{\bar{\phi}, g_1, \kappa_1\} = \Gamma_R\{Z_\phi^{1/2}\bar{\phi}, F(\kappa_2, \kappa_1, g_1), \kappa_2\} \tag{16.10}$$

Since the bare theory is independent of our choice (and changes) of a renormalization scale, it is kept constant under these transformations. This can be expressed by stating that

$$\kappa\frac{\partial}{\partial\kappa}\Gamma^{(N)}(\{p_i\}, \lambda, m_c^2, \Lambda)\big|_{\lambda,\Lambda} = 0 \tag{16.11}$$

as the UV regulator $\Lambda \to \infty$. Consequently, we find

$$\kappa\frac{\partial}{\partial\kappa}\left[Z_\phi^{-N/2}\Gamma_R^{(N)}(\{p_i\}, g_R(\kappa), \kappa)\right]_{\lambda,\Lambda} = 0 \tag{16.12}$$

16.1.1 The RG equations

Therefore, the renormalized N-point vertex functions satisfy the partial differential equation

$$\left[\kappa\frac{\partial}{\partial\kappa}+\bar{\beta}(g_R,\kappa)\frac{\partial}{\partial g_R}-\frac{N}{2}\gamma_\phi(g_R,\kappa)\right]_{\lambda,\Lambda}\Gamma_R^{(N)}(\{p_i\},g_R(\kappa),\kappa)=0 \tag{16.13}$$

where we used the definitions

$$\bar{\beta}(g_R(\kappa),\kappa)=\kappa\frac{\partial g_R}{\partial\kappa}\bigg|_{\lambda,\Lambda} \tag{16.14}$$

$$\gamma_\phi(g_R(\kappa),\kappa)=\kappa\frac{\partial\ln Z_\phi}{\partial\kappa}\bigg|_{\lambda,\Lambda} \tag{16.15}$$

and set $\Lambda\to\infty$.

In general, the coupling constant has dimensions. Let us define a dimensionless bare coupling constant u_0

$$\lambda=u_0\kappa^\epsilon \tag{16.16}$$

and a dimensionless renormalized coupling constant u such that

$$g_R=u\kappa^\epsilon \tag{16.17}$$

where $\epsilon=4-D=D-\Delta_4$, with $\Delta_4=4(D-2)/2$ being the scaling dimension of the operator ϕ^4 at the massless free-field fixed point.

In terms of the dimensionless renormalized coupling constant u, eq. (16.13) becomes the *Callan-Symanzik equation*

$$\left[\kappa\frac{\partial}{\partial\kappa}+\beta(u)\frac{\partial}{\partial u}-\frac{N}{2}\gamma_\phi(u)\right]_{\lambda,\Lambda}\Gamma_R^{(N)}(\{p_i\},u(\kappa),\kappa)=0 \tag{16.18}$$

(again, with $\Lambda\to\infty$), where $\beta(u)$ is the RG beta function, which is defined by

$$\beta(u)=\kappa\frac{\partial u}{\partial\kappa}\bigg|_{\lambda,\Lambda} \tag{16.19}$$

and

$$\gamma_\phi(u)=\frac{\partial\ln Z_\phi}{\partial\ln\kappa}\bigg|_{\lambda,\Lambda} \tag{16.20}$$

Notice that here we defined the sign of the beta function *opposite* to that used in chapter 15. Hence, a positive beta function means that the coupling constant increases as the momentum scale increases, and vice versa.

Equation (16.19) and eq. (16.20) as they stand are somewhat awkward to use, since they involve the bare dimensionless coupling constant u_0 in terms of the dimensionless renormalized coupling constant u instead of the other way around. For this reason, let us use the chain rule to write

$$\kappa\left(\frac{\partial u}{\partial \kappa}\right)_{\lambda}=-\kappa\frac{\left(\frac{\partial \lambda}{\partial \kappa}\right)_{u}}{\left(\frac{\partial \lambda}{\partial u}\right)_{\kappa}} \tag{16.21}$$

where, by dimensional analysis,

$$\lambda=\kappa^{\epsilon}u_0(u,\kappa/\Lambda) \tag{16.22}$$

Using

$$\kappa\left(\frac{\partial \lambda}{\partial \kappa}\right)_{u}=\epsilon\lambda \tag{16.23}$$

we find that the beta function is

$$\beta(u)=\kappa\left(\frac{\partial u}{\partial \kappa}\right)_{\lambda}=-\epsilon\left(\frac{\partial \ln u_0}{\partial u}\right)^{-1} \tag{16.24}$$

In this form, the beta function $\beta(u)$ can be expressed as a power series expansion in the dimensionless renormalized coupling constant u. Each coefficient of this series is a function of $\epsilon=4-D$.

Similarly, we can rewrite the anomalous dimension $\gamma_{\phi}(u)$ as

$$\gamma_{\phi}(u)=\frac{\partial \ln Z_{\phi}}{\partial \ln \kappa}\bigg|_{\lambda}=\kappa\left(\frac{\partial u}{\partial \kappa}\right)_{\lambda}\frac{\partial \ln Z_{\phi}}{\partial u} \tag{16.25}$$

We find

$$\gamma_{\phi}(u)=\beta(u)\frac{\partial \ln Z_{\phi}}{\partial u} \tag{16.26}$$

which also can be written as a power series expansion in the dimensionless renormalized coupling constant u.

16.1.2 General solution of the Callan-Symanzik equations

Let us now solve the Callan-Symanzik equation, eq. (16.18). Let $x=\ln\kappa$, and write the renormalized vertex function as follows:

$$\Gamma_R^{(N)}(\{p_i\},u,\kappa)=\exp\left(\frac{N}{2}\int_{u_1}^{u}du'\frac{\gamma_{\phi}(u')}{\beta(u')}\right)\Phi^{(N)}(\{p_i\},u,\kappa) \tag{16.27}$$

By requiring that this expression satisfies the Callan-Symanzik equation, eq. (16.18), we find that the function $\Phi^{(N)}$ must satisfy the simpler equation

$$\left[\frac{\partial}{\partial x}+\beta(u)\frac{\partial}{\partial u}\right]\Phi^{(N)}(\{p_i\},u,\kappa)=0 \tag{16.28}$$

It is straightforward to see that the solutions to this equation have the general form

$$\Phi^{(N)}(\{p_i\},u,\kappa)=\mathcal{F}^{(N)}\left(\{p_i\},x-\int_{u_2}^{u}\frac{du'}{\beta(u')}\right) \tag{16.29}$$

where $\mathcal{F}^{(N)}$ is a (so far) arbitrary (differentiable) function.

Thus, the Callan-Symanzik equation requires the renormalized vertex functions to have the following form:

$$\Gamma^{(N)}(\{p_i\}, u\kappa) = \exp\left(\frac{N}{2}\int_{u_1}^{u} du' \frac{\gamma_\phi(u')}{\beta(u')}\right) \mathcal{F}^{(N)}\left(\{p_i\}, x - \int_{u_2}^{u} \frac{du'}{\beta(u')}\right) \tag{16.30}$$

where u_1 and u_2 are two integration constants. It should be apparent that the full form *scaling function* $\mathcal{F}^{(N)}$ cannot be obtained in perturbation theory; only its behavior in special asymptotic limits can be determined by these limited means. In other words, the scaling function contains nonperturbative information about the theory.

Let us now rescale all the momenta $\{p_i\}$ by the same scale factor ρ, $p_i \to \rho p_i$. Dimensional analysis implies that the vertex functions should be rescaled by a prefactor of the form

$$\Gamma_R^{(N)}(\{\rho p_i\}, u, \kappa) = \rho^{N+D-ND/2}\Gamma_R^{(N)}(\{p_i\}, u, \kappa) \tag{16.31}$$

Using the form of the general solution, eq. (16.30), of the Callan-Symanzik equation, we find

$$\Gamma_R^{(N)}(\{\rho p_i\}, u, \kappa) = \rho^{N+D-ND/2} \exp\left(\frac{N}{2}\int_{u_1}^{u} du' \frac{\gamma_\phi(u')}{\beta(u')}\right)$$
$$\times \mathcal{F}^{(N)}\left(\{p_i\}, x - \ln\rho - \int_{u_2}^{u} \frac{du'}{\beta(u')}\right) \tag{16.32}$$

where we used that, since $x = \ln\kappa$, the rescaling $\kappa \to \kappa/\rho$ is equivalent to the shift $x \to x - \ln\rho$.

Now we make explicit the notion that the RG induces a flow in the space of coupling constants by introducing a *running* coupling constant $u(\rho)$, which is a function of the scale change ρ. To this effect, define

$$s = \ln\rho = \int_{u}^{u(\rho)} \frac{du'}{\beta(u')} \tag{16.33}$$

such that the running coupling constant $u(s)$ obeys the differential equation

$$\frac{\partial u(s)}{\partial s} = \beta(u(s)) \tag{16.34}$$

with the initial condition for the flow $u(s=0) = u$.

We can rewrite the solution of the Callan-Symanzik equation, eq. (16.32), in terms of the running coupling constant $u(\rho)$ defined in eq. (16.33). Using the form of the general solution, eq. (16.30), of the Callan-Symanzik equation, we obtain

$$\Gamma_R^{(N)}(\{\rho p_i\}, u, \kappa) = \rho^{N+D-ND/2} \exp\left(\frac{N}{2}\left[\int_{u_1}^{u(s)} du' \frac{\gamma_\phi(u')}{\beta(u')} - \int_{u}^{u(s)} du' \frac{\gamma_\phi(u')}{\beta(u')}\right]\right)$$
$$\times \mathcal{F}^{(N)}\left(\{p_i\}, x - \ln\rho - \int_{u_2}^{u(s)} \frac{du'}{\beta(u')}\right) \tag{16.35}$$

which can be written in the equivalent, and more illuminating, form

$$\Gamma_R^{(N)}(\{\rho p_i\}, u, \kappa) = \rho^{N+D-ND/2} \exp\left(\frac{N}{2}\int_u^{u(s)} du' \frac{\gamma_\phi(u')}{\beta(u')}\right) \Gamma_R^{(N)}(\{p_i\}, u(s), \kappa) \tag{16.36}$$

This result implies that a change in the momentum scale in the renormalized N-point vertex function is equivalent to

1) the rescaling of the vertex function by its canonical dimension (i.e., the scaling dimension in the massless free-field theory),
2) the introduction of a running coupling constant (i.e., a flow along an RG trajectory), and
3) an anomalous multiplicative factor associated with the anomalous dimension of the field.

These changes embody the main effects of an RG transformation. It should be evident that the construction used in this chapter has the same physical content as the more intuitive and physically transparent approaches discussed in chapter 15.

16.1.3 Fixed points and scaling behavior

Let us now discuss the consequences of eq. (16.36). This is a rather complex expression. We split the analysis in two steps by (1) first considering its scaling predictions at a fixed point of the RG, and (2) looking at the corrections to the scaling behavior as the fixed point is approached (or departed from).

A: Scaling behavior at a fixed point We begin by looking at the predictions for the behavior of the renormalized N-point vertex functions at an RG fixed point. As in chapter 15, a fixed point of the RG is the value u^* of the dimensionless renormalized coupling constant at which its beta function has a zero: $\beta(u^*) = 0$.

At a fixed point u^*, the integral in the second factor of eq. (16.36) can be computed exactly. The result is

$$\int_u^{u(s)} du' \frac{\gamma_\phi(u')}{\beta(u')} = \int_{s_0}^{s_0+s} \gamma_\phi(u(s'))ds' = \gamma_\phi(u^*)s = \gamma_\phi(u^*) \ln \rho \tag{16.37}$$

Hence, the integral is just the value of the anomalous dimension γ_ϕ *at the fixed point* u^*, multiplied by the logarithm of the scale change. Therefore, at a fixed point, the N-point vertex function obeys the scaling relation

$$\Gamma_R^{(N)}(\{\rho p_i\}, u^*, \kappa) = \rho^{N+D-N\frac{D}{2}-N\gamma_\phi(u^*)/2}\Gamma_R^{(N)}(\{p_i\}, u^*, \kappa) \tag{16.38}$$

In the case of the $N=2$ point function, we obtain

$$\Gamma_R^{(2)}(\rho p, u^*, \kappa) = \rho^{2-\gamma_\phi(u^*)}\Gamma_R^{(2)}(p, u^*, \kappa) \tag{16.39}$$

If we choose the scale change to be $\rho = \kappa/p$, eq. (16.39) implies that

$$\Gamma_R^{(2)}(p, u^*, \kappa) = \left(\frac{p}{\kappa}\right)^{2-\gamma_\phi(u^*)} \Gamma_R^{(2)}(\kappa, u^*, \kappa) \equiv p^2 \left(\frac{p}{\kappa}\right)^{-\gamma_\phi(u^*)} \tag{16.40}$$

since, by definition, the renormalized two-point vertex function of a massless theory is $\Gamma_R^{(2)}(\kappa, u^*, \kappa) = \kappa^2$. Furthermore, this result implies that the renormalized two-point function in real space has the scaling behavior

$$\langle \phi(x)\phi(y)\rangle_R^* = \frac{\text{const.}}{|x-y|^{D-2+\eta}} \tag{16.41}$$

Hence, the scaling dimension of the field ϕ, Δ_ϕ, at the fixed point is

$$\Delta_\phi = \frac{1}{2}(D-2+\gamma_\phi^*) \tag{16.42}$$

from which we deduce that

$$\eta = \gamma_\phi^* \tag{16.43}$$

This explains why $\gamma_\phi(u^*)$ is called the *anomalous* (or fractal) dimension of the field, since it measures the deviation of Δ_ϕ, the scaling dimension of the field ϕ, from its free-field value, $(D-2)/2$.

The main conclusion of this analysis is that, at a general fixed point, the scaling behavior of the observables is different from that of a free-field theory. This result is essentially nonperturbative.

B: Corrections to scaling behavior Most physical systems are not at a fixed point. Therefore, it is important to quantify what corrections, if any, there will be to the predicted scaling behavior away from the fixed point. As we saw in chapter 15, the fixed points can either be stable or unstable, and hence, the flow can be attractive or repulsive. Let us consider both cases. These effects are called corrections to scaling behavior. The types of corrections to scaling depend on the behavior of the beta function near the fixed point. Since, by construction, the beta function is a regular function of the coupling constant, this question depends then on the order of the zero associated with the fixed point.

Let us consider first the simplest case of a linear zero and write

$$\beta(u) = \beta'(u^*)(u-u^*) + O((u-u^*)^2) \tag{16.44}$$

Likewise, we can expand the anomalous dimension function $\gamma_\phi(u)$ near the fixed point u^*:

$$\gamma_\phi(u) = \gamma_\phi(u^*) + \gamma_\phi'(u^*)(u-u^*) + O((u-u^*)^2) \tag{16.45}$$

We have to consider two cases:

1) If $\beta'(u^*) > 0$, then as $\rho \to 0$ (i.e., $s \to -\infty$), the coupling constant u flows in the IR into the fixed point, $u \to u^*$. This is the case of an IR-stable (UV-unstable) fixed point. In other words, this fixed point is stable at long distances (low energies). In this case, the associated operator is *irrelevant*. This is called an IR fixed point.
2) If $\beta'(u^*) < 0$, then as $\rho \to \infty$ (i.e., $s \to +\infty$), the coupling constant u flows in the UV into the fixed point, $u \to u^*$. This is the UV-stable (IR-unstable) fixed point. This fixed point is stable at short distances (high energies). In this case, the associated operator is *relevant*. This is called a UV fixed point.

The theory will generally have several fixed points. As the energy scale is lowered, the RG flow goes from the UV fixed point to the IR fixed point.

Near the fixed point, the integral of eq. (16.37) becomes

$$\begin{aligned}\int_u^{u(s)} du' \frac{\gamma_\phi(u')}{\beta(u')} &= \int_u^{u(s)} \left[\frac{\gamma_\phi(u^*) + \gamma_\phi'(u^*)(u-u^*) + \cdots}{\beta'(u^*)(u-u^*) + \cdots} \right] du' \\ &= \frac{\gamma_\phi(u^*)}{\beta'(u^*)} \ln \left[\frac{u(s) - u^*}{u - u^*} \right] + \frac{\gamma_\phi'(u^*)}{\beta'(u^*)} (u(s) - u) + \cdots \end{aligned} \tag{16.46}$$

Then,

$$\exp\left(\frac{N}{2} \int_u^{u(s)} du' \frac{\gamma_\phi(u')}{\beta(u')} \right) = \exp\left(-\frac{N}{2} \gamma_\phi(u^*) s - \frac{N}{2} \frac{\gamma_\phi(u^*)}{\beta'(u^*)} (u(s) - u) \right) \tag{16.47}$$

In case (1), the IR fixed point, the coupling u flows into the fixed point at u^* in the IR (as $\rho \to 0$). Hence in this case, the scaling form of the N-point vertex function becomes

$$\begin{aligned}\Gamma_R^{(N)}(\{\rho p_i\}, u, \kappa) = \rho^{N+D-\frac{ND}{2} - \frac{N}{2}\gamma_\phi(u^*)} \exp\left(-\frac{N}{2} \frac{\gamma_\phi'(u^*)}{\beta'(u^*)} (u^* - u) \right) \\ \times \Gamma_R^{(N)}(\{p_i\}, u^*, \kappa)\end{aligned} \tag{16.48}$$

For the two-point vertex function, we now find

$$\Gamma_R^{(2)}(p, u, \kappa) = p^2 \left(\frac{p}{\kappa} \right)^{-\gamma_\phi(u^*)} \exp\left(-\frac{\gamma_\phi'(u^*)}{\beta'(u^*)} (u^* - u) \right) \tag{16.49}$$

Hence, the correction to scaling at long distances (low energies) near the IR fixed point amounts to a nonuniversal multiplicative factor correction to the two-point function (and to the other N-point functions).

It is obvious that in case (2), the UV fixed point, we obtain the same result but at short distances (or low energies).

C: Marginality and renormalizability More interesting is the case of a *marginal* operator. In this case, the beta function will typically have a double zero:

$$\beta(u) = A(u - u^*)^2 + \cdots \tag{16.50}$$

This is the case in all theories at their renormalizable dimension (i.e., the critical dimension D_c). In the marginal case, the scaling dimension of the operator associated with the coupling constant u is equal to the dimension of spacetime: $\Delta = D$. This happens for ϕ^4 theory in $D = 4$ dimensions, and also in the case of gauge theories. In the case of the nonlinear sigma model, the critical dimension is $D_c = 2$.

Here too we have two cases: (1) $A > 0$, a marginally stable IR fixed point, and (2) $A < 0$, a marginally stable UV fixed point. The case $A < 0$ is especially important, as it describes asymptotically free theories. The case $A > 0$ is sometimes called asymptotically trivial.

We first need to solve the equation of the beta function:

$$\frac{\partial u}{\partial s} = A(u - u^*)^2 \tag{16.51}$$

The solution is

$$u(s) = u^* - \frac{u - u^*}{A(u^* - u)s + 1} \tag{16.52}$$

We can now solve for s as a function of the running coupling constant $u(s)$:

$$s = -\frac{1}{A}\left[\frac{1}{u(s) - u^*} - \frac{1}{u - u^*}\right] \tag{16.53}$$

As $u(s) \to u^*$, we find

$$s = -\frac{1}{A(u(s) - u^*)} \tag{16.54}$$

Using these results, we get

$$\int_u^{u(s)} du' \frac{\gamma_\phi(u')}{\beta(u')} = \gamma_\phi(u^*)s + \frac{\gamma_\phi'(u^*)}{A} \ln\left(\frac{u(s) - u^*}{u - u^*}\right) \tag{16.55}$$

and, again as $u(s) \to u^*$,

$$\exp\left(-\frac{N}{2}\int_u^{u(s)} du' \frac{\gamma_\phi(u')}{\beta(u')}\right) = \text{const.}\, \rho^{-\frac{N}{2}\gamma_\phi(u^*)} (\ln \rho)^{\frac{N}{2}\frac{\gamma_\phi'(u^*)}{A}} \tag{16.56}$$

In particular, we find that the two-point function near the fixed point is now

$$\Gamma_R^{(2)}(p, u, \kappa) = \text{const}\, p^2 \left(\frac{p}{\kappa}\right)^{-\gamma_\phi(u^*)} \left(\ln \frac{p}{\kappa}\right)^{\gamma_\phi'(u^*)/A} \tag{16.57}$$

Thus, in the marginal cases, we find a logarithmic correction to the fixed-point behavior. This is a consequence of the slow change of the coupling constant near the fixed point. This result applies for the IR fixed point at low energies ($p \to 0$) and the UV fixed point at high energies ($p \to \infty$).

16.2 Perturbative renormalization group for the massless ϕ^4 theory

Our next task is to compute the RG functions for ϕ^4 theory. We only consider the massless theory, and we work to two-loop order.

16.2.1 Renormalization constants to two-loop order

Let us now use dimensional regularization to compute the coefficients of the renormalization constants for the massless theory of eq. (13.65) and eq. (13.67).

We begin with the wave-function renormalization. As we saw in ϕ^4 theory, the first nontrivial contribution appears at two-loop order. The coefficient of this contribution, z_2, is given by eq. (13.65). It can be written as

$$z_2 = \frac{1}{6}\kappa^{2\epsilon} E_3'(\kappa) \tag{16.58}$$

where $E'_3(\kappa)$ is

$$E'_3(\kappa) = \left.\frac{\partial E_3(p)}{\partial p^2}\right|_{p^2=\kappa^2} \tag{16.59}$$

and where

$$E_3(p) = \int \frac{d^D q_1}{(2\pi)^D} \int \frac{d^D q_2}{(2\pi)^D} \frac{1}{q_1^2 q_2^2 (p - q_1 - q_2)^2} \tag{16.60}$$

Using dimensional regularization for the integral in eq. (16.59), one finds

$$z_2 = -\frac{1}{48}\left(\frac{1}{\epsilon} + \frac{5}{4}\right) \tag{16.61}$$

where, as before, $\epsilon = 4 - D$. Notice that z_2 has a simple pole in ϵ. We are keeping only the nonvanishing terms as $\epsilon \to 0$.

Next look at the coefficients λ_2 and λ_3 of the expansion for the coupling constant to two-loop order, eq. (13.67). The computation of the coefficient λ_2 involves only the one-loop integral of eq. (13.29) of the massless theory, evaluated at the symmetric point of the four external momenta with a momentum transfer scale $p^2 = \kappa^2$:

$$I_{SP} = \int \frac{d^D q}{(2\pi)^D} \frac{1}{q^2 (p - q)^2} \tag{16.62}$$

We have already done this calculation in section 13.8 using dimensional regularization. Using these results, one finds that at finite but small $\epsilon = 4 - D$, the coefficient is

$$\lambda_2 = \frac{3}{2}\left(\frac{1}{\epsilon} + \frac{1}{2}\right) \tag{16.63}$$

As shown in eq. (13.67), the coefficient λ_3 involves the one-loop integral I_{SP} of eq. (13.29); the two-loop integrals I_{4SP} of eq. (13.31),

$$I_{4SP} = \int \frac{d^D q_1}{(2\pi)^D} \int \frac{d^D q_2}{(2\pi)^D} \frac{1}{q_1^2 (p_1 + p_2 - q_1)^2 q_2^2 (p_3 - (q_1 + q_2))^2} \tag{16.64}$$

(again, with the external momenta at the symmetry point); and the coefficient z_2 computed above. Using these results and the computation of I_{4SP} at the symmetric point (cf. Amit, 1980), we obtain

$$\lambda_3 = \frac{9}{4\epsilon^2} + \frac{37}{24\epsilon} \tag{16.65}$$

Notice the double pole in ϵ of this two-loop coefficient.

16.2.2 RG functions and fixed points

Our next task is to compute the beta function $\beta(u)$ and the anomalous dimension γ_ϕ in terms of the expressions of eq. (16.2)–eq. (16.4), which that relate the bare and renormalized coupling constant and mass, and the wave-function renormalization Z_ϕ. We will use the two-loop results for ϕ^4 theory given in section 16.2.1. There we wrote an expansion of the dimensionless bare coupling constant u_0 as a power series in the dimensionless renormalized coupling constant u of the form

$$u_0 = u\left(1 + \lambda_2 u + \lambda_3 u^2 + \cdots\right) \tag{16.66}$$

and a similar expansion for the wave-function renormalization Z_ϕ:

$$Z_\phi = 1 + z_2 u^2 + \cdots \tag{16.67}$$

These coefficients can be computed in terms of Feynman diagrams either (1) by using renormalization conditions, or (2) by using the minimal subtraction scheme in dimensionally regularized diagrams. Here we will do it both ways.

The beta function $\beta(u)$, using the result of eq. (16.24), can be written as a series expansion in powers of u:

$$\beta(u) = -\epsilon \left(\frac{\partial \ln u_0}{\partial u}\right)^{-1} = -\epsilon u\left(1 - \lambda_2 u + 2(\lambda_2^2 - \lambda_3)u^2\right) + \cdots \tag{16.68}$$

where λ_2 and λ_3 are given explicitly in eq. (16.63) and eq. (16.65).

Similarly, we can expand the anomalous dimensions γ_ϕ:

$$\gamma_\phi = \beta(u)\frac{\partial \ln Z_\phi}{\partial u} = -2\epsilon\, z_2 u^2 + \cdots \tag{16.69}$$

where z_2 is given by eq. (16.61).

In chapter 13, we used renormalization conditions to calculate the coefficients λ_2, λ_3, and z_2 (eqs. (16.63), (16.65), and (16.61), respectively). Using these results, we find the following result for the beta function $\beta(u)$ and the anomalous dimension γ_ϕ to two-loop order:

$$\beta(u) = \kappa\frac{\partial u}{\partial \kappa} = -\epsilon u + \frac{3}{2}\left(1 + \frac{\epsilon}{2}\right)u^2 - \frac{17}{12}u^3 + O(u^4) \tag{16.70}$$

$$\gamma_\phi = \frac{1}{24}u^2 + O(u^3) \tag{16.71}$$

Notice that although the coefficients λ_2, λ_3, and z_2 have poles in ϵ, the beta function and the anomalous dimension are regular functions of ϵ (as they should be).

The fixed points are the zeros of the beta function, which is a polynomial in the renormalized dimensionless coupling constant u. By construction, it always has a zero at $u = 0$, representing the massless free-field theory. For $D < 4$ (or $\epsilon > 0$), for small enough ϵ it also has a zero at a value u^*, which can be expressed as a power series expansion in ϵ. At two-loop order, the nontrivial (Wilson-Fisher) fixed point is at

$$u^* = \frac{2}{3}\epsilon + \frac{34}{9}\epsilon^2 + O(\epsilon^3) \tag{16.72}$$

Since to any finite order in perturbation theory, the beta function is a polynomial function of u, one may wonder whether there are other possible fixed points. However, only the trivial fixed point at $u = 0$ and the Wilson-Fisher fixed point $u^* = f(\epsilon)$ (where $f(\epsilon)$ is a regular function of epsilon that vanishes as $\epsilon \to 0$) should be considered, since all other fixed points will be zeros at a finite distance from $u = 0$ that is not tuned by ϵ. Furthermore, these additional zeros will change drastically with the order of the expansion of the beta function. In other words, finite fixed points not tuned by ϵ are unphysical consequences of the truncation of the beta function to a polynomial with finite degree.

The anomalous dimension γ_ϕ of eq. (16.71) at this fixed point is

$$\gamma_\phi(u^*) = \frac{\epsilon^2}{54} + O(\epsilon^3) \tag{16.73}$$

Thus, as promised, we see that at two-loop order, the field ϕ at the Wilson-Fisher fixed point has an anomalous dimension.

It is now easy to see that for $D < 4$, the free-field fixed point is unstable in the IR, since the dimension of the operator ϕ^4 is $\Delta_4(FF) = 2(D-2) \leq D$. In contrast, the two-loop beta function, eq. (16.70), tells us that the dimension of ϕ^4 at the Wilson-Fisher (WF) fixed point is $\Delta_4(WF) = 4 - \frac{17}{27}\epsilon^2 > D$. Thus, the same operator is relevant at one fixed point (free field) and irrelevant at the other (Wilson-Fisher). Hence, the IR RG flow is from the free field to the Wilson-Fisher fixed point.

16.3 Dimensional regularization with minimal subtraction

We now use dimensional regularization in the minimal subtraction (MS) scheme to compute the renormalization functions. Once again we consider the massless theory in D dimensions. As we have seen, the Feynman diagrams expressed in terms of integrals develop singularities in the form of poles in $\epsilon = 4 - D$ (as $D \to 4$). In fact, the only vertex functions that have primitive divergences that develop poles in ϵ are $\frac{\partial \Gamma^{(2)}}{\partial p^2}$ and $\Gamma^{(4)}$. The poles in ϵ correspond to logarithmic divergences in the momentum scale in $D = 4$ dimensions.

Once again, let us write the bare dimensionless coupling constant u_0 and the wave-function renormalization Z_ϕ as a series expansion in powers of the dimensionless renormalized coupling constant u:

$$u_0 = u + \sum_{n=1}^{\infty} \lambda_n(\epsilon) u^n \tag{16.74}$$

$$Z_\phi = 1 + \sum_{n=1}^{\infty} z_n(\epsilon) u^n \tag{16.75}$$

The coefficients $\{\lambda_n(\epsilon)\}$ and $\{z_n(\epsilon)\}$ are singular functions of ϵ chosen in such a way that

$$\Gamma_R^{(2)}(p; u, \kappa) = Z_\phi \Gamma^{(2)}(p; u_0, \kappa) \tag{16.76}$$

$$\Gamma_R^{(4)}(\{p_i\}; u, \kappa) = Z_\phi^2 \Gamma^{(4)}(\{p_i\}; u_0, \kappa) \tag{16.77}$$

are finite as $D \to 4$. In the minimal subtraction scheme, the coefficients are determined by imposing the condition that the poles in ϵ found in the bare functions $\Gamma^{(2)}$ and $\Gamma^{(4)}$ are minimally subtracted.

This scheme works as follows. First write the expressions for the bare two-point function to two-loop order:

$$\Gamma^{(p)}(p;u_0,\kappa)=p^2(1-B_2u_0^2) \tag{16.78}$$

where we used the condition that the renormalized theory is massless. Similarly, the bare four-point function is found to be

$$\Gamma^{(4)}(\{p_i\};u_0,\kappa)=\kappa^{\epsilon}u_0\left(1-A_1u_0+(A_2^{(1)}+A_2^{(2)})u_0^2\right) \tag{16.79}$$

where

$$A_1=\frac{1}{2}\left[I\left(\frac{p_1+p_2}{\kappa}\right)+\text{two permutations}\right] \tag{16.80}$$

$$A_2^{(2)}=\frac{1}{2}\left[I_4\left(\frac{p_1}{\kappa},\ldots,\frac{p_4}{\kappa}\right)+\text{five permutations}\right] \tag{16.81}$$

The integrals were given in section 13.2 (cf. eq. (13.36)).

First consider $\Gamma^{(2)}$. To order u^2, we find

$$\Gamma_R^{(2)}(p,\kappa)=p^2(1+z_1u+z_2u^2)(1-B_2u^2) \tag{16.82}$$

$$=p^2\left[1+z_1u+(z_2-B_2)u^2+O(u^3)\right] \tag{16.83}$$

Thus, we set

$$z_1=0,\qquad z_2=[B_2]_{\text{sing}} \tag{16.84}$$

where $[B_2]_{\text{sing}}$ is the singular part (in ϵ) of B_2. Hence,

$$z_2=[B_2]_{\text{sing}}=-\frac{1}{48\epsilon} \tag{16.85}$$

Therefore, in the MS scheme, the wave-function renormalization is

$$Z_\phi=1-\frac{u^2}{48\epsilon}+O(u^3) \tag{16.86}$$

Next we use the same approach for the four-point function. To order u^3, we can write the renormalized four-point vertex function as

$$\begin{aligned}\Gamma_R^{(4)}(p_1,\ldots,p_4;u,\kappa)=&\ \kappa^{\epsilon}(1+2z_2u^2)(u+\lambda_2u^2+\lambda_3u^3)\\&-(u^2+2\lambda_2u^3)\frac{1}{2}\left[I\left(\frac{p_1+p_2}{\kappa}\right)+\text{two permutations}\right]\\&+u^3\left\{\frac{1}{4}\left[I^2\left(\frac{p_1+p_2}{\kappa}\right)+\text{two permutations}\right]\right.\\&\left.+\frac{1}{2}\left[I_4\left(\frac{p_1}{\kappa},\ldots,\frac{p_4}{\kappa}\right)+\text{five permutations}\right]\right\}\end{aligned} \tag{16.87}$$

Collecting terms, we get

$$\Gamma_R^{(4)}(p_1,\ldots,p_4;u,\kappa)=\kappa^{\epsilon}\left\{u+u^2\left[\lambda_2-\frac{1}{2}\left[I\left(\frac{p_1+p_2}{\kappa}\right)+\text{two permutations}\right]\right]\right.$$
$$+u^3\left[\lambda_3+2z_2-\frac{3}{2}\lambda_2\left[I\left(\frac{p_1+p_2}{\kappa}\right)+\text{two permutations}\right]\right.$$
$$+\frac{1}{4}\left[I^2\left(\frac{p_1+p_2}{\kappa}\right)+\text{two permutations}\right]$$
$$\left.\left.\frac{1}{2}\left[I_4\left(\frac{p_1}{\kappa},\ldots,\frac{p_4}{\kappa}\right)+\text{five permutations}\right]\right]\right\} \tag{16.88}$$

We then cancel the singularities by setting

$$\lambda_2=\frac{1}{2}\left(\left[I\left(\frac{p_1+p_2}{\kappa}\right)\right]_{\text{sing}}+\text{two permutations}\right)=\frac{3}{2\epsilon} \tag{16.89}$$

Similarly,

$$\lambda_3=-2z_2+\lambda_2\left(\left[I\left(\frac{p_1+p_2}{\kappa}\right)\right]_{\text{sing}}+\text{two permutations}\right)$$
$$-\frac{1}{4}\left(\left[I^2\left(\frac{p_1+p_2}{\kappa}\right)\right]_{\text{sing}}+\text{two permutations}\right)$$
$$-\frac{1}{2}\left(\left[I_4\left(\frac{p_1}{\kappa},\ldots,\frac{p_4}{\kappa}\right)\right]_{\text{sing}}+\text{five permutations}\right) \tag{16.90}$$

from which we have

$$\lambda_3=\frac{9}{4\epsilon^2}-\frac{17}{24\epsilon} \tag{16.91}$$

We find that to two-loop order, the bare and renormalized dimensionless coupling constants are related by

$$u_0=u+\frac{3}{2\epsilon}u^2+\left(\frac{9}{4\epsilon^2}-\frac{17}{24\epsilon}\right)u^3+O(u^4) \tag{16.92}$$

We can now use these results to obtain the beta function and the anomalous dimension to two-loop order in the MS scheme:

$$\beta(u)=-\epsilon u+\frac{3}{2}u^2-\frac{17}{12}u^3+O(u^4) \tag{16.93}$$

$$\gamma_\phi(u)=\frac{u^2}{24}+O(u^3) \tag{16.94}$$

Notice that the expression for the two-loop beta functions of eq. (16.70) and eq. (16.93) are slightly different. These differences arise from the different renormalization schemes used. Nevertheless, these differences only amount to a redefinition of the location of the fixed point, which does not change the value of the exponents.

16.4 The nonlinear sigma model in two dimensions

In section 12.5, we derived the low-energy effective action of a ϕ^4 theory with an $O(N)$ global symmetry spontaneously broken to its $O(N-1)$ subgroup, the $O(N)$ nonlinear sigma model. The action of the nonlinear sigma model is

$$S=\frac{1}{2g}\int d^Dx\left(\partial_\mu \boldsymbol{n}(x)\right)^2, \qquad \text{with } \boldsymbol{n}^2(x)=1 \tag{16.95}$$

where the constraint is imposed locally. Here g is the coupling constant. In this section, we discuss the renormalizability and the RG for the nonlinear sigma model in $D=2$ and $D=2+\epsilon$ dimensions.

As we saw in section 13.4, by dimensional analysis, we expect the nonlinear sigma model to be renormalizable in $D=2$ spacetime dimensions. Although this theory is nonrenormalizable as an expansion around the trivial fixed point at $g=0$, it is renormalizable around its nontrivial fixed point in $D=2+\epsilon$, which we expect to have the same universal properties as the Wilson-Fisher fixed point in the same dimension.

The nonlinear sigma model is interesting for several reasons. It arises in high-energy physics as the low-energy limit of theories of spontaneously broken chiral symmetry, and hence as a model to describe pions. In that context, the coupling constant g is the inverse of the pion decay constant. It also arises in classical statistical mechanics as the long-wavelength description of the Heisenberg model of a classical ferromagnet, described in terms of an N-component spin degree of freedom with unit length. Here, the coupling constant is T/J, where T is the temperature and J is the exchange constant. This model also arises as the effective action of quantum antiferromagnets. Generalizations of the nonlinear sigma model also play a key role in perturbative string theory, where (supersymmetric) nonlinear sigma models describe contributions to the string amplitudes, and even in the theory of localization of electrons in random potentials.

However, in addition to its relevance to wide areas of physics, this model is of particular interest because it has many analogies with non-abelian four-dimensional gauge theories, while being much simpler.

16.5 Generalizations of the nonlinear sigma model

The $O(N)$ nonlinear sigma model, whose action is given in eq. (16.95), has a simple geometrical interpretation that leads to many generalizations. The field $\boldsymbol{n}(x)$ of the nonlinear sigma model takes values on the $(N-1)$-dimensional sphere S_{N-1} which is the coset of the broken symmetry group $O(N)$ with the unbroken global symmetry $O(N-1)$ (i.e., $S_{N-1}\cong O(N)/O(N-1)$). In this sense, the field can be regarded as a mapping of the Euclidean spacetime $\mathbb{R}^D$ onto the sphere S_{N-1}. We call the sphere S_{N-1} the *target manifold*. Hence, the target space of the nonlinear sigma model is a manifold that is a coset. We will see in chapter 19 that with natural boundary conditions, it can also be regarded as a map from the (Euclidean) spacetime compactified to the sphere S_D onto the sphere S_{N-1}.

16.5.1 The $\mathbb{CP}^{N-1}$ nonlinear sigma model

The first generalization we discuss is the $\mathbb{CP}^{N-1}$ nonlinear sigma model. Here $\mathbb{CP}^{N-1}$ is the complex projective space of $N-1$ dimensions. Let us define a real field with $M=N^2-1$

components, $n^a(x)$ (with $a=1,\ldots,M$) and a field $z_\alpha(x)$, with N *complex* components, such that

$$n^a(x) = \sum_{\alpha,\beta=1}^{N} z_\alpha^*(x)\, \tau^a_{\alpha\beta}\, z_\beta(x) \tag{16.96}$$

with the constraint

$$\sum_{\alpha=1}^{N} |z_\alpha(x)|^2 = 1 \tag{16.97}$$

In eq. (16.96), the $\tau^a_{\alpha\beta}$ are the N^2-1 generators of $SU(N)$, and they satisfy

$$\sum_{a=1}^{N^2-1} \tau^a_{\alpha\beta}\tau^a_{\gamma\delta} = N\delta_{\alpha\gamma}\delta_{\beta\delta} - \delta_{\alpha\beta}\delta_{\gamma\delta} \tag{16.98}$$

It follows that the constraint of eq. (16.97) on the complex field z_α implies that the n^a field is also constrained such that

$$\boldsymbol{n}^2 = N-1 \tag{16.99}$$

In the special case $N=2$, we have a mapping of a constrained two-component complex field z_α to the *three-component* constrained real field n^a. In this special case, the symmetry is $SU(2)$, and the mapping of eq. (16.96) relates the $SU(2)$ symmetry of the field z_α with the group $O(3)$ of the field n^a. For $N=2$, this mapping, known as the Hopf map, relates the fundamental (spinor) representation of $SU(2)$ to the adjoint (vector) representation of $SU(2)$.

In the general case, the N-component complex field is in the fundamental representation of $SU(N)$. With the constraint of eq. (16.97), it describes the breaking of $SU(N)$ down to its $SU(N-1)$ subgroup. However, the real field n^a of eq. (16.96) is invariant under the local $U(1)$ (gauge) symmetry

$$z_\alpha(x) \mapsto e^{i\phi(x)} z_\alpha(x) \tag{16.100}$$

Configurations of the field z_α differing by this $U(1)$ gauge transformation are physically equivalent. Hence, in this case, the target manifold is not the coset $SU(N)/SU(N-1)$. Instead, it is the complex projective space

$$\mathbb{CP}^{N-1} \cong \frac{SU(N)}{SU(N-1)\otimes U(1)} \tag{16.101}$$

The simplest local Lagrangian that has these symmetries is

$$\mathcal{L}[z_\alpha, z_\alpha^*, \mathcal{A}_\mu] = \frac{1}{g}\left|\left(\partial_\mu - i\mathcal{A}_\mu(x)\right) z_\alpha(x)\right|^2, \tag{16.102}$$

with the local constraint $||z||^2=1$. Here $A_\mu(x)$ is a $U(1)$ gauge field. This action is invariant under the global symmetry

$$z_\alpha(x) \mapsto U_{\alpha\beta} z_\beta(x) \tag{16.103}$$

where $U \in SU(N)$, and the local $U(1)$ (gauge) symmetry,

$$z_\alpha(x) \mapsto e^{i\phi(x)} z_\alpha(x), \qquad \mathcal{A}_\mu(x) \mapsto \mathcal{A}_\mu(x) + \partial_\mu\phi(x) \tag{16.104}$$

Notice that the action of the $\mathbb{CP}^{N-1}$ nonlinear sigma model, eq. (16.102), depends on the $U(1)$ gauge field $\mathcal{A}_\mu$ only through the covariant derivative of the complex field z_α. The (Euclidean) partition function for the $\mathbb{CP}^{N-1}$ nonlinear sigma model is

$$Z_{\mathbb{CP}^{N-1}} = \int \mathcal{D}z_\alpha \mathcal{D}z_\alpha^* \mathcal{D}\mathcal{A}_\mu \exp\left(-\int d^D x\, \mathcal{L}[z_\alpha, z_\alpha^*, \mathcal{A}_\mu]\right) \prod_x \delta(||\boldsymbol{z}(x)||^2 - 1) \qquad (16.105)$$

It is apparent from this partition function that the gauge field $\mathcal{A}_\mu$ does not have any dynamics of its own, and furthermore, that the action is a quadratic form in the gauge field. Thus we can integrate out this field explicitly. Indeed, the equation of motion of the gauge field is

$$\frac{\delta S}{\delta \mathcal{A}_\mu} = \frac{\delta}{\delta \mathcal{A}_\mu}\left(|\partial_\mu z_\alpha|^2 + i\left(z_\alpha^* \partial_\mu z_\alpha - (\partial_\mu z_\alpha)^* z_\alpha\right)\mathcal{A}_\mu + \mathcal{A}_\mu^2\right) = 0 \qquad (16.106)$$

which is equivalent to the identification

$$\mathcal{A}_\mu(x) \equiv -\frac{i}{2}\left(z_\alpha^* \partial_\mu z_\alpha - (\partial_\mu z_\alpha)^* z_\alpha\right) \qquad (16.107)$$

If we now substitute this expression for the gauge field A_μ back into the Lagrangian in eq. (16.102), we obtain a nonlinear action for the complex field z_α. In the special case of $N=2$, one obtains the identity

$$\frac{1}{4}\left(\partial_\mu \boldsymbol{n}\right)^2 = \left|\left(\partial_\mu - i\mathcal{A}_\mu\right) z_\alpha\right|^2 \qquad (16.108)$$

and the resulting action is equal to the action of the $O(3)$ nonlinear sigma model. This defines the Hopf mapping of the sphere S_3 of the z field onto the sphere S_2 of the $\boldsymbol{n}$-space.

By inspecting the Lagrangian in eq. (16.102), we see that the coupling constant of the $\mathbb{CP}^{N-1}$ nonlinear sigma model has the same units as the coupling constant of the $O(N)$ nonlinear sigma model. Thus, these models are also expected to be renormalizable in $D=2$ dimensions.

16.5.2 The principal chiral field

Another generalization is the theory of the principal chiral field. Let G be a compact Lie group, and let the principal chiral field be $g(x)$, which takes values on G (i.e., $g(x) \in G$). The Lagrangian is then

$$\mathcal{L} = \frac{1}{2u^2}\text{tr}\left(\partial_\mu g^{-1}(x)\partial^\mu g(x)\right) \qquad (16.109)$$

where u is the coupling constant. This Lagrangian is invariant under global transformations

$$g(x) \mapsto h^{-1} g(x) v \qquad (16.110)$$

where $h \in G$ and $v \in G$. The global symmetry is thus $G_R \otimes G_L$. The nonlinear nature of the field is hidden in the condition that $g(x)$ is a group element. For instance, if $G = U(N)$, then $g(x)$ satisfies $g(x)^{-1} = g(x)^\dagger$. Similar constraints apply more generally.

16.5.3 General nonlinear sigma models

This construction can be made more general. Consider a field $\phi(x)$ whose target space is a differentiable manifold M. The Euclidean Lagrangian is

$$\mathcal{L}[\phi] = \frac{1}{2u} a^{2-D} g_{ij}[\phi(x)] \partial_\mu \phi^i(x) \partial_\mu \phi^j(x) \tag{16.111}$$

where u is the dimensionless coupling constant, and $g^{ij}[\phi(x)]$ is a Riemannian metric on the manifold M, with a being a short-distance cutoff scale. A homogeneous space of the form of a coset $M = G/H$ is a special case. We have already discussed the cases in which M is a sphere and a complex projective space. An example, arising in the theory of Anderson localization, is the case of a coset of the form $O(n+m)/(O(n) \otimes O(m))$ (in the limit n, $m \to 0$). In string theory, the field is the coordinate of the bosonic string in a target space, such as the Calabi-Yau manifolds.

16.6 The $O(N)$ nonlinear sigma model in perturbation theory

All nonlinear sigma models are renormalizable in $D = 2$ spacetime dimensions. Here we focus on the simpler case of the $O(N)$ model, following the work of Polyakov (1975b) and Brézin and Zinn-Justin (1976a). The general case was proven by Friedan et al. (1984). It will be discussed at the end of this section.

To discuss the structure of perturbation theory for the $O(N)$ nonlinear sigma model, we need to choose coordinates on its target space, the sphere. To this end, we decompose the field into a longitudinal field $\sigma(x)$, representing the component along the direction of symmetry breaking, and $N-1$ fields $\boldsymbol{\pi}(x)$, representing the Goldstone bosons, transverse to the direction of symmetry breaking.

Hence, we write $\boldsymbol{n}(x) = (\sigma(x), \boldsymbol{\pi}(x))$, subject to the local constraint $\boldsymbol{n}^2(x) = \sigma^2(x) + \boldsymbol{\pi}^2(x) = 1$. The partition function of the $O(N)$ nonlinear sigma model is the functional integral

$$Z[\boldsymbol{J}(x)] = \int \mathcal{D}\sigma \mathcal{D}\boldsymbol{\pi} \prod_x \delta(\sigma(x)^2 + \boldsymbol{\pi}(x)^2 - 1) \exp\left(-\int d^D x\, \mathcal{L}[\sigma, \boldsymbol{\pi}; \boldsymbol{J}]\right) \tag{16.112}$$

where $\mathcal{L}[\sigma, \boldsymbol{\pi}; \boldsymbol{J}]$ is the Euclidean Lagrangian of the nonlinear sigma model

$$\mathcal{L}[\sigma, \boldsymbol{\pi}; \boldsymbol{J}] = \frac{1}{2g}\left((\partial_\mu \sigma)^2 + (\partial_\mu \boldsymbol{\pi})^2\right) - (J_\sigma(x)\sigma(x) + \boldsymbol{J}_\pi(x) \cdot \boldsymbol{\pi}(x)) \tag{16.113}$$

and where $\boldsymbol{J}(x) = (J_\sigma(x), \boldsymbol{J}_\pi(x))$ is a symmetry-breaking field.

Except for the local constraint $\boldsymbol{n}^2 = 1$, this theory looks like a free field. We can deal with the constraint in one of two ways. One option is to replace the delta function that enforces the constraint by a path integral over a Lagrange multiplier field $\lambda(x)$:

$$\prod_x \delta(\sigma(x)^2 + \boldsymbol{\pi}(x)^2 - 1) = \int \mathcal{D}\lambda \exp\left(-\int d^D x \lambda(x)(\sigma(x)^2 + \boldsymbol{\pi}(x)^2 - 1)\right) \tag{16.114}$$

This leads to a path integral over the fields σ, $\boldsymbol{\pi}$, and λ. The other option is to solve the constraint and work with fewer degrees of freedom. In this section, we use the second option, which, in the end, is a matter of choice. Notice that this issue is very similar to the problem in gauge theory and the role of gauge fixing. There we had the choice of fixing the gauge first and quantizing later, or to quantize first and impose the gauge condition later.

Thus, we will first integrate out the longitudinal component $\sigma(x)$, using

$$\int d\sigma\, \delta(\sigma^2+\boldsymbol{\pi}^2-1)\, F(\sigma,\boldsymbol{\pi}) = \frac{1}{2\sqrt{1-\boldsymbol{\pi}^2}}\, F(\sqrt{1-\boldsymbol{\pi}^2},\boldsymbol{\pi}) \tag{16.115}$$

In other words, the quantity $\mathcal{J}[\boldsymbol{\pi}(x)]$,

$$\mathcal{J}[\boldsymbol{\pi}(x)] = \prod_x \left(2\sqrt{1-\boldsymbol{\pi}^2(x)}\right)^{-1} \tag{16.116}$$

is the Jacobian of the change of variables. In fact, in our choice of coordinates,

$$\mathcal{D}\boldsymbol{\pi}\ \mathcal{J}[\boldsymbol{\pi}] \equiv \frac{\mathcal{D}\boldsymbol{\pi}}{2\sqrt{1-\boldsymbol{\pi}^2}} \tag{16.117}$$

(which uses a shorthand notation for the Jacobian) is the $O(N)$-invariant Haar measure for the sphere S_{N-1}.

Then we can write the partition function as

$$Z[\boldsymbol{J}] = \int \frac{\mathcal{D}\boldsymbol{\pi}}{2\sqrt{1-\boldsymbol{\pi}^2}} \exp(-S_{\text{eff}}[\boldsymbol{\pi};\boldsymbol{J}]) \tag{16.118}$$

where the effective action is

$$S_{\text{eff}}[\boldsymbol{\pi};\boldsymbol{J}] = \frac{1}{2g}\int d^Dx \left[\left(\partial_\mu\sqrt{1-\boldsymbol{\pi}^2(x)}\right)^2 + (\partial_\mu \boldsymbol{\pi}(x))^2\right] - \int d^Dx \left(J_\sigma(x)\sqrt{1-\boldsymbol{\pi}^2(x)} + \boldsymbol{J}_\pi(x)\cdot\boldsymbol{\pi}(x)\right) \tag{16.119}$$

It is important to note that, even though the effective action $S_{\text{eff}}[\boldsymbol{\pi};\boldsymbol{J}]$ seems to have only a global $O(N-1)$ symmetry, the inclusion of the Jacobian factor in the functional integral renders the partition function globally $O(N)$ invariant. The Jacobian factor will play a key role in what follows. Notice that we could have alternatively incorporated the Jacobian factor in the effective action with a contact term of the form

$$S_{\text{contact}} = -\frac{1}{2a^D}\int d^Dx\, \ln(1-\boldsymbol{\pi}^2(x)) \tag{16.120}$$

where a is a short-distance cutoff (where we have implicitly used a lattice regulator).

In what follows, we denote $J_\sigma = H$ and $\boldsymbol{J}_\pi = \boldsymbol{J}$ (which is a vector with $N-1$ components). We work with the partition function in the form

$$Z[H,\boldsymbol{J}] = \int \frac{\mathcal{D}\boldsymbol{\pi}}{2\sqrt{1-\boldsymbol{\pi}^2}} \exp\left[-\frac{1}{g}S[\boldsymbol{\pi},H] + \int d^Dx\, \boldsymbol{J}(x)\cdot\boldsymbol{\pi}(x)\right] \tag{16.121}$$

where

$$S[\boldsymbol{\pi}, H] = \int d^D x \left[\frac{1}{2}(\partial_\mu \boldsymbol{\pi}(x))^2 + \frac{1}{2} \frac{(\boldsymbol{\pi}(x) \cdot \partial_\mu \boldsymbol{\pi}(x))^2}{(1 - \boldsymbol{\pi}^2(x))} - H(x)\sqrt{1 - \boldsymbol{\pi}^2(x)} \right] \tag{16.122}$$

For this form of the action, we see that an expansion of the partition function in powers of the coupling constant g is just the loop expansion for a theory with the action of eq. (16.122).

16.6.1 Ward identities

Ward identities play a central role in the renormalization of the nonlinear sigma model and in the proof of renormalizability. To see how this works, recall that the actual global symmetry is not $O(N)$ but the coset $O(N)/O(N-1)$.

Let us consider a global infinitesimal transformation mixing the fields σ and $\boldsymbol{\pi}$:

$$\delta \boldsymbol{\pi}(x) = \sqrt{1 - \boldsymbol{\pi}^2(x)}\, \boldsymbol{\omega}, \qquad \delta \sqrt{1 - \boldsymbol{\pi}^2(x)} = -\boldsymbol{\omega} \cdot \boldsymbol{\pi}(x) \tag{16.123}$$

where $\boldsymbol{\omega}$ is an $(N-1)$-component constant infinitesimal vector. Assume that the UV regulator of the theory is consistent with the global symmetry. This is the case with dimensional regularization and with lattice regularization and other schemes. Hence, the action, the integration measure of the functional integral, and the regularization preserve the full global symmetry, and the partition function is invariant. Now define

$$F[\boldsymbol{J}, H] = g \ln Z[\boldsymbol{J}, H] \tag{16.124}$$

which is the generating function of the connected correlators of the nonlinear sigma model. Then the invariance of the partition function under the infinitesimal transformation of the fields, eq. (16.123), implies the Ward identity

$$\int d^D x \left[J_i(x) \frac{\delta F}{\delta H(x)} - H(x) \frac{\delta F}{\delta J_i(x)} \right] = 0 \tag{16.125}$$

As in section 12.2, we define $\Gamma[\boldsymbol{\pi}, H]$, the generating functional of the one-particle irreducible vertex functions of the $\boldsymbol{\pi}$ fields, as the Legendre transform of F:

$$\Gamma[\boldsymbol{\pi}, H] = \int d^D x \, \langle \boldsymbol{\pi}(x) \rangle \cdot \boldsymbol{J}(x) - F[\boldsymbol{J}, H] \tag{16.126}$$

Following the same line of argument used in chapter 12, the following identities hold:

$$\begin{aligned} \langle \boldsymbol{\pi}(x) \rangle = \frac{\delta F}{\delta \boldsymbol{J}(x)}, \qquad & \frac{\delta F}{\delta H(x)} = \langle \sigma(x) \rangle \\ \frac{\delta \Gamma}{\delta \boldsymbol{\pi}(x)} = \boldsymbol{J}(x), \qquad & \frac{\delta \Gamma}{\delta H(x)} = \langle \sigma(x) \rangle \end{aligned} \tag{16.127}$$

where

$$\frac{\delta \Gamma}{\delta H} = -\frac{\delta F}{\delta H} \tag{16.128}$$

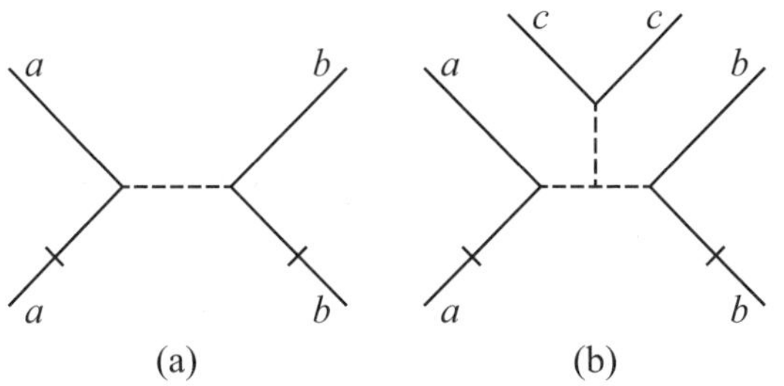

Figure 16.1 Feynman rules for the $O(N)$ nonlinear sigma model: (a) $n=0$ vertex, and (b) $n=1$ vertex. Here $a, b, c=1, \ldots, N-1$. Vertices with $n \geq 2$ have more insertions. The perpendicular dashes represent derivatives on these external legs. The broken lines are shown for clarity.

The Ward identity for the generating functional Γ is

$$\int d^D x \left[\frac{\delta \Gamma}{\delta \boldsymbol{\pi}(x)} \frac{\delta \Gamma}{\delta H(x)} + H(x) \boldsymbol{\pi}(x) \right] = 0 \tag{16.129}$$

where we used the notation $\langle \boldsymbol{\pi}(x) \rangle \equiv \boldsymbol{\pi}(x)$. We will see next that this Ward identity can be used to show that the nonlinear sigma model is renormalizable in $D=2$ dimensions.

16.6.2 Primitive divergences

The first step is to analyze the primitive divergences in the expansion in powers in g. We will see that $D=2$ is special. In two dimensions, the coupling constant g is dimensionless, and so are the fields $\boldsymbol{\pi}$, and $\sigma = \sqrt{1-\boldsymbol{\pi}^2}$. The operator $\left(\partial_\mu \boldsymbol{\pi}\right)^2$ has dimension 2, as do the operator $(\boldsymbol{\pi} \cdot \partial_\mu \boldsymbol{\pi})^2/(1-\boldsymbol{\pi}^2)$ and the symmetry-breaking field H. In particular, if we expand the action of eq. (16.122) in powers of the field $\boldsymbol{\pi}$, the operators obtained in each term of the expansion all have dimension 2, since all of them involve just two derivatives and powers of the field $\boldsymbol{\pi}$. That is to say, all the operators of the expansion are equally relevant (marginal, actually). Thus, we cannot truncate the expansion at any order. Moreover, a truncation of the expansion would break the symmetry, since the terms are related by symmetry. A general term in this expansion is a vertex of the form $(\boldsymbol{\pi} \cdot \partial_\mu \boldsymbol{\pi})^2 \boldsymbol{\pi}^{2n}$ (see figure 16.1a,b).

Let us first determine the Feynman rules for the nonlinear sigma model. This requires that we formally expand the action in eq. (16.122) in powers of the coupling constant g. To make this process more explicit, let us define the rescaled field $\boldsymbol{\varphi}(x) = \boldsymbol{\pi}(x)/\sqrt{g}$ and expand the resulting action. The result is the Lagrangian

$$\begin{aligned} \mathcal{L} = & \frac{1}{2} \left(\partial_\mu \boldsymbol{\varphi}\right)^2 + \sum_{n=0}^{\infty} \frac{g^{n+1}}{2} \left(\boldsymbol{\varphi}^2\right)^n \left(\boldsymbol{\varphi} \cdot \partial_\mu \boldsymbol{\varphi}\right)^2 \\ & - H(x) \left(1 + \sum_{n=1}^{\infty} \frac{(-1)^n (2(n-1))!}{2^{2n-1}(n-1)!n!} g^n \left(\boldsymbol{\varphi}^2\right)^n \right) + \sqrt{g} \boldsymbol{J}_\pi \cdot \boldsymbol{\varphi} \end{aligned} \tag{16.130}$$

The lowest-order vertex, with $n=0$, is shown in figure 16.1a. In momentum space, it carries a weight of $\frac{g}{2} \boldsymbol{q}^{(1)} \cdot \boldsymbol{q}^{(2)}$, where $\boldsymbol{q}^{(1)}$ and $\boldsymbol{q}^{(2)}$ are the momenta on the two external legs.

Consider the Feynman diagrams obtained at one-loop order with $H=0$ and $\boldsymbol{J}=0$. The leading contributions to the propagator of the $\boldsymbol{\pi}$ field are shown in figure 16.2a,b.

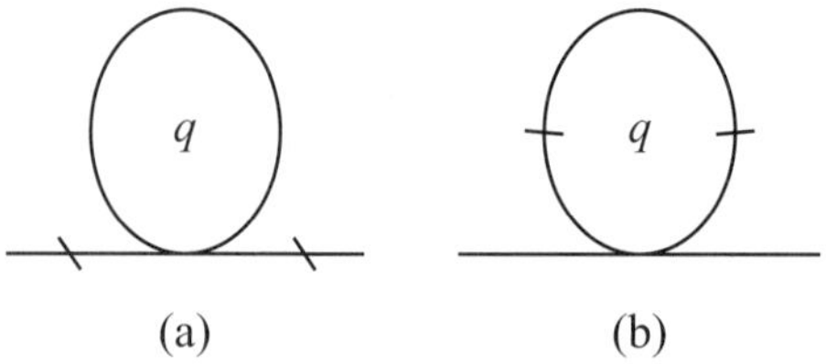

Figure 16.2 One-loop contributions to the two-point function of π. The dashes denote derivatives (a) in the external lines, and (b) in the internal loop.

Figure 16.2a has one derivative on each external leg. The internal loop yields a contribution proportional to the integral

$$\int \frac{d^2q}{(2\pi)^2} \frac{1}{q^2} \propto \ln \Lambda \tag{16.131}$$

and is logarithmically divergent in the UV momentum cutoff Λ. This graph contributes to a term in the effective action, which in momentum space is of the form $p^2|\pi(p)|^2$, and hence contributes to the wave-function renormalization. In contrast, the contribution of the graph in figure 16.2b has the form

$$\int \frac{d^2q}{(2\pi)^2} \frac{1}{q^2} q^\mu q^\nu \propto g^{\mu\nu}\Lambda^2 \tag{16.132}$$

and is quadratically divergent in the UV cutoff Λ and looks like a mass renormalization.

However, the field π is a Goldstone boson, and the Ward identity guarantees that it should be massless at every order in perturbation theory. So what has gone wrong? The remedy to this problem is readily found by noting that the action of the nonlinear sigma model has a contact term, eq. (16.120), arising from the integration measure. The contact term yields a contribution of quadratic order in the field π, which also looks like a quadratically divergent mass term and exactly cancels the offending term. Below we will prove this statement explicitly. However, dimensional regularization is often used (as we will do in section 16.7.1). We have already seen that this regularization replaces a logarithmic divergence with a pole in ϵ. Furthermore, in dimensional regularization, quadratic divergences are regularized to zero, and one does not have to be concerned with this problem. However, in other schemes, such as lattice regularization or Pauli-Villars, these cancellations must be checked at every order in perturbation theory.

16.7 Renormalizability of the two-dimensional nonlinear sigma model

We now examine the restrictions that the Ward identities impose on the possible structure of the singularities in the nonlinear sigma model. By power counting, we only need to worry about operators of scaling dimensions two or less, since (1) any operator with higher dimension would make the theory nonrenormalizable, and (2) such operators will be irrelevant at distance scales large compared to the UV cutoff $a \sim \Lambda^{-1}$.

To this end, let us expand the effective one-particle irreducible action Γ in powers of the coupling constant g:

$$\Gamma = \sum_{n=0}^{\infty} g^n \Gamma^{(n)} \tag{16.133}$$

This is done by organizing the Feynman graphs in powers of g. At the lowest order in g, the tree level, we recover (as expected) the classical action of the nonlinear sigma model:

$$\begin{aligned} \Gamma^{(0)} &= S(\boldsymbol{\pi}, H) \\ &= \int d^2x \left[\frac{1}{2} \left(\partial_\mu \boldsymbol{\pi} \right)^2 + \frac{1}{2} \frac{\left(\boldsymbol{\pi} \cdot \partial_\mu \boldsymbol{\pi} \right)^2}{1 - \boldsymbol{\pi}^2} - H(x) \sqrt{1 - \boldsymbol{\pi}^2} \right] + \text{measure terms} \end{aligned} \tag{16.134}$$

Let us now expand the effective action to order one loop, $\Gamma = \Gamma^{(0)} + g\Gamma^{(1)}$. We use the Ward identity of eq. (16.129) and demand that it hold at every order in g, resulting in the requirement:

$$\int d^2x \left[\frac{\delta \Gamma^{(0)}}{\delta \boldsymbol{\pi}} \frac{\delta \Gamma^{(1)}}{\delta H} + \frac{\delta \Gamma^{(1)}}{\delta \boldsymbol{\pi}} \frac{\delta \Gamma^{(0)}}{\delta H} \right] = 0 \tag{16.135}$$

Next, define the operator

$$\Gamma^{(0)} \star \equiv \int d^2x \left[\frac{\delta \Gamma^{(0)}}{\delta \boldsymbol{\pi}} \frac{\delta}{\delta H} + \frac{\delta \Gamma^{(0)}}{\delta H} \frac{\delta}{\delta \boldsymbol{\pi}} \right] \tag{16.136}$$

in terms of which eq. (16.135) becomes

$$\Gamma^{(0)} \star \Gamma^{(1)} = 0 \tag{16.137}$$

As the cutoff is removed, $\Lambda \to \infty$, the quantity $\Gamma^{(1)}$ will develop a singular part, which we denote by $\Gamma^{(1)}_{\text{div}}$. However, since the divergent and finite parts (the latter one will be denoted by $\Gamma^{(1)}_{\text{reg}}$) have different dependence in the regulator, they must obey eq. (16.137) separately. Hence, $\Gamma^{(1)}_{\text{div}}$ must satisfy an equation of the form of eq. (16.137). The divergence contained in $\Gamma^{(1)}$ can be canceled by adding a counterterm to the action $S(\boldsymbol{\pi}, H)$ of the form $gS_1(\boldsymbol{\pi}, H)$ such that

$$S_1(\boldsymbol{\pi}, H) = -\Gamma^{(1)}_{\text{div}} + O(g) \tag{16.138}$$

In this way, the new, renormalized action $S + gS_1$ satisfies the Ward identity to all orders in g.

By power counting, we know that $\Gamma^{(1)}_{\text{div}}$ is a local function of the field $\boldsymbol{\pi}$ with scaling dimension 2. Since the scaling dimension of $H(x)$ is also 2, it follows that $\Gamma^{(1)}_{\text{div}} = O(H)$. Thus, we can write $\Gamma^{(1)}_{\text{div}}$ as an expression of the form

$$\Gamma^{(1)}_{\text{div}} = \int d^2x \left[B(\boldsymbol{\pi}) + H(x) C(\boldsymbol{\pi}) \right] \tag{16.139}$$

where $B(\boldsymbol{\pi})$ contains at most two derivatives, and $C(\boldsymbol{\pi})$ has no derivatives. The requirement that $\Gamma^{(1)}_{\rm div}$ obeys eq. (16.137) yields the following condition on B and C:

$$0=\int d^2x\left[\frac{\delta\Gamma^{(0)}}{\delta H(x)}\frac{\delta C}{\delta\boldsymbol{\pi}}H(x)+\frac{\delta\Gamma^{(0)}}{\delta\boldsymbol{\pi}}C(\boldsymbol{\pi})+\frac{\delta\Gamma^{(0)}}{\delta H(x)}\frac{\delta B}{\delta\boldsymbol{\pi}}\right] \tag{16.140}$$

From the expression of the tree-level action, $\Gamma^{(0)}$, we find the explicit expressions

$$\begin{aligned}\frac{\delta\Gamma^{(0)}}{\delta H(x)}&=-\sqrt{1-\boldsymbol{\pi}^2}\\ \frac{\delta\Gamma^{(0)}}{\delta\boldsymbol{\pi}}&=-\partial^2\boldsymbol{\pi}+\frac{(\boldsymbol{\pi}\cdot\partial_\mu\boldsymbol{\pi})}{1-\boldsymbol{\pi}^2}\partial_\mu\boldsymbol{\pi}+\frac{(\boldsymbol{\pi}\cdot\partial_\mu\boldsymbol{\pi})^2}{(1-\boldsymbol{\pi}^2)^2}\boldsymbol{\pi}+\frac{H(x)}{\sqrt{1-\boldsymbol{\pi}^2}}\boldsymbol{\pi}\end{aligned} \tag{16.141}$$

By collecting terms, we can write

$$\begin{aligned}&\int d^2x\left[-\sqrt{1-\boldsymbol{\pi}^2}\frac{\delta C}{\delta\boldsymbol{\pi}}+\frac{\boldsymbol{\pi}}{\sqrt{1-\boldsymbol{\pi}^2}}C(\boldsymbol{\pi})\right]H(x)\\ &+\int d^2x\left\{\left[-\partial^2\boldsymbol{\pi}+\boldsymbol{\pi}\frac{\partial^2(1-\boldsymbol{\pi}^2)^{1/2}}{\sqrt{1-\boldsymbol{\pi}^2}}\right]C-\sqrt{1-\boldsymbol{\pi}^2}\frac{\delta B}{\delta\boldsymbol{\pi}}\right\}=0\end{aligned} \tag{16.142}$$

Since $H(x)$ and $\boldsymbol{\pi}(x)$ are arbitrary functions of the coordinates, this integral implies that $C(\boldsymbol{\pi})$ must be the solution of

$$\sqrt{1-\boldsymbol{\pi}^2}\frac{\delta C}{\delta\boldsymbol{\pi}}=\frac{\boldsymbol{\pi}}{\sqrt{1-\boldsymbol{\pi}^2}}C(\boldsymbol{\pi}) \tag{16.143}$$

and that

$$\int d^2x\left\{\left[-\partial^2\boldsymbol{\pi}+\boldsymbol{\pi}\frac{\partial^2\sqrt{1-\boldsymbol{\pi}^2}}{\sqrt{1-\boldsymbol{\pi}^2}}\right]C-\sqrt{1-\boldsymbol{\pi}^2}\frac{\delta B}{\delta\boldsymbol{\pi}}\right\}=0 \tag{16.144}$$

The most general solution is

$$\Gamma^{(1)}_{\rm div}=\lambda S^{(0)}+\mu\int d^2x\left[\frac{(\boldsymbol{\pi}\cdot\partial_\mu\boldsymbol{\pi})^2}{(1-\boldsymbol{\pi}^2)^2}+\frac{H(x)}{\sqrt{1-\boldsymbol{\pi}^2}}\right] \tag{16.145}$$

where $S^{(0)}$ is the classical action, and λ and μ are two singular functions of the regulator.

If we now define $S^{(1)}=-\Gamma^{(1)}_{\rm div}$ and write the field $\boldsymbol{\pi}$ in terms of a rescaled field $Z^{1/2}\boldsymbol{\pi}$ (i.e., this is the wave-function renormalization), we can write the renormalized action as

$$S=\int d^2x\left\{\frac{Z}{Z_1}\left[\frac{1}{2}(\partial_\mu\boldsymbol{\pi})^2+\frac{1}{2}\frac{(\boldsymbol{\pi}\cdot\partial_\mu\boldsymbol{\pi})^2}{\left(\frac{1}{Z}-\boldsymbol{\pi}^2\right)}\right]-H\sqrt{\frac{1}{Z}-\boldsymbol{\pi}^2}\right\} \tag{16.146}$$

where

$$\frac{Z}{Z_1} \equiv 1 - \lambda g + O(g^2)$$

$$Z \equiv 1 - 2\mu g + O(g^2) \tag{16.147}$$

Now define the renormalized dimensionless coupling constant t_R and the renormalized field H_R by the conditions

$$g = t_R Z_1 \kappa^{2-D}$$

$$H = H_R \frac{Z_1}{\sqrt{Z}} \tag{16.148}$$

where κ is an arbitrary renormalization scale. Thus, to one-loop order, it is sufficient to do a renormalization of the coupling constant and a wave-function renormalization.

However, since we rescaled the field, we must modify the transformation laws to read

$$\delta\pi = \sqrt{\frac{1}{Z} - \pi^2}\,\omega \tag{16.149}$$

and the measure $d\pi/\sqrt{1-\pi^2}$ is no longer invariant. To solve this problem, we must replace the measure by $d\pi/\sqrt{1/Z-\pi^2}$, which has an effect that will become apparent at two-loop level. Hence we have succeeded in making the renormalized one-particle irreducible function Γ finite, and it satisfies the Ward identity.

We will now show that the one-loop result implies that the nonlinear sigma model is renormalizable to all orders in the loop expansion. At every order, the key is the Ward identity. Let us use the following induction argument. Assume that to order $n-1$, we have succeeded in renormalizing the theory. Thus, at order n we must satisfy

$$\Gamma^{(0)} \star \Gamma^{(n)} = -\left(\Gamma^{(1)} \star \Gamma^{(n-1)} + \Gamma^{(2)} \star \Gamma^{(n-2)} + \cdots\right) \tag{16.150}$$

However, by hypothesis, the right-hand side contains only renormalized terms. Hence the singular part $\Gamma^{(n)}_{\text{div}}$ must also satisfy the same equation that we solved at one-loop order, that is,

$$\Gamma^{(0)} \star \Gamma^{(n)}_{\text{div}} = 0 \tag{16.151}$$

Hence, $\Gamma^{(n)}_{\text{div}}$ has the same form as $\Gamma^{(1)}_{\text{div}}$. Therefore, to all orders in a expansion in the coupling constant, we can renormalize the nonlinear sigma model by renormalizing the coupling constant and with a wave-function renormalization. This completes the proof of renormalizability.

In summary, in two dimensions, the renormalized action is

$$\frac{S}{g} = \int d^2x \left\{ \frac{Z}{2Z_1 g}\left[\left(\partial_\mu \pi\right)^2 + \left(\partial_\mu \sqrt{1 - Z\pi^2}\right)^2 \right] - \frac{H}{g}\sqrt{\frac{1}{Z} - \pi^2} \right\} \tag{16.152}$$

In general dimension $D > 2$, the result is

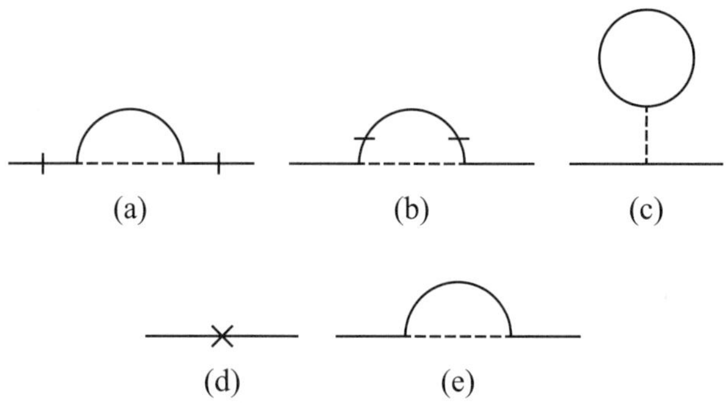

Figure 16.3 One-loop contributions to $\Gamma^{(2)}$. The dashes denote derivatives in the external lines (a), and in the internal loop (b); the tadpole diagram (c) and diagram (e) originate from the symmetry-breaking field H; here we have included the lowest-order (quadratic) term coming from the integration measure (d).

$$\frac{S}{g}=\frac{\kappa^{D-2}}{2Z_1 t}\int d^2x\left\{Z\left[(\partial_\mu \boldsymbol{\pi})^2+\frac{(\boldsymbol{\pi}\cdot\partial_\mu\boldsymbol{\pi})^2}{(\frac{1}{Z}-\boldsymbol{\pi}^2)^2}\right]-\frac{HZ_1}{\sqrt{Z}}\sqrt{1-Z\boldsymbol{\pi}^2}\right\} \tag{16.153}$$

where

$$g=t\kappa^{2-D} \tag{16.154}$$

16.7.1 Renormalization to one-loop order

Let us carry out this program to one-loop order. It will suffice to study the renormalization of the two-point function of the $\boldsymbol{\pi}$ fields. First observe that the bare propagator of the $\boldsymbol{\pi}$ field in momentum space is

$$G_0^{ij}(p)=\delta_{ij}\frac{g}{p^2+\frac{H}{g}} \tag{16.155}$$

where $i,j=1,\ldots,N-1$, and that (except terms coming from the integration measure) every vertex contributes with a weight of $1/g$. The coupling constant g is thus the parameter that organizes the loop expansion.

To zeroth (tree-level) order, the one-particle irreducible two-point function is then

$$\Gamma_{ij}^{(2)}(p)=\frac{1}{g}\left(p^2+\frac{H}{g}\right) \tag{16.156}$$

The Feynman diagrams for the one-loop contributions to $\Gamma^{(2)}(p)$ are shown in figure 16.3. The explicit expression for $\Gamma^{(2)}$ at one-loop order is

$$\begin{aligned}\Gamma^{(2)}(p)=&\frac{1}{g}\left(p^2+\frac{H}{g}\right)\\&+p^2\int\frac{d^Dq}{(2\pi)^D}\frac{1}{p^2+\frac{H}{g}}+\int\frac{d^Dq}{(2\pi)^D}\frac{q^2}{q^2+\frac{H}{g}}+\frac{H}{2g}(N-1)\int\frac{d^Dq}{(2\pi)^D}\frac{1}{q^2+\frac{H}{g}}\\&-\frac{\Lambda^D}{(2\pi)^D}+\frac{H}{g}\int\frac{d^Dq}{(2\pi)^D}\frac{1}{q^2+\frac{H}{g}}\end{aligned} \tag{16.157}$$

Here, the second through the fourth terms on the right-hand side are the expressions for the Feynman diagrams in figure 16.3a–c, and the final two terms correspond to the diagrams in figure 16.3d,e. Notice the negative contribution of the fifth term, which arises from the integration measure.

It is straightforward to see that the following identity holds:

$$\int \frac{d^D q}{(2\pi)^D} \frac{q^2}{q^2 + \frac{H}{g}} - \frac{\Lambda^D}{(2\pi)^D} + \frac{H}{g} \int \frac{d^D q}{(2\pi)^D} \frac{1}{q^2 + \frac{H}{g}} = 0 \tag{16.158}$$

For $D=2$, this identity tells us that the quadratically divergent contributions, which could have induced a mass term (as well as the N-independent logarithmically divergent contributions), cancel exactly. This is a manifestation of the Ward identity.

As a result, the final form of the 1-PI two-point function to one-loop order is

$$\Gamma^{(2)}(p) = \frac{1}{g}\left(q^2 + \frac{H}{g}\right) + \left(p^2 + \frac{(N-1)}{2g} H\right) \int \frac{d^D q}{(2\pi)^D} \frac{1}{q^2 + \frac{H}{g}} \tag{16.159}$$

The divergences in $\Gamma^{(2)}$ can be taken care of by means of a renormalized dimensionless coupling constant t and a wave-function renormalization Z:

$$\Gamma_R^{(2)}(p, t, H_R, \kappa) = Z\Gamma^{(2)}(p, g, H, \Lambda) \tag{16.160}$$

with

$$g = t\kappa^{-\epsilon} Z_1, \qquad H = H_R \frac{Z_1}{\sqrt{Z}} \tag{16.161}$$

where we set $\epsilon = D-2$.

The renormalization constants Z_1 and Z can be expanded in a power series of the dimensionless renormalized coupling constant t_R:

$$\begin{aligned} Z &= 1 + at + O(t^2) \\ Z_1 &= 1 + bt + O(t^2) \end{aligned} \tag{16.162}$$

The renormalized 1-PI two-point function becomes

$$\begin{aligned} \Gamma_R^{(2)} &= \frac{p^2}{t}\kappa^{D-2}\left[1 + t\left(a - b + \kappa^{2-D} \int \frac{d^D q}{(2\pi)^D} \frac{1}{q^2 + \frac{H_R}{t}}\right) + \cdots\right] \\ &+ \frac{H_R}{t}\kappa^{D-2}\left[1 + t\left(\frac{a}{2} + \kappa^{2-D}\frac{(N-1)}{2} \int \frac{d^D q}{(2\pi)^D} \frac{1}{q^2 + \frac{H_R}{t}}\right) + \cdots\right] \end{aligned} \tag{16.163}$$

The coefficients a and b will be chosen so that $\Gamma_R^{(2)}(p)$ is finite. There are many ways of doing this. For instance, we can choose these coefficients to cancel the singular part of the integral in eq. (16.163). If we use dimensional regularization, this procedure is equivalent to the minimal subtraction procedure of 't Hooft and Veltman (1972). Here we will use the following somewhat different choice, which subtracts the expressions at a renormalization scale κ:

$$a = -\kappa^{2-D}(N-1)\int \frac{d^Dq}{(2\pi)^D}\frac{1}{q^2+\kappa^2}$$

$$b = -\kappa^{2-D}(N-2)\int \frac{d^Dq}{(2\pi)^D}\frac{1}{q^2+\kappa^2} \tag{16.164}$$

Hence,

$$a-b = -\kappa^{2-D}\int \frac{d^Dq}{(2\pi)^D}\frac{1}{q^2+\kappa^2} \tag{16.165}$$

With this choice, the renormalization constants Z and Z_1 become

$$Z = 1 - t\kappa^{2-D}(N-1)\int \frac{d^Dq}{(2\pi)^D}\frac{1}{q^2+\kappa^2}$$

$$Z_1 = 1 - t\kappa^{2-D}(N-2)\int \frac{d^Dq}{(2\pi)^D}\frac{1}{q^2+\kappa^2} \tag{16.166}$$

Next we study the behavior of the integral for dimensions $D = 2+\epsilon$ and expand it in powers of ϵ:

$$\int \frac{d^Dq}{(2\pi)^D}\frac{1}{q^2+\kappa^2} = \frac{1}{(4\pi)^{D/2}}\Gamma\left(1-\frac{D}{2}\right)\kappa^{D-2} \tag{16.167}$$

Using the asymptotic expression for the Euler gamma function

$$\Gamma(z) = \frac{1}{z} - \gamma + O(z) \tag{16.168}$$

where γ is the Euler-Mascheroni constant, the integral becomes

$$\int \frac{d^Dq}{(2\pi)^D}\frac{1}{q^2+\kappa^2} = -\frac{1}{2\pi\epsilon} + O(1) \tag{16.169}$$

With these prescriptions, the renormalization constants take the simple form

$$Z = 1 + \frac{(N-1)}{\epsilon}t + O(t^2)$$

$$Z_1 = 1 + \frac{(N-2)}{\epsilon}t + O(t^2) \tag{16.170}$$

Therefore, this procedure is equivalent to minimal subtraction (MS).

Hence, we find that, to lowest order in ϵ, the renormalized 1-PI two-point function is

$$\Gamma_R^{(2)}(p,t,H_R,\kappa) = \frac{1}{t}\left(p^2+\frac{H_R}{t}\right) - \frac{1}{2}\left(p^2+\frac{(N-2)}{2}\frac{H_R}{t}\right)\ln\left(\frac{H_R}{t\kappa^2}\right) + O(t^2) \tag{16.171}$$

Notice that it is not possible to set $H_R \to 0$. In other words, the true IR behavior is not accessible to perturbation theory.

16.7.2 RG for the nonlinear sigma model

Next we compute the beta function for the nonlinear sigma model,

$$\beta(t) = \kappa \frac{\partial t}{\partial \kappa}\Big|_{\text{bare}} \tag{16.172}$$

where, as before, we hold the bare theory fixed. Since the bare coupling constant g and the renormalized dimensionless coupling constant t are related by $g = \kappa^{-\epsilon} Z_1 t$ (with $\epsilon = D - 2$), varying the renormalization scale κ at fixed bare coupling constant

$$\kappa \frac{\partial g}{\partial \kappa} = 0 \tag{16.173}$$

leads to the result

$$0 = -\epsilon t + \beta(t)\left(1 + t\frac{\partial \ln Z_1}{\partial t}\right) \tag{16.174}$$

From eq. (16.170), we find

$$\frac{\partial \ln Z_1}{\partial t} = \frac{N-2}{\epsilon} \tag{16.175}$$

Hence, at one-loop order, the beta function is given by

$$\beta(t) = \epsilon t - (N-2)t^2 + O(t^3) \tag{16.176}$$

Similarly, we find that the anomalous dimension is

$$\begin{aligned}\gamma(t) &= \kappa \frac{\partial \ln Z}{\partial \kappa}\Big|_{\text{bare}} \\ &= \beta(t)\frac{\partial \ln Z}{\partial t} \\ &= (N-1)t + O(t^2)\end{aligned} \tag{16.177}$$

Notice that, contrary to what we found in ϕ^4 theory, the nonlinear sigma model already has an anomalous dimension at the one-loop level.

We can similarly derive the Callan-Symanzik equations to the N-point 1-PI vertex functions of the π fields by requiring that the bare functions $\Gamma_B^{(N)}$ remain constant as the renormalization scale changes:

$$\kappa \frac{\partial \Gamma_B^{(N)}}{\partial \kappa}(p_i y, g, H, \Lambda) = 0 \tag{16.178}$$

resulting in the Callan-Symanzik equation

$$\left[\kappa\frac{\partial}{\partial \kappa} + \beta(t)\frac{\partial}{\partial t} - \frac{N}{2}\gamma(t) + \left(\frac{\gamma(t)}{2} + \frac{\beta(t)}{t} - (D-2)\right)H_R\frac{\partial}{\partial H_R}\right]\Gamma_R^{(N)}(p_i y, t, H_R, \kappa) = 0 \tag{16.179}$$

We will shortly solve this equation along the RG flow near a fixed point.

Next, we need to find the fixed points of this theory (i.e., the values t^* where $\beta(t^*) = 0$). We have two cases.

1) For $D \leq 2$, the only finite fixed point is at $t^* = 0$. This fixed point is IR unstable (or, equivalently, it is the UV fixed point). The case of $D = 2$ is special, and we discuss it in detail below. In this case, the fixed point is marginally IR unstable.
2) For $D > 2$, the fixed point at $t^* = 0$ is IR stable and has a finite basin of attraction in the IR. This fixed point represents the spontaneously broken symmetry state where this theory has $N - 1$ exactly massless Goldstone bosons. At this fixed point, the theory is no longer renormalizable in the sense that it has a large number of irrelevant operators, which dominate the UV behavior. However, for $D = 2 + \epsilon$, a new finite fixed point emerges at

$$t_c = \frac{\epsilon}{N-2} + O(\epsilon^2) \tag{16.180}$$

This fixed point is IR unstable. We will see that this is the UV fixed point of the theory. The theory is renormalizable at this nontrivial fixed point of the $2 + \epsilon$ expansion.

Thus, for $t < t_c$, the theory flows in the IR toward the fixed point at $t^* = 0$, and for $t > t_c$ the theory flows in the IR toward $t \to \infty$. However, the behavior in this phase is not accessible to perturbation theory. Nevertheless, we will be able to infer that in this phase, the correlation length is finite and the symmetry is unbroken. To justify this inference requires a nonperturbative definition of the theory, such as a lattice regularization, where it becomes identical to the $O(N)$ Heisenberg model of classical statistical mechanics. Then this phase is simply the high-temperature phase. Another option, which we discuss in chapter 17, is to define this theory in terms of its $1/N$ expansion.

Let us now explore the structure of the correlators. Dimensionally, in momentum space, the 1-PI vertex function has units $[\Gamma_R^{(N)}] = \kappa^D$. Hence, under a change of momentum scale ρ, the renormalized vertex functions must obey

$$\Gamma_R^{(N)}(\{p_i\}, t, H_R, \kappa) = \rho^D \Gamma_R^{(N)}\left(\left\{\frac{p_i}{\rho}\right\}, t, \frac{H_R}{\rho^2}, \frac{\kappa}{\rho}\right) \tag{16.181}$$

As we saw in section 15.2, the correlation length ξ (i.e., the inverse of the mass) satisfies the Callan-Symanzik–type equation:

$$\left(\kappa\frac{\partial}{\partial\kappa} + \beta(t)\frac{\partial}{\partial t}\right)\xi(t, \kappa) = 0 \tag{16.182}$$

For $t < t_c$, the solution is

$$\xi(t, \kappa) = \kappa^{-1} \exp\left(\int_0^t \frac{dt'}{\beta(t')}\right) \tag{16.183}$$

Since the slope of the beta function at t_c is $\beta'(t_c) = -\epsilon$, we find that the correlation length has the universal form

$$\xi(t, \kappa) = \kappa^{-1}\left|\frac{t - t_c}{t_c}\right|^{-\nu} \tag{16.184}$$

with a correlation length exponent

$$\nu = -\frac{1}{\beta(t_c)} = \frac{1}{\epsilon} + O(1) \tag{16.185}$$

Hence, as $t \to t_c$ from below, the correlation length diverges. In chapter 21, we will see that at t_c, the theory exhibits conformal invariance. Notice that at one-loop order of the ϵ expansion, the value of the exponent ν is the same for all N. This is only correct at one-loop order, and at higher orders, it has N-dependent corrections.

Now let us examine the behavior of the vacuum expectation value of the field σ (i.e., the magnitude of the broken symmetry). At the classical level, $\sigma = \sqrt{1-\boldsymbol{\pi}^2}$. Here we set the symmetry-breaking field to be zero: $H=0$. Under the renormalization procedure that we use, the field σ, just as for the field $\boldsymbol{\pi}$, must be rescaled by the wave-function renormalization:

$$\sigma_R(t,\kappa) = Z^{-1/2}\sigma_B(g,\Lambda) \tag{16.186}$$

Consequently, the renormalized field σ_R obeys the Callan-Symanzik equation

$$\left(\beta(t)\frac{\partial}{\partial t} + \frac{\gamma(t)}{2}\right)\sigma_R(t,\kappa) = 0 \tag{16.187}$$

which has the solution

$$\sigma_R(t,\kappa) = \text{const.}\exp\left(-\frac{1}{2}\int_0^t \frac{\gamma(t')}{\beta(t')}dt'\right) \tag{16.188}$$

For $t \to t_c$ from below, we obtain that $\sigma_R(t,\kappa)$ obeys the scaling law

$$\sigma_R(t,\kappa) \sim |t-t_c|^\beta \tag{16.189}$$

where the exponent β is

$$\beta = -\frac{\gamma(t_c)}{2\beta'(t_c)} = \frac{N-1}{2(N-2)} + O(\epsilon) \tag{16.190}$$

Notice that as $N \to \infty$, $\beta \to 1/2$. We will revisit this result in chapter 17, where we discuss the large-N regime of field theories.

Finally, let us examine the renormalization of the two-point function. Here, too, we set $H=0$. For a change of scale ρ, define the running dimensionless coupling constant $t(\rho)$,

$$t(\rho) = \exp\left(\int_t^{t(\rho)} \frac{dt'}{\beta(t')}\right) \tag{16.191}$$

to find

$$\begin{aligned}\Gamma_R^{(2)}(p,t,\kappa) &= \rho^D\Gamma_R^{(2)}\left(\frac{p}{\rho}, t, \frac{\kappa}{\rho}\right)\\ &= \rho^2\exp\left(-\int_t^{t(\rho)} \frac{(\gamma(t')-\epsilon)}{\beta(t')}dt'\right)\Gamma_R^{(2)}\left(\frac{p}{\rho}, t(\rho), \kappa\right)\end{aligned} \tag{16.192}$$

where we have set $D=2+\epsilon$.

At the fixed point, $t(\rho) = t_c$, we get

$$\Gamma_R^{(2)}(p, t_c, \kappa) = \rho^2 e^{-(\gamma(t_c) - \epsilon)\ln\rho}\, \Gamma_R^{(2)}\left(\frac{p}{\rho}, t_c, \kappa\right)$$

$$\equiv \rho^{2-\eta}\Gamma_R^{(2)}\left(\frac{p}{\rho}, t_c, \kappa\right) \tag{16.193}$$

From the condition that

$$\Gamma_R^{(2)}(\kappa, t_c, \kappa) = \kappa^2 \tag{16.194}$$

we find the standard result

$$\Gamma_R^{(2)}(p, t_c, \kappa) = \left(\frac{p}{\kappa}\right)^{2-\eta}\kappa^2 \tag{16.195}$$

where the exponent η is

$$\eta = \gamma(t_c) - \epsilon = \frac{\epsilon}{N-2} + O(\epsilon^2) \tag{16.196}$$

Thus, we find a nonvanishing anomalous dimension already at one-loop order. In contrast, in ϕ^4 theory, the anomalous dimension only appears at two-loop order.

We now turn to the important case of $D = 2$ dimensions. In $D = 2$ dimensions, $t_c = 0$, and the one-loop beta function reduces to

$$\beta(t) = -(N-2)t^2 + O(t^3) \tag{16.197}$$

and

$$\gamma(t) = (N-1)t + O(t^2) \tag{16.198}$$

From the integral

$$\int_{t_0}^{t} \frac{dt'}{\beta(t')} = \frac{1}{N-2}\left(\frac{1}{t} - \frac{1}{t_0}\right) \tag{16.199}$$

we find that in $D = 2$, the correlation length $\xi(t, \kappa)$ behaves as

$$\xi(t, \kappa) = \left[\kappa^{-1}\exp\left(-\frac{1}{(N-2)t_0}\right)\right]\exp\left(\frac{1}{(N-2)t}\right) \tag{16.200}$$

Hence, at $D = 2$, the correlation length diverges as $t \to 0$ with an essential singularity in t. We will shortly find that the same behavior occurs in $D = 4$-dimensional Yang-Mills gauge theory.

Furthermore, the running coupling constant $t(\rho)$ now obeys

$$\ln\rho = \frac{1}{N-2}\left(\frac{1}{t(\rho)} - \frac{1}{t}\right) \tag{16.201}$$

Hence, as the momentum scale ρ increases, the running coupling constant $t(\rho)$ flows to zero, albeit logarithmically slowly:

$$t(\rho) = \frac{t}{1 + (N-2)t\ln\rho} \approx \frac{1}{(N-2)\ln\rho} \to 0, \text{ as } \rho \to \infty \tag{16.202}$$

In other terms, the effective (running) coupling constant becomes very weak at large momenta (or short distances). This behavior is known as *asymptotic freedom*. In this regime, the renormalized coupling constant is weak, and renormalized perturbation theory works.

The flip side of this result is the behavior at long distances. By inspection of eq. (16.202), we see that there is a scale ρ^* where $t(\rho^*) \to \infty$. It easy to see that this scale is determined by the correlation length, $\rho^* \sim \xi^{-1}$. To be more precise, this momentum scale sets a lower bound on the applicability of renormalized perturbation theory. The physics at length scales longer than ξ (even the existence of a finite correlation length ξ itself!) is beyond the reach of perturbation theory.

This result is most remarkable. In $D=2$ dimensions, the classical action is dimensionless and scale invariant, and the coupling constant is dimensionless. However, as we see, at the quantum level, the theory flows to strong coupling in the IR, and a nontrivial length (and hence a mass) scale appears. This phenomenon is often called *dimensional transmutation*. It is characteristic of asymptotically free theories. This behavior is also found in four-dimensional non-abelian gauge theories and in the Kondo problem.

In this context, it is instructive to compute the explicit form of the two-point 1-PI function $\Gamma_R^{(2)}$. This can be done by evaluating the integrals of eq. (16.192). Here we need to extract the changes in the two-point function due to the effects of the RG flow. We have already discussed this problem in section 16.1.3, where we examined the effects of corrections to scaling. Using the beta function $\beta(t)$ and the anomalous dimensions $\gamma(t)$ in $D=2$, we find

$$\exp\left(-\int_t^{t(\rho)} \frac{\gamma(t')}{\beta(t')} dt'\right) = \left(\frac{t(\rho)}{t}\right)^{(N-1)/(N-2)}$$
$$= (1+t(N-2)\ln \rho)^{-(N-1)/(N-2)} \tag{16.203}$$

Setting $\rho = p/\kappa$, and plugging this result into eq. (16.192), we find that the two-point function has a logarithmic correction to scaling of the form

$$\Gamma_R^{(2)}(p,t,\kappa) = p^2 \left(t(N-2)\ln(p/\kappa)\right)^{-(N-1)/(N-2)} \tag{16.204}$$

This is the behavior of the two-point function at short distances (or high energies): $\Lambda \gg p \gg \rho^*$, where $\rho^* = \xi^{-1}$ is the scale at which $t(\rho) \to \infty$. The same behavior is also found in four-dimensional non-abelian gauge theories.

16.7.3 Renormalization of general nonlinear sigma models

The perturbative renormalization group of the $O(N)$ nonlinear sigma model has been extended to the general case of a nonlinear sigma model whose target space is a manifold M. The Lagrangian of such general models is given in eq. (16.111). This is a theory whose target space M is a manifold with coordinates $\phi_i(x)$, and it is equipped with a positive-definite Riemann metric on M, $g_{ij}[\phi]$, which plays the role of the coupling constants of the nonlinear sigma model. In more conventional models, the target space M is a homogeneous space, the coset of a Lie group G by a compact subgroup H. For example, in the case of the $O(N)$ nonlinear sigma model, whose target space is the sphere $S_{N-1} \cong O(N)/O(N-1)$, the fields are the $\boldsymbol{\pi}(x)$ fields of the Goldstone modes, and the metric is

$$g_{ij}(\boldsymbol{\pi}) = \frac{1}{g}\left(\delta_{ij} + \frac{\pi_i \pi_j}{1-\boldsymbol{\pi}^2}\right) \tag{16.205}$$

which is the Riemann metric for the sphere S_{N-1} in terms of the tangent coordinates, the Goldstone fields. Notice that the Jacobian factor of the integration measure of the path integral is just

$$\frac{1}{\sqrt{1-\pi^2}} = \det(g_{ij}(\pi)) \tag{16.206}$$

Friedan (1985) showed that these general nonlinear sigma models are renormalizable field theories in $D=2$ dimensions. Since the metric g_{ij} on M has the interpretation of a set of coupling constants of the nonlinear sigma model, they must flow under the RG. Friedan computed their beta functions

$$\beta_{ij}(g) = -\Lambda\frac{\partial g_{ij}}{\partial \Lambda} \tag{16.207}$$

and derived the tantalizing result (in $D=2+\epsilon$ dimensions)

$$\beta_{ij}(u^{-1}g) = -\epsilon u^{-1}g_{ij} + R_{ij} + \frac{u}{2}R_{ipqr}R_{jpqr} + O(u^2) \tag{16.208}$$

where R_{ipqr} is the curvature tensor of the manifold M, and $R_{ij}=R_{ipjp}$ is the Ricci tensor of the metric g_{ij}. The result that we obtained for the $O(3)$ model (actually the sphere S_2, the symmetric space $S_2=O(3)/O(2)$) simply means that the coefficient of the one-loop beta function is nothing other than the (Ricci) curvature of the sphere S_2!

A remarkable consequence of this result is that the fixed-point condition (i.e., that the beta function vanishes, $\beta_{ij}=0$) implies that the Ricci tensor on M obeys a generalized form of the Einstein equations of general relativity (Friedan, 1985)

$$R_{ij} - s g_{ij} = \nabla_i v_j + \nabla_j v_i \tag{16.209}$$

where $s=1(-1)$ for manifolds with positive (negative) curvature, or zero for flat manifolds; here v_i is some vector field on M. Nonlinear sigma models with $\alpha=+1$ are asymptotically free: for $\alpha=-1$, they are asymptotically trivial. For $\alpha=0$, the lowest nonvanishing term of the beta function is of order u^2. This result, relating the fixed-point condition of nonlinear sigma models to Einstein's metrics, played a key role in the development of string theory and of its connection with a quantum theory of gravity (Polchinski, 1998). Interestingly, since the RG flow of the nonlinear sigma model is effectively a Ricci flow, it played a significant role in the proof of the Poincaré conjecture, a key problem in mathematics, by G. Perelman (2002).

16.8 Renormalization of Yang-Mills gauge theories in four dimensions

We now discuss briefly the renormalization of four-dimensional gauge theories. This material is discussed extensively in several classic textbooks, particularly in the book by Peskin and Schroeder (1995). Here we highlight the main points and compare them with what we did in ϕ^4 theory and the nonlinear sigma model.

For the sake of definiteness, consider a theory of quarks and gluons. This is a Yang-Mills gauge theory with (color) gauge group $G=SU(N_c)$. The Lagrangian of this theory contains gauge fields (gluons) A_μ^a in the adjoint representation of $SU(N_c)$ (and hence, $a=1,\ldots,$

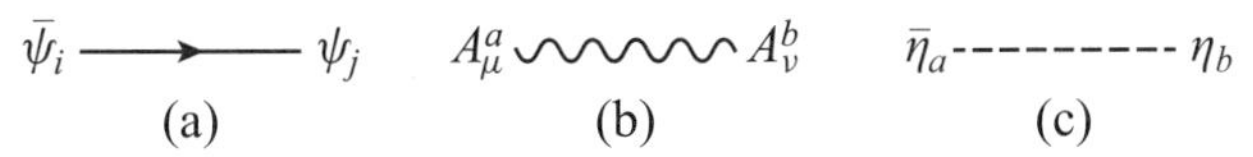

Figure 16.4 QCD propagators: (a) quark propagator, (b) gluon propagator, and (c) ghost propagator.

N_c^2-1), and Dirac fermions (quarks) ψ_i that carry the quantum numbers of the fundamental, N_c-dimensional representation of $SU(N_c)$. Here the color index $i=1,\ldots,N_c$. We omit, for now, the Dirac indices.

16.8.1 Perturbation theory

Here we use the path-integral quantization of Yang-Mills theory coupled to Dirac fermions (i.e., QCD), discussed in section 9.8, in the Feynman–'t Hooft gauges with parameter λ. The Lagrangian of the theory is

$$\mathcal{L}=-\frac{1}{4g^2}F^a_{\mu\nu}F^{\mu\nu}_a+\bar{\psi}_i\left(i\not{D}[A]-M\right)\psi_i+\frac{\lambda}{2g^2}\left(\partial_\mu A^\mu_a\right)^2-\bar{\eta}^a\partial_\mu D^\mu_{ab}[A]\eta^b \tag{16.210}$$

with $a=1,\ldots,N_c^2$, and $i=1,\ldots,N_f$. Here $D^\mu_{ij}[A]=\delta_{ij}\partial_\mu-igA^a_\mu t^a_{ij}$ is the covariant derivative in the fundamental representation of $SU(N_c)$ (with t^a_{ij} being the generators of $SU(N_c)$ in the fundamental representation), and $D^{ab}_\mu[A]=\delta_{ab}\partial_\mu-gf_{abc}A^c_\mu$ is the covariant derivative in the adjoint representation. Here η_a are the ghost fields, and ψ_i are massive Dirac fermions (quarks) with N_f flavors.

The Feynman rules in the 't Hooft–Feynman gauges are as follows. The propagator of the fermions (quarks), represented by a solid oriented line (shown in figure 16.4a), is

$$S_{ij}(p)=\frac{i}{\not{p}-M+i\epsilon}\delta_{ij}, \tag{16.211}$$

with $i,j=1,\ldots,N_c$. The propagator of the gauge field (in the Feynman gauge, with $\lambda=1$) is represented by a wavy line (shown in figure 16.4b) and is given by

$$\mathcal{D}^{ab}_{\mu\nu}(p)=-\frac{i}{p^2+i\epsilon}\delta_{ab}g_{\mu\nu} \tag{16.212}$$

where $a,b=1,\ldots,N_c^2-1$, and $g_{\mu\nu}$ is the metric tensor of Minkowski spacetime. The propagator of the ghost fields (shown as a broken line in figure 16.4c) is

$$\mathcal{C}_{ab}(p)=\frac{i}{p^2+i\epsilon}\delta_{ab} \tag{16.213}$$

This theory has several vertices: a quark-gluon vertex (shown in figure 16.5a) with weight $-ig\gamma^\mu t^a_{ij}$, a trilinear gluon vertex (shown in figure 16.5b) with weight $-g((q_\mu-k_\mu)g_{\nu\lambda}+$ two permutations$)f^{abc}$, a quadrilinear gluon vertex (shown in figure 16.5c) with weight $-ig^2f^{abe}f^{dce}(g^{\mu\gamma}g^{\nu\lambda}-g^{\mu\lambda}g^{\nu\gamma})+$ five permutations, and a ghost-gluon vertex (shown in figure 16.5d) with weight $-gf^{abc}q^\mu$. Notice the important fact that the weights of both the trilinear gluon vertex and the ghost-gluon vertex carry factors that are linear in momentum.

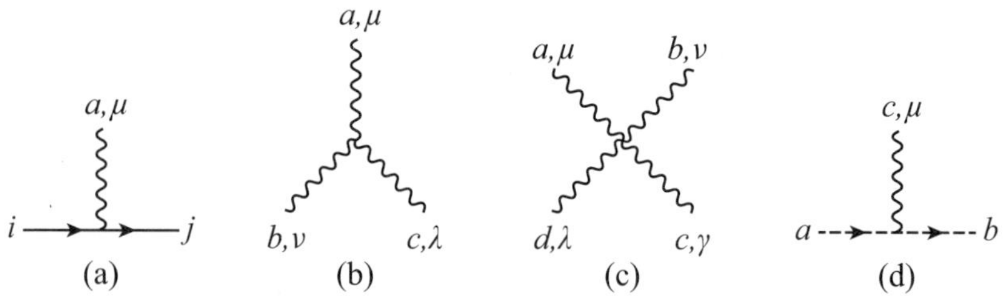

Figure 16.5 QCD vertices: (a) quark-gluon vertex, (b) trilinear gluon vertex, (c) quadrilinear gluon vertex, and (d) ghost-gluon vertex.

At the tree level, the 1-PI gluon propagator in the Feynman gauge ($\lambda = 1$) is

$$\Gamma_0^{(2)ab}{}_{\mu\nu}(p) = -ip^2 \delta_{ab} g_{\mu\nu} \tag{16.214}$$

Now consider its one-loop corrections, $\Gamma^{(2)}_{\text{one loop}}$. These gluon self-energy corrections are represented by the following sum of Feynman diagrams:

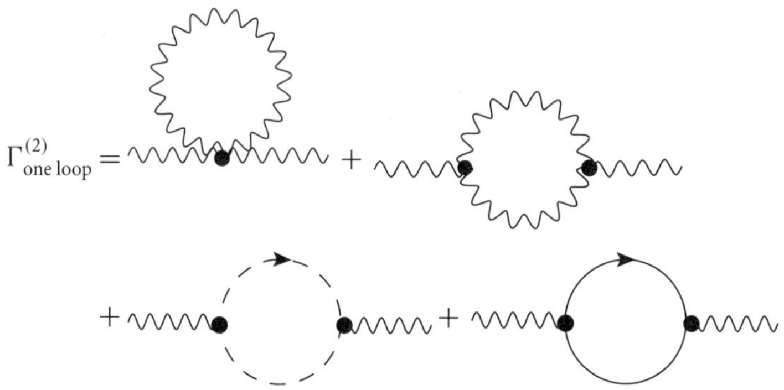

(16.215)

The pure gluon contributions, the first three terms on the right-hand side of this equation, are characteristic of the nonlinearities of Yang-Mills theory and are absent in QED. The last term, the quark loop, however, is similar to the electron loop in QED. If the theory has a flavor symmetry $SU(N_f)$, the quark loop has a coefficient of N_f. Here I will spare you the explicit expressions of these Feynman diagrams, which can be found in many textbooks. Notice that, due to Fermi statistics, both the ghost loop and the quark loop have a minus sign in their weight.

A superficial inspection of these one-loop contributions to the gluon self-energy seem to suggest that the gluons acquire a mass due to these radiative corrections. If this were the case, it would be disastrous, since mass terms for the gauge fields are forbidden by local gauge invariance. A closer examination reveals that this is not the case. The reason lies in the momentum dependence of the trilinear gluon vertex and of the ghost-gluon vertex.

Indeed, due to the momentum dependence of these vertices, there will be an explicit factor of the momentum either on the external legs of the gluon bubble diagram (the second term on the right-hand side of eq. (16.215)) or two momenta on the internal gluon loop. Thus, the gluon bubble diagram is actually a sum of two terms, one in which a factor of the momentum appears on each internal gluon leg and one in which two factors of momentum appear in the internal gluon loop,

(16.216)

where the dashes in eq. (16.216) indicate a momentum factor. The same considerations apply to the ghost loop diagram (the third term on the right-hand side of eq. (16.215)).

Thus, we have three possible contributions to a potential gluon mass term: one coming from the gluon tadpole diagram (this has the quadrilinear vertex), one from the gluon bubble (with two factors of momentum on the internal gluon loop), and one from the ghost loop (again, with two factors of momentum in the internal ghost loop). Fortunately, these three contributions to a potential gluon mass term exactly cancel out!

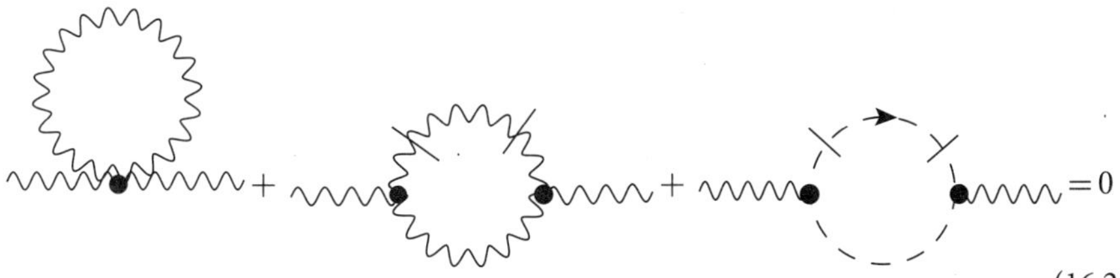

(16.217)

Of course, this cancellation is not an accident. It is a consequence of gauge invariance, and it is a manifestation of the Ward identities of Yang-Mills theory. However, this argument is formal in the sense that the integrals in these Feynman diagrams are divergent. In particular, in $D=4$ dimensions, the contributions of the Feynman diagrams that could produce a gluon mass term are quadratically divergent. Hence, for this cancellation to be obeyed, it is crucial that the regularization used be consistent with gauge invariance. In a non-abelian gauge theory, this is a nontrivial requirement. It is here that the use of dimensional regularization plays a key role for two reasons. One reason is that dimensional regularization is naturally compatible with gauge invariance. The other (and related) reason is that in dimensional regularization, quadratic divergences are regularized to zero.

16.8.2 Renormalization group

Once the cancellation of the quadratic divergences is taken care of, we find that the one-loop contribution to the gluon 1-PI two-point function is the sum of three Feynman diagrams:

$\Gamma^{(2)}_{\text{one loop}} =$ 

(16.218)

These diagrams are logarithmically divergent in $D=4$ dimensions. These three contributions have the same form:

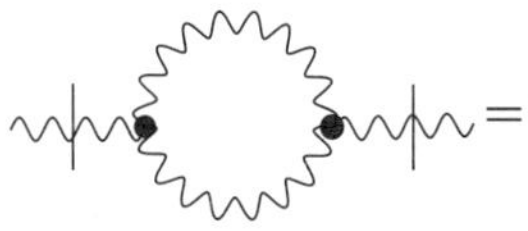

$$= Cg^2\left[\int \frac{d^Dq}{(2\pi)^D}\frac{1}{q^2(p+q)^2}\right]p^2 i(p^2 g^{\mu\nu} - p^\mu p^\nu)\delta_{ab} \qquad (16.219)$$

where C is a constant that differs for the three diagrams. What is important is that this constant depends on the gauge group and on the representation of the fermions. The integral in eq. (16.219) is manifestly logarithmically divergent as $D \to 4$, and hence, it has a pole in $\epsilon = D - 4$. The ghost-loop term yields a contribution of the same form but with a minus sign due to the fermionic nature of ghosts. Finally, the fermion-loop diagram yields an expression of the same form (and also negative). The fermion loop is the only contribution to the photon self-energy, since Maxwell's theory is free.

Putting it all together, we find that the one-loop correction to the 1-PI gluon two-point function in the Feynman gauge ($\lambda = 1$) is (with no quarks)

$$\Gamma^{\mu\nu}_{\text{one loop}\,ab}(p) = -C(N_c)\delta_{ab}\frac{g^2}{16\pi^2}\left(p^2 g^{\mu\nu} - p^\mu p^\nu\right)\frac{5}{3}\left[-\frac{2}{\epsilon} + \ln\left(\frac{p^2}{\mu^2}\right)\right] \tag{16.220}$$

where μ is a renormalization scale. Here we used that for the color group $SU(N_c)$, the constant is the quadratic Casimir operator of the adjoint representation, and it is simply equal to $C(N_c) = N_c$. Also, we used that if the T^a are the generators of $SU(N_c)$ in the adjoint representation, then they obey

$$\text{tr}(T_a T_b) = -C(N_c)\delta_{ab}, \quad (T_a)_{bc} = i f_{abc} \tag{16.221}$$

where f_{abc} are the structure constants of $SU(N_c)$, and $C(N_c)$ is the quadratic Casimir.

Following the renormalization prescription of dimensional regularization with minimal subtraction, the pole in ϵ can be canceled by a wave-function renormalization of the non-abelian gauge field A_μ^a,

$$\Gamma^{(2)\mu\nu}_{R\ \ ab}(p, \mu) = Z_3^{-1}\Gamma^{(2)\mu\nu}_{ab} \tag{16.222}$$

where

$$Z_3 = 1 + g^2\frac{C(N_c)}{16\pi^2}\left[\frac{5}{3} + \frac{1}{2}\left(1 - \frac{1}{\lambda}\right)\right]\frac{2}{\epsilon} \tag{16.223}$$

The renormalizability of Yang-Mills theory implies that the counterterms have the same form as the terms of the bare Lagrangian to all orders in perturbation theory. The renormalized Yang-Mills Lagrangian then has the form

$$\begin{aligned}\mathcal{L}_{\text{YM}} + \delta\mathcal{L}_{\text{YM}} = {} & \frac{1}{2}Z_3\left(\partial_\mu A_\nu - \partial_\nu A_\mu\right)^2 + \frac{\lambda}{2}\left(\partial_\mu A^\mu\right)^2 \\ & - gZ_1\left(\partial_\mu A_\nu - \partial_\nu A_\mu\right)[A^\mu, A^\nu] + \frac{g^2}{2}Z_4[A_\mu, A_\nu]^2 \\ & - \tilde{Z}_3 \partial_\mu \bar{\eta}\partial^\mu \eta + g\tilde{Z}_1 \partial_\mu \bar{\eta}_b A_a^\mu \eta_c f_{abc}\end{aligned} \tag{16.224}$$

where a trace over the color indices is implied. Here we denoted the matrix-valued gauge field $A_\mu = t_a A_\mu^a$. The relation between the renormalized and bare fields is

$$\begin{aligned} & A^\mu = Z_3^{1/2} A_R^\mu, \quad \eta = \tilde{Z}_3^{1/2}\eta_R, \quad \bar{\eta} = Z_3^{1/2}\eta_R \\ & g = Z_1 Z_3^{-3/2} g_R, \quad \lambda = Z_3^{-1}\lambda_R \end{aligned} \tag{16.225}$$

When coupled to fermions (quarks), the renormalized Lagrangian has the following additional terms:

$$\mathcal{L}_{\text{fermions}} = Z_2 \bar{\psi} i \partial\!\!\!/ \psi - Z_2 M \bar{\psi} \psi - i g Z_{1F} \bar{\psi} A\!\!\!/_a t^a \psi \tag{16.226}$$

Clearly, these renormalization constants cannot be independent, since otherwise the renormalized Lagrangian would not be gauge invariant. This condition is enforced by the Ward identities of this theory, known as the Slavnov-Taylor identities. A consequence of these identities is that the renormalization constants satisfy the relations

$$\frac{Z_4}{Z_1} = \frac{Z_1}{Z_3} = \frac{\tilde{Z}_1}{\tilde{Z}_3} = \frac{Z_{1F}}{Z_2} \tag{16.227}$$

Hence, this theory, just as in the case of the nonlinear sigma model, only has a coupling-constant renormalization and a wave-function renormalization. The coupling-constant renormalization is

$$g_R = Z_g g \tag{16.228}$$

where, to one-loop order,

$$Z_g \equiv Z_1^{-1} Z_3^{3/2} = 1 + \frac{g^2}{16\pi^2}\left(\frac{11}{6} C(N_c) - \frac{2}{3} N_f T_f\right) \tag{16.229}$$

where

$$\text{tr}\,(t_a t_b) = -T_f \delta_{ab} \tag{16.230}$$

Here T_f depends on the representation. For the fundamental representation, $T_f = 1/2$.

16.8.3 QCD: Asymptotic freedom

The one-loop beta function is

$$\beta(g) = \mu \frac{\partial g}{\partial \mu} = -\frac{g^3}{8\pi^2}\left(\frac{11}{6} C(N_c) - \frac{2}{3} N_f T_f\right) \tag{16.231}$$

where N_f is the number of fermion flavors. For $SU(N_c)$, $C(N_c) = N_c$ and $T_f = 1/2$. The resulting beta function is

$$\beta(g) = -a g^3 \tag{16.232}$$

where

$$a = \frac{1}{16\pi^2}\left(\frac{11}{3} N_c - \frac{2}{3} N_f\right) \tag{16.233}$$

For the color gauge group $SU(3)$, the quantity in parenthesis is $(33 - 2N_f)/3$.

Thus, provided the quantity in parenthesis in the beta function of eq. (16.231) is positive, the theory is asymptotically free. The solution of this equation is the same as what we found in the nonlinear sigma model in $D = 2$. Thus, if κ is an arbitrary renormalization scale, then the running coupling constant is

$$g^2(\kappa) = \frac{1}{\text{const.} + 2a \ln \kappa} \tag{16.234}$$

For large κ (high energies), the running coupling constant becomes weak:

$$g^2(\kappa) \approx \frac{1}{2a \ln \kappa} \to 0 \text{ as } \kappa \to \infty \tag{16.235}$$

Also,

$$\frac{1}{g^2(\kappa)} - \frac{1}{g^2(\kappa^*)} = 2a \ln\left(\frac{\kappa}{\kappa^*}\right) \tag{16.236}$$

Conversely, in the IR, the running coupling constant flows to large values. Let the renormalization scale $\kappa \sim \Lambda$ (the UV scale); then we can ask: At what scale $\kappa^* = 1/\xi$ does the running coupling constant become strong (i.e., diverge)? Here this scale has the form

$$\xi = \frac{1}{\Lambda} \exp\left(\frac{1}{2ag^2}\right) \tag{16.237}$$

Thus we have that, (1) perturbation theory works on scales shorter than ξ, and (2) the IR, long-distance, behavior is not accessible to perturbation theory. This behavior will persist so long as the coefficient a of the beta function of eq. (16.233) remains positive. Already at the one-loop level, we can see that asymptotic freedom survives unless the number of flavors becomes greater than a critical value $N_f > N_f^{\text{crit}}$. For $SU(3)$, this number is $N_f^{\text{crit}} \sim 16$.

Notice that in the case of QED, whose gauge theory is free, the coefficient is always negative. In this case, the coupling constant flows to zero in the IR but becomes strong in the UV, just as in the case of ϕ^4 theory.

As you can see, there is a close analogy in the behaviors of the two-dimensionial non-linear sigma model and $D = 4$ non-abelian gauge theories. Both theories are asymptotically free in the UV and flow to strong coupling in the IR. Hence, in both theories, the IR behavior is the regime in which the field fluctuations become wild and strongly nonclassical. In chapter 18, we will see that in this regime, the nonlinear sigma model is in its symmetric (unbroken) phase and that the gauge theory is in its confinement regime.

16.8.4 QED: Asymptotic "triviality"

In the case of QED, the gauge sector of the theory is free, and the RG flow is entirely due to the coupling between fermions (electrons and positrons) with the electromagnetic field. Although the renormalization of the theory has the same structure as in the non-abelian theory, it is substantially simpler. Indeed, in this case, only the fermion loop contributes to the photon self-energy. The resulting beta function in $D = 4$ dimensions for QED is

$$\beta(g) = +\frac{N_f}{24\pi^2} g^3 \tag{16.238}$$

Thus, contrary to the case of QCD, the coupling constant becomes weak in the IR (and strong in the UV). This is analogous to what we found for the case of ϕ^4 theory. This behavior is usually referred to as the triviality of QED, in the sense that the effective coupling constant vanishes in the IR and hence the photon-mediated scattering between electrons becomes very weak at very low energies.

There is also a dynamical scale in this theory, and it has the same form as in eq. (16.237). However, since the coupling constant runs to weak coupling in the IR, this scale does not represent in this case a breakdown of perturbation theory or of the vacuum state. The difference is that this scale now represents a short-distance (or high-energy) regime that is inaccessible to perturbation theory.

Exercises

16.1 Renormalization of the $\mathbb{CP}^{N-1}$ nonlinear sigma model

In this exercise, you will work out the program of the perturbative RG in the case of the $\mathbb{CP}^{N-1}$ model in two-dimensional Euclidean spacetime using dimensional regularization and the minimal subtraction scheme (Hikami, 1979). The Euclidean action of the model is (cf. eq. (16.102))

$$S=\frac{1}{g}\int d^2x\left|\left(\partial_\mu - i\mathcal{A}_\mu(x)\right) z_\alpha(x)\right|^2 \tag{16.239}$$

where z_i, with $i=1,\ldots,N$, is an N-component scalar field that satisfies the constraint $\sum_{i=1}^{N}|z_i|^2=1$, $\mathcal{A}_\mu$ is a $U(1)$ gauge field, and g is a dimensionless coupling constant. In this theory, the gauge-invariant bilinear $z_i^* z_i$ (for some i) has an expectation value (but not for the other bilinears) that spontaneously breaks the global $SU(N)$ symmetry to $SU(N-1)$ while respecting the local $U(1)$ symmetry.

1) Check that this theory has a global $SU(N)$ symmetry and a local $U(1)$ symmetry.
2) Integrate out the $U(1)$ gauge field $\mathcal{A}_\mu$, and show that the resulting effective Lagrangian for the z_i complex fields is local and is given by

$$\mathcal{L}=\frac{1}{g}\left(\partial_\mu z_i \partial_\mu z_i^* + (z_i^*\partial_\mu z_i)(z_j^*\partial_\mu z_j)\right) \tag{16.240}$$

Show that this effective Lagrangian respects the same symmetries as in part 1.

3) Assume that the spontaneous symmetry breaking occurs in the Nth component, and work in the unitary gauge, in which $z_N=z_N^*\equiv\sigma$ is real. Furthermore, rewrite the remaining $N-1$ complex fields in terms of $2(N-1)$ real fields as $z_i=\pi_{2i-1}+iz_{2i}$. In terms of these $2(N-1)$ real fields, the constraint is $\sigma^2+\sum_{k=1}^{2N-2}\pi_k^2=1$. It will be convenient to make the following change of variables:

$$\sigma^2=\frac{1}{2}(1+\tilde{\sigma}) \qquad \pi_k=\frac{\tilde{\pi}_k}{\sqrt{2(1+\tilde{\sigma})}} \tag{16.241}$$

Show that in the new variables, the constraint becomes

$$\tilde{\sigma}'^2+\sum_{k=1}^{2N-2}\tilde{\pi}_k^2=1 \tag{16.242}$$

and that the effective Lagrangian for the $2N-2$ component $\tilde{\boldsymbol{\pi}}$ field is

$$\mathcal{L}=\frac{1}{g}\left\{\frac{(\partial_\mu\tilde{\pi}_i)^2}{2(1+\sqrt{1-\tilde{\boldsymbol{\pi}}^2})}+\frac{\frac{1}{2}+\sqrt{1-\tilde{\boldsymbol{\pi}}^2}}{8(1-\tilde{\boldsymbol{\pi}}^2)(1+\sqrt{1-\tilde{\boldsymbol{\pi}}^2})^2}(\partial_\mu\tilde{\boldsymbol{\pi}}^2)^2\right.$$
$$\left.-\frac{1}{4(1+\sqrt{1-\tilde{\boldsymbol{\pi}}^2})^2}\left(\sum_{k=1}^{N-1}\tilde{\pi}_{2k-1}\overleftrightarrow{\partial_\mu}\tilde{\pi}_{2k}\right)^2\right\} \tag{16.243}$$

which now has a structure similar to what we found in the $O(N)$ nonlinear sigma model. Also, as in the $O(N)$ nonlinear sigma model, control the IR divergences with a symmetry-breaking field, which here is the gauge-invariant coupling $H|z_N|^2=H\sigma^2=H(1+\tilde{\sigma})/2$.

4) Show that this theory is renormalizable by defining a renormalized dimensionless coupling constant g_R and the renormalization constants Z and Z_1, such that $\tilde{\boldsymbol{\pi}}=\sqrt{Z}\tilde{\boldsymbol{\pi}}_R$, $g=Z_1 g_R$, and $H/g=\sqrt{Z}H_R/g_R$. Compute the renormalization constants Z and Z_1 to one-loop order. Use these results to compute the beta function for the dimensionless renormalized coupling constant g_R and the anomalous dimension γ.

16.2 Renormalization of perturbative Yang-Mills theory

In section 16.8, we studied the renormalization properties of non-abelian gauge theories. In this problem, you are asked to work out several steps that we discussed summarily in the main text. Work with a Yang-Mills theory in $D=4$ dimensions with gauge (color) group $SU(N_c)$, whose Lagrangian is given by eq. (16.210), in the Feynman gauge with $\lambda=1$. Assume that the matter content is a massive Dirac field with N_f flavors. Focus on the gluon self-energy one-loop diagrams shown in eq. (16.215).

1) Give an explicit expression for each of the four one-loop vacuum polarization diagrams (the gluon self-energy) shown in eq. (16.215). Express your results for each diagram as a sum of two terms, in the form of eq. (16.217) and eq. (16.218). For each diagram, give an expression using dimensional regularization in Euclidean dimension D.
2) Show explicitly the cancellation of the vacuum polarization diagrams shown in eq. (16.217).
3) Give an explicit derivation of the one-loop gluon self-energy correction of eq. (16.220).

16.3 The Wilson loop in QED

In section 9.7, we calculated the expectation value of the Wilson loop operator in quantized Maxwell theory,

$$W_\Gamma=\left\langle\exp\left(ie\oint_\Gamma dx_\mu A_\mu\right)\right\rangle \tag{16.244}$$

where Γ is a closed loop, and used it to show that the effective potential $V(R)$ between two static sources separated a distance R is the familiar Coulomb potential. In this problem, we will do the same calculation for QED at the one-loop level (see Kogut, 1983). We work in $D=4$ dimensions in the Euclidean signature. The

Euclidean Lagrangian is

$$\mathcal{L} = \bar{\psi}(i\not{D} - m)\psi + \frac{1}{4}F_{\mu\nu}^2 \tag{16.245}$$

where $\not{D} = \gamma_\mu(\partial_\mu + ieA_\mu)$, with $\mu = 1, \ldots, 4$. Here γ_μ are the Dirac gamma matrices, which in the Euclidean signature are antihermitian and obey the (pseudo) Clifford algebra $\{\gamma_\mu, \gamma_\nu\} = -2\delta_{\mu\nu}$. Work in the Feynman gauge.

1) Show that to one-loop order in e^2, the expectation value of the Wilson loop operator W_Γ is the same as in free Maxwell theory but with the photon propagator corrected to one-loop order (i.e., by the one-loop vacuum polarization diagram $\Pi_{\mu\nu}$).
2) Show that in momentum space, the vacuum polarization diagram is

$$\Pi_{\mu\nu}(q) = -e^2 \int \frac{d^4p}{(2\pi)^4} \text{tr}\left(\frac{\gamma_\mu(\not{p} + m)\gamma_\nu(\not{p} - \not{q} + m)}{(p^2 + m^2)((p-q)^2 + m^2)} \right) \tag{16.246}$$

3) Now consider the massless theory, and set $m = 0$. Show that to one-loop order, the vacuum polarization leads to a correction to the Euclidean photon propagator (in the Feynman gauge) of

$$G_{\mu\nu}(q) = \frac{\delta_{\mu\nu}}{q^2}\left(1 + \frac{\alpha}{3\pi}\log\left(\frac{q^2}{M^2}\right) + \cdots\right) \tag{16.247}$$

where $\alpha = \frac{e^2}{4\pi}$ is the fine structure constant, and M is a UV cutoff.
4) Show that the result in part 3 implies that to one-loop order the effective potential $V(R)$ becomes

$$V(R) = -\frac{\alpha(R)}{R}\left(1 - \frac{2\alpha}{3\pi}\log(MR)\right) \tag{16.248}$$

with the renormalization condition that for $R = 1/M$, the potential is the bare Coulomb law.
5) Show that $\alpha(R)$ obeys the beta function

$$-R\frac{\partial \alpha(R)}{\partial R} = \beta(\alpha(R)) \tag{16.249}$$

where

$$\beta(\alpha) = \frac{2}{3\pi}\alpha^2 \tag{16.250}$$

17

The 1/*N* Expansions

As we saw in chapter 16, perturbation theory, even when applicable, only describes a regime of a QFT. In the case of the nonlinear sigma model in $D=2$ dimensions, and in Yang-Mills theory in $D=4$ dimensions, the running coupling constant is weak only at short distances, but the long-distance behavior is inaccessible to perturbation theory, since the coupling constant runs to large values that are outside the perturbative regime. So we have theories that at low energies have a vacuum state that is essentially different from the free-field ground state. To understand the physics of the actual ground state requires the use of nonperturbative methods. A key tool in this respect is the study of the generalizations of the theories of interest in their "large-N" limits. Here, N can mean the rank of the symmetry group or the rank of the representation. The behavior is different in each case. We begin by considering first the simpler, and more tractable, case of scalar fields.

There is a long history of studying theories in this limit, both in statistical physics and in QFT. In statistical physics, it goes back to the classic work by Berlin and Kac on what they called a "spherical model" of a phase transition (Berlin and Kac, 1952). This solvable model was later shown to be equivalent to the large-N limit of the classical Heisenberg model (with an N-component order parameter) by Stanley (1968). With the advent of the RG, this limit was studied by S.-K. Ma (1973) in the context of a ϕ^4 theory with a global $O(N)$ symmetry. In QFT, large-N limits became a mainstay tool for studying nonperturbative behavior of asymptotically free theories, such as nonlinear sigma models (Brézin and Zinn-Justin, 1976b), the $\mathbb{CP}^{N-1}$ models, the Gross-Neveu models (Gross and Neveu, 1974), and especially Yang-Mills gauge theory ('t Hooft, 1974; Witten, 1979b). These methods have also been extensively used in theories of the Kondo problem (Read and Newns, 1983), quantum antiferromagnetism (Sachdev and Read, 1991), and the study of quantum phase transitions (Sachdev, 1999). It has also played a central role in the theory of random matrices (Mehta, 2004).

17.1 The ϕ^4 scalar field theory with $O(N)$ global symmetry

Let us begin by considering the $O(N)$ ϕ^4 theory. This theory has a scalar field $\boldsymbol{\phi}(x)$ that is an N-component vector that transforms in the fundamental (vector) representation of the global symmetry group $O(N)$. The (Euclidean) Lagrangian is

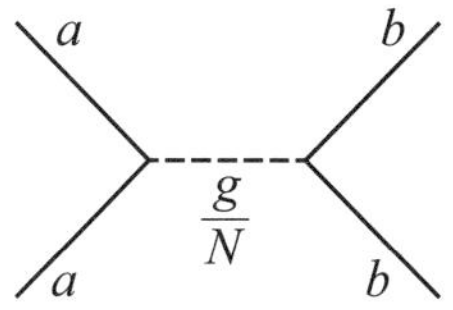

Figure 17.1 The interaction vertex of the $O(N)$ scalar field theory with coupling constant $\lambda = \frac{g}{N}$.

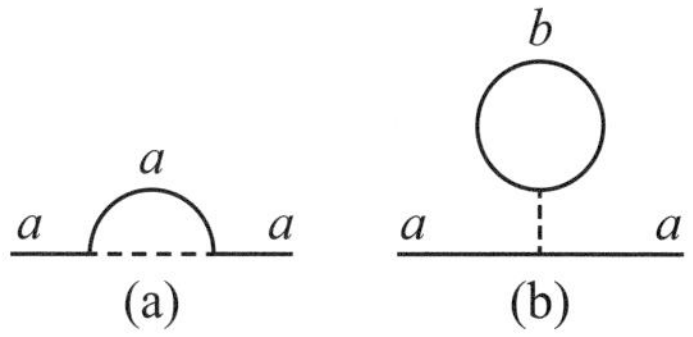

Figure 17.2 One-loop contributions to the 1-PI two-point function of the ϕ field in the $O(N)$ theory: (a) the rainbow diagram and (b) the tadpole diagram.

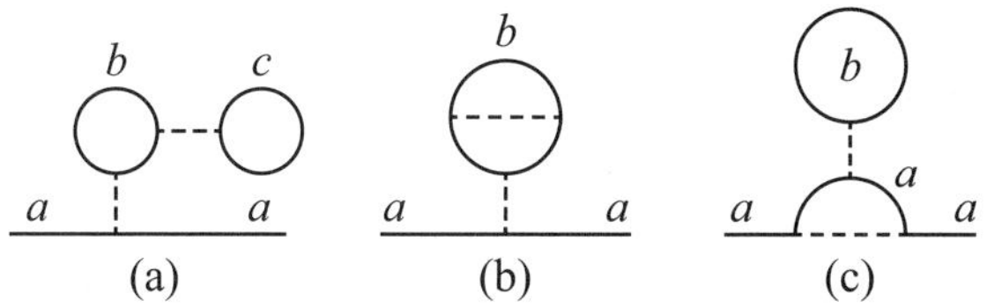

Figure 17.3 Two-loop contributions to the 1-PI two-point function of the ϕ field in the $O(N)$ theory: (a) the two-loop tadpole diagram, (b) the rainbow-tadpole two-loop diagram, and (c) the two-loop rainbow diagram.

$$\mathcal{L} = \frac{1}{2}\left(\partial_\mu \boldsymbol{\phi}(x)\right)^2 + \frac{m_0^2}{2}\boldsymbol{\phi}^2(x) + \frac{g}{4!N}\left(\boldsymbol{\phi}^2(x)\right)^2 \tag{17.1}$$

where, as usual, repeated indices are summed over. Notice that we have made the replacement of the conventional coupling constant $\lambda \mapsto \frac{g}{N}$. We will see shortly the necessity of this replacement. The interaction vertex of the $O(N)$ theory is shown in figure 17.1, where, for clarity, the contact interaction has been formally split.

17.1.1 Diagrammatic approach to the large-N limit

What is the dependence on N of the Feynman diagrams of the 1-PI two-point function (i.e., the ϕ field self-energy)? The diagrams up to one-loop order are shown in figure 17.2. It is easy to see that since each one-loop diagram contributes with a factor of the coupling constant, the rainbow diagram of figure 17.2a contributes with a factor of $\frac{g}{N}$, while the tadpole diagram of figure 17.2b contributes with a factor of $\frac{g}{N}N = g$, where the factor of N comes from the independent sum over the index b running inside the loop. Thus, in the limit $N \to \infty$, the leading term in figure 17.2b and figure 17.2a is a $1/N$ correction.

To see the emerging pattern in the large-N limit, let us look at the two-loop diagrams in figure 17.3. By counting powers of N, we see that the diagram in figure 17.3a contributes with a factor of $\frac{g^2}{N^2}N^2 = g^2$, that the diagram in figure 17.3b contributes with a factor of

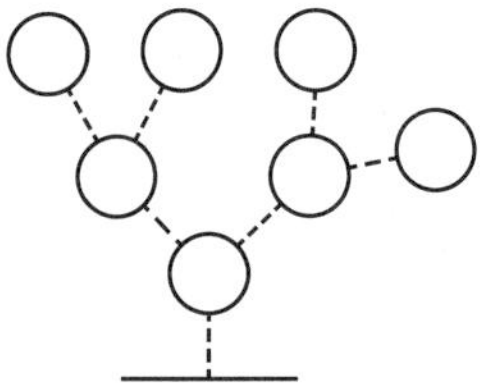

Figure 17.4 A typical contribution to the two-point function of the $O(N)$ scalar field in the $N \to \infty$ limit.

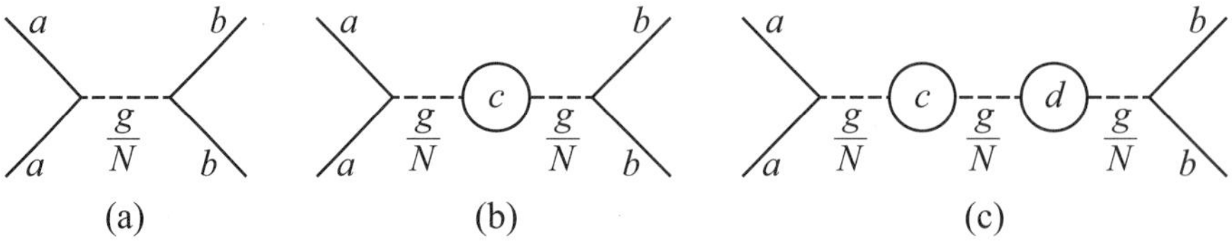

Figure 17.5 Two-loop contributions to the 1-PI four-point function of the ϕ field in the $O(N)$ theory: (a) the bare vertex, (b) the one-loop diagram, and (c) a two-loop bubble diagram.

$\frac{g^2}{N^2}N = \frac{g^2}{N}$, and the diagram in figure 17.3c contributes with a factor of $\frac{g^2}{N^2}$. Hence, only the diagram in figure 17.3a survives in the $N \to \infty$ limit.

We now see the pattern: Diagrams in which the number of independent sums over the internal indices is equal to the order in perturbation theory have a finite limit as $N \to \infty$, whereas the contributions of the other diagrams can be organized as a formal expansion in powers in $1/N$. A typical diagram that contributes to the two-point function in the $N \to \infty$ limit is shown in figure 17.4. In fact, we have already reached this conclusion in section 11.6, where we saw that this sum of diagrams (also known as the Hartree approximation) yields a self-consistent expression for the one-loop self-energy of eq. (11.60).

Moving on to the four-point function, it is easy to see that as $N \to \infty$, the only surviving contribution is the sum of the bubble diagrams shown in figure 17.5, since each new diagram contributes with an extra factor of $\frac{g}{N}$, and a sum over an internal index yields a factor of N.

We can make these arguments explicit by writing the Dyson equation for the two-point function, which symbolically becomes

$$G = G_0 + G_0\, \Sigma\, G \tag{17.2}$$

In the $N \to \infty$ limit, the self-energy Σ is the sum of all tree-like diagrams

$\Sigma =$ 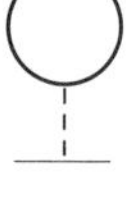 (17.3)

where the internal loop is the full two-point function in the $N \to \infty$ limit. More explicitly, we can write

$$\Sigma(p) = -\frac{1}{6}\left(\frac{g}{N}\right) N \int \frac{d^Dq}{(2\pi)^D} \frac{1}{q^2 + m_0^2 - \Sigma(q)} \tag{17.4}$$

which does not depend on the external momentum p, and the N dependence cancels out. This expression can be recast in terms of an effective mass:

$$m^2 = m_0^2 - \Sigma(p) = m_0^2 + \frac{g}{6} \int \frac{d^D q}{(2\pi)^D} \frac{1}{q^2 + m^2} \tag{17.5}$$

This is just the renormalized mass (squared) at the one-loop level (cf. eq. (11.62), discussed in chapter 11). The only difference is that in the $N \to \infty$ limit, this expression is exact. Thus, in this theory, the large-N limit is equivalent to a self-consistent one-loop approximation.

We now have to consider the 1-PI four-point vertex function $\Gamma^{(4)}_{ijkl}(p_1, \ldots, p_4)$, which by symmetry can be written as

$$\Gamma^{(4)}_{ijkl}(p_1, \ldots, p_4) = \frac{1}{3N} \left[\delta_{ij}\delta_{kl} F_4(p_1 + p_2) + \delta_{ik}\delta_{jl} F_4(p_1 + p_3) + \delta_{il}\delta_{jk} F_4(p_1 + p_4) \right] \tag{17.6}$$

The quantity $F_4(p)$ in the large-N limit is the sum of the bubble diagrams

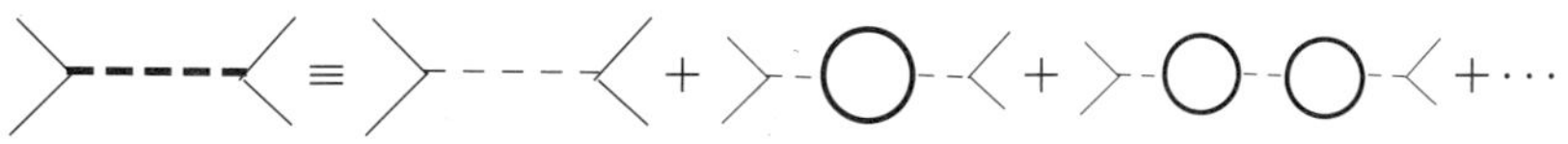
(17.7)

More explicitly, we can write the exact expression (in the $N \to \infty$ limit) for the sum of bubble diagrams $F_4(p)$

$$F_4(p) = \frac{g}{1 + \frac{g}{6} I(p)} \tag{17.8}$$

where $I(p)$ is the one-loop bubble diagram:

$$I(p) = \int \frac{d^D q}{(2\pi)^D} \frac{1}{(q^2 + m^2)((p - q)^2 + m^2)} \tag{17.9}$$

Therefore, the sum of bubble diagrams yields the form of the effective interaction in the large-N limit. The sum of bubble diagrams is reminiscent of the random phase approximation of Bohm and Pines (Pines and Bohm, 1952; Pines and Nozières, 1966), widely used in the theory of screening in electron fluids. In this context, the sum of tadpole diagrams leading to eq. (17.3) is analogous to the Hartree approximation of the electron self-energy in the theory of the electron fluids.

The expression for $I(p)$ is UV finite for $D < 4$ and has a logarithmic divergence as $D \to 4^-$. The methods we introduced to compute these integrals yield the result (in D Euclidean dimensions):

$$I(p) = \frac{\Gamma\left(2 - \frac{D}{2}\right)}{(4\pi)^{D/2}} 2^{4-D} \int_0^1 du((1 - u^2)p^2 + 4m^2)^{(D-4)/2} \tag{17.10}$$

The important observation here is that if the renormalized mass vanishes, then the IR behavior of the bubble leads to an IR singularity of the four-point function $\Gamma^{(4)}(p) \propto p^{D-4}$. As we will see, this result means that the effective coupling constant must have a nontrivial scale dependence, as dictated by the RG flow.

17.1.2 Path-integral approach

The diagrammatic analysis of the large-N limit of ϕ^4 theory is useful, but it has the drawback that it seemingly applies only to the symmetric phase of the theory in which the $O(N)$ symmetry is not spontaneously broken. Although a similar diagrammatic analysis can be done in the broken symmetry phase, there is a more general approach that uses the large-N limit of the path integral for the partition function. To this end, let us introduce an auxiliary (Hubbard-Stratonovich) field $\alpha(x)$ to rewrite the partition function for the theory with the Lagrangian of eq. (17.1):

$$\begin{aligned} Z &= \int \mathcal{D}\boldsymbol{\phi} \exp\left(-\int d^D x \left[\frac{1}{2}(\partial_\mu \boldsymbol{\phi})^2 + \frac{m_0^2}{2}\boldsymbol{\phi}^2 + \frac{g}{2N}(\boldsymbol{\phi}^2)^2\right]\right) \\ &= \int \mathcal{D}\alpha \int \mathcal{D}\boldsymbol{\phi} \exp\left(-\int d^D x \left[\frac{1}{2}(\partial_\mu \boldsymbol{\phi})^2 + \frac{m_0^2}{2}\boldsymbol{\phi}^2 - \frac{N}{2g}\alpha^2 + \alpha\boldsymbol{\phi}^2\right]\right) \\ &= \int \mathcal{D}\alpha \left(\mathrm{Det}\left[-\partial^2 + m_0^2 + 2\alpha\right]\right)^{-N/2} \exp\left(+\int d^D x \frac{N}{2g}\alpha^2\right) \end{aligned} \tag{17.11}$$

After integrating out the $\boldsymbol{\phi}$ fields, we can rewrite the partition function as a path integral over the field $\alpha(x)$,

$$Z = \int \mathcal{D}\alpha \exp\left(-N S_{\text{eff}}[\alpha]\right) \tag{17.12}$$

where the effective action $S_{\text{eff}}[\alpha]$ is

$$S_{\text{eff}}[\alpha] = \frac{1}{2}\mathrm{tr}\ln\left[-\partial^2 + m_0^2 + 2\alpha\right] - \frac{1}{2g}\int d^D x\, \alpha^2(x) \tag{17.13}$$

Therefore, the large-N limit of ϕ^4 theory is the semiclassical limit of the theory of the field α, whose effective action is given by eq. (17.13). Hence, in the $N \to \infty$ limit, the path integral for the partition function is determined by the configurations $\alpha_c(x)$ that make the effective action stationary, that is, the classical field $\alpha_c(x)$ satisfying the saddle-point equation

$$\frac{\delta S_{\text{eff}}}{\delta \alpha} = 0 \tag{17.14}$$

which implies that the gap equation

$$G(x, x; m^2) = \frac{\alpha_c}{g} \tag{17.15}$$

must be satisfied. Here we have set $m^2 = m_0^2 + 2\alpha_c$, and

$$G(x, y; m^2) = \left\langle x \left| \frac{1}{-\partial^2 + m^2} \right| y \right\rangle \tag{17.16}$$

is the propagator of a massive scalar field of mass squared, m^2. Hence, the classical field α_c obeys the equation

$$\frac{\alpha_c}{g} = \int \frac{d^D p}{(2\pi)^D} \frac{1}{p^2 + m_0^2 + 2\alpha_c} \tag{17.17}$$

By analogy with the BCS theory of superconductivity, equations of this type are generally called "gap equations." It should be apparent that the self-energy computed using the Dyson equation and the classical field are related by $\Sigma = -2\alpha_c$.

Returning to the effective action of the field $\alpha(x)$, we see that the 1-PI two-point function of the field $\tilde{\alpha}(x) = \alpha(x) - \alpha_c$ is

$$\Gamma^{(2)}_{\alpha\alpha}(x, y) = 2G(x, y; m^2)G(y, x; m^2) + \frac{1}{g}\delta(x - y) \tag{17.18}$$

which, in momentum space, is easily seen to be the inverse of $F_4(p)/g$. The first term of the right-hand side of this equation is the bubble diagram.

This analysis requires a renormalization prescription. It should also be supplemented by an analysis of the broken symmetry state. For brevity, we will do this only in section 17.2, which is devoted to the large-N limit of the nonlinear sigma model.

17.2 The large-*N* limit of the *O*(*N*) nonlinear sigma model

Let us now turn to the case of the $O(N)$ nonlinear sigma model and its large-N limit. We have already discussed that, in the Euclidean metric, the nonlinear sigma model is the formal continuum limit of the classical Heisenberg model. It has been known since the 1960s (Stanley, 1968) that the large-N limit of this model is equivalent to the "spherical model" proposed by Berlin and Kac (1952). Here we work in D Euclidean dimensions.

As discussed in section 12.5, at the classical level, the $O(N)$ nonlinear sigma model can be regarded as a limit of a ϕ^4 theory in its broken symmetry state. Indeed, we can rewrite the Euclidean Lagrangian of an N-component ϕ^4 theory as

$$\mathcal{L} = \frac{1}{2}(\partial_\mu \boldsymbol{\phi})^2 + \frac{\lambda}{4!}(\boldsymbol{\phi}^2 - \phi_0^2)^2 \tag{17.19}$$

which, in the $\lambda \to \infty$ limit, becomes the Lagrangian of a nonlinear sigma model with an N-component unit-vector field $\boldsymbol{n}(x) = (\sigma(x), \boldsymbol{\pi}(x))$, such that $\boldsymbol{n} = \sigma^2 + \boldsymbol{\pi}^2 = 1$ as a constraint everywhere in D-dimensional Euclidean spacetime. The partition function of the $O(N)$ nonlinear sigma model is the path integral

$$Z[H, \boldsymbol{J}] = \int \mathcal{D}\sigma \mathcal{D}\boldsymbol{\pi}\ \delta(\sigma^2 + \boldsymbol{\pi}^2 - 1)\ \exp\left(-\frac{S}{g}\right) \tag{17.20}$$

where $1/g = \phi_0^2$. Here the δ function acts at all points of D-dimensional Euclidean spacetime. The action S is

$$S = \int d^D x \left[\frac{1}{2}(\partial_\mu \sigma)^2 + \frac{1}{2}(\partial_\mu \boldsymbol{\pi})^2 - H\sigma - \boldsymbol{J} \cdot \boldsymbol{\pi}\right] \tag{17.21}$$

where $(H, \boldsymbol{J})$ is a symmetry-breaking field. We will now see that even though classically (and in perturbation theory for $D > 2$) this is a theory of a broken symmetry state, this theory has (again for $D > 2$) both a broken symmetry phase and a symmetric phase separated by a critical (and nonuniversal) value of the dimensional coupling constant g.

17.2.1 Large-N limit

That the nonlinear sigma model has a finite large-N limit can be gleaned from the perturbative expression for the 1-PI two-point function of the $\boldsymbol{\pi}$ field, presented in eq. (16.157) and figure 16.3. There we saw that as N becomes large, we can have a finite limit, provided the coupling constant g and the field H are scaled by N and the field $\boldsymbol{\pi}$ is also scaled by an appropriate power of N.

Here we will use a functional approach to study the large-N limit. We begin by implementing the constraint by means of an integral representation of the δ function in terms of a Lagrange multiplier field $\alpha(x)$, in terms of which the partition function now reads

$$Z = \int \mathcal{D}\sigma \mathcal{D}\boldsymbol{\pi} \mathcal{D}\alpha \exp\left(-\frac{S}{g} + \int d^D x \frac{\alpha(x)}{2g}(1 - \sigma^2(x) - \boldsymbol{\pi}^2(x)) \right) \tag{17.22}$$

Let us rescale the $(N-1)$-component fields $\boldsymbol{\pi} = \sqrt{g}\boldsymbol{\varphi}$, which in turn can be integrated out to yield

$$\begin{aligned} Z = \int \mathcal{D}\sigma \mathcal{D}\alpha \left(\text{Det}[-\partial^2 + \alpha]\right)^{-(N-1)/2} \\ \times \exp\left(\frac{1}{2} \int d^D x \int d^D y \, G(x-y;\alpha) \, \boldsymbol{J}(x) \cdot \boldsymbol{J}(y) \right) \\ \times \exp\left(-\frac{1}{g} \int d^D x \left[\frac{1}{2}(\partial_\mu \sigma)^2 - H\sigma - \frac{1}{2}\alpha(\sigma^2 - 1) \right] \right) \end{aligned} \tag{17.23}$$

where the kernel $G(x-y;\alpha)$ satisfies

$$(-\partial^2 + \alpha(x)) G(x-y;\alpha(x)) = \delta(x-y) \tag{17.24}$$

for an arbitrary configuration of the Lagrange multiplier field. Hereafter we will drop the explicit x-dependence of the field α. For the time being, let us set the sources $\boldsymbol{J}(x) = 0$ (although later they will be restored).

It is convenient to rescale the bare coupling constant g and the field σ as follows:

$$g = \frac{g_0}{N-1}, \qquad \sigma(x) = \sqrt{g(N-1)}\, m(x) \tag{17.25}$$

The effective action of the rescaled field $m(x)$ and of the Lagrange multiplier field α becomes

$$S_{\text{eff}}(m,\alpha,H) = \int d^D x \left[\frac{1}{2}(\partial_\mu m)^2 + \frac{1}{2}\alpha m^2 - \frac{\alpha}{2g_0} - \frac{Hm}{\sqrt{g_0}} \right] + \frac{1}{2}\text{tr}\ln\left(-\partial^2 + \alpha\right) \tag{17.26}$$

and the partition function is

$$Z[H] = \int \mathcal{D}m \mathcal{D}\alpha \exp\left(-(N-1) S_{\text{eff}}(m,\alpha,H)\right) \tag{17.27}$$

Hence, once again, the large-N limit is the semiclassical limit, in this case of the effective action for the fields m and α coupled to the source H. Therefore, in the large-N limit, the partition function of eq. (17.27) is dominated by the configurations that leave the effective action of eq. (17.26) stationary. By varying S_{eff} with respect to the field $m(x)$, we get

$$\frac{\delta S_{\text{eff}}}{\delta m(x)} = -\partial^2 m(x) + \alpha(x) m(x) - \frac{H}{\sqrt{g_0}} = 0 \tag{17.28}$$

Likewise, by varying S_{eff} with respect to $\alpha(x)$, we obtain

$$\frac{\delta S_{\text{eff}}}{\delta \alpha(x)} = -\frac{1}{2g_0} + \frac{1}{2} m^2(x) + \frac{1}{2}\frac{\delta}{\delta \alpha(x)} \operatorname{tr} \ln \left(-\partial^2 + \alpha\right) = 0 \tag{17.29}$$

where

$$\frac{\delta}{\delta \alpha(x)} \operatorname{tr} \ln \left(-\partial^2 + \alpha\right) = \left\langle x \left| \frac{1}{-\partial^2 + \alpha} \right| x \right\rangle \tag{17.30}$$

Hence, we obtain a second saddle-point equation:

$$\left\langle x \left| \frac{1}{-\partial^2 + \alpha} \right| x \right\rangle + m^2(x) - \frac{1}{g_0} = 0 \tag{17.31}$$

Let us now seek a uniform solution of eq. (17.28) and eq. (17.31) of the form

$$\sigma(x) = M, \quad \alpha(x) = \bar{\alpha} \tag{17.32}$$

with H constant, and $M^2 = g_0 m^2$. Thus, the solution of eq. (17.28) is

$$\bar{\alpha} = \frac{H}{M} \tag{17.33}$$

while eq. (17.31) becomes

$$g_0 \left\langle x \left| \frac{1}{-\partial^2 + \frac{H}{M}} \right| x \right\rangle = 1 - M^2 \tag{17.34}$$

or, what is the same,

$$g_0 \int \frac{d^D p}{(2\pi)^D} \frac{1}{p^2 + \frac{H}{M}} = 1 - M^2 \tag{17.35}$$

17.2.2 Renormalization

Provided $H/M \neq 0$, eq. (17.35) is IR finite. However, for $D \geq 2$, the integral is UV divergent. We will absorb the strong dependence in the UV in a set of renormalization constants. Thus, define a dimensionless renormalized coupling constant t, a renormalized M_R, and a renormalized H_R through the relations (here κ is an arbitrary momentum scale) (see eq. (16.160) and eq. (16.161))

$$\begin{aligned} g_0 &= t\kappa^{-\epsilon} Z_1 \\ M &= Z^{1/2} M_R \\ H &= Z_1 Z^{-1/2} H_R \end{aligned} \tag{17.36}$$

where we have set $\epsilon = D-2$, and Z is the wave-function renormalization. Notice that

$$\frac{H}{M} = \frac{Z_1}{Z}\frac{H_R}{M_R} \tag{17.37}$$

With these definitions, eq. (17.35) becomes

$$t\kappa^{-\epsilon}\frac{Z_1}{Z}\int \frac{d^Dp}{(2\pi)^D}\frac{1}{p^2+\frac{H_R}{M_R}\frac{Z_1}{Z}} = \frac{1}{Z} - M_R^2 \tag{17.38}$$

We have encountered the integral in this equation several times before. It is given by

$$\int \frac{d^Dp}{(2\pi)^D}\frac{1}{p^2+\mu^2} = \frac{1}{(4\pi)^{D/2}}\Gamma\left(-\frac{\epsilon}{2}\right)\mu^{\epsilon} \tag{17.39}$$

Using this result, we can recast eq. (17.38) as

$$t\kappa^{-\epsilon}\left(\frac{Z_1}{Z}\right)^{1+\epsilon/2}\left(\frac{H_R}{M_R}\right)^{\epsilon/2}\frac{1}{(4\pi)^{D/2}}\Gamma\left(-\frac{\epsilon}{2}\right) = \frac{1}{Z} - M_R^2 \tag{17.40}$$

We now use dimensional regularization with (quasi) minimal subtraction by defining Z and Z_1 in such a way that the singular dependence in ϵ is canceled. Thus, we choose

$$Z = Z_1 \tag{17.41}$$

and

$$\frac{1}{Z} = 1 + t\frac{1}{(4\pi)^{D/2}}\Gamma\left(-\frac{\epsilon}{2}\right) \tag{17.42}$$

With these choices, eq. (17.40) becomes

$$1 - M_R^2 = \frac{t}{t_c}\left(1 - \left(\frac{H_R}{\kappa^2 M_R}\right)^{\epsilon/2}\right) \tag{17.43}$$

which is finite as $\epsilon \to 0$. Here we introduced the quantity t_c,

$$t_c = \left(\frac{D-2}{2}\right)\frac{(4\pi)^{D/2}}{\Gamma\left(2-\frac{D}{2}\right)} \tag{17.44}$$

which we will shortly identify with the value of the critical coupling constant. This equation relates M_R, the (renormalized) expectation value of the sigma field, to the renormalized dimensionless coupling constant t (in units of t_c) and the renormalized symmetry breaking field H_R. In the statistical mechanical interpretation, this is the equation of state (in the $N \to \infty$ limit).

17.2.3 Phase diagram and spectrum

Let us now show that, in the $N \to \infty$ limit and for $D > 2$, the nonlinear sigma model has a phase transition between a broken symmetry phase and a symmetric phase. To this end, we seek a solution to eq. (17.43) for $H_R = 0$ and $M_R \neq 0$, that is,

$$1 - M_R^2 = \frac{t}{2\pi\epsilon} \tag{17.45}$$

and notice that, as $M_R \to 0$, t approaches (from below) the value t_c given in eq. (17.44), which defines the critical coupling constant. It is worth noting that, as expected, $t_c \to 0$ as $\epsilon \to 0$. Using this expression for t_c, we can write eq. (17.45) as

$$1 - M_R^2 = \frac{t}{t_c} \tag{17.46}$$

or, what is equivalent,

$$M_R = \left(1 - \frac{t}{t_c}\right)^{\beta} \tag{17.47}$$

where the exponent is $\beta = 1/2$. We already found this result in our study of the perturbative RG in chapter 16, where we noted that $\beta \to 1/2$ as $n \to \infty$. Hence, we expect that there will be corrections to this result if we go beyond the $N \to \infty$ limit. Thus, for $t < t_c$, the $O(N)$ symmetry is indeed spontaneously broken.

Let us now examine the behavior for $t > t_c$. We will now use the definition of t_c in eq. (17.44) to write eq. (17.45) in the simpler form

$$1 - M_R^2 = \frac{t}{t_c}\left(1 - \left(\frac{H_R}{M_R \kappa^2}\right)^{\epsilon/2}\right) \tag{17.48}$$

Now recall that, by dimensional analysis and the Ward identity, the susceptibility χ_R is

$$\chi_R = \kappa^2 \frac{M_R}{H_R} \tag{17.49}$$

in terms of which we can write

$$1 - M_R^2 = \frac{t}{t_c}\left(1 - [\chi_R(t, H_R)]^{-\epsilon/2}\right) \tag{17.50}$$

We now take the limit $H_R \to 0$ and $M_R \to 0$ holding χ_R fixed, and obtain

$$\chi_R = \left(\frac{t}{t_c} - 1\right)^{-\gamma} \tag{17.51}$$

where the exponent is $\gamma = \frac{2}{\epsilon}$ (again in the $N \to \infty$ limit).

Having understood the question of spontaneous symmetry breaking, we will now inquire as to the behavior of the $\boldsymbol{\pi}$ fields. We expect that the $\boldsymbol{\pi}$ fields should be the Goldstone bosons of the broken symmetry state, and hence, should be massless in the broken symmetry phase and massive in the unbroken phase. Moreover, we also expect the σ field to be massive in the broken phase and that all the masses become equal as $H \to 0$.

To this end, we now restore the fields $\boldsymbol{J} \neq 0$, the sources of the $\boldsymbol{\pi}$ fields. Although, in principle, we obtain more complicated saddle-point equations if $\boldsymbol{J} \neq 0$, in practice, we can still use our solutions obtained for $\boldsymbol{J} = 0$ in the regime in which $\boldsymbol{J}$ is infinitesimally small (i.e., a "linear response"). Hence, we can approximate the kernel $G(x-y;\alpha)$ in eq. (17.23), by its approximate form with $\alpha=\bar{\alpha}=\frac{H}{M}=\frac{H_R}{M_R}$. We readily see that, in this limit, the two-point function of the π fields is

$$\langle \pi(x)\cdot\pi(y)\rangle = G\left(x-y;\frac{H_R}{M_R}\right) \tag{17.52}$$

In other words, provided $\frac{H_R}{M_R}\neq 0$, the π fields are massive, and their mass (squared) is

$$m_\pi^2=\frac{1}{\xi_\pi^2}=\frac{H_R}{M_R} \tag{17.53}$$

where ξ_π is the correlation length of the π fields, which, as we see, is given by

$$\xi_\pi=(\chi_R\kappa^{-2})^{1/2}=\kappa^{-1}\left|1-\frac{t_c}{t}\right|^{-\nu} \tag{17.54}$$

where $\nu=\gamma/2=1/\epsilon$ (again, as $N\to\infty$).

Thus, in the symmetric phase, we find that as $H_R\to 0$,

$$m_\pi^2=\chi_R^{-1}=m_\sigma^2 \tag{17.55}$$

where m_σ^2 is the mass (squared) of the σ field. Hence, for $t>t_c$, we have a spectrum of N massive bosons forming a multiplet of $O(N)$ (as we should). However, for $t<t_c$, the symmetry is spontaneously broken, and $M_R\neq 0$ as $H_R\to 0$. Hence, in the broken symmetry state $m_\pi^2=0$, and we have (as we should) a spectrum of $N-1$ massless (Goldstone) bosons.

In the special case of $D=2$, the correlation length $\xi_\pi=\xi_\sigma=\xi$ becomes

$$\xi=m^{-1}=\kappa^{-1}\exp\left(\frac{2\pi}{t}\right) \tag{17.56}$$

This result shows that the dependence of the correlation length on the dimensionless coupling constant t is an essential singularity. Thus, $N\to\infty$ is truly nonperturbative. We will encounter this behavior in all asymptotically free theories.

17.2.4 Renormalization group

Let us now return to the definitions in eq. (17.36) to obtain the RG functions in the large-N limit. We begin with the beta function for the dimensionless coupling constant t,

$$\beta(t)=\kappa\left.\frac{\partial t}{\partial\kappa}\right|_B \tag{17.57}$$

where we hold the bare theory fixed. From the definition $g_0=t\kappa^{-\epsilon}Z_1$, we find

$$\beta(t)\left(1+t\frac{\partial \ln Z_1}{\partial t}\right)=\epsilon t \tag{17.58}$$

Using the expression for Z_1 (and of Z) in eq. (17.42), and the definition of t_c of eq. (17.44), we find that the beta function is exactly given by

$$\beta(t) = \epsilon t - \epsilon \frac{t^2}{t_c} \tag{17.59}$$

Hence, in the $N \to \infty$ limit, the beta function terminates at the quadratic order in t, where it agrees with the one-loop result (cf. eq. (16.176)) in the large-N limit and for $D = 2 + \epsilon$. We see that, for $D > 2$, the beta function has two zeros (or fixed points): (1) a trivial fixed point at $t = 0$ and (2) a nontrivial fixed point at $t = t_c$. Furthermore, the slope of the beta function is negative at the nontrivial fixed point. Hence, this fixed point is unstable in the IR (and hence stable in the UV). Conversely, the trivial fixed point is stable in the IR and unstable in the UV.

In contrast, for $D = 2$, the beta function simply becomes

$$\beta(t) = \kappa \frac{\partial t}{\partial \kappa} = -\frac{t^2}{2\pi} \tag{17.60}$$

which says that the $O(N)$ nonlinear sigma model is asymptotically free in $D = 2$, and the trivial fixed point at $t = 0$ is IR unstable. As we saw, the $N \to \infty$ limit predicts that the theory is in a massive phase with an unbroken global symmetry for all values of the coupling constant.

Similarly, we can compute the RG function $\gamma(t)$,

$$\gamma(t) = \beta(t) \frac{\partial \ln Z}{\partial t} \tag{17.61}$$

and find

$$\gamma(t) = \frac{t}{t_c} \epsilon \tag{17.62}$$

Hence, the anomalous dimension η at the nontrivial fixed point at t_c is

$$\eta = \gamma(t_c) - \epsilon = 0 \tag{17.63}$$

In other terms, the anomalous dimension $\eta = O(1/N)$, and to determine it requires a computation of the leading correction in the $1/N$ expansion. This is consistent with the result we obtained in the $2 + \epsilon$ expansion (cf. eq. (16.196)), where we found that $\eta \propto 1/N$.

An important moral of this discussion is that, while for $D > 2$ the perturbative definition of the theory, based on an expansion in powers of the coupling constant, is nonrenormalizable, here we find that the theory defined around the nontrivial fixed point is renormalizable. Notice that we were able to access this UV fixed point only by using the large-N limit. In principle, the $2 + \epsilon$ expansion could be used to the same end. However, this presents technical difficulties associated with the behavior of the expansion that we will not discuss here.

17.3 The $\mathbb{CP}^{N-1}$ model

We now turn to the $\mathbb{CP}^{N-1}$ model (D'Adda et al., 1978; Witten, 1979b; Coleman, 1985), introduced in section 16.5.1. As shown there, these models can be described by an

N-component complex field $z(x)$ of unit norm, that is, $z(x) = (z_1(x), \ldots, z_N(x))$ with $z_i(x) \in \mathbb{C}$ and $\boldsymbol{z}^\dagger(x) \cdot \boldsymbol{z}(x) = \sum_{i=1}^N |z_i(x)|^2 = 1$, minimally coupled to a $U(1)$ gauge field $A_\mu(x)$. The action of this model is (cf. eq. (16.102)):

$$S[z_\alpha, z_\alpha^*, \mathcal{A}_\mu] = \frac{1}{g} \int d^D x \left| \left(\partial_\mu - i\mathcal{A}_\mu(x)\right) \boldsymbol{z}(x) \right|^2 \tag{17.64}$$

Here g is the coupling constant, which has units of $[L]^{D-2}$, as in the case of the $O(N)$ model. As we saw, this model is invariant under the local $U(1)$ gauge transformations

$$\boldsymbol{z}(x) \mapsto \exp(i\phi(x))\boldsymbol{z}(x), \qquad \mathcal{A}_\mu(x) \mapsto \mathcal{A}_\mu(x) + \partial_\mu \phi(x) \tag{17.65}$$

Notice that the gauge field only enters in the action through the covariant derivative, and that we do not have a separate term for the gauge field, as we would in QED.

In the absence of the $U(1)$ gauge field, this would be a theory of an N-component complex field with unit norm or, what is equivalent, a $2N$-component real vector of unit length. This would be a nonlinear sigma model with $SU(N)$ global symmetry or, equivalently, $O(2N)$. Hence, for $D > 2$, we would expect to have a broken symmetry state with $2N - 1$ Goldstone bosons. However, in the $\mathbb{CP}^{N-1}$ model, a $U(1)$ subgroup of $SU(N)$ has been gauged. In this situation, we expect that in the broken symmetry state, there is a Higgs mechanism (see section 18.11). Consequently, one of the otherwise $2N - 1$ massless Goldstone bosons should be absent, "eaten" by the gauge field, which should be massive in this phase. However, in the symmetric phase, the $\boldsymbol{z}$ fields should be massive, and the gauge field should be strongly fluctuating.

Here we use a functional approach very similar to what we did in the case of the $O(N)$ nonlinear sigma model. Thus, we implement the local constraint, $\boldsymbol{z}^\dagger \boldsymbol{z} = 1$ with an integral representation of the delta function using a (real) Lagrange multiplier field $\alpha(x)$. The partition function of the $\mathbb{CP}^{N-1}$ model is

$$\mathcal{Z} = \int \mathcal{D}\boldsymbol{z}\mathcal{D}\boldsymbol{z}^\dagger \mathcal{D}\mathcal{A}_\mu \mathcal{D}\alpha \exp\left(-\frac{1}{g}\int d^D x \left[|(\partial_\mu - i\mathcal{A}_\mu)\boldsymbol{z}|^2 - \alpha(\boldsymbol{z}^\dagger \boldsymbol{z} - 1)\right]\right) \tag{17.66}$$

To have a well-defined large-N limit, we need to rescale the coupling constant as $g = g_0/N$.

We can now integrate out the N-component complex field $\boldsymbol{z}$ and obtain, after rescaling the $\boldsymbol{z}$ fields by $\sqrt{g_0}$, the partition function in terms of the effective action for the gauge field $\mathcal{A}_\mu$ and the Lagrange multiplier field α:

$$S_{\text{eff}}[\mathcal{A}_\mu, \alpha] = \operatorname{tr}\ln\left(-D_\mu[\mathcal{A}_\mu]^2 + \alpha\right) - \frac{1}{g_0}\int d^D x \alpha(x) \tag{17.67}$$

where $D_\mu[\mathcal{A}] = \partial_\mu - i\mathcal{A}_\mu$ is the covariant derivative. The partition function now has the form

$$\mathcal{Z} = \int \mathcal{D}\mathcal{A}_\mu \mathcal{D}\alpha \exp\left(-N S_{\text{eff}}[\mathcal{A}_\mu, \alpha]\right) \tag{17.68}$$

Notice that, unlike what we did in the $O(N)$ nonlinear sigma model, we are treating symmetrically all the components of the $\boldsymbol{z}$ field. Although we will find a phase transition, the analysis of the broken symmetry state will be easier in a somewhat less symmetric formulation.

Clearly, as $N \to \infty$, this partition function will be dominated by the classical configurations that leave the effective action stationary. The difference between the $O(N)$ nonlinear sigma model and the $\mathbb{CP}^{N-1}$ model is that, in addition to the Lagrange multiplier field α, we now have the gauge field $\mathcal{A}_\mu$. Thus we will have two saddle-point equations, each requiring that the effective action be stationary under separate variations of the gauge field and of the Lagrange multiplier field.

Thus we find two conditions. The first one, obtained by varying S_{eff} with respect to the Lagrange multiplier field $\alpha(x)$,

$$\frac{\delta S_{\text{eff}}}{\delta \alpha(x)} = 0 \tag{17.69}$$

implies that

$$\frac{1}{g_0} = \left\langle x \left| \frac{1}{-D[\mathcal{A}_\mu]^2 + \alpha} \right| x \right\rangle \tag{17.70}$$

The second saddle-point equation, obtained by varying with respect to the gauge field $\mathcal{A}_\mu(x)$,

$$\frac{\delta S_{\text{eff}}}{\delta \mathcal{A}_\mu(x)} = 0 \tag{17.71}$$

is the condition that, at the classical level, the $U(1)$ gauge current vanishes,

$$\frac{\delta S}{\delta \mathcal{A}_\mu(x)} = j_\mu[\mathcal{A}] = 0 \tag{17.72}$$

where S is the action of the $\mathbb{CP}^{N-1}$ model. Thus, in the absence of external sources, we can set the classical configuration of the gauge field $\langle \mathcal{A}_\mu \rangle = 0$, which trivially satisfies the condition of a vanishing $U(1)$ gauge current.

Thus, the only equation to be solved in the large-N limit, eq. (17.70), is the gap equation

$$\frac{1}{g_0} = \left\langle x \left| \frac{1}{-\partial^2 + \alpha} \right| x \right\rangle = \int \frac{d^D p}{(2\pi)^D} \frac{1}{p^2 + \alpha_c} \tag{17.73}$$

Notice that the classical value of the Lagrange multiplier, α_c, plays the role of a mass term for the $\boldsymbol{z}$ fields, and

$$G(x-y|\alpha_c) = \left\langle x \left| \frac{1}{-\partial^2 + \alpha_c} \right| y \right\rangle \tag{17.74}$$

is the propagator for the $\boldsymbol{z}$ field.

We have encountered the integral of the form of the right-hand side of eq. (17.73) several times, most recently in eq. (17.39). This integral is UV divergent for $D > 2$ and logarithmically divergent for $D = 2$. Thus, as before, we need to define renormalized quantities. Here it suffices to renormalize the coupling constant g_0^2. Using dimensional analysis, define a renormalized dimensionless coupling constant t,

$$g_0 = t\kappa^{-\epsilon} Z \tag{17.75}$$

where Z is a renormalization constant, and κ is the renormalization scale. Using eq. (17.39) for the integral, the saddle-point equation of eq. (17.73) becomes

$$\frac{1}{Z} = 1 + t\kappa^{-\epsilon} \frac{1}{(4\pi)^{D/2}} \Gamma\left(-\frac{\epsilon}{2}\right) \alpha_c^{\epsilon/2} \tag{17.76}$$

Once again, let us use the minimal subtraction procedure to cancel the singular dependence of ϵ. To this end, choose Z to have the same form as in the nonlinear sigma model, eq. (17.42). This choice results in the saddle-point equation becoming

$$1 = \frac{t}{t_c}\left(1 - (\alpha_c \kappa^{-2})^{\epsilon/2}\right) \tag{17.77}$$

where t_c is the same value for the critical coupling that we found for the nonlinear sigma model (cf. eq. (17.44)). We can now readily solve for the value α_c:

$$\alpha_c = \kappa^2 \left(1 - \frac{t_c}{t}\right)^{2/\epsilon} \tag{17.78}$$

This solution is only allowed if $t > t_c$. For $t < t_c$, the only possible solution is $\alpha_c = 0$.

In other words, for $D > 2$ and in the limit $N \to \infty$, the $\mathbb{CP}^{N-1}$ model has two phases separated by a phase transition at t_c. For $t < t_c$, the z fields are massless; for $t > t_c$, they are massive with mass $m_z^2 = \alpha_c$. But for $D = 2$, $t_c \to 0$, there is only one phase, and the z fields are massive for all values of the coupling constant. It is easy to see that in $D = 2$, the mass (squared) is

$$m_z^2 = \alpha_c = \kappa^2 \exp\left(-\frac{\pi}{t}\right) \tag{17.79}$$

In other words, in $D = 2$, the $\mathbb{CP}^{N-1}$ model is asymptotically free and exhibits dynamical mass generation.

That the $\mathbb{CP}^{N-1}$ model is asymptotically free in $D = 2$ can be seen by computing the beta function (in the large-N limit)

$$\beta(t) = \kappa \frac{\partial t}{\partial \kappa}\bigg|_B = \epsilon t - \frac{\epsilon}{t_c} t^2 \tag{17.80}$$

which is the same as the $O(N)$ nonlinear sigma model in the $N \to \infty$ limit. Thus, here too, for $D > 2$, we have a UV fixed point at t_c and an IR fixed point at $t = 0$. In $D = 2$ dimensions, this theory is asymptotically free.

It is instructive to compute the leading corrections in the $1/N$ expansion. To lowest order in the $1/N$ expansion, the partition function is

$$\begin{aligned}
\mathcal{Z} &= \exp(-NS_{\text{eff}}[\alpha_c]) \\
&\times \int \mathcal{D}\tilde{\alpha} \exp\left(-\frac{1}{2}\int d^Dx \int d^Dy\, \tilde{\alpha}(x)\Pi(x-y|\alpha_c)\tilde{\alpha}(y)\right) \\
&\times \int \mathcal{D}A_\mu \exp\left(-\frac{1}{2}\int d^Dx \int d^Dy\, A_\mu(x)\Pi_{\mu\nu}(x-y|\alpha_c)A_\nu(y)\right) \\
&\times \left[1 + O\left(\frac{1}{N}\right)\right]
\end{aligned} \tag{17.81}$$

where the kernel $\Pi(x-y|\alpha_c)$ is

$$\Pi(x-y|\alpha_c) = G(x-y|\alpha_c)G(y-x|\alpha_c) \tag{17.82}$$

which is clearly a bubble diagram. Here, $G(x-y|\alpha_c)$ is the propagator of the z fields, given by eq. (17.74). The Fourier transform $\Pi(p)$ is

$$\Pi(p)=\int \frac{d^Dq}{(2\pi)^D}\frac{1}{(q^2+\alpha_c)((q+p)^2+\alpha_c)} \tag{17.83}$$

In contrast, the polarization tensor $\Pi_{\mu\nu}(x-y)$ is the correlation function of the currents of the z fields. By gauge invariance, it must be transverse, that is, it should obey the Ward identity:

$$\partial^x_\mu \Pi_{\mu\nu}(x-y)=0 \tag{17.84}$$

Explicitly, $\Pi_{\mu\nu}(x-y)$ is given by

$$\Pi_{\mu\nu}(x-y)=\partial^x_\mu G(x-y|\alpha_c)\partial^x_\nu G(y-x|\alpha_c)-2\delta_{\mu\nu}G(x,x|\alpha_c) \tag{17.85}$$

In momentum space, this kernel is given by

$$\Pi_{\mu\nu}(p)=\int \frac{d^Dq}{(2\pi)^D}\frac{(2q_\mu+p_\mu)(2q_\nu+p_\nu)}{(q^2+\alpha_c)((q+p)^2+\alpha_c)}-2\delta_{\mu\nu}\int \frac{d^Dq}{(2\pi)^D}\frac{1}{(q^2+\alpha_c)} \tag{17.86}$$

At long distances, $p\to 0$, $\Pi_{\mu\nu}(p)$ behaves as

$$\Pi_{\mu\nu}(p)\simeq \frac{1}{48\pi\alpha_c}(p^2\delta_{\mu\nu}-p_\mu p_\nu) \tag{17.87}$$

where $\alpha_c=m_z^2$ is the mass squared of the z particles. Hence, in the massive phase, $\alpha_c\neq 0$, the low-energy effective action of the gauge field $\mathcal{A}_\mu$ has the Maxwell form

$$S_{\text{eff}}[\mathcal{A}_\mu]=C\frac{N}{4\alpha_c}\int d^Dx\, \mathcal{F}^2_{\mu\nu}(x) \tag{17.88}$$

where $\mathcal{F}_{\mu\nu}=\partial_\mu\mathcal{A}_\nu-\partial_\nu\mathcal{A}_\mu$ is the field strength of this "emergent" gauge field, and $C=\frac{1}{48\pi}$.

The z fields couple minimally to the gauge field $\mathcal{A}_\mu$. Thus, we see that in the symmetric phase (with $t>t_c$), the z particles will experience a long-range "Coulomb" interaction and should form gauge-invariant bound states of the form $z^*_\alpha z_\beta$. This is particularly strong in $D=2$ spacetime dimensions. In two dimensions, as in the case of the $O(N)$ nonlinear sigma model, the theory is asymptotically free, and the coupling constant flows to large values. The theory is in a phase dominated by nonperturbative effects. In the large-N limit, we find that $t_c=0$, and the fields are massive.

However, there is more than that. In $D=2$ dimensions (i.e., $1+1$ - dimensional Minkowski spacetime), the Coulomb interaction is *linear* (i.e., $\propto|x-y|$, where x and y are the spatial coordinates of the z particles). This can be checked by computing the expectation value of the Wilson loop operator, corresponding to the worldlines of a heavy particle-antiparticle pair (see section 9.7). As we know, the Coulomb interaction in $1+1$ dimensions becomes $V(R)=\sigma R$. This is a *confining* potential. Hence, not only are these bound states very tight, but there are also no z particles in the spectrum. We will return to the problem of confinement in gauge theory in chapter 18.

To study the broken symmetry state, with $t<t_c$, we need to modify our approach somewhat. Call the component $z_1=z_\parallel$ and the $N-1$ remaining (complex) components $\boldsymbol{z}_\perp$.

The Lagrangian now is

$$\mathcal{L} = \frac{1}{g}|D_\mu[\mathcal{A}]z_\parallel|^2 + \frac{1}{g}|D_\mu[\mathcal{A}]\boldsymbol{z}_\perp|^2 + \frac{\alpha}{g}(|z_\parallel|^2 + |\boldsymbol{z}_\perp|^2 - 1) \tag{17.89}$$

To have a well-defined large-N limit, set $g = g_0(N-1)$ (as we did in the case of the nonlinear sigma model). We can now integrate out the $N-1$ complex transverse fields, $\boldsymbol{z}_\perp$, and obtain the effective action

$$S_{\text{eff}}[\mathcal{A}, \alpha, \rho] = (N-1)\operatorname{tr}\ln\left(-D[\mathcal{A}]^2 + \alpha\right) - \frac{N-1}{g_0}\int d^Dx\, \alpha(x) + \frac{N-1}{g_0}\int d^Dx \Big[|D_\mu[\mathcal{A}]z_\parallel|^2 + \alpha|z_\parallel|^2\Big] \tag{17.90}$$

This effective action is invariant under $U(1)$ gauge transformations, and requires that we fix the gauge. Here it is convenient to use the unitary gauge, $z_\parallel = \rho \in \mathbb{R}^+$. The path integral for the field ρ (whose action is given in the third term of eq. (17.90)) becomes

$$Z[\mathcal{A}, \alpha] = \int \mathcal{D}\rho\, \exp\left(-\frac{(N-1)}{g_0}\int d^Dx\, \left[(\partial_\mu\rho)^2 + \rho^2\mathcal{A}_\mu^2 + \alpha\rho^2\right]\right) \tag{17.91}$$

The saddle point equation for α now is

$$\frac{1}{g_0} - \frac{\rho^2}{g_0} = G(x, x|\alpha) \tag{17.92}$$

and the saddle point equation for ρ is given by

$$-\partial^2\rho + \rho\mathcal{A}_\mu^2 + \alpha\rho = 0 \tag{17.93}$$

We will see solutions with $\mathcal{A} = 0$ (as before). For $g_0 < g_0^c$ (i.e., $t < t_c$), set $\alpha_c = 0$ and seek a solution with ρ constant,

$$\rho_c = \left(1 - \frac{t}{t_c}\right)^\beta \tag{17.94}$$

with $\beta = 1/2$, in the $N \to \infty$ limit.

Thus, in the broken symmetry in $D > 2$, $\alpha_c = 0$, and the fields $\boldsymbol{z}_\perp$ are massless (i.e., the $2(N-1)$ Goldstone bosons). In this phase, the gauge field $\mathcal{A}_\mu$ is massive, and the mass is ρ_c^2. So, in the broken symmetry phase, the gauge field is "Higgsed." In chapter 18, we will discuss the Higgs mechanism. In chapter 19, we will discuss the role of topology in field theory and return to this model to discuss its instantons (and solitons) and their role.

17.4 The Gross-Neveu model in the large-*N* limit

The Gross-Neveu model is a theory of interacting massless Dirac fermions in $1+1$ dimensions whose Lagrangian (in the Minkowski metric) is (Gross and Neveu, 1974)

$$\mathcal{L} = \bar{\psi}_a i\partial\!\!\!/ \psi_a + \frac{g}{2}\left(\bar{\psi}_a\psi_a\right)^2 \tag{17.95}$$

Here the Dirac fermions are bispinors, and the index $a=1,\ldots,N$ labels the fermionic "flavors." The theory has a global $SU(N)$ flavor symmetry,

$$\psi \mapsto U\psi \tag{17.96}$$

where $U \in SU(N)$. This theory is also invariant under *discrete* chiral transformations

$$\psi \mapsto \gamma_5 \psi \tag{17.97}$$

where γ_5 is a hermitian 2×2 (with $\gamma_5^2 = I$) Dirac matrix that anticommutes with the Dirac matrices γ_μ. Hence, we have the algebra

$$\{\gamma_\mu, \gamma_\nu\} = 2g_{\mu\nu} I, \quad \{\gamma_5, \gamma_\mu\} = 0 \tag{17.98}$$

with I being the 2×2 identity matrix, and $g_{\mu\nu} = \text{diag}(1, -1)$ the metric of $D=$two-dimensional Minkowski spacetime.

The Gross-Neveu model has a $U(N)$ flavor symmetry and a $\mathbb{Z}_2$ chiral symmetry. The fermion mass bilinear is odd under the discrete chiral transformation, eq. (17.97), that is,

$$\bar{\psi}\psi \mapsto -\bar{\psi}\psi \tag{17.99}$$

We will see that in the Gross-Neveu model, the discrete chiral symmetry is broken spontaneously, and there is a dynamical generation of a mass for the fermions.

We can also define a chiral version of the Gross-Neveu model, whose Lagrangian is

$$\mathcal{L} = \bar{\psi}_a i \partial\!\!\!/ \psi_a + \frac{g}{2}\left((\bar{\psi}_a \psi_a)^2 - (\bar{\psi}_a \gamma_5 \psi_a)^2\right) \tag{17.100}$$

which, in addition to the global $SU(N)$ flavor symmetry, now has a $U(1)$ chiral symmetry under the transformations

$$\psi \mapsto e^{i\theta\gamma_5}\, \psi \tag{17.101}$$

The fermion bilinears can be put together in a two-component real vector field $(\bar{\psi}\psi, i\bar{\psi}\gamma_5\psi)$ that transforms as rotation by a global angle 2θ under the global $U(1)$ chiral symmetry. We will see that this symmetry is (almost) spontaneously broken.

We will also see that in $D=2$ spacetime dimensions, the Gross-Neveu models (both with discrete and continuous chiral symmetry) are asymptotically free and exhibit dynamical mass generation. As presented here, the Gross-Neveu model can also be defined in higher dimensions with the proviso that it is no longer renormalizable, meaning that with a suitable regularization it has a phase transition. In $D=4$ dimensions, this model is closely related to the Fermi theory of weak interactions. The chiral version of the model is closely related to the massless Thirring model (in $D=2$) and to the Nambu–Jona-Lasinio model (in $D=4$) (Nambu and Jona-Lasinio, 1961).

17.4.1 The Gross-Neveu model

Both versions of the Gross-Neveu model can be solved in the large-N limit. We begin with the nonchiral model of eq. (17.95), by decoupling the four-fermion interaction term using a real scalar field σ. On scaling the coupling constant $g = g_0/N$, the Lagrangian is now

$$\mathcal{L} = \bar{\psi}_a i\partial\!\!\!/ \psi_a - \sigma \bar{\psi}_a \psi_a - \frac{N}{2g_0}\sigma^2 \tag{17.102}$$

Now we see that the discrete chiral symmetry is equivalent to the $\mathbb{Z}_2$ ("Ising") symmetry $\sigma(x) \mapsto -\sigma(x)$. Moreover, if $\langle \sigma \rangle \neq 0$, then the Ising symmetry would be broken spontaneously, and $\langle \bar{\psi}\psi \rangle \neq 0$ as well. Furthermore, a nonvanishing value for $\langle \sigma \rangle$ means that the fermions have become dynamically massive. Thus, this is a theory of dynamical mass generation by the spontaneous breaking of the chiral symmetry.

We now proceed to study this theory in the large-N limit. To this end, we first integrate out the fermionic fields and obtain the following effective action for the σ field, S_{eff}:

$$S_{\text{eff}} = -i \text{tr} \ln \left(i\partial\!\!\!/ - \sigma \right) - \int d^D x \frac{\sigma^2}{2g_0} \tag{17.103}$$

where the first term now arises from the fermion determinant (i.e., from summing over fermion bubble diagrams). The partition function is now

$$Z = \int \mathcal{D}\sigma \exp(iNS_{\text{eff}}[\sigma]) \tag{17.104}$$

Thus, as in the examples of the nonlinear sigma model and the $\mathbb{CP}^{N-1}$ models, the large-N limit of the theory is the semiclassical approximation of a field that couples to a composite operator, which in the case of the Gross-Neveu model is the fermion mass bilinear, $\bar{\psi}\psi$. The main difference is that the Gross-Neveu model is a fermionic theory, which is the reason for the negative sign in front of the first term in the effective action (aside from the fact that the determinant involves the Dirac and not the Klein-Gordon operator). Notice that we are working in Minkowski spacetime (hence the factor of i in the first term of the effective action).

The stationary (saddle-point) equation is

$$\frac{\partial S_{\text{eff}}}{\partial \sigma(x)} = -i \left\langle x \left| \frac{1}{i\partial\!\!\!/ - \sigma_c} \right| x \right\rangle - \frac{\sigma_c}{g_0} = 0 \tag{17.105}$$

where σ_c is the saddle-point (uniform) value of the field $\sigma(x)$, and we can identify

$$S_{ab}(x-y; m) = -i \left\langle x \left| \frac{1}{i\partial\!\!\!/ - m} \right| y \right\rangle \delta_{ab} \tag{17.106}$$

with the Feynman propagator for a Dirac field with mass $m \equiv \sigma_c$.

Using the momentum space form for the Dirac propagator, we can readily write the saddle-point equation as

$$\text{tr} \int \frac{d^D p}{(2\pi)^D} \frac{-i}{p\!\!\!/ - \sigma_c} = \frac{\sigma_c}{g_0} \tag{17.107}$$

where the usual Feynman contour prescription has been assumed, and the trace runs over the Dirac indices. This result, together with the definition of the propagator, implies that at $N = \infty$, we can identify the chiral condensate with

$$\langle \bar{\psi}\psi \rangle = \frac{\sigma_c}{g_0} \tag{17.108}$$

Hence we have a chiral condensate if $\sigma_c \neq 0$.

We now use the Dirac algebra and a Wick rotation of the integration path into the complex plane, $ip_0 \mapsto p_D$, to write the saddle-point equation as

$$2\int \frac{d^Dp}{(2\pi)^D}\frac{\sigma_c}{p^2+\sigma_c^2}=\frac{\sigma_c}{g_0} \tag{17.109}$$

Therefore, in this theory, either $\sigma_c=0$, and the chiral symmetry is unbroken, or we have a spontaneously broken discrete chiral symmetry state with $\sigma_c \neq 0$. In the latter case, the value of σ_c is the solution of the equation

$$2\int \frac{d^Dp}{(2\pi)^D}\frac{1}{p^2+\sigma_c^2}=\frac{1}{g_0} \tag{17.110}$$

which is identical to the saddle-point equation we found in the $\mathbb{CP}^{N-1}$ model (cf. eq. (17.73)). Therefore it has the same solution.

Here too, we will first define a renormalized dimensionless coupling constant t, such that $g_0=t\kappa^{-\epsilon}Z$, where $\epsilon=D-2$. The resulting beta function for the dimensionless coupling constant t has the same form as in the nonlinear sigma model and the $\mathbb{CP}^{N-1}$ model,

$$\beta(t)=\epsilon t-\epsilon\frac{t^2}{t_c} \tag{17.111}$$

but with a value of t_c that is 1/2 of the value for the nonlinear sigma model, eq. (17.44).

We conclude that, for $t>t_c$, the fermion of the theory has the dynamically generated mass

$$m=\sigma_c=\kappa\left(1-\frac{t_c}{t}\right)^{1/\epsilon} \tag{17.112}$$

and it remains massless for $t<t_c$. Notice that, at $N=\infty$, the dynamically generated mass m and the chiral condensate $\langle\bar{\psi}\psi\rangle$ are related by

$$\langle\bar{\psi}\psi\rangle=\frac{m}{g_0} \tag{17.113}$$

Here too, the case of $D=2$ is special. Indeed, for $D=2$, the theory is asymptotically free and has a dynamically generated mass for all values of the coupling constant, which, as $N\to\infty$, is

$$m=\sigma_c=\kappa\exp\left(-\frac{\pi}{t}\right) \tag{17.114}$$

An alternative way to understand what happens is to compute the effective potential for a constant value of the field σ. At $N=\infty$, the partition function is just

$$Z[\sigma_c]=\exp\left(-iNU(\sigma_c)\right) \tag{17.115}$$

where σ_c is the constant value of the field σ that minimizes the potential

$$U(\sigma_c)=i\,\mathrm{tr}\ln\left(i\partial\!\!\!/+\sigma_c\right)+\int d^2x\frac{\sigma_c^2}{2g_0} \tag{17.116}$$

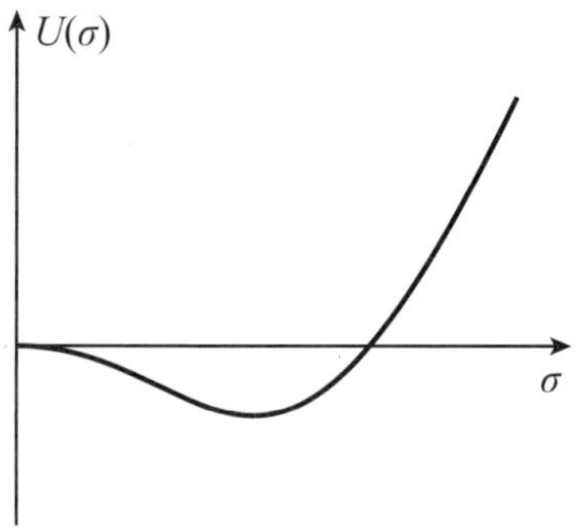

Figure 17.6 Effective potential for the Gross-Neveu model.

Using the properties of the two-dimensional Dirac gamma matrices, we have

$$\operatorname{tr}\ln(i\partial\!\!\!/ - \sigma_c) = \operatorname{tr}\ln(\partial^2 + \sigma_c^2) \tag{17.117}$$

Then, after a Wick rotation, the potential becomes

$$U(\sigma_c) = \operatorname{tr}\ln(-\nabla^2 + \sigma_c^2) + V\frac{\sigma_c^2}{2g_0} \tag{17.118}$$

where $V = L^2$ is the volume of $D = 2$ Euclidean spacetime. Hence, the problem reduces to the computation of the determinant of the Euclidean Klein-Gordon operator. Using the expression for the determinant of the two-dimensional Euclidean Klein-Gordon operator, computed using the ζ-function regularization in section 8.9 (see eq. (8.212)), we can write the potential $U(\sigma)$ as

$$U(\sigma) = V\frac{\sigma^2}{4\pi}\left[\ln\left(\frac{\sigma^2}{\kappa^2}\right) - 1\right] + V\frac{\sigma^2}{2g_0} \tag{17.119}$$

This potential, shown in figure 17.6, has a minimum at the value of σ_c obtained in eq. (17.114).

Eq. (17.108) allows us to identify the fluctuations of the field σ with the fluctuations of the chiral condensate. Thus, the propagator of the σ field is the propagator of the composite operator $\bar{\psi}\psi$. Expanding the effective action to quadratic order in the fluctuations about its expectation value, $\tilde{\sigma}(x) = \sigma(x) - \sigma_c$, we find

$$S_{\text{eff}}[\tilde{\sigma}] = -\frac{N}{2}\int d^2x \int d^2y\, \tilde{\sigma}(x) K(x-y) \tilde{\sigma}(y) \tag{17.120}$$

where the kernel is

$$K(x-y) = \operatorname{tr}\left(S_F(x,y) S_F(y,x)\right) + \frac{1}{g_0}\delta(x-y) \tag{17.121}$$

Here $S_F(x,y)$ is the Feynman propagator for a massive Dirac field in two dimensions, which, on Wick rotation, falls off exponentially at distances long compared to $\xi = \sigma_c^{-1}$. This result also implies that the correlator of the fluctuations of the composite operator $:\bar{\psi}\psi(x):=$

$\bar{\psi}\psi(x) - \langle\bar{\psi}\psi(x)\rangle$, where the chiral condensate $\langle\bar{\psi}\psi\rangle$ is given by eq. (17.113), also falls off exponentially with distance:

$$\langle :\bar{\psi}\psi(x) :: \bar{\psi}\psi(y) :\rangle \sim \exp(-|x-y|/\xi) \tag{17.122}$$

It is straightforward to see that, again in the $N \to \infty$ limit, the composite operator behaves as a scalar bound state with mass equal to twice the fermion mass. Hence at $N=\infty$, the composite operator represents a bound state on threshold (i.e., the binding energy is $O(1/N)$).

17.4.2 The chiral Gross-Neveu model

We close with a brief discussion of the *chiral* Gross-Neveu model, whose Lagrangian is given by eq. (17.100). This Lagrangian is invariant under the global $U(1)$ chiral transformations of eq. (17.101), which leads to important changes. Since we know that in the case of the model with a discrete chiral symmetry, the global discrete chiral symmetry is spontaneously broken, we may suspect that this may also be the case in the chiral Gross-Neveu model as well. We will see that two-dimensional spontaneous breaking of a $U(1)$ symmetry is subtle and, as a result, this claim is almost (but not completely) correct.

To study this theory in its large-N limit (here N is the number of Dirac fermion flavors) we proceed as before and use a Hubbard-Stratonovich decoupling of the quartic fermionic interactions. Since the Lagrangian has two quartic terms, we need two real scalar fields, denoted by $\sigma(x)$ and $\pi(x)$. The partition function for $D=2$ is now

$$Z = \int \mathcal{D}\bar{\psi}\mathcal{D}\psi\mathcal{D}\sigma\mathcal{D}\pi \exp\left(iS(\bar{\psi},\psi,\sigma,\pi)\right) \tag{17.123}$$

where the action is

$$S = \int d^2x \bar{\psi}\left(i\not{\partial} + \sigma(x) + i\pi(x)\gamma_5\right)\psi - \frac{N}{2g_0}\int d^2x\left(\sigma^2(x) + \pi^2(x)\right) \tag{17.124}$$

Under the continuous chiral symmetry of eq. (17.101), the two-component real field (σ,π) transforms as a rotation by a global angle 2θ.

Again, integrate out the fermions to obtain the effective action for the fields σ and π:

$$S_{\text{eff}}[\sigma,\pi] = -iN\text{tr}\ln\left(i\not{\partial} + \sigma(x) + i\pi(x)\gamma_5\right) - \frac{N}{2g_0}\int d^2x\left(\sigma^2(x) + \pi^2(x)\right) \tag{17.125}$$

The $U(1)$ symmetry requires that the effective potential can only depend on $\sigma_c^2 + \pi_c^2$, where σ_c and π_c are the solutions of the saddle-point equations. After tracing over the Dirac indices and a Wick rotation of the integration contours, the saddle-point equations are

$$2\sigma_c \int \frac{d^2p}{(2\pi)^2}\frac{1}{p^2 + \sigma_c^2 + \pi_c^2} = \frac{\sigma_c}{g_0}, \quad 2\pi_c \int \frac{d^2p}{(2\pi)^2}\frac{1}{p^2 + \sigma_c^2 + \pi_c^2} = \frac{\pi_c}{g_0} \tag{17.126}$$

Therefore, if (σ_c, π_c) is a solution of these saddle-point equations, any other uniform configuration obtained by a rotation is also a solution. Hence, any solution will break the continuous chiral symmetry spontaneously. Another way to see that this must be the case is to compute the effective potential $U(\sigma,\pi)$. A simple calculation shows that the result is the same as for the nonchiral Gross-Neveu model of eq. (17.119) with the replacement $\sigma \to (\sigma^2 + \pi^2)^{1/2}$. Thus, the effective potential has the standard "sombrero" shape of a system with a $U(1)$ global symmetry.

In $D = 2$, the chiral Gross-Neveu model is asymptotically free, as is the nonchiral model. In fact, the beta function is the same for both models. In addition, the chiral model also has dynamical mass generation in $D = 2$ dimensions (as it does after a phase transition for $D > 2$). The main difference between the chiral and nonchiral models is the way the chiral symmetry is broken. Since the spontaneously broken chiral symmetry is continuous, we expect that there should be an associated Goldstone boson.

A straightforward way to see this is to rewrite the σ and π fields in terms of an amplitude field ρ and a phase field θ:

$$\sigma(x) + i\pi(x)\gamma_5 = \rho(x) \exp(i\theta(x)\gamma_5) \tag{17.127}$$

Clearly, the amplitude field $\rho(x)$ will be massive and, at $N = \infty$, is going to be pinned at the value $\rho_c = (\sigma_c^2 + \pi_c^2)^{1/2}$, which is the same as the mass for the nonchiral model. However, the phase field $\theta(x)$ must be the Goldstone boson. Hence, the symmetry dictates that the effective low-energy action must depend only on the derivatives of θ. To leading order in an expansion in powers of the inverse mass, the low-energy action is

$$S_{\text{eff}}[\theta] = \int d^2x \frac{N}{4\pi} (\partial_\mu \theta(x))^2 \tag{17.128}$$

which shows that the phase field is indeed massless. The propagator of a massless scalar field in $D = 2$ is

$$G(x - y) = \frac{1}{2N} \ln(x - y)^2 \tag{17.129}$$

which does not decay at long distances. In this sense, the field θ does not describe a physically meaningful excitation, and "does not exist" (Coleman, 1973).

However, it is easy to see that, in the large-N limit, composite operators such as the fermion bilinears

$$\bar{\psi}(1 \pm \gamma_5)\psi(x) = \rho(x) \exp(\mp i\theta(x)) \tag{17.130}$$

exhibit a power-law decay as a function of distance

$$\begin{aligned} \langle \bar{\psi}(1+\gamma_5)\psi(x)\bar{\psi}(1-\gamma_5)\psi(y)\rangle &= \langle \rho(x)\exp(i\theta(x))\rho(y)\exp(-i\theta(y))\rangle \\ &\propto \frac{\rho_c^2}{|x-y|^{1/N}} \end{aligned} \tag{17.131}$$

albeit with a nontrivial exponent $\propto 1/N$. Thus, the correlator of the fermion mass terms does not approach a constant at infinity, and in this sense, there is no chiral condensate. Thus in $D = 2$, although there is dynamical mass generation, the chiral symmetry is almost (but not quite) spontaneously broken. In chapter 19, we will see that this behavior corresponds to a line of fixed points and not to a broken-symmetry state. We will also return to this point in chapter 21, where we discuss conformal field theories. This behavior is a manifestation of the Mermin-Wagner theorem (Mermin and Wagner, 1966; Hohenberg, 1967) (known as Coleman's theorem in high-energy physics (Coleman, 1973)), which states that continuous global symmetries in $D = 2$ classical statistical mechanics and in (1+1)-dimensional QFT cannot be spontaneously broken. But for $D > 2$ and for g_0 larger than a critical value, the dynamical mass generation does correspond to a state with a spontaneously broken symmetry.

17.5 Quantum electrodynamics in the limit of large numbers of flavors

Consider now the large-N limit of QED. The Euclidean action is

$$S=\int d^Dx\left[\frac{1}{4e^2}F_{\mu\nu}^2-\bar{\psi}i(\partial\!\!\!/+iA\!\!\!/)\psi\right] \tag{17.132}$$

This theory has a local $U(1)$ gauge invariance and a global $U(N_f)\times U(N_f)$ flavor symmetry. Here $D=4-\epsilon$. The coupling constant (i.e., the fine structure constant) is

$$\alpha=\frac{e^2}{4\pi}\kappa^{-\epsilon} \tag{17.133}$$

where κ is the renormalization scale. The one-loop beta function for the theory is

$$\beta(\alpha)=-\epsilon\alpha+\frac{2N_f}{3\pi}\alpha^2+O(\alpha^3) \tag{17.134}$$

(actually, the beta function is known to four-loop order). In $D=4$ dimensions in the IR, the theory flows to $\alpha\to 0$, just as in the case of ϕ^4 theory. So it is trivial in the IR. Now, for $D<4$, the beta function has a finite fixed point at

$$e_*^2=24\pi^2\frac{\epsilon}{4N_f}\kappa^{\epsilon} \tag{17.135}$$

Thus, the coupling constant flows to a finite value in the IR, where the theory becomes scale invariant and nontrivial.

To proceed with the large-N_f limit, we integrate out the fermions and write the partition function (in the Euclidean signature) as

$$Z=\int \mathcal{D}A_\mu i\exp\left(-S_{\text{eff}}[A_\mu]\right) \tag{17.136}$$

where

$$S_{\text{eff}}[A_\mu]=\int d^Dx\left[\frac{1}{4e^2}F_{\mu\nu}^2-N_f\text{tr}\ln\left(\partial\!\!\!/+iA\!\!\!/\right)\right] \tag{17.137}$$

On rescaling the coupling constant

$$e^2=\frac{e_0^2}{N_f} \tag{17.138}$$

we obtain, as before, an action of the form

$$S_{\text{eff}}[A_\mu]=N_f\left[\int d^Dx\frac{1}{4e_0^2}F_{\mu\nu}^2-\text{tr}\ln\left(\partial\!\!\!/+iA\!\!\!/\right)\right] \tag{17.139}$$

In the limit $N_f \to \infty$, the partition function is dominated by the semiclassical configurations. The saddle-point equations

$$\frac{\delta S_{\rm eff}[A]}{\delta A_\mu(x)} = 0 \tag{17.140}$$

are trivially satisfied by $A_\mu = 0$. The leading $1/N_f$ corrections are obtained by expanding to quadratic order in A_μ:

$$Z[A] = \left[\operatorname{Det}\left(i\not{\partial}\right)\right]^{N_f} \exp\left(-\frac{N_f}{2}\int d^Dx \int d^Dy\, A_\mu(x) K_{\mu\nu}(x-y) A_\nu(y)\right) \tag{17.141}$$

where the kernel $K_{\mu\nu}(x-y)$ is given by (in momentum space)

$$K_{\mu\nu}(p) = \left[p^2\delta_{\mu\nu} - p_\mu p_\nu\right] K(p) \tag{17.142}$$

where

$$K(p) = \frac{1}{e_0^2} + \frac{D-2}{2(D-1)}\left[b(D)p^{D-4} - a(D)\Lambda^{D-4}\right] + O(\Lambda^{-2}) \tag{17.143}$$

where Λ is the UV momentum cutoff,

$$b(D) = -\frac{\pi}{\sin\left(\frac{\pi D}{2}\right)} \frac{\Gamma^2(D/2)}{\Gamma(D-1)} S_D \tag{17.144}$$

is universal, and $a(D)$ is a correction to scaling that depends on the choice of regularization (and hence is not universal).

The fixed point is determined by canceling out the correction to scaling against the bare Maxwell term. The fixed point for $D<4$ thus determined is located at the value of the charge

$$e_*^2 = \frac{2(D-1)}{(D-2)a(D)} \frac{\Lambda^\epsilon}{N_f} \tag{17.145}$$

The scale-invariant effective action at the fixed point is the nonlocal expression

$$S_{\rm eff}[A_\mu] = \frac{N_f}{4}\frac{(D-2)}{(D-1)} b(D) \int d^Dx \int d^Dy F_{\mu\nu}(x) G_D(x-y) F_{\mu\nu}(y) \tag{17.146}$$

where $G(x-y)$ is given by

$$G_D(x-y) = \left\langle x \left| \frac{1}{(-\partial^2)^{\frac{4-D}{2}}} \right| y \right\rangle \tag{17.147}$$

and its Fourier transform is

$$G(p) = p^{D-4} \tag{17.148}$$

In $D=2$, the model with $N_f=1$ is known as the Schwinger model (Schwinger, 1962). One can check that the large N_f analysis predicts that in $D=2$, the effective action is the same as that of a massive scalar field with mass squared $N_f e^2/\pi$. This result agrees with a direct analysis of the Schwinger model in $D=2$ spacetime dimensions using the chiral anomaly or, what is the same, bosonization (see section 20.9.1). There is a subtlety here that the large N_f approach superficially misses: In addition to the massive scalar, there are also $N_f - 1$ massless scalars. This result, easily found by bosonization (see section 20.3), is important in the realization of chiral symmetry (i.e., in the behavior of fermion bilinears).

17.6 Matrix sigma models in the large-rank limit

The discussion of the previous sections suggests that theories become simpler and solvable in a suitable large-N limit. We will now see that, indeed, in this limit, theories do become simpler, but they are not simple enough to be solvable as in other cases we have discussed. As we will see, the reason for the complexity can be traced back to the fact that in the theories that we have considered, the number of Lagrange multiplier fields (Hubbard-Stratonovich fields) is independent of N, which allows for a simple $N \to \infty$ limit. From a perturbative point of view, we were able to achieve simplicity, since even though the number of diagrams grows with the order of perturbation theory, they do not grow as fast as N. Thus, in the cases that we have examined, the corrections to the large-N limit are down by powers of $1/N$.

We now consider theories in which the scalar field transforms as a rank-N tensor (rather than as a vector). For example, let $\phi_{ij}(x)$ be an $N \times N$ real matrix field (here $i, j = 1, \ldots, N$) that transforms under the global symmetry group $O(N) \times O(N)$. A nonlinear sigma model can be defined by imposing the constraint that the field ϕ is an $O(N)$ rotation matrix. As such, it must obey the local constraint that the inverse must be its transpose (i.e., $(\phi(x)^{-1})_{ij} = \phi(x)_{ji}$). The partition function must again have a local constraint, which is now

$$Z = \int \mathcal{D}\phi \prod_x \delta(\phi_{ij}(x)\phi_{kj}(x) - \delta_{ik}) \exp(-S[\phi]) \tag{17.149}$$

We can now use a representation of the delta function

$$\prod_x \delta(\phi_{ij}(x)\phi_{kj}(x) - \delta_{ij}) = \int \mathcal{D}\lambda_{ij}(x) \exp\left(i \int d^D x \lambda_{ik}(x)(\phi_{ij}(x)\phi_{kj}(x) - \delta_{ik}) \right) \tag{17.150}$$

Hence, the matrix-valued constraint requires that the Lagrange multiplier field should also be a matrix of the same rank as the field itself. Thus the rank of the Lagrange multiplier field diverges as $N \to \infty$. In contrast, in the theories with vector symmetries (that we have discussed in this chapter), the rank of the Lagrange multiplier field is fixed and independent of N. The same considerations apply to all matrix-valued scalar fields (e.g., the principal chiral models on a Lie group G, or general Grassmannian manifolds, and tensor (non-abelian) generalizations of the Gross-Neveu model). We will encounter a similar structure in the case of non-abelian Yang-Mills gauge theory in section 17.7.

Another, and simpler, example is a matrix scalar field theory, in which the field $\phi_{ij}(x)$, with $i, j = 1, \ldots, N$, is a real symmetric matrix, $\phi_{ij} = \phi_{ji}$. In this case, the theory will have a global $O(N)$ symmetry. In another class of theories of this type, the field is an $N \times N$ complex hermitian matrix, $\phi_{ij} = \phi^*_{ji}$, and the global symmetry is $SU(N)$. Finally, a third class consists of a theory on $N \times N$ complex matrices, and the global symmetry is $SU(N) \times SU(N)$. Let us consider a theory for an $N \times N$ matrix field ϕ (here the trace acts after matrix multiplication). The general form of the (Euclidean) Lagrangian is

$$\mathcal{L} = \frac{1}{2}\text{tr}(\partial_\mu \phi \partial_\mu \phi^\dagger) + \frac{m^2}{2}\text{tr}(\phi\phi^\dagger) + \alpha \frac{g_4}{N}\text{tr}(\phi\phi^\dagger\phi\phi^\dagger) + \cdots \tag{17.151}$$

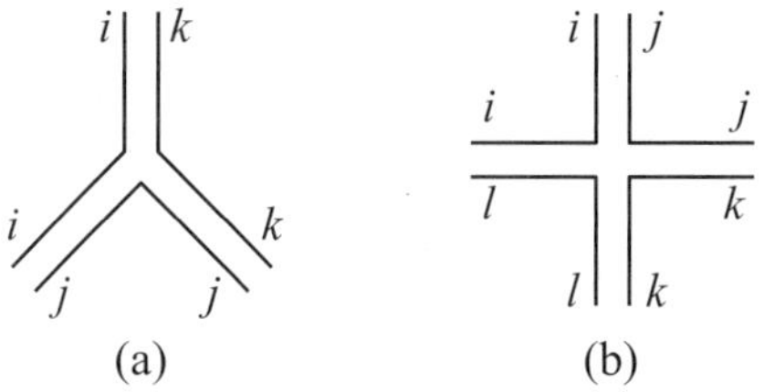

Figure 17.7 (a) The trilinear vertex coupling in $O(N) \times O(N)$ matrix field theory has a weight of $1/\sqrt{N}$, and (b) the quartic vertex has a weight $1/N$.

where $\alpha = 1$ if the field is a real symmetric matrix, $\alpha = 2$ if it is hermitian, and $\alpha = 4$ if it is complex. If the matrix field is real symmetric, a cubic term is also allowed, but not in the other cases. Here g_4 is the coupling constant of the quartic term. Similarly, the coupling constant for an allowed cubic term will be denoted by g_3, and so forth. To obtain a finite large-N limit, we will scale the coupling constant of the trilinear term by $1/\sqrt{N}$, the quartic coupling by $1/N$, and so forth.

The propagator of the matrix field has the form

$$G_{ij|kl}(x-y) = \langle \phi_{ij}(x)\phi_{kl}(y) \rangle \tag{17.152}$$

We use a "double-line" representation to track the propagation of the indices, introduced by G. 't Hooft in the context of Yang-Mills gauge theory ('t Hooft, 1974). In this picture, the two lines of the free propagator are unoriented if the matrix field is real symmetric, and they can be represented as

$G_{ij|kl}(x-y) =$ 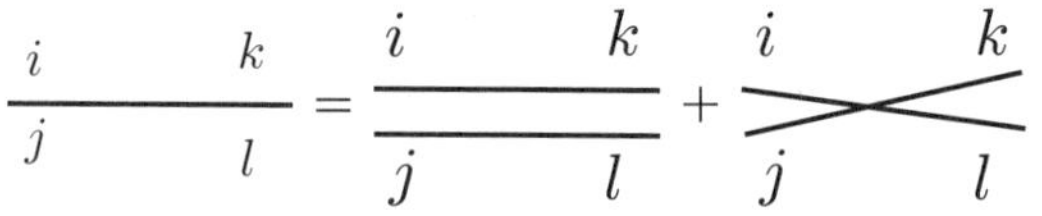 (17.153)

If the matrix field is hermitian, the two lines are oppositely oriented, and the crossed term in eq. (17.153) is absent in this case. In the case of a complex matrix, both lines have the same orientation. The trilinear and quartic vertices are represented in figure 17.7 using the double-line representation.

We will now see that the expansion in Feynman diagrams has a topological character. We follow the work by 't Hooft (1974) (see also the work by Brézin and coworkers (Brézin et al., 1978)). A general Feynman diagram has P propagators, V vertices (of different types), and I closed internal loops. Denote by V_3 the number of trilinear vertices, by V_4 the number of quartic vertices, and so forth, with $g_3, g_4, \ldots$ the associated coupling constants. Then, for a vacuum diagram, we must have

$$2P = 3V_3 + 4V_4 + \cdots \tag{17.154}$$

Each internal loop with a given index is geometrically the face of a polyhedron. The Euler relation then says that

$$V - P + I = \chi = 2 - 2H \tag{17.155}$$

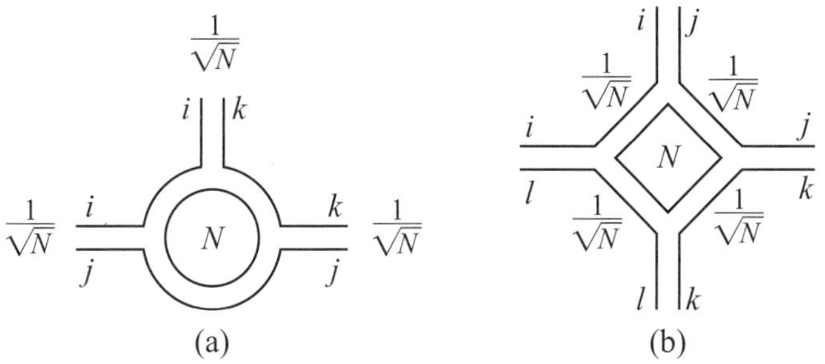

Figure 17.8 (a) The one-loop planar diagram contribution to the trilinear vertex coupling in $O(N) \times O(N)$ matrix field theory has a weight of $(1/\sqrt{N})^3 N = 1/\sqrt{N}$, and (b) the one-loop planar diagram contribution to the quartic vertex has a weight $(1/\sqrt{N})^4 N = 1/N$.

where χ is the Euler character of the surface, and H is the number of holes (the genus) of the surface on which the polyhedron is drawn (zero for a plane or sphere, one for a torus, etc.). With these definitions, since each closed loop contributes a factor of N, the contribution of the diagrams is proportional to

$$g_3^{V_3} g_4^{V_4} \cdots N^I = (g_3\sqrt{N})^{V_3} (g_4 N)^{V_4} \cdots N^{2-H} \tag{17.156}$$

Therefore, provided each coupling is scaled by the appropriate power of N (e.g., $g_r \propto N^{1-r/2}$), the vacuum energy (in units of N^2) has a finite large-N limit given by the diagrams with $H=0$ (no handles; e.g., the sphere). The leading correction, which is $O(1/N^2)$, is given by diagrams drawn on the torus, and so forth. Thus in the $N \to \infty$ limit, the planar diagrams (those with $H=0$) yield the exact answer. This also means that the $1/N$ expansion of these theories is a topological expansion (i.e., an expansion in powers of $1/N^{2H}$, where H is the genus of the surface on which the diagrams are drawn).

Figure 17.8 shows one-loop contributions to the trilinear and quartic vertices. Notice that in these one-loop planar diagrams (i.e., the propagator lines do not cross), the internal loop contributes a factor of N from the summation over the internal index, while each of the three vertex insertions contributes a factor of $g_3/\sqrt{N}$, where g_3 is the coupling constant for the cubic term of the action. Thus, the overall contribution is $(1/\sqrt{N})^3 N = 1/\sqrt{N}$ for the trilinear vertex, and $(1/\sqrt{N})^4 N = 1/N$ for the quartic vertex. So these diagrams are of the same order in N as the bare vertex itself but of order g_3^3 and g_3^4 in the coupling constant, respectively. It is now easy to see that subdividing the internal loop in the diagram by stretching a pair of lines that end at a pair of trilinear vertices leads to a diagram with two loops but *of the same order in* N (i.e., $1/\sqrt{N}$) but of higher order in g_3. The same is true for the quartic vertex.

We can now repeat this process an indefinite number of times, and each insertion gains an extra factor on N for the new loop and a factor of $g_3^2(1/\sqrt{N})^2$ for the two trilinear vertices. Hence, in the large-N limit, we must sum over all planar diagrams of this type of the same order in $1/N$ but of increasing order in the coupling constant g_3. In contrast, if one of the internal propagator double lines were to be crossed (as in the second term of the r.h.s. of eq. (17.153)), then the factor of N will disappear, while the overall factor of $1/N^{3/2}$ will remain. Hence, such a nonplanar diagram is down by one factor of $1/N$ relative to the planar diagrams. The moral is that the leading order in $1/N$ has diagrams of all orders in the coupling constants $g_3, g_4, \ldots$. In this sense, this theory is nonperturbative.

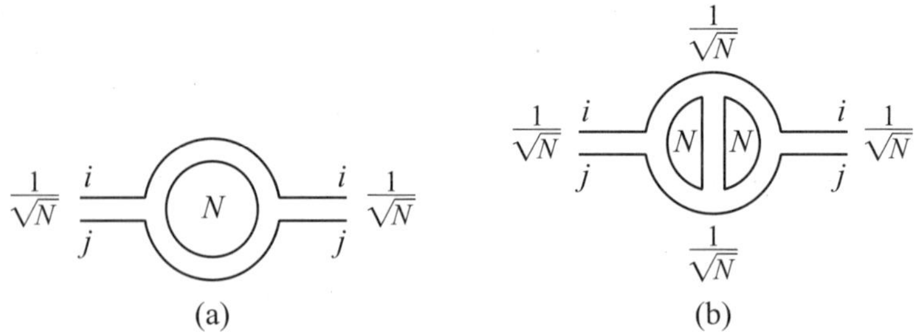

Figure 17.9 Perturbative contributions to the propagator of the $O(N) \times O(N)$ matrix field theory: (a) the one-loop planar diagram, and (b) a two-loop planar diagram.

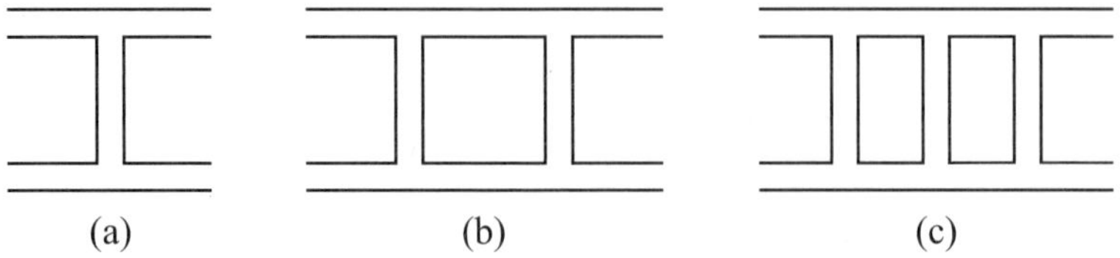

Figure 17.10 Perturbative contributions to the quartic vertex of the $O(N) \times O(N)$ matrix field theory: (a) tree-level, (b) one-loop planar diagram, and (c) two-loop planar diagram.

Let us now examine the corrections to the propagator. Figure 17.9a shows a one-loop contribution. The internal loop has a weight of N and the two vertices contribute with $(1/\sqrt{N})^2$. Thus this diagram has a contribution of $1 = N^0$. In contrast, figure 17.9b has two internal loops and four trilinear vertices. Its weight is $N^2(1/\sqrt{N})^4 = 1$. Hence, it is of the same order as the "leading" diagram. Clearly, here, too, we can continue with this process ad infinitum by inserting inside each closed loop a double propagator line stretched between two trilinear vertices. The resulting diagram is of the same order (1 in this case) in the $1/N$ expansion.

A similar analysis can be made for the four-point function. Some of the contributing planar diagrams are shown in figure 17.10a–c. The tree-level diagram in figure 17.10a is of order $(1/\sqrt{N})^2 = 1/N$. Therefore at $N = \infty$, the particles of this theory are not interacting, and all scattering appears at order $1/N$. The one-loop planar diagram in figure 17.10b also is of order $(1/\sqrt{N})^2 = 1/N$, as is the two-loop planar diagram in figure 17.10c.

As these diagrams show, the large-N limit is, in some sense, a theory in which both ladder diagrams and bubble diagrams are summed over consistently. Even though some significant simplification has been achieved by taking the large-N limit, the theory is highly nontrivial and is still poorly understood.

This analysis leads to a picture for a general planar diagram as a set of nodes (the coordinates of the trilinear vertices) linked to each other (and to the external points) in all possible planar ways by propagator lines. The resulting class of Feynman diagrams have the form of a "fishnet." Such a diagram can also be interpreted as a picture of a tessellated surface, on which the propagators are the edges of polygons meeting at vertices. In the limit where the number of insertions goes to infinity, the surface approaches a smooth surface. Thus, a general one-loop planar diagram is a sum over all possible surfaces anchored on the external loop. We will see in our discussion of gauge theory in the next section that this picture is equivalent to a type of *string theory*.

17.7 Yang-Mills gauge theory with a large number of colors

We now turn to the important case of gauge theories. In the case of Yang-Mills, this theory is non-abelian, and the gauge fields take values in the algebra of a compact Lie group, such as $U(N_c)$, known as the color group. In the large-N_c limit, $SU(N_c)$ and $U(N_c)$ are essentially equivalent. Using $U(N_c)$ as the color group simplifies the analysis. Thus, it is natural to ask: How does this theory behave in the limit of a large number of colors, $N_c \to \infty$? However, when coupled to fermions (quarks), we can also consider the regime in which their flavor symmetry group (a global symmetry of the theory) is $SU(N_f)$, and we can also consider the limit $N_f \to \infty$. These two limits lead to theories with very different characters.

17.7.1 Yang-Mills planar diagrams

Let us consider first pure Yang-Mills theory with a color group $U(N_c)$ in the limit $N_c \to \infty$. The Yang-Mills gauge field is a matrix-valued vector field that takes values on the algebra of $U(N_c)$. As such, it can be written as $A_\mu^{ij}(x) = A_\mu^a(x) t_a^{ij}$, where $i, j = 1, \ldots, N_c$, and t_a^{ij} are the N_c^2 generators of $U(N_c)$ in the fundamental representation. In other words, since the Yang-Mills gauge field takes values in the algebra, it is in the adjoint representation of $U(N_c)$. The Yang-Mills action is (dropping gauge-fixing terms)

$$S_{\rm YM} = \frac{1}{4} \int d^4x \, \left(F_{\mu\nu}\right)_j^i \left(F^{\mu\nu}\right)_i^j \tag{17.157}$$

where

$$\left(F_{\mu\nu}\right)_j^i = i \left(\left[D_\mu, D_\nu\right]\right)_j^i \tag{17.158}$$

is the Yang-Mills field strength, and

$$D_\mu^{ij} = \delta_{ij} \partial_\mu + i \frac{g}{\sqrt{N_c}} A_\mu^{ij} \tag{17.159}$$

is the covariant derivative in the fundamental representation of $U(N_c)$; here $i, j = 1, \ldots, N_c$. Here we have rescaled the Yang-Mills coupling as $g \to g/\sqrt{N_c}$, and g is now called the 't Hooft coupling constant ('t Hooft, 1974).

Since the gauge fields are $U(N_c)$ matrices, it is natural to use a double-line representation for its propagator and to write it in the following form:

$$\langle A_\mu^{ij}(x) A_\nu^{kl} \rangle = \delta_i^l \delta_k^j \mathcal{D}_{\mu\nu}(x-y) \equiv \tag{17.160}$$

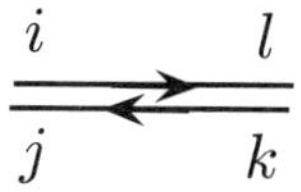

This representation of the propagator simply follows the way the color indices are contracted to each other in the Feynman diagrams.

We can also rewrite the cubic and quartic vertices in the double-line notation, as shown in figure 17.11a,b. Each cubic vertex has a weight of $1/\sqrt{N_c}$, and the quartic vertex also has a weight of $1/N_c$. In contrast, each loop contributes a factor of N_c. It is easy to see that, just as in the example of the matrix-valued scalar field, in the large-N_c limit, the perturbative expansion of the propagator of the gauge field consists of the sum of all possible planar diagrams. To see this, regard each closed loop as a polygon. Then the perturbation theory rules tell us how to fit the polygons together. Let us count the N_c-dependence of a diagram.

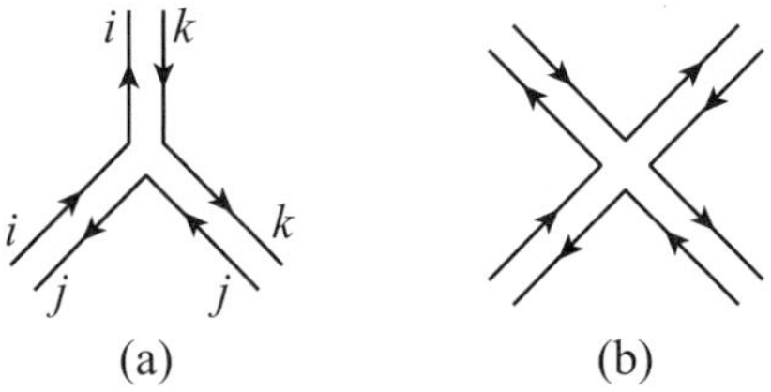

Figure 17.11 Yang-Mills vertices in double-line notation: (a) the cubic vertex, and (b) the quartic vertex.

Each diagram will have V vertices, E edges (the propagators), and F faces (the polygons) and will have an overall weight of

$$N_c^{V-E+F} = N_c^{\chi} \tag{17.161}$$

where

$$\chi = V - E + F \tag{17.162}$$

is a topological invariant known as the *Euler character* of the two-dimensional surface. The diagram, as before, is a tessellation of the surface. Therefore, the weight N_c of a diagram is given by the Euler character of the surface! However, for a connected orientable surface, the Euler character χ is

$$\chi = 2 - 2H - B \tag{17.163}$$

where H is the number of handles of the surface, and B is the number of boundaries (or holes). For an oriented surface without boundaries, $B=0$ and $H=g$, where g (not to be confused with the coupling constant!) is known as the *genus* of the surface. For the vacuum diagrams, which do not have edges, we have

$$\chi = 2 - 2g \tag{17.164}$$

Thus, the $1/N_c$ expansion for the vacuum diagrams is a sum over closed surfaces with increasing genus. The leading term is the sphere, which has no handles, and hence $g=0$. The weight for the sphere is N_c^2. The next contribution is the torus, which has $g=1$, and hence $\chi=0$. Such diagrams scale as $N_c^0=1$. Thus, the $1/N_c$ expansion is a sum over surfaces of different topologies! In contrast, diagrams with quark loops, such as the example shown in figure 17.12, have one boundary (the quark loop) and hence for them, $B=1$. Therefore, compared to the vacuum diagrams, for diagrams with one quark loop, the largest value of χ is 1, and their weight is, at most, N_c.

17.7.2 QCD strings and confinement

In other terms, in the $N_c \to \infty$ limit, only planar diagrams with simple topology contribute to the partition function as well as to the correlators, and therefore, to all physical amplitudes. Although, as we will see, these observations imply that the theory is simpler in this limit, it is by no means as trivial to solve it as in the "vector" large-N limits discussed earlier in this chapter. With some provisos, to date, this problem remains largely unsolved. The only controlled solutions are in 1+1 dimensions, where the dynamics of the gauge field is much more trivial.

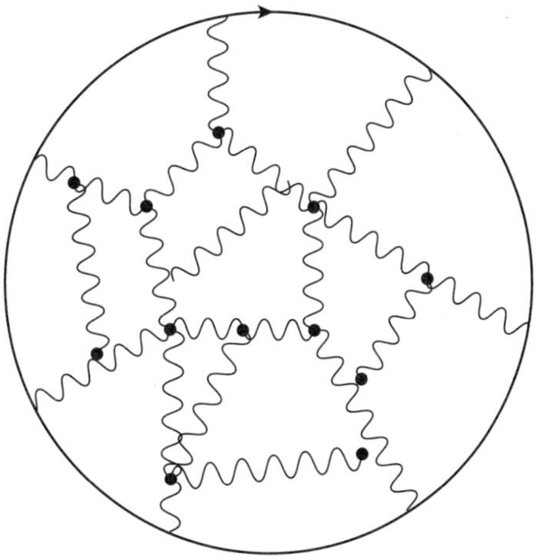

Figure 17.12 A quark loop "fishnet" diagram.

This is the context in which 't Hooft proposed a solution ('t Hooft, 1974). He considered Yang-Mills theory with fermions in 1+1 dimensions in the large-N_c limit, and showed that in this case, it is a confining theory: The energy to separate a quark–antiquark pair over a distance R is $V(R) = \sigma R$, where σ is the string tension, introduced in section 9.7. Hence, quarks do not exist as asymptotic states and are confined, and the spectrum of states consists of color-singlet bound states (i.e., mesons, hadrons, glue-balls, etc). In fact, in 1+1 dimensions, it is possible to solve the theory even at finite N using non-abelian bosonization methods (whose extensions to higher dimensions presently are not known). A similar result is found in the Schwinger model (i.e., QED also in 1+1 dimensions). Another way to address the problem of a strongly coupled gauge theory is lattice gauge theory, which, at the expense of having an explicit Lorentz invariance, can be done in any dimension. We will discuss this approach in chapter 18.

The fact that the $1/N_c$ expansion can be related to a sum over surfaces of different topologies suggests an alternative physical picture in terms of propagating *strings*. In this picture, the total contribution of a single quark-loop diagram is a sum over the contributions of all possible surfaces whose boundary is the loop itself. For example, consider the regime in which the quarks are very heavy. In this case, the diagrams compute the expectation value of a Wilson loop $W[\gamma]$

$$W[\gamma] = \left\langle \text{tr}_F \exp\left(i \oint_\gamma dx_\mu A^\mu \right) \right\rangle \tag{17.165}$$

where γ is the loop, and F tells us that the quarks are in the fundamental representation of $U(N_c)$. A Wilson loop represents a process in which a pair consisting of a heavy quark and a heavy antiquark are created in the remote past and are annihilated in the future. Thus, we can reinterpret the sum over surfaces as the path-integral of a one-dimensional object (i.e., a string), stretching from the quark to the antiquark, which, as time goes by, sweeps over the surface. This picture then suggests that Yang-Mills theory can be understood as a type of string theory known as QCD strings.

The path integral for a particle is a sum over its histories with a weight given in terms of the action. The Feynman path integral for the amplitude of a particle to go from $X^\mu(0)$ at

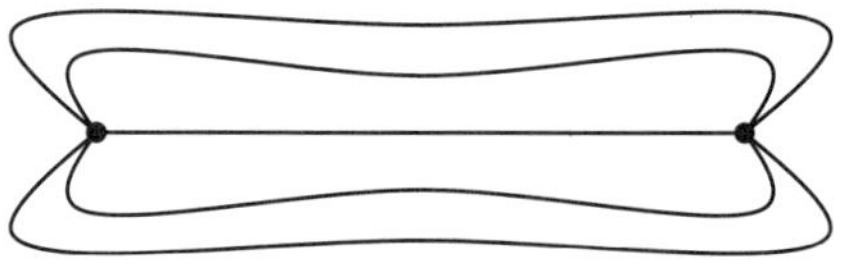

Figure 17.13 Qualitative picture of a QCD string: The quark and the antiquark create a chromoelectric field that is compressed by the Yang-Mills vacuum into a long sausage shape of length R.

time 0 to $X^\mu(\tau)$ at time τ is

$$\langle X^\mu(\tau)|X^\mu(0)\rangle = \int \mathcal{D}X^\mu(t) \exp\Big[- S(X^\mu, \dot{X}^\mu)\Big] \tag{17.166}$$

In the case of a free relativistic particle of mass m, the action is (in the Euclidean signature)

$$S = \int_0^\tau dt\, mc \sqrt{(\dot{X}^\mu)^2} \tag{17.167}$$

and is the proper length of the history of the particle.

A string is a curve in spacetime and is given as a map from a two-dimensional worldsheet labeled by (σ, τ) to Minkowski spacetime of the form $X^\mu(\sigma, \tau)$. The (Euclidean) path integral for a string has the same form as for the particle. For a relativistic string, the action is

$$S = \int d\sigma \int d\tau\, T \sqrt{\det[\partial_a X^\mu \partial_b X_\mu]} \tag{17.168}$$

which is the proper area of the surface swept out by the string. Here, T denotes the string tension, and the worldsheet indices are $a, b = \sigma, \tau$.

Let us assume that the string ansatz is correct and examine the string path integral. In the case of a particle, the path integral is dominated by the history with minimal proper length. This is the classical trajectory. The weight of the path integral is then determined by the action of the classical trajectory, which is proportional to the minimal displacement in spacetime. Likewise, in the case of the string, the path integral is dominated by a history corresponding to a *minimal surface.* It is then obvious, if these assumptions are correct, that the effective potential for a quark-antiquark pair must be linear in their separation and the coefficient T of the string action, eq. (17.168), is indeed the string tension. This picture is expected to hold if Yang-Mills theory in the large-N_c limit is a confining theory.

If this picture is correct, then in physical terms it means that the chromoelectric field configuration created by a static quark-antiquark pair does not have the dipolar configuration of classical electrodynamics. Instead, the field lines are compressed by the actual vacuum state of this strongly interacting theory to a long "sausage," as schematically shown in figure 17.13. In other terms, this picture requires that the Yang-Mills vacuum should *expel* the chromoelectric fields much in the same way that a superconductor expels magnetic fields. In the case of a superconductor, flux expulsion (the Meissner effect) results because the ground state is a condensate of Cooper pairs: an *electric charge condensate.* Hence, in the Yang-Mills case, for the chromoelectric field to be expelled, the vacuum must be a *magnetic condensate*, that is, a condensate of magnetic monopoles (or a "dual superconductor").

17.7.3 The planar limit and the Maldacena conjecture

Although the surviving diagrams of the large-N_c limit suggest the string picture, it is far from obvious how to derive an effective string action from the sum of planar diagrams. A possible solution to this problem has emerged not from directly summing the planar diagrams but, unexpectedly, from the Maldacena conjecture, which is based on string theory. Although string theory is beyond the scope of this book, I will highlight the main arguments that led to a connection to the large-N_c limit of a gauge theory (Maldacena, 1998).

The argument goes as follows. Maldacena considered a ten-dimensional string theory on a ten-dimensional spacetime compactified to $S_5 \times AdS_5$, where S_5 is a five-dimensional sphere, and AdS_5 is a five-dimensional anti–de Sitter (AdS) spacetime, which is a space of constant negative curvature. In $d+2$ dimensions, the metric of AdS_{d+2} spacetime is

$$ds^2 = \frac{R^2}{r^2}\left(-dt^2 + d\boldsymbol{x}^2 + dr^2\right) \tag{17.169}$$

where R is the curvature of AdS spacetime, and $0 \leq r < \infty$ is the "radial" coordinate. An AdS spacetime has a boundary at infinity (along the fifth dimension of AdS_5), which behaves as a flat four-dimensional Minkowski spacetime. Maldacena showed that the classical limit of string theory on this spacetime, which is a supergravity theory, is dual to the strong coupling limit of a version of Yang-Mills theory at the four-dimensional boundary. However, it is not quite a "plain vanilla" Yang-Mills theory but the large-N_c limit of a super-Yang-Mills theory, a supersymmetric version of this theory.

The way this mapping works is as follows. Consider a theory on AdS_5 as a spacetime. In Maldacena's original argument, it was a theory of supergravity, but this was later extended to other (simpler) theories. One solves the classical equations of motion of this theory. This involves, for example, solving classical Yang-Mills theory coupled to classical Einstein gravity, imposing the condition that the spacetime remains asymptotically AdS_5. The Maldacena mapping states that the boundary values of the solutions at infinity yield the expectation values of physical observables in the dual QFT. Thus, the boundary values of the gravitational field yield the expectation value of the energy-momentum tensor operator of the QFT. Similarly, the boundary values of the gauge field lead to the expectation value of the associated conserved currents in the QFT. In this picture, the behavior deep in AdS space is viewed as the IR behavior of the field theory. Conversely, the boundary values yield the UV behavior of the (boundary) theory. In this sense, the fifth dimension of AdS_5 plays the role of the RG flow, and there is a one-to-one correspondence between the gravity theory in the bulk AdS and the strongly coupled gauge field theory defined on the boundary. This correspondence is the *holography principle* of 't Hooft and Susskind ('t Hooft, 1993; Susskind, 1995).

More specifically, consider a theory that has some quantum fields (observables) that we denote by $\mathcal{O}$, defined at the boundary of AdS_5. Let us consider, for simplicity, a scalar field ϕ defined on AdS_5, whose equations of motion are the covariant Laplace equation $D_i D^i \phi = 0$, where D_i is the covariant derivative on AdS_5. Let ϕ_0 be the value of the solution of this Laplace equation at the boundary of AdS_5. Regard ϕ_0 as the source of the field $\mathcal{O}$. Hence, there will be a coupling in the boundary QFT of the form $\int_{\mathcal{M}_4} \phi_0 \mathcal{O}$, where $\mathcal{M}_4$ is four-dimensional Minkowski spacetime. The quantity

$$Z[\phi_0] = \left\langle \exp\left(\int_{\mathcal{M}_4} \phi_0 \mathcal{O}\right)\right\rangle \tag{17.170}$$

is the generating functional of all the correlators of the field $\mathcal{O}$.

Let us now consider the partition function $Z_S[\phi_0]$ in the five-dimensional (supergravity) theory computed with the boundary condition ϕ_0 at AdS_5 infinity. In the limit, the classical partition function is

$$Z_S[\phi_0] = \exp\left(-I_S[\phi]\right) \tag{17.171}$$

where $I_S[\phi]$ is the classical action for the field ϕ with boundary value ϕ_0. In this language, the conjecture becomes the identity (Gubser et al., 1998; Witten, 1998)

$$\left\langle \exp\left(\int_{\mathcal{M}_4} \phi_0 \mathcal{O} \right) \right\rangle = \exp\left(-I_S[\phi]\right) \tag{17.172}$$

where the left-hand side is computed in the strongly coupled QFT, and the right-hand side is computed in the classical theory on AdS_5. This dictionary has been extended (or further conjectured) to hold even in cases where there is no string theory from which it can descend.

If this conjecture is correct (and it has survived many checks since it was formulated in 1998), computations on a classical theory of gravity on AdS_5 can be mapped to the strong coupling limit of a gauge theory in the large-N limit, known as the large-N limit of super-Yang-Mills theory with $\mathcal{N} = 4$ supersymmetries. For technical (but crucial) reasons, the conjecture requires that the theory at the boundary be conformally invariant. In chapter 21, we will discuss conformal invariance in QFT. We will see that conformal invariance requires that the theory should not have any scales. Now, if what we have discussed here is correct, Yang-Mills theory (and QCD) is expected to be a theory with a dynamical scale, the confinement scale, so it is not scale invariant. Furthermore, it is not supersymmetric. Thus, for the Maldacena conjecture to explain confinement, it must also hold in a theory that breaks the supersymmetry and is somehow not conformally invariant. Such deformations of this theory have been constructed but are technically beyond the scope of this book.

Exercises

17.1 Renormalization of a field theory of fermions in the large-N limit

Consider the chiral Gross-Neveu model, which is a simple model of chiral symmetry breaking in particle physics, and of charge-density waves in condensed matter. For simplicity, consider the case of $(1+1)$-dimensional spacetime, although it is easy to work out a generalization to higher dimensions. Most of the problems below are formulated for the theory in Minkowski spacetime. Naturally, you will have to rotate the theory to Euclidean spacetime to do the integrals and to derive the RG equations.

The Lagrangian density of the chiral Gross-Neveu model is

$$\mathcal{L} = \bar{\psi}_a i\partial\!\!\!/ \psi_a + \frac{g_0}{2N}\left(\left(\bar{\psi}_a \psi_a\right)^2 - \left(\bar{\psi}_a \gamma_5 \psi_a\right)^2 \right) \tag{17.173}$$

where ψ_a is a two-component Dirac spinor

$$\psi_a(x) \equiv \begin{pmatrix} R_a \\ L_a \end{pmatrix} \tag{17.174}$$

with R_a and L_a being the amplitudes for the (chiral) right and left fields, respectively, with $a = 1, \ldots, N$. In this exercise assume that N is so large that the limit $N \to \infty$

is a reasonable approximation. We will use the basis for the spinors in which the two-dimensional γ-matrices are given in terms of Pauli matrices: $\gamma_0 = \sigma_1$, $\gamma_1 = i\sigma_2$, and $\gamma_5 = -\sigma_3$, and we use the notation $\partial\!\!\!/ = \partial_\mu \gamma^\mu = \gamma_0 \partial_0 - \gamma_1 \partial_1$. Notice that the usual coupling constant g has been redefined by a scale factor: $g = g_0/2N$.

1) The Lagrangian of this system contains an interaction term that is quartic in the Fermi fields. Instead of using straightforward perturbation theory you will study this system in the large-N limit. To do this, you first need to verify the following Gaussian identity, also known as a Hubbard-Stratonovich transformation:

$$\int \mathcal{D}\sigma(x) \exp\left(-i\frac{N}{2g_0} \int d^2x\, \sigma^2(x) - i\int d^2x\, \sigma(x)\bar{\psi}(x)\psi(x)\right)$$
$$= \mathcal{N} \exp\left(i\frac{g_0}{2N} \int d^2x\, (\bar{\psi}\psi)^2 \right) \tag{17.175}$$

where $\mathcal{N}$ is a suitable normalization constant, and

$$\bar{\psi}\psi \equiv \sum_{a=1}^{N} \bar{\psi}_a(x)\psi_a(x) \tag{17.176}$$

The field $\sigma(x)$ does not carry any indices.

2) Use an identity of the type of the one derived in problem 1, involving two scalar fields $\sigma(x)$ and $\omega(x)$, to write the Lagrangian of the chiral Gross-Neveu model in a form that is *quadratic* in the Fermi fields.
3) This model is invariant under the continuous global chiral transformation $\psi_a = e^{i\theta\gamma_5}\psi'_a$. What transformation law should the scalar fields σ and ω satisfy? How are these fields related to the operators Δ_0 and Δ_5 of the exercises in chapter 12?
4) Integrate out the Fermi fields, and find the effective action for the scalar fields σ and ω. Watch for the factors of N, and be careful with the signs! By an appropriate rescaling of the scalar fields, show that the effective action has the form $S_{\text{eff}} = N\bar{S}$. Determine the form of $\bar{S}$.
5) Now consider the limit $N \to \infty$. Find the *saddle-point equations*, which determine the average values of the scalar fields in this limit. Find the solution of the saddle-point equations with lowest energy. Is the solution unique? Use dimensional regularization. What quantities need to be renormalized to make the saddle-point equations finite? How many renormalization constants do you need? Give your answers in terms of coupling-constant and wave-function renormalizations. Be careful to include the dependence on the dimensionality $2+\epsilon$. Determine the renormalization constants using the minimal subtraction scheme.
6) Compute the β function. Find its fixed points and flows in 1+1 dimensions. Solve the differential equation $\beta(g_0) = \kappa \frac{\partial g_0}{\partial \kappa}$, where κ is a momentum scale. Determine the asymptotic behavior of g_0 in the limit $\kappa \to \infty$. Is the interaction term relevant, irrelevant, or marginal?
7) Use the results of the previous parts to write the saddle-point equations in terms of renormalized quantities alone. In particular, find the dependence of the average values of the scalar fields on the renormalized coupling constant.

8) Consider now the fermion propagator in the $N \to \infty$ limit. Are the fermions massive or massless? If the former is true, what is the value of the fermion mass, and how does it relate to the expectation values of the scalar fields?
9) Find the effective action for the scalar fields to leading order in the $\frac{1}{N}$ expansion (i.e., to order $\frac{1}{N}$). Determine the propagator of the scalar fields at this order. Are the scalar fields massive or massless?
10) Consider now the effect of a field that breaks the chiral symmetry. The extra term in the Lagrangian is $\mathcal{L}_{\text{sources}}$, given by

$$\mathcal{L}_{\text{sources}} = H_0(x)\bar{\psi}_a(x)\psi_a(x) + H_5(x)\bar{\psi}_a(x)\gamma_5\psi_a(x) \tag{17.177}$$

Find the new effective action of this theory in the presence of these symmetry-breaking fields. Derive the modified saddle-point equations. Solve the new saddle-point equations for the case $H_0(x) = H$ and $H_5(x) = 0$.
11) Repeat the renormalization procedure employed for the theory without sources, now for the case with sources present. Be careful to include a wave-function renormalization. Derive the renormalized saddle-point equations. Renormalize the propagators of part 9.
12) By functionally differentiating the path integral with respect to the sources, derive an equation of identities that relates expectation values of the scalar fields σ and ω to expectation values of the fermion bilinears $\bar{\psi}\psi$ and $\bar{\psi}\gamma_5\psi$. In particular, find a formula that relates the propagators of σ and ω to the propagators of the fermion bilinears.
13) Use the Ward identity you derived in the exercises of chapter 12 to derive a relation between the two-point functions of the scalar fields at zero momentum, and the external symmetry-breaking field. Do the results you found in part 9 satisfy these relations?
14) Derive the RG equation (Callan-Symanzik equation (15.79)) satisfied by the scalar fields in the absence of external sources. Solve these Callan-Symanzik equations in terms of a momentum rescaling factor ρ and a *running* coupling constant. *Note*: Unlike renormalized perturbation theory, here you will find a solution of the RG equations that holds for all values of the coupling constant. This is possible because of the large-N limit, which is nonperturbative in the coupling constant.
15) Use the solutions of part 14 to find the asymptotic behavior of the two-point functions of the scalar fields at large momenta.

18

Phases of Gauge Theories

So far we have primarily focused on the behavior of QFTs in perturbation theory in powers of a weak coupling constant g. However, we have also seen that perturbation theory often fails, in the sense that, even in the perturbative regime, the effective (or renormalized) coupling constant runs to strong coupling at low energies (the IR). We should regard this behavior as an indication that the fixed point at $g = 0$ represents an *unstable* ground state, and that the true ground state has very different physics.

In some cases, we were able to use the $1/N$ expansion to investigate the strong coupling regime. Unfortunately, the limit of a large number of colors, N_c, of gauge theories is not solvable, as it reduces to the sum of all the planar Feynman diagrams, which is an unsolved problem. It has been believed since the late 1970s that their ground states represent a phase in which the theories are *confining*. As noted in chapter 17, the $N_c \to \infty$ limit of a supersymmetric version of Yang-Mills theory was solved using the gauge/gravity duality, known as the Maldacena conjecture. Even in that case, confinement could only be proved by breaking the supersymmetry (which breaks the conformal invariance). Although much progress has been made, the existing proof of confinement remains qualitative in character.

In this chapter, we discuss an alternative, nonperturbative approach to QFT, in which the theory is regularized in the UV by placing the degrees of freedom on some (typically hypercubic) lattice. In this language, the Euclidean QFT resembles a problem in classical statistical mechanics.

18.1 Lattice regularization of quantum field theory

The strong coupling behavior of essentially any field theory can be studied by defining the theory with a lattice UV cutoff. In this representation, pioneered by Wilson (Wilson and Kogut, 1974), Kogut and Susskind (1975a), and others, most theories (including gauge theories) can be studied by using methods borrowed from statistical mechanics. In this framework, the theory is defined in a discretized D-dimensional Euclidean spacetime, that is, a D-dimensional hypercubic lattice of lattice spacing a_0 (which, most of the time, we will set to 1). The path integral of the Euclidean QFT becomes a partition function in classical statistical mechanics on a hypercubic lattice. In this version of the theory, one of the lattice directions is interpreted as a discretized imaginary time, while the remaining

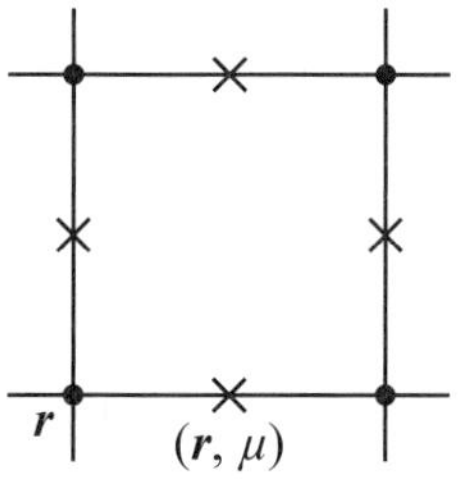

Figure 18.1 A two-dimensional Euclidean lattice. Here $\boldsymbol{r}$ represents the lattice sites, and $(\boldsymbol{r}, \mu)$ are the links.

$D-1$ directions correspond to the (also discretized) spatial directions. Although in lattice theories, rotational invariance (the Euclidean version of Lorentz invariance) is broken by the lattice, the discrete point group symmetries of the hypercubic lattice are preserved. However, rotational invariance is recovered at a nontrivial fixed point, at which the correlation length diverges.

In this perspective, since in the path integral the Planck constant $\hbar$ enters in the same way as the temperature T in classical statistical mechanics, weak-coupling perturbation theory is equivalent to the low-temperature expansion in classical statistical mechanics. Conversely, the strong coupling regime of QFT is identified with the high-temperature regime of statistical mechanics, which is described by the high-temperature expansion.

At least in principle, this connection allows one to access the strong coupling regime but at the price of working with a cutoff. As we saw in chapter 15, the cutoff can only be removed by investigating the behavior of the theory close to a nontrivial fixed point. But to do this, it is necessary to extrapolate these expansions outside their radius of convergence! Another virtue of this approach is that, at least in abelian theories, it is possible to make duality transformations that relate two (generally different) theories: one at weak coupling, and the other at strong coupling!

Finally, once these theories are defined on a lattice, it is possible to use powerful numerical techniques, such as Monte Carlo simulations, to investigate their properties and phase transitions. This numerical approach has become a powerful tool for computing the low-energy spectrum of hadrons, as well as for the study of thermodynamic properties of gauge theory at finite temperatures.

18.2 Matter fields

Let $\{\boldsymbol{r}\}$ be the coordinates of the sites of a D-dimensional hypercubic lattice, and hence, $\boldsymbol{r}$ is a D-tuplet of integer-valued numbers. Let $\mu = 1, \ldots, D$ label the D possible directions on the hypercubic lattice. A link of the lattice is thus labeled by the coordinates of a site and a direction, $(\boldsymbol{r}, \mu)$. Thus, if $\boldsymbol{r}$ is a lattice site, $\boldsymbol{r} + \boldsymbol{e}_\mu$ is the nearest-neighbor site along the direction of the unit vector $\boldsymbol{e}_\mu$ (with $\mu = 1, \ldots, D$). A case in $D = 2$ dimensions is shown in figure 18.1.

The matter fields are defined to be on the lattice sites, $\boldsymbol{r}$. Let us consider first the case in which the matter field is a scalar field. Let $\phi(\boldsymbol{r})$ be a scalar field, transforming in some irreducible representation of a compact Lie group G. For instance, in the case of a principal chiral field, the field $\phi(\boldsymbol{r})$ is a group element and obeys the global transformation law,

$$\phi'(\boldsymbol{r}) = V\phi(\boldsymbol{r})U^{-1} \tag{18.1}$$

The lattice action must be invariant under global transformations. Thus, it must be written in terms of G-invariant quantities. A typical term is

$$\operatorname{tr}\sum_{\boldsymbol{r}} \left(\phi^\dagger(\boldsymbol{r})\phi(\boldsymbol{r}+\boldsymbol{e}_\mu)\right) \tag{18.2}$$

which is clearly invariant under the global symmetry,

$$\begin{aligned}\operatorname{tr}\sum_{\boldsymbol{r}} \left(\phi'^\dagger(\boldsymbol{r})\phi'(\boldsymbol{r}+\boldsymbol{e}_\mu)\right) &= \operatorname{tr}\sum_{\boldsymbol{r}} \left(U\phi'^\dagger(\boldsymbol{r})V^{-1}V\phi'(\boldsymbol{r}+\boldsymbol{e}_\mu)U^{-1}\right)\\ &= \operatorname{tr}\sum_{\boldsymbol{r}} \left(\phi^\dagger(\boldsymbol{r})\phi(\boldsymbol{r}+\boldsymbol{e}_\mu)\right)\end{aligned} \tag{18.3}$$

since $U^\dagger = U^{-1}$, and where we have used the cyclic invariance of the trace. In what follows, we will consider field theories in which the group is (a) $G = \mathbb{Z}_2$, in which case the matter fields are Ising spins, $\sigma(\boldsymbol{r}) = \pm 1$; (b) $G = U(1)$, in which case the matter fields are phases (i.e., XY spins), $\exp(i\theta(\boldsymbol{r}))$, with $\theta(\boldsymbol{r}) \in [0, 2\pi)$; (c) $G = O(3)$, and the degrees of freedom are unit-length real vectors, $\boldsymbol{n}(\boldsymbol{r})$ (such that $\boldsymbol{n}^2(\boldsymbol{r}) = 1$) (i.e., the nonlinear sigma model); and so on.

18.3 Minimal coupling

To promote the global symmetry G to a local symmetry, we need to introduce gauge fields and minimal coupling. The gauge fields are connections (vector fields) and are defined to be on the links of the lattice. In general, the gauge fields will be group elements $\mathcal{U}_\mu(\boldsymbol{r}) \in G$ defined on the links $(\boldsymbol{r}, \mu)$ of the lattice, such that $\mathcal{U}_\mu(\boldsymbol{r}) = \mathcal{U}_\mu^{-1}(\boldsymbol{r}+\boldsymbol{e}_\mu)$. Since they are group elements, they can be written as

$$\mathcal{U}_\mu(\boldsymbol{r}) = \exp(iA_\mu(\boldsymbol{r})) \tag{18.4}$$

where $A_\mu(\boldsymbol{r})$ is an element in the algebra of the Lie group G. Hence, $A_\mu(\boldsymbol{r}) = A_\mu^k(\boldsymbol{r})\lambda^k$, where λ^k are the generators of G. Clearly the vector field $A_\mu(\boldsymbol{r})$ can be interpreted as implementing parallel transport between two lattice sites. Hence

$$A_\mu(\boldsymbol{r}) \sim \int_{\boldsymbol{r}}^{\boldsymbol{r}+\boldsymbol{e}_\mu} dx_\nu A^\nu(x) \sim a_0 A_\mu(\boldsymbol{r}) \tag{18.5}$$

As a result, a local gauge transformation is

$$\mathcal{U}'_\mu(\boldsymbol{r}) = V(\boldsymbol{r})\mathcal{U}_\mu(\boldsymbol{r})V^{-1}(\boldsymbol{r}+\boldsymbol{e}_\mu), \quad \phi'(\boldsymbol{r}) = V(\boldsymbol{r})\phi(\boldsymbol{r}) \tag{18.6}$$

A gauge-invariant term in the action involves the lattice version of a covariant derivative:

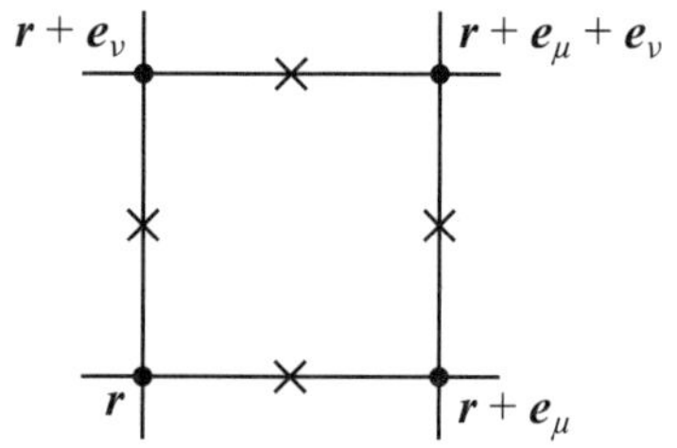

Figure 18.2 An elementary plaquette: The matter fields live on the sites (bold dots) and the gauge fields on the links (crosses).

$$\begin{aligned}\operatorname{tr}\left[\phi'^{\dagger}(\boldsymbol{r})\mathcal{U}'_\mu(\boldsymbol{r})\phi'(\boldsymbol{r}+\boldsymbol{e}_\mu)\right] &= \operatorname{tr}\left[\phi^{\dagger}(\boldsymbol{r})V^{-1}(\boldsymbol{r})\mathcal{U}'_\mu(\boldsymbol{r})V(\boldsymbol{r}+\boldsymbol{e}_\mu)\phi(\boldsymbol{r}+\boldsymbol{e}_\mu)\right]\\ &= \operatorname{tr}\left[\phi^{\dagger}(\boldsymbol{r})\mathcal{U}_\mu(\boldsymbol{r})\phi(\boldsymbol{r}+\boldsymbol{e}_\mu)\right]\end{aligned} \tag{18.7}$$

The Euclidean action for the matter field is

$$-S_{\text{matter}} = \frac{\beta}{2}\sum_{\boldsymbol{r},\mu}\left(\operatorname{tr}\left[\phi^{\dagger}(\boldsymbol{r})\mathcal{U}_\mu(\boldsymbol{r})\phi(\boldsymbol{r}+\boldsymbol{e}_\mu)\right]+\text{c.c.}\right) \tag{18.8}$$

where c.c. denotes, as usual, complex conjugation. It is a simple exercise to show that in the naive continuum limit, $a_0 \to 0$, the Euclidean action becomes

$$S_{\text{matter}} = \int d^Dx\, \frac{1}{2g^2_{\text{matter}}}\operatorname{tr}\left[\left(D_\mu\phi\right)^{\dagger} D_\mu\phi\right] \tag{18.9}$$

where D_μ is the covariant derivative, and $g^2_{\text{matter}} \equiv a_0^{D-2}/\beta$.

The action for the gauge fields must also be locally gauge invariant. Hence, it must be written in terms of *Wilson loops* (Wilson, 1974). Let $(\boldsymbol{r};\mu,\nu)$ define an elementary oriented plaquette of the hypercubic lattice with vertex at the lattice site $\boldsymbol{r}$ and a face (the plaquette) with edges along the μ and ν directions (see figure 18.2). The Wilson loop for an elementary plaquette with boundary γ is

$$W_\gamma = \operatorname{tr}\prod_{(\boldsymbol{x},\mu)\in\gamma_{\boldsymbol{r};\mu\nu}}\mathcal{U}_\mu(\boldsymbol{x}) \tag{18.10}$$

where the product runs on the links of the oriented plaquette $(\boldsymbol{r};\mu,\nu)$.

The gauge-invariant Wilson action for the gauge fields is

$$-S_{\text{gauge}} = \frac{1}{2g}\sum_{\boldsymbol{r};\mu\nu}\operatorname{tr}\left(\mathcal{U}_\mu(\boldsymbol{r})\mathcal{U}_\nu(\boldsymbol{r}+\mathrm{e}_\mu)\mathcal{U}_\mu^{-1}(\boldsymbol{r}+\mathrm{e}_\nu)\mathcal{U}_\nu^{-1}(\boldsymbol{r})\right)+\text{c.c.} \tag{18.11}$$

where g is the coupling constant. It is elementary to show that in the naive continuum limit, $a_0 \to 0$, it becomes the Yang-Mills action,

$$S_{\text{gauge}} = \frac{1}{4\tilde{g}}\int d^Dx\operatorname{tr}F^2_{\mu\nu} \tag{18.12}$$

where we have set $\tilde{g} = a_0^{D-2}g$.

An important physical observable that we will analyze is the expectation value of a Wilson loop operator on a large closed loop Γ of the lattice,

$$\left\langle W_\Gamma \right\rangle = \left\langle \text{tr} \prod_{(\boldsymbol{x},\mu)\in\Gamma} \mathcal{U}_\mu(\boldsymbol{x}) \right\rangle \tag{18.13}$$

which in the naive continuum limit, $a_0 \to 0$, is

$$\langle W_\Gamma \rangle \sim \left\langle \text{tr} \left[P \exp\left(i \oint_\Gamma dx_\mu A^\mu(x) \right) \right] \right\rangle \tag{18.14}$$

It is implicit in this notation that, in a gauge theory with gauge group G, the Wilson loop carries the quantum numbers of a representation of the gauge group. Since the Wilson loop represents the coupling to a static, infinitely heavy matter field, these are the quantum numbers carried by the matter field. We will see that in theories in which matter and gauge fields are both dynamical, the behavior of this observable depends on the representation (i.e., depends on the *charge* of the matter fields). A general compact gauge group G has many such nontrivial representations.

In what follows, we assume, unless stated to the contrary, that the Wilson loop carries the *fundamental* charge of the gauge group (i.e., the quantum numbers of the lowest nontrivial irreducible representation of the gauge group).

18.4 Gauge fields

The partition function of a gauge theory with a general gauge group G is

$$Z = \int \prod_{r,\mu} d\mathcal{U}_\mu(\boldsymbol{r}) \exp\left[\frac{1}{2g} \sum_{r;\mu\nu} \text{tr} \left(\mathcal{U}_\mu(\boldsymbol{r}) \mathcal{U}_\nu(\boldsymbol{r}+\boldsymbol{e}_\mu) \mathcal{U}_\mu^{-1}(\boldsymbol{r}+\boldsymbol{e}_\nu) \mathcal{U}_\nu^{-1}(\boldsymbol{r}) \right) + \text{c.c.} \right] \tag{18.15}$$

where we have used the Wilson action. Here $d\mathcal{U}_\mu$ denotes the invariant (Haar) measure of the group G. As in the continuum case, this partition function requires gauge fixing to have an integration measure that sums over gauge classes of configurations.

Examples of gauge theories are

$\mathbb{Z}_2$: This is the Ising gauge theory, whose gauge group is $\mathbb{Z}_2$ (Wegner, 1971; Balian et al., 1975). We will use the notation $\mathcal{U}_\mu \equiv \tau_\mu$, with $\tau_\mu = \pm 1$. The partition function of the Ising gauge theory is

$$Z_{\mathbb{Z}_2} = \sum_{\tau_\mu=\pm 1} \exp\left(\frac{1}{g} \sum_{r,\mu\nu} \tau_\mu(\boldsymbol{r}) \tau_\nu(\boldsymbol{r}+\boldsymbol{e}_\mu) \tau_\mu(\boldsymbol{r}+\boldsymbol{e}_\nu) \tau_\nu(\boldsymbol{r}) \right) \tag{18.16}$$

where g is the coupling constant.

$U(1)$: This theory is known as compact electrodynamics (Banks et al., 1977; Polyakov, 1977). In this case, $\mathcal{U}_\mu \in U(1)$ and can be written as $\mathcal{U}_\mu(\boldsymbol{r}) = \exp(iA_\mu(\boldsymbol{r}))$, where $A_\mu \in [0, 2\pi)$. The partition function is

$$Z_{U(1)} = \prod_{r;\mu,\nu} \int_0^{2\pi} \frac{dA_\mu(\boldsymbol{r})}{2\pi} \exp\left[\frac{1}{g}\sum_{\boldsymbol{r},\mu\nu} \cos(\Delta_\mu A_\nu(\boldsymbol{r}) - \Delta_\nu A_\mu(\boldsymbol{r}))\right] \tag{18.17}$$

where $\Delta_\mu \varphi(\boldsymbol{r}) \equiv \varphi(\boldsymbol{r}+\boldsymbol{e}_\mu) - \varphi(\boldsymbol{r})$ is the lattice right-difference operator. Here, too, we write $K = 1/g$. Maxwell's theory is both the limit in which $a_0 \to 0$ and in which the fields A_μ are not compact, and hence take values on the real numbers, $\mathbb{R}$.

$SU(N_c)$: This is the lattice version of the Yang-Mills gauge theory with gauge group $G = SU(N_c)$, where N_c is the number of colors. Here the lattice degrees of freedom are group elements, $\mathcal{U}_\mu \in SU(N_c)$. The action is given by eq. (18.11), and the partition function is a sum over gauge field configurations with a Haar invariant measure (and suitably gauge fixed).

We will not consider a detailed version of what is known about these theories (Kogut, 1979). Instead we will discuss their *phase diagrams* and classify their possible ground states and the behavior of physical observables.

18.5 Hamiltonian theory

18.5.1 Global symmetries

In chapter 14, we introduced the concept of the transfer matrix in statistical mechanics and showed that it is the analog of the Euclidean evolution operator. Using the transfer matrix approach, we related the classical Ising model in D dimensions to a quantum Ising model in $D-1$ dimensions. The Hamiltonian of the quantum model has, in addition to the Ising interaction term, a transverse field term that plays the role of a kinetic energy. It is easy to see that this is a general result.

Let us summarize how this works (Kogut and Susskind, 1975a; Fradkin and Susskind, 1978; Kogut, 1979). Define a Hilbert space of states on a hypersurface at a fixed (discrete) imaginary time. The transfer matrix generates a new configuration (a new state) at the next time slice. In this way, a configuration of the degrees of freedom in D dimensions is generated, and the partition function is a sum over such configurations. For instance, in the case of the Ising model, as we showed in section 14.4, the Hilbert space is the tensor product of spin configurations at fixed time, and we found that the resulting quantum Hamiltonian is that of the Ising model in a transverse field, cf. eq. (14.18),

$$H_{\text{Ising}} = -\sum_{\boldsymbol{r}} \sigma_1(\boldsymbol{r}) - \lambda \sum_{\boldsymbol{r},j} \sigma_3(\boldsymbol{r})\sigma_3(\boldsymbol{r}+\boldsymbol{e}_j) \tag{18.18}$$

where $j = 1, \ldots, D-1$, and $\sigma_1(\boldsymbol{r})$ and $\sigma_3(\boldsymbol{r})$ are Pauli matrices acting on the spin states at $(\boldsymbol{r})$. Here, λ is the coupling constant. In this picture, the low-temperature phase of the classical theory is the large-λ regime, and the high-temperature phase of the classical theory is the small-λ phase of the quantum theory.

Likewise, in the XY model, which has a global $U(1)$ symmetry, the degrees of freedom are phases, $\theta(\boldsymbol{r}) \in [0, 2\pi)$, and $L(\boldsymbol{r})$ is the conjugate momentum, such that $[\theta(\boldsymbol{r}), L(\boldsymbol{r}')] = i\delta_{\boldsymbol{r},\boldsymbol{r}'}$. The Hamiltonian now is

$$H_{XY} = \frac{1}{2}\sum_{\boldsymbol{r}} L^2(\boldsymbol{r}) - \frac{1}{g}\sum_{\boldsymbol{r},j} \cos\left(\Delta_j \theta(\boldsymbol{r})\right) \tag{18.19}$$

where, as before, we use the notation $\Delta_j\theta(\boldsymbol{r})=\theta(\boldsymbol{r}+\boldsymbol{e}_j)-\theta(\boldsymbol{r})$, with $j=1,\ldots,d$, for the lattice finite difference, and g is the coupling constant.

This construction generalizes to theories with non-abelian global symmetries, such as the $O(3)$ nonlinear sigma model. In that case, the degrees of freedom are labeled by configurations of vectors of unit length, $\boldsymbol{n}$, at each lattice site. The quantum lattice Hamiltonian is

$$H_{\text{NLSM}}=\frac{1}{2}\sum_{\boldsymbol{r}}\boldsymbol{J}^2(\boldsymbol{r})-\frac{1}{g}\sum_{\boldsymbol{r},j}\boldsymbol{n}(\boldsymbol{r})\cdot\boldsymbol{n}(\boldsymbol{r}+\boldsymbol{e}_j) \tag{18.20}$$

where $J_i(\boldsymbol{r})$ are the generators of $O(3)$, the group of rotations, which satisfy the commutation relations

$$[J_i(\boldsymbol{r}),J_j(\boldsymbol{r}')]=i\epsilon_{ijk}J_k(\boldsymbol{r})\delta_{\boldsymbol{r},\boldsymbol{r}'},\qquad [J_i(\boldsymbol{r}),n_j(\boldsymbol{r}')]=i\epsilon_{ijk}n_k(\boldsymbol{r})\delta_{\boldsymbol{r},\boldsymbol{r}'} \tag{18.21}$$

where $\boldsymbol{J}^2$ is the quadratic Casimir operator of the group $O(3)$.

In all three examples, in general dimension $D>D_c$ (where the critical spacetime dimension D_c depends on whether the symmetry group is discrete or continuous), these theories have a symmetric phase and a phase with a spontaneously broken global symmetry, separated by a phase transition. In each phase, there is a power series expansion with a radius of convergence smaller than the critical value of the coupling constant for the phase transition. Although in all the examples we have discussed, the transition is continuous, this is not generally the case. For example, in the case of discrete symmetry groups (e.g., the permutation group), the transition can be first order.

18.5.2 Local symmetries and gauge theory

Something similar works for a gauge theory as well, with one important difference. In chapter 9, we discussed the canonical quantization of gauge theories. There we saw that the space of states has to be projected onto the physical subspace of gauge-invariant states. This condition was implemented as a constraint that the physical states obey Gauss's law, which plays the role of the generator of time-independent gauge transformations. The lattice theory has exactly the same structure.

We begin with the $\mathbb{Z}_2$ gauge theory. The analog of the temporal gauge is to set the $\mathbb{Z}_2$ gauge fields along the imaginary time direction to 1. Let $\sigma_j^1(\boldsymbol{r})$ and $\sigma_j^3(\boldsymbol{r})$ be two Pauli matrices defined on each link $(\boldsymbol{r},j)$. The Hilbert space is the set of states of the $\mathbb{Z}_2$ gauge fields on each link of the lattice (e.g., the eigenstates of σ_3). The transfer-matrix construction leads to a quantum Hamiltonian that is now a sum of two terms, one on links (the "kinetic energy") and the other on plaquettes (the "potential energy"). It has the form (Fradkin and Susskind, 1978)

$$H_{\mathbb{Z}_2\text{GT}}=-\sum_{\boldsymbol{r},j}\sigma_j^1(\boldsymbol{r})-\lambda\sum_{\boldsymbol{r};jk}\sigma_j^3(\boldsymbol{r})\sigma_k^3(\boldsymbol{r}+\boldsymbol{e}_j)\sigma_j^3(\boldsymbol{r}+\boldsymbol{e}_k)\sigma_k^3(\boldsymbol{r}) \tag{18.22}$$

Let us define at every site $\boldsymbol{r}$ the operators $Q(\boldsymbol{r})$

$$Q(\boldsymbol{r})=\prod_j\left(\sigma_j^1(\boldsymbol{r})\sigma_{-j}^1(\boldsymbol{r})\right) \tag{18.23}$$

where $\sigma_{-j}^1(\boldsymbol{r})=\sigma_j^1(\boldsymbol{r}-\boldsymbol{e}_j)$. It is easy to see that $Q(\boldsymbol{r})$ flips the values of σ^3 on all the links that share the site $\boldsymbol{r}$ and is the generator of local $\mathbb{Z}_2$ gauge transformations. It is also apparent that it commutes with every term of the Hamiltonian, $[Q(\boldsymbol{r}),H]=0$, and that they commute with each other, $[Q(\boldsymbol{r}), Q(\boldsymbol{r}')] = 0$. Hence, $Q(\boldsymbol{r})$ is the generator of time-independent $\mathbb{Z}_2$ gauge transformations and plays the role of a (suitably generalized) Gauss law. Furthermore, since the operators $Q(\boldsymbol{r})$ are hermitian and obey $Q(\boldsymbol{r})^2 = 1$, their eigenvalues are just ± 1. Therefore, the quantum theory is defined on the physical space of gauge-invariant states $|\text{Phys}\rangle$, and obey

$$Q(\boldsymbol{r})|\text{Phys}\rangle = |\text{Phys}\rangle \tag{18.24}$$

which are invariant under local $\mathbb{Z}_2$ gauge transformations. States for which, at some value of $\boldsymbol{r}$, $Q(\boldsymbol{r}) = -1$ will be viewed as having $\mathbb{Z}_2$ sources located at $\boldsymbol{r}$. We will shortly see that, for any spacetime dimension $D > 2$, the $\mathbb{Z}_2$ gauge theory has two phases: a confining phase (for $\lambda < \lambda_c$), and a deconfined phase (for $\lambda > \lambda_c$).

Let us now move on to the case of a gauge theory with gauge group $U(1)$, known as compact electrodynamics. Here we work in the temporal gauge, $A_0 = 0$. In this case, the degrees of freedom are vector potentials defined on the links of the lattice taking the values $A_j(\boldsymbol{r}) \in [0, 2\pi)$. The conjugate momenta are also defined on the links, and are electric fields $E_j(\boldsymbol{r})$, which obey the commutation relations $[A_j(\boldsymbol{r}), E_k(\boldsymbol{r}')] = i\delta_{\boldsymbol{r},\boldsymbol{r}'}$. Thus, in this theory, electric fields behave as angular momenta and their eigenvalues are integer numbers. The Hamiltonian for the $U(1)$ gauge theory is (Fradkin and Susskind, 1978)

$$H_{U(1)\text{GT}} = \sum_{\boldsymbol{r},j} \frac{1}{2} E_j^2(\boldsymbol{r}) - \frac{1}{g} \sum_{\boldsymbol{r};jk} \cos\left(\Delta_j A_k(\boldsymbol{r}) - \Delta_k A_j(\boldsymbol{r})\right) \tag{18.25}$$

This theory is invariant under local gauge transformations. The local generators are written in terms of the lattice divergence operator as

$$Q(\boldsymbol{r}) \equiv \Delta_j E_j(\boldsymbol{r}) \equiv \sum_j \left(E_j(\boldsymbol{r}) - E_j(\boldsymbol{r} - \boldsymbol{e}_j)\right) \tag{18.26}$$

It is easy to see that the unitary transformations $\exp(i \sum_{\boldsymbol{r}} \theta(\boldsymbol{r}) Q(\boldsymbol{r}))$ generate local gauge transformations $A_j(\boldsymbol{r}) \to A_j(\boldsymbol{r}) + \Delta_j \theta(\boldsymbol{r})$. Moreover, the operators $Q(\boldsymbol{r})$ commute with each other, $[Q(\boldsymbol{r}), Q(\boldsymbol{r}')] = 0$, and with the Hamiltonian. Hence, they are the generators of time-independent gauge transformations. We can now define the physical Hilbert space of states, $|\text{Phys}\rangle$, as the eigenstates of $Q(\boldsymbol{r})$ with zero eigenvalue,

$$Q(\boldsymbol{r})|\text{Phys}\rangle \equiv \Delta_j E_j(\boldsymbol{r})|\text{Phys}\rangle = 0 \tag{18.27}$$

which is the analog of the Gauss law constraint of Maxwell's electrodynamics. The main (and important!) difference is that in compact electrodynamics, the operator $Q(\boldsymbol{r})$ has a spectrum of integer-valued eigenvalues. This implies that states that satisfy $Q(\boldsymbol{r})|\Psi\rangle = n(\boldsymbol{r})|\Psi\rangle$ have the interpretation of having sources that carry integer-valued electric charges. Therefore, in this theory, states with fractional charges are not allowed.

The non-abelian gauge theories are constructed in a similar fashion. Here, for simplicity, we consider the case of a theory with the $SU(2)$ gauge group. In this case, the Hamiltonian is (using a conventional notation; Kogut and Susskind, 1975a)

$$H_{SU(2)\text{GT}} = \frac{g}{2}\sum_{r,j} E_j^a(r)E_j^a(r)$$

$$-\frac{1}{2g}\sum_{r;jk}\left[\operatorname{tr}\left(\mathcal{U}_j(r)\mathcal{U}_k(r+e_j)\mathcal{U}_j^{-1}(r+e_k)\mathcal{U}^{-1}(r)\right)+\text{c.c.}\right] \tag{18.28}$$

where, as before, the degrees of freedom are group elements, $\mathcal{U}_j(r) \in G$, and $E_j^a(r)$ (with $a = 1, \ldots, D(G)$) are the generators of the Lie group G. In the case $G = SU(2)$, the group elements $\mathcal{U}_j(r)$ are 2×2 unitary matrices, the generators $E_j^a(r)$ are the angular momentum operators, and $E_j^a(r)E_j^a(r)$ is the quadratic Casimir operator. The generators $E_j^a(r)$ and the group elements $\mathcal{U}_j(r)$ satisfy the commutation relations

$$[E_j^a(r), E_k^b(r')] = i\epsilon^{abc}E_j^c(r)\delta_{r,r'}\delta_{jk}, \tag{18.29}$$

where ϵ^{abc} are the structure constants of $SU(2)$, and

$$[E_j^a(r), \mathcal{U}_k(r')] = \frac{1}{2}\sigma^a\mathcal{U}_j(r)\delta_{jk}\delta_{r,r'}, \tag{18.30}$$

where σ^a are three Pauli matrices (the elements of the algebra of the group $SU(2)$).

The generator of time-independent gauge transformations, $Q^a(r)$, now takes values on the algebra of the group,

$$Q^a(r) = \sum_{j=1}^{d}\left(E_j^a(r) - E_j^a(r-e_j)\right) \tag{18.31}$$

and the Gauss law condition is

$$Q^a(r)|\text{Phys}\rangle = 0 \tag{18.32}$$

and defines the subspace of physical states. Here, too, states that are not annihilated by $Q^a(r)$ are viewed as not being in the vacuum sector but as created by external probe sources.

However, in a non-abelian theory, the Gauss-law generators do not commute with each other. Here one can only specify the eigenvalues of the diagonal generators, those in the Cartan sub-algebra and the eigenvalue of the Casimir operator. For instance, in the $SU(2)$ gauge theory, the sources carry the quantum numbers (j, m), where $j(j+1)$ is the eigenvalue of the Casimir operator, and m is the eigenvalue of the diagonal generator of $SU(2)$. Hence, here the sources are also labeled by a definite set of possible eigenvalues.

18.6 Elitzur's theorem and the physical observables of a gauge theory

There is a fundamental result known as Elitzur's theorem (Elitzur, 1975). It states that in a gauge theory with a compact gauge group, the only operators that can have nonvanishing expectation values must be invariant under local gauge transformation. This theorem always holds, regardless of the phase in which the gauge theory is.

This theorem stands in striking contrast with the case of theories with global symmetries. In that case, we saw that if the global symmetry group is compact, the ground state may break this symmetry spontaneously. The direct implication of Elitzur's theorem is that a local gauge symmetry *cannot be broken spontaneously*. This result may seem surprising, since there are regimes of theories, such as the ones in which the Higgs mechanism is operative, in which it is often stated (incorrectly) that the gauge symmetry is spontaneously broken. We will clarify this apparent contradiction shortly.

To understand the meaning of Elitzur's theorem, let us first revisit the assumptions behind the concept of spontaneous symmetry breaking. In that case, we have a theory with a global compact symmetry group G. We assume that the theory is in a finite but large volume V with an extensive number of degrees of freedom. The action of the theory is invariant under global G transformations.

Now consider a theory in which a symmetry-breaking field, with strength h, is added to the action. The theory now is no longer invariant under the global symmetry transformations G. For the sake of definiteness, let us think of an Ising model, which has a global $\mathbb{Z}_2$ symmetry. In this case, h represents an external uniform magnetic field. Spontaneous symmetry breaking here means that there is a nonvanishing local order parameter. Consider now a *finite* system ($V < \infty$) with a fixed value of the symmetry-breaking field h. The presence of the symmetry-breaking field causes the action to increase for configurations in which the majority of the degrees of freedom are in a direction opposite with respect to the symmetry-breaking field. Their contribution to the partition function is then suppressed.

If we now lower the temperature T (i.e., the coupling constant) below some critical value T_c, two things can happen: (1) if h is reduced all the way to zero at fixed temperature and size, then the misaligned configurations will not be suppressed, the average magnetization will vanish, and the symmetry is not broken; or (2) we take the thermodynamic limit first ($N \to \infty$), which has the effect of imposing an infinite penalty on the misaligned configurations. In this case, the symmetry-breaking field is then turned off, and the local magnetization (which is odd under the symmetry) will have a nonvanishing expectation value whose sign is determined by the now-suppressed symmetry-breaking field h. Therefore, the order of limits between the size of the system and the symmetry-breaking field matters. We saw the same result when we derived Goldstone's theorem in section 12.4.

Another way to understand this result is in terms of perturbation theory: in the broken-symmetry phase, the order in perturbation theory needed to mix two states that are related by the global symmetry is proportional to the size of the system. In that case, if the thermodynamic limit is taken first, the configurations related by the global symmetry will never mix. In other words, spontaneous breaking of a global symmetry is possible, because the states that are related by the action of the global symmetry in the thermodynamic limit are infinitely far apart from each other. A consequence of this fact is that systems with a global symmetry are sensitive to the effects of symmetry-breaking fields (even for infinitesimal ones) in the thermodynamic limit.

The situation is completely different in the case of a gauge theory. Consider, again for simplicity, the $\mathbb{Z}_2$ gauge theory in the Hamiltonian picture. Suppose we begin with some state $|\Psi\rangle$ that is not invariant under gauge transformations (i.e., such that for some sites $\{\boldsymbol{r}\}$, $Q(\boldsymbol{r})|\Psi\rangle = |\Phi\rangle \neq |\Psi\rangle$). Since the operator $Q(\boldsymbol{r})$ (i.e., the Gauss law) is the generator of time-independent local gauge transformations, the states $|\Psi\rangle$ and $|\Phi\rangle$ differ by a local gauge transformation. However, since the gauge group (in this case, $\mathbb{Z}_2$) is compact, the orbit of locally inequivalent states is finite. This implies that after acting a finite number of times on the images of $|\Psi\rangle$, one will find a state that has a finite overlap with $|\Psi\rangle$. Hence,

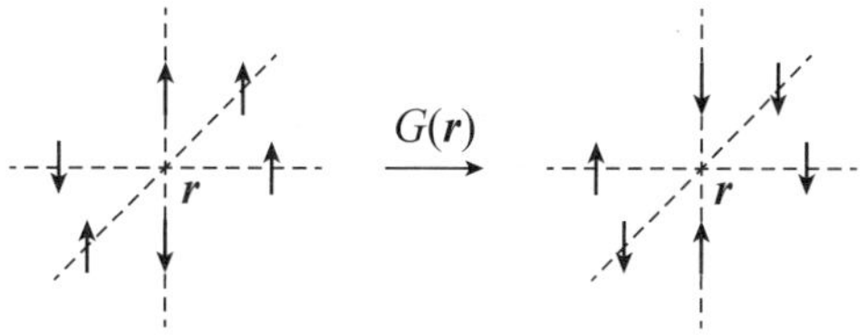

Figure 18.3 A $\mathbb{Z}_2$ gauge transformation at site $\boldsymbol{r}$, denoted by $G(\boldsymbol{r})$: The gauge field variables (denoted by arrows) on all the links attached to the site $\boldsymbol{r}$ are flipped.

this process does not involve acting all the way to the boundaries, and in fact, it holds for an infinitely large system. The argument given above holds also for continuous compact groups (e.g., as $U(1)$, $SU(N)$) since in all cases, the orbit of the gauge group has a finite measure.

Another, more formal way to obtain this result is to consider, again for simplicity, a $\mathbb{Z}_2$ gauge theory on a hypercubic (Euclidean) spacetime lattice in D dimensions. The following arguments can be generalized with minor changes to theories with an arbitrary compact gauge group. Here we consider a modified theory that has a term on the links of the lattice, of strength h, that breaks the local $\mathbb{Z}_2$ gauge invariance. The partition function of the modified $\mathbb{Z}_2$ gauge theory is (in a compact notation)

$$Z[K,h] = \text{Tr} \exp\left(K \sum_{\text{plaquettes}} \tau_\mu \tau_\nu \tau_\mu \tau_\nu + h \sum_{\text{links}} \tau_\mu \right) \tag{18.33}$$

where $K = 1/g$ (g being the coupling constant), and the trace is the sum over all configurations of the $\mathbb{Z}_2$ gauge fields, here denoted by $\tau_\mu = \pm 1$ on each link (of direction μ) of the lattice. The integration measure (i.e., the sum over the gauge field configurations) and the action of this theory, except for the presence of the symmetry-breaking field h, are invariant under local $\mathbb{Z}_2$ gauge transformations.

Let us consider the computation of the expectation value of a local gauge-noninvariant operator, such as $\tau_\mu(\boldsymbol{r})$ located on a single link $(\boldsymbol{r}, \mu)$:

$$\langle \tau_\mu(\boldsymbol{r}) \rangle_{K,h} = \frac{1}{Z[K,h]} \text{Tr} \left[\tau_\mu(\boldsymbol{r}) \exp\left(K \sum_{\text{plaquettes}} \tau_\mu \tau_\nu \tau_\mu \tau_\nu + h \sum_{\text{links}} \tau_\mu \right) \right] \tag{18.34}$$

Let us consider a local gauge transformation $G(\boldsymbol{r})$ of the gauge fields $\tau_\mu(\boldsymbol{r})$ on the links emanating from a site $\boldsymbol{r}$ (shown in figure 18.3):

$$\tau_\mu(\boldsymbol{r}) \mapsto \tau'_\mu(\boldsymbol{r}) = -\tau_\mu(\boldsymbol{r}), \quad \forall \text{ links } (\boldsymbol{r}, \mu) \text{ attached to site } \boldsymbol{r} \tag{18.35}$$

All other gauge fields are unaffected by the local transformation $G(\boldsymbol{r})$. We next perform a change of variables of the gauge fields, $\tau_\mu(\boldsymbol{r}) \mapsto \tau'_\mu(\boldsymbol{r})$, which coincides with the gauge transformation $G(\boldsymbol{r})$ of eq. (18.35) on the links attached to the site $\boldsymbol{r}$, and while the other links are unchanged. On making the change in the expectation value of eq. (18.34), we find

$$\langle \tau_\mu(\boldsymbol{r})\rangle = -\frac{1}{Z[K,h]} \mathrm{Tr}\left[\tau'_\mu(\boldsymbol{r}) \exp\left(K \sum_{\text{plaquettes}} \tau'_\mu \tau'_\nu \tau'_\mu \tau'_\nu + h \sum_{\text{links}} \tau'_\mu \right)\right]$$
$$= \left\langle -\tau_\mu(\boldsymbol{r}) \exp\left(-2h \sum_{\{(\boldsymbol{r},\nu)\}} \tau_\nu(\boldsymbol{r}) \right)\right\rangle_{K,h} \tag{18.36}$$

where $\{(\boldsymbol{r}, \nu)\}$ is the set of links attached to the site $\boldsymbol{r}$, and the expectation value is taken in the theory with couplings K and h.

Next, observe that the following bound on a local change holds:

$$\left| \left\langle \tau(\boldsymbol{r},\mu)\right\rangle_{K,h} - \left\langle -\tau_\mu(\boldsymbol{r},\mu)\right\rangle_{K,h} \right| = \left| \left\langle -\tau_\mu(\boldsymbol{r})\left[\exp\left(-2h \sum_{\{(\boldsymbol{r},\nu)\}} \tau_\nu(\boldsymbol{r})\right) - 1\right]\right\rangle_{K,h} \right| \tag{18.37}$$

where D is the dimension. The expression on the left-hand side of eq. (18.37) has the upper bound

$$\left| \left\langle -\tau_\mu(\boldsymbol{r})\left[\exp\left(-2h \sum_{\{(\boldsymbol{r},\nu)\}} \tau_\nu(\boldsymbol{r})\right) - 1\right]\right\rangle_{K,h} \right| \leq (e^{4Dh} - 1)\left| \langle \tau_\mu(\boldsymbol{r},\mu)\rangle_{K,h} \right| \tag{18.38}$$

Hence, the left-hand side of eq. (18.37) has the same upper bound. However, since as $h \to 0$, this upper bound approaches zero, we obtain

$$\lim_{h\to 0} \langle \tau_\mu(\boldsymbol{r},\mu)\rangle_{K,h} = \lim_{h\to 0} \langle -\tau_\mu(\boldsymbol{r},\mu)\rangle_{K,h} \tag{18.39}$$

which implies that, in the limit $h \to 0$, the expectation value of this gauge-noninvariant operator (and, in fact, of *any* gauge-noninvariant operator) must vanish:

$$\langle \tau_\mu(\boldsymbol{r},\mu)\rangle_K = 0 \tag{18.40}$$

Note that this proof has two key ingredients: (1) that the transformation is local, and (2) that the change in the action due to the gauge transformation is finite. Note that, in contrast, in the case of a global symmetry, the change in the action diverges in the thermodynamic limit.

Therefore, we conclude that only locally gauge-invariant operators can have a non-vanishing expectation value. A direct consequence of this result is that *gauge symmetries cannot be spontaneously broken*. As stated above, this theorem holds for all theories with a local gauge invariance with a compact gauge group.

18.7 Phases of gauge theories

Let us now discuss the phases that can exist in pure gauge theories. For simplicity, we focus on the $\mathbb{Z}_2$ gauge theory and on compact electrodynamics. However, I will comment on which results are generic. Let us consider both the Euclidean and the Hamiltonian versions whenever necessary.

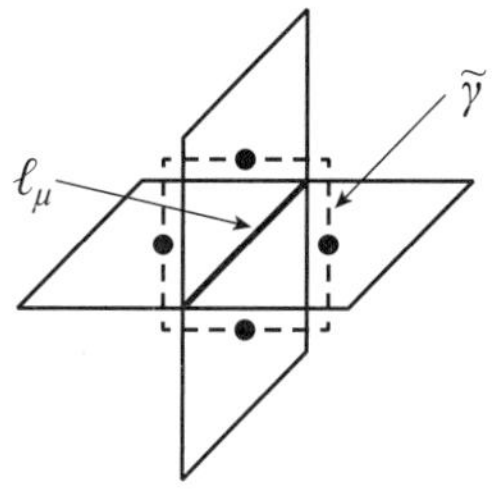

Figure 18.4 The leading term of the weak coupling expansion to the partition function of the $\mathbb{Z}_2$ gauge theory in $D=3$ dimensions is the flipped link ℓ_μ. The four plaquettes (labeled by dots) that share this link are flipped. A flipped link can then be regarded as a closed magnetic loop on the dual lattice (shown as a closed broken curve). In $D=4$ dimensions, the dual magnetic loop is a surface and, more generally, a $(D-2)$-dimensional hypersurface.

18.7.1 Weak coupling: Deconfinement

In the weak coupling limit, $g \to 0$, perturbation theory works. In this limit, the Gibbs weight for each plaquette of the D-dimensional hypercubic lattice (using a synthetic notation),

$$\exp\left[\frac{1}{2g}\operatorname{tr}\left(\mathcal{U}_\mu \mathcal{U}_\nu \mathcal{U}_\mu^{-1} \mathcal{U}_\nu^{-1}\right) + \text{c.c.}\right] \tag{18.41}$$

is dominated by flat gauge field configurations (i.e., such that $F_{\mu\nu}=0$). Hence, in this extreme limit, we can set $\mathcal{U}_\mu = I$ (up to gauge transformations). This is the phase we have studied so far. For simplicity, we will consider the $\mathbb{Z}_2$ gauge theory whose partition function is

$$Z[\mathbb{Z}_2] = \operatorname{Tr} \exp\Big(\sum_{\text{plaquettes}} \frac{1}{g} \tau_\mu \tau_\nu \tau_\mu \tau_\nu\Big) \tag{18.42}$$

where the trace is the sum over all configurations of gauge fields $\tau_\mu = \pm 1$ (modulo gauge fixing).

In the ultraweak coupling regime, $g \to 0$, we can start from the configuration in which $\tau_\mu = 1$ at all links, up to gauge transformations, and neglecting for now interesting topological issues arising on closed manifolds. In this configuration, the product of the four τ_μ gauge fields on each plaquette is equal to $+1$. A configuration with one link variable flipped to the value -1, flips the values of all plaquettes that share this link from $+1$ to -1. This increases the action by $2/g$ times the number of flipped plaquettes N_p (where $N_p = 4$ in $D=3$ dimensions). In $D=3$ dimensions, the flipped plaquettes can be regarded as being threaded by a closed loop on the dual lattice, which we will call a magnetic loop, as shown in figure 18.4. In $D>3$ dimensions, the magnetic loop is a surface in $D=4$ dimensions, and, in general, a $D-2$ hypersurface in D dimensions. Thus, this configuration has a Gibbs weight of $\exp(-\frac{2}{g}N_p)$ times the number of places where we can flip a link. In the $\mathbb{Z}_2$ gauge theory, this expansion has a finite radius of convergence for $D>2$. This expansion is the analog of the low-temperature expansion in a spin system.

For a $U(1)$ theory, the weak coupling regime is described by Maxwell's electrodynamics, whose excitations are photons. In fact, in section 9.7, we computed the expectation value of the Wilson loop operator for Maxwell's theory in Euclidean spacetime. There we found that the expectation value of a Wilson loop on a closed contour Γ, such as the

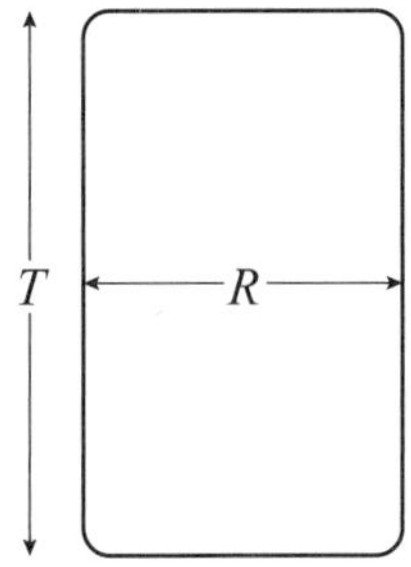

Figure 18.5 A Wilson loop Γ in Euclidean spacetime. Here T is the imaginary time span, and R is the spatial size of the loop.

one shown in figure 18.5, is

$$\langle W_\Gamma \rangle = \exp\left(-TV(R)\right) \tag{18.43}$$

By explicit computation, we showed that the effective potential $V(R)$ is just the Coulomb interaction

$$V_{\text{Maxwell}}(R) = \frac{e^2}{R} \tag{18.44}$$

where we have identified the coupling constant with the electric charge, $g = e$. Notice that, in this case, we could have chosen the charge of the Wilson loop to be ne, with $n \in \mathbb{Z}$, which would have only changed the result by a factor of n^2.

We will call this behavior the Coulomb phase. Note that on a lattice, the Wilson loops are not smooth everywhere and have cusps, which yield singular contributions proportional to $\ln R$. We will ignore these singularities in what follows.

Notice that in $D=4$ dimensions, the charge e is dimensionless, and the ratio T/R is invariant under dilatations of spacetime, $R \to \lambda R$ and $T \to \lambda T$. From this perspective, we should expect that at a fixed point of the RG (i.e., in a scale-invariant theory), the expectation value of the Wilson loop operator should be a universal function $F(x)$ of the scale-invariant ratio $x = R/T$. Hence, in any dimension, at a fixed point of a gauge theory, we expect that the effective potential should obey a $1/R$ law.

In contrast, for the case of Yang-Mills theory in $D=4$ dimensions, this phase, which would have massless quarks and gluons, is asymptotically free and IR unstable, for example, the RG beta function is $\beta(g) = -Ag^2 + O(g^3)$ (where A is a positive constant). In this case, the coupling constant flows to strong coupling in the IR, and weak-coupling perturbation theory breaks down beyond a finite length scale ξ. We will see that this is the confinement scale. The expectation value of the Wilson loop exhibits a crossover from loops smaller than the confinement scale, where they behave much in the same way as in perturbative Yang-Mills theory (with a slowly scale dependent coupling constant), to an area law for large loops (as we will see below). State-of-the-art numerical Monte Carlo simulations in $D=4$ dimensions have established that the area-law behavior holds from the strong coupling regime down to a sufficiently weak coupling regime where perturbative RG calculations are still reliable. These numerical results are strong evidence that confinement holds in the continuum field theory.

Gauge theories with a discrete gauge group exhibit a different behavior. Indeed, in $D > 2$ dimensions, the $\mathbb{Z}_2$ gauge theory has a convergent weak coupling expansion for $g < g_c$. This expansion is reminiscent of the low-temperature expansion in a classical spin system. Indeed, in this theory, $g = 0$ also implies that only flat configurations of the gauge field will contribute. This means that the product of the τ_μ gauge variables on each plaquette should be equal to 1. Up to gauge transformations, this is equivalent to setting $\tau_\mu = 1$ on all the links of the lattice.

Let us compute the expectation value of the Wilson loop operator in the weak coupling phase of the $\mathbb{Z}_2$ gauge theory. This theory has only one nontrivial representation, and hence, there is only one $\mathbb{Z}_2$ charge. Clearly, in this limit, the expectation value of the Wilson loop is trivially $\langle W_\gamma \rangle = 1$. The first excitation above this state consists of flipping the value of τ_μ from $+1$ to -1 on just one link. However, this flips the value of the product of the $\mathbb{Z}_2$ gauge fields on all the plaquettes that share this link. The number of such plaquettes is $2(D-1)$. Therefore, the cost in the action of flipping just one τ variable is $4(D-1)$. For only one flip, for a lattice of N sites and a loop γ of perimeter $L = 2(T+R)$, the expectation value of the Wilson loop is

$$\langle W_\Gamma \rangle = \frac{1 + (N-2L)\exp(-4(D-1)/g) + \cdots}{1 + N\exp(-4(D-1)/g) + \cdots} \tag{18.45}$$

Let us consider the case of n flipped links. Provided g is small enough, we can, to a first approximation, ignore configurations with adjacent flipped links and treat the flipped links as being dilute. In this approximation, the nth-order correction to the numerator of eq. (18.45) is

$$\frac{1}{n!}(N-2L)^n \exp(-4n(D-1)/g) \tag{18.46}$$

In the same approximation, the denominator of eq. (18.45) (i.e., the partition function Z) has a contribution

$$\frac{1}{n!}N^n \exp(-4n(D-1)/g) \tag{18.47}$$

Therefore, in the dilute limit, the partition function is

$$\begin{aligned} Z &\simeq 1 + N\exp(-4(D-2)/g) + \cdots + \frac{1}{n!}N^n \exp(-4n(D-1)/g) + \cdots \\ &= \exp(N\exp(-4(D-1)/g)) \end{aligned} \tag{18.48}$$

To the same approximation, the numerator of eq. (18.45) now becomes

$$\begin{aligned} &1 + (N-2L)\exp(-4(D-1)/g) + \cdots + \frac{1}{n!}(N-2L)^n \exp(-4n(D-1)/g) + \cdots \\ &= \exp\left((N-2L)\exp(-4(D-1)/g)\right) \end{aligned} \tag{18.49}$$

Therefore, to this order of approximation, we find

$$\langle W_\Gamma \rangle = \exp(-2\exp(-4(D-1)/g)L) \tag{18.50}$$

which, as it should, has a finite value in the thermodynamic limit, $N \to \infty$.

In other words, in the limit of g small enough, the Wilson loop obeys a *perimeter law* of the form

$$\langle W_\Gamma \rangle = \exp(-f(g)L) \tag{18.51}$$

To the order of approximation that we have used, we have $f(g) = 2\exp(-4(D-1)/g) + \cdots$, up to exponentially small corrections in the spatial extent R of the loop. In this regime, the effective potential becomes

$$V(R) = \lim_{R/T \to 0} 4\exp(-2(D-1)/g)(1 + R/T) = 4\exp(-2(D-1)/g) + \cdots \tag{18.52}$$

which, up to exponentially small additive corrections, is a constant.

Thus, to this order of approximation, the effective potential $V(R)$ is simply twice the self-energy cost of each source. Since there is no dependence on R, at this order, their effective interaction potential $V_{\text{int}}(R) \approx 0$. One can prove that this is actually the leading behavior (for a very large loop) of an expansion that has a finite radius of convergence. The leading nonvanishing contribution to the interaction potential decays exponentially fast with R: $V(R) \sim \exp(-R/\xi)$, where ξ is the range of the effective interaction. These results also hold for other gauge theories with a discrete gauge group (e.g., $\mathbb{Z}_N$), provided the spacetime dimension is $D > 2$.

18.7.2 Strong coupling and confinement: Hamiltonian picture

A: $\mathbb{Z}_2$ gauge theory We begin the discussion of confinement in the strong coupling regime from the perspective of the Hamiltonian picture. Consider two theories: the $\mathbb{Z}_2$ (Ising) gauge theory and the $U(1)$ (compact electrodynamics) gauge theory.

Consider first the Hamiltonian $\mathbb{Z}_2$ gauge theory in D space dimensions. Write the Hamiltonian of the theory, eq. (18.22), as the sum of two terms, $H_{\mathbb{Z}_2} = H_0 + \lambda H_1$, where $\lambda = 1/g$ is the coupling constant, and where we set

$$H_0 = -\sum_{\boldsymbol{r},j} \sigma_j^1(\boldsymbol{r}), \quad H_1 = -\sum_{\boldsymbol{r};jk} \sigma_j^3(\boldsymbol{r})\sigma_k^3(\boldsymbol{r}+\boldsymbol{e}_j)\sigma_j^3(\boldsymbol{r}+\boldsymbol{e}_k)\sigma_k^3(\boldsymbol{r}) \tag{18.53}$$

The physical Hilbert space consists of the gauge-invariant states that satisfy

$$Q(\boldsymbol{r})|\text{Phys}\rangle = \prod_j \left(\sigma_j^1(\boldsymbol{r})\sigma_{-j}^1(\boldsymbol{r})\right)|\text{Phys}\rangle = |\text{Phys}\rangle \tag{18.54}$$

In the strong coupling limit, $g \gg 1$ (i.e., $\lambda \ll 1$), we can regard H_0 as the unperturbed Hamiltonian and H_1 as the perturbation. In this limit, the Hamiltonian H_0 is diagonal in the basis of the $\{\sigma_j^1(\boldsymbol{r})\}$ operators on the links. In this basis, the generators of local gauge transformations $\{Q(\boldsymbol{r})\}$ are also diagonal. Since acting on gauge-invariant states, $Q(\boldsymbol{r}) = +1$ for all sites $\boldsymbol{r}$, in this basis, gauge-invariant states must have at most an even number of links, where $\sigma_j^1 = -1$ at every site of the lattice.

The unperturbed ground state, $|\text{gnd}\rangle_0$, then must be the state with $\sigma_j^1 = +1$ on all links of the lattice. The energy of the unperturbed ground state is $E_{\text{gnd}_0} = -Nd$. It is easy to see that, since the plaquette operators in H_1 flip the values of σ_j^1 around a plaquette from $+1$ to -1, the excited states are closed loops γ of the lattice, where $\sigma_j^1 = -1$. In particular, for λ small enough (g large enough), only small "flipped" loops are present in the ground state. For larger values of λ (smaller values of g), the ground-state wave function will have contributions from states with loops of increasingly larger size until, at some critical value of the coupling constant, g_c, the size of the loops will diverge: the loops "proliferate."

Let us now consider a different sector of the Hilbert space defined by the condition that the operators $\{Q(\boldsymbol{r})\}$ are equal to $+1$ everywhere except at two sites, which we denote by $\boldsymbol{R}$ and $\boldsymbol{R}'$, where $Q(\boldsymbol{R}) = Q(\boldsymbol{R}') = -1$. This choice is consistent with the requirement that the

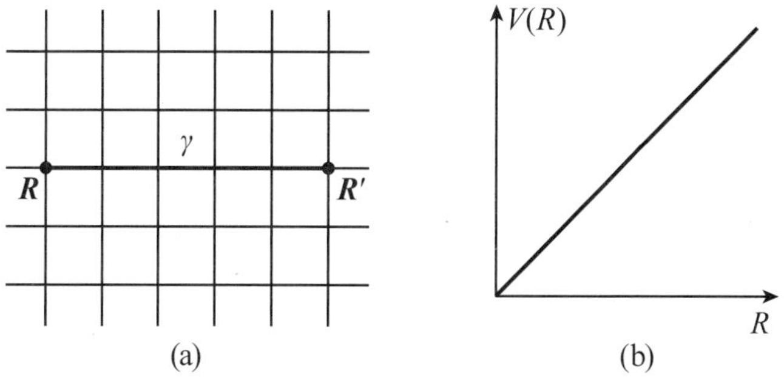

Figure 18.6 a) A string state in the strong coupling regime of the $\mathbb{Z}_2$ gauge theory: $\sigma_j^1 = -1$ on the bold links stretching from $\boldsymbol{R}$ to $\boldsymbol{R}'$ along the path γ. In this state, $\sigma_j^1 = +1$ on all other links. (b) The confining potential of two $\mathbb{Z}_2$ charges separated by a distance $|\boldsymbol{R} - \boldsymbol{R}'|$ in the strong coupling regime.

global $\mathbb{Z}_2$ symmetry must be unbroken, which requires that there must be an even number of $\mathbb{Z}_2$ sources. The eigenvalue of the operator $Q(\boldsymbol{r})$ is the $\mathbb{Z}_2$ "electric" charge of the state.

For simplicity, choose $\boldsymbol{R}$ and $\boldsymbol{R}'$ to be on the same row of the lattice (see figure 18.6a). We say that this sector of the Hilbert space consists of states that have two $\mathbb{Z}_2$ "charges" $Q = -1$: one at $\boldsymbol{R}$, and the other at $\boldsymbol{R}'$. The ground state in this sector must have the smallest number of links with $\sigma_j^1 = -1$ and must obey $Q = -1$ only at $\boldsymbol{R}$ and at $\boldsymbol{R}'$. It has $Q = +1$ on all other sites. It is obvious that the ground state in this sector, which we denote by $|\boldsymbol{R}, \boldsymbol{R}'\rangle$, has $\sigma_j^1 = -1$ on the links of the shortest path γ on the lattice from $\boldsymbol{R}$ to $\boldsymbol{R}'$. In the case shown in figure 18.6a, the shortest path γ is just the straight line (shown as the bold links in the figure).

The operator that creates this state is a *Wilson arc* of the form

$$W_\gamma[\boldsymbol{R}, \boldsymbol{R}'] = \prod_{(\boldsymbol{r},j)\in\gamma} \sigma_j^3(\boldsymbol{r}) \tag{18.55}$$

where γ is an open path on the links of the lattice with endpoints at $\boldsymbol{R}$ and $\boldsymbol{R}'$. It is easy to see that the operator $Q(\boldsymbol{r})$ commutes with the Wilson arc for all $\boldsymbol{r} \neq \boldsymbol{R}, \boldsymbol{R}'$, but it anticommutes if $\boldsymbol{r} = \boldsymbol{R}, \boldsymbol{R}'$. Thus, this operator is not invariant under local gauge $\mathbb{Z}_2$ transformations, and it is not allowed in the vacuum sector of the theory. However, it is allowed in the sector where $Q(\boldsymbol{r}) = -1$ at $\boldsymbol{r} = \boldsymbol{R}, \boldsymbol{R}'$ (i.e., in a sector with two static sources).

The energy of this state is

$$E[\boldsymbol{R}, \boldsymbol{R}'] = E_{\text{gnd}_0} + 2|\boldsymbol{R} - \boldsymbol{R}'| \tag{18.56}$$

In other terms, the energy cost of introducing two $\mathbb{Z}_2$ sources at $\boldsymbol{R}$ and $\boldsymbol{R}'$ is the effective potential $V(|\boldsymbol{R} - \boldsymbol{R}'|)$:

$$V(|\boldsymbol{R} - \boldsymbol{R}'|) = \sigma(g)|\boldsymbol{R} - \boldsymbol{R}'| \tag{18.57}$$

where

$$\sigma(g\lambda) = 2 - O(\lambda^2) \tag{18.58}$$

is the *string tension*. Hence, the energy needed to separate two $\mathbb{Z}_2$ charges grows linearly with their separation $|\boldsymbol{R} - \boldsymbol{R}'|$, and we say that the charges are confined; see figure 18.6b. In other words, the two charges experience a force much like that of a string.

It is easy to see that the expansion in powers of $\lambda = 1/g$ has a finite radius of convergence. These corrections are virtual states on longer strings, and lead to a progressive reduction of the string tension σ as λ increases (g decreases). In other words, the string becomes progressively "fatter," with a transversal thickness ξ, such that $\xi \to 0$ as $\lambda \to 0$, and its energy per unit length progressively decreases. In this picture, the phase transition to the deconfined phase occurs when the string tension $\sigma \to 0$. It turns out (as we will see) that this is a continuous transition, that is, $\sigma(\lambda) \propto |\lambda - \lambda_c|^{\rho}$ (here ρ is a critical exponent) in $D=2+1$ spacetime dimensions, but it is discontinuous in higher dimensions.

However, in the deconfined phases, $g \ll 1$ ($\lambda \gg 1$), the action of the Wilson arc operator is quite simple. In this limit, the plaquette operator, which is a product of the σ^3 link operators around each plaquette, plays the role of the unperturbed Hamiltonian, and the link operator σ^1 is the perturbation. In the extreme limit $g \to 0$, the natural basis (up to gauge fixing) for the Hilbert space consists of the eigenstates of σ^3 on the links. In this limit, the ground state has $\sigma^3 = +1$ on each link (up to local gauge transformations). This state is not gauge invariant. One can fix the gauge and then choose this state as the ground state. In this basis, the Wilson arc operator is diagonal and acts as a c-number, and the effective interaction is short ranged.

Alternatively, one can work in the basis in which the $\mathbb{Z}_2$ charge operator is diagonal. This is the σ^1 basis, which we used in the strong coupling limit. We saw that for the strong coupling regime, in this basis, the vacuum can be viewed as a linear superposition of states, each of the form of a collection of closed strings (loops) on the lattice. In the confining phase, the loops have finite size, typically on the order of the confinement scale. The phase transition to the deconfined (weak coupling) phase can be viewed as a proliferation of the loops, which become arbitrarily large. In the extreme deconfined phase, $g \to 0$, the vacuum state is the equal-amplitude linear superposition of configurations of all loops, regardless of their size. We will see shortly that this state embodies the topological nature of the deconfined phase.

In addition to these "electric excitations," this theory also has "magnetic excitations." In 2+1 dimensions, the operator that creates a magnetic excitation flips the value of the local plaquette operator from $+1$ to -1. In $d=2$ space dimensions, this operator is similar to one that creates a magnetic flux ("monopole") and is defined as

$$M[\widetilde{\boldsymbol{R}}] = \prod_{(\boldsymbol{r},j)\in\widetilde{\gamma}(\widetilde{\boldsymbol{R}})} \sigma_j^1(\boldsymbol{r}) \tag{18.59}$$

Hence, this operator is the product of σ^1 link operators, a seam of links of the lattice pierced by an open curve $\widetilde{\gamma}$ of the dual of the square lattice with an endpoint at the dual site $\widetilde{\boldsymbol{R}}$, as shown in figure 18.7. From its definition, it is clear that in the deconfined phase, this gauge-invariant operator creates a pair of $\mathbb{Z}_2$ "monopoles" at the opposite ends of its "Dirac string" $\widetilde{\gamma}$. These monopole states have finite energy and are deconfined. Thus, in its deconfined phase, the theory has electric and magnetic charges that have a finite energy gap. Nevertheless, both excitations are created by nonlocal operators. In contrast, in the confined phase, $g \gg 1$, this operator acts as a c-number. Hence, it has a finite expectation value in the confined phase. In this sense, the confined phase of the $\mathbb{Z}_2$ gauge theory is a *magnetic condensate*.

With some important modifications, an analog of these operators exists in higher dimensions. While the Wilson arc can be defined in all dimensions, the analog of the magnetic operator changes with dimension. In $D=3+1$ dimensions, the magnetic operator is

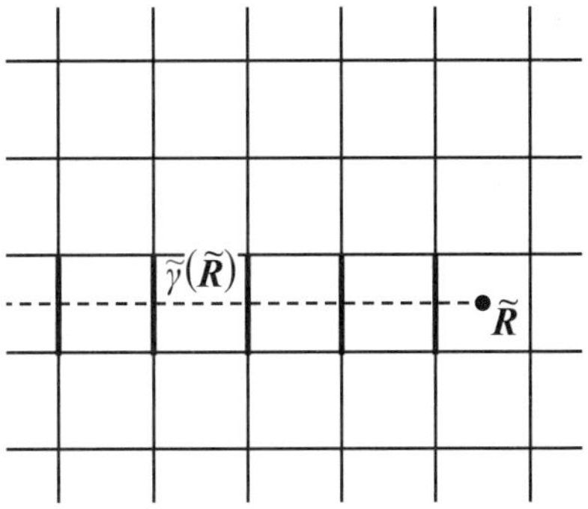

Figure 18.7 A "magnetic excitation" (or $\mathbb{Z}_2$ monopole) is created by a nonlocal operator that flips the value of the plaquette operator at the plaquette labeled by the black dot. This operator is defined on a path $\tilde{\gamma}$ of the dual lattice.

defined on a closed loop $\tilde{\gamma}$ of the dual lattice, and is known as a 't Hooft magnetic loop ('t Hooft, 1978, 1979). The closed loop $\tilde{\gamma}$ on the dual (cubic) lattice is a closed curve on the dual lattice piercing a closed tube of plaquettes on the cubic lattice. The closed curve $\tilde{\gamma}$ is the boundary of an open surface $\tilde{\Sigma}$, with the topology of a disk.

In the $\mathbb{Z}_2$ gauge theory, the 't Hooft magnetic loop operator, which we will denote by $\tilde{W}[\tilde{\gamma}]$, is a product of σ^1 link operators on all the links of the lattice that pierce the plaquettes of the surface $\tilde{\Sigma}$ on the dual lattice. It was originally introduced by 't Hooft as a criterion for confinement in non-abelian gauge theories, where it acts on the center $\mathbb{Z}_N$ of the group $SU(N)$. In two space dimensions, the 't Hooft operator (i.e., the monopole operator $M[\tilde{\boldsymbol{R}}]$ defined above) is labeled by a point, in three space dimensions by a closed loop, in four space dimensions by a closed surface (which is the boundary of a three-volume), and so on. Returning to the case of $3+1$ dimensions, it is straightforward to see that the 't Hooft loop has an *area* law in the *deconfined* phase of the $\mathbb{Z}_2$ gauge theory, and a *perimeter* law in the *confined* phase. In other words, the 't Hooft and the Wilson loop operators exhibit opposite behaviors.

B: $U(1)$ gauge theory It is straightforward to see that the same behavior is seen on all other gauge theories with a compact gauge group G. To clarify how this happens, we will consider briefly the case of a $U(1)$ gauge theory ("compact electrodynamics") also in the Hamiltonian picture. The Hamiltonian of the $U(1)$ gauge theory is given in eq. (18.25). It can also be split into a sum of two terms, an unperturbed Hamiltonian H_0,

$$H_0 = \sum_{\boldsymbol{r},j} \frac{1}{2} E_j^2(\boldsymbol{r}) \tag{18.60}$$

and a perturbation H_1,

$$H_1 = -\frac{1}{g} \sum_{\boldsymbol{r};jk} \cos\left(\Delta_j A_k(\boldsymbol{r}) - \Delta_k A_j(\boldsymbol{r})\right) \tag{18.61}$$

The physical Hilbert space consists of the states that satisfy Gauss's law, which now reads

$$Q(\boldsymbol{r})|\text{Phys}\rangle = \Delta_j E_j(\boldsymbol{r})|\text{Phys}\rangle = 0 \tag{18.62}$$

In the strong coupling regime, $g \gg 1$, H_1 is parametrically small, and we can construct the states in a perturbative expansion in powers in $1/g$. In this regime, it is natural to work in the basis in which the electric field operators on the links, $E_j(\boldsymbol{r})$, are diagonal. Since the gauge fields take values in the compact group $U(1)$ (i.e., $A_j(\boldsymbol{r}) \in [0, 2\pi)$), the eigenstates of the electric fields take values on the integers, $\ell_j(\boldsymbol{r}) \in \mathbb{Z}$, which label the representations of $U(1)$. In this description, the physical Hilbert space of gauge-invariant states is given by the configurations of integer-valued variables $\ell_j(\boldsymbol{r})$ on the links such that $\Delta_j \ell_j(\boldsymbol{r}) = 0$. In other words, we can think of the electric fields as a set of locally conserved, integer-valued "currents."

We can repeat the arguments used in the discussion of $\mathbb{Z}_2$ gauge theory almost verbatim. The (trivially gauge-invariant) ground state, $|\text{gnd}_0\rangle$, is simply the state in which all the electric fields are zero on all links, $\ell_j(\boldsymbol{r}) = 0$, and the unperturbed ground state energy is zero, $E_{\text{gnd}_0} = 0$.

Since the electric fields are integer valued, their sources must also be integers, $q(\boldsymbol{r}) \in \mathbb{Z}$. Hence, states created by a set of static sources $\{q(\boldsymbol{r})\}$, which we label by $|\{q(\boldsymbol{r})\}\rangle$, obey a Gauss law of the form

$$\Delta_j E_j(\boldsymbol{r}) |\{q(\boldsymbol{r})\}\rangle = q(\boldsymbol{r}) |\{q(\boldsymbol{r})\}\rangle \tag{18.63}$$

Let us now consider the case in which we have just two sources, with charges $\pm q \in \mathbb{Z}$, located at sites $\boldsymbol{R}$ and $\boldsymbol{R}'$ of the lattice, just as we did in figure 18.6a. The requirement of global charge neutrality follows from the fact that the global $U(1)$ symmetry is unbroken (it is easy to verify that the energy of states that violate global charge neutrality is divergent). The ground state in this sector of the Hilbert space is the state with $\ell_j(\boldsymbol{r}) = 0$ everywhere, except on the links on the shortest path, γ, stretching from $\boldsymbol{R}$ to $\boldsymbol{R}'$. Once again, the shortest path γ is just the set of links on the straight line between $\boldsymbol{R}$ and $\boldsymbol{R}'$. On each link on γ, the electric field must take the value $\ell_j = q$. Therefore, the energy cost of this state with sources is, once again, a linear function of the separation of the sources, $V(|\boldsymbol{R} - \boldsymbol{R}'|) = \sigma |\boldsymbol{R} - \boldsymbol{R}'|$, where the string tension is $\sigma = \frac{q^2}{2} - O(1/g^2)$. Hence, this theory is confining in the strong-coupling regime. We will see in chapter 19 that the $U(1)$ gauge theory has magnetic monopoles, which play a key role in the confined phase.

C: Non-abelian gauge theories This analysis applies to non-abelian Yang-Mills theories with minor but important changes. Indeed, in the strong coupling limit, the dominant term of the Hamiltonian in eq. (18.28) is the kinetic energy term, which is proportional to the sum of the Casimir operators $E_j^2(\boldsymbol{r})$ on each link. Since the group is non-abelian, the eigenstates are those of the quadratic Casimir and the diagonal (Cartan) generators. For $SU(2)$, the states are $|j, m\rangle$ and carry the quantum numbers of the representations. The ground state is the $SU(2)$ singlet state $|0, 0\rangle$ on every link. The ground-state energy of this trivially gauge-invariant state is zero. We can now consider the Hilbert space for a theory with two static sources. Each must now carry $SU(2)$ quantum numbers, such that the total state remains a singlet. Again, this follows from the fact that the global $SU(2)$ symmetry is unbroken. For the case of two sources separated a distance R (in lattice units) that carry the spinor representation $(1/2, 1/2)$ and its conjugate, we find, once again, that the energy of this state follows a linear potential law, $V(R) = \sigma R$, where the string tension now is $\sigma = \frac{3}{8} g^2 - O(1/g^2)$. Hence, in the strong-coupling regime, the non-abelian theory is confining.

18.8 Hamiltonian duality

Duality is a powerful tool for examining the topological properties of discrete gauge theories. We will now show that a theory with a $\mathbb{Z}_2$ symmetry (be it global or local) has a *dual* theory defined on the dual lattice. This notion generalizes to theories with abelian symmetry groups. The main power of duality transformations is that, in general, they relate weakly coupled theories to strongly coupled theories. In the special case of the symmetry group $\mathbb{Z}_2$, the dual has the same symmetry but is not necessarily realized in the same way. The extension of these concepts to more general theories (e.g., with non-abelian groups) is generally only possible in supersymmetric theories. Modern string theory has greatly generalized the notion of duality across theories and dimensions.

Conventionally, duality is an identification of series expansions of two partition functions and will be discussed briefly in chapter 19. The original notion of duality (and more specifically, Kramers-Wannier duality) is a property of two-dimensional classical Ising models. It was subsequently extended by Wegner to $\mathbb{Z}_2$ gauge theory (Wegner, 1971). Here we use a Hamiltonian approach to duality. In this approach, we define nonlocal operators, such as the kink creation operator of the (1+1)-dimensional quantum Ising model (discussed in chapter 14), the $\mathbb{Z}_2$ monopole, and the magnetic 't Hooft operators, which do not commute with the local degrees of freedom.

The Hamiltonian of the one-dimensional quantum Ising model is

$$H_{\text{Ising}}(\lambda) = -\sum_{n=1}^{N} \sigma^1(n) - \lambda \sum_{n=1}^{N} \sigma^3(n)\sigma^3(n+1) \tag{18.64}$$

Here the lattice is a one-dimensional chain. The dual lattice consists of the midpoints of the lattice (i.e., the set of points $\tilde{n} = n + 1/2$). The operators $\sigma^1(n)$ and $\sigma^3(n')$ satisfy the Pauli matrix (Clifford) algebra (i.e., they commute if $n \neq n'$, and anticommute if $n = n'$, and they square to the identity). Define a dual theory in terms of a new Clifford algebra defined on the sites of the dual lattice as (Fradkin and Susskind, 1978)

$$\tau^3(\tilde{n}) = \prod_{m \leq n} \sigma^1(m), \qquad \tau^1(\tilde{n}) = \sigma^3(n)\sigma^3(n+1) \tag{18.65}$$

Note that the operator $\tau^3(\tilde{n})$ is the kink creation operator defined in chapter 14 (cf. eq. (14.52)). It is trivial to see that the operators in eq. (18.65) obey the same Clifford algebra as the Pauli operators. Moreover, in terms of these operators, the dual Hamiltonian takes the form (up to a term determined by the boundary conditions)

$$H_{\text{Ising}}(\lambda) = -\sum_{\tilde{n}=1}^{N} \tau^3(\tilde{n})\tau^3(\tilde{n}+1) - \lambda \sum_{\tilde{n}=1}^{N} \tau^1(\tilde{n}) = \lambda \tilde{H}_{\text{Ising}}(1/\lambda) \tag{18.66}$$

Hence, the dual Hamiltonian, $\tilde{H}_{\text{Ising}}$, is the same as the original Hamiltonian up to a change $\tilde{\lambda} = 1/\lambda$. This mapping is the equivalent of the famous Kramers-Wannier self-duality of the two-dimensional classical Ising model. We see that in this theory, the dual has the same $\mathbb{Z}_2$ global symmetry.

Duality maps the weak coupling (disordered) phase to the strong coupling (broken symmetry) phase (and vice versa). The phase transition discussed in chapter 14 is at the self-dual point, $\lambda = 1$. Note also that the disordered phase can be viewed as a condensate of kinks, that is, $\langle \tau^3(\tilde{n}) \rangle \neq 0$ is the disordered phase of the theory.

Moving on to $2+1$ dimensions, let us consider the Ising gauge theory whose Hamiltonian is (cf. eq. (18.22)):

$$H_{\mathbb{Z}_2\text{GT}} = -\sum_{\boldsymbol{r},j=1,2} \sigma_j^1(\boldsymbol{r}) - \lambda \sum_{\boldsymbol{r},j,k} \sigma_j^3(\boldsymbol{r})\sigma_k^3(\boldsymbol{r}+\boldsymbol{e}_j)\sigma_j^3(\boldsymbol{r}+\boldsymbol{e}_k)\sigma_k^3(\boldsymbol{r}) \tag{18.67}$$

The lattice now is a square lattice (the set of points labeled by $\boldsymbol{r}$), and the degrees of freedom (the gauge fields) are defined on the midpoints of the links of the lattice. As in the previous case, the operators obey a Clifford algebra, but the theory now has a local (gauge) $\mathbb{Z}_2$ symmetry. The dual of the square lattice is also a square lattice: the set of points on the center of the plaquettes of the square lattice, $\tilde{\boldsymbol{r}} = \boldsymbol{r} + (\boldsymbol{e}_1 + \boldsymbol{e}_2)/2$. Define the operator $\tau^3(\tilde{\boldsymbol{r}})$ to be the $\mathbb{Z}_2$ monopole operator $M(\tilde{\boldsymbol{r}})$ of eq. (18.59),

$$\tau^3(\tilde{\boldsymbol{r}}) = M(\tilde{\boldsymbol{r}}) = \prod_{(\boldsymbol{r}',j)\in\tilde{\gamma}(\tilde{\boldsymbol{r}})} \sigma_j^1(\boldsymbol{r}') \tag{18.68}$$

Recall that the $\mathbb{Z}_2$ monopole operator is defined in terms of a Dirac string along an open path $\tilde{\gamma}$ of the dual lattice. Define the operator $\tau^1(\tilde{\boldsymbol{r}})$ to be the plaquette operator, that is,

$$\tau^1(\tilde{\boldsymbol{r}}) = \sigma_j^3(\boldsymbol{r})\sigma_k^3(\boldsymbol{r}+\boldsymbol{e}_j)\sigma_j^3(\boldsymbol{r}+\boldsymbol{e}_k)\sigma_k^3(\boldsymbol{r}) \tag{18.69}$$

It is straightforward to see that these operators obey the Clifford algebra, since they anticommute on the same site of the dual lattice, commute otherwise, and square to the identity. Notice the important fact that the dual operators are defined on the sites of the dual lattice, whereas the original degrees of freedom are defined on the links of the original lattice.

One can check that these definitions are consistent, provided the gauge theory obeys the Gauss's law constraint, eq. (18.54), which is trivially satisfied by these definitions. It is obvious that, up to the boundary conditions, the Hamiltonian of the dual theory is that of the quantum Ising model in 2+1 dimensions defined on the dual (square) lattice:

$$H_{\mathbb{Z}_2\text{GT}} = \lambda \tilde{H}_{\text{Ising}}(1/\lambda) = -\lambda \sum_{\tilde{\boldsymbol{r}}} \tau^1(\tilde{\boldsymbol{r}}) - \sum_{\tilde{\boldsymbol{r}},j=1,2} \tau^3(\tilde{\boldsymbol{r}})\tau^3(\tilde{\boldsymbol{r}}+\boldsymbol{e}_j) \tag{18.70}$$

Clearly, the theory is no longer self-dual. However, a generalization of this construction to the $\mathbb{Z}_2$ gauge theory with a $\mathbb{Z}_2$ matter field actually is self-dual. Hence, in 2+1 dimensions, duality also maps a weakly coupled theory to a strongly coupled theory, and vice versa. But the theories are no longer the same, as duality now maps *the gauge-invariant sector* of a theory with local symmetry, the $\mathbb{Z}_2$ gauge theory, to a theory with a global $\mathbb{Z}_2$ symmetry, the (2+1)-dimensional quantum Ising model (equivalent to the three-dimensional classical Ising model).

In particular, duality maps the broken symmetry phase of the Ising model, $\lambda < \lambda_c$, to the confined phase of the gauge theory, and the order parameter of the Ising model maps to the $\mathbb{Z}_2$ monopole operator of the gauge theory. Hence, as anticipated, the confining phase can be regarded as a condensate of $\mathbb{Z}_2$ magnetic monopoles. However, the disordered phase of the Ising model maps to the deconfined phase of the $\mathbb{Z}_2$ gauge theory. We will see in section 22.2 that careful consideration of the boundary conditions (which we sidestepped here) shows that, in this deconfined phase, the gauge theory is a topological field theory. In fact, this is true in all dimensions.

Finally, let us consider the $\mathbb{Z}_2$ gauge theory in 3+1 dimensions. As before, the degrees of freedom are defined on the links of the three-dimensional cubic lattice (i.e., they are *vector fields*), and the $\mathbb{Z}_2$ flux is defined on the plaquettes of the lattice. In three dimensions, plaquettes are oriented two-dimensional surfaces. Thus, as expected, the flux is an antisymmetric tensor field. Now, in three dimensions, links are dual to plaquettes, and vice versa. Therefore, the dual of the (3+1)-dimensional $\mathbb{Z}_2$ gauge theory should also be a gauge theory, and as we will see, the dual theory is also the $\mathbb{Z}_2$ gauge theory.

Thus, we once again define a set of Pauli matrices $\tau_{\tilde{l}}^1(\tilde{\boldsymbol{r}})$ and $\tau_{\tilde{l}}^3(\tilde{\boldsymbol{r}})$ (where $\tilde{\boldsymbol{r}}$ is the site at the center of the cube) on the links of the dual lattice such that the dual of the plaquette $\boldsymbol{r}, jk$ is the link $(\tilde{\boldsymbol{r}}, \tilde{l})$, and we identify

$$\tau_{\tilde{l}}^1(\tilde{\boldsymbol{r}}) = \sigma_j^3(\boldsymbol{r})\sigma_k^3(\boldsymbol{r}+\boldsymbol{e}_j)\sigma_j^3(\boldsymbol{r}+\boldsymbol{e}_k)\sigma_k^3(\boldsymbol{r}),$$

$$\sigma_l^1(\boldsymbol{r}) = \tau_{\tilde{j}}^3(\tilde{\boldsymbol{r}})\tau_{\tilde{k}}^3(\tilde{\boldsymbol{r}}+\boldsymbol{e}_{\tilde{j}})\tau_{\tilde{j}}^3(\tilde{\boldsymbol{r}}+\boldsymbol{e}_{\tilde{k}})\tau_{\tilde{k}}^3(\tilde{\boldsymbol{r}}) \tag{18.71}$$

both of which, due to the $\mathbb{Z}_2$ Gauss's law, satisfy the Bianchi identity that the product of the plaquette flux operators on each of the six faces of the cube must be the identity operator. Hence, in 3+1 dimensions, the $\mathbb{Z}_2$ gauge theory is self-dual (Wegner, 1971; Fradkin and Susskind, 1978). Assuming that there is a unique phase transition, it must happen at the self-dual point, $\lambda_c = 1$. It is also straightforward to see that the dual of the Wilson loop operator is the 't Hooft magnetic loop operator defined in section 18.7.2. Here too, duality maps the gauge-invariant sector of one theory at coupling λ to the gauge-invariant sector of the dual at coupling $1/\lambda$, again up to boundary conditions.

We now see the pattern. Duality involves a geometric duality, essentially the duality of forms: in three dimensions, this amounts to stating that sites are dual to volumes, and links are dual to plaquettes. Thus, in 3+1 dimensions, the dual of a gauge theory is a gauge theory. This also implies that the dual of a scalar is an antisymmetric tensor field (a Kalb-Ramond field), and that the dual of a (3+1)-dimensional Ising model is a theory of a $\mathbb{Z}_2$-valued antisymmetric tensor field defined on plaquettes.

However, why does the dual of a $\mathbb{Z}_2$ theory have the same symmetry group? It turns out that if the theory has an abelian symmetry group G, its representations are one-dimensional and form a group, the dual group $\tilde{G}$. For instance, the dual of the group $U(1)$ is the group of integers $\mathbb{Z}$. The group $\mathbb{Z}_2$ is special in that it has only two (one-dimensional) representations, and they form a group isomorphic to $\mathbb{Z}_2$. It is this feature of abelian groups that does not allow for generalizations to non-abelian theories.

18.9 Confinement in the Euclidean spacetime lattice picture

Let us examine the behavior of gauge theory on a Euclidean spacetime lattice in the strong coupling regime. The partition function of a gauge theory with gauge group G has the form (with a suitable gauge-fixing condition)

$$Z[G] = \int \mathcal{D}\mathcal{U}_\mu \exp\left[-\frac{1}{g}\sum_{\text{plaquettes}}\left(\operatorname{tr}\mathcal{U}_\mu\mathcal{U}_\nu\mathcal{U}_\mu^{-1}\mathcal{U}_\nu^{-1} + \text{c.c.}\right)\right] \tag{18.72}$$

Here, "plaquettes" means a sum over all the plaquettes of the lattice. As before, the coupling constant g plays the role of temperature T in statistical mechanics. Therefore, we can examine the behavior of the partition function in the strong coupling regime by an analog of the high-temperature expansion in statistical mechanics. In this limit, we have to expand the exponential in powers of $1/g$ and compute the averages (expectation values) over all configurations of gauge fields, using the Haar measure of the integration as the only weight. Since the action is (by construction) gauge invariant, gauge-equivalent configurations will yield the same value of the average. This amounts to a contribution to the partition function in the form of an overall factor of $v(G)^{ND}$, where $v(G)$ is the volume of the gauge group (which is finite for a compact group G), N is the number of sites (the volume of spacetime), and D is the dimension. It is this divergent contribution that is eliminated by a proper definition of the integration measure, which should sum over gauge classes and not over configurations (i.e., the Faddeev-Popov procedure described in section 9.5).

To simplify the discussion (and the notation), let us describe the strong coupling expansion in the context of the $\mathbb{Z}_2$ gauge theory. The partition function for the $\mathbb{Z}_2$ gauge theory can be written as

$$\begin{aligned} Z[\mathbb{Z}_2] &= \operatorname{Tr} \exp\left(\sum_{\text{plaquettes}} \frac{1}{g} \tau_\mu \tau_\nu \tau_\mu \tau_\nu \right) \\ &= \left(\cosh\left(\frac{1}{g}\right)\right)^{ND(D-1)} \operatorname{Tr} \prod_{\text{plaquettes}} \left[1 + \tau_\mu \tau_\nu \tau_\mu \tau_\nu \tanh\left(\frac{1}{g}\right)\right] \end{aligned} \tag{18.73}$$

where the trace is over the $\mathbb{Z}_2$ gauge fields τ_μ at each link. For a general gauge group, the Gibbs weight for each plaquette is expanded as a sum of characters of the representations of the gauge group. The $\mathbb{Z}_2$ group has only two representations. In what follows, we will drop the overall prefactor, since it does not contribute to expectation values.

Since $\operatorname{Tr} \tau_\mu = 0$, the only nonvanishing contributions must be such that each $\tau_\mu = \pm 1$ gauge field appears twice in the expansion. Thus each link must appear twice. Since each plaquette, by construction, only appears once, the nonvanishing contributions are sets of plaquettes such that all of their edges (links) are glued to other plaquettes in the set. In other words, the plaquettes in the set must cover closed surfaces. An example is presented in figure 18.8, showing the first nonvanishing contribution to the partition function in $D=3$ dimensions. This is the surface of a cube and has six faces. Since each face contributes a factor of $\tanh(1/g)$, this set contributes a factor of $2\tanh^6(1/g)$. This representation holds in all dimensions. It follows that the partition function can be represented as a sum over sets of closed surfaces. In general, the nonvanishing terms of the strong coupling expansion are a sum over closed surfaces Σ of arbitrary genus g. The sum over surfaces of all possible genus g has the form

$$Z = \sum_{\{\Sigma\}} \left(\tanh(1/g)\right)^{\mathcal{A}[\Sigma]} \times \text{entropy factor} \tag{18.74}$$

where $\mathcal{A}[\Sigma]$ is the area of the surface Σ. The sum runs over all closed surfaces of any genus (with self-avoidance and nonoverlapping conditions). The entropy factor counts the number of surfaces with the same area and genus.

A surface can be viewed as the history of a closed string in imaginary time. Figure 18.9a shows a closed surface Σ with genus 0 (i.e., with the topology of a sphere) as being swept in

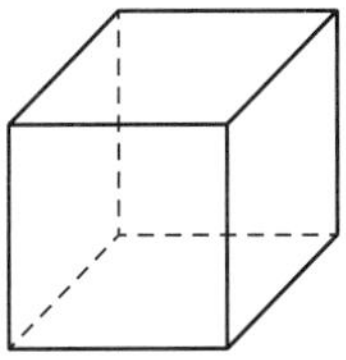

Figure 18.8 The leading nonvanishing contribution to the partition function of a $\mathbb{Z}_2$ gauge theory in $D=3$ dimensions is the surface of a cube.

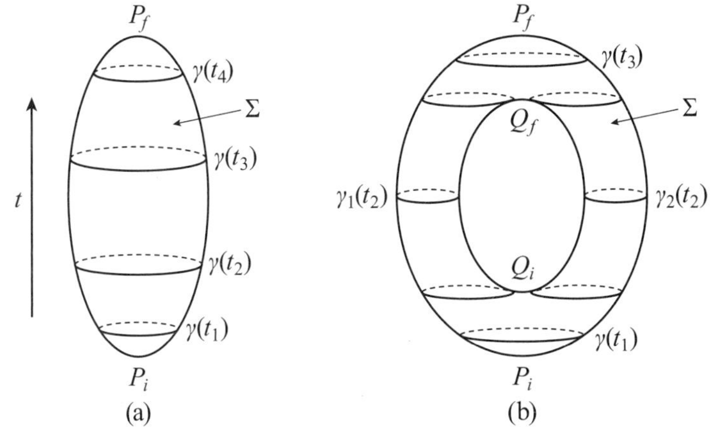

Figure 18.9 A closed surface Σ contributes to the strong coupling expansion by an amount $(2\tanh(1/g))^{A[\Sigma]}$, where $A[\Sigma]$ is its area (in lattice units). (a) A closed surface with genus 0 (with the topology of a sphere); it can be viewed as the history of a closed curve γ (a string) evolving in imaginary time t as it sweeps the surface Σ. Here the closed string is created in the remote past at event P_i, evolves from times t_1 through t_4, and disappears at event P_f. (b) A closed surface of genus 1 (a torus). Here the closed string γ is created at P_i, evolves to Q_i where it splits into two strings γ_1 and γ_2, which after evolving for some time, rejoin at Q_f into the single closed string γ, which eventually disappears at event P_f.

imaginary time t by a string γ after being created at P_i and disappearing at P_f. Figure 18.9b shows the same process for a surface with the topology of a torus (genus 1). In this case, the closed string γ is created at P_i, splits into two closed strings γ_1 and γ_2 at Q_i, which rejoin at Q_f to form the closed string γ, which, in turn, eventually disappears at P_f.

Let us now turn to the behavior of the Wilson loop in the strong coupling limit. It is drastically different from what we found in the weak coupling regime. The result that we will find in the strong coupling regime holds for all gauge theories with a compact gauge group. We will see that the expectation value of the Wilson loop operator for a large loop exhibits an *area law* exponential decay.

Before computing the expectation value of the Wilson loop operator, we will discuss a rigorous result valid for any local theory that satisfies the condition of reflection positivity (i.e., Euclidean unitarity), discussed in section 14.3. This inequality holds for both theories with a global compact symmetry group and theories with a local compact symmetry group. In both types of theories, the Euclidean action is a local function of the fields and has the form $S=-\frac{1}{g}\mathcal{F}[\text{fields}]$. Then, in such theories, the expectation values of correlators of observables are positive, $\langle \mathcal{O} \rangle \geq 0$, where $\mathcal{O}$ is an observable (or product of observables) invariant under the symmetry transformations. Moreover, it has been shown rigorously that the expectation values satisfy a monotonic dependence on the coupling constant g of the theory (Griffiths, 1972):

$$\langle \mathcal{O}[g_1] \rangle > \langle \mathcal{O}[g_2] \rangle, \qquad \text{for} \quad g_1 < g_2 \tag{18.75}$$

This relation is an example of a Griffiths-Ginibre inequality.

In classical statistical mechanics, this inequality says that as the temperature T increases, the correlations decrease. This result holds for systems that obey reflection positivity, such as ferromagnets (or, more generally, unfrustrated magnets). In that context, it implies that the fastest possible decay of a correlation function at long distances is exponential decay.

Since a gauge theory with a compact gauge group satisfies reflection positivity, the inequality eq. (18.75) also applies to these theories. In section 18.6, we saw that in a gauge theory with a compact gauge group, only gauge-invariant observables have a nonvanishing expectation value. Of all such observables, the most relevant one is the Wilson loop operator, and this inequality applies to this operator. It implies that the expectation value of the Wilson loop operator at some coupling constant g_1 is bounded from below by its value at a larger value of the coupling constant g_2. We will now show that, for large enough coupling constant g, the expectation value of the Wilson loop decays as an exponential function of the area bounded by the loop. Thus, in this regime, it obeys an *area law*. It follows that the regimes where the Wilson loop obeys, respectively, area and perimeter laws must be separated by a phase transition.

Let us now compute the expectation value of the Wilson loop operator, using the same geometry as that shown in figure 18.5. Here we will do the computation for the simpler case of the $\mathbb{Z}_2$ gauge theory. However, the calculation works much in the same way for all compact gauge groups, both discrete and continuous. Again using a compact notation, write the expectation value of the Wilson loop operator in the $\mathbb{Z}_2$ gauge theory as

$$\langle W_\gamma[\tau_\mu] \rangle_K = \frac{1}{Z[K]} \operatorname{Tr}\left[W_\gamma[\tau_\mu] \exp\left(K \sum_{\text{plaquettes}} K \tau_\mu \tau_\nu \tau_\mu \tau_\nu \right)\right] \tag{18.76}$$

where

$$W_\Gamma[\tau_\mu] = \prod_{(\mathrm{r},\mu)\in\Gamma} \tau_\mu(\mathrm{r}) \tag{18.77}$$

is the Wilson loop operator.

The computation is elementary. In the strong coupling regime, $K = \frac{1}{g} \ll 1$, we expand the exponential in powers of $\tanh(K)$. The only nonvanishing terms in the trace are such that the $\mathbb{Z}_2$ gauge fields on Γ are matched by gauge fields from the plaquettes in the exponential. It is easy to see that these contributions correspond to the tiling of open surfaces Σ whose boundary is Γ. Each term of the sum contributes a factor of $(2\tanh K)^{\mathcal{A}[\Sigma]}$ times a multiplicity factor. It follows that the leading nonvanishing term is the tiling of the minimal surface Σ_{minimal} with boundary Γ, shown in figure 18.10. In the case of a planar loop, there is only one minimal surface and no multiplicity factor. In this case, the area of the minimal surface is equal to the number of plaquettes enclosed by the loop. For a loop of (imaginary) time span T and spatial extent R (both in lattice units), the area enclosed by the loop is $\mathcal{A}[\Sigma] = RT$. Therefore, to leading order, we find that the expectation value of the

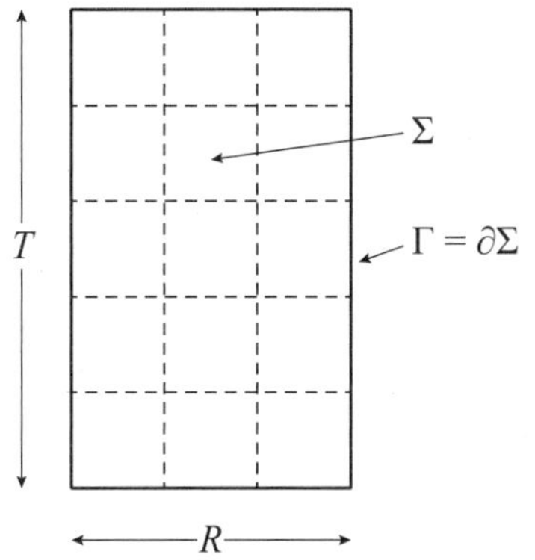

Figure 18.10 The leading contribution to the expectation value of the Wilson loop on Γ is the minimal surface Σ.

Wilson loop decays exponentially with the area enclosed by the loop as

$$\langle W_\Gamma \rangle_K = \exp\left(-\sigma[K]\mathcal{A}[\Sigma]\right) \tag{18.78}$$

where, to leading order, the *string tension* $\sigma(K)$ is

$$\sigma(K) = -\ln\tanh K + \cdots \simeq \ln g + \cdots \tag{18.79}$$

This result, combined with the Griffiths-Ginibre inequality, eq. (18.75), proves that the expectation value of the Wilson loop in a gauge theory with a compact gauge group is bounded by a function that decays exponentially for large loops, following an *area law*. Hence, the area law represents the *fastest* rate at which a Wilson loop can decay for loops of large sizes.

We can now readily obtain the effective potential $V(R)$:

$$\begin{aligned} V(R) &= -\lim_{T/R\to\infty} \ln\langle W_\Gamma \rangle_K \\ &= \sigma(K)R \end{aligned} \tag{18.80}$$

where the string tension, $\sigma(K)$, is given in eq. (18.79).

We have found that the effective potential $V(R)$ grows linearly with separation. Thus the $\mathbb{Z}_2$ sources introduced by the Wilson loop are confined, since the cost of separating them all the way to infinity is divergent. In contrast, in the $g \ll 1$ regime, the energy cost is finite, and equal to self-energy of the sources.

18.10 Behavior of gauge theories coupled to matter fields

In the absence of matter fields, gauge theories have two possible *phases*:

1) a deconfined phase, in which the Wilson loop either obeys a perimeter law or Coulomb behavior (as in the case of Maxwell's theory), and the effective interaction between external charges (sources) either has an exponential decay or a $1/R$ fall off, and

2) a confined phase, in which the Wilson loop obeys an area law, and the interaction between static sources grows linearly with distance.

Confined and deconfined phases are separated by phase transitions, which may be continuous or discrete (first order). If the phase transitions are continuous, they define a nontrivial continuum QFT.

In the case of gauge theories with a discrete gauge group, they may have a deconfined phase if the spacetime dimension $D > D_c = 2$, which is the lower critical dimension. In contrast, the lower critical dimension for a gauge theory with a compact continuous gauge group to have a deconfined phase is $D_c = 4$ spacetime dimensions. This is also the number of dimensions in which their coupling constant is dimensionless and the theory is renormalizable in perturbation theory. This behavior is similar to what we found for matter fields. Matter fields with a discrete global symmetry (e.g., the Ising model) have a lower critical dimension $D_c = 1$, above which the symmetry may be spontaneously broken. The lower critical dimension for matter fields with a continuous global symmetry is $D_c = 2$.

What happens if the gauge fields are (minimally) coupled to dynamical matter fields? In our discussion, matter fields have only entered as static sources (i.e., as infinitely heavy particles that carry the gauge charge). Clearly, if the matter fields are sufficiently heavy and are "uncondensed," the expectation is that their effects should be mild (e.g., a renormalization of the coupling constant of the gauge fields).

However, the behavior of some observables changes, no matter how heavy the matter fields are. Consider a gauge theory deep in its confining phase. We saw that when the matter fields are absent (or rather, are static sources), the Wilson loop operator has an area law, and the energy to separate two charges grows linearly with their separation, as shown in figure 18.6b. We will now see that if the dynamical matter fields carry the fundamental charge of the gauge group, for large enough Wilson loops, the area-law behavior always yields to a perimeter-law behavior. In spite of that, we will see that the theory is still confining.

Consider first the case of a $\mathbb{Z}_2$ gauge theory coupled to an Ising matter field. Let us work with the spacetime lattice picture. The partition function for this theory is (using once again a synthetic notation)

$$Z[\beta, K] = \sum_{\{\tau_\mu\},\{\sigma\}} \exp\left(\beta \sum_{\text{links}} \sigma\tau_\mu\sigma + K \sum_{\text{plaquettes}} \tau_\mu\tau_\nu\tau_\mu\tau_\nu\right) \tag{18.81}$$

We will compute the expectation value of the Wilson loop operator on a closed contour Γ, as defined in eq. (18.77). Notice that, since the $\mathbb{Z}_2$ gauge group has only one nontrivial representation, the Wilson loop has to carry the charge of the representation. So, in this case, only one Wilson loop can be defined.

We are interested in the regime where confinement is strongest (i.e., $K \ll 1$), and the $\mathbb{Z}_2$ matter field is heavy and uncondensed (i.e., $\beta \ll 1$). In this regime, we can expand the Gibbs weight of this partition function in powers of K and β, and determine the leading contribution to the expectation value of the Wilson loop. At $\beta = 0$, we found that in the leading contribution, we had to tile the Wilson loop with plaquettes spanning a minimal surface bounded by the contour Γ. This gave an area-law contribution.

We can similarly set $K = 0$, where the theory should be confining down to the shortest distance scales, and seek the leading-order contribution in β, the coupling to the matter fields. In the definition of the Wilson loop, each link that belongs to Γ has a gauge field

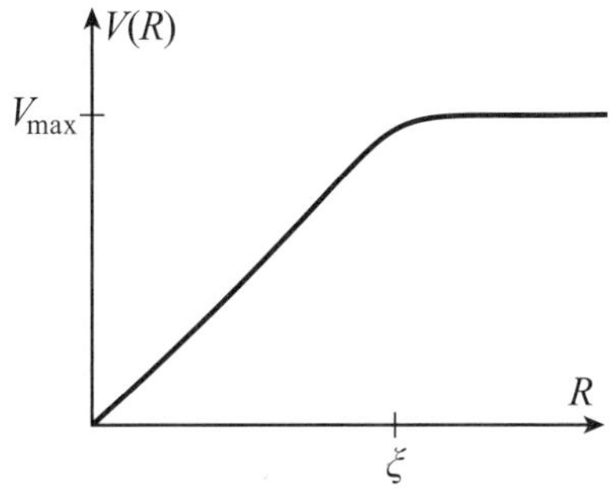

Figure 18.11 The effective potential for the $\mathbb{Z}_2$ gauge theory with dynamical $\mathbb{Z}_2$ (Ising) matter.

variable τ_μ, which gets multiplied around the loop. The leading term that cancels these link variables is the product of the link terms of the gauge-matter term of the action on the links that belong to Γ. In this product, each matter (Ising) variable appears twice (and squares to 1), leaving behind the product of the gauge variables on the links. These, in turn, will square to unity when multiplied by the gauge field on the Wilson loop. Therefore, for any value of β, no matter how small, there will be a perimeter-law contribution.

Thus, in this regime, the lowest-order contributions to the Wilson loop are

$$\langle W_\Gamma \rangle_{K,\beta} = (\tanh K)^{RT} + (\tanh \beta)^{2(R+T)} + \cdots \tag{18.82}$$

where we used that the area of the minimal surface is $A = RT$, and the perimeter of the loop is $L = 2(R+T)$. Clearly, for large-enough loops, the perimeter-law contribution wins over the area law. For loops with $T \gg R \gg 1$, the effective potential will then be

$$V(R) = \begin{cases} \sigma(K)\, R, & \text{for } R \ll \xi \\ 2\rho(k), & \text{for } R \gg \xi \end{cases} \tag{18.83}$$

where, at this order, $\sigma(K) = -\ln\tanh K + \cdots$ is the string tension, $\rho(k) = -\ln\tanh \beta + \cdots$ is the mass (self-energy) of the matter field, and the crossover scale between the two behaviors is $\xi = \sigma(K)/\rho(K)$.

Clearly, these results imply that we have a linear potential until a scale $\xi \simeq 2\rho/\sigma$, where the string breaks by creating a pair of excitations of the matter field (from the vacuum), which screens that test charge of the Wilson loop. Beyond this scale, the potential is essentially the constant value V_{max}, equal to twice the mass of the matter field (see figure 18.11). Physically, the saturation of the linear potential means that the test charges (the Wilson loop) are being exactly screened (or compensated) by the creation of a pair of excitations from the vacuum. This "algebraic screening" behavior does not imply that confinement breaks down, since this excitation is a gauge-invariant state made of two $\mathbb{Z}_2$ matter fields stretching a string of the $\mathbb{Z}_2$ gauge field (see figure 18.12). Furthermore, in this phase, the only local excitations are created by an operator P(plaquette) on small loops of gauge fields around a plaquette (i.e., a "glueball") separated by a distance R, and tightly bound states π(link) of matter and gauge fields ("pions"), also separated by a distance R. An elementary calculation shows that the (connected) correlator of two plaquette operators exhibits the behavior

$$\langle P(\text{plaquette}) P(\text{plaquette}') \rangle_c \simeq \exp(-M_P R) \tag{18.84}$$

Figure 18.12 The effective potential for the $\mathbb{Z}_2$ gauge theory with dynamical $\mathbb{Z}_2$ (Ising) matter.

where, in this limit, the mass of this excitation is $M_P = 4|\ln \tanh K|$. Likewise, the (connected) correlator of two "pions" on links separated by a distance R in this regime is

$$\langle \pi(\text{link})\pi(\text{link}')\rangle_c \simeq \exp(-M_\pi R) \tag{18.85}$$

where, at this level of approximation, the mass is $M_\pi = 2|\ln \tanh \beta| + |\ln \tanh K|$.

Hence, in spite of the perimeter-law behavior, all states are local gauge-invariant states, as required by Elitzur's theorem. Furthermore, the spectrum consists of massive locally gauge-invariant bound states. It is straightforward to show that this behavior holds for any theory with any compact gauge group in a confining phase. In addition, in the confining phase of the $\mathbb{Z}_2$ theory, there are no states in the spectrum that carry the $\mathbb{Z}_2$ charge. We will see shortly that, in the $\mathbb{Z}_2$ theory, such states do exist in the deconfined phase, although the associated operator is nonlocal. This behavior turns out to be because in the deconfined phase, the $\mathbb{Z}_2$ theory is topological. Similar behavior will be found in other theories with a discrete gauge group, such as $\mathbb{Z}_N$.

The "compensation" of the test charge that we found should not be confused with more conventional Debye screening of a test charge in a plasma. Let us call this behavior "algebraic screening." To understand the difference, we need to consider a theory in which more than one charge is possible. One simple example is a theory with gauge group $U(1)$. The partition function is now

$$Z[\beta, K] = \int_0^{2\pi} \frac{d\theta}{2\pi} \int_0^{2\pi} \frac{dA_\mu}{2\pi} \exp\left(\beta \sum_{\text{links}} \cos(\Delta_\mu \theta - qA_\mu) + K \sum_{\text{plaquettes}} \cos F_{\mu\nu}\right) \tag{18.86}$$

where the flux on a plaquette is denoted by $F_{\mu\nu} = \Delta_\mu A_\nu - \Delta_\nu A_\mu$, and $q \in \mathbb{Z}$ is the $U(1)$ charge of the matter field $\exp(i\theta)$. Symmetry allows us to expand the Gibbs weights in the representations of the group $U(1)$:

$$\exp(\beta \cos\theta) = \sum_{\ell \in \mathbb{Z}} I_\ell[\beta] \exp(i\ell\theta) \tag{18.87}$$

This Fourier expansion can be regarded as an expansion in the characters of the representations of the group $U(1)$, which are just $\chi_\ell(\theta) = \exp(i\ell\theta)$, where $\ell \in \mathbb{Z}$. The character expansion applies to any group. In the specific case of the $U(1)$ group, the coefficients of this expansion are the modified Bessel functions $I_\ell(\beta)$, given by

$$I_\ell(\beta) = I_{-\ell}(\beta) \int_0^{2\pi} \frac{d\theta}{2\pi} e^{\beta \cos\theta} e^{-i\ell\theta} \tag{18.88}$$

For $\beta \ll 1$, it is well approximated by $I_\ell(\beta) \simeq (\beta/2)^\ell / \ell!$.

Using these expansions, the partition function of the $U(1)$ theory of eq. (18.86) becomes

$$Z[\beta, K] = \sum_{\{\ell_\mu\},\{m_{\mu\nu}\}} \prod_{\text{sites}} \delta_{\Delta_\mu \ell_\mu, 0} \prod_{\text{links}} \delta_{\Delta_\nu m_{\mu\nu} + q\ell_\mu, 0} \prod_{\text{links}} I_{\ell_\mu}[\beta] \prod_{\text{plaquettes}} I_{m_{\mu\nu}}[K] \tag{18.89}$$

where the variables $\ell_\mu \in \mathbb{Z}$ and $m_{\mu\nu} \in \mathbb{Z}$ are defined, respectively, on the links and plaquettes of the lattice.

The Wilson loop operator with test charge $p \in \mathbb{Z}$ on a closed loop of the lattice Γ is

$$W_p[\Gamma] = \exp\left(ip \sum_{(\boldsymbol{r},\mu)\in\Gamma} s_\mu(\boldsymbol{r}) A_\mu(\boldsymbol{r}) \right) \tag{18.90}$$

where $s_\mu(\boldsymbol{r})$ is an oriented kink variable that is $s_\mu = 1$ on links on the closed contour Γ, and is zero otherwise. The expectation value of this operator is computed by making the following replacement in the partition function of eq. (18.89):

$$\prod_{\text{links}} \delta_{\Delta_\nu m_{\mu\nu} + q\ell_\mu, 0} \mapsto \prod_{\text{links}} \delta_{\Delta_\nu m_{\mu\nu} + q\ell_\mu, p s_\mu} \tag{18.91}$$

Deep in the confined phase, $K \ll 1$ and $\beta = 0$, the link variables are set to zero: $\ell_\mu = 0$. In this regime, we get the area-law behavior for Wilson loops with any test charge p. In this limit, only configurations with $m_{\mu\nu} = p$ on the smallest possible number of plaquettes can contribute (i.e., plaquettes $(\boldsymbol{r}, \mu\nu)$ that tile the minimal surface Σ whose boundary is the contour Γ). Hence, the leading configuration is $m_{\mu\nu}(\boldsymbol{r}) = p\Theta(\boldsymbol{r}, \mu\nu)$, where $\Theta(\boldsymbol{r}, \mu\nu) = 1$ if the plaquette $(\boldsymbol{r}, \mu\nu) \in \Sigma$, and zero otherwise. Their total contribution to expectation value of the Wilson loop of charge p is an area law

$$\langle W_p[\Gamma] \rangle = \left(\frac{I_p(K)}{I_0(K)} \right)^{RT} \tag{18.92}$$

and we find a potential linear in R, with a string tension

$$\sigma(K, p) = -\ln(I_p(K)/I_0(K)) \approx \ln(p!(2/K)^p) + \cdots \tag{18.93}$$

which is finite for all p, and in this limit, is large.

However, for $\beta \ll 1$ but finite, we need to satisfy two constraints, $\Delta_\mu \ell_\mu = 0$ and $\Delta_\nu m_{\mu\nu} + q\ell_\mu = p s_\mu$, while making both $|m_{\mu\nu}|$ and $|\ell_\mu|$ as small as possible. We find several behaviors. If the charge of the Wilson loop is a multiple of the charge of the dynamical matter field, $p = qr$ (for some integer r), we find two leading contributions: (1) $\ell_\mu = 0$ and $m_{\mu\nu} = p$ on the plaquettes that tile the minimal surface Σ, and (2) $m_{\mu\nu} = 0$ everywhere and $\ell_\mu = r$ on links on the contour Γ. The first contribution reproduces the area-law behavior, while the second yields a perimeter law:

$$\langle W_{p=qr}[\Gamma] \rangle = \left(\frac{I_p(K)}{I_0(K)} \right)^{RT} + \left(\frac{I_r(\beta)}{I_0(\beta)} \right)^{2(R+T)} \tag{18.94}$$

Hence, for $p = qr$, the interaction potential saturates to the value $V_{\max} = -2\ln(I_r(\beta)/I_0(\beta)) \approx 2\ln(r!(2/\beta)^r) + \cdots$, and, for sufficiently large loops, we find the same algebraic screening as found before.

In contrast, if the charge of the Wilson loop is smaller than the charge of the dynamical matter field, $p < q$, then the leading contribution comes from configurations with $\ell_\mu = 0$ and $m_{\mu\nu} = p$ on the plaquettes of the minimal surface (and zero otherwise). In other words, in this case, the leading behavior is the area law of eq. (18.92), the effective potential is linear in R, and there is no algebraic screening.

Similarly, if the charges are not multiples of each other, but still $p > q$ (e.g., $p = qr + k$), the solutions to the constraints now are $\ell_\mu = qs_\mu$ (only on the links of Γ), and $m_{\mu\nu} = k$ only on the plaquettes of the minimal surface bounded by Γ. Thus, we now find

$$\langle W_{p=qr+k}[\Gamma]\rangle = \left(\frac{I_k(K)}{I_0(K)}\right)^{RT} \left(\frac{I_r(\beta)}{I_0(\beta)}\right)^{2(R+T)} \tag{18.95}$$

In other words, we recover an area law even if the matter field is dynamical, and the effective potential no longer saturates. Now the unscreened part of the test charge k sees a linear potential with a nonvanishing (but smaller) string tension $\sigma(K, k) = \ln(k!(2/K)^k) + \cdots$, and an offset of the linear potential, $V_{\text{eff}} = 2\ln(r!(2/\beta)^r)$.

These results, which represent the leading behavior of a convergent expansion in powers of K and β, show that when the test charge and the charge of the dynamical matter field are not proportional to each other, the Wilson loop of the test charge has an area-law behavior. Only when proportionality holds do we get a perimeter law and saturation, and hence, algebraic screening. The same behavior is found in more general cases. For example, in a non-abelian gauge theory with matter in the adjoint representation, the Wilson loop for the test charge in the fundamental representation will exhibit confinement.

18.11 The Higgs mechanism

The Higgs mechanism arises in theories of gauge and matter fields with a continuous symmetry in the regime where the matter field spontaneously breaks the global symmetry.

18.11.1 The abelian Higgs model

In the simplest case, the symmetry group is $U(1)$, and the matter field is a complex scalar field ϕ coupled to Maxwell's electrodynamics. The Lagrangian density (in Minkowski spacetime) of this theory, known as the abelian Higgs model, is

$$\mathcal{L} = -\frac{1}{4}F_{\mu\nu}^2 + |D_\mu\phi|^2 - V(|\phi|^2) \tag{18.96}$$

where $D_\mu = \partial_\mu + ieA_\mu$ is the covariant derivative. The potential is, as usual, $V(|\phi|^2) = m^2|\phi|^2 + \lambda|\phi|^4$. If $m^2 < 0$, classically it has a minimum at $|\phi_0| = (|m^2|/2\lambda)^{1/2}$, and the global $U(1)$ symmetry is spontaneously broken.

In the Euclidean domain, the Lagrangian becomes

$$\mathcal{L}_E = |D_\mu\phi|^2 + m^2|\phi|^2 + \lambda|\phi|^4 + \frac{1}{4}F_{\mu\nu}^2 \tag{18.97}$$

which is identical to the Landau-Ginzburg theory for a superconductor, with ϕ being the order-parameter field of the superconductor (the Cooper pair condensate), and the charge is $2e$. The path integral of this model describes the classical partition function of a superconductor interacting with the thermal fluctuations of the magnetic field.

Let us rewrite the complex scalar field in terms of an amplitude and a phase field:

$$\phi = \rho e^{i\theta} \tag{18.98}$$

We will focus on the phase, $m^2 < 0$, in which the scalar field has a broken (global) $U(1)$ symmetry. In this phase, the phase field θ is the Goldstone boson of the spontaneously broken global $U(1)$ symmetry. Deep in this phase, the amplitude field is essentially pinned to its classical expectation value, $\rho_0 = |\phi_0|$. Then the Lagrangian of the abelian Higgs model, eq. (18.96), becomes

$$\mathcal{L} = \rho_0^2 \left(\partial_\mu \theta + eA_\mu\right)^2 - \frac{1}{4} F_{\mu\nu}^2 \tag{18.99}$$

We can now fix the gauge $\theta = 0$, known as the unitary gauge (the same as the London gauge in superconductivity), or equivalently, make a gauge transformation, $A_\mu \to A_\mu - \partial_\mu \Phi$, with $\theta = -e\Phi$. In both descriptions, the phase field θ disappears from the theory (i.e., the Goldstone boson is "eaten" by the gauge field). The Lagrangian now is

$$\mathcal{L} = -\frac{1}{4} F_{\mu\nu}^2 + e^2 \rho_0^2 A_\mu^2 \tag{18.100}$$

Hence, in the broken symmetry phase, the gauge field becomes massive, and in this theory, the mass of the photon is $m = \sqrt{2}e\langle\phi\rangle$. The phenomenon of the gauge field becoming massive upon eating a Goldstone boson is known as the Higgs mechanism. Physically, this phenomenon is equivalent to the expulsion of a magnetic flux in a superconductor, known as the Meissner effect.

Another key feature of the Higgs fields is that they yield masses to Dirac fermions. Thus, consider adding a massless Dirac field ψ to this theory, minimally coupled to the gauge field A_μ, and suppose that the fermions also couple to the Higgs field through Yukawa couplings. The fermionic sector of the theory is now

$$\mathcal{L}_{\text{fermions}} = \bar{\psi} i \not{D} \psi + G \bar{\psi} \psi \phi \tag{18.101}$$

We now see that if the scalar field has a vacuum expectation value, $v = \langle\phi\rangle$, the Yukawa coupling G becomes a fermion mass term with a mass $m = Gv$, or equivalently, that $\langle\bar{\psi}\psi\rangle = G\langle\phi\rangle$. The reader familiar with the BCS theory of superconductivity will recognize that there is a very close parallel between the mechanism of mass generation via the Higgs mechanism and the development of a superconducting gap by Cooper-pair condensation (Schrieffer, 1964).

The analysis that we have done here is classical, and the computation of perturbative corrections is largely straightforward, except for the theory of the phase transition from the symmetric to the Higgs phase. However, the nonperturbative behavior, which we discuss shortly, is subtle, as we shall see.

18.11.2 The Georgi-Glashow model

The Higgs mechanism plays a key role in the theory of weak interactions. An example is the Georgi-Glashow model, which is a theory that unifies the weak and strong interactions with electromagnetism (Georgi and Glashow, 1974).

This theory involves a three-component real scalar field $\boldsymbol{\phi}$ (i.e., a Higgs field with a global spontaneously broken $O(3)$ symmetry) coupled to an $SU(2)$ gauge field, $\boldsymbol{A}_\mu$, which is a matrix with values in the algebra of the group $SU(2)$ (i.e., $\boldsymbol{A}_\mu = A_\mu^a t^a$, where t^a are the three generators of $SU(2)$). The Lagrangian of this theory in the broken symmetry state is (in Euclidean spacetime)

$$\mathcal{L} = \frac{1}{2}\left(D_\mu \boldsymbol{\phi}\right)^2 - \frac{1}{2} m^2 \boldsymbol{\phi}^2 + \frac{\lambda}{4!}(\boldsymbol{\phi}^2)^2 + \frac{1}{4g^2} \mathrm{tr} F_{\mu\nu}^2 \tag{18.102}$$

where, for the gauge group $SU(2)$, the covariant derivative is

$$D_\mu \boldsymbol{\phi} = \partial_\mu \boldsymbol{\phi} + \boldsymbol{A}_\mu \times \boldsymbol{\phi} \tag{18.103}$$

and the field strength is

$$F_{\mu\nu} = i[D_\mu, D_\nu] = \partial_\mu \boldsymbol{A}_\nu - \partial_\nu \boldsymbol{A}_\mu + \boldsymbol{A}_\mu \times \boldsymbol{A}_\nu \tag{18.104}$$

Let us write the scalar field as $\boldsymbol{\phi}^T = (\phi_1, \phi_2, \phi_3)$ and assume that the pattern of spontaneous symmetry breaking is $\boldsymbol{\phi} = (0, 0, m/\sqrt{\lambda})$. This classical expectation value breaks the $O(3)$ symmetry down to $U(1)$. By repeating the line of reasoning that we used in the abelian Higgs model, we find that in the broken symmetry state, the triplet of gauge fields of the Georgi-Glashow model can be rearranged as

$$W_\mu^{\pm} = \frac{1}{\sqrt{2}}(A_\mu^{(1)} \mp i A_\mu^{(2)}), \quad A_\mu = A_\mu^{(3)} \tag{18.105}$$

It is straightforward to see that in this theory, the Higgs mechanism implies that the doublet $W_\mu^{\pm}$ has a mass (squared) $m_W^2 = g^2 m^2/\lambda$, and the field $A_\mu^{(3)}$ is massless. The conclusion is that the massive gauge bosons $W_\mu^{\pm}$ mediate the weak interactions, while the massless gauge fields $A_\mu^{(3)}$ are the photons of the electromagnetic sector. Also, just as in the case of the abelian Higgs model, the two Goldstone bosons of the spontaneously broken $O(3)$ symmetry disappear from the spectrum. However, the longitudinal component, $\sigma = \phi_3 - m/\sqrt{\lambda}$, of the field $\boldsymbol{\phi}$ is massive, with $m_\sigma^2 = 2m_0^2$. This massive excitation is the Higgs particle.

The full theory of weak interactions, the Weinberg-Salam model, is an extension of the Georgi-Glashow model that includes, among other fields, $SU(2)$ doublets of Dirac fermions (electrons and neutrinos). In this theory, the fermions acquire a mass through Yukawa couplings of the fermions with the Higgs fields.

18.11.3 Observables of the Higgs phase

It is tempting to regard the complex scalar fields of the Higgs models as order parameters, as they are if the symmetry is global. However, in a theory in which the symmetry is local (i.e., it is a gauge symmetry), the complex scalar field $\phi(x)$ (and its generalizations) is not gauge invariant. This naturally leads to the question: What is the gauge-invariant meaning of the

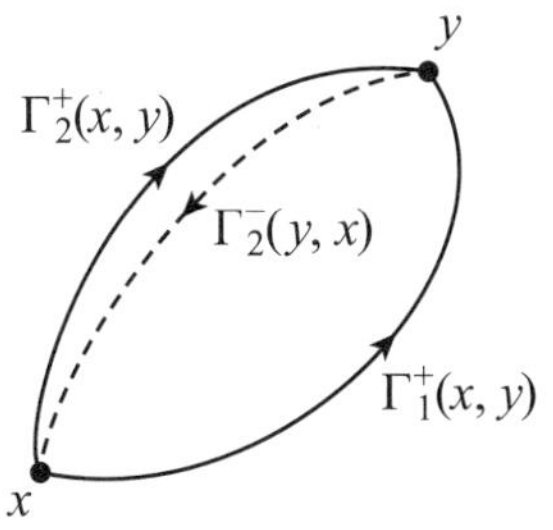

Figure 18.13 Path dependence of the gauge-invariant correlator of eq. (18.106). Here $\Gamma_1^+(x,y)$ and $\Gamma_2^+(x,y)$ are two oriented paths stretching from x to y. $\Gamma_2^-(y,x)$ is the reversed path of $\Gamma_2^+(x,y)$.

concept of spontaneous symmetry breaking when the symmetry is local? In other words, can we make the concept of spontaneous breaking of a local symmetry compatible with the requirements of Elitzur's theorem? For the concept of spontaneous symmetry breaking of a local symmetry to be meaningful, it has to be possible to construct a gauge-invariant order parameter that uniquely distinguishes the Higgs phase from other phases. Otherwise, the Higgs phase would not be a genuinely separate phase of the theory.

Let us discuss this first in the case of a theory in which the gauge group is $\mathbb{R}$, and the theory has a Maxwell gauge field A_μ. A simple way to make the correlators of the Higgs field $\phi(x)$ gauge invariant is to define the nonlocal operator

$$G_\Gamma(x,y) = \left\langle \phi^\dagger(x) \exp\left(ie \int_{\Gamma(x,y)} dz_\mu A^\mu(z) \right) \phi(y) \right\rangle \tag{18.106}$$

where $\Gamma(x,y)$ is an arbitrary path stretching from x to y. This correlator is trivially invariant under the local gauge transformations $A_\mu \to A_\mu + \partial_\mu \Phi$ and $\phi(x) \to e^{ie\Phi(x)}\phi(x)$. Deep in the broken symmetry state, $m^2 < 0$, we can approximate $\phi(x) \simeq \phi_0 e^{i\theta(x)}$, where ϕ_0 is real and constant. In the unitary (London) gauge, $\theta = 0$, the correlator reduces to the computation of the expectation value:

$$G_\Gamma(x,y) \simeq \phi_0^2 \left\langle \exp\left(ie \int_{\Gamma(x,y)} dz_\mu A^\mu(z) \right) \right\rangle \tag{18.107}$$

Since the gauge field is massive, it is easy to check that this correlator decays exponentially fast for large separations $|x-y|$. Therefore, this gauge-invariant operator is not an order parameter of the spontaneously broken symmetry but a test of the massive gauge field. In fact, this path-dependent operator can be defined for any gauge theory, both abelian and non-abelian.

Moreover, the nonlocal gauge-invariant correlator of eq. (18.106) is *path dependent* (as shown in figure 18.13):

$$\begin{aligned} \exp\left(ie \int_{\Gamma_1^+} dz_\mu A^\mu(z) \right) &= \exp\left(ie \int_{\Gamma_2^+} dz_\mu A^\mu(z) \right) \times \exp\left(ie \oint_{\Gamma_1^+ \cup \Gamma_2^-} dz_\mu A^\mu(z) \right) \\ &= \exp\left(ie \int_{\Gamma_2^+} dz_\mu A^\mu(z) \right) \times \exp\left(i\frac{e}{2} \int_\Sigma dS_{\mu\nu} F^{\mu\nu} \right) \end{aligned} \tag{18.108}$$

where $\partial\Sigma = \Gamma_1^+ \cup \Gamma_2^-$. Clearly, the operator depends on the amount of flux in the surface Σ bounded by the two paths. Although the factorization that we used above holds only for abelian theories, the path dependence of the operator also holds for non-abelian theories.

Is there an alternative? Yes, there is, provided the gauge group is $\mathbb{R}$ (i.e., a noncompact gauge theory). In fact, in the context of QED, Dirac observed that it is unphysical (and violates gauge invariance) to define an operator that creates a charged particle without its static Coulomb field (Dirac, 1966), that is, without creating a coherent state of photons. This is done by considering instead the nonlocal operator

$$\phi^\dagger(\boldsymbol{x}) \exp\left(ie \int d^3z \boldsymbol{E}(\boldsymbol{z}) \cdot \boldsymbol{A}(\boldsymbol{z})\right) \phi(\boldsymbol{y}) \tag{18.109}$$

where the integral in the exponential extends over all space, $\boldsymbol{E}(\boldsymbol{z})$ is the classical electric (Coulomb) field created by the charges at $\boldsymbol{x}$ and $\boldsymbol{y}$, and $\boldsymbol{A}$ is the vector potential of the theory. Under a gauge transformation, both the operators $\phi^\dagger(x)$ and $\phi(y)$, and the exponential operator (which creates a coherent state of photons) change under a gauge transformation as

$$\phi(x) \mapsto \phi(x) e^{ie\Phi(x)}, \quad \boldsymbol{A}(x) \mapsto \boldsymbol{A}(x) + \nabla\Phi(x) \tag{18.110}$$

where $\Phi(z) \to 0$ as $|z| \to \infty$. However, gauge invariance is maintained if the gauge-dependent contributions cancel each other out, that is, if

$$e^{-ie\Phi(x)} e^{ie\Phi(y)} e^{i\int d^3z \boldsymbol{E}(z)\cdot\nabla\Phi(z)} = 1 \tag{18.111}$$

This condition can be met if the vector field $\boldsymbol{E}(z)$ satisfies

$$\nabla \cdot \boldsymbol{E}(z) = \rho(z) \equiv e\delta^3(z-x) - e\delta^3(z-y) \tag{18.112}$$

In other words, the vector field $\boldsymbol{E}(z)$ is just the classical Coulomb (electric) field created by the two charges $\pm e$ located at x and y, respectively. Furthermore, the operator

$$\exp\left(ie \int d^3z\, \boldsymbol{E}(z) \cdot \boldsymbol{A}(z)\right) \tag{18.113}$$

creates a coherent state of photons. Also, since we can always solve the Poisson equation, eq. (18.112), in terms of the potential $U(z)$ such that $\boldsymbol{E} = -\nabla U$, with $-\nabla^2 U = \rho$, we see that the exponent in the coherent state operator becomes

$$\begin{aligned}\int d^3z\, \boldsymbol{E}(z) \cdot \boldsymbol{A}(z) &= -\int d^3z\, \nabla U(z) \cdot \boldsymbol{A}(z) \\ &= \int d^3z\, U(z) \nabla \cdot \boldsymbol{A}(z)\end{aligned} \tag{18.114}$$

In particular, in the Coulomb gauge, $\nabla \cdot \boldsymbol{A}(z) = 0$, the coherent state operator reduces to the identity operator. Hence, provided that the Coulomb gauge can be defined unambiguously, the nonlocal operator in eq. (18.109) reduces to the correlation function of the complex scalar fields.

Can we always define the operator in eq. (18.109)? The answer to this question is that it is possible only if the Coulomb gauge is well defined. However, this is only possible for Maxwell's theory, since it is abelian and noncompact. In fact, there is a topological obstruction to the definition of the Coulomb gauge if the group is compact (abelian and non-abelian). We will see in chapter 19 that this obstruction has a topological origin and that it is related to the existence of magnetic monopoles. Therefore, we conclude that for a general compact gauge group, it is not possible to construct an operator that has the properties of an order parameter.

18.12 Phase diagrams of gauge-matter theories

We just showed that in a general compact gauge theory, it is not possible to find a gauge-invariant observable related to the order parameter $\langle\phi\rangle$ of the theory with a spontaneously broken *global* symmetry. These results naturally raise several questions:

1) For a compact gauge group, can the Higgs phase be defined nonperturbatively? In other words, is there a Higgs phase distinct from the other phases of a gauge theory?
2) What is the relation between the Higgs phases and the other phases of the theory?
3) What are the observables of the Higgs phase, and what is their nonperturbative behavior?

In other words, we are asking what the global *phase diagram* is of a theory of gauge fields coupled to matter fields. We will see that the global properties of the phase diagram, which embodies the nonperturbative behavior of the theory, depend on whether the matter carries the fundamental charge of the gauge group. This is the problem we discuss now (Fradkin and Shenker, 1979).

For simplicity (and conciseness), let us consider a theory with a gauge group $U(1)$ and a matter field that is a complex scalar field that carries charge $q \in \mathbb{Z}$. With some exceptions (noted below), our results apply to any theory with a compact gauge group G. Also, we do not consider the case in which the matter field is fermionic. This analysis assumes that all symmetries of the theory are gauged and there are no global symmetries (which may be unbroken or spontaneously broken).

The action of the theory (defined on a hypercubic lattice in D Euclidean dimensions) is

$$-S = \sum_{\text{links}} \beta \cos(\Delta_\mu \theta - qA_\mu) + \sum_{\text{plaquettes}} K \cos F_{\mu\nu} \tag{18.115}$$

Here the complex scalar field is $\phi = e^{i\theta}$, and the gauge field is $\mathcal{U}_\mu = e^{iA_\mu}$. The operator Δ_μ is the right lattice difference, and $F_{\mu\nu}$ is the flux of the gauge field A_μ on each plaquette. As before, K is related to the gauge coupling constant by $K = 1/g^2$.

The phase diagram can be mapped in the $\beta - K$ plane, with $\beta > 0$ and $K > 0$. We can (and will) figure out much of the global features of the phase diagram by considering extreme regimes and then extrapolating to the middle of the diagram. However, the arguments presented below can be shown to be rigorously correct within the radius of convergence of well-defined expansions (of which we will capture only their leading terms).

Consider first the regime $g^2 \to 0$ (or, equivalently, $K \to \infty$) and β finite. In this limit, the gauge fields are weakly coupled and are dominated by the flat-field configurations, $F_{\mu\nu} \simeq 0$

(modulo 2π). In the extreme regime, this corresponds to the pure matter theory with a global $U(1)$ symmetry. For general dimension D, this theory has two phases, separated by a phase transition at some value β_c:

1) For $\beta < \beta_c$, the global $U(1)$ symmetry is unbroken, $\langle e^{i\theta} \rangle = 0$, and the correlation functions of the scalar field are short-ranged and decay exponentially with distance,

$$\langle e^{i\theta(x)} e^{-i\theta(0)} \rangle \sim e^{-|x|/\xi} \tag{18.116}$$

where ξ is the correlation length. Hence, in this phase, the complex scalar field is massive, and its mass is $m \sim 1/\xi$.

2) For $\beta > \beta_c$, the global $U(1)$ symmetry is spontaneously broken, $\langle e^{i\theta} \rangle \neq 0$. If the dimension $D > 2$, then the correlation function has the asymptotic behavior

$$\langle e^{i\theta(x)} e^{-i\theta(0)} \rangle \sim \left| \langle e^{i\theta} \rangle \right|^2 + O(1/|x|^{D-2}) \tag{18.117}$$

In this phase, the global $U(1)$ symmetry is spontaneously broken, and the theory has a massless excitation, the Goldstone boson of the broken symmetry.

This analysis applies to any theory with a continuous global symmetry G, provided $D > 2$. For $D = 2$ (e.g., the nonlinear sigma models), we have already seen that they flow to strong coupling. In that case, the global symmetry is actually unbroken, and hence, $\beta_c \to \infty$. The $U(1)$ case is special in that it has a phase transition at a finite value β_c (known as the Kosterlitz-Thouless transition), and the global $U(1)$ symmetry is also unbroken for $\beta > \beta_c$, but the correlators exhibit power-law decays: This theory has a line of fixed points. For g small but finite, we expect a Higgs phase if $\beta > \beta_c$, and a Coulomb phase (i.e., massive charged matter fields and a massless photon) for $\beta < \beta_c$. We will see shortly whether these expectations are actually met. However, for $D > 1$, in theories with a discrete global symmetry (e.g., the Ising and $\mathbb{Z}_N$ models), the phases with a spontaneously broken symmetry occur for $\beta > \beta_c$.

Let us now consider the regime $\beta \to 0$ and $K = 1/g^2$ fixed. In this regime, the matter fields are massive, with a large mass $m \approx |\ln \beta|$. In this regime, we expect the matter fields to decouple, and we have a pure gauge theory. Again, for a continuous gauge group G, and for $D > 4$, the gauge theory has a Coulomb phase with massless gauge fields for $g < g_c$ ($K > K_c$), and a confined phase in the opposite regime, $g > g_c$. As before, in the confined phase of the pure gauge theory the Wilson loop displays an area-law behavior, and in the deconfined phase it shows a perimeter-law behavior. This phase structure also holds in $D = 4$ dimensions for the $U(1)$ gauge theory. For theories with a discrete gauge group, the same phases arise for $D > 2$, except that in their deconfined phases, all excitations are massive. We will see in chapter 22 that the deconfined phases of discrete gauge theories are topological field theories.

The results of these two regimes, the weakly coupled gauge theory and the heavy matter field, are summarized in figure 18.14.

Let us now focus on the regime in which the gauge theory is strongly coupled, $g \to \infty$ (or $K \to 0$) with β fixed. In this regime, the fields fluctuate wildly. In the regime with β small, we saw that, if the matter field carries the fundamental charge, $q = 1$, the expectation value of the Wilson loop (with the fundamental, and in fact, any charge) has a crossover from an area-law behavior to a perimeter law for large enough loops. This crossover is compatible with the theory being in a confined phase. Furthermore, in the unitary gauge, $\theta = 0$, which

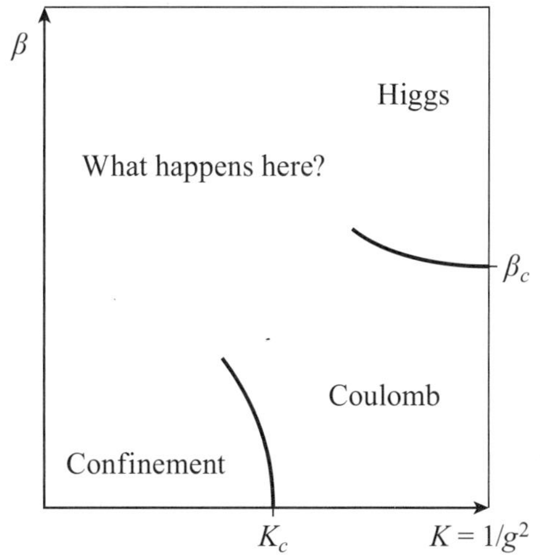

Figure 18.14 Tentative phase diagram depicting the behaviors of the weakly coupled gauge theory and the heavy matter field.

is always globally well defined, as $g \to \infty$, the action reduces to a sum over independent link variables A_μ. The partition function of the theory is free of singularities on a vertical strip parallel to the β axis. Hence, the crossover behavior of the Wilson loop should extend along the entire strip, all the way to $\beta \to \infty$.

Let us now inquire about the behavior of the theory for large β as a function of K. The behavior depends on the charge q carried by the matter field. In the unitary gauge, $\theta = 0$, the action takes the simpler form

$$-S = \sum_{\text{links}} \beta \cos(qA_\mu) + \sum_{\text{plaquettes}} K \cos F_{\mu\nu} \tag{18.118}$$

The behavior for large β now depends on whether $q = 1$ or $q \neq 1$.

In the case of $q = 1$, in the limit $\beta \to \infty$, the link term of the action forces the gauge fields to be $A_\mu = 0$ (modulo 2π) on every link of the lattice. Hence, the partition function is also free of singularities on a strip along the horizontal axis at $\beta \to \infty$. This is a startling result, since this "strip of analyticity" ranges from the Higgs phase at large β and small g all the way to the confining regime at large g and small β. It has been proven rigorously that this strip of analyticity exists for all theories with a compact gauge group, abelian or non-abelian, discrete or continuous, if the matter field carries the fundamental charge of the gauge group (Fradkin and Shenker, 1979; see also Osterwalder and Seiler, 1978). In all such theories, there is no global distinction between a Higgs phase and a confined phase. This result is sometimes called Higgs-confinement complementarity.

However, the Coulomb phase is also stable, since in this phase, there is a finite-energy state (a "particle") that carries the fundamental quantum number of the gauge group. In the case of $U(1)$ theory, the Coulomb phase has a state in the spectrum that carries the $U(1)$ charge, the electron of QED. In the case of the discrete gauge groups (e.g., $\mathbb{Z}_2$), a finite-energy state carries the $\mathbb{Z}_2$ charge, but the operator that creates this state can be shown to be nonlocal. We will see in chapter 22 that this is related to the fact that the deconfined phases of discrete gauge theories are effectively topological field theories.

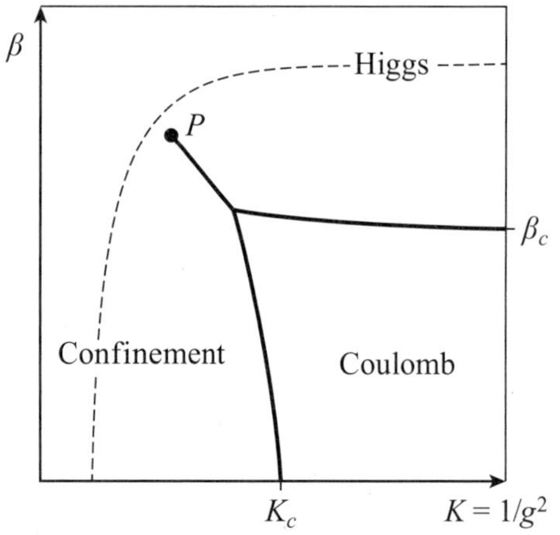

Figure 18.15 Phase diagram for a $U(1)$ gauge-matter theory with a matter field carrying the fundamental, $q = 1$, charge of the gauge field. The dashed curve represents the boundary of the strip of analyticity mentioned in the text. The dot (labeled by P) is a critical endpoint, similar to the critical point of water.

We can now summarize our understanding for the theory with a matter field that carries the fundamental gauge charge in the form of a complete phase diagram, shown in figure 18.15. The theory has two phases: (1) a Coulomb phase, and (2) a confinement-Higgs phase. In this theory, confinement and Higgs are not separate phases: The theory is confining everywhere, except in the Coulomb phase. This phase diagram applies to the case of the compact $U(1)$ gauge theory coupled to a fundamental scalar (i.e., the compact abelian Higgs model) in $D = 4$ dimensions.

The nature of the phase transitions is a more subtle problem. The RG analysis predicts a runaway flow into a regime where the symmetry is broken. This predicts that the Coulomb-Higgs transition is weakly first order. The confinement-Coulomb transition is known to be first order from numerical simulations. The dot labeled by P in figure 18.15 is a critical endpoint, at the end of a line of first-order transitions, similar to the critical point of water. In other words, while there is no global distinction between confinement and Higgs, across the line of first-order transitions, local gauge-invariant observables (e.g., the flux through a local plaquette) will exhibit a jump. Interestingly, part of the phase diagram for hot and dense QCD predicts a jump from nuclear matter to the quark-gluon plasma, and has a structure similar to that in figure 18.15; hence, these are not genuinely separate phases.

In this case of $q > 1$, in the limit $\beta \to \infty$, the link term of the action forces the gauge field to take one of the following possible q values on each link: $A_\mu = 2\pi n/q$, with $n = 0, 1, \ldots, q-1$. Therefore, in the limit $\beta \to \infty$, the theory reduces to a gauge theory with the discrete gauge group $\mathbb{Z}_q$. Thus, in the large-β regime, the $\mathbb{Z}_q$ theory can be regarded as the low-energy limit of the full theory.

If the dimension $D > 2$, the $\mathbb{Z}_q$ gauge theory has a confined phase for $K < K_c[\mathbb{Z}_q]$ and a deconfined phase in the opposite regime. Therefore, as shown in figure 18.16, for general dimension D, the theory now has a large confined phase separate from a Coulomb phase (for β small) and a "Higgs" (deconfined) phase (for β large). Furthermore, using a Griffiths inequality (cf. eq. (18.75)), it is straightforward to see that $K_c[\mathbb{Z}_q] \leq K_c[U(1)]$, as shown in figure 18.16.

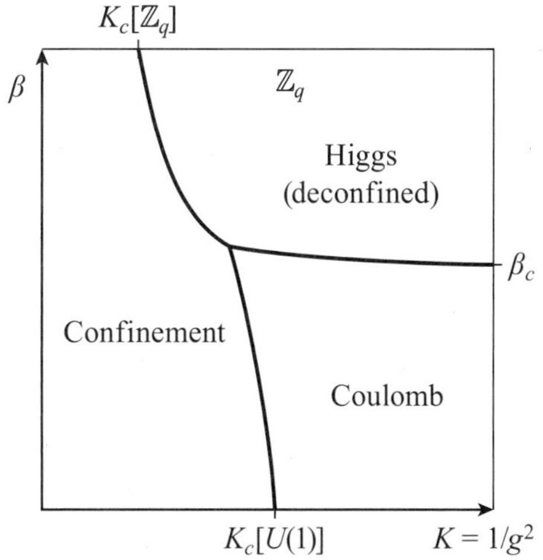

Figure 18.16 Phase diagram for a $U(1)$ gauge-matter theory with a matter field carrying charge $q > 1$ of the gauge field. The top of the phase diagram is the $\mathbb{Z}_q$ discrete gauge theory, which has a confined and a deconfined phase. In this case, there is a global distinction between the Higgs phase and the confinement phase.

The Higgs (deconfined) phase and the confined phase can be distinguished by the behavior of the Wilson loop carrying the fundamental charge (or in fact, any charge $p < q$). As we have seen, in this case, the Wilson loop with a charge $p < q$ is not algebraically screened and exhibits an area law in the confined phase. In the deconfined phase of the $\mathbb{Z}_q$ theory, for $K \gtrsim K_c[\mathbb{Z}_q]$ and large enough β, the Wilson loop exhibits a perimeter law, and the fundamental charge is indeed deconfined in this phase. This phase is a Higgs phase in the sense that the $U(1)$ gauge field is massive, although, as explained above, there isn't a local order parameter. In chapter 22, we will see that this phase is an example of a topological phase, and the low-energy limit of this discrete gauge theory is a topological field theory.

This analysis also holds in the case of a non-abelian gauge theory with a matter field carrying a charge different from that of the fundamental charge. For example, if the gauge group is $SU(N)$ and the matter field carries the *adjoint* representation of $SU(N)$, the "unbroken" sector of the gauge group is discrete gauge group $\mathbb{Z}_N$ which is the center of the group $SU(N)$, and the adjoint representation of $SU(N)$ is invariant under the action of the center, $\mathbb{Z}_N$, of the group.

Exercises

18.1 Lattice Dirac fermions

In this chapter, we have seen how theories of scalar and gauge fields are regularized by being defined on a lattice. In this exercise, you will work with a theory of Dirac spinors also defined on a lattice. For concreteness, the Euclidean theory is used here (see section 8.7). The action of the Dirac theory in $D = 4$ Euclidean dimensions is

$$S = \int d^4x \bar{\psi}(x)(i\partial\!\!\!/ + m)\psi(x) \equiv \int d^4x \bar{\psi}(x)\left(\frac{i}{2}\gamma_\mu \overleftrightarrow{\partial_\mu} + m\right)\psi(x)$$

where γ_μ is the set of four antihermitian matrices that satisfy the (pseudo) Clifford algebra $\{\gamma_\mu, \gamma_\nu\} = -2\delta_{\mu\nu}$. We will discretize the Dirac equation on a $(D+4)$-dimensional hypercubic lattice. The theory should be "naively" discretized by replacing derivatives by lattice differences, resulting in the lattice action

$$S = \sum_{x,\mu} \frac{i}{2}(\bar{\psi}(x)\gamma_\mu\psi(x+e_\mu) - \bar{\psi}(x+e_\mu)\gamma_\mu\psi(x)) + \sum_x m\bar{\psi}(x)\psi(x)$$

where $\{x\}$ is the set of sites of a four-dimensional hypercubic lattice (of lattice spacing $a=1$), and $\{e_\mu\}$ is a set of four unit vectors, with $\mu = 1, \ldots, 4$. The partition function is the Grassmann integral

$$Z = \prod_x d\bar{\psi}(x)d\psi(x) \exp(-S[\bar{\psi}, \psi])$$

Here, and above, $\bar{\psi}(x)$ and $\psi(x)$ are two independent four-component spinors of Grassmann variables defined on each lattice site, and m is the mass.

1) Show that, in the thermodynamic limit of an infinite system with periodic boundary conditions, the lattice Dirac propagator is, in momentum space,

$$S(p) = \frac{1}{\sum_\mu \gamma_\mu \sin p_\mu + m}$$

where the momenta span the first Brillouin zone (BZ), $-\pi < p_\mu < \pi$ (again, with $\mu = 1, \ldots, 4$).

2) Show that in the limit of small momentum ($|p_\mu| \ll \pi$), this propagator reproduces the usual free Dirac propagator with mass m in Euclidean spacetime.

3) Denote by K^i_μ the set of 15 momenta at the BZ points $(\pi, 0, 0, 0)$ (plus three permutations), $(\pi, \pi, 0, 0)$ (plus five permutations), $(\pi, \pi, \pi, 0)$ (plus three permutations), and (π, π, π, π). Show that near the BZ momentum K^i_μ, the lattice propagator has the form (in terms of $q_\mu = K^i_\mu - p_\mu$, with $|q_\mu| \ll \pi$):

$$S(K^i_\mu - q_\mu) = \frac{1}{\sum_\mu \gamma_\mu \cos(K^i_\mu) \sin q_\mu + m}$$

Show that the propagator near these momenta can be brought to the standard form (with mass m) by a similarity transformation $s(K^i_\mu) = \gamma_\mu\gamma_5$. In other words, this lattice theory describes 16 Dirac fermions, not one!

4) A way of eliminating these unwanted 15 "species" of fermions is to add the following (Wilson) mass term to the lattice action:

$$S_W = \sum_{x,\mu} \frac{r}{2}(\bar{\psi}(x+e_\mu)\psi(x) + \bar{\psi}(x)\psi(x+e_\mu) - 2\bar{\psi}(x)\psi(x))$$

This construction is known as the Wilson fermion. Show that the lattice propagator is now

$$S(p) = \frac{1}{\sum_\mu \gamma_\mu \sin p_\mu + m + r\sum_\mu (1 - \cos p_\mu)}$$

with $0 < r < 1$. Show that for $r \neq 0$, the unwanted 15 species of fermions have a large mass $m + 2rk$ (with $k = 1, 2, 3, 4$), while the fermion with $|p| \ll \pi$ is unaffected (to linear order in p_μ).

5) Let us now examine how chiral symmetry is realized. To this end, set the mass $m \to 0$. The chiral transformation is $\psi \to \exp(i\theta\gamma_5)\psi$. Show that the Wilson term is not invariant under a continuous chiral transformation, even if the mass term is $m = 0$.
6) Determine the chiral charge of each of the 16 species of fermions, and show that eight species have chiral eigenvalue $+1$ and that the remaining eight species have chiral eigenvalue -1.

18.2 Wilson loop in $U(1)$ electrodynamics with fermions
In this exercise, consider a lattice $U(1)$ gauge theory coupled to lattice fermions (defined in exercise 18.1) that carry the unit charge of the gauge field. Work on a four-dimensional Euclidean hypercubic lattice, and consider only the strong coupling limit of this theory. The total action is $S = S[A_\mu] + S[\bar{\psi}, \psi, A_\mu]$, where $S[A_\mu]$ is the action of the pure $U(1)$ gauge theory with coupling constant g,

$$S[A_\mu] = -\frac{1}{g}\sum_{x,\mu\nu} \cos(\Delta_\mu A_\nu - \Delta_\nu A_\mu)$$

and $S[\bar{\psi}, \psi, A_\mu]$ is obtained from the lattice action for Wilson fermions by the minimal coupling procedure:

$$\bar{\psi}(x)\psi(x + e_\mu) \to \bar{\psi}(x)\exp(iA_\mu(x))\psi(x + e_\mu)$$

1) Show that the partition function of Dirac fermions in a background gauge field can be expanded in powers of $1/m$ and that this expansion is a sum over closed loops of the lattice. What is the dependence of the contribution of a loop of length L lattice units on the mass m? What is the dependence of a loop of length L on the background gauge fields?
2) Here the gauge fields are dynamical and are in the strong coupling regime. Consider the expectation value of a Wilson loop on a closed contour Γ (a rectangle of $R \times T$ lattice units), $W_\Gamma = \langle \exp(i\sum_\Gamma A_\mu(x))\rangle$ (as in eq. (18.90)). Compute this expectation value to lowest order in the strong coupling expansion in $1/g$ in the limit where the fermion mass is large, $m \gg 1$. Do you find an area or a perimeter law?

19

Instantons and Solitons

In this chapter, we discuss the role of topology in QFT. Topology plays different roles in QFT. One is in identifying classes of configurations of the Euclidean path integral that are topologically inequivalent. Such configurations are called *instantons* and describe *topological excitations*, which play a key role in the nonperturbative definition of the path integral both in theories with global symmetries and in gauge theories, and in the theory of phase transitions. A closely related problem is that of *solitons*, classical configurations (and quantum states) with finite energy and nontrivial topology. Instantons and solitons are classified by a set of topological invariants, known as *topological charges*. When these topological invariants are included in the action of the theory (often as a consequence of quantum anomalies), they change the behavior of the theory in profound ways.

There are many reasons for considering topological excitations. One motivation is the study of mechanisms for quantum disorder, such as the physical origin of phases of a QFT exhibiting confinement in the case of a gauge theory, and/or lack of long-range order in theories with global symmetries. This is also related to the problem of quantum tunneling processes in QFT. In addition, this analysis leads to an understanding of the existence of topological excitations both in QFT and in statistical physics. We begin by defining what is meant by a topological excitation.

19.1 Instantons in quantum mechanics and tunneling

Consider first a very simple problem, the quantum mechanics of a double-well anharmonic oscillator. The potential $U(q)$ for the coordinate q has the form shown in figure 19.1. The Lagrangian is

$$L = \frac{1}{2}\left(\frac{dq}{dt}\right)^2 - U(q) \tag{19.1}$$

where we set the mass $m = 1$. The potential is

$$U(q) = -\frac{\mu^2}{2}q^2 + \frac{\lambda}{4}q^4 \tag{19.2}$$

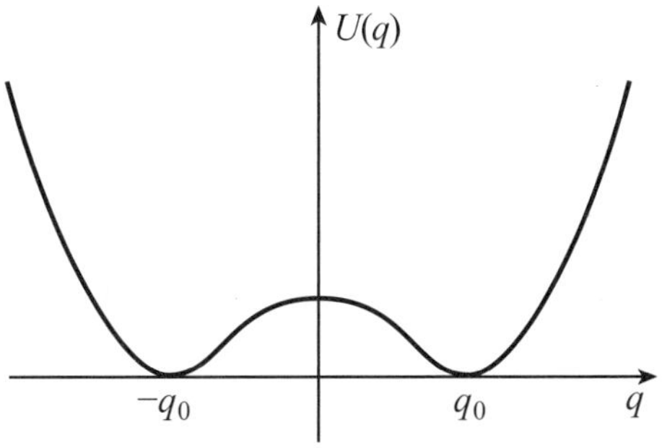

Figure 19.1 The potential $U(q)$ of a double-well anharmonic oscillator.

with λ small. This potential has two classical minima at $\pm q_0 = \pm\sqrt{\frac{\mu^2}{\lambda}}$. At the classical level, the symmetry $q \to -q$ is spontaneously broken. We will see that this symmetry is restored by tunneling processes involving instantons. Here we follow closely the work by Polyakov (1977), and the beautiful lectures by Coleman (1985).

Suppose we want to compute the (imaginary time) time-ordered correlator $C(\tau) = \langle q(0)q(\tau)\rangle$. By expanding in eigenstates, the correlator becomes

$$\langle q(0)q(\tau)\rangle = \sum_n |\langle 0|q|n\rangle|^2 e^{-(E_n - E_0)\tau} \tag{19.3}$$

At long imaginary time, $\tau \to \infty$, due to tunneling processes, the correlator decays exponentially

$$C(\tau) \sim e^{-\tau \Delta E} \tag{19.4}$$

where ΔE is the splitting of the two low-lying states, the symmetric and antisymmetric combination of the states on either well. As usual, the imaginary time correlator is computed by the expression

$$\langle q(0)q(\tau)\rangle = \frac{1}{Z}\int \mathcal{D}q(\tau)\, q(0)q(\tau)\, e^{-\mathcal{E}[q]} \tag{19.5}$$

where $\mathcal{E}[q]$ is the action in imaginary time

$$\mathcal{E}[q] = \int_{-\infty}^{\infty} d\tau \left[\frac{1}{2}\left(\frac{dq}{d\tau}\right)^2 - \frac{\mu^2}{2}q^2 + \frac{\lambda}{4}q^4\right] \tag{19.6}$$

and Z is the partition function

$$Z = \int \mathcal{D}q(\tau)\, e^{-\mathcal{E}[q]} \tag{19.7}$$

Hence, this problem is equivalent to the computation of the correlation function for a problem in classical statistical mechanics in one dimension, where $\mathcal{E}[q]$ is the classical energy.

We will do this calculation using the semiclassical expansion in imaginary time. Let us first seek the extrema of the Euclidean action $\mathcal{E}[q]$, which satisfy $\delta\mathcal{E} = 0$. They are the

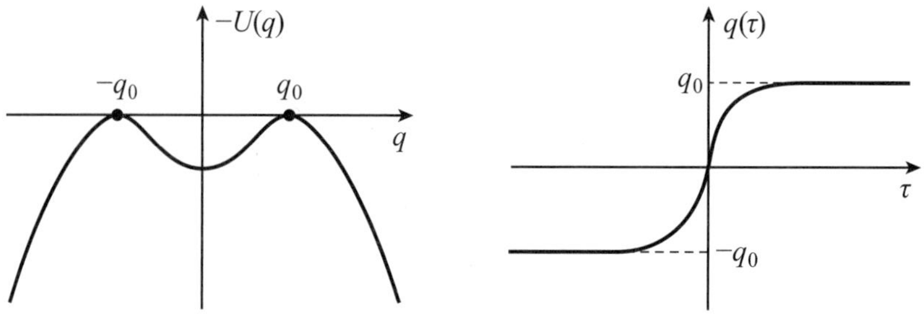

(a) (b)

Figure 19.2 (a) The inverted potential $-U(q)$. (b) A tunneling trajectory.

solutions of the Euler-Lagrange equation

$$\frac{d}{d\tau}\frac{\delta\mathcal{E}}{\delta\dot{q}}=\frac{\delta\mathcal{E}}{\delta q} \tag{19.8}$$

These solutions will be denoted by $\bar{q}(\tau)$. For this system, the Euler-Lagrange equations are

$$\frac{d^2\bar{q}}{d\tau^2}=-\mu^2\bar{q}+\lambda\bar{q}^3 \tag{19.9}$$

which are the same as the classical equations of motion of a dynamical system $q(\tau)$ with the inverted potential, $-U(q)=\frac{\mu^2}{2}q^2-\frac{\lambda}{4}q^4$, shown in figure 19.2a.

This system has a conserved "energy" E

$$E=\frac{1}{2}\left(\frac{d\bar{q}}{d\tau}\right)^2-U(\bar{q}) \tag{19.10}$$

which corresponds to the Lagrangian in real time.

The static solutions of the equation of motion, eq. (19.9), are the classical ground states, $\bar{q}=\pm\sqrt{\mu^2/\lambda}$, for which $\mathcal{E}(\bar{q})=-\frac{\mu^4 T}{4\lambda}$, where T is the imaginary time span, and $T\to\infty$.

In addition, there is also a solution of eq. (19.9) with $E=0$, which obeys

$$\frac{d\bar{q}}{d\tau}=\pm\sqrt{2U[q]} \tag{19.11}$$

The solutions are

$$\bar{q}_c(\tau)=\pm\sqrt{\frac{\mu^2}{\lambda}}\tanh\left(\frac{\mu(\tau-a)}{\sqrt{2}}\right) \tag{19.12}$$

These solutions interpolate between q_0 and $-q_0$ (and vice versa) as τ goes from $-\infty$ to $+\infty$ and represent the tunneling process. One such solution is depicted in figure 19.2b. Their Euclidean action (or energy) is

$$\mathcal{E}[\bar{q}_c(\tau)]-\mathcal{E}[\pm q_0]=2\sqrt{2}\frac{\mu^3}{\lambda} \tag{19.13}$$

which is finite.

Trajectories with finite Euclidean actions are called *instantons*. This trajectory is topological in the sense that it interpolates smoothly between two inequivalent ground states (or vacua) at $\tau \to \pm\infty$, and this trajectory cannot be smoothly deformed to the trivial configurations at $\pm q_0$. In this particular problem, we can either have an instanton (i.e., the trajectory that goes from $-q_0$ to $+q_0$) or an anti-instanton (which executes the reverse trajectory).

The instanton solution has an arbitrary parameter a, and the Euclidean action does not depend on this parameter. Such parameters are called *zero modes* of instantons. This solution has a contribution to the partition function of order $\tau \exp(-2\sqrt{2}\mu^3/\lambda)$. This contribution becomes important at imaginary times long enough so that $\tau > \exp(+2\sqrt{2}\mu^3/\lambda)$.

Let us compute the correlator of eq. (19.5) using the semiclassical expansion. This requires that we evaluate the contribution to the path integral presented in eq. (19.5) at each classical solution. The full result is the sum of their individual contributions, determined separately for both the numerator and the denominator. As we will see, the one-instanton approximation will fail at long times τ, and we will also need to include multi-instanton processes.

Let us compute first the contribution of one instanton, eq. (19.12). Since the instanton is a solution of the Euler-Lagrange equation, the first correction appears at quadratic order in the expansion of the imaginary time action $\mathcal{E}$. Thus, we need to examine the kernel

$$\left.\frac{\delta^2 \mathcal{E}}{\delta q(\tau)\delta q(\tau')}\right|_{\bar{q}_c(\tau)} = \left[-\frac{d^2}{d\tau^2} + (-\mu^2 + 3\lambda \bar{q}_c^2(\tau))\right]\delta(\tau - \tau') \tag{19.14}$$

where $\bar{q}_c(\tau)$ is the instanton solution of eq. (19.12). Let $\{\psi_n(\tau)\}$ be a complete set of eigenstates of this operator,

$$\left[-\frac{d^2}{d\tau^2} + (-\mu^2 + 3\lambda \bar{q}_c^2(\tau))\right]\psi_n(\tau) = \omega_n^2 \psi_n(\tau) \tag{19.15}$$

where $\{\omega_n^2\}$ is the spectrum of eigenvalues. The eigenstates satisfy the boundary conditions $\psi_n(-\infty) = \psi_n(+\infty) = 0$. If the spectrum of eigenvalues of this equation were strictly positive, $\omega_n^2 > 0$, we could then use the eigenstates $\{\psi_n\}$ to parametrize the trajectories $q(\tau)$ and compute the path integral in the semiclassical approximation, as we have done in previous chapters.

However, in this case, this is not possible since, although the spectrum of the fluctuation kernel is nonnegative, its lowest eigenstate has zero eigenvalue (i.e., it is an exact zero mode). The reason is that although the Euclidean action $\mathcal{E}$ is translationally invariant, the instanton solution is not, and it has the parameter a. Since $\mathcal{E}[\bar{q}_c(\tau)]$ does not depend on a, then

$$\frac{\partial \mathcal{E}[\bar{q}_c(\tau)]}{\partial a} = 0 \tag{19.16}$$

Thus,

$$\frac{\delta}{\delta \bar{q}_c(\tau)} \frac{\partial \mathcal{E}[\bar{q}_c(\tau)]}{\partial a} = 0 \tag{19.17}$$

and

$$\frac{\delta^2 \mathcal{E}}{\delta \bar{q}_c(\tau)\delta \bar{q}_c(\tau')} \frac{d\bar{q}_c(\tau)}{da} = 0 \tag{19.18}$$

Therefore, we find that

$$\psi_0(\tau) = \frac{d\bar{q}_c(\tau)}{da} \tag{19.19}$$

is an exact eigenstate of the fluctuation kernel of eq. (19.14) with eigenvalue $\omega_0^2 = 0$. In other words, it is an exact zero mode of the fluctuation kernel, and there is no restoring force along this direction in the space of functions.

The resolution of this problem is to treat the parameter a as a collective coordinate of the solution, and it must be quantized exactly. We have dealt with a similar problem in chapter 9 on the quantization of gauge theory. There, we introduced the Faddeev-Popov gauge-fixing procedure. We will see that it is the way to solve our problem here as well.

The Euclidean action is invariant under translations, $q(\tau) \mapsto q_a(\tau) = q(\tau + a)$, $\mathcal{E}[q] = \mathcal{E}[q_a]$, and the integration measure is also invariant under translations, $\mathcal{D}q = \mathcal{D}q_a$. Let us then follow the Faddeev-Popov procedure and define

$$1 = \int dF\, \delta(F[q_a]) = \int_{-\infty}^{+\infty} da\, \delta(F[q_a])\, \frac{\partial F}{\partial a} \tag{19.20}$$

which we insert into the expression of the partition function to obtain

$$\begin{aligned} Z &= \int \mathcal{D}q \, \exp(-\mathcal{E}[q]) \\ &= \int \mathcal{D}q \int_{-\infty}^{+\infty} da\, \delta(F[q_a])\, \frac{\partial F}{\partial a}\, \exp(-\mathcal{E}[q]) \end{aligned} \tag{19.21}$$

Now perform the change of variables, $q \to q_{-a}$, or $q(\tau) \to q(\tau - a)$, to find

$$Z = \int \mathcal{D}q_{-a} \int_{-\infty}^{+\infty} da\, \delta(F(q))\, D[q_{-a}, a]\, \exp(-\mathcal{E}[q_a]) \tag{19.22}$$

where we defined

$$D[q, a] = \frac{\partial F(a)}{\partial a} \tag{19.23}$$

which plays the role of the Faddeev-Popov determinant.

Using the translation invariance of the measure and of the action, we obtain the result

$$Z = \int \mathcal{D}q \int_{-\infty}^{+\infty} da\, \delta(F(q))\, D[q_{-a}, a] \exp(-\mathcal{E}[q]) \tag{19.24}$$

Let us now compute the Jacobian $D[q_{-a}, a]$ for the specific choice

$$F[q_a] \equiv \int_{-\infty}^{\infty} d\tau\, \frac{\partial \bar{q}_c}{\partial a}\bigg|_{a=0} \left(q(\tau + a) - \bar{q}_c(\tau)\big|_{a=0}\right) \tag{19.25}$$

Then

$$\frac{\partial F}{\partial a}[q_a] = \int_{-\infty}^{\infty} d\tau \left.\frac{\partial \bar{q}_c}{\partial a}\right|_{a=0} \frac{\partial q}{\partial a}(\tau + a) \tag{19.26}$$

Therefore,

$$D \equiv D[q_{-a}, a] = \int_{-\infty}^{\infty} d\tau \left.\frac{\partial \bar{q}_c}{\partial a}\right|_{a=0} \left.\frac{\partial q}{\partial a}(\tau)\right|_{a=0} \tag{19.27}$$

Now parametrize the histories $q(\tau)$ as

$$q(\tau) = \bar{q}_c(\tau, a) + \sum_{n\neq 0} \xi_n \psi_n(\tau - a) \tag{19.28}$$

where $\{\psi_n(\tau)\}$ are the eigenstates of the fluctuation kernel with strictly positive eigenvalues, $\omega_n^2 > 0$, and $\bar{q}_c(\tau, a) = \bar{q}_c(\tau - a)$. Using

$$\frac{\partial \bar{q}_c(\tau, a)}{\partial a} = -\frac{\partial \bar{q}_c(\tau)}{\partial \tau}, \qquad \frac{\partial \psi_n(\tau - a)}{\partial a} = -\frac{\partial \psi_n(\tau)}{\partial \tau} \tag{19.29}$$

we can write

$$\left.\frac{\partial q(\tau)}{\partial a}\right|_{a=0} = -\left[\frac{\partial \bar{q}_c(\tau)}{\partial \tau} + \sum_{n\neq 0} \xi_n \frac{\partial_n(\tau)}{\partial \partial \tau}\right]_{a=0} \tag{19.30}$$

Then the Jacobian D is

$$D = A + \sum_{n\neq 0} \xi_n r_n \tag{19.31}$$

where we defined

$$A = \int_{-\infty}^{\infty} d\tau \left(\frac{\partial \bar{q}_c}{\partial \tau}\right)^2, \qquad r_n = \int_{-\infty}^{\infty} d\tau \left.\frac{\partial \bar{q}_c}{\partial \tau}\right|_{a=0} \left.\frac{\partial \psi_n}{\partial \tau}\right|_{a=0} \tag{19.32}$$

With these results, the integration measure becomes

$$\mathcal{D}q(\tau) = da \left(A + \sum_{n\neq 0} \xi_n r_n\right) \prod_{n\neq 0} d\xi_n \simeq A\, da \prod_{n\neq 0} d\xi_n \tag{19.33}$$

where the approximate form is accurate in the limit $\lambda \ll \mu^3$, with A given in eq. (19.32).

Therefore, to this level of approximation, the one-instanton contribution to the partition function is

$$\begin{aligned} Z_1 = \int \mathcal{D}\tilde{q}(\tau) \int_{-\infty}^{\infty} da\, A\, \exp(-\mathcal{E}[\bar{q}_c]) \\ \times \exp\left(-\frac{1}{2} \int d\tau \int d\tau' \left.\frac{\delta^2 \mathcal{E}}{\delta q(\tau) \delta q(\tau')}\right|_{\bar{q}_c(\tau)} \tilde{q}(\tau)\tilde{q}(\tau')\right) \end{aligned} \tag{19.34}$$

where

$$\tilde{q}(\tau) = q(\tau) - \bar{q}_c(\tau, a) = \sum_{n \neq 0} \xi_n \psi_n(\tau - a) \tag{19.35}$$

which only involves the eigenstates with nonvanishing eigenvalues. By integrating over the nonzero modes, we find that the one-instanton contribution to the partition function is

$$Z_1 = A \left(\int_{-\infty}^{\infty} da \right) \exp(-\mathcal{E}[\bar{q}_c]) \prod_{n \neq 0} \omega_n^{-1} \tag{19.36}$$

while the contribution of the trivial saddle point, $\bar{q} = q_0$, to the partition function is

$$Z_0 = \prod_{n \neq 0} \omega_{n,0}^{-1} \tag{19.37}$$

where $\omega_{n,0}$ are the eigenvalues for the trivial saddle point.

We can similarly compute the contributions to the numerator of the correlation function $\langle q(0)q(\tau)\rangle$, eq. (19.5). Putting it all together, we find that, up to multi-instanton contributions that will be discussed shortly, the correlator is given by

$$\begin{aligned} \langle q(0)q(\tau)\rangle &= \frac{\frac{\mu^2}{\lambda} + A\,K\left(\int_{-\infty}^{\infty} da\, \bar{q}_c(0,a)\bar{q}_c(\tau,a)\right) \exp(-\mathcal{E}[\bar{q}_c(\tau)])}{1 + A\,K\left(\int_{-\infty}^{\infty} da\right) \exp(-\mathcal{E}[\bar{q}_c])} \\ &\simeq \frac{\mu^2}{\lambda} + A \exp(-\mathcal{E}[\bar{q}_c])\, K \int_{-\infty}^{\infty} da \left(\bar{q}_c(0,a)\bar{q}_c(\tau,a) - \frac{\mu^2}{\lambda} \right) \end{aligned} \tag{19.38}$$

where K is the ratio

$$K = \frac{\prod_{n \neq 0} \omega_n^{-1}}{\prod_{n \neq 0} \omega_{n,0}^{-1}} = \frac{\mathrm{Det}'\left[-\frac{d^2}{d\tau^2} + (-\mu^2 + 3\lambda \bar{q}_c^2(\tau))\right]^{-1/2}}{\mathrm{Det}'\left[-\frac{d^2}{d\tau^2} + 4\mu^2\right]^{-1/2}} \tag{19.39}$$

where Det$'$ denotes the determinant without the zero modes. The ratio of determinants K can be calculated using different methods (e.g., see section 5.2.2). In particular, the numerator of K is the determinant for (essentially) the Schrödinger operator for the Pöschl-Teller potential, and the denominator is that for a free particle.

It remains to compute A, given in eq. (19.32), to find

$$A = \mathcal{E}[\bar{q}_c] = 2\sqrt{2}\frac{\mu^3}{\lambda} \tag{19.40}$$

and

$$\int_{-\infty}^{\infty} da \left[\bar{q}_c(0,a)\bar{q}_c(\tau,a) - \frac{\mu^2}{\lambda} \right] = -\frac{2\mu^2}{\lambda} \frac{\tau}{\tanh(\mu\tau/\sqrt{2})} \simeq -\frac{2\mu^2}{\lambda}\tau \tag{19.41}$$

where we used the approximate form for long times, $\tau \gg \sqrt{2}/\mu$.

Therefore, at this level of approximation, the correlator at long times $\tau \gg \sqrt{2}/\mu$ becomes

$$\langle q(0)q(\tau)\rangle = \frac{\mu^2}{\lambda} + KA\exp(-\mathcal{E}[\bar{q}_c])\frac{2\mu^2}{\lambda}\tau \tag{19.42}$$

Thus, we find that, as expected, the instanton contribution is exponentially small. However, as anticipated, this approximation fails at sufficiently long times τ,

$$\tau \gtrsim \exp(\mathcal{E}[\bar{q}_c])/(2KA) \tag{19.43}$$

where $\mathcal{E}[\bar{q}_c] = 2\sqrt{2}\mu^3/\lambda$.

We will now see that the solution of this problem is to include multi-instanton processes, which will amount to showing that the series, whose first two terms are given by eq. (19.42), exponentiates. Note that if λ is small, the failure of the approximation occurs at times much longer than the width of an instanton $\Delta\tau \sim 1/\mu$, so we should be able to regard the sum over multi-instanton processes as a dilute gas of (exponentially) weakly interacting instantons (figure 19.3). In this limit, we can denote a multi-instanton process by

$$\bar{q}_c(\tau) \simeq \sqrt{\frac{\mu^2}{\lambda}}\prod_{j=1}^{N}\text{sign}(\tau - a_j) \tag{19.44}$$

where $\{a_j\}$ are the locations of the instantons (and anti-instantons). The Euclidean action of an N-instanton configuration is

$$\mathcal{E}_c^N \simeq N\frac{2\sqrt{2}\mu^3}{\lambda} \tag{19.45}$$

plus exponentially weak interaction terms.

In this limit, the problem becomes essentially identical to the low-temperature expansion of the classical one-dimensional Ising model. Indeed, we can picture the instantons as the domain walls of the Ising model. In this interpretation, the restoration of the $\mathbb{Z}_2$ symmetry, $q \to -q$, is just the Landau-Peierls proof of the absence of spontaneous symmetry breaking in one-dimensional classical statistical mechanics in systems with a discrete global symmetry (see Landau and Lifshitz, 1959b).

We can now do this calculation explicitly:

$$\langle q(0)q(\tau)\rangle = \frac{1}{Z}\frac{\mu^2}{\lambda}\sum_{N=0}^{\infty} C(\tau)^N \exp(-2\sqrt{2}\mu^3/\lambda) \times \int_{a_1<a_2<\cdots<a_N} da_1\cdots da_N \prod_{j=1}^{N}\text{sign}(\tau - a_j) \tag{19.46}$$

where

$$Z = \sum_{N=0}^{\infty} C(\tau)^N \exp(-2\sqrt{2}\mu^3/\lambda)\int_{a_1<a_2<\cdots<a_N} da_1\cdots da_N \tag{19.47}$$

and

$$C(\tau) = \frac{2AK\tau}{\tanh(\mu\tau/\sqrt{2})} \simeq 2AK\tau \tag{19.48}$$

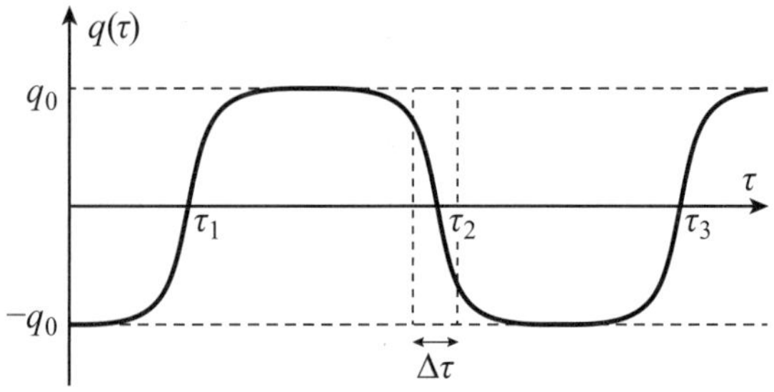

Figure 19.3 A multi-instanton process as a sequence of instantons and anti-instantons. Here $\tau_1, \tau_2, \tau_3, \ldots$, are the locations of an infinite sequence of instantons and anti-instantons in imaginary time τ. Here $\Delta\tau \sim 1/\mu$ is the width of the instanton.

and we have used the behavior at $\tau \gg 1/\mu$. Also,

$$\int_{a_1<a_2<\cdots<a_N} da_1 \cdots da_N = \frac{T^N}{N!} \tag{19.49}$$

where $T \to \infty$ is the total span in imaginary time. On doing the sums, we find

$$\langle q(0)q(\tau)\rangle = \frac{\mu^2}{\lambda} \exp(-(\Delta E)\tau) \tag{19.50}$$

where

$$\Delta E \simeq 2AK \exp(-2\sqrt{2}\mu^3/\lambda) \tag{19.51}$$

Thus, the ground state and the first excited state of the double well are split by an amount ΔE. Notice that eq. (19.51) has an essential singularity in λ, showing its nonperturbative character.

This rather elaborate presentation is the prototype of an instanton calculation. It is a sum of a large number of contributions, each being exponentially small and having a weight with an essential singularity in the coupling constant, λ, in the case at hand. This approach will succeed in theories for which the Euclidean action of the instanton is finite and the instanton solution is labeled only by the location of the instanton. We will see below that we can relax the finite-action condition up to a logarithmic divergence. However, in many cases of interest, such as theories that are classically scale invariant, the instanton has zero modes associated with its scale, which will complicate the analysis.

19.2 Solitons in (1+1)-dimensional ϕ^4 theory

We can similarly look at ϕ^4 theory in (1+1)-dimensional Minkowski spacetime. The Lagrangian density is

$$\mathcal{L} = \frac{1}{2}(\partial_t\phi)^2 - \frac{1}{2}(\partial_x\phi)^2 - U(\phi) \tag{19.52}$$

whose classical Hamiltonian density is

$$\mathcal{H} = \frac{1}{2}\Pi^2 + \frac{1}{2}(\partial_x\phi)^2 + U(\phi) \tag{19.53}$$

We can now consider finite-energy static classical configurations that extremize the Hamiltonian $\mathcal{H}$. These extremal static configurations obey the same equations in the coordinate x as the instanton solutions of the double-well problem in imaginary time, which we just discussed. These finite-energy classical solutions are known as *solitons*. In this theory, the soliton solutions obey *boundary conditions* corresponding to distinct uniform classical solutions. In other words, the theory has a global $\mathbb{Z}_2$ symmetry, and the soliton configuration interpolates between the two classically degenerate states at $\pm\phi_0$. They can be viewed as a domain wall in one space dimension.

Since we are interested in the broken-symmetry state, it will be convenient to use the potential

$$U(\phi) = \frac{\lambda}{4!}(\phi_0^2 - \phi(x)^2)^2 \tag{19.54}$$

for which $m_0^2 = -\lambda\phi_0^2/12 < 0$. Its global minima are at $\pm\phi_0$, and their energy is zero. The finite-energy solution $\phi(x)$ that extremizes the Hamiltonian, and interpolates between the two global minima of the energy, obeys the same equation in the x coordinate as the instanton of the double-well potential in imaginary time τ. The classical soliton solution is

$$\phi(x) = \pm\phi_0 \tanh\left(\frac{(x-x_0)}{\sqrt{2}\xi}\right) \tag{19.55}$$

where $\xi = \left(\frac{\lambda\phi_0^2}{6}\right)^{-1/2} = |m_0|^{-1}$ is the correlation length. The energy of the classical soliton is

$$E_{\text{soliton}} = \frac{2}{3\sqrt{3}}\phi_0^3\sqrt{\lambda} \tag{19.56}$$

Furthermore, it is straightforward to show that there is also a boosted soliton solution of the form $\phi(x \pm vt)$, representing a soliton that moves at speed $v < 1$ (in units of the speed of light, where we have set $c = 1$). Thus, these classical soliton solutions can be regarded as particles that have finite energy.

It is possible to promote these classical soliton solutions to a quantum state using semiclassical quantization. While in quantum mechanics, the semiclassical approximation is WKB, the equivalent (in a broad sense) in QFT is substantially more subtle and technical. This approach is not only viable but also yields the exact answer in a class of special theories that have an infinite number of conservation laws, known as quantum-integrable theories. All known theories of this type are in 1+1 spacetime dimensions. In higher dimensions, the only known theories that are exact at the semiclassical level involve supersymmetry. We will not pursue supersymmetric theories.

It is easy to see that the soliton of ϕ^4 theory is the analog of the kink that we found in the context of the quantum one-dimensional Ising model in chapter 14. Such kinks (or soliton) solutions are domain walls separating regions with distinct broken symmetry states. Solutions of this type exist in the spontaneously broken phase of any theory in one space dimension with a discrete global symmetry group. Section 14.6 introduced a quantum operator that created kinks (see eq. (14.52)).

In the next sections, we will generalize the concepts of instantons and solitons to other theories.

19.3 Vortices

Let us now consider theories with global continuous symmetries. For simplicity, we first discuss the case in which the group is $U(1)$. In this case, the matter field is a complex scalar field $\phi(x) \in \mathbb{C}$. We will be interested in the phase in which the global $U(1)$ symmetry is spontaneously broken at the classical level. Deep in this phase, we can approximate the complex scalar field by a complex field of fixed modulus (i.e., a $U(1)$ nonlinear sigma model), and write

$$\phi(x) = \phi_0 \, e^{i\theta(x)} \tag{19.57}$$

where ϕ_0 is the vacuum expectation value in the broken symmetry phase, and $\theta(x) \in [0, 2\pi)$ is a (compact) phase field, the Goldstone boson of the spontaneously broken $U(1)$ symmetry.

The configurations of systems with compact symmetry groups (global or local) are classified by homotopy groups. This subject can be quite formal and heavily mathematical. We will work at a more physical and intuitive level, at the price of mathematical rigor. There are many excellent reviews and textbooks on the subject, such as the book by Nash and Sen (1983), the book by Coleman (1985), and Polyakov's (1987) book. There are two outstanding reviews, one by Mermin (1979), which focuses on topological defects in ordered matter, and the other by Eguchi, Gilkey, and Hanson (Eguchi et al., 1980), geared mostly toward gauge theory.

19.3.1 Topology in $d = 1$ dimension

Let us consider first the case of a $U(1)$ soliton in $D = 1 + 1$ spacetime dimensions. We will work with periodic boundary conditions in space. Hence, we have effectively compactified (wrapped) the spatial line coordinate onto a circumference of radius $R = L/(2\pi)$, where L is the total length. Thus, we are assuming that the spacetime manifold is $S^1 \times \mathbb{R}$, where the circle S^1 is space and $\mathbb{R}$ is time. The analysis done here also applies to a quantum system (i.e., a field theory in 0 + 1 dimensions, with a $U(1)$ symmetry, such as a planar rigid rotor). In this context, the tunneling process from state $|0\rangle$ to state $|2\pi\rangle$ is described by an instanton.

As in the ϕ^4 soliton, we seek a finite-energy static solution of the classical equations of motion. A classical static configuration of the field $\phi(x)$ is a mapping of every point of the spatial manifold, which we refer to as the *base space*, to the possible values of the phase θ, the *target space*. Since the phase $\theta(x)$ is defined mod 2π, the target space is topologically isomorphic to a circumference S^1. Therefore, in this case, the static classical configurations are maps of the S^1 base space onto the S^1 target space (as shown in figure 19.4):

$$\phi : S^1 \longrightarrow S^1 \tag{19.58}$$

Since the phases $\theta(x)$ are additive (mod 2π), smooth configurations can also be added. Hence, under addition (composition), smooth mappings of S^1 onto S^1 form a group. An example is the constant field configuration, which is a mapping of the entire base space $x \in S^1$ to one particular point of the target space S^1, labeled by the value θ_0 of the phase of

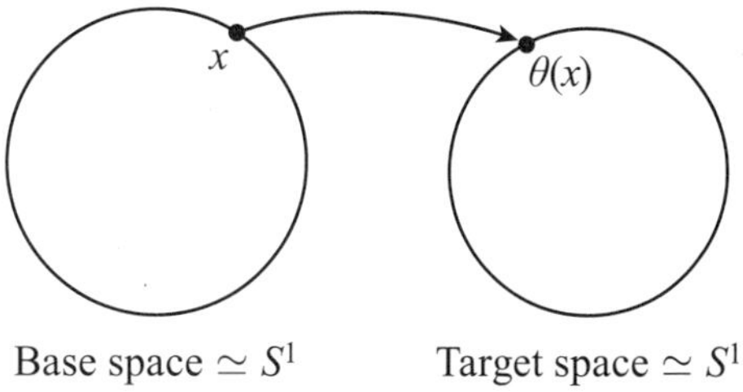

Figure 19.4 A configuration of a complex field as a mapping.

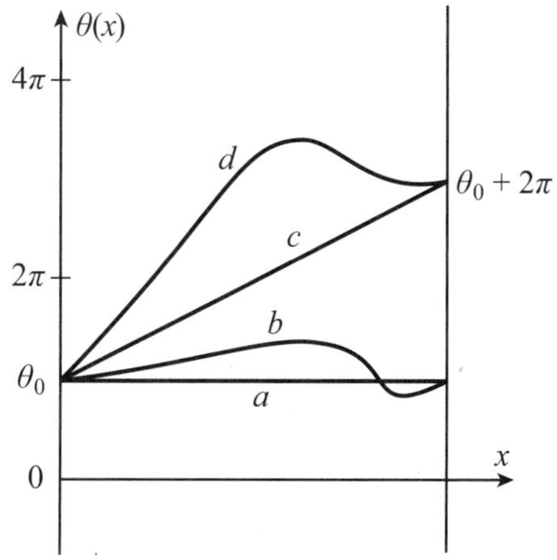

Figure 19.5 Homotopies of $S^1 \to S^1$. (a) A constant configuration, (b) a configuration that is homotopic to a, (c) a configuration with winding number $+1$, and (d) a configuration homotopic to c.

the field. In contrast, many configurations can be obtained by a smooth deformation of the constant field configuration. In terms of the mapping, these configurations can be regarded as closed "curves" on the S^1 target space that begin and end at the same value θ_0 of the phase. Such configurations can be trivially deformed back onto the constant field configuration, θ_0. We will say that such configurations are "contractible" to one another.

In topology, smooth mappings are called *homotopies*. If two configurations (mappings) can be deformed smoothly into each other, we say that they are *homotopic* to each other. In other words, configurations (mappings) that are homotopic to each other are, in a topological sense, equivalent. Therefore, the operation of smooth deformations of mappings defines an equivalence relation between mappings, and we say that mappings that can be deformed smoothly into each other belong to the same *equivalence class*. For example, in figure 19.5, four configurations, labeled a, b, c, and d, are shown. Configurations a and b can be smoothly deformed into each other and, hence, are homotopic to each other. So are configurations c and d. However, configurations a and b are not homotopic to configurations c and d, since a and b are periodic on $[0, L)$, but c and d jump by 2π at L, which is invisible for the complex field ϕ.

The existence of an equivalence relation implies that the mappings can be classified. The question now is: How many distinct equivalence classes are there of mappings of S^1 base onto S^1 target? For any smooth mapping $\theta(x)$, we can define an integer-valued quantity known as the *winding number*, defined as the total change of the phase field across S^1 base (in units of 2π):

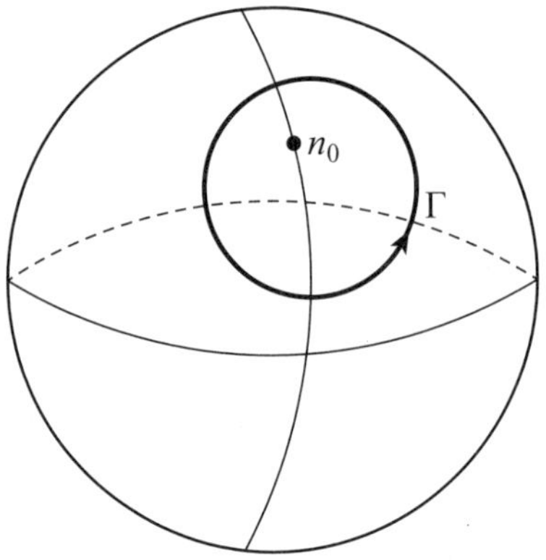

Figure 19.6 The homotopies of S^1 into S^2 are topologically trivial.

$$N = \frac{(\Delta\theta)_L}{2\pi} = \frac{1}{2\pi}\int_0^L dx\, \partial_x \theta(x) = \frac{1}{2\pi}\int_0^L dx\, e^{i\theta(x)} i\partial_x e^{-i\theta(x)} \tag{19.59}$$

where we have used the notation $(\Delta\theta)_L = \theta(L) - \theta(0)$. Since the configurations are required to obey periodic boundary conditions mod 2π on S^1 (i.e., $\theta(x+L) = \theta(x) + 2\pi k$, with $k \in \mathbb{Z}$), it follows that the winding number N is an integer.

Under any smooth deformation of a configuration $\theta(x)$, the value of the winding number cannot be changed continuously, since it is an integer. In this sense, the winding number is a *topological invariant*, which classifies the configurations into classes. Thus, the homotopy classes of mappings of S^1 to S^1 are isomorphic to the integers. This statement means that the homotopy classes form a group, known as the *homotopy group*, which in this case is denoted by $\pi_1(S^1)$. Here the subindex denotes that the base is S^1, and the argument is the target. Let us represent this statement by the equation

$$\pi_1(S^1) \cong \mathbb{Z} \tag{19.60}$$

which states that the equivalence classes are isomorphic to $\mathbb{Z}$.

It is now natural to ask whether (and when) an extension of this analysis always holds for other continuous symmetries and other dimensions. The general answer to this question is no. To see this, let us consider a theory in 1+1 spacetime dimensions, with a classically spontaneously broken $O(3)$ global symmetry, such as the nonlinear sigma model, and ask whether it has classical static soliton solutions.

Since the target space of the $O(3)$ nonlinear sigma model is the two-sphere, S^2, the configurations are now mappings of the S^1 base space to the S^2 target space. Therefore, we can represent all configurations as a closed curve Γ on the two-sphere S^2, shown in figure 19.6. However, all closed smooth curves on S^2 are contractible, meaning that they can be smoothly deformed to a point, such as the arbitrary point $\boldsymbol{n}_0$ shown in the figure. Consequently, all configurations are topologically trivial. The same fact holds for mappings of S^1 to the n-sphere, S^n (with $n > 1$), which are all topologically trivial. We express this statement by saying that their homotopy groups are trivial:

$$\pi_1(S^n) = 0 \tag{19.61}$$

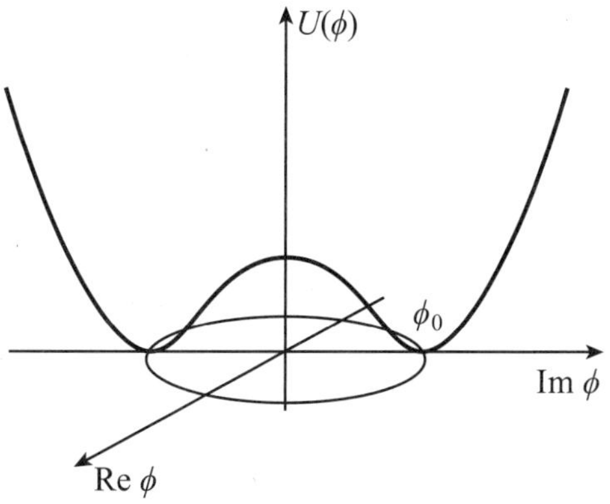

Figure 19.7 The potential $U(\phi)$ in the broken symmetry phase.

Therefore, we conclude that (1+1)-dimensional $O(n)$ nonlinear sigma models, with $n > 1$, do not have topologically nontrivial classical static soliton solutions.

19.3.2 Vortices in $D = 2$ dimensions

The preceding analysis does not imply that the topology is necessarily trivial in $D > 1$ dimensions. Far from it! To see how nontrivial topology arises in $D > 1$, let us consider the case of a complex field $\phi(x)$ with a spontaneously broken global $U(1)$ symmetry in $D = 2$ dimensions. Again, $D = 2$ here will be either the dimension of a Euclidean spacetime (in which case the configurations with nontrivial topology are instantons) or the space dimension of a (2+1)-dimensional spacetime (where the configurations are solitons).

19.3.3 The complex scalar field

For definiteness, consider a theory whose Euclidean action in $D = 2$ spacetime dimensions (or energy in $d = 2$ space dimensions) is

$$S = \int d^2x \left(\frac{1}{2} \left|\partial_\mu \phi\right|^2 + U(\phi) \right) \tag{19.62}$$

where the potential $U(\phi)$ has the form shown in figure 19.7 (e.g., $U(\phi) = u(|\phi|^2 - \phi_0^2)^2$, with $u > 0$). It is invariant under the global $U(1)$ symmetry, $\phi(x) \to \exp(i\varphi)\phi(x)$, and has a minimum at $|\phi(x)| = \phi_0$.

Parametrize the complex field in terms of two real fields, the amplitude $\rho(x)$ and the phase $\theta(x)$, such that

$$\phi(x) = \rho(x) \exp(i\theta(x)) \tag{19.63}$$

Since the $U(1)$ symmetry is broken, assume that, at large distances, the amplitude $\rho(x)$ approaches the limit

$$\lim_{|x| \to \infty} \rho(x) = \phi_0 \tag{19.64}$$

where, as before, ϕ_0 is the vacuum expectation value in the broken symmetry phase.

Let us now consider a circumference $C(R)$ of radius R, large enough so that on it, we can approximate $\rho(x) \simeq \phi_0$ to any desired accuracy. While the amplitude $\rho(x)$ is essentially fixed to the value ϕ_0 for large enough R, the phase $\theta(x)$ can change on $C(R)$. In this limit, the

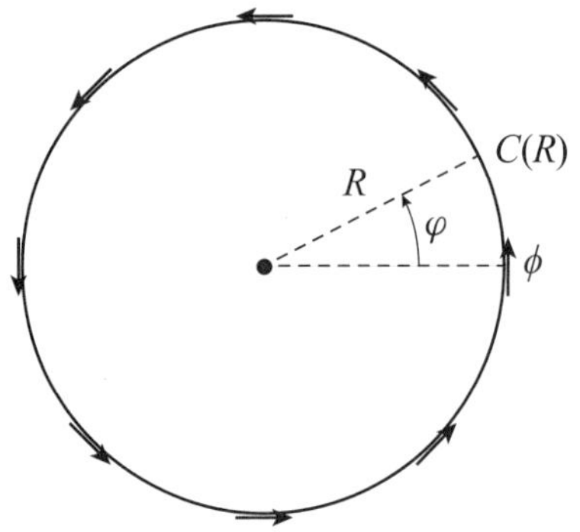

Figure 19.8 A vortex in $D=2$ dimensions. Here R is the radius of the large circumference $C(R)$. The phase θ of the complex field $\phi(x)$ winds as the vector $\boldsymbol{\phi}=(\text{Re }\phi, \text{Im }\phi)$ rotates around the circumference by the azimuthal angle φ.

complex field takes the asymptotic form $\phi(x)\mapsto \phi_0 \exp(i\theta(x))$, where $x\in C(R)$. Therefore, the value of the field $\phi(x)$ on the circumference $C(R)$ defines a map of the S^1 base space (the points on the circumference $C(R)$ labeled by the azimuthal angle $\varphi\in[0,2\pi)$), to the S^1 target space (the values of the phase θ of the field ϕ). From our previous analysis, we see that these mappings can also be classified by the homotopy group $\pi_1(S_1)\cong\mathbb{Z}$, with each class being classified by a winding number called the *vorticity*,

$$n=\frac{(\Delta\theta)_C}{2\pi}=\int_0^{2\pi}\frac{d\varphi}{2\pi}e^{i\theta(\varphi)}i\partial_\varphi e^{-i\theta(\varphi)}. \tag{19.65}$$

where $(\Delta\theta)_C$ is the total change of the phase of the field ϕ on one turn of the circumference $C(R)$.

The configuration defined by this topology is called a *vortex*, schematically shown in figure 19.8. Clearly, if the configuration $\phi(x)$ has a nonvanishing vorticity N_v, the field itself must vanish somewhere in the interior of the circle whose boundary is $C(R)$, say, at the center, since otherwise this configuration will have a singularity. In addition, the amplitude $\rho(r)$ must vanish fast enough as $r\to 0$ so that the gradient terms in the action are small enough that the Euclidean action either does not diverge or diverges, at most, logarithmically.

To see how this works, assume that the potential $U(\phi)$ is sufficiently steep that the amplitude $\rho\simeq\phi_0$ essentially for all values of x except inside a small core of radius a. In that core, it vanishes fast enough that there is no short-distance singularity, at the price of a finite contribution to the action cut off by the size of the core. Thus, for all values of points outside the core, $|x|>a$, we can set $\phi(x)=\phi_0\exp(i\theta(x))$.

What is the action of a configuration of vortices of this type? Outside the core of the vortex, the gradient of the field simply becomes

$$\partial_\mu\phi(x)\simeq\phi_0 e^{i\theta(x)}\, i\partial_\mu\theta(x)\Rightarrow\left|\partial_\mu\phi\right|^2\simeq\phi_0^2(\partial_\mu\theta(x))^2 \tag{19.66}$$

and the Euclidean action of the configuration is

$$S=\frac{\phi_0^2}{2}\int d^2x\left(\partial_\mu\theta(x)\right)^2 \tag{19.67}$$

A configuration of the field $\phi(x)$ has a local current density j_μ, which, in the approximations that we are using, becomes $j_\mu = \phi_0 \partial_\mu \theta(x)$. By analogy with a fluid, define the *vorticity* $\omega(x)$ as the curl of the current,

$$\omega(x) = \epsilon_{\mu\nu} \partial_\mu j_\nu(x) = \phi_0 \epsilon_{\mu\nu} \partial_\mu \partial_\nu \theta(x) \tag{19.68}$$

which vanishes everywhere except at the locations x_j of the vortices (i.e., the singularities of the phase field $\theta(x)$). Let us assume that we have vortices at a set of points $\{x_j\}$ with vorticities $\{n_j\}$, so that the local vorticity is

$$\omega(x) = \sum_j n_j \delta^2(x - x_j) \tag{19.69}$$

where the phase field $\theta(x)$ satisfies eq. (19.68). We will see now that $\theta(x)$ is given by the distribution of vortices as

$$\theta(x) = \sum_j n_j \text{Im} \ln(z - z_j) \tag{19.70}$$

where $z = x_1 + ix_2$ are complex coordinates of the plane. Hence, the phase field is multi-valued and has a branch cut ending at each vortex.

To see how this comes about, define $\vartheta(x)$ as the (Cauchy-Riemann) dual of the phase field $\theta(x)$,

$$\partial_\mu \vartheta(x) = \epsilon_{\mu\nu} \partial_\nu \theta(x) \tag{19.71}$$

which satisfies the Poisson equation

$$-\partial^2 \vartheta(x) = \omega(x) \tag{19.72}$$

In terms of the field $\vartheta(x)$, the action takes the two-dimensional electrostatic form:

$$S = \frac{\phi_0^2}{2} \int d^2x \, (\partial_\mu \vartheta)^2 = -\frac{\phi_0^2}{2} \int d^2x \, \vartheta \, \partial^2 \vartheta = \frac{\phi_0^2}{2} \int d^2x \, \omega(x) \vartheta(x) \tag{19.73}$$

Solving Poisson's equation, eq. (19.72),

$$\vartheta(x) = \int d^2y \, G(x - y) \omega(y) \tag{19.74}$$

in terms of the two-dimensional Green function $G(x - y)$, which satisfies

$$-\partial^2 G(x - y) = \delta^2(x - y) \tag{19.75}$$

whose solution is

$$G(x - y) = \int \frac{d^D p}{(2\pi)^D} \frac{e^{ip\cdot(x-y)}}{p^2} = \frac{\Gamma\left(\frac{D}{2} - 1\right)}{4\pi^D |x - y|^{D-2}} \tag{19.76}$$

Defining

$$G(0) = \lim_{a \to 0} G(a) \tag{19.77}$$

in terms of which, as $D \to 2$, we obtain

$$G(|x-y|) - G(a) = \frac{1}{2\pi} \ln\left(\frac{a}{|x-y|}\right) \tag{19.78}$$

which diverges (logarithmically) as $|x-y| \to \infty$. However, in $D=2$, we find that $G(a) = \frac{1}{2\pi}$ $\ln(L/a)$, where L is the linear size of the system. Hence $G(a)$ diverges logarithmically in the UV ($a \to 0$) and in the IR ($L \to \infty$).

Then we can write

$$\begin{aligned} S &= \frac{\phi_0^2}{2} \int d^2x\, \omega(x) \vartheta(x) \\ &= \frac{\phi_0^2}{2} \int d^2x \int d^2y\, \omega(x)\, G(x-y)\, \omega(y) \\ &= \frac{\phi_0^2}{2} \sum_{i,j} n_i n_j G(x_i - x_j) \\ &= \frac{\phi_0^2}{2} \left(\sum_j n_j\right)^2 G(0) + \phi_0^2 \sum_{i>j} n_i n_j \left[G(x_i - x_j) - G(0)\right] \end{aligned} \tag{19.79}$$

Since $G(0)$ is divergent, the action will diverge unless the total vorticity of the configuration vanishes: $\sum_j n_j = 0$.

Thus, the action of a collection of vortices (with zero total vorticity) is

$$S[n] = \frac{\phi_0^2}{2\pi} \sum_{i>j} n_i n_j \ln\left(\frac{a}{|x_i - x_j|}\right) \tag{19.80}$$

In particular, for a vortex-antivortex pair, with $n_1 = -n_2 = 1$, separated at a distance $R \gg a$, the action is $S[1,-1;R] = \left(\frac{\phi_0^2}{2\pi}\right) \ln(R/a)$. Therefore, the Euclidean action of a set of vortices is the same as the energy of a set of classical electrical charges with a logarithmic interaction.

In effect, we are rewriting the full partition function as

$$Z \propto \sum_{\{n_j\}} \prod_j \int d^2x_j\, \delta\left(\sum_j n_j\right) \exp(-S[n]) \tag{19.81}$$

where x_j are the coordinates of vortices and antivortices. (Here we have neglected an uninteresting prefactor.) In other words, we have mapped the problem to the thermodynamics of a neutral two-dimensional Coulomb gas at temperature $T = \pi/\phi_0^2$.

Hence, we see that individual vortices have a logarithmically divergent action $\frac{1}{2\pi}\ln(L/a)$. Although this is a violation of our criterion that instantons must have finite action, they play a key role in this theory. We will show below that, precisely due to the logarithmic form of the interaction, the partition function of the theory can be recast as a partition function of vortices and antivortices, which exhibit a phase transition.

To see how this works, let us estimate the free energy cost of a vortex. The free energy is $F = U - TS$, where U is the energy, S is the entropy, and $T = \pi/\phi_0^2$. An argument due to Kosterlitz and Thouless (1973) estimates the contribution to the free energy of a single vortex with $n=1$ to be

$$F_{\text{vortex}} = \frac{1}{2\pi}\ln(L/a) - T\ln(L/a)^2 \tag{19.82}$$

where L is the linear size of the system. The first term is the logarithmically divergent self-energy of a single vortex (cut off in the IR by L). The second term is the entropy of a vortex, which counts the logarithm of the number of places where the vortex can be located. For ϕ_0^2 large enough, the energy wins over the entropy, $F_{\text{vortex}} > 0$, and free vortices are suppressed. However, vortex-antivortex pairs (i.e., dipoles with vanishing total vorticity) have a finite energy and logarithmic entropy. Thus, in this regime, there will be a finite density of such dipoles.

But for T large enough (or what is the same, ϕ_0^2 small enough), the entropy will win over the energy, $F_{\text{vortex}} < 0$, and the system becomes unstable against the proliferation of vortices (and antivortices). Hence, there should be a phase transition between a regime in which vortices and antivortices can only occur in bound pairs, to another state in which vortices and antivortices unbind and proliferate, leading to a state best described as a neutral plasma. This simple argument predicts that the critical temperature is $T_c = \frac{\pi}{2}$ (in units of $1/\phi_0^2$). In section 19.9, we derive a more precise version of this Kosterlitz-Thouless transition.

19.3.4 The abelian-Higgs model

Now consider the theory of a complex scalar field minimally coupled to a gauge field with a Maxwell action. In this theory, the abelian-Higgs model (or equivalently, a superconductor coupled to a gauge field), the symmetry is local. This will result in a finite action for the vortex. To this end, consider the Euclidean Lagrangian

$$\mathcal{L} = \frac{1}{2}\left|D_\mu\phi\right|^2 + u(\phi_0^2 - |\phi|^2)^2 + \frac{1}{4}F_{\mu\nu}^2 \tag{19.83}$$

where u is the coupling constant, $D_\mu = \partial_\mu - ieA_\mu$ is the covariant derivative (with $\hbar = c = 1$), and $j_\mu = i(\phi^* D_\mu\phi - (D_\mu\phi)^*\phi)$ is the current.

We seek a vortex solution of the same form as before: $\phi \to \phi_0\exp(i\theta)$. However, now for the action to be finite, we require that $|D_\mu\phi| \to 0$ as $r \to \infty$. In this limit, the covariant derivative becomes

$$D_\mu\phi \to i(\partial_\mu\theta + eA_\mu)\phi_0\exp(i\theta) \tag{19.84}$$

Therefore, as $r \to \infty$,

$$|D_\mu\phi|^2 \to 0 \Leftrightarrow \partial_\mu\theta - eA_\mu \to 0 \tag{19.85}$$

and the gauge field becomes, asymptotically, a pure gauge. Nevertheless, on an arbitrary closed contour Γ that encloses a vortex, the phase of the matter field has a winding number (the vorticity)

$$n = \frac{(\Delta\theta)_\Gamma}{2\pi} = \frac{e}{2\pi}\oint_\Gamma dx_\mu A_\mu(x) \tag{19.86}$$

However, using Stokes' theorem, we have

$$\oint_\Gamma dx_\mu A_\mu(x) = \int_\Sigma dS_\mu B_\mu = \Phi \tag{19.87}$$

where Σ is the region of the plane with boundary Γ, $B(x)$ is the magnetic field, and Φ is the magnetic flux through Σ. Therefore, there is a relation between the vorticity and the flux

$$n = \frac{\Phi}{\phi_0} \tag{19.88}$$

where, upon restoring conventional units, $\phi_0 = 2\pi\hbar c/e$ is the flux quantum (not to be confused with the asymptotic value of the field!), and e is the electric charge. A configuration that obeys this flux-quantization condition is known as an *Abrikosov vortex* (Abrikosov, 1957), which is also known as the Nielsen-Olesen vortex (Nielsen and Olesen, 1973).

Since the gauge field becomes a pure gauge at infinity, the magnetic field must vanish at infinity: $B(x) \to 0$ as $r \to \infty$. Similarly, the complex scalar field must vanish as $r \to 0$, and $B(x)$ remains finite as $r \to 0$.

The vortex solutions (for topological charge n) of the Euclidean equations of motion obey the asymptotic conditions

$$\lim_{r\to\infty} \phi(r,\varphi) = \phi_1, \qquad \lim_{r\to\infty} A_i(r,\varphi) = n\partial_i\varphi \tag{19.89}$$

and

$$\epsilon_{ij}F_{ij} = 4\pi n\delta^2(\boldsymbol{x}), \qquad \frac{1}{4\pi}\oint dx_i A_i = n \in \mathbb{Z} \tag{19.90}$$

Here and below, (r,φ) are the polar coordinates of the plane. The solutions can be found using the ansatz

$$\phi(x_1,x_2) = f(r)g_n(\varphi), \qquad A_i = -ia(r)g_n^{-1}(\varphi)\partial_i g_n(\varphi) \tag{19.91}$$

and obey the boundary conditions

$$f(0) = a(0) = 0, \qquad \lim_{r\to\infty} f(r) = \phi_0, \ \lim_{r\to\infty} a(r) = 1 \tag{19.92}$$

These equations do not have any explicit analytic solutions, and in general are solved numerically. The asymptotic behavior at large distances, $r \to \infty$, of the vortex solutions is

$$a(r) = -\frac{1}{e} + m_v r K_1(m_v r), \qquad f(r) = \phi_0 + O(\exp(-m_\phi r)) \tag{19.93}$$

where $K_1(z)$ is a Bessel function, and the masses m_v for the gauge field and m_ϕ for the complex scalar field are, respectively,

$$m_v = e\phi_0, \qquad m_\phi = 2\sqrt{2u}\phi_0 \tag{19.94}$$

The two scales, m_ϕ and m_v, are, respectively, the inverses of the correlation length $\xi \sim M_\phi^{-1}$ of the scalar field and of the penetration depth $\lambda \sim m_v^{-1}$ (here we are using the terminology of superconductivity) of the gauge field; see figure 19.9. It is well known from the theory of superconductivity that if $\xi > \lambda$ (a type-I superconductor), vortices with a topological charge of the *same sign* have an attractive interaction, and conversely, in the type-II regime, $\xi < \lambda$, they repel each other. The case of a neutral complex scalar field (i.e., without a dynamical gauge field) corresponds to the limit $\lambda \to \infty$, where the vortices repel

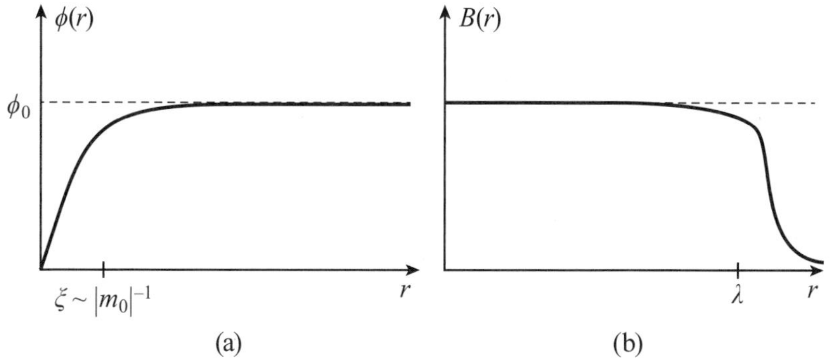

Figure 19.9 Schematic behavior of the vortex solution. (a) Configuration of the amplitude of the complex scalar field, and (b) of the magnetic field $B(x) = \epsilon_{\mu\nu}\partial_\mu A_\nu(x)$. Here $\xi = m_\phi^{-1}$ is the correlation length, and $\lambda = m_v^{-1}$ is the penetration depth.

each other, as was shown in section 19.3.3. It can be shown that the Euclidean action for a vortex solution with topological charge n obeys the Bogomol'nyi bound: $S_n \geq \pi |n| \phi_0^2$.

At the special value $m_v = m_\phi$ (or equivalently, $e^2 = 8u$), the crossover point between type I and type II, the vortices do not interact with one another, at least classically. At this special point (the Bogomol'nyi point), the vortex equations can be reduced to two first-order partial differential equations, which have a similar structure to the instanton self-dual equations that we discuss in section 19.4. In the field-theory literature, these equations are known as the BPS equations, for Bogomol'nyi, Prasad, and Sommerfield (Bogomol'nyi, 1976; Prasad and Sommerfield, 1975). The BPS solutions saturate the Bogomol'nyi bound, $S_n^{BPS} = \pi |n| \phi_0^2$.

The effects of instantons in the two-dimensional abelian Higgs model have been studied in detail (Callan et al., 1976) at the level of the effects of semiclassical fluctuations, including the computation of the fluctuation determinants for vortex solutions (Schaposnik, 1978). The result is that the the partition function of the model can be expressed in terms of the partition function of a dilute gas of vortices and antivortices, of the form

$$Z = \sum_{\{n_j\}} \frac{z^N}{N!} \prod_j \int d^2x_j \exp\left(-\frac{2\pi m_v^2}{e^2} \sum_{i>j} n_i n_j K_0(m_v |\boldsymbol{x}_i - \boldsymbol{x}_j|) \right) \tag{19.95}$$

where $N = \sum_j n_j$ is the total vorticity, z is a fugacity that accounts for effects of short-distance singularities, and $K_0(m_v|x|)$ is a Bessel function.

This expression differs from the one we found in section 19.3.3 in the case of the complex scalar field in several important ways. One difference is that the interaction is proportional to the Bessel function $K_0(m_v|x|)$. This interaction decays exponentially fast at distances longer than the penetration depth $\lambda \sim 1/m_v$, and crosses over to a logarithmic interaction at short distances. Hence, the self-energy of a vortex is finite, instead of being logarithmically divergent. As a result, the free energy of a single vortex is always negative, $F_{\text{vortex}} < 0$, signaling a proliferation instability.

Consequently, the Kosterlitz-Thouless argument now implies that the vortices (and antivortices) always proliferate and are always in a plasma phase. We will see below that in this state in which the instantons proliferate, external test charges are confined. Thus, this example provides a scenario for how instantons can provide a mechanism for confinement. However, it is important to stress that this works because the only zero modes of the instantons (the vortices) are their locations. We will see that classically scale-invariant theories with instantons have additional zero modes related to their scale, which will complicate the analysis. The important result is that instantons lead to an area law for a Wilson loop carrying a charge that is a fraction of the charge of the Higgs field. Hence, these static external sources are confined. But Wilson loops with the charge of the Higgs field have a perimeter law. We have seen in chapter 18 that a lattice version of this theory shows that the theory is confining, because the only allowed states are neutral bound states.

19.4 Instantons and solitons of nonlinear sigma models

Let us now consider the $O(3)$ nonlinear sigma model in two dimensions and show that it has instanton solutions. The Euclidean action of the nonlinear sigma model is

$$S = \int d^2x \, \frac{1}{2}(\partial_\mu \boldsymbol{n})^2, \quad \text{with} \quad \boldsymbol{n}^2 = 1 \tag{19.96}$$

and its partition function is

$$Z = \int \mathcal{D}\boldsymbol{n} \, \exp(-S[\boldsymbol{n}]/g) \tag{19.97}$$

where g is the coupling constant. Here $\boldsymbol{n}$ has three real components and obeys the constraint $\boldsymbol{n}^2 = 1$.

To derive the Euclidean equations of motion, let us implement the constraint using a Lagrange multiplier field $\lambda(x)$, in terms of which the action becomes

$$S = \int d^2x \, \frac{1}{2}\left[(\partial_\mu \boldsymbol{n})^2 - \lambda(x)(\boldsymbol{n}(x)^2 - 1)\right] \tag{19.98}$$

The Euler-Lagrange equations are

$$\frac{\delta S}{\delta n_a(x)} - \partial_\mu \frac{\delta S}{\delta \partial_\mu n_a(x)} = 0, \qquad \frac{\delta S}{\delta \lambda(x)} = 0 \tag{19.99}$$

The second equation simply implies that the constraint $\boldsymbol{n}^2 = 1$ is obeyed everywhere, while the first equation yields the condition

$$-\partial^2 n_a(x) = \lambda(x) n_a(x) \tag{19.100}$$

On taking the inner product with the field $\boldsymbol{n}(x)$, this equation becomes

$$\lambda(x) = -\boldsymbol{n} \cdot \partial^2 \boldsymbol{n} \tag{19.101}$$

where we used the constraint. Plugging this result into the first equation in eq. (19.99), the equation of motion becomes

$$\partial^2 n^a(x) = n^a(x)\boldsymbol{n}(x) \cdot \partial^2 \boldsymbol{n}(x) \tag{19.102}$$

which is nonlinear and nontrivial.

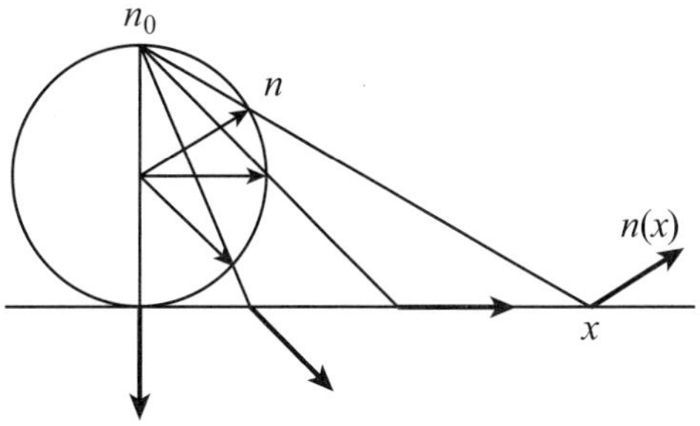

Figure 19.10 The stereographic projection of a field configuration.

To extremize (minimize) the Euclidean action, a smooth configuration $\boldsymbol{n}(x)$, which satisfies the constraint $\boldsymbol{n}^2 = 1$, must be such that

$$\lim_{r\to 0} r||\partial_\mu \boldsymbol{n}||^2 = 0 \tag{19.103}$$

for the action to be finite. Hence, we require that the field $\boldsymbol{n}(x)$ must approach a fixed (but arbitrary) value $\boldsymbol{n}_0$ as $r \to \infty$. Thus, the requirement that Euclidean action is finite implies that, for the allowed configurations, the base space (the plane $\mathbb{R}^2$) has been compactified to a sphere S^2, at least in a topological sense. This can be done, for instance, using the stereographic projection shown in figure 19.10.

However, the target space of the $O(3)$ nonlinear sigma model is also the two-sphere S^2. Thus, the finite-action solutions are smooth maps:

$$\boldsymbol{n} : S^2 \mapsto S^2 \tag{19.104}$$

Let us now show that these maps are classified by the homotopy group $\pi_2(S^2)$. The homotopy classes are labeled by an integer $\pi_2(S^2) \cong \mathbb{Z}$, which we call the *topological charge* Q. Let Q be defined by the expression

$$Q = \frac{1}{8\pi} \int_{S^2} d^2x\, \epsilon_{\mu\nu} \boldsymbol{n}(x) \cdot \partial_\mu \boldsymbol{n}(x) \times \partial_\nu \boldsymbol{n}(x) \tag{19.105}$$

We will show that Q is an integer, a topological invariant that labels the topological class.

Let ξ_1 and ξ_2 be the two Euler angles of the S^2 target sphere. The infinitesimal oriented-area element S_a^{target} (shown in figure 19.11) is given by

$$d\boldsymbol{S}^{\text{target}} = \frac{1}{2}\left(\frac{\partial \boldsymbol{n}}{\partial \xi_1} \times \frac{\partial \boldsymbol{n}}{\partial \xi_2} - \frac{\partial \boldsymbol{n}}{\partial \xi_2} \times \frac{\partial \boldsymbol{n}}{\partial \xi_1} \right) \tag{19.106}$$

We can write Q as follows:

$$Q = \frac{1}{8\pi} \int_{S^2_{\text{base}}} d^2x\, \epsilon_{\mu\nu} \epsilon_{abc} n_a(x) \frac{\partial n_b}{\partial x_\mu} \frac{\partial n_c}{\partial x_\nu} \tag{19.107}$$

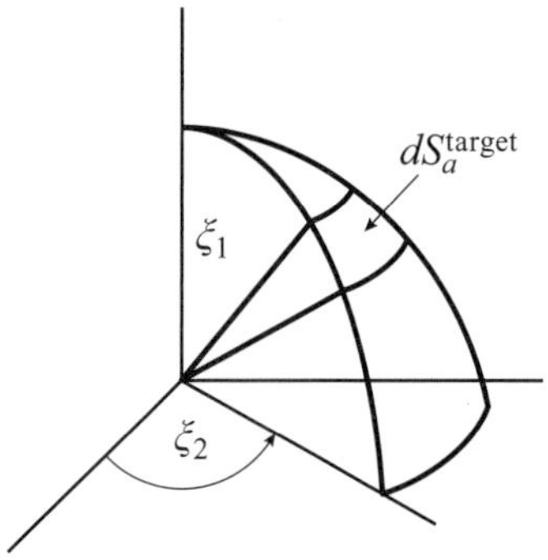

Figure 19.11 The infinitesimal oriented-area element of S^2.

The configuration $\boldsymbol{n}(x)$ maps a point x of S^2_{base} to a point ξ of S^2_{target}. Regarded as a change of variables, the mapping induces the change in the measure (a Jacobian):

$$\epsilon_{rs} d^2\xi = \epsilon_{\mu\nu} \frac{\partial \xi_r}{\partial x_\mu} \frac{\partial \xi_s}{\partial x_\nu} \tag{19.108}$$

On performing the change of variables, Q becomes

$$\begin{aligned} Q &= \frac{1}{8\pi} \int_{S^2_{\text{target}}} d^2\xi \; \epsilon_{rs}\epsilon_{abc} n_a \frac{\partial n_b}{\partial \xi_r} \frac{\partial n_c}{\partial \xi_s} \\ &= \frac{1}{4\pi} \int_{S^2_{\text{target}}} d\boldsymbol{S}_{\text{target}} \cdot \boldsymbol{n} \\ &= \frac{1}{4\pi} \int_{S^2_{\text{target}}} |d\boldsymbol{S}_{\text{target}}| \end{aligned} \tag{19.109}$$

where we used the fact that $\boldsymbol{n}$ and $d\boldsymbol{S}_{\text{target}}$ are parallel. However, the last integral is just the area of S^2, which is equal to 4π.

Therefore, Q is an integer that counts how many times the two-sphere S^2_{target} is being swept as the configuration $\boldsymbol{n}(x)$ spans the entire compactified plane, the base sphere S^2_{base}. This result does not change if the configuration $\boldsymbol{n}(x)$ is changed smoothly. We conclude that $Q \in \mathbb{Z}$ is a topological invariant of the class of mappings, and that the mappings are classified by the homotopy group

$$\pi_2(S^2) \cong \mathbb{Z} \tag{19.110}$$

Now let us show that the topological charge (or, rather, its absolute value) places a lower bound on the Euclidean action of a configuration of the field $\boldsymbol{n}$ in a given topological class, labeled by Q. To see this, consider the trivial identity

$$\left(\partial_\mu \boldsymbol{n} \pm \boldsymbol{n} \times \partial_\nu \boldsymbol{n}\right)^2 \geq 0 \tag{19.111}$$

Since

$$(\boldsymbol{n} \times \partial_\nu \boldsymbol{n})^2 = \boldsymbol{n}^2 (\partial_\nu \boldsymbol{n})^2 - (\boldsymbol{n} \cdot \partial_\nu \boldsymbol{n})^2 = (\partial_\nu \boldsymbol{n})^2 \tag{19.112}$$

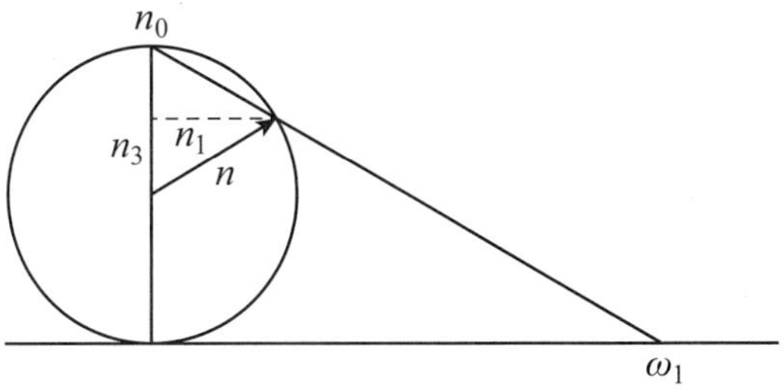

Figure 19.12 Stereographic projection of the target space.

and $\boldsymbol{n}^2 = 1$, and $\boldsymbol{n} \cdot \partial_\mu \boldsymbol{n} = 0$, it follows that

$$(\partial_\mu \boldsymbol{n} \pm \epsilon_{\mu\nu} \boldsymbol{n} \times \partial_\nu \boldsymbol{n})^2 = 2(\partial_\mu \boldsymbol{n})^2 \pm 2\epsilon_{\mu\nu} \boldsymbol{n} \cdot \partial_\mu \boldsymbol{n} \times \partial_\nu \boldsymbol{n} \geq 0 \tag{19.113}$$

which implies the following bound:

$$(\partial_\mu \boldsymbol{n})^2 \geq \epsilon_{\mu\nu} \boldsymbol{n} \cdot \partial_\mu \boldsymbol{n} \times \partial_\nu \boldsymbol{n} \tag{19.114}$$

Hence, the action $S[\boldsymbol{n}]$ of a configuration of the fields has a lower bound

$$S[\boldsymbol{n}] = \frac{1}{2} \int d^2x \, (\partial_\mu \boldsymbol{n})^2 \geq \frac{1}{2} \int d^2x \, \epsilon_{\mu\nu} \boldsymbol{n} \cdot \partial_\mu \boldsymbol{n} \times \partial_\nu \boldsymbol{n} \tag{19.115}$$

which is to say,

$$S[\boldsymbol{n}] \geq 4\pi |Q| \tag{19.116}$$

There is a class of configurations $\boldsymbol{n}$ that saturate the bound, $S[\boldsymbol{n}] = 4\pi |Q|$. To do that, they must satisfy the identity

$$(\partial_\mu \boldsymbol{n} \pm \epsilon_{\mu\nu} \boldsymbol{n} \times \partial_\nu \boldsymbol{n})^2 = 0 \tag{19.117}$$

which implies that these special configurations obey the self-dual (and anti-self-dual) equation

$$\partial_\mu \boldsymbol{n} = \pm \epsilon_{\mu\nu} \boldsymbol{n} \times \partial_\nu \boldsymbol{n} \tag{19.118}$$

together with the constraint $\boldsymbol{n}^2 = 1$. The solutions of the self-dual equation, eq. (19.118), are the instantons (and anti-instantons) of the (1+1)-dimensional nonlinear sigma model. These solutions are also the static configurations of the solitons of the (2+1)-dimensional nonlinear sigma model, where they are known as skyrmions.

Let us solve the self-dual equations using the stereographic projection of S^2 target space onto a (target) plane, $\mathbb{R}^2$, with coordinates (ω_1, ω_2) (see figure 19.12):

$$\omega_1 = \frac{2n_1}{1 - n_3}, \qquad \omega_2 = \frac{2n_2}{1 - n_3} \tag{19.119}$$

It will be convenient to define complex coordinates ω:

$$\omega = \omega_1 + i\omega_2 = 2\frac{n_1 + in_2}{1 - n_3} \tag{19.120}$$

Similarly, we also define $n = n_1 + in_2$. Using

$$\partial_1 \omega = \frac{2}{(1-n_3)^2}\left(\partial_1 n + n \overleftrightarrow{\partial_1} n_3\right) \tag{19.121}$$

we can rewrite the self-dual equations as

$$\partial_1 n = \mp in \overleftrightarrow{\partial_2} n_3, \qquad \partial_2 n = \mp in \overleftrightarrow{\partial_1} n_3 \tag{19.122}$$

or, more compactly,

$$\partial_1 \omega = \pm i \partial_2 \omega \tag{19.123}$$

which is equivalent to saying that ω_1 and ω_2 obey the Cauchy-Riemann equations

$$\frac{\partial \omega_1}{\partial x_1} = \pm \frac{\partial \omega_2}{\partial x_2}, \qquad \frac{\partial \omega_1}{\partial x_2} = \mp \frac{\partial \omega_2}{\partial x_1} \tag{19.124}$$

Therefore, $\omega(z)$ must be an analytic function of $z = x_1 + ix_2$ (but not an entire function). As such, $\omega(z)$ can have zeros and poles but not branch cuts.

In terms of the function $\omega(z)$, the action for an instanton takes the form

$$S = \int d^2x \frac{\left|\frac{d\omega}{dz}\right|}{\left(1 + \frac{|\omega|^2}{4}\right)^2} \tag{19.125}$$

such that

$$|Q| = \frac{S}{4\pi} \tag{19.126}$$

A solution of these equations is an analytic function $\omega(z)$ that has a zero of order $p \in \mathbb{Z}$

$$\omega(z) = \text{const.} \left(\frac{z - z_0}{\lambda}\right)^p \tag{19.127}$$

where z_0 is a complex number, and λ is real. This instanton solution has topological charge $Q = p$. A solution with a pole of order p is an anti-instanton with topological charge $Q = -p$.

In both cases, these solutions have two arbitrary parameters, known as the zero modes of the instanton: the location of the instanton (represented by the complex number z_0 on the plane) and the scale of the instanton (or of the anti-instanton) set by λ. Notice that the vortices discussed in section 19.3 have only one zero mode (their location), while the instantons have, in addition, an arbitrary scale. This is a consequence of the classical scale invariance of the two-dimensional nonlinear sigma model.

The general solution has the form

$$\omega(z) = \prod_i \left(\frac{z - z_i}{\lambda}\right)^{m_i} \prod_j \left(\frac{\lambda}{z - z_j}\right)^{n_j} \tag{19.128}$$

whose topological charge is

$$Q = \sum_i (m_i - n_i) \tag{19.129}$$

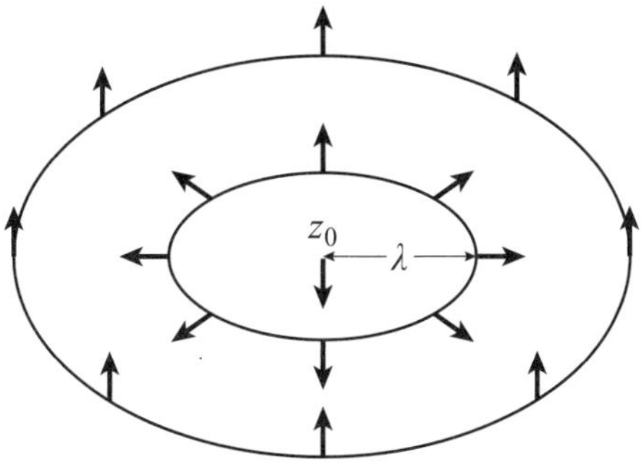

Figure 19.13 Instanton with topological charge $Q = 1$. The origin z_0 and the scale (radius) λ are zero modes of the instanton.

where instantons have topological charges m_i and anti-instantons have $-n_j$.

Since the action for this solution saturates the bound, $S = 4\pi|Q|$, it depends only on the total topological charge Q and not on where the instantons and anti-instantons are located. In other words, the instantons (and anti-instantons) do not interact with each other. We already encountered this feature at the Bogomol'nyi point of the abelian Higgs model. We will find the same feature in other theories with instanton solutions.

The instanton with topological charge $Q = 1$ located at the origin, $z_0 = 0$, is the solution

$$\omega(z) = \frac{z}{\lambda} \tag{19.130}$$

where λ is the scale, or radius, of the instanton. On retracing the steps of this construction, we find that this corresponds to the configuration of the field $\boldsymbol{n}(\boldsymbol{x})$

$$n_3(\boldsymbol{x}) = \frac{\boldsymbol{x}^2 - 4\lambda^2}{\boldsymbol{x}^2 + 4\lambda^2}, \qquad n_i(\boldsymbol{x}) = \frac{4\lambda x_i}{\boldsymbol{x}^2 + 4\lambda^2} \tag{19.131}$$

where $i = 1, 2$ labels the transverse components. This solution is shown qualitatively in figure 19.13.

The construction of the instantons of the $O(3)$ nonlinear sigma model is very elegant and beautiful. However, it poses problems. In the case of the abelian-Higgs (or superconductor) model, the vortex solutions have only one arbitrary parameter, the location of the vortex. We will see in section 19.9, following the work of Kosterlitz and Thouless and its generalizations, that it is possible to compute the partition function of the model, essentially exactly, in terms of a sum over the configurations of vortices and antivortices. In fact, if vortices are ignored, the partition function is trivial.

However, it has turned out to be very hard to recast the partition function of the nonlinear sigma model in terms of a sum over instanton and anti-instanton configurations. As already seen, in addition to the location of the instantons, as a consequence of the scale invariance of the classical theory, the solution has an arbitrary parameter, the scale λ. Thus, in addition to summing over all possible locations of instantons and anti-instantons, one must also sum over instantons of all possible sizes, ranging from the UV to the IR, leading to serious IR problems.

As we know from the RG analysis discussed in chapter 15, the classical fixed point is unstable to quantum fluctuations, and the effective coupling constant flows to strong coupling. One would have expected that the partition function of the instantons would provide a physical picture based on how the $O(3)$ symmetry is restored at long scales (and scale invariance is broken) in terms of a gas (or plasma) of instantons and anti-instantons. To date, this program has only been completed successfully in theories with large enough supersymmetry that the fluctuation determinant for an instanton can be computed exactly and no corrections are needed to the semiclassical approximation.

19.5 Coset nonlinear sigma models

We have focused on the case of the $O(3)$ nonlinear sigma model and showed that it has topologically nontrivial configurations classified by the homotopy group $\pi_2(S^2) = \mathbb{Z}$. However, this is a special property of the group $O(3)$. For the group $O(N)$, with $N > 3$, all configurations are trivial, since $\pi_2(S^N) = 0$. This poses the question: Are there other theories that more generically (for all N) have instantons?

We will now see that there are theories that generically (for "all N") have instantons. They are nonlinear sigma models whose target spaces are cosets of the form G/H, where G is a simply connected Lie group, and H has at least one $U(1)$ subgroup. Since G is simply connected, the configurations of the nonlinear sigma model can be smoothly deformed to the identity $I \in G$. Hence G is topologically trivial: $\pi_2(G) = 0$. Hence, a principal chiral nonlinear sigma model whose field is a group element, $g(x) \in G$, with action

$$S = \frac{1}{2u^2} \int d^2x \, \mathrm{tr}(\partial_\mu g(x) \partial_\mu g^{-1}(x)) \tag{19.132}$$

does not have instantons (here u^2 is a dimensionless coupling constant).

But chiral theories on the coset G/H do have instantons. To see that this is true, let us follow Polyakov's construction and define a field $\varphi_a(x)$

$$\varphi_a(x) = g_{ab}(x) \varphi_b^{(0)} \tag{19.133}$$

where $g(x) \in G$, and $\varphi^{(0)}$ is a constant field that is invariant under the action of the subgroup H (i.e., for all $h \in H$):

$$h_{ab} \varphi_b^{(0)} = \varphi_a^{(0)} \tag{19.134}$$

Note that, with these definitions, the field $g(x)$ does not have to be continuous.

In fact, let $g^N(x)$ be the matrix-valued fields defined on the northern hemisphere of S^2_{base}, and let $g^S(x)$ be defined on the southern hemisphere of S^2_{base}. We require that at the equator, the common boundary of the two hemispheres, which is isomorphic to S^1, the field $\varphi(x)$ is continuous on the whole of S^2_{base}. Consider configurations of the field $g(x)$ that are discontinuous at the equator, so that at points x on the equator of S^2_{base}, we have

$$g^N(x) = g^S(x) \, h(x) \tag{19.135}$$

where $x \in S^1$ and $h(x) \in H$. In contrast, we now show that $\varphi(x)$ is continuous at the equator, since it is restricted to each hemisphere. Thus, we define

$$\varphi^N(x) = g^N(x) \varphi^{(0)}, \qquad \text{and } \varphi^S(x) = g^S(x) \varphi^{(0)} \tag{19.136}$$

Then, at the equator,

$$\varphi^N(x) = g^N(x)\varphi^{(0)} = g^S(x)h(x)\varphi^{(0)} \equiv g^S(x)\varphi^{(0)} = \varphi^S(x) \tag{19.137}$$

where $h(x) \in H$. Hence, the field $\varphi(x)$ is continuous at the equator.

Moreover, its field configurations define a mapping of S^2_{base} to G/H. Such mappings can be classified according to the maps from the equator, S^1, onto H. If the subgroup H contains at least a $U(1)$ subgroup, that is,

$$H = U(1) \times \text{something trivial} \tag{19.138}$$

then we can use the winding number of the maps of S^1 to $U(1)$ to classify the maps of S^2 to G/H. In other words, we have shown that, even though $\pi_2(G) = 0$, we now have

$$\pi_2(G/H) = \pi_1(H) \tag{19.139}$$

Since for $H \simeq U(1)$, $\pi_1(H) = \mathbb{Z}$, we conclude that

$$\pi_2(G/H) = \mathbb{Z} \tag{19.140}$$

19.6 The $\mathbb{CP}^{N-1}$ instanton

One example of a coset nonlinear sigma model is the $\mathbb{CP}^{N-1}$ model, discussed in sections 16.5.1 and 17.3, where we solved it in the large-N limit. The $\mathbb{CP}^{N-1}$ model is a nonlinear sigma model of an N-component complex field $z_a(x)$ (with $a = 1, \ldots, N$), which transforms in the fundamental representation of $SU(N)$, with a gauged $U(1)$ subgroup. The field obeys everywhere the constraint $\sum_{a=1}^{N} |z_a(x)|^2 = 1$. The classical Lagrangian is

$$\mathcal{L} = \frac{1}{g^2}|(\partial_\mu + iA_\mu)z_a|^2 \tag{19.141}$$

where $A_\mu(x)$ is a dynamical $U(1)$ gauge field, and g is the dimensionless coupling constant. In this case, the coset is

$$\mathbb{CP}^{N-1} \cong \frac{SU(N)}{SU(N-1) \otimes U(1)} \tag{19.142}$$

which has instantons for all values of N. Recall that the $\mathbb{CP}^2$ model is equivalent to the $O(3)$ nonlinear sigma model. However, we will see that the $\mathbb{CP}^{N-1}$ model has instantons for all values of N.

At the classical level, this theory breaks the $SU(N)$ symmetry spontaneously down to an unbroken $SU(N-1)$ subgroup and the $U(1)$ gauged subgroup. As in the case of the $O(3)$ nonlinear sigma model, the requirement that the action is finite implies, in this case, that the covariant derivative vanishes at long distances,

$$\lim_{|x|\to\infty} (\partial_\mu + iA_\mu)z_a = 0 \tag{19.143}$$

which can be achieved by the condition

$$\lim_{|x|\to\infty} z(x) = z_0\, e^{i\theta(x)} \tag{19.144}$$

where z_0 is arbitrary and constant (but with $z_0 \cdot z_0 = 1$), and $\theta(x) \in [0, 2\pi)$ is an arbitrary phase. Thus, the winding of the phase θ is an integer-valued topological invariant. In terms of the $\mathbb{CP}^{N-1}$ field, the topological charge is

$$Q = \frac{1}{2\pi i}\int_\Sigma d^2x\, \partial_\mu(\epsilon_{\mu\nu} z_a^* \partial_\nu z^a) = \frac{1}{2\pi}\int_\Sigma d^2x\, \epsilon_{\mu\nu}\partial_\mu A_\nu = \frac{1}{2\pi}\oint_{\partial\Sigma} dx_\mu \partial_\mu \theta \tag{19.145}$$

where Σ is a large disk with boundary $\partial\Sigma$, and where we have used the identification of the gauge field A_μ in terms of the $\mathbb{CP}^{N-1}$ field.

Much as in the case of the $O(3)$ nonlinear sigma model, the topological charge Q of the $\mathbb{CP}^{N-1}$ model sets a lower bound for the action of a configuration. To show this, define the field C_μ^a (with $a = 1, \ldots, N$),

$$C_\mu^a(x) = \partial_\mu z^a(x) - z^a(z_b^* \partial_\mu z_b) \tag{19.146}$$

which obeys the obvious inequality

$$(C_\mu^a(x) \pm i\epsilon_{\mu\nu} C_\nu^a(x))^2 \geq 0 \tag{19.147}$$

which gives the condition that

$$|\partial_\mu z_a|^2 + (z_a^* \partial_\mu z_a)^2 \geq \pm i\epsilon_{\mu\nu}\partial_\mu(z_a^* \partial_\nu z_a) \tag{19.148}$$

(repeated indices are summed over). Here the left-hand side becomes the Lagrangian of the $\mathbb{CP}^{N-1}$ model obtained, after integrating out the gauge field A_μ. Therefore, the action obeys the now-familiar inequality

$$S[z_a] \geq 2\pi |Q| \tag{19.149}$$

The instantons of the $\mathbb{CP}^{N-1}$ model are the configurations that saturate this bound and satisfy the self-dual (and anti-self-dual) equation:

$$C_\mu^a(x) = \pm\, i\epsilon_{\mu\nu} C_\nu^a(x) \tag{19.150}$$

The solutions of these first-order partial differential equations yield the instanton solutions in terms of rational functions of the complex plane, much in the same way as what we did for the $O(3)$ model. In terms of the gauge field A_μ, the $Q = \pm n$ instanton (and anti-instanton) at the origin, $x = 0$, is

$$A_\mu^\pm = \pm n\, \epsilon_{\mu\nu} \frac{x_\nu}{x^2 + \lambda^2} \tag{19.151}$$

where λ is an arbitrary scale.

Naively, we expect that since the instantons have finite Euclidean action, their contribution to the partition function would be exponentially small, on the order of $\exp(-S_E/g^2)$, up to an "entropic" prefactor, which is (naively) a subleading contribution. This expectation is naive since, in addition to summing over the locations of the instantons (which in the case of vortices can be computed), in the nonlinear sigma models, one has to sum (integrate) over instantons of all possible scales. Thus, it is far from obvious that this naive argument actually holds.

19.7 The 't Hooft–Polyakov magnetic monopole

We now turn to the case of non-abelian gauge theories. We begin by constructing the analog of the $U(1)$ vortex. This is the Dirac magnetic monopole.

In section 19.3.3, we showed that the complex scalar field in 1+1 Euclidean dimensions has singular vortex configurations of the phase field $\theta(x)$. These configurations are multivalued and have branch cuts. In 1931, Dirac described a magnetic monopole as being a configuration of magnetic fields created by the current I flowing through an infinitely long and infinitesimally thin solenoid. Let us assume that the infinitesimally thin solenoid, the "Dirac string," runs along the x_3-axis from $x_3 \to -\infty$ and ends at the origin, $\boldsymbol{x} = (0, 0, 0)$. Then, the end of the long solenoid acts as a positive magnetic pole. The magnetic flux through the solenoid is $2\pi q$, where q is the magnetic charge. Outside the solenoid, there is an isotropic magnetic field $\boldsymbol{B}(\boldsymbol{x})$, radiating outward from the end of the solenoid. In other words, the magnetic field is

$$B_i(\boldsymbol{x}) = \frac{q}{2}\frac{x_i}{|\boldsymbol{x}|^2} - 2\pi q \delta_{i,3}\delta(x_1)\delta(x_2)\theta(-x_3) \tag{19.152}$$

The first term is the (magnetic) "Coulomb" field of the magnetic monopole of magnetic charge q at $\boldsymbol{X} = 0$. The second (and singular) term is the solenoid, known as the Dirac string. Notice that, if the solenoid is included, this is an allowed configuration of Maxwell's equations. Thus, this is a magnetic monopole of charge q at the origin with its singular Dirac string attached. However, the solenoid makes this configuration singular, much in the same way as the vortex configuration discussed in section 19.3.3. Indeed, the phase field configuration (in $D = 2$) of a vortex of topological charge n at the origin can be written as

$$\partial_i \theta(\boldsymbol{x}) = n\epsilon_{ij}\frac{x_i}{|\boldsymbol{x}|^2} - 2\pi n \delta(x_1)\theta(-x_2) \tag{19.153}$$

The second term is singular and is similar to a Dirac string. It represents the branch cut of the phase field.

However, we have seen that in the abelian Higgs model, a theory of a complex scalar field coupled to the $U(1)$ gauge field has regular vortex solutions in the classically spontaneously broken phase of this theory. G. 't Hooft (1976a) and A. M. Polyakov (1975a) showed that a regular configuration, which at long distances becomes a Dirac magnetic monopole, exists in the Higgs sector of the Georgi-Glashow model (Georgi and Glashow, 1974), which we discussed in section 18.11.2.

The Georgi-Glashow model has a three-component real scalar field $\boldsymbol{\phi} = (\phi_1, \phi_2, \phi_3)$, which transforms under the adjoint representation of $SU(2)$, and a Yang-Mills gauge field taking values in the algebra of the gauge subgroup $G = SU(2)$ associated with the weak interactions of a grand unified gauge theory with gauge group $SU(5)$. In the spontaneously broken phase, the Lagrangian in $D = (2+1)$-dimensional Euclidean space is

$$\mathcal{L} = \frac{1}{2}(D_\mu \boldsymbol{\phi})^2 - \frac{m^2}{2}\boldsymbol{\phi}^2 + \frac{\lambda}{4!}(\boldsymbol{\phi}^2)^2 + \frac{1}{4}\text{tr}F_{\mu\nu}^2 \tag{19.154}$$

where $D_\mu \boldsymbol{\phi} = \partial_\mu \boldsymbol{\phi} + g\boldsymbol{A}_\mu \times \boldsymbol{\phi}$ is the covariant derivative in the adjoint representation of $SU(2)$. In this phase, the $SU(2)$ gauge symmetry is spontaneously broken down to its $U(1)$ subgroup. As a subgroup of the compact Lie group $SU(2)$, the unbroken $U(1)$ subgroup is compact.

The classical (Euclidean) equations of motion are (with $a=1,2,3$ and $i,j=1,2,3$)

$$D_i F^{aij} = g\epsilon^{abc}(D_j\phi_b)\phi_c \tag{19.155}$$

and

$$D_i D_i \phi_a = -\lambda \boldsymbol{\phi}^2 \phi_a + \lambda f^2 \phi_a \tag{19.156}$$

where $\lambda f^2 = |m^2|$. The Euclidean action of the instanton in three Euclidean spacetime dimensions (or the energy of the soliton in three space dimensions) is

$$S_E = \int d^3x \left[\frac{1}{4} F^a_{ij} F^{aij} + \frac{1}{2} D_i\phi_a D_i\phi_a + \frac{\lambda}{4!}(\boldsymbol{\phi}^2 - f^2)^2 \right] \tag{19.157}$$

Let us first look at the zero-energy solutions—the classical vacuum. They are $A^a_i = 0$, $\boldsymbol{\phi}^2 = f^2$, and $D_i\boldsymbol{\phi} = 0$. In this case, the latter reduces to $\partial_i \boldsymbol{\phi} = 0$, and $\boldsymbol{\phi}$ is a constant vector.

We now seek finite action solutions. As in our analysis of the instantons of the nonlinear sigma model, the finite action requirement implies that the fields should approach the vacuum solution sufficiently fast as $r \to \infty$ (here, $r = |\boldsymbol{x}|$ is the radial coordinate). In three dimensions, the required asymptotic behavior is that, as $r\to\infty$, $r^{3/2} D_i\phi^a \to 0$, and $\boldsymbol{\phi}^2 \to f^2$. In spherical coordinates (r,θ,φ), the θ component of the covariant derivative is

$$D_\theta \phi^a = \frac{1}{r}\frac{\partial \phi^a}{\partial \theta} + g\epsilon^{abc} A^b_\theta \phi^c \tag{19.158}$$

Hence, $D_\theta\phi^a \to 0$ (as $r\to\infty$), provided that $A^b_\theta \sim \frac{1}{r}$, also as $r\to\infty$. The same asymptotic behavior holds for the other components. However, if $A_i \sim 1/r$, then by dimensional counting, the field strength $F \sim 1/r^2$, and $F^2 \sim 1/r^4$ (again, as $r\to\infty$), which is integrable at large r in three dimensions.

This analysis shows that, unlike the case of the $D=2$ nonlinear sigma models, the finite action solutions of the Georgi-Glashow model may have fields ϕ^a that are not equivalent at spatial infinity, since they are allowed to asymptotically point in different directions. In particular, the asymptotic behavior for the scalar field only requires that at spatial infinity (i.e., at the surface of a large sphere S^2 of radius $r\to\infty$), the magnitude of the field must be fixed but not its direction.

Thus, the finite Euclidean action solutions are maps of the base space, the large sphere S^2 with large radius r, to the S^2 target space of the scalar field with fixed norm. In other words, the topology is the same as in the case of the $D=2$ nonlinear sigma model, and the configurations are also classified by the homotopy group $\pi_2(S^2) \cong \mathbb{Z}$. In addition, the classes are labeled by the topological charge $Q \in \mathbb{Z}$ of the $D=2$ $O(3)$ nonlinear sigma model shown in eq. (19.105). For instance, the configuration of the scalar field $\boldsymbol{\phi}$ on the large sphere at spatial infinity S^2 with topological charge $Q=1$ is the hedgehog configuration (a "hairy ball") shown in figure 19.14.

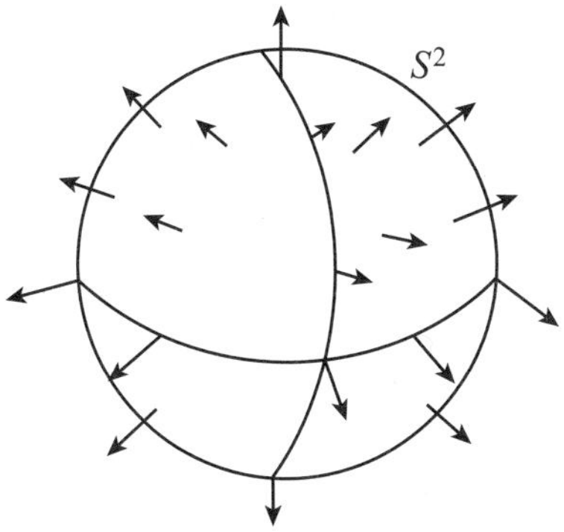

Figure 19.14 Hedgehog configuration of the $O(3)$ field $\boldsymbol{\phi}$ on the large sphere S^2 at spatial infinity.

How is this related to the magnetic monopole? To see the relation, let us use a gauge-invariant formulation due to 't Hooft. Define the gauge-invariant field strength tensor $F_{\mu\nu}$,

$$F_{\mu\nu} \equiv \widehat{\phi}_a F^a_{\mu\nu} - \frac{1}{g}\epsilon^{abc}\widehat{\phi}_a D_\mu \widehat{\phi}_b D_\nu \widehat{\phi}_c \tag{19.159}$$

where $\widehat{\phi}^a = \phi^a/||\boldsymbol{\phi}||$. For a topologically trivial configuration such as $\widehat{\boldsymbol{\phi}} = (0, 0, 1)$, this field strength reduces to

$$F_{\mu\nu} = \partial_\mu A^3_\nu - \partial_\nu A^3_\mu \tag{19.160}$$

In three dimensions, we can define the dual (pseudovector) $F^*_\mu = \frac{1}{2}\epsilon_{\mu\nu\lambda}F^{\nu\lambda}$, and in four dimensions, the dual tensor $F^*_{\mu\nu} = \frac{1}{2}\epsilon_{\mu\nu\lambda\rho}F^{\lambda\rho}$. Then, for topologically trivial configurations, the dual tensors satisfy the Bianchi identities, $\partial^\mu F^*\mu = 0$ (in $D=3$) and $\partial^\mu F^*_{\mu\nu} = 0$, just as in Maxwell's theory. Hence, configurations whose scalar fields have trivial configurations at infinity do not have magnetic monopoles.

However, we can use the definition of the field tensor $F_{\mu\nu}$ in eq. (19.159) to compute the divergence of its dual and find, in $D=4$,

$$\partial^\nu F^*_{\mu\nu} = \frac{1}{2g}\epsilon_{\mu\nu\lambda\rho}\epsilon_{abc}\partial^\nu\widehat{\phi}^a\partial^\lambda\widehat{\phi}^b\partial^\rho\widehat{\phi}^c \equiv \frac{4\pi}{g}j_\mu \tag{19.161}$$

where j_μ is the topological current. In $D=3$, we find

$$\partial^\mu F^*_\mu = \frac{4\pi}{g}j_0(x) \tag{19.162}$$

where j_0 is the topological charge density. Its integral over all space is given by

$$\begin{aligned}\int d^3x\, j_0(x) &= \frac{1}{8\pi}\int d^3x \epsilon_{ijk}\epsilon^{abc}\partial_i\widehat{\phi}^a\partial_j\widehat{\phi}^b\partial_k\widehat{\phi}^c \\ &= \frac{1}{8\pi}\int d^3x \epsilon_{ijk}\epsilon^{abc}\partial_i\left(\widehat{\phi}^a\partial_j\widehat{\phi}^b\partial_k\widehat{\phi}^c\right) \\ &= \frac{1}{8\pi}\int_{S^2} d^2S_i\epsilon_{ijk}\epsilon^{abc}\widehat{\phi}^a\partial_j\widehat{\phi}^b\partial_k\widehat{\phi}^c\end{aligned} \tag{19.163}$$

which is the integer-valued topological charge Q that classifies the configurations of the field $\widehat{\boldsymbol{\phi}}$ on the sphere S^2 at spatial infinity. Therefore, in $D=3$ dimensions, the Bianchi identity now yields the topological charge density $j_0(x)$, whose integral over all space is the topological charge. The same line of argument yields, in $D=4$, the result that the Bianchi identity is equal to the topological current j_μ.

We conclude that in $D=3$, the instantons are pointlike, whereas in $D=4$, they are currents. Also, in $D=4$, the monopoles are pointlike finite-energy topological solitons. It follows that, since the magnetic field is $B_i = \frac{1}{2}\epsilon_{ijk}F_{jk}$, the divergence of the field is

$$\partial_i B_i = \frac{4\pi}{g} j_0(x) \tag{19.164}$$

Hence, the magnetic charge of a monopole of topological charge Q is

$$m = \int d^3x\, \partial_i B_i(x) = \frac{Q}{g} \tag{19.165}$$

Let us work out the monopole with topological charge $Q=1$. Use the spherically symmetric ansatz

$$\begin{aligned} \phi^a(x) =& \delta_{ia}\frac{x^i}{r}F(r), \\ A_i^a(x) =& \epsilon_{aij}\frac{x^i}{r}W(r) \end{aligned} \tag{19.166}$$

where the functions $F(r)$ and $W(r)$ of the radius $r=|x|$ have the asymptotic behavior

$$\lim_{r\to\infty} F(r) = f, \qquad \lim_{r\to\infty} grW(r) = 1 \tag{19.167}$$

In this solution, at large distances $r\to\infty$, the magnetic field points outward and isotropically away from the origin, and it has the asymptotic behavior of a magnetic monopole of charge $1/g$:

$$\boldsymbol{B}(\boldsymbol{x}) \sim \frac{\boldsymbol{x}}{gr^3}, \qquad \text{as } r\to\infty \tag{19.168}$$

On introducing the functions $K(r)$ and $H(r)$,

$$K(r) \equiv 1 - grW(r), \qquad H(r) \equiv grF(r) \tag{19.169}$$

the field equations become

$$\begin{aligned} r^2\frac{d^2K}{dr^2} =& K(K^2-1) + H^2K, \\ r^2\frac{d^2H}{dr^2} =& 2HK^2 + \lambda\left(\frac{H^2}{g^2} - r^2f^2\right)H \end{aligned} \tag{19.170}$$

As in the other cases we have discussed, these equations are in general quite difficult to solve, except in the BPS limit, $\lambda \to 0$, where

$$K(r) = \frac{gfr}{\sinh(gfr)}, \qquad H(r) = \frac{gfr}{\tanh(gfr)} - 1 \tag{19.171}$$

Notice that both $K(r)$ and $H(r)$ are regular functions as $r \to 0$, where $H \to 0$ and $F \to 1$. Hence, the 't Hooft–Polyakov monopole is regular near the origin, and the potential singularity is smeared at distances shorter than the length scale $\xi \sim 1/(gf)$.

It can be shown that these solutions satisfy (and saturate) the Bogomol'nyi bound

$$E \geq 4\pi \frac{Qf}{g} \tag{19.172}$$

and hence, have finite action. They also satisfy the Bogomol'nyi equation

$$F^a_{ij} = \epsilon_{ijk} D_k \phi^a \tag{19.173}$$

In summary, the Georgi-Glashow model has monopole solutions, the 't Hooft–Polyakov monopole, instantons in 2+1 dimensions and solitons in 3+1 dimensions, with a quantized magnetic charge. However, unlike the Dirac monopole, the 't Hooft–Polyakov monopole has finite Euclidean action (or energy) and is not a singular configuration. Shortly we will see that in 2+1 dimensions, these monopole instantons lead to the confinement of static sources charged under the unbroken $U(1)$ subgroup of $SU(2)$.

19.8 The Yang-Mills instanton in $D = 4$ dimensions

Let us now discuss the instanton solutions of pure Yang-Mills gauge theory in $D=4$ (Euclidean) dimensions. Let G be a simple and compact gauge group. The Euclidean Yang-Mills action is

$$S = \frac{1}{4} \int d^4x \, \mathrm{tr} F^2_{\mu\nu} \tag{19.174}$$

As before, we seek finite Euclidean action solutions, $S < \infty$. This requires that, at long distances, the field strength vanishes

$$F_{\mu\nu} \sim O(1/r^2), \qquad \text{as } r \to \infty \tag{19.175}$$

Since the field strength vanishes at long distances, in the same asymptotic limit, the gauge field must approach a pure gauge transformation labeled by $g(x) \in G$, that is,

$$A_\mu \sim g^{-1} \partial_\mu g + O(1/r), \qquad \text{as } r \to \infty \tag{19.176}$$

Therefore, on a sphere S^3 of large radius R, the gauge field configurations A_μ are mapped onto the gauge transformations labeled by the group elements $g \in G$. In other words, the finite action solutions are in one-to-one correspondence with the smooth mappings of the large sphere S^3 onto the gauge group G:

$$g(x) : S^3 \mapsto G \tag{19.177}$$

In the case of $G = SU(2)$, we can write the group elements as

$$g(x) = n_4(x)\, I + i\boldsymbol{n}(x)\cdot\boldsymbol{\sigma}, \qquad g^{-1} = g^{\dagger} \tag{19.178}$$

where I is the 2×2 identity matrix, and σ_i are the three Pauli matrices. Since in $SU(2)$, $\det g = 1$, the four-component real vector $(\boldsymbol{n}, n_4)$ must satisfy the constraint

$$\boldsymbol{n}^2 + n_4^2 = 1 \tag{19.179}$$

Thus, $SU(2) \cong S^3$. Therefore, the maps of $S^3 \mapsto G$ are maps of $S^3 \mapsto S^3$. These mappings are classified by the homotopy group

$$\pi_3(SU(2)) \cong \pi_3(S^3) \cong \mathbb{Z} \tag{19.180}$$

The topological charge, or winding number, that classifies these maps is the Pontryagin index, which counts the number of times that S^3 covers S^3. As in the $\pi_2(S^2)$ case, the Pontryagin index is the Jacobian of the map $g(x)$. Moreover, since the very simple and compact group G has an $SU(2)$ subgroup, the Pontryagin index classifies all these maps, and $\pi_3(G) = \mathbb{Z}$.

In general, for $g \in G$, the Pontryagin index is

$$Q = \frac{1}{24\pi^2}\int_{S^3} \epsilon_{\mu\nu\lambda}\,\mathrm{tr}(L_\mu L_\nu L_\lambda) \tag{19.181}$$

where $L_\mu = g^{-1}\partial_\mu g$, and S^3 is the boundary of four-dimensional Euclidean spacetime. In the case of $SU(2)$, the Pontryagin index is

$$Q = \frac{1}{32\pi^2}\int_{S^3} d^3x\, \epsilon^{abcd}\epsilon_{\mu\nu\lambda}\, n^a\partial_\mu n^b \partial_\nu n^c \partial_\lambda n^d \tag{19.182}$$

After some algebra, it can be shown that the topological charge Q can be written as an integral in four-dimensional Euclidean spacetime Ω, whose boundary is S^3, of a total derivative,

$$Q = \frac{1}{32\pi^2}\int_{\Omega} d^4x\, \epsilon^{\mu\nu\lambda\rho}\,\mathrm{tr}\left(F_{\mu\nu}F_{\lambda\rho}\right) \equiv \frac{1}{8\pi^2}\int d^4x\, \mathrm{tr} F \wedge F^* \tag{19.183}$$

where $F^*_{\mu\nu} = \frac{1}{4}\epsilon_{\mu\nu\lambda\rho}F^{\lambda\rho}$ is the dual tensor. Therefore, the Pontryagin index Q is given by

$$Q = \frac{1}{8\pi^2}\int d^4x\, \mathrm{tr}\left(F^{\mu\nu}F^*_{\mu\nu}\right) \tag{19.184}$$

which labels the topological classes of all compact simply connected gauge groups.

Let us now show that the Pontryagin index places a lower bound on the Euclidean action of gauge fields belonging to a topological class. The argument is similar to the one used for the instantons of the nonlinear sigma model in section 19.4. Thus, rewrite the Yang-Mills

Euclidean action as

$$S = \frac{1}{4g^2}\int d^4x\,\mathrm{tr}F_{\mu\nu}^2$$
$$= \frac{1}{8g^2}\int d^4x\,\mathrm{tr}\left(F_{\mu\nu} - F_{\mu\nu*}\right)^2 + \frac{1}{4g^2}\int d^4x\,\mathrm{tr}\left(F^{\mu\nu}F^*_{\mu\nu}\right)^2$$
$$= \frac{8\pi^2}{g^2}Q + \frac{1}{8g^2}\int d^4x\,\mathrm{tr}\left(F_{\mu\nu} - F_{\mu\nu*}\right)^2 \tag{19.185}$$

Since the last term is manifestly positive, we find the lower bound for the Euclidean action:

$$S \geq \frac{8\pi^2}{g^2}Q \tag{19.186}$$

Once again, we can seek the configurations that saturate the bound, which satisfy the self-duality ("Cauchy-Riemann") equation

$$F_{\mu\nu} = F^*_{\mu\nu} \tag{19.187}$$

for which

$$S = \frac{8\pi^2}{g^2}Q \tag{19.188}$$

are instantons with $Q > 0$.

Similarly, the gauge fields that satisfy the anti-self-dual equation

$$F_{\mu\nu} = -F^*_{\mu\nu} \tag{19.189}$$

also satisfy

$$S = \frac{8\pi^2}{g^2}|Q| \tag{19.190}$$

and are anti-instantons with $Q < 0$.

In addition, if a gauge field is self-dual, $F_{\mu\nu} = F^*_{\mu\nu}$, it also satisfies the equation $D_\mu F^{\mu\nu} = 0$. An example of this solution is the $Q = 1$ instanton of the $SU(2)$ Yang-Mills theory, for which the gauge field is given by

$$A^a_\mu = -\eta_{a\mu\nu}\frac{(x_\nu - a_\nu)}{(\boldsymbol{x} - \boldsymbol{a})^2 + \rho^2} \tag{19.191}$$

where we have introduced the tensor $\eta_{abc} = \epsilon_{abc}$, $\eta_{ab0} = \delta_{ab}$, and so forth. Here ρ is the arbitrary scale of the instanton (reflecting the classical scale invariance of four-dimensional Yang-Mills theory), and a_μ is an (also arbitrary) location of the instanton. The field strength of this solution is

$$F^a_{\mu\nu} = -4\eta_{a\mu\nu}\frac{\rho^2}{(\boldsymbol{x} - \boldsymbol{a})^2 + \rho^2} \tag{19.192}$$

19.9 Vortices and the Kosterlitz-Thouless transition

Let us now discuss two specific examples of theories in which topological excitations (instantons) either drive their phase transitions or lead to a nonperturbative phase at all values of the coupling constant. We will work with the lattice formulation of these theories, where the results can be obtained more simply. The cases that we consider have a (compact) $U(1)$ symmetry, either global or local: the $U(1)$ nonlinear sigma model in $D=2$ dimensions, known in statistical mechanics as the *XY* model, and Polyakov's compact electrodynamics in $D=3$ dimensions. Much of what we do can be extended to theories with more general abelian symmetry groups, both continuous or discrete, but not to non-abelian symmetries.

Consider a system on a two-dimensional square lattice with a $U(1)$ degree of freedom at every site $\boldsymbol{r}$, a two-component vector of unit length, $\boldsymbol{n}(\boldsymbol{r})=(\cos\theta(\boldsymbol{r}), \sin\theta(\boldsymbol{r}))$, where the phase is defined on the interval $0\leq\theta(\boldsymbol{r})<2\pi$. The energy (or lattice Euclidean action) is (here $\mu=1,2$ represents the two directions on the square lattice)

$$E=-J\sum_{\boldsymbol{r},j}\boldsymbol{n}(\boldsymbol{r})\cdot\boldsymbol{n}(\boldsymbol{r}+\boldsymbol{e}_j)=-J\sum_{\boldsymbol{r},\mu}\cos(\Delta_\mu\theta(\boldsymbol{r})) \tag{19.193}$$

where $\Delta_\mu\theta(\boldsymbol{r})=\theta(\boldsymbol{r}+\boldsymbol{r}_\mu)-\theta(\boldsymbol{r})$ is the lattice difference operator, and the $\boldsymbol{e}_\mu$ are the two orthonormal vectors of the square lattice: $\boldsymbol{e}_\mu\cdot\boldsymbol{e}_\nu=\delta_{\mu\nu}$.

The partition function of this model is

$$Z=\prod_{\boldsymbol{r}}\int_0^{2\pi}\frac{d\theta(\boldsymbol{r})}{2\pi}\exp\left(-\frac{J}{T}\sum_{\boldsymbol{r},\mu}\cos(\Delta_\mu\theta(\boldsymbol{r}))\right) \tag{19.194}$$

In the low-temperature regime, $T\ll J$, we can naively take the continuum limit and write

$$Z\simeq\int\mathcal{D}\theta\,\exp\left(-\frac{1}{2g}\int d^2x\,(\nabla\theta)^2\right) \tag{19.195}$$

(with $g=Ta^2/J$, where a is the lattice spacing), which is formally a massless free-field theory if it were not for the fact that the lattice model is invariant under local periodic shifts of the phase:

$$\theta(\boldsymbol{r})\mapsto\theta(\boldsymbol{r})+2\pi n(\boldsymbol{r}),\qquad \text{with } n(\boldsymbol{r})\in\mathbb{Z} \tag{19.196}$$

This local symmetry requires that the only allowed observables obey the periodicity condition. In other words, the field θ, even in the continuum, is compactified, and the global symmetry of the theory is $U(1)$ and not $\mathbb{R}$.

In section 19.3.3, we discussed this continuum theory. There we saw that this theory has singular vortex configurations (instantons). The lattice model also has vortices, but the lattice definition makes them regular. We now consider how the vortices arise in the lattice model and what role they play.

Our main tool will be a *duality transformation*, a generalization of a method first introduced by Kramers and Wannier in the context of the two-dimensional classical Ising model and closely related to topology. In general, this duality transformation has two ingredients. One is a geometric duality, or duality of forms. Geometric duality says that in D dimensions, a p-form is dual to a $D-p$ form. In $D=2$ dimensions, this means that a theory with a global symmetry, and hence defined on the links of the lattice (one-forms), is dual to a theory defined on the links of the *dual* lattice, and hence also on one-forms. Therefore, in two dimensions, a theory with a global symmetry is dual to another theory also with a global symmetry, defined on the dual lattice. We will see next that in three dimensions, the dual is a gauge theory. However, the second ingredient is that the dual of a theory is defined on the representations of the group. If the group is abelian, its representations are one-dimensional and define a group (the dual group). This is where this approach breaks down in non-abelian theories, since their representations are, in general, not one-dimensional and do not form a group.

Let us see how this works out in the context of a theory with a $U(1)$ symmetry. To simplify the notation, from now on we set $J=1$ (or, what is the same, T is now measured in units of J).

By inspection of the partition function of eq. (19.194), we see that the Gibbs weight of a configuration can be expressed as a product over links:

$$\exp\left(-\frac{1}{T}\sum_{r,\mu}\cos(\Delta_\mu\theta(\boldsymbol{r}))\right)=\prod_{r,\mu=1,2}\exp\left(-\frac{i}{T}\cos(\Delta_\mu\theta(\boldsymbol{r}))\right) \tag{19.197}$$

Consider now an expression of the form $\exp(V[\theta])$, which we require to be a periodic function of θ. Therefore, it can be expanded in a Fourier series

$$\exp(V[\theta])=\sum_{\ell\in\mathbb{Z}}V_\ell e^{i\ell\theta} \tag{19.198}$$

since $e^{i\theta}\in U(1)$, and $e^{i\ell\theta}$ are vectors in the ℓth irreducible representation of $U(1)$. Then since $e^{i\theta_1}\in U(1)$ and $e^{\theta_2}\in U(1)$, we have $e^{i(\theta_1+\theta_2)}\in U(1)$. Similarly, if $e^{i\ell_1\theta}\in U(1)$, and $e^{i\ell_2\theta}\in U(1)$, then $e^{i(\ell_1+\ell_2)\theta}\in U(1)$. Then the representations of $U(1)$ form a group under addition. In other words, the representations form a group, the dual group, which is isomorphic to the group of integers, $\mathbb{Z}$.

In particular,

$$\exp(\beta\cos\theta)=\sum_{\ell\in\mathbb{Z}}I_\ell(\beta)e^{i\ell\theta} \tag{19.199}$$

where

$$I_\ell(\beta)=\int_0^{2\pi}\frac{d\theta}{2\pi}\exp(\beta\cos\theta-i\ell\theta) \tag{19.200}$$

is a Bessel function with an imaginary argument. Here we have set $\beta=1/T$.

Let us now modify the Gibbs weight of eq. (19.194) using the more tractable expression (which has the same symmetries)

$$Z=\prod_r\int_0^{2\pi}\frac{d\theta(\boldsymbol{r})}{2\pi}\sum_{\{\ell_\mu(\boldsymbol{r})\}}\exp\left(-\sum_{r,\mu}\frac{\ell_\mu^2(\boldsymbol{r})}{2\beta}+i\sum_{r,\mu}\ell_\mu(\boldsymbol{r})\Delta_\mu\theta(\boldsymbol{r})\right) \tag{19.201}$$

where we have defined a set of integer-valued variables, $\ell_\mu(\boldsymbol{r})\in\mathbb{Z}$, for each link $(\boldsymbol{r},\mu)$ of the square lattice.

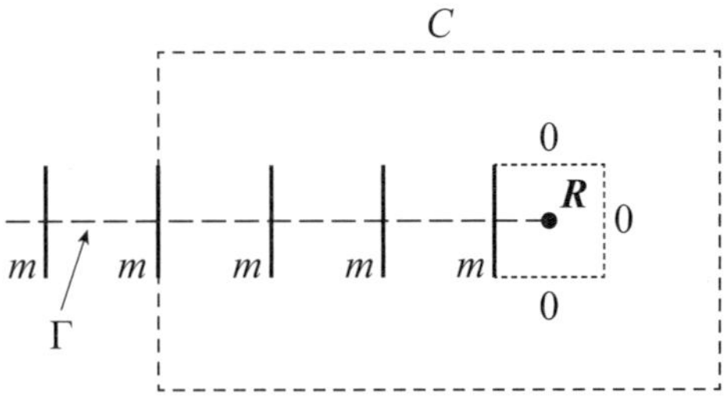

Figure 19.15 A Dirac string Γ is a configuration of the discrete gauge fields ℓ_μ that creates a vortex at the site $\boldsymbol{R}$ of the dual lattice. Here $\ell_\mu = m$ on all the links pierced by the Dirac string Γ, and is 0 on all other links. The circulation of the integer-valued gauge field ℓ_μ on any closed contour C, $\sum_C \ell_\mu = m$ if C encloses site $\boldsymbol{R}$, and equals zero otherwise.

There is an equivalent formulation that is quite instructive. Note that what we have done is equivalent to making the replacement (on every link)

$$\exp(\beta \cos \Delta_\mu \theta(\boldsymbol{r})) \mapsto \text{const.} \sum_{\ell_\mu(\boldsymbol{r})} \exp\left(-\frac{\beta}{2}(\Delta_\mu \theta(\boldsymbol{r}) - 2\pi \ell_\mu(\boldsymbol{r}))^2\right) \tag{19.202}$$

where on the right-hand side, the variable θ now takes all possible real values. Periodicity is recovered by the local discrete transformation

$$\theta(\boldsymbol{r}) \mapsto \theta(\boldsymbol{r}) + 2\pi k_\mu(\boldsymbol{r}), \qquad \ell_\mu(\boldsymbol{r}) + k_\mu(\boldsymbol{r}) \tag{19.203}$$

where $k_\mu(\boldsymbol{r})$ is an integer-valued local gauge transformation. Therefore, the variables $\ell_\mu(\boldsymbol{r})$ play the role of a gauge field whose gauge group is $\mathbb{Z}$. In this picture, the variable θ is not compact, and it does not have any vorticity. The field is compactified by the discrete gauge symmetry. Vortices are created by Dirac strings of the integer-valued gauge fields, such as in the example shown in figure 19.15, which shows a Dirac string on the open path Γ of the dual lattice. On the links pierced by Γ, we have set $\ell_\mu = m$, and ℓ_μ is zero on all other links. The circulation on a closed path C of the discrete gauge field is $\sum_C = m$ if the dual site $\boldsymbol{R}$ is enclosed by C and is zero otherwise. The Dirac string forces a jump of $2\pi m$ on θ across the Dirac string. Thus, this configuration represents a vortex of topological charge m at $\boldsymbol{R}$.

We now return to the duality transformation by integrating out the periodic $\theta(\boldsymbol{r})$ field in the partition function of eq. (19.201). Since

$$\int_0^{2\pi} \frac{d\theta(\boldsymbol{r})}{2\pi} \exp(i\theta(\boldsymbol{r})\Delta_\mu \ell_\mu(\boldsymbol{r})) = \delta(\Delta_\mu \ell_\mu(\boldsymbol{r})) \tag{19.204}$$

where the lattice divergence is defined as

$$\Delta_\mu \ell_\mu(\boldsymbol{r}) = \sum_{\mu=1,2} \left[\ell_\mu(\boldsymbol{r}) - \ell_\mu(\boldsymbol{r} - \boldsymbol{e}_\mu)\right] \tag{19.205}$$

the partition function takes the equivalent form of a sum over configurations on the integer-valued link variables, $\ell_\mu(\boldsymbol{r})$:

$$Z = \sum_{\{\ell_\mu(\boldsymbol{r})\}} \prod_{\boldsymbol{r}} \delta(\Delta_\mu \ell_\mu(\boldsymbol{r})) \exp\left(-\sum_{\boldsymbol{r},\mu} \frac{\ell_\mu^2(\boldsymbol{r})}{2\beta} \right) \tag{19.206}$$

Formally, this partition function is a sum of configurations of closed loops. If $\beta \ll 1$ (i.e., $T \gg J$), the partition function is dominated by the smallest loops: This is the high-temperature expansion.

The constraint that the loops are closed, $\Delta_\mu \ell_\mu(\boldsymbol{r}) = 0$ at every site $\boldsymbol{r}$, is solved by

$$\ell_\mu(\boldsymbol{r}) = \epsilon_{\mu\nu} \Delta_\nu S(\boldsymbol{R}) \tag{19.207}$$

where $S(\tilde{\boldsymbol{r}}) \in \mathbb{Z}$ is a set of integer-valued variables defined on the sites $\boldsymbol{R}$ of the dual of the square lattice (i.e., the set of points at the center of the plaquettes of the square lattice). After solving the constraint, the partition function becomes that of the discrete Gaussian model:

$$Z_{\text{DGM}} = \sum_{\{S(\boldsymbol{R})\}} \exp\left(-\frac{1}{2\beta} \sum_{\boldsymbol{R},\mu} (\Delta_\mu S(\boldsymbol{R}))^2 \right) \tag{19.208}$$

Thus, we have obtained the result that the dual of the XY model, which has a global $U(1)$ symmetry, is a theory defined on the dual lattice with integer-valued degrees of freedom, with a global symmetry, $S(\boldsymbol{R}) \to S(\boldsymbol{R}) + P$, where $P \in \mathbb{Z}$. This model is known as the discrete Gaussian model. As we will see, it is closely related to the vortices of the original problem. The configurations of the discrete variable $S(\boldsymbol{R})$ can be regarded as describing a discretized two-dimensional surface, such as the surface of a three-dimensional cubic crystal.

Therefore, the dual of the theory with a global $U(1)$ symmetry is a theory with a global $\mathbb{Z}$ symmetry. Moreover, under duality, $\beta \to 1/\beta$. Hence, duality exchanges strong coupling with weak coupling. In summary, duality is the set of correspondences

$$U(1) \leftrightarrow \mathbb{Z}, \qquad \beta \leftrightarrow \frac{1}{\beta}, \qquad \text{direct lattice} \leftrightarrow \text{dual lattice} \tag{19.209}$$

It is instructive to determine the dual of the order-parameter field $\exp(i\theta(\boldsymbol{r}))$, and of its correlation function,

$$G(\boldsymbol{r} - \boldsymbol{r}') = \langle \exp(i\theta(\boldsymbol{r})) \exp(-i\theta(\boldsymbol{r}')) \rangle \tag{19.210}$$

Let $n = \pm 1$ at the sites $\boldsymbol{r}$ and $\boldsymbol{r}'$ where the operators are inserted, that is,

$$n(\boldsymbol{x}) = \delta(\boldsymbol{x} - \boldsymbol{r}) - \delta(\boldsymbol{x} - \boldsymbol{r}') \tag{19.211}$$

In the presence of these insertions, the constraint on the discrete gauge fields in the partition function of eq. (19.206) now becomes

$$\Delta_\mu \ell_\mu(\boldsymbol{r}) = n(\boldsymbol{r}) \tag{19.212}$$

The solution is now

$$\ell_\mu(\boldsymbol{r}) = \epsilon_{\mu\nu} \left(\Delta_\nu S(\boldsymbol{R}) + B_\nu(\boldsymbol{R}) \right) \tag{19.213}$$

where $B_\mu(\boldsymbol{R})$ is any integer-valued vector field on the links of the dual lattice that obeys

$$\epsilon_{\mu\nu}\Delta_\mu B_\nu(\boldsymbol{R}) = n(\boldsymbol{x}) \tag{19.214}$$

Hence, under duality, we have the correspondence

$$\text{charge} \leftrightarrow \text{flux} \tag{19.215}$$

which is a discrete version of electromagnetic duality. The correlator then is mapped to a ratio of two partition functions in the discrete Gaussian model

$$\langle \exp(i\theta(\boldsymbol{r}))\exp(-i\theta(\boldsymbol{r}'))\rangle = \frac{Z_{\text{DGM}}[\boldsymbol{r},\boldsymbol{r}']}{Z_{\text{DGM}}} \tag{19.216}$$

where the numerator is the partition function of eq. (19.208) with discrete fluxes ±1 at $\boldsymbol{r}$ and $\boldsymbol{r}'$, respectively. The free-energy cost of the fluxes, $\Delta F[\boldsymbol{r},\boldsymbol{r}']$, is

$$\Delta F[\boldsymbol{r},\boldsymbol{r}'] = -\frac{1}{\beta}\ln\langle \exp(i\theta(\boldsymbol{r}))\exp(-i\theta(\boldsymbol{r}'))\rangle \tag{19.217}$$

Hence, in the symmetric (disordered) phase, where the correlator decays exponentially with distance, $G(\boldsymbol{r}-\boldsymbol{r}') \sim \exp(-|\boldsymbol{r}-\boldsymbol{r}'|/\xi)$, the free-energy cost increases linearly with distance as $\Delta F[\boldsymbol{r},\boldsymbol{r}'] \sim \sigma|\boldsymbol{r}-\boldsymbol{r}'|$, with $\sigma = T/\xi$, and the fluxes are confined.

The discrete Gaussian model of eq. (19.208) is a partition function over the integers. Now let us use the Poisson summation formula to transform the partition function in two different and useful ways. The Poisson summation formula is the following identity for a series of a function $f(n)$, where $n \in \mathbb{Z}$,

$$\sum_{n\in\mathbb{Z}} f(n) = \lim_{y\to 0^+}\int_{-\infty}^{\infty} dx \sum_{m\in\mathbb{Z}} f(x)\, e^{2\pi i m x} e^{-ym^2} \tag{19.218}$$

where $y > 0$ plays the role of a convergence factor. Using the Poisson summation formula, the partition function Z_{DGM} of eq. (19.208) becomes

$$Z_{\text{DGM}} = \sum_{\{m(\boldsymbol{R})\}}\prod_{\boldsymbol{R}}\int_{-\infty}^{\infty} d\phi(\boldsymbol{R})\, \exp(-S_{\text{DGM}}(\phi, m)) \tag{19.219}$$

where

$$S_{\text{DGM}}(\phi, m) = \frac{1}{2\beta}\sum_{\boldsymbol{R},\mu}(\Delta_\mu\phi(\boldsymbol{R}))^2 + y\sum_{\boldsymbol{R}} m^2(\boldsymbol{R}) + 2\pi i\sum_{\boldsymbol{R}} m(\boldsymbol{R})\phi(\boldsymbol{R}) \tag{19.220}$$

We now work with this result to obtain two useful mappings. Since the parameter y suppresses the contributions to the partition function of eq. (19.219) of progressively large values on the variable $|m|$, we will make the approximation of keeping (for now) only the values $m = 0, \pm1$. For y large enough (or $z = \exp(-y)$ small enough), we can approximate the sum over the m variables as

$$\sum_{m(\boldsymbol{R})\in\mathbb{Z}} \exp(-ym^2(\boldsymbol{R})+2\pi i m(\boldsymbol{R})\phi(\boldsymbol{R})) \simeq 1+2z\cos(2\pi\phi(\boldsymbol{R}))+O(z^2)$$

$$\simeq \exp(2z\cos(2\pi\phi(\boldsymbol{R}))+O(z^2)) \qquad (19.221)$$

Then the partition function of eq. (19.219) reduces to

$$Z=\int_{-\infty}^{\infty}\prod_{\boldsymbol{R}} d\phi(\boldsymbol{R}) \exp\left(-\frac{1}{2\beta}\sum_{\boldsymbol{R},\mu}(\Delta_\mu\phi(\boldsymbol{R}))^2+2z\sum_{\boldsymbol{R}}\cos(2\pi\phi(\boldsymbol{R}))\right) \qquad (19.222)$$

which, in the formal continuum limit, $a\to 0$, becomes

$$Z\simeq \in \mathcal{D}\phi \exp\left(-\int d^2x\, \mathcal{L}_{\text{SG}}[\phi]\right) \qquad (19.223)$$

where

$$\mathcal{L}_{\text{SG}}=\frac{T}{2}(\partial_\mu\phi)^2-g\cos(2\pi\phi) \qquad (19.224)$$

and $g=4z/a^2$ (a is the lattice spacing). On the change of variables $\varphi=\sqrt{T}\phi$, the Lagrangian takes the form

$$\mathcal{L}=\frac{1}{8\pi}(\partial_\mu\varphi)^2-g\cos\left(\frac{\varphi}{R}\right) \qquad (19.225)$$

This is the Lagrangian of the sine-Gordon theory in (1+1)-dimensional Euclidean spacetime, using the notation that will be introduced in section 21.6.1, with $R=\sqrt{T/\pi}$.

The sine-Gordon theory is invariant under discrete uniform shifts $\phi(x)\mapsto\phi(x)+2\pi nR$, where R is the compactification radius. This is a theory of a compactified boson, the φ field, with a periodic perturbation. The two-dimensional free compactified boson is a conformal field theory that we will discuss in some detail in section 21.6.1.

For the present purposes, it will suffice to say that the cosine operator is a vertex operator with scaling dimension $\Delta=1/R^2$. In chapter 15, we saw that in two spacetime dimensions, an operator is irrelevant if $\Delta>2$, which means that it is irrelevant if $\frac{1}{R^2}=\frac{\pi}{T}>2$ (or, what is the same, for $T<T_c=\frac{\pi}{2}$). But at this value of T, the cosine operator is not only marginal but is actually marginally relevant. Therefore, the coupling constant g runs to strong coupling under the RG to a regime in which the cosine operator becomes dominant. Furthermore, in this regime, the field φ will be pinned at the minima of the cosine potential, and its fluctuations have a mass gap.

The value $T_c=\frac{\pi}{2}$ that we just obtained is identical to that at which the energy-entropy Kosterlitz-Thouless argument (discussed in section 19.3.3) predicts that vortices will proliferate. We will now see that this is not an accident.

In section 19.3.3 we showed that the partition function for a complex scalar field in two Euclidean dimensions can be rewritten as the partition function of a neutral Coulomb gas, in which the charges are the vortices of the theory. Let us use duality to see how that comes about. Our starting point is the partition function written in the form of eq. (19.219), and the action is given in eq. (19.220). Instead of doing the sums over the integer-valued variables $m(\boldsymbol{R})$, we instead integrate the continuous variables $\phi(\boldsymbol{R})$. The resulting form of the partition function is

$$Z = Z_0 \sum_{\{m(\boldsymbol{R})\}} \delta\left(\sum_{\boldsymbol{R}} m(\boldsymbol{R}) = 0\right)$$

$$\times \exp\left(-\sum_{\boldsymbol{R}} y\, m^2(\boldsymbol{R}) - \frac{1}{2}\sum_{\boldsymbol{R},\boldsymbol{R}'} \left(\frac{2\pi}{T}\right)^2 m(\boldsymbol{R}) G_{\text{reg}}(\boldsymbol{R}-\boldsymbol{R}') m(\boldsymbol{R}')\right) \tag{19.226}$$

where $G_{\text{reg}}(\boldsymbol{R}-\boldsymbol{R}') = G(\boldsymbol{R}-\boldsymbol{R}') - G(0)$, and $G(\boldsymbol{R}-\boldsymbol{R}')$ is the lattice Green function, that is, the solution of the difference equation

$$-\Delta_{\boldsymbol{R}}^2 G(\boldsymbol{R}-\boldsymbol{R}') = \delta(\boldsymbol{R}-\boldsymbol{R}') \tag{19.227}$$

and $G(0)$ is IR divergent. Here Δ^2 is the lattice Laplacian operator (i.e., $\Delta^2 f(\boldsymbol{R}) = \sum_\mu \Delta_\mu f(\boldsymbol{R})$), and Z_0 is the trivial partition function

$$Z_0 = \int_{-\infty}^{\infty} \prod_{\boldsymbol{R}} d\phi(\boldsymbol{R}) \exp\left(-\frac{1}{2\beta}\sum_{\boldsymbol{R},\mu} (\Delta_\mu \phi(\boldsymbol{R}))^2\right) \tag{19.228}$$

At long distances, $|\boldsymbol{R}-\boldsymbol{R}'| \gg a$, the lattice Green function approaches the continuum result

$$G(\boldsymbol{R}-\boldsymbol{R}') \simeq \frac{1}{2\pi} \ln\left(\frac{|\boldsymbol{R}-\boldsymbol{R}'|}{a}\right) \tag{19.229}$$

Putting it all together, we recover the result that up to an uninteresting prefactor, the partition function is that of a gas of particles with both types of charges with a logarithmic interaction: the two-dimensional neutral Coulomb gas of eq. (19.80).

The moral of this analysis is that, in a theory in two dimensions with a $U(1)$ global symmetry, the phase transition to the disordered state is driven by a process of vortex proliferation. In this language, the mass gap results from the fact that at high temperatures, the neutral Coulomb gas experiences Debye screening.

19.10 Monopoles and confinement in compact electrodynamics

We now discuss Polyakov's compact QED in $D = 3$ Euclidean dimensions. This is an abelian lattice gauge theory with a compact abelian gauge group $U(1)$. In section 19.7, we saw that this theory can be regarded as the low-energy limit of the Georgi-Glashow model. The main difference between this theory and Maxwell's electrodynamics is that it has instanton magnetic monopoles. We will see that in this theory, magnetic monopoles play a similar role to the vortices in two dimensions. However, unlike the case of the vortices, the monopoles always proliferate, and this leads to confinement.

Let us consider a $U(1)$ abelian lattice gauge theory on a cubic lattice. Thus, as before, the degrees of freedom are vector fields defined on the links $(\boldsymbol{r}, \mu)$ of the lattice (with $\mu = 1, 2, 3$).

The partition function is

$$Z = \prod_{r,\mu} \int_0^{2\pi} \frac{dA_\mu(r)}{2\pi} e^{-S[A_\mu]} \tag{19.230}$$

where

$$S[A_\mu] = -\frac{1}{2g^2} \sum_{r,\mu\nu} \cos F_{\mu\nu}(R) \tag{19.231}$$

Here

$$F_{\mu\nu}(r) = \sum_{\text{plaquette}} A_\mu(r) \equiv \Delta_\mu A_\nu(r) - \Delta_\nu A_\mu(r) \tag{19.232}$$

is the flux on the plaquette $(r, \mu\nu)$.

Since the action is invariant under periodic shifts, $A_\mu(r) \to A_\mu(r) + 2\pi \ell_\mu(r)$, the Bianchi identity for the field strength $F_{\mu\nu}(r)$ can only be satisfied modulo 2π. However, the operator $\exp(iF_{\mu\nu})$ obeys the Bianchi identity

$$\prod_{\text{cube faces}} e^{iF_{\mu\nu}} = 1 \tag{19.233}$$

Thus, this theory has magnetic monopoles with quantized magnetic charge.

We can analyze this theory by an extension of the duality we just used in $D = 2$. Our first step is to express the partition function as

$$Z = \int_{-\infty}^{\infty} \prod_{r,\mu\nu} \frac{dA_\mu(r)}{2\pi} \sum_{\{m_{\mu\nu}(r)\}} \exp\left(-\frac{1}{2g^2} (F_{\mu\nu}(r) + 2\pi m_{\mu\nu}(r))^2 \right) \tag{19.234}$$

where the gauge fields are no longer periodic, and the $m_{\mu\nu}$ are integer-valued two-form variables (Kalb-Ramond fields) defined on the plaquettes. This expression for the partition function is invariant under the replacements

$$\begin{aligned} A_\mu &\to A_\mu + 2\pi \ell_\mu \\ F_{\mu\nu} &\to F_{\mu\nu} + 2\pi(\Delta_\mu \ell_\nu - \Delta_\nu \ell_\mu) \\ m_{\mu\nu} &\to m_{\mu\nu} + \Delta_\mu \ell_\nu - \Delta_\nu \ell_\mu \end{aligned} \tag{19.235}$$

which enforces periodicity. In this form, $F_{\mu\nu}$ does not have monopoles and satisfies the Bianchi identity. However, the two-form integer-valued fields $m_{\mu\nu}$ do not satisfy the Bianchi identity. In fact, the quantity

$$N(R) = \frac{1}{2} \epsilon_{\mu\nu\lambda} \Delta_\mu m_{\nu\lambda} \in \mathbb{Z} \tag{19.236}$$

defined on the sites of the dual lattice (i.e., the centers of the cubes), measures their integer-valued violation of the Bianchi identity. Thus, the integers $\{m_{\mu\nu}\}$ defined on the plaquettes represent Dirac strings of a configuration of monopoles.

The partition function is now

$$Z = \sum_{\{m_{\mu\nu}(r)\}} \int \mathcal{D}A_\mu \exp\left(-\frac{g^2}{4} \sum_{r,\mu\nu} m^2_{\mu\nu}(r) + i \sum_{r,\mu} A_\mu(r) \Delta_\nu m_{\mu\nu}(r) \right) \tag{19.237}$$

The integral over the gauge fields now leads to the constraint

$$\Delta_\mu m_{\mu\nu} = 0 \tag{19.238}$$

on every plaquette of the cubic lattice. Let us solve this constraint in terms of the integer-valued field $S(\boldsymbol{R})$ defined on the sites of the dual lattice (the centers of the elementary cubes):

$$m_{\mu\nu}(\boldsymbol{r}) = \epsilon_{\mu\nu\lambda} \Delta_\lambda S(\boldsymbol{R}) \tag{19.239}$$

The partition function reduces to the following:

$$\begin{aligned} Z &= \sum_{\{m_{\mu\nu}(\boldsymbol{r})\}} \prod_{\boldsymbol{r},\mu} \delta(\Delta_\mu m_{\mu\nu}) \exp\left(-\sum_{\boldsymbol{r},\mu\nu} \frac{g^2}{4} m_{\mu\nu}^2(\boldsymbol{r}) \right) \\ &= \sum_{\{S(\boldsymbol{R})\}} \exp\left(-\frac{g^2}{2} \sum_{\boldsymbol{R},\mu} (\Delta_\mu S(\boldsymbol{R}))^2 \right) \end{aligned} \tag{19.240}$$

Therefore, in close analogy with what we found for vortices in $D=2$, we find that the dual of compact electrodynamics is the discrete Gaussian model, but now in $D=3$ dimensions!

We can now repeat almost verbatim what we just did in two dimensions. We will conclude that compact electrodynamics is dual to the three-dimensional sine-Gordon theory. It is also dual to the three-dimensional (charge-neutral) Coulomb gas, defined in terms of integer-valued variables $m(\boldsymbol{R})$ that are defined at the centers of the cubes (the monopoles):

$$Z_{\text{CG}} = \sum_{\{m(\boldsymbol{R})\} \in \mathbb{Z}} \exp\left(-\frac{1}{2} \sum_{\boldsymbol{R},\boldsymbol{R}'} \left(\frac{2\pi}{g} \right)^2 m(\boldsymbol{R}) G(\boldsymbol{R}-\boldsymbol{R}') m(\boldsymbol{R}') \right) \tag{19.241}$$

Here, $G(\boldsymbol{R}-\boldsymbol{R}')$ is the three-dimensional Green function, which at long distances, behaves as

$$G(\boldsymbol{R}-\boldsymbol{R}') = \frac{a}{4\pi |\boldsymbol{R}-\boldsymbol{R}'|} \tag{19.242}$$

where a is the lattice spacing. So, now we find a three-dimensional Coulomb gas.

However, the physics is very different. In contrast to the $D=2$ case, the energy (Euclidean action) of a monopole is now *IR finite*. The energy-entropy argument then implies that the entropy (which goes as $\ln L^3$ in $D=3$) always wins over the energy (which is finite). Thus the monopoles always proliferate, and they are in the plasma phase. This conclusion also follows from the sine-Gordon picture, where one finds that the cosine operator is always relevant in all dimensions $D > 2$.

A straightforward calculation shows that the Wilson loop always obeys the area law for all values of the coupling constant g. To see this, we need to find the dual of the Wilson loop operator of a closed loop Γ, which, for simplicity, we take to be planar:

$$W[\Gamma] = \left\langle \exp\left(i \oint_\Gamma A_\mu \right) \right\rangle \tag{19.243}$$

The insertion of this operator modifies the constraint to be

$$\Delta_\nu m_{\mu\nu}(\boldsymbol{r}) = j^\Gamma_\mu(\boldsymbol{r}) \tag{19.244}$$

where j^Γ_μ is the current that defines the Wilson loop, on the closed contour Γ of the lattice (the boundary of an open surface Σ). The solution to this constraint is

$$m_{\mu\nu}(\boldsymbol{r}) = \epsilon_{\mu\nu\lambda}(\Delta_\lambda S(\boldsymbol{R}) + B_\lambda(\boldsymbol{R})) \tag{19.245}$$

where $B_\lambda(\boldsymbol{R})$ is an integer-valued gauge field defined on the links of the dual lattice such that

$$\epsilon_{\mu\nu\lambda}\Delta_\nu B_\lambda(\boldsymbol{R}) = j^\Gamma_\mu(\boldsymbol{r}) \tag{19.246}$$

which has the form of a discrete version of Ampère's law. A solution of this constraint is to set $B_\lambda = 1$ on all links of the dual lattice piercing the surface Σ.

In the dual theory, the expectation value of the Wilson loop is

$$W[\Gamma] = \frac{1}{Z} \sum_{\{S(\boldsymbol{R})\}} \exp\left(-\frac{g^2}{2} \sum_{\boldsymbol{R},\mu} (\Delta_\mu S(\boldsymbol{R}) + B_\mu(\boldsymbol{R}))^2 \right) \tag{19.247}$$

where Z is the partition function in eq. (19.240). This expectation value can be computed whether in the three-dimensional sine-Gordon formulation or in the three-dimensional Coulomb gas. In the latter representation, the expectation value of the Wilson loop is given by

$$W[\Gamma] = W[\Gamma]_{\text{Maxwell}} \times \left\langle \exp\left(4\pi i \sum_{\boldsymbol{R},\boldsymbol{R}'} \Delta_\mu B_\mu(\boldsymbol{R}) G(\boldsymbol{R}-\boldsymbol{R}') m(\boldsymbol{R}') \right) \right\rangle_{\text{CG}} \tag{19.248}$$

where $\langle \mathcal{O}[m] \rangle_{\text{CG}}$ denotes the expectation value of the operator $\mathcal{O}[m]$ in the three-dimensional Coulomb gas (whose partition function is given in eq. (19.241)), and

$$W[\Gamma]_{\text{Maxwell}} = \exp\left(-\frac{g^2}{2} \sum_{\boldsymbol{r},\boldsymbol{r}'} j^\Gamma_\mu(\boldsymbol{r}) G(\boldsymbol{r}-\boldsymbol{r}') j^\Gamma_\mu(\boldsymbol{r}') \right) \tag{19.249}$$

is the expectation value of the Wilson loop in Maxwell's theory (i.e., the abelian gauge theory) without magnetic monopoles, which are represented by the integer-valued degrees of freedom $\{m(\boldsymbol{R})\}$. Recall that in section 9.7, we calculated $W[\Gamma]_{\text{Maxwell}}$ and showed that the effective interaction between two static sources obeys the Coulomb law, $\sim 1/R^{d-2}$, which in 2+1 dimensions is a logarithmic interaction.

The Coulomb gas expectation value shown in eq. (19.248) is the contribution of the monopoles. The quantity $\Delta_\mu B_\mu(\boldsymbol{R})$ is nonzero only on the links of the dual lattice piercing the surface Σ (where $\Gamma = \partial\Sigma$). It is equal to $+1$ on the sites of the dual lattice just above Σ, and is -1 on the sites just below Σ. Hence, $\Delta_\mu B_\mu(\boldsymbol{R})$ is a uniform density of lattice dipoles perpendicular to Σ. In other words, it is essentially the negative of the self-energy of this uniform dipole distribution. A straightforward calculation shows that, for a very large Wilson loop, this expectation value behaves as $\exp(-\sigma \mathcal{A}[\Sigma])$, where $\mathcal{A}[\Sigma]$ is the minimal area spanned by the loop Γ, with a string tension $\sigma \propto g^2$.

Therefore, we conclude that in compact electrodynamics in 2+1 dimensions, large Wilson loops obey the area law, and small loops obey the Maxwell law. The crossover happens at a length scale ξ, the confinement scale, which in the Coulomb gas is the Debye screening length. On scales large compared to the screening length, the discreteness of the Coulomb charges is negligible. In this sense, confinement is the result of the proliferation (or "condensation") of magnetic monopoles.

What is the role of monopoles in higher dimensions and in other theories? In $d=4$ dimensions, compact QED has monopole solitons with finite energy. In the Euclidean path-integral, they are represented by closed loops, much the same way as vortex loops in (2+1)-dimensional theories with a global $U(1)$ symmetry (e.g., the three-dimensional classical XY model). However, while the three-dimensional XY model has a continuous phase transition, the numerical evidence suggests that the compact QED in four dimensions has a first-order confinement transition, and hence is not described by a fixed point. However, although non-abelian theories have monopole instantons in $D=4$ dimensions, as we saw earlier in this chapter, aside from suggestive numerical evidence, it has not been possible to show that confinement is actually driven by a proliferation (or condensation) of magnetic monopoles.

Exercises

19.1 Decay of the false vacuum

Consider a theory of a single real scalar field ϕ with Lagrangian

$$\mathcal{L}=\frac{1}{2}(\partial_\mu \phi)^2 - U(\phi) \tag{19.250}$$

where $U(\phi)=\lambda(\phi^2-\phi_0^2)^2-h(\phi-\phi_0)$. Assume that the theory is in the broken symmetry state with $\phi=-\phi_0$. For $h>0$, however small, this state is in an unstable (metastable) vacuum, and the field will decay by quantum tunneling to the state at $+\phi_0$. Thus the original state will have a lifetime determined by a decay rate Γ. Physically, this means that a bubble of the broken symmetry state $+\phi_0$ should appear and grow. This problem is closely related to the problem of nucleation in the theory of phase transitions.

Being a tunneling problem, the transition will not occur at any order in perturbation theory, and you must use instanton methods in D-dimensional Euclidean spacetime. Thus, consider solutions of the Euclidean classical equations of motion, $\partial_\mu \phi \partial_\mu \phi_c = U'(\phi_c)$, such that $\lim_{\tau\to\pm\infty}\phi(\boldsymbol{x},\tau)=0$, with $\lim_{|\boldsymbol{x}|\to\infty}=0$. Assume that the instanton is a spherically symmetric bubble in D Euclidean dimensions, with $0<r<\infty$ being the radial coordinate in D dimensions.

1) Show that the Euclidean equation of motion for a spherically symmetric instanton is

$$\frac{d^2\phi_c(r)}{dr^2}+\frac{D-1}{r}\frac{D\phi_c}{dr}=U'(\phi_c(r))$$

with the boundary conditions $\lim_{r\to\infty}\phi_c(r)=0$ and $\lim_{r\to 0}\frac{d\phi_c}{dr}=0$.

2) Regarding the variable r as "time" and $\phi_c(r)$ as the coordinate of a particle, find the form of the effective potential acting on the particle. What is the interpretation of the term that is first order in r derivatives?

3) Show that the instanton solution for this problem is the same as for a particle starting at $-\phi_0$ at time $r=0$ and returning to $-\phi_0$ after bouncing off a wall. Show that for h sufficiently small, the action of the bounce is twice the action $S_{\text{instanton}}$ of the one-dimensional instanton solution worked out in section 19.1.
4) Use an analytic continuation argument to show that the lifetime Γ of the unstable state is proportional to $V \exp(-S_{\text{instanton}})$, where V is the volume.

19.2 Nonsingular vortices in two dimensions
In section 19.3.2, we discussed vortex solutions of a complex scalar field in $D=2$ Euclidean dimensions and saw that it has vortex solutions. There we also saw that deep in the broken symmetry phase, where the complex scalar field is equivalent to an $O(2)$ nonlinear sigma model, the energy of a vortex scales at long distances as $\ln R$, where R is the large radius of the $D=2$ system. However, in this regime, the vortices become singular at short distances and depend sensitively on the short-distance cutoff. However, in section 19.4, we saw that the $O(3)$ nonlinear sigma model has finite Euclidean-action instanton solutions. In this problem, you will construct a nonsingular vortex in $D=2$ dimensions by considering a nonlinear sigma model in which the global $O(3)$ symmetry is broken down to $O(2)$. The Euclidean action of this system is

$$\mathcal{S} = \int d^2x \left[\frac{1}{2g} (\nabla \boldsymbol{n})^2 + \Delta (\boldsymbol{n} \cdot \boldsymbol{n}_0)^2 \right] \tag{19.251}$$

where, as before, $\boldsymbol{n}^2 = 1$. Here $\boldsymbol{n}_0$ is an arbitrary unit vector, and Δ is a coupling constant. For simplicity, choose $\boldsymbol{n}_0 = \boldsymbol{e}_z$.

1) Show that the classical scale invariance of the $D=2$ $O(3)$ nonlinear sigma model is broken by the anisotropy term. Show that for anisotropy $\Delta > 0$, the classical ground states lie on the xy plane.
2) In what follows, it will be useful to use the parametrization

$$\boldsymbol{n}(\boldsymbol{x}) = (\cos\phi(\boldsymbol{x}) \sin\theta(\boldsymbol{x}), \sin\phi(\boldsymbol{x}) \sin\theta(\boldsymbol{x}), \cos\theta(\boldsymbol{x}))$$

in terms of two fields: $0 < \phi(x) < 2\pi$ and $0 < \theta(x) < \pi$. Find the expression of the Euclidean action in terms of the fields ϕ and θ.
3) In the two-dimensional plane $\boldsymbol{x} = (r\cos\varphi, r\sin\varphi)$, consider a rotationally invariant field configuration $(\phi(r,\varphi), \theta(r,\varphi))$ with the asymptotic behaviors: (a) $\lim_{r\to\infty} \theta(r,\varphi) = \pi/2$, and (b) $\lim_{r\to 0} \theta(r,\varphi) = 0$, with $\phi(r,\varphi) = \varphi$ for all r. Show that this configuration asymptotically is a vortex and that its topological charge is 1/2 of the instanton charge (known as a "meron").
4) Write the expression of the Euclidean action for this configuration for the field θ, on rescaling the radial coordinate $r = r_0 e^t$ (where $r_0 = 1\sqrt{g\Delta}$). Show that the extremal solution of this Euclidean action obeys the equation of motion of a pendulum (parametrized by $\theta(t)$) in a time-dependent potential. Sketch the qualitative form of the solution, and give an interpretation of the scale r_0.

20

Anomalies in Quantum Field Theory

20.1 The chiral anomaly

In classical field theory, discussed in chapter 3, we learn that symmetries dictate the existence of conservation laws. At the quantum level, conservation laws of continuous symmetries are embodied in Ward identities, which dictate the behavior of correlation functions. However, we will now see that in QFT, there are many instances of symmetries of the classical theory that do not survive at the quantum level due to a quantum *anomaly*. Most examples of anomalies involve Dirac fermions that are massless at the level of the Lagrangian.

Global and gauge symmetries of theories of Dirac fermions involve understanding their currents, which are products of Dirac operators at short distances. Such operators require a consistent definition (and a normal-ordering prescription). Hence, some sort of regularization is needed for a proper definition of current operators. The problem is that not all of the symmetries can simultaneously survive regularization. While in scalar field theories (at least in flat spacetimes), this is not a problem, it turns out to be a significant problem in theories with massless Dirac fermions, in which some formally conserved currents become anomalous.

In path-integral language, a symmetry is *anomalous* if the action is invariant under the symmetry but the measure of the path integral is not. In this sense, quantum anomalies often arise in the process of regularization of a QFT. We will see, however, that they are closely related to topological considerations as well.

The subject of anomalies in QFT is discussed in many excellent textbooks, such as the book by Michael Peskin and Daniel Schroeder (1995). It is also a subject that can be fairly technical and mathematically quite sophisticated. For these reasons, we will keep the presentation as physically transparent and as simple as possible, even at the price of some degree of rigor.

The prototype of a quantum anomaly is the axial (or chiral) anomaly. Classically, and at the free-field level, a theory with a single massless Dirac fermion has two natural continuous global $U(1)$ symmetries: gauge invariance and chiral symmetry. Then, as we saw in chapter 7, Noether's theorem implies the existence of two currents: the gauge current $j_\mu = \bar{\psi}\gamma_\mu\psi$, and the axial (chiral) current $j_\mu^5 = \bar{\psi}\gamma_\mu\gamma^5\psi$. Superficially, in the massless theory at the free-field level, the gauge and the chiral currents are separately conserved:

$$\partial^\mu j_\mu = 0, \qquad \partial^\mu j_\mu^5 = 0 \tag{20.1}$$

A Dirac mass term is gauge invariant, but it breaks the chiral symmetry explicitly. The chiral current is not conserved in its presence. Indeed, using the Dirac equation, one finds that conservation (eq. (20.1)) of the axial current j^5_μ is modified to

$$\partial^\mu j^5_\mu = 2m\, i\bar{\psi}\gamma^5\psi \tag{20.2}$$

where m is the mass.

The anomaly was first discovered in the computation of so-called triangle Feynman diagrams, involved in the study of decay processes of a neutral pion to two photons, $\pi^0 \to 2\gamma$, by Adler (1969), and Bell and Jackiw (1969). They and subsequent authors showed that, in any gauge-invariant regularization of the theory, an extra term appears on the right-hand side of eq. (20.2), even in the massless limit, $m \to 0$. This term is the chiral (or axial) anomaly.

The form of the anomaly term depends on the dimensionality, and it turns out to have a topological meaning. Although the anomaly arises from short-distance singularities in the QFT, it is a finite and universal term. Here "universality" means that although its existence depends on how the theory is regularized, its value depends only on the symmetries that are preserved by the regularization and not on its detailed form.

The important physical implication of the anomaly is that in its presence, the anomalous current is not conserved, and consequently, the associated symmetry cannot be gauged. This poses strong constraints on what theories of particle physics are physically sensible, particularly in the weak-interaction sector.

20.2 The chiral anomaly in 1+1 dimensions

We first discuss the chiral anomaly in (1+1)-dimensional theories of Dirac fermions. For simplicity, consider a theory of a single massless Dirac fermion ψ coupled to a background gauge field A_μ. The Lagrangian is

$$\mathcal{L} = \bar{\psi} i\partial\!\!\!/ \psi + e\bar{\psi}\gamma^\mu\psi\, A_\mu \tag{20.3}$$

The standard derivation of the anomaly uses subtle (and important) arguments about regulators and what symmetries they preserve (or break). Although we will do that shortly, it is worthwhile to use first a transparent and physically compelling argument, originally due to Nielsen and Ninomiya (1983).

20.2.1 The anomaly as particle-antiparticle pair creation

Recall that in 1+1 dimensions, the Dirac fermion is a two-component spinor. Let us work in the chiral basis, which simplifies the analysis. In the chiral basis, the Dirac spinor is $\psi = (\psi_R, \psi_L)^t$ (where the upper index t means transpose). In (1+1)-dimensional Minkowski spacetime, the Dirac matrices are

$$\gamma_0 = \sigma_1, \qquad \gamma_1 = i\sigma_2, \qquad \gamma_5 = \sigma_3 \tag{20.4}$$

where σ_1, σ_2, and σ_3 are the standard 2×2 Pauli matrices. In this basis, the gauge current j_μ and the chiral current j^5_μ are

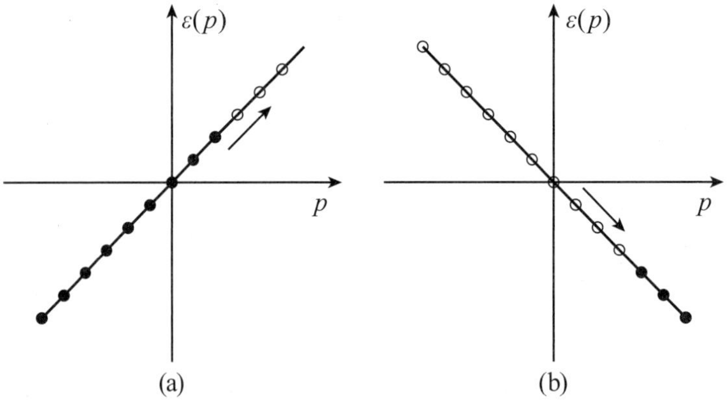

Figure 20.1 Chiral anomaly in 1+1 dimensions: pair creation by a uniform electric field E and the chiral anomaly. (a) The Fermi point of the right-movers increases with time, reflecting particle creation. (b) The Fermi point of left-movers decreases, reflecting the creation of antiparticles (holes).

$$j_0 = \psi_R^\dagger \psi_R + \psi_L^\dagger \psi_L, \qquad j_1 = \psi_R^\dagger \psi_R - \psi_L^\dagger \psi_L$$
$$j_0^5 = \psi_R^\dagger \psi_R - \psi_L^\dagger \psi_L, \qquad j_1^5 = -(\psi_R^\dagger \psi_R + \psi_L^\dagger \psi_L) \tag{20.5}$$

Hence, j_0 measures the total density of right- and left-moving fermions, and j_1 measures the difference of the densities of right- and left-moving fermions. Notice that the components of the chiral current j_μ^5 essentially switch the roles of charge and current. In short, we can write

$$j_\mu^5 = \epsilon^{\mu\nu} j_\nu \tag{20.6}$$

In particular, the total gauge charge $Q = \int dx\, j_0(x)$ of a state is the total number of right- and left-moving particles, $N_R + N_L$, and the total chiral charge of state $Q^5 = \int dx\, j_0^5(x)$ is

$$Q^5 = N_R - N_L \equiv N_R + \bar{N}_L \tag{20.7}$$

where $\bar{N}_L$ is the number of left-moving antiparticles.

In the chiral basis, the Dirac equation in the temporal gauge, $A_0 = 0$, is

$$i\partial_0 \psi_R(x) = (-i\partial_1 - A^1)\psi_R(x), \qquad i\partial_0 \psi_L(x) = (i\partial_1 - A^1)\psi_L(x) \tag{20.8}$$

In the temporal gauge, a uniform electric field is $E = \partial_0 A^1$, and A^1 increases monotonically in time. The Dirac equation, eq. (20.8), states that as A^1 increases, the Fermi momentum $p_F = \varepsilon_F$ (with ε_F being the Fermi energy) increases by the amount

$$\frac{dp_F}{dx_0} = eE \tag{20.9}$$

The total number of right-moving particles is N_R. The density of states of a system of length L is $L/(2\pi)$. So the rate of change in the number of right-moving particles is (see figure 20.1)

$$\frac{dN_R}{dx_0} = \frac{1}{L}\frac{L}{2\pi}\frac{d\varepsilon_F}{dx_0} = \frac{e}{2\pi}E \tag{20.10}$$

Assume that the UV regulator of the theory is such that the total fermion number is conserved, $Q=0$:

$$Q=\int_{-\infty}^{\infty} dx\, j_0(x) = N_R + N_L = 0 \tag{20.11}$$

Therefore, if $N_R > 0$ increases, then $N_L < 0$ decreases. Or, what is the same, the number of left-moving *antiparticles* must increase by the same amount that the N_R *particles* increase. This leads to the conclusion that the total chiral charge, $Q^5 = N_R - N_L = N_R + \bar{N}_L$, must increase at the rate

$$\frac{dQ^5}{dx_0} = \frac{dN_R}{dx_0} + \frac{d\bar{N}_L}{dx_0} = \frac{e}{\pi}E \tag{20.12}$$

But in this process, the total electric (gauge) charge Q is conserved. Notice that the details of the UV regularization do not affect this result, provided the regularization is gauge invariant. Hence, the chiral (axial) charge is not conserved if the gauge charge is conserved (and vice versa).

In covariant notation, the nonconservation of the chiral current j_μ^5 is

$$\partial^\mu j_\mu^5 = \frac{e}{2\pi}\epsilon_{\mu\nu}F^{\mu\nu} \tag{20.13}$$

This is the chiral (axial) anomaly equation in 1+1 dimensions. Notice that it tells us that the total chiral charge is e times a topological invariant, the total instanton number (the flux) of the gauge field!

20.3 The chiral anomaly and abelian bosonization

Let us now reexamine this problem, taking care of the short-distance singularities. Let $|0\rangle_D$ be the vacuum state of the theory of free massless Dirac fermions (i.e., the filled Dirac sea). We will normal-order the operators with respect to the Dirac vacuum state. We need to be careful when treating composite operators, such as the currents, since they are products of Dirac fields at short distances.

To this effect, let us carefully examine the algebra obeyed by the density and current operators. This current algebra leads to the concept of bosonization, introduced by Mattis and Lieb (1965), based on results by Schwinger (1959) and rediscovered (and expanded) by Coleman (1975), Mandelstam (1975), and Luther and Emery (1974) in the 1970s.

The vacuum state should be charge neutral and should have zero total momentum and zero current. Thus both the charge density j_0 and the charge current j_1 should annihilate the vacuum. As usual, the zero current condition is automatic, while the charge neutrality of the vacuum is ensured by a proper subtraction. Since the charge and the current densities are fermion bilinears, they behave as bosonic operators. Naively, one would expect that they should commute with each other. We will now see that this expectation is wrong.

Let us introduce the operators for the right- and left-moving densities, naively defined as $j_R(x_1) = \psi_R^\dagger(x_1)\psi_R(x_1)$, and $j_L(x_1) = \psi_L^\dagger(x_1)\psi_L(x_1)$, where x_1 is the space coordinate. The propagators of the right- and left-moving fields are

$$\langle \psi_R^\dagger(x_0, x_1)\psi_R(0,0)\rangle = -\frac{i}{2\pi(x_0 - x_1 + i\epsilon)}$$

$$\langle \psi_L^\dagger(x_0, x_1)\psi_L(0,0)\rangle = +\frac{i}{2\pi(x_0 + x_1 + i\epsilon)} \tag{20.14}$$

and diverge at short distances. Define the normal-ordered densities by a point-splitting procedure (at equal times)

$$j_R(x_1) =: j_R(x_1) : + \lim_{\epsilon \to 0} \langle \psi_R^\dagger(x_1 + \epsilon)\psi_R(x_1 - \epsilon)\rangle \tag{20.15}$$

and similarly with j_L. Here the expectation value is computed in the free massless Dirac vacuum. From the expressions for the propagators, we see that

$$\begin{aligned} \langle \psi_R^\dagger(x_1 + \epsilon)\psi_R(x_1 - \epsilon)\rangle &= \frac{i}{4\pi\epsilon} \\ \langle \psi_L^\dagger(x_1 + \epsilon)\psi_L(x_1 - \epsilon)\rangle &= -\frac{i}{4\pi\epsilon} \end{aligned} \tag{20.16}$$

Then the equal-time commutator of two right-moving currents is found to be (on taking the limit $\epsilon \to 0$):

$$[j_R(x_1), j_R(x_1')] = -\frac{i}{2\pi}\partial_1\delta(x_1 - x_1') \tag{20.17}$$

Similarly, we find

$$[j_L(x_1), j_L(x_1')] = \frac{i}{2\pi}\partial_1\delta(x_1 - x_1') \tag{20.18}$$

Clearly the right-moving densities (and the left-moving densities) do not commute with each other!

These results imply that the equal-time commutators of properly regularized charge density and current operators j_0 and j_1 are

$$[j_0(x_1), j_1(x_1')] = -\frac{i}{\pi}\partial_1\delta(x_1 - x_1'), \quad [j_0(x_1), j_0(x_1')] = [j_1(x_1), j_1(x_1')] = 0 \tag{20.19}$$

In other words, the currents of a theory of free massless Dirac fermions (which has a global $U(1)$ symmetry) obey the algebra of eqs. (20.17)–(20.19). This is known as the $U(1)$ Kac-Moody current algebra.

The singularities of the commutators of currents are known as Schwinger terms. Recall that we obtained a similar result in section 10.11, where we discussed the local conservation laws of a system of non-relativistic fermions at finite density. This is not an accident, since in one space dimension, at low energy, a system of nonrelativistic fermions at finite density is equivalent to the theory of a massless Dirac field (where the speed of light is identified with the Fermi velocity).

The $U(1)$ current algebra of eq. (20.19) is reminiscent of the equal-time canonical commutation relations of a scalar field $\phi(x)$ and its canonical momentum $\Pi(x)$:

$$[\phi(x_1), \Pi(x_1')] = i\delta(x_1 - x_1') \tag{20.20}$$

Indeed, we can identify the normal-ordered density $j_0(x_1)$ with

$$j_0(x) = \frac{1}{\sqrt{\pi}}\partial_1\phi(x) \tag{20.21}$$

and the normal-ordered current $j_1(x)$ with

$$j_1(x) = -\frac{1}{\sqrt{\pi}}\Pi(x) = -\frac{1}{\sqrt{\pi}}\partial_0\phi(x) \tag{20.22}$$

With these operator identifications, the canonical commutation relations of the scalar field $\phi(x)$, eq. (20.20), imply

$$\frac{1}{\pi}[\partial_1\phi(x_1), \Pi(x_1')] = \frac{i}{\pi}\partial_1\delta(x_1 - x_1') \tag{20.23}$$

which reproduces the Schwinger term of the $U(1)$ Kac-Moody current algebra of eq. (20.19).

The operator identifications of eq. (20.21) and eq. (20.22) can be written in the Lorentz covariant form

$$j_\mu = \frac{1}{\sqrt{\pi}}\epsilon_{\mu\nu}\partial^\nu\phi \tag{20.24}$$

which satisfies the local conservation condition

$$\partial^\mu j_\mu = 0 \tag{20.25}$$

Thus, the regularization we adopted is consistent with the conservation of the $U(1)$ current. The identification (or mapping) of the $U(1)$ fermionic current of eq. (20.24) shows that there can be a mapping between operators of the theory of massless Dirac fermions to operators of a theory of scalar fields, which are bosons. Such mappings are called *bosonization.*

However, is it compatible with the conservation of the chiral current j_μ^5? In eq. (20.6), we showed that the chiral and the $U(1)$ currents are related to each other by $j_\mu^5 = \epsilon_{\mu\nu}j^\nu$. Therefore, the divergence of the chiral current is identified with

$$\partial^\mu j_\mu^5(x) = \epsilon^{\mu\nu}j_\nu(x) = \frac{1}{\sqrt{\pi}}\epsilon^{\mu\nu}\epsilon_{\nu\lambda}\partial^\lambda\phi(x) = \frac{1}{\sqrt{\pi}}\partial^2\phi \tag{20.26}$$

Therefore,

$$\partial^\mu j_\mu^5(x) = 0 \Leftrightarrow \partial^2\phi = 0 \tag{20.27}$$

and the chiral (or axial) current is conserved if and only if the scalar field is free and massless, which has the Lagrangian $\mathcal{L}_B$,

$$\mathcal{L}_B = \frac{1}{2}(\partial_\mu\phi)^2 \tag{20.28}$$

Note also that the Hamiltonian density of the Dirac theory

$$\mathcal{H}_D = -(\psi_R^\dagger i\partial_1\psi_R - \psi_L^\dagger i\partial_1\psi_L) \tag{20.29}$$

must be identified with the Hamiltonian density of the massless scalar field

$$\mathcal{H}_B = \frac{1}{2}\left(\Pi^2 + (\partial_1\phi)^2\right) \tag{20.30}$$

which, after normal-ordering, can be expressed in terms of the $U(1)$ density and current in the (Sugawara) form

$$\mathcal{H} = \frac{\pi}{2}\left(j_0^2 + j_1^2\right) \tag{20.31}$$

To see how this is related to the chiral anomaly, let us couple the Dirac theory to an external $U(1)$ gauge field A_μ. The coupling term in the Dirac Lagrangian is

$$\mathcal{L}_{\text{int}} = eA^\mu j_\mu \tag{20.32}$$

where j_μ is the $U(1)$ Dirac current. In the theory of the massless scalar field, we identify this term with

$$\mathcal{L}_{\text{int}} = \frac{e}{\sqrt{\pi}} \epsilon_{\mu\nu} \partial^\nu \phi \, A^\mu \tag{20.33}$$

which states that the coupling of the fermions to the gauge field is equivalent to the coupling of the scalar field to a source

$$J(x) = \frac{e}{2\sqrt{\pi}} \epsilon_{\mu\nu} F^{\mu\nu} \equiv \frac{e}{2\sqrt{\pi}} F^* \tag{20.34}$$

where F^* is the dual of the field strength $F^{\mu\nu}$, which in 1+1 dimensions is a scalar.

The equation of motion of the scalar field ϕ now is

$$\partial^2 \phi = \frac{e}{\sqrt{\pi}} \epsilon_{\mu\nu} \partial^\mu A^\nu \tag{20.35}$$

Therefore, the divergence of the axial current j_μ^5 does not vanish and is given by

$$\partial^\mu j_\mu^5 = \frac{1}{\sqrt{\pi}} \partial^2 \phi = \frac{e}{2\pi} \epsilon_{\mu\nu} F^{\mu\nu} \tag{20.36}$$

which reproduces the chiral (axial) anomaly of eq. (20.13)!

Let us consider a system with periodic boundary conditions in space, which we take to be a circle of circumference L. The total charge is

$$Q = e \int_0^L dx_1 \, j_0(x_1) \tag{20.37}$$

which is measured relative to the vacuum state, which is neutral, $Q_{\text{vacuum}} = 0$. It then follows that the charge must be quantized and is an integer multiple of the electric charge, $Q = Ne$. However, using the identification of the current, eq. (20.24), results in the condition

$$Q = \frac{e}{\sqrt{\pi}} \int_0^L dx_1 \, \partial_1 \phi(x_1) = \frac{e}{\sqrt{\pi}} \Delta\phi \tag{20.38}$$

where we have defined

$$\Delta\phi = \phi(x_1 + L) - \phi(x_1) \tag{20.39}$$

Therefore, in the sector of the Hilbert space with $N \in \mathbb{Z}$ fermions, the scalar field must obey the generalized periodic boundary condition (here $x = x_1$):

$$\phi(x + L) = \phi(x) + 2\pi N R_\phi \tag{20.40}$$

where

$$R_\phi = \frac{1}{2\sqrt{\pi}} \tag{20.41}$$

Hence, the scalar field is *compactified*, and R_ϕ is the compactification radius.

Therefore the target space of this scalar field is not the real numbers, $\mathbb{R}$, but the circle S^1, whose radius is R_ϕ. This fact, which is a consequence of the charge quantization of the Dirac fermions, restricts the allowed observables of the bosonized side of the theory to be invariant under shifts $\phi \to \phi + 2\pi n R_\phi$, with $n \in \mathbb{Z}$. For example, the *vertex operators* V_α,

$$V_\alpha(x) = \exp(i\alpha\phi(x)) \tag{20.42}$$

are allowed only for the values $\alpha = 2\pi n R_\phi = n\sqrt{\pi}$.

Using methods similar to those we have discussed here, Mandelstam (1975) showed that the bosonic counterpart of the Dirac fermion operator, ψ_R and ψ_L, can be identified as

$$\begin{aligned}
\psi_R(x) &= \frac{1}{\sqrt{2\pi a}} :\exp\left(-\sqrt{\pi}\int_{-\infty}^{x_1} dx_1'\, \Pi(x_0, x_1') + i\sqrt{\pi}\phi(x)\right): \\
&\equiv \frac{1}{\sqrt{2\pi a}} :\exp(2i\sqrt{\pi}\phi_R(x)): \\
\psi_L(x) &= \frac{1}{\sqrt{2\pi a}} :\exp\left(-\sqrt{\pi}\int_{-\infty}^{x_1} dx_1'\, \Pi(x_0, x_1') - i\sqrt{\pi}\phi(x)\right): \\
&\equiv \frac{1}{\sqrt{2\pi a}} :\exp(-2i\sqrt{\pi}\phi_L(x)):
\end{aligned} \tag{20.43}$$

where a is a short-distance cutoff, and $:\mathcal{O}:$ denotes normal ordering. Notice that in terms of the scalar field, the Mandelstam operators are essentially a product of an operator that creates a kink (a soliton), which shifts the field by $\sqrt{\pi}$, and a vertex operator that measures the charge. After some algebra, one can show that these operators create states that carry the unit of charge.

Eq. (20.43) introduces the fields ϕ_R and ϕ_L, the right- and left-moving components of the scalar field, respectively,

$$\phi = \phi_R + \phi_L, \qquad \vartheta = -\phi_R + \phi_L \tag{20.44}$$

where

$$\vartheta(x) = \int_{-\infty}^{x_1} dx_1'\, \Pi(x_0, x_1') \tag{20.45}$$

is called the dual field. The fields ϕ and ϑ satisfy the Cauchy-Riemann equations

$$\partial_\mu \phi = \epsilon_{\mu\nu}\partial^\nu \vartheta \tag{20.46}$$

We also record here the identification of the Dirac mass operator $\bar{\psi}\psi$ and of the chiral mass operator $i\bar{\psi}\gamma^5\psi$:

$$\bar{\psi}\psi = \frac{1}{2\pi a} :\cos(2\sqrt{\pi}\,\phi):, \qquad i\bar{\psi}\gamma^5\psi = \sin(2\sqrt{\pi}\,\phi) \tag{20.47}$$

These identifications can be derived using the operator product expansion (OPE).

In particular, these operator identifications imply that the Lagrangian of the free massive Dirac theory

$$\mathcal{L} = \bar{\psi} i \partial\!\!\!/ \psi - m\bar{\psi}\psi \tag{20.48}$$

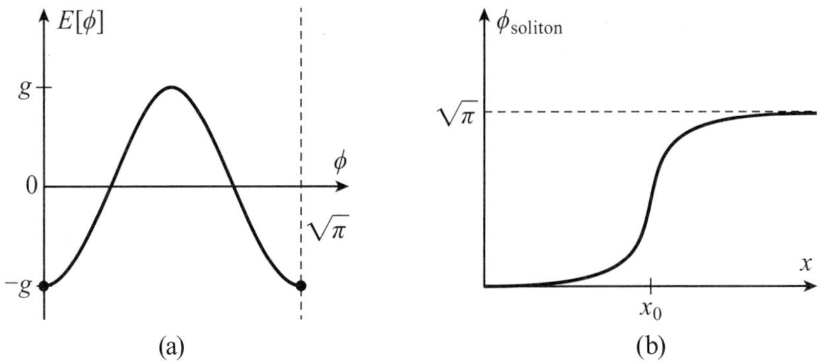

Figure 20.2 (a) The sine-Gordon potential energy of a periodic function of ϕ with period $\sqrt{\pi}$; the black dots are the classical vacua at $\phi = 0, \sqrt{\pi}$. (b) Sketch of the sine-Gordon soliton.

maps to the sine-Gordon Lagrangian

$$\mathcal{L}_B = \frac{1}{2}(\partial_\mu \phi)^2 - g : \cos(2\sqrt{\pi}\,\phi) : \tag{20.49}$$

with $g = m/(2\pi a)$.

20.4 Solitons and fractional charge

We will now see that theories of scalar fields coupled to Dirac fermions have states with fractional charge if the scalar fields have solitons. For simplicity let us focus on (1+1)-dimensional theories.

20.4.1 The sine-Gordon soliton

We already saw a hint of this in the derivation of the connection between the sine-Gordon theory and a massive Dirac fermion. The Lagrangian of the sine-Gordon theory is

$$\mathcal{L}_{\text{SG}} = \frac{1}{2}(\partial_\mu \phi)^2 + g \cos(2\sqrt{\pi}\,\phi) \tag{20.50}$$

This theory is invariant under the discrete shifts of the field, $\phi(x) \to \phi(x) + 2\pi n/\beta$, where $n \in \mathbb{Z}$. The Hamiltonian density

$$\mathcal{H}_{\text{SG}} = \frac{1}{2}\Pi^2 + \frac{1}{2}(\partial_x \phi)^2 - g \cos(2\sqrt{\pi}\,\phi) \tag{20.51}$$

is, of course, invariant under the same global symmetry.

The classical vacua of this theory are $\phi_c = 2\pi N/\beta$. In addition to these uniform classical states, this theory has solitons that connect the uniform states. Let $\phi(x)$ be a static (i.e., time-independent) field configuration (here x is the space coordinate). The total energy of such a configuration is (as shown in figure 20.2a)

$$E[\phi] = \int_{-\infty}^{\infty} dx \left[\frac{1}{2} \left(\frac{d\phi}{dx} \right)^2 - g\cos(2\sqrt{\pi}\,\phi(x)) \right] \tag{20.52}$$

We are interested in nonuniform configurations $\phi(x)$ that extremize $E[\phi]$ and interpolate between two classical vacuum states. These classical solitons are solutions of the Euler-Lagrange equation

$$\frac{d^2\phi}{dx^2} = 2g\sqrt{\pi}\ \sin(2\sqrt{\pi}\,\phi) \tag{20.53}$$

with the boundary conditions

$$\lim_{x\to-\infty} \phi(x) = 0, \qquad \lim_{x\to+\infty} \phi(x) = \sqrt{\pi} \tag{20.54}$$

The classical equation for the static soliton is the same as the equation for a physical pendulum of coordinate ϕ as a function of time. This problem is similar to our discussion of quantum tunneling and instantons in chapter 19. The classical energy of the pendulum is the negative of the sine-Gordon potential, and the classical vacuum states correspond to the top of the pendulum potential. The soliton is the solution that rolls down the hill from the peak at $\phi = 0$ to the valley at $\phi = \sqrt{\pi}/2$ and then climbs to the summit of the next hill at $\phi = \sqrt{\pi}$. The soliton solution shown in figure 20.2b is

$$\phi_{\text{soliton}}(x) = \frac{1}{\sqrt{\pi}} \arccos\left[\tanh(2\sqrt{\pi}\ \sqrt{g}\,(x - x_0))\right] \tag{20.55}$$

where x_0 is a zero mode of the soliton. The total energy of the soliton (measured from the energy of the uniform classical solution) is finite: $E_{\text{soliton}} = 8\sqrt{g}/\beta$. In this soliton solution, the total change of the angle is $\Delta\phi_{\text{soliton}} = \sqrt{\pi}$ which, using eq. (20.38), corresponds to a fermion charge $Q = e\frac{\Delta\phi}{\sqrt{\pi}} = e$. Thus, the fermion of the free massive Dirac theory is the soliton of the sine-Gordon theory.

20.4.2 Fractionally charged solitons

Now consider a theory, again in 1+1 dimensions, with two real scalar fields, φ_1 and φ_2, coupled to a massless Dirac fermion. Assume that the scalar fields have a nonvanishing vacuum expectation value. The Lagrangian is

$$\mathcal{L} = \bar{\psi} i \partial\!\!\!/ \psi + g\bar{\psi}(\varphi_1 + i\gamma^5\varphi_2)\psi \tag{20.56}$$

Note that a constant value of $\langle\varphi_1\rangle$ is the same as a Dirac mass $m = g\langle\varphi_1\rangle$, and that a constant value of $\langle\varphi_2\rangle$ is the same as a chiral mass $m_5 = g\langle\varphi_2\rangle$. In section 17.4, we discussed the chiral and the nonchiral Gross-Neveu models (which we solved in their large-N limit), where these mass terms arise from spontaneous breaking of the chiral symmetry,

$$\psi \to e^{i\eta\gamma_5}\psi \tag{20.57}$$

where η is an arbitrary angle for the chiral model and is equal to $\pi/2$ for the model with a discrete chiral symmetry.

We can make the relation with chiral symmetry more apparent by writing the scalar fields as

$$\varphi_1 = |\varphi| \cos\theta, \qquad \phi_2 = |\varphi| \sin\theta \tag{20.58}$$

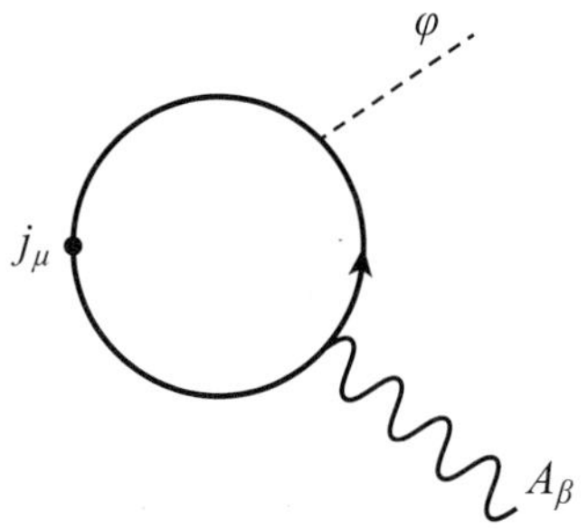

Figure 20.3 Induced current $\langle j_\mu \rangle$ by a small, adiabatic change in the phase of the complex scalar field φ.

where $|\varphi| = (\varphi_1^2 + \varphi_2^2)^{1/2}$. The Lagrangian now is

$$\mathcal{L} = \bar{\psi} i \partial\!\!\!/ \psi + g|\varphi| \bar{\psi} e^{i\theta\gamma_5} \psi \tag{20.59}$$

Thus, a global chiral transformation with angle θ, eq. (20.57), changes the phase field: $\theta \to \theta + 2\eta$.

Now consider the case in which the scalar field (φ_1, φ_2) has a soliton, such that the phase θ winds by $\Delta\theta$ as x goes from $-\infty$ to $+\infty$. Assume that the soliton is static, so that this winding is adiabatic. The question that we consider is: What is the charge of the soliton?

Goldstone and Wilczek used a perturbative calculation (the Feynman diagram of figure 20.3) in the regime where $|\varphi|$ is large compared to the gradients of φ_1 and φ_2. They showed that there is an induced charge current $\langle j_\mu \rangle$ given by (Goldstone and Wilczek, 1981):

$$\langle j_\mu \rangle = \frac{1}{2\pi} \epsilon_{\mu\nu} \epsilon^{ab} \frac{\varphi_a \partial_\nu \varphi_b}{|\varphi|^2} = \frac{1}{2\pi} \epsilon_{\mu\nu} \partial^\nu \tan^{-1}(\varphi_2/\varphi_1) \tag{20.60}$$

This equation implies that the total charge Q is

$$Q = \int_{-\infty}^{\infty} dx\, j_0(x) = \frac{1}{2\pi} \Delta \tan^{-1}(\varphi_2/\varphi_1) \tag{20.61}$$

The same result can be found using the bosonization identities. Indeed, using eqs. (20.47), we can readily see that the bosonized expression of the Lagrangian of eq. (20.56) is

$$\mathcal{L}_B = \frac{1}{2}(\partial_\mu \phi)^2 + \frac{g}{2\pi a} \cos(2\sqrt{\pi}\phi - \theta) \tag{20.62}$$

The minimum of the energy of this sine-Gordon theory occurs for $\phi = \theta/\sqrt{4\pi}$.

A soliton can be represented by a fermion with Dirac mass m, and hence $\varphi_1 = m/g$, and a φ_2 field that approaches the values $\pm v$ as $x \to \pm\infty$. In this case, the phase changes by $\Delta\theta = 2\tan^{-1}(gv/m)$. Therefore, to minimize the energy, the sine-Gordon field ϕ winds by $\Delta\phi = \Delta\theta/\sqrt{4\pi}$. Then the total charge Q induced by the soliton is

$$Q_{\text{soliton}} = e\frac{\Delta\phi}{\sqrt{\pi}} = e\frac{\Delta\theta}{2\pi} = \frac{e}{\pi} \tan^{-1}\left(\frac{gv}{m}\right) \tag{20.63}$$

Of particular interest is the limiting case $m \to 0$, for which the soliton charge approaches the half-integer value:

$$Q_{\text{soliton}} = \frac{e}{2}, \quad \text{and} \quad \Delta\theta = \pi \tag{20.64}$$

This fractionally charged soliton was known to exist in ϕ^4 theory (see section 19.2) coupled to Dirac fermions (Jackiw and Rebbi, 1976), and in the one-dimensional conductor polyacetylene (Su et al., 1979). For example, consider a theory with one massless Dirac fermion coupled to a ϕ^4 field in its $\mathbb{Z}_2$ broken symmetry state with vacuum expectation value $\langle \phi \rangle = \phi_0$. The Lagrangian is

$$\mathcal{L} = \bar{\psi} i \partial\!\!\!/ \psi - g\phi\bar{\psi}\psi + \frac{1}{2}(\partial_\mu \phi)^2 - \lambda(\phi^2 - \phi_0^2)^2 \tag{20.65}$$

This theory has a discrete chiral symmetry, under which $\bar{\psi}\psi \mapsto -\bar{\psi}\psi$ and $\phi \mapsto -\phi$. Now consider a static soliton state of the ϕ field whose configuration is $\phi(x) = \pm\phi_0 \tanh\left(\frac{(x-x_0)}{\sqrt{2}\xi}\right)$ (see eq. (19.55)). The arguments given above predict that the soliton has fractional charge $+1/2$ (or more properly, fractional fermion number).

How does the presence of the soliton affect the fermionic spectrum? The stationary states of the Dirac Hamiltonian for the two-component spinor can be written as

$$H_D[\phi]\psi = \alpha i \partial_1 \psi + \beta g \phi(x_1)\psi \tag{20.66}$$

where α and β are two Pauli matrices, which can always be chosen to be real, say, σ_1 and σ_3, respectively. The Dirac Hamiltonian anticommutes with the third Pauli matrix, σ_2. Thus if ψ is an eigenstate of energy E, then $\sigma_2\psi$ is an eigenstate with energy $-E$. Therefore, the spectrum of states with nonzero eigenvalues is symmetric, and there is a one-to-one correspondence between positive and negative energy states.

However, if the field $\phi(x)$ has a soliton (or an antisoliton), there is a state whose energy is exactly $E=0$, a zero mode. To construct the state, we rewrite the Dirac equation as

$$i\beta\alpha\partial_1\psi(x_1) + g\phi(x_1)\psi(x_1) = 0 \tag{20.67}$$

Note that $\beta\alpha = \gamma_1$, which is an antihermitian matrix, $\gamma_1^\dagger = -\gamma_1$, and hence its eigenvalues are $\pm i$. We can choose a basis in which the zero modes are eigenstates of γ_1, and write them as $\psi_\pm(x_1)\chi_\pm$, where $\chi_\pm = (1,0)^\dagger$ or $(0,1)^\dagger$, respectively. The functions $\psi_\pm(x_1)$ satisfy the first-order differential equation

$$\partial_1\psi_\pm(x_1) = \pm g\phi(x_1)\psi_\pm(x_1) \tag{20.68}$$

whose physically sensible solution must be square integrable:

$$\int_{-\infty}^{\infty} dx_1 \, |\psi_\pm(x_1)|^2 = 1 \tag{20.69}$$

For a soliton, which asymptotically satisfies $\lim_{x\to\pm\infty}\phi(x_1) = \pm\phi_0$, the square-integrable solution for the eigenstate has the spinor solution

$$\chi_- = \begin{pmatrix} 1 \\ 0 \end{pmatrix} \tag{20.70}$$

whose γ_1 eigenvalue is $-i$, and the wave function is

$$\psi_-(x_1) = \psi_-(0)\ \exp\left(-\int_0^{x_1} dx_1'\ \phi(x_1')\right) \tag{20.71}$$

This wave function is even under $x_1 \to -x_1$ since the soliton $\phi(x_1)$ changes sign under this operation. For the soliton profile of eq. (19.55), at long distances, we can approximate $\psi_-(x_1) \simeq \psi_-(0)\ \exp(-\phi_0|x_1|)$, and find an exponential decay (as expected for a bound state). For the antisoliton, the situation is reversed: Now the scalar field obeys $\lim_{x_1 \pm\infty} \phi(x_1) = \mp\phi_0$, and we must choose the spinor with γ_1 eigenvalue $+i$. However, the normalizable wave function is still the same as in eq. (20.71).

20.5 The axial anomaly in 3+1 dimensions

The axial anomaly was discovered in four dimensions in the computation of triangle Feynman diagrams with fermionic loops. Here, too, the axial anomaly is the nonconservation of the axial (chiral) current j_μ^5 in a theory with a gauge-invariant regularization. If the fermions are massless, the chiral anomaly for a $U(1)$ theory is the identity

$$\partial^\mu j_\mu^5 = -\frac{e^2}{8\pi^2} F^{\mu\nu} F_{\mu\nu}^* \tag{20.72}$$

where $F_{\mu\nu}^* = \frac{1}{2}\epsilon_{\mu\nu\lambda\rho}F^{\lambda\rho}$ is the dual field strength tensor. This result is known to hold in QED to all orders in perturbation theory. Just as in the 1+1-dimensional case, the right-hand side is a total divergence, which has a topological meaning.

Let us now see how this result arises using the four-dimensional version of the Nielsen-Ninomiya argument that we discussed in 1+1 dimensions in section 20.2. Consider a theory with a single, right-handed Weyl fermion in a uniform magnetic field B along the direction x_3, whose vector potential is $A_2 = Bx_1$, with all other components being zero. The Dirac equation for the right-handed Weyl (i.e., two-component) spinor ψ_R is

$$[i\partial_0 - (\boldsymbol{p} - e\boldsymbol{A})\cdot\boldsymbol{\sigma}]\psi_R = 0 \tag{20.73}$$

where $\boldsymbol{p} = -i\boldsymbol{\partial}$ is the momentum operator, and $\boldsymbol{\sigma}$ consists of the three 2×2 Pauli matrices. As usual, the solutions to this equation are expressed in terms of the Weyl spinor Φ,

$$\psi_R = [i\partial_0 + (\boldsymbol{p} - e\boldsymbol{A})\cdot\boldsymbol{\sigma}]\Phi = 0 \tag{20.74}$$

leading to

$$[i\partial_0 - (\boldsymbol{p} - e\boldsymbol{A})\cdot\boldsymbol{\sigma}][i\partial_0 + (\boldsymbol{p} - e\boldsymbol{A})\cdot\boldsymbol{\sigma}]\Phi = 0 \tag{20.75}$$

The spinor Φ can be chosen to be an eigenstate of p_2 and p_3. Then Φ is the solution of the harmonic oscillator equation:

$$\left[-\partial_1^2 + (eB)^2\left(x^1 + \frac{p_2}{eB}\right) + p_3^2 + eB\sigma_3\right]\Phi = \omega^2\Phi \tag{20.76}$$

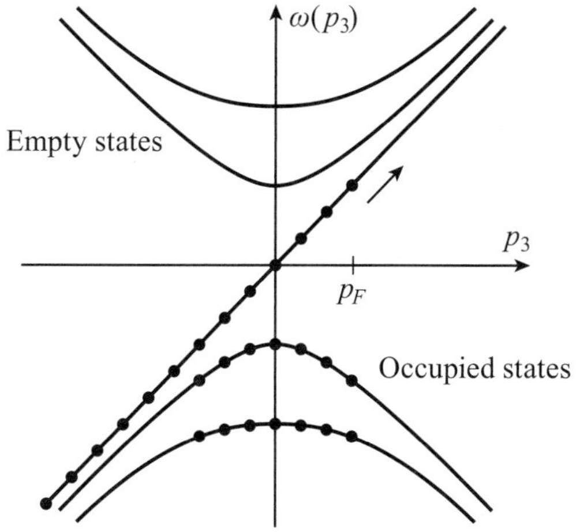

Figure 20.4 The axial anomaly in 3+1 dimensions: spectrum of right-handed Weyl fermions in parallel B and E uniform magnetic and electric fields along the direction x_3. The arrow shows that right-handed fermions are being created, and the Fermi momentum p_F increases with time.

The eigenvalues are the Landau levels

$$\omega(n, p_3, \sigma_3) = \pm\left[2eB\left(n+\frac{1}{2}\right)+p_3^2+eB\sigma_3\right]^{1/2} \tag{20.77}$$

with $n=0, 1, 2, \ldots$, except for the mode $n=0$ and $\sigma_3=-1$, for which

$$\omega(n=0, \sigma_3=-1, p_3) = \pm p_3 \tag{20.78}$$

It is straightforward to find the eigenfunctions. They are, with the exception of the zero modes, the usual one-dimensional linear-oscillator wave functions multiplied by plane-wave states (for each spinor, eigenstates $(1,0)$ and $(0,1)$). The wave function ψ_R vanishes for the zero modes with $n=0$ and $\sigma_3=-1$ and eigenvalue $\omega=-p_3$, and it is nonzero for the zero mode with $n=0$, $\sigma_3=-1$ and for $\omega=p_3$. The spectrum is shown in figure 20.4.

Let us now turn on a uniform electric field E parallel to the magnetic field B. The levels with $n\neq 0$ are either totally occupied or are empty. In fact, there is a one-to-one correspondence between these levels, and their contributions cancel out to the quantity of interest. Only the zero mode matters: $n=0$, $\sigma_3=-1$, and $\omega=p_3$. Except for a density-of-states factor, $LeB/(4\pi^2)$ (where L is the linear size of the system), its contribution is the same as in the (1+1)-dimensional case. Thus, the rate of creation of right-handed Weyl fermions is

$$\frac{dN_R}{dx_0} = \frac{1}{L}\frac{LeB}{4\pi^2}\frac{dp_F}{dx_0} = \frac{e^2}{4\pi^2}EB = \frac{dQ_R}{dx_0} \tag{20.79}$$

Similarly, for left-handed Weyl fermions, ψ_L, the rate is

$$\frac{dN_L}{dx_0} = -\frac{1}{L}\frac{LeB}{4\pi^2}\frac{dp_F}{dx_0} = -\frac{e^2}{4\pi^2}EB = \frac{dQ_L}{dx_0} \tag{20.80}$$

Therefore, the rate of creation of right-handed particles and of left-handed antiparticles is

$$\frac{dN_R}{dx_0}+\frac{d\bar{N}_L}{dx_0}=\frac{e^2}{2\pi^2}EB=\frac{dQ_5}{dx_0} \tag{20.81}$$

which is the axial anomaly in 3+1 dimensions.

20.6 Fermion path integrals, the chiral anomaly, and the index theorem

At the beginning of this chapter, we noted that in the context of the path integral, the axial (or chiral) anomaly arises due to a noninvariance of the fermionic measure under chiral transformations (Fujikawa, 1979). Thus, although the action is invariant, the partition function is not.

We will see how this takes place in a simple theory: a free massless Dirac fermion coupled to a $U(1)$ gauge field. In an even D-dimensional Euclidean spacetime, the Lagrangian is

$$\mathcal{L}=\bar{\psi}\not{D}\psi \tag{20.82}$$

where, as usual, $\not{D}=\gamma^\mu(\partial_\mu+iA_\mu)$, and $\not{D}$ is a hermitian operator.

In the definition of the path integral, one needs to specify a complete basis of field configurations that will define the integration measure. It is natural to use the eigenstates of the operator $\not{D}$. Let $\{\psi_n\}$ be a complete set of spinor eigenstates of $\not{D}$,

$$\not{D}\psi_n(x)=\lambda_n\psi_n(x) \tag{20.83}$$

where the eigenvalues are $\lambda_n\in\mathbb{R}$, and the eigenstates are complete and orthonormal,

$$\int d^dx\,\psi_n^\dagger(x)\psi_m(x)=\delta_{n,m},\qquad \sum_n\psi_n^\dagger(x)\psi_n(y)=\delta(x-y) \tag{20.84}$$

and the eigenstates are functions of the gauge field (since $\not{D}$ depends on the gauge field).

Then, we expand the fields in this basis,

$$\psi(x)=\sum_n a_n\psi_n(x),\qquad \bar{\psi}(x)=\sum_n b_n\psi_n^\dagger(x) \tag{20.85}$$

where the coefficients $\{a_n\}$ and $\{b_n\}$ are independent Grassmann numbers. The fermionic integration measure is defined as

$$\mathcal{D}\bar{\psi}\mathcal{D}\psi\equiv\prod_n da_n db_n \tag{20.86}$$

Under a local chiral transformation

$$\psi(x)\mapsto\psi_n'(x)=e^{-i\alpha(x)\gamma_5}\psi(x),\quad \bar{\psi}(x)\mapsto\bar{\psi}_n'(x)=\bar{\psi}(x)e^{-i\alpha(x)\gamma_5} \tag{20.87}$$

a linear transformation is induced on the coefficients $\{a_n\}$ and $\{b_n\}$,

$$a_n \mapsto a'_n = \sum_m C_{nm} a_m, \quad b_n \mapsto b'_n = \sum_m C_{nm} b_m \tag{20.88}$$

where

$$C_{nm} = \int d^d x \, \bar{\psi}_n^\dagger(x) e^{-i\alpha(x)\gamma_5} \psi_m(x) \tag{20.89}$$

are numbers.

Since $\{\gamma_5, \not{D}\} = 0$, we can choose the states $\psi_n(x)$ to be eigenstates of γ_5 and have definite chirality. Then, for each eigenvalue $\lambda_n \neq 0$, there are two linearly independent states with opposite chirality. In this representation, the measure of the coefficients $\{a_n\}$ and $\{b_n\}$ changes by a Jacobian factor

$$\prod_n da'_n = (\det C)^{-1} \prod_n da_n \tag{20.90}$$

and similarly for the b_n coefficients. For an infinitesimal chiral transformation, the Jacobian is

$$\begin{aligned}(\det C)^{-1} &= \prod_n \left[1 - i \int d^d x \, \alpha(x) \sum_n \psi_n^\dagger(x) \gamma_5 \psi_n(x) \right] \\ &= \exp\left(i \int d^d x \alpha(x) \sum_n \psi_n^\dagger(x) \gamma_5 \psi_n(x) \right)\end{aligned} \tag{20.91}$$

Therefore, under an infinitesimal local chiral transformation, the fermionic measure changes as

$$\mathcal{D}\bar{\psi}\mathcal{D}\psi \mapsto \mathcal{D}\bar{\psi}'\mathcal{D}\psi' = \exp\left(2iN_f \sum_n \psi_n^\dagger(x) \gamma_5 \psi_n(x) \right) \tag{20.92}$$

where we have allowed for the possibility of having N_f flavors of fermions. If it is not equal to 1, the exponential prefactor is the noninvariance of the measure.

The problem now reduces to the computation of the sums in the exponent of eq. (20.92). However, this sum is ambiguous, and its definition requires a regulator. Define the sum to be

$$\sum_n \psi_n^\dagger(x) \gamma_5 \psi_n(x) = \lim_{\varepsilon \to 0^+} \sum_n e^{-\lambda_n^2 \epsilon} \psi_n^\dagger(x) \gamma_5 \psi_n(x) \tag{20.93}$$

where a factor has been introduced in each term of the sum that damps out the contribution of the eigenstates of $\not{D}$ with large eigenvalues. Notice that here we made a choice to regularize the sum using the spectrum of the *gauge-invariant operator* $\exp(-\not{D}^2 \epsilon)$. This is the heat kernel method used by Fujikawa. The detailed computation of the fermion determinants using the ζ-function approach is found in Gamboa Saraví et al. (1984).

Atiyah and Singer proved a theorem, the *Atiyah-Singer theorem*, which relates the computation of this sum to an index of the Dirac operator (Atiyah and Singer, 1968). We will not go through the details of this proof, but simply quote the result of the sum in $D = 2$

and $D = 4$ dimensions:

$$\sum_n \psi_n^\dagger(x)\gamma_5\psi_n(x) = \frac{e}{4\pi}\epsilon_{\mu\nu}F^{\mu\nu}, \qquad \text{for } U(1) \text{ in } D = 2$$

$$\sum_n \psi_n^\dagger(x)\gamma_5\psi_n(x) = \frac{1}{8\pi^2}F^{\mu\nu}F^*_{\mu\nu}, \qquad \text{for } U(1) \text{ in } D = 2$$

$$\sum_n \psi_n^\dagger(x)\gamma_5\psi_n(x) = \frac{e^2}{16\pi^2}\text{tr}(F^{\mu\nu}F^*_{\mu\nu}), \qquad \text{for } SU(N) \text{ in } D = 4 \tag{20.94}$$

Once again, the right-hand side of these equations involves the topological density of the gauge field.

Let us integrate both sides of eq. (20.94) over all Euclidean spacetime. The key observation now is that for the states with eigenvalues $\lambda_n \neq 0$, the contributions of the two degenerate chiral eigenstates cancel each other, and, consequently, only the states with zero eigenvalue, $\lambda_n = 0$ (the "zero modes"), contribute to the sum. Then the result of the sum is finite. Let $n_\pm$ be the number of zero modes with chirality ± 1. Also, the integral of the right-hand side yields the topological charge Q of the gauge field.

The Atiyah-Singer theorem states that if $\mathcal{D}$ is an (elliptic) differential operator, and $\mathcal{D}^\dagger$ is its adjoint, the index $I_\mathcal{D}$ is

$$I_\mathcal{D} = \dim\ker(\mathcal{D}) - \dim\ker(\mathcal{D}^\dagger) \tag{20.95}$$

where $\ker(\mathcal{D})$ denotes the subspace spanned by the kernel of the operator $\mathcal{D}$. Consider now the quantity

$$I_\mathcal{D}(M^2) = \text{Tr}\left(\frac{M^2}{\mathcal{D}^\dagger\mathcal{D} + M^2}\right) - \text{Tr}\left(\frac{M^2}{\mathcal{D}\mathcal{D}^\dagger + M^2}\right) \tag{20.96}$$

As $M^2 \to 0$, only the contribution of the zero eigenvalues of $\mathcal{D}^\dagger\mathcal{D}$ survives. The normalizable zero eigenvalues of $\mathcal{D}^\dagger\mathcal{D}$ (which are the same as those of $\mathcal{D}$) each contribute 1 to this expression. Likewise, the normalizable zero eigenvalues of $\mathcal{D}\mathcal{D}^\dagger$ (which are the eigenvalues of $\mathcal{D}^\dagger$) each contribute -1. Therefore, we can write the index of $\mathcal{D}$ as

$$I_\mathcal{D} = \lim_{M^2 \to 0} I_\mathcal{D}(M^2) \tag{20.97}$$

Then, it follows that the topological charge Q is equal to the index

$$n_+ - n_- = Q \tag{20.98}$$

This result is actually general and also holds for non-abelian gauge theories. It implies that if the gauge field has an instanton, say, with $Q = 1$, then there must be at least one zero mode.

20.7 The parity anomaly and Chern-Simons gauge theory

So far we have considered only even-dimensional spacetimes. On a spacetime of even dimension $D = 2n$, the Dirac spinors have $2n$ components. Thus in 1+1 dimensions, they are two-spinors, in 3+1 dimensions, they are four-spinors, and so forth. Similarly, in 1+1 dimensions, the Dirac operator is a 2×2 matrix-valued differential operator, and the Dirac gamma matrices are 2×2 matrices that can be chosen to be real (in the Euclidean signature). Likewise in 3+1 dimensions, the Dirac operator is a 4×4 matrix, and the Dirac gamma matrices can also be chosen to be real (again, in the Euclidean signature). In both cases, there is a special matrix, γ_5, which anticommutes with the Dirac operator. If γ_5 does not enter in the Dirac operator, then the theory is time-reversal (or CP) invariant.

Let us now consider the Dirac theory in 2+1 dimensions. The Dirac spinors are still two-component, as they are in 1+1 dimensions. In the case of a single massive Dirac field, the Lagrangian is

$$\mathcal{L} = \bar{\psi}(i\not{D} - m)\psi \tag{20.99}$$

where D_μ is the covariant derivative. Although this Lagrangian has the same form as in even dimensional spacetimes, there is the fundamental difference that it involves all three γ_μ Dirac gamma matrices (with $\mu = 0, 1, 2$). In particular, the Dirac Hamiltonian

$$H_{\rm D} = \boldsymbol{\alpha} \cdot \boldsymbol{p} + \beta m \tag{20.100}$$

involves the three Pauli matrices. However, the Hamiltonian is not real, since, for instance, if we chose a representation such that $\alpha_1 = \sigma_1$, $\alpha_2 = \sigma_3$, then the mass term involves the complex matrix $\beta = \sigma_2$. Thus the Hamiltonian is not invariant under time reversal, which will map it onto the complex conjugate and flip the sign of the mass term. Likewise, we could have chosen a representation in which $\alpha_2 = \sigma_2$. In that representation, complex conjugation is equivalent to parity, defined as $x_1 \to x_1$ and $x_2 \to -x_2$. In addition, there is no natural definition of a γ_5 matrix and there is no chirality.

These simple observations suggest that in a theory of massive Dirac fermions in 2+1 dimensions, parity (or time reversal) may be necessarily broken. We will now see that this is a subtle question and that the physics depends on the regularization. Here too, we have a choice of either making the theory gauge-invariant or parity (and time-reversal) invariant.

This phenomenon is known as the parity anomaly (Redlich, 1984a,b). It arises in the computation of the effective action of the gauge field A_μ coupled to a theory of massive Dirac fermions. If gauge invariance is preserved by the regularization, then the effective action at low energies (low compared to the mass of the fermion) must be a sum of locally gauge-invariant operators. In even-dimensional spacetimes, the operator with lowest dimensions is the Maxwell term (or the Yang-Mills term in the non-abelian case). If we define the covariant derivative as $D_\mu = \partial_\mu + iA_\mu$, then the gauge field carries units of momentum (in all dimensions). Thus the Maxwell term (and the Yang-Mills term) is an operator of dimension 4. If this operator appears from integrating out massive fermions, then the prefactor of this effective action in $D = 3$ spacetime dimensions must be proportional to $1/m$, where m is the mass of the Dirac field.

However, in 2+1 dimensions, there is a locally gauge-invariant term with dimension 3 that one can write. It is the Chern-Simons term (Deser et al., 1982a,b):

$$\mathcal{L}_{\rm CS} = \frac{k}{4\pi} \epsilon_{\mu\nu\lambda} A^\mu \partial^\nu A^\lambda \equiv \frac{1}{4\pi} A \wedge dA \tag{20.101}$$

in the abelian theory, and

$$\mathcal{L}_{\rm CS} = \frac{k}{8\pi} {\rm tr}\left(\epsilon_{\mu\nu\lambda} A^\mu \partial^\nu A^\lambda + \frac{2}{3} \epsilon^{\mu\nu\lambda} A_\mu A_\nu A_\lambda \right) \equiv \frac{k}{8\pi} {\rm tr}\left(A \wedge dA + \frac{2}{3} A \wedge A \wedge A \right) \tag{20.102}$$

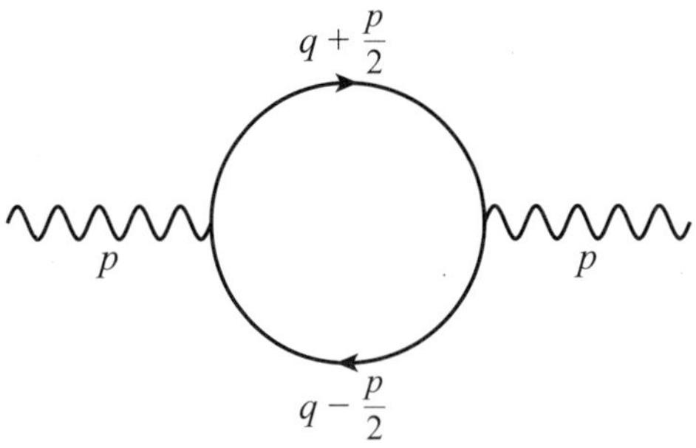

Figure 20.5 One-loop fermion diagram.

in the non-abelian theory. Here A_μ is in the algebra of a non-abelian simply connected Lie group G. Here the Lagrangian has been written in a simpler form using the notation of differential forms. (The factor of 1/2 difference in the prefactors of eqs. (20.101) and (20.102) is due to the normalization of the traces in eq. (20.102).)

The Chern-Simons gauge theory has deep connections with topology, particularly with the theory of knots (Witten, 1989), and it has many applications in physics (e.g., in the physics of the quantum Hall effects; see Fradkin, 2013). We will discuss the Chern-Simons theory in chapter 22. Since it is first order in derivatives and involves the Levi-Civita symbol, the Chern-Simons action is locally gauge invariant, and it is odd under parity and time reversal. In this sense, it is natural to consider it as an effective action for a theory with broken parity and time reversal.

However, the Chern-Simons action is not gauge invariant if the spacetime manifold is open (has an edge). But even if the spacetime manifold is closed (e.g., a three-sphere S^3, a three-torus), the action (or, rather, the weight in the path integral) is not invariant under large gauge transformations, which wrap around the space manifold, unless the parameter k is quantized: $k \in \mathbb{Z}$.

Let us do a one-loop perturbative calculation of the effective action of the gauge fields (Redlich, 1984b). In the abelian theory, this requires the computation of a Feynman diagram with a single fermion internal loop and two gauge fields in the external legs (a polarization bubble diagram), shown in figure 20.5. In the non-abelian theory, there is, in addition, a triangle diagram with three external gauge fields and a single internal fermion loop.

Both the bubble and the triangle diagrams are UV divergent and require regularization. Since we are after an anomalous contribution, we cannot use dimensional regularization. Anomalies are dimension-specific, and dimensional regularization yields ambiguous results, although ad hoc procedures have been devised to solve this problem. The most commonly used regularization is Pauli-Villars. As we have already seen in section 13.7, Pauli-Villars amounts to the introduction of a set of fields (fermions in this case) with very large mass. This procedure will keep the theory gauge invariant, but there will be a finite term that breaks parity (and time reversal)—the parity anomaly.

We only quote the result for the abelian bubble diagram, which is denoted by $\Pi_{\mu\nu}(p)$. Here p_μ is the external momentum (of the gauge fields). The expression to be computed is

$$\Pi^{\mu\nu}(p) = \int \frac{d^3q}{(2\pi)^3} \text{tr}\left[S\left(q+\frac{p}{2}\right)\gamma^\mu S\left(q-\frac{p}{2}\right)\gamma^\nu\right] \tag{20.103}$$

where $S(q)$ is the propagator of a Dirac fermion of mass m.

By power counting, this diagram is superficially linearly divergent as the UV regulator $\Lambda \to \infty$. Lorentz invariance yields an expression of the form (again in Euclidean spacetime),

$$\Pi_{\mu\nu}(p) = \Pi_0(p^2)g_{\mu\nu} - i\epsilon_{\mu\nu\lambda}p^\lambda \Pi_A(p^2) + (p^2 g^{\mu\nu} - p^\mu p^\nu)\Pi_E(p^2) \tag{20.104}$$

where $\Pi_0(p^2)$ contains the linearly divergent contribution, and it is not gauge invariant. The other two terms are finite and gauge invariant.

If we assume that the only effect of the Pauli-Villars regulators is to subtract the linearly divergent (and parity-even) term, the low-energy limit (or what is the same for large fermion mass, the gauge-invariant contribution) yields an effective action for the abelian gauge field A_μ of the form (again in Euclidean spacetime)

$$S_{\text{eff}}[A] = i\frac{1}{8\pi}\text{sign}(m)\int d^3x\, \epsilon_{\mu\nu\lambda}A^\mu \partial^\nu A^\lambda + \frac{1}{4g^2}\int d^3x\, F_{\mu\nu}^2 + \cdots \tag{20.105}$$

where $g^2 = 1/(\pi|m|)$. Thus, we obtain a parity and time-reversal odd contribution (the Chern-Simons term), and a parity and time-reversal even Maxwell term. Clearly these are the first two terms of an expansion in powers of $1/m$.

However, on closer examination, this well-established result has a problem. The coefficient of the Chern-Simons term is (up to a sign) $k = 1/2$. This is a problem, since on a closed manifold, invariance under large gauge transformations requires that it should be an integer. Furthermore, this calculation predicts that this term should be present even in the massless limit, where the theory should be time-reversal invariant, and with this term, it is not.

The solution of this apparent conundrum lies in the fact that the Pauli-Villars regulator is a heavy Dirac fermion, which has two contributions to the effective action. One is the cancellation of the linearly divergent and non-gauge-invariant term in the polarization tensor in eq. (20.104). However, it also has finite contributions. One of them is a Chern-Simons term, of the same form as in eq. (20.105), also with coefficient 1/2 but also with a sign ambiguity. Thus the total contribution to the coefficient of the Chern-Simons term is either $k = 0$ or $k = \pm 1$. In the first case, time-reversal invariance is preserved (and there is no parity anomaly). In the second case, the coefficient is correctly quantized (and time-reversal symmetry is broken), but the sign of k depends not just on the sign of the fermion mass but also on the sign of the regulator mass.

More formally, in the massless case, let us write the fermion determinant as

$$\det(i\not{D}[A]) = |\det(i\not{D}[A])| \exp\left(-i\frac{\pi}{2}\eta[A]\right) \tag{20.106}$$

where $\eta[A]$ is the Atiyah-Patodi-Singer η-invariant,

$$\eta[A] = \sum_{\lambda_k > 0} 1 - \sum_{\lambda_k < 0} 1 \tag{20.107}$$

which is the regularized and gauge-invariant spectral asymmetry of the Dirac operator. In this language, this contribution is often expressed as the $U(1)_{-1/2}$ Chern-Simons action (Alvarez-Gaumé et al., 1985).

A physically more intuitive way to reach the same result is to consider a theory in which the Dirac fermions are defined on a spatial lattice. This is the setting appropriate for investigating the quantum anomalous Hall effect. There is a theorem by Nielsen and Ninomiya which states that it is not possible to have a local theory or chiral fermions on a (any) lattice (and in any dimension). The upshot of this "fermion doubling" theorem is that the number of Dirac fermions must be even. This theorem has been a longstanding obstacle to nonperturbative studies of theories of weak interactions using lattice gauge-theory methods (Nielsen and Ninomiya, 1981).

In 2+1 dimensions, the Nielsen-Ninomiya theorem implies that, at a minimum, there should be two Dirac fermions. So, we see that if the mass of one fermion is much larger than the mass of the other, then the "doubler" plays the role of the Pauli-Villars regulator. The other relevant consideration is that the coefficient of the Chern-Simons term is the same as the value of the Hall conductivity of the theory of fermions. In units of e^2/h, the Hall conductivity is given by a topological invariant of the occupied band of states of the lattice model. This topological invariant is also equal to an integer, the first Chern number, the nontrivial winding number of the Berry phase of the single-particle states on the Brillouin zone.

20.8 Anomaly inflow

In the previous sections, we have discussed chiral anomalies in 1+1 and 3+1 dimensions and the parity anomaly in 2+1 dimensions. We will now see that these anomalies are related. The relation involves considering systems in different space dimensions, where the lower dimensional system is defined on a topological defect of its higher dimensional partner. These topological defects are vortices and domain walls. The connection between anomalies in different dimensions in theories with topological defects is known as the anomaly inflow.

We will see that, in the presence of topological defects, the anomalies of the theory defined on the defect precisely match the anomalies of the theory in the larger system (the "bulk"). The discussion here follows the general construction of the anomaly inflow by Callan and Harvey (1985).

20.8.1 Axion strings

For the sake of concreteness, let us consider a theory in even ($D=4$) dimensions with a γ^5 coupling to a stringlike topological defect, a vortex (or "axion string"). Assume that the Dirac fermions are coupled through mass terms to a complex scalar field, $\varphi=\varphi_1+i\varphi_2$, that has a vortex topological defect running along the x_3 axis, shown in figure 20.6. The Lagrangian in the vortex background is

$$\mathcal{L}=\bar{\psi}i\partial\!\!\!/\psi+\bar{\psi}(\varphi_1+i\gamma^5\varphi_2)\psi \tag{20.108}$$

The vortex has the usual form $\varphi=f(\rho)\exp(i\vartheta)$, where ϑ is the phase of the complex scalar field. Here we use cylindrical coordinates, with $x_1=\rho\cos\phi$, and $x_2=\rho\sin\phi$. The vortex solution behaves at infinity as $\lim_{\rho\to\infty}f(\rho)=\varphi_0$, and it vanishes as $\rho\to 0$, where the vortex has a singularity. In the asymptotic regime, the phase of a single vortex with winding number $N=+1$ is $\vartheta=\phi$. The vortex singularity is an "axion string."

We first show that, in the presence of the vortex, the Dirac operator has zero modes. Let us rewrite $\gamma^5=\gamma^0\gamma^1\gamma^2\gamma^3$ as $\gamma^5=\Gamma^{\text{int}}\Gamma^{\text{ext}}$, where $\Gamma^{\text{int}}=i\gamma^0\gamma^1$, and $\Gamma^{\text{ext}}=\gamma^2\gamma^3$. Here, Γ^{int} is the γ^5 matrix of the (1+1)-dimensional world and measures the chirality of the modes

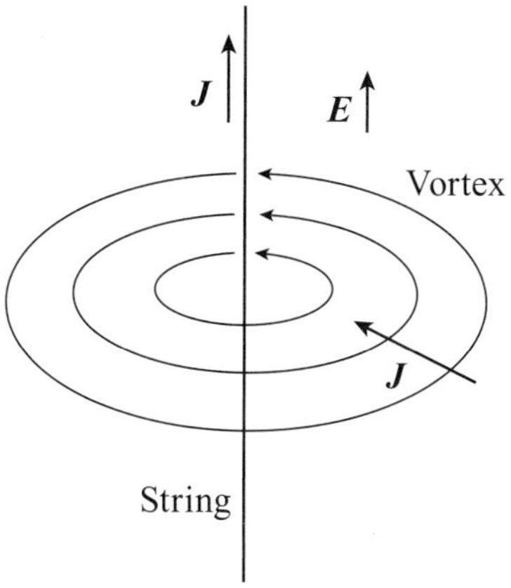

Figure 20.6 The anomaly inflow in the axion string, the singularity of a vortex. In the presence of a uniform electric field $\boldsymbol{E}$ parallel to the axion string, in the "bulk" three-dimensional world, there is a Hall-like current flowing toward the string, and flows in the string, parallel to $\boldsymbol{E}$. The direction of the current on the string is determined by the sign of the vorticity.

of the Dirac equation on the string. Let $\psi_\pm$ be eigenfunctions of γ^5 with eigenvalue ± 1. In this basis, the Dirac equation is

$$\begin{aligned} i\partial\!\!\!/^{\text{int}}\psi_- + i\gamma^2(\cos\varphi + i\Gamma^{\text{ext}}\sin\varphi)\partial_\rho\psi_- &= f(\rho)e^{i\vartheta}\,\psi_+ \\ i\partial\!\!\!/^{\text{int}}\psi_+ + i\gamma^2(\cos\varphi + i\Gamma^{\text{ext}}\sin\varphi)\partial_\rho\psi_+ &= f(\rho)e^{i\vartheta}\,\psi_- \end{aligned} \tag{20.109}$$

The zero-mode solution is

$$\psi_- = \eta(x^{\text{int}})\exp\left(-\int_0^\rho d\rho' f(\rho')\right), \qquad \text{where } \psi_+ = -i\gamma^2\psi_- \tag{20.110}$$

and the spinor η satisfies

$$i\partial\!\!\!/^{\text{int}}\eta = 0, \qquad \Gamma^{\text{int}}\eta = -\eta\,\cdot \tag{20.111}$$

For the antivortex, with winding number $N=-1$, the phase field is $\vartheta = -\phi$, and the fermion zero mode on the axion string is left-moving.

Hence, a massless chiral fermion is traveling along the string, and the direction of propagation is determined by the axial charge of the fermion. Now, in the presence of an external electromagnetic field A_μ, the chiral fermion has a gauge anomaly,

$$\partial^\mu J_\mu = \frac{1}{2\pi}\epsilon_{\mu\nu}\partial^\mu A^\nu \tag{20.112}$$

where now $\mu, \nu = 0, 1$. However, we saw that the anomaly of a chiral fermion, eq. (20.10), means that in the presence of an electric field E parallel to the string, the gauge charge is not conserved. Where is this charge coming from?

To see where this charge may be coming from, we need to consider the rest of the spectrum of the Dirac operator (i.e., the nonzero modes). Thus we need to examine what happens in the higher-dimensional bulk. This can be done using the method of Goldstone and Wilczek discussed in section 20.4 (see figure 20.3). In 3+1 dimensions, the result is

$$\langle J_\mu \rangle = -i\frac{e}{16\pi^2}\epsilon_{\mu\nu\lambda\rho}\frac{(\varphi^*\partial^\nu\varphi - \varphi\partial^\nu\varphi^*)}{|\varphi|^2}F^{\lambda\rho} \tag{20.113}$$

Far from the vortex singularity, we can approximate $\varphi \simeq \phi_0 \exp(i\vartheta(x))$ to find

$$\langle J_\mu \rangle = \frac{e}{8\pi^2}\epsilon_{\mu\nu\lambda\rho}\partial^\nu\vartheta(x)F^{\lambda\rho} \tag{20.114}$$

For a vortex running along the x_3 axis, far from the singularity, we get $\vartheta(x) = \phi$. If the background field is an electric field of magnitude E along the direction x_3, the induced current J_μ has only a radial component,

$$J_\rho = -\frac{e}{4\pi^2\rho}E \tag{20.115}$$

and it is pointing inward, toward the axion string, and is perpendicular to the applied electric field. It is this inward component of the current that compensates for the anomaly on the string. Moreover, using the singularity of the phase, $(\partial_1\partial_2 - \partial_2\partial_1)\vartheta(x_1, x_2) = 2\pi\delta(x_1)\delta(x_2)$, we obtain

$$\partial^\mu\langle J_\mu \rangle = \frac{e}{4\pi}\epsilon^{ab}F_{ab}\delta(x_1)\delta(x_2) \tag{20.116}$$

Hence, the current is conserved away from the singularity, and the nonconservation at the string singularity is matched by the gauge anomaly on the string!

20.8.2 Domain wall in 2+1 dimensions

Now consider the effects of domain walls. We consider first a theory in 2+1 dimensions with the Lagrangian

$$\mathcal{L} = \bar\psi i\not\partial\psi + \frac{1}{2}(\partial_\mu\phi)^2 - g\phi\bar\psi\psi - U(\phi) \tag{20.117}$$

where ψ is a two-component Dirac fermion. Here ϕ is a real scalar field in its broken symmetry state, and we write the interaction in the form $U(\phi) = \frac{\lambda}{4!}(\phi_0^2 - \phi^2)^2$. Consider a static domain wall, that is, a time-independent classical solution of the field ϕ that has a kink (soliton) along the direction x_2 and is constant along x_1. Let $\phi_W(x_2)$ denote the domain wall.

As in the preceding section 20.8.1, we set $\Gamma^{\text{int}} = \gamma^0\gamma^1$, and $\Gamma^{\text{ext}} = \gamma^2$, and the Dirac equation becomes

$$i\not\partial^{\text{int}}\psi - \Gamma^{\text{int}}\partial_2\psi = g\phi_W(x_2)\psi \tag{20.118}$$

The zero mode is

$$\psi = \eta(x_{\text{int}})\exp\left(-\int_0^{x_2} dx_2'\, g\phi_W(x_2)\right) \tag{20.119}$$

with

$$i\not\partial^{\text{int}}\eta = 0, \qquad \text{and } \Gamma^{\text{int}}\eta = \eta \tag{20.120}$$

This is a right-moving Dirac fermion along the wall.

In the broken symmetry state of the scalar field the fermions are massive, and their mass is sign(x_2) ϕ_0. Hence, the sign of the mass term changes across the wall. The contribution of the fermions to the effective action of a gauge field A_μ, far from the location of the wall, is

$$S_{\text{eff}} = \int d^3x \frac{(1+\text{sign}(x_2))}{8\pi} \epsilon_{\mu\nu\lambda} A^\mu \partial^\nu A^\lambda \tag{20.121}$$

where the integral extends over all of 2+1-dimensional spacetime, except for an infinitesimal region surrounding the domain wall.

Under a gauge transformation, $A_\mu \to A_\mu + \partial_\mu \Phi$, this action changes as

$$\begin{aligned}\Delta S_{\text{eff}} &= \int d^3x \, \frac{1}{8\pi}(1+\text{sign}(x_2))\epsilon^{\mu\nu\lambda}\partial_\mu \Phi \, \partial_\nu A_\lambda \\ &= -\frac{1}{2\pi}\int d^2x \, \Phi(x) \epsilon_{\mu\nu} F^{\mu\nu}\end{aligned} \tag{20.122}$$

where an integration by parts has been done. This result cancels the gauge anomaly of the chiral fermion in 1+1 dimensions, carried by the fermion zero modes on the wall. In the theory of quantum Hall effects, these are the chiral edge states.

It is straightforward to see that a uniform electric field parallel to the wall induces a charge current that flows toward the wall. But the electric field will induce a current along the wall, by the pair-creation mechanism already discussed. In each sector, bulk and boundary, there is a gauge anomaly, but these anomalies cancel each other, as we just saw.

20.8.3 Domain wall in 3+1 dimensions

Let us now discuss, briefly, the problem of Dirac fermions coupled to a domain wall in 3+1 dimensions. Physically, this problem arises in the behavior of three-dimensional topological insulators (Boyanovsky et al., 1987; Qi et al., 2008), and as an approach in lattice gauge theory to study theories of chiral fermions (Kaplan, 1992). The Lagrangian is

$$\mathcal{L} = \bar{\psi} i \not{\partial} \psi + g\varphi(x)\bar{\psi}\psi \tag{20.123}$$

Since we are in 3+1 dimensions, the fermions are the usual Dirac four-component spinors. Here, $\varphi(x)$ is a real scalar field. Assume that $\varphi(x)$ is a static classical configuration with a domain wall, that is, $\varphi(x)$ depends only on x_3: $\varphi(x) = \phi_0 f(x_3)$, where $f(x_3) \to \pm 1$ as $x_3 \to \pm\infty$, and ϕ_0 is the vacuum expectation value of $\varphi(x)$. Clearly, $g\varphi(x)$ plays the role of a Dirac mass, $m(x_3)$, which varies from $g\phi_0$ to $-g\phi_0$.

The Dirac Hamiltonian has the standard form,

$$H = -i\boldsymbol{\alpha} \cdot \boldsymbol{\nabla} + m(x_3)\beta \tag{20.124}$$

where $\boldsymbol{\alpha}$ and β are the Dirac matrices. The geometry of this problem suggests that we split the Hamiltonian into on- and off-wall components, $H = H_{\text{wall}} + H_\perp$, where

$$H_{\text{wall}} = -i\alpha_1\partial_1 - i\alpha_2\partial_2, \qquad H_\perp = -i\alpha_3\partial_3 + m(x_3)\beta \tag{20.125}$$

Let $\psi_\pm$ be an eigenstate of the antihermitian Dirac matrix $\gamma_3 = \beta\alpha_3$ with eigenvalues $\pm i$, respectively,

$$\gamma_3\psi_\pm = \pm i\psi_\pm \tag{20.126}$$

We seek a zero-mode solution $H_\perp \psi_\pm = 0$. Hence,

$$\pm\, \partial_3 \psi_\pm + m(x_3)\psi_\pm = 0 \tag{20.127}$$

and a solution of

$$(i\gamma_0 - i\gamma_1\partial_1 - i\gamma_2\partial_2)\psi_\pm = 0 \tag{20.128}$$

In other words, ψ_+ is a solution of a massless Dirac equation in 2+1 dimensions. The requirement that $\psi_\pm$ is an eigenstate of γ_3 reduces the number of spinor components from four to two.

The solution of these equations has the form

$$\psi_\pm = \eta_\pm(x_0, x_1, x_2)\, F_\pm(x_3), \qquad \text{where } \pm\, \partial_3 F_\pm(x_3) = -m(x_3)F_\pm(x_3) \tag{20.129}$$

The solution $F_\pm(x_3)$ is

$$F_\pm(x_3) = F(0)\exp\left(\mp \int_0^{x_3} dx_3'\, m(x_3')\right) \tag{20.130}$$

If the domain wall behaves like $m(x_3) \to g\phi_0$ as $x_3 \to \infty$, then $F_+(x_3)$ is the normalizable solution. This behavior also implies that the spinor $\eta_\pm(x_0, x_1, x_2)$ should be an eigenstate of γ_3 with eigenvalue $+i$. It then follows that the Dirac fermions on the (2+1)-dimensional wall are massless (eq. (20.128)).

We conclude that the Dirac equation in 3+1 dimensions coupled to a domain-wall defect has zero modes that behave as massless Dirac fermions in 2+1 dimensions confined to the wall. Note that, if the Dirac fermions have, in addition to the Dirac mass m, a γ^5 mass m_5, the fermions of the (2+1)-dimensional world of the domain wall now have a mass equal to m_5. Therefore, we expect that the fermions on the wall should exhibit the parity anomaly determined by the sign of m_5!

This question can be answered using, again, an extension of the Callan-Harvey argument. Indeed, instead of coupling the fermions to a real scalar field, they will now be coupled to a complex scalar field $\varphi = \varphi_1 + i\varphi_2$, with the same coupling to the fermions as in the case of the axion string, eq. (20.108), except that now we have a domain wall, with $g\varphi_1(x_3) = m(x_3)$ and $g\varphi_2 = m_5$. Once again, we can use the result of eq. (20.113), and infer that there is a current J_μ induced by the coupling to the electromagnetic gauge field A_μ. Indeed, since the θ angle is a slowly varying function of x_3, an electric field parallel to the wall induces a current also parallel to the wall and the electric field, a result reminiscent of the Hall effect. Likewise, a magnetic field normal to the wall induces a charge on the wall.

In addition, since there is only one Dirac spinor on the wall, the effective action of the gauge field on the wall is the same as predicted from the parity anomaly, with a coefficient of 1/2 the quantized value. So, we seemingly find again the same difficulties.

However, the bulk effective action supplies, in this case, the requisite missing contribution. Hence, here too, the anomaly cancellation is nonlocal and involves the whole edge and the bulk contributions. Indeed, the effective Lagrangian of the gauge field A_μ coupled to a massive Dirac fermion with a γ^5 mass, as in eq. (20.108), has a "θ term" of the form $\frac{\theta}{\rho\pi^2} F^*_{\mu\nu}F^{\mu\nu}$, where $\theta \to \pi$ as $\gamma^5 \to 0$. We will discuss θ terms in section 20.9.

Thus, even in the abelian theory, whose gauge-field sector does not have instantons, the θ terms have implications if the three-dimensional space of the (3+1)-dimensional spacetime

Ω has a boundary $\partial\omega$. Using the divergence theorem, one readily finds that the θ term of the action integrates to the boundary, where it becomes (for $\theta = \pi$)

$$S_{\text{boundary}} = \frac{1}{8\pi} \int_{\partial\Omega\times\mathbb{R}} d^3x\, \epsilon_{\mu\nu\lambda} A^\mu \partial^\nu A^\lambda \tag{20.131}$$

This is a Chern-Simons action with level $k = 1/2$. However, the effective action Dirac fermion at the boundary is also a Chern-Simons gauge there with level $k = 1/2$. We have already discussed that this result is incompatible with the requirement that the level k of the Chern-Simons theory is an integer. We can now see that if the (2+1)-dimensional theory is the boundary of a (3+1)-dimensional theory, the contribution of the bulk to the effective action matches the parity anomaly of the boundary, leading to a consistent theory. This result has important implications in the theory of topological insulators in 3+1 dimensions.

20.9 θ vacua

In this section, we investigate the role of the topological charge in several models of interest. We do this by weighting different topological sectors of the configuration space by their topological charge. In a theory defined in Euclidean spacetime, since the topological charge Q is an integer, let us weight the configurations by a factor $\exp(i\theta Q)$, where θ is an angle in the interval $[0, 2\pi)$. Thus the partition function now depends on the parameter θ:

$$Z(\theta) = \int \mathcal{D}\,\text{fields}\; e^S\, e^{i\theta Q} \tag{20.132}$$

The dependence of the partition function on the angle θ provides a direct way to assess the role of instantons in the physical vacuum of the theory (Callan et al., 1976, 1978).

20.9.1 The Schwinger model and θ vacua

The Schwinger model is the theory of quantum electrodynamics in 1+1 dimensions (Schwinger, 1962). The Lagrangian of the Schwinger model is

$$\mathcal{L} = \bar\psi i\partial\!\!\!/ \psi - m\bar\psi\psi - \bar\psi\gamma^\mu\psi A_\mu - \frac{1}{4e^2}F_{\mu\nu}^2 - \frac{\theta}{4\pi}\epsilon_{\mu\nu}F^{\mu\nu} \tag{20.133}$$

Here we included the last term, proportional to the topological charge and parametrized by the θ angle. In this theory, this term is equivalent to a Wilson loop of charge $\frac{\theta}{2\pi}$ on a closed contour at the boundary of the two-dimensional spacetime. Therefore, it can be interpreted as the effect of the uniform electric field E generated by two charges $\pm\frac{\theta}{2\pi}$ at $\pm\infty$.

In 1962, Schwinger considered the massless theory (with $\theta = 0$) and showed that the spectrum is exhausted by a free massive scalar field, a boson of mass $m_{\text{Schwinger}} = \frac{e}{\sqrt{\pi}}$. Recall that in 1+1 dimensions, the QED coupling constant, the electric charge e, has units of mass. The massless fermions of the Lagrangian are not present in the spectrum of physical, gauge-invariant states. Schwinger further showed that the gauge fields became massive and that their mass is that of the boson.

The world of 1+1 dimensions is quite restrictive and simple. Indeed, in 1+1 dimensions, for obvious reasons, the gauge field does not have transverse components. When coupled to a charged-matter field, it acquires a longitudinal component that is the same as the local charge fluctuations of the matter field. In particular, if the matter field is absent, the Coulomb gauge condition $\partial_1 A_1 = 0$ is solved by $A_1 = \text{const}$. The *confinement* of charge-matter fields is easily seen even classically. Indeed, in one spatial dimension, the electric field created by a pair of static charges is necessarily uniform, and the Coulomb interaction grows linearly with the separation R between the charges, $V(R) = -qR$. Therefore, at least at the classical level, the static sources are confined.

Let us now look at the quantum version of the theory. The simplest way to analyze the quantum theory is to use the method of bosonization, already discussed in section 20.3. There we used bosonization to identify fermionic current (eq. (20.24)), the fermion operators (eq. (20.43)), and the mass terms (eq. (20.47)). The bosonized version of the Lagrangian of eq. (20.133) is

$$\mathcal{L} = \frac{1}{2}(\partial_\mu \phi)^2 - g\cos(\sqrt{4\pi}\phi) + \frac{1}{\sqrt{\pi}}\epsilon_{\mu\nu}\partial^\nu \phi A^\mu - \frac{1}{4e^2}F_{\mu\nu}^2 - \frac{\theta}{4\pi}\epsilon_{\mu\nu}F^{\mu\nu} \tag{20.134}$$

where $g \propto m$. Upon gauge fixing (e.g., the Feynman–'t Hooft gauges), we can integrate out the gauge field and obtain the effective Lagrangian (Coleman et al., 1975; Kogut and Susskind, 1975b):

$$\mathcal{L} = \frac{1}{2}(\partial_\mu \phi)^2 - \frac{e^2}{2\pi}\phi^2 - g\cos(\sqrt{4\pi}\phi + \theta) \tag{20.135}$$

The parameter θ was originally introduced in the operator solution of the massless Schwinger model to ensure consistency of the boundary conditions (Lowenstein and Swieca, 1971). It was later given a physical interpretation in terms of the global degrees of freedom of the gauge theory (i.e., as a background electric field; Kogut and Susskind, 1975b).

Let us discuss the massless case first (Lowenstein and Swieca, 1971; Casher et al., 1974; Kogut and Susskind, 1975b). If the fermion mass is zero, $m = 0$, then the boson ϕ is massive, and its mass squared is $m^2_{\text{Schwinger}} = e^2/\pi$. This is Schwinger's result. After some simple calculations, one can also show that the fermion propagator (in the Coulomb gauge) is zero. Thus there are no fermions in the spectrum (or, more precisely, the fermions are confined). But since the boson is massive, we see that the Dirac mass operator has a finite expectation value, $\langle \bar{\psi}\psi \rangle = \langle \cos(\sqrt{4\pi}\phi) \rangle \neq 0$, since, up to weak Gaussian fluctuations, the boson ϕ is pinned to the classical value $\phi_c = 0$. Therefore, the chiral symmetry is spontaneously broken. In the massless case, the parameter θ is irrelevant in the sense that the vacuum energy is independent of the value of θ. Vacua with different values of θ rotate into one another under the action of the global chiral charge, Q_5, without any energy cost.

Things are different if the fermions are massive ($m \neq 0$). In this case, the cosine term in the action of eq. (20.135) is now present. While here, too, the fermions are confined and the scalar field ϕ is massive, there is a tension between the Schwinger mass term and the cosine term. Although this theory is no longer trivially solvable, we see that the parameter θ will change the vacuum energy. To lowest order in g, the vacuum energy density is a periodic function of θ, $\varepsilon(\theta) = g\cos\theta$. The physical interpretation is that, since the background electric field is quantized in multiples of 2π, it will change abruptly at the value $\theta = \pi$ (mod 2π).

20.9.2 θ vacua and the $\mathbb{CP}^{N-1}$ model

Since the $\mathbb{CP}^{N-1}$ model can be solved in the large-N limit (see section 17.3) and has instantons for all values of N, we may ask: What role do instantons play in the large-N limit?

Following the approach used in section 17.3, let us rescale the coupling constant $g^2 \to g_0^2/N$. In this limit, the bound for the Euclidean action is $S \geq 2\pi N|Q|/g_0^2$, and one would expect that the instanton contribution should be $O(\exp(-cN))$.

We address this problem by modifying the action of the $\mathbb{CP}^{N-1}$ model by adding a term proportional to the topological charge Q of the field configuration. In Euclidean spacetime, the partition function becomes (Witten, 1979b)

$$Z[\theta] = \int \mathcal{D}z\mathcal{D}z^*\mathcal{D}A_\mu \, \exp\left(-\frac{1}{g^2}\int d^2x \, |(\partial_\mu + iA_\mu)z_a|^2 + i\theta \, Q\right) \tag{20.136}$$

The contribution to the weight of the path integral of the term involving the angle θ can be expressed as

$$\exp(i\theta Q) = \exp\left(i\frac{\theta}{2\pi}\int_\Sigma d^2x \, \epsilon_{\mu\nu}\partial^\mu A^\nu\right) = \exp\left(i\frac{\theta}{2\pi}\oint_{\partial\Sigma} dx_\mu A^\mu\right) \tag{20.137}$$

or, what is the same,

$$\frac{Z[\theta]}{Z[0]} = \left\langle \exp\left(i\frac{\theta}{2\pi}\oint_{\partial\Sigma} dx_\mu A^\mu\right)\right\rangle_{\theta=0} \tag{20.138}$$

Therefore, the addition of the term with the θ angle is equivalent to the computation of the expectation value of the Wilson loop operator with a test charge $q = \theta/(2\pi)$ on a contour $\partial\Sigma$ that encloses the full spacetime Σ. Thus, the θ term can be interpreted as the response to a background electric field $E = \theta/(2\pi)$.

The presence of the θ term does not change the asymptotically free character of the theory in the weak coupling regime. Indeed, its beta function predicts that under the RG, the theory flows to strong coupling, a regime inaccessible to the perturbative RG. So other approaches must be used to study this regime. One nonperturbative approach is large N and the $1/N$ expansion.

Since $Q \in \mathbb{Z}$, the parameter $\theta \in [0, 2\pi)$, and one expects the partition function (and the free energy) to be a periodic function of the angle θ. Notice that since the weight of a configuration is now a complex number, the theory is not invariant under time reversal, or equivalent under CP, unless $\theta = 0, \pi \mod (2\pi)$. Terms of this type often arise in theories with Dirac fermions, due to the effects of quantum anomalies.

Since the topological charge Q is the integral of a total derivative, the saddle-point equations of this theory are the same as in the theory with $\theta = 0$. So the θ dependence on physical quantities comes from quantum fluctuations. To leading order in the $1/N$ expansion, the θ dependence of the ground-state energy density, $\varepsilon_0(\theta)$ (i.e., the response to the background electric field E), is computed by

$$\varepsilon_0(\theta) = -\lim_{L,T\to\infty}\frac{1}{LT}\ln\left(\frac{Z[\theta]}{Z[0]}\right) \tag{20.139}$$

The result turns out to be a simple periodic function of θ (shown in figure 20.7),

$$\varepsilon_0(\theta) = \frac{1}{2}\left(\frac{e_{\text{eff}}\theta}{2\pi}\right)^2 + O(1/N^2), \qquad \text{for } |\theta| < \pi \tag{20.140}$$

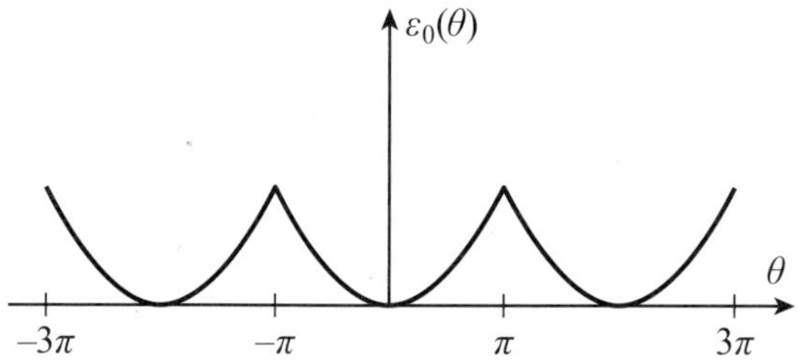

Figure 20.7 Ground-state energy density of the $\mathbb{CP}^{N-1}$ model in the large-N limit as a function of the angle θ. It has cusps for θ, an odd-integer multiple of π, where there is a first-order transition and CP is spontaneously broken.

and is periodically defined in other intervals. Here we defined

$$e_{\text{eff}}^2 = 12\pi^2 \frac{M^2}{N} \tag{20.141}$$

where M^2 is the square of the dynamically generated mass of the $\mathbb{CP}^{N-1}$ model:

$$M = \mu\, e^{-2\pi/g_R^2} \tag{20.142}$$

The fact that $0 < \varepsilon_0[\theta] < \infty$ implies that the test particles of (fractional) charge $\theta/(2\pi)$ are confined.

The function $\varepsilon_0(\theta)$ is a periodic quadratic function of θ with cusp singularities at odd multiples of π. It has a smooth $1/N$ expansion, in clear contradiction with the expected N-dependence from a naive instanton argument, which predicts a $\cos\theta$ dependence. This cusp implies a discontinuity (a jump) in the topological density, $q = \langle Q \rangle = \frac{\partial \varepsilon_0}{\partial \theta}$, for θ an odd multiple of π. Since $\langle Q \rangle \neq 0$ violates CP (or T) invariance, this symmetry is spontaneously broken, at least in the large-N limit.

20.9.3 The O(3) nonlinear sigma model and θ vacua

The $\mathbb{CP}^1$ model is equivalent to the $O(3)$ nonlinear sigma model. Its topological charge is given by eq. (19.105), and the modified Euclidean action is now

$$S = \frac{1}{2g^2} \int d^2x\, (\partial_\mu \boldsymbol{n})^2 + i\frac{\theta}{8\pi} \int_{S^2} d^2x\, \epsilon_{\mu\nu} \boldsymbol{n}(x) \cdot \partial_\mu \boldsymbol{n}(x) \times \partial_\nu \boldsymbol{n}(x) \tag{20.143}$$

In particular, the effective action of eq. (20.143) is the effective field theory of one-dimensional spin-s quantum Heisenberg antiferromagnets, with $\theta = 2\pi s$ and $g^2 \propto 1/s$. Hence, for large values of s, the theory is weakly coupled, and for small values of s is strongly coupled (e.g., for $s = 1/2$). Since s can only be an integer or a half-integer, the value of θ is either 0 (mod 2π) for $s \in \mathbb{Z}$, or π (mod 2π) for $s = 1/2$ (mod $\mathbb{Z}$).

The perturbative RG shows that the theory is asymptotically free at weak coupling regardless of the value of θ. For integer s, where the θ term is absent, the coupling constant g^2 flows under the RG to strong coupling. But, for half-integer values of s, it has been shown that the RG flows to a fixed point with a finite value of g^2 and $\theta = \pi$. Let us explore how this works (Haldane, 1983; Affleck, 1986a; Affleck and Haldane, 1987).

This nonperturbative result is known from a series of mappings between different models. The spin-1/2, one-dimensional Heisenberg quantum antiferromagnet is an integrable system with a global $SU(2)$ symmetry. It is solvable by the Bethe ansatz, and the entire spectrum is known, as are (most of) its correlation functions. In particular, the scaling dimensions of all local operators are known. At the isotropic point, the theory is massless and conformally invariant at low energies. The massless excitations are fermionic solitons. It is a conformal field theory, a subject that we discuss in chapter 21, with central charge $c=1/2$. In particular, the correlation functions of the (normalized) chiral $SU(2)$ (spin) currents $J^{\pm}(x)$ are known to decay as a power law:

$$\langle J^{\pm}(x) J^{\pm}(y)\rangle = \frac{k/2}{|x-y|^2} \tag{20.144}$$

In the case at hand, $k=1$. The exponent of this power law implies that $J^{\pm}$ has scaling dimension 1.

The Wess-Zumino-Witten (WZW) model is an $SU(N)$ nonlinear sigma model in 1+1 dimensions with field $g(x)\in SU(N)$ (and $g^{-1}=g^{\dagger}$). Thus its target space is the group manifold $G=SU(N)$. The Minkowski space action is

$$S[g] = \frac{1}{4\lambda^2}\int_{S^2_{\text{base}}} d^2x \, \text{tr}\left(\partial^{\mu} g \partial_{\mu} g^{-1}\right) + \frac{k}{24\pi}\int_B \epsilon^{\mu\nu\lambda} \text{tr}\left(g^{-1}\partial_{\mu} g\, g^{-1}\partial_{\nu} g\, g^{-1}\partial_{\lambda} g\right) \tag{20.145}$$

The second term of this action is the WZW term. We will see in section 21.6.4 that the level k must be an integer. Much as for other nonlinear sigma models, the WZW model is asymptotically free at weak coupling, and the coupling constant runs to strong coupling under the perturbative RG. However, by explicit computation, it has been shown that, if $k\neq 0$, the beta function has a zero, and the RG has an IR fixed point at a finite value of the coupling constant:

$$\lambda_c^2 = \frac{4\pi}{k} \tag{20.146}$$

At this IR fixed point, the theory is conformally invariant. This conformal field theory (CFT) is known as the $SU(2)_k$ Wess-Zumino-Witten (WZW) model with level k. We discuss this CFT in chapter 21. The case of interest for the $O(3)$ nonlinear sigma model is the $SU(2)_1$ WZW theory.

The $SU(2)_1$ WZW CFT has only one CP-invariant relevant perturbation, the operator $\mu(\text{tr}g)^2$. For $\mu>0$, the theory flows in the IR to a fixed point with $\text{tr}g=0$, where $g\rightarrow i\boldsymbol{\sigma}\cdot\boldsymbol{n}$, with $\boldsymbol{n}^2=1$. In this limit, the WZW term of the action becomes $\pi Q[\boldsymbol{n}]$, where $Q[\boldsymbol{n}]$ is the topological charge of the nonlinear sigma model. Hence, in the IR, the theory flows to the $O(3)$ nonlinear sigma model of eq. (20.143) with $\theta=\pi$. These arguments then imply that the $O(3)$ nonlinear sigma model with $\theta=\pi$ is at a nontrivial fixed point described by the $SU(2)_1$ WZW CFT.

However, for $\theta=2\pi n$ (with $n\in\mathbb{Z}$), the topological term in eq. (20.143) has no effect, since for these values of θ, the topological term of the action is equal to an integer multiple of 2π. Hence, for $\theta=2\pi n$, the theory flows to the massive phase (and CP-invariant phase) described above. This is also the behavior found in spin chains where s is an integer. Results from extensive (and highly accurate) numerical simulations of different quantum spin chains (using the density matrix RG approach) are consistent with this picture.

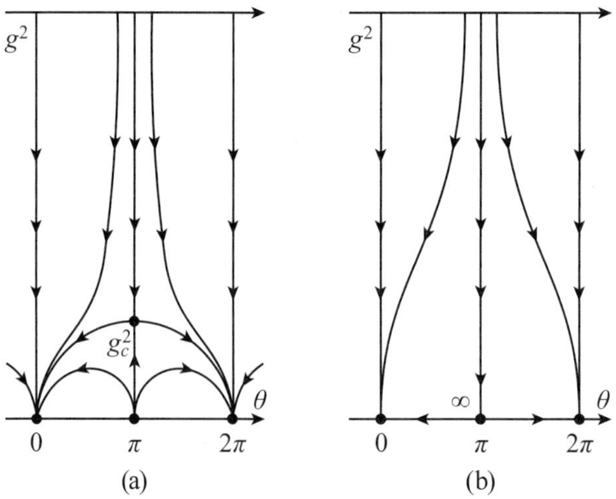

Figure 20.8 Conjectured RG flows of the $\mathbb{CP}^{N-1}$ model as a function of the coupling constant g^2 and of the angle θ. (a) The RG flow for $N=2$ (the $O(3)$ nonlinear sigma model) has a nontrivial finite IR fixed point at $(g^2, \theta) = (g_c^2, \pi)$. (b) The RG flow in the large-N limit describes a first-order transition at $\theta = \pi$ (mod 2π) and a spontaneous breaking of *CP* invariance.

In spite of much analytic and numerical work, what happens for general values of N and θ remains a matter of (educated) speculation. The currently accepted "best guess" for the global RG flow is shown in figure 20.8. Figure 20.8a is the conjectured RG flow for $N=2$ and shows a continuous phase transition at $\theta = \pi$, controlled by the finite fixed point at (g_c^2, π). Figure 20.8b shows the (also conjectured) RG flow for $N \geq 2$. It depicts the case of a first-order transition controlled by a fixed point at (∞, π), where the correlation length vanishes, and the mass diverges (a "discontinuity" fixed point).

Analytically, these flows have been derived by saturating the partition function of the nonlinear sigma model with the instanton (and anti-instanton) contribution (Pruisken, 1985). As noted above, this is problematic, since the integration over instanton sizes is IR divergent. The divergence is suppressed by means of an IR cutoff, which breaks the classical scale invariance of the theory. In turn, this leads to an instanton contribution to the RG. The result is suggestive, and predicts an RG flow of the type shown in figure 20.8. However, this procedure is problematic in many ways. In particular, the θ term of the action is the total instanton number and, in the absence of the IR regulator, it is insensitive to the effects of local fluctuations. Thus, a priori, one would not have expected that θ could be renormalized.

There is, however, numerical (Monte Carlo) evidence of a lattice version of the $\mathbb{CP}^{N-1}$ model showing that, for $N \geq 3$, as $\theta \to \pi$, the topological density remains finite but has a jump at $\theta = \pi$. However, for the $\mathbb{CP}^1$ model, the topological susceptibility appears to be divergent as $\theta \to \pi$, which is consistent with a continuous phase transition. But since the Gibbs weight of a generic configuration is not a positive real number, the Monte Carlo method converges poorly at low temperatures, leading to significant error bars in the results. So these results cannot be regarded as definitive.

20.9.4 θ vacua in four-dimensional gauge theories

Returning to the problem of the Dirac fermions coupled to a background gauge field, whose Lagrangian is given in eq. (20.82), we see that, at least at the formal level, we can integrate out the fermions. The partition function is

$$Z[A] = \int \mathcal{D}\bar{\psi}\mathcal{D}\psi \exp\left(-\int d^4x\, \bar{\psi}\,(i\not{D}[A] + m)\psi \right) = \det(i\not{D}[A] + m) \tag{20.147}$$

The chiral anomaly implies that the low-energy effective action of the gauge field, in an expansion in powers of $1/m$, must contain an extra term in addition to the Maxwell term:

$$S_{\text{eff}}[A] = -\text{tr}\ln(i\not{D}[A] + m) = i\frac{\theta}{32\pi^2}\int d^4x\, \epsilon_{\mu\nu\lambda\rho}F^{\mu\nu}F^{\nu\lambda} + \frac{1}{4e^2}\int d^4x\, F_{\mu\nu}F^{\mu\nu} \tag{20.148}$$

where $\theta = \pi$ for a single flavor of fermions. A similar result is found in the non-abelian case.

Although in four-dimensional gauge theory the θ term is a total derivative (and as such it does not affect the equations of motion), it has profound implications. In non-abelian gauge theories, such as a Yang-Mills, the axial anomaly has far-reaching consequences.

The θ term has strong implications in non-abelian theories, since these theories have instantons. 't Hooft has shown that in an $SU(2)$ Yang-Mills theory with N_f doublets of massless Dirac fermions, the chiral anomaly of the $U(1)$ current j^5_μ is the analogous expression

$$\partial^\mu j^5_\mu = -\frac{N_f g^2}{8\pi^2}\text{tr}(F^{\mu\nu}F^*_{\mu\nu}) \tag{20.149}$$

However, the integral of this expression on the right-hand side over all (Euclidean) four-dimensional spacetime is proportional to the instanton number Q (the Pontryagin index of eq. (19.183)) of the configuration of gauge fields. The total change of the axial charge, ΔQ^5, due to the instanton is ('t Hooft, 1976b)

$$\Delta Q^5 = 2N_f Q \tag{20.150}$$

This result implies that in a theory of this type, the axial symmetry is broken explicitly, and nonperturbatively, by instantons.

Even the abelian $U(1)$ theory is interesting. The Dirac quantization condition requires that the flux $F_{\mu\nu}$ across an arbitrary closed surface of four-dimensional Euclidean spacetime be an integer multiple of 2π. This condition, in turn, requires that the weight of the path integral be a periodic function of θ with period 2π. Let us define the complex quantity

$$\tau = \frac{\theta}{2\pi} + i\frac{4\pi}{g^2} \tag{20.151}$$

The partition function is a function of τ. Periodicity in $\theta \to \theta + 2\pi$ means that the partition function (on a manifold with Euclidean signature) is invariant under the transformation $\mathcal{T}: \tau \to \tau + 1$. Witten has shown that in addition, and quite generally, there is an *S duality* transformation, $\mathcal{S}: \tau \to -1/\tau$, under which the partition function transforms simply. More concretely, the transformations $\mathcal{S}$ and $\mathcal{T}$ do not commute with each other, and they generate the infinite group of modular transformations $SL(2, \mathbb{Z})$. More precisely, the 2×2 integer-valued matrices are elements of a subgroup $\Gamma \subseteq SL(2, \mathbb{Z})$,

$$\begin{pmatrix} a & b \\ c & d \end{pmatrix} \in \Gamma \tag{20.152}$$

such that, under a modular transformation in Γ, a function of the complex variable $F(\tau)$ transforms as

$$F\left(\frac{a\tau+b}{c\tau+d}\right)=(c\tau+d)^u(c\bar{\tau}+d)^v F(\tau) \tag{20.153}$$

where $\bar{\tau}$ is the complex conjugate of τ, and the exponents u and v depend on topological invariants of the manifold (such as the Euler characteristic). Depending on the manifold, the partition function may be invariant under $SL(2,\mathbb{Z})$ or may transform irreducibly as in the above transformation law.

Another example is a theory of two bosons ϕ_I (with $I=1,2$) in two dimensions, each satisfying the compactification conditions $\phi_I \simeq \phi_I + 2\pi$. Therefore, the field (ϕ_1, ϕ_2) is a mapping of the base manifold, which we will take to be the two-torus T^2, to its target manifold, which is also a two torus T^2. These mappings are classified by the homotopy group $\pi_2(T^2)=\mathbb{Z}$. We now consider the (Euclidean) action

$$S=\int d^2x \frac{1}{4\pi g^2}[(\partial_\mu\phi_1)^2+(\partial_\mu\phi_2)^2]+i\frac{B}{4\pi}\int d^2x \epsilon_{IJ}\epsilon_{\mu\nu}\partial_\mu\phi_I\partial_\nu\phi_J \tag{20.154}$$

Actions of this type are often considered in toroidal compactifications of (closed) string theory. It is easy to see that the second term is the analog of the θ angle term discussed above (see eq. (20.148)). Indeed, the last term of eq. (20.154) is equal to $2\pi iBQ$, where $Q\in\mathbb{Z}$ is the topological charge of the field configuration (ϕ_1,ϕ_2). As a result, the partition function is a periodic function of B with period 1. It turns out that it is also invariant under the same duality transformation as Maxwell's theory, and hence is invariant under the extended $SL(2,\mathbb{Z})$ modular symmetry (Cardy and Rabinovici, 1982; Cardy, 1982; Shapere and Wilczek, 1989). As demonstrated in exercise 20.2, this structure leads to a complex phase diagram in which electric-magnetic composites condense. In the four-dimensional case, the composite objects are magnetic monopoles with electric charges, called *dyons* (Witten, 1979a).

Exercises

20.1 Fermion zero modes of a vortex.

In this exercise, consider a theory of a Dirac fermion coupled to a complex scalar field with a vortex (Jackiw and Rossi, 1981; Weinberg, 1981). The Lagrangian of this problem is

$$\mathcal{L}=\bar{\psi}i\not{D}(A)\psi - i\frac{g}{2}\phi\bar{\psi}\psi^c + i\frac{g}{2}\phi^*\bar{\psi}^c\psi$$

Here $D_\mu(A)=i\partial_\mu - qA_\mu$ is the covariant derivative, and ψ^c is the charge conjugate $\psi_i^c = C_{ij}\bar{\psi}_j$, where C is the charge conjugation matrix. Assume that the vortex has topological charge n and has azimuthal symmetry with the z axis. Under these conditions, the problem can be reduced to the study of the Dirac equation in 2+1 dimensions. Work in the representation in which the Dirac fermion is a two-component spinor and the Dirac matrices are given in terms of Pauli matrices: $\gamma_0=\sigma_3$, $\gamma_1=i\sigma_2$, and $\gamma_2=-i\sigma_1$. The complex scalar field and the gauge field configurations are, in polar coordinates $\boldsymbol{r}=(r\cos\varphi, r\sin\varphi)$,

$$\phi(\boldsymbol{r})=f(r)e^{in\varphi}, \quad q\boldsymbol{A}(\boldsymbol{r})=-\boldsymbol{e}_\varphi A(r)$$

where $e_\varphi^2 = 1$, and $q = 2e$ is the charge of the complex scalar field (e.g., a superconducting order-parameter field). The functions $f(r)$ and $A(r)$ obey the asymptotic conditions for a vortex:

$$f(r) \sim f_0 r^{|n|} \text{ as } r \to 0, \ \lim_{r \to \infty} f(r) = f_\infty$$

$$\lim_{r \to 0} A(r) = 0, A(r) \sim -\frac{n}{r} \text{ as } r \to \infty$$

so that the magnetic flux of the vortex is

$$-\frac{1}{2} \int d^2x \, \epsilon^{ij} F_{ij} = \oint_\Gamma d\boldsymbol{r} \cdot \boldsymbol{A}(\boldsymbol{r}) = 2\pi \frac{n}{2e}$$

The precise form of the functions $f(r)$ and $A(r)$ will not be needed.

1) Consider first the theory without a vortex ($\boldsymbol{A} = 0$). Find the spectrum of single-particle states, and show the fermions have a mass $\mu = gf(\infty)$. Show that the Dirac Hamiltonian anticommutes with $\gamma_0 = \sigma_3$ and that it induces a symmetry in the spectrum of states. Find the relation between the spinors for positive and negative energy states.
2) Consider now the case of a vortex with vorticity $n = +1$. Show that in this case, the Dirac equation has a single solution with zero energy, $E = 0$ (a *zero mode*). Find the exact form of the eigenspinor. Show that the zero-mode solution is real and is an eigenstate of σ_3. How does the solution change for a vortex with $n = -1$?
3) Now consider the general case of a vortex with topological charge $n > 0$. Show that the zero modes can be chosen to be eigenstates of σ_3, and that, in Cartesian coordinates, the equation of the zero modes takes the form

$$\mathcal{D} \begin{pmatrix} u \\ v \end{pmatrix} = 0$$

where

$$\mathcal{D} = (-\partial_1 + i\tau_2 \partial_2) + e(-A_2 - i\tau_2 A_1) + g(-\phi_1 \tau_3 - \phi_2 \tau_1)$$

Here the spinor eigenstate of σ_3 is written in terms of two real functions, $\psi = u + iv$, the complex field is $\phi = \phi_1 + i\phi_1$, and τ_i $(i = 1, 2, 3)$ are the three Pauli matrices. Use eq. (20.97) to find the index of the operator $\mathcal{D}$, and show that the index is equal to n, the topological charge of the vortex.

20.2 Physical effects of the θ angle

In this exercise, you will consider the classical statistical mechanical $\mathbb{Z}_N$ model with a θ angle on a square lattice (Cardy and Rabinovici, 1982; Cardy, 1982; Shapere and Wilczek, 1989). Assume periodic boundary conditions. A $\mathbb{Z}_N$ model is a spin system with N states whose degrees of freedom have the form of a clock with N hours. Consider a *doubled* model, in which $\mathbb{Z}_N$ clock degrees of freedom appear *both* on the sites of the square lattice and on the sites of the dual (square) lattice. Use methods similar to the duality transformation discussed in section 19.9 in the context of the two-dimensional classical XY model.

The model is defined as follows. Let $\varphi(\boldsymbol{r}) \in \mathbb{R}$ and $\varphi(\boldsymbol{R}) \in \mathbb{R}$ be two real fields defined on the square lattice of sites $\{\boldsymbol{r}\}$ and on the dual (also square) lattice of sites

$\{\boldsymbol{R}\}$, respectively. Also define an integer-valued variable $s_{\mu 1} \in \mathbb{Z}$ on each link of the direct lattice and $s_{\mu 2} \in \mathbb{Z}$ on each link of the dual lattice (notice that the two types of links intersect). In addition, define an integer-valued variable $n_a \in \mathbb{Z}$ on the sites of the direct and of the dual lattice. The partition function of the model is (Cardy and Rabinovici, 1982; Cardy, 1982)

$$Z_N(g,\theta) = \mathrm{Tr}\exp\Bigg(-\frac{1}{2g^2}\sum_{\mu,a}(\Delta_\mu \varphi_a - 2\pi s_{\mu a})^2 + iN\sum_a n_a \varphi_a + i\frac{N\theta}{32\pi^2}\sum \epsilon_{\mu\nu}\epsilon_{ab}(\Delta_\mu\varphi_a - 2\pi s_{\mu a})(\Delta_\nu \varphi_b - 2\pi s_{\nu b})\Bigg)$$

Here, Tr denotes the sums and integrals over all local degrees of freedom. Also we have omitted the sums over the sites of both the direct and the dual lattices.

1) Show that this model is invariant under the following local symmetries: $\varphi_a \to \varphi_a + 2\pi p_a$, and $s_{\mu a} \to s_{\mu a} + \Delta_\mu p_a$ (on each link). This proves that the fields φ_a are periodic with period 2π, and hence, that the target space is a two-torus T^2.
2) Use the duality transformations introduced in section 19.9 to show that this theory has two types of vortices with integer topological charges m_a (each on the dual and on the direct lattice, respectively), and that (up to an irrelevant prefactor) the partition function is a generalized two-dimensional Coulomb gas of the form
$$Z_N(g^2,\theta) = \mathrm{Tr}\exp(-S[\{m_a\},\{n_a\};g,\theta])$$
where
$$\begin{aligned} S[\{m_a\},\{n_a\};g,\theta] = \\ &= \frac{\pi}{g^2}\Bigg[\sum_{\boldsymbol{R}\neq \boldsymbol{R}'} m_1(\boldsymbol{R})m_1(\boldsymbol{R}')\ln(|\boldsymbol{R}-\boldsymbol{R}'|) + \sum_{\boldsymbol{r}\neq\boldsymbol{r}'} m_2(\boldsymbol{r})m_2(\boldsymbol{r}')\ln(|\boldsymbol{r}-\boldsymbol{r}'|)\Bigg] \\ &+ \frac{Ng^2}{4\pi}\Bigg[\sum_{\boldsymbol{r}\neq\boldsymbol{r}'}\left(n_1(\boldsymbol{r}) + \frac{\theta}{2\pi}m_2(\boldsymbol{r})\right)\left(n_1(\boldsymbol{r}') + \frac{\theta}{2\pi}m_2(\boldsymbol{r}')\right)\ln(|\boldsymbol{r}-\boldsymbol{r}'|) \\ &+ \sum_{\boldsymbol{R}\neq\boldsymbol{R}'}\left(n_2(\boldsymbol{R}) + \frac{\theta}{2\pi}m_1(\boldsymbol{R})\right)\left(n_2(\boldsymbol{R}') + \frac{\theta}{2\pi}m_1(\boldsymbol{R}')\right)\ln(|\boldsymbol{R}-\boldsymbol{R}'|)\Bigg] \\ &+ iN\sum_{\boldsymbol{r},\boldsymbol{R}}(m_1(\boldsymbol{R})n_1(\boldsymbol{R}) - m_2(\boldsymbol{r})n_2(\boldsymbol{R}))\Theta(\boldsymbol{r}-\boldsymbol{R}) \\ &- \frac{1}{8}N^2g^2\Bigg[\sum_{\boldsymbol{r}} n_1(\boldsymbol{r})^2 + \sum_{\boldsymbol{R}} n_2(\boldsymbol{R})^2\Bigg] - \frac{\pi^2}{2g^2}\Bigg[\sum_{\boldsymbol{R}} m_1(\boldsymbol{R})^2 + \sum_{\boldsymbol{r}} m_2(\boldsymbol{r})^2\Bigg] \end{aligned}$$
where $\Theta(\boldsymbol{r}-\boldsymbol{R})$ is the angle subtended by the sites $\boldsymbol{r}$ and $\boldsymbol{R}$ with respect to an (arbitrary) fixed direction. In this form, we have a Coulomb gas of electric charges n_a and magnetic charges m_a. This shows that the effect of the θ angle is to make dyons: composites of electric and magnetic charges.
3) Show that the partition function of the generalized Coulomb gas is a a *periodic* function of the θ angle with period 2π.

4) Show that this generalized Coulomb gas has a dual under the exchange of electric charges with magnetic charges such that the action transforms as follows:

$$S[\{m_a\},\{n_a\};g,\theta]=S[\{n_a\},\{-m_a\};g',\theta']$$

where

$$g'^2=\frac{g^2/(\pi^2N^2)}{(2\pi^2/g^2)^2+(N\theta/2)^2},\qquad \theta'=-\frac{\pi^2N^2\theta}{(2\pi^2/g^2)^2+(N\theta/2)^2}$$

Hint: This result can be derived using the Poisson summation formula, eq. (19.218).

5) Show that, in terms of the complex quantity z,

$$z=\frac{2\pi}{Ng^2}+i\frac{\theta}{2\pi}$$

the electric-magnetic duality transformation, $m\to n$ and $n\to -m$, reduces to

$$z'=\frac{1}{z}$$

6) Denote the duality transformation by $\mathcal{S}$ and the periodicity of the θ angle by $\mathcal{T}$. Show that $\mathcal{S}^2=1$ and that $\mathcal{S}$ and $\mathcal{T}$ do not commute. Also show that an arbitrary sequence of $\mathcal{S}$ and $\mathcal{T}$ is equivalent to the $SL(2,\mathbb{Z})$ transformation

$$z'=\frac{az+b}{cz+d},\qquad ad\neq bc$$

where a, b, c, and d are arbitrary integers.

7) Use a generalization of the Kosterlitz-Thouless energy-entropy argument discussed in section 19.9 to show that a state without electric and magnetic charges (i.e., with $n_a=m_a=0$) is unstable to the proliferation of dyons of electric charge n and magnetic charge m if the following inequality is satisfied:

$$\frac{2\pi m^2}{Ng^2}+\frac{Ng^2}{2\pi}\left(n+\frac{\theta}{2\pi}m\right)^2<\frac{4}{N^2}$$

21

Conformal Field Theory

21.1 Scale and conformal invariance in field theory

In chapter 15, we looked at explicit constructions of RG transformations using ideas largely inspired by the theory of phase transitions. As anticipated, the key concept is that of a fixed point. In particular, we identified a class of fixed points at which the physical length scale, the correlation length, diverges. From the point of view of QFT, these are the fixed points of interest, since their vicinity defines a continuum QFT. Notice that, for example, in eq. (15.88), the UV cutoff a enters only to provide the necessary units to the correlation length. But aside from that, in this limit the UV regulator essentially disappears from the theory.

In this sense, the fixed points associated with continuous phase transitions express the behavior of the theory in the IR, whereas the fixed points that have vanishing correlation lengths define the phases of the theory and are controlled by the behavior of the theory in the UV. Thus, in one phase, the QFT is represented as an RG flow from the UV to the IR in that as a coupling constant is varied, the correlation length grows from its microscopic definition to a behavior largely independent of the microscopic physics.

In this section, we discuss general properties of scale-invariant theories. As such, these theories must be defined as QFTs at a fixed point with a divergent correlation length (representing some continuous phase transition), at which the theory is scale and rotationally invariant (or Lorentz invariant in the Minkowski signature). From a "microscopic" point of view, scale and conformal invariance are emergent symmetries of the fixed-point theory.

A general result (Polchinski, 1988) shows that, under most circumstances, scale-invariant theories have a much larger symmetry: conformal invariance. The general framework for such theories is known as conformal field theory (CFT). This approach allows us to describe not only the CFT but also deformed CFTs by the action of relevant operators. (Polyakov, 1974).

Scale invariance alone implies that observables in this theory should obey scaling. Hence, they should transform irreducibly under scale transformations, dilatations of the form $x' = \lambda x$. Thus, expectation values of a physical observable $F(x)$ must transform homogeneously under dilatations as $F(\lambda x) = \lambda^k F(x)$, where k is called the *degree* of the homogeneous function. Scale transformations are a subgroup of more general transformations known as *conformal transformations*. Conformal transformations are coordinate transformations that preserve the angles (i.e., scalar products) between vectors in a space (or spacetime).

Several excellent treatments of CFT in the literature have inspired the presentation of the topic in this book. Modern general approaches to CFT are the 2015 TASI lectures by David Simmons-Duffin (2017), the 1988 Les Houches Summer School lectures by Paul Ginsparg (1989) on applied CFT, and the lectures by John Cardy (1996) on conformal invariance and statistical mechanics. Other excellent presentations are found in the book *Conformal Field Theory* by Philippe Di Francesco, Paul Mathieu, and David Sénéchal (Di Francesco et al., 1997), and in the two-volume book *String Theory*, by Joseph Polchinski (1998).

21.2 The conformal group in *D* dimensions

In this section, we discuss the general consequences of conformal invariance in a field theory (Ginsparg, 1989; Simmons-Duffin, 2017). Consider a local field theory in a flat D-dimensional spacetime, which will be regarded as $\mathbb{R}^D$. The flat metric will be $g_{\mu\nu} = \eta_{\mu\nu}$, with signature (p, q). The line element is, as usual,

$$ds^2 = g_{\mu\nu} dx^\mu dx^\nu \tag{21.1}$$

The change of the metric tensor under a change of coordinates $x_\mu \mapsto x'_\mu$ is

$$g_{\mu\nu} \mapsto g'_{\mu\nu}(x') = \frac{\partial x^\alpha}{\partial x'^\mu} \frac{\partial x^\beta}{\partial x'^\nu} g_{\alpha\beta}(x) \tag{21.2}$$

Conformal transformations are a subgroup of diffeomorphisms (i.e., differentiable coordinate transformations) that leave the metric tensor invariant up to a local change of scale, that is,

$$g'_{\mu\nu}(x') = \Omega(x) g_{\mu\nu}(x) \tag{21.3}$$

where $\Omega(x)$ is the conformal factor. In this case, if v^μ and w^μ are two vectors whose scalar product is $v \cdot w = g_{\mu\nu} v^\mu w^\nu$, then the quantity

$$\frac{v \cdot w}{\sqrt{|v|^2 |w|^2}} \tag{21.4}$$

is invariant under conformal transformations, which, therefore, preserve angles. Here $|v|^2 = g_{\mu\nu} v^\mu v^\nu$ is the norm of the vector v_μ (an example is shown in figure 21.1). Conformal transformations form a group, known as the conformal group. The Poincaré group, consisting of spacetime translations and Lorentz transformations, is a subgroup of the conformal group.

Let us first look at the generators of infinitesimal conformal transformations. Under an infinitesimal transformation, $x_\mu \mapsto x'_\mu = x_\mu + \varepsilon_\mu$, the line element transforms as

$$ds^2 \mapsto ds^2 + \left(\partial_\mu \varepsilon_\nu + \partial_\nu \varepsilon_\mu\right) dx^\mu dx^\nu \tag{21.5}$$

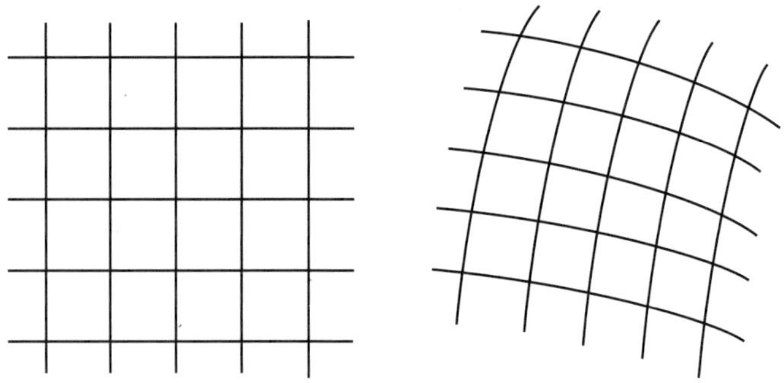

Figure 21.1 A conformal mapping transforms Cartesian coordinates (a) into curvilinear coordinates (b) while preserving angles.

For this mapping to be conformal, we require that

$$\partial_\mu \varepsilon_\nu + \partial_\nu \varepsilon_\mu = \frac{2}{D}\, \partial \cdot \varepsilon \, \eta_{\mu\nu} \tag{21.6}$$

which implies that for an infinitesimal conformal transformation, the conformal factor $\Omega(x)$ is

$$\Omega(x) = 1 + \frac{2}{D}\, \partial \cdot \varepsilon \tag{21.7}$$

Eq. (21.6) implies that $\partial \cdot \varepsilon$ is the solution of the partial differential equation

$$\left(\eta_{\mu\nu}\partial^2 + (D-2)\partial_\mu \partial_\nu\right)(\partial \cdot \varepsilon) = 0 \tag{21.8}$$

The case $D=2$ is special (and especially important). For $D=2$ and for the Euclidean signature, $\eta_{\mu\nu} = \delta_{\mu\nu}$, eq. (21.6) implies that the components of ϵ_μ satisfy

$$\partial_1 \varepsilon_2 = \partial_2 \varepsilon_1, \qquad \partial_2 \varepsilon_1 = -\partial_1 \varepsilon_2 \tag{21.9}$$

In other words, the vector field ε_μ satisfies the Cauchy-Riemann equations. Moreover, the function $f(x_1, x_2) = \varepsilon_1 + i\varepsilon_2$ is an *analytic* function of the coordinates. This feature has powerful implications for the special case of $D=2$ and will be discussed separately.

Returning to the general case $D > 2$, eq. (21.8) implies that third derivatives of ε must vanish and, hence, it can have at most a quadratic dependence on the coordinates x_μ. In general, we have three cases:

1) Zeroth-order in x_μ: $\varepsilon_\mu = a_\mu$, which represents translations.
2) Linear in x_μ: We have two choices, (a) $\varepsilon_\mu = \omega_{\mu\nu} x^\nu$, which represents an infinitesimal rotation (or a Lorentz transformation in the Minkowski signature), and (b) $\varepsilon_\mu = \lambda x_\mu$, which represents an infinitesimal scale transformation.
3) Quadratic order in x_μ: $\varepsilon_\mu = b_\mu x^2 - 2x_\mu b \cdot x$ (or $\frac{x'_\mu}{x'^2} = \frac{x_\mu}{x^2} + b_\mu$), which represents an infinitesimal special conformal transformation.

The infinitesimal generators of conformal transformations are

$$a^{\mu}\partial_{\mu}, \qquad (p+q) \text{ translations,} \tag{21.10}$$

$$\omega^{\mu\nu}x_{\nu}\partial_{\mu}, \qquad \frac{1}{2}(p+q+1)(p+q-1) \text{ rotations,} \tag{21.11}$$

$$\lambda x^{\mu}\partial_{\mu}, \qquad \text{one scale transformation,} \tag{21.12}$$

$$b^{\mu}(x^2\partial_{\mu}-2x^{\mu}x^{\nu}\partial_{\nu}), \qquad (p+q) \text{ special conformal transformations.} \tag{21.13}$$

The total number of generators is $\frac{1}{2}(p+q+1)(p+q+2)$. This algebra is isomorphic to the algebra of the group $SO(p+1,q+1)$.

Finite conformal transformations are

1) Translations: $x'_{\mu}=x_{\mu}+a_{\mu}$ (which has $\Omega=1$).
2) Rotations: $x'_{\mu}=\Lambda_{\mu\nu}x^{\nu}$, with $\Lambda\in SO(p,q)$ (also with $\Omega=1$).
3) Scale transformations: $x'_{\mu}=\lambda x_{\nu}$ (with $\Omega=\lambda^{-2}$).
4) Special conformal transformations: $x'_{\mu}=(x_{\mu}+b_{\mu}x^2)/(1+2b\cdot x+b^2x^2)$ (with $\Omega=(1+2b\cdot x+b^2x^2)^2$).

In $D=p+q$ dimensions, the Jacobian J of a conformal transformation is

$$J=\left|\frac{\partial x'}{\partial x}\right|=(\det g'_{\mu\nu})^{-1/2}=\Omega^{-D/2} \tag{21.14}$$

Translations and rotations have $J=1$, dilatations have $J=\lambda^{D}$, and special conformal transformations have $J=(1+2b\cdot x+b^2x^2)^{-D}$.

21.3 The energy-momentum tensor and conformal invariance

In section 3.8, we discussed the role of the energy-momentum tensor $T^{\mu\nu}$ in the context of classical field theory and, equivalently, the stress tensor in classical Euclidean field theory. We will now see that its counterpart in QFT plays a prominent role in the presence of conformal invariance. In section 3.10, we showed that the energy-momentum tensor can be viewed as the response to an infinitesimal change of the metric (see eq. (3.198)). Furthermore, we showed that in a Poincaré-invariant theory (i.e., invariant under translations and Lorentz transformations), the energy-momentum tensor is locally conserved (i.e., $\partial_{\mu}T^{\mu\nu}=0$) and can always be made symmetric: $T^{\mu\nu}=T^{\nu\mu}$.

The associated Noether charge of the energy-momentum tensor is the total linear momentum 4-vector P^{μ}, which in chapter 3 is defined as the integral of $T^{\mu\nu}$ on a constant-time hypersurface. However, since $T^{\mu\nu}$ is conserved, the quantity defined by the surface integral on any closed hypersurface Σ (the boundary of a region Ω),

$$P^{\nu}[\Sigma]=-\int_{\Sigma}dS_{\mu}T^{\mu\nu} \tag{21.15}$$

does not change under smooth changes of the shape of the boundary Σ or even the size of the enclosed region. This holds provided that no operators are inserted in the bulk of Ω (or become included in Ω as Σ changes).

At the quantum level, the classical conservation law is replaced by the (conformal) Ward identity

$$\partial_\mu \Big\langle T^{\mu\nu}(x)\mathcal{O}_1(x_1)\cdots\mathcal{O}_N(x_N)\Big\rangle = -\sum_{j=1}^{N}\delta(x-x_j)\partial_j^\nu\Big\langle\mathcal{O}_1(x_1)\cdots\mathcal{O}_N(x_N)\Big\rangle \tag{21.16}$$

where $\{\mathcal{O}_j(x)\}$ is an arbitrary set of local operators. Let $\Omega(x_1)$ be a simply connected region of spacetime (which we will take to be Euclidean and flat), and let $\Sigma(x_1)$ be its boundary, which encloses only the operator $\mathcal{O}(x_1)$ but not any of the other operators involved in the expectation values of eq. (21.16). Then the divergence theorem applied to the left-hand side of eq. (21.16) says that

$$\begin{aligned}\int_{\Omega(x_1)} d^Dx\,\partial_\mu\Big\langle T^{\mu\nu}(x)\mathcal{O}_1(x_1)\cdots\mathcal{O}_N(x_N)\Big\rangle &= \\ &= \int_{\Sigma(x_1)} dS_\mu\Big\langle T^{\mu\nu}(x)\mathcal{O}_1(x_1)\cdots\mathcal{O}_N(x_N)\Big\rangle \\ &= -\Big\langle P^\nu[\Sigma(x_1)]\mathcal{O}_1(x_1)\cdots\mathcal{O}_N(x_N)\Big\rangle\end{aligned} \tag{21.17}$$

Performing the same computation on the right-hand side of eq. (21.16), we obtain

$$\begin{aligned}-\int_{\Omega(x_1)} d^Dx\sum_{j=1}^{N}\delta(x-x_j)\partial_j^\nu\Big\langle\mathcal{O}_1(x_1)\cdots\mathcal{O}_N(x_N)\Big\rangle &= \\ &= -\,\partial_{x_1}^\nu\Big\langle\mathcal{O}_1(x_1)\cdots\mathcal{O}_N(x_N)\Big\rangle\end{aligned} \tag{21.18}$$

Therefore,

$$\Big\langle P^\nu[\Sigma(x_1)]\mathcal{O}_1(x_1)\cdots\mathcal{O}_N(x_N)\Big\rangle = \partial_{x_1}^\nu\Big\langle\mathcal{O}_1(x_1)\cdots\mathcal{O}_N(x_N)\Big\rangle \tag{21.19}$$

In the operator language the expectation value of a product of operators is interpreted as the vacuum expectation value of the time-ordered product of the operators. We can now consider two oppositely oriented surfaces Σ_1 and Σ_2 at two different times t_1 and t_2, and the region Ω whose boundary consists of both surfaces. If only one operator (say, at x_1) is included in Ω, it is clear that we can always deform the region to a ball centered about x_1. Conversely, we can deform the simply connected region into the shell just described. Then, using the fact that Σ_1 and Σ_2 are oppositely oriented, the right-hand side of eq. (21.19) can be rewritten as

$$\Big\langle\,(P^\mu[\Sigma_2]-P^\mu[\Sigma_1])\,\mathcal{O}(x)\cdots\Big\rangle = \langle 0|T\{[P^\mu,\mathcal{O}(x)]\cdots\}|0\rangle \tag{21.20}$$

where T denotes the Euclidean imaginary time-ordered product. Hence, we can make the operator identification

$$[P^\mu,\mathcal{O}(x)] = \partial^\mu\mathcal{O}(x) \tag{21.21}$$

(a factor of i is missing, since we are in the Euclidean signature).

Section 21.2 showed that if a theory is conformally invariant, it has more symmetries in addition to translation invariance. Thus, rotational (or Lorentz) invariance implies the existence of a conserved angular momentum tensor $M_{\mu\nu}$, which at the classical level is

$$M_{\mu\nu} = -\int_\Sigma dS^\rho \left(x_\mu T_{\nu\rho} - x_\nu T_{\mu\rho}\right) \tag{21.22}$$

At the quantum level, the action of $M_{\mu\nu}$ on a local operator $\mathcal{O}^a(0)$ at the origin causes it to transform as an irreducible representation of $SO(d)$ (here a labels the representation of $SO(d)$, again using the Euclidean signature). Thus,

$$\left[M_{\mu\nu}, \mathcal{O}^a(0)\right] = \left[\mathcal{S}_{\mu\nu}\right]^a_b \mathcal{O}^b(0) \tag{21.23}$$

where a, b are the indices of the $SO(d)$ representation, and the matrix $\mathcal{S}_{\mu\nu}$ obeys the same algebra as $M_{\mu\nu}$ and acts on an operator at a general location x as

$$\left[M_{\mu\nu}, \mathcal{O}(x)\right] = \left[x_\mu \partial_\nu - x_\nu \partial_\mu + \mathcal{S}_{\mu\nu}\right] \mathcal{O}(x) \tag{21.24}$$

Likewise, the generator of infinitesimal dilatations is expressed in terms of the dilatation operator

$$D = -\int_\Sigma dS_\mu x_\nu T^{\mu\nu} \tag{21.25}$$

and the conserved dilatation current j_D^μ is

$$j_D^\mu = x_\nu T^{\mu\nu} \tag{21.26}$$

An operator $\mathcal{O}(0)$ that transforms irreducibly under dilatations is an eigen-operator of D, that is,

$$[D, \mathcal{O}(0)] = \Delta \mathcal{O}(0) \tag{21.27}$$

where the eigenvalue Δ is the *scaling dimension* (or dimension) of the operator, a quantity that we encountered in chapter 15. We also have the Ward identity:

$$[D, \mathcal{O}(x)] = (x^\mu \partial_\nu + \Delta)\, \mathcal{O}(x) \tag{21.28}$$

Similarly, the generator of infinitesimal special conformal transformations is given by

$$K^\mu = -\int_\Sigma dS_\rho \left(2x_\mu x^\nu T^{\nu\rho} - x^2 T^{\mu\rho}\right) \tag{21.29}$$

The operators D and K_μ generate symmetries (and are conserved), provided the energy-momentum tensor is traceless,

$$T^\mu_\mu = 0 \tag{21.30}$$

The operators P^μ, $M^{\mu\nu}$, D, and K^μ satisfy the conformal algebra

$$\left[M_{\mu\nu}, P_\rho\right] = \delta_{\nu\rho} P_\mu - \delta_{\mu\rho} P_\nu \tag{21.31}$$

$$\left[M_{\mu\nu}, K_\rho\right] = \delta_{nu\rho} K_\mu - \delta_{\mu\rho} K_\nu \tag{21.32}$$

$$\left[M_{\mu\nu}, M_{\rho\sigma}\right] = \delta_{\nu\rho} M_{\mu\sigma} - \delta_{\mu\rho} M_{\nu\sigma} + \delta_{\nu\sigma} M_{\rho\mu} - \delta_{\mu\sigma} M_{\rho\nu} \tag{21.33}$$

$$\left[D, P_\mu\right] = P_\mu \tag{21.34}$$

$$\left[D, K_\mu\right] = -K_\mu \tag{21.35}$$

$$\left[K_\mu, P_\nu\right] = 2\delta_{\mu\nu} D - 2M_{\mu\nu} \tag{21.36}$$

Finally, we note two more Ward identities. The first involves rotations (and Lorentz transformations) and the symmetry of the energy-momentum tensor,

$$\partial_\mu \left\langle \left(T^{\mu\nu}(x) x^\rho - T^{\mu\rho}(x) x^\nu \right) \mathcal{O}_1(x_1) \cdots \mathcal{O}_N(x_N) \right\rangle = \\ = \sum_{i=1}^{N} \delta(x - x_i) \left[\left(x_i^\mu \partial_i^\rho - x_i^\rho \partial_i^\nu + \mathcal{S}^{\nu\rho} \right) \left\langle \mathcal{O}_1(x_1) \cdots \mathcal{O}_N(x_N) \right\rangle \right] \tag{21.37}$$

which, using the conformal Ward identity of eq. (21.16), becomes

$$\left\langle \left(T^{\mu\nu}(x) - T^{\nu\mu}(x) \right) \mathcal{O}_1(x_1) \cdots \mathcal{O}_N(x_N) \right\rangle = \\ = - \sum_i \delta(x - x_i) \mathcal{S}_i^{\mu\nu} \left\langle \mathcal{O}_1(x_1) \cdots \mathcal{O}_N(x_N) \right\rangle \tag{21.38}$$

Hence, the energy-momentum tensor as a local operator is generally symmetric away from other operator insertions.

The other Ward identity involves the dilatation current, j_D^μ, and it is given by

$$\partial_\nu \left\langle T_\nu^\mu(x) x^\nu \mathcal{O}_1(x_1) \cdots \mathcal{O}_N(x_N) \right\rangle = \\ = - \sum_i \delta(x - x_i) \left(x_i^\nu \partial_\nu^i + \Delta_i \right) \left\langle \mathcal{O}_1(x_1) \cdots \mathcal{O}_N(x_N) \right\rangle \tag{21.39}$$

Using eq. (21.16) once again, we obtain

$$\left\langle T_\mu^\mu(x) \mathcal{O}_1(x_1) \cdots \mathcal{O}_N(x_N) \right\rangle = - \sum_i \delta(x - x_i) \Delta_i \left\langle \mathcal{O}_1(x_1) \cdots \mathcal{O}_N(x_N) \right\rangle \tag{21.40}$$

Thus, the symmetry of the energy-momentum tensor holds in expectation values away from operator insertions. It also implies that at the quantum level, the trace of the energy-momentum tensor is the generator of scale transformations. Hence, as an operator, we have

$$\left[T_\mu^\mu, \mathcal{O}_i(0) \right] = \Delta_i \mathcal{O}_i(0) \tag{21.41}$$

from which it follows that, as an operator, the trace of the energy-momentum tensor must be the same as the divergence of the dilatation current,

$$T^{\mu}_{\mu} = \partial_{\mu} j^{\mu}_{D} \tag{21.42}$$

Therefore, scale invariance requires the energy-momentum tensor to be traceless.

21.4 General consequences of conformal invariance

Let us consider now a theory that is invariant under the action of the generators of the conformal group, eqs. (21.31)–(21.36). Thus, the vacuum state $|0\rangle$ is, by definition, annihilated by all the generators.

Consider now the consequences for the two-point function of two operators $\mathcal{O}_i$ and $\mathcal{O}_j$,

$$F_{ij}(x,y) = \langle \mathcal{O}_i(x_i)\mathcal{O}_j(x_j)\rangle = \langle 0|T\left(\mathcal{O}_i(x_i)\mathcal{O}_j(x_j)\right)|0\rangle \tag{21.43}$$

Translation and rotation invariance imply that it must depend only on the distance: $F(x_i,x_j) = F(|x_i - x_j|)$.

To examine the behavior under scale transformations of operators of a theory with conformal invariance, we use the condition that the vacuum must be scale invariant:

$$D|0\rangle = 0 \tag{21.44}$$

Then, we must also have

$$\begin{aligned} 0 &= \langle 0| \left[D, \mathcal{O}_i(x_i)\mathcal{O}_j(x_j)\right] |0\rangle \\ &= \langle 0| \left(\left[D, \mathcal{O}_i(x_i)\right]\mathcal{O}_j(x_j) + \mathcal{O}_i(x_i)\left[D, \mathcal{O}_j(x_j)\right]\right) |0\rangle \\ &= \left(x_i^{\mu}\partial^i_{\mu} + \Delta_i + x_j^{\mu}\partial^j_{\mu} + \Delta_j\right) \langle 0|\mathcal{O}_i(x_i)\mathcal{O}_j(x_j)|0\rangle \end{aligned} \tag{21.45}$$

The solution of this equation shows that the correlator obeys a power law

$$F_{ij}(x_i - x_j) = \frac{C}{|x_i - x_j|^{\Delta_i + \Delta_j}} \tag{21.46}$$

where C is an arbitrary constant.

In addition, the action of special conformal transformations, whose infinitesimal generators are the operators K_{μ}, further restricts the form of the two-point functions of primary operators to obey an "orthogonality condition" to vanish unless the scaling dimensions are equal (see eq. (15.96)):

$$\langle \mathcal{O}_1(x_1)\mathcal{O}_2(x_2)\rangle = \begin{cases} \dfrac{C_{12}}{|x_1 - x_2|^{2\Delta}}, & \text{if } \Delta_1 = \Delta_2 = \Delta \\ 0, & \text{otherwise} \end{cases} \tag{21.47}$$

Let us now turn to the three-point function of primary operators. Invariance under translations and rotations, and covariance under dilatations, require that the three-point functions of primary operators must have the form

$$\left\langle \mathcal{O}_1(x_1)\mathcal{O}_2(x_2)\mathcal{O}_3(x_3)\right\rangle = \sum_{a,b,c} \frac{C_{abc}}{|x_1-x_2|^a|x_2-x_3|^b|x_3-x_1|^c} \tag{21.48}$$

where the sum is restricted to values of a, b, and c such that $a+b+c=\Delta_1+\Delta_2+\Delta_3$. However, covariance under special conformal transformations implies the additional restriction that $a=\Delta_1+\Delta_2-\Delta_3$, $b=\Delta_2+\Delta_3-\Delta_1$, and $c=\Delta_3+\Delta_1-\Delta_2$. Therefore, the three-point function must have the general form (Polyakov, 1970) (see eq. (15.97))

$$\left\langle \mathcal{O}_1(x_1)\mathcal{O}_2(x_2)\mathcal{O}_3(x_3)\right\rangle = \\ = \frac{C_{123}}{|x_1-x_2|^{\Delta_1+\Delta_2-\Delta_3}|x_2-x_3|^{\Delta_2+\Delta_3-\Delta_1}|x_3-x_1|^{\Delta_3+\Delta_1-\Delta_2}} \tag{21.49}$$

where C_{123} is a so-far undetermined constant. In fact, if the operators are normalized such that the coefficient of the two-point function $C_{12}=1$, then the coefficients C_{123} of the three-point function must be universal numbers. We have used these results in sections 15.3 and 15.4, where we introduced the RG and the operator product expansion (OPE).

Finally, let us consider the implications of conformal invariance for N-point functions, with $N \geq 4$. Here the behavior is more complex. For instance, in the case of the four-point function, the general behavior, already presented in eq. (15.98), also depends on the cross ratios

$$\langle \mathcal{O}_1(x_1)\mathcal{O}_2(x_2)\mathcal{O}_3(x_3)\mathcal{O}_4(x_4)\rangle = F\left(\frac{r_{12}r_{34}}{r_{13}r_{24}}, \frac{r_{12}r_{34}}{r_{23}r_{41}}\right) \prod_{i<j} r_{ij}^{-(\Delta_i+\Delta_j)+\Delta/3} \tag{21.50}$$

where $\Delta=\sum_{i=1}^4 \Delta_i$, and $r_{ij}=|x_i-x_j|$. Again, once the two-point function is normalized as before, the prefactor is a universal function of the two cross ratios.

In summary, conformal invariance of the theory restricts the form of the correlation functions and reveals that there are some quantities (such as the scaling dimensions, the coefficient of the three-point functions, and the functions of the cross ratios) that are not determined, unless additional conditions are imposed. We will see that by imposing unitarity (or, equivalently, reflection positivity) and some additional symmetries, we can in some cases fully determine these quantities. This approach yields powerful results in two dimensions (discussed below), and to some extent in three dimensions (which we will not discuss here).

In a physically sensible theory, the correlators must obey cluster decomposition and must decay at long distances. Thus the scaling dimensions of the operators, Δ_i, must be nonnegative real numbers (i.e., $\Delta_i \geq 0$ for all operators). In addition, using the commutation relation of eq. (21.35), we find

$$DK_\mu\mathcal{O}(0) = \left(\left[D, K_\mu\right] + K_\mu D\right)\mathcal{O}(0) = (\Delta-1)\, K_\mu\mathcal{O}(0) \tag{21.51}$$

In other words, the operator K_μ lowers the scaling dimension of an operator $\mathcal{O}$. Hence, if we act repeatedly with the operators K_μ on an operator $\mathcal{O}$, one obtains operators of the form $K_{\mu_1}\cdots K_{\mu_N}\mathcal{O}$, which can have an arbitrarily low dimension. Since the allowed dimensions must be nonnegative, the theory must have a special class of operators such that

$$\left[K_\mu, \mathcal{O}(0)\right] = 0 \tag{21.52}$$

Away from the origin, at a finite location x, the action of K_μ generalizes to

$$[K_\mu, \mathcal{O}] = \left(2x_\mu x_\nu \partial_\nu - x^2 \partial_\mu + 2x_\mu \Delta - 2x^\nu \mathcal{S}_{\mu\nu}\right) \mathcal{O}(x) \tag{21.53}$$

Operators that obey eq. (21.52) (i.e., are invariant under special conformal transformations) are called *primary fields*. Then, given a primary operator $\mathcal{O}$ of scaling dimension Δ, we can construct an (in principle) infinite tower of descendant operators $P_{\mu_1} \cdots P_{\mu_N} \mathcal{O}(0)$, of increasing dimension, $\Delta + N$, since the action of one momentum operator increases the dimension of the operator $\mathcal{O}$ by 1. Hence, the scaling dimensions Δ of the primary operators play the role of quantum numbers of representations of the conformal group, and they label an infinite tower of descendant operators.

We now state the axioms that a conformal field theory must satisfy.

1) The vacuum state $|0\rangle$ of the theory must be invariant under conformal transformation.
2) The theory has a set of primary operators (or fields) that satisfy:
   i) They are eigenoperators of the dilatation operator D with eigenvalue Δ.
   ii) They are eigenoperators of the angular momentum operator $M_{\mu\nu}$ with eigenvalue $\mathcal{S}_{\mu\nu}$.
   iii) They commute with the operator K_μ.
3) Under a conformal transformation, scalar (spinless) primary fields transform as

$$\mathcal{O}_j(x) \mapsto \left|\frac{\partial x'}{\partial x}\right|^{\Delta_j/D} \mathcal{O}_j(x') \tag{21.54}$$

   where, as before (cf. eq. (21.14))

$$\left|\frac{\partial x'}{\partial x}\right| = \Omega^{-D/2} \tag{21.55}$$

   is the Jacobian of the conformal transformation with scale factor Ω.
4) The theory is covariant under conformal transformations in the sense that the correlators of the primary fields satisfy

$$\langle \mathcal{O}_1(x_1) \cdots \mathcal{O}_N(x_N)\rangle = \left|\frac{\partial x'}{\partial x}\right|_{x=x_1}^{\Delta_1/D} \cdots \left|\frac{\partial x'}{\partial x}\right|_{x=x_N}^{\Delta_N/D} \langle \mathcal{O}_1(x_1') \cdots \mathcal{O}_N(x_N')\rangle \tag{21.56}$$

5) The correlators must obey unitarity or, equivalently in a Euclidean theory, must satisfy reflection positivity.
6) All other fields in the theory can be expressed as linear combinations of primary fields and their descendants.

These axioms extend to the case of operators with spin, such as currents. One such example is the energy-momentum tensor, which in dimensions $D > 2$ is a primary field with scaling dimension D. We will see shortly that in two dimensions, it obeys an anomalous algebra and is no longer a primary field.

21.5 Conformal field theory in two dimensions

We now turn to CFT in two dimensions. This case has been studied in greater detail, and it is better understood than the D-dimensional case. It has many physical applications. Originally it was developed to formulate perturbative string theory. It also has direct application to QFTs in 1+1 dimensions and to two-dimensional classical critical phenomena (Belavin et al., 1984; Friedan et al., 1984; Ginsparg, 1989).

21.5.1 Classical conformal invariance in two dimensions

Let us consider theories in two-dimensional flat Euclidean spacetime. Eq. (21.9) shows that two-dimensional conformal transformations ε_μ obey the Cauchy-Riemann equations. Hence, the conformal transformations are analytic (or anti-analytic) functions. Let us write the Euclidean coordinates as complex coordinates $z=x_1+ix_2$, $z\in\mathbb{C}$. Define also

$$\partial_z=\frac{1}{2}\left(\partial_1-i\partial_2\right), \qquad \partial_{\bar z}=\frac{1}{2}\left(\partial_1+i\partial_2\right) \tag{21.57}$$

such that

$$\partial_z z=1,\ \partial_{\bar z}\bar z=1,\ \partial_z\bar z=0,\ \partial_{\bar z}z=0 \tag{21.58}$$

In what follows, we use the notation $\partial_z=\partial$ and $\partial_{\bar z}=\bar\partial$.

For a general vector field v^a $(a=1,2)$, we can also define the complex components

$$v^z=v^1+iv^2,\ v^{\bar z}=v^1-iv^2,\ v_z=\frac{1}{2}\left(v^1-iv^2\right),\ v_{\bar z}=\frac{1}{2}\left(v^1+iv^2\right) \tag{21.59}$$

In Cartesian indices $(1,2)$, the Euclidean metric is the identity, $g_{ab}=\delta_{ab}$, whereas in complex coordinates, the metric tensor is

$$g_{z\bar z}=g_{\bar z z}=\frac{1}{2},\ g_{zz}=g_{\bar z\bar z}=0,\ g^{z\bar z}=g^{\bar z z}=2,\ g^{zz}=g^{\bar z\bar z}=0 \tag{21.60}$$

It is natural to write the conformal transformations as $\varepsilon(z)=\varepsilon_1+i\varepsilon_2$ and $\bar\varepsilon(\bar z)=\varepsilon_1-i\varepsilon_2$. Two-dimensional conformal transformations are the analytic (and anti-analytic) coordinate transformations

$$z\mapsto f(z),\quad \bar z\mapsto \bar f(\bar z) \tag{21.61}$$

under which the Euclidean interval transforms as

$$ds^2=dzd\bar z\mapsto \left|\frac{\partial f}{\partial z}\right|^2 dzd\bar z \tag{21.62}$$

and the Jacobian is

$$\Omega=\left|\frac{\partial f}{\partial z}\right|^2 \tag{21.63}$$

A natural basis for conformal transformations are the functions $\varepsilon_n=z^{n+1}$ and $\bar\varepsilon_n=-\bar z^{n+1}$ (with $n\in\mathbb{Z}$). The infinitesimal generators of two-dimensional classical conformal transformations are

$$\ell_n=-z^{n+1}\partial_z,\quad \bar\ell_n=-\bar z^{n+1}\partial_{\bar z}\quad (n\in\mathbb{Z}) \tag{21.64}$$

which obey the classical local algebra

$$[\ell_n, \ell_m] = (n-m)\ell_{n+m}, \quad [\bar{\ell}_n, \bar{\ell}_m] = (n-m)\ell_{n+m} \tag{21.65}$$

We will see shortly that at the quantum level, this algebra is corrected to include a key new term, the conformal anomaly. Hence, we have two independent algebras. Thus, at the formal level, z and $\bar{z}$ are independent variables, and hence we work not just with complex functions but with those on $\mathbb{C}^2$. We will need to project onto the physical subspace in which $\bar{z} = z^*$.

The classical conformal algebra, eq. (21.65), is called *local*, since the generators are not all well defined on the Riemann sphere, $S^2 = \mathbb{C} \bigcup \infty$. Holomorphic (analytic) conformal transformations are generated by vector fields $v(z)$,

$$v(z) = \sum_{n \in \mathbb{Z}} a_n z^{n+1} \partial_z \tag{21.66}$$

Such vector fields are generally singular as $z \to 0$. The only exceptions are conformal transformations generated by $\ell_{-1}, \ell_0, \ell_1$, and $\bar{\ell}_{-1}, \bar{\ell}_0, \bar{\ell}_1$. From their definitions, we see that ℓ_{-1} and $\bar{\ell}_{-1}$ generate translations, $\ell_0 + \bar{\ell}_0$ generates dilatations (i.e., translations on the radial polar coordinate r), $i(\ell_0 - \bar{\ell}_0)$ generates rotations (i.e., translations on the angular polar coordinate θ), and ℓ_1 and $\bar{\ell}_{-1}$ are the generators of special conformal transformations. The finite form of these transformations is

$$z \mapsto \frac{az+b}{cz+d}, \quad \bar{z} = \frac{\bar{a}\bar{z}+\bar{b}}{\bar{c}\bar{z}+\bar{d}} \tag{21.67}$$

where $a, b, c, d \in \mathbb{C}$ and $ad - bc = 1$. This is the group $SL(2, \mathbb{C})/\mathbb{Z}_2$ (this quotient says that the transformation is unaffected by a change in sign of a, b, c, d). These are the only globally well-defined conformal transformations and the only ones that also exist in dimensions $D > 2$.

Since the transformations generated by $\ell_{-1}, \ell_0, \ell_1$ and $\bar{\ell}_{-1}, \bar{\ell}_0, \bar{\ell}_1$ are globally well defined, we will work in the basis of their eigenstates. Thus, we consider the eigenstates of ℓ_0 and $\bar{\ell}_0$, and denote their eigenvalues, known as the conformal weights, by the real numbers h and $\bar{h}$ (not the complex conjugate). Since $\ell_0 + \bar{\ell}_0$ generates dilatations and $i(\ell_0 - \bar{\ell}_0)$ generates rotations, the scaling dimension Δ and the spin s are given by $\Delta = h + \bar{h}$ and $s = h - \bar{h}$.

21.5.2 Quantization

Let us now turn to the quantum theory, working with Euclidean coordinates, where σ^1 labels space and σ^0 labels imaginary time. Assume that the space coordinate is finite and periodic. Hence, identify $\sigma^1 \cong \sigma^1 + 2\pi$, while the imaginary time coordinate σ^0 can take any real value, positive or negative. In other words, the Euclidean spacetime has been compactified to an infinitely long cylinder of circumference 2π, as shown in figure 21.2a. Let $\zeta = \sigma^2 + i\sigma^1$ be the coordinates on the cylinder, and consider the conformal mapping to the plane with coordinates $z = x^1 + ix^2$,

$$z = \exp \zeta = \exp\left(\sigma^0 + i\sigma^1\right) \tag{21.68}$$

which maps the cylinder with coordinates ζ to the complex plane with coordinates z. Notice that the infinite past on the cylinder, $\sigma^0 \to -\infty$, maps to the origin, $z = 0$, on the plane, and that the infinite future on the cylinder, $\sigma^0 \to +\infty$, maps to the point of infinity, $z \to \infty$,

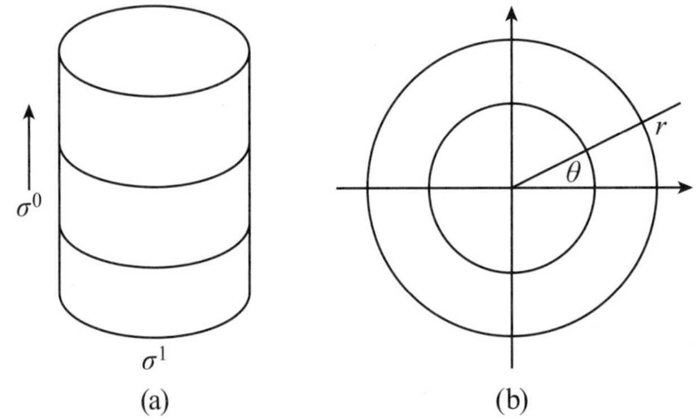

Figure 21.2 A conformal mapping transforms the cylinder to the plane.

on the complex plane. Equal-time surfaces on the cylinder, $\sigma^0 =$ constant, map to a circle of constant radius on the z plane.

To develop a quantum theory, we need the operators that implement conformal transformations on the plane. Thus, a dilatation $z \mapsto \exp(a+z)$ on the plane is a time translation on the cylinder, $\sigma^0 \mapsto \sigma^0 + a$. Therefore, the dilatation generator can be regarded as the Hamiltonian H of the system. In other words, on the cylinder, we have a Hilbert space of states on an imaginary time surface, $\sigma^0 =$ constant. On the plane, the Hilbert space is defined on circles of constant radii, which relate to each other by the action of the dilatation operator D. Likewise, the linear momentum P on the cylinder becomes the generator of rotations on the plane. For obvious reasons, this way of defining the quantum theory is known as radial quantization (Friedan et al., 1984).

As we have seen in chapter 3, symmetry generators are constructed by Noether's prescription. Thus, given a conserved current j^μ, the conserved charge Q is obtained as the integral of the time component of the current on a fixed time surface. The associated symmetry transformations act on a field $\mathcal{O}$ as $\delta_\varepsilon \mathcal{O} = \varepsilon\,[Q, \mathcal{O}]$. Here we are interested in coordinate transformations generated by the energy-momentum tensor $T^{\mu\nu}$. As before, conformal invariance requires that the energy-momentum tensor be conserved, $\partial_\mu T^{\mu\nu}$, and traceless, $T^\mu_\mu = 0$, as operators acting on the Hilbert space.

It will be convenient to have the components of $T^{\mu\nu}$ in complex coordinates:

$$T \equiv T_{zz} = \frac{1}{4}\left(T_{00} - 2iT_{10} - T_{11}\right)$$

$$\bar{T} \equiv T_{\bar{z}\bar{z}} = \frac{1}{4}\left(T_{00} + 2iT_{10} - T_{11}\right)$$

$$T^\mu_\mu \equiv \Theta = T_{00} + T_{11} = 4T_{z\bar{z}} = 4T_{\bar{z}z} \tag{21.69}$$

So, in general, the tensor has three components, T, $\bar{T}$, and Θ. The local conservation of the energy-momentum tensor yields the conditions

$$\bar{\partial} T + \frac{1}{4}\partial\Theta = 0, \quad \partial\bar{T} + \frac{1}{4}\bar{\partial}\Theta = 0 \tag{21.70}$$

Conformal invariance requires that the energy-momentum tensor be traceless: $\Theta = 0$. Then the conservation laws simply become

$$\bar{\partial} T = 0, \text{ and } \partial \bar{T}(\bar{z}) = 0 \tag{21.71}$$

Therefore, the remaining nonvanishing components of $T^{\mu\nu}$ satisfy

$$T(z) \equiv T_{zz}(z), \quad \bar{T}(\bar{z}) = T_{\bar{z}\bar{z}} \tag{21.72}$$

and are, respectively, holomorphic (analytic) and antiholomorphic (anti-analytic). We will see that this property means that expectation values of physical observables factorize into analytic and anti-analytic components. In the Minkowski spacetime signature, they are referred to as the right- and left-moving (or chiral and antichiral) components of the fields.

Given an infinitesimal conformal transformation $\varepsilon(z)$, the conserved charge Q is

$$Q(\varepsilon, \bar{\varepsilon}) = \frac{1}{2\pi i} \oint_{C(r)} \left(dz T(z) \varepsilon(z) + \bar{T}(\bar{z}) \bar{\varepsilon}(\bar{z}) \right) \tag{21.73}$$

where $C(r)$ is a circle of radius r centered at the origin, $z = 0$. The variation of a field $\Phi(u, \bar{u})$ is the equal-time commutator with the charge

$$\delta_{\epsilon,\bar{\varepsilon}} \Phi(u, \bar{u}) = \frac{1}{2\pi i} \oint_{C(r)} [dz T(z) \varepsilon(z), \Phi(u, \bar{u})] + \left[d\bar{z} \bar{T}(\bar{z}) \bar{\varepsilon}(\bar{z}), \Phi(u, \bar{u}) \right] \tag{21.74}$$

A primary field $\Phi(z, \bar{z})$ transforms under a local conformal transformation as

$$\Phi(z, \bar{z}) \mapsto \left(\frac{\partial f}{\partial z} \right)^h \left(\frac{\partial \bar{f}}{\partial \bar{z}} \right)^{\bar{h}} \Phi(f(z), \bar{f}(\bar{z})) \tag{21.75}$$

where h and $\bar{h}$ are the conformal weights. For an infinitesimal transformation, this should be

$$\delta_{\epsilon,\bar{\varepsilon}} \Phi(z, \bar{z}) = \left((h \partial \varepsilon + \varepsilon \partial) + (\bar{h} \bar{\partial} \bar{\varepsilon} + \bar{\varepsilon} \bar{\partial}) \right) \Phi(z, \bar{z}) \tag{21.76}$$

Since equal-time surfaces on the cylinder map onto circles of fixed radius on the complex plane, let us introduce the concept of the *radially ordered product* (analogous to a time-ordered product) for two bosonic operators $A(z)$ and $B(w)$:

$$R(A(z)B(w)) = \begin{cases} A(z)B(w), & \text{if } |z| > |w|, \\ B(w)A(z), & \text{if } |z| < |w| \end{cases} \tag{21.77}$$

(with a minus sign for fermions). Then, the equal-time commutator is the contour integral of the radially ordered product, shown in figure 21.3,

$$\begin{aligned} &\delta_{\epsilon,\bar{\varepsilon}} \Phi(u, \bar{u}) = \\ &\frac{1}{2\pi i} \left(\oint_{|z|>|u|} - \oint_{|u|>|z|} \right) \left(dz \varepsilon(z) R(T(z), \Phi(u, \bar{u})) + d\bar{z} \bar{\varepsilon}(\bar{z}) R(\bar{T}(\bar{z}) \Phi(u, \bar{u})) \right) \\ &\quad = \frac{1}{2\pi i} \oint \left(dz \varepsilon(z) R(T(z) \Phi(u, \bar{u})) + d\bar{z} \bar{\varepsilon}(\bar{z}) R(\bar{T}(\bar{z}) \Phi(u, \bar{u})) \right) \\ &\quad = h \partial \varepsilon(u) \Phi(u, \bar{u}) + \varepsilon(u) \partial \Phi(u, \bar{u}) + \bar{h} \bar{\partial} \bar{\varepsilon}(\bar{u}) \Phi(u, \bar{u}) \end{aligned} \tag{21.78}$$

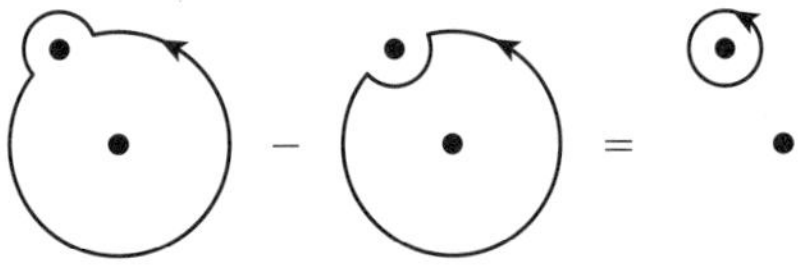

Figure 21.3 Computation of an equal-time commutation relation.

where we imposed consistency with eq. (21.76). This implies that for the last two lines of eq. (21.78) to be consistent with each other, the product of $T(z)$ and $\bar{T}(z)$ with $\Phi(u, \bar{u})$ should have short-distance singularities and obey the OPEs

$$
\begin{aligned}
T(z)\Phi(u,\bar{u}) &= \frac{h}{(z-u)^2}\Phi(u,\bar{u}) + \frac{1}{z-u}\partial_u \Phi(u,\bar{u}) + \cdots \\
\bar{T}(\bar{z})\Phi(u,\bar{u}) &= \frac{\bar{h}}{(\bar{z}-\bar{u})^2}\Phi(u,\bar{u}) + \frac{1}{\bar{z}-\bar{u}}\partial_{\bar{u}} \Phi(u,\bar{u}) + \cdots
\end{aligned} \tag{21.79}
$$

In other words, a field $\Phi(z, \bar{z})$ is a primary field if it has an OPE with the holomorphic T and antiholomorphic $\bar{T}$ components of the energy-momentum tensor of the form of eq. (21.79). Here, the ellipsis represents nonsingular contributions, which drop out of the contour integrals, and hence from the commutators. Eq. (21.79) has to be understood as an operator identity. A more precise way to state this requirement is the following conformal Ward identity:

$$
\begin{aligned}
&\Big\langle T(z)\mathcal{O}_1(u_1,\bar{u}_1)\cdots \mathcal{O}_N(u_N,\bar{u}_N)\Big\rangle \\
&\quad = \sum_{j=1}^{N}\left(\frac{h_j}{(z-u_j)^2} + \frac{1}{z-u_j}\partial_j\right)\Big\langle \mathcal{O}_1(u_1,\bar{u}_1)\cdots \mathcal{O}_N(u_N,\bar{u}_N)\Big\rangle
\end{aligned} \tag{21.80}
$$

This identity requires that the correlation functions be meromorphic functions of z with singularities at the positions of the operators.

In a two-dimensional CFT, the correlator of two primary fields has the form

$$
\langle \mathcal{O}_i(z,\bar{z})\mathcal{O}_j(u,\bar{u})\rangle = \delta_{ij}\frac{1}{(z-u)^{2h_i}(\bar{z}-\bar{u})^{2\bar{h}_i}} \tag{21.81}
$$

As we have already seen in section 15.4, the primary fields obey an algebra known as the operator product expansion (OPE) which has the form

$$
\lim_{z\to u, \bar{z}\to\bar{u}} \mathcal{O}_i(z,\bar{z})\mathcal{O}_j(u,\bar{u}) = \sum_k \frac{C_{ijk}}{(z-u)^{h_i+h_j-h_k}(\bar{z}-\bar{u})^{\bar{h}_i+\bar{h}_j-\bar{h}_k}}\mathcal{O}_k(u,\bar{u}) \tag{21.82}
$$

Consider now an N-point function of primary fields. As before, in addition to power laws determined by the conformal dimensions of the fields, a general correlator is determined by a scaling function of the fields. The OPE then implies that the N-point function can be written as a sum of products of three-point functions, known as *conformal blocks*. Consider now *fusing* pairs of primary fields. Since the choice of which pair of fields is being fused is arbitrary, the algebra encoded by the OPE—the fusion algebra—must be associative.

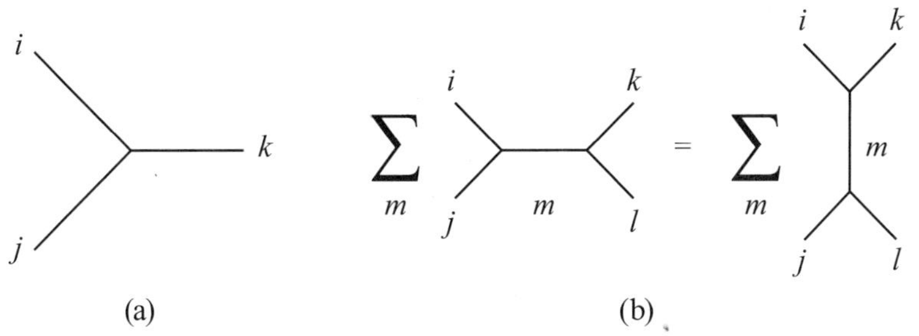

Figure 21.4 The OPE algebra: (a) Fusion of two primary fields. (b) Consistency condition for the fusion algebra.

This requirement leads to a set of consistency conditions for the OPE, which, in turn, implies constraints on the correlators of the CFT, shown in figure 21.4. These consistency conditions, known as crossing and unitarity symmetries, impose severe constraints on the explicit forms of the conformal blocks, which, in specific CFTs, are stringent enough to determine these functions completely.

21.5.3 The Virasoro algebra

In section 21.5.2, we worked out the form of the OPE between the energy-momentum tensor and a primary field. All operators (or fields) in the theory can be classified into families, each labeled by a primary field. The other members of each family are called the descendants of the primary. In this sense, the primary field is the highest weight of the representation.

However, we have not said anything about the energy-momentum tensor and its OPE with itself. We know that in dimensions $D > 2$, the energy-momentum tensor is a primary field. However, this is not the case in $D = 2$ dimensions, where the energy-momentum tensor $T(z)$ is holomorphic and has dimension 2. Thus, it must have conformal weight $(h, \bar{h}) = (2, 0)$. By performing two conformal transformations in sequence, we see that the OPE of the energy-momentum tensor with itself must have the following form:

$$T(z)T(u) = \frac{c/2}{(z-u)^4} + \frac{2}{(z-u)^2}T(u) + \frac{1}{z-u}\partial T(u) \tag{21.83}$$

The first term, proportional to the identity field, is allowed by analyticity, Bose symmetry, and scale invariance. The coefficient c is known as the *central charge* of the CFT.

In left-moving (light-cone) coordinates, x_-, of Minkowski spacetime, the Virasoro algebra is

$$-i[T(x_-), T(x'_-)] = \delta(x_- - x'_-)\partial_- T(x_-) - 2\partial_-\delta\delta(x_- - x'_-) + \frac{c}{24\pi}\partial_-^3\delta(x_- - x'_-) \tag{21.84}$$

The last term on the right-hand side of the algebra, a Schwinger term, is the central extension of the Virasoro algebra. In this case, it is an anomaly of the trace of the energy-momentum tensor.

The OPE implies that the correlator of the energy-momentum tensor must be

$$\langle T(z)T(u)\rangle = \frac{c/2}{(z-u)^4} \tag{21.85}$$

For $\bar{T}$, we have instead

$$\bar{T}(\bar{z})\bar{T}(\bar{u}) = \frac{\bar{c}/2}{(\bar{z}-\bar{u})^4} + \frac{2}{(\bar{z}-\bar{u})^2}T(u) + \frac{1}{\bar{z}-\bar{u}}\partial T(u) \tag{21.86}$$

and the correlator is instead

$$\langle \bar{T}(\bar{z})\bar{T}(\bar{u})\rangle = \frac{\bar{c}/2}{(\bar{z}-\bar{u})^4} \tag{21.87}$$

In principle, the central charges c and $\bar{c}$ can be different. If that is the case, then the CFT is chiral. If an additional symmetry, called modular invariance, is imposed, then the two central charges must be equal, $c=\bar{c}$. This condition follows from the requirement that the gravitational anomaly cancels, and it is usually required. However, there are physical systems, such as the edge states of the fractional quantum Hall fluids, which are chiral theories, and this condition is violated. This is possible since in that case, these states cannot exist by themselves but only as boundaries of a higher-dimensional system.

Equation (21.83) implies that under an infinitesimal conformal transformation, $\varepsilon(z)$, the energy momentum tensor changes as

$$\delta_\varepsilon T(z) = \varepsilon(z)\partial T(z) + 2\partial\varepsilon(z)T(z) + \frac{c}{12}\partial^3\varepsilon(z) \tag{21.88}$$

For a conformal finite transformation $z \mapsto z' = f(z)$, the energy-momentum tensor transforms as

$$T'(z') = (f'(z))^2 T(z') + \frac{c}{12}\{z';z\} \tag{21.89}$$

and similarly for $\bar{T}$ (upon replacing $c \to \bar{c}$). Here $f'(z) = \frac{df}{dz}$, and $\{u;z\}$ is the Schwartzian derivative

$$\{z';z\} = \frac{f'''f' - \frac{3}{2}f''^2}{f'^2} \tag{21.90}$$

The second term in the transformation of eq. (21.89) is known as the *conformal anomaly*. It is an anomaly in the sense that it is absent in the classical theory, where the energy momentum-tensor transforms *homogeneously* (with rank 2), as shown in the first term of the transformation. For this reason, in a two-dimensional CFT, the energy-momentum tensor is not a primary field. The conformal anomaly is a quantum effect that violates the naive homogeneous transformation law.

It is useful to write the Laurent expansion of the energy-momentum tensor

$$T(z) = \sum_{n\in\mathbb{Z}} \frac{L_n}{z^{n+2}}, \quad \bar{T}(\bar{z}) = \sum_{n\in\mathbb{Z}} \frac{\bar{L}_n}{\bar{z}^{n+2}} \tag{21.91}$$

in terms of the operators L_n and $\bar{L}_n$ (the "modes"), which have scaling dimension n that satisfies the hermiticity condition:

$$L_n^\dagger = L_{-n} \tag{21.92}$$

The series expansion of eq. (21.91) can be inverted using contour integrals

$$L_n = \frac{1}{2\pi i}\oint_C dz\, z^{n+1}T(z), \quad \bar{L}_n = \frac{1}{2\pi i}\oint_C d\bar{z}\, \bar{z}^{n+1}\bar{T}(\bar{z}) \tag{21.93}$$

where there are no other operator insertions inside these contours.

The commutation relations for the operators L_n (and similarly for $\bar{L}_n$) are obtained using the OPE for the energy-momentum tensor. The result is

$$[L_n, L_m] = (n-m)L_{n+m} + \frac{c}{12}(n^3 - n)\delta_{n+m,0} \tag{21.94}$$

This is the Virasoro algebra. Similarly, for the mode expansion of $\bar{T}$, we get

$$[\bar{L}_n, \bar{L}_m] = (n-m)\bar{L}_{n+m} + \frac{\bar{c}}{12}(n^3 - n)\delta_{n+m,0} \tag{21.95}$$

Finally, since the OPEs of $T(z)$ and $\bar{T}(\bar{z})$ have no singularities, the generators of the two Virasoro algebras must commute:

$$[L_n, \bar{L}_m] = 0 \tag{21.96}$$

The algebra of eq. (21.94), the Virasoro algebra, differs from the classical conformal algebra of eq. (21.65) by the conformal anomaly term, the term proportional to the identity operator on the right-hand side of eq. (21.94). This term is known as the central extension of the Virasoro algebra. As we will see, in QFT it arises as a Schwinger term that reflects the short-distance singularities of the theory.

The central extension is absent if $n = 0, \pm 1$, for which the operators L_0, L_1, and L_{-1} satisfy a closed subalgebra without a central extension:

$$[L_1, L_{-1}] = 2L_0, \quad [L_{\pm 1}, L_0] = \pm L_{\pm 1} \tag{21.97}$$

These operators generate the global conformal group $SL(2, \mathbb{C})$.

The action of the Virasoro generators on the primary fields is derived from eq. (21.79),

$$\begin{aligned} [L_n, \mathcal{O}] &= z^{n+1}\partial\mathcal{O} + h(n+1)z^n\mathcal{O} \\ [\bar{L}_n, \mathcal{O}] &= \bar{z}^{n+1}\bar{\partial}\mathcal{O} + \bar{h}(n+1)\bar{z}^n\mathcal{O} \end{aligned} \tag{21.98}$$

where, as before, $\Delta = h + \bar{h}$ is the scaling dimension, and $s = h - \bar{h}$ is the spin.

For a theory quantized on a cylinder (see figure 21.2a) the operator $H = L_0 + \bar{L}_0$ is regarded as the quantum Hamiltonian, and $P = L_0 - \bar{L}_0$ is the linear momentum. But for radial quantization (see figure 21.2b), $L_0 + \bar{L}_0$ is the dilatation operator of the Euclidean theory, and $L_0 - \bar{L}_0$ is the angular momentum. Let $|0\rangle$ be the vacuum state of a two-dimensional CFT, defined as the state with the lowest (zero) eigenvalue of L_0

$$L_0|0\rangle = \bar{L}_0|0\rangle = 0 \tag{21.99}$$

The lowering operators of L_0 ($\bar{L}_0$) are L_{-n} ($\bar{L}_{-n}$) with $n > 0$.

Just as in the theory of angular momentum, a state annihilated by the lowering operators is said to be a highest-weight state. The vacuum is a highest-weight state, because it has the lowest eigenvalue of $L_0 + \bar{L}_0$. Similarly, the state $\mathcal{O}|0\rangle$ is an eigenstate of L_0 ($\bar{L}_0$) with eigenvalue h ($\bar{h}$) and is also a highest-weight state. Given a highest-weight state, a tower of states (in principle infinite), the descendants, is constructed by acting on the highest-weight state with lowering operators. The space of states of a CFT is then a sum of irreducible representations of the algebra of the L and $\bar{L}$ each generated from a highest-weight state,

$$\mathcal{O}|0\rangle \equiv |h, \bar{h}\rangle \tag{21.100}$$

resulting from the action of a primary field on the vacuum state. The L_0 and $\bar{L}_0$ eigenvalues of the highest-weight state are, respectively,

$$L_0\mathcal{O}|0\rangle = h\mathcal{O}|0\rangle, \qquad \bar{L}_0\mathcal{O}|0\rangle = \bar{h}\mathcal{O}|0\rangle \tag{21.101}$$

A representation of the Virasoro algebra is built from the highest-weight state by the action of the lowering operators, L_{-n} (with $n \geq 1$). A state is in the nth level if the L_0 eigenvalue is $h+n$. The nth level is spanned by the states $L_{-k_1}\cdots L_{-k_n}\mathcal{O}|0\rangle$, with $k_1 \geq \cdots \geq k_n \geq 0$ and $\sum_j k_j = n$. There are $P(n)$ such states, where $P(n)$ is the number of ways of writing n as a sum of positive integers (partitions). The higher-level states correspond to operators of increasingly higher dimensions obtained by applying products of energy-momentum tensors to the highest-weight state.

In summary, a two-dimensional CFT is characterized by the following data: the central charge c, the conformal weights h of the primary fields $\mathcal{O}$, and the coefficients of the OPEs of the primary fields (i.e., their fusion rules). We discuss below how this works in a few examples of two-dimensional CFTs of interest. However, several questions arise. One is whether there are additional constraints, aside from conformal invariance, that restrict (or even specify) which two-dimensional CFTs are allowed.

One constraint, natural from the point of view of QFT, is unitarity (or its Euclidean version, reflection positivity). This is a powerful constraint. Note that there are many examples of systems in classical statistical mechanics that are conformally invariant but not unitary. So far as what we have discussed is concerned, the number of primaries may well be infinite. It is natural to ask whether there are constraints that will restrict the primaries to a finite number. Such theories are known as rational CFTs. Another way to further restrict the CFTs is to impose the condition that the theory may also have global continuous symmetries and the associated conserved currents. In this case, the algebra of the currents, generally known as a Kac-Moody algebra, combined with the Virasoro algebra, provides a framework to construct CFTs.

21.5.4 Physical meaning of the central charge

The central charge plays a crucial role in a two-dimensional CFT. We will now see that it has a direct physical meaning (Affleck, 1986b; Blöte et al., 1986).

Let Z be the partition function of the theory, which we will regard as either a QFT in $D=2$ Euclidean spacetime or a classical statistical mechanical system in $D=2$ dimensions. The free energy of the system is $F = -\ln Z$. Consider a system of linear size L. Then, in a local theory, we expect the system to have a well-defined thermodynamic limit, $L \to \infty$. In this limit, the free energy can be expressed in terms of finite densities as

$$F = f_0 L^2 + BL + \cdots \tag{21.102}$$

The coefficient of the leading term, f_0, is the free-energy density. It is independent of the boundary conditions and in general, it is not universal. The second term is also generally not universal and it appears if the system has a boundary. Among the correction terms is a term $O(L^0)$, which, as we will see, is universal if the system has no boundaries (i.e., it is defined on a sphere, a torus, or the like) and is conformally invariant.

To understand the origin of these universal corrections, consider (for simplicity) the case of a CFT on a manifold $\mathcal{M}$ without boundaries. Such a manifold is in general curved. Under a global infinitesimal dilatation, $x^\mu \mapsto (1+\varepsilon)x^\mu$, the action changes by an amount determined by the trace of the energy-momentum tensor, Θ,

$$\delta S = -\frac{\varepsilon}{2\pi} \int_{\mathcal{M}} \Theta(x) \sqrt{g} d^2x \tag{21.103}$$

where $g^{\mu\nu}$ is the metric of the manifold $\mathcal{M}$, and $g = \left| \det g^{\mu\nu} \right|$ is the determinant. The factor of $1/(2\pi)$ is introduced for later convenience.

In the RG, the total partition function must be invariant under such a rescaling. Thus it must be true that

$$Z = \exp(-F(L)) = \exp(-F(L+\delta L) - \langle \delta S \rangle) \tag{21.104}$$

so that the change in the free energy is the *negative* of $\langle \delta S \rangle$. Therefore,

$$L\frac{\partial F}{\partial L} = \frac{1}{2\pi} \int_{\mathcal{M}} \langle \Theta \rangle \sqrt{g} d^2x \tag{21.105}$$

We have seen that conformal invariance implies that the energy-momentum tensor is traceless, $\Theta = 0$. This is true only in flat spacetime. If the manifold $\mathcal{M}$ has a scalar curvature, then there is a scale, and the energy-momentum tensor is not traceless. The expectation value of the trace of the energy-momentum tensor is derived by considering a weakly curved system. Such an infinitesimal coordinate transformation is represented by the infinitesimal change in the metric $\delta g_{\mu\nu} = \partial_\nu \varepsilon_\mu + \partial_\mu \varepsilon_\nu$, so that the change in the action is

$$\delta S = -\frac{1}{4\pi} \int T^{\mu\nu} \delta g_{\mu\nu} d^2x \tag{21.106}$$

However, this expression also applies to changes in the metric due to changes in the geometry, not just to coordinate changes. Thus the invariance of the partition function implies

$$\operatorname{Tr} \exp\left(-S + \frac{1}{4\pi} \int T^{\mu\nu} \delta g_{\mu\nu} d^2x \right) \Big|_{\text{new geometry}} = \operatorname{Tr} \exp(-S) \Big|_{\text{old geometry}} \tag{21.107}$$

The computation of the changes of the expectation values for all three components of the energy-momentum tensor requires us to compute integrals of the correlators of energy-momentum tensors. These integrals, in turn, need a short-distance regularization. This calculation leads to the important result that, for closed manifolds,

$$\langle \Theta(x) \rangle = \frac{c}{12} R(x) \tag{21.108}$$

This relation, called the *trace anomaly*, measures the response to a weak deformation of the geometry. Here $R(x)$ is the scalar curvature of the two-dimensional space. Notice that $\langle \Theta \rangle \neq 0$ in flat spacetime if the system is not at a fixed point. Although the term "trace anomaly" is also used in such systems, it is clear that there is nothing anomalous about them.

Another interesting interpretation of the central charge is obtained by considering conformally invariant theories defined on an infinite cylinder, a strip with coordinates $-\infty < u < \infty$ and a periodic coordinate $0 \leq v \leq \ell$. We can think of this geometry in two different ways. One is as a (1+1)-dimensional CFT at finite temperature $T = 1/\ell$ (in units in which the Boltzmann constant $k_B = 1$). The other interpretation is that we are considering a CFT on a circle with circumference ℓ. In both cases the theory on the cylinder is related to the theory on the infinite flat plane by the conformal mapping

$$w = u + iv = \frac{\ell}{2\pi} \ln z \tag{21.109}$$

The energy-momentum tensor on the cylinder and on the plane are related by the transformation law, eq. (21.89), which for this specific conformal mapping yields

$$T(w)_{\text{cylinder}} = \left(\frac{2\pi}{\ell}\right)^2 \left(z^2 T(z)_{\text{plane}} - \frac{c}{24}\right) \tag{21.110}$$

By translational and rotational invariance, the expectation value of the energy-momentum tensor on an infinite plane must vanish. Hence

$$\langle T(w)_{\text{cylinder}} \rangle = -\frac{c\pi^2}{6\ell^2} \tag{21.111}$$

Let us consider first the case of a CFT on a finite spatial interval of length ℓ. In this case, the coordinate u is imaginary time, and the periodic coordinate v is space. In this case the Hamiltonian is

$$\begin{aligned} H &= \frac{1}{2\pi} \int_0^\ell T_{uu} dv = \frac{1}{2\pi} \int_0^\ell \left(T(v) + \bar{T}(v)\right) dv \\ &= \frac{2\pi}{\ell} \left(L_0 + \bar{L}_0\right) - \frac{\pi c}{6\ell} \end{aligned} \tag{21.112}$$

where we used eq. (21.111) and the definitions

$$L_0 = \frac{1}{2\pi i} \oint zT(z) dz, \qquad \bar{L}_0 = -\frac{1}{2\pi i} \oint \bar{z}\bar{T} d\bar{z} \tag{21.113}$$

Similarly, the momentum operator P is

$$P = \frac{1}{2\pi} \int_0^\ell T_{uv} dv = \frac{2\pi}{\ell} (L_0 - \bar{L}_0) \tag{21.114}$$

Eq. (21.112) implies that the energy of the ground state, defined by $L_0|0\rangle = \bar{L}_0|0\rangle = 0$, is

$$E_{\text{gnd}} = -\frac{\pi c}{6\ell} \tag{21.115}$$

where we have set to zero the extensive (and nonuniversal) part of the ground-state energy.

Eq. (21.115) is the Casimir effect, which we have already discussed in chapter 8 for a free massless scalar field. Here we see that it is a general result of a CFT. An important observation is that it is proportional to the central charge c. In this sense, the central charge "counts" the number of degrees of freedom (even though c generally is not an integer).

In addition, we see that the energy E and momentum P eigenvalues of the highest weight state $|h, \bar{h}\rangle$ are

$$E = E_{\text{gnd}} + \frac{2\pi\,\Delta}{\ell}, \qquad P = \frac{2\pi s}{\ell} \tag{21.116}$$

where $\Delta = h + \bar{h}$ is the scaling dimension, and $s = h - \bar{h}$ is the conformal spin of the primary field.

In the other interpretation of the theory on the cylinder, the space coordinate has infinite extent, $-\infty \leq u \leq \infty$, while the periodic coordinate $0 \leq v \leq \ell$ is interpreted as imaginary time with length $\ell = 1/T$, where T is the temperature. Now eq. (21.115) becomes the thermal contribution to the free-energy density (i.e., per unit length) f of a CFT at temperature T,

$$f = \varepsilon_{\text{gnd}} - \frac{\pi}{6} c T^2 \tag{21.117}$$

where the extra factor of T comes from the definition of the free energy, and $\varepsilon_{\text{gnd}} = \lim_{\ell\to\infty} E_{\text{gnd}}/\ell$. Here we have assumed that the speed of the excitations, the speed of light, has been set to 1.

The last term of eq. (21.117) is a generalization of the Stefan-Boltzmann law for blackbody radiation. From here it follows that the specific heat $C(T)/\ell$ of a CFT is

$$\lim_{\ell\to\infty} \frac{C(T)}{\ell} = \frac{\pi}{3} c T \tag{21.118}$$

This result, again, suggests the interpretation of the central charge as a measure of the number of degrees of freedom.

21.5.5 The C-theorem

So far we have discussed general properties of CFTs. However, in general, a theory may not be conformal. We have seen this in detail when discussing the RG in chapter 15, where we considered a theory at a fixed point perturbed by a set of interactions labeled by some coupling constants. In this case, we found that there is a flow in coupling-constant space induced by the RG. Let us now revisit the RG flows in the case of two-dimensional CFTs.

The C-theorem is a deceptively simple yet profound result due to A. B. Zamolodchikov (Zamolodchikov, 1986; Ludwig and Cardy, 1987). Consider a set of two-dimensional (Euclidean) CFTs defined in terms of a set of coupling constants $\{g_i\}$, representing the couplings of a set of primary fields $\mathcal{O}_i$. Consider continuum field theories only. Assume that all possible fixed points are critical fixed points (hence with a divergent correlation length) and are represented by two-dimensional CFTs. Further assume that in this space of coupling constants, all irrelevant operators introduced by the regularization (i.e., lattice effects) have already flown to zero. In this framework, the RG induces a flow in the space of coupling constants that links different CFTs.

The C-theorem is stated as follows: There exists a function C of the coupling constants that is nonincreasing along the RG flows, and it is stationary only at the fixed points. Moreover, at the fixed points, the C function is equal to the central charge of the CFT of the fixed point.

Before presenting the proof, it is worth discussing its meaning and implications. Intuitively, this result makes a lot of sense. Consider two CFTs related by the RG flow. Let us call them theory I and theory II. Since the flow is always from the UV to the IR, what the

RG describes is the gapping-out of a set of degrees present in theory I but absent in theory II by turning on some relevant operators. In this sense, theory II has fewer degrees of freedom than theory I. Therefore, we expect that the central charges of theory I will be larger than the central charges of theory II: $c_I \geq c_{II}$.

The proof goes as follows. Assume that the theories are invariant under translations and Euclidean rotations, and hence have a conserved energy-momentum tensor $T^{\mu\nu}$. Also assume that the theories obey reflection positivity. Away from the fixed points, the energy-momentum tensor has components $T = T_{zz}$, $\bar{T} = T_{\bar{z}\bar{z}}$, and a nonvanishing trace $\Theta = T^z_z + T^{\bar{z}}_{\bar{z}} = 4T_{z\bar{z}}$. Under rotations, T, Θ, and $\bar{T}$ have spins $s = 2, 0, -2$, respectively. Their correlation functions have the form

$$\langle T(z,\bar{z})T(0,0)\rangle = \frac{F(a\bar{z})}{z^4}$$
$$\langle \Theta(z,\bar{z})T(0,0)\rangle = \langle T(z,\bar{z})\Theta(0,0)\rangle = \frac{G(z\bar{z})}{z^3\bar{z}}$$
$$\langle \Theta(z,\bar{z})\Theta(0,0)\rangle = \frac{H(z\bar{z})}{z^2\bar{z}^2} \tag{21.119}$$

where F, G, and H are nontrivial scalar functions. Conservation of the energy-momentum tensor, $\partial_\mu T^{\mu\nu} = 0$, in complex coordinates implies

$$\bar{\partial} T + \frac{1}{4}\partial\Theta = 0 \tag{21.120}$$

Taking correlation functions with $T(0,0)$ and $\Theta(0,0)$, we obtain

$$\dot{F} + \frac{1}{4}\left(\dot{G} - 3G\right) = 0$$
$$\dot{G} - G + \frac{1}{4}\left(\dot{H} - 2H\right) = 0 \tag{21.121}$$

where

$$\dot{F} \equiv z\bar{z}F'(z\bar{z}), \quad \dot{G} \equiv z^3\bar{z}G'(z\bar{z}), \quad \dot{H} \equiv z^2\bar{z}^2 H'(z\bar{z}) \tag{21.122}$$

On eliminating the function G using the conservation laws and defining the function C,

$$C \equiv 2F - G - \frac{3}{8}H \tag{21.123}$$

we obtain

$$\dot{C} = -\frac{3}{4}H \tag{21.124}$$

Now, reflection positivity requires that

$$\langle \Theta(z,\bar{z})\Theta(0,0)\rangle \geq 0 \tag{21.125}$$

which implies that $H \geq 0$. Therefore, C is a nonincreasing function of $R = (z\bar{z})^{1/2}$, and it is stationary (i.e., $\dot{C} = 0$ only if $H = 0$).

Under an RG transformation, the short-distance cutoff is scaled as $a \to a(1 + \delta\ell)$. Since the quantity C is a dimensionless function of R and of the coupling constants $\{g_i\}$, a change in the UV scale is equivalent to sending $R \to R(1-\delta\ell)$ and to a new set of couplings to $\{g_i'\}$ according to the RG equations defined by the beta functions, $\{\beta_i\}$. Therefore, the function $C(R, \{g_i\})$ satisfies the Callan-Symanzik RG equation

$$\left(R\frac{\partial}{\partial R} + \sum_i \beta_i(\{g\})\frac{\partial}{\partial g_i}\right) C(R, \{g\}) = 0 \tag{21.126}$$

where

$$\frac{dC}{d\ell} = -\sum_i \beta_i(\{g\})\frac{\partial C}{\partial g_i} \tag{21.127}$$

is the rate of change of C along the RG trajectory, at fixed R. This result implies that if we define

$$C(\{g\}) \equiv C(1, \{g\}) \tag{21.128}$$

then this quantity is nonincreasing under the RG. Furthermore, it vanishes only if $H=0$, which, by reflection positivity, means that $\Theta = 0$ and the theory is scale invariant (and hence is a fixed point). In addition, at the fixed point, $G = H = 0$ and $F = c/2$ (where c is the central charge). Hence, at the stationary points, the function C is equal to the central charge of the fixed point, $C = c$.

The function C can be computed using the perturbative RG discussed in section 15.6. There we showed that the beta functions have the form

$$\frac{dg_i}{d\ell} = (2-\Delta_i)g_i - \frac{1}{2}\sum_{k,l} c_{ijk}g_j g_k + \cdots \tag{21.129}$$

(no summation in the first term), where the c_{ijk} are the coefficients of the OPEs of the primary fields $\{\mathcal{O}_i\}$. This equation implies that, to this order, the RG equations describe gradient flows,

$$\frac{dg_i}{d\ell} = \frac{\partial}{\partial g_i}\widetilde{C}(\{g\}) \tag{21.130}$$

where

$$\widetilde{C}(\{g\}) = \frac{1}{2}\sum_k (2-\Delta_k)g_k + \frac{1}{6}\sum_{ijk} c_{ijk}g_i g_j g_k \tag{21.131}$$

We can now use the function $\widetilde{C}$ to compute the Zamolodchikov's C function. Since it must have the same stationary points as $\widetilde{C}$, to this order, they must be proportional

$$C(\{g\}) = c + \alpha\widetilde{C}(\{g\}) + O(g^4) \tag{21.132}$$

where α is a constant to be determined. This can be done using perturbation theory.

To this end, consider the case of a CFT that is perturbed by a single relevant operator, a primary field $\mathcal{O}$ of conformal weight (h, h), scaling dimension $\Delta = 2h$, and spin $s = 0$. The perturbed action is

$$S = S_{CFT} - \lambda \int d^2z\, \mathcal{O}(z, \bar{z}) \tag{21.133}$$

For this perturbation to be relevant, we must have $\Delta < 2$, and hence, $h < 1$. The coupling constant λ has dimension $(1-h, 1-h)$.

Let us compute the changes of the correlators with insertions of the energy-momentum tensor. To first order in λ, we get

$$\langle T(z,\bar{z})\cdots\rangle = \langle T(z,\bar{z})\cdots\rangle_{CFT} + \lambda \int \langle T(z)\mathcal{O}(w,\bar{w})\rangle_{CFT} d^2w + \cdots \tag{21.134}$$

From the OPE, we have

$$\begin{aligned} T(z)\mathcal{O}(w,\bar{w}) &= \frac{h}{(z-w)^2}\mathcal{O}(w,\bar{w}) + \frac{1}{z-w}\partial\mathcal{O}(w,\bar{w}) + \cdots \\ &= \frac{h}{(z-w)^2}\mathcal{O}(z,\bar{z}) + \frac{1-h}{z-w}\partial\mathcal{O}(z,\bar{z}) + \cdots \end{aligned} \tag{21.135}$$

Thus, the integral in eq. (21.134) is divergent, and hence, it requires that we introduce a short-distance cutoff (e.g., a cutoff step function that excludes a circle of radius a from the integral). However, the cutoff violates conformal invariance (as before!) and, as a result, $\bar{\partial} T \neq 0$:

$$\bar{\partial} T = \lambda \int d^2w \frac{1-h}{z-w}(z-w)\partial\mathcal{O}(z,\bar{z})\delta(|z-w|^2 - a^2) + \cdots \tag{21.136}$$

Since the energy-momentum tensor must still be conserved, $\bar{\partial} + \frac{1}{4}\partial\Theta = 0$, we see that the energy-momentum tensor now has a nonvanishing trace:

$$\Theta(z,\bar{z}) = -4\pi\lambda(1-h)\mathcal{O}(z,\bar{z}) \tag{21.137}$$

Using eq. (21.124), we find that the C function is

$$C(\{g_i\}) = c(g^*) - 3(2-\Delta_j)g^j g^i + 2c_{ijk}g^i g^j g^k + O(g^4) \tag{21.138}$$

where $g^* = \{g_i^*\}$ is the fixed point.

In the simple case of a single relevant perturbation, this result implies that the central charge at a nearby fixed point $g^* = -(2-\Delta)^2/c_{111}$ is $c' = c - (2-\Delta)^3/c_{111}^2 + O((2-\Delta)^4)$.

The generalization of the C-theorem to dimensions $D > 2$ turns out to be quite subtle and to require concepts from quantum information theory, such as the entanglement entropy. This is the subject of intense research at this time, and we will not discuss this problem here.

21.6 Examples of two-dimensional CFTs

21.6.1 The free compactified boson

Consider a free massless scalar field $\phi(x)$ in two spacetime dimensions. Let us work with the Euclidean signature. We briefly discussed this theory in section 19.9.

The action of the free massless scalar field is

$$S = \frac{1}{8\pi}\int d^2x\, \partial_\mu\phi\partial^\mu\phi \tag{21.139}$$

The two-point function of the field ϕ is

$$\langle \phi(\boldsymbol{x})\phi(\boldsymbol{y})\rangle = -\ln(\boldsymbol{x}-\boldsymbol{y})^2 + \text{const.} \tag{21.140}$$

(which requires a short-distance subtraction that determines the value of the additive constant). In complex coordinates, this function becomes

$$\langle \phi(z,\bar{z})\phi(w,\bar{w})\rangle = -\left[\ln(z-w) + \ln(\bar{z}-\bar{w})\right] + \text{const.} \tag{21.141}$$

The holomorphic and antiholomorphic components are split if we look at the correlators of gradients of the field:

$$\begin{aligned}\langle \partial_z\phi(z,\bar{z})\partial_w\phi(w,\bar{w})\rangle &= -\frac{1}{(z-w)^2}\\ \langle \partial_{\bar{z}}\phi(z,\bar{z})\partial_{\bar{w}}\phi(w,\bar{w})\rangle &= -\frac{1}{(\bar{z}-\bar{w})^2}\end{aligned} \tag{21.142}$$

This implies that the field $\partial\phi$ obeys the OPE

$$\partial\phi(z)\partial\phi(w) = -\frac{1}{(z-w)^2} \tag{21.143}$$

The quantum energy-momentum tensor in complex coordinates is the normal-ordered operator

$$T(z) = -\frac{1}{2} : (\partial\phi(z))^2 : \tag{21.144}$$

Here normal-ordering means the limit

$$T(z) = -\frac{1}{2}\lim_{w\to z}\left(\partial\phi(z)\partial\phi(w) - \langle\partial\phi(z)\partial\phi(w)\rangle\right) \tag{21.145}$$

The OPE of $T(z)$ with $\partial\phi(z)$ can be found using Wick's theorem:

$$\begin{aligned}T(z)\partial\phi(w) &= -\frac{1}{2} : (\partial\phi(z))^2 : \partial\phi(w)\\ &= - : \partial\phi(z)\overbrace{\partial\phi(z) : \partial\phi}(w)\\ &= \frac{1}{(z-w)^2}\partial\phi(z)\end{aligned} \tag{21.146}$$

Hence, the OPE between $T(z)$ and $\partial\phi$ is

$$T(z)\partial\phi(w) = \frac{1}{(z-w)^2}\partial\phi(w) + \frac{1}{z-w}\partial_w^2\phi(w) \tag{21.147}$$

Thus, the holomorphic operator $\partial\phi$ has conformal weight $h=1$.

Another interesting operator is the vertex operator $V_\alpha(z) =: \exp(i\alpha\phi(z)) :$, where $\alpha \in \mathbb{R}$. Here, as before, normal ordering means not contracting the operators inside the exponential when it is expanded in powers. This rule is equivalent to a multiplicative wave-function renormalization of the operator. The OPE of the energy-momentum tensor with the vertex operator is

$$T(z)V_\alpha(w) = \frac{h_\alpha}{(z-w)^2} V_\alpha(w) + \frac{1}{z-w}\partial V_\alpha(w) \tag{21.148}$$

We find that the vertex operator is a primary field of conformal weight h_α:

$$h_\alpha = \frac{\alpha^2}{2} \tag{21.149}$$

We could have deduced this result by computing the two-point function of the vertex operator

$$\langle V_\alpha(z)V_{-\alpha}(w)\rangle = \exp\left(\alpha^2\langle\phi(z)\phi(w)\rangle\right) = \frac{1}{(z-w)^{2h_\alpha}} \tag{21.150}$$

where the conformal weight h_α is given by eq. (21.149).

Similarly, we can also use Wick's theorem to calculate the OPE of the energy-momentum tensor with itself:

$$\begin{aligned} T(z)T(w) &= \frac{1}{4} :(\partial\phi(z))^2::(\partial\phi(w))^2: \\ &= \frac{1/2}{(z-w)^4} + \frac{2}{(z-w)^2}T(w) + \frac{1}{z-w}\partial T(w) \end{aligned} \tag{21.151}$$

This equation implies that the holomorphic component of the energy-momentum tensor $T(z)$ has conformal weight $h=2$ and that the central charge of the free massless scalar field is $c=1$.

In chapter 11, we discussed the concept of spontaneous symmetry breaking in theories with a global symmetry. The massless scalar field ϕ arises formally as the phase field (i.e., the Goldstone boson) of an order-parameter field with a global $U(1)\simeq O(2)$ symmetry in two dimensions. The action of the field ϕ, eq. (21.139), has a global shift symmetry $\phi(x)\to\phi(x)+a$, where a is a real number. The propagator of the field ϕ breaks cluster decomposition, since it grows at long distances (as well as at short distances). In this sense it is not a physical field.

However, the operator $\partial_\mu\phi$ is invariant under such shifts. Likewise, the vertex operators, which exhibit power-law behavior, transform nontrivially under the shift symmetry and are physical. Moreover, since it is a phase field, the physical operators must be invariant under *periodic* shifts,

$$\phi(z,\bar{z}) \mapsto \phi(z,\bar{z}) + 2\pi R \tag{21.152}$$

where R is the compactification radius. In other words, the field is *compactified*. This terminology comes from string theory, in which the field ϕ is the coordinate of a string on a compactified space of radius R. This condition implies that the allowed operators must be invariant under the periodic shifts of eq. (21.152). The compactification condition imposes the strong restrictions on the allowed primary fields.

One operator that is always allowed is the "current" operator $J=i\partial\phi(z)$, since it is automatically invariant under shifts. However, the only vertex operators allowed must be such that $R\alpha\in\mathbb{Z}$. Thus, the allowed vertex operators are

$$V_n(z) =: \exp(in\phi(z)/R): \tag{21.153}$$

with n integer, with conformal weight

$$h_n = \frac{n^2}{2R^2} \tag{21.154}$$

An interesting case has compactification radius $R = 1/\sqrt{2}$. In this case, the operators $J^3(z) = i\partial\phi(z)$ and $J^{\pm}(z) = \exp(\pm i\sqrt{2}\phi(z))$ have conformal weight $(1, 0)$. These operators obey the OPEs

$$\begin{aligned} J^+(z)J^-(w) &= \frac{1}{(z-w)^2} + \frac{\sqrt{2}}{z-w} J^3(w) \\ J^3 J^{\pm}(w) &= \frac{\sqrt{2}}{z-w} J^{\pm}(w) \end{aligned} \tag{21.155}$$

and similarly for $\bar{J}^3$ and $\bar{J}^{\pm}$.

If we define $J^{\pm} = \frac{1}{\sqrt{2}}(J^1 \pm iJ^2)$, then this current algebra can also be written as

$$J^i(z)J^j(w) = \frac{\delta^{ij}}{(z-w)^2} + \frac{i\sqrt{2}\epsilon^{ijk}}{z-w} J^k(w) \tag{21.156}$$

This defines an $SU(2)_1$ Kac-Moody algebra of the $SU(2)$ currents. The first term on the right-hand side represents the central extension of the $SU(2)$ algebra at "level" 1 (the prefactor of the Kronecker delta). In terms of the mode expansions

$$J^i(z) = \sum_{n \in \mathbb{Z}} \frac{J^i_n}{z^{n+1}}, \quad \text{where } J^i_n = \oint \frac{dz}{2\pi i} z^n J^i(z) \tag{21.157}$$

the modes obey the commutation relations

$$\left[J^i_b, J^j_m\right] = i\sqrt{2}\epsilon^{ijk} J^k_{n+m} + n\delta^{ij}\delta_{n+m,0} \tag{21.158}$$

The vertex operators $V_{\pm 1}(z) = \exp(\pm i\phi(z)/\sqrt{2})$ have conformal weight $(1/4, 0)$, and their two-point functions are

$$\langle V_1(z)V_{-1}(w)\rangle = \frac{1}{(z-w)^{1/2}} \tag{21.159}$$

and are double valued. It can be shown that these fields constitute the $j = 1/2$ and $m = \pm 1/2$ (spinor) representation of the $SU(2)$ current algebra.

A more general case of this type occurs when the compactification radius is $R = \sqrt{m}$, with $m \in \mathbb{Z}$. In this case, the conformal weights of the vertex operators are $h_n = n^2/(2m)$. Their two-point functions are

$$\langle V_n(z)V_{-n}(w)\rangle = \frac{1}{z^{n^2/m}} \tag{21.160}$$

In general, these two-point functions are multivalued, since under a rotation by 2π, the argument changes by $2\pi n^2/m$. In other words, these chiral fields represent anyons (or parafermions). Now, among this infinite list, some operators remain local, in the sense that they are single valued. This happens for $n = \sqrt{mp}$ (where p is an integer). However, the nonlocal part of these operators can be classified into $m-1$ sectors, with $n=0, 1, \ldots, m-1$. Each sector is labeled by a primary field, the vertex operator $V_n(z)=\exp(in\phi(z)/\sqrt{m})$. Hence, instead of having infinitely many primaries, these compactified bosons have a finite number (m) of primaries. CFTs with a finite number of primaries are called *rational CFTs* (Ginsparg, 1989).

We close this discussion by noting a case of special interest, $R=1$. In this case, there are only two sectors, one labeled by the identity $V_0=I$ and the other by the primary $V_1(z)=\exp(i\phi(z))$. In this case, the two-point function is

$$\langle V_1(z)V_{-1}(w)\rangle = \frac{1}{z-w} \tag{21.161}$$

which is odd under the exchange $z \leftrightarrow w$, a property that we expect in a fermion. We will see below that this is indeed the Dirac fermion.

21.6.2 The free massless fermion CFT

Consider now a free massless relativistic fermion in two dimensions. As before, set the speed of light to unity. We will work in the Euclidean signature. A relativistic fermion in $D=2$ Euclidean dimensions is a two-component spinor. Let us work in the chiral basis, in which the upper component of the spinor is a right-moving field, and the lower component is a left-moving field. In the Euclidean signature, the upper component of the spinor is holomorphic, and the lower component is antiholomorphic. A chiral (holomorphic) fermion is the $D=2$ version of a Weyl fermion. In addition, the fermionic fields may be complex (Dirac) or real (Majorana).

A: Majorana fermions Let ψ_R denote a real (Majorana) chiral (right-moving) fermion and ψ_L an antichiral (left-moving) Majorana fermion. (The notation is changed here to avoid confusion with the standard notation for relativistic fermions.) We already encountered a theory of Majorana fermions in the context of the solution of the two-dimensional Ising model in chapter 14. We will see shortly that the CFT of the Majorana fermion is closely related to the Ising CFT.

The action for a massless Majorana spinor in the Euclidean signature is (including both chiralities)

$$S=\frac{1}{8\pi}\int d^2x\left(\psi_R\bar{\partial}\psi_R+\psi_L\partial\psi_L\right) \tag{21.162}$$

Here we used that in $D=2$ Euclidean dimensions, the Dirac operator can be represented as

$$\partial\!\!\!/ = \sigma_1\partial_1+\sigma_2\partial_2=\begin{pmatrix} 0 & \partial \\ \bar{\partial} & 0\end{pmatrix} \tag{21.163}$$

The equations of motion of this theory are

$$\bar{\partial}\psi_R=0, \quad \partial\psi_L=0 \tag{21.164}$$

Hence, ψ_R is only a function of z and is a holomorphic field, and ψ_L is only a function of $\bar{z}$ and is an antiholomorphic field.

With the normalization used in eq. (21.162), the OPEs of the Majorana Fermi fields are

$$\psi_R(z)\psi_R(w) = -\frac{1}{z-w}, \qquad \psi_L(\bar{z})\psi_L(\bar{w}) = -\frac{1}{\bar{z}-\bar{w}} \tag{21.165}$$

which also specify the propagators. In this theory, the holomorphic and anti-holomorphic components of the energy-momentum tensor are

$$T(z) = \frac{1}{2} : \psi_R(z)\partial\psi_R(z) :, \qquad \bar{T}(\bar{z}) = \frac{1}{2} : \psi_L(\bar{z})\bar{\partial}\psi_L(\bar{z}) : \tag{21.166}$$

It follows that the OPE of the above energy-momentum tensor $T(z)$ with the chiral Majorana spinor is

$$T(z)\psi_R(w) = \frac{1/2}{(z-w)^2}\psi_R(w) + \frac{1}{z-w}\partial\psi_R(w) \tag{21.167}$$

Hence the chiral Majorana spinor ψ_R has conformal weight $(h, \bar{h}) = (1/2, 0)$, as can also be read off eq. (21.165). Thus, the chiral Majorana fermion has scaling dimension $\Delta = 1/2$ and spin $s = 1/2$ (as it should). Similarly, one finds that the antichiral field ψ_L has conformal weight $(h, \bar{h}) = (0, 1/2)$ and has scaling dimension $\Delta = 1/2$ and (conformal) spin $-1/2$.

Finally, the OPE of the energy-momentum tensors can be computed using Wick's theorem to yield

$$T(z)T(w) = \frac{1/4}{(z-w)^4} + \frac{2}{(z-w)^2}T(w) + \frac{1}{z-w}\partial T(w) \tag{21.168}$$

and similarly for $\bar{T}$. This result implies that the central charge of a chiral Majorana spinor is $c = 1/2$.

Another primary field that can be constructed is the Majorana mass operator $\psi_R(z)\psi_L(\bar{z})$, which mixes the right- and left-moving sectors. It is straightforward to see that this is a primary field with conformal weight $(1/2, 1/2)$ and, hence, scaling dimension $\Delta = 1$ and (conformal) spin $s = 0$.

B: Dirac fermions A Dirac (complex) chiral fermion has the form $\psi_R = \eta_R + i\chi_R$, where η_R and χ_R are chiral Majorana fermions. The Euclidean action of the Dirac fermion is

$$S = \frac{1}{4\pi}\int d^2x \bar{\psi} \partial\!\!\!/ \psi \tag{21.169}$$

which is just the sum of the action of the Majorana fermions η and χ, each of the form of eq. (21.162). It then follows that, since each Majorana field has central charge $c = 1/2$, the central charge of the complex (Dirac) fermion is $c = 1$. The energy-momentum tensor of the Dirac field is just the sum of the energy-momentum tensors of the two Majorana fermions.

Just as in the Majorana case, the Dirac fermions ψ_R and ψ_L have conformal weights $(1/2, 0)$ and $(0, 1/2)$, respectively. Hence their two-point functions are

$$\langle \psi_R^\dagger(z)\psi_R(w)\rangle = -\frac{1}{z-w}, \qquad \langle \psi_L^\dagger(\bar{z})\psi_L(\bar{w})\rangle = -\frac{1}{\bar{z}-\bar{w}} \tag{21.170}$$

The Dirac fermion is a complex field and has a global $U(1)$ symmetry. Consequently, it has a locally conserved current $J^\mu = (J_0, J_1) = \bar{\psi}\gamma^\mu\psi$, which can be decomposed into right- and left-moving (holomorphic and antiholomorphic) components, $J_R = \psi_R^\dagger \psi_R$ and $J_L = \psi_L^\dagger \psi_L$, which have conformal weights $(1,0)$ and $(0,1)$, respectively. Their two-point functions are

$$\langle J_R(z) J_R(w) \rangle = \frac{1}{(z-w)^2}, \qquad \langle J_L(\bar{z}) J_L(\bar{w}) \rangle = \frac{1}{(\bar{z}-\bar{w})^2} \tag{21.171}$$

The $U(1)$ currents J_R and J_L satisfy an OPE of the form

$$J_R(z) J_R(w) = \frac{1}{(z-w)^2}, \quad J_L(\bar{z}) J_L(\bar{w}) = \frac{1}{(\bar{z}-\bar{w})^2} \tag{21.172}$$

The singular term on the right-hand side is known as the Schwinger term of the $U(1)$ currents.

In terms of the mode expansions, we find a $U(1)$ Kac-Moody algebra

$$\left[J_n^R, J_m^R\right] = n \delta_{n+m,0} \tag{21.173}$$

and similarly for the modes of the left-moving currents. The right- and left-moving currents commute with each other.

The expressions of the fermion two-point functions and the Dirac current algebra, along with the results we obtained for a compactified free massless boson with compactification radius $R=1$, suggest that there is a direct connection. This mapping is known as *bosonization*. To see how it works, we assert that the mapping is the following set of relations between the two theories:

$$\begin{aligned} \psi_R(z) &\leftrightarrow e^{i\phi(z)}, & J_R(z) &\leftrightarrow -i\partial\phi(z) \\ \psi_L(\bar{z}) &\leftrightarrow e^{-i\bar{\phi}(\bar{z})}, & J_L(\bar{z}) &\leftrightarrow i\bar{\partial}\bar{\phi}(\bar{z}) \end{aligned} \tag{21.174}$$

where $\bar{\phi}$ is the left-moving (antiholomorphic) component of the field $\phi(z,\bar{z}) = \phi(z) - \bar{\phi}(\bar{z})$.

A proof of the bosonization mapping requires showing that the two theories have identical spectrums. This involves computing the partition functions on a torus (with suitable boundary conditions) for both theories and showing that they agree. This is a technical argument that is not pursued here.

21.6.3 Minimal models and the two-dimensional Ising model CFT

In chapter 14, we looked at the statistical mechanics of the two-dimensional classical Ising model (or, equivalently, the one-dimensional quantum Ising model). There we saw that this is secretly a theory of free relativistic Majorana fermions that become massless at the critical point. Here we will see that this is a special case of a large class of conformal field theories known as the minimal models.

In section 21.5.3, we introduced the Virasoro algebra, eq. (21.94), of the generators L_n and $\bar{L}_n$ of conformal transformations in two Euclidean dimensions. There we showed that the vacuum state of a CFT, $|0\rangle$, is annihilated by both L_0 and $\bar{L}_0$. We also showed that a primary field $\mathcal{O}$ defines a highest-weight state of the Virasoro algebra, $\mathcal{O}|0\rangle = |h, \bar{h}\rangle$, where h $(\bar{h})$ is the L_0 $(\bar{L}_0)$ eigenvalue of the highest-weight state. We also argued that given a highest-weight state, an infinite number of descendant states can be constructed by acting repeatedly with the lowering operators, L_n $(\bar{L}_{-n})$ with $n > 0$, on the highest-weight state.

However, for this scheme to be consistent, all the descendants must be linearly independent. A linear combination of descendants that vanishes defines a null state, which implies that the states are linearly dependent. A representation is constructed by removing all the null states.

This scheme follows closely the construction of the angular momentum representations of the group of rotations. In the case of the group of rotations, the requirement that the representations be unitary (i.e., that the norm of the states be positive) leads to the quantization of the angular momentum eigenvalues. We will now sketch an argument, due to Belavin, Polyakov, and Zamolodchikov (Belavin et al., 1984) that the requirement of unitarity similarly leads to powerful restrictions on the values of the representations of the conformal group (i.e., on the allowed values of the central charge c of the conformal weights $(h, \bar{h})$ and on the fusion rules of the allowed primary fields, which are encoded in their fusion rules).

The null states are found in a straightforward way. For example, at level 1, the only possibility is $L_{-1}|h\rangle = 0$. But then $h=0$, and $|h\rangle = |0\rangle$, the vacuum state. At level two, it may happen that

$$L_{-2}|h\rangle + aL_{-1}^2|h\rangle = 0 \tag{21.175}$$

for some value of a. Acting with L_1 on this equation, we find the consistency condition

$$\begin{aligned}[L_1, L_{-2}]|h\rangle + a[L_1, L_{-1}^2]|h\rangle =& 3L_{-1}|h\rangle + a\left(2\{L_{-1}, L_0\}\right)|h\rangle \\ =& \left(3 + 2a(2h+1)\right) L_{-1}|h\rangle = 0\end{aligned} \tag{21.176}$$

which can only happen if $a = -3/2(2h+1)$. If now we act with L_2 on eq. (21.175), we find the condition

$$\begin{aligned}[L_2, L_{-2}]|h\rangle + a[L_2, L_{-1}^2]|h\rangle =& \left(4L_0 + \frac{c}{2}\right)|h\rangle + 3aL_1L_{-1}|h\rangle \\ =& \left(4h + \frac{c}{2} + 6ah\right)|h\rangle = 0\end{aligned} \tag{21.177}$$

so that the central charge must satisfy $c = 2(-6ah - 4h) = 2h(5-8h)/(2h+1)$. Thus, this will work if the highest-weight state $|h\rangle$ at this value of c satisfies

$$\left(L_{-2} - \frac{3}{2(2h+1)}L_{-1}^2\right)|h\rangle = 0 \tag{21.178}$$

Such a state, with a null descendant at level 2, is said to be degenerate at level 2. It can be shown that this equation implies that the N-point functions of primary fields satisfy a set of differential equations. In the case of the four-point functions, they can be expressed in terms of hypergeometric functions.

Unitarity means that the inner product in the space of states is positive definite. The inner products of any two descendant states can be computed using the Virasoro algebra. A state $|\psi\rangle$ that has negative norm, $\langle\psi|\psi\rangle < 0$, is called a "ghost." In a unitary theory, a ghost should not be found at any level of the representation. Given the collection of descendants of a highest-weight state, one can define a matrix whose elements are their inner products. The determinant of this matrix is known as the Kac determinant. A zero eigenvector of this matrix implies that there are null vectors. Therefore, it will be sufficient to look for zeros of the Kac determinant.

At level two, we have the two-component basis consisting of $L_{-2}|h\rangle$ and $L_{-1}^2|h\rangle$. The matrix is

$$\begin{pmatrix} \langle h|L_2L_{-2}|h\rangle & \langle h|L_1^2L_{-2}|h\rangle \\ \langle h|L_2L_{-1}^2|h\rangle & \langle h|L_1^2L_{-1}^2|h\rangle \end{pmatrix} = \begin{pmatrix} 4h+\frac{c}{2} & 6h \\ 6h & 4h(2h+1) \end{pmatrix} \tag{21.179}$$

The determinant of this matrix can be written as

$$2(16h^3 - 10h^2 + 2h^2c + hc) = 32(h - h_{1,1}(c))(h - h_{1,2}(c))(h - h_{2,1}(c)) \tag{21.180}$$

where

$$h_{1,1}(c) = 0, \qquad h_{1,2}(c) = h_{2,1}(c) = \frac{1}{16}(5-c) \mp \sqrt{(1-c)(25-c)} \tag{21.181}$$

The $h=0$ root is actually due to a null state at level 1, $L_{-1}|0\rangle = 0$, which implies that $L_{-1}^2|0\rangle = 0$ as well. This feature is repeated at all orders.

At level N, the Hilbert space is made up of states of the form

$$\sum_{n_i} a_{n_1\ldots n_k} L_{-n_1}\cdots L_{-n_k}|h\rangle \tag{21.182}$$

with $\sum_i n_i = N$. At level N, we need the determinant of the $P(N)\times P(N)$ matrix of inner products of the form

$$M_N(c,h) = \langle h|L_{m_l}\cdots L_{m_1}L_{-n_1}\cdots L_{-n_k}|h\rangle \tag{21.183}$$

If the determinant vanishes, $\det M_N(c,h)=0$, then there are null states for each c and h. If negative, the determinant has an odd number of negative eigenvalues (at least one). The representation of the Virasoro algebra at those values of c and h is not unitary.

Kac gave an explicit expression for the determinant

$$\det M_N(g,h) = \alpha_N \prod_{pq\le N} (h - h_{p,q}(c))^{P(N-pq)} \tag{21.184}$$

Upon defining

$$m = -\frac{1}{2} \pm \frac{1}{2}\sqrt{\frac{25-c}{1-c}} \tag{21.185}$$

or, equivalently,

$$c = 1 - \frac{6}{m(m+1)} \tag{21.186}$$

the quantity $h_{p,q}$ becomes

$$h_{p,q}(m) = \frac{[(m+1)p - mq]^2 - 1}{4m(m+1)} \tag{21.187}$$

The Kac determinant does not have zeros at any level if $c>1$ and $h\ge 0$. For $c=1$, the determinant vanishes if $h=n^2/4$, but it does not become negative. Thus, there is no obstacle, in principle, to having unitary representations for $c\ge 1$ and $h\ge 0$.

For $0 < c < 1$ and $h > 0$, Friedan, Qiu, and Shenker (Friedan et al., 1984) showed that unitary theories only exist for the set of discrete values of the central charge given by eq. (21.186) for integer values $m = 3, 4, \ldots$. For each value of c, there are $m(m-1)/2$ allowed values of h given by eq. (21.187), where p and q are integers such that $1 \leq p \leq m-1$ and $1 \leq q \leq p$. The set of CFTs that satisfy these conditions is known as the minimal models, and the set defined by eq. (21.186) constitutes their central charges.

Now consider the first minimal model, with $m = 3$. These results imply that it has central charge $c = \frac{1}{2}$. Let us label the primary fields by $\mathcal{O}_{p,q}(z, \bar{z}) = \phi_{p,q}(z)\bar{\phi}_{p,q}(\bar{z})$ with conformal weights $(h, \bar{h})$. For $m = 3$, the allowed values are $h = 0, \frac{1}{2}, \frac{1}{16}$, with primaries

$$\mathcal{O}_{1,1} : (0, 0), \qquad \mathcal{O}_{2,1} : \left(\frac{1}{2}, \frac{1}{2}\right), \qquad \mathcal{O}_{2,1} : \left(\frac{1}{16}, \frac{1}{16}\right) \tag{21.188}$$

This is telling us that we have a theory whose nonchiral primaries have scaling dimensions $\Delta = 0, 1, \frac{1}{8}$. It also has a chiral primary with scaling dimension $\frac{1}{2}$ and spin $s = \frac{1}{2}$.

Chapter 14 presented the solution to the two-dimensional Ising model. In section 14.7, we showed that the critical behavior is described by a theory of Majorana fermions that become massless at the critical point. The minimal model described here with $m = 3$ is consistent with these results from chapter 14. Indeed, the model has central charge $\frac{1}{2}$, as a massless Majorana fermion does. It also has a chiral primary field $(\frac{1}{2}, 0)$ which is a fermion with dimension $\frac{1}{2}$ and (conformal) spin $\frac{1}{2}$ (and similarly for the antichiral field). The Majorana fermion mass $\varepsilon = \psi_R \psi_L$ is a field with conformal weight $(\frac{1}{2}, \frac{1}{2})$, scaling dimension $\Delta_\varepsilon = 1$, and (conformal) spin $s = 0$.

We also find another field, $\mathcal{O}_{2,1}$, which has conformal weight $\left(\frac{1}{16}, \frac{1}{16}\right)$. Hence, it has scaling dimension $\frac{1}{8}$ and (conformal) spin $s = 0$. In chapter 14, we did not compute the spin-spin correlation function. However, it is known that this correlator has a power-law behavior with exponent $\eta = \frac{1}{4}$. Thus the scaling dimension of the spin operator, σ, the order parameter of the Ising model, is $\Delta = \frac{1}{8}$! Thus, we conclude that the CFT of the two-dimensional Ising model is indeed the $m = 3$ minimal model.

We close by noting that the fusion rules (i.e., the OPEs) of the primaries are also determined. In the case of the Ising model, we have three conformal families, each labeled by a primary field, 1, $[\varepsilon]$, and $[\sigma]$, and obeying the fusion rules:

$$[\sigma][\sigma] = 1 + [\varepsilon], \quad [\sigma][\varepsilon] = [\sigma], \quad [\varepsilon][\varepsilon] = 1 \tag{21.189}$$

This approach also allows us to compute the four-point function of the σ field, the order parameter of the Ising model. This is a beautiful but quite technical subject, and we will not pursue it here.

21.6.4 The Wess-Zumino-Witten CFT

The Wess-Zumino-Witten (WZW) model is a $(1+1)$-dimensional nonlinear sigma model whose degree of freedom is an element of a Lie group G. We already discussed this model briefly in section 20.9.3. The action of this model is

$$S[g] = \frac{1}{4\lambda^2} \int_{S^2_{\text{base}}} d^2x \, \text{tr}\left(\partial_\mu g \partial^\mu g^{-1}\right) + \frac{k}{24\pi} \int_B \epsilon^{\mu\nu\lambda} \text{tr}\left(g^{-1}\partial_\mu g \, g^{-1}\partial_\nu g \, g^{-1}\partial_\lambda g\right) \tag{21.190}$$

In the second term of the action, the WZW term, the field $g(x)$ whose base space is S^2_{base}, is extended to the interior of a three-dimensional ball $B \subset S^3$ whose boundary is S^2_{base}, subject to the condition that $g(x) = 1$ (the identity) at the center of the ball B. The extension to the interior of the ball B is arbitrary. However, not all extensions are equivalent, since the mappings of the ball, S^3, to the group manifold G ($SU(N)$ in this case) are classified by the homotopy group $\pi_3(SU(N)) = \mathbb{Z}$, with topological charge

$$Q = \frac{1}{24\pi^2} \int_{S^3} \epsilon^{\mu\nu\lambda} \text{tr}\left(g^{-1}\partial_\mu g\, g^{-1}\partial_\nu g\, g^{-1}\partial_\lambda g\right) \tag{21.191}$$

Thus, the second term of the WZW action has an ambiguity equal to $2\pi kQ$, which is unobservable if $k \in \mathbb{Z}$. This is essentially the same argument that led to the quantization of spin in the coherent-state path integral for spin in section 8.10.

The WZW term was introduced by Witten (1984) in his work on current algebras and non-abelian bosonization and by Polyakov and Wiegmann (1983) as an effective action of (1+1)-dimensional free fermionic theories. This theory has an IR stable fixed point at $\lambda_c^2 = 4\pi/k$, where the theory has full conformal invariance and describes a CFT. The WZW CFT was solved by Knizhnik and Zamolodchikov (1984).

To motivate the structure of the WZW CFT we first consider a theory of N massless Dirac fermions in 1+1 dimensions with $U(N)$ global symmetry. Witten worked with a theory of free Majorana fermions and the $O(N)$ group. Here we are following later results by Affleck (1986b). The Lagrangian for the $U(N)$ group is

$$\mathcal{L} = \bar{\psi}_j i\partial\!\!\!/ \psi_j \tag{21.192}$$

Here $j = 1, \ldots, N$ labels the species of Dirac fermions (not their Dirac components). We will work in the chiral basis, in which, in terms of the Pauli matrices, the Dirac matrices are $\gamma_0 = \sigma_1$, $\gamma_1 = i\sigma_2$, and $\gamma_5 = \sigma_3$. In this basis, the Dirac components of the spinor field ψ_i are, respectively, the right-moving component $\psi_{R,i}$ (with chirality $+1$) and the left-moving component $\psi_{L,i}$ (with chirality -1). This Lagrangian is invariant under global $U(N) \times U(N)$ transformations of the right- and left-moving fields. Each $U(N)$ can be regarded as a direct product of a $U(1)$ group, generated by the identity matrix of the $U(N)$ algebra, and the $SU(N)$ subgroup of $U(N)$. We have a set of right- and left-moving (formally) separately conserved currents

$$J_R = \psi^\dagger_{R,i}\psi_{R,i}, \qquad J_L = \psi^\dagger_{L,i}\psi_{L,i} \tag{21.193}$$

and

$$J^a_R = \psi^\dagger_{R,i} t^a_{ij} \psi_{R,j} \qquad J^a_L = \psi^\dagger_{L,i} t^a_{ij} \psi_{L,j} \tag{21.194}$$

for the $SU(N)$ currents. Here, the $N^2 - 1$ matrices t^a_{ij} are the generators of the group $SU(N)$. The generators are normalized such that $\text{tr}(t^a t^b) = \frac{1}{2}\delta_{ab}$, and they obey the $SU(N)$ algebra $[t^a, t^b] = if_{abc}t_c$, where the f_{abc} are the structure constants of $SU(N)$.

As in the case of the abelian theory, discussed in section 20.3, these currents are affected by the $U(1)$ chiral anomaly, and their commutators have Schwinger terms. The non-abelian currents also have a (non-abelian) chiral anomaly. A careful calculation, using the point-splitting procedure used in section 20.3 (cf. eq. (20.15)), which preserves gauge-invariance, shows that at equal times, the right- and left-moving currents obey the algebras

$$[J_R(x), J_R(y)] = i\frac{N}{2\pi}\delta_{ab}\partial_x\delta(x-y), \quad [J_L(x), J_L(y)] = -i\frac{N}{4\pi}\delta_{ab}\partial_x\delta(x-y) \tag{21.195}$$

for the abelian currents. This is the $U(1)_N$ Kac-Moody algebra. Similarly, for the non-abelian currents, we find

$$\begin{aligned}
[J_R^a(x), J_R^b(y)] &= if_{abc}J_R^c(x)\delta(x-y) + i\frac{k}{4\pi}\delta_{ab}\partial_x\delta(x-y) \\
[J_L^a(x), J_L^b(y)] &= if_{abc}J_L^c(x)\delta(x-y) - i\frac{k}{4\pi}\delta_{ab}\partial_x\delta(x-y)
\end{aligned} \tag{21.196}$$

This is the $SU(N)_k$ Kac-Moody algebra. The Schwinger term in eq. (21.196) is the central extension of the $SU(N)_k$ Kac-Moody algebra.

The second terms on the right-hand side of both sets of equations are Schwinger terms, analogous to the ones we discussed in the abelian theory. Mathematically, the Schwinger terms are called "central extensions," and the parameter k is the level of the Kac-Moody algebra, or its central extension. We already encountered a central extension in the Virasoro algebra, eq. (21.84). In the free Dirac theory that we are considering, the parameter $k=1$. Furthermore, it is known from the mathematical literature that this current algebra has unitary representations only if $k \in \mathbb{Z}$. Therefore, the level k cannot be an arbitrary real number and is quantized.

Let us use light-cone components (i.e., left- and right-moving), $x_\pm = \frac{1}{2}(x_0 \mp x_1)$, and the notation $\partial_\pm = \partial/\partial_{x_\pm}$. The light-cone components T and $\bar{T}$ for the left- and right-moving components, respectively, of the energy-momentum tensor are

$$T = \frac{1}{2}(H-P) = i :\psi_{L,i}^\dagger \partial_- \psi_{L,i}:, \qquad \bar{T} = \frac{1}{2}(H+P) = i :\psi_{R,i}^\dagger \partial_+ \psi_{R,i}: \tag{21.197}$$

where H and P are, respectively, the (normal-ordered) Hamiltonian and linear momentum. A straightforward (but lengthy) calculation leads to the result that the normal-ordered left- and right-moving components of the energy momentum tensor, denoted by T and $\bar{T}$, respectively, are

$$T = \frac{\pi}{N}J_L J_L + \frac{2\pi}{N+1}J_L^a J_L^a, \qquad \bar{T} = \frac{\pi}{N}J_R J_R + \frac{2\pi}{N+1}J_R^a J_R^a \tag{21.198}$$

The $U(1)$ currents are given in abelian bosonization by

$$J_L = \sqrt{\frac{N}{4\pi}}\partial_-\phi, \qquad J_R = -\sqrt{\frac{N}{4\pi}}\partial_+\phi \tag{21.199}$$

where ϕ is a boson (scalar field) with compactification radius

$$R = \frac{1}{\sqrt{4\pi N}} \tag{21.200}$$

Let us now focus on the $SU(N)$ currents. Their right- and left-moving currents obey the conservation laws

$$\partial_- J_R^{ij} = 0, \qquad \partial_+ J_L^{ij} = 0 \tag{21.201}$$

Witten showed that these equations can be solved in terms of a group-valued field, a matrix $g(x_-) \in SU(N)$, in terms of which the chiral currents are (suppressing the group indices)

$$J_R^a(x) = \frac{i}{2\pi}\,\mathrm{tr}\Big(g^{-1}(x)\partial_+ g(x) t^a\Big), \qquad J_L^a(x) = -\frac{i}{2\pi}\,\mathrm{tr}\Big((\partial_- g(x))g^{-1}(x) t^a\Big) \tag{21.202}$$

The conservation of the currents implies that

$$\partial_-(g^{-1}\partial_+ g) = 0, \qquad \partial_+((\partial_- g)g^{-1}) = 0 \tag{21.203}$$

The results of eq. (21.202) provide an operator identification of the non-abelian chiral fermionic currents in terms of the currents of the WZW model. These identifications are the non-abelian version of the identification of the currents in the abelian theory in terms of the compactified boson ϕ.

The next task is to find a local Lagrangian for the matrix-valued field $g(x)$. Since $g(x) \in SU(N)$, it is not a free field. Indeed, this is a principal chiral field. Its natural action is

$$\mathcal{L} = \frac{1}{4\lambda^2}\int d^2x\,\mathrm{tr}(\partial^\mu g \partial_\mu g^{-1}) \tag{21.204}$$

This, however, cannot be the correct answer, since as we have seen, the nonlinear sigma model is asymptotically free with a positive beta function (using the high-energy physics sign convention), while the theory of free massless fermions is a fixed-point theory and, as such, it is scale (and conformally) invariant. In addition, the nonlinear sigma model is not compatible with the conservation equations of eq. (21.203).

So, what operators can be added to the action of the nonlinear sigma model to drive the theory to a fixed point and make it consistent with the conservation laws? It is easy to see that all additional $SU(N)$-invariant local operators are irrelevant, so they cannot do the job. In fact, the only way to drive this theory to a nontrivial fixed point is to add to the action a WZW term

$$S_{\mathrm{WZW}}[g] = \frac{k}{24\pi}\int_B \epsilon^{\mu\nu\lambda}\,\mathrm{tr}\left(g^{-1}\partial_\mu g\, g^{-1}\partial_\nu g\, g^{-1}\partial_\lambda g\right) \tag{21.205}$$

which we discussed briefly in section 20.9.3, eq. (20.145). We will also obtain a consistent representation of the conservation laws.

Indeed, the conservation encoded in the full WZW action, eq. (21.190), is

$$\begin{aligned} 0 &= \frac{1}{2\lambda^2}\partial_\mu(g^{-1}\partial^\mu g) - \frac{k}{8\pi}\epsilon^{\mu\nu}\partial_\mu(g^{-1}\partial_\nu g) \\ &= \left(\frac{1}{2\lambda^2} + \frac{k}{8\pi}\right)\partial_-(g^{-1}\partial_+ g) + \left(\frac{1}{2\lambda^2} - \frac{k}{8\pi}\right)\partial_+(g^{-1}\partial_- g) \end{aligned} \tag{21.206}$$

Therefore, for the conservation law of eq. (21.203) to be satisfied, we require that the coupling constant has to be

$$\lambda_c^2 = \frac{4\pi}{k} \tag{21.207}$$

for $k > 0$ (and the negative of this for $k < 0$). In addition, if the conservation law holds, then it is satisfied by $g(x_+, x_-) = A(x_-)B(x_+)$ (which are arbitrary $SU(N)$ matrices). Then, at this particular value of the coupling, at least classically, the right- and left-moving waves decouple. This is reminiscent of the behavior of free massless fermions. Furthermore, Witten carried out the program of canonical quantization for the WZW theory and showed that at the quantum level, the currents of the WZW model obey the $O(N)_k$ Kac-Moody algebra, provided that the theory is at the value of the coupling constant λ_c. The same result holds for other groups, including $SU(N)_k$.

The perturbative RG also offers a hint that the special value λ_c of eq. (21.207) may be a fixed point. First notice that, since k is an integer, it cannot flow under the RG, and only the coupling constant can flow. A one-loop calculation, which is only accurate in the weak coupling regime, yields a beta function (using the high-energy physics sign convention) of

$$\beta(\lambda) = -\lambda^2\left(\frac{N-1}{4\pi}\right)\left[1-\left(\frac{\lambda^2 k}{4\pi}\right)^2\right] \tag{21.208}$$

which has an IR-stable fixed point at λ_c. Of course, this argument cannot be trusted unless k is very large. However, eq. (21.207) turns out to be the exact answer.

Witten conjectured a set of bosonization identities for the fermion bilinears, mass terms, which mix the left- and right-moving sectors. For a theory with $O(N)$ symmetry (i.e., N free massless Majorana fermions), which is identified with the $O(N)_1$ WZW model (and, hence, with level $k=1$), the identification of the operators is

$$-i:\psi_L^i(x)\psi_{j,R}(x):=Mg_j^i(x) \tag{21.209}$$

where $g_j^i \in O(N)$, and M is a "mass" whose precise form depends on the normal ordering of the operators on the left-hand side of the equation. For free massless Dirac fermions with $U(N)$ symmetry, whose current algebra is $U(1)_N \times SU(N)_1$, they are mapped by bosonization to the $U(1)_N$ free massless compactified boson ϕ, and the $SU(N)_1$ WZW nonlinear sigma model whose degree of freedom is $g \in SU(N)$ (again, with level $k=1$). In the Dirac case, the mass fermion bilinears are identified with group elements as (Affleck, 1988)

$$:\psi_L^{i\dagger}\psi_{R,j}(x):=M\exp\left(i\sqrt{\frac{4\pi}{N}}\phi(x)\right)g_j^i(x) \tag{21.210}$$

The action of the WZW model at its IR fixed point is

$$\mathcal{L} = \frac{k}{16\pi}\int d^2x\,\mathrm{tr}(\partial^\mu g\partial_\mu g^{-1}) + \frac{k}{24\pi}\int_B \epsilon^{\mu\nu\lambda}\mathrm{tr}\left(g^{-1}\partial_\mu g g^{-1}\partial_\nu g g^{-1}\partial_\lambda g\right) \tag{21.211}$$

This is the action of the WZW CFT. Let us now analyze this theory from the CFT point of view.

Knizhnik and Zamolodchikov (1984) solved the WZW CFT for a semi-simple Lie group G. They showed that at the special value of the coupling constant λ_c, the left- and right-moving modes of the WZW model decouple, and the theory has an enhanced $G \times G$ symmetry. At this value of λ, the theory has full conformal invariance. In the rest of this section, we discuss the WZW CFT on the Euclidean complexified plane, with coordinates $z = x_1 + ix_2 \in \mathbb{C}$, as is standard in two-dimensional CFTs (Belavin et al., 1984).

The WZW theory has a Virasoro algebra generated by its energy-momentum tensor, and a Kac-Moody algebra generated by its chiral currents. Focusing on the left-moving (holomorphic) sector, Knizhnik and Zamolodchikov showed that the generators of the two algebras obey the OPEs

$$T(z)T(z') = \frac{c}{2(z-z')^4} + \frac{2}{(z-z')^2}T(z') + \frac{1}{z-z'}T(z') + \cdots \tag{21.212}$$

$$T(z)J_L^a(z') = \frac{1}{(z-z')^2}J_L^a(z') + \frac{1}{z-z'}J_L(z') + \cdots \tag{21.213}$$

$$J_L^a(z)J_L^b(z') = \frac{k\delta^{ab}}{(z-z')^2} + \frac{f^{abc}}{(z-z')}J_L^c(z') + \cdots \tag{21.214}$$

supplemented with the asymptotic conditions, $T(z) \sim z^{-4}$ and $J_L^a(z) \sim z^{-2}$ as $z \to \infty$. Equation (21.212) states that $T(z)$ is the generator of a Virasoro algebra with central charge c, and eq. (21.214) states that the $J^a(z)$ are the generators of a Kac-Moody algebra with level k. Equation (21.213) simply states that the chiral currents $J_L^a(z)$ are fields with dimension $(\Delta, \bar{\Delta}) = (1, 0)$ (see below). The right-moving (antiholomorphic) components $\bar{T}(\bar{z})$ and $J_R^a(\bar{z})$ obey similar equations.

The primary fields of this theory, $\phi_l(z, \bar{z})$, have dimensions $(\Delta_l, \bar{\Delta}_l)$ and obey the OPEs

$$T(w)\phi_l(z, \bar{z}) = \frac{\Delta_l}{(w-z)^2}\phi_l(z, \bar{z}) + \frac{1}{w-z}\partial_z\phi_l(z, \bar{z}) + \cdots \tag{21.215}$$

$$J_L^a(w)\phi_l(z, \bar{z}) = \frac{t_l^a}{w-z}\phi_l(z, \bar{z}) + \cdots \tag{21.216}$$

Equation (21.215) just states that $\phi_l(z, \bar{z})$ is a Virasoro primary field. In eq. (21.216), t_l^a are the generators of the group G for the field ϕ_l.

The energy-momentum tensor $T(z)$ and the left-moving currents $J_L^a(z)$ admit the mode expansions (the Laurent expansion of eq. (21.91)), which here become

$$T(z) = \sum_{n\in\mathbb{Z}} \frac{L_n}{z^{n+2}}, \qquad J_L^a(z) = \sum_{n\in\mathbb{Z}} \frac{J_n^a}{z^{n+1}} \tag{21.217}$$

and similar expressions for the right-moving (antiholomorphic) components. To simplify the notation, the label L has been dropped in the WZW current. The WZW primary fields ϕ_l satisfy the following equations (for $n > 0$):

$$\begin{aligned} L_n\phi_l &= 0, & L_0\phi_l &= \Delta_l\phi_l \\ J_n^a\phi_l &= 0, & J_0^a\phi_l &= t_l^a\phi_l \end{aligned} \tag{21.218}$$

The singular terms in the OPEs of eqs. (21.212)–(21.214) imply that the operators L_n and J_n^a obey the algebra

$$[L_n, L_m] = (n-m)L_{n+m} + \frac{1}{12}c(n^3 - n)\delta_{n+m,0} \tag{21.219}$$

$$[L_n, J_m^a] = -mJ_{n+m}^a \tag{21.220}$$

$$[J_n^a, J_m^b] = f^{abc}J_{n+m}^c + \frac{1}{2}k\delta^{ab}\delta_{n+m,0} \tag{21.221}$$

where eq. (21.219) is the Virasoro algebra, and eq. (21.221) is the Kac-Moody algebra.

The complete set of local fields includes, in addition to the primary fields $\{\phi_l\}$, their descendants under the action of both the Virasoro and Kac-Moody operators, $\{L_{-n}\}$ and $\{J^a_{-m}\}$ (and their right-moving counterparts). The descendant fields constitute the Verma modulus of the primary field ϕ_l and have dimensions $\Delta_l^{\{n,m\}} = \Delta_l + \sum_{i=1}^N n_i + \sum_{j=1}^M m_j$, and similarly for their antiholomorphic components. In this sense, neither $T(z)$ nor $J^a_L(z)$ is a primary field, since $T(z) = L_{-1}\, I$ and $J^a_L(z) = J^a_{-1}\, I$, where here I is the identity field.

What we have described here applies to any CFT with a Virasoro and a Kac-Moody algebra. Let us now apply this formalism to the WZW models, following closely the work (and notation) of Knizhnik and Zamolodchikov (1984). The WZW model is a nonlinear sigma model with a field g that takes values on the group G.

The key property of the WZW model is its conserved chiral currents of eq. (21.202). In complex coordinates, the conserved currents of the WZW model are denoted by $J(z) = J_a(z)T^a$ and $\bar{J}(\bar{z}) = \bar{J}_a(\bar{z})t^a$, and are given by

$$J(z) = -\frac{k}{2}(\partial_z g(z,\bar{z}))g^{-1}(z,\bar{z}), \quad \bar{J}(\bar{z}) = -\frac{k}{2}g^{-1}(z,\bar{z})(\partial_{\bar{z}} g(z,\bar{z})) \tag{21.222}$$

and satisfy the conservation laws, $\partial_{\bar{z}} J = 0$ and $\partial_z \bar{J} = 0$. Notice that we have rescaled the chiral currents relative to the expressions given in eq. (21.202).

The main assumption of Knizhnik and Zamolodchikov is that the set of fields of the WZW theory contains a *primary* field $g(z,\bar{z})$, whose conformal weights (dimensions) are $\Delta_g = \bar{\Delta}_g = \Delta$, and that it satisfies the equations

$$\kappa\, \partial_z g(z,\bar{z}) =: J_a(z)t^a\, g(z,\bar{z}):, \qquad \kappa\, \partial_{\bar{z}} g(z,\bar{z}) =: \bar{J}_a(\bar{z})t^a\, g(z,\bar{z}): \tag{21.223}$$

where κ is given below. Since g is a primary field, we expect that it will have an OPE with chiral (Kac-Moody) currents of the form

$$J_a(w)t^a g(z,\bar{z}) = \frac{c_g}{w-z} c(z,\bar{z}) + \kappa\, \partial_z g(z,\bar{z}) + \cdots \tag{21.224}$$

where c_g, defined by $t^a t^a = c_g I$, is equal to $c_g = 2C(g)$, where $C(g)$ is the quadratic Casimir of the representation, and the ellipsis indicates regular terms as $w \to z$. The normal-ordered product of eq. (21.223) is defined as the limit

$$:J_a(z)t^a\, g(z,\bar{z}) := \lim_{w\to z}\left(J^a(w) - \frac{t^a}{w-z}\right)t^a g(z,\bar{z}) \tag{21.225}$$

The OPE of eq. (21.224) implies that there is a field in this CFT,

$$\chi \equiv (J^a_{-1}t^a - \kappa L_{-1})g = 0 \tag{21.226}$$

which is a null field. Here we used that $\partial_z g = L_{-1} g$. Therefore, χ is a null state, and the representation is degenerate. Consistency then requires that χ be a primary field that satisfies

$$\begin{aligned} L_0 \chi &= (\Delta+1)\chi, \qquad J^a_0 \chi = t^a \chi \\ L_n \chi &= J^a_n \chi = 0, \quad \text{for } n > 0 \end{aligned} \tag{21.227}$$

While the first equation is automatically satisfied, the second equation holds provided

$$c_g + 2\Delta\kappa = 0, \qquad c_V + k + 2\kappa = 0 \tag{21.228}$$

where c_V, defined by

$$f^{acd} f^{bcd} = c_V \delta_{ab} \tag{21.229}$$

is the quadratic Casimir of the adjoint representation. These conditions imply that the scaling dimension Δ of the primary field g must be

$$\Delta = \frac{2C(g)}{c_V + k} \tag{21.230}$$

and that the parameter κ is

$$\kappa = -\frac{1}{2}(c_V + k) \tag{21.231}$$

Another way to reach the same conclusions is to construct the energy-momentum tensor of the quantized theory. Assume that the energy-momentum tensor has the Sugawara form,

$$T(z) = \frac{1}{2\kappa} : J^a(z) J^a(z) :, \qquad \bar{T}(z) = \frac{1}{2\kappa} : \bar{J}^a(z) \bar{J}^a(z) : \tag{21.232}$$

with the same constant κ used above. This structure of the energy-momentum tensor means that the OPE of the currents should be

$$J^a(z) J^a(z') = \frac{kD}{(z-z')^2} + 2\kappa T(z) + \cdots \tag{21.233}$$

where D is the dimension of the group G (i.e., the number of generators). Since the energy-momentum tensor $T(z)$ and the currents $J^a(z)$ must also satisfy the Virasoro and Kac-Moody algebras, eqs. (21.212)–(21.214), these equations will be satisfied only if the central charge c is given by

$$c = \frac{kD}{c_V + k} \tag{21.234}$$

and the constant κ is given by eq. (21.231).

These results also imply that the generators of the two algebras must be related to each other through the expression

$$L_n = \frac{1}{2\kappa} \sum_{m \in \mathbb{Z}} : J^a_m J^a_{n-m} : \tag{21.235}$$

where normal ordering here means that J_n with $n < 0$ are placed to the left of J_m with $m > 0$. Then, if we use this definition for L_{-1} and have it act on the primary field g, we obtain the null state χ of eq. (21.226).

Furthermore, this line of reasoning actually applies to *all* the primary fields in the theory, and not just to the WZW field g. Thus the scaling dimensions of all primary fields are given by

$$\Delta = \frac{2C(\phi_l)}{c_V + k}, \qquad \bar{\Delta} = \frac{2\bar{C}(\phi_l)}{c_V + k} \tag{21.236}$$

where $C(\phi_l)$ is the quadratic Casimir of the representation associated with the primary field ϕ_l. One result that also follows is that the slope of the beta function at the WZW fixed point is

$$\left.\frac{d\beta(\lambda^2, k)}{d\lambda^2}\right|_{\lambda^2=4\pi/k} = \frac{2c_V}{c_V + k} \tag{21.237}$$

which agrees with Witten's one-loop result.

Knizhnik and Zamolodchikov further also showed that the correlators of the g field satisfy the differential equation

$$\left[\kappa \frac{\partial}{\partial_{z_i}} - \sum_{j=1}^{N} \frac{t_i^a t_j^a}{z_i - z_j}\right]\langle g(z_1, \bar{z}_1) \cdots g(z_N, \bar{z}_N)\rangle = 0 \tag{21.238}$$

which is known as the Knizhnik-Zamolodchikov equation. This equation, and its generalization for other primary fields, can then be used to obtain the correlation functions of the WZW theory.

We end our discussion of the WZW CFT by applying these results to the case of the group $G = SU(N)$. In this case, the dimension of the group is $D = N^2 - 1$, and the Casimir of the adjoint representation is $c_V = N$. This implies that for the $SU(N)_k$ WZW CFT, the momentum-energy tensor is

$$T(z) = -\frac{1}{(N+k)} : J^a(z) J^a(z) : \tag{21.239}$$

The central charge is

$$c = \frac{k(N^2 - 1)}{N + k} \tag{21.240}$$

and the scaling dimensions are

$$\Delta_l = \frac{2C_l}{N + k} \tag{21.241}$$

In particular, the dimension of the WZW field g is

$$\Delta = \frac{N^2 - 1}{2N(N + k)} \tag{21.242}$$

Note that the central charges for the WZW theory are, in general, fractional numbers. Thus, these fixed points are, in general, not free-field theories. However, level $k = 1$ theories can represent free fields.

For instance, in the case of $SU(N)_1$, we readily obtain that $c = 1$ and that $\Delta = \bar{\Delta} = (N-1)/2N$. Both results are consistent with the non-abelian bosonization identification of the free-fermion mass terms in eq. (21.210) with a product of the $U(1)_N$ vertex operator $\exp(i\sqrt{4\pi/N}\phi)$ and the WZW field g. Indeed, one can see that the dimensions of these operators add up to $1/2$. Hence, the mass terms have dimension 1 and conformal spin 0, as they should. Moreover, the central charges also add up to the correct value, $c(U(1)_N) +$

$c(SU(N)_1) = N$, the value for N free Dirac fields. In addition, the energy-momentum tensor of the free fermions becomes the sum

$$T_{\text{Dirac}} = \sum_{i=1}^{N} :\psi_i^\dagger(z)\partial_z\psi_i(z): \mapsto -\frac{1}{2N} :J(z)J(z): - \frac{1}{N+1} :J^a(z)J^a(z): \tag{21.243}$$

In other words, the abelian $U(1)$ sector and the non-abelian $SU(N)$ sector factorize, with the central charges and scaling dimensions adding up to their free-field values, and the full Hilbert space decomposes into the tensor product of the individual Hilbert spaces.

Exercises

21.1 CFT on a cylinder

Consider a two-dimensional CFT with central charge c. Let $\phi(z,\bar{z})$ be a primary field of the CFT with scaling (conformal) dimension $\Delta = (h, \bar{h}) \equiv (h, h)$. The two-point function of the primary field on the infinite plane is

$$\langle \phi(0,0)\phi(z,\bar{z})\rangle = \frac{1}{z^h \bar{z}^h}$$

Now consider the same correlator on an infinite cylinder of circumference ℓ. Consider the conformal mapping

$$w = \frac{\ell}{2\pi} \ln z$$

that maps the plane, with coordinate $z = x + iy$, to the cylinder, with coordinate $w = \sigma + i\tau$, where $0 < \tau \leq \ell$.

1) Check that this mapping is conformal.
2) Use the general transformation law eq. (21.75) of a primary field under a conformal mapping $w(z)$ to show that the form of the two-point function of the primary field on the cylinder is

$$\langle \phi(0,0)\phi(\sigma,\tau)\rangle = \frac{(2\pi/\ell)^{2h}}{[2\cosh(2\pi\sigma/\ell) - 2\cos(2\pi\tau/\ell)]^{2h}}$$

3) Now think of the cylinder as the Euclidean spacetime at finite temperature $T = 1/\ell$. Use the general result you derived in part 2 to show that the correlator at equal imaginary time τ decays exponentially at long distances. What determines the value of correlation length $\xi(T)$?
4) Now think of the cylinder as the Euclidean spacetime $S_1 \times \mathbb{R}$, where S_1 is a circle of circumference $\ell = L$, and $\mathbb{R}$ is the imaginary time τ. In other words, we have a theory at zero temperature on a one-dimensional space of length $L = \ell$ with periodic boundary conditions in σ. Find the expression of the two-point function, which is a periodic function of σ with period L at equal imaginary times. How does it depend on the scaling dimension Δ?

21.2 Compactified boson on a cylinder

Consider the theory of the compactified boson $\varphi(x,t)$ of section 21.6.1 defined on a cylinder, $\varphi(x+\ell, t) \equiv \varphi(x,t) + 2\pi m R$, where R is the compactification radius, and where $m \in \mathbb{Z}$ is the winding number. The action of the boson is

$$S = \int_{-\infty}^{\infty} dt \int_0^{\ell} dx \, \frac{1}{8\pi} (\partial_\mu \varphi)^2$$

1) Use canonical quantization to show that the field expansion of the compactified boson is

$$\varphi(x, t) = \varphi_0 + \frac{4\pi n}{R\ell} t + \frac{2\pi Rm}{\ell} x + i \sum_{k \neq 0} \frac{1}{k} \left(a_k e^{2\pi ik(x-t)/\ell} - \bar{a}_{-k} e^{2\pi ik(x+t)/\ell} \right)$$

where φ_0 is the center-of-mass zero mode (and is a periodic degree of freedom restricted to the range $[0, R)$), n/R is the momentum of the center of mass, and the oscillators obey the commutation relations $[a_k, a_p] = k\delta_{k+p}$, $[\bar{a}_k, \bar{a}_p] = k\delta_{k+p}$, and $[a_k, \bar{a}_p] = 0$.

2) Show that the Virasoro generators L_0 and $\bar{L}_0$ are given by

$$L_0 = \frac{1}{2} \left(\frac{n}{R} + \frac{mR}{2} \right)^2 + \sum_{n>0} a_{-n} a_n$$

$$\bar{L}_0 = \frac{1}{2} \left(\frac{n}{R} - \frac{mR}{2} \right)^2 + \sum_{n>0} \bar{a}_{-n} \bar{a}_n$$

in terms of which the Hamiltonian is $H = \frac{2\pi}{\ell}(L_0 + \bar{L}_0)$ and the linear momentum is $P = \frac{2\pi}{\ell}(L_0 - \bar{L}_0)$. This result shows that the Hilbert space of the compactified boson can be classified into sectors, each labeled by the state $|n, m\rangle$, with charge n and winding number m.

3) Compute the ground-state energy and linear momentum of the state $|n, m\rangle$. Relate this result to the scaling dimensions of a *vortex* of topological charge m of a vertex operator of "electric" charge n in the classical two-dimensional XY model.

21.3 CFT with a boundary

In this exercise, consider again a theory of a compactified boson $\varphi(x_1, x_2)$ in 1+1 dimensions. However, now consider the case in which, in the Euclidean signature, the boson lives on a space with a boundary. For simplicity, assume that the Euclidean spacetime is the right half-plane, $x_1 \geq 0$, and x_2 is unrestricted. In this problem, you look at the effects of boundary conditions. Consider two boundary conditions at $x_1 = 0$: (1) Neumann or free ($\partial_1 \varphi = 0$), and (2) Dirichlet or pinned ($\varphi = 0$). The action of the boson is the same as in exercise 21.2, and the compactification radius is R.

1) Compute the explicit form of the two-point function

$$G(\boldsymbol{x}, \boldsymbol{y}) = \langle \varphi(\boldsymbol{x}) \varphi(\boldsymbol{y}) \rangle$$

on the half-space for both Neumann and Dirichlet boundary conditions.

2) Consider now the case in which both $\boldsymbol{x}$ and $\boldsymbol{y}$ are close to the boundary (e.g., $\boldsymbol{x} = (d, x_2)$ and $\boldsymbol{y} = (d, y_2)$). Find the expression for the two-point function in the

limit $d \to 0$ for both Neumann and Dirichlet boundary conditions. How does it differ from the case $d \to \infty$? You may use the regularized expression

$$\ln|\boldsymbol{x}-\boldsymbol{y}| = \lim_{a\to 0} \ln\left(a^2 + |\boldsymbol{x}-\boldsymbol{y}|^2\right)^{1/2}$$

3) Compute the two-point function of two vertex operators,

$$C_n(\boldsymbol{x},\boldsymbol{y}) = \langle \exp(in\varphi(\boldsymbol{x}))\exp(-in\varphi(\boldsymbol{y}))\rangle$$

(where n is an integer) for both types of boundary conditions. Again, the operators are defined to be at $\boldsymbol{x} = (d, x_1)$ and $\boldsymbol{y} = (d, y_1)$. Use your result to compute the scaling dimension of the vertex operators close ($d \to 0$) and far ($d \to \infty$) from the wall for both boundary conditions. Does the criterion for relevance and irrelevance of an operator change in the presence of the boundary? How does it depend on the boundary condition?

4) Consider now the case of an operator that inserts a *vortex* of charge m at some location $\boldsymbol{x}$. The correlator of two such operators is the ratio of two partition functions $Z_m[\boldsymbol{x},\boldsymbol{y}]/Z$, where

$$Z_m[\boldsymbol{x},\boldsymbol{y}] = \int \mathcal{D}\varphi \exp\left(-\int_\Sigma d^2x \frac{1}{8\pi}(\partial_\mu\varphi - A_\mu)^2\right)$$

where the vector potential A_μ has vanishing curl everywhere except at $\boldsymbol{x}$ and $\boldsymbol{y}$, where it is equal to $\pm 2\pi m$ (respectively). The space Σ is again the right half-plane. The denominator Z is the partition function without vortices (i.e., with $A_\mu = 0$). For simplicity, the points $\boldsymbol{x}$ and $\boldsymbol{y}$ are the same as before. Then, the vector potential can be taken to be

$$A_1(z_1,z_2) = 2\pi m\delta(z_1 - R)[\theta(x_2 - z_2) - \theta(y_2 - z_2)], \quad A_2(z_1,z_2) = 0$$

Compute the correlator of two such vortices away from and close to the boundary for both boundary conditions, and find the scaling dimensions in both cases. Compare your result for the scaling dimensions with those you found for the vertex operators.

22

Topological Field Theory

22.1 What is a topological field theory?

Let us now consider a special class of gauge theories known as topological field theories. These theories often (but not always) arise as the low-energy limit of more complex gauge theories. In general, we expect that at low energies, the phase of a gauge theory will be either confining or deconfined. While confining phases have (for really good reasons!) attracted much attention, deconfined phases are often regarded as trivial, in the sense that the general expectation is that their vacuum states will be unique, and the spectrum of low-lying states is either massive or massless.

Let us consider a gauge theory whose action on a manifold $\mathcal{M}$ with metric tensor $g_{\mu\nu}(x)$ is

$$S = \int_{\mathcal{M}} d^D x \sqrt{g}\, \mathcal{L}(g, A_\mu) \tag{22.1}$$

In section 3.10, we showed that, at the classical level, the energy-momentum tensor $T^{\mu\nu}(x)$ is the linear response of the action to an infinitesimal change of the local metric,

$$T^{\mu\nu}(x) \equiv \frac{\delta S}{\delta g_{\mu\nu}(x)} \tag{22.2}$$

That a theory is topological means that it depends only on the topology of the space in which it is defined, and consequently, it is independent of the local properties that depend on the metric (e.g., distances, angles). Therefore, at least at the classical level, the energy-momentum tensor of a topological field theory must vanish identically,

$$T^{\mu\nu} = 0 \tag{22.3}$$

In particular, if the theory is topological, the energy (or Hamiltonian) is also zero. Furthermore, if the theory is independent of the metric, it is invariant under arbitrary coordinate transformations. Thus, if the theory is a gauge theory, the expectation values of Wilson loops are independent of the size and shape of the loops. Whether a theory of this type can be consistently defined at the quantum level is a subtle problem, which we briefly touch on below.

It turns out that, due to the nonlocal nature of the observables of a gauge theory, the low-energy regime of a theory in its deconfined phase can have nontrivial global properties. In what follows, we say that a gauge theory is *topological* if all local excitations are massive (and in fact, we will send their mass gaps to infinity). The remaining Hilbert space of states is determined by global properties of the theory, including the topology of the manifold of their spacetimes. In several cases, the effective action of a topological field theory does not depend on the metric of the spacetime, at least at the classical level. In all cases, the observables are nonlocal objects, Wilson loops and their generalizations.

We focus on two topological field theories: the deconfined phases of discrete gauge theories (particularly the simplest case, $\mathbb{Z}_2$), which exist in any spacetime dimensions $D > 2$, and Chern-Simons gauge theories, which are well understood in 2+1 dimensions.

22.2 Deconfined phases of discrete gauge theories

Consider a simple problem: a $\mathbb{Z}_2$ gauge theory in $D > 2$ spacetime dimensions in its deconfined phase. In particular, we will work in the Hamiltonian formulation (cf. eq. (18.67)), in the extreme deconfined limit $\lambda \to \infty$. In this limit, the ground state must satisfy the condition that each plaquette operator is equal to 1 (i.e., no $\mathbb{Z}_2$ flux). These are the *flat* (i.e., no curvature or flux) configurations of the $\mathbb{Z}_2$ gauge field. In this limit, at the local level, this state is satisfied by the configuration that has $\sigma_j^3 = +1$ at every link.

However, this is not a gauge-invariant state. Furthermore, if the number of sites of bulk of the lattice is N, then there are 2^N gauge-equivalent states, obtained by the action of the generator of local gauge transformations $Q(\boldsymbol{r})$ of the $\mathbb{Z}_2$ gauge theory (eq. (18.54)). If the spatial manifold is open, we can fix the gauge locally (e.g., we fix the axial gauge $\sigma_1^3(\boldsymbol{r}) = 1$ on all links along the x_1 axis). Up to a definition of the gauge fields at the boundary of the manifold, this local gauge-fixing condition removes all redundancies.

However, this local gauge-fixing condition no longer specifies the state completely if the manifold is closed. For example, if the spatial manifold is a two-torus, the axial local gauge-fixing condition does not affect the so-called large gauge transformations, which wrap around a noncontractible loop of the two-torus, such as those shown in figure 22.1a. In fact, on a closed manifold, such as the two-torus, there is a multiplicity of quantum states specified by the solutions of the flat configuration condition. These states are exactly degenerate in the limit where the system has infinite extent.

These degenerate states are specified by the eigenvalues of the Wilson loops along the noncontractible loops of the torus. Indeed, in this ultra-deconfined limit, the eigenvalues of the Wilson loop operators W_{Γ_1} and W_{Γ_2} on the two noncontractible loops Γ_1 and Γ_2 are ± 1. Hence, on a two-torus, there are four linearly independent states. In contrast, a surface of genus 2, a pretzel, has four noncontractible loops (or one-cycles), shown in figure 22.1b. In general, on a two-dimensional closed surface with g handles (which in the continuum limit is equivalent to a Riemann surface of genus g), the deconfined phase of the $\mathbb{Z}_2$ gauge theory has exactly 2^{2g} degenerate ground states. In other words, in the low-energy limit of the deconfined phase, the Hilbert space of the theory has finite dimension 2^{2g}, which grows exponentially with the genus of the surface. This degenerate finite-dimensional Hilbert space has a purely topological origin. The one-cycles shown in figure 22.1, known as the canonical one-cycles, are linearly independent states of this Hilbert space and constitute a basis of the space.

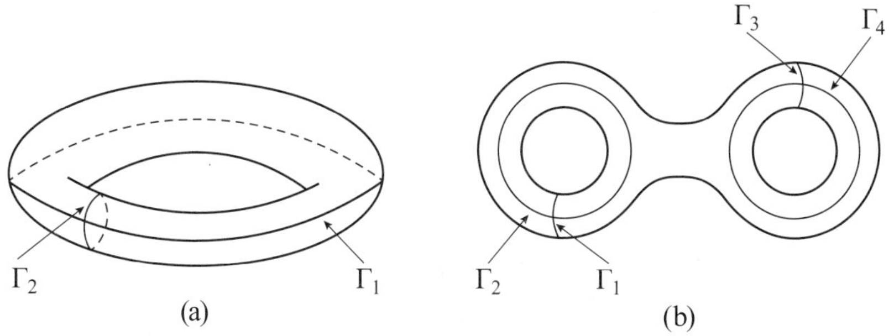

Figure 22.1 (a) A two-torus and its two noncontractible loops, Γ_1 and Γ_2; (b) the four noncontractible loops of the pretzel.

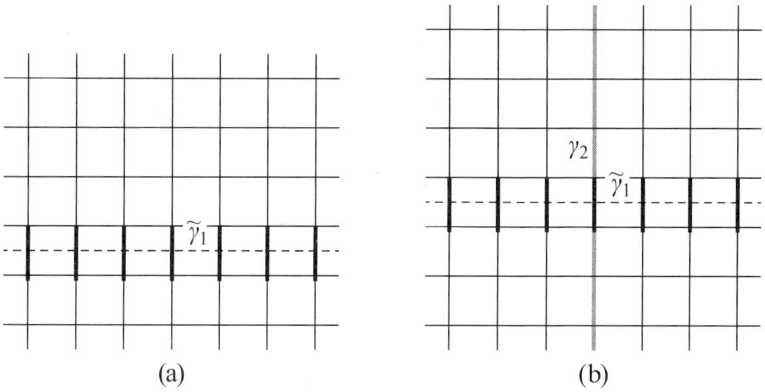

Figure 22.2 (a) A magnetic 't Hooft loop $\widetilde{W}[\tilde{\gamma}_1]$ on the noncontractible cycle $\tilde{\gamma}_1$ of the dual lattice; this operator is a product of σ^1 operators on all the links of the lattice pierced by the loop. (b) A Wilson loop $W[\gamma_2]$ on the noncontractible loop γ_2 and a 't Hooft loop on the noncontractible loop $\tilde{\gamma}_1$. These operators anticommute with each other.

The magnetic 't Hooft loops (defined in section 18.7) and the Wilson loops (both defined on noncontractible loops or cycles) form a closed algebra. Let $\widetilde{W}[\tilde{\gamma}_i]$, with $i = 1, 2, \ldots, g$, be the 't Hooft loops on the canonical one-cycles $\tilde{\gamma}_i$ of a surface of genus g, and let $W[\gamma_j]$, with $j = 1, 2, \ldots, g$, be on the canonical one-cycles γ_j, such as those shown in figure 22.2. Consider a Wilson loop operator, which is a product of σ^3 operators on the loop γ_j, and a 't Hooft loop operator, which is a product of σ^1 operators on the links of the lattice pierced by the loop $\tilde{\gamma}_i$. Since the one-cycles $\tilde{\gamma}_i$ and γ_j cross, they share a link of the lattice, as in figure 22.2b. Hence, these operators anticommute with each other. In general, we have the algebra

$$[W[\gamma_i], W[\gamma_j]] = 0, \quad [\widetilde{W}[\tilde{\gamma}_i], \widetilde{W}[\tilde{\gamma}_i]] = 0, \quad \forall\, i, j \tag{22.4}$$

but

$$\begin{aligned} \{W[\gamma_i], \widetilde{W}[\tilde{\gamma}_j]\} &= 0, \quad \text{if } \gamma_i \text{ and } \tilde{\gamma}_j \text{ intersect} \\ [W[\gamma_i], \widetilde{W}[\tilde{\gamma}_j]] &= 0, \quad \text{otherwise} \end{aligned} \tag{22.5}$$

Also,

$$W[\gamma_i]^2 = \widetilde{W}[\tilde{\gamma}_i]^2 = I \tag{22.6}$$

where I is the 2×2 identity matrix. It is easy to show that the algebra is the same if the loops are smoothly deformed in the bulk of the system. In other words, the algebra only knows whether the loops cross or not. We will see shortly that this property is related to the concept of braiding.

Since the Wilson loop operators commute with each other, and with the plaquette operator of the Hamiltonian, we can choose the basis of the topological space of states to be the eigenstates of the $2g$ Wilson loops, whose eigenvalues are ± 1. Moreover, since Wilson and 't Hooft loops on cycles belonging to different handles commute with each other, it will suffice to consider the algebra restricted to the two one-cycles of one handle. In that case, there are just two states for each Wilson loop, which we label by $|\pm, i\rangle$, where $i = 1, 2$ labels the two one-cycles of that handle. It is straightforward to see that the algebra of the Wilson and 't Hooft loops implies that

$$\begin{aligned} \widetilde{W}[\tilde{\gamma}_2]|+,1\rangle &= |-,1\rangle, \quad \text{and} \quad \widetilde{W}[\tilde{\gamma}_2]|-,1\rangle = |+,1\rangle \\ \widetilde{W}[\tilde{\gamma}_1]|+,2\rangle &= |-,2\rangle, \quad \text{and} \quad \widetilde{W}[\tilde{\gamma}_1]|-,1\rangle = |+,2\rangle \end{aligned} \tag{22.7}$$

With only minor changes, this analysis can be extended to other gauge theories with a discrete abelian gauge group, such as the $\mathbb{Z}_N$ gauge theories. The topological degeneracy is now N^{2g}.

In section 18.8, we discussed the concept of duality. In particular, we showed that the deconfined phase of the $\mathbb{Z}_2$ gauge theory is the dual of the symmetric phase of the (2+1)-dimensional quantum Ising model. The existence of this finite-dimensional topological Hilbert space in the gauge theory seems to contradict the duality mapping. In fact, what we have actually proved is that duality is a relation between local operators of the two theories. The topologically inequivalent subspaces of the gauge theory have the same local content. So, the more precise statement is that duality is blind to the global topology, and the mapping holds for any topological subspace.

So far, we have discussed only the case of a theory in 2+1 dimensions. Let us now consider the role of topology in higher dimensions. We will focus on the extreme deconfined limit of the $\mathbb{Z}_2$ gauge theory in 3+1 dimensions with the spatial topology of a three-torus, T^3, and the spacetime manifold is $T^3 \times \mathbb{R}$, where $\mathbb{R}$ is time. A three-torus has three noncontractible one-cycles. Once again, the flat configurations can be labeled by the eigenstates of the Wilson loops on the noncontractible one-cycles. Hence, the topological degeneracy is now 8. For a spatial manifold of genus g, the degeneracy is 2^{3g}.

Much as in the (2+1)-dimensional case, in 3+1 dimensions, the 't Hooft operators play an important role. The difference is that while in 2+1 dimensions, the 't Hooft loops are strings (essentially, Dirac strings of π flux), in 3+1 dimensions, the 't Hooft operators are defined on surfaces, which on T^3 are two-cycles. There are three inequivalent two-cycles that wrap around the three-torus. Now, a Wilson loop on a γ_1 one-cycle will anticommute with the 't Hooft operator of a two-cycle on a surface Σ_{23} pierced by the Wilson loop. Hence, the algebra of eqs. (22.4)–(22.6) holds here too, and the same result follows.

However, this analysis has a caveat. In 2+1 dimensions, a finite but small amount of the kinetic energy term of the gauge theory Hamiltonian, eq. (18.67), does not affect in an essential way the behavior of Wilson and 't Hooft loops on one-cycles of the torus. But things are different in the deconfined phase in 3+1 dimensions. While the behavior of the Wilson loop is similar, the behavior of the 't Hooft loop is very different. Indeed, even if the kinetic energy term is parametrically very small, a closed 't Hooft loop obeys an area law and, in principle, a 't Hooft operator of a two-cycle will create a state with infinite energy. However, the area law of the 't Hooft loop turns into a perimeter law once the gauge theory is coupled to a dynamical $\mathbb{Z}_2$ matter field, even one that is very heavy. In that case, the analysis is the same.

We will see that the algebra of the operators in eqs. (22.4)–(22.6) is intimately related to the concept of *braiding*, which we discuss next.

22.3 Chern-Simons gauge theories

A particularly important example of a topological field theory is Chern-Simons gauge theory. The main interest in Chern-Simons theory is that it is a topological field theory. In this context, the expectation values of the Wilson loop operators are given in terms of topological invariants associated with the theory of knots, such as their linking numbers and generating functions (known as the Jones polynomial). As we will see, the states created by Wilson loops of Chern-Simons theory are anyons, states that obey fractional statistics. Chern-Simons theories are the low-energy effective field theories of topological phases, such as the topological fluids of the fractional quantum Hall effects.

Chern-Simons gauge theory can be defined on spacetimes of odd dimension D. The simplest case is in $2+1$ dimensions with an abelian gauge group $U(1)$, whose action on a spacetime manifold $\mathcal{M}$ is

$$S=\frac{k}{4\pi}\int^{\mathcal{M}} d^3x\, \epsilon_{\mu\nu\lambda}A^{\mu}\partial^{\nu}A^{\lambda}\equiv\frac{k}{4\pi}\int_{\mathcal{M}} AdA \tag{22.8}$$

where, in the last equality, we introduced the notation of forms.

On a closed manifold $\mathcal{M}$, this action is invariant under both local and large gauge transformations (which wrap around the closed manifold), provided the parameter k, known as the *level*, is an integer. The Chern-Simons action is also odd under time reversal $\mathcal{T}$, $x_0\to -x_0$, and parity $\mathcal{P}$, defined as $x_1\to -x_1$ and $x_2\to x_2$, but it is invariant under $\mathcal{PT}$. The Chern-Simons action is the theory with the smallest number of derivatives that satisfies all these symmetries. For a general non-abelian gauge group G, the Chern-Simons action becomes

$$S=\frac{k}{4\pi}\int_{\mathcal{M}} d^3x\, \mathrm{tr}\Big(AdA+\frac{2}{3}A\wedge A\wedge A\Big) \tag{22.9}$$

Here, the cubic term is shorthand for

$$\mathrm{tr}\Big(A\wedge A\wedge A\Big)\equiv \mathrm{tr}\Big(\epsilon_{\mu\nu\lambda}A^{\mu}A^{\nu}A^{\lambda}\Big) \tag{22.10}$$

for a gauge field A^{μ} that takes values on the algebra of the gauge group G.

A closely related (abelian) gauge theory is the so-called *BF* theory (Horowitz, 1989), which, in a general spacetime dimension D (even or odd), is a theory of a vector field A^{μ} (a one-form) and an antisymmetric tensor field B with $D-1$ Lorentz indices (a $D-1$ form), known as a Kalb-Ramond field. Its action is

$$S=\frac{k}{2\pi}\int_{\mathcal{M}} d^Dx\, \epsilon_{\mu\nu\lambda\ldots}B^{\lambda\cdots}\partial^{\mu}A^{\nu} \tag{22.11}$$

where, once again, k is an integer. We will see shortly that this theory has the same content as the topological sector of a discrete gauge theory, which we discussed in section 22.2.

Chern-Simons theory was studied by mathematicians in the context of the classification of knots and representations of the Braid group. This theory first entered in physics as the effective action of a theory of N Dirac fields with mass m in 2+1 dimensions, coupled to a background gauge field, and the *parity anomaly* of the fermion determinant (Deser et al., 1982a; Redlich, 1984a). Up to subtleties related to regularization, the low-energy effective action of the gauge field is found to be (in the abelian case)

$$S_{\text{eff}}[A^\mu] = -i\text{tr}\ln(i\not{D}[A]+m) = \int d^3x \frac{N}{8\pi}\text{sign}(m) A dA + O(1/m) \tag{22.12}$$

and similar expression for the non-abelian case. The fact that, superficially, the level of this effective action is $k=\frac{N}{2}\text{sign}(m)$ requires special care in the regularization of the fermion determinant. We discussed this problem in section 20.7. This theory has had great success in explaining the phenomenon of statistical transmutation, by which bosons turn into fermions through a process of flux attachment. This mechanism is the basis of the modern theory of anyons, particles with fractional statistics (Wilczek, 1982).

22.4 Quantization of abelian Chern-Simons gauge theory

Here we focus on the simpler case of canonical quantization of the abelian Chern-Simons theory in 2+1 dimensions (Dunne et al., 1989; Elitzur et al., 1989; Witten, 1989). Consider this theory coupled to a set of conserved currents j_μ. Its Lagrangian is

$$\mathcal{L} = \frac{k}{4\pi}\epsilon_{\mu\nu\lambda}A^\mu\partial^\nu A^\lambda - j^\mu A_\mu \tag{22.13}$$

which, in Cartesian components, becomes

$$\mathcal{L} = A_0\left(\frac{k}{4\pi}\epsilon_{ij}\partial_i A_j - j_0\right) + \frac{k}{4\pi}\epsilon_{ij}A^i\partial^0 A^j + \boldsymbol{j}\cdot\boldsymbol{A} \tag{22.14}$$

As usual, the A_0 component plays the role of a Lagrange multiplier field that enforces a constraint. While in Maxwell's theory the constraint is Gauss's law, $\boldsymbol{\nabla}\cdot\boldsymbol{E}=j_0$, in Chern-Simons theory, the constraint implies that

$$\frac{k}{2\pi}B = j_0 \tag{22.15}$$

where $B=\epsilon_{ij}\partial_i A_j$. In other words, this constraint implies a condition that flux and charge must be glued (or attached) to each other. We will see that this is closely related to fractional statistics.

However, the equation of motion of the gauge field is

$$\frac{k}{2\pi}F^*_\mu = j_\mu \tag{22.16}$$

Therefore, in the absence of external sources, $j_\mu=0$, the solutions of the equations of motion are the flat connections, $F_{\mu\nu}=0$, and hence are pure gauge transformations.

At the classical level, the second term on the right-hand side of eq. (22.14) implies that the spatial components of the gauge field, A_i, form canonical pairs. Thus at the quantum level, as operators, they satisfy the equal-time commutation relations

$$\left[A_1(\boldsymbol{x}), A_2(\boldsymbol{y})\right] = i\frac{2\pi}{k}\delta(\boldsymbol{x}-\boldsymbol{y}) \tag{22.17}$$

Finally, the third term of eq. (22.14) implies that the Hamiltonian is

$$H = -\int d^2x\, \boldsymbol{j}(\boldsymbol{x}) \cdot \boldsymbol{A}(\boldsymbol{x}) \tag{22.18}$$

Hence, in the absence of sources ($\boldsymbol{j}=0$), the Hamiltonian vanishes: $H=0$.

The Chern-Simons action is locally gauge invariant, up to boundary terms. To see this, let us perform a gauge transformation, $A_\mu \to A_\mu + \partial_\mu \Phi$, where $\Phi(x)$ is a smooth, twice differentiable function. Then,

$$\begin{aligned} S[A^\mu + \partial^\mu \Phi] &= \int_{\mathcal{M}} (A^\mu + \partial^\mu \Phi)\epsilon_{\mu\nu\lambda}\partial^\nu(A^\lambda + \partial^\lambda \Phi) \\ &= \int_{\mathcal{M}} d^3x\, \epsilon_{\mu\nu\lambda} A^\mu \partial^\nu A^\lambda + \int_{\mathcal{M}} d^3x\, \epsilon_{\mu\nu\lambda}\partial^\mu \Phi \partial^\nu A^\lambda \end{aligned} \tag{22.19}$$

Therefore, the change is

$$\begin{aligned} S[A^\mu + \partial^\mu \Phi] - S[A^\mu] &= \int_{\mathcal{M}} d^3x\, \partial^\mu \Phi F^*_\mu \\ &= \int_{\mathcal{M}} d^3x\, \partial^\mu(\Phi F^*_\mu) - \int_{\mathcal{M}} d^3x\, \Phi \partial^\mu F^*_\mu \end{aligned} \tag{22.20}$$

where $F^*_\mu = \epsilon^{\mu\nu\lambda}\partial_\nu A_\lambda$ is the dual field strength. However, in the absence of magnetic monopoles, this field satisfies the Bianchi identity, $\partial^\mu F^* mu = \partial^\mu(\epsilon_{\mu\nu\lambda}\partial^\nu A^\lambda) = 0$. Therefore, using Gauss's law, we find that the change of the action is a total derivative and integrates to the boundary

$$\delta S = \int_{\mathcal{M}} d^3x\, \partial^\mu(\Phi F^*_\mu) = \int_\Sigma dS_\mu \Phi F^*_\mu \tag{22.21}$$

where $\Sigma = \partial\mathcal{M}$ is the boundary of $\mathcal{M}$. In particular, if Φ is a nonzero constant function on $\mathcal{M}$, then the change of the action under such a gauge transformation is

$$\delta S = \Phi \times \text{flux}(\Sigma) \tag{22.22}$$

Hence, the action is not invariant if the manifold has a boundary, and the theory must be supplemented with additional degrees of freedom at the boundary.

Indeed, the flat connections (i.e., the solution of the equations of motion, $F_{\mu\nu}=0$) are pure gauge transformations, $A_\mu = \partial_\mu \varphi$, and have an action that integrates to the boundary. Let $\mathcal{M} = D \times \mathbb{R}$, where D is a disk in space, and $\mathbb{R}$ is time. The boundary manifold is $\Sigma = S^1 \times \mathbb{R}$, where S^1 is a circle. Thus, in this case, the boundary manifold Σ is isotropic to a cylinder. The action of the flat configurations reduces to

$$S = \int_{S^1 \times \mathbb{R}} d^2x \frac{k}{2\pi} \partial_0\varphi\partial_1\varphi \tag{22.23}$$

which implies that the dynamics on the boundary is that of a scalar field on a circle S^1, and obeys periodic boundary conditions.

Although classically the theory does not depend on the metric, it is invariant under arbitrary transformations of the coordinates. However, any gauge-fixing condition will automatically break this large symmetry. For instance, we can specify a gauge condition at the boundary in the form of a boundary term $\mathcal{L}_{\text{gauge fixing}} = A_1^2$. In this case, the boundary action of the field φ becomes

$$S[\varphi] = \int_{S^1 \times \mathbb{R}} d^2x \frac{k}{2\pi} \left[\partial_0\varphi\partial_1\varphi - (\partial_1\varphi^2) \right] \tag{22.24}$$

The solutions of the equations of motion of this compactified scalar field have the form $\varphi(x_1 \mp x_0)$ (where the sign is the sign of k), and are right- (left-) moving chiral fields, depending on the sign of k. This boundary theory is not topological, but it is conformally invariant.

A similar result is found in non-abelian Chern-Simons gauge theory. In the case of the $SU(N)_k$ Chern-Simons theory on a manifold $D \times \mathbb{R}$, where D is a disk whose boundary is Γ and $\mathbb{R}$ is time, the action is

$$S_{\text{CS}}[A] = \int_{D \times \mathbb{R}} d^3x \left[\frac{k}{8\pi} \text{tr} \left(\epsilon_{\mu\nu\lambda} A^\mu \partial^\nu A^\lambda + \frac{2}{3} \epsilon^{\mu\nu\lambda} A_\mu A_\nu A_\lambda \right) \right] \tag{22.25}$$

This theory integrates to the boundary, $\Gamma \times \mathbb{R}$, where it becomes the chiral (right-moving) $SU(N)_k$ Wess-Zumino-Witten (WZW) model (at level k) at its IR fixed point, $\lambda_c^2 = 4\pi/k$:

$$S_{\text{WZW}}[g] = \frac{1}{4\lambda_c^2} \int_{\Gamma \times \mathbb{R}} d^2x \, \text{tr} \left(\partial_\mu g \partial^\mu g^{-1} \right) + \frac{k}{24\pi} \int_B \epsilon^{\mu\nu\lambda} \text{tr} \left(g^{-1}\partial_\mu g \, g^{-1}\partial_\nu g \, g^{-1}\partial_\lambda g \right) \tag{22.26}$$

Here, $g \in SU(N)$ parametrizes the flat configurations of the Chern-Simons gauge theory. Therefore, the boundary theory is a nontrivial CFT, the chiral WZW CFT (Witten, 1989).

22.5 Vacuum degeneracy on a torus

Let us construct the quantum version of this theory on a manifold $\mathcal{M} = T^2 \times \mathbb{R}$, where T^2 is a spatial torus of linear size L_1 and L_2. Since this manifold does not have boundaries, the flat connections, $\epsilon_{ij}\partial_i A_j = 0$, do not reduce to local gauge transformations of the form $A_i = \partial_i \Phi$. Indeed, the holonomies of the torus T^2 (i.e., the Wilson loops on the two noncontractible cycles of the torus Γ_1 and Γ_2 shown in figure 22.1) are gauge-invariant observables:

$$\int_0^{L_1} dx_1 A_1 \equiv \bar{a}_1, \qquad \int_0^{L_2} dx_1 A_2 \equiv \bar{a}_2 \tag{22.27}$$

where $\bar{a}_1$ and $\bar{a}_2$ are time dependent. Thus, the flat connections now are

$$A_1 = \partial_1 \Phi + \frac{\bar{a}_1}{L_1}, \qquad A_2 = \partial_2 \Phi + \frac{\bar{a}_2}{L_2} \tag{22.28}$$

whose action is

$$S = \frac{k}{4\pi} \int dx_0 \epsilon_{ij} \bar{a}_i \partial_0 \bar{a}_j \tag{22.29}$$

Therefore, the global degrees of freedom $\bar{a}_1$ and $\bar{a}_2$ at the quantum level become operators that satisfy the commutation relations

$$[\bar{a}_1, \bar{a}_2] = i\frac{2\pi}{k} \tag{22.30}$$

We find that the flat connections are described by the quantum mechanics of $\bar{a}_1$ and $\bar{a}_2$. A representation of this algebra is

$$\bar{a}_2 \equiv -i\frac{2\pi}{k}\frac{\partial}{\partial \bar{a}_1} \tag{22.31}$$

Furthermore, the Wilson loops on the two cycles become

$$W[\Gamma_1] = \exp\left(i \int_0^{L_1} A_1 \right) \equiv e^{i\bar{a}_1}, \quad W[\Gamma_2] = \exp\left(i \int_0^{L_2} A_2 \right) \equiv e^{i\bar{a}_2} \tag{22.32}$$

and satisfy the algebra

$$W[\Gamma_1]W[\Gamma_2] = \exp(-i2\pi/k)\, W[\Gamma_2]W[\Gamma_1] \tag{22.33}$$

Under large gauge transformations,

$$\bar{a}_1 \to \bar{a}_1 + 2\pi, \qquad \bar{a}_2 \to \bar{a}_2 + 2\pi \tag{22.34}$$

Therefore, invariance under large gauge transformations on the torus implies that $\bar{a}_1$ and $\bar{a}_2$ define a two-torus target space.

Let us define the unitary operators

$$U_1 = \exp(ik\bar{a}_2), \quad U_2 = \exp(-ik\bar{a}_1) \tag{22.35}$$

which satisfy the algebra

$$U_1 U_2 = \exp(i2\pi k) U_2 U_1 \tag{22.36}$$

The unitary transformations U_1 and U_2 act as shift operators on $\bar{a}_1$ and $\bar{a}_2$ by 2π, and hence generate the large-gauge transformations. Moreover, the unitary operators U_1 and U_2 leave the Wilson loop operators on noncontractible cycles invariant:

$$U_1^{-1} W[\Gamma_1] U_1 = W[\Gamma_1], \quad U_2^{-1} W[\Gamma_2] U_2 = W[\Gamma_2] \tag{22.37}$$

Let $|0\rangle$ be the eigenstate of $W[\Gamma_1]$ with eigenvalue 1, $W[\Gamma_1]|0\rangle = |0\rangle$. The state $W[\Gamma_2]|0\rangle$ is also an eigenstate of $W[\Gamma_1]$ with eigenvalue $\exp(-i2\pi/k)$, since

$$W[\Gamma_1]W[\Gamma_2]|0\rangle = e^{i2\pi/k}W[\Gamma_2]W[\Gamma_1]|0\rangle = e^{-i2\pi/k}W[\gamma_2]|0\rangle \tag{22.38}$$

More generally, since

$$W[\Gamma_1]W^p[\Gamma_2]|0\rangle = e^{-i2\pi p/k}W^p[\Gamma_2]|0\rangle \tag{22.39}$$

we find that, provided $k \in \mathbb{Z}$, there are k linearly independent vacuum states $|p\rangle = W^p[\Gamma_2]|0\rangle$, for the $U(1)$ Chern-Simons gauge theory at level k. This theory is denoted as the $U(1)_k$ Chern-Simons theory. Therefore the finite-dimensional topological space on a two-torus is k-dimensional. It is trivial to show that, on a surface of genus g, the degeneracy is k^g.

We see that in the abelian $U(1)_k$ Chern-Simons theory, the Wilson loops must carry k possible values of the unit charge. This property generalizes to the non-abelian theories, where it is technically more subtle. Here we only state some important results. For example, if the gauge group is $SU(2)$, we would expect that the Wilson loops will carry the representation labels of the group $SU(2)$, and, as such, they will be labeled by (j, m), where $j = 0, \frac{1}{2}, 1, \ldots$, and the $2j+1$ values of m satisfy $|m| \le j$. However, it turns out that $SU(2)_k$ Chern-Simons theory has fewer states, and that the values of j are restricted to the range $j = 0, \frac{1}{2}, \ldots, \frac{k}{2}$.

22.6 Fractional statistics

Another aspect of the topological nature of Chern-Simons theory is the behavior of expectation values of products of Wilson loop operators. Let us compute the expectation value of a product of two Wilson loop operators on two positively oriented closed contours γ_1 and γ_2. We will do this computation in the abelian Chern-Simons theory $U(1)_k$ in (2+1)-dimensional Euclidean space. Note that the Euclidean Chern-Simons action is purely imaginary, since the action is first order in derivatives. The expectation value to be computed is

$$W[\gamma_1 \cup \gamma_2] = \left\langle \exp\left(i \oint_{\gamma_1 \cup \gamma_2} dx_\mu A_\mu\right)\right\rangle_{\text{CS}} \tag{22.40}$$

We will see that the result depends on whether the loops γ_1 and γ_2 are linked or unlinked, as in the cases shown in figure 22.3.

This calculation is simpler than the one we did for Maxwell's theory in section 9.7. As in Maxwell's case, the expectation value of a Wilson loop on a contour (or union of contours, as in the present case) γ can be written as

$$\left\langle \exp\left(i \oint_{\gamma} dx_\mu A_\mu\right)\right\rangle_{\text{CS}} = \left\langle \exp\left(i \int d^3x J_\mu A_\mu\right)\right\rangle_{\text{CS}} \tag{22.41}$$

where the current J_μ is

$$J_\mu(x) = \delta(x_\mu - z_\mu(t))\frac{dz_\mu}{dt} \tag{22.42}$$

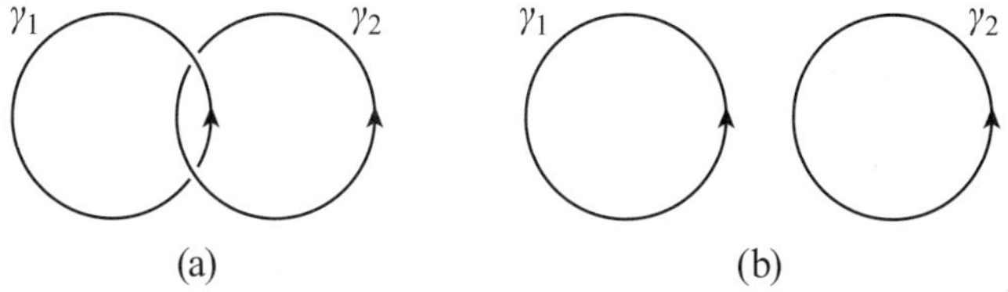

Figure 22.3 (a) Two linked Wilson loops forming a knot; (b) two unlinked Wilson loops.

Here $z_\mu(t)$ is a parametrization of the contour γ. Therefore, the expectation value of the Wilson loop is (Witten, 1989)

$$\left\langle \exp\left(i \oint_\gamma dx_\mu A_\mu\right)\right\rangle_{\text{CS}} \equiv \exp(iI[\gamma]_{\text{CS}})$$
$$= \exp\left(-\frac{i}{2}\int d^3x \int d^3y J_\mu(x) G_{\mu\nu}(x-y) J_\nu(y)\right) \tag{22.43}$$

where $G_{\mu\nu}(x-y) = \langle A_\mu(x) A_\nu(y)\rangle_{\text{CS}}$ is the propagator of the Chern-Simons gauge field. Since the loops are closed, the current J_μ is conserved, and $\partial_\mu J_\mu = 0$, and the effective action $I[\gamma]_{\text{CS}}$ of the loop γ is gauge invariant.

The Euclidean propagator of Chern-Simons gauge theory (in the Feynman gauge) is

$$G_{\mu\nu}(x-y) = \frac{2\pi}{k} G_0(x-y)\epsilon_{\mu\nu\lambda}\partial_\lambda \delta(x-y) \tag{22.44}$$

where $G_0(x-y)$ is the propagator of the massless Euclidean scalar field, which satisfies

$$-\partial^2 G_0(x-y) = \delta^3(x-y) \tag{22.45}$$

As usual, we can write

$$G_0(x-y) = \left\langle x \left| \frac{1}{-\partial^2} \right| y \right\rangle \tag{22.46}$$

Using these results, we find the following expression for the effective action

$$I[\gamma]_{\text{CS}} = \frac{\pi}{k}\int d^3x \int d^3y J_\mu(x) J_\nu(y) G_0(x-y)\epsilon_{\mu\nu\lambda}\partial_\lambda \delta(x-y)$$
$$= \frac{\pi}{k}\oint_\gamma dx_\mu \oint_\gamma dy_\nu \epsilon_{\mu\nu\lambda}\partial_\lambda G_0(x-y) \tag{22.47}$$

Again, since the current J_μ is conserved, it can be written as the curl of a vector field, B_μ:

$$J_\mu = \epsilon_{\mu\nu\lambda}\partial_\nu B_\lambda \tag{22.48}$$

In the Lorentz gauge, $\partial_\mu B_\mu = 0$, we can write

$$B_\mu = \epsilon_{\mu\nu\lambda}\partial_\nu \phi_\lambda \tag{22.49}$$

Hence,

$$J_\mu = -\partial^2 \phi_\mu \tag{22.50}$$

where

$$\phi_\mu(x) = \int d^3y\, G_0(x-y) J_\mu(y) \tag{22.51}$$

On substituting this result into the expression for B_μ, we find

$$B_\mu = \int d^3y \epsilon_{\mu\nu\lambda} \partial_\nu G_0(x-y) J_\lambda(y) = \oint_\gamma \epsilon_{\mu\nu\lambda} \partial_\nu G_0(x-y) dy_\lambda \tag{22.52}$$

Therefore, the effective action $I[\gamma]_{\text{CS}}$ becomes

$$I[\gamma]_{\text{CS}} = \frac{\pi}{k} \oint_\gamma dx_\mu \oint_\gamma dy_\nu \epsilon_{\mu\nu\lambda} \partial_\lambda G_0(x-y) = \frac{\pi}{k} \oint_\gamma dx_\mu B_\mu(x) \tag{22.53}$$

Let Σ be an oriented open surface of the Euclidean three-dimensional space whose boundary is the oriented loop (or union of loops) γ. Then, using Stokes' theorem, we write the last integral in eq. (22.53) as

$$I[\gamma]_{\text{CS}} = \frac{\pi}{k} \int_\Sigma dS_\mu \epsilon_{\mu\nu\lambda} \partial_\nu B_\lambda = \frac{\pi}{k} \int_\Sigma dS_\mu J_\mu \tag{22.54}$$

The integral in the last part of this equation is the flux of the current J_μ through the surface Σ. Therefore, this integral counts the number of times n_γ the Wilson loop on γ pierces the surface Σ (whose boundary is γ), and therefore it is an integer: $n_\gamma \in \mathbb{Z}$. Let us call this integer the *linking number* (or Gauss invariant) of the configuration of loops. In other words, the expectation value of the Wilson loop operator is

$$W[\gamma]_{\text{CS}} = e^{i\pi n_\gamma / k} \tag{22.55}$$

The linking number is a *topological invariant*, since, being an integer, its value cannot be changed by smooth deformations of the loops, provided they are not allowed to cross.

We now show that this property of Wilson loops in Chern-Simons gauge theory leads to the concept of fractional statistics. Consider a scalar matter field that is massive and charged under the Chern-Simons gauge field. The excitations of this matter field are particles that couple minimally to the gauge field. Here we are interested in the case in which these particles are very heavy. In that limit, we can focus on states that have a few of these particles, which will be in their nonrelativistic regime.

Consider, for example, a state with two particles, which in the remote past (at time $t = -T \to -\infty$) were located at two points A and B. This initial state will evolve to a final state at time $t = T \to \infty$, in which the particles either go back to their initial locations (the direct process), or to another one in which they exchange places, $A \leftrightarrow B$. At intermediate times, the particles follow smooth worldlines. These two processes, direct and exchange, are shown in figure 22.4. There we see that the direct process is equivalent to a history with two unlinked loops (the worldlines of the particles), whereas in the exchange process, the two loops form a link. It follows from the preceding discussion that the two amplitudes differ by the result of the computation of the Wilson loop expectation value for the loops γ_1 and γ_2. Let us call

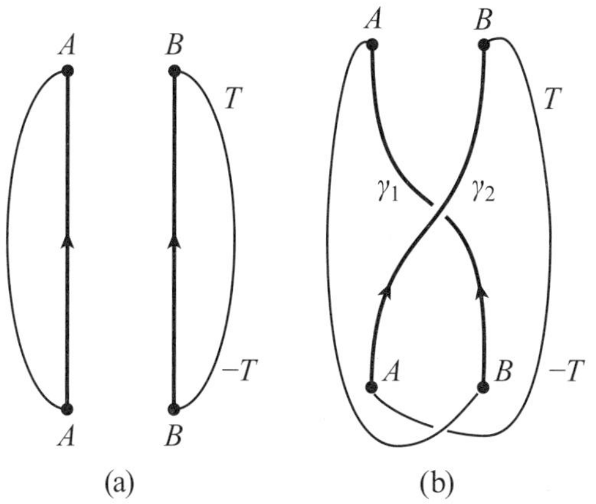

Figure 22.4 (a) A direct process is represented by two unlinked Wilson loops. (b) An exchange process is represented by a topological link of two Wilson loops forming a knot (or braid).

the first amplitude W_{direct} and the second W_{exchange}. The result is

$$W_{\text{exchange}} = W_{\text{direct}} e^{\pm i\pi/k} \tag{22.56}$$

where the sign depends on how the two worldlines wind around each other.

An equivalent interpretation of this result is that if $\Psi[A, B]$ is the wave function with the two particles at locations A and B, the wave function where their locations are exchanged is

$$\Psi[B, A] = e^{\pm i\pi/k} \Psi[A, B] \tag{22.57}$$

Clearly, for $k = 1$ the wave function is antisymmetric and the particles are fermions, while for $k \to \infty$ they are bosons. For other values of k, the particles obey *fractional statistics* and are called *anyons*. The phase factor $\phi = \pm\pi/k$ is called the *statistical phase*.

Notice that, while for fermions and bosons, the statistical phase $\varphi = 0, \pi$ is uniquely defined (mod 2π), for other values of k the statistical angle is specified up to a sign that specifies how the worldlines wind around each other. Indeed, mathematically, the exchange process shown in figure 22.4b is known as a *braid*. Processes in which the worldlines wind clockwise and counterclockwise are braids that are the inverse of each other. Braids can also be stacked on top of each other yielding multiples of the phase φ. In addition to stacking braids, Wilson loops can be fused: seen from some distance, a pair of particles will behave like a new particle with a well-defined behavior under braiding. This process of fusion is closely related to the concept of fusion of primary fields in CFT, discussed in chapter 21.

What we have just described is a mathematical structure called the *Braid* group. The example worked out using abelian Chern-Simons theory yields one-dimensional representations of the Braid group, with the phase φ being the label of the representations. For $U(1)_k$, there are k types of particles (anyons). These one-dimensional representations are abelian in that, in the general case of $U(1)_k$, acting on a one-dimensional representation p (defined mod k) with a one-dimensional representation q (also defined mod k) yields the one-dimensional representation $p + q$ (mod k). We denote the operation of *fusing* these representations (particles!) as $[q]_{\text{mod}\, k} \times [p]_{\text{mod}\, k} = [q + p]_{\text{mod}\, k}$. These representations are in one-to-one correspondence with the inequivalent charges of the Wilson loops and also with the vacuum degeneracy of the $U(1)_k$ Chern-Simons theory on a torus.

A richer structure arises in the case of the non-abelian Chern-Simons theory at level k (Witten, 1989), such as $SU(2)_k$. For example, for $SU(2)_1$, the theory has only two representations; both are one-dimensional and have statistical angles $\varphi = 0, \pi/2$.

However, for $SU(2)_k$, the content is more complex. In the case of $SU(2)_2$, the theory has (a) a trivial representation [0] (the identity, $(j, m) = (0, 0)$), (b) a (spinor) representation [1/2] ($(j, m) = (1/2, \pm 1/2)$), and (c) the representation [1] ($(j, m) = (1, m)$, with $m = 0, \pm 1$). These states will fuse, obeying the following rules: $[0] \times [0] = [0]$, $[0] \times [1/2] = [1/2]$, $[0] \times [1] = [1]$, $[1/2] \times [1/2] = [0] + [1]$, $[1/2] \times [1] = [1/2]$, and $[1] \times [1] = [0]$ (note the truncation of the fusion process).

Of particular interest is the case $[1/2] \times [1/2] = [0] + [1]$. In this case, we have two fusion channels, labeled [0] and [1]. The braiding operations now act on a two-dimensional Hilbert space and are represented by 2×2 matrices. This is an example of a non-abelian representation of the Braid group. These rather abstract concepts have found a physical manifestation in the physics of fractional quantum Hall fluids, whose excitations are vortices that carry fractional charge and anyon (Braid) fractional statistics.

Why this is interesting can be seen by considering a Chern-Simons gauge theory with four quasistatic Wilson loops. For instance, in the case of the $SU(2)_2$ Chern-Simons theory, the Wilson loops (heavy particles!) can be taken to carry the spinor representation, [1/2]. Let us call the four particles A, B, C, and D. We would expect that their quantum state would be completely determined by the coordinates of the particles. This, however, is not the case, since, if we fuse A with B, the result is either a state [0] or a state [1]. Thus, if the particles were prepared originally in some state, braiding (and fusion) will lead to a linear superposition of the two states. This braiding process defines a unitary matrix, a representation of the Braid group. The same is true with the other particles. However, it turns out that for four particles, there only two linearly independent states. This twofold degenerate Hilbert space of topological origin is called a *topological qubit*. Moreover, if we consider a system with N (even) number of such particles, the dimension of the topologically protected Hilbert space is $2^{\frac{N}{2}-1}$. Hence, for large N, the entropy per particle grows as $\frac{1}{2} \ln 2 = \ln \sqrt{2}$. Therefore, the qubit is not an "internal" degree of freedom of the particles but is a collective state of topological origin. Interestingly, there are physical systems, known as non-abelian fractional quantum Hall fluids, that embody this physics and are accessible to experiments! For these reasons, the non-abelian case has been proposed as a realization of a topological qubit (Kitaev, 2003; Das Sarma et al., 2008).

Exercises

22.1 Fractional charge and statistics in $U(1)_k$ Chern-Simons theory

In this exercise consider the $U(1)_k$ abelian Chern-Simons theory with gauge field A_μ coupled to two types of sources, j_μ and an external (background) electromagnetic field $A^{\rm em}_\mu$. The Lagrangian is now

$$\mathcal{L} = \frac{k}{4\pi} \epsilon_{\mu\nu\lambda} A^\mu \partial^\nu A^\lambda + j_\mu A^\mu + \frac{e}{2\pi} A^{\rm em}_\mu \epsilon^{\mu\nu\lambda} \partial_\nu A_\lambda$$

1) Show that the electromagnetic current of this theory is $J^{\rm em}_\mu = \frac{e}{2\pi} \epsilon_{\mu\nu\lambda} \partial^\nu A^\lambda$, and show that this current is locally conserved.

2) Consider now the case in which the current j_μ represents the worldlines of very heavy particles that are charged under the Chern-Simons gauge field. Integrate

out the Chern-Simons gauge field, and find the effective action for the matter currents j_μ and for the electromagnetic gauge field A_μ^{em}.

3) Use the effective action of the previous part to compute the electromagnetic current induced by the external electromagnetic field. Use the result to compute the conductivity tensor σ_{ij} $(i,j=1,2)$, such that $J_i^{\text{em}}=\sigma_{ij}E_j^{\text{em}}$, where $\boldsymbol{E}^{\text{em}}$ is the external electric field.

4) Use the effective action to show that the particles represented by the currents j_μ have *electromagnetic* charge $q=e/k$ and fractional statistics $\theta=\pi/k$.

22.2 *BF* theory

A simple generalization of abelian Chern-Simons theory is a theory with $U(1)\times U(1)$ local gauge invariance. This theory has two $U(1)$ gauge fields, A_μ and B_μ. Its Lagrangian density is

$$\mathcal{L}=\frac{k}{2\pi}\epsilon_{\mu\nu\lambda}A^\mu\partial^\nu B^\lambda$$

where k is an integer. At the classical level, this is a topological field theory.

1) Show that time-reversal acts to exchange the fields A_μ and B_μ.

2) Quantize this theory in the $A_0=B_0$ gauge. Find the canonical momenta for the fields $A_i(x)$ and $B_i(x)$ (with $i=1,2$), and determine the canonical commutation relations. What is the Gauss's law constraint for this theory?

3) Consider a theory of this type quantized on a spatial torus. Let γ_1 and γ_2 be the two noncontractible loops of the torus. Consider now the Wilson loop operators $W_A[\gamma_i]=\exp(i\oint_{\gamma_i}d\boldsymbol{x}\cdot\boldsymbol{A}(\boldsymbol{x}))$ and similarly for the field B_μ. Show that the Wilson loop operators commute with each other if they act on the same gauge field. Show that Wilson loops on noncontractible loops of the torus acting on different gauge fields do not commute with each other and obey the algebra

$$W^A[\gamma_i]W^B[\gamma_j]=\epsilon_{ij}e^{\pm 2\pi i/k}W^B[\gamma_j]W^A[\gamma_i]$$

4) Use this algebra to show that the *BF* theory at level k has k^2 degenerate vacua.

5) Consider now the *BF* theory coupled to two matter currents j_μ^I (with $I=1,2$), representing two types of particles, and two background electromagnetic fields A_μ^{em} and B_μ^{em}. The couplings have the same form as in exercise 22.1. Generalize the methods used in exercise 22.1 to compute the charges and statistics of the heavy particles. How many types of particles do you find?

References

Abrikosov, A. A. 1957. On the magnetic properties of superconductors of the second group. *Soviet Physics JETP*, **5**, 1174. [*Zh. Ekzp. i Teor. Fiz.* **32**, 1442 (1957).]

Abrikosov, A. A., Gorkov, L. P., and Dzyaloshinski, I. E. 1963. *Methods of Quantum Field Theory in Statistical Physics*. Englewood Cliffs, NJ: Prentice-Hall.

Adler, S. E. 1969. The axial-vector vertex in spinor electrodynamics. *Physical Review*, **177**, 2426.

Affleck, I. 1986a. Exact critical exponents for quantum spin chains, non-linear σ-models at $\theta = \pi$ and the quantum Hall effect. *Nuclear Physics B*, **265**, 409.

Affleck, I. 1986b. Universal term in the free energy at a critical point and the conformal anomaly. *Physical Review Letters*, **56**, 746.

Affleck, I. 1988. Critical behavior of $SU(N)$ quantum chains and topological non-linear sigma models. *Nuclear Physics B*, **305**, 582.

Affleck, I., and Haldane, F. D. M. 1987. Critical theory of quantum spin chains. *Physical Review B*, **36**, 5291.

Alvarez-Gaumé, L., Della Pietra, S., and Moore, G. 1985. Anomalies and odd dimensions. *Annals of Physics*, **163**, 288.

Amit, D. J. 1980. *Field Theory, the Renormalization Group and Critical Phenomena*. New York: McGraw-Hill.

Atiyah, M., and Singer, I. 1968. Index of elliptic operators I. *Annals of Mathematics*, **87**, 484.

Balian, R., Drouffe, J. M., and Itzykson, C. 1975. Gauge fields on a lattice. II. Gauge-invariant Ising model. *Physical Review D*, **11**, 2098.

Banks, T., Myerson, R., and Kogut, J. 1977. Phase transitions in abelian lattice gauge theories. *Nuclear Physics B*, **129**, 493.

Becchi, C., Rouet, A., and Stora, R. 1974. The abelian Higgs Kibble model, unitarity of the S-operator. *Physics Letters B*, **52**, 344.

Becchi, C., Rouet, A., and Stora, R. 1976. Renormalization of gauge theories. *Annals of Physics*, **98**, 287.

Belavin, A. A., Polyakov, A. M., and Zamolodchikov, A. B. 1984. Infinite conformal symmetry in two-dimensional quantum field theory. *Nuclear Physics B*, **241**, 333.

Bell, J. S., and Jackiw, R. 1969. The PCAC puzzle: $\pi^0 \to \gamma\gamma$ in the σ-model. *Nuovo Cimento A*, **60**, 47.

Berlin, T. H., and Kac, M. 1952. The spherical model of a ferromagnet. *Physical Review*, **86**, 821.

Berry, M. V. 1984. Quantal phase factors accompanying adiabatic changes. *Proceedings of the Royal Society of London A*, **392**, 45.

Blöte, H. W. J., Cardy, J. L., and Nightingale, M. P. 1986. Conformal invariance, the central charge, and universal finite-size amplitudes at criticality. *Physical Review Letters*, **56**, 742.

Bogoliubov, N. N., and Shirkov, D. V. 1959. *Introduction to the Theory of Quantized Fields*. New York: Interscience.

Bogomol'nyi, E. B. 1976. Stability of classical solutions. *Soviet Journal of Nuclear Physics*, **24**, 449. [*Yad. Fiz.* **24**, 861 (1976).]

Bollini, C. G., and Giambiagi, J. J. 1972a. Dimensional renormalization: The number of dimensions as a regularizing parameter. *Il Nuovo Cimento B*, **12**, 20.

Bollini, C. G., and Giambiagi, J. J. 1972b. Lowest order "divergent" graphs in ν-dimensional space. *Physics Letters B*, **40**, 566.

Boyanovsky, D., Dagotto, E., and Fradkin, E. 1987. Anomalous currents, induced charge and bound states on a domain wall of a semiconductor. *Nuclear Physics B*, **285**, 340.

Brézin, E., and Zinn-Justin, J. 1976a. Renormalization of the nonlinear σ model in $2+\epsilon$ dimensions: application to the Heisenberg ferromagnets. *Physical Review Letters*, **36**, 691.

Brézin, E., and Zinn-Justin, J. 1976b. Spontaneous breakdown of continuous symmetries near two dimensions. *Physical Review B*, **14**, 3110.

Brézin, E., Itzykson, C., Parisi, G., and Zuber, J. B. 1978. Planar diagrams. *Communications in Mathematical Physics*, **59**, 35.

Callan, C. G., and Harvey, J. A. 1985. Anomalies and fermion zero modes on strings and domain walls. *Nuclear Physics B*, **250**, 427.

Callan, C. G., Dashen, R. F., and Gross, D. J. 1976. The structure of the gauge theory vacuum. *Physics Letters B*, **63**, 334.

Callan, C. G., Dashen, R. F., and Gross, D. J. 1978. Towards a theory of strong interactions. *Physical Review D*, **17**, 2717.

Cardy, J. 1996. *Scaling and Renormalization in Statistical Physics*. Cambridge Lecture Notes in Physics. Cambridge: Cambridge University Press.

Cardy, J. L. 1982. Duality and the θ parameter in Abelian lattice models. *Nuclear Physics B*, **205**, 17–26.

Cardy, J. L., and Rabinovici, E. 1982. Phase structure of $\mathbb{Z}_p$ models in the presence of a θ parameter. *Nuclear Physics B*, **205**, 1–16.

Casher, A., Kogut, J., and Susskind, L. 1974. Vacuum polarization and the absence of free quarks. *Physical Review D*, **10**, 732.

Coleman, S. 1973. There are no Goldstone bosons in two dimensions. *Communications in Mathematical Physics*, **31**, 259.

Coleman, S. 1975. Quantum sine-Gordon equation as the massive Thirring model. *Physical Review D*, **11**, 2088.

Coleman, S. 1985. *Aspects of Symmetry*. Cambridge: Cambridge University Press.

Coleman, S., Jackiw, R., and Susskind, L. 1975. Charge shielding and quark confinement in the massive Schwinger model. *Annals of Physics*, **93**, 267.

D'Adda, A., Lüscher, M., and Di Vecchia, P. 1978. A $1/N$ expandable series of non-linear σ-models with instantons. *Nuclear Physics B*, **146**, 63.

Das Sarma, S., Freedman, M., Nayak, C., Simon, S. H., and Stern, A. 2008. Non-abelian anyons and topological quantum computation. *Reviews of Modern Physics*, **80**, 1083.

Deser, S., Jackiw, R., and Templeton, S. 1982a. Three-dimensional massive gauge theories. *Physical Review Letters*, **48**, 975.

Deser, S., Jackiw, R., and Templeton, S. 1982b. Topologically massive gauge theories. *Annals of Physics*, **140**, 372.

Di Francesco, P., Mathieu, P., and Sénéchal, D. 1997. *Conformal Field Theory*. Berlin: Springer-Verlag.

Dirac, P. A. M. 1928. The quantum theory of the electron. *Proceedings of the Royal Society of London A*, **117**, 610.

Dirac, P. A. M. 1931. Quantised singularities in the electromagnetic field. *Proceedings of the Royal Society of London*, **133**, 60.

Dirac, P. A. M. 1933. The Lagrangian in quantum mechanics. *Physikalische Zeitschrift der Sowjetunion*, **3**, 64.

Dirac, P. A. M. 1966. *Lectures on Quantum Field Theory*. New York: Academic Press.

Doniach, S., and Sondheimer, E. H. 1998. *Green's Functions for Solid State Physicists*. London: Imperial College Press.

Dunne, G. V., Jackiw, R., and Trugenberger, C. A. 1989. Chern-Simons theory in the Schrödinger representation. *Annals of Physics*, **194**, 197.

Eguchi, T., Gilkey, P. B., and Hanson, A. J. 1980. Gravitation, gauge theories and differential geometry. *Physics Reports*, **66**, 213.

Elitzur, S. 1975. Imposibility of spontaneously breaking local symmetries. *Physical Review D*, **12**, 3978.

Elitzur, S., Moore, G., Schwimmer, A., and Seiberg, N. 1989. Remarks on the canonical quantization of the Chern-Simons-Witten theory. *Nuclear Physics B*, **326**, 108.

Faddeev, L. D. 1976. Introduction to functional methods. In: *Methods of Field Theory*. Proceedings of the Les Houches Summer School, 1975, session XXVIII, R. Stora and J. Zinn-Justin, eds. Amsterdam: North-Holland.

Faddeev, L. D., and Popov, V. 1967. Feynman diagrams for the Yang-Mills field. *Physics Letters B*, **25**, 29.

Feynman, R. P. 1948. Space-time approach to non-relativistic quantum mechanics. *Reviews of Modern Physics*, **20**, 367.

Feynman, R. P. 1972. *Statistical Mechanics, A Set of Lectures*. Frontiers in Physics. Reading, MA: W. A. Benjamin.

Feynman, R. P. 2005. *A New Approach to Quantum Theory*. Singapore: World-Scientific Publishing. (This is Feynman's 1942 PhD thesis).

Feynman, R. P., and Hibbs, A. R. 1965. *Quantum Mechanics and Path Integrals*. New York: McGraw-Hill.

Fisher, M. E. 1967. The theory of equilibrium critical phenomena. *Reports on Progress in Physics*, **30**, 615.

Fradkin, E. 2013. *Field Theories of Condensed Matter Physics*, Second Edition. Cambridge: Cambridge University Press.

Fradkin, E., and Shenker, S. H. 1979. Phase diagrams of lattice gauge theories with Higgs fields. *Physical Review D*, **19**, 3682.

Fradkin, E., and Stone, M. 1988. Topological terms in one- and two-dimensional quantum Heisenberg antiferromagnets. *Physical Review B*, **38**, 7215(R).

Fradkin, E., and Susskind, L. 1978. Order and disorder in gauge systems and magnets. *Physical Review D*, **17**, 2637.

Friedan, D., Qiu, Z., and Shenker, S. 1984. Conformal invariance, unitarity, and critical exponents in two dimensions. *Physical Review Letters*, **52**, 1575.

Friedan, D. H. 1985. Nonlinear models in $2 + \epsilon$ dimensions. *Annals of Physics*, **163**, 318.

Fujikawa, K. 1979. Path-integral measure for gauge-invariant fermion theories. *Physical Review Letters*, **42**, 1195.

Gamboa Saraví, R. E., Muschietti, M. A., Schaposnik, F. A., and Solomin, J. E. 1984. Chiral symmetry and functional integral. *Annals of Physics*, **157**, 360.

Gell-Mann, M., and Low, F. 1951. Bound states in quantum field theory. *Physical Review*, **84**, 350.

Gell-Mann, M., and Low, F. 1954. Quantum electrodynamics at small distances. *Physical Review*, **95**, 1300.

Georgi, H., and Glashow, S. 1974. Unity of all elementary-particle forces. *Physical Review Letters*, **32**, 438.

Ginsparg, P. 1989. Applied conformal field theory. In: *Champs, Cordes et Phénoménes Critiques/Fields, Strings and Critical Phenomena*. Proceedings of the Les Houches Summer School 1988, Session XLIX, Brézin, E., and Zinn-Justin, J., eds. Amsterdam: Elsevier Science.

Goldenfeld, N. 1992. *Lectures on Phase Transitions and the Renormalization Group*. Reading, MA: Addison-Wesley.

Goldstone, J., and Wilczek, F. 1981. Fractional quantum numbers on solitons. *Physical Review Letters*, **47**, 986.

Gradshteyn, I. S., and Ryzhik, I. M. 2015. *Table of Integrals, Series, and Products*. Eighth ed. Amsterdam: Elsevier Science.

Griffiths, R. B. 1972. Rigorous results and theorems. In: *Phase Transitions and Critical Phenomena*, vol. 1. C. Domb and M. S. Green, eds. New York: Academic Press.

Gross, D. J., and Neveu, A. 1974. Dynamical symmetry breaking in asymptotically free field theories. *Physical Review D*, **10**, 3235.

Gubser, S. S., Klebanov, I. R., and Polyakov, A. M. 1998. Gauge theory correlators from non-critical string theory. *Physics Letters B*, **428**, 105.

Haldane, F. D. M. 1983. Continuum dynamics of the 1-D Heisenberg antiferromagnet: Identification with the $O(3)$ nonlinear sigma model. *Physics Letters A*, **93**, 464.

Hawking, S. W. 1977. Zeta function regularization of path integrals in curved spacetime. *Communications in Mathematical Physics*, **55**, 133.

Henneaux, M., and Teitelboim, C. 1992. *Quantization of Gauge Systems*. Princeton, NJ: Princeton University Press.

Hertz, J. A, Roudi, Y., and Sollich, P. 2016. Path integral methods for the dynamics of stochastic and disordered systems. *Journal of Physics A: Mathematical and Theoretical*, **50**, 033001.

Hikami, S. 1979. Renormalization group functions of $\mathbb{CP}^{N-1}$ non-linear σ-model and N-component scalar QED model. *Progress in Theoretical Physics*, **62**, 226.

Hohenberg, P. C. 1967. Existence of long-range order in one and two dimensions. *Physical Review*, **158**, 383.

Horowitz, G. T. 1989. Exactly soluble diffeomorphism invariant theories. *Communications in Mathematical Physics*, **125**, 417.

Itzykson, C., and Zuber, J. B. 1980. *Quantum Field Theory*. First ed. New York: McGraw-Hill.

Jackiw, R., and Rebbi, C. 1976. Solitons with fermion number 1/2. *Physical Review D*, **13**, 3398.

Jackiw, R., and Rossi, P. 1981. Zero modes of the vortex-fermion system. *Nuclear Physics B*, **190**, 681.

Jordan, P., and Wigner, E. P. 1928. Pauli's equivalence prohibition. *Zeitschrift für Physik*, **47**, 631.

Kadanoff, L. P. 1966. Scaling laws for Ising models near T_c. *Physics*, **2**, 263.

Kadanoff, L. P. 1969. Operator algebra and the determination of critical indices. *Physical Review Letters*, **23**, 1430.

Kadanoff, L. P. 1977. The application of renormalization group techniques to quarks and strings. *Reviews of Modern Physics*, **49**, 267.

Kadanoff, L. P., and Ceva, H. 1971. Determination of an operator algebra for the two-dimensional Ising model. *Physical Review B*, **3**, 3918.

Kadanoff, L. P., and Martin, P. C. 1961. Theory of many-particle systems. II. Superconductivity. *Physical Review*, **124**, 670.

Kaplan, D. B. 1992. A method for simulating chiral fermions on the lattice. *Physics Letters B*, **288**, 342.

Kitaev, A. Yu. 2003. Fault-tolerant quantum computation by anyons. *Annals of Physics*, **303**, 2. (arXiv:quant-ph/9707021).

Klauder, J. R., and Skagerstam, B. 1985. *Coherent States*. Singapore: World-Scientific.

Knizhnik, V. G., and Zamolodchikov, A. B. 1984. Current algebra and Wess-Zumino model in two dimensions. *Nuclear Physics B*, **247**, 83.

Kogut, J. 1979. An introduction to lattice gauge theory and spin systems. *Reviews of Modern Physics*, **51**, 659.

Kogut, J. 1983. The lattice gauge theory approach to quantum chromodynamics. *Reviews of Modern Physics*, **55**, 775.

Kogut, J., and Susskind, L. 1975a. Hamiltonian formulation of Wilson's lattice gauge theories. *Physical Review D*, **11**, 395.

Kogut, J., and Susskind, L. 1975b. How quark confinement solves the $\eta \to 3\pi$ problem. *Physical Review D*, **11**, 3594.

Kosterlitz, J. M. 1974. The critical properties of the two-dimensional *XY* model. *Journal of Physics C: Solid State Physics*, **7**, 1046.

Kosterlitz, J. M., and Thouless, D. J. 1973. Order, metastability and phase transitions in two-dimensional systems. *Journal of Physics C: Solid State Physics*, **6**, 1181.

Landau, L. D. 1937. On the theory of phase transitions. *Zh. Eksp. Teor. Fiz.*, **7**, 19.

Landau, L. D., and Lifshitz, E. M. 1959a. *Classical Mechanics*. Course of Theoretical Physics, vol. 1. Oxford: Pergamon Press.

Landau, L.D., and Lifshitz, E. M. 1959b. *Statistical Physics*. Course of Theoretical Physics, vol. 5. Oxford: Pergamon Press.

Lieb, E., Schultz, T., and Mattis, D. C. 1961. Two soluble models of an antiferromagnetic chain. *Annals of Physics (N.Y.)*, **16**, 407.

Lowenstein, J. H., and Swieca, J. A. 1971. Quantum electrodynamics in two dimensions. *Annals of Physics*, **68**, 172.

Ludwig, A. W. W., and Cardy, J. L. 1987. Perturbative evaluation of the conformal anomaly at new critical points with applications to random systems. *Nuclear Physics B*, **285**, 687.

Luther, A., and Emery, V. J. 1974. Backward scattering in the one-dimensional electron gas. *Physical Review Letters*, **33**, 589.

Ma, S-K. 1973. Introduction to the renormalization group. *Reviews of Modern Physics*, **45**, 589.

Maldacena, J. M. 1998. Large *N* limit of superconformal field theories and supergravity. *Advances in Theoretical and Mathematical Physics*, **2**, 231.

Mandelstam, S. 1975. Soliton operators for the quantized sine-Gordon equation. *Physical Review D*, **11**, 3026.

Martin, P. C. 1968. *Measurements and Correlation Functions*. New York: Gordon and Breach.

Martin, P. C., Siggia, E. D., and Rose, H. A. 1973. Statistical dynamics of classical systems. *Physical Review A*, **8**, 423.

Mattis, D. C., and Lieb, E. H. 1965. Exact solution of a many-fermion system and its associated boson field. *Journal of Mathematical Physics*, **6**, 304.

Mehta, M. L. 2004. *Random Matrices*. Third ed. Pure and Applied Mathematics, vol. 142. New York: Academic Press.

Mermin, N. D. 1979. The topological theory of defects in ordered media. *Reviews of Modern Physics*, **51**, 591.

Mermin, N. D., and Wagner, H. 1966. Absence of ferromagnetism or antiferromagnetism in one- or two-dimensional isotropic Heisenberg models. *Physical Review Letters*, **17**, 1133.

Migdal, A. A. 1975a. Phase transitions in gauge and spin-lattice systems. *Soviet Physics JETP*, **42**, 743. [*Zh. Ekzp. i Teor. Fiz.* **69**, 1457 (1975).]

Migdal, A. A. 1975b. Recursion equations in gauge field theories. *Soviet Physics JETP*, **42**, 413. [*Zh. Ekzp. i Teor. Fiz.* **69**, 810 (1975).]

Nambu, Y., and Jona-Lasinio, G. 1961. Dynamical model of elementary particles based on an analogy with superconductivity. I. *Physical Review*, **122**, 345.

Nash, C., and Sen, S. 1983. *Topology and Geometry for Physicists*. New York: Academic Press.

Nielsen, H. B., and Ninomiya, M. 1981. Absence of neutrinos on a lattice: (I). Proof by homotopy theory. *Nuclear Physics B*, **185**, 20.

Nielsen, H. B., and Ninomiya, M. 1983. The Adler-Bell-Jackiw anomaly and Weyl fermions in a crystal. *Physics Letters B*, **130**, 389.

Nielsen, H. B., and Olesen, P. 1973. Vortex-line models for dual strings. *Nuclear Physics B*, **61**, 45.

Onsager, L. 1944. Crystal statistics. I. A two-dimensional model with an order-disorder transition. *Physical Review*, **65**, 117.

Osterwalder, K, and Seiler, E. 1978. Gauge field theories on a lattice. *Annals of Physics*, **110**, 440.

Parisi, G. 1988. *Statistical Field Theory*. Reading, MA: Addison-Wesley.

Patashinskii, A. Z., and Pokrovskii, V. L. 1966. Behavior of ordered systems near the transition point. *Soviet Physics JETP*, **23**, 292. [*Zh. Ekzp. i Teor. Fiz.* **50**, 439 (1966).]

Pauli, W., and Villars, F. 1949. On the invariant regularization in relativistic quantum theory. *Reviews of Modern Physics*, **21**, 434.

Perelman, G. 2002. The entropy fomula for the Ricci flow and its geometric applications. *arXiv Mathematics e-prints*, math/0211159.

Perelomov, A. 1986. *Generalized Coherent States and Their Applications*. Berlin: Springer-Verlag.

Peskin, M. E., and Schroeder, D. V. 1995. *An Introduction to Quantum Field Theory*. Reading, MA: Perseus Books.

Pines, D., and Bohm, D. 1952. A collective description of electron interactions: II. Collective vs individual particle aspects of the interactions. *Physical Review*, **85**, 338.

Pines, D., and Nozières, P. 1966. *The Theory of Quantum Liquids,*. vol. I. New York: W. A. Benjamin.

Polchinski, J. 1988. Scale and conformal invariance in quantum field theory. *Nuclear Physics B*, **303**, 226–236.

Polchinski, J. 1998. *String Theory*. Cambridge: Cambridge University Press.

Polyakov, A. M. 1970. Conformal symmetry of critical fluctuations. *JETP Letters*, **12**, 381. [*ZhETF Pis. red.* **12**, 538 (1970)].

Polyakov, A. M. 1974. Non-Hamiltonian approach to quantum field theory. *Soviet Physics JETP*, **39**, 10. [*Zh. Eksp. Teor. Fiz.* **66**, 23 (1974)].

Polyakov, A. M. 1975a. Compact gauge fields and the infrared catastrophe. *Physics Letters B*, **59**, 82–84.

Polyakov, A. M. 1975b. Interaction of Goldstone particles in two dimensions. Applications to ferromagnets and massive Yang-Mills fields. *Physics Letters B*, **59**, 79.

Polyakov, A. M. 1977. Quark confinement and topology of gauge theories. *Nuclear Physics B*, **120**, 429.

Polyakov, A. M. 1987. *Gauge Fields and Strings*. London: Harwood Academic.

Polyakov, A. M., and Wiegmann, P. B. 1983. Theory of nonabelian Goldstone bosons and two dimensions. *Physics Letters B*, **131**, 121.

Prasad, M. K., and Sommerfield, C. M. 1975. Exact classical solution of the 't Hooft monopole and the Julia-Zee dyon. *Physical Review Letters*, **35**, 760.

Pruisken, A. M. M. 1985. Dilute instanton gas as the precursor of the integral quantum Hall effect. *Physical Review B*, **32**, 2636.

Qi, X.-L., Hughes, T. L., and Zhang, S.-C. 2008. Topological field theory of time-reversal invariant insulators. *Physical Review B*, **78**, 195424.

Read, N., and Newns, D. M. 1983. On the solution of the Coqblin-Schreiffer Hamiltonian by the large-*N* expansion technique. *Journal of Physics C: Solid State Physics*, **16**, 3273.

Redlich, A. N. 1984a. Gauge noninvariance and parity nonconservation of three-dimensional fermions. *Physical Review Letters*, **52**, 18.

Redlich, A. N. 1984b. Parity violation and gauge noninvariance of the effective gauge field action in three dimensions. *Physical Review D*, **29**, 2366.

Sachdev, S. 1999. *Quantum Phase Transitions*. Cambridge: Cambridge University Press.

Sachdev, S., and Read, N. 1991. Large-*N* expansion for frustrated and doped quantum antiferromagnets. *International Journal of Modern Physics B*, **5**, 219.

Schaposnik, F. A. 1978. Pseudoparticles and confinement in the two-dimensional abelian Higgs model. *Physical Review D*, **18**, 1183.

Schrieffer, J. R. 1964. *Theory of Superconductivity*. Reading, MA: Addison-Wesley.

Schulman, L. S. 1981. *Techniques and Applications of Path Integration*. New York: John Wiley & Sons.

Schultz, T. D., Mattis., D. C., and Lieb, E. H. 1964. Two-dimensional Ising model as a soluble problem of many fermions. *Reviews of Modern Physics*, **36**, 856.

Schwinger, J. 1959. Field theory commutators. *Physical Review Letters*, **3**, 296.

Schwinger, J. 1962. Gauge invariance and mass. *Physical Review*, **125**, 397.

Shapere, A., and Wilczek, F. 1989. Self-dual models with theta terms. *Nuclear Physics B*, **320**, 669.

Simmons-Duffin, D. 2017. The conformal bootstrap. In: *TASI 2015: New Frontiers in Fields and Strings*, J. Polchinski, P. Vieira, and O. DeWolfe, eds. Singapore: World Scientific.

Simon, B. 1983. Holonomy, the quantum adiabatic theorem, and Berry's phase. *Physical Review Letters*, **51**, 2167.

Stanley, H. E. 1968. Spherical model as the limit of infinite spin dimensionality. *Physical Review*, **176**, 718.

Stone, M., and Goldbart, P. 2009. *Mathematics for Physics: A Guided Tour for Graduate Students*. Cambridge: Cambridge University Press.

Su, W. P., Schrieffer, J. R., and Heeger, A. J. 1979. Solitons in polyacetylene. *Physical Review Letters*, **42**, 1698.

Susskind, L. 1995. The world as a hologram. *Journal of Mathematical Physics*, **36**, 6377.

't Hooft, G. 1974. A planar diagram theory for strong interactions. *Nuclear Physics B*, **72**, 461.

't Hooft, G. 1976a. Magnetic charge quantization and fractionally charged quarks. *Nuclear Physics B*, **105**, 538.

't Hooft, G. 1976b. Symmetry breaking through Bell-Jackiw anomalies. *Physical Review Letters*, **37**, 8.

't Hooft, G. 1978. On the phase transition towards permanent quark confinement. *Nuclear Physics B*, **138**, 1.

't Hooft, G. 1979. A property of electric and magnetic flux in non-abelian gauge theories. *Nuclear Physics B*, **153**, 141.

't Hooft, G. 1993. Dimensional reduction in quantum gravity. In *Conference on Highlights of Particle and Condensed Matter Physics (SALAMFEST), Trieste, Italy, March 8–12, 1993*, A. Ali, D. Amati, and J. Ellis, J., eds., vol. C930308. Singapore: World-Scientific.

't Hooft, G., and Veltman, M. 1972. Regularization and renormalization of gauge fields. *Nuclear Physics B*, **44**, 189.

Tyutin, I. V. 2008. Gauge invariance in field theory and statistical physics in operator formalism. *arXiv e-prints*. Preprint of P.N. Lebedev Physical Institute, no. 39, 1975.

Wegner, F. J. 1971. Duality in generalized Ising models and phase transitions without local order parameters. *Journal of Mathematical Physics*, **12**, 2259.

Weinberg, E. J. 1981. Index calculations for the fermion-vortex system. *Physical Review D*, **24**, 2669.

Weinberg, S. 2005. *The Quantum Theory of Fields*. First ed. Cambridge: Cambridge University Press.

Widom, B. 1965. Equation of state in the neighborhood of the critical point. *Journal of Chemical Physics*, **43**, 3898.

Wiegmann, P. B. 1989. Multivalued functionals and geometrical approach for quantization of relativistic particles and strings. *Nuclear Physics B*, **323**, 311.

Wilczek, F. 1982. Magnetic flux, angular momentum, and statistics. *Physical Review Letters*, **48**, 1144.

Wilson, K. G. 1969. Non-Lagrangian models of current algebra. *Physical Review*, **179**, 1499.

Wilson, K. G. 1971. Renormalization group and critical phenomena. I. Renormalization group and the Kadanoff scaling picture. *Physical Review B*, **4**, 3174.

Wilson, K. G. 1974. Confinement of quarks. *Physical Review D*, **10**, 2445.

Wilson, K. G. 1983. The renormalization group and critical phenomena. *Reviews of Modern Physics*, **55**, 583.

Wilson, K. G., and Fisher, M. E. 1972. Critical exponents in 3.99 dimensions. *Physical Review Letters*, **28**, 240.

Wilson, K. G., and Kogut, J. B. 1974. The renormalization group and the ϵ expansion. *Physics Reports C*, **12**, 75.

Witten, E. 1979a. Dyons of charge $e\theta/2\pi$. *Physics Letters B*, **86**, 283.

Witten, E. 1979b. Instantons, the quark model, and the $1/N$ expansion. *Nuclear Physics B*, **149**, 285.

Witten, E. 1984. Non-abelian bosonization in two dimensions. *Communications in Mathematical Physics*, **92**, 455.

Witten, E. 1989. Quantum field theory and the Jones polynomial. *Communications in Mathematical Physics*, **121**, 351.

Witten, E. 1998. Anti de Sitter space and holography. *Advances in Theoretical and Mathematical Physics*, **2**, 253.

Wu, T. T., and Yang, C. N. 1976. Dirac monopole without strings: Monopole harmonics. *Nuclear Physics B*, **107**, 365.

Zamolodchikov, A. B. 1986. "Irreversibility" of the flux of the renormalization group in a 2D field theory. *JETP Lett.* **43**, 730, 1986 [*Pis'ma Zh. Eksp. Teor. Fiz.*, **43**, 565].

Zinn-Justin, J. 2002. *Quantum Field Theory and Critical Phenomena*. fourth ed. International Series of Monographs in Physics. Oxford: Oxford University Press.

Index